Integrated Water Resources Planning *for the 21st Century*

Proceedings of the 22nd Annual Conference
Cambridge, Massachusetts
May 7-11, 1995

Sponsored by the
Water Resources Planning and Management Division
American Society of Civil Engineers

Co-sponsored by the:
Boston Society of Civil Engineers
American Consulting Engineers Council of New England
New England Water Environment Association
New England Interstate Water Pollution Control Commission
Massachusetts Assoc. of Land Surveyors & Civil Engineers
Massachusetts Municipal Engineers Association
Massachusetts Society of Professional Engineers
Society of Women Engineers - Boston Section
Water Environment Federation
New England Water Works Association
Boston Water and Sewer Commission
Massachusetts Water Resources Authority
Cambridge Water Department
Massachusetts Port Authority
South Essex Sewerage District
Tufts University
Harvard University
Northeastern University
Massachusetts Institute of Technology
Boston University
Wentworth Institute of Technology

Corporate Sponsors:
Camp Dresser & McKee
CH2M Hill
Metcalf & Eddy
Parsons Brinckerhoff
Rizzo Associates
Stone and Webster
Whitman & Howard

National Chair: Larry A. Roesner, PhD., P.E.
Edited by: Michael F. Domenica, P.E.

Published by the
American Society of Civil Engineers
345 East 47th Street
New York, New York 10017-2398

ABSTRACT

This proceedings, *Integrated Water Resources Planning for the 21st Century,* contains papers presented at the 22nd Annual Conference of ASCE's Water Resources Planning and Management Division held in Cambridge, Massachusetts on May 7-11, 1995. The major theme of the conference was the challenge facing water resource professionals to develop and implement decision-making approaches that integrate the numerous objectives and constraints in reaching balanced water management strategies. Much attention was given to the principles of watershed-based planning and sustainable development as underlying foundations for sound water resources protection and development. International events included two sessions covering water resources issues in Pacific Rim countries and a Symposium on the water resources management aspects of the North American Free Trade Agreement (NAFTA). Papers presented at the conference included covered such topics as: 1) urban drainage and stormwater; 2) water rights and policy; 3) watersheds and wetlands; 4) water pollution control; 5) water supply planning and management; 6) economics, flood control and risk assessment; 7) water conservation and stochastic hydrology; 8) information resources; and the North American Free Trade Agreement (NAFTA).

Library of Congress Cataloging-in-Publication Data

Water Resources Planning and Management Conference
(22nd: 1995: Cambridge, Mass.)
Integrated water resources planning for the 21st century: proceedings of the 22nd Annual Conference, Cambridge, Massachusetts, May 7-11, 1995 / sponsored by the Water Resources Planning and Management Division American, Society of Civil Engineers; co-sponsored by the Boston Society of Civil Engineers...[et al.]; edited by Michael F. Domenica.
p. cm.
Includes index.
ISBN 0-7844-0081-4
1. Integrated water development—Congresses. 2. Sustainable development—Congresses. 3. Free trade—North America—Congresses. I. Domenica, Michael F. II. American Society of Civil Engineers. Water Resources Planning and Management Division. III. American Society of Civil Engineers. Boston Society of Civil Engineers Section. IV. Title.
TC409.W3692 1995 95-15111
333.91'15—dc20 CIP

Library of Congress Catalog Card No: 95-15111
ISBN 0-7844-0081-4
Manufactured in the United States of America.

FOREWORD

This volume contains papers presented at the 22nd Annual Conference of ASCE's Water Resources Planning and Management Division, held in Cambridge, Massachusetts, May 7–11, 1995. The conference theme, Integrated Water Resources Planning for the 21st Century, identifies the overall challenge facing water resources professionals as we approach the next century—the integration of water resources planning, both internally as we address quantity and quality issues for surface water, groundwater, estuaries and oceans, and externally as we integrate water resources priorities with pressing societal needs for transportation, food, land use management, health care and other human necessities. The sessions of the conference embodied these themes as engineers, planners and scientists dealt with such issues as watershed management, sustainable development, risk assessment, pollution trading, aesthetics in water resources development, decision support systems, water resource economics and internetworking as a vehicle for integrating planning efforts around the world. As much as providing solutions to water resources management problems, the conference focused on new ways of thinking about solving problems. Concepts and approaches such as those described above are not intended to give readers or listeners ready answers to their problems, but to provide guidance from experience on how to approach developing site-specific solutions based on the unique balance of water resources issues that prevail in each setting.

The conference consisted of over 370 presentations, papers and panel discussions covering a wide range of water resources issues. There were two multiple-session discussions of international water issues: one covering Pacific Rim water resources development and one addressing the implications of the North American Free Trade Agreement (NAFTA) on water resources management. In addition, a special symposium covering NAFTA in more detail was held on the last day of the conference. The Task Committee on Water Supply and Conservation sponsored a two day symposium on water conservation covering such topics as demand management, water pricing, demand forecasting, and reuse. There was a special symposium on stochastic hydrology, moderated by Dr. Peter Rogers of Harvard University, in honor of the late Dr. Myron Fiering. Several sessions were dedicated to results of the Demonstration Erosion Control (DEC) Project of the U.S. Army Corps of Engineers. Also, in an effort to increase student participation in ASCE activities, Dr. Robert Traver and Dr. Robert Montgomery coordinated presentations of student design projects——from both the students and professors perspectives.

Papers included in these *Proceedings* were accepted after review of abstracts submitted by authors. All papers directly related to the main topics of conference were accepted. Papers are eligible for nomination for ASCE award and for discussion in the Journal of the Water Resources Planning and Management Division.

The above symposia and special sessions were complemented by individual papers covering a wide range of topics, including uban stormwater management; water rights and policy; watersheds and wetlands; water pollution control, especially CSO control; groundwater and conjunctive use; and computer technology in water resources planning, including the use of electronic journals and internetworking tools. Speakers and presenters at the 96 session conference represented over 20 countries and every region of the United States.

Two technical tours were offered to conference participants. A boat tour of Boston Harbor focused on the numerous water-related facilities serving the metropolitan area, including very new and very old sewerage facilities. Later in the week, attendees traveled to western Massachusetts to visit two water resources projects upon which much of the state and region depend: the Quabbin Reservoir, which is the primary water supply source for the metropolitan Boston area, and the Northfield Mountain Hydroelectric Pumped Storage Project, which provides peaking power to the New England electricity network.

The conference was held in conjunction with the annual meeting of the Boston Society of Civil Eigineers, which served as a co-sponsor and assisted in many details of conference planning. The BSCES was responsible for organizing the western Massachusetts technical tour. The New England Water Environment Federation organized the Boston Harbor tour.

Co-sponsors and corporate sponsors of the conference included:

Boston Society of Civil Engineers
American Consulting Engineers Council of New England
New England Water Environment Association
New England Interstate Water Pollution Control Commission
Massachusetts Asso. of Land Surveyors & Civil Engineers
Massachusetts Municipal Engineers Association
Massachusetts Society of Professional Engineers
Society of Women Engineers—Boston Section
Water Environment Federation
New England Water Works Association
Boston Water and Sewer Commission
Massachusetts Water Resources Authority
Cambridge Water Department
Massachusetts Port Authority
South Essex Sewerage District
Tufts University
Harvard University
Northeastern University
Massachusetts Institute of Technology
Boston University
Wentworth Institute of Technology

Corporate Sponsors:

Camp Dresser & McKee
CH2M HILL
Metcalf & Eddy
Parsons Brinckerhoff
Rizzo Associates
Stone and Webster
Whitman & Howard

The photograph on the cover of these proceedings is the winner of the fifth annual Water Resources Planning and Management Division photography contest. Members of all ASCE Student Chapters and clubs as well as Younger Members were invited to submit original work photographs consistent with the theme of this year's conference. Judging was by the sponsoring Membership Development Committee.

The winning photograph shows Modified Roosevelt Dam near Globe, Arizona under construction in May, 1994. Modifications include the addition of 23.5 Meters (77 feet) to the height of the existing cyclopean masonry dam for the purposes of improving the dam's safety, adding flood control space, increasing conservation storage, and expanding public recreational opportunities. Its an excellent example of the theme of this year's conference: "Integrated Water Resources Planning for the 21st Century."

The photograph was taken and submitted by Todd Ingersoll, a civil engineering graduate student at Arizona State University, who received the first place prize of a complementary three-day conference registration as well as hotel accommodations and travel expenses.

Amy Behrman, an ASCE Student Chapter member at Ohio Northern University received honorable mention.

CONTENTS

Chapter A—Urban Drainage and Stormwater
Session A-1
Approaches to Regulatory Objectives
Moderator—Jim Heaney

Session A-2
Urban Stormwater Computer Modeling
Moderator—Wayne Huber

Session A-3
Stormwater Pollution Prevention
Moderator—John J. Warwick

*Manuscript not available at time of printing.

Session A-4
NPDES Related Stormwater Monitoring Needs
Moderator—Ben Urbonas

Session A-5
How Has The "Environmental Regulations Pendulum" Swung in The Past Year?
Moderator—Jonathan Jones

Session A-6
Water Management Issues in The Pacific Rim—I
Moderator—Shaw L. Yu

---

*Manuscript not available at time of printing.

Session A-7
Water Management Issues in the Pacific Rim—II
Moderator—Wayne Huber

Session A-8
Preservation/Restoration of Riverine Corridors in Urban Areas
Moderator—Stuart G. Walesh

Session A-9
Urban Stream Protection and Enhancement
Moderator—Jonathan Jones

Session A-10
Stormwater Modeling
Moderator—Igor Runge

Session A-11
Effects of Urbanization on Urban Water Resources
Moderator—Libby Blank

Chapter B—Water Rights and Policy
Session B-1
Watershed-Based Permitting in Massachusetts
Panel Discussion: Moderator—Arlene O'Donnell; Panelists—Andrew Gottlieb, Dean Spencer, Paul Hogan, Pam Harvey

Session B-2
Sustainability Criteria in Water Resources Planning—I
Moderator—Barbara Lence

*Manuscript not available at time of printing.

Session B-3
Sustainability Criteria in Water Resources—II
Moderator—Charles Howard

Session B-4
Issues in Water Policy and Rights
Moderator—Mike Domenica

Session B-5
Interdisciplinary and Diversity Issues in Water Resources Problems of Today
Panel Discussion: Moderator—Neil S. Grigg; Panelists—Daniel P. Loucks, Amanda Renshaw, Rosanna Hoffman, Robert C. Hunt, Gerald Greene

Session B-6
Strategies for Effective Water Resources Planning—I
Moderator—Jim Heaney

*Manuscript not available at time of printing.

*Manuscript not available at time of printing.

*Manuscript not available at time of printing.

*Manuscript not available at time of printing.

Session C-6
Watershed Planning Approaches
Moderator—C. T. Bathala

Session C-7
Wetlands Restoration
Moderator—Edwin E. Herricks

Session C-8
Wetlands Treatment
Moderator—Bethany Eisenberg

*Manuscript not available at time of printing.

*Manuscript not available at time of printing.

Session D-2
Alternative Wastewater Systems
Moderator—Glenn Haas

Session D-3
Computerized Information Systems for Wastewater Management
Moderator—John Sullivan

Session D-4
Permitting of Wastewater Treatment Facilities
Moderator—Peter Von Zweck

*Manuscript not available at time of printing.

Session D-5
Uncertainty and Risk in Point Source Control
Moderator—George Hecker

Session D-6
Improving Water Quality of Water Supply Reservoirs
Moderator—Joseph M. McGinn

Session D-7
CSO Control Strategies
Moderator—Patrick Hughes

*Manuscript not available at time of printing.

Session D-8
Design Strategies for Controllng CSO's
Moderator—Erez Sela

Session D-9
Impacts and Management of Industrial Discharges
Moderator—Virginia Roach

Session D-10
Wasteload Allocations
Moderator—Dominique Brocard

---

*Manuscript not available at time of printing.

Session D-11
Pollution Control Around the World
Moderator—Katherine Hon

Chapter E—Water Supply Planning and Management
Session E-1
Water Resources Planning Approaches
Moderator—Jery Stedinger

Session E-2
Water Resources Planning Case Studies—I
Moderator—Ali Diba

---

*Manuscript not available at time of printing.

*Manuscript not available at time of printing.

Session E-6
Decision Support Systems in River Basin Management
Panel Discussion: Moderator—Paul Kirshen; Panelists—Richard B. Allen, John Labadie, Daniel P. Loucks, Quentin W. Martin, Arland Whitlock

Session E-7
Water Treatment Needs and Response
Moderator—Douglas Reed

Session E-8
Storage and Transport Management—I
Moderator—Allison Keyes

Session E-9
Storage and Transport Management—II
Moderator—Alan Silbovitz

Session E-10
Planning for Metropolitan Boston's Water Supply System
Moderator—John Shawcross

Session E-11
Water and Wastewater Utility Management
Moderator—Ronald Sharpin

Chapter F—Economics, Flood Control, and Risk Assessment
Session F-1
Optimal Risk-Based Decision Making—I
Moderator—Timothy Jacobs

*Manuscript not available at time of printing.

Session F-2
Optimal Risk-Based Decision Making—II
Moderator—Jay Lund

Session F-3
Probabilistic Weather Forecasts in Reservoir Operations at TVA
Moderator—Janet C. Herrin

*Manuscript not available at time of printing.

Session F-4
Water Resources Economics I—Environmental Economics in Water Resources Planning
Moderator—Robert Hamilton

Session F-5
Water Resources Economics II—Water Policy and Pricing Issues
Moderator—Duane J. Rosa

Session F-6
Flood Plain Management: Policy and Case Studies—I
Moderator—David D. Dee, Jr.

*Manuscript not available at time of printing.

Session F-7
Flood Plain Management: Policy and Case Studies—II
Moderator—David D. Dee, Jr.

Session F-8
Flood Control and Flow Forecasting I
Moderator—Paul Kirshen

Session F-9
Flood Control and Flow Forecasting II
Moderator—Laurel Saito

*Manuscript not available at time of printing.

Session F-10
Flood Control and Flow Forecasting III
Moderator—Uli Kappus

Session F-11
Reservoir Management and Hydropower
Moderator—Richard A. Westmore

Chapter G—Groundwater and Conjunctive Use
Session G-1
Groundwater Management and Modeling—I
Moderator—Richard J. Hughto

Session G-2
Computational Issues in Aquifer Remediation—I
Moderator—William W.G. Yeh

Session G-3
Computational Issues in Aquifer Remediation—II
Moderator—David Dougherty

Session G-4
Computational Issues in Aquifer Remediation—III
Moderator—Christine A. Shoemaker

*Manuscript not available at time of printing.

Session G-5
Computational Issues in Aquifer Remediation—IV
Moderator—David Ahlfeld

Session G-6
Conjunctive Water Use Modeling and Management
Moderator—Miguel Marino

Session G-7
Groundwater Management and Modeling—II
Moderator—Keith A. London

*Manuscript not available at time of printing.

Session G-8
Miscellaneous Groundwater Management—Case Studies
Moderator—Peter Silbermann

Session G-9
Inter-American Water Management and Sustainable Development—I
Moderator—David Eaton

Session G-10
Inter-American Water Management and Sustainable Development—II
Moderator—David Eaton

---

*Manuscript not available at time of printing.

Session I-4
DEC Project: Design and Construction Techniques
Moderator—Charles Little, Jr.

Session I-5
DEC Project: Performance of Grade Control Structures
Moderator—Roger H. Smith

Session I-6
Student Multidisciplinary Design Projects
Panel Discussion: Moderator—Robert G. Traver; Student Panelists

Session I-7
Student Multi-disciplinary Design Projects: Faculty Perspective
Moderator—Robert J. Montgomery

*Manuscript not available at time of printing.

*Manuscript not available at time of printing.

*Manuscript not available at time of printing.

Session J-4
Information Sharing and Intersociety Cooperation Among Canada, Mexico, and the USA
Panel Discussion: Moderator—Walter Lyon

A WATERSHED APPROACH TO URBAN STORMWATER MANAGEMENT INTEGRATING FEDERAL, STATE AND LOCAL OBJECTIVES

Thomas M. Grisa, P.E.[1]
Member, ASCE

Abstract

Federal, state and local agencies all have specific and sometimes contradictory objectives for managing urban stormwater runoff, making it difficult to address local interests when federal and state agencies loom overhead with their requirements and regulations. For some, the tendency would be to resist these mandates as they are perceived to be heavy-handed. But, local municipalities and counties are finding that it is sound engineering practice and wise policy to incorporate all of these objectives when developing a stormwater management plan to avoid costly capital expenditures or fines in the future. This paper will present the cooperative approach taken by the City of Wauwatosa to integrate federal mandates, state goals, and local concerns into a comprehensive stormwater management plan which will accommodate the stormwater requirements of the 1990's and needs of the 21st century.

Introduction

Recent federal mandates have required municipalities and counties to improve the quality of stormwater discharges from municipalities and counties, from industrial facilities and from construction sites.

---

1) Assistant City Engineer, City of Wauwatosa, 7725 W. North Avenue, Wauwatosa, WI 53213

Furthermore, reauthorization of the Clean Water Act will expand on this first mandate to include even smaller communities and more industries in an effort to further improve the water quality of our receiving streams.

In addition, states like Wisconsin, which depend substantially on tourism for their economic viability, have exceeded federal requirements by developing stormwater quality programs such as Wisconsin's Priority Watershed Program. This program was conceived of in the 1960's and established over 15 years ago, long before stormwater quality became a popular national issue. The program addresses stormwater quality on a watershed basis, to improve stormwater quality where water quality of streams and lakes is below desired standards. Unlike the unfunded federal mandates, the Wisconsin Department of Natural Resources (WDNR) encourages municipalities to participate in this voluntary program by providing funding for planning, operations and construction activities. The City of Wauwatosa responded to this opportunity by spearheading a project to address stormwater concerns in the Lower Menomonee River watershed which includes the northwest portion of the city and portions of two other municipalities.

Developing a Team Approach

Like most municipalities, located in Milwaukee County, the City of Wauwatosa has historically addressed only stormwater drainage issues and flooding concerns with management practices focused primarily on infrastructure improvements within the City's corporate limits. Certainly, cooperation with neighboring communities has occurred, but most solutions to stormwater problems have not been addressed on a watershed basis.

However, to properly address the City's stormwater needs and evaluate stormwater quality issues pertinent to the priority watershed, municipal boundaries would need to be crossed to consider upstream conditions. The City of Wauwatosa approached the City of Brookfield and the Village of Butler, who also received grants from the DNR for stormwater management, to participate in the project with Wauwatosa. At first, there was reluctance on everyone's part since the project would involve three municipalities located in two counties, a state environmental regulatory agency and the unknown extent of future federal involvement.

Furthermore, some high ranking officials were not fully convinced that the goal of the project, addressing stormwater quality in addition to stormwater quantity, was a valid one. As one engineer put it, "All I need is a larger diameter storm sewer to convey my stormwater to your city, and let you take it from there since you're downstream." Despite these initial concerns, an agreement was ultimately reached for proceeding with the project.

The project area consists of approximately 921 hectares (2275 acres) with a land use distribution of approximately 37 percent industrial uses, 17 percent commercial/business development, 21 percent residential and 25 percent parks and open space concentrated primarily along the Menomonee River. The industrial uses include manufacturing, warehousing and distribution centers and transportation facilities (railroads, freeways and local roads).

Because of the complexity of the project and our lack of specific expertise in the area of stormwater quality issues, we retained the engineering firm, Woodward-Clyde Consultants, to bring the needed technical background to the project. The first activity performed by the project team was to develop a stormwater management committee made up of representatives from each municipality, the DNR, the County Parks Department, the Department of Transportation, a Railroad Company, a Country Club, owners of vacant land, and other major industries in the study area currently or likely to be affected by stormwater discharge permit rules.

The intent of this committee was to review the overall project approach and to provide review comments on the proposed alternative stormwater management strategies. Furthermore, committee members were asked to help educate and inform their respective entities about the project and general stormwater issues.

Maintaining Unity Despite Conflicting Interests

As expected, each committee member brought his own agenda to the meetings. As the project progressed and as the committee discussed the findings of the project, these agendas became very evident. Industry was concerned with potential permit requirements and source controls, the municipalities were interested in proposed structural and source controls, the DNR wanted to see the goals of the Priority Watershed Program achieved and everyone was interested in the location of proposed ponds.

Since the study area is nearly 100 percent developed, open spaces became prime targets for location of stormwater retention facilities. These open spaces consisted of small parcels of undeveloped land zoned for industrial use, a Country Club and County owned land. The not-in-my-backyard (NIMBY) principle became each agency's battle cry as all saw value in their own property and found reasons to propose the pond elsewhere.

In depth discussions were held for each proposed site identifying the reasons for selecting the site including location in the watershed, effectiveness of the proposed facility to improve stormwater runoff quality and economic viability. These reasons were not always convincing and some ponds were reduced in size or eliminated altogether with the approval of the DNR. In two cases, the proposed ponds were located within a County golf course. The project was expanded at the request of Milwaukee County Parks to include preliminary engineering for specifically locating and configuring the ponds to enhance several holes on the golf course.

Considerable effort was expended to ensure all the concerns and interests were addressed by the committee for the final report involving give and take by all parties involved. However, the effort was worthwhile since it resulted in recommendations that all could accept and established positive working relationships for future projects.

Cost-Benefit Analysis of Urban Stormwater Controls

Thomas A. Blood[1], Richard L. Schaefer[2]

Abstract

In response to flooding and stream channel erosion problems resulting from urbanization, a jurisdiction in Washington's Puget Sound region proposes to revise its required methods of hydrologic analysis and performance standards for on-site runoff controls. The jurisdiction encompasses a large metropolitan area that includes significant spawning and rearing habitat for anadromous fish. The proposed revisions would better protect beneficial uses of area receiving waters, but would also increase the required size of on-site runoff flow controls for some types of new development. We prepared an analysis of the relative economic costs and benefits to be derived from implementing the proposal to assess the cost effectiveness of the proposed regulation.

Introduction

Hydrologic analyses of watersheds within the jurisdiction showed that stormwater control facilities, particularly on-site flow control facilities, do not meet intended or required levels of performance. Various means of resolving the discrepancies between performance goals and actual performance were studied, and specific revisions to the drainage regulations were recommended.

The jurisdiction recommended that the hydrologic design methods for on-site flow control facilities be changed from an event-based method to a continuous model approach. It was also recommended that flow control performance for a site be determined by the nature of the receiving water body; that is, where discharge is to a sensitive water body, such as a salmon spawning stream channel, flow controls would be more stringent. Based on case

[1] Analyst, R.W. Beck, 2101 Fourth Avenue, Suite 600, Seattle, Washington 98121.

[2] Director, R.W. Beck, 2101 Fourth Avenue, Suite 600, Seattle, Washington 98121.

studies that applied the revised regulations to a series of development scenarios, it was expected that the changes to drainage regulations will generally require larger flow control facilities, and these facilities will consequently consume a greater percentage of the developable land area.

Because of the potential for these proposed changes to economically affect development in the jurisdiction, the relative costs and benefits from the changes were considered. The cost-benefit analysis compared the quantifiable economic costs and benefits expected from continued development under both current and proposed requirements. Other, non-quantified benefits of the proposed revisions were also noted in the analysis.

Analysis Structure

Two watersheds were studied to evaluate the effects proposed requirements would have upon land development and receiving streams. The first watershed is lightly developed and contains a high-quality resource stream system. This watershed was selected as representative of rural areas expected to develop to densities of 0.2 to 1.0 dwelling units per acre (0.5 to 2.5 dwelling units per hectare). The second watershed was selected as a typical partially urbanized drainage basin containing lesser-quality resource stream systems and zoned for urban densities of 3.5 to 20 dwelling units per acre (9 to 50 units per hectare). The remaining available land area in each watershed was assumed to build out to its zoned densities over 20 years. The resulting stream flows were modeled for each watershed using continuous hydrologic simulation techniques. The hydrologic model results were then used to estimate the relative stability of the stream channels over time, the disturbance of anadromous fish spawning and rearing habitat in the stream channels, and the consequent impacts to production of anadromous fish in the streams.

Benefits

Maintaining fish species in the watersheds—specifically steelhead and coho, sockeye, and chinook salmon—was identified as a major quantifiable benefit of the proposed runoff flow control standards. The goal of this part of the benefit analysis was to derive the approximate dollar value of the fish produced in these two watersheds each year. This data would then be applied to the fish population survival levels associated with the current and proposed flow control standards. The differences between the fish population values in each watershed over time were considered the marginal benefits to be derived from implementing the proposed standards. These marginal benefit values were then input as the benefit into the cost-benefit equation.

Fisheries data for the two modeled watersheds was obtained from the Washington Department of Fish & Wildlife (WDFW). Because these fish are tagged, WDFW data include estimates of the annual sport and commercial catch for these fish runs. Based on a review of the existing literature, fish dollar values were determined to consist of recreational, commercial and existence values. These values are described briefly below.

Recreational Values: Based on the existing literature, recreational values for a fishery are comprised of money spent by sport anglers for fishing licenses, tackle, guiding fees, travel expenses, and other related items. Several existing economic contingent and travel valuation surveys conducted on the Washington and Oregon coasts to measure these values were analyzed, and recreational values were standardized and averaged. Using WDFW data (i.e., for angler days spent-to-catch ratios) for Puget Sound, per fish recreational values were calculated. Based on WDFW tag return program data, the number of fish reared in the modeled watersheds and later caught by anglers was determined, and the per fish recreation value was then applied to these fish.

Commercial Values: Commercial values for salmon were calculated by using 1994 retail salmon prices. Based on WDFW tag return program data, the number of commercially caught fish reared in the modeled watersheds was determined, and the per fish commercial value was then applied to these fish.

Existence Values: Existence values are defined (and have been theoretically upheld by economists) as values that individuals may be willing to pay for the knowledge that a resource exists. For example, a Chicago resident may contribute to a conservation organization to preserve grizzly bears in Yellowstone Park even though the individual may never visit Yellowstone. The Corps of Engineers conducted a definitive study to determine the existence values to Pacific Northwest residents of salmon in the Columbia River watershed based on a survey of 700 households in four states. Our analysis applied these per fish existence values to salmon and steelhead reared in the modeled watersheds.

Once recreational, commercial and existence values were determined, the annual benefit to the fishery to result from the proposed standards was estimated:

1. Habitat degradation indices (resulting from the different standards) were matched with fish productivity curves to show the percentage of each fish population that would survive under stream conditions resulting from each standard.

2. Appropriate dollar values were applied to the numbers of each fish population that would survive (benefit) each year from each standard.

3. Total fish benefit values were projected over a 50-year period at two discount rates for each standard, and the results were included in the cost-benefit equation.

Costs

The costs of on-site runoff controls will increase with larger facility volume requirements. Cost increases will result primarily from the greater land area required to site the facilities. To estimate the cost impacts of the proposed change in standards, a financial model and case studies were prepared in cooperation with development community representatives. Example proposals for single-family residential, multifamily, and commercial developments meeting current standards were redesigned to conform to the proposed requirements. The original proposals and the modified designs were used to estimate differences in property yields and costs. The costs of facilities meeting current and proposed standards were compared considering the purchase cost of the land occupied by the facilities, the lost yield or profit on the occupied land (opportunity cost), and construction and maintenance costs.

For each development type, unit area costs were computed for facilities constructed under current and proposed standards. The unit costs were applied to the developable land area in each study watershed. Capital costs were spread over the 20-year buildout period, and the facilities were estimated to have an average service life of 40 years. Hence, a 50-year life cycle was set for the analysis, the median end of the useful life of the facilities.

Findings

Because of significant prior degradation in the partially urbanized basin, the current anadromous fish production in the watershed is already depressed. Because the fish population base upon which future benefit value was measured was already reduced, the computed benefit value, particularly over the short term, was relatively small. Further, because of the large quantity of development to occur in the urban basin that would be affected by the proposed standards, the computed costs were high when compared to the benefit to the marginal fish population. It may be warranted in such instances to associate existence value, either in whole or in part, to the number of residents in the watershed holding the existence value, rather than with the number of fish in the stream.

In the rural basin, the large existing fish population and the smaller projected development growth will likely produce economic benefits that equal or slightly exceed the increased costs of development from implementing the proposed standards.

The contrasted findings for the two basins would indicate that, in terms of economic benefit to fisheries, it is cost effective to concentrate urban growth in those watersheds that have already sustained in-stream resource damage to the point of severely depressing fish population. The findings also argue for rigorous control of runoff in watersheds having high-quality fish resources to avoid possible degradation of the resource.

Although not included in this analysis, other economic factors affected by runoff control requirements, can also be appropriately evaluated using a cost-benefit approach. These factors include relative measures of flooding damage, water quality, other in-stream resources and uses, and legal and regulatory liabilities.

Alternative Methodologies for Numerical Regulatory Effluent Limitations for Storm Water Discharges to Ocean Waters

L. Donald Duke [1]
Michelle H. Wilhelm [2]

Abstract

Rational numerical effluent limits for pollutants in storm water discharges may not be developed as directly or readily as for industrial discharges and other point sources because of variability in sources and uncertainty in quantifying pollutant discharges. The current most widely-used method for setting standards for discharges to ocean waters, the dilution factor approach, may not be appropriate for storm water discharges. This research identifies three categories of alternative methods that may be applied, and then attempts to develop numerical limits using existing data for one particular storm water conveyance discharging into Santa Monica Bay, California. Existing information about current pollutant discharges and environmental impacts of storm water pollutants does not support development of limits using any of the three alternative methods. Instead, this research applies four specific statistical methods derived from modifications to existing methods.

Introduction

Storm water discharges are subject to NPDES water quality regulations. To date, regulations for municipalities and industrial facilities, designed to protect receiving water quality, have specified Best Management Practices (BMPs) rather than setting numerical effluent limits for specific pollutants in storm water discharges. This research evaluates various methodologies that could be used to develop such pollutant-specific numerical limits for specific storm water discharges using information that is currently available.

---

[1]Assistant Professor, Environmental Science and Engineering Program, School of Public Health, University of California, Los Angeles, CA 90095-1772
[2] Radian Corp., 300 N. Sepulveda Blvd., El Segundo, CA 90245

Developing Numerical Limits: Current Approach; Three Potential Alternatives

At present, water quality limits required by NPDES regulations for point source discharges to ocean waters in California are determined by specifying water quality objectives for the receiving waters, and then calculating a dilution factor for the discharge based on outfall design and other factors that govern mixing characteristics. This form of limit is less applicable to urban storm water discharges for several reasons, including: variability of discharge flow regimes leading to uncertain and highly variable mixing characteristics at the outfall; outfalls not designed to promote efficient mixing, compared to long tubular dispersal devices and similar mechanisms routinely installed at point source outfalls; uncertain pollutant sources in a large urban watershed, and accompanying difficulty in requiring control actions by responsible parties; and extreme variability over time in pollutant discharge concentration, with attendant uncertainty about pollutant loads and predicted environmental effects. Three alternative methods are suggested (Duke and Wilhelm, 1994).

Concentration-based limits are well-suited to avoid adverse impacts from potential short-term increases in pollutant concentrations near storm water outfalls. Unlike effluent limits for point sources with continuous flow and relatively uniform characteristics, effluent limits for storm water need to consider the variable nature of flow and of pollutant concentration. Statistical analysis of pollutant concentration, storm water flow, and pollutant mass are necessary to characterize existing discharges because of variability over time. The form of these limits would be maximum allowable concentration averaged over a short period such as one hour. Concentration-based limits are best suited for volatile organic compounds and other pollutants that do not persist in the environment; and for other substances which have the potential to cause adverse effects on biota or the ecosystem after short-term acute exposure to the pollutants. If such limits are to be based on rational analysis designed to prevent adverse effects of pollutant discharges, significant additional research is necessary to demonstrate relationships between concentration of specific pollutants and environmental effects in receiving waters. An alternative is to develop limits which prohibit pollutant loads from increasing beyond existing conditions. These limits may be specified simply by monitoring and statistical analysis of existing discharge concentrations.

Mass-based limits related to receiving water pollutant concentration are well-suited to regulate long-term accumulation of pollutants in receiving waters. Mass-based limits are most appropriate for pollutants that degrade slowly in the environment, bind to suspended sediments, or bioaccumulate in aquatic organisms. Such pollutants may include: metals; chlorinated organic compounds such as pesticides; and some other organic substances which are relatively slow to volatilize. Additional research on the relationship between mass or concentration of pollutants and adverse effects on the receiving water ecosystem would be necessary to make the policy decisions about allowable mass loads needed to develop such limits for individual pollutants.

The third form of limit would specify mass discharge based on long-term accumulation of pollutants in sediments of the receiving water, which is the form in which pollutants are most likely to persist in the environment. Such limits are the form recommended in the Restoration Plan for Santa Monica Bay promulgated by the Santa Monica Bay Restoration Foundation (1994). However, existing data are far from adequate to rationally develop such limits. Additional necessary research would be necessary in areas including: relationships between sediment concentrations of pollutants and adverse effects on biota and the ecosystem; relationship between pollutant concentrations in sediments and in the water column; and relationships between pollutant discharge loadings, including variations over time, and steady-state or dynamic concentrations of pollutants in receiving water sediments.

A full plan for protection of the receiving water quality may integrate all three forms of numerical limits. Appropriate limits for multiple specific pollutants may be developed in any one or all three of these general forms.

Application of Methods to Ballona Creek, Santa Monica Bay, California

Existing information about Ballona Creek (Stenstrom and Strecker, 1993) was used to develop preliminary numerical limits for discharge of two pollutants, the metal lead and the chlorinated organic pesticide dieldrin. Four specific methodologies are applied. Three are modifications of the dilution factor approach, commonly applied to point source discharges to ocean waters. The fourth statistically evaluates existing pollutant discharge data with the intent to set limits that will prohibit any increase in pollutants over existing levels.

The first method computes the average dilution factor for Ballona Creek discharge throughout the wet-weather season, assuming average flow to be a good predictor of mixing behavior. The results fail to reflect either impacts during a storm or local impacts during low flow, both periods when mixing is not well-described by this method.

The second method combine the average dilution approach with statistical anaiysis of pollutant concentration in discharges. The method is demonstrated not to be valid with existing data, because the flow rate of Ballona Creek that corresponds to each sample data point is too low to assume mechanical mixing, an assumption that is necessary when developing the dilution-factor equation.

The third method uses data collected only during storm discharges, when flow in Ballona Creek is sufficiently high to predict mechanical mixing. However, existing monitoring data for Ballona Creek do not include such storm discharge, either for concentration of pollutants or for storm water flow.

The fourth approach uses existing storm water concentration data for Ballona Creek to statistically predict behavior of pollutant discharge under existing conditions. The 95th, 90th and 50th quantile concentrations are estimated utilizing the available wet weather concentration data. One or more of these statistically derived concentrations can be used as discharge limits, not to be exceeded a specified number of times during a season. The approach does not use health or ecosystem end points, but sets limits designed simply to prohibit discharge at higher concentrations than currently experienced.

Conclusions

Existing methods for developing numerical water quality limits for ocean discharges may not be appropriate for storm water conveyances discharging urban runoff. However, the current state of knowledge does not support any methods for developing rational numerical discharge limits designed to be protective of adverse environmental impacts in receiving waters. Only minor modifications to existing methods are possible if numerical limits are to be developed in the immediate future.

References

Stenstrom, Michael K., and Strecker, Eric, 1993. Assessment of Storm Drain Sources of Contaminants to Santa Monica Bay. UCLA School of Engineering and Applied Science Report No. ENGR 93-62.

SantaMonica Bay RestorationProject,1994.Santa Monica Bay Restoration Plan. Monterey Park, CA: Santa Monica Bay Restoration Foundation.

Duke, L. Donald, and Wilhelm, Michelle H., 1994. Draft: Investigation of Feasibility of Numerical Effluent Limits for Storm Water Discharges to Ocean Waters. Report to State Water Resources Control Board, Sacramento CA.

Acknowledgment

This research was funded by California State Water Resources Control Board, Ocean Waters Unit; and Santa Monica Bay Restoration Foundation.

USING SWMM IN URBAN STORMWATER MASTER PLANNING

By Glenn A. Bottomley[1]

ABSTRACT

A computer simulation model has been developed for the watershed study area of an eastern coastal city with flat low-lying terrain whose outfall is subject to tidal influence. The U.S. Environmental Protection Agency Stormwater Management Model (SWMM) computer simulation model was selected to analyze the existing complex urban storm drainage network. The SWMM model was selected for its capability to simulate; surcharged conditions, split flows, flow reversals and dynamic flow routing to account for the volume of stormwater runoff in the storm drain conduits; as well as to identify time duration of flooding, under capacity conduits and water surface elevation. The detailed existing storm drain network coded into the model challenged hardware and software limitations. Methodology and computer processor comparison is discussed. A comparison is made of the SWMM peak discharges and other hydrologic models.

INTRODUCTION

The watershed area consists of 150 acres of fully-developed residential neighborhoods with some light business and industrial land use nearly fully-developed. The slope of the area consists of flat to mild slopes normally encountered in low-lying coastal areas. The outfall for the study area is an expansive creek which is a tributary to a major tidal river. Normal high tides do not significantly influence the storm drain system. The existing residential neighborhoods experience frequent flooding during rainfall events and ponding after rainfall subsides. The drainage area and storm drain system is characterized by; roadside ditches with lack of positive grade, residential lot grading below street grade, major ditches ponding, residential structures in low-lying areas, and retrofit drainage improvements

---

[1]Civil Engineer, Parsons Brinckerhoff Quade & Douglas, Inc.
11 Koger Executive Center, Suite 100, Norfolk, VA 23502

that result in split-flow and branching configurations. The objective of the study is to analyze the existing system, identify deficiencies and recommend and analyze improvements. The objective of this paper is to report on the case study and the use and methodology of applying SWMM in analyzing the existing system and planning the improvements.

SWMM OVERVIEW

The EPA SWMM version 4.21 was utilized for the analysis of the existing system and the proposed improvements. SWMM is probably the most popular of urban runoff models and, as such, there is a large amount of literature and experience to refer to (Nix, 1994). The program is executed on a DOS-based personal computer and occupies over 3 megabytes of memory. SWMM simulates storm events based on rainfall hyetographs and drainage area hydrologic data and then routes the stormwater runoff through the conveyance system to the points of outfall. The simulation period may be for a single event or continuous. Data input consists of a batch file created in free format input not restricted to fixed columns. It is beneficial to create your data file set with the appropriate template date files, provided with recent versions of SWMM, and organize data in columns to facilitate manipulating and troubleshooting the simulation data. The DOS editor or a word processor can be used to construct the input file as well as to review and print the output. Proprietory "add-on" software is available to assist in developing the batch files and to provide logic assistance in creating the model and generating output. SWMM is capable of operating in metric units.

SWMM consists of four computational blocks: RUNOFF, TRANSPORT, EXTRAN and STORAGE/TREATMENT. The RUNOFF block represents the total drainage area by idealized subcatchments and simulates the rates of runoff using input rainfall hyetographs and accounting for losses due to infiltration, evaporation and surface storage. Basic input parameters are rainfall intensities and for subcatchments; width, area, percent directly connected impervious area, overland flow slopes, depression storage and infiltration rates. RUNOFF will then compute inflow hydrographs for the specified inlets. It should be noted that the RUNOFF and TRANSPORT blocks are capable of routing runoff for a simple conduit network with known gradients and flow directions. (CDM, 1993)

The TRANSPORT block routes flows through the conveyance system; however, it cannot simulate surcharged conditions or "pressure flow.

The EXTRAN block represents the conveyance system as a series of links and nodes, or conduits and junctions, in the form of pipes and channels and performs a dynamic routing of flows through the conveyance system. EXTRAN is capable of simulating the following conditions: backwater, branched or looped system, surcharged, flow reversals, weirs, orifices, pumping facilities, on-line and off-line storage facilities (CDM, 1993). Detailed input for the conduits, junctions and ground elevations is required.

PROJECT APPROACH

The project's objective was to analyze the existing system and identify deficiencies and analyze improvements. RUNOFF and EXTRAN were used to model the surcharged and flooding conditions of the network. The City provided a detailed inventory and map of the existing storm drain and drainage area subdivides. Due to the nature and extent of flooding, the availability of data and the perceived number of potential alternatives, a very detailed conveyance system was coded into the model maintaining the City's drainage area subdivides. The resulting model consisted of 106 subcatchments, hydrographs stored for 93 inlets and 121 conduits. The developer of a model should consider the benefits and cost of constructing a large comprehensive model, as it requires a significant investment of manhours and computer time.

ANALYSIS

Inflow hydrographs were developed utilizing a design hyetograph specifying an incremental depth of rainfall at half-hour increments for a duration of six hours for a single event simulation. The time step specified for EXTRAN is an important element in model stability and must accommodate the shortest conduit in the system. The detailed SWMM model network had several short conduits that violated time step criteria and a parameter was set in EXTRAN to automatically create equivalent conduits by adjusting the length and roughness coefficient. The storm drain volume increased 1.8% as a result of the additional conduit length. Specifying a short time step to ensure stability and accuracy also increases computation time. A five second time step over a six hour duration resulted in 4320 steps. The following figure demonstrates the requirement for a high speed processor.

PC Processor /Speed	Run Time
386/20 MHz	57 min.
486/33 MHz	13 min.
486/66 MHz	10 min.

Systems in surcharge require a special iteration loop which significantly increases the run time of a model.

EXTRAN's ability to balance the hydraulic head allowed for the analysis of several split-flow junctions in the network and also demonstrated significant flow flooding out of a junction which was also observed in the field. In addition, potential split-flow solutions were able to be analyzed. A storage alternative was evaluated by assigning a constant storage volume per foot of depth to a junction; however, evaluation of flooding junction durations demonstrated a minimal impact due to lack of readily available land area. Tidal tailwater conditions were simulated with the boundary condition parameters by specifying a stage-history of water surface elevations based on the graphical sinusoidal tide curve for Hampton Roads with a linear adjustment for evaluation of storm tides. Beneficial output for presentation of results from EXTRAN includes maximum junction water surface elevation, time duration of junction flooding, computed flow and velocity, time of occurrence and ratio of maximum flow to theoretical conduit design flow.

To ensure the model was computing a reasonable peak discharge, the results were compared to peak discharges arrived at by different methods as summarized below.

Method	Discharge
SWMM Baseline	3.1 cms
SWMM Improvements	4.2 cms
Rational	4.8 cms
TR-55	5.1 cms

It should be noted that as a result of increasing the size of the downstream conduits in the SWMM model the peak discharge flow rate approached the discharge quantities of the different methods.

REFERENCES

Camp Dresser & McKee, Inc., *Runoff, Transport and Extran Applications, Workshop for the Center for TREEO*, March, 1993.

Huber W.C. and R.E. Dickinson, *Storm Management Model, Version 4, User's Manual*, U.S. Environmental Protection Agency, Athens, GA 1988.

Nix S.J., *Urban Stormwater Modeling and Simulation,* 1994.

Roesner L.A., A.A. Aldrich and R.E. Dickinson, *Stormwater Management Model, Version 4, Part B: EXTRAN Addendum,* U.S. Environmental Protection Agency, Athens, GA June 1988.

A Comparison of Unsteady Flow Routing Models

John F. Dummer[1] SMASCE and Dr. Wayne C. Huber[2] MASCE

Abstract

Current engineering design practices require that, in some situations, unsteady flow routing be modeled. In these cases a high level of modeling sophistication is required, and selection of the proper model for the given situation is critical. The model selected must adapt to available input data, properly model the given situation, and deliver the results in a format useful for the intended purpose. Three U.S., nonproprietary computer models capable of unsteady flow routing are: The Environmental Protection Agency's (EPA) Storm Water Management Model (SWMM), the National Weather Service (NWS) Dynamic Wave Operational Model (DWOPER), and the U.S. Army Corps of Engineers Hydrologic Engineering Center (HEC) model UNET. This paper compares these models with the intent of aiding in the selection of the proper model for a desired application.

Introduction

Computer models capable of unsteady flow routing have a wide range of application. Real-time flood forecasting is one situation requiring such a model. Planning and design of floodplain structures such as levees and dikes is another application for unsteady flow models, as it is often desired to determine the effects of encroachment of development on a floodplain. Another use for unsteady flow models is the determination of the predicted elevation of a flood with a given recurrence interval such as a 100-year flood. HEC-2, which is a steady state model, is currently used for this purpose. Use of a dynamic model in this situation is a newer alternative that allows the total hydrograph to be examined rather than just the maximum flow, providing for a more complete analysis. Another application of unsteady models is sewer system analysis and design. For the uses discussed above it is desirable that any model intended for widespread acceptance

[1] Graduate Student, Department of Civil Engineering, Oregon State University, Apperson Hall 202, Corvallis, Oregon 97331

[2] Professor and Head, Department of Civil Engineering, Oregon State University, Apperson Hall 202, Corvallis, Oregon 97331

simulate flow in natural channels as well as man-made conveyances, including urban drainage systems. The desirable unsteady flow model must also be capable of modeling typical floodplain structures as well as components of sewerage systems.

The present research focuses on three models: The Environmental Protection Agency's (EPA) Storm Water Management Model (SWMM), the National Weather Service (NWS) Dynamic Wave Operational Model (DWOPER), and the U.S. Army Corps of Engineers Hydrologic Engineering Center (HEC) model UNET. Each of these computer models is nonproprietary, capable of unsteady flow simulation (solves the Saint Venant equations), and is available in the United States. This paper evaluates each of the three models describing the model in terms of the method used to solve the Saint Venant equations, the range of application, and the special features included with each model. A comparison of the models based on the model characteristics and conclusions based on the comparison are also included.

DWOPER

Motivation for the creation of DWOPER came from the National Weather Service need to enhance their capabilities in the area of flood forecasting. The main use for the model is day to day operational forecasting; however, the model is also applicable for simulating unsteady flows for engineering planning, design, or analysis (Fread, 1987).

A weighted four-point nonlinear implicit finite difference scheme is used to obtain solutions to the Saint Venant equations via a Newton-Raphson iterative technique (Fread, 1987). This solution technique allows for unequal distance steps and large time steps, facilitating the longer simulation periods desired for flood forecasting. Combining the finite difference equations with the boundary conditions at the upstream and downstream edges of the study area yields a sufficient number of equations to solve for the water surface elevation and the flow rate. The Newton-Raphson method is then used to obtain the solution. This solution technique is unique to the models compared in that the equations are not linearized.

The effects of lateral flows, wind, lock and dam systems, dendritic river systems, off-channel storage and flow diversions can be accounted for in this model. Automatic calibration of the roughness parameter in the friction slope term of the momentum equation is another feature of this model. The roughness parameter often varies with discharge or stage and with distance along the river. The DWOPER model contains an internal feature that can be accessed to automatically determine the optimum roughness parameter that will minimize the difference between computed and observed values (Fread, 1987).

SWMM

The EXTRAN block of SWMM was also selected for comparison. EXTRAN was developed in the early 1970's for the City of San Francisco as an analysis tool for the sewer system there. EXTRAN addresses branched or looped networks, backwater from tidal or nontidal conditions, free surface flow, pressure flow, flow reversals, flow transfer by weirs, orifices and pumping facilities and storage at on or off-line facilities (Roesner et al., 1992).

Solution of the Saint Venant equations, or gradually varied, one-dimensional unsteady flow equations for open channels is accomplished using an explicit solution employing a modified Euler method. This solution technique is unique to the models compared in that it is the only one to employ an explicit solution. The momentum equation is applied to discharge in each link and the continuity equation to head at each node, with implicit coupling during the time-step. The relative ease in computation associated with the explicit solution comes at the expense of overall model stability. Stability constraints and recommendations regarding time steps are well documented (Roesner et al., 1992).

EXTRAN is conceptually made up of a system of conduits which are connected by nodes. The conduits represent the pipes which would make up a sewer network and the nodes the manholes between them. Eight types of conduit can be modeled in EXTRAN including natural channels of irregular cross-section. Diversion structures including orifices, transverse weirs, and side-flow weirs can be modeled in EXTRAN, as can pump stations and storage basins. Six choices for outfalls are also included in the model including outfalls with and without tide gates.

UNET

UNET is the third model chosen for comparison, and can be described as a one-dimensional unsteady flow program that can simulate a complex network of open channels with off-stream storage (Barkau, 1993). It is part of the HEC River Analysis System. The model is linked with the HEC Data Storage System (DSS), allowing use of that database as a convenient and efficient database and graphing tool.

Solution of the unsteady flow equations is accomplished through a four-point implicit finite difference scheme, also known as the box scheme. The equations are first linearized and then solved using finite difference approximations of the equations of unsteady flow along with the boundary conditions (Barkau, 1993).

Several boundary conditions are specifically addressed by the model; provision is made in UNET for modeling levee failures, and the resulting storage interactions that they create. Also included is the capability to model gated spillways and weir overflow structures. Bridge and culvert hydraulics are given special consideration as are pumped diversions, storage areas and ice covered streams. UNET only considers subcritical flow. The computer code for this model, as with DWOPER and EXTRAN, is in the computer language FORTRAN.

Conclusions

Each of the models investigated provides a unique perspective from which to solve unsteady flow routing problems. Much of the reason for the differences in the models arises from the different intended purposes in mind when the models were created. DWOPER was created for flood forecasting on large rivers and that is the perspective from which it is applied. The use of an implicit solution technique, which allows longer simulations, is evidence of this. UNET also uses an implicit solution technique, but special consideration is given to culverts and the solution of a network of open channels. This indicates an emphasis on smaller waterway situations. EXTRAN, while fully capable of river modeling, was designed for sewer flow analysis and is set up with that objective in mind. It is important that modelers understand this difference in initial model objective. The time investment required to efficiently set up one of these models is significant and initial investment into finding the model with the desired attributes would be well worth the time required.

One interesting point when comparing these models arose when considering the units used by each model. The most comprehensive treatment of units was found to be in EXTRAN where U.S. customary or metric units can be used. DWOPER is geared toward the use of U.S. customary units with some provision for metric units, and the UNET model only uses U.S. customary units. This could be a significant point when selecting a model.

The present research is ongoing and simulations using each of these models is currently being undertaken. A reach of the Peace River in Florida between Bartow and Ft. Meade has been selected for comparative analysis of the models. Discharge information from gaging stations at Bartow and Ft. Meade along with cross-section information for the reach is being analyzed. This will allow comparison of actual discharges at the downstream end with simulated discharges generated by each of the models.

References

Barkau, Robert L., UNET One-Dimensional Unsteady Flow Through a Full Network of Open Channels User's Manual, CPD-66, Version 2.1, U.S. Army Corps of Engineers, Davis, California, 1993.

Fread, D.L., "National Weather Service Operational Dynamic Wave Model," Hydrologic Research Laboratory, National Weather Service, NOAA, Silver Springs, Maryland, 1987.

Roesner, Larry A., John A. Aldrich, and Robert E. Dickinson, Storm Water Management Model User's Manual Version 4: EXTRAN Addendum, Third Printing EPA/600/3-88/001b (NTIS PB88-236658/AS), Environmental Protection Agency, Athens, GA, 1992.

Evaluation of the Wilmette Runoff Control Program

Eric D. Loucks[1] and Michael C. Morgan[1]

Introduction

The east side of Wilmette, Illinois experienced basement back-ups from the combined sewer system serving the Village during heavy rainfall. Regulatory constraints required that the solution include temporary storage of flows in addition to relief sewers. Temporary ponding in street rights-of-way was found to be the least cost means of providing the required storage. The east side runoff control plan includes 17,900 m. (58,000 ft.) of relief sewer in the 510 ha. (1260 acre) combined sewer area. The plan also provides nearly 20,000 m^3 (700,000 ft^3) of surface storage in street rights-of-way. This storage will be implemented through construction of 325 street berms and accompanying inlet restrictors designed to retain water on the streets. Stormwater catch basins in the area are restricted such that the inflow to the combined sewer system cannot exceed its full flow capacity.

Runoff control in conjunction with street storage is being provided in Wilmette for two reasons. First, the storage reduces peak discharges in the system thus reducing the need for additional relief sewer capacity. In Wilmette, between 15 and 20 million dollars in relief sewer construction costs will be saved due to the use of street storage. The second reason is to reduce the peak discharge from Wilmette to facilities of the Metropolitan Water Reclamation District of Greater Chicago (MWRDGC). The MWRDGC is responsible for transportation, treatment and eventual discharge of sewerage flows in Wilmette and 185 other communities. Approval from MWRDGC is required for all system modifications and new discharge connections. Projects consisting only of relief sewers are generally

----

[1]Water Resources Engineers, RUST Environment & Infrastructure, Inc., 1501 Woodfield Rd, Schaumburg, IL 60173

unacceptable. Hydraulic simulations were used to demonstrate a reduction of 28 percent in the peak discharge from the Village. A maximum total peak release rate was negotiated with MWRDGC which serves as a constraint on the system design.

System Analysis

Analysis of the street storage facilities was a major challenge in the design of the overall runoff control system. Cost effectiveness depended on optimal operation of the storage facilities which were constrained by a strict set of performance criteria. A computer modeling strategy was formulated for the proposed system using the USEPA Stormwater Management Model (SWMM) (Huber and Dickinson, 1992). This strategy consisted of three linked components:

- hydrologic model using the RUNOFF module of SWMM
- street storage model using the EXTRAN module
- combined sewer system model using the EXTRAN module

Three innovations were required in the development of the street storage model. First, the EXTRAN code was modified to accept input of stage-storage relations for storage junctions and to generate descriptive storage junction output summaries of storage junction levels and outflows. Second, it was determined that the standard EXTRAN orifice formulation did not adequately represent field conditions for flow through a catch basin restrictor. An alternate equivalent pipe formulation was developed for use in the EXTRAN model. Third, the EXTRAN weir code was used to model flow overtopping of the berms into adjacent street storage sites.

The configuration of a typical street storage facility is illustrated in Figure 1. Typical maximum allowable ponding depths are 30 cm (12 in.) at the gutter flow line and 15 cm (6 in.) at the crown. The flow is retained in the street by berms or by the longitudinal slope of the roadway. The lateral extent of the street ponding is restricted to the street right-of-way which can, in turn, be the limiting factor in setting the design depth. The release from each ponding area is regulated by flow restrictors placed in catch basins located at the low point. These can be tied together so that only one restrictor per storage area is necessary.

Street Storage

Standard input of variable storage junction data in EXTRAN is accomplished by specifying the relationship between depth and surface area. While depth versus

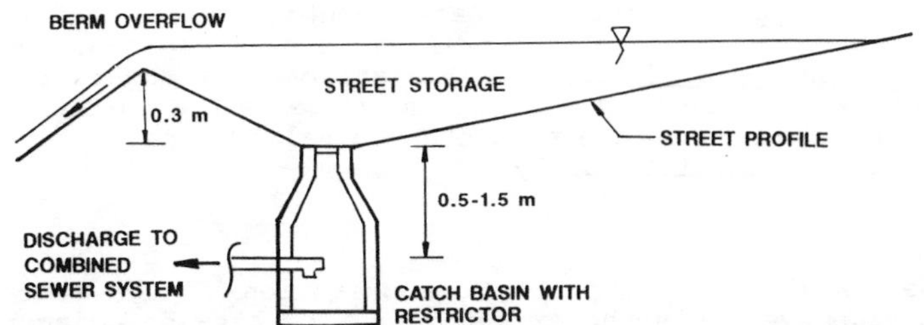

Figure 1. Profile of a street ponding area along the gutter flow line (not to scale).

surface area ties in nicely with the EXTRAN solution technique, it is difficult to obtain directly for street storage sites. Available street storage volumes are computed from street cross-sections using the end area method. Software is available to compute street storage at depth intervals of 0.1 feet. The EXTRAN code was modified to accept stage versus storage volume input and to print an enhanced summary of storage junction results.

The summary provides the maximum depth, storage and discharge for each storage site and identifies whether an overflow from the junction occurred.

Catch Basin Restrictors

The configuration of a catch basin restrictor differs greatly from the situation assumed when an EXTRAN orifice is employed. In EXTRAN, orifice flow is represented using an equivalent pipe. Inherent assumptions in the equivalent pipe method are that the depth of water in the upstream junction exceeds the orifice diameter and that the upstream and downstream junction elevations are approximately the same. The catch basin restrictors installed in Wilmette are 0.5 to 1.5 meters below street grade and the receiving sewer system is one to three meters lower than the restrictors. EXTRAN offers no dependable method to place an orifice invert far below the storage junction invert.

Laboratory tests by Spring (1984) demonstrated that a PVC tee-restrictor in a catch basin behaves as a classical orifice for a wide range of heads. The flow for a particular orifice area and head is given by formula $Q=C_d a(2gh)^{1/2}$, where g is acceleration due to gravity and C_d is a discharge coefficient found to be 0.6 to 0.65. In the context of the EXTRAN model, this formula is much better suited to the EXTRAN weir code rather than an equivalent pipe representation. The EXTRAN weir code was modified to accept a new type of weir representing a catch basin restrictor. Data inputs are the orifice diameter, the depth of the orifice below the ground, and the discharge coefficient. This approach is superior as long as there is no downstream submergence. Even then it is still more accurate than the equivalent pipe, but not as stable computationally.

Berm Overflow

The third issue affecting the operation of the street storage system is the exchange of flow between adjacent ponding areas due to berm overflow. Optimization of the use of street storage, available combined sewer capacity and relief sewer capacity requires detailed accounting of flow and volume. Berm overflow is employed to fully utilize available storage and to convey stormwater to relief sewer locations from individual ponding areas, which may not have sufficient storage volume. Simulation of berm overflow has been implemented in the EXTRAN model using the standard transverse weir input. Thus, the EXTRAN model of the runoff control system consists of an above-ground network of storage junctions and berm overflows connected to the underground combined sewer and relief sewer system.

Conclusion

The Wilmette runoff control program is being implemented in five phases. SWMM simulations have been used for feasibility studies, system design and post-construction verification. In the feasibility analysis, storage sites were grouped together using a single storage junction to represent ten or more ponding areas. During design and construction, the planning level models were refined to support and verify the design of street storage locations, relief sewer configurations, relief sewer connections to existing combined sewers, and restrictor sizes. These models stretched the traditional data limits of EXTRAN. Current models representing the two completed phases feature over 250 pipes and 350 junctions including more than 100 storage junctions.

References

Huber, W.C. and R. Dickinson, "Stormwater Management Model Version 4 User's Manual," 2nd Printing, October, 1992.

Spring, B.H., "Hanging Trap and Scepter Orifice Tests", Valparaiso University, December, 1983.

Stormwater Retrofitting Techniques For Water Quality Benefits

Gordon England, P.E.

With the advent of NPDES Stormwater Permits and increased environmental awareness, many municipalities are confronted with the daunting task of retrofitting existing developed areas which provide no water quality treatment for the stormwater runoff. For most areas of the Country, cleaning stormwater is a new concept. This paper will cover traditional as well as new techniques that can be used by engineers to design retrofit projects.

Many studies have been performed documenting the pollutant loadings in stormwater runoff. The sources of this pollution are many including greases, oils and heavy metals from cars, yard clippings, fertilizers, pesticides, pet wastes from yards, dirt and trash from streets and parking lots, dirt and erosion from construction sites and improper storage and use of toxic material at industrial sites.

Florida has instituted stormwater treatment methods such as retention ponds for approximately the last 15 years. These regulations apply to new development where ample land is available for treatment methods such as retention ponds. These ponds are designed to provide 80%-90% pollutant removal.

The first step in pond design is an analysis of rainfall records for the area. In Florida we get about 50 inches of rainfall a year and 90% of the rainfall events are 1" or less. In other words, we get a great number of small storms which tend to occur almost daily during the summer. Therefore, the standard for treatment is to create pond volumes which will hold the runoff from 1" of rainfall, although many municipalities require ponds to hold the volume equivalent to 1" of runoff over the drainage area as a higher standard. Systems discharging to protected waters such as shell fish areas or drinking water supplies are required to retain 1.5 inches of runoff or more.

Depending on soil and groundwater conditions, various designs can be used. The most efficient design is an offline dry retention pond. The pond diverts the first flush of rainfall, usually 1" or more of runoff, which has the majority of the pollutants, into a dry retention pond. After the pond fills, the remaining rainfall bypasses the pond and flows to a second detention pond if flood control is desired. The water in the dry pond percolates into the ground allowing the pollutants to filter out.

Retention ponds trap a certain volume of water which does not leave the site. These ponds are generally used for stormwater treatment. Detention ponds temporarily store runoff and then slowly bleed that volume down via an orifice or weir over several days. Detention ponds can be used for stormwater treatment, for flood attenuation or both.

If the groundwater is low but the soils do not percolate well, an offline dry detention pond may be used which slowly bleeds the detention volume down over several days allowing pollutants to settle out. If the groundwater is not at least 2 feet below the pond, there may end up being cattail problems since these ponds stay wet so long due to the time required for the groundwater mound to dissipate through lateral percolation.

If the grades do not allow an offline system, an online dry pond can be used where a designated volume will be stored below a weir and excess water will flow over the weir at a designed rate. Online ponds are not as efficient as offline ponds since pollutants can be intermingled with excess flows leaving the pond.

Often a dry pond is not feasible due to soil or groundwater conditions. In those cases wet ponds are used which have a permanent pool volume. These ponds can be offline or online. They are designed to detain a certain volume and slowly bleed that volume down via an orifice or weir over several days. This allows the suspended pollutants to settle out and biological reaction to occur to remove dissolved nutrients.

The challenge in retrofitting built out areas is that usually there is little or no land available for ponds. There are various options in these areas depending on available revenues and the desired treatment level. Since these methods are newly developed, there is little data on removal efficiencies and maintenance intervals play a crucial role in the treatment effectiveness.

Underground vaults can be used for detention ponds which allow normal usage over their tops. Large access ways must be provided for maintenance. These can be nearly as efficient as ponds, except for biological treatment, but are extremely expensive and difficult to maintain. They are rarely used except in commercial or downtown areas where land values are extremely high.

Exfiltration pipes may be used where porous soil and low groundwater conditions occur. These are perforated pipes in gravel beds which percolate a desired retention volume into the ground. They are highly efficient when used as offline systems but are not very cost effective. These systems have higher maintenance and limited lifespans, just like septic tank laterals, since the soil pores will eventually clog. For this reason, they should not be used under pavement. In order to extend their lives, sediment sumps should be used at all inlets to keep dirt out of the pipes and skimmers should be used at pipe openings to keep oils and trash out of the pipes. Since these systems are not designed for conveyance capabilities, skimmers over pipe openings do not impede significant flows.

Retention volumes can be created in roadside swales by constructing concrete ditch blocks [weirs] in the swales. This is feasible when there is a low groundwater table, permiable soils, and ditch capacity to raise the water surface during storms.

Baffle boxes are another technique used to trap sediments, floating trash and yard clippings. They are basically large septic tanks constructed in-line with existing pipes and therefore require no new right-of-way. Most heavy metals attach to suspended solids so they are also trapped in the box.

Baffle boxes are our most commonly used method for retrofitting. The weirs in the box are set at pipe invert elevations so as to not impede flows in high flow storms. The weirs create multiple chambers which slow the water down and allow sediments to fall out and be trapped in the chambers. Trash screens catch floating debris but are hinged at the top to swing out of the way in high flow situations. Manways are built over each chamber which allow for access for cleaning with a vacuum truck. The manways need to be within 15' of pavement for a heavy vacuum truck to reach them. The removed sludge is not considered hazardous and may be dumped anywhere vacuum trucks normally dump. Accumulation rates vary with each location and depend on many factors such as rain intervals and intensities, yard mowing practices, drainage basin size, abundance of trees which drop leaves in streets, land use, soil characteristics, landscape practices, flow velocities, etc. Some baffle boxes will fill every few weeks, collecting over 60,000 pounds of dirt a month, while others fill only twice a year.

It is important to have a monthly inspection and cleaning program for baffle boxes as well as other devices. They should be cleaned before the chambers become full or a large storm will resuspend some of the dirt and floating trash and carry it out to the receiving water body. Monthly cleanouts are recommended for this reason, as well as to remove stagnant water before it turns anaerobic with odor problems.

While engineers like to have simple numbers for removal efficiencies for treatment facilities, the complexity of the situation makes this difficult. The variables of loading rates, cleaning intervals, rainfall intensities and pipe velocities have so far defied simple analysis. In addition, the main pollutants trapped, such as large grit rolling along the bottom of the pipe, yard clippings and floating trash, are invisible to most testing techniques.

We are currently undergoing research to determine optimum box sizing and shapes as well as quantifying the pollutant load of grass clippings and how they release their nutrient load in water with time.

The costs of a baffle box vary depending on pipe size, utility relocation and pavement repair. Often an existing inlet is replaced with a baffle box but the box may also be placed behind the inlet to preserve the street and/or utilities. The cost of most installations will be between $15,000 and $30,000.

For locations with small flow rates, a fiberglass inlet weir has been developed which fits inside of existing curb inlets and manholes. This weir has a trash screen allowing it to trap dirt, yard clippings and floating trash. Flows under 4 cfs will flow through the weir while higher flows will flow over the weir to minimize upstream flooding. The fiberglass weirs are purposely designed to leak water through cracks so as to slowly release trapped water but not the sediments. Original designs used a concrete weir but the orifices kept clogging leading to septic, smelly conditions after a few weeks of dry weather. This effluent then washed out in high flows. Also grass clippings washed over the top every time sprinklers drained into the street and through the inlet. The fiberglass weir solves these problems.

These installations cost only $500 - $600 and are nondestructive to existing systems. The tradeoff is that they need frequent maintenance or the trapped pollutants will be washed out in high flows. Cleaning out can be performed with a vacuum truck, a small pump and storage unit on a truck, or by hand if the manholes are dry. Once again efficiency depends on rain intensities and cleanout intervals.

For grated inlets another device called a grate inlet trash box has been developed which drops into the existing inlet and traps dirt, trash and oils. The box has filter cloth covered drain holes in the bottom allowing it to dry out between storms. The specially designed lid acts as a skimmer keeping floating trash in the box as well as holding an oil absorbent pad which removes oil in the runoff. The box is designed as an orifice having flow capacities similar to the grated inlet. During high flows, the water will flow out bypass holes and still retain trash. The main concern in determining the box size is to keep the bottom of the box above outflow pipes. No tools or materials are needed, making for a two minute installation at a cost of approximately $500. These boxes come in a wide variety of common sizes.

Another inexpensive treatment method is to construct an inlet with no outflow pipe to be used as a sediment trap in a gutter line. It will fill up with water and dirt and then be bypassed by gutter flows. When the inlet is above the water table a weep hole can be used to keep the inlet dry.

At outfalls or in canals, trash screens can be constructed out of fencing to trap floating debris. As long as velocities are low, head losses should be minimal. The fence bottom should be above the ditch bottom for better hydraulics. Vinyl coated fences are recommended to minimize corrosion.

Erosion in open channels is a major source of sediment deposition. This erosion can be reduced by stabilizing the slopes or by piping the channels. Sediment sumps can be placed in channels to collect dirt for regular removal.

In summary, retrofitting existing storm drain systems is a new field of civil engineering which will present many challenges. Traditional designs and attitudes will need to change with these new demands on our engineering skills. Environmental mandates will force us to work more closely with governmental entities to create cost effective stormwater treatment methods. A willingness to experiment will lead to the development of many new treatment methods. Along with the construction of these systems must come a commitment to maintain them or they will be useless.

Blue-Green Revisited: Integrating Stormwater Management Into the Urban Planning and Development Process

William N. Lane,[1] M. ASCE

Abstract

"Blue-green" stormwater management, based on open and natural drainage systems integrated into multipurpose open space/environmental corridors or greenways, is the most promising approach in newly developing or urbanizing areas. In complex metropolitan areas, fragmented governmental jurisdiction, and the incremental site-by-site nature of the urban development process, are the main obstacles to achieving environmentally sound and cost-effective urban stormwater management on a watershed or metropolitan scale. A planning and coordinated implementation process can be developed to overcome these obstacles.

Introduction

Watershed-based water quality planning and the NPDES stormwater permit program are focusing greater attention on managing urban stormwater runoff for water quality as well as quantity. Opportunities for retrofitting stormwater management practices into existing development are often constrained and expensive. The most promising approach to achieving environmentally sound and cost-effective stormwater management is to integrate stormwater management as a basic consideration in the urban planning and development process.

The "blue-green" approach to stormwater management, based on natural drainage systems integrated into multipurpose open space/environmental corridors or greenways, is a popular and practical approach in newly developing or urbanizing areas. The trend toward blue-green stormwater management corresponds with an increasing emphasis on open space corridor and greenway systems in regional and metro-

[1]Deputy Director, Dane County Regional Planning Commission, Madison, WI

politan land use planning. The blue-green corridors approach addresses several functions: (1) protecting water resources, drainage and hydrologic functions; (2) pollution control; (3) protecting public health, safety and property; (4) outdoor recreation and education; (5) wildlife habitat; and (6) enhancing scenic beauty and shaping urban form.

If metropolitan/regional open space corridor networks incorporating blue-green drainage systems are such a good idea (having been advocated for at least thirty years), why are there so few successful examples? The answer lies in the fragmented planning and decisionmaking authority in most metropolitan areas, and in the incremental, site-by-site nature of the urban development and implementation process.

To be successful, a planning and coordination structure needs to be developed for an urban watershed (or even better, for the entire urban area) that: (a) provides the proper geographic scale and institutional level for coordinated planning that involves all responsible governmental units and agencies; (b) takes a comprehensive and multipurpose approach to corridor/stormwater planning and management, involving staff and agencies responsible for all major functional areas; and (c) focuses on developing continuing coordinated implementation procedures and programs to translate plans into action, primarily by involving implementation agencies and staff in the planning, by incorporating stormwater plans in community land use and comprehensive plans, and by integrating stormwater management into day-to-day urban development control and infrastructure management programs of local communities.

Geographic Scope and Institutions for Integrated Stormwater Planning

Stormwater management is best approached on a watershed basis; however, jurisdictions of urban governments and management agencies rarely correspond to watershed boundaries. The degree of jurisdictional fragmentation and overlap, and the number of responsible agencies involved, pose formidable obstacles to achieving necessary coordination and agreement in complex urban and metropolitan areas.

In a few instances, a suitable institution for watershed-wide or areawide urban stormwater planning and management may already exist--a large city or urban county government, a regional drainage or flood control district, a regional service or utility authority or water resources agency. (Using, or modifying, existing agencies and institutions will almost always be easier and quicker than creating new ones.)

Where, as is usually the case, a single suitable institution doesn't exist, it is often possible to develop an areawide or watershed framework for coordinated stormwater planning and implementation. Most of the larger, more complex urban areas have regional or metropolitan comprehensive planning agencies or councils of government responsible for areawide planning and intergovernmental coordination. These agencies can be useful vehicles for structuring areawide or watershed stormwater planning and intergovernmental coordination and implementation.

It is also possible to develop an areawide or watershed approach based on either formal or ad hoc intergovernmental agreements or arrangements between the responsible governments or agencies. These special arrangements often lose effectiveness in implementation, unless they are formally structured as a continuing institution.

Comprehensive Multipurpose Approach

The "blue-green" approach to stormwater planning and management--integrating drainageways, streams and stormwater facilities into multipurpose environmental corridors or greenways--requires a comprehensive perspective. Conflicts and competing needs can only be balanced and resolved in a planning and coordinating framework which reflects and represents all the important interests.

First, the framework must include the agencies and staff responsible for the major functions--stormwater management, environmental protection, land use/urban development control, and open space/recreation.

Second, the planning must include both source control and drainage system management. Hydrologic and environmental changes resulting from urban development are so drastic and irreversible that either approach alone will be inadequate to prevent degradation of urban water resources.

Effective source control often includes regulations limiting increases in erosion and runoff from individual construction or development sites. Source control ordinances have been criticized as being ineffective, expensive, cumbersome or even counterproductive. Since source control regulations usually only limit **increases** in impacts (erosion and runoff) resulting from land development and disturbance, they cannot be expected to fully correct either hydrologic or water quality impacts, or to have much effect on major floods. Source control also includes off-site or public stormwater management facilities--detention/retention basins, infiltration practices, vegetated drainage swales and buffer strips along channels--that need to be planned and managed on a watershed basis. Realistically, however, source control must include construction and development site erosion and runoff controls to be sufficiently effective.

Translating Plans Into Action

Most urban development, and development and expansion of stormwater drainage systems, occurs on a piecemeal, site-by-site incremental basis. The value of a comprehensive watershed plan for stormwater management is to provide a framework and context for individual day-to-day public and private decisions and actions, so that a coherent and functional overall stormwater management system results. Without a plan, the result is often a fragmented hodgepodge of conflicting and uncoordinated reactions and solutions to problems created by earlier decisions (or mistakes).

To perform its function, the plan must be directed at implementation processes and programs. First, the planning process itself must include those actually involved in plan implementation decisions and processes--particularly agencies and staff responsible for land use regulation, development site reviews, stormwater management, environmental regulation, and recreation and open space programs. Their involvement accomplishes two ends: (1) a more practical and implementable plan; and (2) a greater likelihood that the plan's concepts and proposals will be carried over to basic and routine operational decisions and programs.

Secondly, the planning process should specifically outline and design plan implementation procedures and programs. The plan should include policies and criteria or guidelines for day-to-day administrative procedures, such as site review or environmental permit issuance; financing and budgeting proposals for inclusion in annual budgets and capital improvement programs; a detailed program for ongoing maintenance and operations needs; and a structure or process for continuing intergovernmental or interdepartmental coordination of plan implementation.

Despite the obstacles, coordinated "blue-green" stormwater management on a watershed or areawide basis is possible even in complex urban areas, if proper attention is paid to the institutional structure for planning and implementation.

Development of a SWPPP for a Multi-Component Plant

Erik P. Drake[1], A.M. ASCE, Steven H. Wolf[1], M. ASCE,
Donald P. Galya[1], M. ASCE, and Roy E. Piatelli[2]

Abstract

As part of their National Pollutant Discharge Elimination System (NPDES) Permits, facilities must develop and implement Storm Water Pollution Prevention Plans (SWPPPs). One such plan, designed to lessen adverse environmental impacts by reducing pollution at its source, was developed for General Electric River Works in Lynn, Massachusetts, a large multi-component plant. To minimize pollutants in storm water discharges, the facility specific plan was designed to identify potential sources of these pollutants and to implement physical controls and management practices.

Introduction

Studies have shown that storm water runoff from industrial and urbanized areas is a major source of water pollution (U.S. EPA, 1992). Under the Clean Water Act, the Environmental Protection Agency has developed the National Pollutant Discharge Elimination System (NPDES) to manage discharges to the waterways of the United States and its territories. Recent revisions to the NPDES Permit for Storm Water Discharges Associated with Industrial Activity, now require the development and implementation of a Storm Water Pollution Prevention Plan (SWPPP). This plan is designed to lessen adverse environmental impacts by reducing storm water pollutants at their sources. The plan assesses specific facility activities and identifies potential sources of pollutants which have the possibility to affect the quality of storm water discharges. Based on the assessment of pollutant sources, appropriate Best Management Practices (BMPs) are implemented in order to prevent or reduce pollution in discharged storm water runoff. BMPs are practices and measures designed to minimize the risk of discharging pollutants to the environment.

[1] ENSR Consulting and Engineering, 35 Nagog Park, Acton, MA 01720.
[2] General Electric River Works, 1000 Western Avenue, Lynn, MA 01910.

A SWPPP was prepared for General Electric (GE) River Works, located in Lynn, Massachusetts approximately seven miles northeast of the City of Boston. The River Works is a multi-component plant that develops, manufactures, assembles, and tests aircraft engines and marine drive systems for both military and commercial applications. The 46 building complex covers over 200 acres and is over 100 years old. The facility is divided into several business units, each with its own operations, goals, and restraints. Approximately 4000 feet of the facility borders the Saugus River or an adjacent tidal marsh. Storm water drainage from the facility is collected by a network of catch basins and eventually discharged to the Saugus River. The River Works facility is currently permitted for 10 storm water discharges.

Methods

In order to evaluate potential pollutant sources at GE River Works, a comprehensive site survey was performed over a two month period. Building exteriors and grounds for the entire GE River Works facility were inspected for potential pollutant sources. The challenges to such a survey included the size and age of the facility along with its multi-component nature. The plants Environmental, Health and Safety Department facilitated this evaluation through existing pollution prevention programs, an informed staff, effective plant communication, and thorough record-keeping and documentation.

The inspections included identification of the general level of housekeeping, material and waste storage areas, loading/unloading areas and doors, open-ended pipes extending from buildings, catch basin locations, and any indications of exposed materials (such as discoloration or residues around pipes or vents). A site summary description and list of questions were then created for each building/area. Based on this information a second reconnaissance was performed with plant personnel familiar with the specific building(s) in order to fill in any information gaps.

Based on the size of the facility, it was necessary to establish a ranking system to assign potential risk levels to significant materials identified at the facility during the reconnaissances. The relative ranking of a source was determined by applying the following criteria:

- the presence of significant materials exposed to storm water;
- the quantity of an exposed material;
- the potential of exposure through spills, ruptures or leaks;
- the mobility of the materials;
- the distance to the storm water collection system (i.e., possible containment and/or clean up of a released material prior to its entrance into the storm water system);
- the existence of structural controls (e.g., berms and covering);

- the applicability of existing management practices to limit the potential for a spill to occur or to contain a spill should it occur; and
- evidence of historical spills (e.g., residues, discoloration, structural damage to pavement or catch basins, or past spill records).

The amount of material required to be considered a significant quantity is related to its potential to contribute to storm water pollution and is dependant on the type of material. A limited quantity of material would not cause a measurable increase in concentration of a given pollutant in the storm water but could contribute to such an increase if additional sources existed. Based on these definitions and the above criteria, the following five ranking levels were developed:

Negligible
- pollutant is not present in significant quantities; or
- significant quantities of pollutant are present but are not exposed to storm water and no potential for leaks or spills exists.

Low
- significant quantities of pollutant are only exposed to storm water by leaks or spills and structural controls are adequate for containment; or
- limited quantities of pollutant could be exposed to storm water (either by spills or handling practices), and existing structural controls are ineffective in preventing contamination of the storm water.

Low+
- limited quantities of pollutant are continually exposed to storm water, and existing structural controls are ineffective in preventing contamination of the storm water.

Medium
- significant quantities of pollutant are continually exposed to storm water, but existing structural controls are adequate to prevent contamination of the storm water; or
- significant quantities of pollutant are only exposed to storm water through leaks or spills, and existing structural controls are ineffective in preventing contamination of the storm water.

High
- significant quantities of pollutant are exposed to storm water, and existing structural controls are inadequate to prevent contamination of storm water; or
- significant quantities of pollutant are not currently exposed to storm water, but storage practices and spill data indicate that the potential for such exposure is high and existing structural controls are inadequate.

Results

Based on the results of the surveys, a facility specific plan was designed to: 1) identify potential sources of pollutants to storm water; and 2) implement physical controls and management practices to minimize pollutants in storm water discharges. Potential pollutant sources were grouped in terms of their risks. Action items were proposed for all potential risks which were ranked as "High", "Medium", or "Low+". Sources that received a "High" potential risk received immediate attention. Sources that received a "Medium" potential risk were addressed within a six month implementation period. Sources that received a "Low+" potential risk are being addressed as regularly scheduled maintenance and improvements are performed.

As part of the implementation of the SWPPP, BMPs tailored to prevent storm water pollution were developed for the GE River Works Facility. These management practices are designed to minimize the potential for spills or other material exposure that could adversely impact storm water. Because of the multitude of different operations performed at the facility, broad based BMPs needed to be developed which were both practical and workable.

Regularly scheduled inspection and compliance programs were also established to reevaluate potential risks and to assure continued preventative practices are maintained. All previously identified potential risks are reexamined as part of these evaluations. Sources which were determined to have "High", "Medium", or "Low+" potential risks are reassessed to determine if the risk has been eliminated or remains. Sources that previously received a "Low" or "Negligible" potential risk will also be reviewed to determine if their potential risks have changed. During each inspection any new potential risks will be identified, and the necessity for emending any BMPs will be evaluated.

Conclusion

SWPPPs are implemented to minimize the presence of contaminants in storm water discharges from a facility. The development and implementation of a SWPPP for a large multi-component facility is a complicated task. This task was made manageable at GE River Works by making use of facility resources, evaluating the facility on a building by building basis, and ranking the different risks relative to their potential for impact to the local waters. This methodology allowed GE River Works to formulate a comprehensive SWPPP which effectively minimizes the risk of storm water pollution from the facility.

References

U.S. EPA (1992). Storm Water Management for Industrial Activities: Developing Pollution Prevention Plans and Best Management Practices. EPA/832/R-92-006, September, 1992.

Control and Treatment of Coal Pile Stormwater Runoff at the University of Massachusetts, Amherst

Kenneth W. Carlson, David H. Knowlton, and Leonard H. Gillan[1]

Abstract

The goal of this project is to abate both groundwater and surface water degradation resulting from acid drainage and leachate generated from the coal handling and storage facility at the Amherst campus of the University of Massachusetts. Stormwater runoff and leachate from the coal pile have been degrading the local aquifer and Taylor Brook, a state classified anti-degradation stream, for twenty years. Taylor Brook is part of the Connecticut River watershed, which is a major water supply source for the region.

Introduction

The University of Massachusetts at Amherst uses coal to operate it's central steam generation plant. The steam plant provides heat, steam, and hot water and also cogenerates a portion of the electricity for the campus facilities. A coal storage and handling facility was constructed in 1965 to ensure a continuous supply of coal to the steam plant.

The facility is located at the end of Eastman Lane, east of Tillson Farm and East Pleasant Street in Amherst, Massachusetts. The pile is triangular in shape and is bounded on the east by the Central Vermont Railroad tracks, on the southwest by a spur of the track, and by a densely wooded area zoned as residential and mostly undeveloped, to the north. The end of Eastman Lane forms the western corner of the pile. The site slopes from west to east, with the highest spot at the end of Eastman Lane. East of the railroad tracks, the terrain drops off at a very steep slope (6 meter drop in 36 meters) towards a wetland and Taylor Brook.

Taylor Brook is classified as a Class B warm-water fishery. It originates in wetlands and flows in a north to south direction, passing within 90 meters of the coal pile site as it makes its way to its confluence with Adams Brook. Adams Brook then

[1]Associate, Senior Project Engineer, and Associate Project Engineer, Weston & Sampson Engineers Inc., 5 Centennial Drive, Peabody, MA 01960-7985

flows to the Fort River, which ultimately flows to the Connecticut River. Prior to 1982, a number of independent studies were undertaken to determine why water quality in Taylor Brook had deteriorated to the point that local farm animals allegedly refused to drink from Taylor Brook.

One study concluded that the deterioration of water quality in Taylor Brook was due to leachate released from the coal pile. Leachate was reportedly discharged to Taylor Brook as a non-point source, through the wetlands area between the coal pile and the brook. The analysis of samples from monitoring wells, coal pile runoff, and surface water from Taylor Brook showed levels of pH, total dissolved solids (TDS), sulphate, aluminum, iron, and manganese beyond the limits set by the EPA Drinking Water Standards. In addition, levels of acidity, specific conductance, hardness and chemical oxygen demand (COD) were above values typically seen in groundwater.

In the late 1980's, the University began purchasing pre-washed coal and applying a surfactant, as necessary, to the stockpiled coal to limit the amount of coal fines at the site. In addition, the University began collecting the coal pile runoff and treating it prior to discharging to the town of Amherst sewer system. The treatment system consisted of a headwall, soda ash neutralization chamber, pump station to pump drainage into a holding lagoon, and a pump station to the town's sewer system.

In 1984, the owners of land adjacent to the site brought legal action against the University and the state Division of Capital Planning and Operations (DCPO), claiming that runoff and coal dust from the pile had contaminated land they intended to develop. This legal action resulted in a monetary settlement to the land owners and a renewed interest by DCPO and the University to abate groundwater and surface water contamination. A consultant was retained to recommend remediation measures for the site, based on the results of the earlier sampling program. The recommendations were the following 1) construct an impervious clay/soil liner under the coal pile to collect leachate and prevent infiltration into groundwater, 2) pump and treat groundwater underneath the pile to remove the contaminated plume beneath and downgradient of the site, 3) construct a sedimentation/equalization basin to settle out coal fines and sediment as well as contain runoff from the pile, and 4) install a package treatment plant to treat runoff before discharge to Taylor Brook. Before final construction documents were completed, the consultant went out of business and Weston & Sampson Engineers, Inc. (Weston & Sampson) was retained to finish the project.

Weston & Sampson's Project Approach

Weston & Sampson's approach to the project was to 1) verify the previous groundwater and surface water treatment system design by conducting hydrogeological and treatability studies, 2) assist DCPO and the University with permitting issues and 3) evaluate and recommend improvements to the previous design.

Weston & Sampson performed a hydrogeological study to support the design of a pump-and-treat groundwater treatment system as recommended by the previous consultant. Four groundwater monitoring wells were installed as part of the study. With the exception of pH, iron, manganese and aluminum, the water quality found in the wells met EPA drinking water standards. Surface water samples were taken from Taylor Brook during dry and wet weather periods at locations upstream and downstream of the coal pile. The quality of surface water was the same at both locations during both events. Stormwater runoff samples were collected from the influent pipe to the equalization chamber of the stormwater treatment system in the northeast corner of the site. Total suspended solids concentrations were above the NPDES permit limits and manganese was detected above the EPA secondary drinking water standards.

The hydrogeological report concluded that the adverse effect of the coal pile on water quality has decreased significantly over the past decade due to interim pollution control measures at the site. Purchasing pre-washed low sulfur coal, spraying the pile with a surfactant, and pumping the treated runoff to the town sewer system have all played a role in the improvement of the water quality. The surface water of Taylor Brook does not appear to be significantly impacted and groundwater quality degradation appears to be limited in degree and extent. Accordingly, the current plan to install a liner and treat stormwater runoff appears to be adequate and groundwater treatment is not required at this time.

The results of the hydrogeological study were used to prepare applications for the various permits required for this project. A Notice of Intent was prepared and submitted to the Amherst Conservation Commission and DEP division of Wetlands and Waterways. An Order of Conditions was obtained that included the standard erosion protection requirements (hay bales and silt fences) and specific requirements relating to effluent, ground and surface water monitoring. These monitoring requirements were later included as part of NPDES discharge permit.

The NPDES permit application was prepared in accordance with the Massachusetts and Federal Clean Water Acts. The permit was issued jointly by USEPA and DEP, and included effluent monitoring requirements. The permit also required implementation of ground and surface water monitoring. Surface water monitoring conditions included biological monitoring, as well as monitoring of inorganic parameters. As part of the NPDES permit application, a variance was requested and received from the anti-degradation provisions of the Massachusetts Water Quality-Standard 314 CMR for Taylor Brook. Limited degradation may be allowed as a variance from these regulations if the discharge will not cause any significant lowering of water quality. In addition, completed plans and specifications were submitted for DEP's approval as part of NPDES process. Compliance with MEPA process was assured by submitting an ENF. As a result of ENF review, the Secretary of EOEA determined that an EIR is not required.

Two laboratory characterization/treatability studies were performed on composite samples of coal pile leachate. The first study, completed by a package treatment plant manufacturer, concluded that clarification with alum will meet all

EPA storm water discharge NPDES General Permit requirements and filtration, although not necessary, can be used if a safety factor is required. Lime precipitation was compared to alum coagulation, but proved to be less effective. A second treatability study, performed by a subconsultant confirmed the results of the first study.

The liner to be installed under the coal pile will be an unreinforced polypropylene material. A bentonite liner was not recommended by the manufacturer for this application after careful review of leachate parameters. The liner will extend to the perimeter of the site to include the access roadways as well as the coal pile, and cover a total area of approximately 0.8 hectares. The synthetic liner was selected after considering chemical compatibility with leachate, heat resistance, ease and speed of installation and cost.

The storage volume of the basin was based on containing the peak design (100-year, 24-hour) storm runoff volume of 1440 cubic meters (m^3) while treating the incoming runoff at a rate of 0.2 cubic meter per minute. Therefore, the required storage volume was reduced by 288 m^3. Due to the need to provide sufficient freeboard in the basin, the selected storage volume is 1,272 m^3. Should the treatment process be interrupted for any reason, the storage/equalization basin will still provide sufficient capacity to contain the peak design storm runoff. The basin will be constructed of poured-in-place reinforced concrete protected by a corrosion resistant coating. The storage/equalization basin will have provisions for overflow when the maximum capacity is reached. For safety reasons, the perimeter of the basin will be surrounded with a handrail and chain link fence.

Collected leachate and runoff will flow by gravity from a sump in the basin to the pump vault. The submersible pump station will provide flow to the raw water controller, inside the treatment plant building, at the mixing tank. The level controls will be set so that the pump station and treatment plant will operate continuously for a minimum of 9 hours before shuttling off automatically.

The designed treatment process is based upon the Infilco Degremont, Inc. (IDI) Accelator Treatment Plant followed by the IDI Accelapak sand filter. The final design is based on meeting the discharge limits set forth by the NPDES discharge permit. A mixing tank will be utilized for initial pH adjustment of the influent with caustic soda. The influent will be treated with Alum (Aluminum Sulfate) and an Anionic Flocculant (Polymer). The treatment tank will provide contact clarification of the influent and produce sludge consisting of chemical floc materials. Sludge will be automatically discharged into a sump, while the clarified effluent will be directed to the gravity filter. Sludge will be pumped from the sump to a holding tank located outside the building. A hose will also be provided for use in applying the sludge back onto the coal pile. The IDI Accelapak Filter will provide final polishing of the effluent prior to disposal to Taylor Brook. The filter will be equipped with all appurtenances required for fully automatic operation, including a backwash cycle. Automatic sampling of the effluent will also take place to conform with the NPDES permit.

Summary of NPDES Monitoring Needs

Harry Torno, M.ASCE[1], John Warwick, M.ASCE[2], and Ben Urbonas, M.ASCE[3]

Abstract

The Urban Water Resources Research Council of ASCE sponsored an Engineering Foundation Conference on Stormwater NPDES-Related Monitoring Needs, which was held August 7-12, 1994, in Crested Butte, Colorado. The Conference was prompted by concerns that, while we as a nation are spending millions of dollars on stormwater monitoring, we still are not able to predict the effects of stormwater discharges on the environment, particularly in the long term.

The Conference explored and available technology associated with stormwater monitoring, and brought together regulators, the regulated community and their consultants, and academicians, to determine what must be done to achieve the various goals set forth for stormwater in the Clean Water Act, and the municipal and industrial NPDES regulations that implement that act. This paper summarizes the key points raised in the Conference presentations, and related discussions.

Introduction

200 or so cities and counties in the U.S. have been required, by Part II of the U. S. Environmental Protection Agency's municipal NPDES separate stormwater discharge permit application, to spend large sums of money (estimated by some at about $60 million) to collect stormwater data. Many questions have been raised about both the validity of the requirement to collect these huge amounts of data, and of the data itself, particularly in terms of evaluating the impacts on the

---

[1] Consultant, 2880 Seapoint Drive, Victoria, British Columbia, Canada V8N 1S8

[2] Director, Graduate Hydrology Program, University of Nevada-Reno, 1000 Valley Road, Reno, NV 89512

[3] Chief, Master Planning & South Platte River Programs, Urban Drainage and Flood Control Dist., 2480 W. 26th Avenue, Suite 156B, Denver, CO 80211

environment and the usefulness of required corrective measures which might be required. For example, the Dallas-Fort Worth area spent approximately $2.9 million to satisfy Federal requirements, and industry in that area may have had to spend even more.

These costs seem to be due largely to a profound lack of understanding, by both the regulators and the regulated community (including their consultants), of the cost and complexity of obtaining meaningful stormwater quantity and quality data. The goal of the Conference, therefore, was to explore where monitoring technology stands at this time, and to determine separate stormwater systems monitoring needs.

Thoughts and Observations

Current water quality standards tell us little about the environmental health of the nation's receiving waters. End-of-pipe measurements of pollutant discharges, used in the implementation of these water quality standards, tell us nothing about the impact of those discharges on the environment, and may, in fact, be a significant detriment to the improvement of environmental quality. They drain resources (time, energy, money), and most stormwater professionals cannot support their cost effectiveness. Furthermore, the standards impair effective communication between regulators and the regulated community, and confuse issues like what research should be done.

Effective stormwater monitoring, whether physical, chemical or biological, is very difficult and expensive. Currently, NPDES regulations (and the individual regulators, whose interpretations of those regulations may vary considerably) require permittees to collect vast amounts of stormwater monitoring data. Much of the data collected appears to be inappropriate and of little value, either locally or nationally, either to the regulator or the permittee.

Individual regulators charged with writing regulations and/or permits, and with their implementation, frequently do not have enough training and experience, considering the technical and administrative complexities involved. Regulated communities and industries are concerned that permit writers will continue to use the results of current monitoring, whether meaningful or not, to justify additional monitoring and additional control requirements, and that the costs involved will be prohibitive.

Toxic concentrations of certain pollutants are sometimes found in stormwater, but there is a question about the degree of stress they cause. This is due to both the short-term exposure of the biota to these toxicant concentrations (there is, in fact, little data available on the duration of the exposure), and the form of the constituent that is measured. Most measurements are of total recoverable concentrations of a pollutant, whereas, in reality, the dissolved or ionic form is the real toxicant.

Very little meaningful monitoring is being directed toward measuring the actual effect of stormwater discharges on the short- or long-term health of the environment. Furthermore, there is no consensus on how this monitoring should best be accomplished.

Research continues into the use of biosensors to identify short-term instream toxic effects, biological inventory techniques to assess cumulative long-term impacts, and of the integrated application of event-based monitoring and computer simulation to estimate the short-term impacts of BMP implementation. To date, however, we have generally been unable to identify reliable indicators (which can substitute for the monitoring of a wide range of pollutants or toxic effects) of pollution or of environmental degradation. We have a long way to go in establishing these indicators, if indeed it can be done at all.

Cause and effect relationships in urban ecosystems are very difficult to determine, and experiments must be designed with each specific site in mind. Rigid national protocols will not help, and may in fact hinder sound local experimental design. It will nonetheless be necessary to have general scientific guidance at the national level so as not to sacrifice good science for regulatory expedience.

Some leading biologists, biochemists, ecologists, etc. have undertaken the study of the effects of urban stormwater discharges on the environment, particularly in England and the U.S. This involvement, which is critical if the programs alluded to here are to be successful, has already served to broaden the knowledge base, and to facilitate communication between disciplines which have, in the past, not had much interaction.

There is a clear need for large, nationally-funded investigations that can direct sufficient resources, provide sound experimental design, and provide adequate quality control, to, over time, permit environmental scientists and engineers to draw conclusions concerning environmental health, to evaluate alternatives for controlling or improving stormwater quality, and to determine how best to utilize our limited resources to achieve the greatest benefits.

Excellent progress has been made in developing techniques for locating illicit connections, which are a significant source of stormwater contamination. This progress is reported in the Conference Proceedings (Torno, 1994), and in several EPA documents. Despite this progress, however, further work is still necessary.

Source control of pollutants is a critical fundamental need. All projects which reported lead concentrations from urban runoff, for instance, clearly indicated that lead concentrations are lower now than they were, say, ten years ago, and that these reductions can be directly attributed to the use of lead-free gasoline. Diazinon, on the other hand, continues to be a ubiquitous urban environmental pollutant.

Virtually all of the studies reported at the Conference found diazinon in urban runoff from residential areas. Some even suggested that diazinon could be used as an indicator of urban runoff. Its importance as an environmental pollutant at the detected levels may deserve re-evaluation. Similarly, automobile brake pads may be a significant source of copper in stormwater. Limiting its use in brake pads may be most cost-effective if the control of copper, at the levels found in stormwater, is deemed necessary.

Funding constraints at the local and state level are impediments to reaching the goal of a better environment. EPA's estimates of funding requirements have, for the most part, been too low. Local administrators frequently have difficulty in justifying the cost of stormwater monitoring programs to their constituents because there is no perceived benefit - stormwater is not viewed as a serious environmental problem. At the same time, although less federal funds are available, local governments still must comply with NPDES regulations.

Despite the many ongoing projects which are evaluating best management practices (BMPs) for the enhancement of stormwater quality, there is a lack of consistency in the way monitoring data are being collected, the type and form of information being reported, and the methods for drawing conclusions (and reporting those conclusions) from the data. Most such evaluations appear to lack a design engineer's perspective, in that they do not permit the use of the information gathered to develop more effective and reliable designs. As a result, the data base available for BMP design is very limited.

There is a consensus that the meaningful monitoring of receiving waters is the most difficult technical problem currently facing us. If these monitoring programs are to provide us with the requisite information on the health of the environment, and on the impacts that stormwater discharges will have on that environment, they must be very carefully designed - both the programs and their physical, chemical and biological components - by environmental scientists in a wide variety of disciplines working in concert with engineers and planners.

The Conference Proceedings (Torno, 1994) contain several excellent examples of well-designed and managed programs (see papers by Livingston, and by Shaver and Maxted), so it is clear that such projects are possible. They typically require, however, the co-operation of various levels of government, and of industry, and must, in addition, involve the best experts in the engineering and scientific community.

Reference

Torno, Harry C. (Ed.), "Stormwater Related NPDES Monitoring Needs," Proceedings of an Engineering Foundation Conference, August 7-12, Mount Crested Butte, Colorado, ASCE, New York, 1995.

Watershed/Ecosystem Issues in Urban Runoff Monitoring and Management

Edwin E. Herricks (Member ASCE)[1]y

Abstract

At the Engineering Foundation Conference "Stormwater NPDES Related Monitoring Needs" (NPDES Monitoring Conference) critical issues concerning impervious area, biological and ecological integrity/health and watershed integration were a focus for formal and informal discussions. Ecosystem and watershed issues have been regularly addressed in Engineering Foundation Conferences sponsored by the Urban Water Resources Council (UWRRC), but a new urgency is suggested if findings presented at this conference are confirmed in other regions with other development characteristics. The following review of formal and informal conference discussions emphasizes issues of monitoring and analysis of bioassessment and toxicity testing data generated for compliance monitoring and watershed management.

Introduction

Although the primary focus in stormwater and urban runoff research in the past 20 years has been chemical water quality alterations and physical effects, there is a body of information on biological effects of stormwater and urban runoff. Reports and publications deal with the effects of stormwater and urban runoff on receiving water biota that include sediment quality impacts. A typical finding is that a number of biological organism groups (fish, benthic macroinvertebrates, algae, rooted macrophytes, etc.) in addition to studies of water and sediment quality are needed.

When viewed from an ecological perspective the results from biological studies suggest that stormwater and urban runoff affect both habitat stability and

[1]Professor of Environmental Biology, Department of Civil Engineering, 3215 NCEL, MC-250, University of Illinois at Urbana-Champaign, 205 N. Mathews St, Urbana, IL 61801

chemical toxicity. Habitat instability can be expected to alter the composition of aquatic communities by favoring those species capable of withstanding continuous changes in habitat. Aquatic communities can also be expected to change composition and abundance reflecting the influence of runoff characteristics, not natural variability in environmental conditions. Chemical toxicity can be expected to differentially affect less tolerant species. The community present may be specific for the characteristic contaminants in urban runoff. The combined effect of habitat instability and chemical toxicity will be the absence of valued species.

In summary, stormwater and urban runoff can be expected to produce short-term and long-term changes in receiving waters. When physical conditions are altered in a stream, there will be corresponding alteration of habitat for stream biota. Watershed development will also alter water quality in receiving streams. The combination of changed physical habitat and altered water quality must be recognized as the major environmental consequences of stormwater and urban runoff.

Monitoring Trends

Several papers at the NPDES Monitoring Conference (Torno 1994), Waller, et al., Herricks, et al., Livingston, et al. and Shaver, et al. provide a new synthesis of impact issues. Waller presented an analysis of episodic toxicity, including the use of a Toxicity Identification Procedure, that assists in the focus of specific contaminants. Herricks discussed an ongoing project that is intended to identify test systems for episodic event, time scale, toxicity. He indicated that a battery of tests have been used, and consistent with the results Waller presented from Ft. Worth, indicated infrequent, and low levels of toxicity in stormwater. The discussions led by Livingston focused on a measure of the long-term consequences of stormwater and urban runoff, the effects on sediment. In addition, Livingston reviewed Florida's Stormwater Bioassessment Project, a comprehensive, ecoregion-based approach to ecosystem assessment based on the adoption of a standardized metrics set of biological measures. Shaver presented a watershed approach that couples modeling with biological and habitat metrics.

It was these presentations that provided the core of evidence for the discussion of ecosystem effects. Preliminary assessments of toxicity testing indicate limited single effect toxicity, yet there is clear ecosystem degradation associated with watershed development. Shaver has been able to quantify a threshold that became a primary focus of discussion. He noted that with an increase of impervious area to a range of 10 to 20% there is a corresponding decline in biological integrity and ecosystem health.

Ecosystem Issues

A new synthesis on ecosystem issues is beginning to emerge as watershed considerations effectively integrate ecosystem considerations in more general analyses of stormwater and urban runoff effects. Since the early 1980's biological integrity has been defined in terms of chemical variables, flow regime, habitat conditions, biological energy sources, and biotic interaction in communities (Karr 1981). The picture is complex, but clear. An emphasis on hydrology, without a recognition of the importance of flow regimes and habitat structure important to biota, leaves a tremendous gap in effective management. Considering chemical variables alone, particularly depending on measurements that do not address real concentration and duration of exposure characteristics /that assess toxicity or consider the long-term consequence of sediment accumulation and biotic effect, does little to expand our understanding of stormwater and urban runoff effect in receiving systems. Finally, addressing runoff issues without a watershed context leads to a failure to consider a range of factors the control biological communities and contribute to system integrity.

Discussions at the NPDES Monitoring Conference have led to a consolidation of thinking that is encouraging new initiatives to address critical environmental management needs. One particularly exciting endeavor is a regional study to confirm development thresholds that appear to significantly impair environmental integrity. This study will build on advances in Florida and Delaware reported by Livingston, Shaver, and their colleagues to make similar assessments nationwide. The developing understanding of time-scale toxicity attacks questions of cause and effect directly. Waller's reporting that it is possible to identify specific toxicants that make a major contribution to observed effects can begin to focus attention on important chemical issues in a more general watershed/ecosystem management effort. The recognition that test systems are available that can assess single events, and procedures are available to address contaminant accumulation and associated environmental effects also provide direction for better use of toxicity testing and bioassessment activities, particularly when those activities are coordinated in an integrated watershed/ecosystem management approach.

References

Karr, J. R. 1981. Assessment of biotic integrity using fish communities. Fisheries 6(6):21-27.

Torno, H. C. 1994. *Stormwater NPDES Related Monitoring Needs* American Society of Civil Engineers, New York. 675 pp.

Monitoring of Best Management Practices

Eric W. Strecker[1], M. ASCE, and Ben R. Urbonas[2], M. ASCE

Abstract In order to improve our knowledge of how to identify and design the most effective Best Management Practices (BMPs) for stormwater water quality enhancement, BMP studies should be based on consistent stormwater monitoring techniques and provide consistent reporting information for a variety of physical, chemical, watershed, biological and climatic data. This paper presents a summary of the recommendations on consistent data for BMP effectiveness monitoring studies made during the recent Engineering Foundation Conference on Stormwater NPDES Related Monitoring Needs . Other considerations that affect data transferability, such as effectiveness estimations, statistical testing, etc., are also discussed.

Introduction Many studies have assessed the ability of stormwater treatment BMPs (e.g., wet ponds, grass swales, stormwater wetlands, sand filters, dry detention, etc.) to reduce pollutant concentrations and loadings in stormwater. However, in reviewing these individual BMP evaluations, it is apparent that inconsistent study methods and reporting make assessments of overall BMP effectiveness difficult, if not impossible. Often studies included the analysis of different constituents and utilized different methods for data collection and analysis. These differences alone contribute significantly to the range of BMP effectiveness reported. This makes it almost impossible to assess the factors that may have contributed to the variation in performance (such as watershed differences, climate, design, size, and configuration, etc.). If more consistent and complete data were collected, the data would likely be much more useful for improving BMP design.

Constituents for Assessing BMP Performance The choice of constituents to include is subjective. To include a parameter in a recommended list of monitoring constituents, Strecker (1994) considered the following criteria: the constituent is prevalent in typical urban stormwater at concentrations that may cause water quality impairment; analytical test can be related back to potential water quality impairment; constituent sampling methods are straightforward and reliable for a moderately careful investigator; laboratory analysis is economical on a widespread basis; and the constituent is one where treatment is a viable option. Not all of the constituents recommended fully "meet" all of the criteria listed above; however, the factors were considered in the

---

[1] Director, Water Resources Engineering, Woodward-Clyde Consultants, 111 S.W. Columbia, Suite 990, Portland, Oregon 97201.

[2] Chief, Master Planning & South Platte River Programs, Urban Drainage and Flood Control District, 2480 West 26th Avenue, Suite 156, Denver, Colorado 80211.

recommendations. When developing a list of constituent analyses for an individual BMP evaluation, it is important to consider the upstream land use activities. The constituents recommended in Table 1 are generally present and of concern in "typical" urban stormwater as identified by the Nationwide Urban Runoff Program (NURP) (EPA, 1983) and by the Federal Highway Administration (Driscoll et al., 1990). These parameters were reconfirmed during the recent municipal NPDES monitoring programs which analyzed for a large list of constituents (Torno, 1994).

TABLE 1
Recommended Standard Analytical Tests for Urban Stormwater BMP Assessments
Adapted from Strecker (1994)

Lab Analyses	EPA Method Number	Detection Limit (mg/l)
Conventional		
TSS	160.2	4
TDS	160.1	10
TOC	415.1	
COD	410.1	1
Total Hardness	SM314	1
Nutrients		
(NH_3 - N)	350.2	0.1
Total phosphorus (as P)	25401	0.05
Ortho-phosphate (as P)	365.2	0.05
Nitrate + nitrite (NO^-_3 + NO^-_2 as N)	353.1 or .2	0.05
Total Metals		
Pb (lead)	239.2	0.001
Cu (copper)	220.2	0.001
Zn (zinc)	289.2	0.001
Dissolved Metals (sample filtered with 0.45 μm)		
Cd (cadmium)	213.2	.0002
Pb (lead)	239.2	0.001
Cu (copper)	220.2	0.001
Zn (zinc)	289.2	0.001

Total suspended solids (TSS) can indicate problems for biota and are also often used as a surrogate for other contaminants which bind or adsorb to them. Although TSS is often correlated with some of the other parameters, it is typically not a strong enough correlation to eliminate other parameters. However, TSS is a good indicator of pollutant removal efficiency and should be included in any evaluation of BMP performance. Chemical Oxygen Demand (COD) provides a measure of oxygen demand. The test is used to measure the organic content of the sample, as is Total Organ Carbon (TOC). Both tests are indicators of the potential for oxygen depletion. As the most important nutrients in algal production are nitrogen and phosphorus compounds; three forms of nitrogen and two forms of phosphorus are recommended. Heavy metals such as copper (Cu), lead (Pb), zinc (Zn), and cadmium (Cd) have been identified as the most significant and commonly occurring metals in urban stormwater. It is recommended that both the total and dissolved form of each metal be analyzed to enable comparisons with water quality criteria; therefore, hardness should be measured for each sample. Pesticides and herbicides and volatile and semi-volatile organics have not generally been detected at a high frequency and in quantities that exceed available criteria. Accurately measuring oil and grease is very difficult. There is much disagreement over the form of bacteria to utilize for studies; therefore, none was recommended.

BMP Effectiveness Data Reporting A wide variety of methods for calculating the effectiveness of a BMP to remove pollutants have been used. Water quality samples should be collected (or compiled from discrete data) as a flow-weighted composite sample. Urbonas (1995) recommended calculating a percent removal of load on a storm-by-storm basis and then averaging the removal rates. Strecker (1994) suggested that, for BMPs that have wet storage, it may be more appropriate to develop a statistical characterization of inflow and outflow concentrations and compare the difference. This would eliminate the problem of smaller storms failing to displace the storage volume, so that the inflow and outflow are not from the same event. However, both approaches have merit.

General Parameters to Report with Effectiveness Urbonas (1995) summarized the information that should be recorded regarding the physical, climatic, and geological parameters which likely affect the performance of a BMP. Table 2 presents these parameters as modified during the conference. Readers are asked to comment on the ASCE JWRPM article by Urbonas by July 1, 1995.

Other Considerations Also discussed was the need for good data reporting formats, the use of standard quality assurance/quality control procedures, sample collection techniques and their potential effect on reported pollutant removals, and statistical considerations (sample size). Oswald (1994) discussed the need to collect samples from at least 80 percent of the storm and the need to target from 10 to 20 storms. Green, et. al., (1994) recommended the use of a model to design the sampling program and also questioned the need to standardize methods used for flow estimation. Roa-Espinosa and Bannerman (1994) are testing the use of a source area sample which may enable the analysis of individual source control measures in the future.

APPENDIX

Driscoll, E.D., P.E. Shelley, and E.W. Strecker. 1990. "Pollutant Loadings and Impacts from Stormwater Runoff, Volume III: Analytical Investigation and Research Report." FHWA-RD-88-008, Federal Highway Administration.

Green, D., T. Grizzard and C. Randall. 1994. "Monitoring of Wetlands, Wet Ponds, and Grassed Swales." Proceedings of the Engineering Foundation Conference on Stormwater Monitoring Related Monitoring Needs. Aug. 7-12, Crested Butte, Colorado.

Oswald, G., and R. Mattison, 1994. "Monitoring the Effectiveness of Structural BMPs." *Ibid.*

Roa-Espinosa, A. and R. Bannerman, 1994. "Monitoring BMP Effectiveness at Industrial Sites." *Ibid.*

Strecker, E.W., 1994. "Constituents and Methods for Assessing BMPs." *Ibid.*

Torno, H.C., Editor. 1994. *Ibid.*

U.S. Environmental Protection Agency. 1983. "Final Report on the National Urban Runoff Program. Water Planning Division, U.S. EPA." Prepared by Woodward-Clyde Consultants.

Urbonas, B.R. 1995. "Recommended Parameters to Report with BMP Monitoring Data." J. Water. Resour. Plng. and Mgmt., ASCE, 121(1), 23-24.

TABLE 2
Parameters to Report with Water Quality Data for Various BMPs

Parameter Type	Parameter (1)	Retention (Wet) Pond (2)	Extended Detention Basin (3)	Wetland Pond Basin (4)	Grass/Swale Wetland Channel (5)	Sand/Leaf Compost Filter (6)	Oil & Sand Trap (Vault) (7)	Infiltration and Percolation (8)
Tributary Watershed	Tributary watershed area	•	•	•	•	•	•	•
	Total tributary watershed impervious percentage	•	•	•	•	•	•	•
	Percent of impervious area hyd. connected	•	•	•	•	•	•	•
	Gutter, sewer, swale, ditches in watershed?	•	•	•	•	•	•	•
	Land use types (res, comm, ind, open) and acreages	•	•	•	•	•	•	•
General Hydrology	Average storm runoff volume	•	•	•	•	•	•	•
	50th percentile storm runoff volume	•	•	•	•	•	•	•
	Coefficient of variation of runoff volumes	•	•	•	•	•	•	•
	Average daily base flow volume	•	•	•	•	•	•	•
	Average runoff interevent time	•	•	•	•	•	•	•
	50th percentile interevent time	•	•	•	•	•	•	•
	Coefficient of variation of interevent times	•	•	•	•	•	•	•
	Average storm duration	•	•	•	•	•	•	•
	50th percentile storm duration	•	•	•	•	•	•	•
	Coefficient of variation of storm durations	•	•	•	•	•	•	•
	2-year flood peak velocity				•		•	
	Depth high groundwater or impermeable layer		•	•				•
Water	Water temperature	•	•	•	•	•	•	•
	Alkalinity, hardness and pH	•	•	•	•	•	•	•
	Sediment settling velocity distribution, when available	•	•	•	•	•	•	•
	Facility on- or off-line?	•	•	•	•	•	•	•
	If off-line, amount of flow bypassed annually	•	•	•	•	•	•	•
General Facility	Type and frequency of maintenance	•	•	•	•	•	•	•
	Inlet and outlet dimensions and details	•	•	•	•	•	•	•
Wet Pool	Solar radiation, when available	•		•	•			
	Volume of permanent pool	•		•		•	•	
	Permanent pool surface area	•		•		•	•	
	Littoral zone surface area	•						
	Length of permanent pool	•		•		•	•	
Detention Volume	Detention (or surcharge) volume	•	•	•		•	•	•
	Detention basin's surface area	•	•	•		•	•	•
	Length of detention basin		•	•		•	•	•
	Brim-full emptying time	•	•	•		•	•	•
	Half-brimful emptying time	•	•	•		•	•	•
	Bottom stage volume		•					
	Bottom stage surface area		•					
Pre-Treatment	Forebay volume	•	•	•		•	•	•
	Forebay length	•	•	•		•	•	•
	Other BMPs upstream?	•	•	•	•	•	•	•
Wetland Plant	Wetland type, rock filter present?			•	•			
	Percent of wetland surface at $P_{0.3}$ and $P_{0.6}$ depths			•	•			
	Meadow wetland surface area			•	•			
	Plant species and age of facility	•	•	•	•			

Adapted from Urbonas (1995)

Fecal Coliform Standards and Stormwater Pollution

Brent McCarthy[1]
Gary Mercer[1]

I. Abstract

This paper presents fecal coliform water quality standards for six eastern states. It then presents preliminary information from NPDES stormwater permit applications on fecal coliform levels in the eastern part of the U.S. Next, fecal coliform levels in stormwater are compared with state standards. The appropriateness of applying fecal coliform standards for non-point source pollution control for cleaning up stormwater are discussed, followed by a proposed approach for controlling fecal coliform pollution from stormwater systems.

II. State Water Quality Standards for Fecal Coliform

Table 1 shows coliform standards in 6 eastern states for the most sensitive riverine use. Standards for Connecticut are for total coliform, while the standards are for fecal coliform in the other 5 states.

TABLE 1

Connecticut -	Monthly moving average for 12 months ≤ 100 total coliform/100 ml; no values shall exceed 500 total coliform/100 ml.
Illinois -	Over 30 day period, geometric mean ≤ 200 fecal coliform/100 ml (minimum of 5 samples); no more than 10% of samples in any 30 day period can exceed 400 fecal coliform/100 ml. For Lake Michigan, based on a minimum of 5 samples over not more than a 30-day period, fecal coliform can not exceed a geometric mean of 20/100 ml.
Massachusetts -	Arithmetic mean in any representative set of samples can not exceed 20 fecal coliform/100 ml; no more than 10% of samples can exceed 100 fecal coliform/100 ml.
Michigan -	Over 30 day period, the geometric mean of any series of 5 or more representative samples must be less than 200 fecal coliform/100 ml, for all waters of the state. This concentration may be exceeded if due to uncontrollable non-point sources. (Storm drainage outfalls would be considered point sources, according to Michigan's definition).

1 Principal Engineers, Camp Dresser & Mckee Inc., Ten Cambridge Center, Cambridge, Massachusetts 02142

New Jersey - Geometric mean for representative set of samples can not exceed 200 fecal coliform/100 ml; no more than 10% of samples can exceed 400 fecal coliform/100 ml.

Pennsylvania - Geometric mean of 5 samples taken on 5 consecutive days can not exceed 200 fecal coliform/100 ml; no value can exceed 2,000 fecal coliform/100 ml.

For less sensitive uses, the range in state standards is wider. For example, waters are exempt from fecal coliform standards entirely in Illinois for waters not subject to primary contact recreation and/or not flowing through or adjacent to parks or residential areas. Massachusetts' lowest class (Class C) allows a geometric mean of 1,000 fecal coliform/100 ml, with less than 10% of samples greater than 2,000 fecal coliform/100 ml. (Actually, since all riverine waters in Massachusetts are classified as A or B, for practical purposes, the standards are stricter than this). Connecticut has as a guideline a geometric mean of 200 fecal coliform/100 ml in any group of samples with no more than 10% of samples exceeding 400 fecal coliform/100 ml.

III. Levels of Fecal Coliform in Stormwater Systems

Table 2 summarizes fecal coliform levels in storm drainage systems from 10 cities in the eastern and central United States. The data was taken from EPA Part II Municipal Stormwater Permit applications and stormwater sampling as part of facility planning efforts.

TABLE 2
Fecal Coliform Concentrations in Storm Drains

	No. Samples	Range	Median	Geometric Mean	Mean
Northeast	8	80-5,000	175	234	773
Central	27	0-4,300,000	22,000	3,699	303,701
Midwest	32	50-160,000	6,050	5,868	25,143
South Central	35	0-280,000	4,800	3,096	20,548
Southeast	10	68-3,800	365	339	801
Southeast	31	120-70,000	3,100	3,418	12,559
Southeast	24	1,600-1,900,000	55,000	34,203	151,060
Southeast	27	0-300,000	11,000	7,357	39,005
Southeast	12	11-16,000	105	229	3,105
Southeast	31	100-104,000	3,950	4,675	15,809
All 10 Cities	237	0-4,300,000	4,800	2,618	64,197

As can be seen in the table, the geometric mean of fecal coliform concentrations from storm drains in wet weather for the 237 samples is 2,618 fecal coliform/100 ml. (For the purpose of computing the geometric mean, 0 values were assumed to be 1). Comparing Tables 1 and 2, it is clear that water from most storm drainage systems will violate water quality standards in these 6 states most of the time.

Contamination in storm water systems during dry weather is also a problem. For example, in one northeastern city, 56 of 71 major outfalls had measurable flow during dry weather periods. Surfactants were common in the dry weather flow, and in 5 of 56 of the outfalls, there was visual or odor evidence of sewage contamination (samples were not tested for fecal coliform). This is not surprising considering storm drainage systems in the northeast often are located below the water table and often contain culverted portions of natural streams.

III. Sources of Fecal Coliform in Stormwater

The sources of fecal coliform in stormwater are well documented. They include improper sanitary connections, failed septic systems, and exfiltration from sanitary systems to storm drainage systems. It is important to remember other common sources of contamination as well: warm blooded animals including pets, and other wildlife common in urban and suburban settings, including raccoons, skunks, pigeons, seagulls, and geese. There is also some circumstantial evidence that fecal coliform "grow" outside of warm blooded animals, including in the sediment of stream beds.

The average human discharges about 4 billion fecal coliform per day. Neglecting fecal coliform die-off, that is enough to exceed the most common standard of 200 fecal coliform/100 ml for over one-half million gallons of water per day (or over 5 million gallons where the standard is 20 fecal coliform/100 ml). Assuming warm blooded animals discharge an equivalent number of fecal coliform as humans on a weight basis, discharge in stormwater from either improper disposal of human waste or from uncontrolled discharge of bird and animal wastes into storm drainage systems will easily cause water quality standards violations.

IV. Meeting Water Quality Standards

For traditional point sources (treatment plants), meeting water quality standards has been straight-forward. Wastewater treatment facilities chlorinate their discharge to kill pathogens. The technology is well documented, and contact time and dosage to destroy fecal coliform are well known. Also, as the impact of chlorine on desirable life forms in receiving waters has become of increasing concern, dechlorination facilities are now common processes at wastewater treatment plants.

For storm drainage systems, meeting water quality standards is much more difficult. End-of-pipe chlorination at every outfall with fecal coliform contamination is impractical. End-of-pipe facilities are costly and difficult to site. As opposed to traditional wastewater treatment plants, stormwater flow rates are highly variable. Because of the variability in flow, achieving the proper match between chlorine dosage and contact time is difficult. With widespread end-of-pipe facilities, the necessity to control discharge of chlorine would grow. Another option, building infrastructure for collection, storage, and treatment of stormwater, is also extremely costly. Cities across the nation face multi-billion dollar solutions for control of combined sewer systems. The areal extent of stormwater systems greatly exceeds combined systems, and the cost of collection, storage, and treatment would also be much greater.

A concept used for point sources to help meet water quality standards is the mixing zone. This concept is not important at treatment plants for fecal coliform because disinfection is designed to remove the problem completely, an option probably not available for storm drainage systems. However, mixing zones are important for other contaminants at wastewater treatment plants, where technology alone is not sufficient to reduce contaminants to the point where water quality standards are met.

To meet a 200 fecal coliform/100 ml standard, the geometric mean value presented in Table 2 (2,618 fecal coliform/100 ml) would have to be reduced by over 90%. In view of the varied sources of fecal coliform and the sheer quantity of fecal coliform discharged from individual sources, it will be almost impossible to achieve such a large reduction without disinfection. Removal efficiencies will have to be even higher for states with stricter standards and when maximum allowable fecal coliform counts are considered.

V. A Proposed Approach

To control sanitary sources of pollution in stormwater, the best method includes a staged approach beginning with non-structural best management practices (BMPs) and progressing to more intensive removal methods as required.

Non-Structural BMPs include:

- leash laws
- "pooper-scooper" ordinances
- efforts to discourage "unnatural" levels of bird life, including geese, seagulls, and pigeons.
- street sweeping
- catch basin cleaning

Structural BMPs include:

- removal of illicit sanitary connections
- removal of sewer blockages, repair of sewer breaks
- repair of septic systems
- sewer system infiltration/inflow rehabilitation

These measures will significantly reduce the levels of fecal coliform in receiving waters. However, in many cases, these efforts will not be sufficient to achieve compliance with water quality standards. In these cases, the concept of a mixing zone should be investigated, to allow for a zone around the storm drain outfalls that exceeds water quality standards. When considering the extent of mixing zones, the location of the outfall relative to sensitive uses can be considered, as well as fecal coliform die-off rates.

This approach will (a) reduce the maximum amount of fecal coliform pollution at minimum costs; and (b) focus attention and money on the most severely contaminated storm drainage systems: those that continue to violate water quality standards even after non-structural BMPs, structural BMPs, and mixing zones are applied.

REVISITING THE 1994 NATIONAL ENVIRONMENTAL PENDULUM CONFERENCE

By: Jonathan E. Jones

On January 31 and February 1, 1994, the American Society of Civil Engineers (ASCE) Urban Water Resources Research Council (UWRRC) along with approximately 20 other cc-sponsors, held a two-day conference at the Georgetown University Conference Center entitled: *National Water Resources Regulation - Where is the Pendulum Now?* The purpose of the conference was to foster lively discussion and debate about contemporary water resources management issues. We were fortunate to assemble a slate of outstanding speakers including lawyers, regulators, politicians, engineers, scientists, and others. These people were selected "from both sides of the aisle" and they provided strong and compelling talks on such subjects as wetlands regulation, endangered species, preservation of minimum flows, risk assessment, standards-setting, and other subjects.

There have been many significant developments regarding water resources regulations in 1994 and the early part of 1995, and this session of the May, 1995 WRPMD conference in Boston provides an opportunity on this subject. Our speakers have been asked to address such topics as:

- The likely nature of the new Clean Water Act.
- The significance of the United States Senate vote on the Safe Drinking Water Act reauthorization in 1994, which had a strong emphasis on risk assessment and benefit-cost analysis.
- How concern with private property rights is likely to affect the nature of new laws and the interpretation of existing laws.

[1] Professional Engineer and Vice President with Wright Water Engineers, Inc. in Denver, Colorado.

- The probable nature of stormwater quality management and regulation, on both the municipal and industrial side.

- Multi-media regulation and regulation of sediment.

- Delegation of power between states and the federal government.

- Providing flexibility for municipal and industrial NPDES permit holders on monitoring programs.

We have assembled a group of excellent speakers to address these and other questions in Boston and we look forward to learning their projections for the future.

Combining Water Resource and Environmental Plans

William Whipple, Jr., F ASCE[1]

Abstract

A proposal for a new national planning approach to cope with environmental problems.

Inevitably in a major drainage basin both engineering and environmental problems demand attention. The Corps of Engineers (or Bureau of Reclamation) can take the major role in analyzing and modeling the quantitative relationships involved, determining costs, and suggesting alternative engineering solutions. State and local participants can outline the environmental problems; but the decision as to how they are to be handled should have the benefit of advice of the appropriate federal environmental agency or agencies, perhaps the EPA, Fish and Wildlife Service, and/or National Park Service. In these cases, the appropriate federal environmental agency should participate in the hearings and working sessions, along with the states. The working arrangements can be similar to those developed by the Corps of Engineers for the Drought Preparedness Studies. In the end, when the Corps of Engineers (District Commander) makes his report, it should be accompanied by a report from the appropriate federal environmental agency and a preliminary view from the state (or states). It is important that the participating environmental agency should be authorized (by Congress) to agree to water quality criteria and other regulatory restrictions other than those previously developed as objectives for general use. If such an arrangement were not made, the multi-agency planning would be ineffective, because states and other stakeholders would not be willing to waste their time planning if there were a foregone conclusion that the initially stated environmental requirements could not be modified. The costs of such comprehensive river basin planning should be paid by the federal government, except for normal costs of non-federal participation.

---

[1]Principal, Greeley Polhemus Group, 395 Mercer Road, Princeton, NJ 08540

There remains, of course, the problem of how to coordinate at Washington level any differences of opinion which cannot be negotiated between the field agencies and the states. It would greatly facilitate the process if the national Water Resources Council or some equivalent coordinating agency were to be revived. The Water Resources Council provided a means for coordinating water resources action between the federal agencies, which had been recognized as a serious problem ever since World War II. The council required that each agency conform to general government policy rather than some agency viewpoint. Although congressional committees and the Executive Office of the President exercise powers of overall coordination, the growing difficulties in coordinating water resources planning and environmental regulation show that a more intensive analysis is required. The council would provide the means for such coordination in the establishment of policy and procedures, and possibly in the actual review, where necessary, of plans submitted. It is true that the council did not generally perform this review function. However, since the necessity for such coordination of state-federal and of interagency differences is now so manifest, it seems that either the new council should be utilized for review, or the difference should be referred directly to congressional committees as a preliminary authorization. For major comprehensive plans, the issues are sufficiently important that the ultimate decision should be by Congress (similar to the congressional approval required now for federal construction projects). The revived Water Resources Council would receive recommendations from the agencies, hold a public hearing, consider state views and make a recommendation to the appropriate congressional committee. While it existed, the Water Resources Council consisted of the Secretaries of the departments concerned, to which should be added the EPA.

The above procedure would be appropriate for comprehensive planning reviews of major river basin systems. However, there are other problems, particularly those of runoff pollution control, for which it would not be appropriate. In these cases, the EPA is the federal agency mainly concerned, but the long-range stated objectives will be generally costly to major cities and to industries; and it will be difficult to convince those who have to pay for the work of the value of the environmental benefit, particularly as that benefit characteristically occurs mainly outside of the jurisdiction of the municipality (or industry). In this case, the EPA should be authorized to conduct open planning to solve the problem.[2] In this planning, like the multi-agency planning, the EPA should not be bound by previously established water quality criteria. The EPA would examine the existing state of pollution, the improvement expected from various alternative courses of action, and the environmental improvements expected downstream. If the state agreed with the conclusions of the study, the requirements for runoff pollution control would be imposed. If the state did not agree, the decision could be made by the Water Resources Council. If the governor of the state disagreed with the findings of the council, he could always arrange for his views to be incorpo-

[2]It is probable that legislative authority for such planning already exists under section 208 of the Clean Water Act.

rated in special legislation, which congressmen from the state could introduce if they felt a mistake was being made.

However, with the requirement of absolute mandatory criteria removed and an open planning process, it seems probable that the states and the EPA could come to an accord in most pollution control cases. Although the municipalities and industries themselves might prefer to have no runoff pollution control measures at all, the state is in a better position to take a balanced position; and the EPA would doubtless be reluctant to take many dissenting states before the Water Resources Council for a decision. It is anticipated that such studies would usually show very sound grounds for requiring general use of BMPs for runoff pollution control in future construction, but would recommend relatively few major reductions in runoff pollution from existing urban areas, except where special problems exist.

The Swinging Pendulum: A Receiving System Perspective

Edwin E. Herricks (Member ASCE)[1]

Abstract

Receiving systems have been the focus of regulation and management. Past emphasis has been on chemical conditions, but clean water may not be enough to protect the environment. This paper reviews some of the issues raised at the National Water Resources Regulation: Where is the Environmental Pendulum Now? Conference and the progress, ar at least movement, in addressing these issues since the conference.

Introduction

The swinging pendulum was the paradigm for discussion at a Urban Water Resources Research Council (UWRRC) conference in early 1994 (Holme, 1994). As presented in the purpose statement for the conference, there is conflict among regulatory, the regulated community, advocacy groups, and the public. "Regulators must set standards and demand compliance to conform with the law, the regulated community seeks relief when over-regulation is perceived, environmental advocates seek full protection of the environment, and the public demands quick and cost-effective responses to environmental problems." That forward went on to state, "A balance in environmental regulation must be reached among multiple competing interests to avoid over-emphasizing any single factor such as economics, legal process, political "fads," scientific testing, or engineering feasibility...With more than two (competing) interests the path of the pendulum is erratic, swinging across a field of influence rather than a single path." Receiving systems were an issue at the conference because the laws and regulations, which were the focus of primary attention, have the intent of protecting the environment. A specific focus on receiving systems was provided

[1]Professor of Environmental Biology, Department of Civil Engineering, 3215 NCEL, MC-250, University of Illinois at Urbana-Champaign, 205 N. Mathews St, Urbana, IL 61801

by papers presented by Jim Karr (Karr, 1994) and Ed Herricks (Herricks, 1994). Karr argued that clean water wan not enough. He noted that habitat modifications, exotic species, poor management of sport and commercial stocks and contamination by chemicals have all taken their toll on biotic integrity. He argued for a new, integrated water resources management approach that is focused on protecting ecological health of aquatic ecosystems. Herricks addressed criteria development, particularly the development of biocriteria, with particular attention to wet weather issues. He argued that we must recognize several factors that guide our selection of management approaches. First, we must develop an understanding of physical, chemical, and biological/ecological processes that include an understanding of scale, particularly watershed influences. We must have an awareness of fundamental change produced by existing development. Next we should take advantage of natural restorative processes when attempting to tip the balance to integrity. Finally, we need ecological criteria rather than narrowly focused regulation and engineering practice to achieve stated Clean Water Act goals.

A State-of-the-Art

An underlying theme of Karr's and Herricks' presentations and papers was a call to action, suggesting the need for new approaches to environmental management. Interestingly, the twelve months between the Pendulum Conference and the preparation of this paper has seen additional alarming syntheses regarding the protection of biological integrity in the nation's waters, the continuing development of new techniques to support scientific management, and a potential sharp swing in the pendulum with the election of November 1994. At an Engineering Foundation Conference in August 1994 a watershed-focused study led by Earl Shaver (Shaver et al. 1994) indicated that when development in a watershed crosses a threshold where 10% of the are is impervious that system health is degraded. He felt that development producing greater than 20% impervious area was likely to compromise integrity significantly. This finding, when tied to Karr's clear warning about existing losses of natural resources, raises new concerns about environmental protection.

At the same time, science is providing us with more information about fundamental processes, information that may lead to techniques to stem the loss of integrity. In the past, the emphasis has been on control and management of continuous discharges, and a complex structure of toxicity testing has supported regulation of effluents. Now that we have point sources controlled, there has been a shift to nonpoint sources, particularly stormwater. Unfortunately, in this arena the support for regulation is weak. At the same Engineering Foundation Conference where Shaver presented his findings on watershed degradation Waller (1995) found mixed toxicity in storm event analysis but was able to direct attention to critical issues through the use of Toxicity Identification Evaluation (TIE) procedures. Herricks, et al. (1994) at the same conference reported on a test battery selection procedure that recognizes the unique character of time-scale, episodic event toxicity. He, too, found mixed and low toxicity during storm events. Clearly, we are on the path to a better scientific understanding of both cause and effect when operating at a scale where our science and experimentation has proven effective in the past.

In another arena the foundations of regulation are being explored. One of the underlying themes of the Pendulum Conference was the Clean Water Act reauthorization. The reauthorization did not materialize in the last Congress and is likely to take on a vastly different character with the new Congress. At the same time initiatives are underway that are suggestive of how the pendulum might swing. A group working thorough the Institute for Water Resources is proposing a change to the way money is being spent on monitoring. Their idea is that some portion of the monitoring funds that would normally be spent on NPDES stormwater monitoring should be diverted to a national fund to support integrated analysis of stormwater effects. Clearly, there is a recognition that funds are limited, mandates must have flexibility, and common sense should prevail.

Possible Futures

This brief review of history and an assessment of a state-of-the-art can support some speculation on the swing of the pendulum. First, another quote from the Pendulum Conference purpose is warranted. "With more than two interests, the path of the pendulum is erratic, swinging across a field of influence rather than a single path, pulled this way and that by each competing interest. If the pendulum storms moving, the stop may be temporary as new forces align to pull the pendulum in a new direction. Paths, stopping points, and expected influences of competing forces are unpredictable and regularly result in gridlock!" We are clearly in a situation where prediction is foolish. I can identify some trends and effects that possible outcomes of the changing environmental scene. There is a clear call for control to become more local, reducing the costly mandates from on high that are so much a part of the last twenty years of regulation. Will a local control succeed? I have seen local decision making and the results are mixed. Like it or not, technical decision making requires some technical expertise! Local decisions without the support of the needed expertise are likely to sharpen the lack of integration recognized by Karr, which is still a problem with the old, global, approach to management. Thus I see a trend to local control that in some instances will benefit the environment and in others, lead to severe degradation because the thresholds of effect (20% impervious area) are easily crossed.

I also see incremental improvement in management as we adopt an ecosystem approach. The ecosystem approach did not fair well in the Clean Water Act reauthorization, but at least a popular view of ecology and ecosystems is pervasive. The U.S. Environmental Protection Agency's biocriteria are a clear step forward in establishing a foundation for better management. At the same time, we are learning more about event-related toxicity and identification of procedures for cause and effect determination (e.g. TIE). We have made tremendous advances in the last 20 years in toxicology, environmental toxicology, and ecotoxicology. Those advances will continue! The clouds on the horizon are funding and a continuing lack of integration fostered by disciplinary segregation in training and problem solving. If we can do as Karr suggests "...stop talking about water quality and start formulating policy focused on protection the ecological health of aquatic resource systems" we may be able to tame the erratic swings of the pendulum with a common interest, and need, for ecosystem health protection.

References

Herricks, E. E. 1994. Wet Weather Issues in Receiving Systems: Can a Pendulum Swing Upstream. In H. Holme (ed) *National Water Resources Regulation: Where is the Pendulum Now?* American Society of Civil Engineers, New York. p.168 -176.

Herricks, E. E., I. Milne, and I. Johnson. 1994. Time-Scale Toxic Effects in Aquatic Ecosystems. In H. C. Torno (ed) *Stormwater NPDES Related Monitoring Needs* American Society of Civil Engineers, New York. p. 353 - 374.

Holme, H. 1994. *National Water Resources Regulation: Where is the Pendulum Now?* American Society of Civil Engineers, New York. 275 pp.

Karr, J. R. 1994. Clean Water is not Enough. In H. Holme (ed) *National Water Resources Regulation: Where is the Pendulum Now?* American Society of Civil Engineers, New York. p.79 - 95.

Shaver, E, J. Maxted, G. Curtis, and D. Carter. 1994. Watershed Protection Using an Integrated Approach. In H. C. Torno (ed) *Stormwater NPDES Related Monitoring Needs* American Society of Civil Engineers, New York. p. 435 - 459.

Waller, W. T., M. F. Acevedo, E. L. Morgan, K. L. Dickson, J. H. Kennedy, L. P. Annann, H. J. Allen, and P. R. Keating. 1994. Biological and Chemical Testing in Stormwater. In H. C. Torno (ed) *Stormwater NPDES Related Monitoring Needs* American Society of Civil Engineers, New York. p. 177 - 193.

Integrated Water Resources Planning of Quantity and Quality in Taiwan

Chian Min Wu[1]

ABSTRACT

Water Resources Management (WRM) has evolved into a decision making in complex situations, where conflicting interests demanding limited natural and economic resources. Important issues in this respect are the analysis of natural system and the development and evaluation of strategies adopted to meet the demands of society. Thus, planning studies have become an indispensable tool in WRM.

To facilitate the water resources planning process in Taiwan, the Water Resources Planning Commission (WRPC), ROC, has devoted much manpower and funding to cooperate with many research institutes to study the water resources problems. In past years, the team for National Master Plan has completed the three area studies, which are Chianan, Kaoping and Tanshui, Choshui river basins. In the future, more areas will be implemented. Thus, the strategies for WRM can be considered as a whole in Taiwan. This paper describes the overall setups and the integrated analysis of the water resources master plan in Taiwan, ROC.

INTRODUCTION

The growth of Taiwan's economy in the past three decades was rapid and has induced social and institutional changes in many respects. Though it is expected that the growth will continue at a lower rate, the available water resource is insufficient to meet the rising demand for different water uses.

From a supply oriented viewpoint, the development of water resources, the

[1]Chairman, Water Resources Planning Commission, Ministry of Economic Affairs,Taiwan, ROC

construction of major infrastructures, is considered to be the first measure to meet water demand. But the most feasible reservoir sites have been developed and the cost of new reservoirs will be high, therefore more attention is now given to the demand oriented consideration. The finding is that the possibility of demand reduction seems to exist for all major water user.

Whether the water demand should be governed by a supply oriented or be transferred to a demand oriented basis is the basic issue and it may be decided by some definite analyzing procedure.

THE COMPUTATIONAL FRAMEWORK

In order to facilitate the various analysis steps, a number of different computational tools(Figure 1) have been developed. All modeling developments have been taken place on the personal computer. The following models and modules have been distinguished:

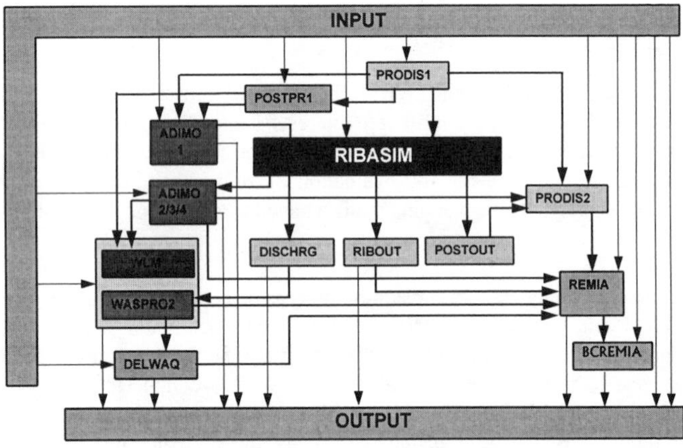

Figure 1. Computational Framework for the Analysis of the River Basins

- PRODIS (PROgram for the Domestic and Industrial Sector) : computation of present and future water demand for households and industry, both from PWS and self-supply, and computation for water shortages and cost of water supply. PRODIS has a request and an allocation stage, which are two separate models: PRODIS1 and PRODIS2.
- Agricultural District Model (ADIMO) : ADIMO computes agricultural water demand from surface water and groundwater (request mode). In the allocation mode, the model computes water shortages in fields and indicates the damages that result from these shortages.

- River basin Distribution Model (RIBASIM) : RIBASIM determines the overall surface water balance in the river basin, including the operation of the reservoirs and the distribution of water in the river network. The model tries to match supply and demand in the system.
- DISCHRG, RIBOUT and POSTOUT : the processing program for the output of RIBASIM model.
- TAIwan WAter Quality model (TAIWAQ) : for water quality calculations in estuary, river and reservoir system under both steady and unsteady conditions. It is a modified version of DELWAQ program from Delft Hydraulics supplemented with the waste production model WASPRO, and the flow distribution model, RIBASIM.
- WASte PROduction model (WASPRO) : a model to calculate waste loads from domestic, industrial, non-point (agriculture) and livestock sources, where possible associated with water flows in the RIBASIM network schematization.
- Regional Model for Impact Assessment (REMIA) : REMIA processes selective outputs from PRODIS, ADIMO and RIBASIM in order to provide a comprehensive report of physical quantities and socioeconomic consequences for different cases. Such cases consist of a combination of a water resources management strategy and so-called scenario specification, reflecting a set of assumptions about economic developments or conditions that affect water demands or supply. These cases are always compared with a base case, in order to determine the specific effects of the strategy under investigation.

The computational framework was used to analyze the present and future water resources management in terms of water quantity and quality, such as the management of existing or additional reservoir, or scenarios for future water demands by industry and municipal.

SETTING UP THE ANALYSIS

The overall aim of the master plan study is to prefer quantitative analysis in support of planning and decision-making for water resources management on a national level. From this general aims, some more specific objectives have emerged, e.g. the development of a computational framework, the execution of integrated analysis on river basin level and on the national level, and the preparation of a national master plan.

The country has been subdivided into four regions (North, Central, South, and East), that coincide with the economic regions used in the national planning. Within these regions, subregions have been distinguished which coincide with one or more separate river basins. There are total 18 subregions for water resources development and utilization. For the purpose of the national master plan study, the river-basin-oriented division is selected for planning and modeling. Each planning area composes one or more river basins. Within Choshui basin, ten irrigation districts have been distinguished. In addition to those tributaries of Choshui river, Tali creek, Peikang creek, have been considered as part of Choshui river basin. The area for Choshui river is about 6178 square kilometer.

Problem identifications of the pilot area show that serious water shortage problems should be expected in the relatively near future of Choshui river basin. The main future problems will be the deficits for the public water supply, PWS(leading to deficits for industrial and domestic use), lowering of groundwater levels from over withdrawal of groundwater sources, especially in the downstream area. In addition, the low flow situation and water pollution problem, which is already an issue in the existing situation will continue to become worsen.

CONCLUSION

In view of the need to support water resources planning and decision making on both the regional and the national level, the following outputs from the joint water resources planning efforts are suggested.

- A national master plan, say once every five years, or at a frequency which fits the planning of other sectors of the economy, formulates an overall strategy of the water resources sector.
- River basin development plans on a pre-feasibility level identify and select water resources development strategies and projects.
- Simulation models have been successfully used by various practitioners. However, in recent years, a tendency has been toward incorporating an optimization scheme into a simulation model to perform a certain degree of optimization.
- Water resources planning should not be carried out in isolation and should be seen as a key part of overall water services planning which itself forms part of national economic planning. Flexibility is essential and alternative strategic must be subjected to rigorous sensitivity analysis.

REFERENCES

Delft Hydraulics (1987). *Interim Report I-V: Master Plan Study; National Master Plan for Water Resources Management, Taiwan.*

Delft Hydraulics (1985-1987). *Master Plan Study; National Master Plan for Water Resources Management, Taiwan.* Final report , Volume I-II.

Delft Hydraulics (1985-1987). *Master Plan Study; National Master Plan for Water Resources Management,* Working Document 1-3

Delft Hydraulics (1993). *The Computational Framework for Water Resources Management,* Volume 1.

Delft Hydraulics (1994). *The Computational Framework for Water Resources Management.*

Modeling of Nonpoint Source Pollution in the Feitsui Reservoir Watershed

Jan–Tai Kuo[1], M. ASCE, Shaw L. Yu[2], M. ASCE, Chian–Min Wu[3]
Shih–Chuan Lee[4] and Ming–Chieh Kuo[4]

Abstract

The VirginiA STorm model (VAST) is a watershed nonpoint source pollution model which is capable of simulating rainfall-runoff processes, estimating the amount of pollutant accumulation and washoff, and generating the watershed hydrograph and pollutograph for a given storm event. The objective of this paper is to describe the application of VAST for modeling nonpoint source pollution in the Feitsui Reservoir Watershed in Northern Taiwan. Data taken between 1993 and 1994 were used to calibrate and verify the model. The results show that VAST and its modified version, VANTU, can be used as a planning tool for nonpoint pollution control in Taiwan. However, the model calibration/verification and application can be quite challenging due to Taiwan's environmental conditions and its physical and climatological characteristics.

Introduction

The nonpoint pollution sources for lakes and reservoirs in Taiwan have been identified as urban and recreational development, and cropping and livestock activities in the watersheds. In an effort to characterize nonpoint source pollution (NPS), to examine the NPS impact on reservoirs and lakes and to plan control strategies such as the use of best management practices (BMP), the Water Resources Planning Commission (WRPC) contracted the National Taiwan University in 1993 to initiate an NPS study for the Feitsui Reservoir Watershed. One of the key work elements is to test the application of a watershed NPS model.

Feitsui Reservoir is located in Northern Taiwan, on the Pei-Shih Creek which is a tributary to the largest river in Northern Taiwan, the Tamsui River. Feitsui (Emerald in Chinese) is a show-case reservoir in Taiwan which supplies a

[1]Prof., Dept. of Civil Engineering, National Taiwan University, Taipei, Taiwan
[2]Prof., Dept. of Civil Engr. & Appl. Mech., Univ. of Virginia, Charlottesville, VA
[3]Chairman, Water Resources Planning Commission, Taipei, Taiwan
[4]Graduate Students, Dept. of Civil Engineering, National Taiwan University

majority of water for the Taipei Matropolitan Area. There were some aborigin people living in the watershed before the dam's construction and continued to live there. In recent years urban development and in particular the many cropping activities such as tea gardens, and occasional camping crowds have caused concerns about the reservoir's water quality. Excessive nutrients and pesticides from cropping areas can be very problematic for the reservoir's health.

The VAST Model and Its Modifications

VAST (Tisdale et al., 1993) is an event watershed nonpoint pollution model. It uses algorithms for simulating watershed hydrology and pollutant accumulation and washoff similar to those found in such major models as HEC-1 and STORM. VAST uses a binary-tree format for the river-tributary arrangement and has pre- and post- processors for ease in data input and output (Kuo et al., 1994). The model is of intermediate complexity and data requirement and can be considered as a "black box-type" of model.

The original VAST uses the unit-hydrograph approach for computing runoff from rainfall. In order to improve its application in Taiwan, a physically-based routine for estimating watershed loss parameters which was developed in Taiwan was used to replace the rainfall loss simulation routine in VAST. The modified version was named VANTU. Other modifications, such as the addition of a Muskingum-Cunge procedure for flood routing, is currently being worked on.

Results and Discussion

Some stormwater sampling were carried out in the Feitsui Reservoir Watershed during 1993 and 1994. Data are still limited due to logistic and other difficulties encountered during the sampling effort. The data, togethger with another set of data collected through a sister-project (Lo et al. 1994) were used in testing VAST and VANTU. The results are presented and discussed below.

1. Hydrograph Simulation

Figure 1 depicts results of VAST modeling of the hydrograph from a storm event of 10/29-30/94. The results suggest that VAST could simulate satifactorily the rainfall-runoff process in the Feitsui Reservoir Watershed.

2. Nonpoint Pollution Simulation

Figure 2 shows an example of VAST model calibration results. The data were those collected for a tea garden in the Watershed. Pollutans examined include total suspended solids (TSS), biochemical oxygen demand (BOD), total phosphorus (TP), and total nitrogen (TN). Calibration paramenters include prior dry days, the washoff coefficient (K), and the fraction of TSS as a specific pollutant. Results indicate that the important parameters are the washoff coefficient, K, and the fraction of TSS as a specific pollutant.

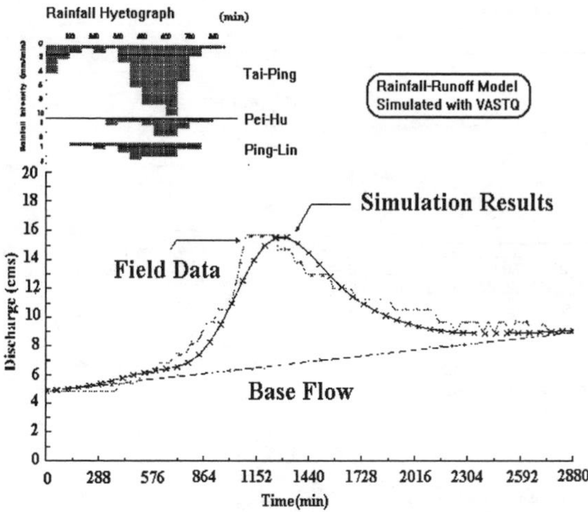

Figure 1. The Hydrograph Simulated with VASTQ

From the quality calibration results, one can derive an observation which is of special interest to water quality modellers. Because of Taiwan's reservoir watersheds are mostly in mountainous areas with very steep slopes, and because of Taiwan's storms are of quite high intensity, the washoff coefficient may be much higher than those reported in the US literature. Also, there seems to be two different K values which would be needed for a single runoff event, rather than a single K that is commonly used in the literature. A lower K value, followed by a higher K value, apper to fit the data well. Furthermore, it was noticed that the total nitrogen concentration was very high at the beginning of the runoff event and then gradually tapered off. This could be due to the excessive application of fertilizers such as chicken manure, which is a common practice in Taiwan.

Conclusions

Modeling nonpoint source pollution for reservoir watershed in Taiwan requires special attention to Taiwan's environmental conditions and its physical and climatological characteristics. The steep slopes and high-land farming, and lack of conservation practices make the washoff phenomenon much more severe than in other parts of the developed world. The farming practices, e.g., excessive use of fertilizers, also contribute to the unique nature of the NPS pollution problem in Taiwan.

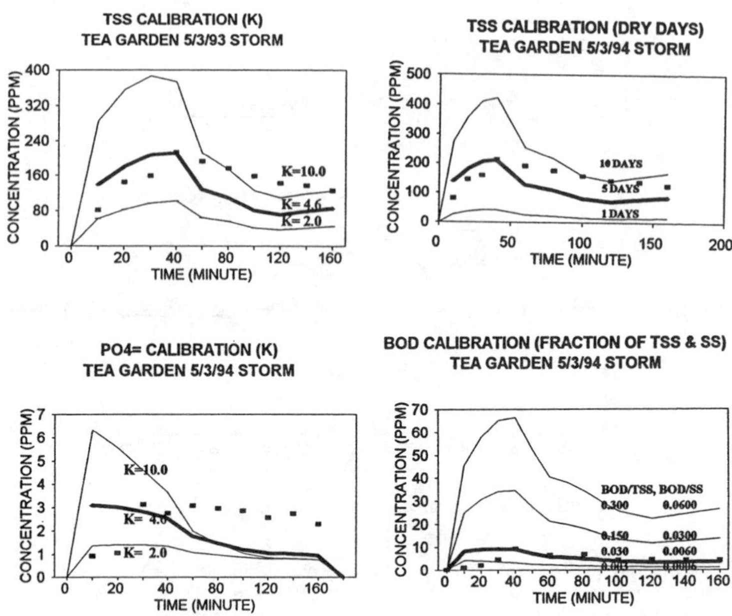

Figure 2. Tea Garden Calibration for Storm of 5/3/94

References

Kuo, J. T., S. L. Yu, S. C. Lee, M. C. Wu, and M. C. Kuo, A Study on Non-point Source Pollution for Reservoirs and Lakes (I). Technical Report 83EC2A371004, Dept. of Civil Engr., National Taiwan University, Taipei, Taiwan, June, 1994.

Lo, S. L., and Yu, S. L., A Survey of Critical Areas – Land Use Effect on Water Quality in the Water Source Quantity and Quality Protection Zones. Report No. 361, EPA–83–E103–09–08. Graduate Institute of Environmental Engr., National Taiwan Univ., Taipei, Taiwan, June, 1994.

Tisdale, T. S., R. J. Kaighn, and S. L. Yu, The Virginia Storm (VAST) Model for Stormwater Management – User's Guide. University of Virginia, December, 1993.

Water conservation through reclamation of sewage for reuse

Kee Kean Chin[1] & Say Leong Ong[2]

Abstract

In major urban centers of ASEAN there have been water shortages. Water conservation through reclamation and reuse is essential. A 45,000 cu m per day treatment plant was built to reclaim sewage for industrial uses. The reclaimed water in general meets with the quality requirements for general washing and process applications of industries such as paper and textile. As demand of ultrapure water increases advanced treatment processes to reduce dissolved ions, silica, organic and inorganic particles and other impurities become necessary. This paper presents data of field studies of the performances of a RO plant and an electrodialysis reverse (EDR) plant respectively producing 30 and 170 cu m/hr ultrpure water for their intended use.

Introduction

The demand of ultrapure water is increasing at a fast pace due to the expansion of the chemical, petrochemical, pharmaceutical, biotechnological and electronic industries. Water purity required by each industry for its manufacturing process, however, differs significantly. Specifications for high pressure bioler feed water although not as stringent as those specified for the electronic industry requires conductivity of 2 uS/cm or lower. The Jurong Industrial Water Work which uses a treatment train consists of (1) screening and prechlorination, (2) chemical clarification, (3) rapid gravity filtration, (4) aeration and (5) post-chlorination is unable to produce ultrapure quality water without further treatment. Treatment processes to produce ultrapure water include: chemical coagulation and sedimentation, multimedia sand filtration (F), cartridge filtration (CF), reverse osmosis (RO) or

[1]Professoral Fellow, Department of Civil Engineering, National University of Singapore, Kent Ridge, Singapore 0511.
[2]Senior Lecturer, Department of Civil Engineering, National University of Singapore, Kent Ridge, Singapore 0511.

the electrodialysis reverse (EDR) and deminerization (DM) with cation and anion exchange beds.

Field Studies

Field studies were carried out at a RO plant (48 cu m/hr) and a water reclamation plant using electrodialysis reverse system. Table 2 shows the characteristics of the raw waters . The Presumptive coliform is around 3 million per 100 ml of secondary treated sewage and that of the faecal coliform 5 hundred thousand per 100 ml. Pre- and post-chlorination at the Jurong Industrial Water Work have eliminated practically all the coliform organisms. There is, however, regrowth of microorganisms in the industrial water distribution pipelines and the inclined plate settling tank as the plate counts of microorganisms are high. Total organic halogen (TOX) including chloroform (3 ppb), 1,1,1-trichloroethane

TABLE 1 Water characteristics

Parameters		Treated sewage	Effluent			
			JIWW	EDR	RO	DM
Colour	HU	30	20	5	-	-
Turbidity	JTU	9	2	2	1	1
pH	Unit	7.1	7.1	6.0	5.8	5.8
BOD5	mg/l	26	2	-	-	-
COD	mg/l	78	32	16	10	8
TOC	mg/l	17	9	-	3	-
Total hardness	mg CaCO3/l	125	119	-	1	2
Total alkalinity	mg CaCO3/l	101	87	-	8	7
TSS	mg/l	37	2	-	-	-
TDS	mg/l	610	565	52	21	14
Chloride	mg/l	20	202	14	0.6	0.4
Conductivity	uS/cm	946	934	-	36	38
Detergent	mg LAS/l	0.13	0.12	-	-	-
Phospate	mg p/l	3.1	0.7	1.4	0.1	nd
Ammonia-N	mg N/l	12.6	11.9	1.1	1.7	1.2
Nitrate-N	mg N/l	18.4	15.6	2.9	4	4
TN	mg N/l	-	-	8	-	-
Floride	mg F/l	0.8	0.6	-	-	-
Sulphate	mg S/l	102	77	3	0.6	3.9
Ca	mg Ca/l	32	25	2.3	0.2	0.9
K	mg K/l	18	16	0.5	1.4	0.8
Mg	mg Mg/l	12	9	0.8	0.1	0.1
Na	mg Na/l	126	120	8	3.2	5
Al	ug Al/l	12	5	-	4	3
Cu	ug Cu/l	140	7	-	7	2
Fe	ug Fe/l	2,300	37	100	48	11
Mn	ug Mn/l	440	64	3	0.02	1
Ni	ug Ni/l	230	36	-	7	4
Pb	ug Pb/l	86	8	-	2	0.5
Zn	ug Zn/l	95	71	-	8	6

(3 ppb), trichloroethane (2 ppb), 1,2-dichlorobenzene (2 ppb) and trihalomethane (3ppb) are observed in the secondary treated sewage effluent. Dibromomethane, bromodichloromethane and bromoform not observed at the secondary treated sewage effluent are formed upon chlorination even though the residual chlorine concentration is lower than 0.2 mg/l in most cases.

At the water reclamation plant using RO system polyaluminum chloride is the primary chemical used in coagulation and flocculation. Without a clarification step mud balls were observed to form in the multimedia sand filter which gave rise to considerable operating problems (Chin & Ong 1992). Filter backashing cycle was around 24 hours and downstream cartridge filter (0.45 um) life averaged only around 22 days. With a clarifier to remove flocs the backwash cycle of sand filter is at present more than 80 hours and the cartridge filter change-out time extended to around 42 days. Cartridge filter effluent has a turbidity of lower than 2 JTU.

TABLE 2 TTHMS in waters (ug/l)

Effluent	TOC	CHC13	CHBrC12	CHBr2C1	CHBr3	TTHMs
Sewage	17	3	ND	ND	ND	3
JIWW	9	10	6	4	3	24
Carbon ad	9	11	7	5	4	28
Cartr fil	8	11	7	5	3	26
RO	1	9	7	5	3	24
DM	0.9	11	8	6	4	28

Spiral wound cellulose acetate RO unit was used for advanced treatment of treated sewage. RO membrane fouling was observed to be a problem in sewage reclamation. High nutrient concentrations in treated sewage provide an ideal condition for biogrowth. The coliform count of the effluent of the Jurong Industrial Water Work was zero after chlorination. The 24 hour plate count of the same water 5 km downstream at the distribution pipe line was around 6,000/100 ml. The count of the influent to the RO unit was 990,700/100 ml and that of the RO effluent was 84,700/100 ml. Replacement of the RO membrane was necessary after 8 months of operation although a 2-year life span is specified by the manufacturer. Conductivity of the RO product water was around 36 uS/cm and RO product water yield was between 60 to 70 % of the feed water (48 cu m/hr). The RO membrane was not able to removed the TTHM's formed during chlorination at the Jurong Industrial Water Work (Table 2).

When the electrodialysis reverse (EDR) process was used the Jurong industrial water was often blended with PUB water. The TDS of the feed and the product waters are given in Table 3. The ratio of JIW/PUB water averaged around 1/1.6. The product water yield was around 85 % and the conductivity of the product water ranged

Table 3 TDS of the feed water and product water of EDR

Operation time, day	TDS	
	Feed water	Product water
0	990	161
15	813	89
30	774	82
45	862	74
60	467	66
75	331	51
100	289	49

from 41 to 180 uS/cm. The EDR uses a series of membranes made from ion exchange resins. The membranes are arranged in stacks with a porous spacer inserted between two ion exchange membranes. When a DC field is imposed in this array cations will pass through the cation selective membrane and the anions through the anion-selective membrane resulting in ions being removed from one channel and concentrated in adjoining channel. The limitation of the EDR system is the possibility of 'hot spots' resulting from extremely high voltage occurring at the membrane boundry layer. This layer tends to ionize water into hydrogen ion and hydroxide ion and thus causes the precipitation of metal hydroxides that foul the EDR membranes.

Conductivity of both the RO product water and the EDR product water is high even for high pressure boiler feed water. The polishing stage often includes demineralization using cation and anion ion exchange beds to be followed by cartridge filtration (< 1 um), ultrafiltration and ultra violet light disinfection. The conductivity of the demineralized water is less than 1 uS/cm.

Conclusion

With limited water resources reclamation of sewage for industrial water is a viable alternative. Both the RO and the EDR systems were able to produce ultrapure water for boiler feed water. The cost of production of ultrapure water for high pressure boiler feed water is high at around US$ 2.0/cu m of product water. The cost will be considerably higher if electronic grade water and ultrapure water for the pharmaceutical industry are produced with more advanced steps of treatment.

References

Chin K K and Ong S L (1992). Experiences of non-potable reuse of wastewaters. Wat. Sci. Tech. Vol 26, No 7-8, pp 1565-1571.

Developing a Wasteload Allocations Protocol for Streams in Taiwan

Wu-Seng Lung[1], M. ASCE, Han-Keng Lee[2], Jan-Tai Kuo[3], and Jehng-Jung Kao[4]

Abstract

To assist the local regulatory agencies in developing wasteload allocations (WLA), the Taiwan Environmental Protection Administration (TEPA) has launched a program to develop protocols for WLA. Funded by the TEPA, this project was designed to standardize water quality models for stream wasteload allocations. STREAM, ESTUARY, and HAR03 are three modeling frameworks selected for adoption and standardization. While these models from the U.S. served as a good starting point for this project, characteristics of local streams required modifications to accommodate the hydraulic and kinetic conditions germane to these receiving waters.

Introduction

Under the current technology-based approach to water pollution control, water quality of many streams in Taiwan has not been improving significantly, due to increasing wasteloads from both municipal and industrial discharges. Trying to reverse this trend, the Taiwan Environmental Protection Administration (TEPA) has developed a water quality-based approach with wasteload allocations (WLA). Section 9 of the Water Pollution Control Regulations (TEPA, 1991) stipulates: Local regulatory agencies may develop wasteload allocations for a receiving water or portions of the receiving water which falls into either of the following categories:

1. Water quality standards are being violated under effluent limitations for point source discharges.
2. Receiving waters require special protections.

Although local regulatory agencies have the mandate and responsibility of conducting a WLA, a protocol must be developed and water quality models must be standardized. As part of the protocol development, a case study is

1. Professor of Civil Engineering, University of Virginia, Charlottesville, VA 22903, USA

2. Associate Professor of Hydraulic Engineering, Feng Chia University, Taiwan

3. Professor of Civil Engineering, National Taiwan University, Taiwan

4. Associate Professor of Environmental Engineering, National Chiao Tung University, Taiwan

being performed to demonstrate how to adopt, configure, calibrate, and verify a water quality model. Wusi (i.e., Wu Creek), a stream in central Taiwan was selected for this purpose. Three water quality models are being configured for a major tributary of Wusi. This paper reports the progress of this effort in Taiwan.

BOD Wasteload Allocations

One of the first water quality problems addressed in this project is BOD wasteload allocations. Although the BOD concentrations in many streams in Taiwan are high, their impact on the dissolved oxygen (DO) levels is relatively small. In fact, many streams exhibit near or over saturation in DO levels. The typical DO sag curve is rarely observed in these streams. It should be pointed out that streams in Taiwan are relatively short and have very steep bottom slopes. A combination of short travel times and significant reaeration results in little depression in DO levels. Therefore, instead of a DO standard, a BOD standard has been promulgated by regulatory agencies for receiving streams. Such a unique characteristics could potentially simplify the modeling work for BOD wasteload allocations.

While the DO component in a stream BOD/DO model could be decoupled from the BOD component with the justification of this unique stream hydraulic characteristics, another local situation would significantly complicate the modeling task. Wastewater discharges in Taiwan are quite intensive and located on drainage pipes far away from the receiving water. A typical wastewater discharge may travel a good distance before it reaches the receiving stream. Attenuation of the waste is therefore significant and must be accounted for in the determination of wasteloads. The complicate network of intensive wastewater discharges simply makes this task of estimating the delivery ratio extremely difficult, if not impossible. Another factor contributing to this difficulty is data deficiency. In fact, a companion project which is currently underway is to develop a data base for wasteloads and delivery ratios

While the BOD deoxygenation is not an important process in terms of the DO budget in these streams, it is still a key to modeling the BOD concentrations, particularly in light of the fact that BOD standards must be met under the water quality-based approach. In addition, industrial wastewaters have a wide range of BOD decay rates which must be independently derived for the modeling work, albeit another difficult task due to data deficiency.

Watershed Hydrology and Design Flow

Another important factor in wasteload allocations is the design flow which is closely related to the watershed hydrology. The prolong period of dry months in many watersheds results in no flow in the streams. For example, 7-day 10-year low flow which are normally used as the design flow in WLA in the U.S. are zero. The wastewater flow is the stream flow during the drought season. For practical purpose, the Q_{75} flow is being considered as the design flow for WLA.

Water Quality Models

The first model adopted is STREAM, which is based on the analytical solution to the steady-state, one-dimensional mass transport equations for CBOD, NBOD, and DO in a stream (Thomann and Mueller, 1987). The program code was initially developed by Manhattan College for their Summer Institute in

Water Pollution Control (see Lung, 1987). The following kinetic processes are included: CBOD decay and deoxygenation, NBOD decay and deoxygenation (nitrification), benthic oxygen demand, algal photosynthetic oxygen production and respiration, and stream reaeration. The Tsivoglou's Equation is selected to quantify the reaeration coefficient since it is particularly appropriate for shallow, rapid moving, and low flow streams.

The ESTUARY model was developed by Manhattan College for one-dimensional estuaries. Again, it is based an analytical solution to the steady-state, one-dimensional mass transport equation for estuaries. It is basically a mass transport model, using salinity as a natural tracer to determine the longitudinal dispersion coefficient. In estuaries, this dispersion coefficient is primarily dominated by tidal actions. In tidal fresh rivers, vertical gradients of the longitudinal velocity make up the bulk of dispersion. As in STREAM, first-order kinetics are formulated for CBOD and NBOD reactions in the water column.

The third model selected for standardization is HAR03, based on the finite segment approach, for steady-state, multi-dimensional analyses of mass transport and water quality constituents characterized with the first-order kinetics. It can be configured to simulate CBOD, NBOD, chloride, salinity, specific conductivity, and DO deficit.

All three models operate on PCs with integrated screen and hardcopy graphics to visualize the model results for comparison with field data (Lung, 1987). Effort to develop a visualization package in Chinese is underway. This is an important aspects because the modeling frameworks will be used by local regulatory agencies in Taiwan.

The Wusi Watershed and Wasteloads

The Wusi watershed (Figure 1) has an area of 2,026 km^2, the 4th largest in Taiwan. The watershed covers the Taichung County, the City of Taichung, and part of the ChangHwa and NanTao Counties. Land use in the mid and lower portions of the watershed is dominated by industries. Both industrialization and urbanization have significantly impact the water quality of the lower Wusi.

Major sources of pollution in the Wusi watershed include industrial, poultry and hog farming, domestic, and tourism. Figure 1 shows the distribution and quantity of BOD loads in the watershed. The total BOD_5 load from domestic wastewater in the watershed is about 49,180 kg/day, from a population of 1.8 million. The total domestic wastewater flow is 5.42 m^3/sec. There are over 200,000 hogs in the drainage area, contributing a BOD_5 load of 20,470 kg/day.

The Wusi Water Quality

Figure 2 shows the BOD and DO concentrations in the main stem of Wusi in recent years. The significant increase in BOD concentrations in the lower river reflects wasteloads from industries and municipalities in the lower watershed. As indicated earlier, their impact on the DO levels is not pronounced (Figure 2) as the organic materials are transported to the sea very rapidly.

References

Lung, W.S. (1987). Water Quality Modeling Using Personal Computers. *Journal of Water Pollution Control Federation*, 59(10):909-913.

Taiwan EPA (1991). Regulations of Water Pollution Control. Taipei, Taiwan.

Taiwan EPA (1993). Wusi River Planning for Pollution Control. Taipei, Taiwan.

Thomann, R.V. and Mueller, J.A. (1987). *Principles of Surface Water Quality Modeling and Control*. Hapers & Row, Publishers, New York, NY.

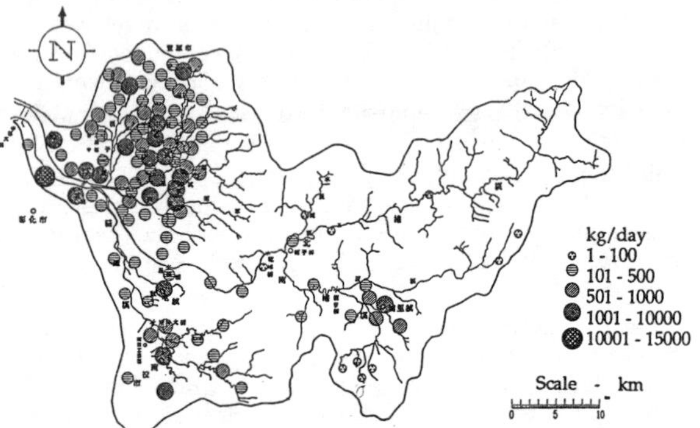

Figure 1. Wusi Watershed and Point Source BOD Loads

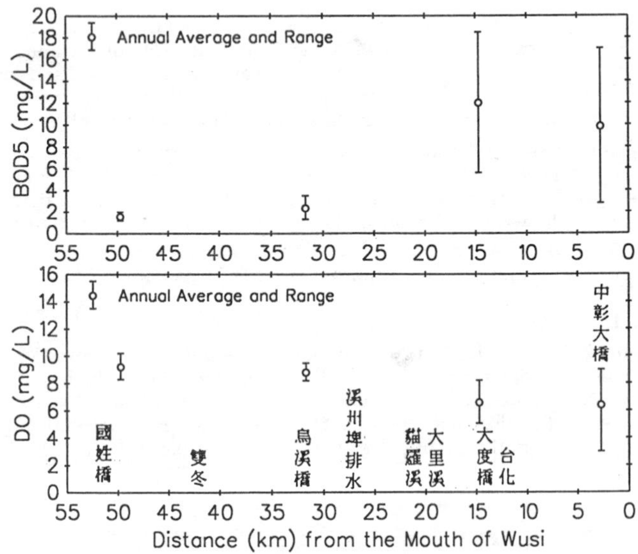

Figure 2. Measured BOD_5 and DO Concentrations in Wusi

Development of A Nonpoint Source Pollution Control Strategy for Taiwan

Shang-Lien Lo[1], M. ASCE, Shaw L. Yu[2], M. ASCE
Jan-Tai Kuo[3], M. ASCE, M.C. Wu[4] and K. Y. Lin[4]

Abstract

Many reservoirs in Taiwan, especially those in mountainous, upland areas, are suffering from pollution coming from their watersheds which are subject to enormous pressure for development. The climate and physical features of the watersheds, as well as the social-economical situations in Taiwan are quite different from those found in developed countries such as the United States. This paper presents a strategy for nonpoint source control in Taiwan which is based on a thorough review of available literature and considerations of the specific environmental conditions in Taiwan.

Introduction

In recent years many reservoirs in Taiwan have been found to either have eutrophication problems or be in danger of becoming eutrophic. Most of the reservoirs in Taiwan are sources of public water supply and the protection of their water quality is of critical concern. The rivers in Taiwan are mostly short with steep slopes. When it rains the stormwater runoff flows quickly in the rivers and out to the sea in a matter of hours to a few days. On the other hand stormwater runoff entering the reservoirs will stay a much longer time. Therefore reservoirs are especially vunerable to pollution from nonpoint sources which are mostly storm-induced.

---

[1]Prof., Grad. Inst. of Envir. Engr., National Taiwan Univ., Taipei, Taiwan
[2]Prof., Dept. of Civil Engr. & Appl. Mech., Univ. of Virginia, Charville, VA
[3]Prof., Dept. of Civil Engr., National Taiwan Univ., Taipei, Taiwan
[4]Grad. Student,Grad.Inst.of Envir.Engr.,National Taiwan Univ.,Taipei,Taiwan

Aside from eutrophication, many reservoirs are also suffering from siltation. A recent report (Water Resources Planning Commission, 1993) indicates that the storage capacity of the island's reservoirs is decreasing at an astounding annual rate of 14.7 million cubic meters due to siltation ! The total capacity estimated in 1994 was 1.8 billion cubic meters, which is only about 76% of the original design capacity. The problem was thought to be Taiwan's geography and climate. The steep mountain terrain, loose soil, intense storms, and earthquakes all contribute to high potential of erosion. But man's activities in the watersheds also have been found to be very much responsible for erosion. Extensive high-land farming, road construction and community development all destroy the vegetation of watershed areas. Excessive use of fertilizers and pesticides, and the seasonal use of river beds as cultivating grounds bring nutrients and other pollutants to the streams and reservoirs. Many laws and regulations have been established to protect water-supply watershed areas. The effectiveness of these laws and regulations has been small because of meaninglessly low fines and panalties and implementation problems.

Other challenges Taiwan must deal with in terms of nonpoint source pollution control include the lack of basic data, institutional and organizational problems, public awareness and commitment, a stable and continuing funding source, and most importantly, how to find a way to balance environmental protection against economic development and industrialization.

A "Three-Prone" Control Strategy

In a previous study (Yu et al. 1993), a thorough review of nonpoint source control literature was made and discussions were made with water quality control officials, environmental practitioners and scholars in Taiwan were held . The purpose was to develop a consensus for establishing a nonpoint source pollution control strategy for Taiwan. The following observations were made after the literature review and the discussions:

1. **Taiwan is a small island country with a very large population. The "economic miracle" has brought to Taiwan an enormous foreign reserve and one of the fastest-growing economy in the world. On the other hand, the environmet has suffered due to years of neglect. It is time for Taiwan to pay much more attention to protecting the environment and to enhance the quality of life for its people and future generations.**

2. **Economic growth must continue. Development in the watersheds**

cannot and should not be prohibited. However, the "best" technology should be used and strictly enforced upon developers (public or private) so that all potential "added-on" pollution is controlled or elimilated. Therfore, a " limited" development strategy should be promoted instead of "no development".

In order to ensure an effective control of nonpoint source pollution, a "three-prone" approach is proposed. The first is to start immeadiately an extensive collection of basic data, especially on the quality side. Stormwater runoff data will provide the basis for nonpoint pollution model calibreation and verification. Data on the performance of certain best management practices (BMPs) will allow BMP selection to be made and design guidelines to be developed to suit Taiwan's conditions. The second is to initiate demonstration studies in selected watersheds on pollution source identification and quantification, model application, and BMP effectiveness. The third, and certainly very important, is to examine the institutional, organizational and legal needs for implementing a full-scale nonpoint source control program in Taiwan. It is suggested that the three elements of the nonpoint control strategy be initiated at the same time and as soon as possible. Eventhough in Taiwan point sources of pollution have not been very well controlled, we cannot wait until point sources are controlled and then begin the nonpoint control effort because the former effort may take many years to complete. In the meantime the reservoirs may render useless because of nonpoint pollution and siltation.

A Case Study

A case study was made (Lo and Yu, 1994) on characterizing nonpoint pollution in a reservoir watershed. The Feitsui Reservoir, which supplies 2 million cubic meters of water per day for 4 million people in the Taipei metropolitan area, was chosen for the study. The reservoir watershed include mostly wooded mountain areas with some cropping (mostly tea gardens and rice fields), recreational and small towns. A sampling program was implemented during the rainy months (March through May) in 1994 at several sampling sites representing tea and fruit gardens and rice fields. Data collected include runoff rate, concentrations of biochemical oxygen demand (BOD), total suspended solids (TSS), total phosphorus (TP), and total nitrogen (TN) during eight storm events.

Pollutant loadings for the storm events were computed and an estimate was made of the total annual unit areal loading for each pollutant, based on information on rainfall and runoff for the watershed studied. Table 1 is a summary of the results.

Table 1. Nonpoint Source Pollutant Loadings for Different Land Uses in the Feitsui Reservoir Watershed

	Loadings in kg/ha/yr			
Land Use	BOD	TSS	TP	TN
Tea Gardens	5.5	104.6	1.71	2.57
Fruit Gardens	3.2	129.4	0.40	0.63
Rice Fields	7.4	144.2	0.24	6.02

Comparing to literature information, the limited data obtained in Taiwan show a high phosphorus loading for tea gardens, high nitrogen loading for rice fields, and generally low loadings for organic content. Much more data are needed before any definitive conclusions can be made.

Conclusions

Nonpoint pollution control in Taiwan is as urgently needed as point source control. Efforts should begin as soon as possible for basic data collection, tests of control technologies, and formulating an institutional framework for implementation of control strategies.

References

Lo, S. L., and Yu, S. L., " A Survey of Critical Areas-- Land Use Effect on Water Quality in the Water Source Quantity and Quality Protection Zones", Report No. 361, EPA-83-E103-09-08. Graduate Institute of Environmental Eng'g, National Taiwan Univ., Taipei, Taiwan, June, 1994, 79p.

Water Resources Planning Commission, " Annual Water Resources Report - 1993", Water Resources Planning Commission, Taipei, Taiwan, 1993.

Yu, S. L., Lo, S. L., J. T. Kuo, " Developing A Nonpoint Pollution Control Strategy for Reservoirs and Lakes in Taiwan ", Report No. 354, NSC82-0410-E-002-332. Graduate Inst. of Envirn. Engr., National Taiwan Univ., Taipei, Taiwan, Nov., 1993, 109p.

State of the Art on Runoff Control Facilities in Japan

Ichiro Tanaka[1], Mitsuyoshi Zaizen[2]

Abstract

Stormwater runoff control has two methods. That is to say; stormwater storage and infiltration.
Concerning on storage facility, it is used not only to reduce runoff volume also to use the facility such as playground on fine weather. By means of stormwater infiltration, we will be able to solve several problems at one time, such as prevention of inundation, groundwater cultivation, CSO problem and so on. But, there are two problems left in future; contamination of groundwater, decline of subgrade's strength.

Introduction

In Japan, the objective of stormwater runoff control is prevention of inundation. Stormwater runoff control has two methods. That is to say; stormwater storage and infiltration. Recently stormwater storage facilities are used as park, tennis court, parking lots on fine weather. Meanwhile in case of infiltration the change from a quick disposal of stormwater to runoff control will be accelerated by the Ministry's revision of a government ordinance for drainage systems on private properties on July 1, 1994. Besides the recognition that the best way to reduce non point pollution would be runoff control is rising up. This report describes an overview of usual method and technical standards, purposes, examples and future of the stormwater runoff control.

----

[1]Senior Researcher, [2]Research Engineer, Japan Institute of Wastewater Engineering Technology (JIWET), Ikebukuro Chitose Bldg., Nishiikebukuro 1-22-8, Toshima-ku, Tokyo, 171 Japan

Usual Method on Planning and Design of Runoff Control

Total volume of rainfall and rainfall distribution (hyetograph) are more important than peak discharge when we design runoff control facilities. Therefore, we usually have adopted alternating-block design storm. In case of short rainfall duration (< 3 hours), rainfall intensity-duration curves for individual frequencies are approximated by Talbot expression. In case of long rainfall duration (> 6 hours), we use Cleveland expression. Design storm return period depends on the scale of facility. In case of on-site facility it is 3-10 years, about off-site facility it is 30-50 years.

Transition of Stormwater Storage Facilities

During the 1960s and the early 1970s, when large-scale developments, such as large residential area or golf course, were constructed, they built stormwater storage facilities to reduce runoff increased by developments. Most of the facilities in the times are off-site facilities and the main purpose is storage by using large area. From 1975, the rising cost of land has brought with it multipurpose stormwater storage facilities of various kinds. Most of them are on-site facilities and they are used as park, tennis court, parking lots on fine weather. After Ministry of Construction established the Technical Standards for Multipurpose Stormwater storage facilities in 1986, they have been constructed more progressively. It is one of interesting parts of the Standards that the critical depth of each facility is stipulated, as shown in Table 1.

Table 1. Critical Depth of facility

The name of facility	Critical Depth (meter)
Green lot (Housing)	0.3
Parking Lot	0.1
Playground (School)	0.3
Child park	0.2
Park	0.3

Rainwater Use

Recently how to restore rainfall to use is one of the most emerging issues in Japan. A terrible shortage of water hit Japan last year, more than 10% people of the total population was obliged to endure suspension of water supply. It is an ironical story that detention tanks constructed for the purpose of runoff reduction have come to draw public attention due to the shortage of water. But we must admit there is an incompatible point between runoff reduction and rainwater use. Because detention tanks for runoff reduction should be empty before it rains, to the contrary detention tanks for rainwater use should be full of water before it rains. Therefore we do not count the volume of detention tanks for rainwater use in the volume of runoff reduction when we plan river improvement or sewer system.

Underground Storage Facility of Ready-made Concrete Blocks

In recent, among the constructing technology of underground storage facilities, the underground storage facility of ready-made concrete blocks has come to be in footlight for its some merit. For example, the concrete blocks being made in factory, we can shorten a period of construction, and being made under ground, we can use the upper land of it ,such as parking lot, playground, park and so on. JIWET has researched and developed the design method of the facility in cooperation with a company, and it is drawing public attention.

Characteristic of Infiltration Facilities

The quantity of stormwater that can infiltrate into soil is dependent on not only the size of facility but also the permeability of soil, groundwater conditions and others. Therefore it is very difficult to estimate the effect at the planning and designing stage unlike storage facilities. But we have many results that infiltration facilities reduce the peak-flow by 40%-50% under the condition of small rainfall, proper installation and suitable maintenance. Moreover infiltration facilities have sub-effects on groundwater cultivation, preservation of spring water, recovery of urban river and so on.

Structure of Infiltration Facilities

There are two structures how to flow into infiltration facilities. One of them is base-cut method, and the other is peak-cut method. Base-cut method means that at first stormwater inflows to infiltration facility and secondly excess water flows into inlet for discharge. In the base-cut method facilities, infiltration starts at the beginning of rainfall so that they have many kinds of effects like prevention of inundation, CSO problem, and groundwater cultivation. Peak-cut method means that at first stormwater flows into inlet for discharge to some extent, and then excess water flows into infiltration facility. The peak-cut method puts emphasis on prevention of inundation.

Actual Results of Infiltration Facility

The typical usage of infiltration is to prevent inundation of city where urbanization is rapidly advancing. By the installation of infiltration facilities, inundation of this area has dramatically disappeared. Table 2 shows some actual results of infiltration facilities in Japan.

Table 2. Results of Infiltration Facility in Japan

The name of city	Sapporo	Tokyo	Hamamatsu
Return Period	10 years	3 years	3 years
Soakaway	476	32,835	1,930
Infiltration Trench	1,629m	211,862m	-
Infiltration Curve	-	69,610m	-
Permeable Pavement	-	483,722m^2	-
Coverage Area	39ha	1,357ha	796ha

(as of March 1994)

Besides these, infiltration has been promoted by many local governments as follows; Kawasaki, Tiba, Yokohama, Shiogama, Amagasaki, Fukuoka and others.

Problems of Infiltration

In Japan, the needs of prevention of inundation have made some of local governments to decide installing infiltration facilities on road. In this case there are important problems; That is to say, contamination of groundwater and decline of subgrade's strength by influence of the infiltration. We will have to research and solve this problem as soon as possible.

Future of Runoff Control Facilities

Runoff control facilities are now being adopted in various locations throughout Japan. By means of them, we will be able to solve some problems at once, such as; improvement of water environment and runoff control.

Urban Wastewater Management in Southern Jiangsu

Michael C. Lee, Ph.D., P.E.[1] and Hua Ju[2]

Abstract

To address the environmental and public health problems caused by increasing pollution of water courses in Southern Jiangsu, China, the Jiangsu Provincial Environmental Protection Project Office (JEPPO) has undertaken a comprehensive environmental strategy and action program with support of World Bank loans. The urban wastewater management component includes pollution abatement in targeted industries, expansion of municipal sewage treatment systems, construction of sewers, integration of municipal and industrial sewage treatment systems, and development of standards to enable municipal governments and agencies of Changzhou, Wuxi and Zhenjiang to manage these projects effectively. In view of the number and complexity of contracts to be let under the project, the three municipalities and JEPPO appointed Camp Dresser & McKee International Inc. (CDM) to act as Project Implementation Consultant to assist the municipalities with implementation management.

Introduction

Increased growth in the Southern Jiangsu region through rapid industrialization, urbanization as well as population growth have contributed to increased pollution and have put a strain on the natural capacity of the environment to pollutants. A large portion of industrial and municipal wastewater an sewage is directly discharged without treatment to watercourses. Because of this, water pollution is especially problematic in Southern Jiangsu causing serious environmental and public health problems. The problem is further compounded due to the heavy dependence on surface water resources for public potable water supply, as well as for us for agriculture, aquaculture and transportation purposes.

---

[1]Associate, Camp Dresser & McKee International Inc., 18881 Von Karman, Suite 650, Irvine, California 92715

[2]Deputy Director, Jiangsu Provincial Environmental Protection Project Office for World Bank Loan, Nanjin, China

The World Bank has assisted the provincial and municipal authorities in the formulation of a comprehensive environmental strategy and action plans for Southern Jiangsu. As part of this program, the World Bank will provide loans for the construction of some 63 municipal facilities for urban wastewater management in Changzhou, Wuxi and Zhenjiang. CDM has been selected as Project Implementation Consultant to provide these authorities with assistance to implement these programs. The key elements of these projects include:

- More than 100 km of trunk sewer, branch, gravity and pressure sewer, including 2 km of submarine pipeline, and 293 km of reticulation sewer
- Some 26 sewage pumping stations of varying capacities
- New and extended sewage treatment works in Changzhou and Wuxi, including a 100,000-m^3/day primary treatment plant, two 50,000-m^3/day secondary treatment plants, and two 5,000-m^3/day secondary plants
- A pilot waste stabilization ponds system of 10ha in area for Zhenjiang, as well as a 400m river outfall and 1.5km of river bank protection

<u>Project Objectives</u>

The overall objective of the project will be to improve the environmental quality and condition of the region. This will be accomplished by not only constructing the necessary treatment works and infrastructure, but to develop sound policies that provide a proper balance between economic development and environmental protection in Southern Jiangsu. Specific objectives of the project include:

- Improve the formation, monitoring and enforcement of environmental policies and regulations
- Improve and later maintain the water quality in the major water bodies in Southern Jiangsu
- Construct infrastructure to effectively transport, treat and discharge industrial wastewater and municipal sewage generated in the three cities
- Establish standardized design procedures, specifications and documents for the efficient design of projects to be funded by World Bank
- Develop standard bid documents, procurement systems, contract selection criteria, quality control measures and project and construction monitoring and management procedures.
- Provide the municipalities and local agencies with the funding and technical assistance to be able to manage and implement the projects

funded by World Bank in an efficient, environmentally sound and cost-effective manner

- Provide sound transfer of technology to local municipalities, agencies and design institutes

<u>Project Approach</u>

CDM has identified a number of challenges associated with the implementation of this project. Because these challenges can all be overcome through good management and thoughtful planning, CDM has already developed an action plan to assure success.

- **DESIGN STANDARDIZATION** Although, the three cities may differ in their desires for standardized equipment, consideration should be given to standardizing either on a city-wide basis or multi-city basis. CDM will propose design standards to facilitate checking and cross-referencing of the contracts. CDM will propose and modify, if necessary, an appropriate standard to best suit the contracts under the project. The proposed 63 projects can be categorized into 7 major groupings: pipelines, pumping, sludge handling, equipment procurement, outfalls, and submarine pipelines. These categories lend themselves very well to standardized design procedures.

- **TIMELY REVIEW OF DETAILED DESIGNS** The contracts under this project cover a wide range of engineering disciplines, therefore, specialists from various fields will be required to review the detailed designs. CDM proposes to convene Technical Review Committees (TRCs) to review the major groups of projects in accordance with each project schedule. These TRCs could be composed of members of JEPPO and the three cities staffs as well as CDM's local associate firm. Technical presentations would be made by design institute staff and views exchanged on the technical approach to each project.

- **TENDER DOCUMENT PREPARATION** A standard tender document format for all the projects will save time required for the preparation of documents and facilitate cross-references between contracts. The standard bid document will also facilitate comparison of bids and shorten the time required to evaluate each bidders submission.

- **PRE-QUALIFICATION OF POTENTIAL BIDDERS** Pre-qualification of potential bidders will ensure that only experienced bidders capable of achieving the high quality standards set by World Bank and the three Municipalities are invited to tender. The Municipalities may depend on CDM's knowledge and experience with contractors around the world and in China during the evaluation of the pre-qualification submissions.

- **CONSTRUCTION MANAGEMENT** Meeting target completion dated for each contract will be critical to the overall progress of the project. CDM will identify the types of elements of the critical path that risk delaying major portions of the project and determine earliest and latest possible dates for completion. Once these sectional completion dates are known, CDM will recommend the most cost- and time-efficient construction packages to meet deadlines.

- **CONSTRUCTION AND EQUIPMENT PROCUREMENT** Procurement is on the critical path of any project because late delivery of equipment or materials can seriously delay or halt construction altogether. CDM will assist the Municipalities by identifying potential suppliers and manufacturers with good records for delivering quality goods on time. If procurement does cause a delay, CDM will revise the remainder of the program to minimize the effect to the entire project.

- **OPERATIONS AND MAINTENANCE** O&M consideration should form an integral part of the Project from initial planning, through design and construction phases. O&M training is an on-going requirement to assure that line staff and managers are regularly updated and reviewed on their procedures. Operations and Maintenance training will form an integral part of the in-city training seminars. CDM will assist the municipalities in the preparation of specifications which will require contractors and equipment manufacturers to submit complete, technically correct and easy to follow O&M manuals and training plans.

Conclusion

For 63 municipal facilities for urban wastewater management in Wuxi, Changzhou and Zhenjiang, good engineering practices combine with local design standards will be performed by local design institutes and municipalities. CDM, as Project Implementation Consultant, will review detailed designs, standard specifications, and bid documents. Modifications or additions will be made to conform with World Bank guidelines and international practices. Unique or not readily available materials or equipment suppliers or contractors will be prequalified. Standard procedures for competitive bidding and award will be established. Finally, monitoring and reporting procedures on quality control and construction management will be established.

Effects of Whole Depth Aeration on Water Quality in Two Hypereutrophic Reservoirs

Chi-Chin Hwang[1] and Ching-Gung Wen[2]

Abstract

Chen-Ching Lake and Fen-Shang Reservoir are two hypereutrophic water supply reservoirs. The reservoirs have been artificially aerated using diffuser systems since Sept. 1992. A post-aeration assessment program was conducted from Feb. 1994 through Jan. 1995. Results of water quality monitoring indicated that artificial aeration significantly reduced ammonia (NH3) concentration in Fen-Shang Reservoir, where culture media for nitrifying bacteria were provided together with the aeration system. Nitrification was insignificant in Chen-Ching Lake despite improved dissolved oxygen (DO) concentrations. Phosphorus removal was minimal in both reservoirs. No definitive reduction in chlorophyll *a* (chl-*a*) concentration was observed in either of the two reservoirs.

Introduction

Artificial aeration has been used to improve water quality in lakes and reservoirs (e.g., Garrell, et al. 1977; Taggart and McQueen 1981; Osgood and Stiegler 1990). In deep lakes, where significant stratification often leads to oxygen depletion in the hypolimnion, a hypolimnetic aeration can raise DO concentration and minimize the adversary effects of bottom anoxia. Due to depth constraint and the absence of an apparent thermocline, whole depth aeration is usually used in shallow lakes and reservoirs. However, a whole depth aeration not only raises dissolved oxygen concentrations, it also induces vertical mixing in the lake. Nutrient rich sediment can be brought up to the overlaying water column and utilized by algal cells. On the other hand, the decreased water clarity and increased algal mixing depths in aerated water bodies can lead to low primary productions (Lind et al. 1992; MacIntyre 1993). The contradictory effects of artificial aeration on algal growth explain the usually uncertain outcomes of lake management using such practices (Fast 1973; Toetz 1990; Osgood and Stiegler 1990). In this study,

[1] Associate Professor, Department of Civil Engineering, Kaohsiung Polytechnic Institute, Kaohsiung, Taiwan

[2] Professor, Department of Environmental Engineering, National Cheng-Kung University, Tainan, Taiwan

we compared the effects of artificial aeration on the water quality of two hypereutrophic reservoirs.

Study Sites Description

Chen-Ching Lake is a shallow (water depth 3-4 m), off-channel water supply reservoir with a surface area of 103 ha. It receives a 310,000 m^3 daily inflow from the near-by Kaoping River. Heavy pollution in source stream caused profound eutrophication in the reservoir. In 1994, the reservoir had a total phosphorus (TP) concentration of 210 mg/m^3, total nitrogen (TN) 2770 mg/m^3, and chl-*a* 24.8 mg/m^3. High ammonia concentration in the raw water dramatically raised chlorine consumption and the cost of water treatment. In addition, filter clogging algae degenerated infiltration efficiency, leading to frequent backwash of filtration basins. Algae and odor producing bacteria also caused fouling smell in the water supply. An aeration system was installed hoping to reduce ammonia concentration through promoting nitrification and to inhibit algal growth by increasing algal mixing depths. The system consisted of five air compressor units and a total of 168 air diffusers spreading on the floor of the reservoir.

Fen-Shang Reservoir has a surface area of 75 ha and a maximum depth of 15 m. It receives a 350,000 m^3 daily inflow from nearby Tong-Kang River. Averaged TP, TN, and chl-*a* concentrations were 1140 mg/m^3, 6430 mg/m^3, and 20.6 mg/m^3 respectively in 1994. An aeration system was installed and in operation since Sept. 1992. The system consisted of 200 air compressor units delivering air to 400 diffusers from five different locations along the shoreline. Floating bacteria culture curtains were installed on 15 cross-sections along the main channel to provide growth media for nitrifying organisms.

Methods

Water quality in both reservoirs was continuously monitored from Feb. 1994 through Jan. 1995. Samples were collected at six sampling stations, starting with Station A near the inlet through Station F near the outlet of each reservoir. In each station, water was withdrawn from 3 different depths: surface, mid-depth, and bottom, using a weighted plastic hose. Surface samples were collected at a depth of 30 cm. Bottom samples were collected 50 cm from the floor. Each sample was analyzed for DO, ortho-phosphorus (ortho-P), TP, NH3, nitrite nitrogen and nitrate nitrogen (NO2-N + NO3-N), total Kjeldahl nitrogen (TKN), chl-*a*, and pH. Secchi disc transparency was also measured at each sampling station.

Results and Discussion

Both reservoirs have relatively short hydraulic detention time (approximately 10 days for Chen-Ching Lake and 7 days for Fen-Shang Reservoir). Therefore, in-lake water quality is greatly affected by the inflow. Any improvement, or otherwise, of lake water quality tends to be masked by the quality of inflow. Comparisons between pre- and post-aeration water quality will not provide decisive conclusions unless the reservoirs have received relatively consistent inputs. In this study, a prolonged draught covered the year before, as well as the fist 3 months of the assessment program. It was followed by an unusually wet season. Therefore, it is more appropriate to evaluate the effectiveness of aeration based on in-lake variations of related water quality parameters.

Chen-Ching Lake

Dissolved oxygen concentration in Chen-Ching Lake held constantly above 5 mg/L during the study period as illustrated in Fig. 1a. Photosynthesis production of oxygen and active vertical mixing have contributed to a relatively high DO concentration in this shallow water body. In fact, dissolved oxygen had never been a problem even before aeration. Despite favorable DO concentrations and warm water temperatures, nitrification in Chen-Ching Lake was minimal judging from the relatively constant NH3-N and NO3-N concentrations shown in Fig. 1b. Nearly unchanged ortho-P and TP concentrations along the main channel as shown in Fig. 1c. reveal a rather insignificant phosphorus removal in Chen-Ching Lake. Declining phosphorus concentrations immediately after the inlet is believed to be caused by the settling-out of particle-bound inorganic phosphorus. The oxic water column might have suppressed sediment phosphorus release. However, continuous mixing generated by aeration have also prevented algal phosphorus from settling and

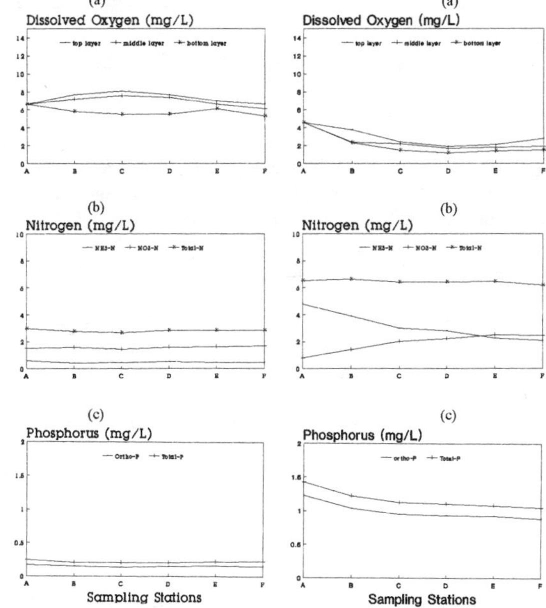

Fig. 1. Variations of dissolved oxygen, nitrogen, and phosphorus along the main channel of Chen-Ching Lake.

Fig. 2. Variations of dissolved oxygen, nitrogen, and phosphorus along the main channel of Fen-Shang Reservoir.

eventual burial. Chl-*a* resembled pre-aeration levels with an average concentration of 24.8 mg/m^3 and a maximum of 48 mg/m^3 measured in June 1994.

Fen-Shang Reservoir

Despite vigorous aeration, DO concentrations in Fen-Shang Reservoir averaged only 2.2 mg/L during the study period (Fig. 2a.). Active nitrification transformed appreciable amount of NH3-N to NO3-N as shown in Fig. 2b. The process consumed large amount of oxygen. Bio-filters installed across the channel provided nitrifying bacteria with growth media, contributing to a nitrification rate much greater than that in Chen-Ching Lake. Phosphorus removal was insignificant as indicated in Fig. 2c. Chl-*a* concentration averaged 20.6 mg/m^3, resembled pre-aeration levels.

Results of this study indicated that aeration had no discernible effect on suppressing algal growth in both reservoirs. Phosphorus removal was minimal because strong turbulence induced by aeration has prevented algal cells and particulate phosphorus from settling. Lacking bacteria growth media, nitrification was insignificant in Chen-Ching Lake. Since dissolved oxygen had not been a problem even before aeration, the prudence of aerating Chen-Ching Lake is questionable. Without aeration, a better phytoplankton settling rate would improve the removal of particulate phosphorus. Nitrification depleted large amount of dissolved oxygen in Fen-Shang reservoir. The entire reservoir would have turned anoxic without aeration.

References

Fast, A.W. (1973)."Effects of artificial destratification on primary production and zoobenthos of El Capitan Reservoir, California. ***Wat. Resour. Res.*** 9,607-623

Fast, A.W., Moss, B., and Wetzel R., (1973)."Effects of artificial Aeration on the chemistry and algae of two Michigan lakes." ***Wat. Resour. Res.*** 9,624,647

Garrell, M.H., Confer, J.C., Kirschner, D., and Fast, A.W., (1977). "Effects of hypolimnetic aeration on nitrogen and phosphorus in a eutrophic lake." ***Wat. Resour. Res.*** 13,343-347

Lind, O.T., Doyle, R., Vodopich D.S., Trotter, B.C., Limon, J.G., and Davalos-Lind, L., (1992)."Clay turbidity: regulation of phytoplankton production in a large, nutrient-rich tropic lake." ***Limnol. Oceanogr.***, 37(3)549-565

MacIntyre, S., (1993)."Vertical mixing in a shallow, eutrophic lake: possible consequences for light climate of phytoplankton." ***Limnol. Oceanogr.***, 38,798-817

Osgood, R.A., and Stiegler, J.E. (1990). "The effects of artificial circulation on a hypereutrophic lake." ***Water Resources Bulletin*** 26,209-217

Taggart, C.T., and McQueen, D.J., (1981). "Hypolimnetic aeration of a small eutrophic kettle lake: physical and chemical changes." ***Arch. Hydrobio.*** 91,150-180

Toetz, D.W.(1981)."Effect of whole lake mixing on water quality and phytoplankton." ***Water Research*** 15,1205-1210

Case Studies in Stream Preservation/Restoration

Ronald L. Rossmiller, PhD, PE, Member[1]

Abstract

This paper presents an overview of what cities are doing to preserve and restore riverine corridors in the United States. This is done by citing case studies from the cities of Scottsdale, Arizona, Englewood, Colorado, and Milwaukee, Wisconsin. To be successful, these programs must have participation from a wide variety of public and private interest groups. A number of objectives are contained in these urban stream corridor programs. The extent to which some or all of these objectives are included in the citys' preservation/restoration programs is discussed.

Introduction

Many cities throughout the United States are recognizing that the river corridors within their boundaries are deteriorating into single purpose flood control channels. Aquatic and wildlife habitat are disappearing, recreational opportunities are nonexistent, access for maintenance is missing, and the corridor is physically unsightly. To combat this trend, many cities are beginning to develop programs that address these problems in a comprehensive manner.

A number of objectives are contained in these urban stream corridor programs, i.e., floodplain habitat for birds and animals, active and passive recreation, flood control for a wide range of rainfall events, aquatic habitat, historic and cultural values, water quality enhancement for urban storm runoff, maintenance trails, a buffer between land uses, and control of land uses within the floodplain through zoning.

----

[1]National Program Director for Stormwater Management, HDR Engineering, Inc., 500 108th Avenue NE, Bellevue, WA 98004

Scottsdale, Arizona

The City of Scottsdale has grown from a sleepy town of some 2,000 people in 1950 to a vital, booming city of well over 100,000 today (Communications and Public Affairs, 1985). Problems with runoff from the infrequent rains in this desert community have also become progressively more destructive during this time as development began to encroach into the floodplain of Indian Bend Wash.

Indian Bend Wash cuts through the middle of Scottsdale and during flood events isolated one side of the city from the other. In 1961, the Corps of Engineers (COE) developed a plan to construct a seven mile long, trapezoidal concrete channel 23 feet deep and 170 feet wide.

Scottsdale's citizens thought that further dividing the city by building a concrete channel was not the solution. In 1964, a committee called the Scottsdale Town Enrichment Program (STEP) Committee was formed. "Turn the problem into an opportunity," the people said. "Make our eyesore a beauty spot of open space that will at the same time alleviate our flood problem."

A study verified the feasibility of this idea - to turn the entire Indian Bend Wash into a greenbelt of recreation running through the heart of the city; create a series of badly-needed recreational parks and lakes which could also serve as an effective way to control floods.

The COE had to be convinced that a greenbelt concept, as opposed to the traditional method of using a concrete channel, would work. This was accomplished by the combined efforts of citizens, unceasing congressional support, an innovative city staff, a dedicated city council, the STEP Committee, and city commissions.

In 1973, on the heels of the worst flood on record, the citizens approved a $10 million bond issue for flood control. By 1976, the city had entered into a contract with the COE to finance parks and recreation over a five-year period. Gradually, parks and lakes began to take shape along the seven-mile reach. Debris and wild underbrush were removed and special grasses with deep roots were planted in their place. Tennis court fences were hinged to swing with the flood waters. Golf courses were also designed to endure floods.

The project now provides a string of parks, golf courses, swimming pools, and fishing and boating lakes within walking distance of more than 60,000 residents. Of the nearly 1,200 acres in the greenbelt, more than 300 are devoted to city parks.

Englewood, Colorado

The city of Englewood has historically been affected by flooding of Little Dry Creek. Pictures from a 1917 flood show several feet of water on Broadway, the city's main street through its retail area. Major floods have also occurred in 1935 and 1973 (Ragland and Pflaum, 1989).

In the mid-1960s Little Dry Creek was enclosed in a box culvert west of Broadway to allow the construction of a 1.5 million square foot shopping center. In addition to existing older businesses along Broadway experiencing a decline in revenue, the culvert entrance was found to be unable to pass the 100-year flood peak. Residential development was also occurring southeast of Englewood along Little Dry Creek.

During 1974 to 1981, the following studies were made: (1) detain a portion of the flow on school recreation property upstream; (2) revitalize the downtown area between Broadway and the shopping center by removing old residences and businesses and building new businesses; and (3) remove this area from the 100-year floodplain by constructing an aesthetic multi-purpose open channel.

Funds for the project came from four sources: (1) tax-increment bonds sold by the Urban Renewal Authority; (2) private moneys from redevelopers; (3) Use Tax revenues from the city; and (4) funds from the Denver Urban Drainage and Flood Control District. Flood control and redevelopment planning and design began in 1981 and construction was completed in 1988. During this period, the developer went bankrupt and negotiations with a new one resulted in further extension of the box culvert to allow for more buildable land. Total cost of the project was over $60 million for the redevelopment portion and over $12 million for flood control work.

The upstream detention basin contains two soccor fields and a baseball diamond. The improved channel is a two-stage system consisting of a low-flow channel with a 40 cfs capacity that meanders through a grass-lined major channel designed for the 100-year peak flow of 3,650 cfs. The low-flow channel consists of a concrete invert with large even boulders along each side.

The channel improvements are enhanced by abundant landscaping, including irrigated sod along the channel corridor and extensive tree and shrub planting. A 10-foot wide maintenance road and recreational path traverses the entire length of the channel. Six new major tenants have been added to the area and sales tax revenues have improved the financial condition of the city.

Milwaukee, Wisconsin

City of Milwaukee personnel are taking a different approach to the restoration of Lincoln Creek. Over the years Lincoln Creek has been turned into a concrete channel. A preliminary study is being completed to remove the concrete lining from a 10-mile reach of the creek and restore it to a natural channel.

The preliminary study report had not been released at the time that this paper was written, so the following comments can only be surmised from the intent of this project.

A concrete channel serves only three land uses, two of which are unintended. One intended use is to convey flood waters in a non-damaging manner, non-damaging to real property, to personal property, and to human life. The two unintended uses are as a receptacle for all types of urban litter and trash and as a transportation route for those who can find access to the channel.

The second intended use is to free up a portion of the floodplain so that it can be utilized for urban development purposes. This use of the floodplain eliminates its use as a temporary storage area for flood waters, as a wildlife habitat, and as a location for filtering out some of the pollutants created by the urbanization process.

Restoration of the creek by removing the concrete channel and replacing it with natural materials will create a waterway with a more pleasing appearance. Restoration of the entire floodplain could be a much more challanging and expensive endeavor, depending on the magnitude of previous floodplain encroachment.

What is done to restore the Lincoln Creek channel and floodplain and how it is accomplished will be interesting to follow in the years to come. Will the people of the city of Milwaukee take a page from the history of Scottsdale, Arizona and Englewood, Colorado, or will they find new paths to take to achieve their objectives?

Appendix 1. References

Communications and Public Affairs. (1985). "Indian Bend Wash." City of Scottsdale, Arizona.

Ragland, Densel A. and John M. Pflaum. (1989). "Little Dry Creek Flood Improvements, Englewood." City of Englewood, Colorado.

PRESERVATION AND ENHANCEMENT OF RIVERINE CORRIDORS IN SOUTHEASTERN WISCONSIN: HISTORIC OVERVIEW AND CURRENT PRACTICE

Robert P. Biebel,[1] Member, ASCE

Abstract

The Southeastern Wisconsin Regional Planning Commission's regional land use and comprehensive watershed and floodland management programs initiated in 1961 have been directed toward the protection and enhancement of riverine environmental corridors within the context of a rapidly urbanizing Region. This paper describes the methods used to quantitatively define and categorize these corridors based upon natural resource base values and discusses the programs used to preserve and enhance these corridors.

Introduction

In Southeastern Wisconsin, the protection of riverine environmental corridors has been carried out through an integrated program involving comprehensive regional land use and watershed planning. The underlying concept includes a quantitative process to identify those natural resources that should be protected and preserved as environmental corridors, defined as linear areas in the landscape containing concentrations of natural resource and related amenities. These corridors often lie along the major stream valleys and are, in effect, a composite of the most important individual elements of the natural resource base in Southeastern Wisconsin.

Although recognizing the importance of the watershed as a rational planning unit for water resources within the Region, the regional planning program also recognizes the need to conduct individual watershed planning programs within the broader framework of comprehensive regional planning. This approach allows for consideration of the regional significance of environmental corridor areas which may in part be related to the linkage of corridors in adjacent watersheds and categorization with regard to regional significance in terms of size and resource values.

---

[1]Chief Environmental Engineer, Southeastern Wisconsin Regional Planning Commission, P.O. Box 1607, 916 N. East Avenue, Waukesha, Wisconsin 53187-1607.

Process for Delineation of Environmental Corridors

The most important environmental areas are placed into one of three categories--primary environmental corridor, secondary environmental corridor, and isolated natural area using a point system for rating the various elements of the resource base (see Table 1 and Map 1). To qualify for the primary environmental corridor category, an area must exhibit a point value of 10 or more and be at least 161.9 ha (400 acres) in size, be at least 3.2 km (two miles) long, and have a minimum width of 61 m (200 feet). The secondary environmental corridors, while not as significant as primary environmental corridors in terms of overall resource values, should be considered for preservation as urban development proceeds. To qualify as a secondary environmental corridor, an area must exhibit a point value of 10 or more and have a minimum area of 40.5 ha (100 acres) and minimum length of 1.6 km (one mile). Isolated natural areas generally consist of those natural resource base elements that have "inherent natural" value, and must be at least two ha (five acres) in size. Of the 6,964.5-square-kilometer (2,689-square-mile) Southeastern Wisconsin Region, the primary environmental corridors, secondary environmental corridors, and isolated natural areas comprise about 18 percent, 3 percent, and 2 percent of the area, respectively.

Table 1. Point Values Assigned to Natural Resource Features

Resource Base or Related Element	Point Value	Resource Base or Related Element	Point Value
Shoreland		Prairie	10
Lake or Perennial		Existing Park Site . .	5
Stream	10	Potential Park Site	
Intermittent Stream .	5	High Value	3
Floodland	3	Medium Value . . .	2
Wetland	10	Low Value	1
Poor Soils	5	Historic Site	
Woodland	10	Structure	1
Wildlife Habitat		Other Cultural . .	1
High Value	10	Archaeological . .	2
Medium Value	7	Scenic Viewpoint . .	5
Low Value	5	Scientific Area	
Steep Slope		State Significance.	15
20 Percent or More .	7	County Significance	10
13-19 Percent	5	Local Significance.	5

Environmental Corridor Protection Strategy

The principal programs designed to protect and enhance the riverine corridors in Southeastern Wisconsin include the development and dissemination of basic inventory data; the preparation of comprehensive watershed water quality and floodland management plans; and the integration of regional planning recommendations and concepts into local land use planning, regulation, and policy.

Map 1. SAMPLE ENVIRONMENTALLY SIGNIFICANT LANDS DELINEATION

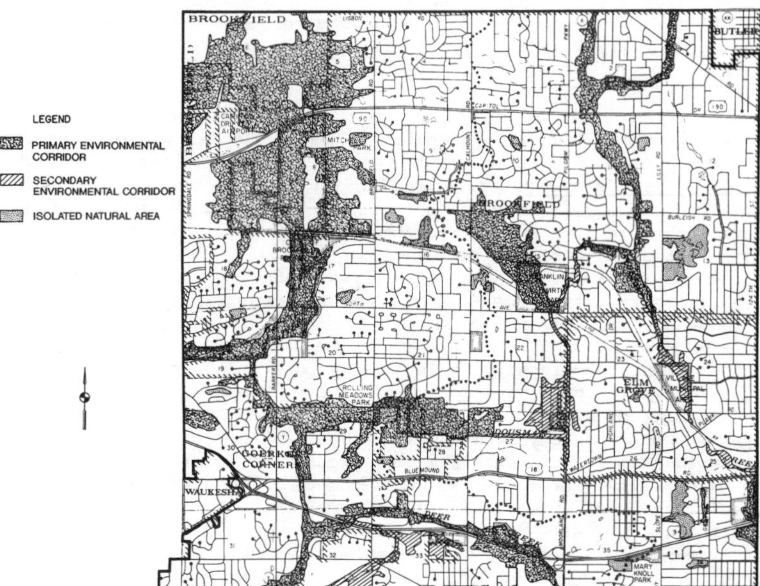

Source: SEWRPC.

Development and Dissemination of Basic Inventory Data: The ability to provide accurate planning information related to environmental corridors is essential to corridor preservation and enhancement. Local plans, regulations, and development-related decisions can be influenced if sound information regarding the resource features can be provided in a timely way. In Southeastern Wisconsin, as noted above, the environmental corridor boundaries are defined and mapped. Most of the flood hazard areas have been delineated through the Commission's comprehensive watershed planning programs, or other programs, and are available on large-scale topographic and cadastral maps. Because corridor delineations have been made based largely on aerial photograph interpretation, corridor boundaries are subject to field interpretation. Thus, the Commission staff often delineate corridor areas by field survey on a proposed development site as part of initial site planning, using the corridor criteria set forth above.

Comprehensive Watershed-Floodland Management Programs: The comprehensive watershed-floodland programs carried out in Southeastern Wisconsin include an integrated planning program directed toward meeting land use, park and open space, floodland management, and water quality considerations. In developing alternative plans for mitigating current and avoiding future floodplain problems, emphasis in the comprehensive plans is placed upon the use of nonstructural measures, such as floodplain preservation and structure removal, and "soft" structural solutions, such as minor channel modification resulting in both improved efficiency and enhanced environmental conditions which tend to allow for enhancement of the riverine corridor areas, rather than less preferable "harder" structural measures, such as channel enclosure or lining.

Implementation of comprehensive watershed plan recommendations is largely dependent upon a cooperative and collegial planning effort involving representatives of government, the private sector, and the public. This is typically accomplished through a public involvement program including direct involvement of advisory committees which review the planning work for technical, economic, environmental, and political soundness. To be effective, this program must serve as a forum for resolving conflicting interests and, if properly done, can significantly assist in building a consensus on plan objectives and recommendations.

Integration of Regional Plan Recommendations and Concepts into Local Planning, Regulation, and Policies: The work and recommendations of the Regional Planning Commission are, by law, strictly advisory. Thus, it has been extremely important to build consensus on Commission recommendations during the planning process if local governments are to incorporate the concepts and recommendations into local plans, regulations, and policies. The Commission maintains a community assistance planning function to assist county and local units of government in planning and plan implementation efforts. It thereby promotes coordination of local and regional plans and plan implementation actions. For example, in many cases, local detailed land use planning programs will include a component providing for the development and delineation of flood hazard areas along minor streams and tributaries to supplement data on the major streams covered by the regional plans and thus provide for effective street and lot layouts.

Current Practice and Expected Future Programs

The major change which has occurred at the areawide systems level of comprehensive regional and watershed planning relates to the current need to address detailed environmental issues to the point of constituting a full environmental impact assessment. More of the design considerations, such as the nature of fish and aquatic plant amenities, types of proposed vegetation, specific locational alignment in selected areas of existing wetland and woodland areas, and adjacent recreation and trail details are now provided at the systems level planning stage in the project, rather than in the initial design phase. This approach has evolved in order to build support for the recommendations during the project formulation stages. This tends to burden the planning phase of the project to demonstrate the positive impacts of potential projects beyond the flood control or other primary functions, but should result in an easier path through the subsequent implementation and regulatory permitting processes.

Evaluating the Efficacy of Low Cost Wetlands to Improve Lower Truckee River Water Quality

Daniel Spinogatti, Jr.[1] and John J. Warwick, member ASCE[2]

Abstract

An artificial wetland was created to receive effluent containing relatively low concentrations of nitrogen from the Numana Fish Hatchery, located in a semi-arid environment near Pyramid Lake, Nevada. Measurement devices are being used to obtain detailed records of total inflows and outflows. Conservative salt tracer tests are being conducted to determine travel times through the wetland. The planned water quality sampling scheme is Lagrangian in nature and will therefore be based on the travel time estimates. Multiple synoptic water quality surveys will be conducted, and laboratory analysis will include all nitrogen species, dissolved oxygen, temperature, conductivity, total suspended solids, and dissolved organic carbon. Project objectives include: 1) investigating the effectiveness of the artificial wetlands in removing nitrogen from the effluent waters; and 2) using the collected data to parameterize (calibrate and verify) a hydrodynamic model (RIVMOD) (Hosseinipour and Martin, 1990) and a water quality mathematical model (WASP) (Ambrose et al., 1991). Alterations to the WASP code will facilitate model simulation of all pertinent aspects of the constructed wetland, allowing for design evaluation and extrapolation to other applications that also involve low concentration effluents (e.g., agricultural and stormwater non-point pollution).

Introduction

Since 1981, the Pyramid Lake Paiute Tribe (PLPT) has utilized the Numana Hatchery, situated between Nixon and Wadsworth, Nevada, as a major supplier of Lahontan Cutthroat trout to Pyramid Lake. The effluent from the hatchery discharges onto the ground-surface of land adjacent to the Lower Truckee River. This low-topographic area, a former oxbow of the Truckee, has developed over time into a "natural" wetlands area, complete with surface water and hydrophytic vegetation. Water samples obtained from effluent and wetlands surface water indicated the presence of low-to-moderate levels of nitrogen (TKN > 2.8 mg-N/L) and total phosphorus (TP > 0.4 mg-P/L) in the waters discharging into the lower Truckee. With support from the Environmental Protection Agency (EPA) Clean Lakes Program, and PLPT work-in-kind funds, an artificial wetlands was constructed at the Numana Hatchery in an attempt to provide low-cost and low-maintenance treatment.

---

[1]Graduate Student and [2]Director, Hydrology/Hydrogeology Program, University of Nevada - Reno, 1000 Valley Road, Reno, NV 89512. Tel.# (702) 784-6250

Constructed (or artificial) wetlands have proven to be a reliable and cost-effective method for removing biochemical oxygen demand (BOD), total suspended solids (TSS), and nutrients, including nitrogen, from secondary wastewater effluent (Reed and Brown, 1992). Many recent studies have involved high-concentration municipal effluents (TN= 20-30 mg/L). A more recent concern involves wetland treatment of low-concentration effluents (TN <3 mg/L), such as agricultural and stormwater runoff (Olson, 1993; Reuter et al., 1992). In both natural and constructed wetlands, the major mechanisms of nitrogen transformation and removal appears to be the aerobic process of nitrification ($NH_4^+ \rightarrow NO_2^- \rightarrow NO_3^-$) followed by the anerobic denitrification process ($NO_3^- \rightarrow N_2O$ and/or N_2 gas) (Gale et al., 1993).

Site Description

The Numana Hatchery is located adjacent to the Lower Truckee River, approximately six miles upstream (south) of the confluence with Pyramid Lake, and within the confines of the PLPT Reservation. Water is supplied to the hatchery from 2 on-site groundwater wells. One well is utilized at any given time, pumping continuously at a rate of approximately 300 gallons per minute. The total capacity of the hatchery's recirculating water system is about 500,000 gallons. All water is circulated through a series of 8 biofilters, which contain a natural ion-exchange resin (clinoptiloite) to remove ammonium. On Monday through Thursday, 2 biofilters per day are cleaned and emptied. On these days, approximately 50,000 gpd is generated from biofilter cleaning and up to 382,000 gpd results from overflow water. On Fridays, Saturdays, and Sundays, the total discharge is composed of overflow water plus 20,000-30,000 gallons from the cleaning/emptying of outdoor pools.

Design of Artificial Wetlands

Construction of the free-water surface (FWS) wetlands system at the Numana Hatchery was completed in September, 1994. A FWS wetland, similar to a natural marsh, is characterized by a soil bottom, emergent vegetation, and a water surface exposed to the atmosphere (Reed and Brown, 1992). The cost of earth-moving required for the wetlands construction was mitigated by donated equipment and labor from Granite Construction of Reno/Sparks, NV. A site map showing the constructed wetlands location is presented in Figure 1.

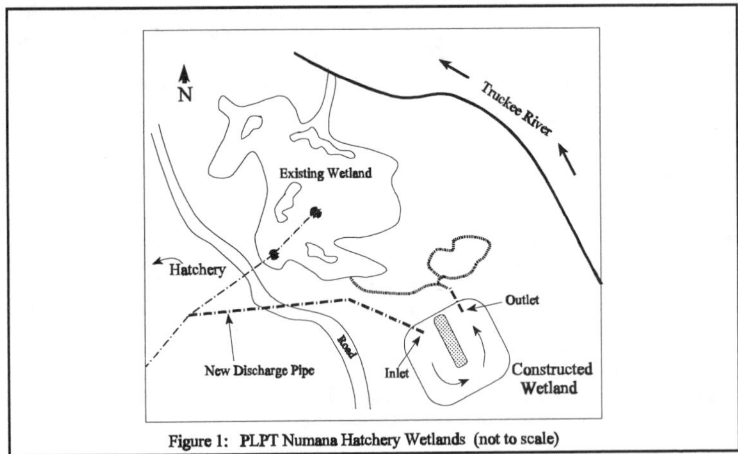

Figure 1: PLPT Numana Hatchery Wetlands (not to scale)

The final design of the wetland allowed for a surface area of approximately 2 acres. The wetland is equipped with a standpipe outlet structure enabling effective control of water stage in the pond. Due to this flexibility of water level control, various retention times can be achieved. Approximately 1.4 days of retention time is gained with each foot of water. A 4 ft. water level yields a maximum theoretical retention time of 5.8 days.

Due to the permeable soils located at the site, a 30 mil. polyethylene liner was utilized to contain the effluent waters within the bermed wetland basin. A soil layer, approximately 3 ft. thick, was backfilled over the installed liner. Emergent-macrophytic vegetation, consisting of hard-stem bulrush (*Scirpus acutus*), was planted within the wetland to maximize microbial growth and treatment potential. Bulrush was obtained from the adjacent natural wetlands area. Planting was completed in October, 1994.

Methodology

Continuous inlet/outlet flow monitoring will be conducted in the constructed wetland. Flow rates in the inlet pipe will be obtained via a pipe-insert propeller flow meter. On the outlet pipe, a Palmer-Bolus flume (with a pressure transducer cavity) is being used to obtain continuous outflow measurements. Figure 2 shows wetland outflow data from Sunday Dec. 11, 1994 (12:00 a.m., day 0) through Sunday Dec. 18, 1994 (12:00 a.m., day 7). The outflows plotted in Figure 2 represent flow values averaged over 15 minute time intervals. The typical weekly outflow variation can be observed from the graph. The four large outflow pulses occurring once per day on Monday (day 1) through Thursday (day 4) represent increased flow due to biofilter backwashing.

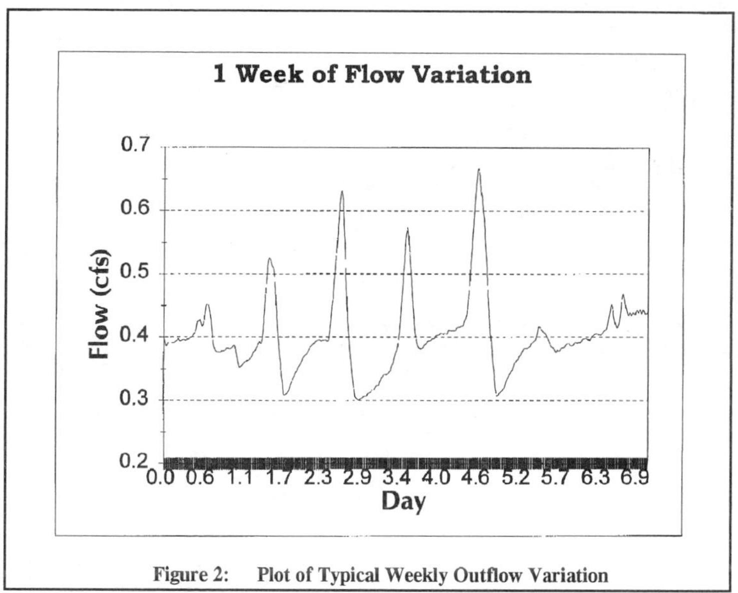

Figure 2: Plot of Typical Weekly Outflow Variation

Due to the "pulse loading" characteristic of this system, an accurate estimation of dispersion is of key importance. Conservative salt tracer tests will be conducted to determine the travel time and dispersion rate at a chosen flow regime (2 ft. water level). A Lagrangian sampling scheme will be arranged based on the results of the travel time studies. This scheme will allow monitoring of essentially one large parcel of water. Three sampling locations will be established, one each at the pond inlet and outlet locations, and one at an intermediate location. Samples will be subjected to laboratory analysis for the following parameters: Total Kjeldahl Nitrogen (TKN), ammonium, nitrate, total suspended solids (TSS), dissolved organic carbon (DOC), total phosphorus, and ortho-phosphate. Field measurements will include: dissolved oxygen, temperature, conductivity, pH, and alkalinity.

Modeling

The constructed wetland system will be modeled much like a "slow moving stream". A hydrodynamics model (RIVMOD) will be calibrated to the observed travel time by adjusting the Manning's roughness coefficient and the longitudinal dispersion rate. RIVMOD will be linked to the Water Quality Analysis Simulation Program (WASP). WASP will be utilized to model the water quality transformations occurring within the wetland.

The WASP model is designed for application to lakes, rivers/streams, and estuaries. The model code will be modified to account for additional nitrogen uptake and/or transformations by vegetation. Literature values will be used in the model for rates of plant uptake of ammonium (NH_4) and nitrate (NO_3). Rates of mineralization, nitrification, and denitrification will be arrived at through the calibration process. The fraction of NH_4 contributed by sediments will be measured in laboratory experiments. Ammonia stripping (volatilization) is assumed to be negligible.

Following calibration, the model will be verified with site data. After model calibration/ verification has been successfully completed, the model will be used to investigate optimal management and operational strategies for the system.

Literature Cited

Ambrose, R.B., T.A. Wool, J.L. Martin, J.P. Connolly, and R.W. Schanz. 1991. WASP5.x, A Hydrodynamic and Water Quality Model: Model Theory, User's Manual, and Programmer's Guide (draft). U.S. EPA. Environmental Research Laboratory, Office of Research and Development. Athens, GA.

Gale, P.M., K.R. Reddy, and D.A. Graetz, 1993. Nitrogen removal from reclaimed water applied to constructed and natural wetland microcosms. Water Environ. Res. 65(2): 162-168.

Hosseinipour, E.Z. and J.L. Martin. 1990. RIVMOD: A One-Dimensional Hydrodynamic and Sediment Transport Model; Model Theory and User's Manual (draft). U.S. EPA. Center for Exposure Assessment Modeling, Environmental Research Laboratory. Athens, GA.

Olson, R.K. (ed.) 1993. Created and Natural Wetlands for Controlling Nonpoint Source Pollution. U.S.EPA, Office of Research and Development, Office of Wetlands, Oceans, and Watersheds. C.K. Smolley; CRC Press Inc., Boca Raton, FL.

Reed, S.C. and D.S. Brown. 1992. Constructed wetlands design: the first generation. Water Environ. Res. 64(6): 776-781.

Reuter, J.E., T. Djohan, and C.R. Goldman. 1992. The use of wetlands for nutrient removal from surface runoff in a cold climate region of California: results from a newly constructed wetland at Lake Tahoe. J. Environ. Mngt. 36:35-53.

ALTERNATIVES FOR TREATING ROADWAY STORMWATER RUNOFF

by Jorge Romero-Lozano[1]

ABSTRACT: Due to public concern regarding environmental protection for the nation's waterways, more stringent water quality control regulations have been developed. This is most obvious in the area of non-point sources of pollutants, such as stormwater drainage. The National Pollutant Discharge Elimination System (NPDES) permitting requirements of the Clean Water Act have led to the required establishment of Best Management Practices (BMP) and/or stormwater runoff management and treatment facilities. This paper presents a case study on the roadway drainage design process of a transportation corridor currently under design, and how it relates to stormwater runoff treatment facilities.

INTRODUCTION

The case study described in this paper is a 35.2-kilometer corridor. The primary focus of the drainage design effort was to develop the most efficient stormwater management system. Efficiency was rated based on the system's effectiveness in treating pollutants in stormwater runoff, as well as providing an effective drainage design to convey stormwater off the roadway for the purpose of enhancing motorist safety. Stormwater treatment became a high priority because the mitigation measures, as set forth by the Environmental Impact Report, required that stormwater runoff be treated in order to comply with local, state, and federal regulations, as initiated by the NPDES requirements for non-point sources of the Clean Water Act.

MITIGATION MEASURES

The mitigation measures were the impetus for the drainage design, which addressed the implementation of stormwater treatment facilities to handle all runoff within the project limits. Some of the more salient issues included: 1) prevention of corridor pollutants from reaching improved and unimproved downstream drainages, 2) maintenance of post-project peak flow rates and

---

[1]Project Engineer, Greiner, Inc., 7310 N. 16th Street, Suite 160, Phoenix AZ 85020

velocities equal to pre-project conditions, 3) maintenance of the same amount of water supply flowing to wetland habitats, and 4) monitoring of the treatment facilities, and their success in eliminating pollutants, throughout the life of the roadway.

DESIGN APPROACH

The corridor was divided into nine design sections, each section ranging in length from 1.6 to 4 kilometers. The nine design sections were divided among five design firms to perform the roadway design. Because of this, careful coordination among the designers was necessary to ensure a congruous design approach throughout the corridor. This coordination effort became indispensable when dealing with drainage issues, since catchment boundaries overlapped design section limits and stormwater treatment had to be in place throughout the corridor. Moreover, even though each design section had its own unique set of issues, it was important that a standardized design effort be maintained so as to realize the construction and maintenance cost effectiveness of one common treatment facility for all design sections.

To minimize the amount of runoff reaching the treatment facilities, it was proposed that the stormwater runoff from the cut and fill slopes be kept separate from that of the paved sections. The reason for this was that the largest volume of pollutants, and those having the greatest potential to degrade the receiving watershed, would originate from vehicular travel on the paved roadway. Therefore, the runoff conveying these pollutants would be the one to target for treatment.

STORMWATER TREATMENT FACILITIES

The primary task at hand was the development of acceptable and efficient stormwater treatment facilities. The most feasible and viable alternatives were identified as follows:

1. Detention Basins with Oil Separators: These facilities would remove the majority of pollutants from the "first flush" storm, namely, the first 1.27 cm of storm runoff. In order to make the first 1.27 cm of runoff amenable to calculation by common methods, it was defined to be the equivalent of a 2-year frequency, 60 minute duration storm. The concept was that this "first flush" would remove the majority of the pollutants from the roadway. Treatment of the captured stormwater runoff would be through the use of oil separators only. The "first flush" would be collected in a detention basin and kept separate from the remainder of the storm runoff by using split flow structures. Under a controlled effluent line, the captured runoff would be run through the oil separators which would remove oil and grease. The primary disadvantage of this facility was that it would not treat all of the stormwater runoff, but rather only a portion thereof.

2. Infiltration Basins: These facilities consisted of graded ponds which would collect all the runoff from a 100-year frequency, 24-hour duration storm. The basin would have a 1.5-meter layer of permeable filter material where, through the mechanics of sorption, precipitation, trapping, straining and bacterial removal, pollutants would be removed. Theoretically, the filtered runoff would percolate through the insitu materials at the bottom of the permeable filter material layer to ground water. However, it was determined by the geotechnical engineer that the native materials did not possess the required percolation rate of 0.0014 cm/sec to drain the basins quickly enough so that, in the case of successive storms, overtopping could occur. In fact, the native materials, which are predominantly clay and silts, have a percolation rate in the range of 10^{-6} cm/sec to 10^{-8} cm/sec, which is essentially impervious. It was also feared that long periods of exposed standing water would create "mosquito ponds." Another concern with this alternative was the lack of space needed to accommodate the basins.

   The size of the basins was based on a 10:1 ratio as specified in the project's Runoff Management Plan, where for every ten hectares of impervious area, one hectare of basin would be required to effectively treat the stormwater runoff. However, the available amount of open area due to right-of-way constraints was not sufficient to accommodate the location of these basins. Another concern was the fact that if runoff remained stored for a significant amount of time, it might saturate the cut, fill and natural slopes and potentially cause them to fail.

3. Detention Basins with Compost Stormwater Filters: These facilities consisted of basins that would detain water to control the influent flow discharging into the filters. The filters would consist of a layer of compost material designed to remove floating surface scums, oil, grease, metals and insoluble chemicals from the stormwater runoff. According to the manufacturer of these systems, CSF Treatment Systems, Inc., these filters have demonstrated that they are capable of removing in excess of 85% of the oil and grease entering the filter, and 82% of the heavy metals (CSF Treatment Systems, Inc., 1992). The compost filter would allow a high release rate of effluent and influent flow. The effluent flow would theoretically consist of "cleaned water" which would then be discharged into a parallel system designed to solely convey effluent runoff from these filters. No untreated storm runoff would be discharged into these lines. The parallel system would ultimately discharge into the nearest stream. Since some of the basins were to be located adjacent to the roadway, particularly within the infield areas of loop ramps, it was determined that an overflow spillway to safely pass flows exceeding the design flows be provided. This system would have an inlet located at an elevation below the top of the basin and would bypass the influent system and the filter, but would then connect directly to the effluent line.

   This alternative was chosen to be the proposed stormwater treatment facility to be implemented corridor-wide because of its treatment capabilities, flexibility in application to the corridor, and also, because it precludes the detaining of water for long periods of time which might cause aesthetic and maintenance problems. Moreover, the estimated cost of the filters made this alternative very feasible: the cost is estimated at $30,000.00 to $100,000.00 for a filter capable of treating flows ranging from 0.03 cms to 0.23 cms.

CONCLUSIONS

Although the success of the stormwater treatment/drainage facilities proposed for this project will not be proven until after they have been put in place and the first significant storm occurs, the overall approach and theoretical analysis appears sound and correct. This type of design, mandated by regulatory measures, urges engineers to move beyond drainage systems designed to remove stormwater from the roadway in the most cost effective manner. Designers must combine this requirement with the need to provide water quality control measures in terms of best management practices and structural methods.

ACKNOWLEDGEMENTS

The writer wishes to express his sincere appreciation to the various reviewers of this paper.

REFERENCES:

CSF Treatment Systems, Inc. (1992). "Technical Summary - Methods & Results". CSF Treatment Systems, Inc., Portland, OR.

Effectiveness of Detention Based Stormwater Management on The Brandywine

Robert G. Traver, M. ASCE[1]

Abstract

This paper presents preliminary results of an investigation into the impact of urbanization on the flows of the Brandywine River. Of specific interest is the effectiveness of detention based stormwater management efforts first introduced to southeastern Pennsylvania in the late 1970's.

Introduction

The impact of urbanization on the flows of the Brandywine River has been of concern for some time. Located in southeastern Pennsylvania, the Brandywine has been designated as one of Pennsylvania's "scenic rivers." Due to it's proximity to both Philadelphia and Wilmington, the watershed has been the subject of continuing development. Legislation was passed in the late 1970's to reduce the impact of urbanization on downstream flooding. These efforts focused on the reduction of peak outflows through the use of stormwater detention basins (Brandywine Conservancy 1980). This study attempts to use existing streamgage and raingage records to quantify the effect of urbanization and detention based stormwater management on the flows of the Brandywine.

Study Focus

This study is currently in the middle of the second of three phases. The initial phase consisted of a computer model study on the effectiveness of point based detention stormwater controls (Traver and Chadderton 1992). The results of this first phase indicated as suspected that point based detention was only partially effective at reducing increased flood flows due to urbanization. The current phase focuses on examining available data for

[1] Assistant Professor, Department of Civil Engineering; Villanova University, Villanova, PA 19085.

statistically significant changes in the minimum, mean, and maximum stream flows (Traver 1994). The third phase will consist of relating regional historic uses to the data set to further understand the cause-effect relationship.

Increased peak flows have been recorded in areas that develop without management controls (ASCE 1992). This has been well documented by the profession. What has not been so well studied is the effects of urbanization on low and average flows, and the effectiveness of detention based storm water management on a regional basis. Experience tells us that as the population of an area increases, more water is consumed from surface or ground water sources. Water supply projects such as dams are constructed to increase the amount of water available for "mining." The effect of these changes should be reflected in the available flow history. The purpose of this second phase then is to examine low, mean, and high flow records for statistically significant evidence of the effect of urbanization on the Brandywine.

The Brandywine River Valley

The Brandywine River Valley in southeastern Pennsylvania has several large population centers, a growing suburbia, as well as large areas that retain their rural atmosphere. The river itself is dendritic in nature with two major branches, known as the East and West Branch. Eight streamgages of varying record lengths are currently in operation on the river, with three on the East Branch and four on the West Branch. All gages are presently in operation, with some going back to the early 1960's. The station at the lowest point of the watershed (Chadds Ford) incorporates a drainage area of 287 square miles and has records dating back to 1911. Monthly streamflow records were collected and converted to inches per unit area to allow for comparison of watersheds with different sizes.

Five of the area raingages were found suitable for use based upon length of precipitation record. Missing records were estimated from the surrounding raingages. The thiessen polygon method was used to associate the raingages to the contributing drainage area for each streamgage (Viesmann et. al. 1989). The data was then imported into a spreadsheet for manipulation and plotting. A five year moving average was used to emphasize trends due to the high variability of the raw data.

Flow Plots

Figure 1 presents the relationship between monthly rainfall volume and low flow observed. Note the corresponding drop in rainfall and low flow from 1975 to present. It is interesting to note the difference in low flows per unit area observed at different points on the stream channel. The stream gages at the upper portion of the East and West Branches recorded a smaller flow per unit area then that at the bottom of the watershed (Chadds Ford). It is also interesting to note the divergence of the low flow curves for the West and East Branch.

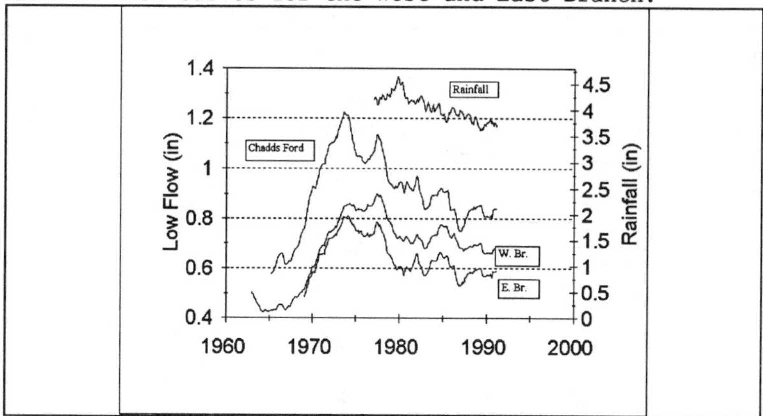

Figure 1. Monthly Low Flows (5 Year Moving Average).

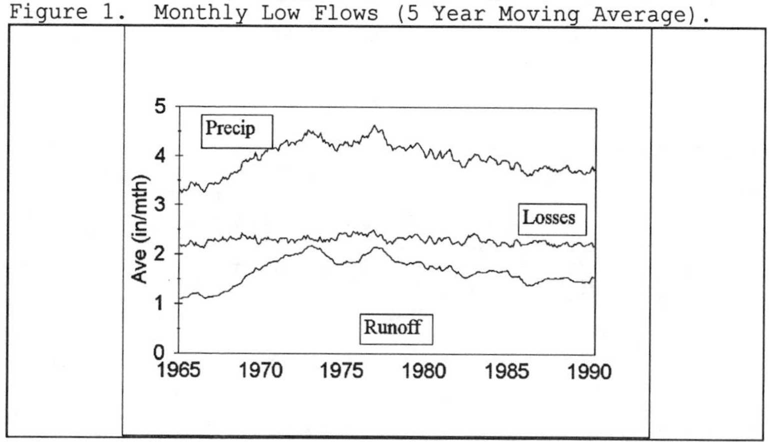

Figure 2. Mean Flows at Chadds Ford.

Mean Flows

Representative plots of monthly average flows are shown in Figures 2 and 3. When observing the flows at Chadds Ford, both rainfall and runoff quantities appear to be dropping while the losses appear linear. The loss rate is calculated as the difference between raingage and streamgage data, and should include water lost to deep aquifers, evapotranspiration, and consumptive surface water withdrawals. Further upstream in the watershed above

Downingtown (West Branch), the losses appear to start to increase in the early 1980's. It is also apparent that there is a relationship between loss rate per unit area and the size of the contributing drainage area.

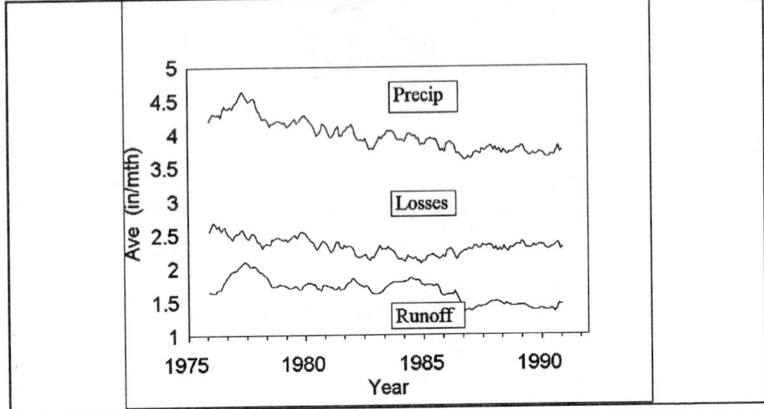

Figure 3. Mean Flows near Downingtown (West Branch).

Summation

Though some interesting observations can be made from the collected data, much more work is needed before any conclusions can be drawn. Reservoir storage, consumptive withdrawals, and population change must be factored into the process. Early Results have demonstrated the importance of including both rain and stream gages to avoid misleading conclusions. Low flows have been dropping in the Brandywine since 1970, but to what extent this is due to urbanization is unknown.

Bibliography

ASCE, Design and Construction of Urban Stormwater Management Systems, ASCE, NY, NY, 1992.

Brandywine Conservancy, Stormwater Management: A Review of Municipal Practices and Regulations. Pa. Dept. of Env. Res., April 1980.

Traver, Robert G., and Chadderton, Ronald A. "Accumulation Effects of Stormwater Management Detention Basins", Water Forum '92, Conference proceedings, ASCE, NY, NY, August 1992.

Traver, R. G., and Chadderton, R. A. "Investigation of Urbanization on the Brandywine River", **Stormwater Runoff and Quality Management**, Proceedings, The Penn. State Univ., University Park, PA, 1994.

Viessman, Warren Jr., and Lewis, Gary L., and Knapp, John W., Introduction to Hydrology, 3rd Ed., Harper & Row, New York, 1989.

TREATMENT OF AN URBAN RIVER USING ROTATING BIOLOGICAL CONTACTORS AND SAND FILTERS

John P. Scaramuzzo[1], Frederic C.Blanc[2], Constantine J. Gregory[2] and John P. Elwood[3]

INTRODUCTION

Urban runoff, combined sewer overflows, agricultural runoff and other non-point source discharges to our waterways represent one of the pollution control challenges of the next two decades. Such discharges often cannot be effectively treated at the discharge points and therefore result in degraded water quality in streams and lakes. Organics, nutrients, bacteria, suspended solids, and other pollutants make it impossible to achieve reasonable stream water quality in many instances. For many smaller lakes, ponds and streams in critical areas, sidestream treatment can effectively raise the level of water quality and change a stream from an eyesore to an asset.

Sidestream treatment is a relatively new concept which involves the improvement of polluted surface waterways by withdrawing a portion of the flow, subjecting it to treatment processes and returning it to the waterway. Sidestream treatment offers the most cost effective method of upgrading the water quality in a number of such streams, ponds or lakes.Such treatment could be seasonally implemented, during the critical period, from spring through fall in a climate such as New England's. Evaluation and design of such sidestream treatment options requires performance data of unit processes at pollutant concentrations lower than the concentrations experienced in municipal wastewater treatment plant effluents. This paper presents some of the results experienced in a pilot study in which a rotating biological contactor and sand filter were used to treat the water from the Muddy River in Boston's Back Bay area.

THE MUDDY RIVER

The Muddy River, a minor tributary of the Charles River, encompasses a drainage area of 8.6 square miles located in Boston, Brookline, and Newton. The Muddy River is an integral part of the "Emerald Necklace", an interconnected series of parks designed by the famous 19th Century landscape architect Frederick Law Olmstead. The major reasons for the present water quality problems in the Muddy River are low flow, stormwater runoff, combined sewer overflow, illegal cross connections, sedimentation, and invasive riparian growth.

---

1. Engineer, Foster Wheeler Environmental Corp., 470 Atlantic Ave. Boston, MA. 02210.
2. Professor, Department of Civil Eng., Northeastern Univ., Boston, MA. 02115.
3. Consultant, 305 County Road, Bourne, MA. 02532.

The average flow of the Muddy River is 12 cfs (7.7 MGD) with flows as low as 1 cfs (0.65 MGD) in the dry summer months. Table 1 shows a comparison between the average water quality in Jamaica Pond which is the headwater of the Muddy River and the Muddy River at the pilot plant site some 2 miles downstream. During and after summer storm events the quality deterioration is much greater than the average numbers indicate.

Parameter	Study	Jamaica Pond (mg/l)	Muddy River (Average)	Change (%)
NH4	NU	0.056	0.626	1000
Organic N	NU	0.638	0.866	35
TKN	NU	0.694	1.492	115
BOD	MDWPC	0.90	3.20	255
DO	USCOE	8.9	1.9	- 69
Total Phos.	USCOE	0.015	0.6	3900

Table 1, Comparison between Jamaica Pond and Muddy River Water Quality

During the late spring and early summer of 1993, a pilot facility was constructed on a site near the river provided by the City of Boston, Parks and Recreation Department. A 55 gpm influent pump, submersed in the Muddy River, fed a 4000 gallon equalization tank which in turn was used to supply the river water to three parallel pilot units one of which was a rotating biological contactor (RBC). Suspended solids, biochemical oxygen demand, organic-nitrogen and ammonia-nitrogen constitute a major portion of the pollution problem in the Muddy River and many other watercourses. The RBC removed BOD, and biologically converted organic-nitrogen and ammonia nitrogen to nitrate. Elimination of BOD and ammonia nitrogen from the water results in a lowering of the oxygen demand in the stream, increasing the level of dissolved oxygen in the stream.

The rotating biological contactor [RBC] pilot plant unit was a four stage unit with 385 square feet of media per stage and 1548 square feet for the unit.The total liquid tankage volume for the four stages was approximately 190 gallons. This resulted in total hydraulic retention times of 54 minutes at a flow rate of 3.5 gpm, 38 minutes at a flow rate of 5 gpm and 27 minutes at a flow rate of 7 gpm. For this experiment, the rotational velocity of the RBC shaft was maintained at a constant 4 rpm which resulted in a peripheral velocity of approximately 0.83 feet per second.

Figure 1 illustrates the average total BOD results for RBC influent (stage 0) and the BOD concentrations in the 4 RBC stages. The increase in the first stage BOD is indicative of the solids accumulation and decomposition problem in RBC pilot units. There were five dates on which total and filtered BOD analyses were performed on the same samples. The data from the five dates indicate that the filtered BOD is typically approximately 43 percent of the total BOD.

FIGURE 1: BOD REMOVAL IN RBC STAGES

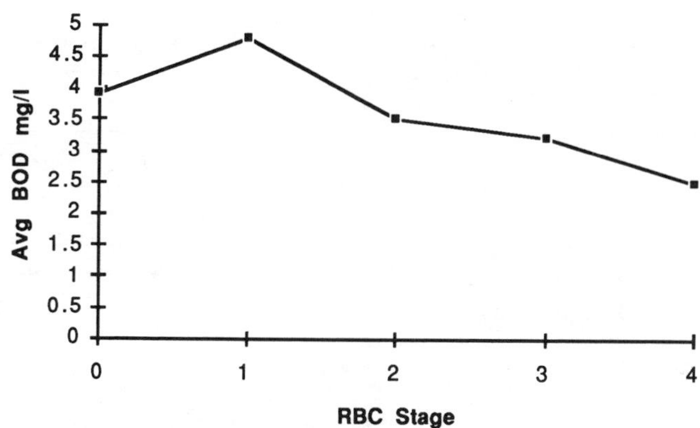

BOD removal for all of the filtered ("soluble") BOD data indicated an average 40 percent removal. Figure 2 shows the average chemical oxygen demand (COD) concentration in the RBC by stage. Process performance indicated 48 to 50 percent soluble COD removal for the RBC process.

FIGURE 2: AVERAGE SOLUBLE COD IN RBC STAGES

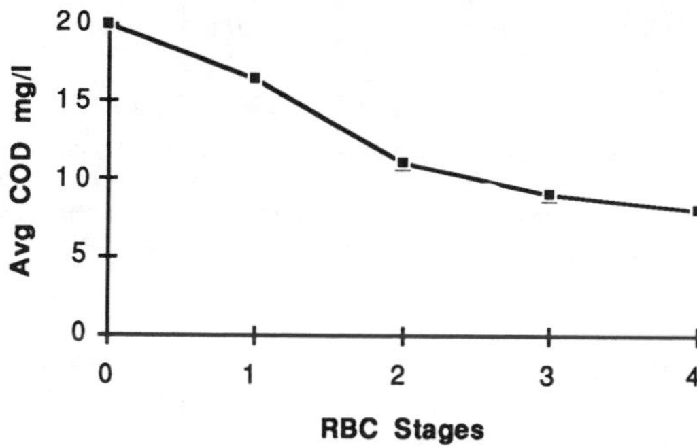

Figure 3 indicates the ammonia nitrogen influent and effluent concentrations for the RBC pilot unit.For a good portion of the study the effluent concentrations were under 0.2 mg/l. Such low concentrations limit both the possible nitrification rate and percentage removals. Ammonia removal for all data over all temperatures was 71 percent from an average influent.

FIGURE 3

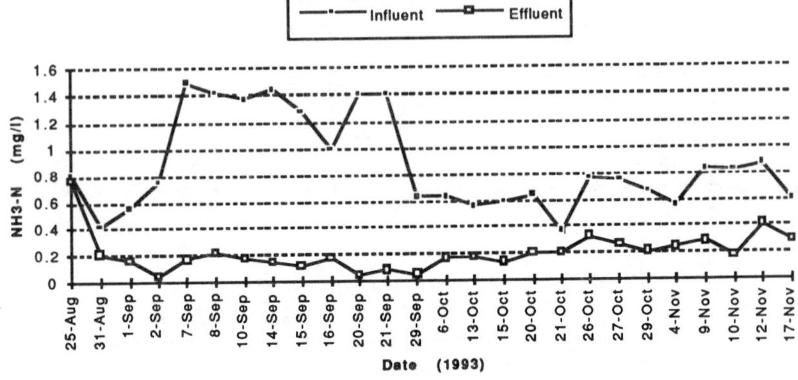

concentration of 0.869 mg/l to an average effluent concentration of 0.210 mg/l.

Filtration is a method of capturing suspended solids from an effluent provided the suspended solids concentrations are sufficiently low. To obtain information relative to filtration of the RBC effluent a sand filter was used during the latter part of the study. The sand filter used was 7 feet tall and 6 inches in diameter containing 12 inches of stone base, 28 inches of coarse sand. The filter was operated at a flow rate of 5 gpmsf for a run time of 72 hours routinely producing a final effluent with a suspended solids concentration of 2.4 mg/l which represents a 45 percent TSS removal. In the total pilot study [references 1,2] fine sand filters exhibited greater suspended solids removal percentages

CONCLUSIONS

Based on the complete study results it can be concluded that the combination of the RBC and a sand filter could treat the waters of a stream such as the Muddy River and achieve removals of 72 percent for total suspended solids, 50 percent for soluble BOD and 85 percent for ammonia nitrogen.

REFERENCES

1. Blanc, F.C., Gregory, C.J. and Elwood, J.R., (1994), "Treatment of an Urban Waterway as a Stormwater Mitigation Measure", a report to U.S.E.P.A, Northeastern University, Boston, MA.
2. Scaramuzzo, J.P.,(1994), "Treatment of an Urban Waterway Using Rotating Biological Contactors, M.S. Thesis, Northeastern University, Boston, MA.

TREATMENT OF AN URBAN RIVER BY SIDESTREAM NITRIFICATION

Adam A. Scott[1], Frederic C. Blanc[2], and Constantine J. Gregory[3].

INTRODUCTION

The Muddy River is an historic urban waterway located in Boston. The Muddy River suffers extensively from the problems of an urban waterway: low flow, sedimentation, and high pollution load from stormdrains and combined sewers. The average water quality levels in the Muddy River are BOD = 5.0 mg/l, NH4 = 0.8 mg/l, and TKN = 1.5 mg/l.

One way to improve water quality levels in a polluted stream is sidestream treatment. In order to test the feasibility of sidestream treatment, a pilot plant was constructed on the banks of the Muddy River. This treatment plant used attached growth processes to remove both carbonaceous nutrients (BOD,COD) and Nitrogen (NH4, Org N, and TKN).

Ammonia and organic Nitrogen are treated by the process of nitrification. Three treatment processes were used for the nitrification of Muddy River water: a rotating biological contactor (RBC), a trickling filter, and a coarse sand filter. All three processes achieved some degree of nitrification.

NITRIFICATION RESULTS

While nitrification has been well studied, it has been mainly in the treatment of primary and secondary wastewaters. The average ammonia levels in urban rivers are considerably lower than these wastewaters. The objective of this research was to evaluate the effectiveness of attached growth processes to reduce nitrogen levels on the low, variable strength water in a polluted stream. The attached

----

1 Environmental Engineer, LEA Guertin & Associates, Inc., 75 Kneeland Street, Boston, MA 02111

2 Professor of Civil Engineering, Northeastern University, 120 Huntington Ave, Boston, MA 02115

3 Professor of Civil Engineering, Northeastern University, 120 Huntington Ave.,

growth processes included a rotating biological contactor, a trickling filter, and a coarse sand filter.

THE ROTATING BIOLOGICAL CONTACTOR

The RBC achieved the highest consistent nitrification and the lowest effluent ammonia levels of all the treatment processes studied. This was due to the relatively low hydraulic loading rate, the large surface area, and the natural mechanical aeration of the RBC. Figure 1-1 shows the % Ammonia and TKN conversion. Relevant factors, including hydraulic loading rate, water temperature, pH, ammonia concentration, and nitrification, are shown in Table 1-2.

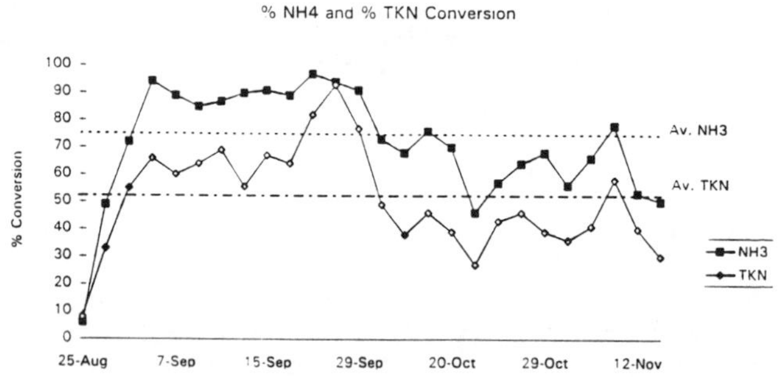

Figure 1-1, % Ammonia and TKN Conversion in RBC

Date	Hydraulic Load	Temp	pH	Inf. NH4	Eff. NH4	% Nit NH4	% Nit TKN	Nit. Rate
	l/min-m^2	Deg C		mg/l	mg/l			g/day-m^2
Aug 25-Sep 2	0.08	22	X	0.632	0.293	55	41	0.058
Sep 3-Sep 20	0.08	20	X	1.344	1.344	90	66	0.151
Sep 21-Sep 29	0.11	16	6.6	1.015	1.015	93	85	0.215
Sep 30-Oct 21	0.11	13	6.5	0.606	0.606	72	43	0.124
Oct 22-Nov 17	0.15	10	6.4	0.703	0.703	60	40	0.169

Table 1-2, Performance Data for RBC

THE TRICKLING FILTER

The trickling filter had the least consistent nitrification and the highest effluent ammonia levels of all the treatment processes studied. This was due to the relatively high hydraulic loading rate and small surface area/volume ratio of media. Figure 1-3 shows the % Ammonia and TKN conversion. Relevant factors, including hydraulic loading rate, water temperature, pH, ammonia concentration, and nitrification, are shown in Table 1-4.

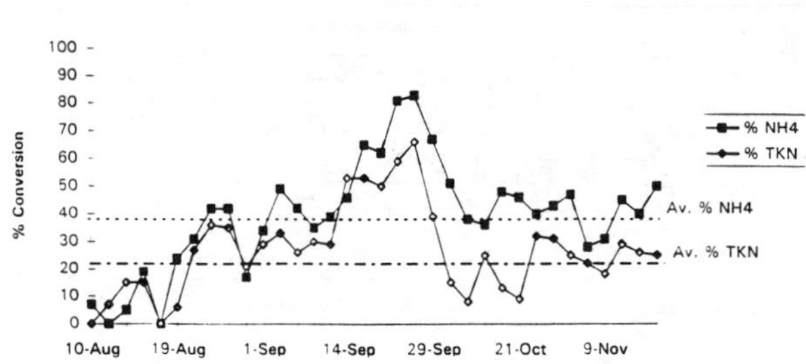

Figure 1-3, % Ammonia and TKN Conversion in TF

Date	Hydraulic Load	Temp	pH	Inf. NH4	Eff. NH4	% Nit NH4	% Nit TKN	Nit. Rate
	l/min-m^3	Deg C		mg/l	mg/l			g/day-m^3
Aug 10-Aug 19	37	X	X	0.273	0.257	9	7	2.4
Aug 20-Sep 13	37	21	X	0.888	0.534	37	30	24.4
Sep 14-Oct 1	9	17	6.7	1.191	0.388	67	53	13.6
Oct 2-Nov 17	18	11	6.4	0.674	0.386	42	21	8.1

Table 1-4, Performance Data for Trickling Filter

THE COARSE SAND FILTER

The coarse sand filter had the highest nitrification rates and moderate effluent ammonia levels. The high nitrification rate was a result primarily of the high surface area/volume ratio of the coarse sand media. The coarse sand filter operated efficiently without any natural or mechanical aeration to the influent river water. Figure 1-5 shows the % Ammonia and TKN conversion. Relevant factors, including hydraulic loading rate, water temperature, pH, ammonia concentration, and nitrification, are shown in Table 1-6.

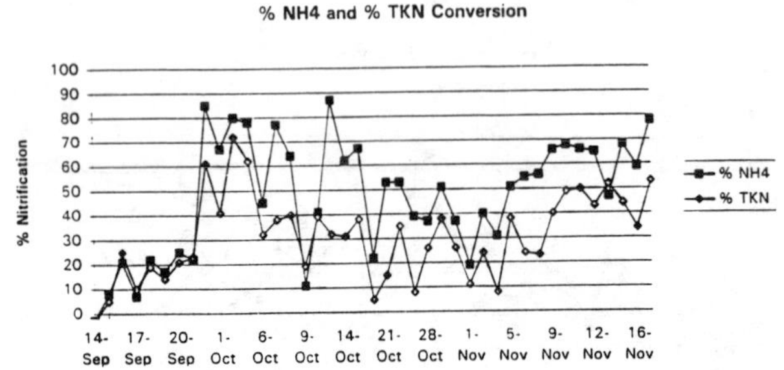

Figure 1-5, % Ammonia and TKN Conversion in CSF

Date	Hydraulic Load	Temp	TSS Removal	Inf. NH4	Eff. NH4	% Nit NH4	% Nit TKN	Nit. Rate
	l/min-m^3	Deg C		mg/l	mg/l			g/day-m^3
Sep 14-Sep 21	106	18	52	1.243	1.057	15	14	32.2
Sep 30-Oct 3	106	14	70	0.871	0.199	78	59	130
Oct 6-Oct 14	266	12	35	0.392	0.178	55	33	130
Oct 15-Nov 8	360	10	X	0.528	0.295	44	23	137
Nov 9-Nov 17	266	9	X	0.755	0.268	64	46	207

Table 1-6, Performance Data for CSF

REFERENCES

1 Antonie, R; "Nitrogen Control with the RBC"; Advances in Water and Wastewater Nutrient Removal; Ann Arbor Press; 1979

2 U.S. Army Corps of Engineers; Reconnaisance Report on the Muddy River; 1992

3 US EPA; Process Design Manual for Nitrogen Control; 1975-603-902; 1975

Urban River Aquatic Resources Enhancement

David C. Nyman, P.E., Member ASCE[1]

Abstract

Urban development has altered the alignments and habitat structures of many rivers and streams. One such river, found in a highly urbanized New England community, flows through a former industrial mill complex in an extensively modified channel. The mill complex is the site of a major commercial redevelopment project which further alters portions of the channel alignment. Project design includes restoration of fisheries habitat along several thousand feet of the affected water course. Habitat restoration involves application of "ecological engineering" techniques to provide biologically appropriate improvements within challenging engineering constraints.

Introduction

The aquatic resources enhancement program comprises three major components: (1) providing in-stream structural improvements of fisheries habitat within the disturbed channel; (2) augmenting riparian vegetation along this channel; and (3) replacing an existing culvert with a 335 meter long, three-barrel box culvert designed to facilitate the passage of a variety of fish species.

The stream restoration effort adapts habitat improvement measures from fisheries management techniques typically practiced in rural areas on relatively low-discharge streams with adjacent natural floodplains. For this project, application of these measures must address more demanding constraints. The design must accommodate high hydraulic energy locations below bridge crossings

[1]Senior Civil Engineer, Fugro East, Inc., 6 Maple Street, Northborough, MA 01532

and existing dam spillways. Stream habitat must be developed within a confined channel with high flood velocities and limited floodplain. Selected areas within the channel require coordination of habitat modification with the removal or capping of hazardous materials releases. The development strategy for the adjacent property also affects the extent and character of in-stream improvements. Therefore, the project team must integrate fisheries management practices, wetland and aquatic vegetation restoration techniques, and hydraulic and structural engineering measures with project development goals to achieve a successful design.

Streambed Improvements

Physical stream bed modifications include a selection of structural measures designed to improve pool-riffle ratio, establish a thalweg ("thread of stream" channel) with velocity/sedimentation conditions supportive of benthic communities, and provide bottom and bank conditions suitable for fish habitat. Typical measures include the following:

Wing deflectors. Low profile V-shaped jetties fabricated of large boulders with cobble infill will be installed on alternate sides of the river at intervals of 5-7 channel widths along portions of the channel. These deflectors direct the river current at low and normal flows, with several benefits, including deepening and narrowing of the channel thalweg, enhancing pool-riffle ratios, protecting stream banks from erosion, removing silt from areas for benthic invertebrate production, and encouraging development of riparian vegetation by silt bar formation (Wesche, 1985).

Low-head rock weirs. An alternating pool-riffle structure is desirable for fisheries habitat. The existing urbanized channel is primarily riffle structure. In order to improve the pool-riffle ratio, the wing deflectors will be augmented by low rock weirs. These are essentially low-head dams constructed of large boulders to create pools at intervals along the channel. Gaps provided at strategic locations between boulders maintain fish passage during low flow periods.

Boulder placement. Boulders set in random locations within the channel provide in-stream cover, break up uniform currents and thus dislodge fine sediments, and provide feeding lies and resting areas for fish (Ontario MNR, undated). Placement of boulder and cobble rubble along the toe of each river bank provides shelter and shade.

Bottom substrate. Channel segments disturbed by realignment and by removal of contaminated sediments will be reconstructed using granular material, cobbles, stones, and boulders similar to the substrate materials of adjacent channel reaches. Field observations, coupled with gradation analyses of existing substrate samples, provide data for specifying the new material.

Energy dissipation aprons. Stone channel protection at high energy locations downstream of bridges, culverts, and existing spillways will stabilize the stream bed. Stepped pools formed with the armor stone will facilitate fish passage at these locations.

Riparian Vegetation

Riparian vegetation provides fisheries and wildlife cover, moderates stream temperatures, helps stabilize embankments, and serves as a source of food, among other benefits (Welsch, 1992). Many sections of the existing river within the project area contain little or no vegetation. To enhance fisheries habitat, plantings are proposed along the margins of the river bed, on the channel embankments, and immediately adjacent to the channel.

Planting design includes a selection of indigenous plants, based on species observed in the vicinity of the project, tolerant of the conditions prevalent in the watercourse. Because of the high energy flows anticipated, especially within confined channel reaches with limited or no adjacent floodplain, plantings may be lost during flood events. In these areas, plantings include shrub and herbaceous species that can self-propagate from stem and root remnants left by a flood event. Large woody species are not proposed within the floodway, as they can become floating debris and cause downstream damage. Bioengineering products allow for additional plantings in channel areas subject to high energy flows; wetland plants bound in a natural fiber matrix are highly resistant to erosive forces. Outside the floodway, shrubs and trees planted along the upper edge of the river embankment provide shade and other riparian benefits.

Culvert Replacement

The reconstruction of an existing box culvert to allow fish passage poses a special challenge. One barrel of the culvert will have a lower invert than the other barrels, and will carry most of the low and normal flows.

Low flow baffles are designed to produce a series of stepped pools within this culvert barrel, to restrict flow velocities to levels conducive to fish passage, without at the same time forming vertical obstacles to their passage. Cobble/stone/gravel substrate will provide resting areas for fish and structure for development of some incidental benthic food source.

The design also considers the introduction of light sufficient for fish passage. The low flow barrel will have a series of light wells with grated openings at grade, to allow daylight penetration into the culvert. The project team adapted daylighting engineering methodology (IES, 1972) to develop a design for lighting the culvert. The grates set at finished grade essentially serve as culvert sky-lights, allowing illumination levels at the bottom of the culvert to meet agency criteria for fish passage. The adapted design methodology allows estimation of light levels considering light transmission through the grates, transmission losses in the light well risers, and transmission losses within the culvert and the water column.

Conclusion

Project design introduces a variety of fisheries habitat improvements into a highly altered, degraded river channel. The habitat restoration techniques are adapted to a high-energy hydraulic setting. Monitoring of the constructed project will provide valuable insight into the value of these aquatic enhancement measures for future stream restoration efforts.

Appendix. References

IES Lighting Handbook. (1972). Illuminating Engineering Society, Fifth Edition, J.E. Kaufman, ed., New York, NY.

Ontario MNR (Ministry of Natural Resources). (Undated). *Community Fisheries Involvement Program: Field Manual, Part 1: Trout Stream Rehabilitation*. Ontario, Canada.

Welsch, D.J. (1992). *Riparian Forest Buffers: Function and Design for Protection and Enhancement of Water Resources*. US Department of Agriculture, Forest Service, Radnor, PA.

Wesche, T.A. (1985). "Stream Channel Modifications and Reclamation Structures to Enhance Fish Habitat." In *The Restoration of Rivers and Streams: Theories and Experience*, J.A. Gore, ed., Butterworth Publishers, Boston, MA.

WATERFILM FLOW DEPTH AND HYDRAULIC RESISTANCE FROM RUNOFF EXPERIMENTS ON PORTLAND CEMENT CONCRETE SURFACES

By James B. Stong[1], A.M. ASCE, and Joseph R. Reed[2], M. ASCE

Abstract

Uniform flow experiments were conducted on Portland cement concrete (PCC) surfaces with variable mean texture depths (MTD) and slopes (S). Uniform flow rates ranged from very low flows up to approximately the largest flow rates possible without the development of roll waves. A regression analysis of the measured sheet flow depths was combined with a Darcy's *f* type expression to define hydraulic frictional resistance. Results appear to indicate that *f* is inversely proportional to Reynold's number, to an exponent other than one (1), and with slope dependency.

Introduction

It is commonly accepted among transportation professionals that the hydroplaning potential of vehicles is a function of waterfilm thickness depths on pavement surfaces. Hence, the prediction of sheet flow depths becomes an important issue in roadway safety.

Reliable empirical expressions have been developed for the calculation of pavement waterfilm thickness depths while a rainfall event of uniform intensity is in progress. However, for that period of time directly following a rainfall event but during which runoff continues, considerably less effort has been made and consequently less is known about the hydraulic runoff conditions.

Woo and Brater (1961) tried to describe flow resistance upon roughened wide rectangular channel surfaces, for sheet flow conditions, by using a Darcy's *f* laminar flow expression:

---

[1] Asst. Prof., Dept of Civ. Engrg., Gonzaga Univ., Spokane, WA 99258

[2] Prof, Dept of Civ. and Envir. Engrg., The Pennsylvania State Univ., University Park, PA 16802

$$f = \frac{C}{N_R} \qquad (1)$$

where: f = Darcy's friction factor
C = Darcy's intercept
N_R = Reynold's number

Woo discovered, among other things, that as the drainage surface slope (S) increased the C value also increased. Surface roughness did not appear to be significant.

Although Woo's work added insight to the complex nature of uniform sheet flow runoff upon roughened surfaces, of itself it was incomplete. Consequently, Stong (1992) performed a more in-depth experimental study. Stong's aim was to more exactly determine the relationship between the hydraulic variables of pavement surface roughness, drainage surface slope, flow rate, and Reynold's number for the determination of both waterfilm flow depths and hydraulic resistance values.

Approach

The approach taken is the same as that taken by Reed (1991); namely that theoretical equations, one of which contains a resistance variable, are algebraically combined with a regression equation containing only experimentally measured variables. The resulting equation is solved for the resistance variable and would have the same unbiased correlation coefficient of the original regression equation, given acceptance of the theoretical equations.

Experimentation

Research was performed by discharging uniform flows, via a distribution box arrangement, onto Portland cement concrete surfaces. The surfaces were 30.48-centimeters wide and 7.31-meters long, bounded by channels so as to create a hydraulically wide rectangular channel situation. Three surface roughnesses defined by mean texture depths (MTD) of 0.025, 0.069, and 0.112 centimeters were tested at each of three slopes (S) of 0.005, 0.015, and 0.025. These values form typical ranges for runoff from roadway or runway surfaces. Keeping slopes under 3% helps insure prevention of roll waves which are not intended to be part of the study. The runoff flow rates (q) used ranged from very low flows up to approximately the largest flow rates possible without the development of roll waves. A total of 73 flow-depths, the consequence of approximately 3300 flow depth measurements by point gage (nearest 0.010 centimeter), was determined from uniform runoff testing.

Results

Of the 73 flow depths measured, 3 were eliminated as

being outside the laminar flow range. It might be noted that the laminar flow range for open channel flow is similar to that defined by the familiar Moody Diagram, ie, Darcy's *f* values vs Reynold's number values plot as a straight line upon log-log paper. The remaining 70 flow depths were placed into a multiple regression routine. The resulting regression with R^2=0.979 was:

$$Y = 75.73\frac{q^{0.315}MTD^{0.117}}{BPN^{1.20}S^{0.227}} \qquad (2)$$

where:
- Y = hydraulic flow depth (cm)
- q = flow rate per unit width (m^3/sec·m)
- MTD = mean texture depth (cm)
- BPN = British Portable Number
- S = drainage surface slope (m/m)

Since the BPN values varied only in a narrow range between 80 and 90, a new regression executed without BPN and with R^2=0.973 was:

$$Y = 0.319\frac{q^{0.313}MTD^{0.0708}}{S^{0.230}} \qquad (3)$$

Still another regression was run without MTD since it's exponent in eq (3) is very small. That regression with R^2=0.963 resulted in:

$$Y = 0.255\frac{q^{0.308}}{S^{0.231}} \qquad (4)$$

One might perceive from this result that perhaps roughness of the surface, as defined by MTD, is not important. However, MTD defines the hydraulic flow depth datum, and in that sense is still included in Y.

The accepted theoretical expressions for Darcy's *f* and Reynold's number, formulated for flow within a wide rectangular channel, are:

$$f = \frac{8gRS}{V^2} = \frac{8gYS}{\left(\frac{q}{Y}\right)^2} = \frac{8gY^3S}{q^2} \qquad (5)$$

$$N_R = \frac{VR}{\nu} = \frac{VY}{\nu} = \frac{\left(\frac{q}{Y}\right)Y}{\nu} = \frac{q}{\nu} \qquad (6)$$

where:
- g = acceleration of gravity (m/sec^2)
- R = hydraulic radius (m)
- Y = hydraulic flow depth (m)
- V = mean flow velocity (m/sec)

ν = *kinematic viscosity* (m^2/sec)

with f, s, q, and N_R as previously defined.

When equation (4), (5), and (6) are combined, and introducing the value of kinematic viscosity at 65°F which is the water temperature at which all experimental testing was conducted, the resultant expression for Darcy's f is:

$$f = 282\frac{S^{0.307}}{N_R^{1.076}} \qquad (7)$$

This equation depicts Darcy's f for the PCC surfaces tested. When a log-log plot of Eq (7), an unbiased expression, is superimposed on the calculated and biased f-N_R values which were determined using the experimental data, the regression equation fits amazingly well.

If one were to compare Eq (7) with the commonly accepted form of Eq (1), one might conclude that C appears to be dependant on flow slope and possibly channel shape. Also, N_R is taken to the power of 1.076 and not 1.00.

It is recommended that additional surfaces, especially asphaltic surfaces, be tested individually due to the different nature of their texture as compared to that for PCC.

Acknowledgement

The authors wish to thank the Mid-Atlantic Universities Transportation Center for their support of this study through grant money from the Federal Highway Administration. The grant was administered by the Pennsylvania Transportation Institute.

References

Reed, J.R. (1991). Manning's n for rainfall-runoff from pavement surfaces. Proceedings, International Conference on Computer Applications in Water Resources, pp. 429-435. Tamsui, Taiwan: Tamkang University, July.

Stong, J.B. (1992). Resistance Variables for sheet flows on Portland Cement Concrete Surfaces. Ph.D. Dissertation, Department of Civil Engineering The Pennsylvania State University, University Park, August.

Woo, D.C., & Brater, E.F. (1961). Laminar flow in rough rectangular channels, Journal of Geophysical Research, 66(12): 4207-4217, December.

Using Remotely Sensed Data To Estimate Rates Of Sedimentation

Ronald E. Fix, P.E., Member A.S.C.E.[1]

Abstract

Traditional methods of measuring lake sedimentation are complicated, slow and expensive. This paper addresses an innovative method to estimate the rates of sedimentation in a reservoir using remotely sensed data including aerial photography and satellite images. Satellite images are available on an approximately two week basis since the early 1970s. The images are interpreted with the aid of a Geographic Information System (GIS), which provides a relatively easy method to monitor changes not only in the reservoir, but also in its upstream drainage basin. The technique, as proposed, can be used to estimate the expected useful life, and to help to identify sources of upstream erosion. Once an erosion source is located, it may be possible to take action to reduce or eliminate the source of lake sedimentation, thereby extending its expected life.

Introduction

The objective of this study is to simplify estimating the rate of sedimentation of a water supply reservoir with a minimal cost to the end water user. Satellite images of the globe are available from at least two international suppliers, and are available on an approximately a fourteen day basis since the early 1970s. The data analysis can be developed using a GIS software which can be run on a personal computer (PC). The image processing can also be processed on a PC. A significant effort was made in establishing ground control for the images using the Global Positioning System (GPS). One of the principle attributes of a water storage reservoir, as pointed out in the 1991/1992 Report of the Institute of Hydrology (1992) is its ability to act as a sediment trap.

---

[1]Graduate Research Assistant, University of Oxford, U.K.

This characteristic has both positive and negative impacts. One of the positive impacts is the reduction of suspended solids in the water discharged from the reservoir. One of the major negative impacts is the loss of storage capacity due to trapped sediment. As time passes water demand increases, while the capacity of the reservoir is decreased.

Methods

With conventional methods the measurement of the impact of sedimentation is very expensive, and time consuming . One method is to completely drain the basin and allow it to dry long enough for the bottom to be surveyed with conventional methods. This method may be practical for very small reservoirs and ones that are not the sole source of water for its users. Another method is to make a bathymetric map of the bottom, using a boat with some form of a depth gauge. For this method, the location of the boat has to be determined at each point a depth measurement is made. For a small study area the boat location can be made by triangulation or with a theodolite/electronic distance meter combination. For larger areas this method is very time consuming and labour intensive. A newer method to determine the boat location is through the use of a Global Positioning System (GPS) system.

The reservoir history, catchment features, hydraulic characteristics, climatological data, inflow, outflow and topological maps of the original basins and the basins as they exist today have to be collected (Anderson and Burt 1990, Labadz et. al. 1991).

The processing of spatial and digital data have been enhanced with the introduction of Geographic Information Systems (GIS). Once the study area has established ground control and digital maps, a GIS system can be programmed to make sophisticated spatial analysis. With the advent of more powerful personal computers GIS systems can analysis large spatial data sets like the Landsat images. Remotely sensed data includes aerial photography, which can be digitised, geo-referenced and used within a GIS system to compare the results with aerial photography (Institute of Hydrology 1993). Landsat (American Earth Resources Satellite) image data have been used to map water turbidity and temperature and to measure the area of surface water (EOSAT 1990, Institute of Hydrology 1993) in relation to areas flooded by rivers and the change of a lake surface over time. There are 97 Landsat 5 images with cloud cover less than ten percent available for the study site for the period since October, 1982 (Arp, 1994).

Case Study

The principal study site is the water supply reservoir system for the City of Tyler, Texas in the United States. The system consists of two adjacent reservoirs located to the southeast of Tyler, in Smith County near the cities of Whitehouse and Troup. There is an abundance of data and engineering reports (Fix 1978, Steele 1983, dePamphilis and Young 1990) for this site. The first reservoir, known as Lake

Tyler, was constructed in the late 1940s and early 50s. During the drought of the early 1960s the water level of Lake Tyler declined to a dangerous level. To relieve this situation, a second reservoir, known as Lake Tyler East, was constructed in mid 1960s on the drainage basin directly to the east of Lake Tyler with channel connecting the reservoirs allowing them to operate as a single unit. The two lakes have approximately the same surface area and storage capacity, but the drainage area of Lake Tyler is "too small" for optimal reservoir operation, and the drainage area of Lake Tyler East is "too large". The combined lake and catchments are near optimal. The lake parameters are given below:

	Lake Tyler	Lake Tyler East
Drainage Area	117 square kilometres (45 square miles)	161 square kilometres (62 square miles)
Surface Area at NWL*	9.57 square kilometres (2,365 acres)	10.15 square kilometres (2,507 acres)
Storage at NWL*	0.053 cubic kilometres (40,553 acre feet)	0.050 cubic kilometres (43,146 acre feet)

*Normal Water Level

In order to reconstruct the hydrological data for the reservoir system, daily reservoir operation data have been collected and entered into computer data files. These data include rainfall, evaporation, raw water pumped and lake levels. These digitised data were used to estimate the reservoir inflows and outflows (Loucks et. al. 1981, da Costa and Loucks 1989, Karamouz 1990). The reservoir operation model has been written in the C++ programming language and compiled for use on a PC. The end result of the model was a 'water budget' for the reservoir system showing the inflows for the system using optimisation techniques for the period of record 1951-1993. The water budget has provided an estimate of the annual reservoir inflows, and has been compared to the changes in deposition within the lake system (Labadz et. al. 1991, Sly 1994).

Conclusion

The satellite images have been timed so that there will be two images for the reservoir system when the lake is at the same level with a few years between the images. More than 120 control points were observed encircled the study area using both Leica™ and Trimble™ GPS receivers at points easily identified on the image. The images were first geo-referenced using the ERDAS™ and Intergraph™ image processing software. Once the images were justified, they were sub-sampled to include the drainage area and then those images were sub-sampled to include only the area immediately surrounding the two lakes. EOSAT™ provided a grant in the form of 6 scenes of the study area. The surface area of the lake was estimated by counting the number of water surface pixels for each image. The surface area for images made with the lake at the same elevation are compared and used to estimate the loss of surface area to sedimentation.

References

Anderson, M.G. And T. P. Burt (1990). *Process Studies In Hillslope Hydrology.* New York: John Wiley & Sons.

da Costa, J.R. and D. P. Loucks (1989). "Computer-Aided Planning and Decision Support." in *Water Resources Planning and Management - Proceedings of the 16th Annual Conference*, New York, NY: American Society of Civil Engineers pp 717-720.

dePamphilis, Patrick D., and Joncie H. Young (1990). *City of Tyler: Lake Palestine Utilization Study*. Tyler, Texas: Wisenbaker, Fix, and Associates.

Fix, Ronald E. (August, 1994). "Global Positioning System An Effective Way to Establish Ground Control for a Small Watershed." Texas Civil Engineer.

Institute of Hydrology (1992). *Report of the Institute of Hydrology 1991/1992.* Oxfordshire, UK.

Institute of Hydrology (1993). *Report of the Institute of Hydrology 1992/1993.* Oxfordshire, UK.

Karamouz, Mohammad (1990)."Uncertainty and Imprecision in Water Resources Systems Operation: A Fuzzy Set Approach." in *Optimizing the Resources for Water Management.* New York, NY: American Society of Civil Engineers pp 449-453.

Labadz, J. C., T. P. Burt, and A. W. R. Potter (1991). "Sediment Yield and Delivery in the Blanket Peat Moorlands of the Southern Pennines." Earth Surface Processes and Landforms, 16, pp 255-271.

Loucks, Daniel P., Jery R. Stedinger, and Douglas A. Haith (1981). *Water Resource Systems Planning and Analysis*, Englewood Cliffs, New Jersey: Prentice-Hall Inc.

Sly, P. G. (1994)."Sedimentary processes in lakes." in *Sediment Transport and Depositional Processes,* ed. Kenneth Pye. Oxford: Blackwell Scientific Publications, pp 157-191.

Steele, Robert R. (1983). *Report on Water Supply, Population and Water Use in Relation to Lake Tyler Water Rights Adjudication.* Austin, Texas: URS Engineers.

Holistic Stormwater Master Planning for Cape Coral, Fl.

William C. Pisano[1], Steve Kiss[2], David Connelly[3], Tom Giles[4]

Abstract

This paper provides an overview of the ongoing stormwater master planning efforts in Cape Coral, Florida. The goal of the plan is to integrate and balance conflicting present and future needs with ever increasing regulatory requirements. These needs include improving drainage service, proper maintenance of canals, roads and streets, enhancing water supply (domestic and nonpotable irrigation), maintaining levels of recreational navigation and fishing, control of stormwater and nonpoint pollution, and preservation of fragile aquatic and terrestrial ecosystems. The effort can be best described as holistic water resource planning. Initial inventory efforts, development of maintenance programs and design, and construction of drainage capital improvement within critical drainage problem areas are underway.

Description of Study

The area of Cape Coral is 288 sq. km with a present population of 93,000 and an estimated population of 374,000 at full buildout. The City owns and maintains 644 km canals (74% are freshwater and the balance are saline), 605 km drainage pipes with 45,000 inlets, 3900 km of roadside swales which collectively comprise the surface water system. This system collects and conveys stormwater flows, provides recreational boating and fishing opportunities, and provides a source of water for the world's largest "secondary" water system for household irrigation and fire protection (canal water mixed with treated sewage).

----

[1] Vice President, Montgomery Watson, Boston, MA
[2] Director of Engineering, City of Cape Coral, FL
[3] Vice President, Montgomery Watson, Boston, MA
[4] President, Avalon Engineering, Cape Coral, FL

Cape Coral's unique amenity is the canal system allowing boating access throughout the City and to the open waters of the Caloosahatchee River and the Gulf of Mexico. Fifty eight weirs control water levels in the upper or freshwater canal system. Two underlying fragile aquifers lie under the surficial aquifer which provide sources of domestic water. Pumped flow from the lowest aquifer is treated by the world's largest reversed osmosis water treatment system. Freshwater "spreader" canals along the City's north-south perimeters confine future growth and "spread" stormwater from the developed areas into the remaining mangrove swamps.

The "holistic" planning approach derives from recognition of the City's need to develop broad scale integrated infrastructure solutions for all of its water related problems. A partial list of problems and concerns is the following.

Since the City's infrastructure was fully developed in the 50's for the sale of lots, there are huge sparsely populated areas throughout the City with only a few fully developed portions in the south east marine areas. However, the road, canal and minor piping drainage systems were completely constructed which result in monumental maintenance requirements.

Localized flooding occurs in a number of the fully developed areas. The capability of the drainage system during extreme events for most areas is unknown. Much of the minor system piping has deteriorated and is being replaced. Portions of the freshwater canal system are choked with weeds. Utilization of concrete vertical sidewalls in canals versus "natural" vegetated sloped sidewalls with docks for boat use is a significant community issue needing resolution.

There is uncertainty whether available canal freshwater (replenished by the surficial aquifer) in combination with future treated sewage volumes can fully support the secondary water system at complete buildout. Long term hydrologic monitoring and regional modeling of the surficial aquifer yield is necessary. Portions of the freshwater spreader canals on the City's perimeter could be used to augment these sources, but these canals are frequently topped with seawater during severe storms. Several of the spreader canals have been breached with resulting saltwater intrusion. Fresh water inputs from upstream wildlife areas could be used for augmentation but such release rates are restricted by regulatory considerations.

Large developed areas are still served by septic systems, but sewers are presently being constructed to alleviate nonpoint pollution. Increasing weir levels in the freshwater area would enhance supply to the secondary water system but would aggravate drainage service in some areas as well as adversely impact septage leaching fields and limit recreational boating and fishing opportunities.

Deteriorated canal water quality has already been experienced in areas with full buildout and in canal with "dead-ends". Interconnection of marine canals to improve tidal flushing in problem areas is a potential control measure.

The historic practice of using groundwater from the mid level aquifer for household irrigation has caused problems to the underlying aquifer (domestic water use). This has prompted the need for the secondary irrigation system. Portions of the new secondary water system entail significant pumpage and costly conveyance facilities. Locating redundant canal intakes and quantifying potential alternative sources are critical operational concerns, especially with regard to future canal water quality degradation resulting from increased urban runoff and other nonpoint pollution sources.

Presently roadside swales are the only practical water quality BMP for most of the residential area as ground water levels are near the ground surface. Over the years much of the early swale system has been modified by homeowners and design standardization is paramount. For some areas purchase of large tracts for creation of stormwater holding and treatment areas may be necessary as the degree of stormwater quality control is becoming more stringent. To confound the planning effort, construction during the original development significantly changed the natural terrestrial systems with considerable and often unknown alteration of upper soil column and vegetation.

A multi-year project is being conducted by Montgomery Watson Americas, Inc. and Avalon Engineering, Inc. The study elements include areawide topological, soils, vegetation and hydrographic surveying, GIS mapping, telemetered hydrologic/meteorlogic and water quality canal network monitoring, monitoring of various stormwater BMPs, areawide canal network hydrologic and hydraulic simulation modeling (and water quality as needed), formal multi- objective economic optimization (using simplified simulation model with objective functions and constraints developed using localized costs and preferences defined via public action groups). Important controls include regional drainage improvements, new canal storage and conveyance configurations to ensure continued secondary system supply with future population growth, additional stormwater BMPs, and operational rules for setting fresh water weir level settings (adjustable gates). The entire canal network together with significant "minor" drainage systems and surficial aquifer linkages will be modeled by Ghioto Associates, Orlando, Florida using the Aquarian Watershed Simulation Model (AWSM).

One example depicting a potential scheme for increasing freshwater canal storage north of Pine Island Road for the secondary water supply system is shown in Fig. 1. The idea is to use movable canal barriers at key freshwater (east-west) canal entry points into the spreader canal system (north-south) that would permit recreational boat traffic, but at the same time prevent entry of brackish water intrusion from the spreader canals. Freshwater volumes on the east side of spreader canal could then be "reclaimed" and held in reserve for potential use in the secondary water system.

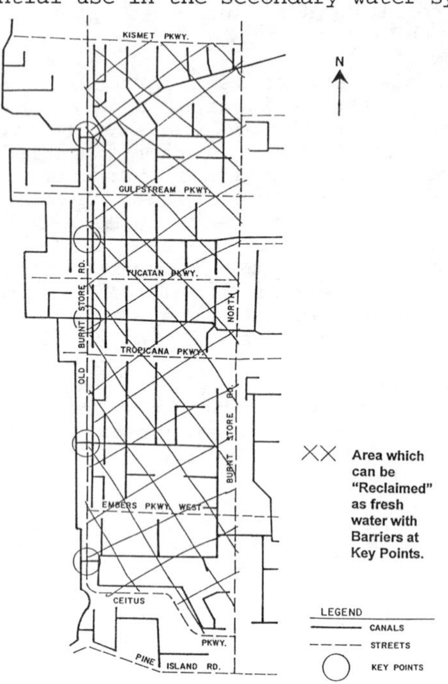

Fig. 1 Freshwater Canal Storage Alternative

Model Validation for Runoff Pollution from Urban Watersheds

Rizwan Hamid, Vassilios A. Tsihrintzis,* and Hector R. Fuentes[1]

Abstract

A methodology for modeling urban runoff and pollutant loadings is presented using data collected and compiled by the Department of Natural Resources Protection (DNRP) at various sites in Broward County, Florida, as part of their National Pollution Discharge Elimination System (NPDES) permit application. The SCS hydrology method was used to calculate flow hydrographs. EPA's formulation was used to simulate the following four quality constituents: Biochemical Oxygen Demand (BOD_5), Total Suspended Solids (TSS), Total Kjeldahl Nitrogen (TKN), and Lead (Pb). Calibration and verification was completed for different storm events at three sites. Hydrographs and constituent loading predictions are presented to illustrate results.

Introduction

Urbanization increases pollutant loads by at least one order of magnitude over natural watershed conditions. Typical pollutants include: suspended sediments from increased construction activities that adversely impact biotic communities; nutrients, such as phosphorous and nitrogen, that cause algal blooms (eutrophication); high BOD levels that lead to anoxic conditions; and trace metals from the urban landscape, such as lead, copper and zinc, that impact aquatic life and contaminate drinking water supplies.

Three watersheds representing residential, commercial and highway land uses, located in Broward County, Florida, were used to predict hydrographs with associated pollutant loadings. The first watershed has a predominantly residential land use, is

---

[1]Respectively, Graduate Assistant, Assistant Professor and Associate Professor, Department of Civil & Environmental Engineering and Drinking Water Research Center, Florida International University, VH 160, Miami, FL 33199, USA.

*Corresponding author, Member ASCE

0.77 acres in area, and is located in a highly dense residential neighborhood of the City of Lauderhill (population density of 11 persons/acre). The effective impervious area is 65% with an average slope of 0.001. There is one catch basin which collects the runoff from the drainage area and discharges it through a single 18-inch pipe into a lake. The second watershed has mainly a commercial land use, is 1.84 acres in size, and is located in the City of Lauderdale Lakes (population density of 12.2 persons/acre). The effective impervious area is 45% with an average slope of 0.001. There are three catch basins which collect the runoff from the drainage area and discharge it through a single 15-inch pipe into a canal. The third watershed drains a highway, is 1.50 acres, and is located in the City of Coconut Creek with a population density of 20.3 persons/acre. The effective impervious area is 98% with an average slope of 0.005. Runoff is collected by three catch basins and is discharged into a small lake. Data obtained by DNRP include precipitation, flow rates at outfalls, and concentrations of the quality constituents from the composite of samples collected during the storm event.

Methodology

The standard SCS hydrology method was used to estimate hydrographs, while formulation by EPA (1979) was used to estimate the pollutant loadings from each watershed. This formulation uses the pollutant buildup and washoff equations, which were programmed into a spreadsheet to obtain initial concentrations. For each site, one rainfall event was used for calibration and another independent event was used for verification of the methodology. Calibration parameters for the runoff include: the curve number (CN), the initial abstraction (I_a), and the antecedent moisture conditions (AMC I, II, or III). Parameters for pollutant loadings include: pollutant removal or decay rates, given by EPA (1979); pollutant loading rates, obtained via a literature search on previous urban nonpoint source pollution studies (Novotny and Olem 1994); washoff coefficients, read from charts based on physical characteristics of the watershed (EPA 1979); and population densities, calculated from population statistics and city maps of the area. These parameters were varied until predicted quantities showed good agreement with those measured in the field. Table 1 presents values used for each watershed. To verify these parameters, an independent storm event was used, keeping all the parameters unchanged.

Results and Conclusions

The results of the calibration and verification for all three watersheds are presented in Table 2. Figure 1 presents measured and predicted hydrographs for the calibration of the watershed with highway land use; the precipitation used is also shown. Figure 2 presents the same information for verification of this watershed. Agreement overall is good. BOD_5 and the TSS concentrations were relatively higher in comparison to TKN and Pb. The highway land use shows the highest amounts of BOD_5 and TKN concentrations, whereas the commercial land use shows the highest amounts of TSS concentrations; all areas show a nearly zero amount of lead.

The methodology described in this paper shows an efficient way of obtaining initial estimates of runoff and pollutant loadings from stormwater runoff, which can be used to support regulatory and decision making purposes within the NPDES program.

Acknowledgements

Mr. John Foglefong, P.E., of the Water Resources Division of the Broward County DNRP provided necessary data and technical assistance.

Appendix. References

Environmental Protection Agency (EPA). (1979). *1978 Needs Survey: Continuous Stormwater Pollution Simulation System - Users Manual.* Office of Water Programs Operations, Washington DC.

Novotny, V. and Olem, H. (1994). *Water Quality - Prevention, Identification, and Management of Diffuse Pollution.* Van Nostrand Reinhold, New York.

Table 1. Quantity and Quality Parameters Used in Simulations

Land use	Quantity Parameters		Quality Parameters			
	CN	I_a (in.)	BOD_5 (lb/ac)	TSS (lb/ac)	TKN (lb/ac)	Pb (lb/ac)
Residential	74	0.09	0.122	0.366	0.007	0.0004
Commercial	63	0.29	0.220	0.880	0.012	0.0004
Highway	90	0.09	3.660	4.880	0.049	0.0004

Table 2. Summary of Results: Calibration and Verification

Land Use (event no.)	Flow				Quality Parameters							
	Measured		Predicted		Measured (mg/L)				Predicted (mg/L)			
	R (in)	Q_p (cfs)	R (in)	Q_p (cfs)	BOD_5	TSS	TKN	Pb	BOD_5	TSS	TKN	Pb
Res. (1)	0.020	0.009	0.020	0.055	1.40	0.90	0.01	0.000	1.40	4.2	0.08	0.004
Res.(2)	0.001	0.006	0.003	0.010	3.20	11.00	0.58	0.000	2.20	6.5	0.13	0.007
Comm.(1)	0.020	0.023	0.020	0.066	8.70	104.00	0.78	0.000	18.90	75.6	1.00	0.040
Comm. (2)	0.002	0.004	0.003	0.010	9.70	15.00	0.00	0.000	4.00	16.0	0.22	0.001
Hwy. (1)	0.180	0.270	0.610	1.850	7.20	19.00	1.67	0.000	7.80	10.4	0.10	0.001
Hwy. (2)	0.030	0.085	0.020	0.089	45.00	43.00	0.45	0.000	34.30	45.7	0.46	0.004

R = Runoff, and Q_p = Peak Flow

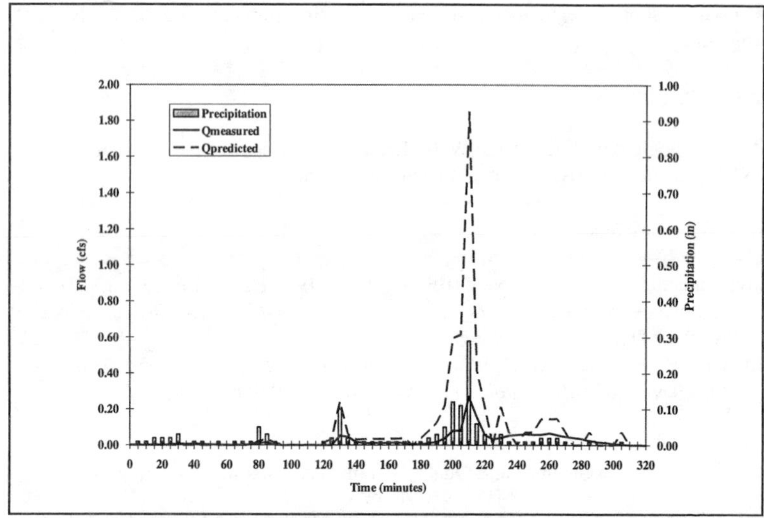

Figure 1. Predicted and Observed Hydrographs from Calibration Run

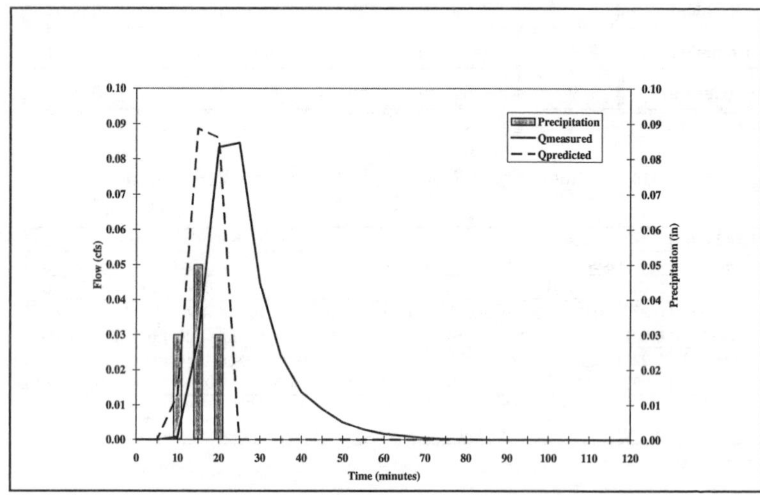

Figure 2. Predicted and Observed Hydrographs from Verification Run

THE EFFECT OF BENTHIC DEPOSITS ON AN URBAN WATERWAY

Philip B. Pedros[1], Frederic C. Blanc[2] and Constantine J. Gregory[2]

Introduction

Streams and rivers provide drainage for the runoff of large areas. Inherent in this natural runoff is a quantity of sediments. Most of this material and the additional contamination from sewered areas and domestic and industrial wastes contributed by man, settles to the bottom of sluggish rivers and lakes. These deposits build up over time and vary in composition, depth and location. If the water immediately above the sediments contains dissolved oxygen than a thin aerobic layer will exist at the sediment water interface; underneath will be the remaining anaerobic portion of the deposits. The thickness of the aerobic layer is determined by the depth that oxygen can diffuse to; normally, only a few millimeters[1] (Fillos, et al, 1972). From the underlying anaerobic layer, products of decomposition diffuse upward into the aerobic layer. Here they may be oxidized, utilized for food, or released directly to the water.

However, should the overlying water be completely exhausted of dissolved oxygen than no aerobic layer would exist. The entire system, the benthic sludge and supernatant water would be anaerobic. The products of decomposition in the lower layers of sediments would diffuse upward and be released directly to the water column. Under such conditions it has been demonstrated that the release of some contaminants is greatly increased.

This research had three principal objectives: to measure the concentration of organics and nitrogen in the river and interstitial waters, to estimate the potential release of these contaminants from the sediments and to verify that high concentrations of contaminants are found in the river after periods of precipitation.

Comments on the Release of COD

The interstitial water COD values ranged from 20.8 mg/L to 132.2 mg/L. Figure 1 shows the average concentration of COD in both the river water and interstitial water. It is apparent that an accumulation of organic material resulted in the benthic deposits. The COD concentration in the interstitial water was 3.5 times greater than in the river water. The average interstitial water concentration was 66 mg/L, while the average river concentration was 19 mg/L.

With oxygen in the overlying water, the biologically oxidizable component of the COD is utilized by the bacteria present in the biofilm. The refractory organics, (i.e. the chemically oxidizable portion of the COD,) are subject to aerobic stabilization. Both the bacterial respiration and the aerobic stabilization require oxygen which is removed from the overlying water.

If the concentration of COD in the anaerobic layer is high, than the quantity of organics that diffuses upward maybe in excess of the amount that can be oxidized within the aerobic layer. This will result in the direct release of the excess organics

1. Philip B. Pedros, Engineer, CDM Federal Programs, 160 N. Washington St. Suit 400, Boston, MA.02146
2. Frederic C. Blanc, Constantine J. Gregory Professors, Department of Civil Engineering ,Northeastern University, Boston, MA. 02115

into the supernatant water. If the overlying water column is aerobic, the biologically oxidizable material will be consumed by bacteria in the water. However, the refractory organics may be oxidized within the water, or may be carried away by the flowing water, if the residence time is too short for oxidation to occur.

Therefore, the oxygen demand exerted on the overlying water column results from the sediment oxygen demand and from the oxidation of materials released directly into the water from the benthic deposits.

Results of Nitrogen Analysis

In this study the concentrations of: ammonia nitrogen, (NH_3), organic nitrogen and total Kjeldahl nitrogen of the interstitial water were analyzed at the same six locations as the organics. The values of total nitrogen, (TKN), ranged from 11 mg/L to 57 mg/L. Figure 2 presents the botton sediment interstitial water nitrogen concentrations.

As shown in Figure 2, all six sampling locations had much higher values of ammonia nitrogen than organic nitrogen, on average, 32 times greater. At site 5, the concentration of ammonia was 50 times greater than the concentration of organic nitrogen.

Nitrogen in The River Water

The concentrations of ammonia nitrogen, organic nitrogen and TKN in the river water are shown on Figure 3. The NH_3 concentration varied from 0.466 mg/L to 0.682 mg/L and averaged 0.56 mg/L. The maximum and minimum organic nitrogen concentrations were 0.71 mg/L and 0.861 mg/L respectively. The overall average value was 0.788 mg/L. The organic nitrogen concentration was found to be, on average, 1.4 times greater than the concentrations of NH_3. This was opposite to what was found in the interstitial water. A comparisaon of Figures 2 and 3 shows that the ammonia nitrogen concentration in the interstitial water averaged more than 60 times the concentration in the river water.

Summary and Conclusions

The concentration of readily biodegradable organics, (i.e. BOD_5,) in the interstitial water was equivalent to the concentration in the river water. The average BOD_5 concentration in the interstitial water was 5.2 mg/L.

The concentration of NH_3 ranged from 10 mg/L to 52 mg/L, which suggests that the sediments, of the Muddy River, have become veritable repositories of ammonia.

The average NH_3 concentration in the interstitial water was 32 times greater than the soluble organic nitrogen concentration. Equivalent low concentrations of organic nitrogen were found in both the river and interstitial waters.

The estimated release rates for an overlying water DO concentration of 0.0 - 0.5 mg/L, varied from 0.5 mg/m^2-hr at site 4 to 8.3 mg/m^2-hr at site 5. The release rates for an overlying water DO between 8.0 - 8.7 mg/L ranged from -1.5 mg/m^2-hr at site 4 to 1.0 mg/m^2-hr at site 5. The negative value indicates that there was no net release of NH_3 from the aerobic layer to the overlying water. The NH_3 that diffused up from the anaerobic layer, at the negative release rate sites was completely utilized by the bacteria within the biofilm. At the sites with negative release rates, NH_3 may have also diffused into the aerobic biofilm layer, from the overlying water, to satisfy the needs of the nitrifying bacteria.

Figure 1

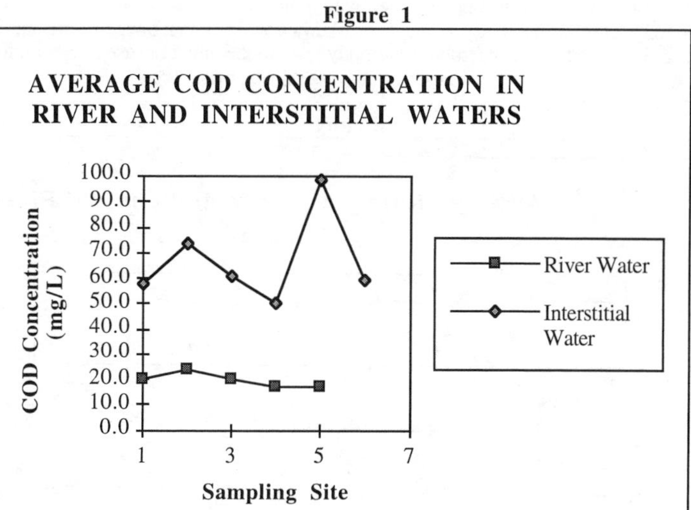

Figure 2

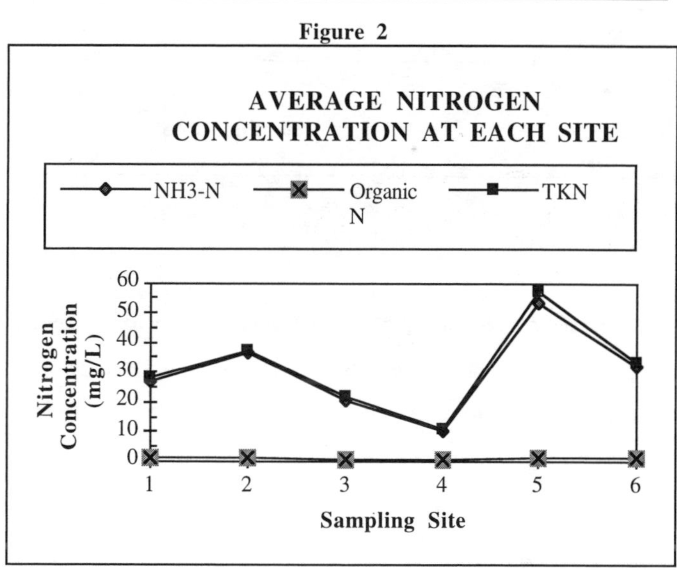

The oxygen demand on an aerobic waterbody may be exerted from two sources. The bacterial respiration, due to consumption of NH_3 within the aerobic layer of the benthic deposit, will cause oxygen to diffuse into the bottom from the overlying water. If the concentration of NH_3 is sufficiently high than an excess amount will be released directly into the waterbody where its breakdown will exert an oxygen demand. The oxygen demand on an anaerobic waterbody is do to the direct release of NH_3 into the water from the benthic deposit.

Figure 3

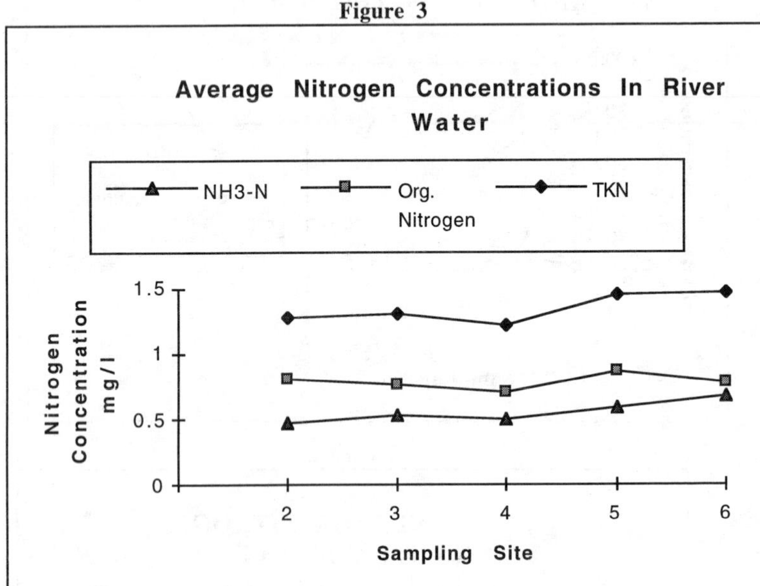

REFERENCES

[1] Fillos, John and Alan H. Molof, "Effect of Benthal Deposits on Oxygen and Nutrient Economy of Flowing Waters," *Water Pollution Control Federation,* 44, No.4 (April 1972), p. 644-662.

Effects of Urbanization on Low Streamflows in North Carolina

Jack B. Evett[1], M. ASCE

Abstract

Historical low-streamflow data were analyzed for a number of gaging stations on streams in and around various urban areas in North Carolina in an attempt to find and document effects of urbanization on low streamflows. Records for streams within each urban area were compared with streams outside the area by two statistical methods. It was concluded from the study that there is some support for the premise that urbanization causes a decrease in streamflows over time, but statistically the results are inconclusive. It appears that most small streams in North Carolina--both urban and rural--are experiencing decreasing flows over time.

Introduction

The hypothesis for this project was that low streamflows in urban areas in North Carolina would exhibit a decreasing trend as urbanization progressed, with the proviso that a similar decreasing trend would not be present for nearby rural areas. Thus, the overall objective of this project was, in effect, to prove or disprove this hypothesis.

Procedures

The project was initiated by selecting key U.S. Geological Survey continuous-record stream-gaging stations

----

[1]Professor, Department of Civil Engineering, The University of North Carolina at Charlotte, Charlotte, NC 28223

for which low-streamflow data were available for a substantial period of time. More specifically, stations were selected both within and outside each of six "urban areas"--Asheville, Greensboro, Raleigh, Charlotte, Goldsboro-Kinston, and Rocky Mount-Tarboro.

For each station, annual "7Q values" were evaluated. (7Q denotes the minimum average value of mean-daily streamflows for any seven consecutive days.) Successive 5-year averaged 7Q values were then plotted on a time graph for all stations in each respective urban area. A "best straight line," as determined by regression analysis, was placed on each graph to indicate trends over time. These lines are referred to as "trend lines." Stations in each urban area were then analyzed to try to detect effects of urbanization on low streamflows.

Statistical Methods of Analysis

The One Sample Run Test for Randomness (OSRTR) (McCuen 1993) was used to test for randomness. Simply stated, it gives a judgment as to whether a series of events was random or exhibited a trend. To try to determine whether there was a significant difference between streamflow trends for urban versus rural stations, with the assistance of Tiwari (1992) a nonparametric method for testing equality of slopes in a simple linear model--the Test for Equality of Slopes (TES)--was developed and applied. This method compares the slopes of the trend lines for two comparative stations (urban versus rural) to see if they have, statistically, the same slope or differing slopes. The project's hypothesis would be proved if the slopes of trends line for urban stations are significantly less (statistically) than those for corresponding rural stations. Otherwise (i.e., if the slopes for urban stations are greater than or equal to those for corresponding rural stations), the hypothesis would be disproved.

Applications and Results

Both the OSRTR and the TES were applied to the stations selected from each of the six urban areas noted previously. Space limitations do not permit presentation herein of specific details of these applications; only a summary of the results is possible.

In the Asheville area, two rivers were considered--one urban, one rural. Results of the OSRTR were that both stations exhibited no significant trend. The TES indicated there was only a 65.8 percent significance that the rural station had a greater slope than the urban one. Hence, it was concluded that, although both stations exhibited a negative (downward) trend with the urban station's trend being more negative, statistically they did not differ significantly; and the project's hypothesis was not proved.

In the Greensboro area, two of the three urban stations had positive slopes; the third was negative, but virtually zero. All three rural stations exhibited negative slopes. This outcome in itself is contrary to the project's hypothesis (that urban slopes should be less than rural ones). The OSRTR indicated that one urban and one rural station exhibited trends while others did not. According to the TES, two comparisons of rural-urban stations indicated a significant difference in slopes at a statistical significance of at least 90 percent, with a third at only 44.5 percent. The two significant at the 90 percent level were, however, contrary to the project's hypothesis, because the rural slopes were less than the urban ones.

In the Raleigh area, slopes of all stations were negative, but the lone urban station had the most negative slope. The OSRTR gave mixed results, with the urban station resulting in "no trend," while the rural stations had two indicating "trend" and two, "no trend." Of the three urban-rural comparisons by the TES, two indicated a significant difference in slopes at a statistical significance of at least 90 percent, with the third at only 15.9 percent. Thus, while the results of the OSRTR were not entirely favorable to the project's hypothesis, it was felt that the overall results did tend to uphold the hypothesis.

In the Charlotte area, slopes of all stations except one rural one were negative. Results of the OSRTR were again inconclusive, with one of the three rural stations and one of the five urban stations indicating "trends" and others, "no trend." Results of the TES were somewhat conclusive. Three of five comparisons indicated a significant difference in slopes at a statistical significance of at least 90 percent. Hence again, while results of the OSRTR were not entirely favorable to the project's hypothesis, the overall results did tend to uphold the hypothesis.

In summary, two of the four urban areas tended to uphold the project's hypothesis, while two did not. (The other two areas considered initially--Goldsboro-Kinston and Rocky Mount-Tarboro--were deleted because of problems with the data.) It was noted, however, that of all stations examined, including the Goldsboro-Kinston ones, 21 out of 24 (8 of 10 urban and 13 of 14 rural) exhibited downward trends over the approximately 30 years of record. Hence, it appears that almost all streams--urban and rural alike--are exhibiting downward trends. The fact in itself that most urban stations exhibited downward trends suggests that the effect of urbanization is to decrease low streamflows over time. However, the fact that most rural stations in the same general areas also exhibited downward trends over the same time period suggests that perhaps it was not urbanization that caused the decrease in low flows. In other words, for whatever reasons, almost all small streams in North Carolina seem to be decreasing.

Conclusions

It is concluded from this project that there is a tendency to support the premise that urbanization causes a decrease in low streamflows in North Carolina, but statistically the results are inconclusive. It appears that most small streams--urban and rural--are experiencing decreasing flows over time, more so than would result from decreasing trends in precipitation alone.

References

McCuen, R. H. 1993. Statistical Hydrology. Prentice Hall, Englewood Cliffs NJ.

Tiwari, R. C. 1992. The University of North Carolina at Charlotte, personal communication.

Notes

(1) The research on which this paper is based was financed in part by the Department of the Interior, U.S. Geological Survey, through the Water Resources Research Institute of The University of North Carolina. The contents of this publication do not necessarily reflect the views and policies of the Department of the Interior, nor does mention of trade names or commercial products constitute their endorsement by the United States government.

(2) Because of space limitations, details regarding all parts of the project presented in this paper could not be included herein. All details, including an extensive bibliography, are included in Effects of Urbanization and Land-Use Changes on Low Stream Flow, Report No. 284, published by the Water Resources Research Institute of The University of North Carolina, June 1994.

Sustainability Criteria in Project Planning

D. P. Loucks, F. ASCE[1]

Abstract

Sustainability emphasizes the need for us to consider the impacts of what we do in our generation on those who follow us in furure generations. While we cannot know with certainty what these impacts may be, or what future generations will want or value, we can attempt to include what we think they will want or value in our current planning and management models. In this way we may at least estimate what they would like us to do today to allow them to better satisfy their needs and desires in the future. This then becomes a multiple-objective planning problem, where if conflicts exist, tradeoffs can be identified and debated. This paper outlines how sustainability can be included among these multiple objectives using economic efficiency as an example framework.

Introduction

ASCE leadership has made it clear that they want this professional organization to be on the forefront with respect to promoting and practicing sustainable development. This is a challenge to each of us involved in designing, planning and/or managing infrastructure and environmental resources. It forces us to not only define what that term means, but also to devise criteria by which we can measure the extent to which we are fulfilling this objective. This Division's Task Force Committee on Sustainability Criteria is one of many efforts underway to do this and to get it accepted in the profession.

This paper is not a report of the Committee, but rather some thoughts on the subject that may provoke some discussion and progress toward accomplishing the task given to us - not only to those of us on the Committee, but to all of us as members in ASCE.

There seem to be, at least, three different arguments pertaining to sustainability objectives and the use of traditional cost - benefit analyses for project planning and evaluation. Some suggest that in order to achieve sustainability - i.e., to insure that present needs are met without compromising those of future generations - we must use a relatively low discount rate when computing present values and comparing alternative investment decisions based on those values. Otherwise, any costs or benefits accrued to future generations will be insignificant when brought to the present.

---

[1]School of Civil and Environmental Engrg., Cornell University, Ithaca, NY 14853

For example, the present value of an irrigation development project might be maximized by practices that maximize immediate crop yields, but that over time degrade the soil to the extent that some 100 years from now it would not be suitable for growing crops. At a high discount rate, say even 10% per year, we might not even notice the difference when those diminishing values are discounted to the present. At a low discount rates, say 2% per year, the impact of such a decision will have a much greater impact on the present value.

Others would argue that the use of low discount rates will favor projects that are economically less attractive, having rates of return less than alternative investments, thereby reducing the wealth that could be passed on to future generations. This in turn could reduce their ability or willingness of those future generations to allocate resources to activities such as environmental protection that might increase not only their quality of life but those of their future generations as well.

Still others argue that such benefit-cost analyses are only suitable when dealing with easily quantifiable expenditures and returns over time. They cannot be applied when dealing with environmental issues and resources that are not priced in the market place, that are variable and uncertain, and involving major costs now but major benefits only far into the future. They argue that when it comes to setting the discount rate for project planning, long-term investments to protect our water and other environmental resources must be treated differently. When it is a question of equity and resource ownership, the argument goes, we should be obligated to bequeath to posterity not only a certain stock of wealth including a certain fraction of that wealth in natural capital - i.e., environmental resources.

It seems to me each of these positions is arguable. But do we need to argue them? In the remaining space available here I will try to outline an approach that takes a multiobjective view of project planning, where at least one of the objectives has to do with sustainability. That objective will include what our best guess is of what our future generations would like us to do for them as they maximize their present value, just as we maximize our present value, from the use of environmental resources. The extent to which we consider those future "needs" relative to our own present "needs", assuming there is a conflict or tradeoff, is a political decision.

A Multiobjective Approach

To introduce the general modelling approach, assume the conditions or resources available to us who live today are denoted by R_0. Also assume our overall objective is to maximize the present value of our total "needs" or welfare, $W_0(X_0 \mid R_0)$, however measured, given the conditions (resources), R_0, we currently have. The decisions we make, X_0, will impact or determine, in part, not only our ability to meet our own needs, i.e., our own welfare, but also the conditions or resources, R_y, available in the future periods (years) y. Like us, our descendents in future periods y may want to maximize the present value of their welfare, $W_y(X_y \mid R_y)$, given their current conditions, R_y. Their current conditions, R_y, are dependent on the resource management decisions of those who preceded them.

Letting a_y be a relative weight given to a future welfare function $W_y(X_y|R_y)$, in period y, a multiple-objective function can be defined that includes the welfare of a series of periods beginning now and terminating at some time in the future:

$$\text{Maximize } \Sigma_y \; a_y W_y(X_y \mid R_y)$$

where each set of future conditions, R_y, depends on the current conditions, R_0, and on the sequence of decisions, X_0, X_1, X_2, X_3, ..., X_{y-2}, and X_{y-1} up to year y.

This multiple-objective modelling approach can be used to determine tradeoffs among various future generations by varying the values of the relative weights a_y. More importantly, however, this modelling approach can help us identify today what we think future generations would like us to do for them. This does not mean we must or will necessarily do what our descendents wish, only that we want to identify what those wishes might be. We can then examine the range of tradeoffs between what we want to do and what we think those who follow us in the future may want us to do.

To implement this modeling approach is to quantify those "welfare" functions which in fact may be a vector of objectives, not all expressible in the same units. To estimate what future generations "needs" might be we can try to place ourselves in our descendents shoes and assume their objectives could be the same as ours are today. For objectives measurable in economic terms, this might be the maximization of the net present value of our income.

Given a constant market-based annual discount rate of r, which includes the inflation rate, and project net incomes, NI(y), in successive years y, the beginning-of-period present value of net benefits n periods from now (based on the following Y periods after period n) is:

$$PVI(n) = \sum_{y=n}^{Y+n} NI(y) / (1+r)^{y-n+1} \qquad \text{for } n = 0, 1, 2, ..., Y \text{ periods}$$

The objective function for the present time might be some weighted combination of successive future present values:

$$\text{Maximize} \quad \sum_{y=0}^{Y} a_y \cdot PVI(y)$$

where the sum of the weights, a_y, could equal 1. Clearly if a_0 is 1 and all the other weights are 0, then we are placing all the weight on the present, as in traditional benefit-cost analyses.

Using the present values defined by the above equations, one could maximize the weighted sum of those present values subject to constraints specifying that the successive present values, PVI(y), for increasing future periods y, do not decrease. This enforces an economic surrogate of sustainability.

Some sensitivity experiments were made using the above modeling approach in some specific project planning exercises. The results were as follows:

• Decreasing the discount rate alone will not always result in the selection of a sustainable policy. Some economically marginal, possibly more polluting (if not constrained), projects may become economically attractive under low discount rates.

• Imposing constraints insuring non-decreasing present values may not always insure fully sustainable management practices given certain values of the discount rate, price and costs.

• Maximizing the present value, today, of future net income will not necessarily result in a sustainable policy, i.e., one which will ensure future generations at least the same level of return from the resources used.

• As the weights assigned to future objectives increase relative to those assigned to present objectives, the current optimal management policy may shift from one that is non-sustainable to one that is sustainable. This occurs as the total net present value of income at the current time decreases. This is the tradeoff that may have to be considered.

Some Conclusions

Economic efficiency objectives are not the only ones that should be considered in any project evaluation, but they have been used here as a means of illustrating how sustainability objectives can be included within an economic efficiency framework. One can only guess at what future generations will select for their objectives, but among them might be objectives similar to those we use today, including economic efficiency. After attempting to define the economic and other objectives of future generations, we can use them in models for estimating the tradeoffs between what we would like to do today for ourselves and what our descendents would like us to do today for them tomorrow.

Since we cannot look into the future with much precision, and since our models require us to make guesses about the future, this planning process should be a sequential and adaptive one. The assumptions concerning the future should be re-evaluated each time planning or decision making takes place. Our estimates of future desires, technology, and economic and environmental conditions will always need updating based on the most recent information we can use to guess about the future. All we can do is to make our best estimates at the time we must make these estimates. Then based on these estimates, generate information useful to planners and decision makers and allow the political process to determine the appropriate tradeoffs that must be made between present and future generations.

While models can help identify and estimate tradeoffs among multiple objectives of present and future generations, it is not a trivial matter to introduce this information and get it used in any decision making process. The actions of interest groups or lobbies are often what determine the relative importance placed on various multiple objectives. There are few if any lobbies representing the interests of future generations. There are many representing the interests of present generations. It is difficult to see a way to change this situation except through continued education, debate, and public discussion of sustainability issues.

Sustainability in International Water Resources
Water Supply and Sanitation in Developing Countries

Philip Roark[1]

Although the term development has always implied sustainability, it is only in the last few years that this relationship has been highlighted and sustainability has come to the forefront of development thinking. USAID, for instance, now describes sustainability, as applied to developing countries, as the ultimate test of development efforts. In spite of agreement that sustainability should be the goal of development assistance there continues to be many projects undertaken by international aid organizations which are clearly unsustainable.

There is wide divergence on the meaning of sustainability, what is needed to achieve it, and how to measure it. This divergence of views is at least partially attributable to the wide range of researchers who have studied the subject. Economists, ecologists, agronomists, and social scientists, to name a few, have each produced definitions to explain the perspective of their own disciplines. This paper provides a perspective of sustainability as applied to the water supply and sanitation (WSS) sector in developing countries, and containing implications for other disciplines as well.

Factors in the equation of sustainability

At one level "sustainability" is simple to define. "Sustainable" means to endure, to last, and to keep in being. Sustainable development is about marshalling resources through projects to ensure that some measure of human well being is maintained over time. **Sustainability may be defined as the outcome of a development project whereby an institutional capacity is**

[1]Project Manager, Chemonics International, 2000 M Street NW, Washington, DC 20036

established to maintain, or expand, a flow of benefits at a specified level for a long period of time after project inputs have ceased.

Defining sustainability is comparatively easy but determining what causes sustainability is much more difficult. The equation of sustainability is complex as shown in figure 1. At the top of the picture are **institutions**. Institutions are the key to sustainability as they provide the permanent guidance on how the WSS sector functions. Institutions may range from water and sewer utilities or agencies at the national level of government to community organizations at the village level. Other institutions at the intermediate level include regional government agencies and the private sector. The private sector includes the entrepreneurial service industry and non profit societies. All exist to provide specific services to the public.

FIGURE 1

FACTORS IN THE SUSTAINABILITY EQUATION

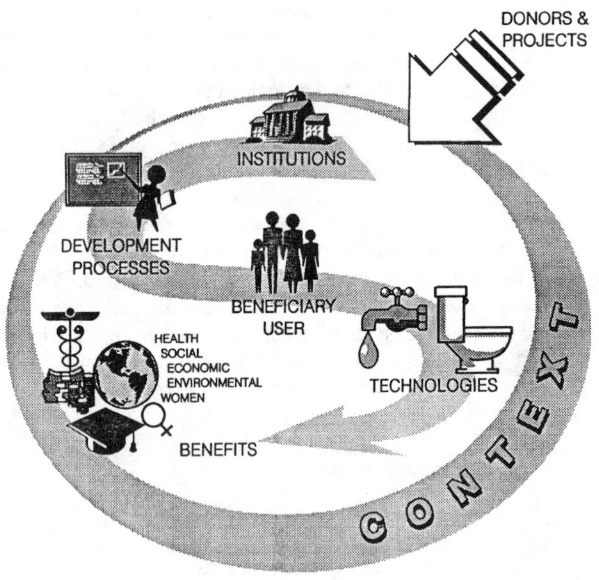

SUSTAINABILITY = (INSTITUTIONS + DEVELOPMENT PROCESS + TECHNOLOGIES) + (CONTEXT) – (DONORS + PROJECTS)

Such institutions utilize various **development processes** to involve, influence, educate, and modify the behavior of a targeted population whose decisions, as **users and beneficiaries** of WSS systems, are critical to sustainability. The process of participation, for example, is meant to spread a net of inclusion for a wide set of beneficiaries so that they may determine the kind and magnitude of benefits aimed at them. Health education, as a training process, is meant to instill a better understanding of the connection between disease and user behavior. Another development process is engineering and construction which may, as an example, design a water supply system and utilize heavy drill rigs for constructing wells in combination with local laborers to dig trenches for pipe lines.

Technologies are the most visible output of the WSS sector and represent the equipment, machines and related services of providing potable water and disposing of waste. Historically, development agencies have placed much emphasis on technology transfer, taking technologies from a wealthy country setting and placing them in countries with poor economies. To be successfully utilized, however, technologies must be appropriate to the skills and resources of the users and operators responsible for maintaining them. An assured supply of spare parts is also critical to their continued use. Technologies alone, however, do not provide benefits. Proper use and maintenance of the technologies is also required which is the focus of many of the development processes related to training and changing user behavior.

The combination of the proper use and management of WSS facilities, along with the cumulative effect of educative and participatory processes, results in a number of **benefits** for the involved population. The most important include improvements in health, economic, social, and environmental status. Health benefits, usually considered the most prominent, include reductions in morbidity and mortality from water related diseases---especially diarrhea, which is the leading cause of mortality among children of developing countries. Economic benefits are seen where water is used to grow or produce tangible products; indirect economic benefits are also derived from increased productivity by healthier workers or students. Environmental benefits occur when WSS facilities eliminate vectors or habitats of disease. It is, of course, fundamental to sustainability that watershed protection measures must be in place to avoid depleting water sources and degrading water quality. Finally, social benefits include the general value of improved knowledge derived from participating in the various development processes.

However, unless these benefits are accepted and truly valued by the intended population, sustainability is still at risk. Here women may play a particularly important role because they often are the traditional decision makers in where to get water, whether to contribute to maintenance once they select their source, and primary users of the water once it enters the home. This broad involvement directly impacts on sustainability of benefits. For example, because many women who live in rural areas and urban slums spend a large part of their day fetching water, they will often necessarily place a higher value on time savings and reliability of the water supply, as opposed to the health benefits of improved water quality. Taking women's circumstances and priorities into account then, is a critical factor of the sustainability equation.

Surrounding the factors already discussed in figure 1 are additional factors that may be classified as **contextual**. These are pervasive and effect not only the WSS sector but the nation at large and are immune to institutional efforts to change them. These factors include the economy, societal norms and attitudes, and the physical environment. For example, project plans for the importation of water pumps can go awry with falling commodity prices and resulting lack of hard currency; or natural disasters such as drought can place high stress on water sources and institutional resiliency to manage them; or cultural attitudes (e.g. selling water is considered taboo in some societies) can have unanticipated impacts. Ignoring the vagaries of contextual factors places the gains made in project benefits at risk and imposes the need for contingency planning to assure sustainability.

Entering into the sustainability picture, external to local institutions are **donors**, and their mechanism of influence, **projects**. Donors, and those responsible for carrying out projects, must consider all the factors described earlier related to contexts, benefits, technologies, targeted beneficiaries, development processes, and institutions in order to design and implement a project. Their careful consideration of these factors and ability to influence most of the factors plays a major role in determining the sustainability of project benefits. Donors must, however, withdraw support at some point in time, thus requiring the local institutions to provide all ongoing support to intended beneficiaries. Donors and projects are thus temporary entities in the sustainability picture and eventually must be subtracted from the sustainability equation.

Local institutional capacity is the key

In terms of development philosophy, it is the role of projects to assist local institutions to build their skills and resources so that they will be able to sustain benefits. While this concept was implicitly presumed in most development efforts, it has only been recently that this concept was stated explicitly as a part of project goals. The ability of the local institution to at least maintain, and preferably increase, the benefit stream is the ultimate test of sustainability. Future development efforts must place increased emphasis on local institution building as the foundation of sustainability.

Sustainability: A Perspective from the South

Claudio I. Meier[1], Student Member ASCE

Abstract

The widespread acceptance of the sustainability concept has not been matched by changes in attitudes and behaviours. The cause might lie in the failure to distinguish a new emerging paradigm from yet another plan to 'solve' the current environmental crisis. Stressing some ethical implications, the concepts of sustainability and sustainable development are reviewed. Perspectives on applying these ideas to developing countries are discussed. It is concluded that these should not be following the same resource-intensive development paths taken decades ago by today's developed nations. Instead, humbler but also more socially and environmentally acceptable alternatives must be found. For this task, disinterested help from Northern countries will be needed and most appreciated.

Introduction

There is currently much debate about the idea of sustainability and the corollary concept of sustainable development. Notwithstanding the fact that many usually antagonistic voices (from multinational corporations to environmental groups) have all willingly embraced these concerns the expected change in attitudes and behaviours has been slow to come. Such seeming consensus results from each group having its own ad-hoc interpretation of what 'sustainable' means. A deeper understanding of the meaning, scope and far-reaching implications of these issues would stir far more controversy, as any paradigm shift should. This article reviews the concept of sustainability, as perceived by a civil engineer from a developing country, who teaches environmental science as well as water-related subjects to engineering students.

---

1 Instructor Professor, Civil Engineering Dept., Universidad de Concepcion, Chile. Present Address: Civil Engineering Dept., Colorado State U., Fort Collins, CO 80523.

Sustainability

An early proponent of this concept, from whom I will borrow heavily, is Herman Daly with his book 'Steady-State Economics'(1977). The sustainable worldview arose in response to the currently widely held and pursued progrowth position, which states that not only is economic growth good per-se (an ethical assumption), but it also can keep going forever (a wrong physical assumption). For standard economics -the theoretical support behind the latter position- the economy is an isolated system in which money as well as products circulate between firms and households. Means are assumed to be unlimited, and maximising economic growth is the final goal. No mention is made of fundamental limits imposed by the environment, wherefrom all resources needed to manufacture products are extracted, and where all manufacturing wastes and obsolete products are dumped back. This linear approach to resource use can not be the rule in a world that has a finite capacity for producing resources (matter and energy) and for absorbing wastes. The progrowth view does not consider physical limits, nor does it recognise ethical ends. When life's basic needs are still to be met, economic growth is obviously required. But beyond that level, do more things solve man's existencial problem? This ethical dilemma is avoided by assuming that 'more is best', no questions asked.

For a sustainable worldview, on the other hand, the economy is part of, and depends on a finite environment. Thus it must stop growing at a certain point,i.e., reach a steady state. A continuously growing per-capita consumption and a continuously growing population are clearly impossible under this scenario. Resource use and waste are minimised by increasing product durability, decentralising, and recycling or reusing as much as possible. The economy is set at the desired, constant level of physical wealth that is deemed sufficient (an ethical decision). Thus, it is assumed that 'enough is better'. Glasby(1988) gives an excellent review on the physical limits to continuous growth, that includes a quantitative case-study for New Zealand.

It is important to realise that sustainability represents far more than a technological goal. It is indeed a whole new way of perceiving our relationship with our planet. Now, any serious attempt at sustainability will require a major redirection of technology. Our profession, with its broad scope covering the decision-making process, as well as the planning, design, construction and operation of infrastructure works, will have a most important role to play in this change. We need to rise to this challenge: recent leading proposals from the American Society of Civil Engineering's Board of Direction, or the chance to discuss these issues at meetings point the right direction.

Sustainable Development

This concept was proposed in 1987 by the World Comission on Environment and Development, in the so-called 'Brundtland Report'(WCED, 1987). It is defined as: development that meets the needs of the present without compromising the ability of future generations to meet their own needs. Development is taken here in its broadest and correct sense to include not only wealth but also equity and environmental quality. The idea of physical limits on the environment's capacity to meet demands is evident, whilst the ethical component is implicit in two words: needs and future. In effect, quantifying needs above the essential necessities (whims?) relies on ethical decisions.Future needs are even more problematic because they are unknown. In theory, the only alternative would be to keep all options open, which would severely constraint our ability to meet present needs; thus, another ethical choice has to be made in this case. We see that this blueprint for a sustainable world is clearly based upon the sustainable worldview, but takes a more pragmatic stance by considering the actual environmental degradation/development crisis that affects most of the world's population. Thus the recommendation for achieving maximum growth in areas in which basic needs are not being met. The following priorities are emphasised: stabilising population; sustaining, enhancing and minimising impacts on Earth's life-supporting systems(atmosphere, waters, soils and species); ensuring equitable access to resources; preserve certain ecosystems, use others on a renewable basis; reduce depletion rates of non-renewables; conserve biodiversity; reorient technology and risk management; and merge ecology and economics in decision-making.

Concerns for Developing Countries

Industrialised countries account for less than 25% of the world's population, but consume about 80% of all resources whilst generating 70% of the world's pollution. Developing countries, with three quarters of the world's population must fare on what's left. Moreover, the gap between the two groups is increasing steadily (MacNeill, 1990). According to WCED(1987), a five to tenfold increase in economic activity will be required over the next 50 years to meet the needs of an increased population and to begin reducing mass poverty in the world. It is said that failing to reduce poverty would cause the world's natural resource capital to collapse. One might also ask how a 5 to 10 times increase in output will be achieved without sacrificing the whole planet... Now, it has been repeatedly shown that overpopulation (mainly in the South) causes mostly local problems, making things even worse for those already poor, whilst it is overconsumption (mainly in the North) that causes most global problems,with effects suffered by all(Parykh and Painuly,1994). The data not only point out the North's responsability for worldwide resource depletion and pollution, they also indicate that in developing countries, the control of consumtion patterns is more important than the control of population to ensure sustainability. Problem is that most developing countries are opening their markets in an effort to attract foreign capital. This creates an ever increasing pressure to drop traditional customs, that consume little, and replace them by westernised, mass-consuming lifestyles. This process is fostered by multinational corporations, eager to penetrate emerging markets; by small national elites, already leading westernised lifestyles, that usually concentrate most economic and political power, and also control mass-media; and by a new trend towards an increasing assertion of national or regional dominance(e.g., Chile and NAFTA). To grow, pay their huge debts, or maintain a semblance of developed world that only a few enjoy, most developing countries are rapidly liquidating their natural assets, causing tremendous social and environmental impacts. Prices paid are usually low, because needy countries are not in a good position to negotiate.

Conclusions

The sustainable worldview seems to be based on sound physical and ethical considerations, and should be embraced. Nonetheless, the agenda of sustainable development seems to be losing ground to private and national economic interests. A commitment from the heart towards disinterested cooperation between all countries, but mainly between the developed North and developing South, will be needed in order to reverse present trends. Developing countries should find alternate development pathways, instead of imitating Westernised ways.

References

Daly, H.E.1977. Steady-State Economics. Freeman.

Glasby, G.P. 1988. Entropy, pollution and environmental degradation. Ambio XVII(5), pp. 330-335.

MacNeill, J. 1990. Strategies for sustainable economic development. In:"Managing Planet Earth: Readings from Scientific American". Freeman.

Parykh, J. and J.P. Painuly. 1994. Population, consumption patterns and climate change: A socioeconomic perspective from the South. Ambio XXIII(7): 434-437.

WCED (World Comission on Environment and Development). 1987. Our Common Future. Oxford University Press.

SUSTAINABLE STORMWATER MANAGEMENT

James P. Heaney[1]
Member, ASCE

ABSTRACT

This paper reviews the general area of sustainability and its interpretation by the worldwide engineering community. Then, an on-going national assessment of natural hazards is described briefly along with a general taxonomy of the four stages of hazards management. The implications of these emerging paradigms on stormwater management are discussed. Finally, a request for involvement of interested people is presented.

INTRODUCTION

The purpose of this paper is to describe the general area of sustainability and to translate this philosophy into an operational definition for water resources in general and for stormwater management in particular. The 1993 Great Flood of the Upper Mississippi River and Lower Missouri River will be used as the case study.

GENERAL PERSPECTIVES ON SUSTAINABILITY

A popular definition of sustainability was developed by the World Commission on Environment and Development (1987).

"Sustainable development...meets the needs of the present without compromising the ability of future generations to meet their own needs."

[1]Professor, Dept. of Civil, Environmental, and Architectural Engineering, U. of Colorado, Boulder, CO

80309-0421

Engineers throughout the world have recognized the vital importance of engineering for sustainable development by forming the World Engineering Partnership for Sustainable Development (WEPSD). The official WEPSD vision is presented below:

"Engineers will translate the dreams of humanity, traditional knowledge, and the concepts of science into action through creative application of technology to achieve sustainable development. The ethics, education, and practices of the engineering profession will shape a sustainable future for all generations. To achieve this vision, the leadership of the world engineering community will join together in an integral partnership to actively engage all disciplines and decision makers to provide advice, leadership, and facilitation for our shared and sustainable world."

Hatch (1994) suggests a set of nine design concepts for sustainable engineering which provides an operational framework:

1. Engineers of all specialties must be educated to understand environmental and economic issues, problems, and especially the risks and potential impacts of their actions.
2. Engineers must adopt the notion of ecosystems thinking which focuses on synthesis or the combining of separate elements to form an integrated and whole coherent system.
3. Engineers need to more carefully examine the aggregate long-term consequences of decisions, both temporally and spatially.
4. Engineers must acquire environmental economic tools to integrate the environmental and social conditions into market economics.
5. Engineers must search more thoroughly for sustainable alternatives. Total life-cycle analysis should be done and include potential impacts beyond the service life of the engineering system.
6. Engineers must continue to develop and apply new technologies to serve sustainability.
7. Engineers must listen to our direct customers but also other people who are affected by our actions.
8. A multidisciplinary team approach should be used to incorporate all perspectives.
9. Education in sustainability must also be extended to the general population so that they have a more complete understanding of these concepts.

SUSTAINABILITY AND HAZARDS MANAGEMENT

White and Haas (1975) summarize the results of the first national assessment of research needs for managing natural disasters. This work laid an important foundation for a broader conceptual view of hazards management. Mileti et al. (1994) provide a current overview of the components of hazards management as an initial part of the second national assessment of hazards management. They partition the management process into four components: preparedness, response, recovery, and mitigation as shown in Table 1. This broader perspective provides a valuable framework for examining the subset of hazards that we deal with in water resources.

Table 1. Societal Adjustment Cycle for Natural and Technological Disasters (Mileti et al. draft, 1994).

A. Preparedness-building and emergency response and management in advance of a disaster to facilitate an effective societal response.

B. Response-societal actions taken immediately before, during, and after a disaster in order to save lives, minimize damage to property, and enhance the effectiveness of recovery.

C. Disaster Recovery-short-term activities to restore vital life support systems to minimum operating standards and long-term activities to return life to previous or improved quality of life.

D. Mitigation-structural and nonstructural policies and activities that will reduce vulnerability to damage from natural disasters.

SUSTAINABILITY AND STORMWATER MANAGEMENT

The 1993 Great Flood in the Upper Mississippi and Lower Missouri River Basins has rekindled a nearly two centuries old debate about stormwater management in the United States. Over the past two centuries, stormwater management has evolved from almost total reliance on structural measures to mitigate the effects of flooding to a wide variety of programs devoted to hazards management at the four levels described in Table 1. Two national studies provide a current summary of emerging paradigms for stormwater management (Federal Interagency Floodplain Management Task Force 1994, and Interagency Floodplain Management Review Committee 1994).

How can we synthesize the strong interest in sustainability by the worldwide engineering community, a more generic framework for hazards management in general, and the results of two recent national commission reviews of current stormwater management policies into an action agenda in water resources engineering? In many ways, we have already come a long way in this direction with the long-standing focus of taking the systems view in water resources engineering (Mays and Tung 1993).

The Natural Hazards Center at the University of Colorado is in the first year of a three year NSF sponsored study to outline future directions of research in the hazards field in general. During our initial meeting with experts in a variety of areas, we attempted to develop operational definitions of sustainability within the context of hazards management. While there seemed to be general concurrence that sustainability was a good general philosophical theme, there was little agreement regarding how it can be operationalized.

We welcome input from the water resources community on how to more effectively integrate this important notion into analysis of water resource systems. Interested people are encouraged to contact me for more information.

APPENDIX. REFERENCES

American Association of Engineering Societies. 1994. The Role of Engineering in Sustainable Development. AAES, Washington, D.C.

Federal Interagency Floodplain Management Task Force. 1994. A Unified National Program for Floodplain Management. Federal Emergency Management Agency FEMA 248 (Draft).

Interagency Floodplain Management Review Committee. 1994. Sharing the Challenge: Floodplain Management into the 21st Century. Report to the Administration Floodplain Management Task Force, Washington, D.C.

Hatch, H. J. 1994. Accepting the Challenge of Sustainable Development. in American Association of Engineering Societies. 1994. The Role of Engineering in Sustainable Development. AAES, Washington, D.C.

Mays, L.W. and Y.K. Tung. 1993. Hydrosystems Engineering and Management. McGraw-Hill, New York.

World Commission on Environment and Development. 1987. Our Common Future. Oxford University Press, Cambridge.

Measuring Sustainability in a Humid Region

Hal Cardwell[1], Assoc. Member ASCE, David Feldman[2] and James Kahn[3]

Abstract

Conflict over water in a humid region such as the southeastern United tates reminds us that issues of sustainable development of water resources are ot confined to the developing world or to arid regions. While sustainable evelopment has become a popular concept since the 1992 Earth Summit in Rio, efinitions of and ways to measure sustainability are just evolving. As part of an ngoing study of policies for sustainable development of water resources in the outheast, we identify five topic areas which need measures of sustainability: astitutional structures, economic aspects, social structures, biological resources, nd physical attributes. These measures need to be developed by considering ctors such as policy relevance, aggregation level, and potential for incorporation an accounting structure. Different spatial and temporal scales need to be ncluded, as do measures of both quantity and quality of the resource.

Introduction

Although a humid area, the southeastern United States has begun to xperience serious conflict over water use and supply. These conflicts threaten ustainability throughout the region - the ability to maintain economic activities hile simultaneously protecting the natural environment. Three example of these nflicts are contemplated interbasin water transfers (*e.g.*, from the Roanoke basin the Virginia Beach area); reservoir operations for multiple use in major eveloped river systems (*e.g.*, TVA's recreational use reservoir plan); and the

---

[1]Post-Doctoral Fellow, P.O. Box 2008, MS-6036, Oak Ridge National Laboratory, Oak idge, TN 37831-6036, managed by Martin Marietta Energy Systems Inc., for the U.S. epartment of Energy under contract DE-AC05-84OR21400.

[2]Senior Research Associate, Energy, Environment, and Resources Center, 327 South tadium Hall, University of Tennessee, Knoxville, TN 37996-0710.

[3]Associate Professor, Department of Economics, University of Tennessee, Knoxville, N 37996.

and the perceived inequity of water pricing mechanisms that penalize small communities undergoing economic revitalization. These conflicts cross local, sta and cultural boundaries, and produce economic inefficiencies, environmental degradation, and costly lawsuits. If they did not cross boundaries perhaps there would be no conflicts (Matthews 1994). Because of the diversity of water management problems and the common geographies and cultures of this humid region, the southeast serves as an excellent focal point for studying competing water uses and divergent management issues affecting these problems.

Sustainability has three dimensions: economic development; maintenanc of ecological values (e.g. sufficient instream flow for flora and fauna, preservatio of aesthetics); and equitable allocation (e.g. the ability to accommodate multiple stakeholder needs, regardless of size or character of a community). The links between these dimensions combine in myriad ways to produce a satisfactory lev of social welfare (Smith *et al.* 1993). In considering ways to measure sustainability, we choose not to use a unified measure or single index (Liverma *al.* 1988) but instead propose five topic areas for indicators and propose desired characteristics.

3. Topic Areas

To develop measures of sustainability we can categorize aspects of water conflict into five topics: 1) institutional structures for water management and conflict resolution, 2) economic aspects of water resources management, 3) socia structures, 4) affected biological resources, and 5) physical attributes of the watersheds. While many water issues and sustainability measures affect more than one of these topic areas, this categorization allows us to identify the eleme of sustainability in water resources development.

Institutional structures. Institutional structures for water management include water law and property rights, water markets, land use regulations, and the responsibilities and effectiveness of private and public water supply and management agencies. We also include legal rights to water and the institution that enforce them. The institutions include numerous federal, state, local, intergovernmental and private agencies.

Economic aspects. Economic aspects of water resources development include pricing structures, private versus local or national funding for water projects, ar the effects of water quality and water shortages on economic development and industry. These aspects include monetary and non-monetary values for clean fr flowing rivers and the value of water diverted for municipal, industrial and agricultural uses.

Social structures. Social structures include the relative value of small versus la communities and their rights to water, the equity of interbasin transfers or changes of water use, and the cultural heritage of a region. Behavior patterns with regard to water may be considered, as well as the ethical and cultural perceptions of water, both as a human resource and as an integral component o the natural landscape.

Biological resources. Biological resources are directly affected by water resourc decisions. Native and exotic fisheries obviously depend on surface water and sediments, as do other aquatic plants and animals, riparian vegetation, and errestrial plants and animals. The southeast, with its plentiful water supply, is home to an extraordinary number of threatened and endangered species.

Physical attributes. The physical aspects of a stream control its potential uses. Physical aspects include hydrology, water quality, channel morphology, aquifer size and quality, geology and soils, and existing water resources developments in the region.

4. Characteristics of Indicators

Indicators of sustainable development are aggregated measures that must be useful to policy makers. In considering indicators, we must consider at least three factors: 1) relevance of the indicator to the policy makers, 2) its effectiveness as a surrogate for a complete description of the sub-system it represents, and 3) the potential of the indicator for aggregation into an accounting structure.

Relevance to Policy. Without relevance to policy options, an indicator has very little value (Verbruggen and Kuik 1991). While researchers can collect volumes of data on individual processes or events, the data must be concise and useful to the decision maker. Examples of indicators that may be especially relevant include dollars lost in shellfish productivity, number or years in water rationing, or difference in water prices over time over space or per unit.

Effectiveness as a Surrogate. By definition an indicator represents an entire system; the choice of indicator can have major influence on the determination of the sustainability of a system. As such it must be of a high enough level of aggregation to capture the net effect on a biological, physical or social system.

Aggregation Potential. The third characteristic to consider about an indicator of development is its ability to be aggregated with other measures of sustainability. Methods of aggregation of these indicators include incorporation in a relative manner as measured against a predetermined standard (ten Brink 1991), and inclusion in a national accounting measure to mimic the Gross National Product indicators (Daly 1989). This implies the need for the indicator to be measurable.

Finally, indicators must span spatial and temporal scales, and address both water quality and quantity issues. Regional, basin-wide, and local spatial scales will all be appropriate. Likewise, political, biological, economic and social systems all operate on multiple time spans. This means measures must address both short-term (monthly or seasonal), mid-range (annual), and long-term (decades or longer) planning horizons. Indicators also need to reflect both the pressure exerted by people on the water resource in terms of demands, and the state of the resource in terms of quality. While these aspects are interrelated, indicators of acceptable quantity lose their value without corresponding measures of quality. We note that these characteristics span the five topic areas identified in Section 3. Table 1 shows sample indicators for the five proposed topic areas. These would need to be developed at local, state and regional levels.

5. Conclusions

Measuring sustainable development involves multiple ways of viewing sustainability. Five topic areas are identified, and multiple criteria are used to evaluate a potential indicator. Multiple spatial and temporal scales must be developed. The issue of sustainability is complicated, precisely because of the inter-relationships between components. The challenge is to identify measures which can be used to move away from conflict, and towards sustainable water resources policies.

Table 1. Sample indicators of sustainability in water resources

physical	• consumptive use - average rainfall (in volume/area/time) • percent of river miles with unfishable, unswimmable water
social	• frequency of rationing • population growth in a region over time • rural vs. urban population change in a region over time
economic	• relative water cost: urban vs. rural, industrial vs. municipal, over time • lost revenue from declines in shellfish productivity over time
institutional	• percent of communities with water plans and change over tim • speed of water conflict resolution
biological	• species abundance or decline • increase or decrease of number of threatened species

6. References

Brink, B. ten (1991). "The AMOEBA approach as a useful tool for establishing sustainable development?" *In search of indicators of sustainable development*, O. Kuik and H. Verbruggen, eds., Kluwer Academic Publishers, Norwell, MA, 1-6.

Daly, H.E. (1989). "Towards a measure of sustainable social net national produ *Environmental accounting for sustainable development*, Y. J. Ahmad S. E Serafy, and E. Lutz, eds., The World Bank, Washington, DC, 8-9.

Liverman, D. M., Hansen, M. E., Brown, B.J., and Meridith, R. W. Jr. (1988). "Global sustainability: toward measurement." *Environ. Manag.*, 12, 133-143.

Matthews, O. P. (1994). "Judicial resolution of transboundary water conflicts." *Water Resour. Bull.*, 30(3), 375-383.

Smith, L. G., Carlisle, T. J. and Meek, S. N. (1993). "Implementing sustainabili the use of natural channel design and artificial wetlands for stormwater management." *J. Environ. Manag.*, 37, 214-257.

Verbruggen, H., and Kuik, O. (1991). "Indicators of sustainable development: a overview." *In search of indicators of sustainable development*, O. Kuik an H. Verbruggen, eds., Kluwer Academic Publishers, Norwell, MA, 1-6.

Sustainability Criteria: An Application to the Hydropower Industry

Slobodan P. Simonovic[1], Barbara J. Lence[2], and Donald H. Burn[2]

Abstract

This project determines key sustainability criteria that should be considered in the process of project selection. Sustainability, as used throughout this work, is defined as the ability to meet the needs of the present without compromising the needs of future generations. The key sustainability issues identified are formalized through the establishment of operational definitions for these issues. These definitions are in the form of sustainability measures, or metrics, that can be used to evaluate a given project. These measures are formally evaluated through their application to project selection for alternatives associated with the North Central Project, an electricity supply problem, in Manitoba, Canada. Results obtained from this evaluation may be used to subsequently refine and revise the sustainability measures that are derived.

Introduction

Applying principles of sustainability requires major *changes in the objectives* on which decisions are based and an understanding of the complicated interrelationships between existing ecological, economic and social factors. The broadest objectives for achieving sustainability are: (i) environmental integrity; (ii) economic efficiency; and (iii) equity (Young, 1992). The second important aspect of sustainable decision making is *the challenge of time* (long term consequences). Sustainable development requires forms of progress that meet the needs of the present without compromising the needs of future generations (WCED, 1987). The third aspect of the sustainability context is *the change in procedural policies* (implementation). Pursuing sustainable project selection will require major changes in both substantive and procedural policies (Jordaan et al., 1993).

---

[1]Professor, Department of Civil and Geological Engineering, The University of Manitoba, Winnipeg, MB, Canada, R3T 2N2

[2]Associate Professor, Department of Civil and Geological Engineering, The University of Manitoba, Winnipeg, MB, Canada, R3T 2N2

Three main criteria for addressing sustainability in project selection have been identified. They are: (i) equity; (ii) reversibility; and (iii) risk. This research develops sustainability measures for these issues that can be used to assist in making project decisions. A detailed case study of the North Central Project, an electricity supply system for seven remote communities in Manitoba, Canada, is used to demonstrate these measures. The project selection process associated with this case study involves analysis, sponsorship, and benefits at the local, provincial, and national level. Since all three sustainability issues identified above are relevant for this case study, it affords a practical opportunity to explore the application of these measures.

Case Study of North Central Project

The North Central Project is the title given to the central electricity supply system for seven North Central communities in Manitoba. Currently, the electrical energy for the communities is generated by means of local diesel generating plants located in each community and distributed through local, independent distribution systems. The use of electrical energy in each of the communities is limited, and customers requiring unlimited energy can obtain this service at considerable charges that vary from time to time and from community to community. In 1984, a study to identify feasible alternatives for the supply of economical, unlimited electrical energy to the customers in the seven communities was initiated. This study led to extensive analysis of the electrical energy supply problem and identification of alternative solutions. The solutions that were identified include: central system supply; local hydro generation systems; and alternative energy systems, such as diesel, wood-fired, wood gasification dual-fuel engine, solar energy, wind energy, peat-fired steam turbine generation, and hybrids of these.

An extensive and systematic analysis of the primary impacts of the various electricity supply options has been undertaken. These include analysis of the impacts on: the biophysical environment such as wildlife habitat and vegetation; the traditional Aboriginal way-of-life; the people in their communities, including ability to pay monthly electrical bills, rewiring costs, and changes in quality of service; and legal and jurisdictional issues including economic development and related benefits. Information is available regarding the distribution of impacts within and between communities, and among different electricity user groups. This information provides the data base for the analysis of the applicability of the sustainability measures for equity, reversibility, and risk, and facilitates the investigation of the implications for sustainable development for other projects of this type, and for general project selection decision-making.

Sustainability Measures

The overall analytical approach for this work consists of the formulation of the sustainability measures for the issues of equity, reversibility, and risk, and the evaluation of the effectiveness of these measures with the application to the case study. A discussion of the key sustainability issues follows.

Equity

Equity is concerned with the distribution of financial and other impacts, both benefits and costs, associated with a given project (Musgrave, 1959). As an example, the distribution of access to affordable electricity in a region has important equity implications in terms of the current production capabilities and growth potential of the various communities. Sustainability requires the identification of such considerations and of methods of analyzing the distribution of impacts within present and between present and future generations. For present generations, the distribution of impacts may be analyzed for all of the current users, or social groups, within a system. An evaluation of equity considerations between generations requires longer time periods of analysis, analysis of impacts between users in each generation, and analysis of impacts from one generation to the next. Analysis of equity for future generations is complicated by generational changes in preferences that may result from changing economic, social, technological, or environmental conditions. For example, demographic shifts in a region may result in both different parties that are experiencing impacts and in different desired use levels for the available resources. Implementing equity in practice requires the definition of groups of equal users, both within and between generations, and the determination of what constitutes equal treatment.

Reversibility

Reversibility, in the context of sustainability, can be viewed as a measure of the degree to which a system can be returned to its original form following the implementation of a development plan. One ideally wants the reversibility of a proposed action to be high so that any negative implications that may result from present decisions can be mitigated in the future. Reversibility recognizes that all development decisions have consequences, some of which will not be immediately apparent. Unanticipated consequences that lead to negative impacts will have to be addressed by future generations. A useful operational concept for reversibility is that decisions that are highly reversible should result in the users of a system still being able to do what they have always been able to do. This essentially implies that sustainability leads to unaltered opportunities for both present and future generations. A quantification of reversibility might be obtained through a comparison of predicted future conditions with the project in place and predicted future conditions should the project not proceed.

Risk

Expanded spatial boundaries, a lengthened time scale, comprehensive multi-objective analysis, and other issues related to sustainability place immense demands on science. A number of questions raised by the sustainability perspective reveal major deficiencies in the knowledge of the behaviour of a wide range of natural and human systems under consideration. Recognizing that many of these deficiencies cannot be eliminated immediately makes it evident that risk and uncertainty are inherent concepts related to sustainability.

To determine the acceptable level of environmental and social protection, the environmental and social risk associated with economic development must be assessed. Risk is a measure of the probability and severity of adverse effects (Lowrance, 1976). Risk addresses three questions: (i) What can go wrong?; (ii) What is the likelihood that it will go wrong?; and (iii) What are the consequences? (Kaplan and Garrick, 1981). Thus, the incorporation of risk assessment and management constitutes another component of sustainable project decision-making. To manage risk over time, the following three questions must also be addressed (Haimes, 1992): (i) What can be done?; (ii) What options are available and what are their associated trade-offs in terms of all costs, benefits, and risks?; and (iii) What are the impacts of current management decisions on future options?. Sustainability thus requires innovative approaches for the derivation of appropriate risk measures since the use of the expected value of risk, and/or the conditional expected value, are no longer sufficient. Knowledge-based risk management is necessary and will be examined in this research.

Summary

This research formalizes the definition of a new set of measures for project evaluation and selection. The integration of sustainability issues into this analysis entails a different way of thinking about the consequences and implications of such decisions. This research promotes an expanded view of the project evaluation and selection process. It is also an integral part of the development of appropriate measures for implementing sustainability principles into decisions for a variety of problems for the power industry. Such decisions may include evaluating the trade-offs between: hydropower and thermal power options; demand side management and supply alternatives; coal and gas generation options; and hydropower production at different sites.

References

Haimes, Y.Y., 1992. "Sustainable development: A holistic approach to natural resource management", *Water International*, Vol. 17, No. 4, pp. 187-192.

Jordaan, J., E.J. Plate, E. Prins, and J. Veltrop, 1993. " Water in our common future", *UNESCO Committee on Water Research COWAR,* UNESCO International Hydrological Programme, Paris.

Kaplan, S., and B.J. Garrick, 1981. "On the quantitative definition of risk", *Risk Analysis,* Vol. 1, No. 1.

Lowrance, W.W., 1976. *Of Acceptable Risk,* William Kaufman, Inc., Los Altos, CA.

Musgrave, R. A., 1959. *The Theory of Public Finance*, McGraw-Hill, 160-161. World Commission on Environment and Development (WCED), 1987. *Our Common Future*. Oxford University Press.

Young, M.D., 1992. *Sustainable Investment and Resource Use*. UNESCO, Man and the biosphere series, Volume 9, The Parthenon Publishing Group, Casterton Hall, Carnforth, UK.

"The Last Dam in the West": Is the Western Water Project Really an Endangered Species?

Uli Kappus, P.E.[1], Ken Steele, P.E.[2], and Dick Westmore, P.E.[3]

Introduction

The settlement and subsequent economic development of the Western United States was largely the result of harnessing the region's water resources. Federal programs to settle the West were supported by efforts to develop its water, energy and mineral resources. Since the mid 1970s, water agencies and environmental interests have discussed a movement away from large dam projects. Concurrently, federal and state governments have implemented new policies that have reduced funding for water development and strengthened regulations designed to protect and enhance the environment. For example, the U.S. Bureau of Reclamation, historically the region's primary agency for water development, has changed its mission from water developer to water manager.

Despite these challenges, agencies that recognize the need for balancing water supply security and environmental values, continue to be successful in implementing major water resources projects. In this paper, several large dam projects that are moving positively toward implementation are discussed. These projects, and many others, are being implemented despite the challenges noted above. This paper discusses the reasons why such projects are moving forward, and presents conclusions about how future water development projects in the Western U.S. will be implemented in light of a continued need for enhancing environmental and recreational values.

Historical Perspective

Historically, water supply in the West has been reliant on dams. The West's highly variable precipitation levels require that water be stored in times of plenty for use in times of drought. The importance of water in the Old West was characterized by physical and legal wars over water rights. Water rights issues continue to become more complex, partly because the western prior appropriation doctrine of "first in time/first in right" was not developed with environmental values in mind. Its focus was on

[1] F. ASCE, Director of Water Resources Services, GEI Consultants, Inc., 1925 Palomar Oaks Way, Suite 300, Carlsbad, CA 92008-6526

[2] M. ASCE, Emergency Storage Project, Project Manager, San Diego County Water Authority, 3211 Fifth Avenue, San Diego, CA 92103-5718

[3] M. ASCE, Water Resources Division Manager, GEI Consultants, Inc., 5660 Greenwood Plaza Blvd., Suite 202, Englewood, CO 80111-2418

quantity of water, not quality or environmental need. With today's regulatory environment and societal needs, there is a strong call to strike a balance between water development and environmental needs. When this balance is struck, a successful project often results.

Part of the impetus for water developers to form a partnership with the environmental community was spawned by the National Environmental Policy Act (NEPA). While these regulations make it more challenging to permit major projects as rapidly as developers might wish, in the long run, the projects that are permitted will better serve all of society. This partnership between water developers and the environmental community has led to permitted projects, and reminds us of Henry Ford's observation that "Coming together is a beginning. Keeping together is progress. Working together is success."

It is clear that, with a rapidly growing population in the west, tensions between historical water users and demands, such as the environmental and recreational interests, will continue at a high level. Not only will water users face the classic hydrologic drought, but the enforcement of regulations, such as the endangered species act, will superimpose a "regulatory drought". Classic examples of this redistribution are the San Francisco Bay/Delta, Columbia River and Colorado River Systems.

Case Studies

The table provides a summary of key information for selected dams in the Western U.S. that recently have been, or are currently being, permitted. These selected dams represent only a small cross-section of planned projects that number well into the many dozens. Three of the projects shown in the table are discussed in more detail below.

Dam Name	State	Structural Height (ft)	Storage Volume (acre-ft)	Dam Type	Estimated Construction Cost ($mill)	Status
Emergency Storage Project Component						
Moosa South	CA	341	68,000	CFR(1)	260	Permitting
San Vicente		302(2)	90,100(3)	RCC(4)	125	
Olivenhain		320	24,000	RCC(4)	120	
New Los Padres	CA	274	24,000	RCC(4)	87	Permitting
Seven Oaks	CA	551	Flood control	E(5)	168	Construction
New Elmer Thomas	OK	113	8,000	RCC(4)	44	Constructed
Greybull	WY	150	30,000	E(5)	43	Permitting
Domenigoni	CA	280	800,000	E(3)	1,500	Design
Ritschard	CO	145	60,000	E(5)	45	Construction

(1) Concrete-face rockfill
(2) Enlargement of existing dam from 220 ft high to 302 ft high
(3) Enlargement of existing reservoir from 90,200 af to 180,300 af
(4) Roller-compacted concrete
(5) Earthfill

Emergency Storage Project

The San Diego County Water Authority (Authority) is developing the Emergency Storage Project (ESP) to ensure the availability of a sufficient water supply for the county in the event of a natural disaster such as a major earthquake. Studies indicate that such an event could sever the imported water pipelines for up to six months. These pipelines deliver 90 percent of the water used by the over 2.5 million people within the Authority's service area.

Current estimates indicate such an event would create a water shortage of over 90,000 acre-feet during the time that would likely be required to repair damaged aqueducts. Economic losses due to a six-month, 60-percent reduction in water supplies are estimated to be $25 billion, according to studies prepared for the Authority. In 1992, GEI Consultants, Inc. was retained by the Authority to develop the engineering studies to support permitting this project.

Siting of major pipelines and reservoirs in San Diego has historically resulted in a high level of public and agency interest. Concerns exist regarding growth inducement, potential water rate impacts, environmental and endangered species issues, and impacts associated with relocating people. To balance these issues, the Authority implemented a systematic decision making process that resulted in a logical and orderly alternatives analysis that fully met permitting requirements. The Authority first identified 32 systems that could meet the ESP need. These systems were then scored for their performance in meeting five primary goals using a decision model. The weights for these goals were recommended by the consensus decision of a 27-member panel comprised of representatives of a wide range of interest groups in the community. Thus, the public actively participated in the decision making process. From the screening efforts, four final systems have been selected for inclusion in the EIR/EIS document.

The ESP system which will be selected is estimated to cost between $480 and $600 million. The Moosa South, San Vicente enlargement, and Olivenhain reservoir options are key components of the final four systems Completion of the EIS/EIR is scheduled for mid-1995, with construction beginning in 1997.

New Los Padres Dam

The New Los Padres Dam is sponsored by the Monterey Peninsula Water Management District. It is located about 25 miles southeast of Monterey, California, on the Carmel River. The $87 million, 274-foot-high RCC dam has been in the permitting stage for over a decade. Its purpose is primarily for domestic water supply for the Monterey Peninsula, but the project also provides for restoration of year-round flows in the Carmel River to benefit the steelhead resources and riparian habitat. Much of the Carmel River is dry during the summer months. The proposed dam is a main-stem storage facility that will provide modern fish transportation facilities around the structure.

The District went through an extensive environmental process including a major public involvement program. As part of the water supply alternatives considered, voters chose not to approve building a desalination plant on Monterey Bay, which is a National Marine Sanctuary. The key environmental issue is that, when releases are made from the reservoir, steelhead are able to fully utilize the river for all stages of their life cycle. The U.S. Army Corps of Engineers will make a decision on the 404 permit application in the spring of 1995.

Ritschard Dam

Ritschard Dam, forming Wolford Mountain Reservoir, is presently under construction and is located in Grand County, Colorado, near Kremmling on Muddy Creek, which is a tributary to the Colorado River. The project owner is the Colorado River Water Conservation District. The dam will provide storage for 60,000 acre-feet for both the Western Slope of Colorado and the Denver Metropolitan area. The final EIS was issued in 1990. The dam will be an earth embankment structure, 1,900 feet long and 140 feet high, with a total cost of $45 million.

The project succeeded in being permitted because it meets the water supply needs of both the Front Range and Western Slope. Extensive mitigation measures were required, including the mitigation of 286 acres that were classified as wetlands.

Conclusion

It is clear that the common pessimistic statement -- "large dams are a thing of the past" -- needs to be corrected. The types of siting and permitting processes used in the past for large dams are now ineffective. The successful water resources manager for a modern dam must show responsible stewardship for fiscal and environmental resources. However, the need for water supply, flood control, hydroelectric power, recreational facilities, and the necessary environmental enhancement opportunities provided by dams will mean that new dam construction will continue to be vital in the West and in all parts of the Country. The key elements shared by the successful projects include: (1) pressing need for additional water, (2) willingness of the owners to recognize, understand and follow regulatory requirements, (3) strong public involvement, and (4) a commitment to effective environmental enhancement and mitigation measures. The practice of dam engineering has always adjusted to the emergence of new technical concepts. The practice is now responding to the societal pressures for addressing past environmental abuses. The key for new water projects is to strike a balance between water and environmental needs, leading to preservation of our quality of life. The modern dam must be considered as one component of an integrated water resources management system to meet long-term water supply and environmental objectives.

WOLFORD MOUNTAIN RESERVOIR
SUCCESSFUL NAVIGATION OF THE REGULATORY, ENVIRONMENTAL AND FINANCIAL IMPEDIMENTS TO IMPOUNDMENTS

David H. Merritt[1], M. ASCE

Abstract

The Wolford Mountain Project, a reservoir currently undergoing construction on Muddy Creek near Kremmling, Colorado is a case study in successfully addressing the conflicts among multi-basin demands for water, environmental issues, and economic constraints in developing water resource projects.

Originally conceived as mitigation for the water supply related impacts of a trans basin diversion from the headwaters of the Colorado River, it represented the first real foray into the project construction arena for the Colorado River Water Conservation District, with the concomitant responsibility to satisfy constituents supporting and opposing the project. The subsequent history of the project has been an effort to avoid impacts wherever possible, mitigate where needed, and then negotiate when necessary.

The two biggest lessons learned from the project development are:
1)The ability to separate proponent's "needs" from "wants" is key to successfully navigating the permitting channels.
2)Complete mitigation of all "impacts" is impossible in resource development.

Introduction

The Wolford Mountain Project, a reservoir project nearing completion in the mountains of Western Colorado, is an example of successful permitting of a reservoir, in the midst of what was widely believed to be the most difficult period in which to bring a water development project to fruitition. The project involved transbasin diversions,

---

[1]Project Manager, Colorado River Water Conservation District, Glenwood Springs, Colorado 81602

endangered plants, four species of endangered fishes, and hundred of acres of wetlands. The result, although long and protracted, has resulted in a project which has been widely hailed as an excellent compromise.

The Wolford Mountain Project had its origins in 1968, when the Northern Colorado Water Conservancy District (Northern) filed for a water right to develop a water project to divert water from the confluence of the Colorado and Fraser Rivers to municipal uses in Northeastern Colorado. Opposition by the CRWCD resulted in a 1985 agreement whereby the CRWCD assumed the responsibility of building a reservoir for Western Colorado, in exchange for a payment of $10.2 million dollars. The CRWCD proceeded with an application to construct a reservoir on Rock Creek, with an alternative being a reservoir on Muddy Creek. This proved to be critical to the success of the entire project. We did not just file for the project which we wanted, but at the very beginning of the project also identified an alternative which would fulfill the needs of the CRWCD.

Permitting

The National Environmental Policy Act (NEPA) compliance process started with the development of an Environmental Impact Statement (EIS), through the associated studies. In the case of the Wolford Mountain Project, this was triggered by the filing of an application for a Special Use Permit with the Routt National Forest for a construction of a reservoir on Rock Creek. At this time, the Lead Agency was selected, the Cooperating agencies identified, the principal contact personnel identified, and the consultant for the preparation of the EIS was selected.

It is extremely important to establish a management structure which will work toward the expeditious and open assessment of the proposed action. It is necessary to identify responsible personnel at the Lead Agency, consultant and proponent who can provide the needed information in a timely fashion and can get the critical decisions made when needed. It is important to recognize that schedules will slip, and that while it is very frustrating to the proponent when this happens, it is more important in the final analysis that the integrity of the process be preserved. Ultimately, what is desired is a project which is permitted, accepted and constructed, not one which has merely survived the NEPA process.

The Projects

Rock Creek - When the CRWCD reached an accord with Northern, the desired project was a reservoir on Rock Creek. This site had had reconnaissance level engineering studies performed, and represented a very good reservoir site. Located on the Routt National Forest, at elevation

8600' above MSL, the dam site was in a narrow canyon situated in pre-Cambrian granites. Other than the US Forest Service (USFS) lands, there was a small tract of private land and a tract of land owned by the Colorado Division of Wildlife (which had been purchased with the intent of constructing a reservoir). There was a "no fee" USFS campground, consisting of a single pit toilet, and a nearly dilapidated stage stop dating from the late 1800's. The wetlands were considered to be the major impediment, with 486 acres of high quality wetlands. The area was not a critical range for deer and elk, and was not generally considered to be a high quality fishery.

Muddy Creek - For over 30 years the CRWCD and others had looked at reservoir sites along Muddy Creek, a tributary to the Colorado River near the town of Kremmling. In the early 1980's the CRWCD had conducted reconnaissance level studies at four sites. However, we had been steered away from Muddy Creek by Division of Wildlife personnel, due to the substantial big game populations which had been recently established. However, we included this site in the NEPA analysis as an alternative site. The reservoir was at lower elevation (7500' MSL), and the bottomlands were operating ranches. The private lands amounted to over 1800 acres. The only public lands involved were Bureau of Land Management(BLM) lands at the dam site, and on the upland shores of the reservoir basin. There were approximately 896 acres of wetlands or wetland-type habitat, albeit in poor to fair condition. The dam site was located in Cretaceous shales, in a wide valley which required the design of an earth embankment, at a much higher cost than Rock Creek.

<u>The Studies</u>

The environmental and engineering studies started in May 1985. Full environmental and feasibility engineering studies were conducted at both sites. The original time frame of one year for completion of the EIS (summer 1986) became elongated to two years with the Draft Environmental Impact Statement (DEIS) being published in August 1987. By this time, significant issues of the trout fishery in Rock Creek had developed, along with concerns about the loss of the high quality, high mountain wetlands. However, while recognizing that "the purpose and need" were valid, and thus the "No Action" alternative was not the preferred Federal action, the USFS and the BLM were not comfortable at that time with designating either project as the Preferred Alternative. Rather, they came out with an EIS without an action identified, and with a request for public comment to assist in the final decision.

The public comment was overwhelmingly not only against a reservoir at Rock Creek, but in favor of a reservoir at

Muddy Creek. As it turned out, Rock Creek was a favorite "secret fishery" of a number of individuals, including the then head of the Division of Wildlife. Additionally, the town of Kremmling was very interested in having a reservoir with recreation potential nearby, and there was minimal opposition by the three ranches involved to a sale of the property.

Based upon public comment, an analysis of the environmental trade-offs, and concurrence of the CRWCD, the Lead Agencies decided that the Federally Preferred Alternative should be a reservoir at Muddy Creek. That was when the real negotiating commenced. There are four native species of fish in the Colorado River basin that are listed as Threatened and Endangered. At the time, they were the subject of a fledgling Recovery Program, and there was question as to how "sufficient progress" would be obtained. While it went beyond the guidelines of the new program, the CRWCD agreed to commit releases from the reservoir toward flows for these fish, until such time as other supplies could be determined.

The wetlands were the last major hurdle to overcome. Studies and negotiations on the final jurisdictional determination took two years. During that time the Supplemental Draft EIS came out (August 1988), and we worked toward the Final EIS (February 1990), the final Mitigation Plan (March 1991), and the "404" Permit (May 1991). Again, the key was the maintenance of a working relationship with the Lead Agencies, and a dedication on the part of the EIS consultants toward producing "good science".

The final engineering design commenced in August 1989, prior to issuance of either the FEIS or the "404" permit. While risky, this resulted in the plans and specifications being submitted for approval to the Office of the State Engineer less than one month after the issuance of the "404" permit, and, one year later, the project was put out to bid. Delayed again by a water rights/water quality dispute, construction on the project started in August 1993. Negotiations on the part of the CRWCD and the Denver Water Board resulted in the project proceeding to construction without the water rights issue being fully resolved. It is likely that the final court decree will be entered prior to completion of the dam, but it has been only through an aggressive continuing attitude of cooperation and negotiation that the project has proceeded through the permitting, design and construction process in the past 10 years.

Interaction Between Professionals and the Public

Stuart G. Walesh[1]

Introduction

One challenge facing today's water professionals is to communicate effectively with the public recognizing both the public's increasingly elevated goals relative to water and the public's growing understanding of water science and technology. A premise of this paper is that citizens want to be involved and should be involved in urban water management.

Unfortunately, some engineers and other professionals with urban water management responsibilities fail to appreciate the importance of the challenge, or they recognize the challenge but are not prepared to meet it. Herrin and Whitlock (1992) somewhat harshly, but perhaps accurately, suggest that the cause of this failure lies with the engineer's formal and informal education. According to them, "Engineers are taught very few skills in interpersonal relationships, much less those of public interface and involvement. We spend little, if any, time addressing it at our conferences and conventions. We then spend thousands of hours and millions of dollars defending our projects when threatened by delays and possible blockage by public intervention."

Need for Interaction Between Water Professionals and the Public

Successful urban water management programs must include meaningful involvement of and interaction with elected and appointed government officials and the public. An urban water planning effort that fails to include a public interaction program plans to fail. Although said over a century ago, and in an entirely different context, the following words of U.S. President Abraham Lincoln are appropriate to modern urban water planning: "With public sentiment, nothing can fail; without it nothing can succeed.

----

[1]Consultant and Professor of Civil Engineering, College of Engineering, Valparaiso University, Valparaiso, IN 46383

Consequently, he who molds sentiment goes deeper than he who enacts statutes or pronounces decisions (Helweg, 1985)." A public interaction program, or lack thereof, is often the principal reason for the successful implementation of an urban water planning program or the failure to implement it (Kurz, 1973).

Definition of the Public

Many subgroups with very different, often competing agendas typically constitute "the public." All should have an opportunity to interact with the water professionals during the urban water management process. Examples of subgroups are environmental organizations, recreation clubs, service groups, professional societies, business associations, educators, students of all ages, individual citizens, and appointed and elected government officials at all levels. The success of a public interaction program is determined more by the number of different "publics" that participate than by the total number of individuals involved. Breadth of public representation and involvement is crucial. Water professionals should be especially wary of the temptation to exclude what they regard as "extremist" elements from the urban water management process. These individuals and groups have a right to be part of the process and to express their views. Attempts to exclude them are likely to aggravate matters and precipitate or elevate conflict.

Objectives of Interaction Program

The interaction program paradigm presented in this paper has three objectives (Walesh, 1989). The first objective is to demonstrate to the public and their appointed and elected representatives that the water resource professionals are aware of the problems, at least in a general sense; want to learn more about them; and want to seek solutions. In other words, water resource professionals need to demonstrate empathy and concern.

The second objective of a public interaction program is to gather supplemental data and information pertinent to the urban water management effort. The third and final objective of public interaction is to build a base of support for rapid plan implementation.

Programs and Events

- **Advisory Committees:** Examples of committee functions are providing data and information, raising issues and concerns, establishing goals, conducting public meetings, reviewing draft reports or report chapters and helping to interpret the planning process and the plan recommendations to committee members' associates and formal or informal constituencies.

- **Public Meetings:** An introductory meeting should be held early in the planning process primarily to achieve the first of the three public interaction objectives, that is, earning the public's trust. One or more intermediate public meetings can focus on status reports and presentation of alternatives that are under consideration. Draft recommendations of the urban water plan can be presented at a final public meeting. To the extent feasible, these meetings should be conducted within affected areas to facilitate ease of access and as a symbol of concern.

- **Contacts With Engineering Firms, Land Developers, and Professional Societies:** Because of their expertise and influence, such groups can provide valuable insight, useful data and information and offer support during plan implementation.

- **Presentations to Service Clubs and Other Community Groups:** Knowledgeable and influential community leaders are typically members of one or more civic organizations such as service clubs, environmental groups, and business and professional associations. Such presentations can help to expand knowledge of and support for an urban water management effort.

- **Field Reconnaissance and Contacts:** Field reconnaissance provides opportunities for informal, one-on-one interaction with inquisitive members of the public.

- **School Programs:** By educating school children about urban water issues, a two-fold result can be achieved. Students gain understanding and, to the extent they share what they learned with their parents, the knowledge is disseminated.

- **Guided and Self-Guided Tours:** Guided and self-guided tours enhance understanding of the location and severity of pollution, flooding, aesthetic, and other water-related problems and provide an opportunity for the water professionals to explain remedial and preventive measures that are under consideration in the planning program.

- **Briefings for Newly-Appointed or Elected Public Officials:** By being introduced to issues and being provided with basic information on on-going or completed urban water planning efforts, new public officials are more likely to be supportive.

- **Workshops:** These events provide an opportunity for in-depth exploration of substantive topics such as issues, findings, alternatives, recommendations, and operations.

- **Telephone-Based Access and Input:** Herrin and Whitlock

(1992) suggest telephone access, preferably toll-free, to resource people or recorded messages offering reports on the status of facilities (e.g., reservoir water levels, operating hours of water-based recreation facilities) or updates on the status of planning efforts (e.g., availability of draft reports, date and location of next public meeting). Voice and electronic mail offer exciting possibilities.

- **Clean-Up Projects:** Government agencies or units that are responsible for urban water management can draw attention to and obtain support for the effort by encouraging and supporting the organization of shoreline and other clean-up efforts by environmental groups or individual citizens.

- **Negotiated Conflict Resolution Among Contending Interests:** Contending interests which might include environmentalists, citizens, land developers, and engineering firms can be provided with an opportunity to negotiate an acceptable solution provided it satisfies applicable rules and regulations.

Supporting Devices

Having selected a set of programs and events to encourage effective interaction between water professionals and the public, attention typically turns to various supporting devices. These are tools and techniques intended to enhance the success of the interaction programs and events. Examples of supporting devices are: newsletters, brochures, fact sheets, and other special publications; press releases and the public media; audio-visual materials; lay-oriented graphics; and physical models.

References

HERRIN, J. C. & A. W. WHITLOCK, "Interfacing With the Public on Water-Related Issues: What TVA is Doing," Saving a Threatened Resource, ASCE, 1992, New York, pages 293-298.

HELWEG, O. J., Water Resources Planning and Management, John Wiley & Sons, New York, 1985.

KURZ, J. W., "Transformation of Plans Into Actions," Urban Planning and Development Division - ASCE, 1973, 99 (2), pages 184-191.

WALESH, S. G., Urban Surface Water Management, John Wiley & Sons, New York, 1989.

GUIDELINES FOR INTERSTATE WATER COMPACTS

Ray Jay Davis,[1]

Abstract

Interstate compacts are a common mechanism for allocating American transboundary water resources among states. A task committee of the Water Regulatory Standards Committee is developing guidelines for consideration by governmental entities that may enter into future compacts. Drafters of compacts should consider compact adoption provisions, parties to compacts, waters covered, compact administration, water resources allocation, water quality, and changes in compact terms.

Introduction

Many streams and lakes are shared by more than one state. Under the American constitutional system, there are three means for allocating the quantity and regulating the quality of these transboundary water resources: federal legislation, Supreme Court litigation among states, and interstate compacts (Grant, 1991; Getches, 1990). In a seminal article, Felix Frankfurter and James Landis (1925) advocated interstate compacts as the preferred approach, and there now are about two dozen allocation compacts (Grant, 1991), as well as other compacts involving pollution control, planning and flood control, and multipurpose regulation (Muys, 1976).

A task committee on shared use of transboundary water resources of the Water Regulatory Standards Committee has undertaken to prepare guidelines to which the states might look for information about issues that should be covered in compacts, commentary about the guidelines, and references to compact provisions and

---

[1]Professor, J. Reuben Clark Law School, Brigham Young University, Provo, UT 84602-8000.

literature about them. This paper introduces topics which these guidelines will address.

Adoption Provisions

The usual steps for compact formation involve authorization by Congress for the states to negotiate the compact, negotiation and ratification by the states, and Congressional consent for the compact (Getches, 1990). The language of compacts then references the authorization and sets forth both the facts concerning the compact negotiations and the text.

Parties

Most compacts are between two states or among more than two. The compact text lists the parties. In recent years, the federal government has become a party in addition to the states in some water allocation compacts. This has been extolled as an idea whose time has come (Sherk, 1994). Whether it is a party or not, the federal government participates in authorization, approval, and (usually) negotiation of the compacts (Newsome, 1994).

Waters Covered

Interstate water sharing agreements should describe accurately and fully the waters covered. Failure to do so in the Colorado River Compact has led to extensive litigation which has not yet answered all pertinent questions about coverage by the compact (Dennett, 1994). The compact should note what part of the main stem of the stream is covered and which tributaries; it should indicate whether atmospheric and/or ground waters are included since they are intimately connected with surface waters. Thus far, none cover atmospheric waters and few deal with ground water (Boyce, 1994).

Administration

Most, but not all, compacts provide for a commission to administer their provisions. They stipulate composition of the commission, means of selection of commissioners, and voting procedures and powers. Employment of staff, means of funding for administration, and power to expend such monies should also be included. Commission powers include making by-laws, stream development, operations, and management, and planning authority. It has been suggested that compacts typically do not grant broad enough authority (Muys, 1976). Dispute resolution, enforcement powers, and sanctions also are not now well articulated in many compacts (Simms et al, 1988).

Allocation

Allocation compacts include one or more formulas for sharing transboundary waters. They can be written in terms of withdrawal or depletion of stated amounts or percentages of flow (McCormick, 1994). They can deal differently with different reaches of the stream. They should be flexible enough to incorporate matters such as storage, climate change, conservation, augmentation, reuse, importation, etc. They need to consider operation of federal laws. A wide variety of legal, hydrologic, and management factors should go into developing the allocation formulas (R. Davis, 1995).

Quality

Not only compacts motivated by quality concerns (Williams, 1994), but also those in which allocation affects quality should address water quality.

Changes

Laws other than compacts which impact their operation change, climate and other environmental parameters change, and water utilization and usability also change. Accordingly, there should be provision in compacts for change -- periodic review provisions (Boyce, 1994) and termination provisions outlining succession to water rights, obligations of compact parties, and distribution of funds. Compact amendments must be instituted following the same procedure as original adoption.

Conclusion

This listing of topics for interstate compact guidelines is not necessarily complete. Other provisions may be required, and perfectly good compacts may not include all topics noted here. But compact drafters will be able to benefit from the ASCE Model Interstate Water Compact documents. They provide a menu from which selections may be made.

References

Boyce, J. (1994). "Wrestling With the Bear: Round Three." Student thesis, J. Reuben Clark Law Library, Provo, UT.

Davis, E. (1994). "Interstate Compacts That Are for the

Birds: A Proposal for Reconciling Federal Wetlands Protection With State Water Rights Through Federal-Interstate Compacts." Student hesis, J. Reuben Clark Law Library, Provo, UT.

Davis, R. (1995). "Principles for Shared Use of Transboundary Water Resources." *Conf. on Water Resources Planning and Management at Cambridge, MA*. ASCE, New York, NY.

Dennett, R. (1994). "Las Vegas and the Virgin River: Cashing In On An Unclaimed Jackpot in the Southern Desert." Student thesis, J. Reuben Clark Law Library, Provo, UT.

Frankfurter, F. & J. Landis (1925). "The Compact Clause of the Constitution -- A Study in Interstate Adjustment," *Yale Law J., 34*, 685.

Getches, D. (1990). *Water Law In a Nutshell*, ch. 10, West Publ, St. Paul, MN.

Grant, D. (1991). "Water Apportionment Compacts Between States," ch. 46 in R. Beck, *Waters and Water Rights*, Michie Co., Charlottesville, VA.

McCormick, Z. (1994). "Interstate Water Allocation Compacts in the Western United States -- Some Suggestions." *Water Resources Bull., 30*, 385.

Muys, J. (1976). "Allocation and Management of Interstate Water Resources: The Emergence of the Federal-Interstate Compact." *Den. J. Int'l Law & Pol'y, 6*, 307.

Newsome, A. (1994). "Calling a Truce in Southeastern Water Wars: Proposal to Adopt a Federal-Interstate Compact." Student thesis, J. Reuben Clark Law Library, Provo, UT.

Sherk, G. (1994), "Resolving Interstate Water Conflicts in the Eastern United States: The Re-emergence of the Federal-Interstate Compact." *Water Resources Bull., 30*, 397.

Simms, R., L. Rolfs & B. Spronk (1988). "Interstate Compacts and Equitable Apportionment." *Rocky Mtn. Mineral Law Inst., 34*, 23-1.

Williams, C. (1994). "Interstate Water Compact for the Potomac River Basin." Student thesis, J. Reuben Clark Law Library, Provo, UT.

The Effects of Water Resources Planning on Land Development Projects

Thomas J. Olenik, Member, ASCE (1)

Abstract:

Land development regulations in the state of New Jersey have been utilized to control development since the mid-1970's. The demand for affordable housing in the state is reaching a critical period since the median price of a single-family dwelling in New Jersey currently exceeds $170,000. At the present time a land development project faces independent action from the local and county planning boards as well as a long list of individual permits required by the New Jersey Department of Environmental Protection (NJDEP). These permits amount to such items as stream encroachment, stormwater management, sanitary sewer extension permits, freshwater wetlands, and restrictions on the use of septic systems or the flow of sewage to existing treatment plants (micromanagement of a watershed). Since all the above items are treated as independent and theoretically unrelated review procedures, the time it takes for a developer to obtain the necessary approvals prior to construction has risen to an average of two to three years. At the present time, the NJDEP is moving towards a watershed-based permitting process that will examine the cumulative effects of all the previously cited permits.

I. Introduction

The development of land (residential, commercial or industrial) in the state of New Jersey illustrates how cumbersome and costly such a process can be when public policies try to steer and regulate all aspects of a project. Since 1975 when New Jersey enacted the Municipal Land Use Law (MLUL), the process has

----

(1) Associate Professor of Civil & Environmental Engineering, Dept. of Civil & Environmental Engineering, New Jersey Institute of Technology, 323 M.L. King Blvd., Newark, NJ 07102 (201) 596-5895; Senior Design Consultant, Semester Consultants, Inc., 112 Lincoln Avenue, Dunellen, NJ 08812

evolved into a complex system of regulatory bodies, engineers, planners and lawyers that has become complicated and costly. Few visitors to New Jersey realize that over half the state's land mass is undeveloped property. The congested northeastern part of the state adjacent to New York City is the one which is most familiar to people from out of state. The pressure for development away from the built-up areas has increased in the last twenty years. Due to the enormous complexities in developing such properties, the related cost of raw land and professional services have also risen markedly, resulting in the median price of a home increasing to $170,000.

The objective of the remainder of this paper is to examine the regulatory process and to discuss the typical problems that must be overcome prior to the successful completion of a project. In this case, 'successful completion' does not necessarily mean economic viability for the developer and/or the public.

II. Project Development Process

The enactment of the MLUL (1975) signified the beginning of the regulatory process in New Jersey. This law established local and county planning boards, along with rules to which these boards must adhere during the planning and review process. These types of boards are not uncommon throughout the United States and do serve a necessary public function. However, numerous state-wide regulations have been instituted since 1980, resulting in the development process outlined below.

1. Zoning Regulations

The MLUL established planning and zoning boards to review a project on a local and county-wide basis. The major changes in these bodies have been the protectionism practiced by rural communities with regard to any development, and the inversion of the planning process by the NJDEP. For example, the undeveloped municipalities have established zoning requirements of one housing unit for 1.2 to 2.0 ha (three to five acres). This low-density zoning aspect coupled with increased infrastructure requirements artificially raises the cost of housing in developed areas. In addition, the requirements of the NJDEP listed below have reduced the importance of the planning board process. That is, due to the state review and permit requirements, the planning board process is no longer the most important aspect of a project's approval procedure.

2. Water Resources Regulations

a. Sewer Extension Permit

This permit is utilized as a planning tool by restricting the extension of sewers and/or treatment plans to areas the state believes should be of low density or not developed at all. Therefore, even though an active sewer line is accessible to a project, a permit to connect new sewers to the existing system may not be issued by NJDEP.

b. Stream Encroachment Permit

As previously discussed, restrictions existed with regard to development within floodplains. These regulations have been greatly expanded to restrict fill in flood hazard areas, and expand hydrological and hydraulic data requirements for all projects that contain streams with drainage areas of 2 ha (50 acres) or more. In addition, a new, more stringent category was established known as a Project of Special Concern. A Project of Special Concern is one in which a project proposed channelization, removal of woodlands near a stream, and those projects that would likely produce serious adverse effects on water resources.

c. Freshwater Wetlands

In 1987 the State of New Jersey passed the Freshwater Wetlands Act to restrict development in these areas, in addition to all other local or federal laws. A field investigation by the developer's environmentalist and state review official establishes a line which severely restricts development. In addition, a 15.2 to 45.6 meter (50 to 150 foot) buffer area is added on as a factor of safety.

d. Other NJDEP Permits or Requirements

1. Dam Safety Permit (retention/detention ponds)
2. Subsurface disposal of sanitary wastes (septic system)
3. Environmental Cleanup Responsibility Act (ECRA) (hazardous wastes)
4. Coastal Area Facilities Review Act
5. Water Quality Permit
6. Statewide Stormwater Permit Program

3. Other Permits

a. Soil Erosion and Sediment Control Permits. The county-wide soil conservation districts (SCD) have greatly expanded their review authority.

b. Local sewerage authority approval (sewer extensions)

c. Local or county board of health (septic systems)

4. Statewide Master Plan (pending)

5. **Federal Permits** (as applicable)

6. **Watershed Management**

Irrespective of the previous section, the NJDEP is embarking upon a watershed-based permitting system. A review panel studying an environmentally sensitive area known as the Great Swamp has recommended that permits issued for a project within a certain watershed consider the cumulative and secondary impacts the permits would have on the entire watershed. A similar but narrower approach was taken by NJDEP in the 1970's through limitation placed upon the siting and expansion of public wastewater treatment facilities and the number of subsurface (septic) disposal systems on a site. The proposed changes in the permitting system will also consider the effects of traffic, water consumption, open space, zoning, etc., on the overall watershed. The result of these considerations will be to develop models of a particular area and to create standards for performance for water and air quality, toxic and hazardous waste generation, and quality of life issues. Based upon the as yet undefined performance standards cited above, an assessment of each project (and related permits) would be made as to its effect on the watershed as a whole.

These regulations and permits are all required in addition to the local technical requirements, which can be in conflict with the above. Also, as the NJDEP expands it regulations under current laws, projects are forced to meet new requirements in the middle of the review process. All of the abovementioned state permits are processed on a singular basis, resulting in severe time delays and policy disputes.

III. Conclusions

The perspective discussed above clearly indicates the complexity and cost involved in the development of a residential or commercial property in the state of New Jersey. In recent years the state's economy has suffered due to the fact that less regulated and more agreeable jurisdictions exist in other states. New Jersey state officials should be alarmed when, 1) the median cost of a house rises to $170,000 2) the budget of the NJDEP is seventy-percent based upon application fees and fines, and 3) the time it takes to bring a project to the construction stage averages between two to three years.

All of the above factors exist prior to the future implementation of a watershed management program. At some point in the near future, New Jersey will regulate itself into a steeper economic decline.

Water Resources Development in a Coastal Environment

Dan Conaty, Parsons Engineering Science, Inc.[1]
David W. Connally, Parsons Engineering Science, Inc.[2]

Abstract

The San Mateo Basin Groundwater Conjunctive Use Project is an example of the present-day challenges involved with developing a utility improvement project in a sensitive coastal environment of California. As sensitive coastal and biological resources experience intensive development pressure, the need to design utility projects to avoid or minimize environmental degradation becomes increasingly important. To facilitate approval of projects located in sensitive coastal areas, a cooperative effort among the project proponents, engineering design and environmental consultants, and the environmental regulators is emphasized.

Project Background

The San Mateo Basin Groundwater Conjunctive Use Project would involve a lease agreement between the U.S. Marine Corps Base, Camp Pendleton (Base) and Tri-Cities Municipal Water District (TCMWD or "District) to manage groundwater resources at the northern portion of the military base. The affected groundwater basin is located at Camp Pendleton, the Marine Corps' amphibious training center for the west coast. This 51,800-hectare (ha) military base is located in northern San Diego County, California. The Base is bordered by the City of San Clemente to the northwest, the Cleveland National Forest and Fallbrook to the north and east, the City of Oceanside to the south, and the Pacific Ocean to the west.

The watershed for the San Mateo drainage basin encompasses an area of about 35,400 ha. The Lower San Mateo Basin is a two-aquifer system,

---

1 Principal Scientist, Parsons Engineering Science, Inc., 9404 Genesee Avenue, Suite 140, La Jolla, California 92037

2 Principal Scientist, Parsons Engineering Science, Inc., 199 S. Los Robles Avenue, Pasadena, California 91109

consisting of an upper unconfined aquifer and a deeper confined aquifer. The upper aquifer is divided into an overlying alluvium and the Upper San Mateo Formation Aquifer. The alluvium extends from ground surface to a depth of 30.5 to 46 meters (m) below ground surface (bgs). The Upper San Mateo Formation Aquifer exists below the alluvium to a depth of 98 to 114 m bgs. The confined aquifer, referred to as the Lower San Mateo Formation Aquifer, exists between 128 to approximately 198 m bgs, and is separated from the upper aquifer by an aquitard. The total groundwater storage capacity of the unconfined aquifer is 58,300 acre feet (af). The confined aquifer has a total storage capacity of 22,100 af.

The groundwater conjunctive use lease agreement would allow TCMWD and the Base to increase production of groundwater resources in the San Mateo groundwater basin and permit the District to store or "bank" imported water to augment local supplies. For its part, the Marine Corps Base would be connected to a municipal water supply system and could relinquish water conveyance and storage system maintenance responsibilities within a portion of it's service area. Conjunctive use of the basin is intended to allow TCMWD and the Base additional flexibility for meeting water demand and increase reliability of those deliveries.

Proposed project facilities consist of wells, transmission pipelines, pressure regulating facilities, and chlorination facilities to provide: potable water to the existing TCMWD regional system; treated imported water to the Base local potable water system, if necessary; and, imported water to recharge local groundwaters from time to time. Project facilities would be designed with the capability to transport basin groundwater and banked water to TCMWDs' service area and to deliver imported water to the North Base Service Area.

The management scheme for this project includes establishment of a Basin Committee. The Committee will set objectives for TCMWD groundwater production and banking, and prepare an Annual Report analyzing the effects of the project on the basin. The Basin Committee objective will be to maximize the sustained yield of groundwater from San Mateo Basin consistent with minimum disturbance to the environment.

Parsons Engineering Science, Inc. is under contract to prepare an environmental document to satisfy requirements of the National Environmental Policy Act (NEPA) and the California Environmental Quality Act (CEQA). Project design would be conducted by another consulting firm; however, the conceptual design process has been substantially influenced by environmental issues.

Major Environmental Issues

Major challenges with the San Mateo Basin Groundwater Conjunctive Use Project involve balancing the need for water storage with several varying environmental issues, including:

- Potential loss of sensitive biological habitats and species during construction;
- Potential impacts to a National Register-eligible archaeological site;
- Potential indirect biological impacts due to groundwater drawdown; and
- Land use compatibility issues associated with developing utilities at a major military installation, in the coastal zone, and on land being leased for state park and agricultural uses.

For the San Mateo Basin Groundwater Conjunctive Use Project, these environmental concerns largely dictated the proposed location of pipelines and well sites. The pipeline alignment location was selected after detailed field studies were conducted for biological and archaeological resources. The pipelines would be situated along an agricultural service road at the margin of a wetland, and will be configured such that permanent loss of wetland habitat and disturbance of an archaeological site are avoided. The pipeline corridor width must be constricted to approximately 8 m at some locations to avoid temporary loss of sensitive riparian woodland habitat.

Project construction activities may also be constrained by environmental considerations. Construction timing may be an issue if it coincides with nesting activities of sensitive bird species in the vicinity. Project construction would also involve the temporary closure of a coastal access route used for various forms of recreation and camping at a San Onofre Beach State Park. The state park has one of the most popular surfing beaches in southern California. This coastal access will likely have to be maintained during construction; however, options are constrained by a freeway on one side an archaeologically sensitive area on the other side of the access road.

Another issue addressed in the environmental document is the operational effect of groundwater drawdowns on sensitive habitats and species located within the basin. The basin is home to several riparian woodland habitats and associated wetlands, as well as several sensitive animal species that depend upon these habitats. Groundwater conditions in the basin were modeled by another consulting firm. Detailed biological studies were conducted under contract to Parsons ES.

Options for reducing potentially significant impacts associated with groundwater drawdown are being reviewed. These could involve restrictions on groundwater pumping and irrigation of vegetation during the dry season. Operation of the project would almost certainly be associated with detailed in-field monitoring to better establish baseline conditions and project conditions, especially the connection between the groundwater and surface water flows in the lower basin along the San Mateo Creek. The groundwater needs of each habitat type are also not well understood and should be better defined. This monitoring, should it be required, would be costly to implement.

Constraints Analysis

As part of the Clean Water Act, Section 404 permitting process, the U.S. Army Corps of Engineers requires that an Alternatives Analysis be prepared to evaluate the relative impacts of each alternative and to determine the Least Environmentally Damaging Practicable Alternative (LEDPA). In its evaluation of a project, the Corps of Engineers must first determine whether avoidance of the wetland resource is possible. If not, resource impacts must then be minimized and mitigation of the lost wetlands is required.

Timing of Alternative Analysis preparation is important because the results can affect the project design process. When designing in a sensitive wetland environment, the LEDPA should be considered as part of the project design prior to or as part of the environmental document preparation process. Under NEPA, project alternatives are assessed with the same level of detail as the proposed project alternative. If prepared properly, this would fulfill Corps of Engineers requirements for Alternatives Analyses under Section 404 of the Clean Water Act.

Under CEQA, alternatives do not have to be analyzed with the same level of detail as the proposed project. However, if it is known at the outset that an Alternatives Analysis will ultimately be required, a thorough analysis of alternatives should be conducted as part of the CEQA process. If the CEQA process is done correctly, the LEDPA would be adopted during lead agency project approval and the Corps of Engineers is more likely to concur with the lead agency's selection of the proposed project alternative.

For the San Mateo Basin Groundwater Conjunctive Use Project, a simple constraints analysis was conducted to determine the LEDPA for analysis in the environmental document. As described above, the proposed project was selected based upon avoidance of impacts to sensitive habitats and a significant archaeological site. For the District, the LEDPA alternative to route the pipeline along the roadway and across an agricultural field was not substantially more expensive than an engineer's preferred-alignment option.

Conclusion

In the 1990s, environmental concerns are largely dictating the location of water utility projects located in sensitive coastal environments. It is no longer adequate to dismiss alternatives based solely on cost, land ownership or engineering considerations. Resource agencies are obliged by law to prioritize wetland avoidance when reviewing projects. Cost and engineering practicality are only two of the factors the Corps of Engineers considers when determining the LEDPA.

It is critical that project designers and planners evaluate all options to avoid permanent and temporary wetland impacts during the pre-design and environmental document phases. For the San Mateo Basin Groundwater Conjunctive Use Project, Parsons Engineering Science has worked with the District, Base, and design engineers to conceptually locate project facilities to avoid or minimize impacts to sensitive environmental areas.

ISSUES IN WATER RIGHTS MODELING

Ralph A. Wurbs[1], Member, ASCE

Introduction

During the 1970's and 1980's, the state of Texas consolidated surface water rights into a prior appropriation permit system. Water rights have become an important consideration in water management in the state. This motivated development of the Water Rights Analysis Package (WRAP), which is a generalized model for simulating river basin management within the framework of a priority based water allocation system (Wurbs et al. 1993). The monthly time step model is used to evaluate reliabilities in supplying permitted water rights, with alternative reservoir/river system management strategies, based on a hydrologic period-of-record simulation. Application of the model in a study of the Brazos River Basin focused on a system of 12 reservoirs owned by the Brazos River Authority and Corps of Engineers, but reflects the fact that over a thousand entities, owning about six hundred reservoirs, hold permits to use the waters of the Brazos River and its tributaries (Wurbs et al. 1994). The impact of natural salt contamination on water supply reliabilities was another major concern in the study.

The simulation study included identification and examination of key issues and considerations in river basin management and associated water availability modeling. The several concerns noted here are illustrative of the numerous complexities of both modeling and administering a water rights system.

Allocation of Shortages Between Competing Water Users

What happens during severe drought conditions when insufficient water is available to satisfy all water rights? The WRAP model simply allocates water based on the priority dates specified in the water rights permits. In the model, no diversions are allowed which adversely impact diversions by more senior rights. A

[1] Associate Professor, Civil Engineering Department, Texas A&M University, College Station, Texas 77843

junior right may be completely curtailed, to allow a more senior right to receive its full permitted diversion amount. In an actual real-world situation involving insufficient water supply, the shortages will likely be shared, to some degree, by water users regardless of the relative seniority of their rights, and temporary demand management measures will be implemented. However, detailed drought contingency plans do not exist at this time for most of Texas. The water rights system has not yet been tested and refined by a major drought comparable to the 1930's and 1950's.

Reservoir Storage Priorities

Reservoir operation in Texas is based on providing long-term storage as protection against infrequent but severe droughts. The right to store water is as important as the right to divert water. If junior appropriators located upstream of a reservoir diminish inflows to the reservoir when it is not spilling, reservoir dependable yield is adversely affected. Each drawdown could potentially be the beginning of a several-year critical drawdown which empties the reservoir. Thus, protecting reservoir inflows is critical to providing a dependable water supply. On the other hand, forcing appropriators, with rights junior to the rights of the reservoir owner, to curtail diversions to maintain inflows to an almost full, or even an almost empty, reservoir is difficult and not necessarily the optimal use of the water resource. If junior diversions are not curtailed, the reservoir will likely later refill anyway, without any shortages occurring.

Handling of the storage aspect of water rights is not yet precisely defined in Texas. Water rights grant both the right to divert and to store water. Water rights permits, in Texas, include a single priority date. Priorities are not specified separately for diversions and storage. Reservoir storage priorities have a very significant effect on simulation results. The effects are primarily reflected in tradeoffs between individual water rights rather than basin totals.

The following scheme was somewhat arbitrarily adopted in the simulation study as being a reasonable approach. The major reservoirs are filled to 80% of capacity with priorities associated with the water rights and then to 100% capacity with priorities junior to all diversion rights in the basin.

Multiple-Reservoir System Operations

Water rights permits in Texas are granted for individual reservoirs rather than multiple-reservoir systems. However, 12 reservoirs containing 63% of the conservation storage capacity in the Brazos River Basin are operated by the Brazos River Authority (BRA) and Corps of Engineers as a system. A large portion of the water use from these reservoirs involve withdrawals from the lower Brazos River which are supplied by unregulated streamflows and releases from multiple reservoirs. The BRA holds an excess flows permit which allows diversion of unregulated streamflows from the lower Brazos River, in lieu of reservoir releases, as long as no other water users are adversely affected. The BRA permits have been modified to allow multiple-reservoir releases as long as the total of the diversions specified in the individual reservoir permits are not exceeded. Flexibility has also been provided for shifting between types of water use as well. The various aspects of multiple-reservoir system operation addressed by Wurbs *et al.* (1994) include: use of excess flows in combination with reservoir releases; balancing multiple-reservoir releases; effects of tributary versus main-stem reservoir releases on salinity in the lower Brazos River; and balancing local versus system diversion reliabilities. The simulation study demonstrates that multiple-reservoir system operations, particularly use of excess flows in combination with reservoir releases, are very beneficial in maintaining water supply reliabilities in the Brazos River Basin.

Return Flows

Simulation results are sensitive to assumptions regarding return flows. Although some recently issued water rights have addressed return flows, most permits do not specify the amount of the diversion to be returned to the stream. Return flows can significantly impact the availability of water to downstream users. The WRAP input file includes a return flow location and factor, reflecting the percentage of the diversion returned to the stream, for each diversion right. Limited historical data compiled by the Texas Water Development Board and Texas Water Commission were used to estimate return flow factors as a function of type of use for different regions of the basin.

Salinity Constraints

In the past, the quality of water supply diversions has not been explicitly considered in the process of evaluating water rights applications and issuing permits. Water quality considerations have typically focused on including restrictions on new water rights permits to maintain instream flows. However, salinity is widely recognized as being an important concern in the Brazos and other river basins. Salt concentrations exhibit great variability, both temporally and spatially. In the Brazos River Basin study, salinity is treated as a constraint to the use of diverted water for off-stream uses.

The modeling study included alternative WRAP runs with maximum allowable total dissolved solids, chloride, and sulfate concentration levels specified for each type of water use. Shortages were declared for diversions any time the specified limits were exceeded. Tolerable or acceptable concentration limits for various types of water use are difficult to precisely define. The study did not include establishment of acceptable salt concentration limits for particular water uses. Rather alternative model runs were made to demonstrate the sensitivity of simulation results to a range of assumed maximum allowable concentrations. Depending on diversion location and salt concentrations for a particular water use, water supply reliability may be controlled by water quality rather than quantity.

Requirements for Hydrologic and Other Data

WRAP input data is voluminous. A major part of the study effort consisted of developing complete sets of unregulated (naturalized) monthly streamflow and salt load sequences covering the 1900-1984 period of analysis at all pertinent locations. Reservoir storage volume versus water surface area data are also difficult to obtain for the numerous smaller reservoirs and had to be approximated. Net reservoir evaporation rates are available from an existing database. Water rights data are available from the Texas Natural Resource Conservation Commission, which is the agency responsible for administering the water rights system.

Interpreting Water Supply Reliability

Water management decisions necessarily require qualitative judgement in determining acceptable levels of reliability for various situations. Tradeoffs occur between the amount of water to commit for beneficial use and the level of reliability that can be achieved. Reliabilities are not very sensitive to changes in diversion amounts. Conversely, yields change greatly with relatively small changes in reliability. The amount of water supplied from the Brazos River Basin can be increased significantly by accepting somewhat higher risks of shortages or emergency demand reductions. Yield versus reliability estimates are not highly precise and can vary significantly with incorporation of different but yet still reasonable assumptions in the model.

Acknowledgements

The research on which this paper is based was financed in part by the United States Department of Interior, U.S. Geological Survey, through the Texas Water Resources Institute. The Texas Water Development Board co-sponsored the research and provided nonfederal matching funds. Contents of this publication do not necessarily reflect the views of the research sponsors, nor does mention of trade names or commercial products constitute their endorsement by the government agencies.

References

Wurbs, R.A., D.D. Dunn, and W.B. Walls (1993). "Water Rights Analysis Package (TAMUWRAP), Model Description and Users Manual," Technical Report 146, Texas Water Resources Institute, College Station, Texas.

Wurbs, R.A., G. Sanchez-Torres, and D.D. Dunn (1994). "Reservoir/River System Reliability Considering Water Rights and Salinity," Technical Report 165, Texas Water Resources Institute, College Station, Texas.

Applying a Prescriptive Model to Help Resolve Conflict Over Alamo Lake, AZ

Kenneth W. Kirby[1]

Abstract

Conflict over reservoir operation increases the need for analytical tools to help reevaluate reservoir system operation. This paper presents an analytical approach combining prescriptive and descriptive models to develop a region of feasible reservoir operation alternatives to help advocacy groups in conflict more quickly find a suitable compromise. This paper presents the methodology used for the prescriptive analysis on Alamo Reservoir, including penalty function generation and operation rule inference from prescriptive model results. The conclusions show the usefulness of applying prescriptive models for solving water use conflicts on small reservoir systems.

Introduction

Alamo Lake is a multiple purpose reservoir owned and operated by the U.S. Army Corps of Engineers (Corps). Alamo Dam is located in Arizona on the Bill Williams River approximately 40 river miles upstream of the confluence with the Colorado River. During the late 1980's, resource agencies responsible for managing the Bill Williams River and Alamo Dam and Reservoir experienced increasing conflict between their individual missions and perspectives. Much of the disagreement stemmed from how the Corps was operating the water conservation pool at Alamo Lake.

In August of 1990 the agencies instituted an interagency planning team to develop a comprehensive water resource management plan for the Bill Williams River corridor. As part of their work, the Bill Williams River Corridor Technical Committee used a descriptive model (HEC-5) to evaluate several operational strategies to meet competing needs more effectively. Beyond the descriptive study, the Hydrologic Engineering Center (HEC) used data from the Bill Williams River Corridor Technical Committee's experience to develop and test a strategy taking advantage of prescriptive models to help resolve conflict over reservoir use.

---

[1] Hydrologic Engineering Coop, Hydrologic Engineering Center, U.S. Army Corps of Engineers, 609 Second Street, Davis, CA 95616

Modeling Strategy

A reservoir system modeling effort is more likely to be successful if conducted systematically. The approach taken for the Alamo study consisted of the seven steps shown in Figure 1. Although this process does not address all aspects of conflict resolution, the structured analytical strategy can help focus activities or provide a common point of reference for the stakeholders in the conflict. This focus can help start the resolution process and help keep it moving.

In general, the strategy applies a prescriptive model to screen alternatives and then uses a descriptive model to test and refine the more promising alternatives. No attempt is made to discover a globally "optimum" solution, but rather to efficiently produce a region of viable alternatives to help those in conflict reach an acceptable compromise.

Gather Data

The data relevant to applying these models generally includes streamflow hydrology, water management objectives, and constraints on the system. The constraints can consist of physical, legal, or economic limits.

Quantify Objectives

Quantifying values of different water uses is essential for prescriptive modeling, and is usually the most difficult step. Previous studies performed by the Hydrologic Engineering Center using prescriptive models employed economic objective functions. This approach is effective for traditional uses such as flood control, hydropower, irrigation, and recreation where widely accepted economic methods exist. However, economic valuation is more controversial for environmental objectives such as endangered species preservation or riparian habitat restoration. Furthermore, the availability of economic expertise can limit economic valuation of uses for small reservoir systems. Therefore, this study formulated and tested a systematic approach to construct penalty functions based on verbal statements of preference. The method establishes quantitative statements of value for each use relative to one another. Unit costs are assigned according to broad preference categories to define the penalty function for each use. The method is referred to as Relative Unit Cost (RUC).

Set up Prescriptive Model

This study applied the Hydrologic Engineering Center's Prescriptive Reservoir Model (HEC-PRM) for the prescriptive phase of analysis. HEC-PRM is a deterministic optimization model with a minimum cost network flow structure and an optimal sequential algorithm for incorporating hydropower.

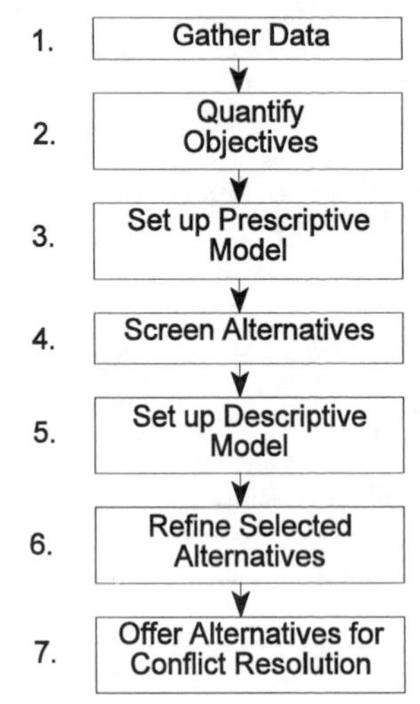
1. Gather Data
2. Quantify Objectives
3. Set up Prescriptive Model
4. Screen Alternatives
5. Set up Descriptive Model
6. Refine Selected Alternatives
7. Offer Alternatives for Conflict Resolution

Figure 1 Steps in Modeling Strategy

Screen Alternatives

The purpose of the prescriptive model is to generate promising alternatives that could improve system operation with regards to current water management objectives. The Relative Unit Cost methodology produces quantitative value statements that are relative rather than absolute, therefore more than one set of penalty functions will likely be required to consider the range of promising alternatives for different relative values between uses. The Relative Unit Cost methodology is well suited for use with some of the multiobjective algorithms. Results of the prescriptive model offer optimal time series for system storages and releases according to the hydrology and penalty functions specified. These results can be analyzed to deduce a descriptive operating strategy.

Set up Descriptive Model

The descriptive modeling phase of analysis is necessary to test and refine results from the prescriptive model. Descriptive models are generally more flexible and can incorporate more of the real system details. The particular modeling software chosen should allow for extremely flexible operating rules. Since the goal is to deduce an operating rule that can mimic the prescribed results, the rule form often varies for different systems.

Refine Selected Alternatives

After screening alternatives with the prescriptive model and setting up the descriptive model, the analyst can simulate reservoir operations on the time scale required and refine the operating rules to improve system performance. This step can be very time consuming and costly. By screening alternatives using a prescriptive model, the most promising alternatives are theoretically available for simulation testing directly, thereby reducing the need to continue to try new alternatives.

Offer Alternatives for Conflict Resolution

The purpose of the modeling exercise is to identify a region of viable alternatives that can help advocacy groups in conflict find an acceptable compromise. The alternatives should be presented along with information regarding how well each alternative satisfies the various objectives.

Findings from the Alamo Study

The process of constructing penalty functions for the different water use objectives proved very useful in understanding interactions between different uses of the system. Verbal statements of preference compiled by the Bill Williams River Corridor Technical Committee were translated to quantitative representations of value using the Relative Unit Cost approach. Solving the Alamo HEC-PRM model for each individual objective provided a useful look at how the reservoir would be operated if it focused on benefitting a single objective. Figure 2 shows the preferred storage time series for riparian habitat restoration, lake fishery, lake recreation, and flood control objectives. This plot clearly demonstrates that the greatest conflict occurs between the riparian habitat and flood control objectives, while the fishery and recreation objectives produce very similar operational schemes.

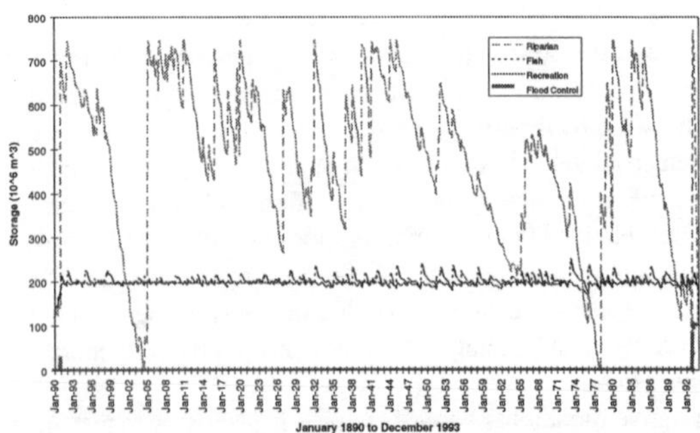

Figure 2 Comparison of Preferred Storages for Each Use

Conclusions

The proposed modeling strategy using an optimization model to screen alternatives and a simulation model to refine and test operation alternatives can be an effective way to discover innovative strategies to more closely meet conflicting demands. By intelligently screening alternatives with the optimization model, the simulation modeling effort can be more focused and efficient. One of the problems encountered when trying to apply this type of approach for reservoir systems today is the difficulty in quantifying objectives for environmentally related uses. The Relative Unit Cost Method developed for the Alamo study helps mitigate this problem. The RUC Method allows the use of optimization models to analyze reservoir systems even where economically based penalties are not practical. Even if most water management objectives can be accurately represented using economic techniques, the RUC Method can still be used to quantify those objectives that cannot be suitably represented using economic techniques. The value functions produced from both methods can then be used in conjunction with multiobjective programming techniques to generate promising alternatives.

Acknowledgements

The Corps' Los Angeles District and the Bill Williams River Corridor Technical Committee provided the data and many useful suggestions and ideas for this study.

References

Bill Williams River Corridor Technical Committee (1994). *Proposed Water Management Plan for Alamo Dam and the Bill Williams River*, Draft June 13, Eric Swanson - Technical Committee Coordinator, Arizona Game and Fish Department.

Kirby, K. (1994). *Resolving Conflict Over Reservoir Operation: A Role for Optimization and Simulation Modeling*, Master's Thesis, University of California, Davis.

PRINCIPLES FOR SHARED USE OF TRANSBOUNDARY WATER RESOURCES

Ray Jay Davis,[1] A.M. ASCE

Abstract

When political entities share water resources, timing and magnitude of water use have been sources of tension. Defined standards for reasonable and equitable water sharing have evolved from this conflict. Guidelines have developed in the United States through Supreme Court doctrine of "equitable apportionment," by compacts among states, and by acts of Congress. Also, on an international level, customary principles of international law and treaty practices have established certain guidelines. This paper summarizes these legal principles.

Introduction

Because water may flow across and along political boundaries jurisdiction over such shared water resources may be exercised by more than one governmental entity. In the event of multiple regulation, the basic rule of transboundary water law is that all involved jurisdictions must respect the rights of all others to a "reasonable" and "equitable" share in beneficial use of the water (Grant, 1991; Utton, 1991).

Through consideration of a variety of factors, decision makers have given content to the terms "equitable" and "reasonable." These factors have been identified in studies of interstate (Simms et al, 1988; Tarlock, 1985) and international water allocation law (Committee, 1966; Griffin, 1958). They may be classified under the following headings: (1) The legal regime, (2) water basin resources, and (3) resource

[1]Prof. J. Reuben Clark Law School, Brigham Young University, Provo, Utah 84602-8000

management. Each group of principles will be discussed in turn.

Legal Regime

Law regarding use of multijurisdictional water bodies may be made locally, on a state basis, nationally, and internationally. Sources for such law stem from agreements between and among governmental entities, adjudication of disputes over water use, and statutes (Getches, 1980). Consideration of the resulting "law of the river" is the starting point for establishing rules about the shared use of such transboundary waters. Accordingly, according to Tarlock (1985), if the jurisdictions involved follow the prior appropriation doctrine, it is likely that the multijurisdictional allocation will also adhere to the temporal priority of beneficial use concept, and if they are riparian jurisdictions, presumptively the law of riparian rights will be applied.

The law of the river includes not only allocation agreements, but also those involving water quality, flood control, planning, and multipurpose regulation (Muys, 1976). Environmental law may affect it (Davis, 1994). Existing law governing utilization may shift toward legal rules for integrated management and optimized water use (Utton, 1991).

Water Basin Resources

In addition to the legal regime, hydrology of the water resource is a place to start consideration of sharing (Rice & White, 1987). Analysis of the water resource includes the annual and seasonal flow in various reaches of the stream and its tributaries, variation (flood and drought potential), ground water availability, water contributions from each jurisdiction, dependency of each entity on the water resource, and availability of water from other sources.

In addition to the water resource, decision makers should consider land resources (geographic conditions in each jurisdiction), atmospheric resources (weather and climate in the river basin), and human resources -- population, economic needs and abilities, including capacity to develop the water and alternative sources and to compensate for use, and sociological considerations.

Resources will change. For example, consideration should be given to population change (Dennett, 1994). Shared water resources and affected other resources are not static.

Resource Management

Utilization and usability of transboundary water resources should be weighed in determining fair and equitable allocation. Among these management factors are timing of use (priority and seasonal timing), nature of use (location, purpose, consumptive character), amount of use (measured quantity and flow, variation, diversion, depletion, percentage of flow), use efficiency (waste, conservation, reuse), effects of use including substantial injury from use (Sherk, 1989) and water quality impact (Harlow, 1994), and flexibility in use (export from and import into the basin, augmentation and depletion of the resource, transfer, and storage capability).

Management of transboundary water resources is subject to change. Indeed the prospect of change may be the reason for making a decision about future allocation. For example, construction of new storage facilities may be related to agreement on water sharing (Cannon, 1994).

Conclusion

Although water basins have similarities, every one differs from every other one. To allocate transboundary water resources fairly and equitably, those differences must be carefully weighed in a delicate balance of factors. Those principles noted here provide guidelines for such reasonable and equitable sharing.

References

Cannon, B. (1994). "The Snake River Compact: Facing an Uncertain Future." Student thesis, J. Reuben Clark Law Library, Provo, UT.

Committee on the Uses of Waters of International Rivers (1966). International Law Ass'n Report, art. IV & V, Helsinki, Finland.

Davis, E. (1994). "Interstate Compacts That Are for the Birds: A Proposal for Reconciling Federal Wetlands Protection with State Water Rights Through Federal-Interstate Compacts." Student thesis, J. Reuben Clark Law Library, Provo, UT

Dennett, R. (1994). "Las Vegas and the Virgin River: Cashing In On An Unclaimed Jackpot in the Southern Desert." Student thesis, J. Reuben Clark Law

Library, Provo, UT.

Getches, D. (1990). *Water Law In a Nutshell*, ch. 10. West Pub., St. Paul, MN.

Grant, D. (1991). "Water Apportionment Compacts Between States," ch. 46 in R. Beck, *Waters and Water Rights*. Michie Co., Charlottesville, VA.

Griffin, W. (1958). "Legal Aspects of the Use of Systems of International Waters." *S. Doc. No. 118*, 85th Cong., 2d Sess. 63.

Harlow, C. (1994). "If At First You Don't Succeed: Resolving Water Quality Issues in the Lake Tahoe Basin Through Interstate Allocation and Regional Development Compacts." Student thesis, J. Reuben Clark Law Library, Provo, UT.

Muys, J. (1976). "Allocation and Management of Interstate Water Resources: The Emergence of the Federal-Interstate Compact." *Den. J. Int'l Law & Pol'y, 6*, 307.

Rice, L & M. White, (1987). *Engineering Aspects of Water Law*, ch. 5 Wiley & Sons, NY, NY.

Sherk, G. (1989), "Equitable Apportionment After Vermejo: The Demise of a Doctrine." *Natural Res. J., 29*, 565.

Simms, R., L. Rolfs & B. Spronk (1988). "Interstate Compacts and Equitable Apportionment." *Rocky Mtn. Min. Law Inst., 34*, 23-1.

Tarlock, A. (1985). "The Law of Equitable Apportionment Revisited, Updated, and Restated." *Univ. Colo. Law Rev., 56*, 381.

Utton, A. (1991). "International Waters," part V in R. Beck, *Waters and Water Rights*. Michie Co., Charlottesville, VA.

Agreements for Sharing Transboundary Water Resources: Hydrologic Information Requirements

William E. Cox[1]

Abstract

Agreements for sharing transboundary water resources must be based on adequate hydrologic information consisting of definition of resource characteristics and the human demands and impacts affecting the resource. Management programs generally are forced to operate without complete hydrologic information, but remedying such defects should be a basic objective of management operations.

Introduction

The need for interjurisdictional agreements for sharing of transboundary water resources arises whenever interference with natural hydrologic processes on one side of a political boundary produces adverse effects on the other side of the boundary. These interferences often take the form of water withdrawal or wastewater discharges but can also include other interventions such as artificial precipitation augmentation. Agreements over use of transboundary water resources generally prescribe rules for sharing use of the resource to reduce conflict.

Such agreements can involve a variety of parties and take many forms, but they must be based on adequate understanding of the resource. Agreements not based on accurate resource characterization cannot produce intended results but will fail as expectations based on the agreement are not met.

Factors Affecting the Scope of Information Requirements

Hydrologic information requirements associated with interjurisdictional agreements tend to have broad scope. One factor responsible for this broad scope is the tendency for water management agreements to be implemented through governmental regulation of water use, either by an interjurisdictional management body or by the individual units of government involved. Regulatory approaches

[1] Professor, Civil Engineering Department, Virginia Tech, 200 Patton Hall, Blacksburg, VA 24061-0105.

require an omniscient agency if the net social benefit produced by the water resource is to be maximized. To achieve this result, a management agency must have substantial information about resource capabilities and limitations as well as about the benefits and costs of alternative water uses.

The scope of the hydrologic information requirement for transboundary water management is similar to that associated with single jurisdiction water management. In general, an individual governmental unit totally encompassing a river basin or ground water system needs the same hydrologic information to make management decisions as would an interjurisdictional management body. The management agency in the interjurisdictional case may need somewhat more precise information about the location of potential water development impacts in areas near boundaries in order to ensure compliance with agreements; however, this difference is primarily one of detail and not in type of information needed.

The transboundary case does differ from the single jurisdiction case with respect to complexity of data collection and management. In the transboundary case (especially where different nations are involved), data collection and management practices may be substantially different. Some jurisdictions may have little hydrologic information because of lack of monitoring or other program deficiencies. Among those with information, differences in form and storage technology may produce incompatibilities requiring substantial effort to mitigate.

The requirement for hydrologic information associated with transboundary water management generally encompasses two information categories: (1) information for defining the dimensions and characteristics of the resource and (2) information for determining the demands and impacts of human activities on the water resource. This second category may not appear to fit within a strict definition of "hydrologic information," but water use and other human activities that affect the behavior and availability of water are just as important as natural phenomena in determining future resource potential and problems. These two categories are considered separately in the next two sections.

Definition of the Resource

The viability of a transboundary water-management agreement depends directly on an adequate definition of the resource. General climatic data are part of the definition. Precipitation is a direct input into activities such as rainfed agriculture and is the basic source of runoff to streams and recharge to ground water systems. Evaporation is an important mechanism for loss of water from its liquid form. Other climatic conditions such as temperature and winds are important because they affect hydrologic processes such as evaporation and also are determinants of water demand.

Streamflow is a basic source of supply in most areas that must be defined in terms of flow rates and their variability over time. While the total amount of streamflow available in the long term is an important parameter, its variability is a key factor in assessing the resource. The occurrence and duration of low flows limits the water-supply potential of a stream, or at least indicates the need for investment in storage facilities. The magnitude and duration of high flows indicates the flooding threat and the need for flood damage reduction measures. Quality of surface water can limit its utility and therefore also must be determined.

Definition of the ground water resource requires determination of the amount of water in storage and the rate at which it is recharged. While the recharge rate limits the long-term rate of withdrawal, the existence of large amounts of water in storage creates the potential for use in excess of recharge for a finite time period. Aquifer characteristics such as depth below the surface, thickness, and transmissivity also are necessary to allow determinations of desirable well locations and pumping rates. Since ground water quality can vary significantly with location and depth, resource definition also requires quality information. Collection of ground water data is generally difficult and costly because of the need for well drilling and installation of specialized equipment.

Data for defining climatic and water resource conditions must be collected through networks of monitoring stations designed to reflect spatial variability and operated over time periods that adequately indicate temporal variability. Because of variability over time, data becomes meaningful for management purposes only after statistical analysis for determining probabilities of occurrence of different events such as rainfall amounts or streamflows. The reliability of statistical indicators of hydrologic phenomena depends not only on adequacy of data collection but also on the extent to which past events are an accurate indication of the future. The possibility of climatic change as a result of changes in atmospheric gases or other factors could invalidate the assumption that past patterns of variation can be used to approximate future occurrences.

Determination of Water Demand

Just as important as defining the natural characteristics of the water resource itself is determination of resource use. Water problems are the result of interaction between supply and demand, and management requires information about both natural water availability and water use. The definition of offstream water demand has several dimensions. The quantity of water withdrawn and the timing of withdrawals are basic data needs, but also important are return flow amounts, locations, and quality. Water-use data also should include information about water-use activities since management programs may impose water-use criteria or other measures that discriminate among different types of water use. Information about land use is also necessary to define baseline conditions since changes can affect runoff and other hydrologic parameters.

Water-demand data also must include instream water use such as recreational activities and maintenance of fisheries. These demands historically have been overlooked in assessing demand but have increasingly been recognized as valid water uses for which minimum flows must be maintained.

Consideration of Future Conditions

Essential hydrologic data is not limited to current conditions but includes both water supply and demand conditions anticipated in the future. Current water supply conditions are usually considered to be an indication of future conditions, but potential modifications over time must be considered. Changes in land use can alter ground water recharge and runoff patterns. Changes in water quality can also impact supply availability. Particularly significant is aquifer contamination with potential to eliminate ground water supplies for substantial periods of time. Water supply projections must evaluate the potential for such supply losses.

Water demand changes also must be projected. Past trends in demand were once considered to be reliable indicators of future demand, but the generally accepted approach at present does not extrapolate past trends. Instead, consideration is given to possible changes in underlying factors influencing demand, including changes in policies intended to reduce demand.

Conclusion

Adequate hydrologic information is a fundamental prerequisite for development and implementation of effective transboundary water resource agreements. Information needs for interjurisdictional water management are not significantly different than they are in the case of management by a single governmental unit although the mechanics of collecting and making use of the information in the transboundary case are more complex. Necessary information includes data that defines the water resource and its natural behavior but also encompasses present and projected water-use data. Management of the water resource must focus on actual operation of the hydrologic system as impacted by human activities.

But the lack of complete data should not serve as an excuse not to develop transboundary management programs. Data collection is an ongoing process that is never complete. Identification of major information deficiencies and the initiation of necessary remedial actions are themselves important water management activities. In the case of transboundary water resources, an assessment of information adequacy on a hydrologic-system basis may await the creation of interjurisdictional agreements and implementing organizations. Development of transboundary agreements therefore is itself an important part of the process for addressing hydrologic information needs.

HYDROLOGY, ENVIRONMENT AND THE SHARED USE OF WATER

Nathaniel D. McClure IV[1] M.ASCE

Abstract. Hydrology, (streamflows, stages, lake levels) is related to environmental factors (fish and wildlife habitat, biodiversity, spawning and propagation of species). Methodologies have been developed to portray these interrelationships. Increased emphasis on the "environment" has elevated its importance in the analyses of alternatives in water resources decisionmaking related to the shared use of water.

Introduction. Since 1970 environmental awareness and concern has grown in sophistication and intensity. Federal laws such as the National Environmental Policy Act and the Endangered Species Act mandate informed decisionmaking and environmental protection. Litigation has reinforced the need for sound technical evaluations, appropriate coordination and adequate documentation.

Previously, environmental implications of water resources activities received limited consideration. Other uses such as flood control, hydropower, navigation, water supply, and irrigation were the principal focus. Many of the complex interrelationships between hydrology and the environment were not well understood. Increased environmental emphasis promoted the need for better ways to analyze and display environmental ramifications. Improved technology and scientific knowledge led to the development of methodologies including sophisticated computer models which simulate, predict, measure and promote a fuller comprehension of the consequences and

---

[1]Chief, Planning and Environmental Division, Mobile District, U.S. Army Corps of Engineers, P.O. Box 2288, Mobile, AL 36628-0001

tradeoffs involved. The focus is now moving toward a more holistic approach to addressing water resource issues. The State of Florida, for example, adopted "Ecosystem Management" to guide environmental planning and regulatory activities (Minasian 1994). Watershed planning, environmentally sustainable development, ecosystem analyses and system-wide management are now common terms.

Two studies by the US Army Corps of Engineers provide insight into hydrologic and environmental relationships and illustrate various methodologies being used. The Missouri River Division (MRD) is conducting a study to review and update the Master Water Control Manual for the Missouri River (MRD 1994) and the Mobile District, in partnership with the States of Alabama, Georgia and Florida, is conducting a comprehensive water resources study of the Alabama-Coosa-Tallapoosa (ACT) and the Apalachicola-Chattahoochee-Flint (ACF) basins. An interdisciplinary approach synthesizing engineering and life/natural sciences skills, in a collaborative fashion, creates mutual understandings and tools that support more informed and balanced water resources planning and management. The two studies illustrate the breadth and complexity of the issues and highlight methodologies being applied.

MRD Study. The report containing output data from models on the array of alternatives considered in the Draft Environmental Impact Statement states:

"Understanding system hydrology, defined in terms of total system storage volume, lake levels and water releases, is crucial for evaluating the impacts that the operational alternatives will have on system beneficial uses such as fish, navigation and recreation." (MRD 1994)

Results are presented in the following resource categories: system hydrology (lake levels and river flows), wetland habitat, riparian habitat, tern and plover nesting habitat, young fish production in mainstem lakes, lake coldwater fish habitat, river coldwater fish habitat, river warmwater fish habitat, and native river fish physical habitat.

The hydrology and predicted responses to alternative water control plans were analyzed using the Long Range Study (LRS) Model. The LRS model was supplemented and revised to aid in the study (Patenode and Wilson 1994). An Environmental Impacts Model assesses key environmental resources for each alternative (MRD 1994). For example, wetlands and riparian habitats are influenced by reservoir and river hydrology. The model evaluates vegetative response to changes in monthly water tables for 13 cover types. Predictions of the general trends in endangered least tern and threatened piping plover nesting habitat employed a special model which tracked annual acreage of nesting habitat. The relationship between acres of habitat and flow was estimated using field verified aerial video-tapes. (Tressler etal 1994).

The analysis of young fish production in the main-stem lakes used a statistical regression model relating production and abundance indices of young fish to hydrologic factors, i.e., lake elevation, inundated area, inflow, discharge and flushing rate (MRD 1994). The analyses of lake cold water fish habitat was based on the extent of cold water retained in the reservoirs through the summer and fall seasons. A temperature regression model predicted effects on cold water river habitat downstream of the mainstem lakes. The habitat was measured in number of miles of river with a suitable temperature regimen. Warm water fish habitat was predicted in a similar fashion. Native river fish evaluation utilized an examination of physical habitat based on cross-sectional flow pattern comparisons.

ACT/ACF Study. The ACT is in Georgia and Alabama and the ACF also involves Florida, terminating in Apalachicola Bay. (McClure and Griffin 1993). The hydrological analyses will use HEC-5 models developed for each basin. The models predict the hydrologic responses associated with reservior operational scenarios and predicted water demands. Environmental analyses address wetlands and riparian water needs, protected species, riverine fauna water needs, reservoir fisheries water needs, and water needs of fish and wildlife management facilities.

Research conducted in the MRD study was utilized, especially in the fishery area. Wetlands and riparian resources are being evaluated using an ecological characterization of the resource (biological, physical and hydrological). Potential effects to protected species will be evaluated by identifying potentially limiting ecological criteria and habitat requirements, followed by predicting potential impacts from stage/flow alterations based on the known life histories and habitat requirements.

The Freshwater Needs Assessment of the Apalachicola River and Bay is designed to determine what is needed to preserve the existing environmental diversity and productivity of this valuable ecological resource (TCG 1993). An interdisciplinary team is conducting physical, biological and chemical studies to assess the system's needs. A three dimensional model will be used to determine circulation, flushing and transport (nutrients) characteristics of the bay.

Conclusions. Water use and management strategies affect the hydrology and the environment. Analytical tools used to predict the biophysical ramifications of these activities are evolving toward a more holistic/ecosystem approach. Decisionmaking should utilize this evolving technology in establishing environmentally sustainable and balanced water management strategies.

References.

McClure, N.D. and R.H. Griffin, "A Partnership Approach to Address and Resolve Water Resource Conflicts," Proceedings of Conserve 93, ASCE, AWRA, AWWA, Las Vegas, Nevada, December 1993.

Minasian, L., "Interpreting and Applying Ecosystem Management Principles," Environmental Exchange Point, Florida Department of Environmental Protection, Tallahassee, Florida, November 1994.

MRD, Master Water Control Manual, Missouri River, Review and Update, Volume 1: Alternatives Evaluation Report, Missouri River Division, US Army Corps of Engineers, Omaha, Nebraska, July 1994.

Patenode, G.A. and K.L. Wilson, "Development and Application of Long Range Study (LRS) Model For Missouri River System," Proceedings of ASCE Water Resources Planning and Management 21st Annual Conference, Denver, Colorado, May 1994.

Technical Coordination Group, "ACF/ACT Comprehensive Study Newsletter" Alabama, Georgia, Florida and US Army Corps of Engineers, Mobile, Alabama, Summer 1993.

Tressler, R., R.Fairbanks, D.Latka, J.Glassley, K.Engel and J.Rude, "Missouri River Least Tern and Piping Plover Model Development," Proceedings of ASCE Water Resources Planning and Management 21st Annual Conference, Denver Colorado, May 1994.

Economic Considerations Relevant For The Shared Use Of Transboundary Resources

Ronald G. Cummings[1]

Abstract

In assessing alternatives for managing transboundary water resources, it is often desirable to estimate economic values that result from water use. Since most water "prices" are cost-based, economic values must be estimated. A brief sketch is given of the strengths and weaknesses of three methods for estimating economic values. Questions are raised concerning the potential viability of water markets in eastern, riparian states.

Introduction

The concern of this paper is with economic considerations relevant for assessing the value of alternative uses of water supplies by parties to any dispute regarding transboundary water resources. For such assessments, we may think of any alternative water use in terms analogous to a "good" that provides utility or satisfaction to individuals. There are basically two classes of water "goods" that are relevant for our valuation purposes: priced and non-priced goods. A priced good is one for which prices are established in markets; in some cases, a priced good can be one for which a market value can be inferred from prices for other goods. A *non*-priced good is not traded in markets and, therefore, does not command a market price. Obviously, there are many instances where clear distinctions do not exist for goods that are of the priced and non-priced genre.

What makes the task of assigning values to goods--mostly services--associated with water use is that there are very few such goods that are traded in markets and therefore have market prices that can be used to value them. Water used for municipal uses is typically priced at the cost of service--for obvious reasons, such cost underestimate the value of water put to this use; this is particularly the case in instances where water prices are based on declining or uniform block rates. It then follows that observed differences in water prices in

[1]Professor of Economics, Georgia State University, Atlanta, GA 30303.

different municipalities cannot be taken as reflecting differences in economic value. Cost-based "pricing" of water is also common for most other water uses. Irrigators typically pay only the costs of facilities required to acquire water. For those who obtain water from federal projects--navigation, recreation, hydropower, irrigation, and recreation as examples--payment for water is limited by statue to the recuperation of project cost. All of this is to say that in general observed "water prices" are *not* prices for water *per se*--prices which might reflect values derived from the use of this resource--they are prices that reflect costs of acquiring water. Exceptions are seen in prices for water resulting from the exchange of water rights in water rights markets that function in many Western States.

If economic values are to be used in the process of assessing the implications of alternative policies or strategies for managing transboundary water resources it then becomes necessary to estimate such values.[2] It is generally the case that one or a combination of three methods are used for this purpose: the hedonic price method; the travel cost method; and the contingent valuation method. In what follows I will briefly describe each of these methods and comment on their relative strengths and weaknesses.

Measuring Economic Values Associated With Water Use

The hedonic price method attempts to derive a price-quantity--demand--relationship for water from the demand relationship that applies to the goods or services that are produced with water. Thus, for example, suppose that we wish to value water quality in a lake. We identify several lakes that differ in terms of water quality. They may also differ in any number of other ways, such as size, climate, and proximity to urban areas. We have information on homes on each of these lakes that have been sold in the past. Differences in prices for which these homes have sold could be expected to reflect many different things; examples include lake differences noted above; size of the home; number of bedrooms, and, most importantly for our purposes, water quality. With the hedonic price method, one uses statistical methods in an effort to attribute portions of a house's value to each of these characteristics. The idea here is that the value of the house is the sum of the values associated with each of these characteristics of the house. Of course, our interest is with the value attributed to the characteristic "water quality." Given sufficient variation in water quality,

[2]Basic to negotiations concerning mechanisms that are to guide the shared use of transboundary water resources are comparisons of values generated by water uses among the affected parties. It should be clear that if such comparisons are to be meaningful there cannot be substantial differences in the extent to which the affected parties are committed to principles of conservation. The failure of any party to conserve water would have the effect of rewarding waste--the non-conserving party's water "needs" are inflated by waste.

a demand relationship can be sketched out and economic value can then be estimated. The strength of this method is that it makes use of observed market values and, all else equal, can be thought of as providing defensible measures of maximum willingness to pay. Its major weaknesses or limitations include: it is often difficult to obtain the variation in the characteristic of interest necessary to trace out a complete demand relationship; its use is predicated on there being a market-priced good that has as one of its characteristics the resource in question-- in our case, water quantity or quality.

The travel cost method is based on the notion that the value of a resource is implied by costs incurred by individuals in order to have access to the resource. In the simplest terms, the method surveys resource users to the end of determining such things as trip expenditures and travel distance. Given sufficient variation in visitor's place-of-origin, one can estimate the relationship between travel costs and visitation. This relationship forms the basis for the estimate of a demand relationship and, therefore, for the estimate of economic value. The primary strength of the method is that resulting measures are indeed market based in the sense that they reflect observable expenditures by the surveyed population. Its major weakness or limitations are: a host of potential biases can arise from the manner in which such things as multiple-purpose trips, substitute sites, and travel time are treated in the analysis; and applications of the method are limited to resource uses that require visitation. In the water resources area, it's use is seen primarily in efforts to estimate recreation values.

Finally, like the travel cost method, the contingent valuation method also makes use of surveys. However, with this method surveyed individuals are directly asked their maximum willingness to pay for the resource. Thus, in the crudest terms, if we want to know the value that individuals place on a particular wetlands area, we simply survey a sample of individuals in the relevant population and ask them their maximum willingness to pay (*via*, e.g., higher taxes) to improve, maintain, or prevent the degradation of the wetlands area. The major strengths of this method are that, all else equal, one can indeed obtain the economic values of interest for our purposes, and the method does not have the application limitations noted for the hedonic price and travel cost methods. The method has a number of *arguable* weaknesses or limitations: values derived with this method *may be* susceptible to substantial biases that arise from (as but a few examples) the hypothetical nature of "payment" and the failure of subjects to adequately consider budgetary substitutes. Some insight as to contemporary concern with the reliability of values estimated with this method is seen in the recommendation by the Department of the Interior and the National Oceanic and Atmospheric Administration that one "calibrate" such values by multiplying them by the factor .5.

Concluding Remarks

In concluding my very brief comments regarding economic issues relevant for assessing the impacts of alternative methods for managing transboundary water resources, I wish to leave you with two (of many possible) thoughts regarding water markets which I hope you may find provocative. Let me make clear at the outset that in posing these questions I do *not* intend to advocate the use of markets; the questions are posed simply for the purpose of stimulating debate along lines that I feel might be productive.

A number of writers, primarily legal scholars, have noted the extent to which evolving water laws in the riparian East and the prior-appropriation West appear to be converging. With the "regulated-riparian" system evolving in Eastern States, we then see the evolution of circumstances which *could* provide one of the basic requisites of an effective market system: fungible water rights. While, unlike conditions extant in a prior appropriation system, a regulated-riparian system does not establish a system of property rights in water, its water use permits could indeed be accorded the status of a marketable usufructuary right. This being the case, I submit the following questions as warranting thought and debate. In doing so, I emphasize the need for debate; I am not so sanguine that I fail to recognize the myriad regulatory and distributive questions that arise in considerations of water markets. First, as an alternative to (e.g.) two states quarreling over the division of an interstate stream, can we think of a *regulated* water market system that effectively allocates water to highest valued uses for both states? Second, it is often the case that the source of a large part of contention arises from the question as to how water is to be allocated during periods of drought; could such contention be ameliorated, if not eliminated, by something akin to a futures market wherein particularly vulnerable entities (such as municipalities and industry) can purchase calls on water resources during such periods from willing, less vulnerable holders of water use permits?

A REVIEW OF WESTERN WATER SHARING AGREEMENTS

OLEN PAUL MATTHEWS[1] and ZACHARY McCORMICK[2]

ABSTRACT

The twenty-two interstate water compacts allocating water between western states provide background information needed for devising future allocation strategies. Three aspects of allocation compacts are significant: 1. Scope--which water and what parties are included? 2. Allocation method--who bears the risk of shortage? and 3. Management system--how are decisions made and disputes resolved?

INTRODUCTION

In the western United States 22 compacts set up a mechanism for allocating surface water. The lack of bargaining incentive and fears of lost state sovereignty, limit the nature and scope of these agreements. Examples of limitations include ignoring the hydrologic connection between groundwater and surface water as well as omitting the relationship between water quality and quantity. In spite of the limited nature of these compacts(or perhaps because they are limited), the western compacts provide a manageable case study on what works and what does not in such agreements. The scope, allocation method, and management system will be examined below.

SCOPE

To prevent conflicts on an intergovernmental level, the scope of compacts should be comprehensive. All the water resources in the river basin and all claims to that water must be included. Major omissions in existing

---

[1]Director, Environmental Institute, Oklahoma State University, Stillwater, OK 74075
[2]Consultant, Albuquerque, NM 87111

western compacts include ignoring hydrologically connected groundwater, failing to include federal and Indian claims, overlooking basin exports.

Most western compacts focus on allocating surface waters. Only the compacts on the Big Blue, Upper Niobrara, and Bear Rivers even consider groundwater (McCormick, 1994). Of these, only the Big Blue Compact (86 Stat. 193, 1972) specifically limits groundwater pumping if surface flow is affected. Because overdrafts of groundwater can result in reductions in surface flow, compact allocations which ignore groundwater can result in unintended consequences. In such situations the upstream state has the advantage as can be seen in the dispute between Colorado and Kansas on the Arkansas River. In litigation before the U. S. Supreme Court, a Special Master's report states that groundwater pumping in Colorado has an impact on water allocated under the compact and recommends that Kansas prevail on this issue (Kansas v. Colorado, No. 105 Original). Including groundwater withdrawals in compacts may not always be technically easy, but leaving groundwater out can cause problems.

Existing western compacts generally omit any quantification of federal and Indian claims, but several do try to bind the federal government or charge potential federal rights to a state. Federal and Indian reserved water rights, federal navigable servitudes, and federal regulatory needs, including environmental ones, all influence the amount of water available for allocation. Many of these rights have not been quantified or even asserted. Because state water rights are often subservient to federal and Indian rights, compact allocations may fail to achieve their intended purpose. If these rights could be quantified in advance of compact negotiation, a degree of certainty could be introduced. Even if this is done however, Congress is not precluded from passing new regulations or asserting new rights. Another, perhaps better, approach is that taken in compacts on the Belle Fourche, Republican, and Klamath Rivers. In these compacts, Congress has acquiesced to conditions binding it to the rights established by compact allocation. If these rights are impaired, compensation will be paid. In a few compacts federal water use will be charged against the state where the use occurs. These states assume the full risk of future federal assertions.

The Yellowstone River Compact (65 Stat. 663, 1950) bans exports outside the basin unless all states party to the compact agree. Because water is considered an article of commerce, out of state export bans are generally precluded by the constitution. An exception to this general prohibition occurs with congressional consent. Consent is given when Congress approves the compact.

ALLOCATION METHOD

Four major methods of allocation are found in western compacts. The major methods are storage allocation and three kinds of flow allocation including models, percentage of flow, and guaranteed quantities. Choice of allocation method depends on the state's goals and how they want to divide the risk of shortage.

Storage allocation is the simplest method with each state being limited to the amount which can be stored. The downstream state receives only what is in excess of the upstream state's storage allowance, plus water originating between the storage facility and the state boundary. In this case, the downstream state assumes the risk that water volumes in excess of the upstream storage allowance will occur. This method seems easy to monitor and enforce, but the compact language may not always be clearly drafted and can cause litigation.

Although hydrologic models have the aura of scientific certainty, they have been less than successful. Models fail to describe the hydrology of a river accurately and comprehensively enough and fail to account for changes resulting from natural causes. In theory, a model will describe the conditions of the river at the time the compact is made and allow the risk of dry years to be divided between the states. Model inaccuracy makes other methods more practical at this time.

Allocating water based on the percentage of flow allows the states to share the risk of dry years in proportion to their allocated percentage. This method is easy to monitor and seems a fair way to divide risks between states, but it may interfere with rights of an individual appropriator. If compacts are established in advance of a water crisis, then this problem should not occur. Pre-compact rights could be protected by limiting the allocation between the states to post-compact diversions and uses.

Guaranteed quantities require a fixed minimum be delivered by the upstream state. In this allocation method the entire risk of annual variation falls on the upstream state. Although monitoring is simple the system may not be fair during times of severe drought. In modified form this method might be more acceptable. On the South Platte River, Colorado has agreed to deliver a minimum quantity of water as long as sufficient volumes are available. If a shortage occurs, Colorado will take steps to limit consumption. This splits the risk of shortfall between the two states.

MANAGEMENT SYSTEM

The third major variable relates to how decisions are made and includes the power of compact commissions (if they exist), voting structure, and how disputes are resolved if parties disagree. Problems managing rivers which cross state boundaries are usually not related to a single event or point in time. Continuing management is required because the river is an interrelated system. But western compact commissions are created by political compromises which very seldom allow the commission powers which would detract from state sovereignty. This is done by limiting the subject matter over which a commission has power. Even when exercising what power they have, a unanimous vote is often required. Even when less than a unanimous vote is allowed, the scope of authority is limited and generally excludes the ability to change the method of allocation. Two compacts allow a federal representative to cast tie breaking votes, but federal representatives are loath to take the side of one state against another. Arbitration and litigation are of course possibilities, but these are not generally mentioned in western compacts.

CONCLUSION

The experimentation western states have done with different methods of allocation provides good information for the future, especially when put in terms of risk allocation. But, western compacts are limited because their scope is not comprehensive and the management structure is weak. This is a result of political reality and the desire for states to retain sovereignty over "their" water. Even with these limitations, existing compacts can provide useful examples, good and bad for future allocations.

REFERENCES

McCormick, Zachary (1994), "Interstate Water Allocations in the Western United States--Some Suggestions", Water Resources Bulletin 30: 385-395.

A Current Case Study: The Rio Grande and the El Paso Tri-state Area

Conrad G. Keyes, Jr., F.ASCE, Boyle Engineering, El Paso
Edmund G. Archuleta, PE, Gen. Mgr. El Paso Water Utilities (EPWU)
J. Frederick Burns, M.ASCE, Boyle Engineering, Albuquerque

ABSTRACT

The presentation will discuss the history and working arrangements of the New Mexico / Texas Water Commission (Commission), a new regional entity which has been created to address the water resource management issues that have been evolving in a U.S./Mexico border area of the arid Southwest. This historic region is on the Rio Grande where three state boundaries intersect: Chihuahua, Texas, and New Mexico. The Rio Grande normally forms the boundary between three burgeoning cities in the region: Ciudad Juarez, El Paso, and Las Cruces, respectively. The flood plain of the river has been long developed as a rich agricultural area under the auspices of one of the earliest reclamation projects to be constructed, The Rio Grande Project, which is known to have "paid off" all of their obligations to the U.S.. Their exists a maze of local, state, national, and international regulatory authorities with overlapping jurisdictions over water resources with which the new Commission has to deal in their efforts to address the issue of meeting changing water demands for an urban population which is rapidly growing on both sides of the border. The status of the Commission's efforts to address the physical requirements as well as the resource management hurdles are discussed.

INTRODUCTION

Around 1980, EPWU determined that inter-state wells could be constructed in New Mexico to supplement their available water resources for municipal and industrial (M&I) uses by the year 2000. This was challenged by New Mexico on the basis that water could not be legally exported across state lines. After the Supreme Court over-turned this premise in their finding that water was legally a commodity of inter-state commerce, the State Engineer of New Mexico, after lengthy, contentious hearings in 1987 and 1988, denied the well permits on the basis that El Paso had not demonstrated the need for additional water supplies for the next 40 years. He also declared the two aquifers of the region as "closed basins" in the State of New Mexico. A prolonged litigation ensued.

In 1989, EPWU in response to Mr. Reynolds findings, embarked on a major water resource management study effort. The EPWU Water Resource Management Plan, resulting from the study, identified that the Hueco Bolson would be effectively depleted of usable water around 2020. New sources of water would be required, even with rigorous conservation and water reclamation and reuse. Accordingly, EPWU and the other parties to the litigation, in an historically unprecedented move, negotiated a litigation settlement agreement that provided a mechanism for the water resources, both surface and ground water, to be considered on a regional basis, according to needs.

The settlement agreement provided that EPWU was to first reduce their water demand by implementing a rigorous conservation program. El Paso has responded by acting into law one of the most comprehensive conservation ordinances in the nation. In the two years it has been in effect, the results have been good. The target set by the EPWU's governing board is a 20% reduction by the year 2000.

The second priority stipulated in the agreement was to use available surface water of the Rio Grande Project for municipal purposes while maintaining the viability of the existing agricultural users. As agricultural lands are urbanized, water usage will be transferred to M&I purposes. The third priority of water resources is to be ground water.

One of the most profound conditions of the settlement agreement was the establishment of the Commission to deal with water issues on a regional basis. In the short period since organization by parties to the litigation, the Commission has embraced participation by all the major water using agencies of the region. Under their auspices, studies have been on-going on the methods and means of increasing efficiencies of water allocations toward the end of increasing the availability of water for all uses. Conjunctive use of all water sources are being studied, along with construction of major conveyance facilities which will enhance water availability quantitatively and qualitatively.

THE RIO GRANDE PROJECT

The Rio Grande Project was authorized by the U.S. Congress around 1915. The keystone of the project was Elephant Butte Dam on the Rio Grande built in 1916, with a reservoir storage capacity of over 2.6 million acre-feet. Six downstream diversion dams were constructed in the 130-mile downstream reach of the river: Percha, Leasberg, Mesilla, American, Internacional, and Riverside. Canals, drains, laterals, and wasteways were constructed to serve about 160,000 acres in the United States and over 20,000 acres in Mexico. Later, in 1937, Caballo Dam was built to provide regulatory capability of power generation releases and flood control. The operations of the project was delegated to two irrigation districts, the Elephant Butte Irrigation District (EBID) in New Mexico and the El Paso County Water Improvement District No. 1 (EPCWID#1) in Texas. The Mexican allocation was specified by a 1906 treaty to be 60,000 acre-feet per annum delivered at the Acequia Madre at the Presa Internacional.

Up until the middle of the century, the natural water resource evolution was that all surface water resources would be used for agricultural purposes, supplemented by shallow well pumping in periods of drought. The surface water resources of the Rio Grande are the only truly renewable resources in the region. Ground water became an attractive alternative for M&I uses because of the relatively high quality, availability near the point of use, and the requirement of minimal treatment. The municipalities have utilized ground water pumped from the Mesilla Bolson in the Rio Grande Valley of New Mexico and the Northwest area of El Paso County, Texas, and the Hueco Bolson in Texas and Mexico. The Mesilla Bolson is a prolific aquifer underlying the Rio Grande from Caballo Dam, downstream passed the Mexican Border. It is hydraulically linked to the river which provides automatic recharge to maintain a basically full condition except in areas of concentrated well pumping where cones of depression occur due to stressing the transmissivity of the aquifer. Two such major cones of depression are now existing in the area of the Las Cruces well field and in the area of the Canutillo well field in West Texas.

The Hueco Bolson underlies a natural basin east of, and parallel to, the Rio Grande Rift. There is no defined watercourse overlying this underground basin, thus the natural recharge is limited to mountain front recharge. The aquifer is being mined by the cities of El Paso and Juarez at an ever increasing rate. Projections now indicate that at present trends all water of usable quality will be used before the year 2025. This reality is what lends the urgency to the development of alternative water resources for the region.

In the 1940's, El Paso initiated their first use of surface water by the construction of the Canal Street Water Treatment Plant, which was later expanded to a capacity of 40 MGD. In 1994, another plant, the Jonathan W. Rogers Water Treatment Plant was placed in service with 40 MGD capacity. The water allocated to operate these plants are under agreement with the EPCWID#1 to use water of lands in the district which has been urbanized. This is only minimally satisfactory because water with which to operate these plants is only available during the normal seven-month irrigation season. To truly convert the municipalities in the region to the use of a sustainable, renewable water source, present operational policy and regulations will have to be changed to provide year-round availability along with water quality protection from contamination by municipal wastewater effluent and agricultural return flows.

THE POLICY MAZE

The Rio Grande watershed encompasses three states in the U.S. and one foreign country, being Colorado, New Mexico, Texas and the Republic of Mexico. In the U.S., the interstate policy issues are governed by the Rio Grande Compact Commission, with representation from each state and one federal member. The International Boundary and Water Commission deals with policy along the reach of the river which forms the Mexico / U.S. border. The laws and policies which govern water rights policies in the two U.S. states are the State Engineer in New Mexico, and the Texas Natural Resources Conservation Commission in that state. Water resource planning in Texas is under the jurisdiction of the Texas Water Development Board, and in New Mexico by the State Engineer and the Interstate Streams Commission. The Bureau of Reclamation is the operating agency of the Rio Grande Project. The two irrigation districts, EBID and EPCWID#1 operate the respective irrigation works.

The newly formed Commission is not intended to add to the confusion of jurisdictions, but rather be a representative of the user agencies of the project water and of the various ground water resources. It is their intent to proceed, with proper study, to develop and build consensus of methods and means by which the finite water resources of the region may be enhanced in both quality and quantity to the benefit of all user agencies. It is their intent to protect the public's interest by providing thoughtful consideration of environmental issues effected by the project operations. The Commission has formed two standing committees: the Management Advisory Committee, and the Legal and Environmental Advisory Committee. These have proven to be hard-working groups which are making significant progress toward solutions to improve the overall water resource availability in the region.

REGIONAL POPULATION GROWTH

The tri-state area around El Paso has been experiencing rapid population growth, much of which is attributable to the border economy. The Maquila program has fostered migration within Mexico to Ciudad Juarez. The population of that city is near 1.5 million. El Paso has experienced a 2.5% growth, and Dona Ana County, New Mexico a 2.8% growth. The stimulus of NAFTA is unknown, but it is anticipated to further population growth.

FINDINGS OF STUDIES TO DATE

The engineering studies performed under the auspices of the Commission to date, along with the efforts of the standing committees have been significant. The studies have been progressively more detailed in nature, the first phase of which was to identify the potential for surface water savings by constructing structural improvements to the antiquated conveyance system and by operational changes. These studies identified the potential of creating over 180,000 acre-feet per annum more usable surface water. Secondly, conceptual works have been identified to serve multiple delivery scenarios, dependent upon what groups opted to participate in the project, and the costs associated with participation. A study has been performed to further enhance water resource availability through the mechanism of aquifer storage and recovery of surface water to utilize spillage losses presently encountered in abnormally wet years. Independently, the EPWU has initiated pilot plant studies of desalinization of the abundant moderately salty water in the Hueco Bolson.

Many of these strategic changes in water resource development will require fundamental statutory changes to implement. The Commission's legal and environmental committee is evaluating the impacts and requirements of such changes if they are deemed to be in the interest of the public to implement. Formal discourse with agencies of Mexico may be initiated pending identification of a final strategic policy consensus by the U.S. user agencies.

PROJECT STRUCTURAL NEEDS IDENTIFIED

The improvements identified on a conceptual basis consist of improvements to the existing canal system by concrete lining, shaping, connecting for continuity, and increasing flow capacity to convey surface water releases from Caballo Dam. Metering, control and telemetry would be included to better meet the agricultural users calls for water as well as the increased M&I demands. Protection of the relatively high water quality upstream can be achieved by isolation from contamination by municipal waste water, drain and return flows from agricultural operations. The dramatic reduction of transport time could eliminate spills due to weather changes between calls and delivery. Two major water treatment plant facilities are envisioned in the Mesilla Valley to serve this rapidly urbanizing area.

The construction costs for the entire project range upward from one-half billion dollars. Funding at this scale is thought to be impracticable for a single project; therefore, the Commission is considering a phased and staged approach concomitant with the population growth projections. To maximize the efficacy of each phase, it is perceived that the project should logically be commenced near American Dam, in El Paso, and proceed upstream. The works required downstream of the American Dam consist of channel improvements and extension of the American Canal between the American Dam and the Riverside Dam. This is a project already funded by Congress, with construction started in 1994.

Presently, the Phase I project could be subdivided into four stages of construction to logically fit the irrigation schedule of the existing canal. Stage A of Phase I would consist of a buried conduit approximately five miles in length through the most urbanized and industrialized reach of the project. The reach has drain inflows with a high salt load and the conduit would connect directly to the headworks of the American Canal. The American Canal is the direct feed to the EPWU's Canal Street Water Treatment Plant, and with the canal improvements in process will also feed the new Jonathan W. Rogers Water Treatment Plant near Riverside Dam.

SCHEDULE AND FINANCIAL ARRANGEMENTS FOR COMPLETION

The environmental and feasibility studies could begin in 1995. For adequate financial backing, the complete engineering planning report needs to be accomplished in 1995. The preliminary design is anticipated to begin sometime in 1996, if the agency approvals are in order for such to occur.This way, application can be processed in the 104th U.S.Congress, with the Border Environment Cooperation Commission, and with federal and state agencies in the tri-state region around El Paso. Obviously, environmental assessments, preliminary engineering with all parties, public involvement, and negotiations with Mexico are high priority items toward successful funding arrangements with all interested parties along the U.S./Mexico border. Any statutory changes by the two U.S. states involved will be identified and inititated in the respective legislative bodies. It may be appropriate to obtain local, state, federal and international financing for all or part of the conveyance system and water treatment of the Rio Grande waters for use by all populations in the El Paso tri-state area.

STANDARDIZING THE SHARED USE OF TRANSBOUNDARY WATER RESOURCES

Stephen E. Draper,[1] Member, ASCE

The availability of reliable sources of water has been a necessity for successful economic and social advancement of the nation-state. When water sources have been shared by two or more political entities the nature, timing and magnitude of the water use by the different parties have been a continual source of conflict. Yet, with all the conflict that has been a part of American history, no defined standards for equitable sharing of water have been developed. Had such standards been available, economic progress of the affected regions and the nation itself may have been better served.

In 1994, the American Society of Civil Engineers created a committee to develop the principles and guidelines for shared use of water. Through the ASCE Special Standards Division's Water Regulatory Standards Committee a special task force was commissioned to establish equitable principles and standards to optimize shared use of water resources.

The need for these guidelines are clear. The history of economic development in the American West is as much about transboundary water sharing as it is mining or raising cattle. Conflict over the waters of the Colorado River still reverberates today in new water-sharing battles. Recently a major dispute has emerged between Alabama, Florida and Georgia over transboundary water sharing of the Alabama-Coosa-Tallapoosa and Appalachicola-Chattahoochee-Flint River Basins. These conflicts are not restricted to the United States. The deficiency of adequate supplies of water is evident throughout much of the world and trans-

[1] The Draper Group, 1401 Peachtree St, NE, Atlanta, GA 30309

boundary sharing of water resources is a major source of conflict. Conflict exist between Turkey, Iraq and Syria in the Tigris Euphrates basin; between Jordan and Israel regarding their opposite bank sharing of Jordan River; and between nations in the Nile River Basin. The breakup of the Soviet Union has caused conflict between former Republics, especially in the arid regions east of the Caspian Sea.

The ASCE program is to approach development of standards for shared use of transboundary water resources (SUTWR) in a systemic manner, integrating surface and groundwater sharing, water quantity and water quality, and the disciplines of law, engineering, ecology and economics. The official ASCE Committee Purpose is straightforward:

> To establish equitable principles and standards to manage shared use of water resources for the purposes of minimizing transboundary conflicts. These principles and standards will be developed for adoption in international agreements, interstate compacts or state-tribal agreements for regulatory purposes along intergovernmental boundaries.

Development of a definitive, specific standard to be applied in all hydrologic circumstance, between any legal or cultural system, anywhere in the world is clearly unachievable. Consequently the Committee has drafted a document outline which is formulated to be descriptive of an ASCE "guideline standard." Such a standard focuses on the *process* of creating or modifying a transboundary water sharing agreement. It sets *guidelines* to ensure all pertinent factors are included in formulating the agreement. As was pointed out so eloquently in the early stages of committee organization, different political systems and/or different cultures and/or different water use customs must be accommodated by the "standard." Otherwise the "standard" would gather dust on the shelves of a few libraries.

This ASCE standard is based on three precepts.

> (1) <u>SIMPLICITY:</u> The final product of our efforts should be as simple as possible (ie, be direct and in language both engineers and lawyers understand).

(2) SUSTAINABILITY: The final product should focus on the principle of sustainability as outlined in the United Nations Environmental Policy & Law Paper No. 27, *Agenda 21: Earth's Action Plan*, Oceana Publications, Inc., 1993.

(3) CONFORMITY: The final product should CONFORM to the individual circumstances of the transboundary shared use so that it can be applied to all circumstances (ie. be "generic").

This last requirement (conformity) arises from a quote having to do with engineers from European water-rich traditions trying to optimize water usage in African and Near-Eastern dryland farming.

> .. Farmers practising [sic] these [dryland farming] techniques know more about their problems and their solutions than scientists ever will. For these reasons dryland farmers need "not messages but methods, not precepts but principles, not a package of practices but a basket of choice, not a fixed menu *table d'hote* but a choice *a la carte*.
>
> - Chambers, Robert, "Farmers First," International Agricultural Development, Nov/Dec 1985.

ASCE intends to set a standard for the *process* that all states and/or nations must use when creating or modifying a transboundary water sharing agreement. The ASCE standard will assert what must be considered prior to actually formulating the agreement. The first step is an **assessment** of the resources impacted by the shared water resource: the sources and uses of the water resource, the ecological matters affected by the transboundary use and the economic units, to include commercial transfers, that may be affected. The ASCE standard will discuss each element that should be considered. The next step is to begin development of the agreement. In that task the drafters must analyze the choices available for **allocation** of the water resource among the parties, with special emphasis on the extreme events (droughts or floods). The ASCE standard will discuss various means to do so. The drafters may choose the method of allocation that best fits their situations needs. The drafters are provided options of **water quality** standardization. The drafters must provide in their agreement a means to ensure the continued quality of waters from the transboundary source. The standard will describe different choices for the **administration** apparatus to supervise implementation. Finally the ASCE standard will focus on the most critical part of the agreement, the **dispute resolution** mechanism. Possible options will be provided and a recommended practice articulated for specific ecologic, hydrologic and economic conditions.

The ASCE standard concludes with **recommended agreements** that optimize shared use. For interstate purposes, the committee feels there is enough data and experience on interstate water sharing agreements so that a comprehensive **Model Interstate Water Sharing Agreement** can be drafted. That Model Agreement would be included in the document. For purposes of the North American countries the same amount of data and experience is not available so a **North American Model Water Sharing Agreement** would not be as detailed. Such a Model Agreement would, however, also be included. In the final phase a "generic" model agreement, useful throughout the world, will be developed. However, this **Universal Model Water Sharing Agreement** would be even less detailed and would most likely consist of nothing more than a sample framework of an agreement.

The necessity of three different guidelines is mandatory in order to allow unrestricted "repetitive" use of the ASCE Standard. Because the legal structure of all states within the United States are similar, there are specific options to the drafters of such an agreement that might not be available to the drafters of an agreement between nation-states (e.g. Israel & Jordan, Mexico & U.S., Paraguay & Brazil, etc.). Because most of the lawyers on the committee are quite familiar with the basics of interstate agreements, they will focus on drafting a detailed model of interstate water sharing. Because Mexico, Canada and the United States have different legal systems, a North American agreement will necessarily be different than the interstate agreement. The drafting of this agreement can, however, be an adaption of the interstate agreement. The general international agreement can then be adapted from the NAFTA agreement.

The final draft of the standard should be ready for review by December 1997. The SUTWR Standard should provide a standard approach to helping eliminate one of the world's most pervasive sources of conflict. By consistent application of the *process*, the drafting of agreements to use transboundary water sources can assure the proper integration of all aspects of water use: ecological, economic, hydrologic and legal.

URBAN STRATEGY FOR THE TRUCKEE RIVER NON-PROFIT INSTREAM FLOW BANK

John W. Fordham, Member ASCE[(1)]
Lori Carpenter[(2)]
W. Alan McKay[(1)]

Introduction

The Truckee River which rises along the eastern slope of the Sierra Nevada mountains in the Lake Tahoe basin and flows eastward into the Great Basin terminating in one of the remnants of the ancient Lake Lahontan, Pyramid Lake, has been subjected to considerable manmade manipulation for the past 120 years. The changes imposed through structural controls in the headwater areas and diversions downstream have resulted in a flow regime which has little resemblance to that which would be seen in a natural state. As with most other western rivers, the control and use of the Truckee's waters has in the past been, and still is, the subject of considerable controversy with many parties vying for the use of that water. This competition for the limited resource has only accelerated in the past two decades as urban demands have increased and environmental needs have been identified. Changes in use from primarily agriculture to allow increased urban uses and to recognize environmental needs of the natural system has resulted in significant legal maneuvering and lawsuits and counter suits filed naming thousands of individuals and entities in both California and Nevada. The most recent development which will have a major effect on river operations is the development of the Truckee River Operating Agreement (TROA). This agreement currently being worked out as a part of Public Law 101-618 is to provide operating criteria which will provide for existing rights and to maximize flows for Pyramid Lake and its threatened and endangered fisheries.

In the process of developing the TROA, most river interests were represented; however, there was no real advocate for instream flows or water rights to provide for the storage and release of flows to enhance the riverine environment during times of low flow. Since the river is fully allocated, flows for this purpose can only be obtained through a change in use or coincident with releases for other purposes. In order to obtain some minimal amount of rights for such purposes, a concept is being pursued, referred to here as the Urban Strategy, to obtain old rights which were never legally transferred from agricultural lands which have since undergone urbanization and are now served with other water rights.

Background

The Truckee River rises along the eastern slopes of the Sierra Nevadas and flows northeast to its terminus Pyramid Lake. Under natural conditions, the river was free flowing

(1)Water Resources Center, Desert Research Institute, University and Community College System of Nevada, P.O. Box 60220, Reno, Nevada 89506
(2)Huffman & Associates, 3969 So. McCarran Blvd., Reno, Nevada 89502

exhibiting high flows during the spring and early summer from melting snow and low flows during late summer and fall. There was some natural control in the upper river basin at Lake Tahoe due to a restricted outlet which served to prolong the high flow period. Along its course below Lake Tahoe it is joined by several major tributaries before entering Nevada. It flows eastward through the Truckee Meadows which was a large natural meadow prior to agricultural and urban development, and further east through a semi-arid western portion of Nevada to Pyramid and Winnemucca Lakes. The unregulated river and lakes supported an extensive fishery and provided a significant wetland for the Pacific flyway, the shallow Winnemucca Lake to the east of Pyramid Lake which has since dried up.

As the Europeans came to the area, drawn primarily by the gold rush to California and later the silver discoveries of the Comstock Lode, there was settlement in the valleys along the Truckee and adjacent Carson Rivers which brought about pressures to manipulate natural flows of the rivers and tributary streams. Agriculture developed in the Truckee Meadows and the Lahontan Valley near the terminus of the Carson River during the 1860s and 1870s brought about large flow diversions. This together with the lumbering operations in the Lake Tahoe basin and along the upper reaches of the Truckee resulted in a significant decline in the river and lake fisheries. As the mining industry began to decline in western Nevada, there was an increase in interest in developing new agricultural areas, both along the Truckee and adjacent Carson Rivers. This interest fit with the emerging federal interest in reclamation and resulted in the Newlands project which includes significant diversions of Truckee River water to the Carson River basin.

By 1905, Derby Dam on the Truckee River was completed and diversions to the Carson basin began. Lahontan Dam and Reservoir were completed in 1914 on the Carson River to provide storage of water from both the Carson and Truckee Rivers. The diversions had an immediate downstream impact, reducing flows in the lower river and inflows to Pyramid and Winnemucca Lakes and further limiting river access for fish spawning. Pyramid Lake began to decline reflecting reduced inflows with a parallel decline in Winnemucca Lake which was completely dry by 1939. Pyramid reached a minimum level in 1967, some 93 feet lower than its historic maximum. In addition to the major change brought about by the Newlands project, there have been several other changes which have effected the amount of and timing of flows in the Truckee River as well as the water quality in the Nevada portion of the river. The Truckee Storage Project resulted in the construction of Boca Reservoir on the Little Truckee River to irrigate about 29,000 acres in the Truckee Meadows. Much of the Truckee Meadows irrigated acreage has now been converted to urban uses. The Washoe Project resulted in two additional upstream reservoirs, Prosser and Stampede which are both multipurpose facilities providing flood control, fishery water, and some M&I storage. These reservoirs together with the low dam on Lake Tahoe control much of the natural flow of the river and provide carryover storage from year to year. The urban growth in the Truckee Meadows has resulted both point and nonpoint discharges to the river resulting in decreased water quality in the river through and downstream of the urbanized area.

Beyond the pure physical control of the river afforded by the dams, reservoirs, and diversion structures, there has been an evolution of legal and institutional control. The impact of these has been to regulate the river to meet diversion rights of various parties based on water right priority and water in storage. Each agreement or decree was based on the then prevailing demands and acknowledged beneficial uses. For the most part, instream flow requirements for a healthy stream were not considered with the result being even more water being controlled by upstream reservoirs for diversion for M&I and agriculture uses. This has resulted in additional lake level declines for Pyramid Lake and reaches of the river with severely depleted flows for longer periods of time and on a more frequent basis than would occur naturally. The river depletion has been most severe during the summer months through the urbanizing Truckee Meadows where both M&I and agricultural diversions are made. The current water rights allocation and resultant storage and diversions from the river have resulted in little or no flow in some reaches during summer and fall resulting in problems related to water quality, fisheries and the loss of a potentially significant aesthetic and financial resource for the tourist based economy of the Reno–Sparks area. During recent years, there has been a significant civic push and monies spent to enhance the river corridor through the urban area to make the riverine environment a part of the tourist draw. To help in this effort and to enhance river water quality, a project has been undertaken to develop a strategy which would provide some additional flow through the urban reaches of the river during low flow periods as explained below.

Objectives and Approach

The objective of the project supported by the Nevada Division of Environmental Protection through the 319(h); 604(b) Clean Lakes funding is to establish a long-term urban strategy for the Truckee Meadows area to achieve water quality standards and a healthy attractive riparian corridor. The strategy, if successful, will obtain water rights which would be available for controlled release to maintain the river through the urban corridor which will be beneficial to all interests.

The Urban Strategy consists of two phases which will be performed concurrently. The first phase consists of attending the TROA meetings and to the extent possible negotiate through the TROA process adequate flows and upstream storage rights for instream flow beneficial uses. To the extent that adequate instream flows cannot be negotiated in the TROA process due to prior existing water rights, etc., the non–profit instream flow bank will be established.

In the second phase *"adequate instream flows"* will be defined. The basis will be water quality criteria established in the Truckee River at the Truckee Meadows Water Reclamation Facility and hydraulic principles such as: bedload transport, bankfull and floodprone channel dimensions, and width/depth ratios. "Adequate instream flows" will be defined comparing the existing conditions to the projected conditions under TROA. First, adequate instream flows will be determined using existing cross–sections of the Truckee River through the greater Truckee Meadows area. Using these existing cross–sections, permanent station will be established where river stage will be determined for monthly low flow frequencies under existing conditions. Results of the above data analysis will define the "natural" or existing flow regime - given man's intervention on the river system. Results for the aforementioned variables will then be compared against the recommended seasonal flows defined under TROA and by other agencies. If the recommended river flows under TROA are found to be lower than the flows defined as adequate to support the instream goals/objectives then those time periods will be deemed to be critical flow periods. These critical low flow periods will define release periods for waters obtained under a nonprofit "flow bank" which would be stored in upstream reservoirs.

The creation of a non–profit instream flow bank whereby private citizens, public agencies, and other community organizations donate water rights under the existing water rights adjudication is essential to the health of the river during low flow periods. The flow bank is to consist of the following: the bank by–laws and operating criteria for water release, criteria for absolution of the bank and trust responsibilities of bank, criteria for use of water, short–term goals for a three–year period, long–term operating criteria and goals, acquisition program, coordination of donated work by civic minded participants, and the flow bank's relationship to the Truckee River Operating Agreement and its release criteria.

The instream flow bank water will release water at low flow periods to help achieve water quality criteria and meet the previously defined minimum stages at set river stations. The timing of these water releases will be in conjunction with irrigation releases and diversions at Derby Dam. The intent of the instream flow bank released flows is that they will also flow to Pyramid Lake in support of the Cui–ui Recovery Plan and water quality objectives in the lower Truckee River below Derby Dam. The non–profit instream flow bank will be held in–trust by an independent organization such as the Trust for Public Lands, Truckee River Yacht Club, or The Nature Conservancy. The non–profit instream flow bank water rights will be held upstream in Stampede Reservoir until flows in the river fall below adequate flows needed to support a healthy riparian habitat and facilitate meeting water quality objectives below or at the Truckee Meadows Water Reclamation Facility.

Current Status

The project to develop the flow bank as described is in its initial stages. Several blocks of currently unused water rights have been identified which could be donated to the bank. These are now held by public agencies or as water rights tied to now urbanized land currently served by other water rights. In the latter case, transfer will require significant legal and administrative efforts. Discussions have been held with both Reno and Sparks city officials, the major urban water purveyor and a number of nonprofit organizations. Some of these entities have committed to donate resources and time to minimize the costs of obtaining and transferring water rights for the flow bank.

Beyond the already identified blocks of water rights and the fractional water rights tied to property now dedicated to the cities, i.e., street rights–of–way, it is envisioned that private citizens will be encouraged to dedicate existing unused rights as the benefits to the river are demonstrated. It will be necessary to provide a mechanism by which the donated rights are guaranteed to be restricted to enhance the river in perpetuity and to that the transfer be of no cost to the donor.

References

Public Law 101-618 Title II "Truckee-Carson-Pyramid Lake Water Rights Settlement Act," 1990.

FISH PASSAGE IN URBAN STREAMS

John R. Genskow, P.E.[1], Member ASCE

Abstract

This paper discusses fish passage criteria for culverts and how these criteria were applied to planning a culvert rehabilitation project along an urban creek. It also discusses how fisheries issues impact urban "in-stream" drainage system maintenance and facility rehabilitation projects.

Introduction

Pigeon Creek No. 1 is an urbanized stream located in southwestern Everett, Washington with base flows of approximately 3 cubic feet per second (0.1 cubic meters per second). This creek discharges to Port Gardner and Possession Sound, a body of water connected to Puget Sound. The lower reaches of this 2-mile (3.2 kilometers) long creek run through a deep ravine in Everett's Forest Park. Land uses in the upper basins include typical urban development including commercial, light industrial, and residential areas. A neighborhood elementary school has adopted the stream, and efforts have been made over the last several years to enhance fisheries habitat by cleaning up the lower creek and installing notched log weirs. Several schools plant salmon in the stream each year.

The ravine is intercepted by Mukilteo Boulevard approximately 4,500 feet (1.4 kilometers) from the mouth. The roadway was constructed on approximately 70 feet (21 meters) of unclassified fill. Stream conveyance is through a 36 inch (0.9 meter) diameter concrete pipe culvert. The culvert presents a blockage to fish passage because of its length, steep slope, and drop at the outlet. The city of Everett has been monitoring the roadway embankment and has recorded movement on the downstream slope in the vicinity of the culvert. The timber bulkhead at the outlet has also deteriorated and is no longer providing slope stabilization benefits. Inspection reports point to leakage from the failing culvert as the primary cause for the slope stability problem and the deteriorated bulkhead.

---

[1]Senior Project Manager, HDR Engineering, Inc., 500 - 108th Avenue N.E., Suite 1200, Bellevue, WA 98004-5538

Approximately 3,000 feet (915 meters) of potential fisheries habitat exist upstream of the culvert. Permit conditions require the city of Everett to provide for fish passage to the potential spawning and rearing habitat upstream as part of the project to rehabilitate the culvert and stabilize the roadway fill. Funding is provided by the City of Everett and the Boeing Company, with financing assistance from the State of Washington Public Works Trust Fund.

Fish Passage Criteria

Fish passage criteria are based on numerous literature references. (A thorough list of the more significant references can be found in the "Stormwater Management Manual for the Puget Sound Basin, The Technical Manual", 1992). The criteria are based on the limits of athletic ability for the weakest fish of each species with no factor of safety built in. Natural channels may exceed these criteria, but the diversity of natural channel beds and formations provide routes with only brief exposure to excessive conditions. Culverts do not provide this diversity. Full bridging structures, incorporating natural bottoms, are the most preferred and should be used whenever possible. The criteria listed in Table 1 are applied by the Washington Department of Fish and Wildlife when establishing permit requirements.

TABLE 1
FISH PASSAGE HYDRAULIC DESIGN CRITERIA

Criteria	Juvenile Salmon	Adult Trout, Pink, And Chum	Adult Chinook Coho, Sockeye Steelhead
Maximum Velocity	na[1]	2 fps (.6 mps)	3 fps (.9 mps)
Minimum Flow Depth	0.3 feet (9 centimeters)	0.8 feet (24 centimeters)	1.0 feet (30 centimeters)

1. An excessive risk of passage failure exists for juvenile salmonids for culverts longer than 60 feet (18.3 meters).

When sufficient data are available, the high flow discharge to determine velocity should be the flow that is not exceeded more than 10% of the time during the months of adult migration. When flow data is not available, the high flow design discharge should be based on the 2-year flood event. The low flow discharge to determine depth should be based on the 2-year, 7-day low flow discharge or the 95% exceedance flow for migration months of the fish species of concern.

When these criteria are exceeded, the slope and diameter of the culvert may be adjusted. Pigeon Creek with 21 feet (6.4 meters) of fall over 270 feet (82.3 meters) through the roadway fill presents an extreme situation. In this case a weir and pool configuration was used to maintain minimum depths and dissipate energy. The maximum drop allowed is 12 inches (30 centimeters) for adult fish and 9 inches (23 centimeters) for juveniles. The volume of the pool behind each weir is based on the following equation:

$$\text{Volume} = 16 * Q_{2\text{-yr}} * \text{feet of drop}$$

16 is a constant with units to give a volume in cubic feet.

Alternative Selection

Three alternatives were evaluated to meet both the culvert rehabilitation (slope stabilization) and fish passage goals: 1) Add a second culvert at a flatter slope for fish passage, install a fish ladder to the second culvert invert, and line the existing culvert to convey flood flows; 2) Line the existing culvert, regulate the outlet with a control structure to meet fish passage criteria for flow depth and velocity, and install a fish ladder to the top of the outlet control structure; and 3) Replace the existing culvert with a large diameter culvert, incorporating a series of weirs along the invert for fish passage. Fish ladder options for alternatives 1 and 2 include a Denil steep pass fishway, weir and pool ladder, and slotted fishway. Analysis of the alternatives included comparison of fish passage criteria, slope impacts, constructability, construction access requirements, environmental impacts, permit requirements, long term maintenance, and cost.

Alternative 3 was the selected alternative. The larger 10 foot (3 meter) diameter with low flow weirs along the bottom provided the best chance for fish passage. Studies utilizing the facilities at The Dalles Dam on the Columbia River (Slatick, Trans. Amer. Fish. Soc., 1971. No. 3) have shown configurations similar to alternatives 1 and 2 can work; however, field experience shows a better success rate for alternative 3 (Powers, Washington Department of Fish and Wildlife, communication, 1994). An alignment was selected that placed the proposed culvert in native materials, allowing for construction of a steel plate culvert by hand tunneling and reducing the environmental damage compared to providing access for large jacking and boring equipment and segments of large pipe. Figure 1 shows a section and profile of the selected alternative.

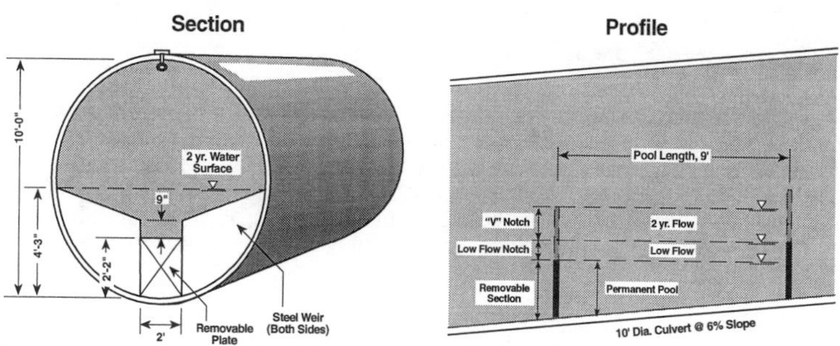

Figure 1. Fishway Pier and Pool

The culvert is sized based on pool volume, construction methodology, and maintenance access requirements. Conveyance capacity exceeds the 100-year design storm event. The weirs will provide greater risk of debris snags, but the larger opening will provide additional area for flow bypass. Removable plates are provided to flush trapped sediments from the pools. Flushing will be done using stream flows during non-migratory months.

Cost/Benefit

The opinion of probable construction cost for the project is approximately $600,000. The portion of cost to stabilize the slope and to convey the 100-year design storm event is approximately $150,000. The remaining $450,000 is to install a larger facility to provide fish passage. This is approximately 10 percent more than alternatives 1 and 2.

The benefit of providing fish passage is creating an opportunity for fish to access approximately 3,000 feet (915 m) of potential habitat upstream. The benefits of the selected alternative over the other alternatives are: 1) improved fish passage conditions, 2) reduced annual maintenance, and 3) reduced permitting costs. The cost per lineal foot of newly accessed habitat is approximately $150 ($492 per meter).

Other Considerations

The potential upstream habitat is within a forested public park. The lower 7,000 feet (2.1 kilometers) of the 11,000 foot (3.4 kilometer) stream is located within this park. However most of the tributary area is urban, including residential, commercial, and manufacturing land uses. Even if fish do get past the culvert, water quality and flow regimes become important variables in the spawning process and juvenile stages. Other local streams also provide habitat potential above fish blockages. This project is expensive because of difficult construction access; environmental issues of the park, deep ravine, and steep slopes; and its location under a major arterial. Consideration should be given to habitat banking or trading. This would involve investing the extra cost of fish passage facilities in locations where the cost per lineal foot of habitat is lower.

Summary

The Pigeon Creek Project needed to be done quickly to prevent structural failure of the 70 foot (21 meter) embankment. The environmental, safety, and socioeconomic impacts of a failure would be significant. Requirements for providing fish passage through the embankment have increased the cost of the project by over 300 percent. The benefit of the added cost is approximately 3,000 feet (914 meters) of potential habitat approximately $150 per lineal foot ($492 per meter) of stream.

However, this is a unique urban stream in that the lower two-thirds is protected and buffered by a woodland park. Other streams may appear more cost effective in terms of blockage removal costs versus feet of potential habitat accessed, but the long term viability of the urbanizing habitat may be questionable. Water quality and flow regimes remain concerns, as they are in all urbanizing areas.

The Pigeon Creek Project, with the support and ownership of the local community and nearby schools, is an example of how a project formulated to address a specific concern can provide opportunity to meet multiple objectives and provide multiple benefits. However, applying these criteria at other culvert locations will significantly increase project and maintenance costs. A balance must be found between the viability of the habitat and the cost of providing fish passage.

Fee-Collection Program To Improve Water Quality

James L. Smyth, M. ASCE[1]

Abstract

Many water utilities are finding themselves in the dilemma of having a water supply reservoir surrounded by increasing urbanization which can increase the salt load or introduce other contaminants to a water supply reservoir. The Sweetwater Authority has two water supply reservoirs, Sweetwater and Loveland, in the 182-square-mile watershed of the Sweetwater River in San Diego County, California. These two reservoirs and the groundwater basins are vital in storing water for approximately 164,000 customers. A pro-active fee-collection system was established in 1985 with the intent to have new development contribute toward the cost of constructing a runoff diversion system. The fee-collection area was limited to only 20% of the watershed area. Sweetwater Authority is now in the process of expanding the fee-collection program to encompass the entire watershed. This paper discusses the strategies and documentation proposed to convince the local land regulating agency to accept and approve the expansion of this fee-collection program.

Water System Background

The Sweetwater Authority is a public water agency which operates a 106-year-old water system. Included with the distribution system is a water filtration plant at Sweetwater Reservoir. Sweetwater and Loveland Reservoirs are almost equal in volume with a combined storage of 54,400 acre-feet, representing nearly a two-year supply of water to the service area. If local water is not available, water can be purchased from the local wholesaler, the San Diego County Water Authority (SDCWA),

---

[1]Chief Engineer, Sweetwater Authority, 505 Garrett Ave., Chula Vista, CA 91910

and stored on a seasonal basis in Sweetwater Reservoir should excess water be available. This process has a regional benefit under local drought conditions. Historically, 10% of the annual demand has been met using local supplies; 90% has been met from the imported supply.

Problems

Upon acquiring the water system in 1977, the State Health Department (SHD) informed Sweetwater of its concern that continued urbanization in the watershed will adversely affect the water quality of Sweetwater Reservoir. Urbanization creates a potential impact to water quality in the watershed due to land development. The contaminants brought about by urbanization must be considered by Sweetwater for proper management of Sweetwater Reservoir. The major concern was the increasing salt loads in dry-weather flows and sewage spills. The State Health Department and the SDCWA have provided support that Sweetwater and Loveland Reservoirs are too valuable as regional storage facilities for both runoff and imported water to allow the quality of stored water to degrade. It is also imperative that local water be utilized to its fullest use due to the uncertainty and costs of imported water.

Initial Project

In response to the SHD's concern and the established policy to utilize Sweetwater Reservoir as a drinking water reservoir, Sweetwater embarked on a study to determine what could be done to mitigate the adverse effects of urbanization on the water quality in the watershed. By 1982, it was concluded that passive watershed management (i.e., street cleaning and controlling use of chemicals by the public) was not sufficient to effectively protect the quality of water in Sweetwater Reservoir. Therefore, Sweetwater decided to pursue the concept of an interceptor system that would collect and divert first-flush wet-weather flows, dry-weather flows and sewage spills around the reservoir. This system, called the Sweetwater Reservoir Urban Runoff Diversion System (URDS) consists of a series of dikes, gates, pipelines, ponds and a pump station to collect and divert poor-quality water. After several storms, or when the runoff quality reaches acceptable levels, a by-pass gate is utilized to allow runoff to flow to the reservoir. This allows the maximum of captured local runoff at an improved quality in the reservoir. Since the watershed area lies outside the service boundary, Sweetwater has no legal ability to assess fees. In 1985, Sweetwater staff convinced the local agency governing the land use, the County of San Diego, to implement a policy requiring future development to pay for a portion of the proposed URDS facilities. The reasoning was that new development would substantially contribute to the poor quality of runoff. At that time the intent was to assess the land with the highest potential for immediate development in the watershed, or approximately 36 square miles; therefore, no consideration was given to charging a fee in the entire watershed.

Current Project

In 1993 staff commenced a project to consider expanding the existing fee-collection area to include the entire watershed area. Montgomery-Watson, San Diego, CA., has been selected as the consultant to assist staff in this project. Processing this change through the County of San Diego may prove to be more difficult than the original assessments approved in 1985. The strategy that will be used is quite similar to that originally used back in the early 1980's. The goals and objectives of this project are: (1) Support the need for fee collection to fund the expansion of the runoff protection system through additional sampling of dry and wet weather runoff, (2) To quantify the impacts of urbanization on water quality, and (3) To estimate the ultimate development in the watershed, and determine the rate at which growth will take place.

For runoff sampling, Sweetwater is using the services of Kinnetic Laboratories/Texscan, Inc., Carlsbad, CA. They will collect samples by remote controlled methods from both developed and undeveloped areas. Sampling will be done during the winter seasons of 1993-94 and 1994-95. Because measuring the amount and quality of all of the sub-area drainage to the reservoirs would entail a large amount of effort, a comparative study of smaller sub-drainage areas characterized by land use is proposed. This method is the one used on most stormwater monitoring studies conducted by the Environmental Protection Agency. Essentially, the amount and quality of stormwater runoff is measured from typical sub-basins which are selected based upon land-use. As in the initial project, we are comparing developed (urban) and undeveloped (pristine) sub-basins. Pollutant loading from these sub-basins are determined and event mean concentrations are determined for each pollutant parameter. Emphasis is given on sampling for significant storms after a dry period. These data can be utilized for three purposes: (1) The event mean concentration data can be compared to regional data bases for similar land use categories, (2) The relative pollutant loadings from each of the two land use categories within the watershed can be directly compared with each other to assess the differences in runoff pollutant loadings, and (3) Utilizing the combined land use, drainage, and population map, data can be used in a simple spreadsheet model to assess the total pollutant loading to both reservoirs due to runoff in the watershed, and the total relative contributions due to various land uses can be assessed. The constituents to be analyzed are: TOC, BOD, TSS, TDS, total Kehldahl nitrogen, total phosphate, total coliform, and pH. In addition to the winter sampling program, Sweetwater Authority staff will collect grab samples of dry-weather flows between March and November, in 1994 and 1995 at key discharge points into Sweetwater Reservoir. The information ultimately generated can then be presented to the County staff to present an overall picture of continued pollutant loading and impacts expected as development progresses within the entire watershed. The institutional arrangements required will be lengthy and involved. Governmental and quasi-governmental groups include the staff of the County of San Diego Planning Department, the County of San Diego Planning Commission, ten various planning groups within the entire watershed (each planning group is comprised of citizens that live within the planning areas), and the County Board of Supervisors for final approval. Sweetwater proposes to initiate this process by fall of 1995.

Conclusions

With the increasing concerns on reliability and cost of imported water, it is imperative that Sweetwater manage and protect its reservoirs. The method of deriving fee-collection areas is an attempt to fund the URDS and URDS II projects in an equitable manner to meet the goals of protecting Sweetwater Reservoir.

Appendix: References

Kinnetic Laboratories, August 31, 1993. Fee Collection Study, Sweetwater Authority Runoff Monitoring Study, Preliminary Alternative Study Designs.

Luke-Dudek Civil Engineers, 1982. Urban Runoff Study, Middle Sweetwater Hydrologic Subunit, San Diego, California.

Montgomery-Watson, 1994. Sweetwater Authority Watershed Sanitary Survey.

Moser, William, Vice-President, Montgomery-Watson Engineers.

Reynolds, Richard A., General Manager, Sweetwater Authority.

Whitewater Investigation of Existing and Proposed Flood Control Reservoirs

Jerry W. Webb[1] and Stephen S. Stout[2]

Abstract.

The economic market associated with whitewater rafting has increased to such a degree that efforts are currently being made in Kentucky and Virginia to find a better way to promote and stimulate growth in this recreational industry. Congressional concern brought on by public interest has resulted in the involvement of the Corps of Engineers in evaluation of augmentation flows for whitewater purposes. Currently two reservoirs within Huntington District are operated to provided whitewater during the fall drawdown. Whitewater releases from existing reservoir projects may be appropriate and in the public interest as long as the change does not significantly affect other project purposes or cause other adverse effects. In order to expand the industry and enhance the quality of the whitewater experience, expansion of a proposed flood control reservoir near Haysi, VA is being considered to provide storage allocation specifically for whitewater recreation. This paper will explain how Huntington District determined the optimal operational approach for the existing J.W. Flannagan and proposed Haysi Reservoirs in southwestern Virginia to provide reliable releases for whitewater use on Levisa Fork.

Methodology of Reservoir Simulation Studies.

The study of droughts and the statistical analysis of whitewater supply can involve many variables, both known and unknown, dependent and independent, with the measure of severity changing with the amount of rainfall deficiency and the duration of time without adequate moisture replenishment. Droughts are serial in nature. A six month drought cannot occur without first experiencing

[1]Supervisory Hydraulic Engineer, USACE, Huntington District, Hydrology Section, 502-8th St., Huntington, WV
[2]Hydraulic Engineer, USACE, Huntington District

a moderate to severe one month, two month, etc. drought. It follows that a multi-season drought may begin with that same one month drought. Therefore, the severity of the total drought changes as the duration increases. This factor must be considered in the analysis of whitewater supply from reservoirs where sufficient storage can only be provided by storing water over multiple seasons.

Analysis of the dependability of whitewater can be performed using many methods from simple mass flow curves to sophisticated statistical analysis of historic flow records. Because of the complexity of basin response, a statistical analysis of the results of a sequential reservoir simulation using an adequate period of historical flow records represents the best approximation of probable yield from a specified reservoir storage increment. A reservoir simulation model can incorporate the variance of streamflow, change in evaporation due to changes in lake size, various demand schedules for whitewater or other purposes and produce results that can readily be analyzed to determine effects of whitewater withdrawal on other project purposes, both upstream and downstream from the reservoir. The computer program **(RESOP)** used for the reservoir simulation studies described in this report was developed by the Huntington District Corps of Engineers. The program uses daily inflow values, monthly evaporation data distributed to daily values, downstream control point flows and a series of operator supplied control parameters to simulate the daily operation of the reservoir. Lake storage and whitewater or other low flow purposes are defined by upper and lower limiting guide curves. Daily losses from storage due to evaporation are computed based on the area of the lake surface for each day. The program also allows the reservoir to be operated for a downstream control point if a minimum flow at the downstream location is desired. Program output includes daily inflow, outflow, storage, lake elevation, evaporation expressed as flow and several control parameters. Summary tables of annual and/or monthly minimum and maximum values are provided.

Hydrologic Data Base.

The reliability of reservoir simulation analysis is enhanced when a lengthy flow record that accurately represents flow characteristics of the basin covering more years than the desired extreme return interval value can be assembled. Systematic flow records have been collected for the Levisa Fork and its tributaries since about 1925 although some station records are not continuous and other stations were established at a much later date. Figure 1 shows a general vicinity map and points of interest.

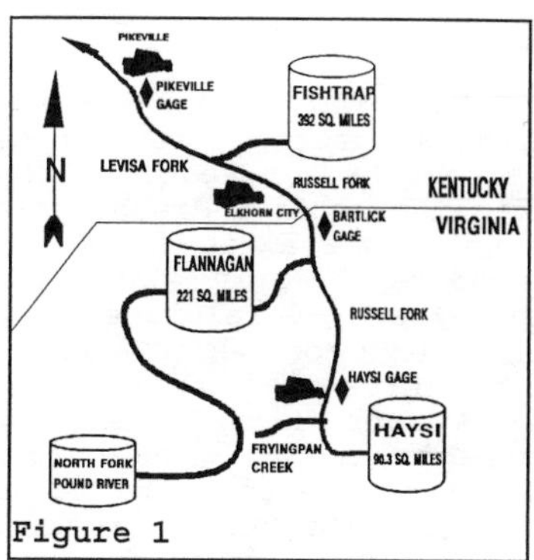

Figure 1

Table 1 provides a summary of the flow records available in this study.

TABLE 1. Gaging Station History Within the Study Area

GAGE	DRAINAGE AREA (Square Miles)	PERIOD OF RECORD	AVERAGE ANNUAL DISCHARGE CFS--- (Inches of Runoff)
Pound River at Flannagan Dam	221	1927-1964[1] 1965-Present[2]	265 --- (16.28)
Russell Fork at Haysi, Va.	286	1927-Present[1]	417 --- (19.80)
Russell Fork at Bartlick, Va.	526	1963-1992[3]	UNAVAILABLE

1. Unregulated Flows
2. Regulated Flows
3. Continuous Record (1963-1983); Broken Record (1983-1992)

Average daily discharge values for 64 years of record were calculated for Haysi, Flannagan and Bartlick using USGS historical records and drainage area ratio concepts. This combination of control points defines a system that is capable of assessing the whitewater capability of the existing Flannagan and the proposed Haysi reservoirs. The critical discharge control point for this whitewater study is located at Bartlick, VA.

It was necessary to compare the actual historic flows to those generated with **RESOP**. **RESOP** was utilized to generate the existing condition modified flows at Bartlick. These discharges were compared to the recorded USGS gage data at Bartlick to assure that a reasonable correlation existed before proceeding with the study.

Summary and Conclusions.

Flannagan and Fishtrap reservoirs are currently used to augment the minimum water quality discharge of 190 cfs at Pikeville. Whitewater alternatives were evaluated by increasing this minimum control discharge at Bartlick for each of the alternatives evaluated. Operational simulations were performed representing base condition releases and potential increases in releases from Flannagan to support additional days of whitewater. In order to evaluate potential whitewater from the proposed flood control reservoir at the Haysi Damsite, base condition releases from Flannagan including the currently authorized 8 days of whitewater at 800 cfs were incorporated into a range of operational simulations requiring additional storage at Haysi.

The procedure used in analyzing results of the simulation at J.W. Flannagan reservoir and the proposed Haysi dam site consisted of selecting the lowest elevation attained for each year, ranking the elevations from lowest to highest and plotting the results using median plotting positions. Best fit curves were then drawn between the plotted points. Figure 2 provides a summary of the simulations for whitewater demands furnishing 800 cfs for periods ranging from 8 to 24 days at Flannagan. Figure 3 provides a summary of wet dam and whitewater simulations which, subject to normal operational contraints, store flows larger than 40 cfs at Haysi.

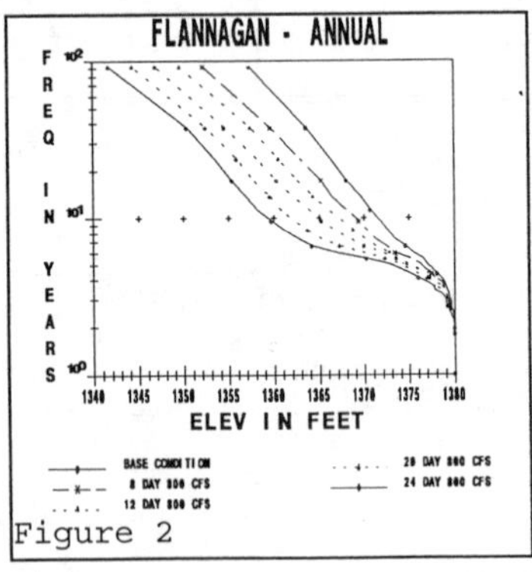

Figure 2

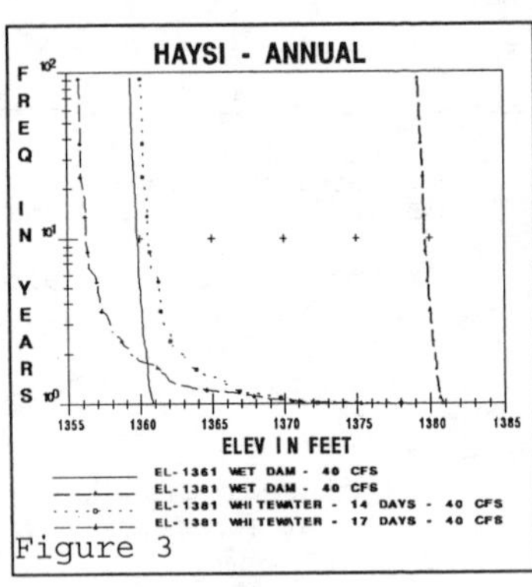

Figure 3

In summary, it is recognized that the procedure of developing synthetic historic records has limitations from a true statistical point of view. However, the relative impacts associated with a consistent water balance reservoir system simulation is considered appropriate in determining potential impacts on other project purposes.

Prior-Spring Lakes Watershed Improvement Project

Paul A. Nelson[1], Daniel M. Parks[2], Elizabeth Erickson[3] and Steven M. Kloiber[2]

Abstract

The Prior Lake-Spring Lake Watershed District (PLSLWD) is currently implementing a six-year water quality improvement project for Spring, Upper Prior, and Lower Prior Lakes. The three lakes are important recreational resources for the community and the focal point of the District. The project was developed as part of a U.S. EPA Clean Lakes Project and partial funding for plan implementation has been received under the EPA Section 319 Program. This paper presents the results of the study and the first year of project implementation.

Introduction

A diagnostic-feasibility study of the lakes was completed in August 1993. The study included a 12-month monitoring program and land use assessment that characterized the sources of water quality problems. Water quality problems on Spring and Upper Prior Lakes are primarily due to excessive algal blooms caused by high phosphorus concentrations. Lower Prior Lake is in good condition but is threatened by future development. Average total phosphorus (TP) concentrations in Spring, Upper Prior, and Lower Prior Lakes are 149 μg/l, 82 μg/l, and 46 μg/l, respectively. The median TP concentration for other regional lakes in Minnesota is 60 μg/l.

Phosphorus loading sources identified in the contributing watersheds included cropland, feedlots, past watershed drainage alterations, urban runoff, and wetland drainage. A large portion, approximately 40% of the Spring Lake TP budget comes from the County Ditch 13 subwatershed. Preliminary surveys indicate that hundreds of restorable wetlands are located in this subwatershed. In addition to watershed

---

[1] Manager, Water Quality Division, Montgomery Watson, Waterford Park, 505 U.S. Highway 169, Suite 555, Minneapolis, MN 55441

[2] Engineer, Montgomery Watson, Waterford Park, 505 U.S. Highway 169, Suite 555, Minneapolis, MN 55441

[3] President, Prior Lake Spring Lake Watershed District, 16670 Franklin Trail SE, Suite 110, Prior Lake, MN 55372

source areas, a significant portion (33%) of the TP load to Spring Lake is from internal recycling.

Internal phosphorus recycling in Spring Lake plays an important role in determining the water quality of Spring Lake as well as Upper Prior Lake immediately downstream. The internal recycling in Spring Lake creates an excess of soluble phosphorus. Approximately 60% of the TP in Spring Lake is in soluble form, the most readily available form for algal uptake. This soluble phosphorus is then discharged to Upper Prior Lake, which in turn discharges to Lower Prior Lake.

Goals for the lakes include improving Spring Lake to partially supporting swimming, Upper Prior Lake to borderline fully supporting/partially supporting swimming, as well as protecting and improving the quality of Lower Prior Lake.

Improvement Plan

The Improvement Plan focuses resources on the identified problems. The plan is composed of two, three-year phases. The first phase focuses on controlling watershed phosphorus sources, particularly in the County Ditch 13 subwatershed. The first phase also includes efforts to improve water quality protection regulations. The second phase of the project focuses on controlling internal phosphorus recycling in Spring Lake as well as continuing and expanding the watershed efforts started in the first phase.

Recommended actions for the first phase includes a public education program, revised stormwater ponding criteria for new development, farm fertilizer management demonstrations, promotion of no-till farming, wetland restoration, and promotion of shoreline aquascaping. Phase 2 actions will include new sediment basins, improvements to existing basins, expansion of the farm fertilizer demonstrations to an incentive program, promotion of no-till farming, and control of internal phosphorus recycling in Spring Lake.

A consortium of local sponsors was developed to investigate funding sources and implement the plan. The development of this consortium with its wide range of expertise was instrumental in initiating implementation and successfully obtaining financial support. Project sponsors included the PLSLWD, the City of Prior Lake, Scott County, Scott Soil and Water Conservation District, the local Extension Service, the U.S. Fish and Wildlife Service, the Prior Lake Association, the Spring Lake Association, and the Prior Lake Lions Club.

Implementation

Implementation of Phase 1 began in May 1994. Initial efforts were focused on public education/information, no-till farming, wetland restoration, aquascaping, farm fertilizer management, and acquiring permits for the Spring Lake chemical treatment system.

Public Education/Information

A project pamphlet, newsletter, and exhibits were developed as part of the education program. A project booth was used to distribute pamphlets and display exhibits at local community events. Newsletters were inserted into the local weekly paper and mailed to households within the watershed.

A soil testing program for urban fertilizer management was also initiated. Residents receive free soils tests for determining lawn fertilizer needs by bringing soil samples to the PLSLWD office. To increase the success of this effort, the program may be revised as a school science project. Students would collect soil samples from their lawns, evaluate the results, and determine actual lawn fertilizer needs. The project would pay for soil testing and provide results to the students for their analysis.

The project also helped coordinate a class on yard waste management. To encourage attendance, $100 composting bins were give out to attendees for $10. Approximately 50 people attended the class. Another class will be offered in the Spring of 1995. This class will focus on residential fertilizer use and will be developed by the local Extension office.

No-Till Farming

Interest in no-till farming in the area is high. Demonstrations sponsored by the local Soil and Water Conservation District in previous years have attracted over 150 attendees. However, no-till drills are expensive, and farm operators cannot purchase equipment based on interest alone. An investigation of rental sources for no-till equipment showed only one drill available for rental, located 15 miles from the project area. In spring 1994 the project purchased a no-till drill and made it available to area farmers at a reduced rental rate to encourage utilization.

The drill was purchased in May 1994 just as tillage season was beginning. No prior advertising of the drill's availability was possible, however, it's availability was soon spread by word of mouth. First-year results exceeded expectations and over 400 acres were planted with the drill. Most farm operators liked the drill and expressed interest in future usage. One surprise was that farm operators tended to use the drill as a planter rather than for no-till farming. Without prior advertising farmers had already tilled the previous fall and did not need the no-till capabilities. In addition, the farmers were cautious and only did test plots of 5 to 10 acres as no-till. In response to this finding the project is lowering the rental rates for 1995 for farmers who use the drill for no-till planting.

Wetland Restoration

A large effort was spent during the first year promoting wetland restoration. Restoration is being promoted by offering additional financial incentives for property owners to participate in existing state and federal restoration programs.

First year efforts include the formation of an advisory group composed of local, state, and federal personnel involved in wetland restoration, inventorying and prioritizing the restorable wetlands, and contacting landowners. The first landowners being contacted are those categorized as high priority restorations.

Prior to personal contact, landowners were mailed a letter describing the project along with a project newsletter. There was limited results and comments from the mailings. This was not surprising since personal contact was always considered necessary. The mailing provided general project information before personal contacts were made.

Personal contact has been more time consuming than originally thought. Numerous unanticipated questions and issues have had to be resolved. Issues have ranged from the replacement of cattle crossings, concerns over potential impacts to well water quality, future tax assessment rates, and landowners desire to excavate organic soils in the drained wetland to sell as topsoil and create open water areas. Another important issue was the transfer of development density credits. Property owners wanted assurance that the sale of wetland easements would not decrease the number of future lots developable on their property. These issues have created the need for project personnel to investigate zoning, taxing, and development regulations and permits not under the jurisdiction of the sponsors.

Aquascaping

Aquascaping is being promoted as an aesthetically pleasing alternative for shoreline management. The purpose is to establish a landscaped shoreline with native aquatic plants instead of groomed lawns. The president of one of the lake associations has already incorporated aquascaping and is very pleased. This serves as a demonstration to other shoreline owners and has created increased interest in this practice. Incentive payments are available in 1995 for the creation of five additional aquascaping demonstrations.

Farm Fertilizer Management Demonstrations

The Extension Service is soliciting farmers for participation in fertilizer management demonstrations. Farmers will receive a payment of $20/acre to work with Extension and develop farm-specific fertilizer management plans. A fertilizer management plan will ensure proper fertilizer application and be more cost-effective for the farmer.

Chemical Treatment

A chemical treatment system was proposed as part of the feasibility study. The purpose of the system was to flocculate pollutants prior to discharge of County Ditch 13 runoff into Spring Lake. Ferric chloride is the proposed flocculant. This technology has been used before on an experimental basis in Minnesota, however, a standard approach for permitting and implementing this technology has not been developed. The first year project efforts for this task focused on developing an acceptable and environmentally sensitive permit together with the agencies. These efforts are complete and the project has received a permit that is based on monitoring and maintaining iron concentrations below the ambient water quality criteria.

Summary

Sponsors of the Prior Lake-Spring Lake Improvement Project have completed the first year of implementation. Many of the elements have been successfully initiated, however, all of the elements have required minor adjustments to improve their effectiveness. On-going evaluation, improvement of project elements, and motivated public participants will be the key to a successful project.

Richmond Creek Drainage Plan

Dana Gumb
NYC Department of Environmental Protection

Michael DeNicola[1]
Sandeep Mehrotra
Robert D. Smith
Hazen and Sawyer, P.C.

Introduction

New York City Department of Environmental Protection (NYCDEP) conducted this project to develop a stormwater management plan which protects and enhances the natural areas of the Richmond Creek watershed. Specifically the objectives of this project were to (1) develop a hydrologic and hydraulic model for the Richmond Creek watershed to evaluate the hydraulic structures and local flooding, (2) establish engineering solutions to meet the 5-year design storm conveyance requirements for each hydraulic structure, and (3) analyze through model simulations the viability of wetlands along the stream and the requirements to maintain these wetlands while meeting the 5-year design storm requirements.

Richmond Creek was recently acquired by NYCDEP as one component in a system of wetlands on Staten Island preserved for stormwater management, called the Bluebelt system. The creek is a natural channel used for stormwater drainage and watershed management in central and western Staten Island. The section of Richmond Creek studied in this project is located between the intersection of Meisner and Rockland Avenues and extends approximately 2.40 km (1½ miles) to Arthur Kill Road. Figure 1 presents the project study area. The watershed encompasses approximately 372 ha (920 acres) of predominantly undeveloped land. Through most of the project area, Richmond Creek is a nontidal, freshwater stream bordered by wetlands. A total of nine hydraulic structures interconnect the surrounding residential areas in the watershed. Currently, this section of Richmond Creek experiences local flooding, stream bank erosion and sediment deposition due to excessive stormwater discharges.

(1) Environmental Engineer, Hazen and Sawyer, P.C., 730 Broadway, New York, New York 10013

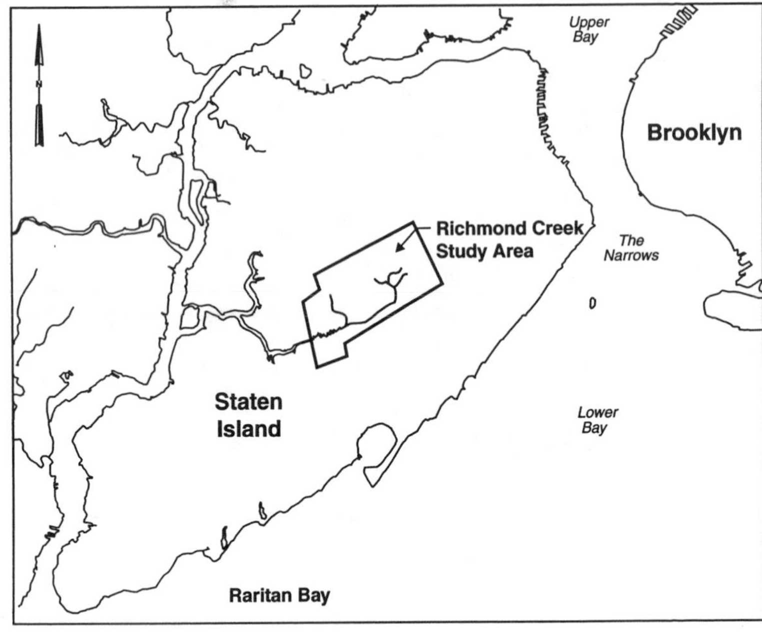

Figure 1: Project Study Area

Results and Discussion

Detailed information concerning the precipitation, topography, hydrology, hydraulics and stormwater management components of the Richmond Creek watershed were compiled from existing reports and information, topographic surveys and field investigations. This information was used to develop a modeling framework to evaluate the effectiveness of the existing hydraulic structures in conveying peak discharges and determine solutions to control peak discharges, eliminate flooding and maintain the surrounding wetlands.

The evaluation of Richmond Creek required hydrologic analyses to determine runoff rates and peak flows and hydraulic analyses to evaluate conveyance capacities and flow elevations. The HEC-1/HEC-2 integrated model was selected for the Richmond Creek watershed based on the characteristics of the system, available data and project objectives. Peak flows at each hydraulic structure, computed by the HEC-1 model for given design storms, were used in the HEC-2 model to calculate steady-state, non-uniform water surface profiles along the Creek. The conveyance capacity of each hydraulic structure was evaluated for the 1, 2, 5, 10, 25, 50 and 100-year design flows under existing development and full development in the watershed.

It was determined that the Lighthouse Avenue culvert, Arthur Kill Road culvert and the Mill Pond weir were not able to convey peak flows for a 5-year design storm without backwater elevations exceeding roadway elevations (refer to Table 1). Engineering alternatives to relieve conveyance restrictions were developed and their effectiveness was evaluated using the HEC-1/HEC-2 model simulations.

Location	Top of Roadway Elevation (m)	HEC-2 Computed Water Surface Elevations (m) for Design Storm Flows						
		1yr	2yr	5yr	10yr	25yr	50yr	100yr
Arthur Kill Culvert	2.3	1.92	2.14	2.57	2.76	2.80	2.97	3.24
Mill Pond Weir *	1.7	2.13	2.34	2.64	2.79	2.83	3.00	3.25
Richmond Hill Bridge	3.5	2.41	2.51	2.72	2.84	2.89	3.07	3.86
Lighthouse Culvert	3.5	3.58	3.61	3.64	3.68	3.73	3.83	4.11
Aultman Culvert	5.0	4.52	4.68	4.94	5.04	5.09	5.13	5.25
St. George Bridge	5.9	5.32	5.40	5.76	5.94	6.04	6.11	6.23
Eleanor Culvert	16.0	14.68	14.77	15.00	15.16	15.46	16.03	16.43
Meisner Culvert	33.7	31.97	32.05	32.25	32.39	32.70	33.08	33.85
Rockland Culvert	34.1	31.36	31.38	31.48	31.58	31.97	33.06	34.18

* The maximum water surface elevation to protect the surrounding area at the Mill Pond Weir was determined to be 2.07m.

Shaded areas indicate where water surface elevations exceed roadway elevations.

Table 1: HEC-2 Results Summary

The recommended improvements for the Richmond Creek watershed were as follows:

- Lighthouse Avenue
  - Increase the effective flow area of the culvert to 3.9 m^2 (42 square feet).
  - Direct a portion of stream flow through the wetlands upstream of Lighthouse Avenue culvert to attenuate the peak discharge.
- Arthur Kill Road/Mill Pond
  - Increase existing weir length at Mill Pond to 12.2m (40 feet).
  - Install a 25 foot side weir at Mill Pond for stormflow diversion downstream of the Arthur Kill Road culvert via a 1.5 m (5 foot) diameter sewer.

The implementation of these improvements will allow the peak flow generated by a 5-year design storm under full development (full buildout) in the watershed to be conveyed using the Richmond Creek natural channel (refer to Table 2). These improvements utilize the upstream wetlands to reduce the effective peak discharge for the 5-year design storm by approximately 25 percent.

In addition, these improvements are consistent with existing flow patterns and hydraulic structures, maintain the existing wetlands and floodplain delineations and minimize construction activities at major arteries.

Location	Top of Roadway (ft)	Creek Invert (ft)	HEC-2 WATER SURFACE ELEVATIONS (m) FOR A 5-YR DESIGN STORM FULL BUILD-OUT	
			Existing Culverts	Recommended Improvements
Arthur Kill Culvert	2.3	0.2	2.57	1.85
Mill Pond Weir	1.7	0.4	2.64	2.07
Richmond Hill Bridge	3.5	1.4	2.72	2.59
Lighthouse Culvert	3.5	2.5	3.64	3.41
Aultman Culvert	5.0	3.8	4.94	4.94
St. George Bridge	5.9	4.5	5.76	5.76
Eleanor Culvert	16.0	14.2	15.00	15.00
Meisner Culvert	33.7	31.3	32.25	32.25
Rockland Culvert	34.1	31.0	31.48	31.48

Shaded areas indicate where water surface elevations exceed roadway elevations.

The recommended improvements include new structures at Arthur Kill Road, Mill Pond and Lighthouse Avenue.

Table 2: HEC-2 Results for the Recommended Improvements

Conclusion

The HEC-1/HEC-2 integrated model proved useful in assessing the existing hydraulic structures and developing improvements to increase the conveyance capacity of Richmond Creek. Based on the computer model simulations, the use of the natural channel was maximized for stormwater conveyance and the surrounding wetlands were maintained. This provides a major component for developing a comprehensive stormwater management plan for the Richmond Creek watershed. In establishing an effective stormwater management plan, stormwater control and treatment issues must be further analyzed. Sediment deposition, erosion, floatables control and stormwater and septic pollution are all issues which need to be addressed if an effective stormwater management plan is to be established. Enhancement of the wetlands and natural areas as well as water quality improvements require solutions such as stormwater BMPs, erosion control programs, dredging and public education programs.

Restoring the South Florida Ecosystem

Adil J. Salem,[1] M. ASCE

ABSTRACT

Restoration of the South Florida Ecosystem is a complex undertaking by many Federal, state, and local agencies. Hydrologic restoration will be the first step in restoring this ecosystem. In this regard, a number of restoration efforts are currently underway by the Corps of Engineers. Several projects are in final design such as the Kissimmee River Restoration project, the Modified Water Deliveries to Everglades National Park project, the Canal C-111 project. In addition, a comprehensive review of the entire C&SF Project to develop ways to restore the south Florida ecosystem while providing for other water-related needs is now underway. This paper describes these efforts and how they relate to the overall goal of restoring more natural hydrology to the Everglades.

INTRODUCTION

The historic Everglades is part of an 18,000 square mile ecosystem which extends from Orlando to Florida Bay. Major regions of this ecosystem include the Everglades, Big Cypress, Lake Okeechobee, the Kissimmee River, Florida Bay, Biscayne Bay, and the Florida Reef Tract.

Since the 1800s, there have been a number of efforts by state and local governments to drain the Everglades. In 1948, the Central and Southern Florida Federal Project (C&SF) was authorized following severe droughts in the 1930s and hurricanes and floods in the 1940s. Today the project consists of about 1,000 miles of primary canals and 1,000 miles of levees, 16 major pump stations, and 150 water control structures. The Federal project is

[1]Chief, Planning Division, U.S. Army Corps of Engineers, Jacksonville District, P.O. Box 4970, Jacksonville, Florida 32232-0019.

supplemented by several hundred miles of secondary canals and levees and many structures constructed by local governments.

The C&SF Project serves many purposes - such as flood control, water supply, water deliveries to Everglades National Park, and water management in general. This project has allowed urban and agricultural development in areas previously considered unsuitable. As a result, this has fostered tremendous population growth in central and southern Florida to include cities such as West Palm Beach, Ft. Lauderdale, and Miami. Agriculture in the region has thrived since drainage and irrigation improvements were advanced by the C&SF Project. But the Project also had adverse impacts on the Everglades and Florida Bay.

THE DEGRADED ECOSYSTEM

The pre-drainage wetlands of southern Florida covered an area estimated at approximately 8.9 million acres. This region was a complex system of hydrologically interrelated landscapes, including expansive areas of sawgrass sloughs, wet prairies, cypress swamps, mangrove swamps, and coastal lagoons and bays. Prior to drainage, the characteristics of this network of wetland landscapes could be described by a set of physical and ecological features that were present across a broad area, and which gave definition to these ecosystems.

As a result of land use and water management practices during the past 100 years in southern Florida, the defining characteristics of the regional wetlands either have been lost or have been substantially altered. There has been increasing concern over the decline of the south Florida ecosystem due to continued rapid growth in the south Florida area and the increasing demands for water for municipal, industrial, agricultural, and environmental needs. This has resulted in a concerted local, State, and Federal effort to restore this ecosystem. At the federal level, a South Florida Ecosystem Restoration Task Force was established in September 1993 to foster Federal agency cooperation in addressing and solving ecosystem problems. In March 1994, the Governor of Florida established the Governor's Commission for a Sustainable South Florida.

THE CORPS OF ENGINEERS RESPONSE

Restoration of the South Florida Ecosystem is a complex undertaking by many Federal, state, and local agencies. Hydrologic restoration will be the first step in restoring this ecosystem. In this regard, a number of restoration efforts are currently underway by the Corps of Engineers. Several projects are in final design such as the Kissimmee River Restoration project, the Modified Water Deliveries to Everglades National Park project, the Canal C-111 project. In addition, a comprehensive review of the entire C&SF Project to develop ways to restore the South Florida while providing for other water-related needs is now underway.

The purpose of the Review Study is to reexamine the C&SF Project to determine the feasibility of structural or operational modifications to the project essential to restoration of the Everglades and Florida Bay ecosystems while providing for other water-related needs. The objective of the Review Study is to identify problems and opportunities, formulate alternative plans, evaluate conceptual alternative plans, and recommend, if feasible, further detailed studies.

Given the complexity of the problems to be considered in the study and the desire to use the skills of specialists in other agencies, a multi-agency approach was developed to complete the formation of the study team. Multi-agency staffing was essential in order to facilitate the flow of needed information among agencies, and, more importantly, to achieve buy-in and ownership by the key public agency stakeholders. The study team included personnel from other agencies such as the South Florida Water Management District, the National Park Service, the Fish and Wildlife Service, and the National Marine Fisheries Service.

ECOSYSTEM RESTORATION

The fundamental tenet of south Florida ecosystem restoration is that hydrologic restoration is a necessary starting point for ecological restoration. Water built the south Florida ecosystem and water management changes have adversely affected this ecosystem. Restoration begins with the reinstatement of the natural distribution of water in space and time. The focus is on the wetlands because the greater part of the predrainage south Florida ecosystem was wet.

The ecological goal of the south Florida ecosystem restoration is to recreate, on a somewhat smaller scale, a healthy ecosystem large enough and diverse enough to survive the natural cycles of droughts, floods, and hurricanes and to support large and sustainable communities of native vegetation and animals. An important lesson from history is that, in this ecosystem, any successful restoration plan developed must encompass the whole regional system, not geographic areas in isolation.

COMPREHENSIVE REVIEW STUDY

Alternative plans were formulated with the premise that the hydrological restoration would create a system that could function similarly to the way it functioned under pre-drainage conditions. In formulating alternative plans, several general principles were developed to guide the formulation effort. Plans should encompass an ecosystem approach instead of being formulated to only maximize local environmental objectives. Where possible, excess water currently being discharged to tide should remain in the system. The quantity, quality, timing, and distribution of water should be restored to more natural conditions where possible. Natural areas should be linked and connected where possible to reduce habitat fragmentation. Barriers should be minimized or removed. Transition zones should be created to reduce seepage losses and conflicts between natural and developed areas. Finally, operational flexibility should be a part of the modified C&SF Project to improve the ability to make operational changes that may become necessary in the future.

To evaluate the effect of hydrology in terms of restoration, an environmental evaluation methodology was developed that would measure environmental outputs of alternative plans. For each alternative plan, there had to be a description of hydrological conditions in order to evaluate environmental outputs. Model output was compared to Natural System Model (NSM) output as a target for restoration. Although a day-by-day, grid-by-grid comparison would not be wise, the trends, magnitudes, and fluctuation ranges are important hydrologic targets. NSM outputs represent the best hydrologic description of pre-drainage conditions available for the Review Study.

During the Review Study, numerous ideas were identified and grouped into basic plans to be evaluated. These basic plans are not final, nor is the Corps recommending or selecting a plan to be implemented. There are additional components and combinations of components that the Corps will study before a final plan is selected.

CONCLUSION

The restored natural areas of south Florida will be smaller in area than the historic ecosystems. Within this framework of change in spatial scales, it is impossible to predict with certainty how and when fish and wildlife populations will respond to improved hydrological conditions. More needs to be known about inter-relationships between life forms and their environment in the south Florida wetlands in order to predict responses.

Integrated Stream Management in the Longwell Run Watershed Carroll County, Maryland

Barbara A. Schauer[1], Kristin D. Barmoy[2]

Abstract

The Longwell Run is located in an urbanized watershed in central Carroll County, Maryland. This stream has experienced degradation typical of waterways in watersheds which encompass a large percentage of commercial and industrial development. With the assistance of a state grant, Carroll County government recently contracted with Black and Veatch to conduct a feasibility study to investigate options for restoring Longwell Run.

The Longwell Run feasibility study includes prioritized recommendations which integrate environmental site planning, sediment and stormwater control, riparian restoration, wetland creation, urban forestry, and pollution prevention on a watershed-wide basis. The restoration plan will be implemented, beginning in 1995, with the financial and technical assistance of various federal, state and local government agencies. The plan proposes use of a variety of techniques including structural stormwater management retrofits, in-stream habitat restoration, riparian reforestation, elimination of fish barriers, wetland creation, stream stewardship projects, and watershed-wide commercial/industrial pollution prevention practices.

Restoration of streams such as the Longwell Run must be approached from a watershed perspective. The Longwell Run Feasibility Study represents a dynamic blueprint to guide the process, provide illustration of concepts, and maintain continuity over the many years that may be necessary to implement the plan.

Introduction

The Longwell Run watershed encompasses one of the most urbanized areas of Carroll County, containing shopping centers, commercial areas, schools, and many areas of

---

[1]Project Engineer, Black & Veatch, 18310 Montgomery Village Avenue, Suite 500, Gaithersburg, Maryland 20879

[2]Chief, Carroll County Bureau of Stormwater Management and Sediment Control, 225 N. Center Street, Westminster, Maryland 21157

light industry. The watershed is approximately 45 percent impervious (460 acres of the 985-acre watershed). Development has caused a significant increase in stormwater runoff to Longwell Run which has, in turn, contributed to the destruction of in-stream and riparian habitat. This is evidenced by extremely eroded and undercut banks, tree falls, reduced sinuosity, increased sedimentation, and loss of in-stream habitat. Urbanization has also meant deforestation and loss of riparian vegetation within the watershed, resulting in a lack of stream shading and bank stabilization.

Large increases in stormwater discharges to the Longwell Run are clearly the critical factor that led to the degradation of the stream. Post-development storm events in urban areas yield an increased volume of runoff that occurs more rapidly and with a greater peak discharge. Impervious surfaces associated with urbanization decrease infiltration and depression storage within the landscape. This typically causes increased frequency and severity of flooding, accelerated channel erosion, and altered natural characteristics of streambeds.

Urbanization can also negatively impact stormwater runoff quality. Compared to natural conditions, stormwater runoff from urbanized areas frequently contains high concentrations of sediment, organic matter, nutrients, heavy metals, pesticides, hydrocarbons, bacteria, trash and debris. In addition, runoff from an urban landscape is warmer leading to increased thermal impacts on the stream community.

The restoration plan for Longwell Run focuses on mitigating the impact of stormwater flow from the urbanized watershed by reducing the frequency and magnitude of storm flow events in the stream, while proposing measures to restore damaged or lost in-stream and riparian habitat. This approach, which combines an urban stormwater management retrofit process with in-stream and riparian restoration methods, has a good chance of creating permanent and measurable improvements in the stream. Key elements of the Longwell Run restoration plan are discussed in the following sections.

Stream Assessment

The restoration plan began with a baseline stream assessment designed to provide a comprehensive picture of the physical, chemical, and biological conditions present within the stream and the adjacent riparian areas. The intent of the assessment was to look closely at the stream conditions to determine the need and optimal locations for retrofit and restoration projects and to identify appropriate restoration techniques for various stream reaches. The assessment contained two primary components, an analysis of physical habitat and water quality indicators, and an analysis of biological indicators, including benthic organisms and fish. Chemical analysis was also conducted on two water and sediment samples from locations in the stream that exhibited visual signs of chemical pollution.

Stormwater Retrofit Projects

Eighteen existing stormwater management facilities control runoff from various sites within the watershed. Most of these facilities, however, do not provide water quality management and are inadequate by today's design standards. Facility retrofits are, thus, critical to the stream restoration plan.

An inventory of potential stormwater management retrofit projects in the watershed was prepared following a watershed-wide investigation. A total of 20 possible retrofit projects were identified during the assessment. The number of feasible retrofit options was double the total initially expected. The County and the consultant had both expected fewer options, because of constraints including the proximity of local businesses to the stream, and the location of potential retrofit sites on private property.

The retrofit projects proposed include structural modifications to existing stormwater management ponds, underground detention tanks, and infiltration basins. Proposed new facilities include constructed wetlands, grassed swales with check dams, and perimeter sand filters for parking lots. In addition, construction of a parallel pipe system for stormwater diversion and modification of a fish barrier were identified as recommended projects. Watershed-wide pollution prevention was also cited as a critical "retrofit" opportunity.

The projects were organized into an inventory and mapped to define the percent of the watershed they effectively control. Figures were also prepared illustrating each proposed retrofit. The projects were then evaluated and prioritized to help the County formulate an action plan. The evaluation criteria chosen were: the ability to remove pollutants before they reach the stream; the direct value in improving habitat quality; construction cost; ease of implementation; and public benefits.

Stream Restoration Projects

Approximately 75 percent of the Longwell Run was determined to be impaired. Poor habitat and physical conditions along the stream were caused largely by severe bank erosion and siltation of stream substrate. The stormwater management retrofit projects and new facilities identified will mitigate the impact of stormwater on the stream, reducing in-stream erosion and siltation.

Erosion and siltation may be further reduced by regrading and revegetating the most severely damaged areas of the stream. The combined effect of reduced flows, resulting from the stormwater management retrofits, and restoration of the most severely damaged sections of the stream will allow the less severely damaged stream reaches to stabilize and revegetate naturally.

An increase in riparian vegetation resulting from stream restoration efforts will provide stream shading and lessen the effects of high summertime temperatures. The combined effect on improving the water quality of Longwell Run by reducing siltation, providing additional in-stream habitat, and stabilizing temperatures is, then, expected to improve the diversity of the benthic community. Ultimately, this increase in favorable food sources and supplies coupled with an increase in habitat variability should result in a more diverse and abundant fish population throughout the entire stream system.

Based on the stream assessment results, six areas were selected to apply stream restoration techniques. Restoration projects are recommended in stream reaches where the habitat is severely impaired and where the County has access to the property needed to complete the restoration. Restoration techniques selected include in-stream fish habitat creation, bank regrading, wetland creation, stream bank stabilization, riparian re-vegetation and reforestation, stream rechannelization.

A restoration concept was prepared for each of the six areas selected which includes a combination of the restoration techniques listed above. The projects were evaluated using criteria similar to that used to evaluate the stormwater management projects which include: direct value in improving in-stream and riparian habitat; ability to mitigate the effects of stormwater flow; cost; ease of implementation; and public benefit.

Public Information and Education

Establishment of a public information and education program is crucial for successful watershed management. This is especially true in the Longwell Run watershed, because even if every structural retrofit recommended is constructed, slightly less than one half of the impervious acres in the watershed can be effectively treated. The restoration plan included possible components of public outreach including the formation of a citizen advisory committee, publicizing watershed location, watershed eduction, stewardship projects, and a listing of sources of technical assistance.

Pollution Prevention

Proactive pollution prevention is far more cost effective than controlling and cleaning up pollution. Pollution may be prevented through careful use of local regulations and educational projects. The Longwell Run watershed contains many business and industries. The stream restoration plan recommends that pollution prevention activities, which include a public education campaign to gain support, be targeted at the business/industry segment of the community.

Interagency Cooperation

A feasibility study like that provided by Black and Veatch is critical to the success of a comprehensive urban stream restoration project. The cooperation of various government agencies and coordinated implementation of watershed management related programs, however, is another critical factor in the likely success of the Longwell Run restoration project. The County, in pursuing their stream restoration goal, will receive financial and technical support from seven other local, state, and federal government agencies.

Carroll County will be establishing an in-stream sampling station just downstream of the Longwell Run project to help fulfill the sampling requirement in the County's municipal National Pollutant Discharge Elimination System permit (NPDES). This station will help the County to track the progress of their efforts by collecting long term data relating to the quality of the stream. In addition, the continuation of the overall Longwell Run restoration project is included as a condition in the County's draft NPDES permit.

Moving the NPDES Program to a Watershed Approach[1]

Gregory W. Currey, Associate Member, ASCE
William E. Hall
Jeffrey L. Lape[2]

Abstract

The National Pollutant Discharge Elimination System (NPDES) program is the principal means of abating and controlling point sources of pollution to waters of the United States. A major challenge facing the NPDES program is managing this effort within the context of both limited resources and environmental impacts and priorities that vary from location-to-location. In order to meet this challenge, the U.S. Environmental Protection Agency (EPA) has developed the NPDES Watershed Strategy. The purpose of the Strategy is to integrate NPDES program functions into a broader watershed management framework and to support development of statewide watershed management approaches. The watershed management approach reflected in the NPDES Watershed Strategy provides a framework for efficiently and effectively addressing point source environmental impacts and supporting other surface water and ground water protection activities at both the national and state levels.

The Need for a Watershed Strategy

Traditionally, the NPDES program has focused on chemical-specific technology-based and water quality-based permit limits and requirements for industrial and municipal wastewater discharges. More recently, NPDES permits have included whole effluent toxicity monitoring requirements and limits. The scope of the NPDES program also includes newer initiatives to address storm

---

[1]The views presented in this paper are solely of the authors and do not necessarily reflect the views of the U.S. EPA or any other Federal agency.

[2]Gregory W. Currey, Civil Engineer; William E. Hall, Environmental Protection Specialist; Jeffrey L. Lape, Wet Weather Programs Manager; U.S. EPA, 401 M Street, SW (4203), Washington, DC 20460

water; sewage sludge; combined sewer overflows (CSOs); and incorporation of sediment criteria and biocriteria in point source discharge permits. The baseline requirements and new initiatives cover hundreds of thousands of point source dischargers including approximately 70,000 industrial and municipal wastewater dischargers; 15,000 CSO points in 1,100 communities; 150,000 industrial storm water dischargers; and 800 municipal separate storm sewer systems.

EPA is promoting a watershed-based approach to addressing the myriad of issues and requirements facing water quality agencies. As a result, EPA released the NPDES Watershed Strategy in March 1994 as a framework for addressing point source permitting issues within the context of limited resources and spatially varying environmental impacts and priorities. The purpose of the Strategy is two-fold: 1) to integrate NPDES program functions into a broader watershed management framework and 2) to support development of statewide watershed management approaches (Office of Wastewater Management, 1994).

Integrating Program Functions

Integrating NPDES program functions into a broader watershed management approach is critical to successfully managing the program within the context of limited resources and spatially varying environmental impacts and priorities. Traditionally, the NPDES program has addressed individual dischargers in isolation, usually in response to a permit application or required inspection. This method of operating makes it difficult to place individual discharger activities and impacts in a larger context and distinguish priorities for program resource allocations. As a result, most large dischargers receive the same amount of attention from permitting and compliance authorities regardless of their environmental significance while smaller, though not necessarily insignificant, dischargers receive less attention. In addition to the potential discrepancies caused by setting priorities based on discharger size, handling facilities located in the same watershed at different times often results in inconsistent management. For example, different staff may develop permit requirements using different sources of information and methods for analysis. Thus, one discharger may be required to implement new forms of treatment before another discharger located just upstream or downstream because their permits were issued at different times.

Overlaying stream hydrography on individual point source dischargers, the relationships among point sources become apparent. Decisions made for one discharger may impact decisions for another; the need for coordinated management is clear. Broadening the scope of the analysis, one can view the stream and the point sources in the context of the entire watershed, including land use, nonpoint sources of pollution, and even demographic characteristics.

Understanding the relationship of point source discharges to other watershed characteristics offers several opportunities for more efficient and environmentally focused program management. For example, a permitting authority can organize permit issuance so that all permits within a watershed expire and are reissued at roughly the same time, thus allowing for more consistent and equitable permit requirements among dischargers. Permit requirements can be developed in the context of basin plans that consider a range of management options to achieve water quality objectives. The permitting authority may then place increased emphasis and resources on discharges with the greatest environmental impact and decrease emphasis on those with low impact.

Supporting Statewide Watershed Management Approaches

A method for integrating and coordinating NPDES program functions with other water management activities that has been developed by a number of states is a statewide watershed management approach. Under this approach, a state is divided into geographic management units drawn around river basins. Activities such as monitoring, planning, assessment, and implementation of management controls are conducted within each basin according to a set schedule. The large basins can be further divided into smaller sub-basins, watersheds, water bodies, or stream reaches to provide greater flexibility and higher resolution for targeting program resources to address specific problems. While basin delineations based on surface waters may not coincide with the boundaries of groundwater aquifers, they still provide a method for coordinating efforts when surface water and ground water management issues overlap.

While basin management units provide a geographic focus for water management, activities also must be coordinated over time to ensure integrated and consistent program implementation within the basin. A statewide watershed management approach can achieve this temporal focus through watershed management cycles. A management cycle has three features that create an orderly system for coordinating and regularly evaluating resource protection activities:

1) a specified time period for completing all elements of the management cycle (e.g., monitoring, environmental assessment, priority setting, management strategy development, basin plan preparation, and plan implementation)

2) a sequence for addressing basins to balance workload from year to year (during any given year, some of the basins would be in the intensive monitoring phase, some in the assessment and prioritization phase, etc.)

3) a schedule of management activities for each basin; this schedule provides a long-term reference and coordinating plan for all participating water management programs, resource protection agencies, the regulated community, and public interest groups.

The NPDES Watershed Strategy represents a commitment by EPA and, in particular, the NPDES program to support statewide watershed management. At least 17 states are developing or implementing a framework to sequence monitoring, assessment, permitting, and other activities within geographic management units. The scope, complexity, and maturity of these programs vary from state to state.

For example, the State of Michigan began operating its permitting program on a watershed basis in 1983, citing a need to balance workload and address clusters of discharges at the same time. Michigan developed a five year schedule for regular permit issuance that delineated 77 basins in the State and subdivided this group into five groups of approximately 15 basins each. Permitting in the basin groups is repeated every five years in the same order. This approach has allowed Michigan to coordinate monitoring and modeling for pollutant load allocations by basin prior to and in support of point source permit issuance. Michigan has not yet fully integrated other control activities, such as nonpoint source control, or compliance and enforcement into this process (Surface Water Quality Division, 1983).

The Commonwealth of Massachusetts, on the other hand, has a relatively young statewide watershed management approach and is scheduled to begin its first round of basin permitting in 1995. Massachusetts, like Michigan, has scheduled monitoring and modeling prior to permit issuance and plans to cycle through all of its 27 basins every 5 years. The Commonwealth also plans to use the same five-year schedule for issuing water withdrawal permits and eventually hopes to include other programs, such as compliance, sludge management, and ground water permitting (Bureau of Resource Protection, 1993).

EPA and the NPDES program are dedicating time and money to the tasks of integrating the permitting program into a broader watershed management framework and supporting the efforts of states like Michigan and Massachusetts that wish to adopt a statewide watershed management approach.

Appendix: References

Bureau of Resource Protection. (1993). "Proposal: Watershed 2000." Massachusetts Department of Environmental Protection, Boston, MA.

Office of Wastewater Management. (1994). *The NPDES Watershed Strategy.* U.S. Environmental Protection Agency, Washington, DC.

Surface Water Quality Division. (1983). "Procedure No. 2: NPDES Permit Issuance Schedule." Michigan Department of Natural Resources, Lansing, MI.

WATERSHED MANAGEMENT PLANNING APPROACH

William P. Boucher[1]
David R. Bingham[2]
Daniel Murray[3]

ABSTRACT

Prevention and control of pollution from various sources are increasing concerns for municipal officials. To effectively develop and provide the required level of pollution control, municipalities must integrate their various existing pollution control programs into an overall watershed management approach. This paper outlines an approach to develop such a watershed management plan designed to provide prevention and control of various pollution sources.

INTRODUCTION

Historically, water pollution control has been approached through the permitting of specific types of discharges, such as industrial point sources, CSOs, and storm water runoff. However, this approach focuses on individual sources rather than an overall watershed/receiving water system. Therefore, the Environmental Protection Agency (U.S. EPA) is developing and implementing a national watershed protection strategy. This strategy addresses integration of existing water pollution control functions into a broader watershed context. This will allow more flexibility in developing pollution prevention and control strategies on a watershed basis, while cost-effectively integrating diverse pollution prevention and control activities.

This paper describes an overall approach to develop a watershed management plan. The approach stresses implementing low cost nonstructural development regulations and municipal source controls. Through proper use of local regulations, a community can implement elements of a watershed plan gradually as development and redevelopment occur.

---

[1]Senior Engineer, Metcalf & Eddy, Inc., 30 Harvard Mill Sq., Wakefield, MA 01880.
[2]Vice President, Metcalf & Eddy, Inc., 30 Harvard Mill Sq., Wakefield, MA 01880.
[3]U.S. EPA, Center for Environmental Research Information, Cincinnati, OH 45268.

THE PLANNING APPROACH

The planning approach outlined in this paper is intended to guide municipal officials through a systematic approach to developing a watershed-wide pollution prevention and control plan and is based on the process discussed in "Handbook: Urban Runoff Pollution Prevention and Control Planning" (U.S. EPA, 1993). The planning process, shown in Figure 1, includes the following steps: 1) Initiate Program; 2) Define Existing Conditions; 3) Collect and Analyze Additional Data; 4) Assess and Rank Problems; 5) Screen and Select BMPs; 6) Implement Plan. These steps apply no matter what source types (e.g., point sources, SSOs, CSOs, nonpoint sources) exist in the watershed.

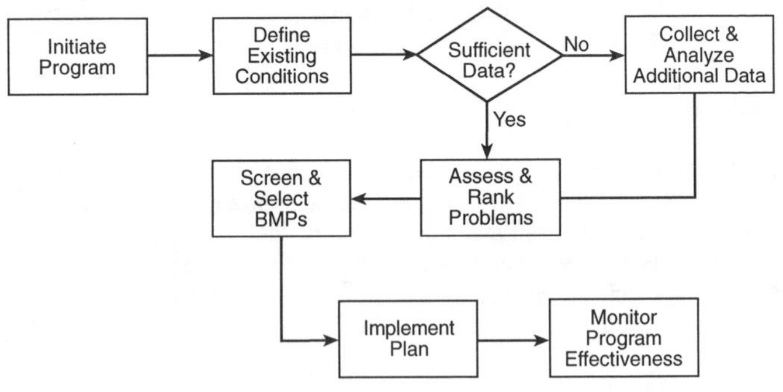

Figure 1. Watershed Planning Approach

Initiate Program

As a first step in the planning process, it is important for municipal officials to develop an overall program structure. Since municipal boundaries often do not coincide with watershed boundaries, it is necessary for individuals from additional impacted communities as well as state and federal officials to be involved in the program. Setting goals is a key aspect of initiating the program, and refinement of goals is an ongoing consideration. Successful implementation of a watershed management program depends on establishing clear goals and objectives. Setting goals is an iterative process that moves from less to more specificity as additional information on the watershed and water resources is obtained. The types of goals set usually depend on the receiving water, regulatory and political requirements, and the public's input on the affected water body.

Determine Existing Conditions

After the program has been initiated, an understanding of existing watershed characteristics and water resource conditions must be developed. Often, extensive data are available which can greatly reduce program costs. The required research is typically done by gathering existing available watershed information, including environmental, infrastructure, municipal, and pollution sources, as well as receiving water data, including hydrologic, chemical, and biological data, and water quality standards and criteria.

In determining existing conditions, a comprehensive review of the local regulatory structure and municipal control practices are particularly important. Improving local development and redevelopment controls or improving maintenance of the municipal infrastructure can often be cost-effective methods for protecting water resources. A review of these practices serves to indicate specific aspects that can be strengthened to improve runoff pollution control. Through the proper use of land use regulations (e.g., zoning ordinances, subdivision regulations, site plan review procedures, and natural resource protection) and comprehensive runoff regulation controls, proper watershed management can be cost-effectively implemented over a number of years as development and redevelopment occur. All regulations impacting water quality should be reviewed in this process. They should be investigated to determine their impact on runoff quantity control (e.g., preservation of open space, control of post-development flow, and requirements for runoff recharge), solids control (e.g., erosion control plan development requirements), and other pollution control (e.g., nutrient limits, aquifer protection requirements, etc.). Specific information related to developing this type of regulatory review can be obtained in Handbook: Urban Runoff Pollution Prevention and Control Planning" (U.S. EPA, 1993). Once these data are gathered, the information is organized into a coherent description of existing conditions to determine the gaps in knowledge.

Collect and Analyze Additional Data

Even in the best of circumstances, municipalities usually will not have all the information required to describe adequately the watershed/receiving waters. The program team, therefore, may have to gather additional information. This additional information, in conjunction with existing data, will help evaluate the existing conditions of the watersheds and water resources of concern more fully. Because of the cost involved in gathering these data, the benefits of additional data collection will have to be weighed against using funds to develop and implement the plan.

Assess and Rank Problems

Once sufficient data have been collected and analyzed, they are then used to assess and rank the pollution problems. A list of criteria (based on data gathered in earlier steps) can be developed to assess problems. These can be either pollutant source criteria, resource criteria, institutional criteria, or goals and objectives criteria. These criteria are used in conjunction with water quality assessment methods and models to determine current impacts and future desired conditions. The sources of pollution can be ranked according to impacts. In this step, it is also important to rank water bodies in order to provide pollution prevention, where appropriate. Often, simple pollution prevention techniques can reduce the future likelihood of problems. The emphasis on ranking problems is central to watershed management since it allows resources to be focused on targeted areas or sources, which improves the likelihood of achieving water resource improvement.

Screen and Select Best Management Practices

Once the water resource problems have been prioritized, efforts can be focused on specific water resource problems and the specific sources of those problems. At this time, a list of various pollution prevention and treatment practices should be compiled and reviewed for their effectiveness in solving the prioritized problems. This list should include regulatory, source control, and structural management practices. Initially, obviously inappropriate practices are eliminated from further consideration. This is an initial screening, and at the end of this step, the shorter list of feasible practices is evaluated further. The analytical tools developed during the assessment of problems and decision factors, such as cost, program goals, environmental effects, public acceptance, and others, as appropriate, are used to make the final selection.

Define and Implement Plan

Once pollution prevention and treatment practices have been chosen, the program can be advanced from planning into implementation. Often, this implementation will be conducted with a phased approach. Solutions that are cost-effective or that have major impact might be implemented early in the program. Pilot or demonstration studies can be used, with the results influencing further implementation. All involved entities must be familiar with and accept their role in funding, implementing, and enforcing the plan. In addition, continuing activities should be clearly defined. For example, maintenance programs should be developed so that structural practices will continue to operate as they were intended.

REFERENCES

U.S. EPA, 1993. U.S. Environmental Protection Agency. Handbook: Urban Runoff Pollution Prevention and Control Planning. EPA/625/R-93/004. Office of Research and Development.

Assessment of Needs for Management of Nonpoint Source Pollution in the United States

T. Dabak,[1] Member ASCE, R. Xue,[2] S. Pett,[1] and C. E. Gross[3]

Abstract

Analytical methodologies were developed to model national needs for controlling nonpoint source (NPS) pollution from agricultural lands, confined animal facilities, and silvicultural activities. The modeled estimates included activities to develop and implement NPS management programs to control runoff from these sources. Only capital investment or initial implementation of NPS control costs were captured.

Introduction

NPS are defined as sources of water pollution that do not meet the legal definition of "point source" in section 502(14) of the *Clean Water Act* (CWA). NPS pollution is not regulated by National Pollutant Discharge Elimination System (NPDES) permits.

The 1987 amendments to the CWA allow the use of State Revolving Funds (SRF) to fund selected non-Federal NPS control activities that are contained in approved Section 319 NPS Management Plans. An effort to report needs for selected aspects of NPS control that are potentially eligible for SRF funding was initially undertaken within the scope of the Environmental Protection Agency's (EPA) 1992 Needs Survey (EPA, 1993). Since few states have developed comprehensive estimates for NPS control, analytical methodologies (or models) were developed to estimate national costs.

The modeled estimates included activities to develop and implement NPS management programs to control pollution from agriculture (cropland, pastureland, and rangeland), confined animal facilities with fewer than 1,000 animal units, and silviculture.

---

[1] Tetra Tech, Inc., 10306 Eaton Place, Suite 340, Fairfax, VA 22030

[2] South Florida Water Management District, 3301 Gun Club Road, West Palm Beach, FL 33416

[3] Environmental Protection Agency, Washington, DC 20460

Database Development

Two national databases were used for cropland, pastureland, and rangeland, namely, the United States Department of Agriculture's (USDA) 1987 National Resources Inventory (NRI) and the USDA's Fiscal Year Statistical Summary data published under the Agricultural Conservation Program by the Agricultural Stabilization and Conservation Service (ASCS). The 1987 NRI was used to develop universal BMP (or Soil Conservation Service's conservation practice) selection criteria and to identify land requiring erosion control as well as irrigation water management. ASCS's statistical summary data reports for the years 1989, 1990, and 1991 were used to develop state-specific cost and erosion reduction efficiency data.

For confined animal facilities, model feedlots developed for the economic analysis for the Coastal Zone Act Reauthorization Amendments (CZARA) were used to represent the most typical facility configurations. The number of livestock operations in each model feedlot were assembled from 1987 Census of Agriculture database for each state. Cost data developed for each model facility in the economic analysis for CZARA (DPRA, 1992) were used to estimate costs of NPS pollution from confined animal facilities.

The 1987 U.S. Forest Service publication *Forestry Statistics of the United States* was used to estimate the area of timberland harvested on non-Federal land in each state annually. The distribution of the timberland area in northern, eastern, and southern states in relation to the type of terrain and presence of streams was developed by considering the geographical characteristics of each state. Cost of BMP implementation for these states were obtained from the economic analysis conducted for CZARA (RTI, 1992). For western states, cost data were approximated by using information obtained from past studies and state forest service officials.

Methodology Development

The development of the methodology for cropland, pastureland, and rangeland was based on a concept called "best management system." A best management system is a combination of conservation practices or management measures that, when applied, will achieve desired NPS pollution control through reduced transport of sediment, nutrients, and chemicals into surface and ground water. Erosion control was addressed by implementation of USDA's soil conservation BMPs. Water quality management was addressed by applying additional control measures, such as nutrient management, pesticide management, and irrigation water management. A computer program was developed to facilitate the computation of cost estimates and employed a decision support system to control the proper execution sequence of various decision processes associated with the cost-estimating methodology.

The methodology developed for the control of pollution from confined animal facilities used model facilities to represent typical facility sizes within each livestock category.

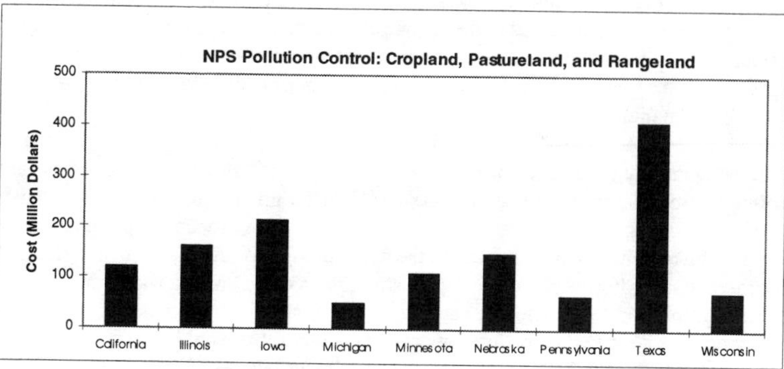

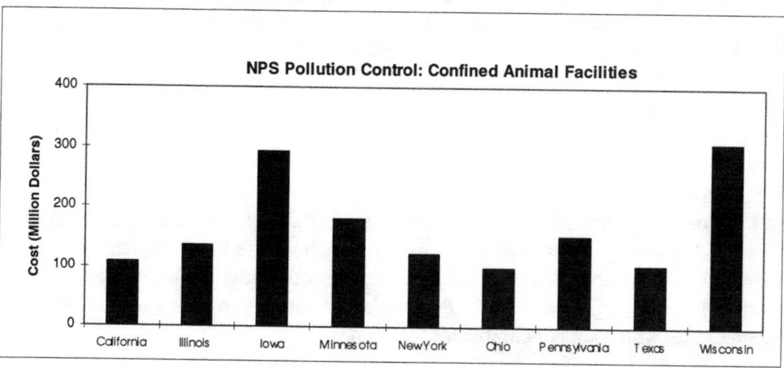

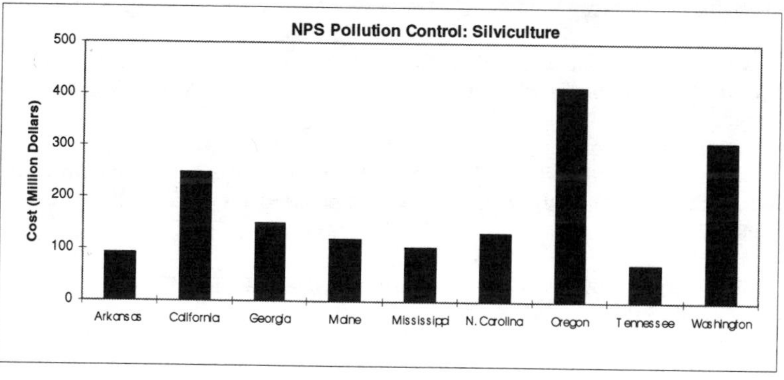

Figure 1 Results of Modeling NPS Pollution Control for Selected States

Livestock categories considered were beef feedlots, dairies, swine feedlots, and broiler and layer houses. The approach used was similar to that used in the economic analysis for CZARA. It was assumed that facility runoff would be controlled through diversions for runoff containment and channeling of on-site effluent to the ultimate control structures. All runoff collected in these control structures were assumed to be used for irrigation. The methodology also assumed that no controls were in place and that management measures and requirements developed under CZARA would be applicable to the entire nation.

The methodology developed for silviculture attempted to estimate cost of controlling pollution from silvicultural operations carried out on private timberland nationwide. Because of the complex and variable nature of forestry, a reasonable representation of silvicultural activities was attempted in the methodology for the northern, eastern, and southern states. For western states, past studies and consultations with local forestry officials were used for developing estimates. The cost estimates attempted to capture those activities identified in the forest management plans that would be eligible for SRF funding. Costs such as the "lost value of timber" (timber left as buffer areas between the stream and the operation) in streamside management areas and maintenance were not included in these estimates.

Results and Conclusions

Results obtained from the application of the cost-estimating methodologies for selected states are shown in Figure 1. These estimates represent an initial effort to assess needs on a national level for selected aspects of NPS control, therefore, the USDA and EPA generally concur on national totals, however, they both caution using individual state estimates. The estimates are for capital investment or initial implementation of NPS controls, activities eligible to receive SRF funds, but not ongoing costs of operation and maintenance, which are not eligible for SRF funds.

Appendix. References

DPRA. (1992). *Economic Impact Analysis of Coastal Zone Management Measures Affecting Confined Animal Facilities*. Report prepared for USEPA, NPS Control Branch.

Research Triangle Institution (RTI). (1992). *Economic Analysis of Coastal Nonpoint Source Pollution Controls: Forestry*. Report prepared for USEPA, NPS Control Branch.

USEPA. (1993). *1992 Needs Survey. Report to Congress*. Office of Water (WH-547). EPA 832-R-93-002.

A Survey of Urban and Agricultural Watershed Management Practices

Barbara A. Schauer[1] and Wayne H. Jenkins[2]

Abstract

Watershed management techniques utilized in urban and agricultural areas differ in a number of ways. These differences are due, in part, to different land uses and combinations of pollutants. However, successful watershed stewardship must combine the management of both types of areas in order to meet pollution reduction and stream protection goals. This paper reviews watershed management and nonpoint source (NPS) pollution reduction literature and discusses both traditional and innovative management techniques.

Introduction

Increasingly, environmental regulators are turning to watersheds as the primary focus for efforts to reduce NPS pollution. Large watersheds are composed of both urban and rural areas, and both types of areas are important contributors of NPS pollution. Urban and rural areas possess unique features which are reflected in the types and concentrations of the NPS pollution found in each area. As a result, different types of watershed management techniques are applied.

Urban Management Practices

Urban stormwater management programs were originally developed to address flood control, and over time evolved to address water quality problems as well. Most programs sought to limit post development peak flows to pre-development levels and to control water quality by treating the "first flush" of runoff. However, projects were dealt with on a case-by-case basis, and the cumulative impacts to receiving streams were ignored. This approach has led to the wide-scale deterioration of stream habitat and a decrease in biological diversity in impacted streams. In addition, large numbers of structural best management practices

---

[1] Project Engineer/Scientist, Black & Veatch, 18310 Montgomery Village Ave., Gaithersburg, Maryland 20857

[2] Water Resources Engineer, Maryland Department of the Environment, Chesapeake Bay and Watershed Management Administration, 2500 Broening Highway, Baltimore, Maryland 21224

(BMPs) have been constructed and now require regular maintenance and inspection. The costs associated with these activities can be substantial. The realization that the current approach to stormwater management is not working has fueled an interest in lessening the reliance on traditional structural BMPs and finding innovative, cost-effective ways to protect the physical integrity and biological health of receiving streams. An overview of selected urban BMPs, both structural and nonstructural, is provide here.

Structural BMPs rely on three basic mechanisms to treat runoff: **infiltration**, **filtration**, and **detention** (U.S. EPA, 1993). Traditional *infiltration* practices include *infiltration trenches and basins, dry wells, and porous pavement*. These practices rely on the absorption of runoff to treat runoff and have often been used to provide water quality benefits. The usefulness of infiltration BMPs is limited by soil permeability, groundwater level, and the location of drinking water aquifers. The innovative practices of bioretention basins use a combination of infiltration, retention, native vegetation, and soil conditioning to treat first flush of runoff from impervious areas (Coffman, et al., 1993)

Traditional **filtration** practices include *filter strips, grassed swales, and sand filters*. These BMPs treat sheet flow by using vegetation or sand to filter and settle particles and have been found to be very effective in the short term with shallow flow (Dillaha, 1990). Problems with effectiveness typically arise as these BMPs become clogged with sediment or are inundated with high flows. *Peat-sand filters* are an innovative practice that combine the high phosphorus and biological oxygen demand (BOD) removal capabilities of peat with a nutrient-removing grass cover crop and a subsurface sand layer to achieve high pollutant removal efficiency (Galli, 1992).

Detention practices temporarily impound runoff to control runoff rates and allow suspended solids and associated pollutants to settle out. *Detention ponds, extended detention ponds and wet ponds* fall into this category. Poor vegetative stabilization, clogging, and excessive sediments build up are problems commonly associated with these practices. *Constructed wetlands* are engineered systems designed to simulate natural wetlands. These innovative BMPs combine infiltration, filtration, and detention with vegetative uptake to treat runoff. Many innovative practices use BMP systems which combine several structural and non-structural BMPs to attenuate, convey, treat, and polish runoff (Schueler, et al., 1991).

In addition to structural BMPs, a variety of **non-structural practices** have been developed. These include *integrated pest management, street sweeping, and waste oil collection*. Non-structural typically work by reducing pollutants at their source before they can be transported by stormwater runoff.

Agricultural Management Practices

The sources of agricultural nonpoint source pollution are erosion from cropland, confined animal facilities, fertilizers, pesticides, and herbicides, grazing, and irrigation. The pollutants contained in these sources are sediments, nutrients, pesticides, herbicides, animal wastes, and salts. Agricultural activities also have the capacity to impact aquatic habitats through physical disturbance caused by animals, equipment, and water management practices (Dillaha, 1993).

Management techniques may be discussed in terms of practices that reduce the quantity of pollutant flow, practices that reduce the concentration of pollutants, and structural measures that may mitigate pollutant concentrations and/or the quantity of flow. The techniques are described in the following paragraphs.

Measures to **reduce the quantity** of flow of NPS pollutants include conservation tillage, contouring plowing, terracing, vegetative filter strips, and the use of cover crops. Conservation tillage is the fastest growing practice in the history of U.S. agriculture. It is considered to be one of the most effective agricultural BMPs. This method leaves at least 30 percent of the soil surface covered with crop residue after planting. Conservation tillage decreases soil erosion and surface runoff, increases infiltration, and reduces incorporation of agricultural chemicals. *Vegetative filter strips* are bands of planted or indigenous vegetation situated between pollutant sources and receiving waters. Filter strips remove sediment and other pollutants from surface runoff and are most effective with shallow flow. *Grassed waterways* are useful for preventing gully formation and result in mitigating the downslope transport of sediment. *Contour planting* configures furrows perpendicular to the direction of the slope and can reduce soil loss 0 to 85 percent, depending on slope, depth of plow, and rainfall. *Terracing* is a very effective technique for reducing NPS pollution in surface runoff. Level terraces are reported to reduce soil loss by 95 percent, nutrient losses by 56 to 92 percent, and runoff by 73 to 88 percent. These reductions are the result of runoff storage and allowing sediment to deposit and water to infiltrate. *Close grown and winter cover crops* such as small grains and grasses are effective in reducing erosion with 70 percent reductions in soil loss and 11 to 96 percent reduction in runoff volume. This technique is effective because it provides a protective canopy and root system to hold soil in place during the winter months. By spring winter cover crops are well-established and will hold soils in place during spring rain storms. *Protective covers* such as straw, gravel, wood chips, and mulches protect soil surfaces from the impact of rain and reduces soil erosion.

Measures to **reduce pollutant concentrations** in runoff include nutrient, pesticide and herbicide management, integrated pest management, and chemical use control and hazardous waste collection. *Chemical management* can be improved to minimize losses in runoff by adjusting the quantity, placement, application, and timing of chemical use. This is effective for reducing the quantity of chemicals in runoff. *Integrated pest management* is the use of management practices for pest control that eliminates unnecessary applications of pesticides and replace pesticide use with biological and cultural controls where possible. The basic premise is that chemicals should not be applied unless absolutely necessary. This is a departure from traditional farming practices that emphasize the use of pesticides as a routine preventive measure. *Chemical use control* can be effective for reducing road salts and motor oil.

Many of the **structural practices** are similar to those used in urban areas. These include *sediment traps and barriers, detention basins and ponds, and infiltration trenches*. *Sediment traps* are small temporary structures used to trap sediments by reducing flow velocities and promoting deposition. As previously described, *detention basins and ponds* are large structures designed to reduce peak runoff rates and remove sediment in stormwater.

Conclusion

Different land uses can produce widely different pollutants and combinations of pollutants. Most watersheds contain different land uses, including urban and agricultural areas. As a result, effective watershed management requires employing a variety of practices and techniques to address specific land use characteristics and pollutants.

References

1. Coffman, Larry, et al., "Design Considerations Associated with Bioretention Practices." Presented at the 20th Anniversary Conference of the Water Resources Planning and Management Division of the ASCE, Seattle, WA, May, 1993.

2. Dillaha, Theo A., "Role of Best Management Practices in Restoring the Health of the Chesapeake Bay: Assessment of Effectiveness." Perspective on the Chesapeake Bay, 1990, M. Haire and E.C Krone, ed., Chesapeake Research Consortium, Gloucester Point, VA, 1990.

3. Galli, John, Analysis of Urban BMP Performance and Longevity in Prince George's County, Maryland - Final Report, Metropolitan Washington Council of Governments, 1992.

4. Schueler, Thomas, et al., "Developing Effective BMP Systems for Urban Watersheds." Nonpoint Sources Watershed Workshop - Nonpoint Source Solutions, September, 1991.

5. U.S. Environmental Protection Agency, Guidance Specifying Management Measures for Sources of Nonpoint Pollution in Coastal Waters. January, 1993.

MASSACHUSETTS' QUABBIN RESERVOIR
A MODEL FOR WATERSHED MANAGEMENT

David A. Oberlander[1], Associate Member

Abstract

The Quabbin reservoir, located 105 kilometers (65 miles) west of Boston, supplies most of the drinking water for approximately 2½ million people in the greater Boston area. The Quabbin was constructed approximately 50 years ago to serve the increasing water demand of the growing metropolis of Boston. The importance of the reservoir system cannot be overstated. Without a clean source of drinking water, residents, businesses, and industries would not be able to thrive in the Boston area. In order to curtail the chance of accidental contamination of the Quabbin system, a comprehensive watershed protection plan has been designed and implemented by the officials in charge of the water supply. This paper provides a brief overview of the watershed protection plan and highlights some of the important aspects of the plan.

Introduction

The responsibilities for providing potable water to the citizens and businesses in the greater Boston area are shared by two independent but related state agencies--the Massachusetts Water Resources Authority (MWRA) and the Metropolitan District Commission (MDC). The MWRA maintains responsibility for treating and distributing a potable water supply to the users while the MDC ensures the watershed areas are protected from potential pollution sources.

Potential pollution sources are ubiquitous, a few examples include: on-site subsurface disposal systems, underground storage tanks for petroleum products and other hazardous chemicals, improper disposal of industrial wastes, certain recreational activities, dumps and improperly maintained sanitary landfills, pesticides, herbicides, high population density (animals and human), and construction activities. The potential pollution sources increase as an area undergoes development. As an area becomes more congested, watershed protection efforts

[1]Senior Engineer, Environmental Division, Maguire Group, Inc., 225 Foxborough Boulevard, Foxborough, Massachusetts 02035-2500.

must increase because of the additional pressure to develop protected watershed areas. Development also results in higher water demand. Historically, Boston's water officials have moved westward from Jamaica Pond, to Sudbury and Framingham, to Wachusett and finally to the Quabbin in search of larger potable water supplies. At present, the option of moving further westward is no longer viable; therefore, watershed protection has become more critical. In addition, water quality regulations are becoming increasingly stringent. A pristine watershed helps insure a quality drinking water source and may allow waivers to some of the requirements of the regulations. Watershed protection is a practical means to restrict the activities within sensitive watersheds in order to protect the water supply. The MWRA/MDC have implemented a watershed protection plan for the three major watersheds--the Quabbin and Ware River watersheds are covered by one plan and the Wachusett watershed has a separate plan.

Watershed Protection

The watershed protection plans for each area have common components with the ultimate goal of protecting the water supply. Components of both plans include: land preservation, control of subsurface disposal systems, inspection of watershed areas, control of development, land use restrictions, handling and cleanup of oil and hazardous waste spills in the watershed, wildlife management, and control of access to sensitive areas.

Quabbin and Ware Watersheds: The MDC owns about one third of the watershed for the Ware River. At present, the area is moderately developed; however, it has few industrial or commercial threats to water quality. Nearly 70% of the Quabbin watershed is owned by the MDC. The remaining watershed area is quite rural with low population density and no significant industrial or commercial areas. Much of the area has not been developed. Recreational use has always been considered a high risk at the Quabbin. At one time, nobody was allowed in the reservoir area. Today, passive recreation such as hiking, bicycling (on paved ways only), fishing, and bird watching is allowed throughout most of the watershed. Other recreational activities--all terrain vehicles, horseback riding, walking dogs, swimming, wading, and camping--are prohibited due to the potential deleterious impacts on the water source.

However, the leading cause of poor water quality at the Quabbin is not recreation, overloaded septic systems, or over development--it is sea gulls. Sea gulls enjoy feeding at landfills and wastewater treatment facilities in neighboring towns. At night, they prefer to roost on the open water away from danger. A favorite spot in past years has been near the Quabbin intake structures. The problem accelerates in the late fall as other area water bodies begin to freeze. The gulls head for the open water of the Quabbin which freezes later than most of the smaller, shallower water bodies. The gulls leave in late January, after the reservoir has frozen over completely. While resting at the Quabbin, the sea gulls clean themselves and leave their droppings in the reservoir. The result is high coliform counts.

The MDC has recognized this problem and now implements a gull harassment program. The program involves scaring gulls from their roosting areas with loud noises and boats. The gulls are driven further up the reservoir--away from the intake areas--which results in lower coliform counts at the intakes. The program has also expanded to area landfills. The idea is to prevent the gulls from feeding at the local landfills. Without a food supply in the area, the gulls are forced to find alternate food supplies--hopefully far away from the reservoir.

A land management plan developed for the Quabbin watershed identifies two important goals to help maintain water quality. The short term goal of the land management plan promotes forest regeneration while the long term goal encourages the establishment of a diverse, multi-layered forest. Regeneration is fostered by cutting out undesirable growth in a conservative manner. Two significant threats to forest growth include whitetail deer and hurricanes. Whitetail deer have impacted regeneration by feeding on the tender stems of young trees. The ideal density of deer for the area is 26 deer per square kilometer (10 deer per square mile). The population density has been estimated to exceed 155 deer per square kilometer (60 deer per square mile). Intense browsing associated with the elevated deer population results in the demise of the lower layer of young trees. Efforts to reduce the damage caused by the deer include two procedures. The first is a closely monitored and controlled hunting season implemented four years ago to reduce the deer population. The second involves the use of solar powered electric fences to discourage deer from entering certain areas. Hurricanes, the other threat to the forest, have the potential of wiping out large tracts of trees. If a multilayered forest is maintained, the chance for major hurricane destruction diminishes. In addition, damaged forest areas will regenerate more rapidly.

Wachusett Reservoir Watershed: Ten percent of the Wachusett watershed is owned by the MDC. Compared to the former two watersheds, the Wachusett watershed is highly developed. The MDC does not have direct control of the watershed area; thus, the watershed is subject to increased development and undesirable land uses. The principal threats to water quality at Wachusett are due to the extensive development and industrial and commercial uses near the reservoir and tributaries.

In an effort to curtail the unwise development of the watershed, MDC has assembled a guide to bylaw adoption for towns within the watershed. The guide outlines tools to control growth and unregulated activities. The tools available in most towns include zoning bylaws, Board of Health regulations, conservation commissions, fire departments (for fuel storage), and planning boards. MDC has produced model bylaws to address aquifer protection districts, soil erosion and sediment control, open space zoning, site plan review, and environmental analysis of subdivisions.

In addition to the achievements previously mentioned--the gull harassment program, land management and acquisition plans, and the generation of model bylaws, the MWRA/MDC have accomplished several other gains. The Cohen Bill also known as the Watershed Protection Act was passed in 1992 to protect the Quabbin, Ware River, and Wachusett watersheds. It regulates certain land uses and activities in two buffer zones. The first buffer includes areas within 120 meters (400 feet) of the reservoirs and within 60 meters (200 feet) of tributaries and surface waters. Any alterations in the first zone are prohibited. The second area consists of land between 60 meters and 120 meters of tributaries and surface waters, wetlands, floodplains, and some aquifers within the watersheds. In the second buffer zone, certain activities such as storage of gasoline, road salt, or other potentially hazardous materials are prohibited.

A comprehensive water sampling and analysis program has been implemented. Bacteriological monitoring of the reservoirs and tributaries is implemented to protect public health. Algae populations are monitored to help prevent algal blooms that can cause taste and odor problems. Storm event sampling is conducted to help detect failed septic systems and other potential pollution sources. Sanitary surveys are routinely done in watershed subareas.

The agencies have taken steps to mitigate some of the existing problems with subsurface disposal systems. For example, a community outreach program is in effect in the Quabbin area to teach communities about the proper operation and maintenance of their septic systems. In the Wachusett area, studies have been conducted to identify sewerage needs. In order to mitigate the problems identified in the study, monies have been appropriated to fund the design and construction of sewer system extensions and innovative subsurface disposal systems.

Conclusion

The MWRA and MDC have been committed to watershed protection for many years. Over the years, as the search for a potable water supply moved westward, water officials realized the importance of protected watershed areas. With each new development of a water supply, additional protection measures were achieved. In addition, the Safe Drinking Water Act (SDWA) served as a further impetus to protect the water supply. Recent amendments to the SDWA, particularly the Surface Water Treatment Rule (SWTR), have caused water agencies to study the idea of watershed protection in greater detail. The SWTR requires large surface water suppliers to filter their water unless a waiver of the rule is granted. Because of the rural nature of the Quabbin and Ware watersheds, the pristine nature of the reservoir and watershed has been easier to control. It is likely that the Quabbin will be issued a filtration waiver because of the strong watershed protection plan, the percentage of land controlled by State agencies, and the quality of the source water. Wachusett Reservoir, on the other hand, will require a much stronger effort to receive a waiver of the filtration requirement. The watershed is more densely populated, less land is owned or controlled by state agencies, and commercial and industrial users are located within the watershed. Nonetheless, a strong watershed protection plan and the careful monitoring of potential threats to water quality will be beneficial to the Wachusett watershed regardless of filtration requirements.

A Wastewater Management Plan for the Wachusett Reservoir Watershed Holden and West Boylston, Massachusetts

Joseph M. McGinn, Kenneth W. Carlson, Peter J. Goodwin[1]

Abstract

The Metropolitan District Commission (MDC) Division of Watershed Management (DWM) is responsible for the management of watersheds and reservoirs that supply drinking water to a large population in Massachusetts. Since the Wachusett Reservoir is a critical part of the reservoir system, the MDC-DWM undertook the preparation of a comprehensive wastewater management facilities plan for the towns of West Boylston and Holden to address water quality and public health concerns from failing on-site wastewater disposal systems in the watershed. The plan recommends both a sewering program and a remediation program for on-site wastewater disposal systems.

Introduction

The MDC-DWM is responsible for the management and operation of a system of watersheds and reservoirs located in central Massachusetts 96.5 km (60 mi) west of Boston, which serve as the source of public water supply for over 2.5 million people in the Boston metropolitan area. Within this system, the 246 x 10^6 m^3 (65 x 10^9 gal) Wachusett Reservoir serves as the primary source reservoir prior to distribution. In order to protect water quality, DWM performs water quality monitoring and sanitary surveys every three years for the sanitary districts established within the watersheds.

In addition to the performance of sanitary surveys, the DWM, in conjunction with the Massachusetts Water Resources Authority (MWRA), completed the development of Watershed Resource Protection Plans for each major watershed in conformance with the requirements of the Massachusetts Department of Environmental Protection (MDEP), established pursuant to the Surface Water

[1]Director, Division of Watershed Management, Metropolitan District Commission, Boston, Massachusetts; Associate and Senior Project Engineer, Weston & Sampson Engineers, Inc., Peabody, Massachusetts

Treatment Rule provisions of the 1986 Safe Drinking Water Act Amendments.

Based on the sanitary surveys and the Wachusett Reservoir Watershed Protection Plan, it was concluded that inadequate or failing on-site wastewater disposal systems (septic systems) were responsible for significant water quality degradation in several tributary streams, a number which discharge directly into the Wachusett Reservoir. Relative to all threats to water quality, septic systems were ranked as the highest existing threat to water quality within the watershed.

The consistent recommendation of every watershed survey and protection plan was that a wastewater facilities plan was required to determine where sanitary sewers might be considered the most appropriate and cost-effective approach to resolve the most severely impacted streams and watercourses. It was further recommended that the facilities plan should address the issues of preventing and remediating problems of failed systems in areas of the watershed which were determined to be inappropriate for sewering and reliant on septic systems for wastewater disposal.

The DWM developed a scope of work for the development of a facilities plan for the towns of West Boylston and Holden, within which the water quality degradation associated with failed septic systems was most clearly evident and pronounced. These two towns are just north of and adjacent to Worcester, the largest city in central Massachusetts, and a major portion of the Wachusett Reservoir lies within the town of West Boylston.

Part A - Sewerage Facilities Plan

Twenty-six potential sewer service areas were identified in the two towns and evaluated for wastewater disposal need. The needs analysis concluded that twenty-three areas should be sewered and the remaining areas in both towns could continue with on-site wastewater disposal systems. The areas recommended for sewering were then prioritized, based upon the severity of existing wastewater disposal problems and proximity to the reservoir and tributaries. Two areas were determined to require a fast-track schedule to alleviate existing water quality and public health problems.

Wastewater flows were developed in order to size and cost the proposed wastewater collection facilities. Flow from most of the sewered area will be conveyed to the Upper Blackstone Water Pollution Abatement District (UBWPAD) plant in Millbury via the city of Worcester sewer system and the MDC Rutland/Holden Interceptor and Relief Sewers. Average wastewater flows for the sewered area were determined to be 44 L/s (1.0 mgd) in the initial year, 66 L/s (1.5 mgd) in the design year 2020 and 75 L/s (1.7 mgd) with future full development.

Three sewering configuration options are evaluated for West Boylston using a conventional/pressure sewer alternative. A "decentralized" option was selected, placing several smaller pumping stations in the system versus one larger station adjacent to the reservoir. The areas in Holden would be serviced by extensions of their existing conventional sewer system, some of which have already been designed.

The plan recommends that wastewater collection system components be designed and constructed with a high degree of watershed protection to mitigate potential adverse environmental impacts. Recommended measures include standby pumps and power; wireless telemetry systems; wastewater storage provisions; bypass connections for force mains and pumping stations; emergency overflow containment, treatment and disinfection; and sleeved stream crossings.

The plan recommends a six phase implementation plan to complete construction within seven years of initiation, consistent with MDC's desire to address EPA Safe Drinking Water Act requirements. The two fast-track projects would be designed and constructed within the first two years, addressing the most critical problem areas. Design and construction of the remaining improvements would be accomplished under two concurrent master schedules, one for each town, over the seven year period.

The recommended wastewater collection system includes approximately 94,483 m (310,000 ft) of gravity sewer, 1,524 m (5,000 ft) of pressure sewer, 24,383 m (80,000 ft) of force main, twenty-three pumping stations and two community wastewater collection, treatment and disposal systems. The estimated capital cost for the recommended project is $50 million ($14,000/lot). Of this, over $5 million ($1,200/lot) in added capital costs can be attributed to watershed protection measures to be incorporated in the project. The estimated capital cost of repairs and remediation for on-site systems in the proposed sewered areas is over $41 million ($10,000/lot), without most of the repaired or replaced systems meeting new state standards.

Legal and institutional requirements for the recommended plan include the negotiation and execution of a number of intermunicipal and interagency agreements. Other actions include filing for necessary environmental and technical reviews with state and local agencies including compliance with the Massachusetts Environmental Policy Act (MEPA) and MDEP regulations where the critical issues will be interbasin transfer of water and environmental review by appropriate state and local agencies. Financial requirements for the recommended plan include determination of financial responsibility between the state and the towns; and authorization and appropriation of funds by the state and towns to finance the plan.

Part B - On-Site Remediation

The focus of Part B included on-site system problem identification and prioritization for the entire study area; evaluation of both short- and long-term remediation measures for use in proposed sewered and non-sewered areas; and a recommended repair/remediation plan.

Both short- and long-term on-site system remediation measures are evaluated for use in the study area. Short-term remediation measures, such as water conservation and system modifications, are proposed to prolong the use of problem on-site systems in areas proposed for sewering, without having to meet all applicable state and local regulations. This would improve existing water quality and public health at a reasonable expense to owners of problem systems that eventually are to be sewered. The plan estimates that 765 (24%) of the systems in the proposed sewered areas would require short-term remediation during the seven-year sewering program at a cost of $7.65 million.

Long-term remediation measures, intended for use in areas where sewers are not being proposed for long term wastewater management included short-term measures and a number of innovative/alternative on-site disposal systems that could meet all local and state regulations. The plan estimates that 1,422 (50%) of the systems in the non-sewered areas would require long-term remediation during the twenty-five year planning period at a cost of $17 million.

The plan recommends that a pilot demonstration project be undertaken for five selected on-site system alternatives. The estimated cost of the pilot program is $160,000 with sampling and analyses for performance monitoring performed by MDC. The capital cost of the recommended repair/remediation program, including short- and long-term remediation; an on-site system inspection and maintenance program; public information program; and pilot demonstration project is estimated at $25 million, with an annual O&M cost of $260,000.

The plan identifies financing mechanisms for remediation of on-site systems including a recently enacted state law allowing local boards of health to provide loans to homeowners to repair on-site systems through a betterment assessment program. Legal and institutional arrangements to allow remediation of on-site systems with state or local funds are also identified, including the establishment of a wastewater management district specifically for areas with on-site systems.

Public Participation

A comprehensive public participation program was an integral part of the facilities planning process. A Wastewater Management Committee was formed with representatives from the two communities; local and state agencies; and the MDC. The committee provided valuable input during the planning process while conducting eight (8) public meetings. The MDC conducted a formal public hearing to receive testimony on the draft plan.

Vulnerability Zone Identification: A Watershed Management Tool for Protecting Reservoir Water Quality

Carol Ruth James[1], Karen E. Johnson, and Edward H. Stewart

Abstract

As part of an extensive watershed management project, an approach was developed for integrating the potential impacts of watershed physical characteristics on raw water quality. This approach uses the capabilities of a geographic information system (GIS) database to produce water quality vulnerability zone maps which, in turn, can be used to develop watershed management strategies. Therefore, these vulnerability zone maps can be used by utilities to proactively manage the watersheds so that source protection becomes an effective first barrier to water quality degradation and to ultimately improve the water quality. An overview of the conceived approach used to produce the water quality vulnerability zone maps is documented below.

Introduction

With the ever increasing regulatory emphasis on raw water quality and corresponding treatment recommendations, surface water utilities need management tools which enable them to strengthen their first barrier to water quality degradation -- their reservoir watersheds. Because of the resultant loss of reservoir volume, utilities have historically viewed the minimization of sediment loading as their primary watershed management strategy. From a water quality perspective, algal blooms and taste and odor concerns have driven the use of in-reservoir treatment controls. Over the last five to ten years, however, because of the emphasis on trihalomethanes (THMs) and now other disinfection by-products (DBPs), utilities have begun to look at ways of controlling THM and other DBP precursors in the watersheds. This is the first proactive movement towards controlling physical characteristics of the watershed to specifically enhance water quality.

This article highlights an approach developed to relate physical characteristics of watersheds to reservoir water quality in an effort to develop better watershed management tools. As part of a project conducted for the San Francisco (Calif.)

---

1 Carol Ruth James (Principal Engineer) and Karen E. Johnson (Supervising Planner) - Montgomery Watson, 355 Lennon Lane, Walnut Creek, CA 94598, (510) 933-2250.

Edward H. Stewart, Watershed Resource Manager, San Francisco Water Department, 1000 El Camino Real, Millbrae, SF 94030, (415) 872-5931.

Water Department, a four-step approach was conceived.

Step 1 Develop relationships between each selected watershed physical characteristic and each water quality parameter group of concern

Step 2 Map these relationships for each water quality parameter group

Step 3 Composite these relationship maps into a vulnerability zone map for each water quality parameter group

Step 4 Composite the individual vulnerability zone maps into one general water quality vulnerability zone map

Step 1 - Develop Relationships Regarding Physical Characteristics

Following the approach outlined above, the physical characteristics of the five 63,000-acre local watersheds were analyzed for source and transport vulnerability for five groups of water quality parameters -- particulates, THM and other DBP precursors, microorganisms (e.g. *Giardia* and *Cryptosporidium*), nutrients (nitrogen and phosphorus), and synthetic organic chemicals. The results of these analyses were entered into a GIS database. Relationships were developed between the five physical characteristics -- soils, slope, vegetation, wildlife concentration, and proximity to water -- and the five water quality parameter groups. For example, soils were evaluated according to their density, organic carbon content, moisture content, nutrient content, and adsorption capacity, respectively. A similar type of analysis was conducted in order to develop relationships between the other four watershed physical characteristics and the five groups of water quality parameters.

One of the more challenging relationships to develop was that regarding proximity to water. This physical characteristic integrates information pertaining to overland flow including precipitation and land surface characteristics (surface roughness, distance, and slope). A minimum distance of 300 feet from the high water line was established for the "high" proximity to water vulnerability zone. This distance was established for the watershed with the least precipitation and was in turn modified to reflect the increased precipitation and relative differences in land surface characteristics associated with the different watersheds. A widely used time of concentration formula for ungauged watersheds, developed by C.E. Ramser (1927), was used.

Tc = [2.2 n (Lo) / (So)^0.5]^0.467

where: n is the Manning's roughness coefficient
Lo is the overland distance from point of farthest contribution in meters
So is the average slope

Solving this formula for Lo resulted in high proximity to water vulnerability zones ranging from 300 to 1,300 feet from the high water lines for each water body (reservoir and tributaries) in each watershed.

Step 2 - Map the Physical Characteristic and Water Quality Relationships

The results of the analyses conducted in Step 1 were entered into an extensive project GIS database. Relationship or vulnerability maps were then prepared for each physical characteristic for each water quality parameter group. These maps identify the areas -- high, medium, and low -- of each watershed which, if ineffectively managed, could adversely affect the water quality of the corresponding receiving waterbody(ies).

Step 3 - Develop Composite Water Quality Vulnerability Zone Maps

Once Step 2 was completed a compositing approach was defined in order to combine the five relationship maps for each water quality parameter group into one vulnerability zone map for each group. The composite approach was based in part on information gathered from existing watershed management plans. In general, the compositing was based on the approach summarized below.

- The high vulnerability areas defined for the "proximity to water" layer take precedence and, therefore, are defined as high vulnerability zones on all water quality composite maps.

- For other areas to be defined as a "high", slope must be high and either soils or vegetation must be high.

- For areas to be defined as a "low", two of either slope, soils, or vegetation must be low.

- All other areas not already defined based on the three points above are defined as medium.

Verification of the composite approach was achieved through the use of the agricultural industry's Universal Soil Loss Equation (USLE). As yet a technique has not been developed for the water industry. However, the USLE closely relates to water quality concerns because it places a higher weighting on the soil types (clays, loams) which are more easily transported and ultimately can become a water quality/treatment problem. Once verified, high, medium, and low vulnerability zones were defined and mapped for each water quality parameter group.

Step 4 - Develop One General Water Quality Composite Map

A general water quality vulnerability zone map was developed for each watershed by combining the vulnerability zone information reflected in each of the five water quality parameter group maps. Where conflicts occurred between individual maps the more restrictive vulnerability zone rating was used. For example, if an area was rated as high vulnerability for particulates but medium vulnerability for microorganisms' contribution, then a high vulnerability rating was assigned in the general water quality vulnerability zone map. The general water quality vulnerability zone maps are, in turn, being used for developing general management strategies as they pertain to water quality protection. In the event a specific water quality issue needs to be addressed then the individual water quality parameter group maps can be accessed. Figure 1 depicts the composite general water quality vulnerability zone map for three watersheds.

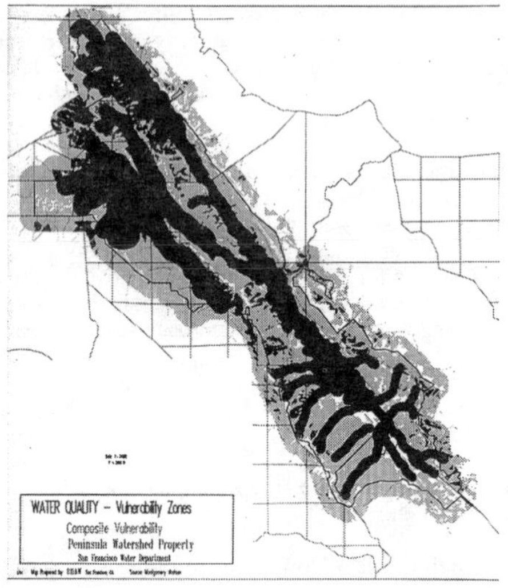

Figure 1. Water Quality Vulnerability Zone Map

Conclusions

Information regarding the vulnerability zones is being used as part of a comprehensive watershed management plan being developed by EDAW, Inc., which also addresses issues such as fire management and wildlife protection management. Identification of the water quality vulnerability zones leads to the identification of management strategies regarding future activities in the watersheds and corresponding mitigation activities which must be pursued to maintain or enhance the current reservoirs' water qualities. For example, knowing that an area in the watershed is a high vulnerability zone with regards to an increased particulate contribution targets that area for "no or low" soil disturbance activities and the implementation of erosion control practices as needed.

In addition, the development of the vulnerability zones highlights the need for additional quantitative information regarding watershed physical characteristics as well as corresponding water quality. Future monitoring programs can be developed based on the identified data gaps and the data produced can further supplement the existing GIS database and ultimately refine the vulnerability zone maps and corresponding watershed management programs.

The steps taken to develop the vulnerability zones, and ultimately the watershed management strategies, can be adapted to other watersheds and serve as a basis for utilities to effectively use their watersheds as the first barrier in protecting their water supplies, and an opportunity to improve their water quality.

Kensico Reservoir Water Pollution Control Study

Andrew J. Sharpe, P.E.[1]
Charles V. Beckers, Jr., P.E.[2]
David Parkhurst, PhD.[3]

Abstract

This paper summarizes the work currently being performed by the New York City Department of Environmental Protection (NYCDEP) and their contractors Roy F. Weston of New York, Inc. (Weston) and Lawler, Matusky & Skelly Engineers (LMS) to identify and quantify coliform sources within the Kensico Reservoir Watershed, assess the impact of selected coliform mitigation strategies on water quality leaving the Reservoir, and develop computer models of the watershed and Reservoir.

Introduction

High quality water from both the Delaware and Catskill watersheds is transported to the Kensico Reservoir in Westchester County, New York via a system of aqueducts and high pressure tunnels. The reservoir serves to balance flows and settle suspended solids present in the water. The combined Delaware and Catskill waters are chlorinated immediately after leaving the reservoir for delivery to the City of New York. The Kensico Reservoir is considered New York City's source water for purposes of compliance with the turbidity and

---

[1]Project Manager, Roy F. Weston of New York Inc., 465 Columbus Avenue, Valhalla, NY 10595

[2]Project Manager, Lawler, Matusky & Skelly Engineers, One Blue Hill Plaza, Pearl River, NY 10965

[3]Assistant Chief, Division of Water Quality Control, New York City Department of Environmental Protection, 465 Columbus Avenue, Valhalla, NY 10595

coliform requirements of the Surface Water Treatment Rule (SWTR).

The NYCDEP officially began compliance monitoring in April 1991. Since that time, effluent-water turbidity has generally been well below the SWTR limit of 5 ntu. On-going studies performed by the NYCDEP and Weston indicate that the Kensico Reservoir acts as a settling basin for removal of suspended solids (and hence turbidity). These same studies show that the watershed and Reservoir appear to act as sources of fecal coliform bacteria (FCB). In spite of this finding, water leaving the Kensico Reservoir has always met the SWTR criterion that no more than ten percent of fecal coliform samples can exceed 20 cfu/100ml in any six-month period in an unfiltered supply. Nevertheless, the NYCDEP recognizes the need for vigilance and is developing infrastructure management programs to sustain and improve upon this level of performance well into the 21st century. This is especially true in light of development pressures within the watershed and the desire to provide water that must meet increasingly stringent quality criteria.

Importance of Fecal Coliform Bacteria

Fecal coliform bacteria, because of both their origin from the intestinal tracts of warm blooded animals (including man) and their abundance and relative ease of detection, are used as indicators of potential contamination of water by human and other animal excreta. FCB *per se* are usually harmless to man, but their presence is taken to indicate potential contamination by a variety of pathogenic organisms that may pose risks to human health, including other bacteria, protozoa and viruses. It follows that controlling FCB sources will also tend to control these pathogen sources as well. For this reason the USEPA has established stringent criteria for FCB in potable water. This principle is at the heart of the work done by NYCDEP from 1991 to 1993 and by NYCDEP and its contractors during 1994. To put this work in context, a single person discharges from 100 to 400 billion coliform organisms per day and the SWTR requires FCB to be less than 20 cfu/100ml in ninety percent of the samples taken during any six month period. Kensico Reservoir's continuing compliance with this criterion demonstrates how successful the current watershed management practices and controls have been.

Conceptual Model for the Kensico Reservoir

The only sources of FCB and associated pathogens are human, animal and bird excreta. FCB or pathogens can reach the reservoir either in surface water or groundwater flows, as is the case with human, domestic pet, or terrestrial wildlife excreta, or by direct discharge, as is the case with gulls, waterfowl, or aquatic mammals. After discharge, this fecal material and its contained organisms are subject to various natural and manmade removal mechanisms, so that not all of the discharged FCB and pathogens necessarily reach the reservoir. For example, coliforms from human sources are controlled and treated either by septic systems or by sewers and publicly owned treatment works. While poorly maintained septic systems and leaking sanitary sewers have the potential to release FCB and pathogens into the surrounding soil and groundwater, the soil structure may partially or totally remove FCB from the flow path to the Reservoir. The FCB and pathogens deposited on the ground surface with the feces of terrestrial wildlife such as raccoons, deer, and shore-grazing geese or of domestic pets can be washed into the streams or directly into the Reservoir as a result of stormwater run-off. However, even in this flow there are mechanisms that retard the movement of bacteria and pathogens and reduce their numbers. Vegetation removes a portion as will the passage of water through wetlands and retention basins. Exposure to light, predatory organisms and high temperatures also reduces the number of FCB.

When the FCB and pathogens enter the Kensico Reservoir, they are subject to additional removal mechanisms. One mechanism is the settling out of bacteria and suspended solids with which they are primarily associated. A second removal mechanism is the die-off of the FCB and pathogens that do not survive and reproduce easily in the open environment. The effectiveness of these mechanisms depends primarily on the residence time of the reservoir and secondarily on the physical, chemical and biological properties of the water body. Seasonal differences in the temperature, and therefore density and stratification of the water, influence the extent of mixing and dilution of the influent water as well as the rate and extent of suspended solids settlement. FCB and pathogens that do survive tend to be held in the sediments and deep water layers, and are unlikely to contribute significantly to the FCB and pathogen levels in the Reservoir effluent waters during periods when the water column is strongly stratified.

Kensico Reservoir Water Pollution Control Study (KRWPCS)

The mechanisms discussed above can be thought of as representing the buffer capacity of the Reservoir with respect to FCB and pathogens. The purposes of the KRWPCS have been to assess the degree to which these mechanisms are effective, to look for the conditions under which this buffer capacity is exceeded or bypassed, and to examine means for controlling these exceedances. Exceedances could happen following a storm event, the direct discharge of fecal material close to the effluent chambers by gulls and geese, or short circuiting of the flow leading to reduced settling and incomplete mixing of the influent water.

The KRWPCS performed by Weston and its subcontractor LMS studied the following FCB sources within the watershed: Sanitary sewers and septic systems, streams, stormwater, gulls, geese and groundwater. The results of the sanitary sewer gauging, internal television inspection, review of septic system failures, and dye studies did not strongly support the notion that these facilities are major sources of FCB through exfiltration or leachfield failure. Only infiltration to the sanitary sewers was observed but under appropriate groundwater conditions exfiltration may occur. Therefore, continued monitoring, inspection and surveillance of these sanitary facilities was recommended. Data gathered and reviewed during the study strongly supported the idea that stormwater, gulls and geese were important FCB contributors and that gulls were at least as important as geese in FCB loadings. Groundwater monitoring wells were installed at a number of locations around the Reservoir. The locations were selected based on geology and proximity to the Reservoir or to potential pollution sources. Analysis of groundwater samples indicated that coliform contamination occurred in some shallow wells, but not in the deep bedrock wells.

Reservoir bathmetry was mapped and dye tracer studies performed to increase our understanding of water movement within the Reservoir. These studies showed that the Reservoir exhibits a three-layer flow pattern when stratified. Water from the aqueducts seeks its own density level between the colder (denser) hypolimnion and the warmer (less dense) epilimnion. Also the dye front traveled from the aqueduct influent chambers to the effluent chambers in 3 to 5 days, but the average detention time during the dye studies was about equal to the gross detention time of 15 to 20 days. Sediment analysis indicated that the bottom was generally free of biological and chemical contamination with the exception of some sites near streams where coliform levels were elevated.

A number of structures for control of FCB in stormwater were recommended. They consisted of detention basins to settle out turbidity and FCB, diversion dams to divert storm flows to adjacent wetlands, stilling basins to reduce erosion at storm sewer effluents, water quality inlets to remove sediment from street runoff, and stream bank stabilization using native vegetation to reduce erosion.

The KRWPCS also examined the relocation of the effluent chambers, and mixing of the Reservoir water to reduce FCB leaving the reservoir. Review of the water quality data indicated that improvements would be marginal at best, especially in light of the high cost of these facilities.

Watershed and reservoir computer models (SWMM and RMA-10 respectively) were developed to determine the impact of the various FCB mitigation strategies and to assist the NYCDEP in reservoir operations.

Preliminary Conclusions

The various reports resulting from the KRWPCS are currently under review by the NYCDEP, and conclusions stated herein are therefore preliminary in nature. The studies performed to date indicate that harassment of the gulls, geese and other waterfowl as well as stormwater control structures, have, or are likely to have a substantial impact on FCB levels. Beyond this, the most promising coliform mitigation strategies in descending order of impact are construction of a wall diverting streams away from the effluent chambers, continued internal television inspection of sanitary sewers, and surveillance of the watershed to identify and correct leaking septic systems and other pollutant sources.

Through the efforts expended in this study and the on-going studies and facility design being performed by the NYCDEP, Kensico Reservoir source water can be expected to continue to meet SWTR criteria.

CAN WATERSHED MANAGEMENT MAKE A DIFFERENCE

Ronald Sharpin(1), Al Dion(2), and Rick Stevens(3)

The city of Groton, Connecticut depends primarily on surface water from the Great Brook and Billings Avery watersheds to supply drinking water to its users. The 40 square kilometer (15.4 square miles) watershed includes the town of Ledyard as well as the city of Groton and provides an estimated yield of 43,532 cubic meters per day (11.5 million gallons per day). The city owns and controls 11.7 square kilometers (4.5 square miles) of the watershed which consists of a series of impounding reservoirs ranging in size from 4 hectacres (10 acres) to 125 hect acres (300 acres) and depths of 3 to 20.4 meters (10 to 67 feet). The raw water is treated by conventional treatment processes including multi-media rapid sand filtration. The treated water is of high quality but the raw water has been exhibiting increasing color of the last few years due to increasing vegetative growth in the reservoir. Large numbers of Canadian geese also inhabit the watershed. The Groton Utilities Department is concerned about the wildlife impact on the Giardia and Cryptosporidium requirements of the Safe Drinking Water Act's emerging Surface Water Treatment Rules.

The city currently administers a watershed protection program that includes annual sanitary surveys but believes an enhanced program could improve raw water quality and minimize the need for treatment plant up-grades and additional costs. An evaluation was done of the proposed enhanced Watershed Protection Program and the relationship between raw water quality and treatment costs. The program follows the schematic shown in Figure 1.

(1) Environmental Consultant, 17 Wessex Road, Newton, MA
(2) Superintendent, Department of Utilities, Groton, CT
(3)Chief Engineer, Department of Utilities, Groton, CT

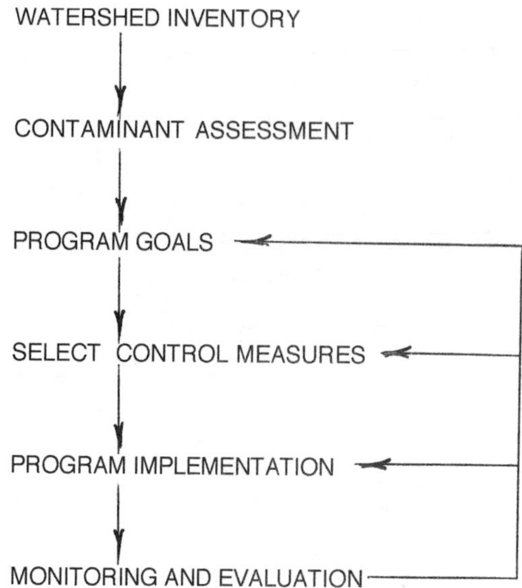

Maps have been prepared and contaminant threats identified. Program goals of the Watershed Protection Plan include:

* Enhancing the raw water quality
* Meeting the existing SWTR and the proposed rules for microcontaminants and disinfection by-products
* Containing costs of treatment
* Developing an integrated protection plan that includes the town of Ledyard and various interest groups

Control measures are selected for the identified contaminant threats based on effectiveness, costs, and compatibility. The plan prefers to utilize land-use controls for protection from potential contaminant threats and structural controls for existing threats. Major contaminant threats identified in the planning and suggested control measures are shown in Table 1.

TABLE 1 THREATS-CONTROLS

ACTIVITY	THREAT	CONTROL
Timber Harvesting	Sediment, Nutrients	Erosion Controls Wet Weather Restrictions Avoid Sensitive Areas On-Site Management
Stormwater Runoff	Sediment Nutrients Vegetative Growth	Limit Input to Reservoir Mechanical Harvesting Grass Carp
Hazardous Materials	UST Leaks Hydrocarbon Spills	UST Restricted Zones UST Containment Vaults UST Management Emergency Response
Raised Reservoir	Drowned Vegetation - Nutrient Input	Clear Shoreline Vegetation

WATER QUALITY PROTECTION MEASURES RELATED TO ROADWAY CONSTRUCTION IN A WATER SUPPLY WATERSHED

William R. Arcieri, Dr. William J. Barry,[1] Roger L. Sanborn, P.E., Donald Lyford, P.E.,[2]

Abstract

Throughout the country highway runoff has received increased scrutiny as a potential contamination source to surface water resources. The recent court ruling in California provides a dramatic example in which a state transportation agency was found to be in violation of the Clean Water Act for not implementing adequate Best Management Practices (BMPs) to reduce water quality impacts resulting from highway runoff. Previous studies have demonstrated that under certain conditions, highway runoff can contain elevated levels of various pollutants(Gupta et al, 1981; Rexnord, 1985). The pollutant sources include exhaust emissions, oil drippings, inadvertent spills, and the wear and tear of moving parts (i.e., brake pads, tires). Particularly, when public water supplies are involved, a combination of BMPs may be needed to obtain adequate treatment. This paper describes the development of an innovative highway drainage system designed to protect a municipal water supply in southeastern New Hampshire. The system relies on various BMPs to not only enhance highway runoff quality but to also provide initial containment in the event of an accidental spill. Maintaining a sufficient balance between the water quality, safety and wetland protection goals was an important aspect in receiving final agency approval for this relatively unique highway drainage design.

---

[1]Senior Water Resource Scientist and Project Manager, respectively. Normandeau Associates. 25 Nashua Rd. Bedford, NH 03110.

[2]Project Manager and Design Engineer, respectively. New Hampshire Department of Transportation, 6 Hazen Drive, Concord, New Hampshire 03302

Project Overview

The New Hampshire Department of Transportation (NHDOT) is currently in the preliminary design phase of a proposed widening of Route 101, a major east-west, two lane highway in southeastern New Hampshire. Approximately 1.2 miles of the expanded highway will traverse through the watershed of Dearborn Brook, the principal feeder stream to a municipal drinking water reservoir located approximately 0.5 mile downstream(Figure 1). Given the relatively short time of travel, the potential impacts from an inadvertent hazardous spill along this travel corridor has always been a serious concern. In addition, pollutant loading estimates indicated that heavy metal concentrations in runoff could pose a long-term threat to the water supply. Since there is no reasonable means to divert runoff out of the watershed, various measures will be incorporated to protect this drinking water source.

Design Objectives

The primary goals of the proposed highway drainage system are to reduce pollutant levels in highway runoff below aquatic life and human health criteria and to provide initial hazardous spill containment. To this end, it was considered critical to be able to collect and control runoff from the entire roadway within the watershed as well as avoid direct discharges to Dearborn Brook. Furthermore, since the groundwater recharged within the project corridor also flows toward the reservoir, preventing groundwater contamination was another important factor.

Drainage System Features

To provide the desired high level of protection, it was recognized early on that a combination of various types of BMPs(i.e., grass swales, oil and grease traps, detention basins, etc.) would be required. In selecting a location for a detention basin, the Dearborn Brook crossing represents the general low point along the right-of-way. However, the construction of a detention basin in this steeply sloped area would require major amounts of earthwork to create a large enough berm around the proposed basin. This amount of disturbance to this relatively stable wooded buffer along the stream corridor would, in itself, jeapordize the water quality of Dearborn Brook due to the potential sedimentation and erosional losses.

As a result, a closed drainage system with curbing and guardrails along the edge of pavement was initially considered; however, it conflicted with the desired safety objectives. The added expense and maintenance as well as the lack of pretreatment typically provided by

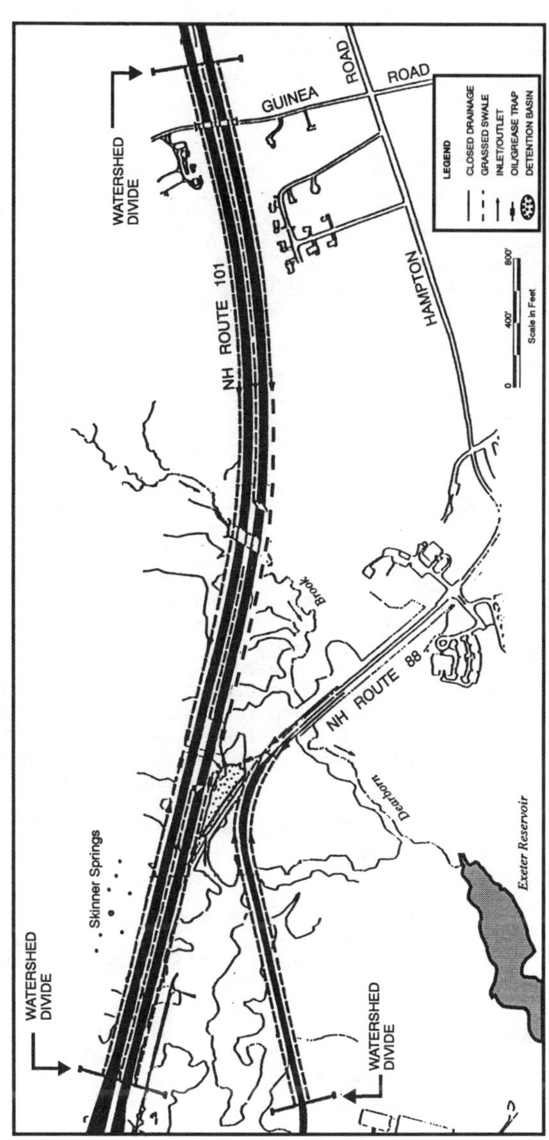

Figure 1. Proposed Dearborn Brook Runoff Containment System

grass swales were also undesirable aspects. In addition, concerns were raised about the ability to contain spills that occurred along the shoulder where truck overturns are likely. Guardrails are generally not considered adequate to prevent trucks from leaving the pavement.

A design alternative was developed by incorporating 40-foot wide vegetated shoulder areas that would not only accomodate grass swales for stormwater conveyance but also function as vehicle recovery areas enabling vehicles to regain control after leaving the pavement. This design alternative combines the attributes of both open and closed drainage designs by having complete runoff containment as well as enhance runoff quality through particulate trapping and vegetative uptake. The widened shoulder area will be designed with gradual 10:1 front and backslopes. Groundwater protection objectives are also met by incorporating an impervious borrow layer beneath the widened shoulder area. In addition, with a 2:1 side slope outside the vehicle recovery area, the redesigned highway width is only slightly larger than that with the original 4:1 side slope embankemt, thus minimizing added wetland impacts and ROW aquisitions.

By extending the length of the bridge structure at the Dearborn Brook crossing, grass swales at the roadbed grade can convey runoff across the brook to a centrally-located detention basin. All runoff will be routed through large oil and sediment traps prior to entering the detention basin. The trap design will be similar to the three-chambered, water quality inlets presented by Schueler (1987). The detention basin will provide extended detention for the 2-year, 24 hour storm event and will include a sediment forebay, a permanent wetland area near the outlet, and a manual shutoff valve on the outlet pipe to contain an inadvertent spill. The basin will outlet to an intermittent stream that eventually empties into the reservoir. A more detailed discussion on detention basin design for water quality purposes can be found in Schueler (1987) and Yu (1993).

References

Gupta, M.K., R.W. Agnew, N.P. Kobringer. 1981. Constituents of highway runoff; FHWA/RD-81/042.

Rexnord, Inc. 1985. Effects of highway runoff on receiving waters. FHWA/RD-84/065.

Schueler, T.R. 1987. Controlling urban runoff: A practical manual for planning and designing urban BMPS. Metro. Washington Council of Governments.

Yu, S.L. 1993. Stormwater management for transportation facilities. Transportation Research Board. NCHRP Synthesis 174. ISBN 0-309-04923-7.

Role of Engineers in Multidisciplinary Hydropower Planning: A Case Study

S. F. Railsback[1], M. ASCE, and M. J. Sale[2]

Abstract

Licensing multiple hydropower projects in the upper Ohio River basin typified situations where engineers work on an multidisciplinary team to support regulatory decisions. Engineers have unique capabilities for quantifying competing objectives and their tradeoffs. It is crucial that input from all professions and parties be given due consideration in the decision process. Engineers can enhance the decision process by soliciting and quantifying the objectives of all participants, designing alternatives that manage the system best for these multiple objectives, and quantifying the tradeoffs between these alternatives.

Introduction

The objective of this paper is to illustrate, using a hydropower licensing example, how water resource engineers can promote effective watershed management decisions. This illustration is made by defining who the decision-makers were, what the technical role of engineers and other professionals were, what non-technical information was important, and examining how engineers interacted with others in decision-making. The case studied concerns environmental impact analysis of large-scale hydropower development but the conclusions are applicable to many complex water resource management processes.

From 1986 through 1988, staff of Oak Ridge National Laboratory assisted the Federal Energy Regulatory Commission (FERC) in preparing an environmental impact statement (EIS), the main planning document for licensing new hydropower projects at 19 existing navigation dams in the upper Ohio River basin (FERC 1988, Railsback et al. 1989). The primary water resource concern with these projects was the potential for cumulative decreases in dissolved oxygen (DO) concentrations; at the dams water is aerated as it spills over crests or through gates, but the hydro

---

[1] Consulting engineer, Lang, Railsback & Assoc., 250 California Avenue, Arcata, CA 95521

[2] Hydrosystems Group Leader, Environmental Sciences Division, Oak Ridge National Laboratory, P.O. Box 2008, Oak Ridge, TN 37831-6036. Research sponsored by the U.S. Department of Energy under contract DE-AC05-84OR21400 with Martin Marietta Energy Systems, Inc.

projects route water through non-aerating turbines instead. Models indicated that the projects as proposed would cause cumulative decreases in DO to levels below state water quality standards under critical conditions. Mitigation in the form of spill flows can increase DO concentrations but reduce power production. For the EIS, we developed a basin-wide resource tradeoff model that selected spill flows at each dam to optimize basin-wide hydropower production while meeting a specified DO criterion. The EIS considered alternative tradeoffs between DO and power production that resulted from the alternative management objectives of maximizing power production, meeting state DO standards (5 mg/L), and (the alternative implemented by FERC) avoiding any water quality impacts to fisheries by providing 6.5 mg/L of DO. Such other environmental impacts as turbine mortality to fish and water level decreases in wetlands were also important.

Who were the decision makers?

The proposed Ohio River basin hydropower projects were non-federal and therefore required licensing by FERC. The licensing process and its EIS define nearly all the environmental mitigation requirements for projects. The FERC commissioners are politically appointed and, typical of such commissions, base their decisions on analyses and recommendations prepared by staff. Although FERC's licensing authority has historically superseded other regulatory authority over non-federal hydropower, they must consult with and give full weight to recommendations made by fisheries and water quality management agencies. States also have water quality certification authority for FERC-licensed projects. Although FERC has greater authority to implement its decisions than most interstate water management agencies, this case study is representative of large-scale water resource decisions made by commissions and their staffs with input from other agencies.

What was the technical role of water resource engineers?

Water resource engineers for several parties had technical roles in the Ohio River Basin case. Hydropower developers' engineers designed the proposed projects. To meet their clients' objectives, these engineers appeared to design the projects to maximize power production or profitability of an individual project, while meeting what were perceived as minimum environmental standards. Engineers for the Army Corps of Engineers and other agencies were involved to protect the existing water resource benefits, most importantly existing river water quality, navigation, and flood control. Hydropower development was not an objective of these agency engineers.

We participated in the decision-making process as staff engineers for FERC by helping assess the social and environmental impacts of the projects, and by conducting the quantitative analyses that allowed comparison and balancing of the power and non-power benefits of alternative system management objectives. Our final recommendations in the EIS were in fact adopted by FERC in hydro licenses.

What other professions made important technical input?

The FERC decision-making process, like many other watershed management processes and all decisions for which an EIS is prepared, guarantees important roles to numerous other professions. Biologists for the resource agencies and FERC examined potential impacts on aquatic and wetland resources and recommended mitigation. Likewise, specialists in recreation, social, and cultural resources examined project effects and mitigation needs.

What non-technical input had a role?

Engineers are trained to base decisions on technical data and analyses. However, the following kinds of non-technical input were important in the Ohio River basin hydro case and almost all other watershed management decisions.

Laws and policies. Watershed management decisions are constrained in many ways by laws and agency policies. Individual laws and policies often seem ill-adapted to the decisions being made, but must be complied with to avoid court challenges. The Federal Power Act and FERC's regulations under this act govern how the licensing process proceeds and requires that water management decisions be coordinated with natural resource agencies. One of the most challenging laws and polices in the upper Ohio River case was state water quality standards, especially the non-degradation policies applied by some states to the Ohio River. We found the policy of preserving existing water quality in a river as heavily modified as the Ohio to be a source of great controversy and difficult to implement. In the absence of a workable definition of non-degradation, we our own, which was to provide DO concentrations (6.5 mg/L) that avoid adverse effects to fisheries. Other important laws and policies included the agencies' fisheries management objectives and the authorized purposes of federal water projects, which limited the Army Corps of Engineers to supporting only certain management objectives.

Values. As is typical, different parties in the case had different opinions concerning the relative value of power (and profit) versus environmental, social, and cultural resources. There were also different values applied to such economic resources as power production, commercial navigation, flood control, and recreation. Laws specify minimum levels of protection for many non-power resources, but decisions on mitigation above these minimum levels are based on values.

Public perceptions. There was surprisingly little participation by the general public or activist organizations in the upper Ohio River licensing case, although one river conservation group participated in an agency-led court challenge of the licensing decision process. For other hydro projects (especially in less disturbed environments) such participation is intense, with the public providing input based on values, emotions, and (sometimes) valuable technical information. As a relatively obscure, presidentially appointed, federal body, FERC often appears less subject to public and political persuasion than other water resource decision-making commissions.

How did engineers contribute to sound decision-making?

As their staff, we contributed the following kinds of information to FERC in support of their decision of what mitigation (especially spill flows) to require at the hydro projects that were licensed.

Synthesizing input from participants in the decision. In preparing an EIS, a scoping process is used to solicit both technical and non-technical input from other agencies and the public. In hydro licensing, mitigation recommendations and supporting technical information are solicited from resource agencies. The role of commission staff engineers in this process is to solicit this input and, giving due weight to the recommendations, concerns, and expertise of professionals in other fields, incorporate it into the decision process.

Synthesizing input includes critically reviewing its technical validity (with qualified assistance from aquatic biologists and other relevant professionals), but not rejecting input because it conflicts with the engineers' objectives or values. Engineers have a responsibility for ensuring our decision-making tools are based on credible science. However, under the laws governing hydro licensing and environmental impact assessment, engineers also have a responsibility to give full weight to the laws, policies, values, and concerns of other professions and the public in the decision-making process. In our opinion, engineers also have an ethical responsibility to not discount the concerns of other professions solely because it conflicts with our objectives or values; we are simply not qualified to judge the work of many other professions.

The history of FERC provides a good example of how excluding the objectives and values of other professions can damage a decision-making process. Although now recognized for having rapidly turned around and become a leader in restoring the environment at water projects, many in natural resource agencies and in the public still perceive FERC as biased towards power development; a lack of confidence in FERC's decision process appeared in this and other licensing cases to make some resource agencies more rigid in their consultations and mitigation recommendations. In fact, whether FERC's decisions in this case were based on sufficient unbiased technical information was challenged by some agencies in federal court. Although our analyses were upheld in court, avoiding even the perception of advocating particular objectives or values is important in a regulatory process.

Quantifying project effects for decision making. We find that in hydro licensing, as in most watershed management decisions, engineers are unique among participants in being equipped (and willing) to quantify the effects of proposed development on the environment and other resources. Such quantification is needed for the tradeoffs between resources to be compared and the results of alternative management objectives to be examined. In the upper Ohio River basin case, a system-wide DO model was developed to predict the effects on water quality of alternative spill flows. Another model predicted power production as a function of spill flows, so that the tradeoffs between alternative DO criteria and power production could be explicitly defined.

Designing good operating alternatives from alternative management objectives. Another contribution we made in the upper Ohio River basin case, which we consider very important but see in surprisingly few other water management decision-making processes, is applying engineering methods to design alternative operating policies that each are an optimal way to implement a different specific management objective. We used an optimization model to maximize power production under alternative DO criteria; in other words we designed system management alternatives (spill flows at dams) that each provided the most power production while meeting alternative DO objectives of 5 and 6.5 mg/L (Chang et al. 1989). Far too often EISs and other decision processes quantitatively compare alternatives that were arbitrarily chosen; there is no reason to believe that any of such alternatives is the best way to meet any of the management objectives.

Conclusions

The upper Ohio River basin case (as well as several other more recent watershed-scale hydropower licensing cases) illustrates how water resource engineers can facilitate sound decision-making by a commission. Part of our success may have been due to FERC's unusual regulatory authority, but several conclusions from the case are widely applicable to engineers' participation in watershed management.

Engineers have to function as part of multidisciplinary teams. Since watershed management decisions are based on many considerations that engineers are not trained in, we must give full consideration to input from other professions.

Engineers have unique capabilities for quantifying objectives and tradeoffs. Engineers are more likely than others to be trained in the systems science techniques needed to quantify and compare environmental, economic, and other objectives, and so often take the lead in developing decision-making tools.

Decision-making agency engineers should not advocate specific objectives or values. To avoid undermining the credibility of the decision-making agency and process, engineers should give full credence to the credible technical input, and to the non-technical concerns, of other parties in the decision process.

Engineers can enhance the impact assessment and decision process through the following steps. First, scoping and other means should be used to obtain concerns and information from all parties. The objectives of participants should be specified (modeled) quantitatively, with assistance from professionals in the appropriate fields. (It may be appropriate to specify the objectives of the decision-making body itself.) Next, in a step we feel is crucial but often neglected, alternatives that "optimally" implement each such objective should be designed. Finally, the tradeoffs, in economic, environmental, and other relevant currencies, between the alternative objectives should be modeled.

References

FERC (1988). "Hydroelectric Development in the Upper Ohio River Basin, FERC Docket No. EL85-19-114, Final Environmental Impact Statement." FERC/FEIS-0051. Federal Energy Regulatory Commission, Washington, D. C.

Railsback, S. F., M. J. Sale, and S-Y. Chang (1989). "Development of Flow Releases for Water Quality Protection at Hydroelectric Plants at Ohio River Basin Navigation Dams". In: *Water Resources Planning and Management, Proceedings of the 16th Annual Conference*. S. C. Harris, ed. American Society of Civil Engineers, New York.

Chang, S-Y., S-L. Liaw, M.J. Sale, and S.F. Railsback (1989). "Methods for Generating Hydroelectric Power Development Alternatives". In: *Proceedings of the Third Scientific Assembly of the International Association of Hydrologic Sciences, Baltimore MD, 10-19 May, 1989.*

Watershed Restoration in the Northern Sierra Nevada: A Coordinated Resource Management Approach

Donna S. Lindquist[1] and Larry L. Harrison[2]

Abstract

A collaborative erosion control program was initiated in California's northern Sierra Nevada in 1985 to reduce erosion and restore the health of the upper North Fork Feather River watershed. Pacific Gas and Electric Company (PG&E) and seventeen public and private sector groups joined forces to develop a watershed restoration strategy. Watershed degradation is a major contributor of sediments affecting operation of PG&E's downstream hydroelectric facilities, and the health and sustainability of environmental resources. To date, a comprehensive erosion control strategy was completed and the Coordinated Resource Management (CRM) process was successfully applied to implement 33 improvement projects. Committees were identified to address technical, financial and policy decisions. Both scientists and engineers play a key role in forming multi-disciplinary technical teams to plan and implement projects. Ecosystem management approaches to watershed restoration can be effectively facilitated by the CRM process.

Introduction

The management of California's water supply has presented challenges and growing controversy over the past century. Population growth, agricultural demand, land development and recreational use have placed heavy demands on a limited water supply, creating conflicts between water users. New water quality standards at both the State and Federal levels are currently being enacted to protect water quality and ensure that both environmental and economic issues are considered. As demand for high quality water increases, watersheds in the northern Sierra Nevada, where much of the State's water supply is produced, are receiving more attention. Many of these watersheds have been degraded by intense use. Once pristine meadows have been down-cut and drained; gullies replace once stable stream systems; sagebrush invades areas once dominated by sedges and willows; and the productivity and biodiversity of fish and wildlife populations have become greatly diminished. Resulting erosion and instability indicate poor resource conditions prevalent throughout the region, which has created political, socio-economic and environmental concerns.

---

[1] Senior Research Scientist, Pacific Gas & Electric Co., 2303 Camino Ramon, San Ramon, CA 94583

[2] Consulting Engineer, 209 Matsqui Road, Antioch , CA 94509

Building a Collaborative Framework

Funds from both government and private sectors to support watershed restoration are limited. Many rural communities have initiated grassroots coalitions to address watershed issues and to implement workable solutions. A notable example is found in Plumas County, California, located in the northern Sierra Nevada. There, more than 140 years of mining, grazing, timber harvesting, and road building have degraded watershed resources and increased soil erosion to seriously threaten the environment and economy of the region. By the mid 1980's at least 60 percent of the watershed had been damaged resulting in decreased soil productivity, poor water quality, frequent damaging floods, and loss of wildlife habitat. (Clifton, 1994).

A local coalition, later called the Feather River Coordinated Resources Management (CRM) Committee, was formed in 1985 to address these issues. Pacific Gas & Electric Company and seventeen public and private sector groups joined forces to develop a collaborative erosion control plan for the upper North Fork Feather River watershed. The primary goal was to improve water quality and quantity in order to enhance environmental resources and stabilize the declining economy of the region. The CRM Committee initiated its first project on a one mile reach of Red Clover Creek to demonstrate new, cost effective erosion control measures, and to test the ability of the group to cooperatively plan, fund and implement projects (Lindquist and Bowie 1988; Harrison 1991).

Coordinated Resource Management Process

The CRM process has been used successfully in many areas of California to provide a framework for locally-driven resource enhancement efforts. Since the CRM's authorization by the Federal and State governments in the early 1980's, approximately 200 CRM groups have been formed in California to cooperatively address resource management problems (Wills and Schramel 1992). Complex issues that extend over large geographic areas, and across multiple jurisdictions and land ownerships can be effectively resolved with this approach (Anderson and Baum 1987; Harrison 1991). Major resource uses are integrated into action plans that reflect the needs of individual stakeholders, minimize conflicts and are consistent with the capabilities of the land and associated resources. Decisions are made by consensus, so the needs of each participant are addressed in development of the final plans. These basic principles have been adopted by the Feather River CRM to organize the program.

It is essential that all parties with a stake in the land and resource issues be involved in the process. Since participation is voluntary, participants must recognize that the value of the benefits they will receive will outweigh the value of their contributions. The contributions of each participant are leveraged by the contributions of the others to provide cost effective benefits. Since the approach is widely used and accepted by land management agencies, CRM enhances credibility, visibility, and funding opportunities. In addition, the CRM process promotes an integrated approach to watershed restoration that embodies the principles of ecosystem management. Ecosystem management is emerging as a new framework for achieving harmonious and mutually dependent sustainability of society and the environment. It focuses on human and natural systems at regional scales over time. Current trends suggest that government agencies, regulators and academics recognize the need for resource management on a holistic level, which can be facilitated with the CRM process.

Feather River CRM Structure and Function

The Feather River CRM was formed to 1) maximize local initiative and control over watershed management issues, 2) to coordinate requests for Federal and State technical and financial assistance, and 3) to promote the development of inter-disciplinary technical teams to plan and manage projects. The Feather River CRM consists of four committees, which are composed of representatives from the eighteen CRM signatory organizations (Clifton 1994). The structure and functions of committees are described below.

The Executive Committee is responsible for policy guidance and dispute resolution, and support in the political arena. The Management Committee is responsible for administration of the program, including policy and budget decisions, approving new projects, identifying financial support opportunities, tracking required monitoring and approving project designs. The Steering Committee is composed of representatives from each contributing organization who review the program status, critique new projects, troubleshoot issues, and interact with landowners. And lastly, the Technical Assistance Committees (TACs), which are inter-disciplinary technical teams formed of volunteers from among the participants to plan and execute each project. The committees and overall program are managed and coordinated by Plumas Corporation, a non-profit economic development company. Plumas Corporation maintains full-time staff to prepare applications and reports, supervise workers, bid and administer contracts, administer funds and provide technical expertise to the project TACs. They have played a significant role contributing to the success of the program.

Technical Assistance Committees: the Role of the Engineer

Inter-disciplinary TAC groups are identified for each CRM project. Since each project has a unique set of challenges and technical needs, the composition of each TAC may vary based on the expertise needed. Generally, disciplines represented on TACs include terrestrial and aquatic biologists, soil scientists, hydrologists, botanists, range scientists, economists and engineers, and to a lesser extent, expertise in geology, geomorphology and cultural resources. The effectiveness of these teams will determine the success of the individual projects, and, ultimately, the success of the whole watershed program. TACs also provide a valuable forum for creative thinking, where technical concepts can be shared with inter-disciplinary critiques.

The TACs provide a unique opportunity for engineers who are often solely responsible for the design of hydraulic structures to work in inter-disciplinary groups to broaden their views. Watershed restoration work frequently requires that degraded stream channels be redesigned and constructed to a state that will again provide the benefits of a healthy "natural stream," providing for fish habitat, sediment transport and a healthy riparian zone as well as channel flow capacity and resistance to erosion. Designing a natural stream to achieve these objectives is much more complex than the usual challenges for hydraulic engineers of designing dams, canals, gate structures and revetments. There are no engineering codes governing the design of natural streams. Few engineering curricula include course-work needed to redesign streams to meet Nature's specifications. Therefore, the engineer must depend on the other specialists as needed to assemble a specification for stream restoration.

The design of natural streams must be coordinated by a liberal dose of practical experience and empirical observations of natural stream behavior. The design process will not be perfect or consistent because every stream and location provides unique design challenges. It is to be expected that portions of reconstructed streams may suffer damage when tested by high flows. These damages should not be seen as project failures, but rather as learning experiences and opportunities to test new solutions. It is only through such trial and error that engineers will learn the intricacies of natural stream design. When problems occur, they are systematically analyzed by the interdisciplinary team and repairs are made incorporating new ideas. Latter projects, consequently, experience much less damage upon the first test with a significant flood flow due to lessons learned.

When it comes to the design of natural channels the hydraulic engineer must understand that it is nature that has the advanced degrees, and to be competitive with nature's credentials, the engineer must recruit all the help possible from the other disciplines in the design of gradients, meanders, floodplains, revetments, bank vegetation, and enhancements for fish and wildlife habitat. Similarly, the other disciplines need to respect what the hydraulic engineer can bring to the table in designing channels to mimic nature, as well a providing engineered hydraulic control structures such as dams and sills and rip-rap if a more natural solution will not work or is too costly to implement.

Conclusions

- CRM provides a framework to organize multiple agencies and landowners to address complex watershed issues that cross jurisdictional and ownership boundaries.
- TACs composed of interdisciplinary teams are necessary to address the complex design issues of stream and watershed restoration projects.
- Project success depends on close interaction and cooperation between engineers and scientists on interdisciplinary teams. Integrating technical knowledge, hands-on experience, understanding lessons from nature, and professionalism are key ingredients for success.
- Ecosystem management approaches to watershed restoration can be effectively facilitated by the CRM process.

References

Anderson, E.W. and R.C. Baum. 1987. Coordinated Resource Management Planning: Does it Work ? *Journal of Soil and Water Conservation 42(3): 161-166.*

Clifton, C.C., (1994), "East Branch North Fork Feather River Erosion Control Strategy," *Plumas National Forest, prepared for the EBNFFR CRM Group.*

Harrison, L.L., (1991), "North Fork Feather River Erosion Control Program," In: *ASCE Waterpower '91 Conference Proceedings, 199-208.*

Lindquist, D.S. and L.Y. Bowie. 1988. "Riparian Restoration in the Northern Sierra Nevada: A Biotechnical Approach." In: *California Riparian Systems Conference Proceedings*, Davis, CA, September 1988.

Wills, L. and J. Schramel. 1993. "A Grass Roots Perspective: Feather River Coordinated Resource Management. In: *Watershed Management Council Proceedings: Overcoming Obstacles, Fourth Biennial Conference*, So. Lake Tahoe, CA: pg. 53-63.

Klamath Basin: Integrated Water Resources Management

Daniel D. Heagerty *

Abstract

As conflicts in watershed resource use increase, water resource management in the Klamath River Basin grows increasingly complex. Pressure to meet growing and often conflicting demands has led to water resource actions that compromise the biological processes in the basin. Historical patterns of watershed management are proving to have limited success in the face of changing demands. However, new management principles are emerging that preserve the health of the ecosystem while protecting long-term human uses within the basin. These principles suggest that the most effective way to achieve sustainable watershed uses will be to consider the basin and its sub-watersheds in a landscape context. This new approach—integrated resource planning and management (IRP)—protects the watershed ecosystem while providing for a sustainable output of resources.

The basin provides an ideal opportunity for a fully integrated landscape-level IRP program to address major water quantity and quality issues similar to those facing basins and watersheds throughout the United States. This paper describes overall program management opportunities in the Klamath River Basin. It also reviews the environmental issues that precipitated program planning, such as the numerous problems with threatened and endangered species; significant water quality degradation and non-compliance in rivers and lakes; water rights issues; and escalating water use conflicts among tribes, federal refuges, irrigators, and fish species.

Overview

The Klamath River Basin encompasses 10.5 million acres of land in southern Oregon and northern California. The U.S. Bureau of Reclamation Klamath Project was one of the earliest federal reclamation projects in the United States. It was established in the early 1900s to provide a reliable water supply for agricultural uses across 233, 625 acres of land, as well as to meet water supply requirements of the Klamath, Tule Lake, and Clear Lake National Wildlife Refuges. Over time,

*CH2M Hill, 825 N.E. Multnomah, Suite 1300, Portland, OR 97225

however, competing water demands have developed that require the project's water supplies to support numerous other uses, such as requirements imposed by the federal Endangered Species Act (ESA); water quality control; flows for hydroelectric power facilities; and Tribal Trust requirements. Today, limited water supplies cannot meet these competing demands.

Stakeholder interest in the basin's management increased dramatically in 1993, when representatives from local, state, federal, and tribal governments, along with private interests, established the Klamath Basin Ecosystem Restoration Office (ERO), a cooperative program to restore the form and function of the Klamath River Basin ecosystem (Lewis, 1995). The ERO was charged with assessing the wide range of basin issues, and particularly with stemming the growing tide of threatened and endangered species crises. There are 113 species of plants and animals currently on the "sensitive species" list for the basin, imposing difficult and often unknown limitations on management decisions. In addition, the Oregon Biodiversity Project, a privately sponsored effort to identify biodiversity issues in Oregon and provide management options for long-term sustainability, has identified the Klamath River Basin as potentially the most important of Oregon's 12 physiographic provinces with regard to biodiversity.

Major Stakeholders

The numerous laws, directives, and orders that have evolved over this century to manage the water supplies of the Klamath River Basin have identified several major stakeholders. Each has interests that must be considered in developing an IRP program:

U.S. Fish and Wildlife Service. The National Wildlife Refuge System includes four refuges located in the Klamath River Basin. These areas have not received full water allocations in some years because of diversions to protect fish. USFWS also manages the Ecosystem Restoration Office in Klamath Falls, and is responsible for non-anadromous threatened and endangered species.

Bureau of Reclamation—Klamath Project Authorization of 1905. This federal project encompasses 233,625 acres.

Federal Energy Regulatory Commission (FERC) and PP&L. Five hydroelectric projects on the Klamath include operating regimes, instream flow requirements, and other provisions in the basin.

Tribes. The Klamath and Modoc Tribes in Oregon, and the Hoopa, Karuk, and Yurok Tribes in California, maintain that their rights to hunt, fish, and gather treaty-protected resources in the basin were guaranteed by the treaties of the 1860s between the tribes and the United States.

Oregon Department of Environmental Quality (DEQ), California Regional Water Quality Control Board—Water Quality. The DEQ has initiated modeling and assessment to determine total maximum daily loads (TMDLs) for point source dischargers on the Klamath River. Nonpoint source control is a significant issue for Oregon and California, and environmental groups are threatening to file suit to force a regulatory crackdown on nonpoint sources of water quality degradation.

Irrigation Districts. The federal project currently provides water to 203,000 acres of land for the production of grains, irrigated pasture, potatoes, onions, and other special crops, via four irrigation districts and numerous individual agricultural users.

U.S. Forest Service and U.S. Bureau of Land Management. The U.S. Forest Service manages seven national forests in the basin. The U.S. Bureau of Land Management manages several thousand acres of range and forest lands. These agencies are developing forest and range management plans aimed at ecosystem management and sustainable development, according to the President's forest management program.

Endangered Species Act (ESA). Competing downstream demands for water to assist anadromous fish movements have come into direct conflict with water requirements for resident fish. Further, in 1995 the National Marine Fisheries Service (NMFS) anticipates listing Coho salmon and steelhead trout on the Klamath as threatened.

IRP Program

Integrated Resource Planning (IRP) consists of principal concepts, demonstrated in and borrowed from the electric utility industry, that have been expanded and adapted to natural resource parameters and characteristics. The central features of IRP include:

- Looking at multiple options for matching resource supply to consumer demand.
- Opening the decision-making process to new ideas and interests.
- Recognizing that uncertainty and change are endemic to long-range planning.
- Evaluating planning decisions from a regional and societal perspective as well as an agency perspective.
- The supply-demand aspect of IRP does not dictate that resources be consumed in the traditional sense of the word.

The Klamath IRP program will need the increasingly sophisticated analytical tools becoming available for watershed and basin-wide assessments, restoration plans, and integrated resources planning. Hydrological and water quality models must be linked with GIS mapping systems to adequately match hydrologic and water quality issues and predictions with locations. Biodiversity and GAP assessments across the basin landscape will be integrated with resource management options development. Ecological health assessment methodologies and applications will be used to address target species and habitat issues. Specific remedial actions taken to address immediate problems with threatened and endangered species will be reviewed, and the short- and long-term requirements for systems integration and monitoring will be assessed. Multi-species habitat protection plans can be developed for long-term management of threatened and endangered species. Partnerships and consensus-building among landowners and agencies have begun, and will remain essential to the program's success.

Conclusion

For resource owners expecting to get a yield or product from their property, predictability in the management of key resources is critical. To avoid a continual "revolving door" in which landowners are never sure what issue or species will be the next to prevent their economic use of the land, multi-species habitat protection plans rapidly are becoming the norm. Under these plans, land managers will have a predictable, sustainable yield of resources, while publicly valued resources remain protected.

A successful Klamath Basin IRP program incorporates the following key premises:

- Watershed restoration is controlled not by a central authority, but by several public and private programs and partnerships.
- Local political and opinion leaders are involved in a coordinated strategy.
- Public and private agencies abandon some of their traditional management focus and share resources, experience, and data.
- Agencies take the opportunity to develop new project delivery systems to ensure that work is done in a timely manner and that local skills are developed and used in practical ways to find management solutions.

Resource decisions are shaped by both technical considerations and individual or collective values. The concept of balancing technical and social considerations is central to integrated resource management. The most viable management concepts are those that have successfully balanced technical and economic criteria with local, regional, and national values.

Bibliography

Lewis, Steven A., project supervisor. Ecosystem Restoration Office, Klamath Falls, Oregon. Letter dated January 12, 1995.

U.S. Department of Interior. Bureau of Reclamation. Klamath Project, Oregon-California. October 1988. U.S. Government Printing Office: 1992-686-850.

Freas, Kathy, Daniel D. Heagerty, Janet Senior, CH2M Hill. "Biodiversity Considerations in Watershed Management." Watersheds Conference 1994, Washington, D.C., October 1994.

Rogers, John W., and B. Fritts Golden, CH2M Hill, *Developing and Implementing Multi-Interest Ecosystem and Watershed Management Programs.* To be published January 1995.

A METHODOLOGY FOR THE ASSESSMENT OF CUMULATIVE IMPACTS OF SECTION 404 CLEAN WATER ACT PERMITTING: THE SANTA MARGARITA WATERSHED CASE STUDY

Eric D. Stein[1]

Abstract

Assessing the cumulative impact of prior activities provides a baseline for effective watershed planning and highlights opportunities for improvement of existing practices. This study used GIS to assess the cumulative impact of projects permitted under Section 404 of the Clean Water Act (CWA) and to provide recommendations for development of the Santa Margarita Watershed Plan.

Introduction

Watershed planning is increasingly utilized to manage aquatic resources on a holistic basis. Assessment of the cumulative impacts of previous activities can be used to provide a framework on which to guide planning and policy decisions, thereby serving as an important component of watershed planning. Such assessments must be scientifically rigorous, yet easy to implement by regulators and planners. The goal of this research was to develop and implement a methodology to assess the cumulative impacts of activities permitted under Section 404 of the CWA on the ecology of the upper Santa Margarita Watershed.

Background

The Santa Margarita River is the second largest river basin on the Southern California coastal plain. It is one of the few remaining free-flowing river systems and provides one of the most expansive, unspoiled riparian habitats in Southern California. Rapid development of the upper watershed beginning in the mid 1980s has been accompanied by disruption of riparian systems, increased runoff and sedimentation, and floodplain encroachment. There is concern that the cumulative impacts associated with development will degrade the ecologic and hydrologic integrity of the entire watershed.

---

[1]U.S. Army Corps of Engineers, Los Angeles District, Regulatory Branch. 300 N. Los Angeles, Street, Los Angeles, CA 90012

Methodology

The complexity of ecological systems compels the use of measures of habitat composition and structure as surrogates for direct measurement of ecological function. The choice of surrogate measures is motivated by the need to balance ecological applicability against ease and expediency of measurement (Kentula et al, 1992). In this study, a semantic categorization was used to assign impact scores based on the following six ecological evaluation criteria: endangered species habitat, structural diversity of habitat, spatial diversity of habitat, open space habitat, linear contiguity of habitat, and adjacent habitats. Project locations were then color-coded based on degree of impact, plotted on a watershed map using ARC/INFO GIS, and analyzed for spatial associations.

Site specific impacts were assessed by comparing the conditions present at each project site prior to issuance of the Section 404 permit to conditions present after issuance of the permit. Each project site was given a pre- and post-project rating of A, B, C, D, or E for each evaluation criterion, with A representing site conditions similar to those present at pristine reference sites and E representing the most degraded condition. Impact scores, ranging from -4 to +4, were based the change in the number of rating categories at each site. See Table 1 for a sample impact evaluation.

Table 1: Sample Impact Evaluation

CRITERION	PRE PROJECT RANK	POST PROJECT RANK	IMPACT SCORE
Endangered Species Habitat	D	D	0
Structural Diversity of Habitats	C	D	-1
Spatial Diversity of Habitats	E	D	+1
Open Space Habitat	B	C	-1
Adjacent Habitats	B	E	-3
Linear Contiguity of Habitats	A	E	-4

Assigning numerical values, rather than letters, to the categories within a criterion would imply a defined arithmetic proportionality between ratings. Justification of such a numerical scale would require knowledge of the weighting and the mathematical relationship between categories. Scientifically defensible data to support such mathematical relationships do not exist. Therefore, I focused on the direction and the number of categories changed, and did not attempt to arbitrarily quantify the magnitude of this change.

Results

The Clean Water Act goal of restoring and maintaining the chemical, physical, and biological integrity of the nation's waters has not been met in the upper Santa Margarita Watershed. Of the activities authorized by Section 404 permits since 1986 (the earliest issuance of permits in this area), 96% have resulted in overall adverse or substantial adverse impacts to aquatic resources. The temporal distribution of impacts has been strongly correlated with increases in population and urban development.

Analysis of impacts based on the six evaluation criteria revealed differences between criteria: Impacts scores for the endangered species habitat criterion were concentrated in the minimal impact category; impact scores for the structural diversity of habitats and spatial diversity of habitats criteria were concentrated in the adverse impact category; and impact scores for the linear contiguity of habitats, adjacent habitats, and open space habitats were concentrated in the substantial adverse impacts. Less than 1% of the acreage impacted received impact scores of enhancement or substantial enhancement. The differences between impacts based on evaluation criteria is related to the degree of review each criterion received during permit evaluation. The criteria which were directly evaluated during permit review; endangered species habitat, structural diversity of habitats, and spatial diversity of habitats; were subject to less severe impacts than those not specifically evaluated; linear contiguity of habitats, adjacent habitats, and open space habitats.

Cumulative impacts to landscape level processes were amplified by the spatial distribution of permitted projects. Impacts were concentrated in the Temecula Valley, where the urban centers are located, fragmenting two relatively undisturbed habitats and possibly hindering interaction and migration between faunal populations. The landscape position of impacts decreased the value of neighboring habitats, may have contributed to species becoming endangered, and may interfere with the recovery of currently endangered species.

In addition to habitat fragmentation, the most egregious impact to the upper watershed has been the loss of riparian floodplains due to urban encroachment and stream channelization. Constriction of streams within steep sided channels isolates them from adjacent uplands, limiting dynamic riparian processes such as overbank seed dispersal, and precluding free movement of organisms between upland and riparian habitats. Furthermore, disconnecting rivers from their floodplains reduces their ability to attenuate flood peaks, limits natural sediment deposition and water quality enhancement, and disrupts downstream successional processes and scour cycles (Harris and Gosselink, 1990).

Discussion

The Santa Margarita Watershed is at a critical juncture in terms of the future health of its aquatic resources. The rapid growth of this region of Southern California is expected to continue, and the conflict between urban development and resource protection is expected to escalate. Because a large portion of the resources of the watershed are relatively intact, the opportunity still exists to develop a management plan which will allow urban development to proceed in a manner which is compatible with protection of the unique resources present in the watershed. However, as this study has shown, the current strategy of managing cumulative impacts on a project by project basis has been ineffective at preventing significant degradation of landscape level functions.

If the current trend of increasing cumulative impacts is to be curtailed before irreversible degradation of aquatic resources has occurred, a holistic watershed management plan must be adopted which incorporates the following strategies: First, project specific impacts must be considered in the context of past actions and landscape level processes. This can be facilitated by developing GIS maps, with layers depicting ecologically important areas, urban zones, levels of current degradation, and risk of future degradation. Second, floodplain integrity must be maintained by explicitly considering impacts to the riparian zone during permit review. In addition, traditional structural channelization should be replaced by wider, more natural stream corridors which utilize riparian vegetation to stabilize slopes, control erosion, and reduce flow velocities (Warner and Hendrix, 1985). Third, permitting and mitigation requirements should be commensurate with the permanence of impacts. In addition, preservation of functionally valuable portions of the landscape should be accepted as mitigation for permitted impacts to less valuable portions of the landscape.

This study has shown that when numerous small decisions on related environmental issues are made more or less independently, the combined consequences of the decisions are not addressed (Odum, 1982). However, modifications to existing permitting procedures and continuation of the proactive planning process may be successful at reducing future cumulative impacts.

References

Harris, L.D. and J.G. Gosselink, Cumulative Impacts of Bottomland Hardwood Forest Conversion on Hydrology, Water Quality, and Terrestrial Wildlife, in *Ecological Processes and Cumulative Impacts: Illustrated by Bottomland Hardwood Wetland Ecosystems*, J.G. Gosselink et al (eds.). Lewis Publishers, Chelsea, MI, 1990

Kentula, M.E., R.P. Brooks, S.E. Gwin, C.C. Holland, A.D. Sherman, and J.C. Sifneos, *An Approach to Improving Decision Making in Wetland Restoration and Creation.* United States Environmental Protection Agency, Environmental Research Laboratory, Corvallis, OR, 1992

Odum, W.E., Environmental Degradation and the Tyranny of Small Decisions. *BioScience*, Vol. 32(9), pp. 728-729, 1982

Warner, R.E. and K.M. Hendrix, *Riparian Resources of the Central Valley and California Desert.* California Department of Fish and Game, Sacramento, CA, 1985

ASSESSING COMMUNITY ATTITUDES AND PERCEPTIONS: METHODS EVERY PROFESSIONAL SHOULD KNOW

by

Roger Durand, Ph.D.[1]
Richard C. Allison, Ph.D., M. ASCE[1]
Judith Durand, M.S.[2]

ABSTRACT

In this paper three alternative methods for assessing community attitudes and perceptions are discussed: surveys (including telephone, in-person, and mail surveys); focus groups; and Delphi techniques. Each method is presented from the viewpoint of the water resources professional who desires -- either "in-house" or through an outside consultant -- to gauge community thinking.

INTRODUCTION

The need for assessing community attitudes and perceptions in water resources planning and decision making has never been more important. A recent survey conducted by the American Water Works Association (AWWA) Research Foundation indicates that the majority of today's public wants more information and more opportunities for involvement in decisions. As one water professional recently stated, "Not as many people want to leave it to the experts; they want to speak and be heard; they want their recommendations turned into action." These desires for more information, involvement, and opportunities to express opinions mean, in turn, that water resources managers and planners need to know more about methods for tapping the public's thinking (AWWA, 1994).

SURVEY METHODS

Surveys, probably the most widely used and well-known methods for assessing community attitudes and perceptions, are generally distinguished by the means utilized to elicit information from subjects: in-person (or "face-to-face"); by telephone; or by

[1] Professor, University of Houston-Clear Lake, Houston, Texas

[2] Principal, Durand Research and Marketing, Houston, Texas

mail. As indicated in Chart 1 (below), these means of eliciting information differ from each other (as well as from the other methods discussed in this paper) by the time required to conduct them, by their costs, by their specific advantages, and by their weaknesses. What the three types of surveys share is that, typically, they all involve gathering cross-sectional data from a sample of subjects selected by random procedures to be representative of a larger population. (Note: This "typical" survey is the one characterized in Chart 1.) Usually, interest lies in generalizing conclusions to the population from which the sample was drawn. And usually, a standardized questionnaire is designed and administered to all subjects (Rubenstein, 1995). Examples of the use of surveys in water resources management and planning abound. They include planning for water conservation through residential retrofit (Neighbors, et al., 1993); managing the recreational uses of water (Durand, Allison, and Hill, 1991); and understanding the risks to human health of water reuse (Durand and Schwebach, 1989).

FOCUS GROUP

The typical focus group involves the interviewing of 8 to 12 individuals who jointly discuss a predetermined topic under the direction of a moderator. The moderator, who must be trained both in interviewing as well as in group dynamics, promotes interaction among the individuals in the group and assures that the discussion does not stray from the topic of interest. A typical discussion session will last from an hour and a half to 2 and 1/2 hours. While such sessions are often held in facilities especially design for focus groups (e.g., one-way mirrored viewing rooms), they can be conducted in offices or homes (Sewart and Shamdasani, 1990). As with surveys, the focus group is compared to other methods in Chart 1 (below). No examples of the use of this method in water resources management and planning have been identified. But the following examples from other areas are illustrative of the method: D. Brinberg and J. Durand, 1983; Axelrod, 1975.

DELPHI

The Delphi is a method whereby the opinions of experts are elicited and combined so that collective knowledge and judgment can be brought to bear to project future trends (Helmer, 1968). In the classsic Delphi a relatively small group (10-15) of experts is first posed an open-ended question. The answers obtained to this question are then presented to the group of experts in a second round. In this second round, each expert is asked to "vote" on the items obtained from answers to the open-ended question in order to identify the most important trends in the minds of the experts. The results of this second round are then tabulated. Successive rounds can be conducted until a desired level of consensus is reached. Of course, it is possible that a consensus is never achieved. The unique feature of a classic Delphi is that the group of experts never meets together. Rather, the entire process is generally conducted by mail so that personalities do not bias the results. Delphis have been employed to forecast trends in uses of the Galveston Bay as part of the Federal government's National Estuaries Program (Durand, Allison, and Hill, 1991). Once again, this method is compared to the others discussed in this paper in Chart 1.

Chart 1: A Comparison of Models

Method	Time Required	Costs	Principal Advantage (s)	Principal Weakness (es)	Statistical Qualities
Survey: In- Person	Considerable	High	Yields in-depth information; data on diverse types of people	Difficult to gather data on physically dispersed people	Accurate estimates of population parameters from known statistics
Survey: Telephone	Small	Moderate	Yields data on diverse types of people	Lack of in-depth information	Accurate estimates of population parameters from known statistics
Survey: Mail	Moderate	Low	Yields data on physically dispersed people	Low response rate; often misses important groups	Accurate estimates of population parameters from known statistics
Focus Groups	Small	Low	Yields in-depth information; useful for initial discovery	Difficult to summarize and interpret results	Results cannot be generalized; statistical tests often not valid
Delphi	Small to Moderate	Low	Yields qualitative forecasts of discontinuous nonlinear trends	Biased results are common	Results cannot be generalized; statistical tests often not valid

CHOOSING A METHOD...

How should water resource professionals choose among methods to assess community attitudes and perceptions? Chart 1 provides criteria that the professional needs to weigh carefully. But the foremost criterion must always be that the method needs to be appropriate to the planning or management decision at hand -- a judgment call that is best decided through experience or with professional advice.

REFERENCES

American Water Works Association White Paper, "Principles of total water management outlined," AWWA Mainstream, November, 1994.

Axelrod, M., "Marketers Get an Eyeful When Focus Groups Expose Products, Ideas, Images, Ad Copy, Etc., to Consumers," Marketing News, February 1975.

Brinberg, D. and J. Durand, "Eating at Fast Food Restaurants: An Analysis Using Two Behavioral Intention Models," Journal of Applied Social Psychology, 1983.

Durand, R., R.C. Allison, and R. Hill, A Socioeconomic Characterization of Galveston Bay, Report to the Galveston Bay National Estuary Program of the United States Environmental Protection Agency and the Texas Water Commission, 1991.

Durand, R. and G. Schwebach, "Gastrointestinal Effects of Water Reuse," American Journal of Public Health, December 1989.

Helmer, Olaf, "Analysis of the Future: The Delphi Method," in James R. Bright (ed.), Technological Forecasting for Industry and Government. Englewood Cliffs, NJ: Prentice-Hall, 1968.

Neighbors, R. et al., "Effectiveness of Retrofit in Water Conservation: Results from a Study of Single Family Residences," in Proceedings of CONSERV93, sponsored by the American Water Works Association, American Society of Civil Engineers, and the American Water Resources Association, December 1993.

Rubenstein, S.M., Surveying Public Opinion. Belmont, CA: Wadsworth, 1995.

Stewart, D.W. and P.N. Shamdasani, Focus Groups. Newbury Park: Sage, 1990.

The Role of Alternatives Analysis in the Permitting of Water Resources Projects

Richard Pyle, P.E[1]., Lee Judd, P.E[2]., and Ken Steele, P.E.[3]

Abstract

Historically, engineers have taken the responsibility for identifying water resources problems, and deciding on solutions for these problems prior to announcing the solution to the public. In recent years however, engineers have faced increasing difficulty in defending these decisions when faced with the broad range of public opinion typically associated with major water resources projects. As a part of its permitting efforts for the Emergency Storage Project (ESP), the San Diego County Water Authority (Authority) chose to move away from the traditional "Decide-Defend-Announce" approach, to an approach that involved a thorough and open alternatives analysis process. During the alternatives analysis phase of the project, the Authority made use of a quantitative decision model to evaluate a broad range of alternative solutions. The use of this model also enabled the Authority to invite the public to be active participants in the selection of the solution to San Diego's emergency storage needs.

Project Context

Approximately 90 percent of the water used in San Diego County is imported from outside the region. The Authority is developing the ESP to protect the 2.5 million people in the Authority service area against the severe economic loss that would

---

[1] M. ASCE, Senior Civil Engineer, San Diego County Water Authority, 3211 Fifth Avenue, San Diego, CA 92103-5718

[2] M. ASCE, Senior Water Resources Engineer, GEI Consultants, Inc., 1925 Palomar Oaks Way, Suite 300 Carlsbad, CA 92008-6526

[3] M. ASCE, Emergency Storage Project Manager, San Diego County Water Authority, 3211 Fifth Avenue, San Diego, CA 92103-5718

occur due to a long term interruption of imported water supplies such as would be caused by a major earthquake. It is projected that the county will need an extra 111×10^6 m^3 (90,100 acre-feet) of emergency storage by the year 2030.

The Authority has implemented a thorough alternatives analysis that will yield an environmentally sound, cost-effective project that meets all legal requirements. The alternatives analysis has considered the use of a full range of possible solutions, including: new pipelines to allow existing reservoirs to be used more effectively; ground water basins; new or expanded reservoirs; and alternative water supplies such as desalination or reclaimed water.

Alternatives Analysis

Due to the magnitude of the proposed project, it was anticipated that permitting would be required under several state and federal regulations, including; the National Environmental Policy Act (NEPA), the California Environmental Quality Act (CEQA), and the Clean Water Act Section 404(b)(1) regulations. Of these regulations, the 404(b)(1) guidelines are the most explicit regarding their requirements for a thorough and complete analysis of all "practicable" alternatives. Specifically, these regulations require that a first and second stage screening process be conducted. The screening process should include reservoir and non-reservoir alternatives as well as combinations of alternatives. The screening process should also screen out alternatives clearly not meeting the project objective and should identify the alternatives to be carried forward into the EIR/EIS.

For the Authority, the purpose of the alternatives analysis was to provide a means for selecting four emergency storage system alternatives from the many possible solutions available. The selected four alternative systems were then studied in detail and included in the draft EIR/EIS documents. The alternatives analysis procedure the Authority selected consisted of a three step screening process. The first step reduced the vast array of possible solutions to 32 alternative systems. The second step reduced these 32 alternative systems to 13 systems, and finally the third step reduced these 13 systems to the final four systems.

Enhancements to the Alternatives Analysis Process

In addition to wanting to meet the requirements of the regulatory agencies, the Authority chose to utilize the alternatives analysis process as the focal point where the engineering and public involvement efforts would meet. The Authority accomplished this by using a decision model, by using this model to allow collaborative public involvement to occur, and by using the model as an aid in managing data collection efforts.

Decision Modeling: Decision modeling is a well recognized approach that has been effectively used by decision makers in the past. While the process goes by several different names, it is essentially founded on a participatory polling (group based decision) process developed by Rand Corporation after World War II for use in making strategic military decisions. Over the past 20 years, the process has been refined to apply to decisions concerning siting studies for major projects.

Decision models are constructed by identifying planning values that are considered to be important when considering the project. For the ESP, five major goals were identified, these being to: 1) Minimize Environmental Concerns, 2) Minimize Social Impacts, 3) Maximize System Implementability, 4) Maximize Operational Effectiveness, and 5) Minimize overall project cost. These goals were then further subdivided into sub-goals, objectives and criteria. Criteria were defined so as to be able to be measured directly. Each goal, sub-goal, objective and criteria is then weighted to reflect the relative importance of that item relative to the complete decision being considered. Using this methodology, a score can be determined for each alternative being considered. This score will reflect the preference for the alternatives being considered, and allows direct comparison between alternatives. This methodology has been widely applied for permitting major projects, and is invaluable in allowing a defensible and thoroughly documented decision that balances competing objectives.

As noted above, this methodology requires that weights be assigned as to the relative importance of the goals, sub-goals, objectives and criteria. This decision is largely subjective, and therefore requires an orderly, interactive examination of numerous project-related factors by an interdisciplinary group of individuals who are knowledgeable about the project as well as the local setting. For the ESP, the first stage of screening was completed by a screening panel made up of an interdisciplinary team of Authority staff and consultants. However, when screening from 13 system alternatives to the final four, the weighting of goal and sub-goals was accomplished by a panel made up of representatives of a broad cross-section of the general public. This panel is discussed in more detail below.

The Emergency Storage Working Committee: While the Authority had implemented a full range of public involvement programs, it was the formation of the Emergency Storage Working Committee (ESWC) that raised the level of public involvement to a new level. The 27-member committee was comprised of a broad cross-section of the community; those who lived and worked in the affected areas, business and environmental leaders, educators, technical specialists, disaster preparedness experts, representatives of key industries and political representatives. Groups were invited to designate a representative to bring their perspectives to the committee.

The committee's work culminated in a confidential voting session in which a consensus recommendation to the Authority regarding the weights to be placed on the goals and sub-goals in the decision model was decided. Each committee members vote carried equal weight. The final "deliverable" of the committee was a consensus report which provided direction to the Authority staff regarding the views of the committee on a range of project issues.

Management of Data Collection Efforts: As discussed above, for the ESP there were many hundreds of possible combinations of solutions that could have been included in the alternatives analysis. The Authority therefore chose to screen alternatives several times, each time applying a greater level of detail when making comparisons. Preliminary screening was completed using basic "common sense" engineering and cost information, and resulted in the selection of 32 system alternatives that met the projects purpose and need. Coarse screening was then completed using regional GIS data and preliminary engineering costs estimates. Coarse screening reduced the number of alternatives being considered to 13, therefore allowing the Authority to focus its study efforts and spend its valuable budget on gaining data for and evaluating in more detail only those alternatives solutions that were reasonably expected to be under final consideration. During fine screening, the 13 system alternatives were reduced to four using more detailed biologic and engineering data.

The alternatives analysis can therefore be used as a management tool, in that the level of data to be gathered and the level of engineering detail required at each point of the project is dictated by the input requirements for the decision model. Therefore, as the schedule for screening activities is determined, the schedule for data acquisition and engineering design is also determined. Although the level of detail and accuracy of data being input into the decision model is less detailed for preliminary screening than for latter screening, it is essential that the level of detail and accuracy is consistent between all alternatives being considered, each time that screening occurs.

Conclusion

The ESP experience has shown that a through alternatives analysis can be the nexus at which preliminary engineering and environmental data, public involvement and decision modeling converge. Each of these three aspects of alternatives analysis are vital, and since alternatives analysis is a major part of the national and state permitting laws, the alternatives analysis process seems destined to be an increasingly important part of water resources projects.

Development of an Integrated Decision Support System for the Guarapiranga Reservoir

by B. Jacobs[1] , D. Agostini[2], L. Oliveira[3], A. Bittencourt[3] and C.A. Pereira[3]

ABSTRACT

CDM and COBRAPE have developed DYNCASS, a decision support tool for watershed management. This application incorporates GIS, data management and modeling functions. Its principal design objectives were to provide an integrated tool with minimal training requirements. DYNCASS was developed for the Guarapiranga water supply reservoir in São Paulo, Brazil to assess water quality problems and develop watershed restoration strategies.

INTRODUCTION

The Guarapiranga Program is a $250 million dollar World Bank watershed restoration project. The program is aimed at improving water quality in the Guarapiranga Reservoir, the source of drinking water to more than 4 million people in the city of São Paulo, Brazil. DYNCASS was developed as a tool to support the analysis and evaluation of the array of possible watershed management activities.

OBJECTIVES

The main DYNCASS design goal was to integrate into one software package the three basic functions for which computers are typically used in the field of water resources:

1) GIS	Used to identify critical areas within the watershed and establish spatial relations between the natural and cultural features.
2) Data Management	Used to quantify pollution levels and identify relations among the water quality indicators.
3) Modeling	Used to explore possible restoration alternatives and development scenarios.

[1] Camp Dresser & McKee International, 10 Cambridge Center, Cambridge, MA 02142. Currently at the Massachusetts Institute of Technology, 77 Massachusetts Avenue, Cambridge, MA 02142

[2] Camp Dresser & McKee International, 10 Cambridge Center, Cambridge, MA 02142

[3] Companhia Brasileira de Projetos e Empreendimentos, 1405 Rua Pinheiros, 1405, I Andar, São Paulo, SP, BRAZIL

In addition to the integration of the three functions listed above, the code's development pursued a set of operational objectives:

- To minimize training requirements, so as to maintain accessibility to both technical and non-technical personnel.
- To allow the user to easily access during an analysis all three of the software's basic functions (GIS, data management, modeling).
- To provide the ability to use the same tool throughout all project phases, from the planning to the detailed phase, allowing the user to review his original assumptions at any time during the project.

GIS AND PRELIMINARY SCOPING

In the Guarapiranga watershed the GIS functions were used to:

- Locate the areas of urban development, the greatest source of potential phosphorus loading, by displaying urban land use coverage on the screen.
- Identify areas not served by sewers, and of major concern, by superimposing on the urban use coverage areas the network of sewers and interceptors.
- Assess the extent to which the critical un-sewered urban areas were being monitored, by superimposing on land-use and sewer maps the locations of in-reservoir and in-stream water quality monitoring stations.

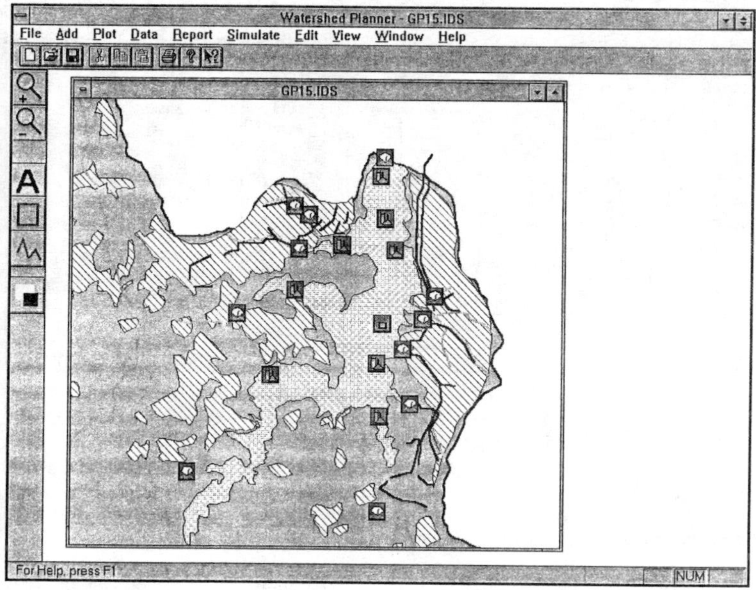

Figure 1 -- GIS module: urban areas, sewers and monitoring network overlays.

DATA MANAGEMENT AND VISUALIZATION

To date, in the Guarapiranga watershed tabular and graphic data visualization tools have been used to:

- Establish a relation between phytoplankton blooms and phosphorus concentrations in the water column.
- Characterize the frequency and severity of hypoxic episodes at the bottom of the reservoir.
- Estimate the effectiveness of algicide application by plotting algicide application rates versus phytoplankton population.
- Characterize rainfall climate using the rainfall statistics modules; parameters computed include average and standard deviation for interstorm intervals, storm depth, storm intensity, storm duration.

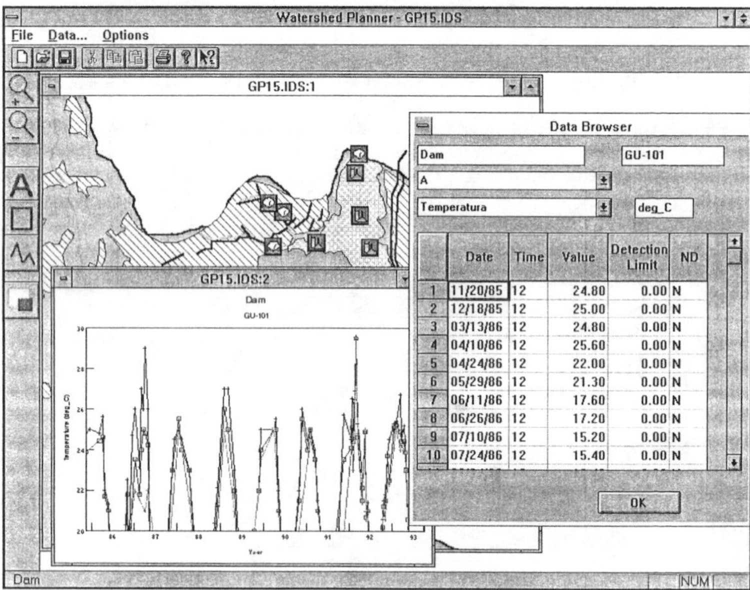

Figure 2 -- Data Visualization: browsing and plotting field data

MODELING THE WATERSHED

DYNCASS provides the user the ability to perform both preliminary and detailed modeling. The package currently incorporates two preliminary models: a land-use spreadsheet model and a Vollenweider eutrophication model. In the Guarapiranga watershed the simple land-use spreadsheet model was used to:

- Rapidly estimate the sensitivity of expected total P loads to parameters such as total population, population sewered, land-use areas and P generation coefficient.

- Establish value ranges to be explored during the detailed modeling phases of the project for specific parameters such as the P generation coefficient.
- Explore possible what-if or best/worst case scenarios.
- Facilitate discussions during project meetings.

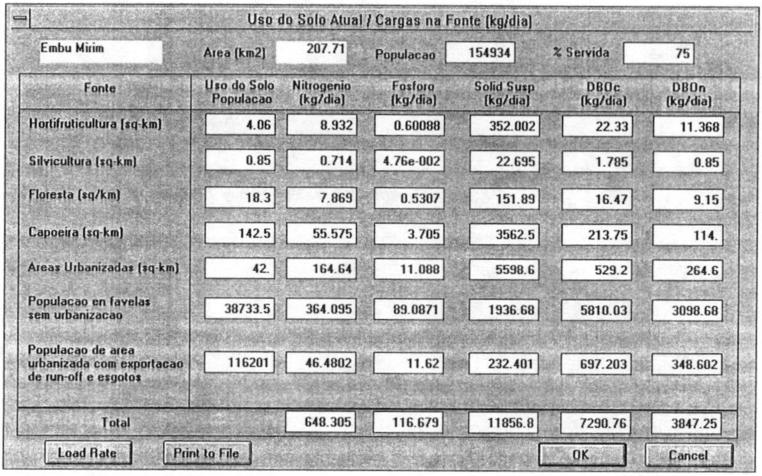

Uso do Solo Atual / Cargas na Fonte (kg/dia)

Embu Mirim | Area (km2) 207.71 | Populacao 154934 | % Servida 75

Fonte	Uso do Solo Populacao	Nitrogenio (kg/dia)	Fosforo (kg/dia)	Solid Susp (kg/dia)	DBOc (kg/dia)	DBOn (kg/dia)
Hortifruticultura (sq-km)	4.06	8.932	0.60088	352.002	22.33	11.368
Silvicultura (sq-km)	0.85	0.714	4.76e-002	22.695	1.785	0.85
Floresta (sq/km)	18.3	7.869	0.5307	151.89	16.47	9.15
Capoeira (sq-km)	142.5	55.575	3.705	3562.5	213.75	114.
Areas Urbanizadas (sq-km)	42.	164.64	11.088	5598.6	529.2	264.6
Populacao en favelas sem urbanizacao	38733.5	364.095	89.0871	1936.68	5810.03	3098.68
Populacao de area urbanizada com exportacao de run-off e esgotos	116201	46.4802	11.62	232.401	697.203	348.602
Total		648.305	116.679	11856.8	7290.76	3847.25

Load Rate | Print to File | OK | Cancel

Figure 3 -- Preliminary Modeling: the Land Use Model

Detailed modeling activities include using an integrated version of the U.S. Army Corps of Engineers runoff model STORM. To date DYNCASS-STORM has been used to:

- Explore the impact on total watershed contaminant loads of the contaminant surface accumulation rates in some of the smaller urban watershed.
- Explore the potential benefits associated with diverting and treating polluted stream flows.
- Explore the potential benefits associated with adding to the diversion scheme outlined above storage facilities to better capture daily flow variations and storm high flows.
- Simplify model calibration activities by allowing a seamless superposition of field and simulated data within the same package.

CONCLUSIONS

DYNCASS has proven to be of great benefit in arriving at restoration strategies for the Guarapiranga Reservoir. The future addition to DYNCASS of other simulation tools, such as QUAL2 or WASP, will strengthen its ability to analyze and predict water quality conditions in a watershed.

The Role of GIS in Watershed Protection[1]

Gregory W. Currey, Associate Member, ASCE[2]
Sumner Crosby III[3]

Abstract

Watershed protection is a geographically-based approach to water management adopted by a number of state water management agencies and promoted on a national level by the U.S. Environmental Protection Agency (EPA), among other Federal government agencies. A key feature of a successful watershed protection approach is the ability to identify, visualize, prioritize, and target water quality problems for watershed management actions. These activities require numerous types of environmental and related data. Geographic information systems (GIS) are a tool for accessing, managing, and analyzing these data on a geographic basis. EPA and state water management agencies have successfully applied GIS to water management projects, but there are additional steps these agencies may take to realize the potential of GIS as a tool for watershed protection.

Introduction

A key element of a successful watershed protection approach is the ability to identify, prioritize, and target water quality problems geographically. These activities require access to numerous types of data from a variety of sources including an assortment of historic scientific reports available in manual files, current ambient water quality data, hydrologic data, or source data, other state and Federal data bases, or the results of other monitoring activities (e.g, sediment

---

[1]The views presented in this paper are solely of the authors and do not necessarily reflect the views of the U.S. EPA or any other Federal agency.

[2]Civil Engineer, U.S. Environmental Protection Agency, 401 M Street, SW (4203), Washington, DC 24060

[3]GIS Specialist, U.S. Environmental Protection Agency, Region 3, 841 Chestnut Building (3WM12), Philadelphia, PA, 19107

or biological samples). Some of the key national databases for water quality management are STORET (including Reach File), the Permit Compliance System (PCS), the Toxics Release Inventory (TRI), and the Waterbody System (WBS) (Office of Information and Resources Management, 1992). GIS provides a framework for spatially linking sources of water quality data to one another and linking these data sources to other geographically defined data such as census data and local land use data.

Prioritizing and Targeting Resources Using GIS

State and Federal water quality management agencies prioritize and target resources for watershed protection at several levels. Strategic planning defines broad initiatives that become the basis for determining where to invest time and financial resources. GIS can be used in the strategic planning process to paint a national or regional picture of critical water quality problems. For example, the WBS can be used in conjunction with thematic mapping features of a GIS to indicate the major source of water quality problems, such as point source discharges of toxics, pesticide runoff from agricultural areas, or habitat destruction caused by urban storm water discharge, in each basin or watershed of a given region. Prioritizing at the watershed or river basin level is important as well. Water bodies or water quality problems in a watershed may be ranked according to such factors as severity of risk to human health and the aquatic community, impairment of designated uses of a water body (actual or potential), and resource value of the water body to the public. Based upon an analysis of spatially integrated water quality data, a water quality agency can determine where to target its resources.

Developing Control Strategies Using GIS

After selecting water bodies and water quality problems for management attention, a water quality agency develops a pollution source control strategy. Developing a control strategy requires clear goals for the stream segment, water body, watershed, or a larger management unit, such as an entire river basin. After setting goals, the agency must specify how they will be achieved, who is responsible for implementation, on what schedule, and how the effectiveness of the strategy will be assessed. GIS could spatially integrate water quality data for the watershed, data on ambient levels of toxic pollutants, PCS data indicating the location of point source dischargers, census data from the Department of Commerce, and land use data from local planning authorities. These data could be used in conjunction with water quality modeling to identify ambient conditions and specific pollutants and stressors in the watershed and determine the most effective and efficient control strategy.

Future Direction for Water Quality Agencies

EPA has made tremendous progress in the past several years in assessing the overall utility of GIS in the environmental field and developing guidelines for GIS acquisition and application. Information management personnel within EPA have developed a thorough knowledge of GIS and are available as resources for water program staff. A number of state water quality agencies are making significant progress in GIS development and application. There are several steps that EPA and state water programs should take, however, in order to fully utilize the capabilities of GIS in watershed protection.

GIS Access and Education: GIS capabilities in EPA and state agencies often are housed in a specialized information resources department. As a greater number of "user friendly" GIS products, such as the ARC/VIEW system, become widely available, the demand for and accessibility of GIS to program staff should increase. EPA and states should develop GIS expertise in their water programs. But, even having a GIS expert in the water program office is not sufficient if program staff and management are not educated about GIS. Introducing program staff and management to GIS workstations and software and educating them about potential applications in the water program would increase the effectiveness of this tool for watershed protection efforts.

Data Access and Monitoring Strategies: Often, data collected by state and Federal agencies and local governments are viewed as the "property" of the program that collected them. For effective watershed protection, data must be viewed as a corporate resource and be maintained on systems throughout each agency. This is generally not the case within EPA and between EPA and state agencies. For example, many states do not enter all of the data elements in the PCS data base and maintain their detailed permit tracking information on a separate state database; therefore, EPA and other states with an interest in a particular watershed have no easy access to a key source of data concerning point sources in the watershed. Also, these data are not always referenced to a commonly identified watershed or river basin. Data should be referenced to US Geological Survey hydrologic unit codes or another standard watershed code as well as to any state defined river basins. Furthermore, data often are collected or requested by one program with little consideration of how monitoring or data reporting could be modified to meet the needs of other programs. Collecting data for purposes with no obvious connections to their primary functions is not a high priority for government programs (Arbeit, 1993). Possible solutions include statewide monitoring strategies covering all water management programs within a state, representation by all key EPA or state program offices on work groups charged with setting individual program monitoring requirements, and better communication among water quality agencies and with local governments concerning data needs.

Standardized Location Identifiers: If national databases pertaining to water quality management are to be used in the context of a GIS system for watershed protection, there must be standardized location identifiers in each database. Stream segment identifiers in data bases such as WBS are largely a matter of state preference. Point data usually are located by latitude and longitude, but the precision of these measurements varies from facility to facility and database to database. Reporting location with a standardized identifier and precision would facilitate data use in a GIS and across agencies. Also, if the watershed is to become the basic management unit for water quality management, each data base containing water quality or pollutant source data should locate the point or line by watershed. Addition of a standardized watershed code in the required elements of each database should meet this requirement.

Integrated Modeling and Data Systems: Implementing a control strategy usually requires entering data to a model and testing several control options. EPA supports numerous DOS-based hydrologic and water quality models that could be integrated with GIS to create a complete data display, analysis, and modeling package. Most modeling applications that have been integrated with a GIS are for local projects and many use simplified versions of standard EPA or other agency models. On more broad national or state level, integrating data modeling functions with GIS is a complex task and will require a substantial commitment by EPA to focus the efforts of both GIS and modeling experts on this work.

In summary, GIS provides a framework for spatially linking sources of water quality data to one another and to a watershed. GIS can be used for prioritizing and targeting resources for watershed protection and for developing pollution control strategies for a watershed, river basin, or an entire region. In order to fully utilize the power of GIS for watershed protection, EPA and state water quality agencies should take steps to educate and provide GIS access to program staff, promote data sharing and standardization of location identifiers, and focus efforts on integrating GIS and modeling applications.

Appendix: References

Arbeit, David. (1993). "Resolving the Data Problem: A Spatial Information Infostructure for Planning Support." *Proceedings - Third International Conference on Computers in Urban Planning and Urban Management*, Richard E. Klosterman and Steven P. French, ed., Atlanta, GA., Vol. I, pp. 3-26.

Office of Information and Resources Management. (1992). *Geographic Information Systems (GIS) Guidelines Document*. U.S. Environmental Protection Agency, OIRM 88-1, Washington, DC.

Evaluation of Wetland Functions and Values: An MCDM Technique

J.E. Yost[1], A.M. ASCE, M.F. Dahab[2], M. ASCE, and R.H. Hotchkiss[3], M. ASCE

Abstract

This paper describes a model for wetland evaluation that allows the user to input parameters concerning one or several wetland sites and develop a numerical ranking of each site for comparison. Multiple Criteria Composite Programming provided the mathematical structure for making a quantitative evaluation of wetlands based on their qualitative and quantitative characteristics. Fifteen basic wetland functions and values related to the natural freshwater wetlands typically found in the Central United States were used to describe the aspects of the wetland system deemed most important and a methodology to normalize these basic indicators was developed. Basic indicators were grouped into composite sets, and weights and balancing factors were given to each indicator and group thus representing a double-weighting scheme. Weights were given to a basic indicator to show the importance of the indicator in relation to others. Balancing factors were given to groups of indicators to reflect the importance of deviations between a basic indicator value and its ideal value. The evaluation resulted in a final wetland site composite value.

Introduction

Scientists, engineers, lawyers, and regulators are finding it necessary to become knowledgeable in wetland ecology and management in an effort to better understand, preserve, and reconstruct these critical ecosystems. Wetlands are now recognized as valuable natural resources that, in their natural state, provide benefits to humans and their environment. Wetlands supply both value and function. While used interchangeably, these terms have different meanings. A wetland function describes what a wetland area does, irrespective of its beneficial worth; however, a wetland value is a subjective explanation of the relative worth of some wetland process or product. The value associated with a wetland function may be high or low depending upon one's interpretation or position in the overall ecosystem. Flood storage capacity upstream from a city may have a high value, yet the same

----

[1]Civil Engineer, EA Engineering, Science, and Technology, Inc. 121 S. 13th Street, Suite #701, Lincoln, NE 68508

[2]Associate Professor, Department of Civil Engineering, University of Nebraska-Lincoln, W349 Nebraska Hall, Lincoln, NE 68588

[3]Assistant Professor, Department of Civil Engineering, University of Nebraska-Lincoln, W348 Nebraska Hall, Lincoln, NE 68588

wetland downstream might have a much lower value to the city. At the same time, the downstream wetland could provide water purification, wildlife, recreational, or other functional values and be of a high value to city residents.

Efforts to manage wetlands, given a limited set of resources, could potentially be improved by allocating resources for wetlands protection according to the relative values of different wetlands. To do this, wetlands can be ranked in some fashion according to their relative importance. The evaluation of wetlands for this purpose is complicated by comparing, with some common denominator, the many functional values of wetlands against human economic systems and needs. This paper describes a model for wetland evaluation that allows the user to develop a numerical ranking for several wetlands as a means of comparison.

Multiple Criteria Composite Programming

Multiple criterion decision making (MCDM) techniques allow for the evaluation on a common, consistent basis, of various social, economic, and environmental factors. MCDM also allows both technicians and decision-makers equal access to the results in a manner that may facilitate the project evaluation and selection process. Multiple Criteria Composite Programming (CP) was selected to provide the mathematical structure for making a quantitative evaluation of wetlands based on both their qualitative and quantitative characteristics. CP, such as the Joint Ecological-Socioeconomic Evaluation of Water Resources Systems (JESEW) developed by Bardossy and Bogardi (Torno, 1988), offers a method for integrating different types of information to arrive at one numerical index. Using a multi-tiered approach, CP is a mathematical programming tool that aggregates and evaluates system indicators to detect how the system would perform under various conditions. The technique is graphically-based in that it identifies solutions that are closest to an ideal solution as determined by a measure of graphical distance.

Fifteen basic wetland functions and values related to the natural freshwater wetlands typically found in the Central United States were selected for use to describe the aspects of the wetland system deemed most important. Beginning with two principal divisions, environmental and socio-economic factors, six composite sets were established. The environmental division consisted of ground water, surface water, and water quality factors, as well as habitat functions. The socio-economic factor division consisted of both social and economic factors. Within the environmental factors, the first ten basic indicators are as follows. Ground water factors consisted of ground water recharge and ground water discharge. Surface water factors consisted of floodflow conveyance/storage/mitigation and erosion control/stabilization. Water quality factors consisted of sediment trapping, toxicant retention, nutrient attenuation/removal, and production export. Habitat functions were divided into aquatic abundance/diversity and wildlife abundance/diversity. The other five basic indicators represented the socio-economic factors. Social functions consisted of recreation, open space/aesthetic quality, and heritage/ education/research. Economic factors were divided into consumptive use value and opportunity cost or alternative use value.

A methodology to normalize these basic indicators was then developed for integration into the evaluation model. With basic indicators grouped into the six composite sets, weights and balancing factors were given to each indicator and group thus representing a double-weighting scheme. Weights were given to each basic indicator to show the importance of that indicator in relation to others. Balancing factors were given to groups of indicators to reflect the importance of deviations between a basic indicator value and its ideal value. The evaluation resulted in a final wetland site composite value. Additional information regarding this methodology and case studies can be found in Yost (Yost, 1994).

Case Study

A case study was performed in which data sets from several wetland sites were compared using the evaluation model. Because the weights and balancing factors were subjective and depended on the program user's judgement, a sensitivity analysis was also conducted to evaluate which indicators, and the weights assigned to them, tended to most strongly influence the results. The process consisted of incrementally changing the assigned values of each weight and balancing factor, and analyzing the effects or results from these alterations. Differing results may in turn impact which wetland site is evaluated highest and/or lowest. While the case study was used to demonstrate the evaluation methodology, it was the execution of the sensitivity analysis that resulted in meaningful comparisons of wetland sites and therefore offers the most assistance in the actual wetland management process.

A second case study involved comparing the results of the first to results obtained using the U.S. Army Corps of Engineers Wetland Evaluation Technique (WET, 1987). The results of this comparison indicated that the proposed model compared favorably to the WET technique. It should be noted that for the goals and objectives of this study, WET literature served as a guideline or accepted standard for creation of the evaluation methodology.

Discussion and Conclusions

In this study, a model for initial wetland evaluation is presented. The model was intended for planning and education purposes and not for detailed impact analysis. The level of this evaluation was structured such that detailed physical, chemical, and biological monitoring data from a wetland site were not required. An evaluation can be conducted using information gathered during an on-site survey along with readily available information resources. While some familiarity with wetlands is beneficial, it does not necessarily require a wetland specialist. Also note that no attempt has been made to address unique wetland types that would require special considerations involving differing methodology for evaluation purposes.

A multi disciplinary strategy allows for a detailed analysis of ecosystem components that in turn can lead to analysis of product outputs and subsequent estimate of socio-economic values on a broad scale. The advantage to using a multiple criterion approach based on CP, the basic structure of this model, is that indicators with different units of measure could be combined into a single value or index. CP uses a double-weighting scheme that gives the user an opportunity to emphasize specific indicators, or groups of indicators, to strengthen and enhance the evaluation process.

One must recognize that the methodology developed in this study to estimate numerical index values for each basic indicator represents a snapshot, or one moment in time, of the wetland site. After executing the case study used to illustrate the wetland evaluation methodology, it was recognized that a "noteworthiness" statement should be added at the end of each basic indicator numerical index question set. This would be used to advise the evaluator to add *bonus points* for intuitive factors or characteristics about a particular wetland site that may have not been given adequate consideration (points) in the question sets.

In summary, based on the study outlined in this paper, the following conclusions can be made:

1. Literature reviewed during this research suggested that wetland functions and values are an important component in the wetland management/decision-making processes.

2. Composite Programming (CP) can be an effective means of making a quantitative evaluation of wetland functions and values that consist of both qualitative and quantitative characteristics related to the natural wetlands typically found in the Central United States.

3. The case study indicated that the methodology and corresponding question sets developed as a means for estimating the numerical index values for each basic indicator favorably agreed with an established and well documented wetland evaluation technique.

4. The sensitivity analysis confirmed that the weights and balancing factors used in the CP structure can influence and/or alter the results (composite values) and thus careful consideration should be given to their selection.

5. While the case study was used to demonstrate the evaluation methodology, it was the execution of the sensitivity analysis that resulted in meaningful comparisons of alternatives (wetland sites) and therefore offered the most assistance in the actual wetland management/decision-making process.

References

Torno, H. C. *Training Manual for the Environmental Evaluation of Water Resources Management and Development*, (Unesco, Paris, 1988).

U.S. Army Corps of Engineers. *Wetland Evaluation Technique (WET) Volume II: Methodology, Operational Draft*, (Vicksburg, MS: Waterways Experiment Station, 1987).

Yost, Jonathan E. 1994. *Assessing Wetland Functions and Values with Multiple Criteria Composite Programming Methodology.* Masters of Science Thesis, University of Nebraska-Lincoln Libraries, Lincoln NE.

Water Management for Wetlands Restoration in Everglades National Park, Florida

By
Thomas Van Lent[1], M. ASCE
Robert Johnson[2]

1 Introduction

The Everglades of today are approximately half of their former extent of 1.2 million hectares, largely due to drainage and flood control activities. The drainage has also been responsible for a loss of landscape heterogeneity. Smith *et al.* [1989] have estimated that the Florida Bay estuary, which receives water from the Everglades, has fresh water inflows averaging 60% less than the pre-drainage condition. Generally speaking, alterations to the landscape by manipulations of the hydrologic regime have lead to an unraveling of ecosystem function [Davis and Ogden, 1994b].

There have been several attempts at reversing the ecological decline by managing the inflows of surface water into Everglades National Park. In 1970, Congress approved a minimum delivery schedule whereby the Park was guaranteed a fixed minimum volume of water. In 1983, the minimum delivery schedule to the Park was abandoned because deliveries were substantially different from natural marsh response to rainfall [Wagner and Rosendahl, 1987; Johnson and Van Lent, 1994], causing damage to nesting alligators and wading birds.

In place of the minimum delivery formula, Congress approved modifications to the schedule of deliveries to the Park and authorized a program of experimental water deliveries for Everglades National Park. The Corps of Engineers, the South Florida Water Management District, and the Department of the Interior have implemented a number of changes in an attempt to improve the viability of the marshes in and near the Park, but some of these modifications have had negative effects on Everglades National Park [Van Lent *et al.*, 1993].

In response to these Congressional directives, managers and researchers at Everglades National Park have begun to address issues related to the restoration of the Park's wetlands. This has necessitated the development of a water management proposal that would promote ecological restoration. This investigation represents an incremental step in an iterative, experimental program

[1]Assistant Professor, Department of Civil and Environmental Engineering, South Dakota State University, Brookings, SD

[2]Supervisory Hydrologist, South Florida Natural Resources Center, Everglades National Park, Homestead, FL

for the restoration of Everglades National Park. We have developed a water management proposal for the Park which attempts to define the hydrologic conditions which likely existed around the time of the Park's establishment. We then use these management criteria to develop proposals for the hydrologic restoration.

2 Water Management Objectives

Wetland restoration is not a fully developed science, nor are there large numbers of case studies upon which to draw. The most important element to a successful project, according to Coat and Williams [1990], is the clear definition of objectives. The National Park Service has established the following objectives for hydrologic improvements for Taylor Slough:

- Reproduce the hydrologic conditions (water depths and inundation patterns) in the Taylor Slough basin which existed at the time of the Park's establishment.
- Reestablish the inter- and intra-annual variations in water levels in response to rainfall variability.
- Restore the historical fresh water flows from Taylor Slough into Florida Bay. The inflows should largely match the historical volumes as well as temporal distribution.
- Modify over time the water management approaches to foster ecosystem productivity and biodiversity.

From a hydrologic perspective, each of these represents a limitation on traditional views of a water delivery scheme, since the above objectives focus on water levels rather than on surface water inflows. In this broad, shallow, and grassy wetland, flows are difficult to measure and manage, while stages are relatively easy to measure and directly determine marsh viability. Managing water levels in the marsh represents a more reasonable and ecologically sound approach because water levels and their resulting seasonal inundation patterns are the primary hydrologic parameters controlling wetland habitats. Moreover, the surface water flow patterns are determined by water levels and surface water gradients.

3 Methodology

As is often the case in water resources management, the alternatives for future direction are constrained by historical policies and actions. Radical departures from the status quo are difficult to implement, particularly when a wide variety of interest groups jointly determine water management policy. Currently, flow to Everglades National Park is determined by a stage regulation schedule. The schedule, developed by the South Florida Water Management District [MacVicar, 1989], is based upon a linear regression. The independent variable is weekly flow, and the dependent variables are weekly total rainfall, weekly estimated evapotranspiration, and previous week's flow. This linear regression was based upon the observed hydrologic data from 1933–1947. As part of the Congressionally-mandated testing program, the next step should be an incremental modification to the existing schedule.

We propose to modify the existing water management criteria in two ways. First, we use marsh water levels as the primary water management index, rather than flow. In the Everglades, management strictly on the basis of flow may not be the best approach if ecological viability is a goal. Everglades biota are tuned primarily to hydroperiod, which is in turn determined by stage and land surface elevation. The C&SF Project canals make it possible to manipulate stages to levels far below pre-Project conditions while maintaining relatively high flows through the marsh. Managing water levels to correspond to rainfall is expected to produce significant improvement in ecosystem performance.

We also propose to improve the methodology for analyzing the historical information. Rather than use the highly serially-correlated flow data in a linear regression, we apply time series to obtain a linear transfer function between rainfall and water level. The historical, observed data is then compared to an Everglades hydrologic simulation model [MacVicar *et al.*, 1984; Fennema *et al.*, 1994]. Using a time series approach to account for the serial correlation leads to a model that is more responsive during the extreme periods, particularly during droughts.

4 Summary

We have developed an incremental modification to the current water management scheme for delivery of water to Everglades National Park. The essential feature of the scheme is a linear transfer function relationship between rainfall and water surface elevation in the marshes of Everglades National Park. The management plan will result in an increase in inundation frequencies in the central Everglades sloughs, a decrease in the frequency of severe dry-downs in the central sloughs, and an increase in the flood risk for some of the basins controlled by the C&SF Project. The proposal is forwarded as an experimental water management plan under the Congressionally-authorized program for experimental water deliveries to Everglades National Park.

References

Coat, R. and P. Williams. (1990). Hydrologic techniques for coastal wetland restoration illustrated by two case studies. In ed. Berger, J., *Environmental Restoration*, pages 236–246. Island Press, Washington, DC.

Davis, S. and J. Ogden, eds. (1994a). *Everglades: The Ecosystem and its Restoration.* St. Lucie Press, St. Lucie, FL.

Davis, S. and J. Ogden. (1994b) Toward ecosystem restoration. In ed. S. Davis and J. Ogden. *Everglades: The Ecosystem and its Restoration*, page 769–796.

Fennema, R., C. Neidrauer, R. Johnson, T. MacVicar, and W. Perkins. (1994). A computer model to simulate natural Everglades hydrology. In ed. S. Davis and J. Ogden. *Everglades: The Ecosystem and its Restoration*, page 249–290.

Johnson, R. and T. Van Lent. (1994). Restoring flows to the Shark Slough basin, Everglades National Park. In *Effects of Human-Induced Changes on Hydrologic Systems*, Amer. Water Resour. Assoc. Bethesda, MD, page in press.

MacVicar, T., T. Van Lent, and A. Castro. (1984). South Florida Water Management Model: A documentation report. Technical Report 84-3, South Florida Water Management District, West Palm Beach, FL.

MacVicar, T. (1989). A rainfall-based plan of water deliveries to Everglades National Park. In C. Neidrauer and R. Cooper. *A two year field test of the Rainfall Plan: A management plan for water deliveries to Everglades National Park*, Appendix A. Technical Report 89-3. South Florida Water Management District, West Palm Beach, FL.

Smith, T., J. Hudson, M. Robblee, G. Powell, and P. Isdale. (1989). Freshwater flow from the Everglades to Florida Bay: A historical reconstruction based upon fluorescent banding in the coral solanastra bournoni. *Bul. Marine Sci.*, 44(1):274–282.

Van Lent, T., R. Johnson, and R. Fennema. (1993). Water management in Taylor Slough and effects on Florida Bay. Technical Report SFRC 93-03, South Florida Research Center, Everglades National Park, Homestead, FL.

Wagner, J. and P. Rosendahl. (1987). History and development of water delivery schedules for Everglades National Park through 1982. South Florida Research Center, Everglades National Park, Homestead, FL. Draft Report.

KAWAINUI MARSH RESTORATION

James Pennaz[1], M.ASCE
and
Margo Stahl[2], C.E.P.

ABSTRACT

Kawainui Marsh is Hawaii's largest remaining coastal wetland system. Located on the Island of Oahu, at the base of the Koolau Mountains, the marsh contains nationally recognized resources including four species of federally listed endangered waterbirds, and is eligible for listing on the National Register of Historic Places due to its archaeological/cultural significance. The marsh has been spared from development over the years as a result of its designation as a flood control basin in 1966 by the U.S. Army, Corps of Engineers for a flood control project.

On January 1, 1988 the project failed to protect the residents of the community of Kailua when floodwaters moving through the marsh overtopped the project levee and flooded homes. Proposed improvements to the flood control project will require filling of some wetland bordering the project levee. Mitigation for this wetland fill will result in the restoration and creation of wetland in the marsh previously degraded during the earlier project construction. Wetland hydraulic design methods used to meet flood control and marsh restoration requirements will be discussed. The authors discuss planning methods and procedures utilized to bring community and agency groups together to plan the levee improvements and marsh restoration effort.

INTRODUCTION

Kawainui Marsh serves as one of the most important ecological, cultural, educational, and recreational resources in the State of Hawaii for residents and non-residents alike, and as an aesthetic buffer to the generally urban character of the rest of the low-lying Kailua area. Interest in wetland protection has become widespread in Hawaii as elsewhere in the country. Hawaii has lost most of its wetlands to development over the years. In recent decades the periphery of Kawainui Marsh has been altered by agricultural and residential development and its waters enriched by years of sewage effluent resulting in a dramatic increase in an exotic floating mat of

[1]Hydraulic Engineer, US Army, Corps of Engineers, Pacific Ocean Division, Bldg. T-1, Fort Shafter, Hawaii 96858 USA
[2]Supervisory Fish and Wildlife Biologist, US Fish and Wildlife Service, Honolulu Office, PO Box 50167, Honolulu, Hawaii 96850 USA

vegetation. Kawainui Marsh represents one of a small number of these floating marshes in the world (Sasser, Gosselink and Shaffer, 1991). The floating marsh rises and falls with changes in water level. Sewage discharges are now diverted to a deep ocean outfall in the northeast area of Kailua Bay.

HISTORY OF KAWAINUI MARSH

Kawainui Marsh is thought to have once been a marine embayment open to the sea. About three thousand years ago the Kailua coastal strand barrier began to form and the bay became a lagoon. The present day marsh may have remained a lagoon until about five hundred years ago when continual organic filling and isolation from the sea transformed the lagoon into a freshwater marsh. The extent that the early Hawaiians contributed to this infilling is a source of professional debate (Athens and Ward, 1991). The marsh remains freshwater except at the upper end of the discharge channel. The water levels in the marsh normally lie several feet above mean sea level and resist saltwater intrusion. The marsh contains resources of national recognition. It provides habitat to four species of federally listed endangered waterbirds and serves as a migratory pathway for diadromous native stream fish and shrimp. The marsh is believed to be the site of an early Hawaiian fishpond and taro pond culture with a complex series of ponds and terraces covering the upper southern end of the marsh. Several heiau (Hawaiian platform temple) and other historic sites fringe the boundary of the marsh.

EXISTING FLOOD CONTROL PROJECT

The existing flood control project includes an earth levee 1,932 meters long with a design crest elevation of 2.9 meters above mean sea level (msl) and an outlet channel (Oneawa Channel) 2,896 meters long flowing northeasterly from the northern end of the marsh into Kailua Bay (Figure 1). The project was designed to accommodate the Standard Project Flood with a peak inflow to the marsh of 512 cubic meters per second (cms) (5.8 million cubic meter volume) and a peak discharge in the Oneawa Channel of 190 cms. Three square kilometers of marsh land were acquired by the local sponsor for temporary storage of flood flows as part of the flood control project. The existing project was authorized by the Flood Control Act of 17 May 1950 and was completed in August 1966. On January 1, 1988 (New Years Flood), the project was overtopped by a flood with an estimated peak inflow to the marsh of 538 cms and a volume of 9.6 million cubic meters.

The original project design concept was to have the marsh act as a natural flood storage basin. Floods from streams entering the southern end of the marsh would flow to an outlet channel at the northern end. This concept did not work during the New Years Flood because of the larger than design volume of floodwater and because of changed conditions within the marsh resulting from a massive increase in vegetation. The increase in vegetation was caused by discharge of sewage effluent into streams which are upstream of the marsh. Between 1966 and 1988, the marsh was rapidly overgrown by vegetation making it substandard as a flood control basin and reducing habitat for endangered waterbirds.

FLOOD CONTROL PLANNING

During project planning, it became clear that flood control must take into account the effect of the vegetation on flooding, the national significance of the wetland and surrounding area, and the desires of the community, local sponsor, and the State of Hawaii. Computer simulations showed that marsh vegetation would cause floods greater than the 10 year event to accumulate at the southern end of the marsh and eventually overtop the levee. Many alternatives were investigated including the removal of vegetation and sediments within the marsh, pumping stations, water diversions, and various combinations of levee raises. Removal of vegetation and sediments within the marsh was favored by local environmental groups because this alternative would restore the marsh to its former condition and greatly enhance the habitat of endangered waterbirds.

The final plan selected was a combination earth levee raise and concrete floodwall which best met engineering, economic, and environmental concerns. Factors against restoring the marsh to its former condition included potential water quality problems in the ocean receiving waters after the vegetation buffer was removed, vegetation and sediment disposal, long term maintenance of open water areas, and life cycle cost.

ENGINEERING DESIGN

The recommended plan consists of a 1,920 meter long, 1.2 meter high concrete floodwall on top of a raised levee. The engineering design analysis was enhanced by using the RMA-2V finite element computer model. This model was used for project design because the irregular combination of thick mats of marsh plants (up to 1.2 meters thick) and open water areas complicates the prediction of water surface elevations, currents and flows throughout the marsh. Traditional level pool routing methods fail to reproduce the actual hydraulic characteristics of flood events within the marsh. The RMA-2V model was used to explain the hydraulic characteristics of the 1988 flood event and evaluate various alternatives considered including vegetation removal. The model was calibrated using observed depth and duration of levee overtopping from the 1988 flood. Additional calibration data was obtained from an April 1989 event that was monitored by continuous recording stage gages located in the marsh. We recommend use of the RMA-2V computer model for wetland simulations because of the models' ability to simulate entire inflow and outflow hydrographs, marsh porosity and roughness characteristics, and varying water surface elevations.

ENVIRONMENTAL ISSUES

Water quality in Kawainui Marsh is dependent upon a multitude of factors including stream water base flows, stormwater runoff, growth and decay of vegetative material, solar heating, water percolation from springs, and livestock grazing. The marsh serves as both a source and a sink of nutrients. It is one of the primary sources of nutrient and sediment pollution to Kailua Bay according to a study by the University of Hawaii Water Resources Research Center.

Open water areas provide valuable habitat to the endangered waterbirds and enhance water quality in the marsh by enabling sunlight to penetrate. However, the removal of extensive amounts of vegetation to create open water may reduce the water purification services that the marsh presently provides to the receiving waters of Kailua Bay.

CONCLUSIONS

Wetland restoration efforts in Kawainui Marsh must take into consideration a number of variables including flood control, water quality, and wildlife habitat.

Restoration efforts, therefore, involve an interdisciplinary team working together to accommodate all concerns.

REFERENCES

Athens, J. Stephen and Jerome V. Ward. 1991. Paleoenvironmental and Archaeological Investigations, Kawainui Marsh Flood Control Project, Oahu Island, Hawaii. Report prepared for U.S. Engineer Division, Pacific Ocean. International Archaeological Research Institute, Inc. for Micronesian Archaeological Research Services, Guam.

Kawainui Marsh Technical and Policy Advisory Committee. 1983. Resource Management Plan for Kawainui Marsh. Prepared for Hawaii Coastal Zone Management Program, Department of Planning and Economic Development, State of Hawaii.

Sasser, Charles, James G. Gosselink and Gary P. Shaffer. 1991. Distribution of nitrogen and phosphorus in a Louisiana freshwater floating marsh. Aquatic Botany 41: 317-331.

U.S. Army Corps of Engineers. 1957. Design Memorandum, Kawainui Swamp. Oahu.

U.S. Army Corps of Engineers. 1992. Final Detailed Project Report and Environmental Impact Statement for Kawainui Marsh Flood Control Project. Jointly prepared by U.S. Army Engineer District, Honolulu and City and County of Honolulu Department of Public Works.

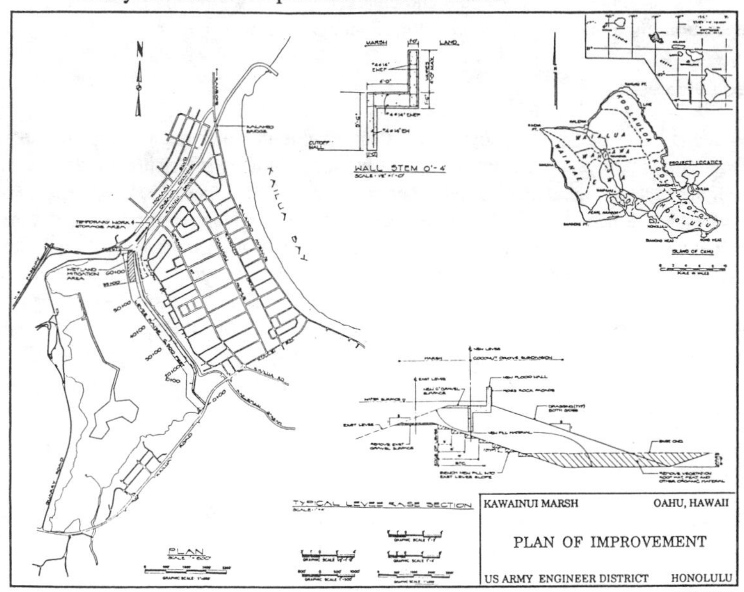

FIGURE 1

Hydraulics of Shallow Water Flow in a Marsh Flowway

C. Charles Tai,[1] M. ASCE and Chou Fang,[2] A. M. ASCE

Abstract

A demonstration project was constructed to test restoration of water quality in Lake Apopka by using a marsh flowway. The marsh flowway acts as a filter to separate sediments and nutrients from the water column. The present project showed that the flow was not uniformly distributed and channels formed within the flowway. Dense vegetation might have contributed to the channelized flow.

Introduction

Wetlands have long been recognized for their value in water quality improvement. They provide physical treatment processes for removing nutrients in the water column. Lake Apopka, located in the central Florida (Fig. 1), is a 12,422-ha (30,671-acre) lake. Since the 1920s, waste water effluent and agricultural waters were discharged into the lake which led to severe eutrophication condition of the lake. Circulation of lake water through reclaimed wetlands was recommended as one of the feasible approaches to restore water quality. The St. Johns River Water Management District (SJRWMD) initiated a 284-ha (700-acre) demonstration marsh flowway project to test this concept. The flowway has been in operation since August 1991. This paper will present data collected and discuss the hydraulics of the project.

Site Description

The demonstration flowway project is located in the north-west corner of lake Apopka (Fig. 1). The system consists of two units connected by a conveyance channel. Only one unit, the 81-ha (200-acre) south marsh flowway was used for a

[1]Director of Engineering and [2]Project Engineer, St. Johns River Water Management District, P.O. Box 1429, Palatka, FL 32178-1429

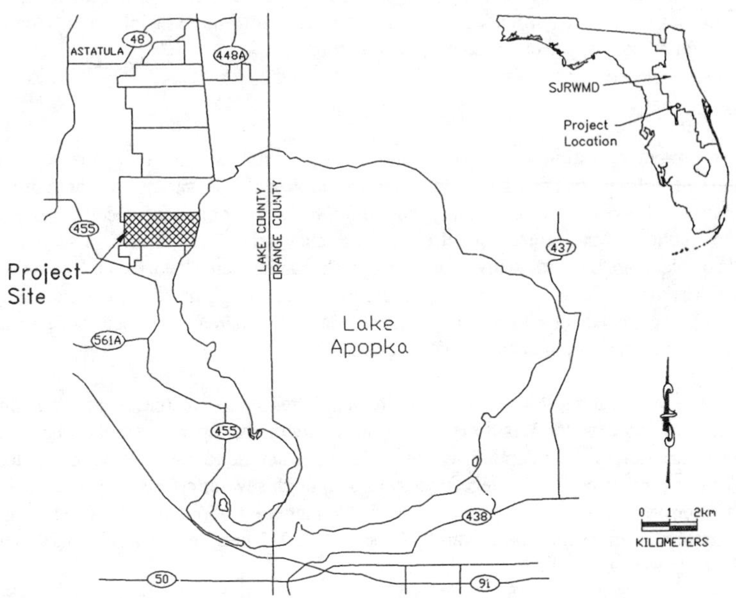

Fig. 1. Location Map.

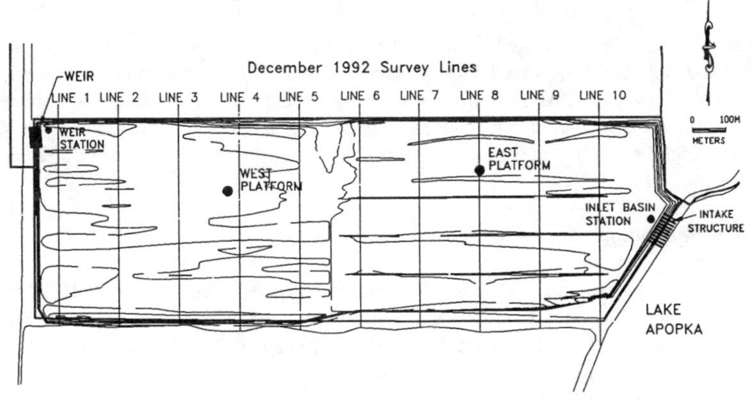

Fig. 2. South Marsh Flowway Site Plan.

detailed study. The south marsh flowway is about 487 m (1,600 ft) wide and 1,610 m (1 mile) long (Fig. 2). This flowway is restored from a farm land and the terrain is flat. Lake water is introduced to the south marsh from the east inlet and released through an overflow weir at the northwest corner.

Experiment

The project operation started in August 1991. Within a short time, the soil and nutrient-rich lake water caused rapid vegetation growth. A variety of plant species emerged; however, the cattail (*Typha Latifolia*) is the dominant species. Before construction of the project, small drainage ditches existed in the eastern section of the project area. Eventually, these ditches became open channels and conveyed majority of the flow in this part of the flowway. Although no functioning drainage ditches remained in the western section, internal channels formed because of variation in vegetation density and soil conditions.

As the vegetation grew thicker, the hydraulic resistance increased as well, and therefore more water was forced to concentrate in the open channels where the resistance is less. The depth of water thus has to increase to compensate for the loss of flow conveyance in the vegetated area. Although several efforts have been made to improve sheetflow condition, water depth increased about 0.36 m (1.18ft) and hydraulic gradient increased from 0.00006 to 0.00013 between November 1991 and September 1993.

Dye tests were conducted to investigate flow path in the flowway. Dye was injected at the inlet and time lapse pictures were taken from a helicopter. Dye concentrations were measured at selected time periods (Fig. 3). Observations confirmed that most of the flow was channelized and water was not evenly distributed across the flowway.

For a uniformly distributed flow in the marsh flowway, detention time can be calculated as $T = V/Q$, where T is the detention time, V is the flowway volume, and Q is the flow rate. As flow becomes concentrated in the open channels, volume available for detention time is reduced. Fig. 3 shows a reduction of detention time T and therefore a reduction of nutrient removal efficiency between October 1992 and September 1993.

Flow Simulation

A two-dimensional finite element flow model based on RMA2 (USACOE, 1993) was developed to study the flow regime considering the effects of plant density and channelization. The calibration involved fine-tuning a distribution of roughness coefficients, namely, Manning's n (Chow, 1959), in the modeling domain. The calibrated model was used to calculate Manning's n for different time periods of

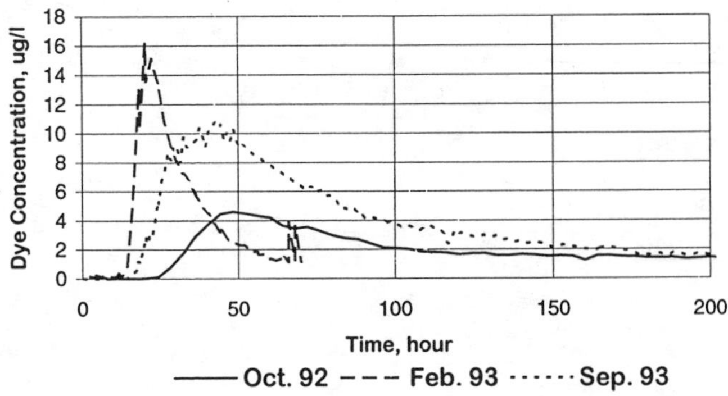

Fig. 3. Dye Concentration at Outlet Weir.

plant development. Using n=0.08 for the channel flow, the n values for vegetated marsh areas increased from 1.0 to 4.0 between November 1991 and September 1993.

Conclusion

Vegetation resistance has been combined with the bed resistance in most models. In the present project, vegetation has dominated everywhere in the flowway except in the open channels. The dense vegetation resulted in a large Manning's n value. Obviously, Manning's n should be estimated based on the vegetation parameters such as stem density, stem diameter, submerged depth, leaf area, etc. However, there is no such model available.

The SJRWMD will construct a 2,025-ha (5,000 acre) full-scale flowway project soon. There is a pressing need to develop acceptable design criteria and concepts. A wetland hydraulic experiment project is underway. The vegetation will be controlled to ensure uniform growth. Vegetation parameters will be monitored and the flow rate and water elevation recorded to develop a procedure for estimating Manning's n values from vegetation and hydraulic parameters.

References

Chow, V.T. Open Channel Hydraulics, 1959, McGraw-Hill Book Company, New York, New York, pp. 115-123.

U.S. Army Corps of Engineers Waterways Experiment Station, RMA2 version 4.27, 1993, 3909 Halls Ferry Road, Vicksburg, MS 39180.

Use of Constructed Wetlands for Upgrading an Advanced Wastewater Treatment Plant for Zinc Removal

Robert Butterworth P.E.[1], Donald L. Ferlow F.A.S.L.A[1], Elizabeth C. Moran Ph.D[1]
Maria J. Aridgides P.E.[1], Wayne Hazelton, Director, Microbiology[2]

Abstract

Stiefel Laboratories, Inc., Oak Hill, New York, is under an Order on Consent issued by the New York State Department of Environmental Conservation (NYSDEC) to upgrade its advanced wastewater treatment plant to remove zinc. The treated effluent must meet water quality standards for streams with intermittent flow suitable for trout spawning. Stiefel's zinc discharge limit is 3.2 grams/day (0.04 mg/L at a design flow of 76 m^3/day). Conventional technologies, such as activated carbon removal, zinc hydroxide precipitation and zinc sulfide precipitation were not able to achieve consistently these stringent effluent limits. Wetlands treatment technology was recommended by Stearns & Wheler as an additional treatment step due to its high potential to remove metals to very low levels, its ability to minimize the amount of toxic waste products (sludges) as well as its cost-effectiveness. Construction of the wetland cells was completed in July 1994. Data obtained during the initial five months of the wetlands operation indicate that the wetland cells can reduce soluble zinc in the influent to the cells from 0.25mg/L to 0.01mg/L.

Project Background

Stiefel Laboratories, Inc., a soap, detergent, and pharmaceutical manufacturing facility located in Oak Hill, New York, discharges process wastewater containing organics, high concentrations of BOD and nitrogen, and metals to an advanced wastewater treatment plant owned and operated by Stiefel. The treatment plant, constructed in 1990, consists of tankage for influent equalization, extended aeration, secondary settling, and aerated sludge storage. Secondary effluent passes through sand filters and carbon columns to further remove organics and suspended solids. The final effluent is aerated before being discharged to an intermittent stream.

1. Stearns & Wheler, 1 Remington Park Drive, Cazenovia, NY 13035
2. Stiefel Research Institute, Oak Hill, NY 12460

During the initial two years of operation, several biological upsets resulted in numerous discharge permit violations. As a result, Stiefel was issued an Order on Consent by the NYSDEC to evaluate and recommend improvements to their wastewater treatment plant. In December 1992, Stearns & Wheler identified a combination of factors that caused the permit violations, including: influent loadings in excess of design values; equipment failures; seasonal low temperatures; and zinc toxicity. Stiefel implemented measures to control shock zinc loadings to the treatment plant by modifying the manufacturing facility's cleaning operations to recover zinc from their waste stream and constructing an additional influent equalization tank which significantly reduced the frequency of biological upsets due to zinc toxicity. However, further removal of zinc was required to meet the stringent effluent discharge limits of 3.2 grams/day (0.04 mg/L at a design flow rate of 76 m^3/day).

Several alternatives were evaluated to reduce the zinc concentration in the plant effluent, including: activated carbon adsorption, zinc hydroxide precipitation and zinc sulfide precipitation. However, these alternatives were not proven to be able to consistently meet Stiefel's zinc discharge limit. In addition, these alternatives would produce a significant amount of chemical sludge containing zinc requiring disposal. Wetlands treatment is an innovative technology capable of reducing metals, including zinc, to very low levels. Based on the review of the available alternatives, Stearns & Wheler recommended construction of wetland cells as a final treatment step to achieve effluent limits for zinc because of its reliability and low operating and maintenance costs.

Wetlands Cell Design

Both natural and constructed wetlands have been demonstrated to effectively remove metals in a number of research and full scale projects. The majority of data on metal removal by wetlands describe wetland remediation of acid mine drainage. Other data present biologically treated wastewater enriched with metals and landfill leachate. Frequently investigated metals include iron, copper, zinc, nickel and cobalt.

A variety of processes are responsible for metal removal in wetlands. They include a mixture of chemical, biological, and microbiological reactions which occur in the aerobic and anaerobic zones. The aerobic and anaerobic zones support different removal processes, resulting in different removal efficiencies. Understanding the chemistry of the predominant removal mechanisms under both conditions is, therefore, necessary to optimize a wetlands system design with respect to metal removal. Hydraulic detention time, wetlands media materials, and type of vegetation should be evaluated and selected to allow the specific removal mechanisms to develop and dominate in the process of metal removal.

Anaerobic wetlands with submerged beds were selected for removal of zinc from the Stiefel effluent. Under anaerobic conditions the mechanisms predominantly responsible for zinc removal are adsorption, hydroxide precipitation and neutralization, and sulfide precipitation of zinc.

It was anticipated that upon start-up of the constructed wetlands, the adsorption of dissolved zinc onto organic matter associated with the wetlands media would be most important. Over time, after development of sulfate-reducing bacterial population, sulfide precipitation would become a more dominant mechanism.

Wetlands Design for Stiefel

The system was constructed in two lined biofilter basins. The total active volume of the biofilters was based on a hydraulic detention time of 15 days, which corresponds to a total surface area of approximately 240 m^2, at a 40 percent media porosity and active bed depth of 0.6 m. The biofilters were designed to operate in series or in parallel. The active media of the biofilters consists of a mixture of coarse sphagnum moss and agricultural limestone. Surface layer of the biofilters contains cattails. Perforated PVC pipes were placed at the influent and effluent end of each biofilter to uniformly distribute and collect the flow.

Zinc Removal

Zinc removal by the constructed wetlands was evaluated from the start-up of the wetlands operation in July 1994 until the middle of December 1994. The wetlands influent and effluent zinc concentrations observed during this period are shown in the figure below. The total and soluble zinc concentration in the effluent from the wetlands was less than the detection limit of 0.01 mg/L. Thus, the zinc discharge was below the SPDES discharge limit of 3.2 grams/day. Data shows that complete zinc removal was achieved immediately after the start-up, despite a very short acclimation period and limited growing season of the wetlands vegetation.

To verify the role of sulfide precipitation, sulfate and sulfide concentrations were measured in the influent and effluent from the wetlands. The results of the sulfate reduction are shown in the figure. The data shows that sulfate reducing conditions were established within three weeks after the wetlands start-up. During this period, the sulfate concentration decreased from an average of 43 mg/L to less than 2 mg/L. However, as the effluent temperature decreased to 55 F in October, the effluent sulfate concentrations started increasing and reached the influent values in November. It is anticipated that the increased sulfate concentrations are due to slowing down of the microbial activity associated with the temperature decrease. Despite the loss of the sulfate reducing conditions the zinc removal was unaffected. The above results suggest that both adsorption and sulfide precipitation were important in zinc removal by the wetlands.

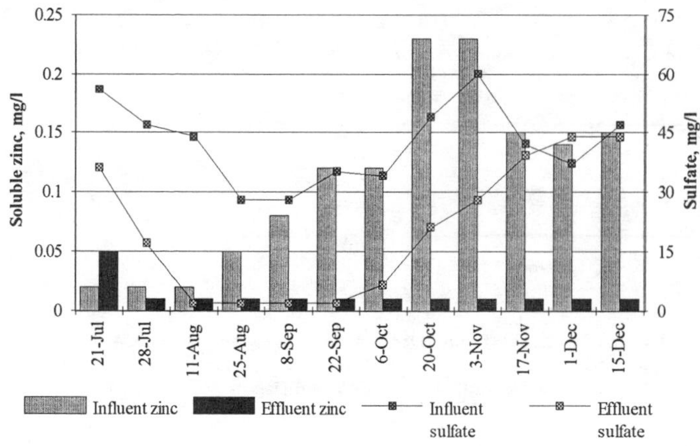

WETLANDS PERFORMANCE

The effect of wetlands on conventional pollutants, including BOD_5, TSS, TKN and NH_4^+ was monitored by bi-weekly analysis of the wetland influent and effluent. Since the influent values of these parameters are very low due to the advanced treatment preceding the wetlands, it was important to monitor possible changes in concentrations due to the wetlands. Slight increases in concentrations of these parameters were measured upon system start-up and lasted approximately 3 to 4 weeks. In the following weeks the impact of the wetlands on the conventional pollutants was minimal.

Conclusions

Based on our knowledge, Stiefel's wetlands system is the first full-scale system in New York State designed for metal removal. It's initial results obtained during start-up indicated that the wetland cells can reduce zinc in the influent to the cells from 0.25mg/L to 0.01 mg/L. In addition, excluding the initial four weeks after start-up, the wetlands did not show significant impact on the conventional pollutants in the plant effluent.

CONSTRUCTED WETLANDS: PERSPECTIVES AND APPLICATIONS

Andrew Dzurik[1], Member,ASCE **and Danuta Leszczynska**[2]

Introduction

Constructed wetlands (CW) for wastewater and stormwater treatment provide an option using natural systems and processes that have come into wider use in the past decade. With new federal NPDES requirements for stormwater runoff, and increasingly stringent water effluent standards by the EPA and state environmental agencies, CWs are likely to become more attractive as a relatively low cost option for treatment. Among the major problems currently confronting wastewater treatment are the high costs of constructing, operating and maintaining a conventional facility. At the same time, there is continuous damage to and destruction of wetlands stemming from the expansion of agriculture, urban growth and construction projects into wetland areas. The growing interest in using natural and constructed wetlands can help to address both of these issues by providing a low cost treatment alternative, and by adding to the inventory of wetlands.

This paper provides a summary of the CWs currently in use in the United States for different purposes. The primary categories are agricultural, municipal, industrial wastewater,and stormwater runoff. Small individual systems occasionally use CWs, but our focus is on larger scale systems that have general applicability to meet wastewater and stormwater treatment needs. Information is given on their use, design features, cost and performance.

System Types

The two primary types of constructed wetlands in use are: freewater surface (FWS) and submerged flow (SF) systems. The FWS wetlands are designed to imitate natural wetlands. They usually have soil bottoms, emergent vegetation, and water exposed to the atmosphere. The vegetation is planted in the shallow basins or channels, with relatively low water depth. The type of the soil ranges from gravel to clay or peat.

SF wetlands are design to maintain the water (or wastewater) level below the surface of the media (rocks, gravel), with no free opening to the atmosphere. They also might be known as vegetated submerged bed (VSB) flow, root zone method (RZM), vegetated rock-reed filter, microbial rock filter, or hydrobotanical systems.

The decision on whether to use a CW system may depend on several factors, including climate, wastewater flow, available sites, and budget. The decision on the

---

[1]Dzurik, Department of Civil Engineering, Florida State University, P.O.Box 2175, Tallahassee, FL. 32316, Tel:904-487-6124; E-mail: Dzurik@evax.eng.fsu.edu.

[2]Leszczynska, Department of Chemistry, Jackson State University, P.O.Box 17910, MS 39217-0510, Tel:601-973-3492; E-mail: Danuta@tiger.jsums.edu

specific type of constructed wetlands to use may depend on the source, type and flow of wastewater and on physico-chemical properties of its contaminants.

Wastewater Pollutants fate Pathways

The fate of any pollutant entering a wetland is affected by a variety of factors as (1) physical, chemical and biological characteristic of the wetland, (2) physico-chemical characteristic of the wetland's water, (3) physico-chemical characteristic and the concentration of particular pollutant, (4) the ratio between pollutants, their mixing pattern and possibility of chemical reactions between them, and (5) weather conditions - temperature and rainfall pattern.

While passing through a wetland cell, each pollutant has a limited number of fate options. Metals, nonmetals and other inorganic and organic pollutants, depending on their character, may be accumulated in biological tissue, transformed through biodegradation, removed from solution through precipitation, adsorption, photolysis, oxidation, hydrolysis, and volatilization, or may remain in soluble form and pass through the wetland system without being influenced by any removal mechanism.

Applications

Studies sponsored by the US EPA indicate a continuing increase in the construction and use of CWs for various purposes -- there are now over 200 CWs in the U.S. (Knight et al.1992). Because the development of this technology is relatively recent, the sizes, designs, operations, applications and effectiveness vary widely. Guidelines, standards, and literature on their design is growing, but the amount of information available is still limited. In order to develop a database of information on CWs in the U.S., the EPA's Office of Research and Development and Office of Water has undertaken a search of information on CW systems that generally treat more than 100,000 gpd (378 m3/d). The existing systems identified in the database include 127 systems with 215 individual treatment cells. Of the total cells, 71% are FWS and the others are SF or hybrid systems. Their sizes extend up to 442 acres (1093 ha), but plans are currently in process for FWS systems of several square miles for treatment of agricultural stormwater runoff in Florida's Everglades region. The list of systems includes 89 municipal, 22 industrial, 5 stormwater, and 11 undefined. Although agricultural use of CWs is common, these are generally small and, therefore, do not appear among the list of larger systems (Reed, 1992).

The following subsections give examples and summary information on various applications of CW systems.

Municipal. Two substantial municipal CWs for wastewater treatment are in Orlando, Florida. One system was designed and constructed for Orange County, and has been in operation since 1987 (CDM, 1887). The relatively simple design consists of 300 acres (120 ha) of created wetlands planted with selected herbaceous vegetation, and integrated with natural wetlands. It was designed as a receiver of secondary treated wastewater and was initially operated at 3 mgd (0.13 m^3/s), with an ultimate design capacity of 6.2 mgd (0.27 m^3/s) . The constructed wetlands, adjacent to natural marshes, serve as major treatment areas and buffer zones at the same time. The recycled wastewater ultimately discharges into a small creek (Best, 1987, 1994, Leszczynska and Dzurik, 1992). The second system, designed for the City of Orlando, about 1200 acres (485 ha) and ultimately designed to treat 16 - 24 mgd (0.17 - 1.1 m^3/s), is divided into three functional classifications. The overall CW system is named "Orlando Wilderness Park" and serves as a major wetland and recreational facility while meeting all regulatory limits for discharge set by the state.

As examples of the cost-effectiveness of a small municipal system, Union, Mississippi, a small rural town of about 2,000, has installed a CW facility for about $500,000, or less than half the cost of building a new treatment plant (Moss, 1993). One community in Kentucky constructed a wetland wastewater treatment facility at a cost of $300,000 -- considerably less than a traditional wastewater treatment plant that would have cost 3 to 4 million dollars (Hemra, 1992). A Tennessee Valley Authority demonstration project at Benton, Kentucky cost about $260,000, or $0.26 per gallon per day ($ 0.07 per liter per day) of design capacity - - only 10% of the estimated cost of a conventional system upgraded to meet federal requirements (Steiner).

Agriculture. Numerous recent examples of CW systems can be found in agriculture. They range widely in size and cost, but they can be especially effective because of the relative abundance of land, and the comparatively low cost of installation. They have a variety of applications, such as animal wastes (e.g., feed lots and dairy farms), and cropland stormwater runoff contaminants (e.g., fertilizer and pesticides). One study found that agricultural CW systems are generally cost effective in comparison with alternative treatment methods that give similar water quality results (Martin, 1993).

Although most agricultural CWs are small, plans underway in South Florida's agricultural area will result in the largest CWs in the country. The sugar industry has major sugar cane crops and processing plants in South Florida between Lake Okeechobee and the Everglades, largely due to the rich soils, tropical climate, and development of a canal system. However, high concentrations of phosphorus in water that flows into the Everglades are causing serious degradation in certain areas. It was determined that "protection of the Everglades system, and the ability to meet the total phosphorus limits and levels, will require installation and use of stormwater treatment areas (STAs)...." (SFWMD, 1992). The land needed for effective treatment will be massive, requiring about 32,600 acres (13,200 ha) of effective basin treatment areas. Total cost will be over 400 million dollars.

Industrial. Industrial applications of CW systems are increasing as federal NPDES standards now include stormwater runoff as well as plant discharge. One example is the Amoco refinery located at Mandan, North Dakota, where a study was undertaken to evaluate alternative wastewater treatment systems to comply with stringent federal standards. The plant occupies 300 acres (122 ha) surrounded by 660 acres (267 ha) dedicated to wastewater treatment and wildlife management. Alternatives evaluated were carbon adsorption, activated sludge, and expansion of the existing biooxidation system. Comparison of the alternatives showed expansion of the biooxidation system to cost $250,000 compared with costs of the other systems ranging from $1 to $3 million. Clearly a substantial savings accrued by use of the expanded CW system (Litchfield, 1989).

After the Logan Aluminum manufacturing plant in Kentucky experienced problems with its effluent spray irrigation and soil infiltration system, an overland flow-wetland system was built to treat the plant's wasteweater and to recycle the effluent back to the plant. Built on 1991 at a cost of 1.4 million dollars, the main objective of the system was to reduce effluent ammonia concentrations and reduce suspended solids (Reily and Wojnar, 1992).

Stormwater Runoff. In 1990 the U.S. EPA published final regulations for the National Pollution Discharge Elimination System (NPDES) stormwater discharge permits, thereby implementing requirements of Section 405 of the Water Quality Act of 1987. The regulations require certain industries and municipalities to obtain NPDES stormwater permits for all storm sewers that drain into public waterways. Although the municipal permits apply only to communities of over 100,000 population, it is conceivable that at some future time, the threshold level would be lower. Aside from specific legal requirements, it makes sense for communities of all sizes, as well as agricultural and industrial areas, to treat stormwater runoff, for this non-point contaminant is now the major source of water pollution, especially from urban areas.

Many constructed wetland systems for stormwater runoff have been built in Florida in the past ten years (Livingston, 1989). The most common type is a wet detention system with a permanent water pool, temporary stormwater storage area above the permanent pool, and a littoral zone planted with native aquatic plants. An outstanding example of a CW for stormwater runoff treatment is the Lake Jackson Restoration Project in Tallahassee, Florida, an experimental regional stormwater facility with major funding from the U.S. EPA. The facility covers about 30 acres, treating stormwater from nearby shopping centers and other commercial activity, highway I-10 and city streets, and low density residential areas. The system includes a 5 acre (2 ha) pond with sand filter, a 2 acre (0.8 ha) detention pond, and three wetland cells in series, approximately 3 acres (1.2 ha) each, with sheet flow between each cell. The constructed wetland system has been quite effective in improving the water quality of Lake Jackson, into which it flows.

Summary and Conclusions

Constructed wetlands are becoming increasingly important as a technology for improving water quality -- for small communities with low budgets, for agricultural areas with waste and runoff problems, for industries with potential contamination from process water, for communities and industries that must manage and treat stormwater runoff, and for mining areas. As the public and private sectors meet more stringent water quality standards for both wastewater and stormwater runoff, CWs can provide an attractive option at a reasonable cost. At the same time, a constructed wetland can be a visual and recreational asset by incorporating good landscape design. In its various applications, effluent from CWs is cleaner than the influent and can be used for recreation, agricultural irrigation, industrial processes, groundwater and stream augmentation, and as a supply of drinking water with additional treatment. CWs can also serve as park areas with habitats for fish and migratory birds, as was done in a number of communities. It appears that interest in constructed wetlands for wastewater treatment is growing in all sectors, and will continue to increase as greater demands are put in place for improved water quality in the face of limited financial resources.

References

Best, G.R., 1987, "Natural Wetlands - Southern Environment: Wastewater to Wetlands, Where Do We Go From Here?" Aquatic Plants for Wastewater Treatment and Resource Recovery, Magnolia Publishing, Inc.

Best, G.R., 1994, "Wetlands Ecological Engineering: An Approach for Integrating Humanity and Nature through Wastewater Recycling through Wetlands," Proceedings, 7th NABS Workshop.

Camp Dresser and McKee, Inc., 1987, "Addendum to December 1985 Wetland Exemption Background Monitoring Report, Eastern Service Area Wastewater Treatment Facility, Phase III," Orange County Public Utilities Division

Hembra, R.L., 1992, "Alternative Strategies Needed to Reduce Wastewater Treatment Costs," U.S. General Accounting Office, August 4, GAO/T-RCED-92-84.

Knight R.L, R.W. Ruble, R.H. Kadlec, and S. Reed, 1993, "Wetlands for Wastewater Treatment: Performance Database" in "Constructed Wetlands for Water Quality Improvement" ed. by G.A.Moshiri, Lewis Publishers.

Leszczynska, D., and A. Dzurik, 1992, "Tertiary Wastewater Treatment through Constructed Wetland Ecosystems," Environment Protection Engineering, 18(1- 2).

Litchfield, D.K. and D.D. Schatz, 1989, "Constructed Wetlands for Wastewater Treatment at Amoco Oil Company's Mandan, North Dakota Refinery," in D.A. Hammer(ed.), Constructed Wetlands for Wastewater Treatment, Lewis.

Livingston, Eric, 1989, "Use of Wetlands for Urban Stormwater Management," in D. Hammer (ed.), Constructed Wetlands for Wastewater Treatment, Lewis.

Martin, J.B., and C.E. Madewell,1971, "Environmental and Economic Aspects of Recycling Livestock Wastes," Southern J. Agric. Econ.3:137-142.

Moos, Shawna, 1993, "More Than Just Sewage Treatment," Technology Review, August.

Reed, Sherwood and Donald S. Brown, 1992, "Constructed Wetland Design - The First Generation," Water Environment Research, 64(6).

Reily, J.M. and H.A Wojnar, 1992, "Treating and Reusing Industrial Wastewater: Constructed Wetland Reduces Ammonia, Solids and Metals," Water Environment and Technology, 4(11).

South Florida Water Management District, 1992, "Surface Water Improvement and Management Plan for the Everglades," Volume I, West Palm Beach, FL.

Steiner, G.R., "Constructed Wetlands -- Low Cost, Simple Sewage Treatment for Rural Communities," unpublished paper, Tennessee Valley Authority, Chattanooga.

CSO TREATMENT USING CONSTRUCTED WETLANDS

RICHARD A. JUBINVILLE, P.E., M, ASCE[1]
MICHAEL J. TOOHILL[2]
ROBERT H. KADLEC, PhD.[3]

ABSTRACT

A CSO study completed in 1988 recommended that the City of Chicopee, MA eliminate a total of 42 CSO discharges by utilizing conventional treatment technology at a projected cost of $98.7 million. The results of a feasibility study commissioned by the City indicated that constructed wetlands could provide a cost-effective alternative to conventional CSO treatment. A pilot-scale facility, using both FWS and SSF wetlands, was recommended to be constructed and operated for a 1- to 2-year period to gather operational data for design and construction of full-scale facilities.

INTRODUCTION

The City of Chicopee, MA has an extensive combined wastewater/stormwater collection system, with a total of 42 combined sewer overflows (CSOs) that discharge directly into the Connecticut and Chicopee Rivers during storm events. A 1988 CSO study (Metcalf & Eddy, 1988) recommended construction of conventional treatment facilities at a projected construction cost of $98.7 million.

In 1993, the Pioneer Valley Planning Commission (PVPC), with participation by the City of Chicopee and the Massachusetts Department of Environmental Protection (DEP), commissioned a study to investigate the feasibility of using constructed wetlands to treat CSO discharges. Manmade wetland treatment systems have proven effective at treating both wastewater and stormwater. As of 1993 their use in treating CSOs had been limited to one case, in Houten, the Netherlands.

Initially five CSOs were selected by the City and the PVPC for evaluation. After completion of a review by the study team, four sites were eliminated due to cost and/or siting limitations. One, CSO 009 (Paderewski Street Pump Station), was selected for further study based on its size and location: close to the Chicopee WWTF, and to a large tract of land at the junction of the Chicopee and Connecticut Rivers. It discharged a

[1]Vice President, Whitman & Howard, Inc., 45 William Street, Wellesley, MA 02181
[2]Environmental Scientist, Whitman & Howard, Inc.
[3]Principal, Wetland Management Services, Inc.

reasonably sized flow for testing a pilot scale constructed wetland, and its location will minimize conveyance costs. Proximity to the WWTF means ready access by the City's operations personnel who will be operating and maintaining the pilot facility.

WETLANDS TREATMENT TECHNOLOGY OVERVIEW

Many constructed and natural wetland systems have been used for wastewater treatment for at least the past 20 years. Constructed systems are either free water surface (FWS) or subsurface flow (SSF) wetlands. FWS wetlands have open water and submergent (pond weeds) or floating plants (lily pads, duck weed), while SSF wetlands contain gravel beds with water below the ground surface and emergent plants. Both types of systems are typically planted with robust, aggressive species such as cattails, bulrushes or reeds.

Comprehensive overviews of a number of constructed wetland systems currently in operation throughout the world are given by Hammer (1989, 1992), the USEPA (1987, 1993), and Morshiri (1993). Most of these systems were constructed for wastewater treatment, with systems created for stormwater treatment being a newer variation.

Test data from the FWS CSO wetland facility in Houton-OOST, Netherlands (van Oorschot, 1990) and other wetlands treatment systems currently in operation indicated that acceptable reductions in BOD, pathogens, and suspended solids levels could be achieved using this technology. Based on the data produced by the literature review, it was decided that a constructed wetland could be a feasible alternative to more conventional CSO technologies for the City.

CHARACTERISTICS OF THE SELECTED CSO

CSO 009 - Paderewski Street Pumping Station - is located approximately 450 meters (1,500 feet) north of the Chicopee WWTF at the end of a 1.5-meter (60-inch) diameter combined sewer and is separated from the Chicopee WWTF by the Massachusetts Turnpike. See Figure 1.

A regulator chamber directs dry weather flow to the pump station for discharge to the Connecticut River Interceptor which flows to the WWTF. The Pump Station has a capacity of 0.041 m^3/S (0.95 mgd).

The design parameters for CSO 009 for a 3-month, 2-hour design storm are 0.029 m^3/S (0.66 mgd) and a peak discharge rate of 1.06 m^3/S (24.3 mgd). The discharge has an average BOD of 68 mg/l, a TSS concentration of 165 mg/l, and a fecal coliform level of 120,000 per 100 ml (Metcalf & Eddy, 1988).

The area proposed to site the wetland contains both upland and wetland vegetation, and is impacted by the 100 year and 500 year floodplain boundaries. A suitable area was selected that avoids the wetland area and is outside of the 100-year floodway zone, where higher water velocities would be a threat to the project. The site is also accessible to the public, which is an advantage as public education is one of the project's priorities.

RECOMMENDED PLAN

Several pumping options to convey CSO 009 to the constructed wetland site were evaluated. The recommended plan is to construct a new dedicated CSO Pump Station with two 0.547 m^3/S (12.5 mgd) submersible pumps. Flows would be transported via a 0.51 meter (20-inch) force main to a vortex degritting unit at the existing WWTF for preliminary treatment. A gravity sewer would discharge flow to the constructed wetland: parallel treatment units consisting of (1) a 0.8 ha (2 acre) SSF wetland, and a 0.6 ha (1.5 acre) FWS wetland. Wetland cells would have a berm height of 1.5 meters (5 feet). The project would occupy approximately 2 ha (5 acres).

A 2-year monitoring period would follow construction in order to provide operational data on both types of wetland systems, and to evaluate vegetation types, wildlife forage value and aesthetic considerations. The ability of the wetland to reduce bacteria to acceptable levels without the use of chlorine or other disinfection measures is a key objective of the pilot facility. The results of the pilot scale operation will provide design parameters for full scale constructed wetland systems for other CSO discharges at other sites or by expanding the pilot facility to accept additional CSO flows from the Chicopee River Interceptor.

Chicopee representatives had three major concerns about the FWS system: (1) the potential for odors; (2) mosquito breeding; and (3) public safety. It is felt that operation of both types wetlands simultaneously will provide valuable operating experience and data to answer these concerns. The vortex treatment unit was added to the plan at the City's request to keep gross solids out of the wetland. Although wetlands can achieve excellent solids removals, the City did not want to have to perform solids removal activities at a site remote from the WWTF.

The project costs for this pilot scale system are estimated to be $1.85 million with over $1.0 million attributed to conveyance and preliminary treatment costs.

CONCLUSIONS

This use of a constructed wetland for CSO treatment in Chicopee appears to be feasible, however, the costs must be compared against those of more conventional CSO treatment technologies, as costs for conveyance can be significant.

PROJECT FOLLOW-UP

Since completion of the feasibility study in April, 1994. The City of Chicopee has asked the project team to evaluate less expensive alternatives to the constructed wetland scheme. In addition, the DEP has relaxed its requirement that an entire CSO flow be captured for treatment.

The project team is currently considering the use of portions of the City's WWTF to provide CSO treatment for flows within the Chicopee River interceptor. A wetland system constructed as a pilot scale facility remains a priority of the project.

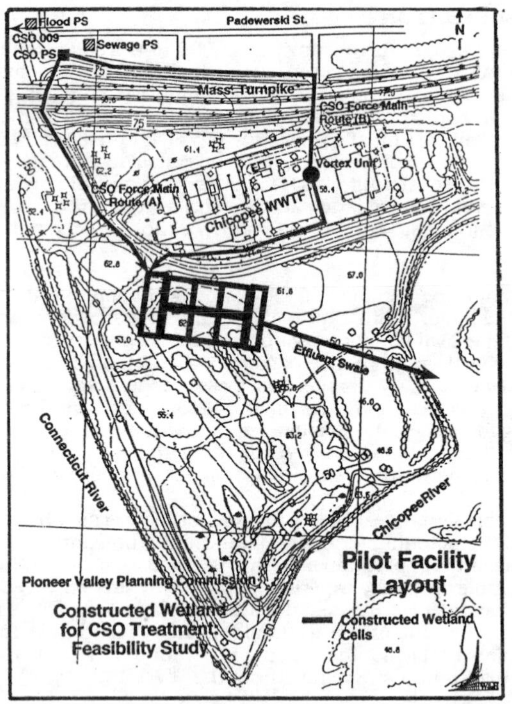

FIGURE 1

APPENDIX

Hammer, D. A. (ed). Constructed Wetlands for Wastewater Treatment Municipal, Industrial, and Agricultural. Lewis Publ., Chelsea, MI, 1989.

Hammer, D. A. Creating Fresh Water Wetlands, Lewis Publ., Chelsea, MI, 1992.

Metcalf & Eddy, Inc. Lower Connecticut River Combined Sewer Overflow Study, 1988.

Moshiri, G. A. in press. Constructed Wetlands for Water Quality Improvement. Lewis Publ., Chelsea, MI, 1993.

U.S. Environmental Protection Agency (USEPA). Report on the Use of Wetlands for Municipal Wastewater Treatment and Disposal. EPA 430/09 88-005, U.S. EPA, Washington, DC, 1987.

U.S. Environmental Protection Agency (USEPA). Created and Natural Wetlands for Controlling Nonpoint Source Pollution. Lewis Publ., Chelsea, MI, 1993.

vanOorschot, M. M. P. De Biezenvelden bi j HouTen-Oost. The Utrecht Plant Ecology News Report, Wetlands for the Purification of Water, No. 11:247-269. Utrecht; The Netherlands, 1990.

WETLAND BIOTRANSFORMATION OF ATRAZINE

K.S. Ro, Dept. of Civil and Environmental Engineering, Louisiana State University Baton Rouge, LA 70803
K.H. Chung, College of Pharmacy, Sung Kyun Kwan University, Suwon, Korea

Natural wetlands are known to be able to decontaminate many of the conventional pollutants such as biochemical oxygen demands (BOD) and total suspended solids (TSS) (U.S. EPA, 1988). The wetlands support a diversity of microorganisms which contributes to the wetlands' ability to degrade many bioresistant chemicals such as atrazine, the most widely used herbicide in the U.S. (Chung et al., 1994, Ro and Chung, 1994). The objectives of this study are 1) to investigate the fate of atrazine under aerobic and anaerobic conditions in laboratory-scale batch reactors containing wetland water and sediment, and 2) to observe the effects of a nitrogen-rich nutrient medium on atrazine biotransformation.

MATERIALS AND METHODS

The wetland water and sediment were obtained from a diked wetland receiving local sugar mill wastewater in Louisiana. The wetland sediment and water compositions are shown in elsewhere (Chung et al., 1994). Two sets of sample batch reactors (500 mL amber bottles) containing initially 10 mg/L atrazine were incubated with 12.5 g dry sediment/L under both aerobic and anaerobic conditions. The aerobic conditions were provided by saturating the water with air previously and incubating the reactors without lids. An anaerobic chamber (Labline model 655) maintained strict anaerobic conditions by continuously purging with Ar gas. All sample batch reactors were incubated in duplicate or triplicate at room temperature (22 + 2 °C) and the average values were reported.

Atrazine, its well-known dealkylated metabolites (deethylatrazine, deisopropylatrazine, and deethyldeisopropylatrazine), and their hydroxy analogs were analyzed by solid phase extraction and HPLC analysis. The sediment-phase concentrations of atrazine and its metabolites and hydroxy analogs were estimated by subtracting aqueous concentration from the total concentrations (both aqueous and sediment phases). The samples were also analyzed with a GC/MS (HP 5890 II) in order to confirm the identity of atrazine and its metabolites. The concentrations of NH_3-N in the sample batch reactors at the end of incubation period were determined according to the

Standard Methods. A detailed description of the experimental methods and materials is shown elsewhere (Chung et al., 1994; Ro and Chung, 1994).

RESULTS AND DISCUSSION

Figure 1 shows the atrazine profile during 70 days of incubation under aerobic conditions. None of the dealkylated metabolites and the hydroxy analogs were observed both in water and the sediment phases. After about 42 days of a stationary phase, atrazine (initially 10 mg/L) disappeared to less than a detectable level (10 ug/L) within 21 days. Approximately 10% of total atrazine was adsorbed on the sediment with a distribution coefficient of 9.979 mL/g (Chung et al., 1994). In order to see the effects of nitrogen-rich nutrient medium on the biotransformation rate of atrazine, we also incubated the sample batch reactors in basal salt media (BSM) (Ro and Chung, 1994). BSM appeared to hinder the aerobic biotransformation rate, removing only 25% of atrazine during 126 days of incubation. Cook hypothesized that the nutrient media without a nitrogen source would force microorganisms to attack atrazine in order to acquire nitrogen, and thus increase the biotransformation rate. Conversely the high level of a nitrogen source in the BSM might have slowed down the biotransformation rate. However, Bernal-Cespedes (1990) observed that an additional nitrogen source [$(NH_4)_3PO_4$] increased the aerobic atrazine degradation rate.

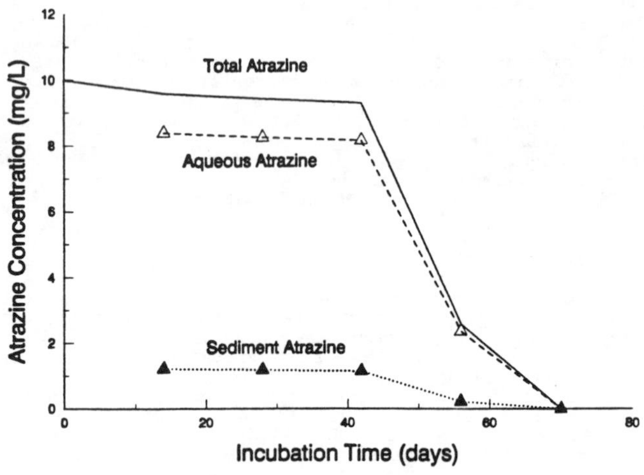

Figure 1 - Atrazine Profile During Aerobic Incubation

In contrast to the aerobic biotransformation, we detected a significant level of hydroxyatrazine in the sample batch reactors during 38 weeks of anaerobic incubation. However, none of the dealkylated metabolites were observed. Figure 2 shows the fate of atrazine and hydroxyatrazine in both aqueous and sediment phases during the incubation period. More than 50% of total hydroxyatrazine was found in the sediment phase, compared to about 10% for atrazine. Although the initial 10 mg/L of atrazine was reduced to 4.2 mg/L in the water phase after 38 weeks of incubation, one must be cautious not to interpret this result as 58% biodegradation of atrazine. Considering the atrazine adsorbed on the sediment and the total hydroxyatrazine, the total triazine species concentration was 7.3 mg/L.

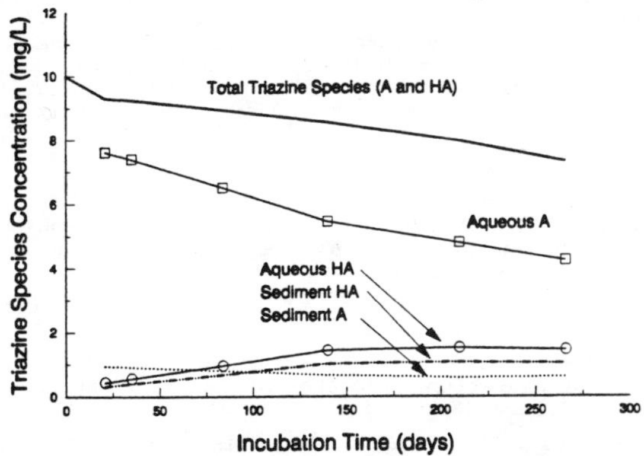

Figure 2 - Triazine Species During Anaerobic Incubation
(A: Atrazine, HA: Hydroxyatrazine)

Of interest is the fact that BSM greatly improved the biotransformation rate under anaerobic conditions. Atrazine completely disappeared from both in water and sediment phases and only about 2.7 mg/L of the total hydroxyatrazine was present after 38 weeks of incubation. This observation contradicted Cook's hypothesis. The decrease in both atrazine and hydroxyatrazine without producing known dealkylated metabolites and the hydroxy analogs suggested a complete mineralization of atrazine to its end products, such as CO_2 and NH_3. At the end of the incubation, we measured the values of mg of total NH_3-N produced per mg atrazine disappeared (Y_{TAN}) in the sample batch reactors. Assuming all atrazine-N has been converted to NH_3-N, a theoretical maximum Y_{TAN} is 0.33. The observed Y_{TAN} for the wetland-water sample batch reactor was 0.28, which was slightly less than that of the theoretical maximum value. The absence of the dealkylated metabolites and their hydroxy analogs and a comparable Y_{TAN} to the theoretical maximum value strongly suggest the mineralization of the disappeared atrazine and hydroxyatrazine to NH_3 and CO_2.

CONCLUSIONS

Atrazine biotransformation characteristics under both aerobic and anaerobic conditions were investigated using laboratory-scale batch reactors containing wetland water and sediment. After 38 weeks of anaerobic incubation, atrazine in the water phase dropped from the initial 10 mg/L to 4.2 mg/L. However, a significant portion of the disappeared atrazine existed as hydroxyatrazine. In contrast to the anaerobic conditions, atrazine disappeared rather rapidly within 21 days of incubation after 42 days of stationary phase under aerobic conditions. None of the dealkylated metabolites and their hydroxy analogs, including hydroxyatrazine, which was prevalent under anaerobic conditions, were observed under aerobic conditions. BSM increased the atrazine biotransformation rate under anaerobic conditions, but decreased the rate under aerobic conditions.

REFERENCES

Chung, K.H., Ro, K.S., and Roy, D. (1994) Fate and Enhancement of Atrazine Biotransformation in Anaerobic Wetland Sediment. submitted for publication, *Water Res.*

Cook, A.M. (1987) Biodegradation of s-Triazine Xenobiotics. *FEMS Microbiol. Rev.*, **46**, 93-116.

Bernal-Cespedes, G.I. (1990) Evaluation of Liquid Solid Contact Reactor Bioremediation Approaches for Atrazine Degradation. M.S. Thesis, Louisiana State University, Baton Rouge.

Ro, K.S. and Chung, K.H. (1994) Atrazine Biotransformation in Wetland Sediment under Different Nutrient Conditions: Aerobic. in press *J. Environ. Sci. Health Part A*.

U.S. EPA (1988) Report on the Use of Wetlands for Municipal Wastewater Treatment and Disposal. EPA 430/09-88-005.

Canal Multiple Use in Metropolitan Phoenix Area

Thomas G. Sands, P.E., M. ASCE[1]

Abstract

Since the late 19th century, Phoenix area canals have delivered water to farms, businesses and residences in the Salt River Valley. Water delivery remained the sole purpose of the canals until recently. Now, the canals are recognized as an aesthetic asset to the community, and proposals for various recreational and commercial uses of the canals are on the table. This paper briefly describes the concept and benefits of canal multiple use in metropolitan Phoenix.

Introduction

The Salt River Project (SRP) was formed in 1902 to provide agricultural water to the Salt River Valley in central Arizona. Over the past 30 years, the area has changed from a small agricultural community to a major metropolitan area. It now includes ten cities and towns, including Phoenix, Scottsdale, Tempe and Mesa and has a total population of approximately 2.5 million. Growth of this magnitude has brought tremendous changes (Sands, 1987). These changes also have required modifications to the infrastructure supporting the area. One of the most fundamental infrastructure networks is the SRP canal system, which supplies water to the Valley's arid desert lands.

The canals originally were designed to transport water from the Salt River to local farms. However, with urbanization, new ideas for uses of the canal have emerged that could result in major changes. As the Valley's population increases, the need for open space and recreational outlets within

[1] Principal Engineer, Salt River Project, P.O. Box 52025, Phoenix, AZ 85072-2025, Phone (602) 236-2371

cities has expanded. Furthermore, the aesthetic qualities of the dirt and concrete canals also has been a topic of discussion. This has led to discussions of beautifying the canals and developing them to provide additional recreation space.

Canal Multiple Use Concept

The aesthetic value of flowing water historically has been ignored. In fact, homes and business typically were constructed to back up to the canals, which were considered nothing more than basic utility corridors. However, the water's aesthetic and recreational potential now is being reconsidered in order to maximize the water's benefit to citizens *before* the water reaches its end destination. The name given to this concept is "canal multiple use." Local supporters believe that the benefits of recreational enhancements to the canals, in the long term, could far outweigh the costs of development.

Regional Framework Proposed

Recently, regional design guidelines were proposed by a consortium led by Arizona State University's College of Architecture (Fifield, 1990). The document's purpose is two-fold. First, it proposes and defines a new urban role for the canals, that of a public amenity. Second, it views the canals as a regional public system and explores their connection with other open spaces, city centers, neighborhoods and public attractions. The stated purpose of the guidelines is to identify, preserve and enhance the regional character of the Valley's major canal systems such that:

- any portion of the canal system is in some way identifiable with the whole.
- the system remains amenable to regional users.
- the system creates a unique image for the whole Metropolitan Phoenix Area.

Expected Benefits

Aesthetic and recreational enhancement of the canals can yield multiple benefits. First, the value of the canal system to the community should increase as it becomes more useful and purposeful. Second, it can create open-space opportunities for private, municipal and commercial purposes. Third, the canals would become open-space links to other underutilized open spaces. Fourth, because most historical discussion of the Valley includes the canals and their vital link to successful agriculture and subsequent urbanization, enhancement of the canals can assist in promoting community awareness of their history.

Challenges

Many challenges present themselves in utilizing the canals for more than their original use of transporting water for irrigation use. One of the more significant challenges is assessing and resolving modifications to SRP's historic operation and maintenance (O&M) practices. However, SRP's O&M procedures are evolving (Salt River Project, 1989) and new methods of maintaining the canals are continually being developed. With cooperation from all parties involved, the aesthetic uses of the canals can be increased while still allowing SRP to fulfill its mission of providing a reliable, adequate and economic supply of high quality water to its shareholders.

Projects Now Underway

Several canal multiple use projects currently are underway. The largest undertaking is the Scottsdale Downtown Waterfront Project, which will incorporate a half mile stretch of canal into a major retail/commercial project. Another is the Sunnyslope Demonstration Project in north central Phoenix, which will create a mile-long linear park in a predominately residential area.

References

Fifield, Pihlak, Cook, and Southerland, 1990. "Metropolitan Canals, A Regional Design Framework."

Salt River Project, 1989. "Canal Multiple Use Guidelines."

Sands, T.G., 1987. "The Urbanization of the Salt River Project," ASCE Irrigation and Drainage Conference Proceedings.

Impacts of Highway Runoff on Surface Water Drinking Supplies

Richard A. Moore, P.E.[1]
Brian T. Butler[2]

Abstract

This paper provides an overview of three potential issues that are frequently the subject of discussion between state transportation officials and environmental regulators. The issues include the impacts of deicing chemicals, conventional pollutants, and spills on public surface water drinking supplies.

Introduction

The watersheds of most public surface water drinking supplies are criss-crossed by local roads and state and interstate highways. Over the past ten years many state transportation agencies and highway departments have begun to recognize the sensitivity of these water resources and have begun to emphasis environmental management of highway activities. The main focus of these activities has been to control deicing chemicals, especially sodium; conventional stormwater pollutants, typically, solids, nutrients, metals and petroleum products, and spills of hazardous materials. However, often these efforts are not judged adequate by regulatory authorities.

The remainder of this paper summarizes the authors current thinking on the management of deicing chemicals, conventional pollutants and spills and how a balanced approach might be reached to meet the needs of both transportation and environmental officials.

---

[1] Vice President, Rizzo Associates, Inc., 235 W. Central St., Natick, MA 01760

[2] Environmental Scientist, Rizzo Associates, Inc., 235 W. Central St., Natick, MA 01760

Deicing Chemicals

Rock salt (sodium chloride) has long been the preferred deicing chemical for highway departments due to both its low relative cost and its effectiveness at breaking the ice/pavement bond at most commonly encountered winter storm temperatures. However, sodium has been regulated in drinking water for many years at concentrations above 20mg/l.

To reduce sodium concentrations in drinking water supplies many highway departments have partially or completely substituted alternative chemicals for rock salt. These include products such as calcium chloride and calcium magnesium acetate (CMA). In all cases costs increase and deicing operations change.

Currently many states have or are considering changing the sodium limits by either increasing the maximum concentration or changing enforceable standards to guidelines or both. For example, Massachusetts has replaced a sodium maximum contaminant limit (MCL) of 20mg/l with a guideline of 28 mg/l.

This reduction in the level of concern for sodium in drinking water supplies represents an opportunity to rethink the use of sodium as the primary highway deicing agent. The seven practices listed below are intended to reduce total rock salt usage while at the same time maintain safe operating conditions during snow and ice events.

- Applying a rock salt brine solution prior to storm onset
- Pre-wetting solid rock salt for application during the storm
- Applying rock salt in windrow on roadway crown or high-side
- Optimize rock salt specifications
- Tailoring rock salt application to weather conditions, equipment, and personnel limitations
- Maximizing plowing operations
- Eliminating rock salt applications below 20°F

Such practices represent a good balance of environmental, cost and safety concerns.

Conventional Pollutants

Conventional pollutants typically associated with highway runoff include solids, nutrients, metals and petroleum products. There is a good deal of data describing the discharge characteristics of these parameters as summarized in Table 1.

Table 1 Highway Runoff Discharge Characteristics

	Runoff Concentrations, mg/l			
	All Studies (1)		Federal Highway Administration	
Parameter	Range	Median	Spring Runoff Median	Winter Runoff Median
Total Suspended Solids	42-204	120	93	204
Oil and Grease	4-10	7.4	10	-
Ortho-Phosphate	0.05-0.65	0.35	0.293	0.57
Total Copper	0.034-0.091	0.047	0.039	0.091
Total Zinc	0.053-0.42	0.24	0.217	0.42

U.S. Department of Transportation (FHWA)
U.S. Environmental Protection Agency (USEPA)
U.S. Geological Survey (USGS)
Texas Department of Transportation (TxDOT)
Florida Department of Transportation (FDOT)
Washington State Department of Transportation (WDOT)
City of Seattle (METRO)

Unlike discharge data there are limited data describing the effects of highway runoff on receiving water bodies, especially large reservoirs. The best available data generally focuses on the physical changes in stream morphology that occurs with increased flows and sedimentation.

Due to the lack of data describing water quality and aquatic impacts, statistical models are often used to predict conditions. These models are by necessity simplifications of very complex systems and often produce conservative results that can not be easily verified in the field.

In general, the interpretations of both field data and model results is not straight forward and there is little regulatory guidance that address stormwater runoff controls. Nevertheless over the past twenty years there has been considerable information developed on the control of conventional stormwater pollutants. These controls can generally be divided into two categories: source controls, or best management practices (BMPs) and discharge controls. BMPs are typically practices, such as street sweeping and catchbasin cleaning, that prevent pollutants from entering the drainage system. Discharge controls are typically end-of-pipe structures such as detention basins.

There is an emerging consensus in the technical and regulatory communities that most BMPs are cost effective, good for the environment, and should be implemented in some form regardless of the local conditions.

There is less agreement concerning when specific discharge controls should be used to control highway runoff. However, two general guidelines can help in their selection. First, discharges should be located as far away form open water as possible, thus maximizing buffer zones to allow for natural settling, filtering, and dissipation of stormwater velocities. Second, discharge controls are most appropriately considered for high quality receiving waters such as drinking water supplies.

Based on these guidelines, increased buffer zones and/or discharge controls should be considered for highway drainage in public surface water supplies. In addition to controlling conventional pollutants this approach can also serve as part of a spill control program as described below.

Spill and Emergency Response

An effective emergency response program is a critical element of any stormwater management plan designed to protect surface public water supplies.

A good emergency response program has a number of components that need to be evaluated and adapted for each particular watershed. In addition to buffer zones and discharge controls previously mentioned, these include:

- Identification and correction of unsafe roadway conditions.
- Education of hazardous materials transporters to avoid sensitive routes.
- Communication among responsible parties (e.g., police, fire, HazMat teams, local emergencies/planning committees and highway departments). This includes response alerts as well as ongoing planning.
- Availability of equipment including absorbent materials, booms, vacuum equipment, and boats, all stored at convenient locations.
- Training for response personnel in chemical identification, health and environmental hazards, deployment, and use of equipment. This includes emergency response exercises.
- Mapping of the drainage system so that the transport pathways and discharge locations can be quickly determined and secured.
- Preparation of a written plan with an ongoing planning system with periodic updates.

In most cases spill and emergency response should be implemented watershed wide with highway departments playing a supporting role with other responsible parties.

Summary

Protecting surface water supplies from adverse roadway impacts requires a common sense balance in managing and controlling deicing chemicals, conventional pollutants and spills.

Grassroots Watershed-Based Wetlands Planning

John E.Murphy, P.E.[1], Mary L. White[2], Alex Strysky[3]

Abstract

This paper discusses recent watershed-based planning efforts involving community organizations, municipal, state, and federal agencies. The purpose is to begin to address the restoration of functions to a highly stressed wetlands system.

Introduction

The Alewife Brook watershed is an urbanized, highly altered sub-basin of the Mystic River watershed, draining into the Boston Harbor, lying primarily in the densely developed communities of Arlington, Belmont, and Cambridge, Massachusetts. A history of urbanization, re-channelization, wetlands filling and pollution, and combined sewer overflow (CSO) discharges have caused extensive impacts to wetlands, water quality, fisheries and wildlife habitat, and flood storage capacity. The goal of restoring functions to this system will require a multi-disciplinary approach, as discussed in this paper.

Wetland & Open Space Loss and The Effect On Flood Levels

While the highest rate of wetlands filling in the Alewife watershed occurred prior to this century,

----

[1]Belmont Conservation Commission, and Associate, Hayden-Wegman, Inc., 214 Lincoln Street, Boston, MA, 02134

[2]Belmont Conservation Commission, and Project Manager, MWRA, 100 First Avenue, Boston, MA 02129-2024

[3]Director, City of Cambridge, Conservation Commission, 57 Inman Street, Cambridge, MA 02139

development impacts to the floodplain have continued. Over the last 60 years or so, a significant portion of Metropolitan District Commission (MDC), Alewife Reservation land and open space abutting the park land has been lost to development of transportation infrastructure, residential complexes, commercial uses, and other urbanization pressures.

In Belmont, for example, the total acreage of MDC park land alone decreased by 30%, from 24 acres in 1931 to only 16 acres today. A similar rate of open space loss has also been experienced in abutting areas of Arlington and Cambridge.

The loss of wetlands and open space has impacted the ability of the Alewife flood plain to retain flood waters as evidenced by a progressive rise in flood levels of tributary water bodies and their low lying flood plains. Storm flood elevations of several water bodies in the Alewife Brook watershed, including Clay Pit Pond, Blair Pond and Little Pond, have all shown a general increasing trend. For example, flood elevation measurements (Betts, 1994) taken at Little Pond over the last 60 years show a gradual rise from an average of about 7 feet above mean sea level (AMSL) back in 1932 to over 9 feet (AMSL) in recent years. This 2-foot rise in flood level can be directly attributed to the loss of wetlands and open space and their effectiveness of absorbing and attenuating storm runoff. Local homeowners situated in low lying areas surrounding these water bodies have experienced a greater frequency and severity of flooding. This loss of flood storage capacity due to development in the flood plain compounds the hydraulic problems downstream of the area which also has been altered by brook crossings and the channelization of Alewife Brook for most of its length.

Stormwater Flow & Water Quality At Alewife Reservation

Water quality monitoring data collected over the last several years at the Alewife Reservation has documented the severe impacts to water quality caused by the urbanization of this watershed. The Massachusetts Water Resources Authority's (MWRA) monitoring program for Boston Harbor and its tributary rivers include sampling sites along Alewife Brook and Little River, which flows through the MDC Reservation and joins Alewife Brook at the eastern end of the reservation.

Monitored water quality parameters showing the greatest change with measured rainfall are bacteria counts of Fecal Coliform and Enterococcus, indicators of sewage pollution. The primary cause of this pollution is combined sewer overflows (CSOs) and stormwater runoff. MWRA sampling data from 1989 through 1991 reveal that rainfall accumulations of as little as 0.10 in. cause escalation of bacterial counts by a factor of 8 over dry weather conditions (MWRA, 1991, 1993). Fecal Coliform and Enterococcus levels of several thousand to over 10,000 colonies per 100 ml have been reported during periods of heavier rainfall. Normal dry weather counts are generally in the 200 to 500 range. In comparison, the State designated water quality standard of Alewife Brook is Class B (fishable/swimmable) with maximum allowable Fecal Coliform of 200 organisms per 100 ml (geometric mean).

In addition to bacterial contamination, Alewife Brook/Little River water quality monitoring data show other pollutant deterioration related to urban development (Metcalf & Eddy, 1994). Elevated levels of oil and grease, nutrients and toxic pollutants have been seen in Alewife Brook. These pollutants have a direct correlation with urban runoff from roads, parking lots and developed land areas. Further evidence of stormwater flow impacts on Alewife Brook is the high degree of turbidity as a result of sediment bearing stormwater.

Restoration Activities

Recognizing that restoration of the Alewife system requires a watershed-wide approach, the conservation commissions of the three watershed communities have begun a coordinated stewardship and restoration effort. In addition to protecting the floodplain and water quality through their authority under the Massachusetts Wetlands Protection Act, they have initiated elements of a comprehensive watershed study.

The cornerstone of the watershed initiative will be the completion of a U.S. Army Corps of Engineers Floodplain Management Study. The MDC, with assistance from the conservation commissions and the State's Wetlands Restoration and Banking Program, has arranged for the Army Corps to begin this study in 1995. This work is expected to yield important information on the hydrology of the watershed, with particular attention focussed on opportunities to enhance the natural flood valley storage potential of the watershed. Local commissions can then determine best management practices to employ for this enhancement. From this data, local by-laws may be drafted to give communities further enforcement abilities to ensure that appropriate management practices for the area are carried out.

The conservation commissions are sponsoring citizen-participation activities in the watershed to not only gather data, but to increase community involvement and interest in the Alewife watershed. Activities planned for 1995 include citizen stormwater monitoring and anadromous fish run assessment.

A number of restoration activities are now or soon to be underway. These include the repair of eroded banks at Clay Pit Pond in Belmont caused in part by rapid water level fluctuations, a restoration plan being developed by the MDC for Blair Pond in Cambridge and Belmont, and planned wetlands restoration of a portion of Alewife Brook in Cambridge.

Another important milestone expected to be reached in 1995 includes the formation of an Alewife watershed association. A strong foundation for this organization already exists among existing community groups and the conservation commissions. The watershed association will serve an important role in filling the fund raising, education, and community involvement needs that are critical to this long-term restoration effort.

Conclusion

The restoration of an urbanized watershed is a complex task that requires a clear understanding of the relationships between the man-made and natural aspects of the environment. It is also important that realistic goals and priorities be set by the watershed communities to advance the restoration effort. It is clear that close coordination among the three conservation commissions, the MDC, state agencies, and community groups is necessary to commence and maintain the long-term investment in time and energy necessary to restore the Alewife system.

Appendix

Betts, Richard B., former Town Engineer, Town of Belmont, provided water elevation records of Clay Pit Pond, Blair Pond, and Little Pond.

MWRA, Tech Reports No. 91-2 and 93-4, Combined Sewer Overflow Receiving Water Monitoring Boston Harbor and Tributary Rivers, June 1989-October 1990, October 1991; and October 1990-September 1991, January 1993.

Metcalf & Eddy, Baseline Water Quality Assessment Report, MWRA, August 1994.

Integrated Planning for Flood and Pollution Control

Rita Fordiani, P.E., BSCES Member,[1] and Melodie Esterberg, P.E.[2]

Abstract

As part of the City of Portland, Maine, Combined Sewer Overflow (CSO) Abatement Study, a master plan was developed which highlighted stormwater management as a key factor in controlling chronic flooding and CSOs throughout the city. This is especially true in the 1,800-acre urban watershed tributary to Capisic Brook where five CSOs discharge approximately 30 million gallons per year of combined sewage to Capisic Brook. The master plan for the Capisic Brook watershed is based on controlling the quantity and quality of stormwater entering the Capisic Brook while integrating natural storage and treatment techniques, park and trail design, and enhancement of wildlife habitat and recreational use opportunities.

Introduction

The Capisic Brook in Portland, Maine, is tributary to Casco Bay which is included in USEPA's National Estuary Program. At present, five CSOs discharge approximately 30 million gallons of combined sewage into the brook. Capisic Brook receives stormwater drainage from the surrounding 1,800-acre urban watershed. The degraded water quality of the Capisic Brook is of principal concern due to the present public health hazard. USEPA awarded the City of Portland a grant to perform a study under the Clean Water Act Section 104(b)(3) to develop an integrated watershed management plan for the Capisic Brook watershed. The goal of the plan is to control chronic flooding and eliminate the five CSOs by managing stormwater in the watershed and integrating natural storage and treatment of stormwater, park and trail design, and enhancement of wildlife habitat and recreational use opportunities. In conjunction with this study,

----

[1]Environmental Engineer, CH2M HILL, 50 Staniford Street, Boston, MA 02114
[2]Project Engineer, City of Portland, 55 Portland Street, Portland, ME 04101

the USDA Natural Resources Conservation Service is performing hydrologic and hydraulic modeling of stormwater flows in the watershed and the brook.

Chronic flooding areas and the five CSOs in the watershed are identified on Figure 1.

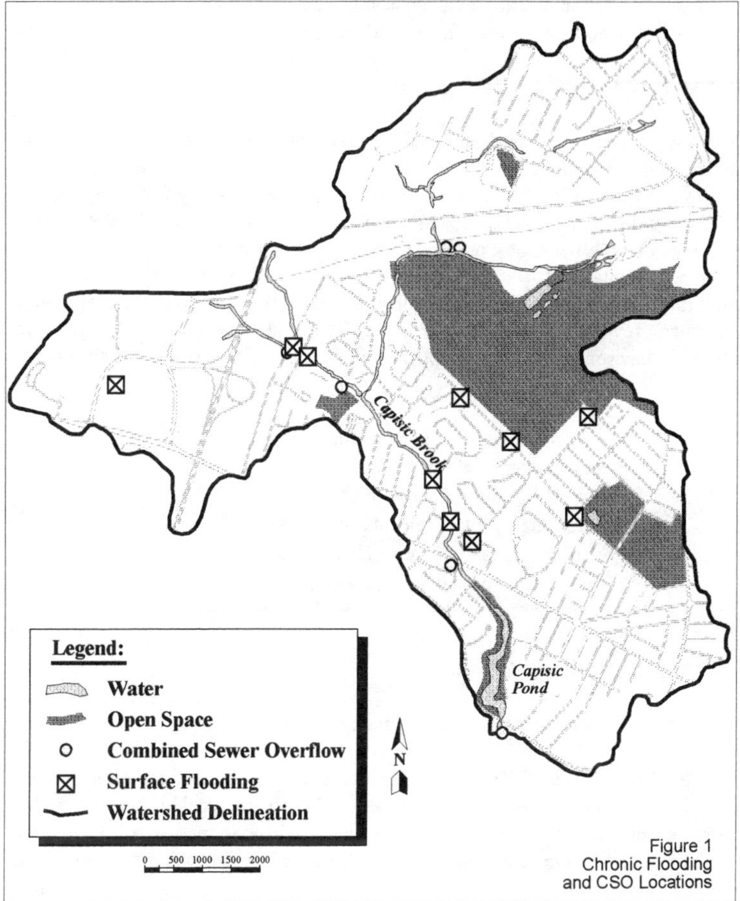

Figure 1
Chronic Flooding
and CSO Locations

Project Tasks

A 60-member Task Force of regulatory agencies (city, county, state, and federal), environmental groups, and public interest groups was developed to facilitate meeting the multiple objectives of the study. Project tasks include the following:

- Alternatives evaluation
- Cost estimating
- Environmental assessment
- Public involvement
- Conceptual design
- Permit and approval application requirements and schedule
- Stormwater management utility program assistance

Evaluation of Alternatives and Cost Estimating

Three systemwide technical alternatives have been developed for alleviating surface flooding and CSOs: 1) direct stormwater flows to the brook; 2) direct stormwater flows o the Portland Wastewater Treatment Facility (WWTF); and 3) a combination of approaches 1 and 2. Due to the limitations of the carrying capacities of both the brook and the sewer system, all of the approaches rely on detaining stormwater for some time in the watershed, either along a street, in a detention site, or along the brook corridor. Potential stormwater storage sites have been previewed with task force participants. Site selection will be based on technical, environmental, and implementation evaluation criteria, such as the following:

- Results of hydrologic and hydraulic modeling of the watershed
- Pollutant removal efficiency
- Equivalent uniform annual costs
- Water quality improvements
- Surrounding land use and enhancement potential
- Construction period and operating impacts
- Design, construction , and operational complexity
- Reliability
- Land requirements
- Public acceptance
- Institutional and development time requirements

Once the flooding and CSO issues are addressed, other equally important issues, such as streambank erosion, litter control, other specific water quality improvements, preservation and increase of wildlife habitat, increase of open space for passive and active recreation, and potential trail design will be incorporated into the conceptual design of the specific sites and into the overall integrated watershed management plan.

Environmental Assessment

Detailed habitat mapping within the brook corridor has been performed using aerial photographs. The mapping was followed by an inspection of the existing brook corridor to highlight areas for wildlife and wetland enhancements. A continuous water quality data collection program is being established, and a site walk with regulatory agencies has been performed to discuss specific concerns and requirements. A principal objective of the environmental assessment is to reach regulatory consensus on approaches to integrate wetland and habitat enhancement with water quality objectives and flood control.

Public Involvement Program

Alternatives and watershed enhancement concepts have been presented to the public. Public involvement activities, to date, include the following:

- Public meetings discussing issues and preferred uses in the watershed
- Capisic Brook Cleanup Day to promote resource awareness and litter removal
- Lawn care seminar on ecological methods of lawn care
- Community newsletters
- Press releases of events

Newsletters are sent to approximately 3000 property owners in the watershed. Attendance at public meetings has been approximately 100 citizens. Public concerns and issues have been collected through newsletter feedback questionnaires, public meetings, and through direct communication with project staff from the City of Portland and the consulting team.

Remaining Tasks and Progress

Once the USDA Natural Resources Conservation Service has completed the hydrologic and hydraulic modeling of the watershed, specific sites will be sized and evaluated. Conceptual designs of integrated technical and environmental plans and cost information will be presented to the Task Force and public for comment. A permit and approval implementation plan and schedule will be developed for selected sites. Areas which may be designated as part of a potential stormwater management utility program will be identified. The remaining tasks are scheduled to be complete approximately 6 months after the receipt of modeling data.

Integrated Water Quantity-Quality Modeling

J. G. Lang[1] and D. P. Loucks, F. ASCE[2]

Abstract

An interactive water quantity-quality simulation model was used to evaluate the impacts of alternative river basin land uses and water quantity and quality control measures. The river basin of interest was the Grand River Basin in Ontario, Canada. Flow and quality conditions in this river promote the growth of benthic plants and algae, which in turn have significant impacts on the nutrient and dissolved oxygen concentrations in the river. The results of this exercise demonstrate that a simplified yet flexible model, that can simultaneously simulate time and space varying flows, storage volumes and qualities, can be an effective tool in basin-wide water quantity and quality planning, especially when in use together with more complex (realistic), but less flexible, water quality models.

Introduction

The Grand River Basin is a narrow 300-km long area of over 6960 km^2 in southwestern Ontario. Long-term mean annual precipitation is about 86 cm, of which about 30 cm runs off to the Grand River. This runoff is the source of non-point pollutants from extensive urban and agricultural development. Point sources include wastewater treatment plants that add to the pollutant loading in the river. High nutrient concentrations in the river, in concert with ideal habitat, lead to prolific growth of rooted aquatic macrophytes and attached algae. These plants significantly influence the dissolved oxygen concentrations in the river. Four upstream multipurpose reservoirs control much of the flow in the river, and are used, in part, as sources for downstream flow augmentation.

Water quality deterioration, potential drinking water shortages, and ever-present flooding concerns motivated a basin-wide comprehensive water management planning program in the late 1970s. Various management alternatives, including the construction of additional reservoirs and improved wastewater treatment, were analyzed using mathematical models. This modeling included the development of the Grand River simulation model, GRSM (Ontario Ministry of Environment, 1982). This model is capable of performing long-term simulations of two-hour time steps, steps short enough to capture the within-day variations that occur within the river. River flows are inputs to the model, and a Monte-Carlo approach accounts for the uncertainty of incoming concentrations.

----

[1]Klohn-Crippen Consultants, Ltd., Richmond, BC, V6X 2W7 Canada
[2]School of Civil and Environmental Engrg., Cornell University, Ithaca, NY 14853

Of interest in this study was whether or not a much simpler water quality model, incorporated into a water quantity simulation model, would be sufficient to examine various water quantity and quality management issues more efficiently and effectively than what could be done using several more complex and less interactive models, including GRSM, used previously. The interactive river-aquifer simulation (IRAS) model (Loucks, French, and Taylor, 1995) was selected as a model that could easily examine and display the results of different assumptions regarding inputs, parameter values, and management options. The smallest time steps this model can simulate are days, and hence it can only attempt to predict daily average water quantity and quality conditions. While IRAS can facilitate the examination of water quantity and quality impacts of numerous system design, operation, inputs, and parameter value alternatives, it can not model adequately the attached plants and their effects on the oxygen budget of the river. Nor can IRAS model the reaeration as a function of flow and reach depth. Hence the need for a comparison between IRAS model and the larger and more complex GRSM model before considering IRAS useful for analyzing different options in the Grand River Basin.

Quality Constituents Modeled

Of interest in the Grand River is the impact of any management option on the oxygen budget of the river. To address this issue, the quality constituents modeled included the carbonaceous biochemical oxygen demand (CBOD), the nitrogenous biochemical oxygen demand (NBOD) which includes organic and ammonium, nitrate nitrogen (NIT), and phosphate phosphorus (P). Sediment oxygen demand and plant (including benthic) biomass were treated as model inputs, although they need not all have been.

Model Calibration and Verification

Using the already calibrated and verified GRSM as a measure of "truth", the IRAS model was calibrated and verified based on the GRSM output. Recognizing that it is impossible to build numerical models that will predict with perfect precision the behavior of open stochastic natural systems, and that GRSM is in itself only an approximation of reality, the goal was to see if IRAS could reproduce GRSM results sufficiently well to be useful in planning. Users of any model are the appropriate ones to judge whether or not the model is sufficiently verified, not the model developers.

Input data were available for the Grand River Simulation Model for twenty periods of four months each: June through September for 20 years. From this data set, two distinctly different periods of 60 days each were chosen for IRAS calibration and verification. These periods were chosen to capture some of the most critical annual water quality conditions. Calibration was first accomplished by trial and error and later by applying genetic algorithms (see Mohan and Loucks, this volume).

The calibrated and verified parameter values included the CBOD decay rate; the NBOD oxidation rate, matching the nitrate concentrations; the NBOD uptake rate, so that the NBOD concentrations matched; the P uptake rate; the aeration rate; and the sediment oxygen demand rate, including the effects of benthic plants.

A comparison of the resulting parameter values in the GRSM and IRAS models showed the impact of different modeling approaches, computational techniques, and simulated time period durations. Considering the differences in the two water quality models, comparisons of the time series of calibrated results showed, perhaps surprisingly, close agreements, with IRAS tending not to reproduce as much variability in the concentrations as predicted by GRSM. Some of this is due to the longer time periods and hence the averaging used by IRAS. These peak concentrations of certain constituents may be important. Hence if simplified models are to be used to examine numerous management options, it may be appropriate to further analyze those alternatives that appear most attractive using a more realistic, and thus often more complex, water quality model.

Management Scenarios

The calibrated IRAS model was used to examine human-induced changes in the Grand River Basin. These involved the quantity as well as quality impacts. The resulting water quality changes predicted with IRAS were compared against results of the same scenarios run through GRSM. Two of these scenarios involved the addition of an industrial wastewater, and the withdrawal of water from the river.

The addition of industrial wastewater was simulated by adding 500 mg/l of carbonaceous BOD to the daily wasteload at Waterloo. Effluents rich in oxygen demanding carbonaceous material may come from a variety of industrial sources, including pulp mills, from which nitrogenous wastes are minimal. The performance of the model was judged from the concentrations downstream of the discharge point. The IRAS and GRSM results are comparable. The results suggest that IRAS, in spite of its relative simplicity, is robust enough in this case to examine quite different concentration regimes.

During June and July, the dry weather flow of the Grand River at Waterloo is maintained by upstream reservoir releases to about 10 m^3/s. To simulate the effect of potential changes in reservoir release policy, or perhaps irrigation or drinking water demand, a constant value of 2.8 m^3/s was withdrawn from river flow upstream of Waterloo. At the same time the industrial discharge of 500 mg/l CBOD was maintained. The resulting plot of IRAS and GRSM results are shown in Figure 1. Again, the results are comparable. The real issue here, however, is the prediction of the dissolved oxygen concentrations, since the 28% lower flow regime has some effect on the atmospheric reaeration rate, not captured automatically in IRAS. When benthic plants were not considered, the oxygen predictions of both models were much the same, even when including sediment oxygen demand. On average, IRAS predicted that the effect of CBOD addition and flow withdrawal would reduce the dissolved oxygen at most by about 0.9 mg/l, while GRSM predicted just over a 0.7 reduction. No model is this precise, so essentially both predict about a 1 mg/l reduction.

GRSM showed that reducing river flows stimulated the growth of all plant communities, since lower flow depths allow more light to penetrate to the bottom of the river. IRAS accounts for this increase in uptake by increasing the concentration of nutrients when the loadings are constant but the flows are reduced. This increases the magnitude of the plant uptake term in the NBOD balance equation, indirectly simulating the increased plant growth. Comparisons of GRSM and IRAS NBOD results for the flow withdrawal scenario are shown in Figure 1.

In summary, IRAS seemed to predict CBOD and nutrient concentrations well, even when plant growth was altered. However the effects of plants on dissolved oxygen were overpredicted by IRAS by about half a mg/l. The within-day effects of plants on dissolved oxygen cannot be captured by IRAS. Integrated water quantity and quality planning can be aided by the combined and effective use of both simple and more detailed simulation models.

References

Lang, J.G., 1994, A Comparison of Simplified and Complex Water Quality Models of a Eutrophic River. IHE, Delft, NL, July

Loucks, D.P., French, P.N. and Taylor, M.R., 1995, Interactive River-Aquifer Simulation Model: Program Description, Cornell University, Ithaca, NY

Mohan, S. and Loucks, D.P., 1995, Genetic Algorithms for Estimating Model Parameters, this volume.

Ontario Ministry of Environment, 1982, Dynamic Water Quality Computer Simulation Model GRSM, Water Resources Branch, internal report.

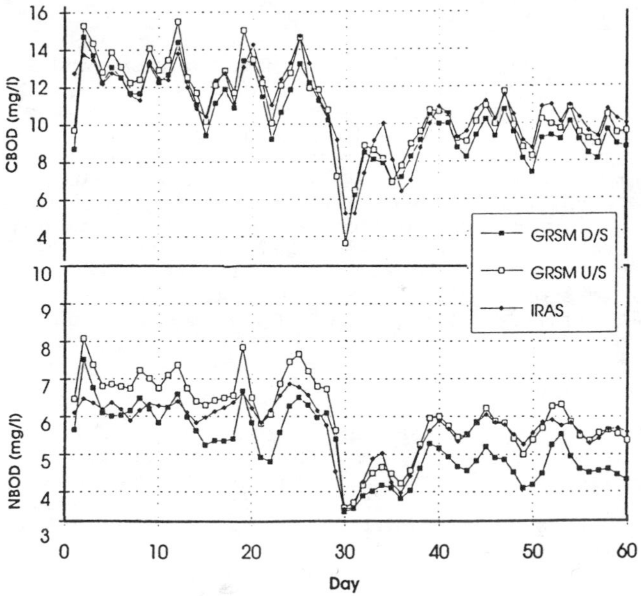

Figure 1. Predicted upstream (GRSM U/S), downstream (GRSM D/S) and average (IRAS) CBOD and NBOD concentrations in a reach of the Grand River with reduced flows and added industrial discharges.

Water Quality Modeling of the Rio Tiete System in Sao Paulo, Brazil

Amy L. Shallcross[1] and Gary W. Mercer, P.E.[2]

Abstract

Water quality modeling of the Rio Tiete and Billings Reservoir system was performed with USEPA's QUAL2EU and WASP4 to evaluate the alternatives of the wastewater master plan for Sao Paulo, Brazil. Planned wastewater collection and treatment systems (primary) throughout the system result in slightly greater overall water quality improvement than those concentrated in specific areas, such as downtown or upstream. Due to the large amount of uncollected and untreated sewage throughout the basin, either greater levels of treatment (secondary, nitrification), or instream aeration will be needed to elicit further water quality improvements throughout the system.

Introduction

The metropolitan area of Sao Paulo, with a population of over 15 million inhabitants, discharges primarily untreated domestic and industrial wastewater to the Rio Tiete System. The receiving waters are anaerobic much of the time, resulting in poor river quality and odors in the downtown area. As a result of high loads of CBOD, nitrogen, and phosphorus discharged from the Rio Pinhieros, Billings Reservoir is in a hypereutrophic condition with frequent phytoplankton blooms and an extensive anaerobic area that releases sulfuric gas.

---

[1]Water Resources Engineer, Camp Dresser & McKee Inc., Raritian Plaza I, Edison, NJ 08818

[2]Water Resources Engineer, Camp Dresser & McKee Inc., Ten Cambridge Center, Cambridge, MA 02142

The State of Sao Paulo, with funding from the World Bank, has taken efforts to plan, design, and construct wastewater collection and treatment systems to abate these problems. Water quality modeling and analyses of the Rio Tiete and Billings Reservoir System were conducted to compare the water quality benefits of five alternatives of wastewater system improvements, under different flow regimes and planning horizons. This modeling study was conducted to further evaluate the benefits of the plan and justify the associated costs.

Study Area and Existing Conditions

The State and City of Sao Paulo are located in the eastern portion of Brazil, near the Atlantic Ocean. Figure 1 presents the Rio Tiete system in Metropolitan Sao Paulo. The system modeled includes 226 km of Rio Tiete, 12 km of Rio Tamanduatei, 26 km of Rio Pinheiros, and 24 km of Billings Reservoir. Also, three dams were

FIGURE 1. Study Area

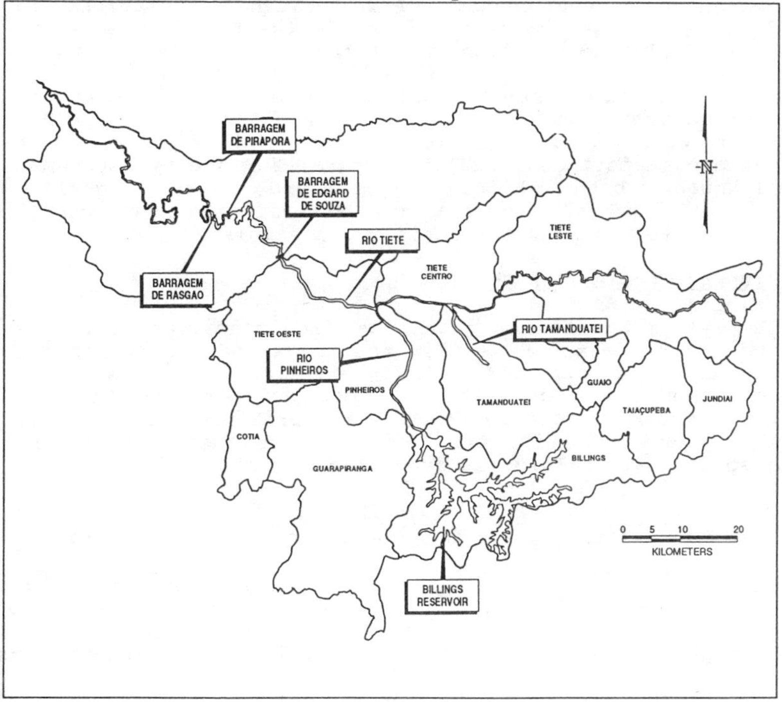

located in the system, Edgard de Souza, Pirapora, and Rasgao. In the 1950's, people in Sao Paulo used the river for recreation. However, accelerated population and industrial growth polluted the river system preventing plant and animal life. Today, the river receives sewer and industrial flows of 40 m^3/s, roughly 60 percent of the flow during dry periods.

Technical Approach

USEPA's QUAL2E model was chosen to simulate riverine and reservoir water quality. QUAL2E is a one-dimensional steady-state water quality model that simulates major water quality interactions, including the nitrogen cycle, the phosphorus cycle, algal production, sediment oxygen demand, atmospheric and dam reaeration, and fecal coliform. Hydraulically, QUAL2E is limited to the simulation of time periods when both stream flow and input wasteloads are essentially constant. As a steady-state model, QUAL2E can be used with field sampling data to evaluate the magnitude and quality characteristics of non-point source loads. Also, diurnal variations in dissolved oxygen concentrations due to photosynthesis can be investigated by using QUAL2E for dynamic simulations.

Both QUAL2E and WASP4 models were developed to examine the water quality of Billings Reservoir. WASP4, EPA's Water quality Analysis Simulation Program, is a dynamic compartment modeling program for aquatic systems. Although both hydrodynamics and water quality (conventional and toxics) can be evaluated with WASP4, only conventional water quality was simulated.

Alternatives Evaluated

The system was modeled under two flow conditions (average and 90% exceedence). 3 flow routing scenarios (balanced, sanitary, hydropower) also were evaluated. These scenarios represent different options for operation of the river/reservoir system and are presented in Table 1. The 5 implementation strategies of the Master Plan were also simulated. These strategies are outlined in Table 2.

TABLE 1. FLOW SCENARIOS MODELED

Flow Scenarios	At the Confluence of Rios Tiete anc Pinhieros
Balanced Flow	Flow divided equally between Rios Tiete and Pinhieros
Sanitary Flow	All flow diverted to Rio Tiete
Hydropower Flow	All flow diverted to Rio Pinhieros anc Billings Reservoir for the operation o: hydropower facilities.

TABLE 2. MASTER PLAN IMPLEMENTATION STRATEGIES MODELED

Alternative	Description
1	Capture, route and treat flows to Upper Rio Tiete through Barueri Treatment Plant
2	Increase capture and treatment of discharges in Suzano, Sao Miguel, and Cotia
3	Distribute improvements throughout all basins (recommended in the Master Wastewater Plan)
4	Increase capture and treatment of discharges to treatment area
5	No improvements

Findings

All alternatives, but #5, will produce a small improvement in the dissolved oxygen concentrations. Distributing the improvements throughout the basins has the largest water quality benefits for the overall system. However, all alternatives, except #5, result in similar benefits. Although the flow scenarios do no affect the upper Rio Tiete, the hydropower flow scenario results in the most benefit for the lower Tiete below Rasgao Dam. The sanitary flow scenario indicates improvement of water quality in Billings Reservoir and a worsening of water quality along the lower Tiete. None of the alternatives significantly reduce the concentrations of fecal coliform, ammonia, or phosphorus.

Additional sampling was recommended to further understand water quality interactions throughout the system. To achieve comparable U.S. water quality standards, additional levels of treatment, such as nitrification, will be required. Less costly alternatives to enhanced treatment should be considered in the short-term. One such alternative includes mechanically increasing instream aeration on the Rio Tiete and Rio Pinhieros below their confluence. Instream aeration has worked successfully to increase dissolved oxygen concentrations in the Chicago Waterways.

UNCERTAINTY ANALYSIS IN NON-POINT SOURCE WATER QUALITY MANAGEMENT DECISIONS

J. S. Yulianti[1], Barbara J. Lence[2], and G. V. Johnson[3]

Abstract

The uncertainty in nature, and therefore in the input data for predictive models, may lead to the question: How warranted are the management decisions that are based on such input data?. This research identifies the significant input parameters for estimating erosion and sedimentation and for selecting non-point source water quality management decisions. It also evaluates which management policies are the most sensitive to these parameters.

Introduction

Non-point source management decisions based on output from rainfall-runoff simulation models may be uncertain due to incomplete or unreliable input data, to inappropriate parameter estimates, or to an inappropriate choice of model. Estimates of economic costs are additional important information in environmental management. Management decisions based on unreliable input data may be questionable. This paper develops approaches that answer two research questions related to environmental management of agricultural land: Which uncertain parameters are significant in terms of the non-point source management model output?, and Which management policies are sensitive to uncertainty in the input parameters?. These approaches are demonstrated for a case study for the Highland Silver Lake Watershed in Illinois.

---

[1] Graduate Student, Department of Civil and Geological Engineeering, 15 Gillson Street, University of Manitoba, MB, Canada, R3T 5V6.

[2] Associate Professor, Department of Civil and Geological Engineeering, 15 Gillson Street, University of Manitoba, MB, Canada, R3T 5V6.

[3] Associate Professor, Department of Agricultural Economics, 66 Dafoe Rd., University of Manitoba, MB, Canada, R3T 2N2.

Identification of Important Input Parameters

A modified Generalized Sensitivity Analysis (GSA) is applied to identify important parameters in the linked process of water quality simulation and water quality management optimization for non-point source pollution control. GSA was originally developed to identify key input parameters in water quality simulation models (Spear and Hornberger, 1980). The GSA approach uses Monte-Carlo simulation to generate various model outcomes, a classification algorithm to separate the simulation results into categories, and a goodness-of-fit test to determine whether each input parameter significantly affects the separation of the simulation results. The classification algorithm can be chosen to fulfil some objectives such as minimizing the difference between the simulated and the actual peak discharge for a given watershed. For example, one category of simulations may represent results that have a peak discharge value that is similar to the actual value for a given rainstorm for the watershed, and another category of simulations may represent cases where the peak discharge value is not similar to actual value. In this case, the GSA results would show which parameters are important for simulating the actual peak discharge.

Lence and Takyi (1992) develop a modified GSA method for identifying important streamflow and temperature information in the linked process of water quality simulation and optimal water quality management for point source pollution. When optimization models are linked with simulation models, the GSA is applied to determine how sensitive the optimization model results are to uncertain simulation model input parameters. In this case, the classification algorithm cannot be based on some actual field conditions. This is because the actual system management solution is not determined by nature. Lence and Takyi (1992) address this issue by basing the classification algorithm on comparisons with the optimization results for the case where the inputs to the simulation model are chosen, or estimated, from the history of record for the system. In this study, the modified GSA (Lence and Takyi, 1992) is applied for non-point source pollution management. The classification algorithm for the linked process of non-point source simulation and optimization is based on comparisons with the non-point source management solution that is determined for a known set of hydrologic and agricultural production parameters. The approach is applied to an intensive crop agriculture area in the Highland Silver Lake Watershed in Illinois. The study area is 256.3 ha and has 37 land management units. Five crop rotations, three tillage systems, and two mechanical control practices are considered. This gives a combination of 30 different management options for each land management unit in the watershed.

The Sediment Delivery Economic (SEDEC) model which incorporates the rainfall, erosion and sediment estimates, and economic considerations in the selection of sediment control management practices is used in this study. This research investigates whether certain parameters in the SEDEC model significantly affect the management decision. In the SEDEC model, erosion is simulated based on the Universal Soil Loss Equation (USLE) and an optimization subroutine determines the optimal choice of management practices that meet a sediment load constraint and a given management objective, e.g., minimizing the total production cost. Fifty-six parameters in the SEDEC model are considered to be uncertain. They are: the rainfall-runoff factor (R), the yield production per unit area for five variety of crops for each of ten soil types, Yci for corn , Ysi for soybean , Ywi for wheat, Ydi for doublecrop soybeans, and Yvi for cover, where i denotes the soil type, and the crop prices for five variety of crops, Pc for corn, Ps for soybean, Pw for wheat, Pd for doublecrop soybeans, and Pv for cover. These are randomly generated in the Monte-Carlo simulation. For this analysis, 1200 Monte-Carlo realizations of the SEDEC model output are determined for a constraint on the sediment load. The stream sediment load ranges from 0 ton/annum to 450 ton/annum in increments of 25 ton/annum. R, Yci, Ysi, Ywi, Ydi and Yvi affect the sediment load, as well as the economic cost and benefit. The distributions of these parameters are used to simulate the erosion and sediment delivery of given management practices in the watershed and the crop prices as well as production costs determine the total cost of a management solution.

The results of the Monte-Carlo simulation are then classified into two categories, the behaviour and the non-behaviour classes. If a realization results in a total cost that is similar to the deterministic case, it is classified as a behaviour(B), and if the total cost is different from the deterministic case it is considered a non-behaviour(NB). For each parameter, the cumulative distribution functions (CDFs) for parameter values in the B and NB classes are evaluated and the Kolomogorov-Smirnov Two Sample Statistic test (K-S test) is used to investigate the degree of separation of the CDFs. For a given input parameter, if the CDFs of that parameter in the B and NB classes are significantly different, then the potential simulation of realizations in either of the two classes can be considered sensitive to that parameter. The K-S test Statistic, the dmn value (where m and n are the number of realizations in the B and NB class, respectively), may be used as a sensitivity index, or an indicator of the importance of the parameter interest.

The dmn values for parameters Ysi, Yvi, Ps, and Pv are not significant; this indicates that 34 out of 56 parameters are significant for the Highland Silver Lake Watershed in terms of the management model output. The dmn values for R are the largest among these so R appears to be the most significant parameter influencing the model output.

Sensitivity of Management Policies

Three management policies are analyzed to determine their sensitivity to the SEDEC model input: the Least Cost policy, the Erosion Standard policy, and the Erosion Tax policy. The Least Cost policy minimizes the total cost that results in a total sediment load at the end point of a given watershed. The Erosion Standard policy minimizes total cost while meeting a limitation on erosion from each farm. The Erosion Tax policy minimizes total cost while imposing tax charges on each farm for any amount of erosion greater than the allowable erosion rate. Monte-Carlo simulation is used to evaluate the sensitivity of the policies to uncertain model input. The 56 parameters in the SEDEC model are considered to be uncertain and are randomly generated in the Monte-Carlo simulation.

For each Monte-Carlo realization, the total cost for maintaining a stream sediment load under each of the three policies is determined for a range of sediment loads up to 450 ton/annum. For the Erosion Standard policy, an erosion limit of 1 ton/acre is considered. The distributions of total costs under each sediment load constraint for each policy are determined from the Monte-Carlo simulation results. For this analysis, 400 Monte-Carlo realizations for each increment stream sediment were performed. For each management policy, the trade-offs between the stream sediment load maintained and the total cost are determined. The sensitivity of the model output due to uncertainty in the input parameters can be estimated by analyzing these trade-off. Small differences between the 5th and the 95th percentile of these trade-off curves indicate that changes in the input parameter values do not significantly influence the model output, while large differences indicate that the model output is sensitive to changes in the input parameters. The Least Cost approach resulted in the smallest difference between the 5th and 95th percentile trade-off curves. This indicates that the Least Cost policy is less sensitive to uncertain input parameters than the other two policies. The Erosion Standard approach, which shows the largest differences between the 5th and 95th percentile trade-offs curves, is the most sensitive management policy.

References

1. Lence, B. J., and A. K. Takyi, Data requirements for seasonal discharge programs: An application of a regionalized sensitivity analysis, *Water Resour. Res.*, 28, 7, 1781-1789, 1992.
2. Spear, R. C., and G. M. Hornberger, Eutrophication in Peel Inlet-II. Identification of critical uncertainties via generalised sensitivity analysis, *Water Res.*, 14, 43-49, 1980.

Regulation of Private Sewage Disposal Systems

Daniel Holzman, P.E.[1]

Abstract

The EPA has identified private sewage disposal systems (septic systems) as major contributors to water pollution. Effective regulation of these systems is extremely difficult due to conflicts between private landowners and public regulatory bodies. Issues of cost and landowner rights versus public health concerns dominate discussion, but technical deficiencies in our understanding of how septic systems actually work also contribute to confusion regarding appropriate regulations. This paper examines elements of current design procedures in Massachusetts, modifications to the existing procedures incorporated in new regulations, and presents areas where the author believes additional research is warranted.

Introduction

Massachusetts regulates private sewage disposal systems under Title 5, Commonwealth of Massachusetts Regulations. The current Title 5 has been in place since 1978; a new Title 5 set of regulations is scheduled for implementation in 1995. The new Title 5 is considerably more detailed than the old, and incorporates stricter regulations in a variety of areas.

My company has represented the Town of Carlisle as the Board of Health consultant dealing with septic system issues for the last three years. Carlisle has no public water or sewer systems, and has experienced considerable growth in the last three years. Some of the new 2 acre minimum lots have been developed on marginal sites, with typical problems including high groundwater, presence of ledge, and adjacency to wetlands. Many of the existing systems in Carlisle predate even the old Title 5, and these systems fail on a regular basis. Carlisle provides a microcosm of septic problems found around Massachusetts.

[1] Civil Department Manager, Barrientos & Associates Inc., 529 Main Street, Charlestown MA 02129, Member ASCE

System Performance and Failures

One of the outstanding difficulties in evaluating septic system performance is the lack of research on actual groundwater pollution caused by systems in place. The purpose of a septic system is to prevent pollutants from entering the groundwater, yet only in research projects do we actually know if the system is performing its function. Obviously, it would be prohibitively expensive to instrument the average $5-10 thousand septic system to determine if pollutants are actually reaching the groundwater, so even answering a basic question such as "Is the system working?" becomes a subjective matter. We believe that one of the most important areas of research in on site disposal is the cause of failure of systems. If we know why they fail, we should be able to improve design procedures.

It is very important to distinguish between evidence of failure and the *cause* of failure. Typically, evidence of failure includes one or more of these effects:

1. Sewage breaking out onto the surface of the ground.
2. Backing up of sewage into the owner's home.
3. Persistent unpleasant odor in the area of the system.
4. Evidence of seepage near or on the septic field, including growth of wet soil species such as moss or ferns.
5. Soft, spongy ground on top of the septic field.

Most septic systems typically stop working properly long before visual or other evidence confirms the failure. Our experience suggests that by the time the average homeowner reports a failure, the system is often completely beyond rehabilitation.

Determining the cause of failure of a system is often extremely difficult. Few if any homeowners have the desire to perform an autopsy on their system, and in many cases the system is abandoned in place. Most of the time, we have only circumstantial evidence available to determine cause of failure. Following are some specific causes of failure which we have been able to determine:

1. About half the failed systems in Carlisle seem to be in areas of high groundwater. Many of the systems were installed before 1978, hence were built with no Title 5 guidelines at all. We have noted that many of the failed systems are in low, wet areas, and we believe that most of these failed systems would not meet the required 4' separation currently in force between the bottom of the system and the maximum anticipated groundwater level.

   In addition, we have noted that many systems installed before 1991 were designed based on a single groundwater reading taken during the Spring season. Records indicate that groundwater fluctuations in a typical Carlisle well are at least 10' between extreme high and low levels. Very limited groundwater level corrections were applied to Carlisle septic design prior to 1991, hence systems designed during unusually dry spring periods are likely to be designed with insufficient offset between the bottom of field and high groundwater levels.

2. Several systems have failed due to use of inferior construction materials. In one case, Orangeburg pipe (a bituminous fiber pipe) was crushed by a vehicle, leading to failure of the system. Several septic tanks have cracked presumably due to settling. At least one distribution box failed due to cracking.

3. At least two systems failed due to improper installation procedures. In one case, the pipe leading from the septic tank to the distribution box was either installed uphill, or settled unevenly. Prior to 1991, Carlisle had no requirement for as built drawings prepared by the engineer of record, hence there is no accurate field record of the installations of most systems in Town.

4. A number of systems have failed due to introduction of inappropriate materials into the effluent. One system apparently failed due to disposal of instant potatoes into the drain. We believe that other systems have failed due to flushing of chemicals or paint down the drain.

5. In a number of cases, we have evidence that the septic tank was infrequently or never pumped. We assume that in those cases, bypass of solids into the field was the proximate cause of failure.

New Regulations

Title 5 has been rewritten in an attempt to resolve some of the outstanding issues of septic design, and tighten inspection, design and monitoring procedures. The new Title 5 significantly tightens procedures in the following areas:

1. For the first time, requires that individual systems be inspected by a licensed inspector at time of transfer of title. If the system fails, requires repair within nine months.

2. Requires that a licensed site evaluator be present during deep hole testing at all sites. System design will still be performed by a licensed professional engineer or sanitarian.

3. Provides regulations governing use of alternative technologies, and encourages use of shared systems for the first time. Strictly regulates use of tight tanks.

4. Allows for three different methods of determining high groundwater at a site, including high groundwater observation, adjustment using USGS or local wells, and high groundwater estimation using soil mottling.

The new Title 5 is much more detailed regarding design procedures, evaluation of requests for variances, methods of installation, and tracking of system performance. Nevertheless, there are still a number of technical areas which are subject to considerable interpretation, and which require clarification and more research.

Areas Requiring Further Research

The technical design of septic systems relies on adequate information about the site, including groundwater levels, percolation rates, soil characteristics, topographic data, and the location of permanent features. Based on my experience, I believe that the following areas require further research and design clarification:

1. Determination of maximum high groundwater. The old Title 5 was very vague on procedures for developing a design high groundwater level at a given site. The new Title 5 is more explicit, but still allows three different methods to be used to determine high groundwater, while providing no guidance on cross verification procedures. I believe that a working paper should be developed which defines the design return frequency high groundwater event, and offers guidance on the relationship of the results of the three methods (mottling, Spring water level observations, and use of indicator wells).

2. Actual procedures for performing percolation tests need to be tightened up. At present, there is no standard tool for preparing the percolation hole, and actual procedures for performing the test vary widely. Specific procedures should be established which will standardize the percolation test procedure, similar to the "standard penetration test" used in soil borings. Highly specific procedures, with a standardized installation tool, could allow the percolation test to function as a soil index property.

3. The use of perimeter drains to reduce the water table adjacent to septic systems seems to be getting much more popular in Carlisle. Typical proposals include use of gravel underdrains with perforated pipe set below the design high groundwater level. The design basis of these systems is poorly documented, in part due to extremely complex mathematical analysis of unconfined drainage into the systems, and the difficulty obtaining reliable permeability coefficients for soil near the drains. A well documented design manual for these drains would be very valuable.

4. Much more research is needed on alternative design technologies for individual users, including sand filters, composting toilets, biologically active sludge beds, and numerous other treatment techniques. Rigorous field testing and design procedures for new systems are sorely needed.

Conclusion

New Title 5 regulations promise to improve on site sewage disposal design and installation. History suggests that many failures could have been avoided through tighter regulation, better enforcement, and improved design in the past. Even with the new regulations, further research is still needed on affordable, effective alternatives to currently used systems, and standardization of design procedures for "standard" systems is also needed.

Environmental Benefits of Wastewater Facilities Planning in a New Hampshire Coastal Community

William R. Hall, Jr. P.E.[1]

Abstract

The Town of Seabrook, New Hampshire has invested more than 15 years in the planning and development of a town-wide wastewater collection and treatment system. Sited on the New Hampshire coast and with a significant portion of the community identified as salt and fresh water wetlands, the environmental benefits of centralized collection and proper treatment of wastewater have been long recognized. Failed on-site subsurface disposal systems, direct discharges and restrictive soils have allowed inadequately treated wastewater to be disposed throughout the community. A primary benefit of the project is providing a collection system to properly handle wastewater for treatment and disposal. Secondary benefits to the project include the opportunity to reclaim a portion of a valuable salt marsh. The result of the planning and design have allowed the Town to correct wastewater disposal problems, reclaim a valuable salt marsh and control surface drainage runoff to rare shellfish areas in Hampton Harbor.

Introduction

The Town of Seabrook is located in the southeastern corner of New Hampshire bordering Salisbury, Massachusetts to the South, Hampton Falls and Kensington to the west and North Hampton to the north. The Town has approximately 1 mile of shoreline bordering the Gulf of Maine in the Atlantic Ocean. Approximately 60 % of the land area in the Town is available for development, with the remaining land identified as unbuildable due to soil and wetlands constraints (Normandeau Associates 1989). Land use throughout the Town is mixed residential and

[1]Principal, Stearns & Wheler, Two Commerce Drive, Bedford, NH 03110

commercial with some light industrial uses primarily in the Route 1 corridor. The Town is bisected by Interstate Route 95 which has limited access and Route 1 (Lafayette Road) offering the greatest concentration of commercial development.

The Town has been actively involved in the development of a wastewater master plan for more than 18 years. The first study, initiated in 1977, recommended that a core area of sewers and a central wastewater treatment plant be constructed. Treated water would be discharged to the Atlantic Ocean through an outfall. A second study in 1986 reviewed different treatment and disposal options in an attempt to identify the best method of wastewater disposal throughout the Town. Both studies recognized the problems of failing septic systems and poor soil types for the siting of new or replacement systems. Soil types in the area of the beach, while permeable, are occasionally saturated with salt water, and the housing density is very high. The original systems were designed for seasonal use and are now used to meet year round demands(Normandeau Associates 1989). The beach area is directly adjacent to Hampton Harbor which has long been a source of shellfish for harvesting.

The final recommendation of the 1986 report was for centralized treatment followed by ocean disposal. The location of the treatment facility was recommended to be on Wright's Island at the New Hampshire/ Massachusetts border with a dedicated ocean outfall handling the treated effluent.

In 1987, the Town retained the joint venture of Jones & Beach/ Stearns & Wheler to review the previous studies, develop recommendations and initiate the design of the wastewater collection system. The basis of design was developed in conjunction with the joint venture, the State and the local planners. The initial concept was to develop a plan which would allow the critical areas of Town to connect to sewers first. This plan would be expanded to include all roads in Town having sewers installed over a six year period. Key elements in the basic design criteria included a 20 year design life for the wastewater treatment plant in accordance with EPA guidelines and a 50 year design life for the collection system. Expandability and process reliability were also key issues in the development of the master plan. Ultimate disposal of treated effluent was to be into the Atlantic Ocean and all water quality criteria needed to be met consistently and continuously. No combined sewer flows or stormwater was to be discharged to the collection system or the treatment facility.

In 1988, the joint venture issued a confirmation report which spelled out the design criteria as well as the plan for sewering the community. The ultimate design population of the Town was considered when developing initial and ultimate flows based upon maximum density and buildout on buildable land within the community. The report recommended the design and construction of approximately 37 miles of sewers with 17 pump stations. Location of the treatment facility to provide secondary levels of treatment was recommended for Wright's Island with ultimate disposal of the treated effluent through a dedicated ocean outfall(Jones & Beach/Stearns & Wheler 1988). The Wright's Island site was selected due to the limited impact to existing wetlands and the ability to improve the environment by controlling access.

Wright's Island had an access road constructed through the salt marsh many years ago and the site had uncontrolled access. Evidence of illegal dumping and disposal activities including demolition debris and appliances was prevalent. In addition, several species of nuisance grasses were encroaching on the salt marsh due to the restricted flushing because of the access road to the island (IEP 1990). In 1989, application was made for the ACOE permit, the USEPA discharge permit and the State water quality and Coastal Zone Management permits. The applications covered all fresh and salt water impacts for the treatment facility, collection and outfall.

The concern over the existing degradation of the salt marsh in the area of Wright's Island was noted by the reviewing agencies and in response, a salt marsh enhancement plan was developed for the project by IEP in 1990. This plan called for the construction of two large box culverts to greatly increase the flushing action of the salt marsh. In addition, the salt marsh had several old flow channels to aid in draining the marsh for salt marsh harvesting. The plan called for the construction of new connecting channels to increase the flow of tidal waters to the existing saltmarsh channels. Additionally, a re-vegetation plan and saltmarsh monitoring program were developed as part of island improvements. The purpose of this effort is to allow for natural flushing of the saltmarsh to reoccur and improve the quality of the salt marsh by re-establishing saltmarsh species in the area(IEP 1990).

Recognizing the efforts that go into constructing a new collection system to deal with wastewater, the Town elected to commission a study addressing the issue of drainage from surface runoff throughout the community. An area that faces repeated drainage problems after precipitation events is the Seabrook Beach area. The Town realized that construction of a wastewater collection system would temporarily impact the residents of the community. For this reason and knowing that stormwater runoff problems exist, the Town elected to have drainage system improvements constructed in conjunction with the sanitary sewers. In the beach area, the drainage problems were compounded by a fluctuating groundwater table and the close proximity of the beach, the saltmarsh and the Hampton Harbor. In discussions with State and Federal regulators, the impact of surface water to the beach area, the salt marsh and Hampton Harbor needed to be minimized. As a result, the drainage system is constructed with perforated pipe to allow for water to flow from the pipe into the ground when the groundwater is low. In addition, oil/water separators and grit traps are provided at key discharge locations to minimize the amount of deleterious material being discharged during storm events.

In conclusion, the community's planning efforts have developed a long range plan to address many facets of the environment of the New Hampshire seacoast. The reconstruction and restoration of the salt marsh is a direct benefit due to the project. The availability of sanitary sewers throughout the entire community will help to alleviate known surface and groundwater contamination problems in many areas throughout the town. A centralized collection and treatment system with process redundancy will provide for continuing water purification under excessively adverse conditions. The shell fishing areas of Hampton Harbor will be protected and enhanced by limiting the discharge of raw sewage and minimizing the drainage water runoff. In short, the direct and indirect benefits of an 18 year effort and a $42 million budget are numerous to the residents of this and surrounding communities.

References:

1. IEP, Inc.(1990), Wright's Island Wastewater Treatment Facility Salt Marsh Enhancement Project, Portsmouth, NH

2. Jones & Beach/Stearns & Wheler(1988), Confirmation Report Wastewater Facilities, Town of Seabrook, NH, Stratham, NH

3. Jones & Beach/Stearns & Wheler(1989), Report on the Site Selection Study for the Town of Seabrook, NH Wastewater Treatment Facility, Stratham, NH

4. Normandeau Associates, Inc.(1989), Seabrook Wastewater Treatment Facilities Summary of Technical Studies, Bedford, NH

GIS in Automated Optimal Design of Sewer Networks

Newland Agbenowosi[1]
R.G. Greene[2]
G.V. Loganathan[3]

Abstract

In the planning and design of sewer networks, most of the decisions are spatially dependent because of the right of way considerations and the desire to have flow by gravity. An optimization pipe design routine coupled with a GIS software is used for selecting alternative network layouts. The GIS also determines the locations of various pump stations. This information is used for determining optimal manhole depths, lifting pump capacities, pipe diameters, and costs.

Introduction

Geographic information systems (GIS) capable of storing, retrieving and managing spatial database are ideally suited for topological network design. GIS help in the integration of various types of maps such as street maps, plat or building lot maps, utility maps, and topographic maps which are all used in the decision making process. Blacksburg, Virginia with an area of 18.8 sq. miles, has a sewer system that includes 17 pump stations because of the hilly terrain. With the increasing development, some pump stations are being upgraded while others are being diverted for better efficiency. A tool that is capable of efficiently designing sewer networks, pipes, and selecting pumps and their locations is desired in planning the new development. Programs NETSEL, LINK, PUMPSEL, and WETWELL (Agbenowosi, 1995) select near optimal pipe network layout, pump station location and pump configuration with the appropriate wetwell design. A FORTRAN code GSDPM3 (Gray et al., 1992) provides the pipe design for the selected network.

Layout Generation Using GIS

The layout generation and subwatershed delineation are performed with the aid of a GIS package, ARC/INFO (ESRI, 1992), coupled with NETSEL and LINK routines (Agbenowosi, 1995). The user provides a map showing the locations of

---

[1]Grad. Student,[2]Assist. Prof.,[3]Assoc. Prof., Civil Engr., Dept., Patton 200, Virginia Tech. Blacksburg, VA 24061.

the proposed manholes, map of proposed building lots and the topography of the area. With the aid of AML (ARC MACRO LANGUAGE) in ARC/INFO (ESRI, 1992), the map of the manhole locations (nodes) is used to generate potential links (pipes) without consideration of right of ways. The building lots map is then used to perform an overlay to delete unacceptable links in the network. The resulting network contains all the allowable links considering right of ways.

The topographic map of the area is converted to a TIN (triangulated irregular network) to determine the various manhole elevations and to generate profiles between two or more manholes. The network containing all the acceptable links and the manhole elevation data are passed to the NETSEL program (Agbenowosi, 1995). The program NETSEL seeks to achieve gravity flow from a maximum number of manholes to the chosen outfall. When gravity flow is not attainable from certain nodes, the program selects the node with the lowest elevation among those that cannot be drained by gravity to the main outfall. It then tries to drain a maximum number of nodes to this node. This node is then designated as a possible outfall where a pump station may be located and all nodes draining into this new outfall constitute a subwatershed. This procedure is repeated until all nodes are being drained into an outfall.

The delineated subwatersheds are then connected to the main outfall using the LINK routine. The LINK routine connects the unconnected subwatersheds based on the distance between their outfalls and nodes in previously connected subwatersheds.

The subwatersheds are numbered WP(1),WP(2),...,WP(N) as follows:

1. The most upstream subwatershed is allocated the number 1 (m=1).
2. Proceed downstream numbering subwatersheds consecutively.
3. When a junction is encountered, check if there are any unnumbered subwatersheds upstream of it. If there exists some unnumbered subwatersheds, follow the unnumbered path to its most upstream subwatershed and start consecutive numbering downstream until return to junction is established. If any other unnumbered path from the same junction exists, repeat step 3. Else go to step 2. At the end of the numbering procedure, the highest number which is N must be assigned to the subwatershed with the main outfall, WP(N).

Sewer Design Procedure

The design procedure after the network has been generated is outlined in the flowchart (See Figure 1). In Figure 1, the following steps are performed.

Step 1: Set m=1.

Step 2: Estimate flows to each pipe in subwatershed WP(m). The flow estimates, together with the elevations of manhole locations and pipe lengths are passed to GSDPM3 for the design of pipe diameters and slopes. Other information required for the design such as Manning n, desired pipe material, and allowable infiltration which are not GIS related are entered by a data file into GSDPM3 (Gray et al., 1992). Design of each pipe is accomplished using the Manning equation and cumulative flow into the pipe. A minimum size of 8 inches is used.

Step 3: If m = N, end. Else go to step 4.
Step 4: Determine the subwatershed WP(k) into which WP(m) drains. Transfer total flow from WP(m) to WP(k). Use programs WETWELL and PUMPSEL (Agbenowosi, 1995) to design wetwell, sump, force main and pump station for WP(m). Set m=m+1, go to step 2.

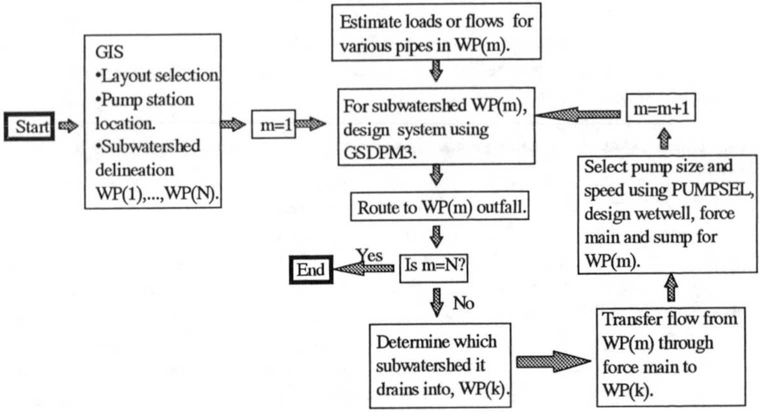

Figure 1. Flow Chart For Design Procedure.

Results

The method has been tested for different scenarios. In each case, the package generated the appropriate layout(s) and subwatersheds. One such scenarios is described in Figures 2 and 3.

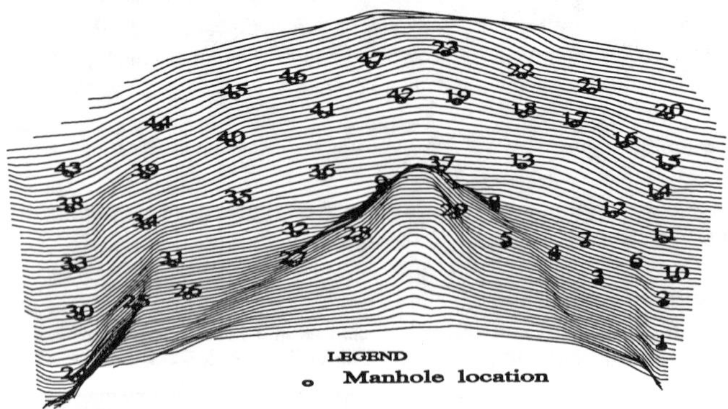

Figure 2. Three Dimensional View of Terrain

Figure 2 represents the three dimensional view of the terrain of the area with the circular spots as the proposed manhole locations. Figure 3 shows the selected layout. In this case, the main outlet has chosen to be manhole number 1. Because of the terrain, one half of the watershed will be drained into manhole number 24 since it is the lowest point on that side of the hill. A pump station will then be placed at number 24 to pump the water to manhole number 9 and then it will flow by gravity to the main outlet.

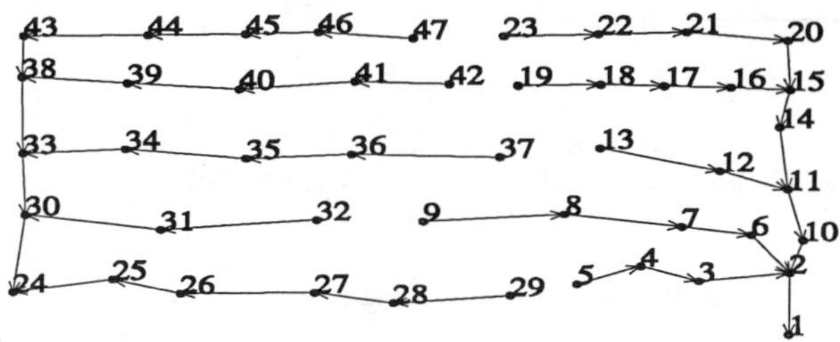

Figure 3. Selected Network Layout

Summary

The major benefit of using GIS is the ability to quickly evaluate costs, and technical objectives in the design process and to eliminate a large number of alternative configurations. The graphics capability of the GIS also enhances the comprehension of the design procedure and makes it easier to determine existing structures or features that may be obstacles to the design of the system. With the completion of the project, city engineers and planners should have a versatile tool for design of sewer networks and pump stations.

References

Agbenowosi, Newland K. "GIS Based Optimal Design of Sewer Networks and Pump Stations." M.S. Thesis, Virginia Tech. Blacksburg, 1995.

Gray, Donald D., Mark A. Pacheco, Richard L. Coffman, and John D. Quaranta. Gravity Sewer Design Program Version 3.0 M: User's Guide. West Virginia University. Morgantown, 1992.

ARC/INFO User Guides. ESRI (Environmental Systems Research Institute). Redlands, 1992.

SANITARY SYSTEM MODELING
AUTOMATED MAPPING/FACILITY MANAGEMENT
FORT BENNING, GEORGIA

Robert W. Carr[1] , M. ASCE and Dale J. Hexom[2]

Abstract

In recent years, many communities and military installations have used databases to help manage sewer systems. The Automated Mapping/Facility Management (AM/FM) database allows the user to keep sewer system data and maps up to date. Studies to determine the capacity of an existing sanitary sewer system can be conducted using the AM/FM data.

The Fort Benning Military Reservation began developing an AM/FM database in 1985 using aerial photography and planimetric mapping. Data from available maps and plans were used to develop the database information, such as manhole inverts, pipe types, pipe lengths, etc. In 1990, a capacity analysis of the Fort Benning sanitary sewer system was initiated. The primary purpose of this study was to provide a sanitary sewer master plan for the major interceptors in the Main Post and Custer Road Family Housing Area. The manhole rim and invert elevations, sewer sizes and sewer lengths were taken from the existing AM/FM database. The analysis showed that the existing data had large amounts of conflicting data. A manhole inspection program was conducted to obtain actual rim and invert data, pipe sizes and lengths. Following the completion of the manhole inspection program, the capacity analysis was completed. The results of the analysis showed those existing sewers which will need to be upgraded to handle the future estimated flows.

This paper will review the methodology used to develop the AM/FM database and to conduct the Fort Benning Sanitary Sewer Capacity Study. Particular attention is given to the interaction between the AM/FM database and the software used for the capacity analysis.

[1]Water Resources Project Manager, RUST Environment & Infrastructure, 1020 N. Broadway, Suite 400, Milwaukee, WI 53202
[2]AM/FM Project Manager, RUST Environment & Infrastructure,

Fort Benning Overview

Fort Benning is an Army Training and Doctrines Command (TRADOC) installation, located just outside Columbus, Georgia, approximately 160 kilometers (100 miles) southwest of Atlanta. The 730-square kilometer (282 square mile) facility is the proud home of the U.S. Army Infantry Center. Over 45,000 civilian and military personnel work and live at Fort Benning.

Development of AM/FM at Fort Benning

Facility maps and information are critical to an installation like Fort Benning in the same manner as they are to any utility system operation. A military facility has to manage not only water, sanitary sewer, and storm drainage maps and data; but also buildings, roads and railroads, electrical distribution and lighting, natural gas, cable television, land use, telephone, heating and cooling, and emergency monitoring control systems. To allow the systematic management of the above data, the Fort Benning Master Planning Branch began to implement an AM/FM/GIS system in 1985.

An accurate base map was developed, complete with stereo-digitized two foot and five foot topographic contours from aerial photography. Using available utility maps, field survey crews located and either panelled or marked with white spray paint the ground level utility features which were then captured in the aerial photography. During stereo-digitizing, each unique utility system feature was digitized into a separate file.

An attribute database structure based on Intergraph Corporation's Data Management and Retrieval System (DMRS) was developed for each utility. Utility system maps, utility studies, and personal interviews with Fort Benning staff were used to gather information that was used to populate the utility system database. During the initial AM/FM system development, little field data verification was performed. Field work essentially consisted of ground control survey for the aerial photography.

Sanitary Sewer System Modelling Preparation

The Fort Benning sanitary sewer system consists of approximately 217 kilometers (135 miles) of sewers, 22 lift stations and two secondary sewage treatment plants. The capacity analysis for Main Post and Custer Road Family Housing areas included all large diameter interceptor sewers and selected trunk sewers in the model. The HYDRA computer program, developed by Pizer, Inc., was used to model the selected sewers. Unfortunately, the Fort Benning GIS database, DMRS, was not directly compatible with the PC based computer model.

Two tasks were required before the AM/FM system maps and database could be useful in the computer model development. First, the digital map files had to be skeletonized down to produce a digital file with the selected sewers. During this task, the issue of pipe connectivity surfaced. The model required that each pipe had to be connected graphically and digitally with no breaks or discontinuities. Unfortunately, there were numerous occurrences in the digital files where discontinuities existed. A data validation program was developed which identified those occurrences and allowed a CADD technician to find the problems in the graphic files and correct them.

The second major task in preparing the GIS database for use in the capacity analysis was to translate data from DMRS to dBASE and then organize the data to the format required by the modelling program. During this task, a major data discrepancy problem was discovered. Sanitary sewer system maps, complete with manhole number, rim and invert elevations, and pipe sizes had been generated from the GIS database. This data was used in the field by survey crews to spot check and validate the data. The field review identified prblems which included incorrect invert elevations, differing pipe sizes, and system configurations which did not match the maps. An additional problem identified was that of duplicate manhole numbers present on the same map sheet. These problems had not previously surfaced since the GIS maps had been developed based on the hard copy maps and sewer system information that the Fort Benning Maintenance Division had available.

The duplicate manhole numbering problem was easily solved with the AM/FM system. When the digital maps were prepared, the CADD system generated a unique identification number for each graphic file element. This number was used to produce a new, unique manhole number. The unique numbers were re-plotted on sewer system maps, and a major field verification effort was undertaken. This effort included a complete survey and inspection of approximately 1,400 manholes. Data gathered for each manhole included: manhole rim and invert elevations; pipe sizes and materials; manhole condition; evidence of infiltration and inflow; etc. The field data was entered into the translated dBASE database. The dBase database was used by HYDRA without further translation.

Sanitary Sewer Capacity Analysis

While the field verification and sewer database update were being conducted, the sanitary system flows for the study area were developed. Sanitary flows were developed from population and employment projections provided by the Fort Benning Master Planning Branch.

During the field verification work, a four-week flow monitoring program was conducted at twenty locations. The selected areas were either residential, commercial or a combination of residential/commercial areas. Based on the flow monitoring data, diurnal curves were developed for each type of area.

The residential sanitary flows are those flows from housing units. Each housing unit was assigned 3.6 persons per unit and 100 gallons per capita (person) per day (gpcd). For barracks and Bachelor Officer Quarters (BOQ), a flow rate of 50 gpcd was used. The Fort Benning building database was used to obtain the number of number of housing units, barracks and BOQs. Each housing unit within the study area is tributary to a modeled interceptor. HYDRA uses the number of units, number of persons per unit and the per capita flow rate to determine the peak flow to each interceptor. These peak flows are then fitted into the diurnal curve from that area and routed downstream. The non-residential sanitary flows are those flows from offices, gyms, laundries, warehouses and other non-residential buildings. Sanitary flows were estimated on a gallons per day per square foot basis. The Fort Benning building database was used to obtain the number of non-residential buildings. Each building within the study area is tributary to a modeled interceptor. HYDRA uses the building type, square footage of the building and the unit flow rate to determine the peak flow to each interceptor. These peak flows are then fitted into the diurnal curve for that type of building and routed downstream.

Inflow and infiltration allowances were estimated and added to the base flow because these clear water flows affect the capacity of the existing sewer system. Inflow was estimated using the flow monitoring data from the four-week flow monitoring program. The infiltration flow allowance of 1,000 gallons per inch-mile of sewer per day was taken from existing Technical Manuals.

The results of the capacity analysis showed that most of the modeled interceptors in the study area have sufficient capacity to carry the estimated flows. Alternatives were developed for any interceptors that were under capacity and a system improvement plan was recommended.

Conclusions

The Fort Benning AM/FM/GIS system allowed for the easy preparation of system study area maps and provided valuable information on data that was missing and needed to be field verified to produce a meaningful system model. The data shortfall in the GIS was a result of the lack of field verification during the early development stages of the AM/FM/GIS database. If more verification of the data reliability been done when the GIS was initially created, less time and expense would have been required during the modelling project. Clean graphics with full connectivity is also important in modelling applications. Being aware of the requirements for potential user applications during GIS development can result in a more fully functional GIS for the end users.

SWIMS: A Computerized Information Management System for NPDES Stormwater and Wastewater Discharge Compliance

Joshua Lieberman[1]

Abstract

A computerized SWIMS (surface water information management system) has been designed and developed to meet the needs of NPDES (National Pollutant Discharge Elimination System) permit compliance. Such permits may involve the collection and reporting of a large volume of chemical and physical parameters. SWIMS is intended to facilitate the three major tasks faced by any NPDES permit holder:

1) Collection and validation of required discharge data and documentation.
2) Production of required periodic discharge monitoring reports (DMR).
3) Warning and analysis of non-compliance with permit requirements.

The SWIMS data model emphasizes generality and flexibility, in order to encompass as wide a range of potential permit requirements as possible. Implementation of SWIMS in Microsoft Access™ has resulted in a graphical application which shares data readily with other Windows applications such as spreadsheets and word processor. This forms a complete and easy-to-use compliance tracking system useful to both the facility operator and the permitting consultant.

Introduction

NPDES (National Pollutant Discharge Elimination System) permits for stormwater and wastewater discharges to the environment require the collection, processing, and reporting of a number of physical and chemical parameters. The number

[1] Senior Hydrologist, ENSR Consulting and Engineering, Inc., 35 Nagog Park, Acton, MA 01720

of parameters may range from a handful for small facilities to hundreds for larger facilities with a number of outfalls. These parameters may include flow, pH, temperature, and other physical characteristics; as well as concentrations of a wide variety of conventional and priority pollutants. Monitoring frequencies may vary from quarterly to continuous. In the case of stormwater, monitoring frequencies may even be determined by the weather. The work involved in submitting a correctly written DMR (discharge monitoring report) may be a significant burden for facility operators who are primarily concerned with what happens inside rather than outside their facility.

Such a task naturally lends itself to computerization, and such applications as spreadsheets have commonly been used to track the necessary monitoring data. However, as permit requirements have become more voluminous and more complicated, spreadsheets have become less and less effective for managing all the various aspects of compliance. This report details the design and development of a compliance tracking system (SWIMS) dedicated especially to meeting the needs of NPDES permit compliance.

Needs Analysis

The initial design step involved examining the specific tasks which SWIMS should be able to accomplish. The first capability required by any NPDES permit is the collection of specified monitoring data, either analyses of water samples for compounds of potential concern, specific in-situ tests such as pH or temperature performed at the outfall, or the output of continuous monitoring instruments such as recording current meters. SWIMS should assist in the compilation and validation of monitoring data, but should also be useful for the planning of monitoring efforts by incorporating information about sampling effort schedules and methods.

The second capability involves summarizing the monitoring data at the appropriate intervals and generating DMR's which are formatted correctly for a given permit. SWIMS should also perform a validation function here, providing some assurance that only validated data is reported.

The third capability is not strictly required by a NPDES permit, but could be crucial to its continued validity! SWIMS should be able to provide a evaluation of compliance over any selected time period and flag particular permit requirements which are not being met. This should include results compliance, such as with pollutant levels, as well as method compliance, such as with monitoring intervals. This capability, augmented by graphical plots and statistics, is particularly important for helping the facility operator and permitting consultants to work together in meeting and/or negotiating permit requirements.

A number of other capabilities are useful but not essential for NPDES permit management, such as automatic sample label generation, or waste stream statistical analysis. The option for future development of such capabilities was incorporated to the extent possible in the initial design.

Design

A flexible and scaleable data model was devised in order to handle the wide variety of possible permit requirements and facility characteristics. This data model is expressed in the following entity-relationship diagram:

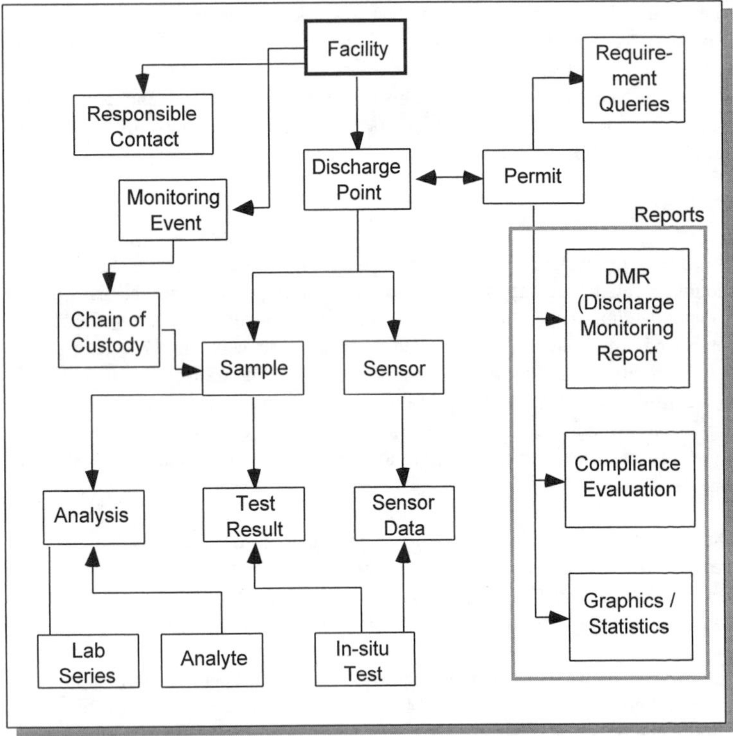

Three basic forms of monitoring data are contained in this design, including analytical laboratory results, results of in-situ tests such as salinity, and the results of continuous monitoring at a particular discharge locality. The actual monitoring is assisted by event and chain-of-custody tables, as well as a separate method table containing reference information for all test and sampling protocols. An important element of the design is the contacts table, which provides contact information (name, address, phone number, etc.) not only for a facility representative, but for each person carrying out sampling, testing, or analysis.

Permit requirements themselves are stored in sql query form. A particular requirement may be either a monitoring requirement (e.g. analyze for TCE), a reporting requirement (e.g. 3-month mean temperature), or a compliance requirement (e.g. minimum allowed dissolved oxygen over 24 hours is 5 mg/l). A collection of requirement queries belonging to a particular permit can then be triggered to generate any of a number of formatted and/or graphical reports. The formats of DMR's required for each NPDES permit can also be stored as part of the system data so that correctly formatted reports may be triggered and printed out automatically at the appropriate intervals.

Implementation

SWIMS has been implemented using the Windows rdbms (relational database management system) Microsoft Access™, although the design itself could be implemented in any other reasonably capable rdbms. Operator experience with earlier implementations showed that a user-centered graphical interface with easily available help functions is vital to the application's usefulness and user-friendliness. Users were particularly frustrated with fixed hierarchies of data entry screens where viewing another screen might involve a time-consuming march through five or six other screens in the hierarchy. The present implementation allows the user to view any of the screens in independent windows at any time, or even compare two views of records in the same screens. The capacity for generating trend and scatter plots is build directly into the SWIMS application and transfer of data to and from other Windows applications is particularly straightforward. For particularly large or complex facilities, the present implementation can also be used as a client application to data tables held remotely on a more powerful database server.

Conclusions

The NPDES permit management system SWIMS provides a permit holder or permitting consultant with powerful tools to manage discharge monitoring and monitoring report generation, as well as graphical and statistical tools for analyzing facility compliance. SWIMS is not intended to supplant expert assistance in obtaining and complying with NPDES permits. It is a cost-effective means of facilitating the burdensome administrative aspects of permit compliance, and can also be helpful in solving or avoiding permit compliance problems.

Water Quality Data Management-Survey of US Agencies

Laurel Saito[1], Associate Member, ASCE
Neil S. Grigg[2], Fellow, ASCE

Abstract

A survey of 200 water quality agencies in the United States was undertaken in 1991 which solicited information from each agency regarding the methods of data storage and management used, the activities involving water quality data performed by each agency, the types of water quality data used, the sources of data, and interagency activities involving water quality data. The results of the survey showed that almost all agencies responded that their water quality data are being used by others, and that the majority of agencies also used data obtained from other agencies. This indicates a high demand for data sharing, and the potential is especially great because of the predominant use of IBM-compatible personal computers, ARC/INFO geographic information systems and dBASE database software.

Introduction

This paper presents the results of a 1991 nationwide survey of water quality agencies in the United States regarding their water quality data management efforts. The survey was performed as part of the first author's thesis work to provide a nationwide assessment of water quality data management activities. The results presented here are discussed in greater detail in Saito, et. al. (1994) and Saito (1992).

Survey development and distribution

The survey was developed to obtain six general categories of information: 1) type of agency (e.g., federal, state or other); 2) type of water quality involvement (e.g., ground water, surface water or other); 3) description of data storage and management (e.g., software such as STORET, WATSTORE, geographic information systems (GIS's), other computerized software or manual data

[1] Assoc. Engineer, Boyle Engineering Corp., 165 S. Union Blvd., Suite 200, Lakewood, CO 80228
[2] Dept. Head and Prof., Dept. of Civil Engineering, Colorado State Univ., Fort Collins, CO 80523

management, and hardware such as personal computers, mainframes, minicomputers or workstations, and other computers); 4) type of activities involving water quality data (e.g., federal standards compliance, state standards compliance and research and development activities); 5) types of data used (e.g., discharge, temperature, pH, dissolved oxygen, major cations, nitrogen, phosphorus, suspended sediments or solids, biochemical oxygen demand (BOD) or chemical oxygen demand (COD), trace metals, pesticides and herbicides, volatile organic compounds (VOCs), bacteriological or viral, chlorophyll *a* or algae, radiological, and other data types); and 6) sources of data and interagency activities.

Approximately 600 surveys were sent out in April and May of 1991 using mailing lists obtained from the U.S. Environmental Protection Agency's (USEPA) Office of Regulations and Standards and the NAWDEX database of the U.S. Geological Survey's (USGS) Water Resources Division. The mailing lists contained agencies from all states and Puerto Rico. dBASE IV was used to generate mailing labels and track responses.

Survey results

Of the surveys sent out, 226 were returned. Although follow-up telephone calls were made to about one-third of the respondents to clarify and verify responses, eight of the returned surveys were still unusable due to incompleteness and difficulty in contacting the responding parties. These responses were not included in the analysis of the survey.

The 218 usable surveys consisted of responses from a total of 200 agencies in 48 states, one territory, and the District of Columbia. 189 of these agencies reported that they were involved with surface water quality, while 144 of the 200 agencies work with ground water quality. "Other" water quality, or water quality activities that did not fall under the general surface or ground water classifications, involved 36 of the agencies. Responses were also broken down according to agency type. Table 1 shows a breakdown of the responding agencies by agency type and kind of water quality data involvement.

Table 1. Responding agencies by agency type and water quality type

Water quality type	Federal agencies	State agencies	Other agencies	All agencies
Surface water	60	74	55	189
Ground water	45	57	42	144
Other water	12	15	9	36

Table 2 summarizes the survey analysis with percentages of agencies that indicated positive responses to survey questions. Only those selections which received the highest percentage of positive responses are shown for each category. It should be noted that the summation of percentages vertically in a section of the table do not necessarily add up to 100 percent because agencies often selected more than one option and all options are not shown.

Table 2. Summary of survey analysis

	Federal agencies	State agencies	All agencies
Surface water quality agencies			
Total number of agencies	60	74	189
Most-used data management	WATSTORE (57%) Manual (52%)	Computer software (92%) STORET (73%)	Computer software (75%) Manual (59%)
Most-used GIS[a]	ARC/INFO (90%)	ARC/INFO (90%)	ARC/INFO (83%)
Most-used computer software[b]	dBASE (36%) QuattroPro (14%)	dBASE (72%) Lotus (60%)	dBASE (55%) Lotus (47%)
% using personal computers	65	92	82
% using mainframes	53	58	50
% using minicomputers/ workstations	50	32	37
Most-frequent activities	Baseline/trend anal. (78%) Cause/effect studies (63%)	NPDES permitting (77%) Baseline/trend anal. (74%)	Baseline/trend anal. (72%) Public inquiries (63%)
Most-used data types	Susp. sed./solids (92%) Discharge (90%)	pH (92%) Temperature (89%)	Temperature (88%) pH (88%)
Most-used data sources	Agency itself (93%) WATSTORE (52%)	Agency itself (95%) Other agencies (66%)	Agency itself (93%) Other agencies (53%)
% whose data is used by others	87	95	90
% with cooperative agreements	78	73	71
Ground water quality agencies			
Total number of agencies	45	57	144
Most-used data management	WATSTORE (73%) Manual (52%)	Computer software (79%) STORET (73%)	Computer software (67%) Manual (53%)
Most-used GIS[a]	ARC/INFO (96%)	ARC/INFO (100%)	ARC/INFO (92%)
Most-used computer software[b]	dBASE (29%) Lotus (29%)	dBASE (60%) Lotus (51%)	dBASE (46%) Lotus (43%)
% using personal computers	62	91	82
% using mainframes	64	58	56
% using minicomputers/ workstations	51	42	42
Most-frequent activities	Baseline/trend anal. (64%) Cause/effect studies (60%)	Fed. drkg. wtr. stds. (72%) Sta. drkg. wtr. stds. (70%)	Baseline/trend anal. (60%) Sta. drkg. wtr. stds. (58%)
Most-used data types	Major cations (89%) Trace metals (87%)	Nitrogen; VOCs (81%) Trace metals (81%)	Trace metals (83%) Major cations (81%)
Most-used data sources	Agency itself (93%) WATSTORE (62%)	Agency itself (95%) Other agencies (72%)	Agency itself (94%) Other agencies (56%)
% whose data is used by others	89	98	93
% with cooperative agreements	82	77	74

[a]Percent of agencies using GIS using this system

[b]Percent of agencies using computer software using these programs.

Conclusions

Three general conclusions were drawn from the results of the water quality data management survey: First, there is a high *demand* for data sharing. At least 90% of the responding agencies indicated that their water quality data is used by others. In addition, over half of the agencies cited other agencies as data sources.

Second, there is a *potential* for data sharing. Over one-third of all surface and ground water agencies were using GIS, and an overwhelming number of these agencies were using ARC/INFO. In addition, the predominant computer software used by all agencies was dBASE. The compatibility of dBASE with ARC/INFO and other GIS's indicates an encouraging potential for data sharing.

Because of the high demand and potential for data sharing, the third conclusion is that there is a need to *enhance* data sharing. From the technical side, this implies that there is a need to develop software that makes data compatible between agencies. In addition to being compatible with ARC/INFO and dBASE, the survey results suggest the importance of developing and enhancing data management software that is appropriate for IBM-compatible personal computers, since most state agencies and more than half of all federal agencies were using personal computers, and almost all of these agencies were using IBM-compatibles.

However, perhaps the more difficult side of enhancing data sharing comes from the organizational perspective. In order for data sharing to be effective, organizations must make adequate arrangements for data coordination with other agencies, both in terms of the activity of coordinating data and in staffing such efforts. However, the organizational forms of public agencies are often driven by programmatic factors such as legislation, appropriated budgets, civil service positions,and political appointments. In this climate, it is difficult to coordinate management issues such as data sharing between different agencies. Issues such as conflicting missions between regulatory agencies and resource development agencies, and conflicts between federal, state,and local agencies can impede data sharing.

Acknowledgments

This research was supported by the Colorado Water Resources Research Institute and the Colorado Commission on Higher Education. This study would not have been possible without the tremendous cooperation of water quality data managers throughout the United States. Their willingness to donate time and knowledge in completing the surveys as well as to talk about water quality data management in follow-up conversations is greatly appreciated.

References

Saito, L. (1992). "Water quality data management." *Tech. Rep. No. 59*, Colorado Water Resources Research Institute, Colorado State University, Fort Collins, Colorado. (July).

Saito, L., Grigg, N.S. and Ward, R.C. (1994). "Water-quality data management: Survey of current trends." *Journal of Water Resources Planning and Management*, Vol. 120, No. 5. (September/October), pp. 587-612.

STREAMLINED PERMITTING FOR SECONDARY WASTEWATER TREATMENT FACILITIES

Jane Wheeler, Raymond Masak, and Bryon Clemence[1]

Introduction

The South Essex Sewerage District (SESD) in Massachusetts is under a federal Consent Decree to bring its wastewater facilities into compliance with the federal Clean Water Act, which requires secondary treatment. After extensive investigations of various sites across the five-community District, the site of the existing primary treatment facilities in Salem was selected as the location of the new 30-mgd secondary plant. Even though the site currently hosts wastewater facilities, the project required numerous environmental approvals from a number of federal, state, and local agencies. To ensure that the demanding Consent Decree schedule was met, it was imperative that permitting for construction and operation of all components be expedited as much as possible.

This paper focuses on the approvals and permitting strategy required for construction of the land-based secondary facilities. Additional permits were also required for construction of a new ocean outfall diffuser, expansion of a landfill for disposal of residuals, and for operation of all facilities.

Develop the Permitting Strategy

With the increase in environmental awareness and regulatory requirements in recent years, a project that may have breezed through the permitting process 10 years ago may find itself lost in the permitting maze today. As a result, it is extremely important to establish and follow a carefully-planned permitting strategy even before project design begins. The permitting strategy should incorporate a number of "rules of thumb" as outlined below. The remainder of this paper relates each of the those steps to the SESD secondary facilities.

1 Wheeler/Clemence, Camp Dresser & McKee Inc., Cambridge, MA
Masak, South Essex Sewerage District, Salem, MA

- Consider permits from day one
- Use "smart" design
- Coordinate with regulatory agencies early and often

Consider Permits from Day One

The permitting process for the secondary facilities actually began long before preliminary design was initiated. SESD and Camp Dresser & McKee Inc. (CDM) conducted an exhaustive siting search covering all five communities within the District and fully evaluated the most feasible sites in terms of a number of environmental, technical, and institutional factors, including permitting. Permitting thresholds were even incorporated in the siting criteria where appropriate. For example, recognizing that 5,000 square feet of wetland disturbance would require a time-intensive Variance from the state Wetlands Protection Act, this threshold was used as one factor to screen out sites from further consideration. By factoring in the permitting issues early-on, potential roadblocks were avoided later.

SESD and CDM also developed a permitting plan early in the process to identify specific permits and permit interrelatonships and to define the overall approach for obtaining permits within the Consent Decree schedule. The plan also established the links between specific project activities and permit requirements (i.e., permit triggering activities). Providing this information in a simple matrix form allowed easier identification of design refinements that could be adopted to simplify permitting (described below).

Use "Smart" Design

Once the site for the secondary facilities was selected, preliminary design commenced. From a permitting perspective, this stage is very important since it sets the framework for permit requirements and provides an opportunity to influence the course of design. In developing the facility layout, for example, sensitive environmental features and regulatory jurisdictions at the site (e.g., wetlands) were avoided as much as possible so that some permitting requirements could be eliminated.

At the outset of preliminary design, in spite of efforts to avoid sensitive site features that could trigger permits, it was determined that 12 environmental permits would be required for construction alone. However, re-evaluation of the specific activities that triggered the permits revealed that some modifications in design or construction of the plant would greatly simplify the permitting requirements and/or reduce the time required to obtain necessary approvals. In the end, by making minor revisions to design, and consolidating construction phases and activities, only five permits were necessary for plant construction. This streamlining was a result of several design modifications, and extensive agency negotiations, which are summarized below.

Using existing storm drain outfalls at the site (instead of constructing new drains) avoided construction within federal wetlands and therefore eliminated the need for an Army Corps of Engineers Section 404 permit, 401 Water Quality Certification, and Coastal Zone Management Consistency Determination. Initially, SESD planned to discharge both stormwater and groundwater from construction dewatering through these existing storm drains, and to apply for a National Pollutant Discharge Elimination System (NPDES) General Stormwater Permit to cover these discharges. However, EPA and the Massachusetts Department of Environmental Protection (MDEP) were concerned about possible contaminants in the groundwater and indicated that an NPDES Individual Permit would therefore be required, which is a far more complicated permitting process than the NPDES General Permit process. Following extensive negotiations with both EPA and MDEP, it was agreed that stormwater could be discharged through the existing storm drains and that groundwater from dewatering activities could be discharged through the existing effluent outfall. The stormwater discharge required a new NPDES General Permit from EPA under the 1992 stormwater regulations. The discharge of groundwater through the existing effluent outfall was permitted under the current NPDES Individual Permit for the existing primary treatment plant at the site. This approach eliminated the need for a new NPDES Individual Permit or offsite disposal of the groundwater. This strategy also avoided the need for a Coastal Zone Management Consistency Determination, which is required for new NPDES Individual Permits.

In additon, combining the two phases of construction (site preparation and plant construction) allowed permit consolidation. As a result, only one Order of Conditions was required instead of two.

The outcome of making a few seemingly-simple changes in preliminary design was to keep the project on schedule, allowing construction to commence on time and in compliance with the Consent Decree. The five permits that were ultimately necessary for construction include:

- Order of Conditions under the State Wetlands Protection Act for work in buffer zone (wetland resource areas were successfully avoided, but the site is within the 100-foot buffer zone, which required Conservation Commission approval)

- National Pollutant Discharge Elimination System (NPDES) Stormwater General Permit from EPA for stormwater discharges during construction

- Approval under Section 106 of the National Historic Preservation Act for archaeological resources found on site

- Chapter 91 Waterways License for construction in historic (i.e., filled) tidelands

- Conditional Plant Approval/Permit to Construct for air emissions

Coordinate with Regulatory Agencies Early and Often

Although many regulatory agencies require detailed design information to be included in the permit application, the aggressive Consent Decree schedule required that SESD submit applications before design completion. Consequently, most applications were submitted during preliminary design and updated as necessary as design progressed. In addition, to ensure that agencies were well-informed about the project, meetings were held with several key agencies in advance of application submittal to answer questions and obtain early agency feedback.

At the outset of preliminary design, there were two different plant designs under consideration, which complicated the permitting process further. A decision was made to proceed with some of the upfront permitting on both plant layouts - in particular, Conservation Commission approval under the Wetlands Protection Act. A Notice of Intent was submitted addressing both plant layouts and the Commission was asked to issue an Order of Conditions that would address either layout. SESD and CDM also prepared some of the draft conditions and reviewed them with the Commission so that all parties were in agreement prior to formal issuance of the Order. One of the conditions suggested by SESD and CDM was to present additional information to the Commission later in the design process, so that the Commission would have an opportunity to review and comment on detailed design information that was not available at the time the Notice of Intent was filed. By approaching the Commission early and participating in the preparation of the permit conditions, the permitting schedule was significantly shortened.

Another approach that was used to streamline permitting and avoid construction delays was to negotiate a special review procedure with the Massachusetts Historical Commission so that construction could commence before archaeological reviews were completed. There were several areas on the plant site that were identified as archaeologically sensitive, and a comprehensive testing program was required in those areas. MHC agreed to review and approve preliminary testing results, rather than waiting several months for laboratory analyses to be completed. Without this shortened process, construction would have been delayed.

Conclusion

All permits for the construction of the secondary plant were obtained on schedule. The success of the permitting effort is attributed to the simplified permitting approach that was established at the project outset and followed over the course of the project. A similar approach would be useful for any project that is driven by a demanding schedule and/or has complex permitting requirements.

Application of CSO-Related Bypasses in Maine

Steven D. Freedman[1]
Mark C. Jordan[2]

Abstract

The recently-published U.S. Environmental Protection Agency (EPA) "Combined Sewer Overflow (CSO) Control Policy" (*Federal Register* 1994) provides an opportunity for communities with CSOs to utilize primary treatment capacity for all or a portion of their wet-weather flows. This concept has been recommended for, and in some cases already successfully employed by, several communities in the State of Maine and elsewhere.

This paper discusses the manner in which primary treatment followed by disinfection has been either proposed for, or is currently employed at, secondary wastewater treatment facilities (WWTF) in Augusta, Bangor, Bath, Portland and South Portland, Maine.

Overview of CSO Control Policies

In addition to the EPA control policy, each state was required to develop and implement its own CSO control policy. Collectively, these policies establish standards and/or guidelines for (1) planning (2) alternative abatement evaluations (3) scheduling (4) minimum levels of controls, or best management practices (5) attainment of water quality standards and (6) minimum treatment levels. Relative to the latter, many states, and EPA, are requiring that CSO discharges receive, as a minimum, the equivalent of primary treatment relative to floatable and settleable solids removals.

One of the most revealing features of the EPA policy, and the focus of this paper, is contained in Section II.C.7, "Maximizing Treatment at the Existing POTW Treatment

[1] Senior Associate, Whitman & Howard, Inc., 500 Southborough Drive, South Portland, Maine 04106-3209

[2]Design Engineer, Portland Water District, 225 Douglass Street, Portland, Maine 04104-3553

Plant". This section allows permittees to utilize excess primary treatment capacity, followed by a CSO-Related Bypass, if it can be shown that the secondary portion of the WWTF is used to its maximum capabilities. The justification for the CSO-Related Bypass is that without such, "severe property damage" would result. In this instance, the property in question is the mixed liquor which would otherwise have been washed-out from the aeration basins had all of the primary effluent been allowed to enter the secondary treatment system.

Case Studies

The application of the CSO-Related Bypass provision of the EPA CSO Control Policy was a key feature in the CSO abatement planning for Augusta, Bangor, Bath, Portland and South Portland, Maine. These five communities applied the provision in a variety of ways for both primary settling and disinfection.

Augusta. The Augusta WWTF is a pure-oxygen facility designed for an average-daily flow (ADF) of 0.35 m^3/s (7.95 mgd) with a peak flow of 0.73 m^3/s (16.7 mgd). Current ADF ranges between 0.15 and 0.17 m^3/s (3.5 and 4 mgd). The connected capacity of the interceptor sewers entering the WWTF site is approximately 1.27 m^3/s (29 mgd). An overflow weir, located downstream of the degritting units, begins to bypass at flowrates in the 0.39-0.44 m^3/s (9- to 10-mgd) range. At full interceptor flow, approximately 0.53 m^3/s (12 mgd) receives primary and secondary treatment; the remainder, up to 0.74 m^3/s (17 mgd), is bypassed directly to the chlorine contact tank.

The CSO facilities plan recommended the following modifications to the WWTF: new headworks building; new bypass structure following primary treatment (CSO-Related Bypass); separate high-rate disinfection facilities for the bypassed flow; and miscellaneous hydraulic and process improvements. These improvements will allow all flow, up to 1.27 m^3/s (29 mgd), to receive full preliminary and primary treatment and disinfection. Flows up to 0.53 m^3/s (12 mgd) will receive secondary treatment. The rate of bypass will be controlled by the pumping rate of the primary effluent to the aeration tanks; the flow not pumped to secondary treatment will be bypassed. These improvements are currently under design and are expected to be operational in 1997.

As a preliminary step to the approval of the CSO facilities plan, the Augusta Sanitary District, the owner and operator of the WWTF, is negotiating an Administrative Order with EPA to establish discharge limits for the CSO-Related Bypass.

Bangor. Bangor's primary WWTF was upgraded to secondary treatment in 1992 using a biofilter-coupled process. The WWTF has a design ADF of 0.79 m^3/s (18 mgd) and a peak wet-weather design flow of 1.31 m^3/s (30 mgd).

As part of the City's CSO planning, it was recommended to dedicate one of the three existing primary settling tanks for wet-weather flows greater than the 1.31-m^3/s

(30-mgd) peak design flow. The remaining two tanks will be used for dry-weather flows. This arrangement will allow the WWTF to handle all 1.31 m^3/s (30 mgd) through the preliminary, primary and secondary treatment units with an additional 0.57 m^3/s (13 mgd) receiving primary treatment and disinfection. The dedicated wet-weather primary settling tank will be used for chlorination and dechlorination.

Bath. The Bath WWTF is a contact-stabilization facility designed for an ADF of 0.07 m^3/s (1.64 mgd) and peak wet-weather flows of 0.55 m^3/s (12.6 mgd). Actual ADF is 0.12 m^3/s (2.8 mgd) with peak wet-weather flows as high as 0.61 m^3/s (14 mgd).

The CSO facilities planning, performed concurrently with planning for the WWTF upgrade and expansion, recommended the following: construction of a new headworks structure; conversion of the existing final clarifiers to primary settling tanks; construction of a new bypass structure following primary treatment (CSO-Related Bypass); conversion from mechanical aeration to fine-bubble aeration; construction of new final clarifiers; construction of new disinfection facilities for both dry- and wet-weather flows -- conventional system for dry-weather flows and high-rate for wet-weather flows; and miscellaneous improvements. The rate of bypass will be controlled by a gate located in the aeration tank splitter box/flow diversion structure.

The upgraded and expanded WWTF will be designed for an ADF of 0.15 m^3/s (3.5 mgd) with a sustained secondary flow of 0.31 m^3/s (7 mgd). All flow entering the WWTF, approximately 0.61 m^3/s (14 mgd), will receive full preliminary and primary treatment. Flows in excess of 0.31 m^3/s (7 mgd) will be bypassed around the secondary units and receive high-rate disinfection. Instantaneous peak or "first-flush" flows in the order of 0.44 m^3/s (10 mgd) will receive secondary treatment. These improvements are currently under design are expected to be operational in 1997.

Portland. The Portland WWTF is a conventional activated-sludge facility which has been operating what is now referred to as a CSO-Related Bypass since it was constructed in 1979. The WWTF is designed to provide preliminary and primary treatment to peak wet-weather flows of 2.63 m^3/sec (60 mgd), secondary treatment to an ADF of 0.87 m^3/sec (19.8 mgd) and peak flows of 1.61 m^3/sec (36.8 mgd). Flow in excess of 1.62 m^3/sec (37 mgd) are diverted from the primary clarifier effluent channel to the primary effluent chlorine contact tank. The bypass is controlled by a flow meter and control valve. The National Pollutant Discharge Elimination System Permit requires the combined effluent meet secondary effluent standards.

As part of the CSO abatement planning, the Portland Water District, the owner and operator of the WWTF, has proposed increasing the peak wet-weather flows to 3.5 m^3/sec (80 mgd) while maintaining peak secondary flows at 1.62 m^3/sec (37 mgd). As a preliminary step to the approval of the Portland CSO abatement plan, the Portland Water District is negotiating an Administrative Order with EPA to establish separate limits for the primary and secondary effluents. The WWTF's State discharge license currently recognizes separate limits for the primary and secondary effluents.

South Portland. The South Portland WWTF is currently upgrading its modified-aeration facility from an ADF of 0.24 to 0.41 m^3/s (5.5 to 9.3 mgd). New headworks, final clarifiers and chlorination facilities are being constructed. The existing final clarifiers are being converted to dedicated wet-weather primary settling tanks. The upgraded and expanded WWTF will be able to handle up to 1.01 m^3/s (23 mgd) through secondary treatment. An additional 1.45 m^3/s (33 mgd) of wet-weather flows will receive preliminary and primary treatment followed by disinfection.

Disinfection for the wet-weather flows will be accomplished by adding the chemical ahead of the settling tanks with dechlorination occurring at the effluent end. These improvements are currently under design and are expected to be operational in 1996.

Other Facilities. In addition to the five facilities discussed above, other communities which are at varying stages of their CSO abatement planning are also considering CSO-Related Bypasses. These include Rockland and Skowhegan, each with existing conventional activated-sludge WWTFs.

Conclusions

The use of the CSO-Related Bypass provision of the EPA control policy is an effective, and often cost-effective wet-weather control measure. It can be applied at WWTFs that either have existing primary settling tanks, or where tankage can be converted for that purpose, such as outdated final clarifiers. The provision also allows for a variety of disinfection options: conventional; high-rate; and concurrent with primary settling.

Appendix I. References

Federal Register. (1994) 59 (No. 75; April 19), 18688.

Appendix II. Notation

The following symbols are used in the paper:

ADF	=	average daily flow;
CSO	=	combined sewer overflow;
EPA	=	United States Environmental Protection Agency;
m^3/sec	=	cubic meters per second;
mgd	=	million gallons per day;
POTW	=	publicly-owned treatment works
WWTF	=	wastewater treatment facility.

Development of a Spatial Decision Support System for Assessment and Management of Combined Sewers

Lindell Ormsbee[1]

Abstract

A spatial decision support system is developed for use in assessing and managing CSO sites for the state of Kentucky. The proposed system contains four separate components: 1) a graphic development component, 2) a graphic display component, 3) a database component, and, 4) a model component. The proposed environment has been developed to provide decision makers with a PC-based platform for CSO management as well as an interface with the national EPA PCS (Permit Compliance System) Database. It is expected that the developed environment will provide the Kentucky Division of Water with a comprehensive framework for the assessment and management of CSO discharges in the state of Kentucky.

Introduction

Kentucky currently contains approximately 341 CSOs associated with 27 separate wastewater treatment systems. The majority of the CSOs are located in those cities which are located along the Ohio River. In 1990, the Kentucky Division of Water (DOW) prepared and submitted to EPA the Kentucky Combined Sewer Overflow Control Strategy, which established a uniform, statewide approach to developing and issuing KPDES permits for Combined Sewer Overflows (CSOs). Since that time, the program has been implemented as each municipality and sanitation district renews their old KPDES permits. The purpose of the permit program is to 1) insure that any CSO is a result of wet weather flow only, 2) bring all wet weather CSOs into compliance with technology-based requirements and applicable water quality standards, and 3) minimize water quality, aquatic biota and human health impacts due to wet weather overflows.

[1]Associate Professor, Department of Civil Engineering
University of Kentucky, Lexington, Kentucky 40506-0281

The CSO permit program involves two separate phases. The first phase of the program requires each municipality or sanitation district to 1) identify the receiving water for each overflow, 2) update the existing sewer use ordinance so as to prohibit the construction of any new combined sewers and minimize any new flows into existing combined sewers, and 3) develop a comprehensive Combined Sewer Operational Plan (CSOP) for the system. The second phase of the program involves the implementation of a Combined Sewer Operational Plan (CSOP) so as to reduce the total loading of pollutants entering receiving streams from the combined sewer system.

Continued maintenance of a CSO permit will require a significant amount of data. The associated data will include: 1) physical and geographical data associated with the CSOs, 2) data with regard to the CSOP, and 3) data resulting from required stream and CSO sampling.

National PCS Database

As part of the regular KPDES permit, a permittee is required to submit monthly or quarterly reports to the Kentucky Division of Water. The frequency of the reporting is dependent upon the size of the system. These reports have been customized for each system and include a wide range of water quality parameters. Upon receipt of the report, the data is processed by Kentucky Division of Water personnel and entered into the National PCS (Permit Compliance System) Database. PCS is a national database developed and managed by the Environmental Protection Agency for use in storing and reviewing all data associated with the NPDES permit program.

With the advent of the CSO permit program, it is anticipated that the PCS database will be expanded to accommodate the additional data associated with the CSO permits. Data associated with individual CSOs will be entered into the database using two different types of ASCII data records. The first data record will be used to record physical and geographical data associated with the CSO. The second data record will be used to record the results from water quality monitoring.

At the present time it is anticipated that data associated with the CSO permits will be processed in much the same way as the regular KPDES permits. That is, each permittee will submit a written report containing the relevant data for the associated system. Once received the data will be evaluated and then input manually into the PCS database.

Kentucky DOW CSO Database

In order for the Kentucky Division of Water to better manage and analyze the resulting CSO data, a decision support system is desired that can provide a link between a relational database and a graphics display environment. In addition to providing the Kentucky DOW personnel with a comprehensive planning and analysis tool, the associated environment could also be structured to provide a convenient link with the existing PCS Database. Given such a database, the Kentucky DOW could maintain a more powerful and comprehensive CSO database in house while at the same time having a link with the PCS environment.

Decision Support System

In order to manage and analyze the data associated with the CSO permit program, a decision support system is being developed for use by the Kentucky Division of Water Personnel. Because of the spatial nature of the collected data, a decision was made to integrate a Geographic Information System (GIS) into the framework of the Decision Support System. The proposed framework is illustrated in Figure 1. As can be seen from the figure, the system may be accessed from windows and provide a direct linkage to the PCS environment through a common ASCII file database. It is recommended that the actual environment contain four separate components: 1) a graphic development component, 2) a graphic display component, 3) a database model, and 4) an analysis model.

The graphic development component is used to create the initial map images and associated database linkages. Based on an examination of several GIS programs, it was recommended that ArcCAD™ be used for the graphic development package. ArcCAD™ is a geographical information system engine that runs within AutoCAD. One of the main advantages of ArcCAD™ is that it can create GIS coverage files that are compatible with ARC/INFO™.

The graphic display component is used to view the spatial coverages of CSO locations and provide graphic access to the associated relational database. Based on an examination of several packages a decision was made to recommend the use of ArcVIEW™ II for the graphic display program. ArcView™ II is a program for use in displaying and querying the contents of a spatial database. The program can be used to display all or part of the database contents and perform some simple spatial analysis functions such as spatial measurement and spatial query.

The database model recommended for use in the decision support system is dBASE IV™. This program was recommended because it can be used to edit and update the database files associated with ArcCAD™ and ArcView™ II. In addition to being able to examining the existing database, dBASE IV™ permits the construction of reports or ASCII files of selected database parameters for use with linkage to the PCS environment.

In order to provide the decision maker with some capability to investigate the impacts of particular storm events on CSO events, some type of analysis tool is recommended. One of the most useful tools would be a simple spreadsheet program. By use of a spreadsheet, general ASCII files could be read in or exported from the ASCII file database for receipt or transmittal to the other system components. Data imported into the spreadsheet program could be analyzed using the built in functions of the spreadsheet and then displayed graphically using the associated graph utilities.

Final Operating Environment

As currently envisioned, data to be submitted for use in monitoring compliance with each CSO permit will be prepared in a specified format and sent to the state EPA on a standard computer diskette. Once received, the information from the diskette will be processed and uploaded to the decision support system. One in this environment, the information may be stored, analyzed and displayed, Once the information has been analyzed, it may then be transferred back across the ASCII file database and uploaded in the PCS system.

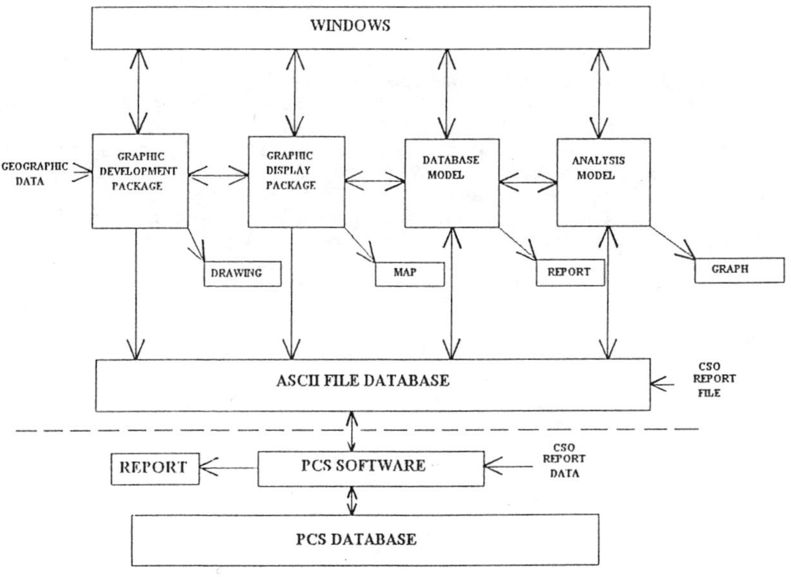

Figure 1. CSO Decision Support System

USE OF UNCERTAINTY ANALYSIS IN WATER QUALITY MODELS

Amy L. Shallcross[1], Gary W. Mercer[2] M.ASCE,
Linfield C. Brown[3], and Patrick J. Reidy[4]

Abstract

Uncertainty analysis was performed on the water quality of the Chicago Waterways and Upper Illinois River. The uncertainty analysis options provided by the UNCAS portion of EPA's water quality model, QUAL2EU, was used. Uncertainties generated by the variability of model inputs were compared with the standard deviations of instream water quality measurements. The uncertainty analyses indicated which inputs had the greatest effect on predicted water quality and that the uncertainty in the input variables create about the same variability in predicted water quality as that of sampling data.

Introduction

Uncertainty analysis is used to examine the variability of predicted water quality values from models such as QUAL2EU.

[1] Water Resources Engineer, Camp Dresser & McKee, Edison, New Jersey.

[2] Senior Water Resources Engineer, Camp Dresser & McKee, Cambridge, Massachusetts.

[3] Professor of Civil and Environmental Engineering, Tufts University, Medford, Massachusetts.

[4] Water Resources and Hazardous Waste Engineer, Camp Dresser & McKee, Cambridge, Massachusetts.

The complex interactions and stochastic nature of water quality variables prevent complete confidence in deterministic solutions of a system's water quality. Variabilities or errors in model inputs contribute to uncertainty of calculated output variables. Analysis of the uncertainty of predicted values is necessary to determine the reliability with which planners can use predicted water quality values. QUAL2E-UNCAS (QUAL2EU) provides three means to evaluate the uncertainty of model results: sensitivity analysis, first order error analysis (FOEA) and Monte Carlo simulations.

Study Area

To examine the uncertainty options available through QUAL2E-UNCAS, a model of the Chicago Waterways and Upper Illinois River Systems were evaluated. The model was developed by the Metropolitan Water Reclamation District of Greater Chicago (MWRDGC)and Camp Dresser and McKee, Inc. (CDM), in 1991. The models simulate a total of 200 miles of riverine water quality. Water quality samples, for calibration and verification of the model, were obtained by the Illinois State Water Survey (ISWS). A total of six 48-hour sampling rounds of data, 2 sampling rounds in three seasons (fall, spring, summer), were collected by ISWS.

First Order Error Analysis

FOEA uses a first order approximation to the relationship for computing variances in multivariate systems. Inputs are assumed independent and the model is assumed to behave linearly. QUAL2EU requires the specification of coefficients of variation and a percentage input perturbation. Coefficients of variation were calculated from field data, where possible (concentrations, flows). Default or literature values were used for rate parameters. A reasonable input perturbation for sensitivity coefficients was chosen to be 5 percent. QUAL2EU outputs normalized sensitivity coefficients (representing percent change in output resulting from one percent change in input) and components of variance (percentage of output variance due to each input

Sensitivity coefficients from the FOEA of the Chicago Waterways model indicated that water quality predictions were most sensitive to temperature, flow depth, and velocity. The Upper Illinois River model indicated that water quality predictions were sensitive to headwater loads, temperature, depth and velocity. Coefficients of variation for dissolved oxygen predicted by the FOEA are lower than those from sampling data, indicating that the model predicts dissolved oxygen with less variability than encountered from sampling. Several constituents had coefficients of variation that differed. Predicted coefficients of variation for orthophosphate were lower for the Chicago Waterways and higher for the Upper Illinois River. The over prediction on the Upper Illinois was associated with the orthophosphate settling rate, which was set higher than usual to account for phosphorus settling. Ammonia and Nitrate coefficients of variation were under predicted by FOEA for the Upper Illinois River. Higher observed coefficients of variation for ammonia and nitrate were due to low concentrations (essentially at detection limits of analytical test procedures) of ammonia in the Upper Illinois River. Actual coefficients of variation for chlorophyll-a (algae) are higher than predicted. The dynamic nature of algal growth is affected by sunlight variability, which is not completely simulated in QUAL2E.

Monte Carlo Simulations

Monte Carlo simulations allow the examination of changes in output with multiple changes of input parameters. Given the variability of input parameters, determined by coefficients of variation and a probability distribution function (p.d.f. - uniform, normal, lognormal), many different combinations of model input are sampled and used to generate water quality predictions. Thousands of simulations are performed in which the model samples all parameters from their respective distributions and recalculates the water quality for several locations of interest in the system.

In the Chicago Waterways and Upper Illinois River model, lognormal p.d.f.s were assigned to all but two input parameters to limit the chances of sampling a negative input value in the Monte Carlo simulations. For dissolved oxygen, the water quality constituent generally of most concern , Monte Carlo analysis indicated that model output variability was comparable to the sampling variability. Error of the predicted dissolved oxygen concentrations was approximately +/- 0.75 mg/l, while error of measured dissolved oxygen concentrations was about +/- 0.5 mg/l. Variances derived from Monte Carlo simulations were generally greater than instream concentration variances. Monte Carlo simulations produced higher variances because (1) the waterways behave more as a dynamic system, rather than steady-state, as modeled; (2) "snapshot" sampling may more closely reflect error in sampling techniques or experimental protocols than the system's true variability; (3) grouped coefficients of variation, required by QUAL2EU, did not reflect the small variability of water reclamation plant loads, which contribute the majority of flow to the waterways and (4) improbable combinations of model input can be sampled (i.e. the inputs are not independent).

Conclusions

The uncertainty analysis options, provided in QUAL2E-UNCAS, provide a valuable tool for assessing the predictive capability of QUAL2E. First order error analysis enables the modeler to determine which input variables have the greatest effect on predicted water quality. The modeler, with estimates of input parameter uncertainty, can use Monte Carlo simulations to determine the range of uncertainty of predicted water quality.

Sediment Quality-Based Point Source Permitting[1]

Gregory W. Currey, Associate Member, ASCE[2]

Abstract

Point source permits developed under the National Pollutant Discharge Elimination System (NPDES) program are a principal means of abating and controlling sources of sediment contamination of waters of the United States. The U.S. Environmental Protection Agency (EPA) is investigating uses of sediment assessment methods in the NPDES program as part of a strategy for addressing contaminated sediments. EPA has conducted preliminary field work to validate procedures for developing chemical-specific, sediment quality-based waste load allocations and permit limits for non-ionic organic compounds and heavy metals. These procedures are based on the same assessment methodologies used to develop proposed chemical-specific sediment quality criteria.

Contaminated Sediment Management Strategy

EPA released its proposed Contaminated Sediment Management Strategy on August 31, 1994 (59 Federal Register 44880). The Strategy describes EPA's understanding of the extent and severity of sediment contamination; the cross program and inter-agency framework for addressing sediment contamination; and actions needed for consideration and reduction of risks posed by contaminated sediments. One of the Strategy's primary goals is to prevent ongoing sediment contamination that may cause unacceptable environmental or human health risks (EPA, 1994a).

Point source permits developed under the NPDES program are a key method for preventing sediment contamination. The Contaminated Sediment

---

[1]The views presented in this paper are of the author and do not necessarily reflect the views of the U.S. EPA or any other Federal agency.

[2]Gregory W. Currey, Civil Engineer, U.S. Environmental Protection Agency, 401 M Street, SW (4203), Washington, DC 20460

Management Strategy outlines actions that the NPDES program may take to address abatement and control of point sources of sediment contamination. One key action is to develop a sediment quality-based permitting procedure that applies the aquatic life sediment quality criteria (SQC) being proposed by EPA for the purpose of sediment quality assessment (EPA, 1994a).

Sediment Quality Criteria

EPA's aquatic life SQC are the Agency's best recommendation of the concentration of a substance in sediment that will not unacceptably affect benthic organisms. EPA published proposed criteria in the Federal Register for five non-ionic organic compounds (acenaphthene, dieldrin, endrin, fluoranthene, and phenanthrene) in February 1994 and will use an Equilibrium Partitioning Approach (EqP) to develop additional SQC for non-ionic organics. EPA is proposing an EqP approach for establishing sediment quality criteria for five heavy metals (cadmium, copper, nickel, lead, and zinc) as well (EPA, 1994b).

It will likely be several years before most states begin adopting SQC into their water quality standards. EPA's water quality regulations, however, clearly authorize states and EPA to implement permit limits based on an interpretation of state narrative criteria (e.g., no discharge of toxics in toxic amounts, no discharge resulting in objectionable sediment deposits). Thus, if a permitting authority finds that a narrative state water quality standard is exceeded in whole or in part due to a particular point source, the permitting authority has a basis for controlling a discharge of the problem pollutant using NPDES permit limits. In some cases, these limits could be based on proposed SQC.

Sediment Quality-based Permitting

The ability to identify point sources of sediment contamination and to determine what decreases in pollutant loadings are needed to protect sediment quality is essential to meeting sediment quality criteria through effluent limitations. A number of existing water quality models, such as WASP4/TOXI4, include a sediment component and, with the proper ambient and loading data, can estimate toxic chemical concentrations in sediments. These models typically require a thorough knowledge of water quality modeling and substantial amounts of data to use. Permit writers generally lack the modeling experience and data necessary to successfully develop wasteload allocations using complex water quality models. To be successful in implementing sediment quality-based permitting, permit writers must have at their disposal a somewhat simplified approach to sediment quality modeling.

To meet the need for a sediment quality-based modeling procedure that is sophisticated enough to be scientifically sound, yet simple enough for routine use by permit writers, EPA is supporting test applications of its SMPTOX3E model. SMPTOX3E is a one dimensional, steady-state, mass balance model that predicts particulate- and dissolved-phase toxic chemical concentrations in the water column and bedded sediments based on loadings data.

Initial Test Application of SMPTOX3E

EPA applied the SMPTOX3E model to an industrial site near Lake Charles, Louisiana. A facility at the site discharged organic chemicals (hexachlorobenzene, hexachlorobutadiene, hexachloroethane, and 1,2,4-trichloroethane) through an effluent/cooling water canal into a bayou. The purpose of the field application was to test the ability of SMPTOX3E to describe chemical concentrations in the water column and bedded sediments resulting from effluent discharges. The model required site-specific information, including flow and loading input, physical transport parameters, and information on kinetic processes. Field data were available for effluent and ambient toxic chemical concentrations, although these data were not acquired specifically to support the model test application.

Given data uncertainties and assumptions in the modeling framework, the results were reasonably consistent with both water column and sediment observed concentrations in the bayou and with water column observed concentrations in the canal. There were, however, large inconsistencies between model results and observations of toxic chemical concentrations in the canal sediments (LTI, 1993). Two hypotheses, based on anecdotal data, may explain the discrepancy between the model results and observed concentrations in the canal sediments.

The first hypothesis is that there may be an undocumented ground water source of chemicals to the canal. Anecdotal reports indicate that historically large quantities of wastes were landfilled near the canal. Given that large amounts of water are flushed through the relatively small canal, a significant ground water loading could elevate sediment concentrations substantially without noticeably affecting water column concentrations. An analysis of the model showed that water column concentrations in the canal would be quite insensitive to chemical concentrations in the bedded sediments (LTI, 1993).

The second hypothesis for explaining the disequilibrium in the canal is that the observed sediment concentrations in the canal have not had sufficient time to recover from historical wastewater loadings, which reportedly were orders of magnitude higher than present levels. This hypothesis could not be confirmed because no historical loadings data were available (LTI, 1993).

Conclusions from the Test Application

EPA drew two major conclusions from the SMPTOX3E test applications at the Lake Charles site. First, any future test applications and accompanying data collection should focus on attempting to validate the model process mechanisms. For example, emphasis should be placed on validating that the model correctly reflects the controlling sediment partitioning mechanisms and provides a reasonable representation of water column responses to a range of hydraulic conditions. Characteristics of the site and available data should determine whether steady-state and/or time variable models are most appropriate for this validation. A time-variable model is appropriate when there is temporal disequilibrium between external loadings and sediment concentrations.

Second, in the Lake Charles test application, steady-state model results were consistent with observed data for areas where chemical concentrations were assumed to be in equilibrium with present loads. Consequently, steady-state models, such as SMPTOX3E, are well suited for effluent permitting purposes where the objective is to determine the long-term average effluent loading required to maintain sediment quality. Any steady-state model used for permitting purposes should be based on the same scientific principles and should contain the same process mechanisms as those models field-validated in more extensive test site applications.

EPA wants to continue field validation of SMPTOX3E by incorporating metals partitioning equations into the model and testing it against field measurements of metals in sediments in the Blackstone River in Massachusetts and Rhode Island. Preliminary model results using existing data from the Blackstone River were generally consistent with observed water column concentrations for five heavy metals during three dry-weather surveys. Once SMPTOX3E is properly validated, it should become a valuable tool for permit writers who want to ensure that NPDES permits prevent ongoing sediment contamination.

Appendix: References

LTI-Limno-Tech, Inc. (1993). *Summary of Conclusions and Recommendations from Test Site Application of SMPTOX3E Model to Hydrophobic Organic Chemicals at PPG Industries, Lake Charles, Louisiana.* LTI-Limno-Tech, Inc., Ann Arbor, MI.

U.S. Environmental Protection Agency. (1994a). *Proposed Contaminated Sediment Management Strategy.* U.S. Environmental Protection Agency, Washington, DC.

U.S. Environmental Protection Agency. (1994b). *Briefing Report to the EPA Science Advisory Board on the Equilibrium Partitioning Approach to Predicting Metal Bioavailability in Sediments and the Derivation of Sediment Quality Criteria for Metals: Volume 1.* U.S. Environmental Protection Agency, Washington, DC.

Water Quality Modeling and Biological Impact Assessment in Support of NPDES Permitting

Donald Galya[1],M.ASCE, Steven Wolf[1],M.ASCE, James Bowen[1],M.ASCE, Kevin Johnson[1], Steven Truchon[1], and Roy Piatelli[2]

Introduction

The General Electric Company facility in Lynn, Massachusetts (GE River Works) is a manufacturing plant that produces aircraft engines and submarine gears. Several of the plant operations require non-contact cooling during engine and gear testing activities. Also, in order to the facility's electric power demands, GE River Works operates a steam-electric power plant with a once-through non-contact cooling system. GE River Works maintains a total of 13 outfalls with thermal, non-contact cooling discharges to the estuarine portion of the Saugus River. Other discharges from the facility to the Saugus River consist primarily of storm water runoff.

Since the Saugus River is a designated Massachusetts Outstanding Resource Waterbody, recent National Pollutant Discharge Elimination System (NPDES) repermitting activities included the completion of an Antidegradation Study in accordance with state regulations. The purpose of the Antidegradation Study was to demonstrate that the facility discharges are provided with the best practical method of treatment, and the level of treatment at the facility is adequate to protect and maintain the existing uses and applicable water quality standards for the river. The study considered potential impacts of facility discharges on both water quality and aquatic biota.

Discharge Reductions

Pollutant loadings from GE River Works to the Saugus River have been reduced over the past several years due to

[1]ENSR Consulting and Engineering, 35 Nagog Park, Acton, MA
[2]General Electric Company, 1000 Western Avenue, Lynn, MA

the installation of wastewater treatment systems; and the reduction, elimination and consolidation of facility operations. Wastewater treatment improvements at the facility consist of the installation oil/water separation systems on outfall drain lines with stormwater runoff and the installation of closed loop cooling systems on three outfall drain lines. The closed loop cooling systems effectively eliminate cooling water discharges from these three outfalls. In addition, several other wastewater streams have been eliminated or reduced by changes in facility operation.

Impact Studies

The thermal component of the GE River Works facility discharges have the most potential for aquatic effects on the Saugus River and therefore the impact analysis focused on these discharges. The impact studies consisted of thermal studies, followed by a biological impact analysis. Detailed, intensive studies were performed at five outfalls where discharge temperatures greater than 32.2°C (90°F) were proposed. The thermal studies included field surveys, and far-field and near-field predictive thermal modeling.

Thermal Surveys

Field thermal plume surveys were performed for the five outfalls selected for detailed analysis. The purpose of the surveys was to delineate the three-dimensional characteristics of the thermal plumes and to obtain sufficient data such that near-field plume models could be developed and calibrated. The surveys consisted of temperature and salinity measurements at approximately one meter depth intervals at several locations within the discharge plumes. Measurements were obtained at high water slack and low water slack tides in order to cover potentially critical tidal phases. Concurrent information on facility operating and discharge characteristics, such heat load, discharge flow rate, and inlet and outfall temperatures were also obtained.

Far-Field Modeling

Far-field modeling was performed in order to evaluate the overall temperature increase in the Saugus River due to the thermal discharges from the GE River Works facility and other heat sources to the river. This modeling study was performed using a one-dimensional, finite difference, transient heat balance model. The model incorporates consideration of freshwater flushing, tidal transport and flushing, and atmospheric heat loss.

Input parameters for the far-field model include the geometric characteristics of the Saugus River, tidal height variations at the open ocean boundary, freshwater inflows at the upstream boundary, heat (or mass) source discharges, ambient temperatures (or concentrations) at the upstream and downstream boundaries. General geometric information on the Saugus River was obtained from a National Oceanic and Atmospheric Administration chart. Calibration of the model was performed using salinity and other data from studies performed by the U.S. Army Corps of Engineers.

A series of model simulations representing average to worst case conditions were performed using a range of estuary hydraulic conditions and discharge thermal loadings. Average estuary hydraulic conditions consisted of average freshwater inflow and tidal amplitude. Worst case conditions consisted of minimum values for inflow and tidal amplitude. Discharge conditions ranged from (1) simulation of only the continuous GE River Works discharges at average heat loading to (2) simulation all thermal discharges on the river with both intermittent and continuous GE discharges at maximum heat loading. The results of these simulations indicate that the maximum temperature increase in the river would range from 0.6°C (1.0°F) for average conditions to 0.8°C (1.5°F) for worst case conditions.

<u>Near-Field Modeling</u>

Near-field plume modeling was performed in order to predict the detailed three dimensional temperature distribution for the five outfalls selected for detailed analysis. Two primary models were used for this study. For the power plant discharge, a hydrothermal jet plume model was used. This model considers jet momentum, plume buoyancy, dilution due to entrainment of ambient water, and plume bending due to ambient currents. The ability of the model to accurately simulate the system was verified using two field plume survey data sets. Conditions existing at the time of the surveys, such as the ambient water temperature and salinity, discharge flow rate, discharge temperature, tidal phase and water level, discharge aspect ratio (width/depth), discharge/ambient velocity ratio, and densimetric Froude number were input to the model.

For the remaining four outfalls selected for detailed analysis, the hydrothermal jet model was not appropriate due to the relatively low densimetric Froude number of the discharges. Instead, a three dimensional, semi-analytical advection-dispersion was used. This model considers advection, parameter decay or heat transfer, and three dimensional dispersion. By incorporating a source term area, the initial mixing of low momentum discharges can be simulated. For this study, the dispersion coefficients and source term area characteristics were calibrated and verified using two independent data sets for each outfall. Subsequently, the calibrated models were used to predict the three dimensional temperature distribution for each outfall for high water slack and low water slack tidal conditions.

The near-field models provided detailed plume temperatures above background temperatures. In order to determine the total increase in temperature above ambient, the far-field temperature increase was added to the near-field plume temperature values. Then, worst-case absolute temperature plumes were predicted by adding the estimated maximum summertime ambient temperature to the total temperature increase values. The study results indicate that, even during maximum thermal discharge and worst-case conditions, only a relatively small area of the Saugus River would exceed the Massachusetts water quality standard for maximum temperature.

Biological Impact Analysis

A biological impact analysis was conducted to evaluate the potential thermal effects of facility discharges on the biological community of the Saugus River. This analysis considered 14 representative taxa/lifestages (5 planktonic, 6 benthic, 3 pelagic). Both localized thermal tolerance and large scale effects were assessed. The analysis indicated that there is little potential for thermal tolerance impacts within the main body of the Saugus River. There is a small potential for limited and localized effects within a salt marsh channel for one of the outfalls but the results of biological field surveys did not indicate the presence of any measurable effects in the vicinity of the outfall.

Large scale thermal effects were evaluated by determining the potential for impacts on fish migration. Rainbow smelt and alewifes were considered in this analysis. The results of the analysis indicated that predicted temperatures in the river are not high enough to adversely effect migration and that migration is presently occurring, apparently unaffected by the thermal discharges.

Conclusion

An Antidegradation Study performed as part of the NPDES repermitting of the GE River Works facility consisted of thermal and biological impact analyses of facility discharges. The study results indicated that the facility discharges do not have any significant thermal or biological impacts. In addition, in recent years the facility has substantially reduced pollutant loadings, thus minimizing the potential for water quality or biological impacts.

Risk Management Analysis for Discharge of Polluted Overflows Into Receiving Waters

David L. Westerling, Member[1]

Abstract

Capital expenditures associated with industrial waste treatment works may be a significant portion of a company's financial liabilities. To date there has not been any significant advances in establishing a comprehensive model for evaluating significantly different industrial waste treatment alternatives. This paper attempts to lay the foundations for a comprehensive evaluation model which establishes and evaluates all the factors involved with the selection of a given technology or processing system. The model encodes the corporate risk profile and indexes the regulatory severity to establish the base rate evaluations of the different alternatives. The model allows the user to treat all inputs as variables and evaluate ranges of options both singularly and in combination to provide adequate confidence in selected values and to establish the output sensitivity of single and groups of variables. The systematic approach the model provides can be applied to a variety of industries and with simple modifications could most likely handle any industrial requirement.

Introduction

The intent of this paper is to provide a risk assessment format which can be used to evaluate a variety of industrial treatment alternatives. The specific industry chosen to illustrate these progressive techniques was a Rolling Mill Industry located on the Blackstone River, Massachusetts.

The model developed evaluates treatment systems with

---

[1]Associate Professor, Department of Civil Engineering, Merrimack College, North Andover, Massachusetts 01845

respect to capital, treatment efficiency, regulatory restrictions, compliance and deviant effluent detection probabilities. The model illustrates monetary exposure based on capital expenditures and level of treatment for given effluent volumes. This approach allows the company to tailor the model to provide a complete compliance cost structure with there own risk preference encoded into the data evaluation. This is a comprehensive model which establishes potential cost/exposure associated with a firms waste stream within a given environmental reporting system. This technique provides the assessment of several treatment alternatives with varying annual costs, ranges of treatment levels, differing points of discharge and different receiving systems. The model uses mass balance, effluent concentrations, treatment capabilities, regulations, historical environmental conditions and liability exposure to compute associated costs.

Background

The industrial waste stream which will be dealt with in this paper is from a non-ferrous rod rolling mill. The figures were adapted from Rod Mill in Chile. A rod rolling mill or rolling mill processes metal as the name suggests by rolling the product between carbide or steel rolls into the desired shape (usually round). Billets of metal are heated to a prescribed temperature and driven through a series of rolling stands each one further reducing the size of the billet until it reaches the desired rod or wire size. The rolling process generates heat as the metal is worked and as the cross section is reduced the product traveling speed increases. This requires rapid cooling of both the product traveling through the mill as well as the processing rolls and mill equipment.

The cooling process historically uses surface water from a river or lake and more recently municipal water to cool the product and equipment. The combined process water flow from this representative mill was 26,300 l/min. containing 28.5 kg/hr of fine scale. The Temperature ranged from 43 degrees C to 49 degrees C with a PH from 7.0 to 9.0 and a FOG measure 25 mg/l. When Copper is run through the mill the process waters carry the copper in the form of dissolved and suspended solids. These process waters must be treated before discharge to remove the elevated levels of copper.

Probabilities

The probability of polluting the receiving waters is a combination of the concentration of effluent from the mill and the volume of the receiving water. In the case of a moving stream or river this volume or flowrate is dependent on the natural phenomena of rainfall and runoff.

The first attempts to investigate probabilities of the natural flood events were made in the area of rainfall data. The U.S. Department of Commerce Weather Bureau (now known as the National Office of Oceanic and Atmospheric Information (NOAA) published a series of technical papers (TP-29,TP-40) on Rainfall Intensity-Frequency Regimes. There remained the problem of applying this data to our own situation on the Blackstone River.

A second attempt to quantify probabilities was to look at runoff data from streamgages. This data provides the probability of a certain rate of runoff for a given frequency. The "Runoff Volume-Duration-Probability Analysis of selected watersheds in Massachusetts was used. Given that our drainage area was 139 square miles, the data from a neighboring watershed at Westville, Massachusetts with a drainage area of 93.8 square miles was used and applied by direct proportion.

With the runoff volume probability data a cumulative distribution function (CDF) was constructed using a Monte Carlo Simulation. The discharge probability data was entered and a random number was assigned using the RAND function in an Excel spreadsheet. For example:

```
IF(RAND()<=0.01,1600,
IF(RAND()<=0.04,1361,
. . .
3IF(RAND()<=0.99,299,299)))))))
```

The Excel quickgraph was used to plot the sorted values.

A mass balance was then conduced on the limits of the range of probabilities (0.01-0.99). For example the mass balance from the treatment plant was:

$$C_{TP}=\frac{C_{OV}Q_{OV}+C_{RS}Q_{RS}}{Q_{TP}}$$

where flow is in million gallons per day (MGD) and

Ctp = concentration from treatment plant (varies)
Qtp = discharge from treatment plant (55 MGD)
Cov = concentration from overflow (varies)
Qov = discharge from overflow (9.5 MGD)
Crs = concentration from rest of system (0.1 ppm

Cu)

Qrs = discharge from rest of system (45.5 MGD)

The concentrations were calculated for each scenario and input into the decision tree with a limit of copper at 12 ug/L or 0.012 parts per million (ppm).

Costs

To evaluate the costs of the alternatives available for disposal of plant effluent, two cost components were used:

A. the initial construction costs,

B. the annualized cost of construction and the annual operating and maintenance (O&M) costs.

A secondary cost is the disposal fee charged by a receiving publicly owned treatment works (POTW), if the effluent is directed there.

Results and Conclusion

The model discussed in this paper allows the integration of dissimilar information from varying sources into a complete and comprehensive analysis package. The spread sheet format provides simple variable adjustment through six major sensitivity nodes: the capital cost, governing body, level of treatment, effluent concentration, likelihood of discovery and corporate, civil, personal and punitive fines and damages. The model results were presented at the conference and are available from the author.

References

Clemen, Robert T., Making Hard Decisions, PWS Kent, 1990.

Economist, Vol.315 Apr.90 p.38. (Ashland)

Means, Site Work Cost Data, R.S. Means Company, 1990.

U.S. Department of Commerce Weather Bureau, Rainfall Intensity-Frequency Regime, Technical Paper No. 29.

U.S. Soil Conservation Service, Runoff Volume-Duration-Probability Analysis, Central Technical Unit, Hydrology Branch, July 1965.

WATER QUALITY MANAGEMENT OF AMMONIA UNDER UNCERTAINTY

Corinne L. Wotton[1] and Barbara J. Lence[2]

Abstract

This paper investigates the importance of uncertain input parameters for ammonia management models and different ambient ammonia criteria. The approach uses a modified Generalized Sensitivity Analysis to determine the significance of each uncertain parameter in a simulation and optimization model. The analysis is demonstrated for the management of ammonia on the White River in Washington State. Results indicate that, for this case study, the river pH and temperature, the nitrification rate, and the midstream flow are the most important parameters under the current recommended criterion. The significance of the nitrification rate and the upstream flow are shown to be dependent on the value of the ammonia criterion.

Introduction

Uncertainties in environmental models are due to poor mathematical representation of the natural system and to inaccurate input data. For an ammonia management model, management decisions are also affected by uncertain ammonia criteria or stream standards. The ammonia criteria recommended by the United States Environmental Protection Agency (USEPA) in recent years have undergone significant changes. There are currently two criteria, one for acute toxicity and one for chronic exposure, that are empirically based functions of the pH and temperature of the water body. Due to their empirical basis, these criteria are expected to change in the future as new information regarding the hazardous effects of ammonia is collected. This work acknowledges that random variables such as streamflow, pH, temperature, and reaction rates complicate the application of ammonia management models and that the ammonia criteria are also uncertain. A modified Generalized

---

[1] Graduate Student, Department of Civil and Geological Engineering, 15 Gillson Street, University of Manitoba, MB, Canada, R3T 5V6.

[2] Associate Professor, Department of Civil and Geological Engineering, 15 Gillson Street, University of Manitoba, MB, Canada, R3T 5V6.

Sensitivity Analysis (GSA) is applied to an ammonia management model to evaluate which input parameters are important in terms of their effect on the management decision. The modified GSA is conducted using a range of ammonia criteria to evaluate whether the significance of the parameters change under different criteria.

Generalized Sensitivity Analysis (GSA) for Ammonia Management

GSA was developed by Spear and Hornberger (1980) to investigate the sensitivity of water quality system behaviour to uncertain input parameters. The GSA developed by Spear and Hornberger is composed of three elements, namely, a Monte Carlo Simulation (MCS), a classification algorithm, and a statistical analysis. The MCS randomly draws input parameter values from statistical distributions based on literature and preliminary field data. These randomly drawn parameter values are combined with other known information and are input into a simulation model. The classification algorithm determines whether or not a MCS response mimics the known system behaviour, or actual water quality conditions, determined from sampling programs. If a MCS response mimics the known system, then the output and corresponding parameter set is classified as a Behaviour. Conversely, if the MCS response does not mimic the known system, then it is classified as a Non-behaviour. The final component of the GSA is a statistical analysis of the Cumulative Distribution Functions (CDFs) of the Behaviour and Non-behaviour classes for each parameter. If the CDFs are significantly different, then the parameter is considered important in simulating acceptable system responses. The Kolmogorov-Smirnov test statistic, the $d_{m,n}$, measures the maximum separation of the CDFs for a given parameter and indicates the sensitivity of the system response to that parameter.

Lence and Takyi (1992) modify the GSA technique for combined simulation and optimization models. The simulation model evaluates water quality in a river based on random values of uncertain parameters and known information. The optimization model determines the optimal allocation of waste while ensuring that a water quality criterion is met. The modification to the original GSA occurs in the classification algorithm. The modified classification algorithm classifies the optimization model output either as Behaviours or Non-behaviours depending on whether the model objective function value is similar to that which would result if the input parameters were known with certainty, i.e., based on a deterministic scenario of the input conditions.

This work demonstrates the modified GSA technique for the management of ammonia in a river. The simulation model evaluates the impact on instream ammonia concentration at locations along the river caused by a mass of ammonia released from each discharger. The optimization model evaluates the minimum uniform removal level required to meet the chronic un-ionized ammonia criterion. A Behaviour results if the MCS response yields a uniform removal level which is greater than the uniform removal level evaluated using the deterministic input data.

Case Study

The GSA for ammonia management is demonstrated for the control of ammonia along a 25 mile section of the White River in Washington State. Eight point source dischargers are located on the river along with a diversion, located at river mile 24.5, and two tributaries, located at river miles 23.3 and 3.5. The diversion withdraws a significant portion of the White River flow for use in power generation. Water quality data, discharger data, and flow data were collected from the Washington State Department of Ecology and the United States Geological Survey. Fourteen years of daily flow records from four gauging stations, one upstream of study area, one mid-stream in study area, and one on each tributary, are used. Water quality data consists of monthly and bimonthly measurements of temperature and pH for four monitoring stations. Period of records for these data range between one and 16 years. Ammonia loads vary between a maximum of 38.4 mg/l, from a municipal waste treatment plant, and a minimum of 0.41 mg/l, from a industrial waste treatment plant. Parameters considered uncertain include low flows at each gauging station, temperature of the main stem of the river and tributaries, pH along the main stem of the river, and the nitrification rate.

Inputs for the deterministic model include the seven-day-ten-year low flow, the 7Q10 flow, at each gauging station, the highest monthly-averaged temperature and pH for the river, and the nitrification rate of 0.45 day^{-1} taken from Pelletier (1993). For the MCS, probability distributions were fit to seven-day low flow data for each gauging station and to temperature values measured in August and pH levels recorded in October, which are the months with the highest monthly average of these values. These parameters fit two-parameter lognormal distributions. The nitrification rate was assumed to follow a uniform distribution with a maximum value of 9.0 day^{-1} and a minimum value of 0.0 day^{-1} based on literature cited in USEPA, 1985.

The modified GSA is applied for several scenarios for the water quality goal. The first scenario uses the chronic un-ionized ammonia criterion currently recommended by the USEPA. The remaining scenarios use a criterion that ranges from 60% to 200% of the current un-ionized ammonia criterion. This represents the feasible range of water quality for the White River.

Results

The magnitude of the $d_{m,n}$ for each parameter represents the degree of sensitivity of that parameter. The larger the $d_{m,n}$, the more sensitive the model response is to that parameter. In order to compare analyses under different ammonia criteria, the $d_{m,n}$ values are normalized by the 95% $d_{m,n}$ for each analysis. These normalized $d_{m,n}$s are the sensitivity indices. A sensitivity index greater than 1 means that the parameter is important in simulating acceptable system responses at 95% confidence. The relative values of the normalized $d_{m,n}$ for different parameters are used to rank the relative importance of the parameters.

Under the current un-ionized ammonia criterion, the pH, temperature, midstream gauging station low flow, and the nitrification rate are sensitive parameters. The pH has the highest sensitivity index (20.50), followed by the temperature (16.40), the midstream gauging station low flow (8.30), and the nitrification rate (1.05). It is expected that pH and temperature affect the model response since the concentration of un-ionized ammonia in the river is dependent upon the temperature of the water and its pH level. Interestingly, the sensitivity of the pH and the temperature rank higher than that of the flows. Historical records prove that much more time and resources have been spent recording flow data than water quality data along the White River. The midstream gauging station is the most significant flow input and is located after the major diversion along the White River. This indicates that the amount of diverted water is more important to water quality along the river than the upstream flow or the flow from the tributaries. Consequently, critical water quality sections exist along the White River after the diversion withdrawal point.

By investigating the sensitivity indices of the parameters under varying un-ionized ammonia criteria one can evaluate how the importance of the parameters may change when the criteria are changed. As the criterion is increased, the sensitivity indices of the pH, the temperature and the flow at the midstream gauging station remain approximately constant and the sensitivity index of the nitrification rate decreases and changes from significant to insignificant at a criterion level slightly higher than the current criterion level. This indicates that the nitrification rate is not important when the allowable instream un-ionized ammonia concentration is less constrained than the current criterion. The sensitivity of flow at the upstream gauging station usually remains insignificant, however there is one criterion level for which it becomes significant. Flows at the gauging stations on both tributaries remain insignificant in all scenarios.

References

Lence, B.J. and A.K. Takyi, Data Requirements for Seasonal Discharge Programs: An Application of a Regionalized Sensitivity Analysis, *Water Resources Research*, 28(7), pp. 1781-1789, 1992.

Pelletier, G.J., Puyallup River Total Maximum Daily Load for Biochemical Oxygen Demand, Ammonia, and Residual Chlorine, Washington State Department of Ecology, 1993.

Spear, R.C. and G.M. Hornberger, Eutrophication in Peel Inlet--II. Identification of Critical Uncertainties Via Generalized Sensitivity Analysis, *Water Research*, 14, pp. 43-49, 1980.

United States Environmental Protection Agency, Rates, Constants, and Kinetics Formulations in Surface Water Quality Modelling (2nd ed.), EPA/600/3-85/040, 1985.

United States Environmental Protection Agency, Memorandum: Revised Tables for Determining Average Freshwater Ammonia Concentrations, 1992.

Genetic Algorithms for Estimating Model Parameters

S. Mohan and D. P. Loucks, F. ASCE[1]

Abstract

Genetic Algorithms (GAs) are search procedures that can be used to find solutions to a variety of hydrologic, hydraulic and water resources problems, including those involving streamflow routing and water quality prediction. From a population of random possible solutions, the algorithm generates a succession of improved solutions using processes observed in natural genetics. These processes include probabilistic reproduction, crossover and mutation schemes. An elitist strategy passes the best solutions of one population generation on to the next generation. This paper briefly outlines a GA procedure and reports on its use for estimating parameter values of some linear and non-linear flow routing and water quality prediction models. The results suggest that genetic algorithms may be an efficient and robust means of calibrating a variety of models where measured time-series data are available.

Introduction

Being able to predict flows and quality constituent concentrations within surface water systems is an essential feature of many simulation models, especially those models that involve small time steps often needed when predicting the impacts of flood events or pollutant concentrations. There are a large number of hydrologic and hydraulic flow-routing and water-quality prediction models that can be used, and each have parameters whose values must be estimated for each specific river reach. The Muskingum method is one of the simplest and most commonly used routing methods, and usually requires the estimation of only two parameters. A more versatile routing model to be defined later requires the estimation of four parameters. Even the simplest water-quality models typically include numerous rate constant parameters for constituent growth and/or decay, "constants" that are dependent on the particular constituents being modeled, the temperature, and in some cases the flow and river-reach characteristics.

The reliability of these flow-routing and water-quality models is dependent, in part, on the accuracy of the parameter values of the models. These parameter values can be estimated by trial and error procedures (e.g., Cudsworth, 1989) if the parameters are few or by some fairly involved statistical procedures (e.g., Beck, 1983) if the parameters are many.

----

[1]School of Civil and Environmental Engrg., Cornell University, Ithaca, NY 14853

This paper reports the use of genetic algorithms (GAs) for estimating values of the parameters for a two- and nine-parameter Muskingum routing model and a four-parameter routing model together with a nine-constituent water-quality model, the later two included within an interactive river-aquifer simulation (IRAS) model (Loucks, French, and Taylor, 1995). The algorithms are robust in that they are independent of the particular model being calibrated, but they themselves have parameters whose values need to be identified.

Genetic Algorithms

Genetic Algorithms (GAs) are adaptive algorithms, used to solve optimization problems. They provide a means of finding (at least close to) the maximum or minimum values of some function. This objective function can be quite nonlinear, discontinuous, and even definable only by simulation models). Genetic algorithms (GAs) have been applied successfully to a variety of optimization and machine-learning applications. The basic principles of GAs are well described in the current literature (e.g., Goldberg, 1989). An early practical engineering example of the use of a GA is Goldberg's study of a system of 10 pipes and 10 pumping stations. Wang (1991) has used a GA for calibrating a conceptual rainfall-runoff model. Simpson et al. (1994) have used a GA for pipe network optimization and reported improved results in comparison with the traditional optimization techniques.

There are many variations of genetic algorithms. They all strive to find the set of variable values that produces the best (maximum or minimum) value of the chosen objective function. One of the more common variations of the algorithms proceeds as follows. A population of sets of randomly-generated variable values for the given problem is created. (The size of this population is one of the GA parameters.) The variable values contained in each of these sets are then substituted into the objective function, and the value of that objective function is computed for each set of variable values. Clearly some sets of variable values will be better than other sets of values. Next, a new population of sets of variable values are "reproduced" or copied from the original population. The probability that a set from the original population will be reproduced or copied into the new population is dependent on how well the variable values satisfied the objective function, at least compared to the best objective function value that was found. Hence the new population will have more of the better sets and less of the worst sets of the old population.

The variable values in each set of the new population are combined into a string (a succession of numbers), and these series of numbers are usually encoded into a binary (0,1) form. Each binary string is analogous to a chromosome containing genes (0,1 bits). These sets of strings are then paired, and each pair may then exchange and/or alter some of their bits so that their values are changed. They do this using crossover and mutation operations. (The probabilities of a crossover and a mutation are two additional parameters of the GA). Crossover involves splitting a string into two parts at a randomly generated crossover point and recombining it with another string that has been split at the same crossover point. Thus for example, a pair of strings having two numbers each, say 4 and 2 and 3 and 5, represented as 01000010 and 00110101, after crossover of the last two bits would be 01000001 and 00110110, or, in decimal values, 4 and 1 and 3 and 6. Mutation is the random alteration of a bit in the string, e.g., the second string would become 00100110 (2 and 6) after the 4th bit is mutated.

Mutation, the probability of which is typically low, assists in keeping diversity in the population. There is always the possibility of finding a better solution where one may not be expected.

A fourth parameter in GA is the length of each binary variable value, i.e., the interval between successive possible discrete values for each variable. Obviously, longer string lengths are required for increased precision).

The new population of variable values are evaluated. Using this information a third population is reproduced, and so on. The process repeats itself until no significant change in the best set of values can be found. The efficiency of this process depends on the GA parameter values.

Genetic algorithms have some potential advantages for some types of problems over three other search methods: calculus-based, enumerative, and random. Calculus based methods are local in scope; that is, if a function has multiple local optimum, these methods may drive toward one of these local values without approaching the global optimum. Calculus-based methods are also dependent on the existence of derivatives. Many real-world functions are discontinuous or, as most simulation models, discrete, and therefore do not work well with a method that prefers smooth and continuous functions. Enumerative methods are very simple but one may have to carry out a large number of iterations to locate the best. Even though simple, enumerative methods are often inefficient, and their execution times can become excessive when the search space is large or the "function" (or simulation program) is complex. Random search methods can perform as poorly as the enumerative methods in many respects. Selecting feasible random solutions and evaluating them can become time-consuming as the space becomes large.

Parameter Estimation

The objective function we selected for the calibration of the flow routing and water quality models was the minimization of the sum of squared differences between the computed and observed time-series of flow and concentration data. Both input and output time-series data are required for each river reach. The flow routing model was calibrated first, and then using the calibrated flow routing model, the water quality model was calibrated.

The Muskingum method assumes that the storage, S, within a reach is a linear function of its inflow, I, and outflow, O. $S = K[\ xI+O(1-x)]=KO+Kx(I-O)$, where K and x are the routing parameters (see any hydrology text or handbook). Where the two-parameter model is inadequate, Cudsworth (1989) proposed a nine-parameter version. These nine parameters account for the weights placed on past inflows at different time intervals. Alternatively,the IRAS routing model assumes N subreaches within a reach. The outflow, O_i, from each subreach i is a function of its initial storage, S_i, and inflow, Q_i: $O_i = (A \cdot S_i + B \cdot Q_i)^C$. Adding the subscript t for time, each subreach's initial storage volume, $S_{i,t+1}$, in the following time period results from a mass balance: $S_{i,t+1} = S_{it} + Q_{it} - O_{it}$. The inflow to the next downstream subreach i+1 is the outflow from the immediate upstream subreach, $Q_{i+1,t} = O_{it}$. The four parameters are A, B, C, (all ranging between 0 and 1) and the positive integer N. The values of these parameters can vary in each reach.

The water quality model was a traditional first-order model, including dispersion, advection, growth, decay and transformation of, in this case, nine water quality constituents. The parameters included the dispersion parameter and the rate constants for each of those constituents in the reaches being modeled (see Lang and Loucks in this volume).

Using flood data on three increasingly complex rivers, (Hoggen, 1989; Cudsworth, 1989; and Pradeep, personal correspondence, 1994) sensitivity analyses were carried out on the GA parameters. For the two and nine-parameter Muskingum model and the four-parameter IRAS routing model, the best ranges of parameter values were identified as: population size =100-250; cross-over probability = 0.6-0.9; and mutation probability = 0.02-0.1. Using flow data reported by Hoggen,, the computed outflow hydrograph using the values obtained from GA were essentially the same as the reported data. There was a slight improvement in prediction using the IRAS model. Predicting the flows of Columbia River, the nine-parameter Muskingum model was calibrated by Cudsworth. His results were compared to the nine calibrated values found using GA. The computed outflows based on GA-derived calibration values were closer to the actual observed values. Both the nine-parameter Muskingum and the four-parameter IRAS routing models were calibrated for the Darling River in Australia, using GA. The four-parameter IRAS GA-calibrated model provided the better fit.

Calibration of water quality models involved similar GA parameter sensitivity studies and included constraints on the reasonable ranges of each of the parameter values. Additional details including graphs showing the results of flow and water quality model calibrations using GAs are shown elsewhere in this volume (Lang and Loucks, 1995).

References

Beck, M.B. and van Straten, G., 1983, Uncertainty and Forecasting of Water Quality, Springer-Verlag, Berlin, Germany

Cudsworth, A.G., 1989, Flood Hydrology Manual, A Water Resources Publication, USDI, Bureau of Reclamation, Denver, CO

Goldberg, D.E., 1989, Genetic algorithms in Search, Optimization and Machine Learning,. Addison Wesley, Reading, MA

Lang, J.G. and Loucks, D.P., 1995, Integrated Water Quantity-Quality Modeling, this volume.

Loucks, D.P., French, P.N. and Taylor, M.R., 1995, Interactive River-Aquifer Simulation Model: Program Description, Cornell University, Ithaca, NY

Simpson, A.R., Dandy, G.C., and Murphy, L.J., 1994, Genetic Algorithms compared to other techniques for Pipe Optimization, Journal. of Water Resources Planning and Management, ASCE, 120(4), 423-443.

Wang, Q.J., 1991, The Genetic Algorithm and Its Application to Calibrating Conceptual Rainfall-Runoff Models, Water Resour. Res., 27 (9), pp 2467-2471.

Determination of Acute Dilution Factors Based on a 1-Hour Exposure of a Drifting Aquatic Organism

Christopher C. Obropta[1], Member ASCE
Scott T. Taylor[2], Associate Member ASCE

Abstract

The EPA UM initial mixing model was used to simulate an ocean outfall off Seven Mile Beach in Cape May County, New Jersey. The UM model was verified via a dye study conducted during the summer of 1992. The verified model was used to develop dilution factors for both the spatially and temporally defined acute mixing zones. A comparison of the methodologies for defining the dilution factors for the spatially defined and temporally defined acute mixing zones illustrates that the temporal method for defining the acute mixing zone and the acute dilution factor yields less stringent effluent limitations for the discharger while still protecting the marine environment.

Introduction

As the focus on surface water quality changes from conventional to toxic pollutants, the importance of accurately defining mixing zones and dilution factors for these mixing zones becomes crucial. A mixing zone is an allocated impact zone where water quality criteria can be exceeded as long as acutely toxic conditions are prevented. An initial mixing model such as UM, developed and distributed by the EPA, can be used to simulate the initial mixing of discharged effluent and the receiving water. After the model is verified via a dye study, it can be used to predict dilution factors at the edge of

----

[1] Associate Project Manager, Omni Environmental Corp., 211 College Road East, Princeton, NJ 08540-6623.

[2] Senior Engineer, Omni Environmental Corp., 211 College Road East, Princeton, NJ 08540-6623.

the acute and chronic mixing zones. These dilution factors can be used to develop acute and chronic wasteload allocations, which are then used to determine average monthly and maximum daily effluent limitations for the discharge.

Theory

The Criterion Maximum Concentration (CMC) is the water quality criteria to protect against acute effects. Although the CMC is derived from 48- to 96-hour exposure tests, the EPA recommends no more than a one-hour exposure period to these acute concentrations for a drifting organism in order to prevent acute toxic effects. The EPA describes several methods for defining the acute mixing zone (also referred to as the CMC mixing zone). Of the methods available, only one defines the acute mixing zone on a temporal basis. Since the acute criteria is a time based criteria, this method is the most logical, but is often overlooked in favor of the more straightforward methods which define the mixing zone on a spatial basis. This temporal method is based on the calculation of the average dilution in the region through which a drifting organism passes during a one-hour period.

Modeling

A field verification study was performed for the Seven Mile Beach outfall by continuously injecting Rhodamine WT red dye into the effluent and measuring dye concentrations in the vicinity of the ocean outfall. These data were used to verify that the UM model was appropriate for simulating the initial mixing of the Seven Mile Beach ocean discharge. The UM model was then used to simulate initial mixing of the discharge at the higher permitted effluent flow.

For modeling future conditions, long term ambient receiving water data were statistically analyzed to determine a critical current speed and density stratification. These critical receiving water conditions were then used with the UM model to determine dilution factors for both the spatially defined acute mixing zone and the temporally defined acute mixing zone.

Using the method recommended by EPA for spatially defining the CMC mixing zone, a distance of 2.53 meters (m) from any diffuser port is obtained. At the permitted flow, a dilution factor of 19.1 was determined at the edge of the 2.53 m mixing zone using the UM model with critical conditions.

To temporally define the acute mixing zone, the EPA recommends that the length of the CMC mixing zone be established such that a drifting organism would not be exposed to one-hour average concentrations exceeding the CMC, or would not receive harmful exposure when evaluated by other valid toxicological analysis. Using the critical receiving water conditions, the UM model was used to predict effluent concentrations for a one-hour period. These effluent concentrations were converted to dilution factors and are plotted against time in Figure 1. The area under the curve in Figure 1 divided by 60 minutes yields a dilution factor of 103, which is the average dilution factor across the one-hour mixing zone. This dilution factor corresponds to the dilution factor predicted at the edge of a 70 m mixing zone. It is important to note that the time of exposure of a drifting oranism using the spatially defined CMC mixing zone is approximately 5 seconds.

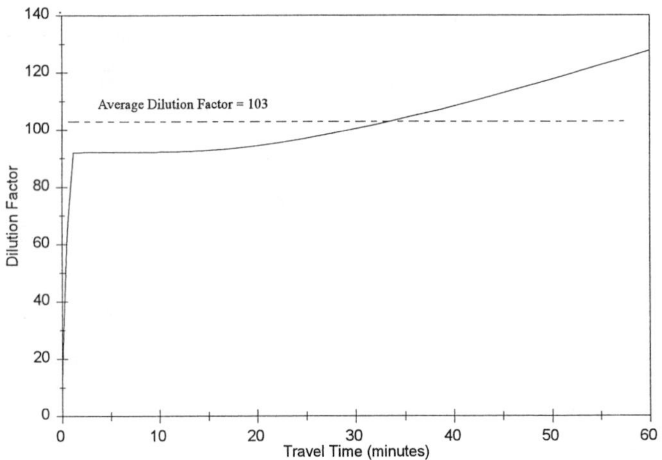

Figure 1: Predicted Dilution Factor vs. Time

Application of Dilution Factors

The dilution factors determined for the Seven Mile Beach outfall were applied to the State of New Jersey acute water quality criteria for chlorine produced oxidants (CPO) in order to assess the potential limitations for that parameter. The reader is referred to EPA's Technical Support Document for Water Quality Based Toxics Control for a detailed description of the methodology involved. Briefly, this methodology includes calculating a wasteload allocation (WLA) based upon the dilution factor and water quality criterion. The WLA is then used to calculate the allowable long term average (LTA) discharge concentration based upon the variability of the data and the number of samples collected per month. The LTA is then utilized to calculate average monthly and maximum daily discharge limitations. These calculations are illustrated in Table 1. Note that applying the spatial mixing zone results in limitations which are one fifth of those calculated by applying the temporal mixing zone.

TABLE 1: CALCULATION OF POTENTIAL CPO LIMITATIONS

New Jersey Acute Criterion (ppb)	13.0
Acute Dilution Factor	19.1/103.0
Acute WLA (ppb)	248.3/1339
Acute LTA 99%ile (ppb)	79.7/429.8
Maximum Daily Limitation 99%ile (ppb)	**248/1337**
Average Monthly Limitation 95%ile (ppb)	**86/462**

Rows with two values separated by a slash indicate:
- Spatially Based CMC Mixing Zone/Temporally Based CMC Mixing Zone

Conclusion

The dilution factor defined for a temporally defined acute mixing zone is much higher than the dilution factor defined for a spatially defined mixing zone. The spatially defined mixing zone only allows for exposure of a floating organism for less than five seconds, while the temporally defined mixing zone is based upon one-hour exposure. Since the acute criteria are based on time of exposure, the temporal definition is a much more appropriate means of developing dilution factors for these acute criteria.

Appendix I

Baumgartner, D.J., et al., "Dilution Models for Effluent Discharges (Second Edition)", United States Environmental Protection Agency, EPA/600/R-93/139, July, 1993.

Omni Environmental Corporation, "Cape May County Municipal Utilities Authority Effluent Plume Study for the Ocean City and Seven Mile Beach Outfalls", December, 1992.

USEPA, "Technical Support Document for Water Quality-Based Toxics Control", March, 1991.

How to Evaluate and Cost UV/Oxidation Systems

Robert Notarfonzo and Wayne McPhee
Solarchem Environmental Systems
130 Royal Crest Court
Markham, Ontario L3R 0A1
905-477-9242

Ultraviolet Oxidation is a proven technology for ground and wastewater treatment. Learn to identify when to consider it and how much it will cost.

Introduction

We are all aware of the fact that no single technology offers the solution for every water treatment problem. Although the conventional technologies of air stripping and activated carbon have proved adequate and cost effective, continued advances in UV/Oxidation have made it the EPA proven technology of choice in an ever increasing number of groundwater and process water applications. The last 5 years has seen the number of full scale UV/Oxidation installations increase from a handful to over several hundred.

This article focuses on the use of UV/Oxidation systems to treat various organic contaminants in ground and process water. Methodology and treatment parameters for estimating preliminary capital and operating costs for UV Peroxide systems (the most common UV/Oxidation arrangement) is provided along with contaminant specific cost comparisons between UV Peroxide and liquid phase activated carbon treatment .

Principles of UV/Oxidation

In the UV/Oxidation process, a high powered lamp emits UV radiation through a quartz sleeve into the contaminated water. An oxidizing agent, typically hydrogen peroxide, is added which is activated by the UV light to form oxidizing hydroxyl radicals:

$$H_2O_2 + UV \rightarrow 2 \bullet OH$$

These radicals indiscriminately destroy the toxic organic compounds in the water. Depending on the nature of the organic species, two types of initial attack are possible: it can abstract a hydrogen atom to form water, as with alkanes or alcohols or it can add to the contaminant, as is the case for olefins or aromatic compounds. The following equation represents the simplified general oxidation process:

$$\text{Chlorinated Organic Molecule} \xrightarrow[\bullet OH]{O_2} \text{Oxygenated Intermediates} \xrightarrow[\bullet OH]{O_2} CO_2 + H_2O + Cl^-$$

The attack by hydroxyl radicals, in the presence of oxygen, initiates a cascade of reactions leading to mineralization (ie. CO_2 and H_2O). In certain applications, catalysts, which are photo active and non-toxic, are added to significantly enhance the system's performance. A UV/Oxidation system can be designed to treat to any discharge requirement.

UV/Oxidation's key advantage is its inherent destructive nature; contaminated water is detoxified with no requirement for secondary disposal. There is no transfer of contaminants from one medium to another. Furthermore, UV systems in combination with hydrogen peroxide have no vapor emissions, hence no air permit is required. The equipment is quiet, compact, and unobtrusive, and preventative maintenance and operating requirements are low in a carefully designed system.

UV/Oxidation System

In a typical UV/Oxidation system reagents are injected and mixed using metering pumps and an in-line static mixer. The contaminated water then flows sequentially through one or more UV reactors where treatment occurs. Pretreatment such as solids removal, pH adjustment, and oil and grease removal is sometimes required. In practice, if the UV system is designed carefully with provision for automated cleaning of the quartz sleeve which surrounds the UV lamp, pretreatment can often be avoided, thus reducing both the capital investment and the ongoing maintenance costs.

The UV lamp inside the reactor is operated at high voltage, typically between 1000 and 3000 volts. Safety interlocks are fitted to protect personnel from both the UV radiation and the high voltage supply. These interlocks are usually linked to a Programmable Logic Controller (PLC), which can be used to control the whole installation including feed pumps, the UV lamps and the reagent delivery systems. Key advantages of a PLC are that it can be accessed via a modem to facilitate diagnostics for easier servicing, and can be reprogrammed to accommodate changes in operation throughout the remediation cycle.

In most groundwater applications, the material specified for the UV reactor is 316L SS, which protects against the oxidants and the UV light, while providing excellent resistance to corrosion.

Electrical Energy per Order (EE/O) and UV Dose

The key design variables are the exposure to UV radiation and the number of orders of magnitude of contaminant concentration removed. These two variables are combined into a single function, the Electrical Energy per Order or EE/O. The EE/O is a powerful scale-up parameter and is a measure of the treatment obtained in a fixed volume of water as a function of exposure to UV light. It is defined as:

"The kilowatt hours of electricity required to reduce the concentration of a compound in 1000 gallons by one order of magnitude (or 90%). The unit for EE/O is kWh/1000 gallons/order".

For example, if it takes 10 kWh of electricity to reduce the concentration of a target compound from 10 ppm to 1 ppm (1 order of magnitude or 90%) in 1000 gallons of groundwater, then the EE/O is 10 kWh/1000 gal/order for this compound. It will then take another 10 kWh to reduce the compound from 1 ppm to 0.1 ppm, and so on.

The EE/O measured in a design test is specific to the water tested and to the compound of interest, and it will vary for different applications. Typical EE/Os for a range of organic contaminants are provided in Table 1. With the EE/O determined, either through design tests or by using Table 1, the UV dose (ie. the amount of electrical energy required to treat 1000 gallons) required to treat a specific case is simply calculated using the following equation (note: for streams with several contaminants the required energy is not additive but determined by the contaminant requiring the greatest UV dosage):

(1) UV Dose (kWh/1000gal) = EE/O*log(initial/ final), where initial is the starting concentration (any units), and final is the anticipated or required discharge standard (same units as initial).

Table 1. Typical EE/Os for Contaminant Destruction

Compound	EE/O (kWh/1000gal/order)		Compound	EE/O (kWh/1000gal/order)
1,4-Dioxane	6		NDMA	5
Atrazine	30		PCE	5
Benzene	5		PCP	10
Chlorobenzene	5		Phenol	5
Chloroform	15*		TCE	4

DCA	15*		Toluene	5
DCE	5		Xylene	5
Freon	10*		TCA	15*
Iron Cyanide	40		TNT	12
Vinyl Chloride	3			

* Reduction catalyst required

Operating Costs

Once the required UV dose is known, the electrical operating cost associated with supplying the UV energy can be calculated using the following equation:

(2) Electrical Cost ($/1000 gal) = UV Dose (kWh/1000gal) * Power Cost ($/kWh)

The second key parameter from the design test is the concentration (ppm) of chemical reagents used, specifically hydrogen peroxide and any catalyst added to improve performance. The peroxide dose is based on the UV absorbance and COD of the water, and is typically in the range 50 to 200 ppm (mg/L). For the purpose of a preliminary cost estimate, the simplest rule of thumb for estimating the amount of peroxide necessary is the greater of 25 ppm or twice the COD concentration. Cost of the hydrogen peroxide varies from $0.005 to $0.008 per ppm concentration/1000 gal. If a catalyst is required, its selection and concentration will vary with the target compound and must be based on design test results. Lamp replacement costs typically range between 40% to 50% of the electrical cost.

Therefore,

(3) Total Operating Cost ($/1000 gal) = 1.45 * Electrical Cost + Peroxide Cost,
where Peroxide Cost = (H_2O_2 conc. in ppm) * ($0.005/ppm/1000 gal).

Capital Costs

Capital cost is a function of system size which is a function of the UV power required to destroy the selected contaminants. Using the EE/O's provided in Table 1, the following equation is used to determine the total UV Power (kW) required:

$$\text{(4) UV Power (kW)} = \frac{\text{EE/O} * 60 * \text{Flow(gpm)} * \log(\text{initial/final})}{1000}$$

$$= \frac{\text{UV Dose} * 60 * \text{Flow (gpm)}}{1000}$$

Once the required UV power is known, Figure 1 can be used to look up the associated capital cost in $US. The capital costs are given as ranges to allow for the actual number of discrete reactors which will be required along with any additional system options required.

The total UV power varies proportionally with flow rate and orders of magnitude of concentration of contaminant removed. For example, doubling the flow rate, or treating from 10 ppm to 0.1 ppm instead of down to 1 ppm, will both double the UV Power required. The equation can theoretically be used to obtain total UV Power for any combination of flow rate or concentration, but its accuracy depends on the EE/O which for a single contaminant can vary significantly from those listed depending on the water matrix and concentration.

It is important to note that the total UV Power, and hence the capital and operating cost, is influenced much more by a change in flow rate than by a similar change in concentration because of its logarithmic dependence on concentration through the log (initial/final) term.

These are the most accurate cost figures possible without performing actual design tests. The design tests consist of batch treatment runs of sampled water while varying UV dosage and reagent concentrations. Capital and operating costs are optimized over the lifetime of the remediation project by selecting the combination of total UV power and reagent concentration where the most economical treatment is obtained.

Cost Comparisons

The unit treatment costs of UV/Oxidation vs. liquid phase activated carbon as a function of influent concentration are compared in Table 2. The following conclusions can be drawn:

Figure . UV Oxidation Capital Cost as a Function of UV Power

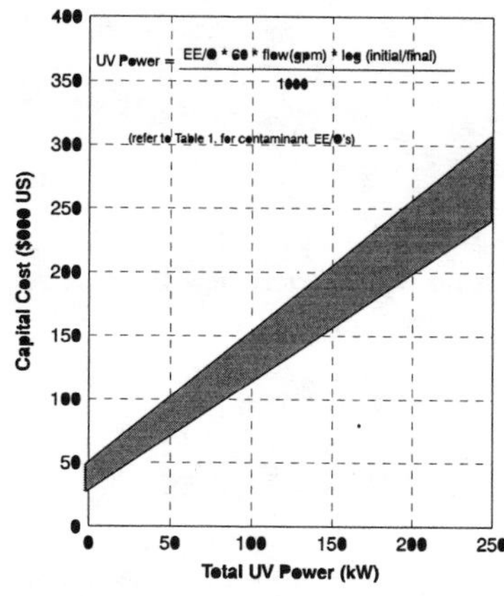

- unit operating costs for UV/Oxidation increase much more slowly with increasing influent concentrations than activated carbon. At concentrations above 50 ppm even good carbon adsorbers become uneconomical.
- UV/Oxidation treatment costs are almost always less for those contaminants which load poorly on carbon regardless of concentration.
- UV/Oxidation treatment costs are competitive for most of the average loading compounds but a definitive answer can only come from the results of a design test.

Since UV/Oxidation capital costs are typically 2 to 3 times that of carbon for the same size flow, longer term projects favor UV/Oxidation where the cumulative savings in operating costs offset the higher capital expense.

Table 2. Operating Cost Comparison between UV/Oxidation and Activated Carbon

Contaminant Concentration (ppm)	UV Peroxide ($/1000 gal)	Poor Carbon Adsorbers ($/1000 gal)	Average Carbon Adsorbers ($/1000 gal)	Good Carbon Adsorbers ($/1000 gal)
0.1	$1 - $2.5	>$3	$0.5 - $3	<$0.5
1	$1.5 - $4	>$10	$0.7 - $10	<$0.7
10	$2 - $6	>$50	$1 - $50	<$1
Typical Contaminants	aromatics, chloroalkenes, nitroaromatics, free and complexed cyanides, chloroalkanes*	vinyl chloride, NDMA, 1,4-dioxane, chloroform, TCA, DCA, methylene chloride	benzene, DCE, phenol, carbon tetrachloride, TCE	PCP, TNT, lindane, chlordane, xylene, PCE, PCBs

* based on use of a proprietary reduction catalyst

The Approach to CSO Abatement for Manchester, New Hampshire

Gary Mercer[1]
Tom Seigle[2]

Abstract

This paper presents the approach undertaken for a CSO abatement study for the City of Manchester, New Hampshire. Manchester has a population of approximately 90,000 inhabitants and a combined sewer area of over 5400 acres. The City, working with New Hampshire Department of Environmental Services and the USEPA, undertook a study to determine the impacts of CSO discharges and plan CSO abatements. The study encompassed flow and water quality monitoring, SWMM modeling, abatement evaluation, and the development of a recommended plan.

Flow and water quality monitoring of the combined sewer system was performed over a two year period of time to: develop an understanding of the system, to develop model parameters, and to characterize the quality of the CSO discharges. The RUNOFF and EXTRAN blocks of SWMM were calibrated and verified to the flow monitoring data. The model was than used to evaluate CSO abatement alternatives, including, storage, treatment, and separation. The benefits/impacts of each control was assessed to develop the recommended plan.

Introduction

The City of Manchester is in the south central part of the state of New Hampshire. The City has a population of approximately 90,000 inhabitants and a combined sewer area of over 5400 acres. The Merrimack River divides the city into an eastern and western half. The combined sewer system conveys a base sanitary flow of 22 mgd

[1] Water Resources Engineer, Camp Dresser & McKee Inc, Cambridge, MA.

[2] Chief Sanitary Engineer, City of Manchester, Manchester, NH

to a waste water treatment plant, which discharges to the Merrimack River. The treatment plant has a secondary capacity of 50 mgd and a primary capacity of approximately 70 mgd. A 18 mgd pump station conveys sewer flows from the west side of the City to the east side where the treatment plant is located. The combined sewer system has 23 overflow structures. All but two of the regulators are located at the downstream end of trunk lines, which control flow into the interceptor system. A regulator is located at the westside pump station and upstream of the waste water treatment plant, both of these relive excess flows in interceptors.

The City undertook a Phase I CSO study in 1992 and determined that CSO discharges contributed to exceedences of water quality standards in the receiving waters. For the Merrimack River, aesthetics and bacteria standards were determined to be exceeded, and for the Piscatquog River, aesthetics, bacteria, and several metal standards were exceeded. With these findings, the City was directed by the New Hampshire Department of Environmental Services and Region 1 USEPA to undertake a CSO Facilities Plan. The development of the CSO Facilities Plan follows the approach presented in the USEPA CSO Policy and New Hampshire CSO Strategy.

Data Collection

A fundamental approach undertaken for this effort was to develop a clear understanding of the combined sewer system and how the system responds to rainfall events. Plans and profiles of the interceptor system were collected and field verified. All regulators and overflow structures were inspected and surveyed. A flow monitoring program was implemented over a two year period. The monitoring program of the combined sewer system collected flow data at 20 locations, including flow measurements at 80 percent of drainage areas, and on all interceptors. For each monitoring location, three to seven storms were monitored. Additionally, a block program was implemented, which placed a wooden block on the overflow weir or pipe at each of the 23 regulators. Regulators were inspected after each rainfall event for a period of 6 months to determine the activation rainfall intensity. In addition to the flow monitoring, water quality samples were collected at three CSOs for two storms and analyzed for conventional and priority pollutants. Water quality sampling and dye studies were also performed in the Merrimack and Piscatquog Rivers to determine the river response to CSO discharges.

Modeling

A computer model of the combined sewer system was developed which simulated the existing conditions and allowed for "what if" simulations of potential CSO abatement alternatives. The RUNOFF and EXTRAN blocks of the USEPA SWMM model were used to simulate the combined system. The RUNOFF model simulates the rainfall and the runoff response of the combined drainage areas. The combined drainage area was divided into over 200 subareas in the RUNOFF model. Calibration and verification of the RUNOFF model was made using the flow monitoring data described above. The EXTRAN model simulated the regulators, overflow pipes, interceptors, and pump stations of the combined sewer system. This model simulates the hydraulics of the combined sewer system and predicts the depth and velocity of flow. The EXTRAN model was calibrated and verified to the flow monitoring data taken in the interceptors and to the block program data. The EXTRAN model calibration and verification encompassed over 50 gage-storms.

The model was applied. Design events were simulated using the calibrated RUNOFF and EXTRAN models to predict the overflow volumes and the peak discharge from each CSO. Rainfall hyetographs for seven design storm events, 2 week, 1 month, 3 month, 6 month, 1 year, 2 year and 5 year, were selected from a 43 year rainfall record for the simulations.

A STORM computer model was developed to predict the annual overflow volume for existing and for each CSO abatement alternatives. STORM is a continuous model, whereas RUNOFF and EXTRAN were run as event models. The control storage and treatment levels determined from the RUNOFF/EXTRAN simulation were used in the STORM model to predict annual overflow volumes at each regulator. For example, the 2 week level of storage determined by the RUNOFF/EXTRAN model was set as storage in the STORM model, and the annual overflow volume at each regulator for a 2 week storage was determined.

Evaluation

The effectiveness of each increment of additional storage or treatment can be evaluated by plots of overflow volume vs. treatment and overflow volume vs. storage. Plots were developed for each regulator and for the system. Plots for the system are presented in Figure 1 and 2. Evaluation of alternatives then proceeded by developing the cost of CSO abatement technologies at each regulator for each level of control. CSO abatement technologies included in the evaluation were: screening and disinfection, swirls, sedimentation tanks, storage tanks and separation. Based on the findings from the alternative evaluation, a reduced set of alternatives were evaluated that focused on the most effective controls at each design storm level.

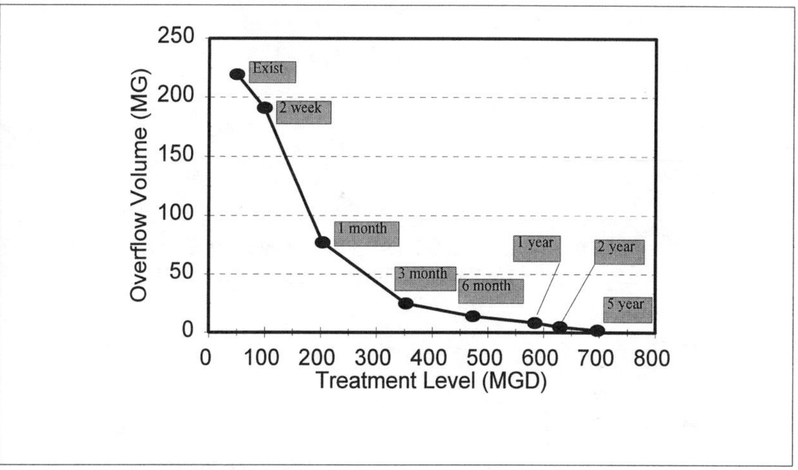

Figure 1. Overflow Volume versus Treatment

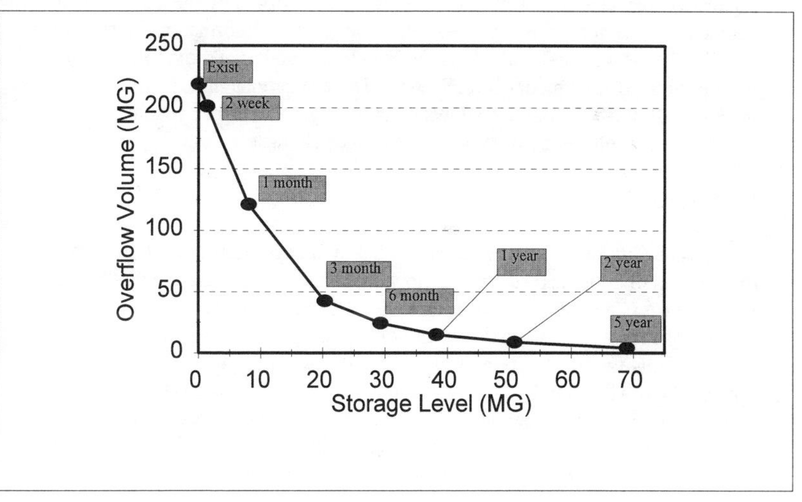

Figure 2. Overflow Volume versus Storage

HYDROLOGIC ANALYSES FOR CSO STORAGE TUNNELS IN FALL RIVER, MASSACHUSETTS

Peter von Zweck[1], and Terry Sullivan[2]

Abstract

An in-line storage tunnel is being designed to control combined sewer overflows in the City of Fall River, Massachusetts. The hydrologic design of the tunnel and its ancillary structures have included use of a wide variety of design tools. The evaluations were performed to ensure that the proposed design will meet both combined sewer overflow performance objectives established by regulatory agencies and operational objectives established by the City. The design coupled data from an existing ARC/INFO Geographic Information System, the SWMM RUNOFF model, the SWMM EXTRAN model, and the fully dynamic transient mixed flow model MXTRANS. The evaluations indicate that the proposed design will limit overflows to an average of four per year at a single location while improving in-system conveyance characteristics.

Introduction

The City of Fall River has embarked on an aggressive program to mitigate the impacts of combined sewer overflows (CSOs). The program will allow the City to meet CSO policies developed by the U.S. Environmental Protection Agency (EPA) and the Commonwealth of Massachusetts. The cornerstone of the program calls for construction of over 10 kilometers of 2.4- to 6.1-meter-diameter tunnel which will drain by gravity to the City's existing wastewater treatment plant. The tunnel will operate as an interceptor during dry weather relieving the existing system which frequently surcharges. The tunnel will also provide 2.0 x 10^5 cubic meters of storage for combined sewer flows during wet weather. The

[1] Water Resources Project Manager, CH2M HILL, 50 Staniford Street, 10th Floor, Boston, Massachusetts 02115, 617/523-2260

[2] Director of Water Pollution Control, City of Fall River, One Government Center, Department of Engineering Services, Fall River, Massachusetts, 02722, 508/324-2320

near surface/gravity concept will result in a savings of approximately \$15 million in construction costs and \$600,000 per year in operating and maintenance costs relative to the deep tunnel concept studied previously.

Database Development

An ARC/CADD database was created to store and manipulate data used in the hydrologic evaluations. This database was created from an existing ARC/INFO database obtained from the City of Fall River Planning Department. The City's database contains topologically correct geographic data layers for a variety of features. As additional data were acquired for CSO design, they were added to the GIS database. The data layers contained in the model GIS are shown in Table 1.

Table 1. - GIS Database

Existing Database	Data Added
Streets	Pump station locations
Street and landmark names	Storm sewers
Water bodies	Catch basins
Railroads	Drainage basin boundaries
Zoning map	Green space map
Combined sewers	1980 census map
Sewer lengths and diameters	USGS topography

Hydrologic Model Development

SWMM models were developed to simulate the performance of the City of Fall River's sanitary sewer system. The SWMM RUNOFF model was used to compute wet weather inflow to the system. Hydraulic analyses of the existing interceptor system and proposed tunnel were performed using the SWMM EXTRAN model. The SWMM models were used to balance the amount of flow directed to the existing interceptor system and the proposed tunnel.

The City's collection system includes over 280 kilometers of combined sewers, 11 pump stations, 1 treatment plant providing 1.9×10^5 cubic meters per day of primary and secondary treatment, and 19 CSO locations throughout the City which discharge into Mount Hope Bay, the Taunton River, and the Quequechan River. Figure 1 identifies the locations of the principal interceptor sewers and pump stations and the CSO locations.

The first task performed during the model development was to convert SWMM version 4.2 models created during the 1993 CSO Facilities Plan Review into a format compatible with the Storm Water Management Model with XP

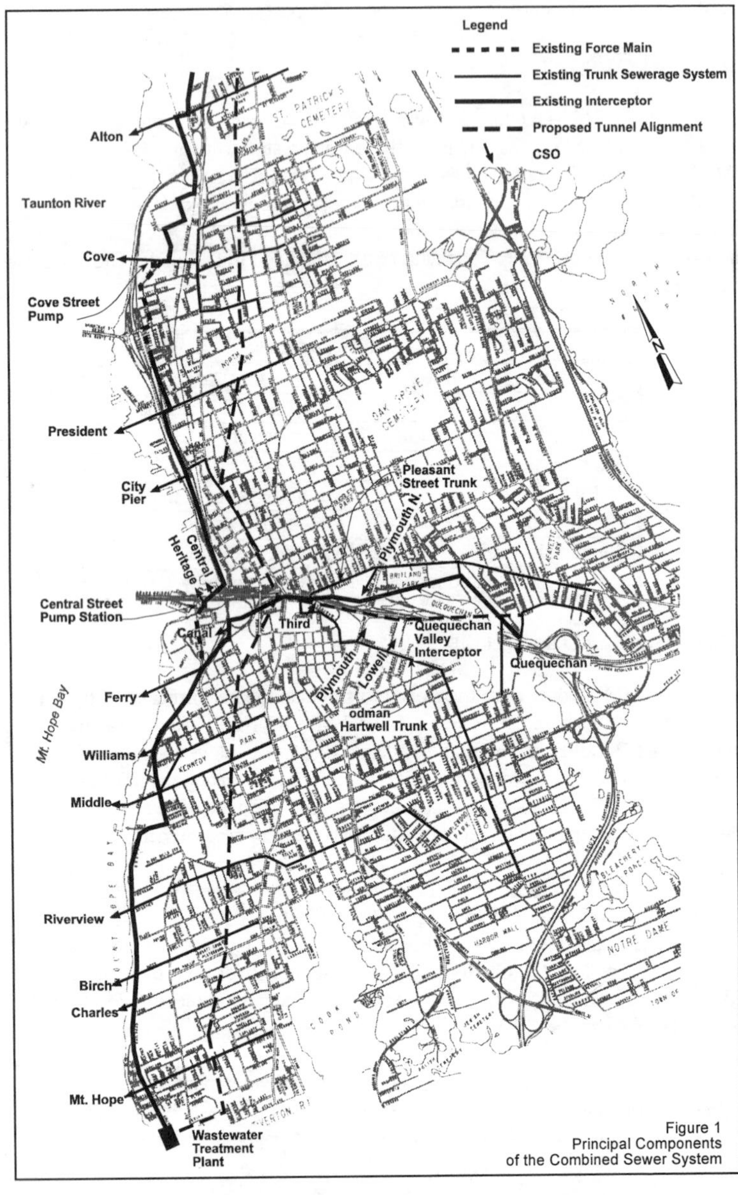

Figure 1
Principal Components
of the Combined Sewer System

graphical interface (XP-SWMM). The conversion was made to facilitate work required throughout the design effort and to provide the city staff with a user-friendly data management tool. The XP-SWMM program allows modelers to quickly obtain and or modify input data. The program's on-screen displays also facilitate review of the large amount of output data produced by the EXTRAN program.

Other tasks performed during the model development included increasing the limits of the model to include all areas impacted by the proposed CSO controls, and refinement of the level of detail used to describe the existing facilities. In most cases, the existing models ended one conduit upstream of each regulating structure. To facilitate modeling of the proposed controls, several additional conduits were added, and the drainage basins tributary to each regulator were sub-divided.

Hydraulic Transient Modeling

Because the EXTRAN model cannot accurately determine the magnitude or velocity of transient pressure waves in tunnels, they were evaluated using the fully dynamic mixed flow model MXTRANS. Results from the MXTRANS model were used to refine designs for the size and alignment of the tunnel, and the elevation of the extreme event overflow weir. These studies were directed by Dr. Charles C. S. Song of the St. Anthony Falls Hydraulic Laboratory.

Summary of Modeling Results

The XP-SWMM models were used to evaluate the performance of the proposed controls for a 3-month design storm. This evaluation indicated that the tunnel system can fully capture and treat runoff from a 3-month storm.

The XP-SWMM model was also run in continuous mode for precipitation data recorded during 1968. This simulation indicated that four events would exceed the tunnel system's capacity. This simulation also indicated that more than 85 percent of the combined sewage flow generated in the City will be captured for treatment.

References

He, J., Song, C.S., and Liu, Y. (1994). "Hydraulic Transient Study of the Fall River Tunnel System." University of Minnesota St. Anthony Falls Hydraulic Laboratory, Minneapolis, MN.

Phase 2A Preliminary Design Report Preliminary Design Documents. (1994). CH2M HILL/PBG&S Team, Boston, MA.

New York City's CSO Facility Planning Program

Peter J. Young, P.E.[1]
Robert D. Smith, P.E.
Hazen and Sawyer, P.C.

Robert Gaffoglio, P.E.
Vincent DeSantis, R.A.
NYC Department of Environmental Protection
Bureau of Environmental Engineering

Abstract

This paper gives an overview of the City of New York's current effort to develop CSO controls through its City-wide CSO Facility Planning Program.

Introduction

The City of New York is served primarily by a combined sewer system with 4,800 miles of combined sewers within its five Boroughs. This large and complex system has close to 450 CSO locations spread throughout the City which discharge to area receiving waters.

To address the water quality impacts due to combined sewer overflow (CSOs), the New York City Department of Environmental Protection (NYCDEP) initiated its current City-Wide CSO Facility Planning Program in 1985. The Program consists of eight planning areas which together cover the entire New York Harbor (see Figure 1). There are four area-wide projects (East River, Jamaica Bay, Inner Harbor and Outer Harbor) and four tributary projects (Flushing Bay, Paerdegat Basin, Newtown Creek and Jamaica Bay Tributary).

While each facility planning project is conducted independently and is specific to the water body in question, the program goals are uniform; compliance with State water quality standards for dissolved oxygen and coliform bacteria.

(1) Environmental Engineer, Hazen and Sawyer, P.C., 730 Broadway, New York, New York 10013

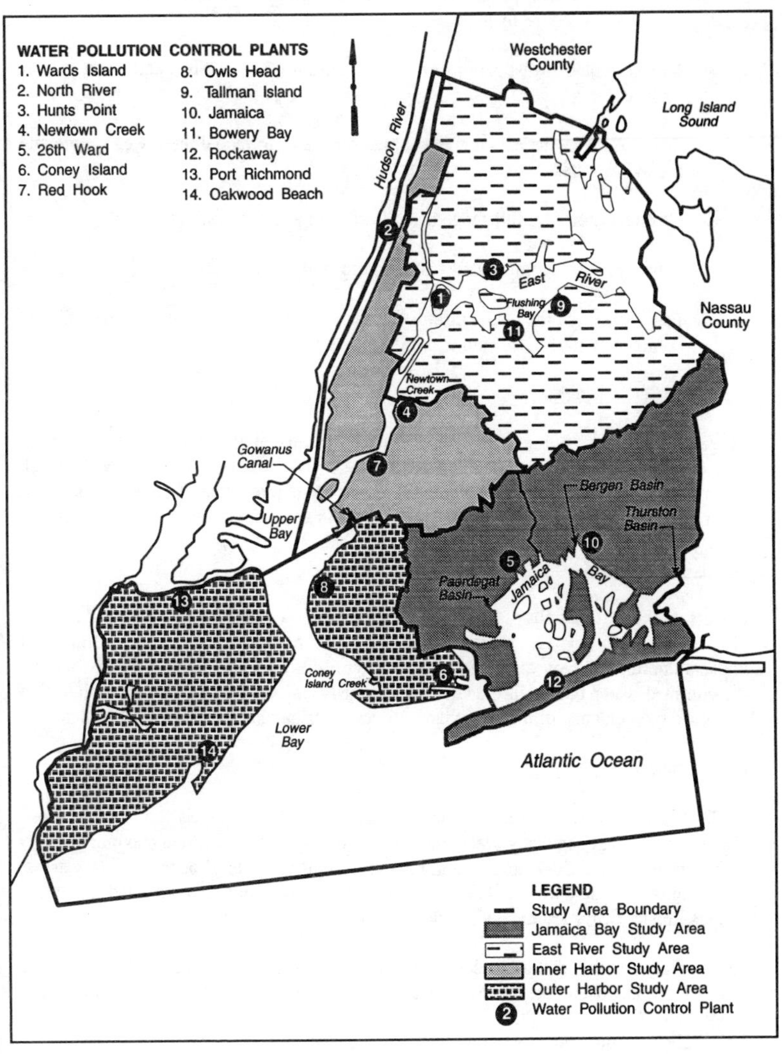
WATER POLLUTION CONTROL PLANTS
1. Wards Island
2. North River
3. Hunts Point
4. Newtown Creek
5. 26th Ward
6. Coney Island
7. Red Hook
8. Owls Head
9. Tallman Island
10. Jamaica
11. Bowery Bay
12. Rockaway
13. Port Richmond
14. Oakwood Beach
Westchester County
Long Island Sound
Hudson River
East River
Flushing Bay
Nassau County
Newtown Creek
Gowanus Canal
Upper Bay
Bergen Basin
Thurston Basin
Paerdegat Basin
Jamaica Bay
Coney Island Creek
Lower Bay
Atlantic Ocean
LEGEND
Study Area Boundary
Jamaica Bay Study Area
East River Study Area
Inner Harbor Study Area
Outer Harbor Study Area
Water Pollution Control Plant

CSO Study Area Locations

CSO Facility Planning

NYC's program is comprehensive and was conceived recognizing the site specific nature of CSO impacts. For each project area, the same approach to selecting CSO controls is employed. This procedure includes:

- Intensive field investigations to collect data on CSO loadings and water quality impacts.
- Detailed mathematical computer-based programs to model the sewer system and receiving waters.
- Consideration of a full range of CSO control alternatives.
- Developing a recommended CSO plan to mitigate impacts.
- Participation of the public throughout the planning process to involve the effected community in the decision-making process.

Status of NYC's Program

Currently, facility planning as described above has been completed for seven of the eight project areas. The facility planning process has given the City not only a thorough understanding of the CSO problem but also an important tool, state-of-the-art mathematic water quality models which can be used for other NYC water quality planning initiatives.

Water Quality Impacts

Results of the CSO Facility Planning Program has shown that for the open water areas of New York Harbor (i.e. Hudson River, East River, Upper Bay), there are minimal impacts due to CSOs on dissolved oxygen and coliform levels. However, in contrast, the confined water bodies in the City are seriously impacted by CSO with violation of dissolved oxygen and coliform standards directly attributable to CSOs.

Recommended Plans

Table 1 summarizes the recommended plans for each of the eight planning areas. Generally, for the open water areas, the CSO recommendations include maximizing flow to the plants, in-line CSO storage, and regulator improvements. For the tributary areas, more capital intensive off-line CSO storage facilities are being recommended due to the serious impact CSOs are having on these areas.

Full implementation of these recommendations will take several years. However, the groundwork has been laid for the future with these recommendations. The NYC CSO Program will eliminate CSO related water quality violation and advance the mission of NYCDEP to protect local receiving waters and restore a great natural resource to the people of NYC.

Table 1	
Summary of Facility Plan Recommendations	
Project Area	**Recommendations**
Paerdegat Basin	• $1.1 \times 10^5 m^3$ (30 mg) off-line CSO storage tank • $7.6 \times 10^4 m^3$ (20 mg) in-line CSO storage • Dredging of waterbody
Flushing Bay	• $1.0 \times 10^5 m^3$ (28 mg) off-line CSO storage tank • $5.7 \times 10^4 m^3$ (15 mg) in-line CSO storage
East River	• $2.6 \times 10^4 m^3$ (7 mg) off-line CSO storage tank (Hutchinson River) • $4.5 \times 10^4 m^3$ (12 mg) off-line storage tank (Westchester Creek) • $2.6 \times 10^4 m^3$ (7 mg) off-line storage tank (Little Neck Bay) • $1.9 \times 10^4 m^3$ (5 mg) in-line storage
Jamaica Bay	• $1.2 \times 10^5 m^3$ (31 mg) off-line CSO storage tank • $1.1 \times 10^4 m^3$ (3 mg) in-line CSO storage • Dredging of waterbody
Inner Harbor	• Regulator improvements • Maximize wet weather flow to plant • $1.6 \times 10^5 m^3$ (41 mg) in-line CSO storage
Outer Harbor	• Regulator improvements • Maximize wet weather flow to plant • $3.8 \times 10^4 m^3$ (10 mg) in-line CSO storage
Newtown Creek	• Maximize wet weather flow to plant • In-stream aeration • Dredging of waterbody
Jamaica Tributaries (in progress)	• (Under development)

A Modeling Strategy for a National Assessment of Water Quality-Based Combined Sewer Overflow Controls

A. Stoddard[1], M. Morton[1], T. Dabak[1], A. Eralp[2]

Abstract

A modeling strategy was developed and implemented for a nationwide assessment of water quality-based combined sewer overflow (CSO) controls. The methodology relied on an automated approach using the 6-digit USGS hydrologic accounting system to construct individual river basin models. Information from EPA data bases (STORET, PCS, IFD, REACH, and NEEDS SURVEY) were used to develop the watershed-based models. An iterative scheme was used to perform waste load allocations to determine the required pollutant load reduction from CSO discharges needed to achieve compliance with water quality standards.

Introduction

The main objective was to determine the level of pollutant reduction needed at CSO discharge points to achieve water quality standards. The information was planned for use in calculating a national cost for controlling CSO discharges. Although models were developed and used to evaluate individual discharge points and receiving water segments, the aggregate results were the main interest. Therefore many simplifying assumptions were made. The main parameters considered were dissolved oxygen (DO) to evaluate overall water quality of the receiving water, total copper to evaluate aquatic toxicity, and fecal coliform bacteria to evaluate use impairment.

About 1300 facilities nationwide use combined sewer systems to carry sanitary and industrial wastewater as well as storm water. These facilities are located mainly in older cities in the Northeast, the mid-Central states, and along the West Coast. Combined sewer overflows (CSOs) occur when the capacity of the combined sewer system is exceeded during a storm event. During these events, part of the combined

[1] Tetra Tech, Inc., 10306 Eaton Place, Suite 340, Fairfax, VA 22030
[2] U.S. Environmental Protection Agency, Washington, DC

flow from the collection system is discharged untreated into receiving waters, which may result in a failure to achieve designated uses of the water body or a violation of water quality standards.

To cover all the CSO facilities in the U.S., over 250 individual models of various water bodies were constructed which represented 158 watersheds, 43 estuaries, and 98 large lakes and open coastal water systems. Since this was a national needs estimate, it was assumed that the water quality impact of intermittent discharge events from CSOs could be conservatively represented using a steady-state modeling approach. Three different water quality models were used depending on the water body type: RIVHW (based on Thomann and Mueller 1987) was used for the riverine analysis; HAR03 (Nossa and Chapra 1974) was used for the estuarine analysis; and CFM, the Coastal Farfield Model (Tetra Tech 1985) was used for the analysis of open waters at coastal and lake sites.

Data Needs and Simplifying Assumptions

Even with the above simplified models, a large amount of data were required to develop model input files. Data were obtained from several EPA databases to characterize various water bodies according to: (a) physical transport and reach geometry [EPA's hydrologic REACH File system]; (b) background water quality and tributary pollutant loads [EPA's STORET system]; (c) pollutant loads from NPDES municipal and industrial dischargers [EPA's Permit Compliance System--PCS]; and (d) CSO runoff and pollutant loads for the design storm event. Data were also obtained from the scientific literature to characterize physical, chemical, and biological parameters needed to specify the modeling framework.

As the main interest was to develop an aggregate national cost estimate, several simplifying assumptions have been incorporated in the water quality analysis to maintain a feasible approach. (1) Uniform water quality standards for DO, total copper, and fecal coliform bacteria were applied to the receiving water of each urban CSO using EPA's standards. It was beyond the scope of the study to incorporate a database to characterize the complex set of water quality standards established by states and local agencies to reflect the designated uses of all waterways in the U.S. (2) The water quality models were not calibrated to baseline conditions. Because of the magnitude of the overall task, kinetic rate constants are based on a global set of model parameters and on best estimates computed as a function of background water quality compiled from STORET. (3) Water quality problems from CSO discharges are highly transient due to the storm event nature of the overflows. In order to facilitate this national assessment, it was assumed that water quality impacts of intermittent discharge events could be conservatively represented with a steady-state modeling approach. By using a pulse load based on the runoff volume and pollutant load of the 5-yr, 6-hr design storm event, the mass balance impact of each CSO was evaluated from (a) the steady-state baseline water quality conditions resulting from average annual tributary flows and municipal and industrial NPDES discharges within a watershed, and (b) the incremental steady-state water quality response to a single design storm runoff volume and pollutant load averaged over a typical time interval between storms.

Example River Application

An example river network created from the REACH file for the Lower Connecticut River (6-digit watershed code 010802) is shown in Figure 1. The boxes represent river reaches. The locations of CSO discharges are marked "CSO" and the municipal/industrial discharges are marked "PCS". Only tributaries having CSO outfalls are shown, however, other tributaries are included in the model to provide boundary conditions and flow balance. The lower portion of the network from reach 029 to the mouth is tidal and was not included in the river model (RIVHW). The geometry, flow, and CSO/PCS locations for watershed 010802 were placed in a computer file designated 010802.RCH. A series of preprocessors create the various data groups for the RIVHW input file by accessing (1) the RCH file to create the segment geometry data group, (2) the STORET data to create tributary boundary conditions, (3) the PCS file for point source load data from municipal/industrial discharges, (4) the IFD file to determine the SIC code for each point source discharge, (5) a master file of CSO facilities for the BOD and copper loads, and (6) a file with default values for various kinetic rates needed for the RIVHW model.

The model was initially run with design rainfall CSO loads. A post-processing program determined which model segments, if any, violated water quality standards. If violations occurred, CSO loads were reduced 5% and the model was run again. This process continued until either no more water quality violations existed or until the CSO loads were reduced to zero. A table was created showing each CSO facility and the percent reduction in BOD and copper required to meet water quality standards.

Conclusions

Study results indicated that 78% of the CSO facilities in the U.S. do not need to reduce BOD loads since the receiving water quality meets the DO standard. However, 19% of the facilities need to reduce their loads 100% in an effort to meet the DO standard. The remaining 3% required reductions of between 5% and 95%. For dissolved copper, 30% of the facilities required no reduction in copper load, 40% required a 100 percent reduction, and 30% required some reduction. Since effluent levels of coliform bacteria in CSO discharges are very high, disinfection would likely be required in all cases to achieve water quality standards.

References

Thomann, R.V. and J.A. Mueller. 1987. *Principles of Surface Water Quality Modeling and Control.* Harper & Row Publishers, New York, NY, 644 pp.

Nossa, G.A. and S.C. Chapra. 1974. Documentation for HAR03: a computer program for the modeling of water quality parameters in a steady state multi-dimensional natural aquatic systems. 2nd Edition. U.S. Environmental Protection Agency, Region II, New York, NY.

Tetra Tech. 1985. Coastal Farfield Model for computing centerline dilutions of a pollutant as it travels away from the zone of initial dilution. Unpublished.

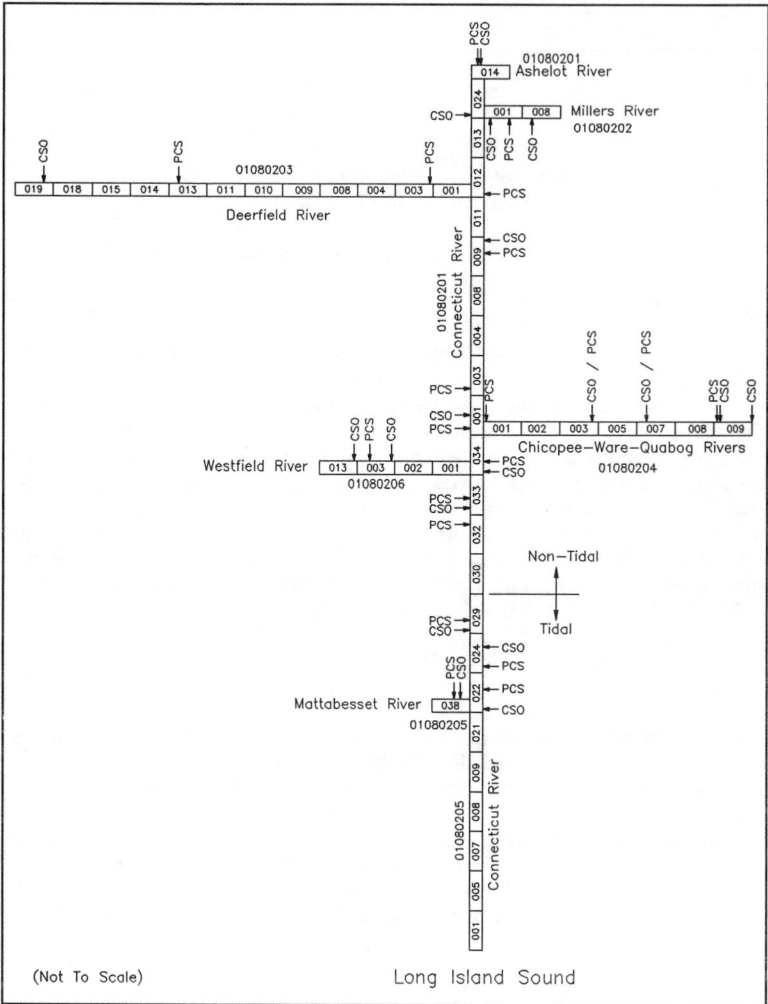

Figure 1. River basin network for the Lower Connecticut River (010802).

Continuous Simulation Approaches for CSO Characterization and Alternatives Analysis

Edward H. Burgess[1], James T. Smullen[2], and Larry A. Roesner[3]

Introduction

The recently released U.S. EPA Combined Sewer Overflow (CSO) Control Policy (FR 18688; April 19, 1994) requires cities with combined sewer systems to characterize the occurrence of CSOs in terms of average annual statistics, and defines control targets in the same manner. In fact, the occurrence and control of CSOs have long been evaluated using average annual statistics as measures of performance (*e.g.*, number of occurrences per year, volume of overflow per year, mass of various pollutant constituents per year, etc.). In order to generate average annual statistics, continuous simulation models - rather than single-event models - are typically applied to simulate the response of the combined sewer system to a precipitation time series. The advantages of continuous simulation for CSO analysis have been described in the literature by Roesner *et al* (1974; 1994), Freedman *et al* (1994) and others.

A number of computer models capable of continuous simulation of combined sewer systems have been developed and applied to characterize CSOs and to support CSO control planning. The most commonly applied models are the U.S. Army Corps of Engineers' **S**torage, **T**reatment, **O**verflow, **R**unoff **M**odel (STORM; Hydrologic Engineering Center, 1977) and the U.S. EPA's **S**torm **W**ater **M**anagement **M**odel - Version 4 (SWMM; Huber and Dickinson, 1988). It should be noted that some CSO planning efforts have attempted to use single-event simulations to characterize CSOs and to support plan development. However, single-event simulation approaches are inherently unable to properly characterize CSOs in terms of long-term statistics (e.g. average annual frequency and volume, etc.), and antecedent conditions and inter-event relationships in the system response cannot be accounted for with these approaches.

---

[1] Principal Engineer, Camp Dresser & McKee, 1811 Losantiville Avenue, Suite 350, Cincinnati, Ohio 45237

[2] Principal Engineer, Camp Dresser & McKee, Inc.

[3] Chief Technical Officer, Camp Dresser & McKee, Inc.

Practical Trade-offs in Continuous Simulation

Although there is general agreement as to the benefits of continuous simulation for CSO analysis, significant differences have evolved in the manner in which continuous simulation models are applied to combined sewer systems. There are generally two approaches:

(1) Simulation of a complete (*i.e.,* 10- to 40-year) long-term precipitation record with simplified characterization of the hydrologic abstractions (and external simulation of the interceptor and regulator hydraulics); and

(2) Detailed continuous simulation of the hydrology and hydraulics of the combined sewer system (including the interceptor sewers and regulators) using an abbreviated precipitation time series (*e.g.,* a "typical year," which replaces the complete time series with synthetic rainfall records for a one-year period).

The fundamental trade-off in these alternative approaches involves either:

(I) simplification of the hydrology of the system (the first approach) or ;

(ii) simplification of the precipitation time series (the second approach).

This trade-off exists because, despite rapid advances in computer technology, it remains impractical to simulate the response of a combined sewer system to a long-term precipitation record using a detailed hydrologic and hydraulic model due to excessive computational runtimes.

Typical Year Analysis

While continuous simulation with a detailed model of the combined sewer system is appealing, the selection of a typical year is generally necessary to reduce computational run-times and model output analysis/debugging to practical levels. Within the framework of continuous simulation modeling of combined sewer systems, probabilities of occurrence of relevant hydraulic conditions (e.g. surcharge, overflows, etc.) are the desired output. These probabilities are the decision variables for evaluation of CSO control alternatives.

While typical-year analysis is a reasonably simple concept, the development (or selection) of a precipitation time series representative of the typical year is not at all simple. First, it is not possible to select from precipitation records alone a typical year precipitation time series that will produce statistically valid probabilities **for the desired output**. The reason that typical year analysis must consider more than the precipitation data alone is that the desired output (the relevant hydraulic conditions) is a function of the rainfall-runoff relationship (a weakly non-linear process) and the hydraulic response of the sewer network to runoff (flow rate and hydraulic grade line) is a strongly non-linear process, as revealed by examination of EXTRAN's solution of the Saint Venant equations. These non-linearities dictate that the hydraulic response, not the precipitation input, must be used as the basis for developing the typical year.

Since hydraulic response data for a period of sufficient length (i.e. ideally 20 years or more) is virtually never available, it is necessary to use a model of the system to generate a simulated hydraulic response time series (e.g. with EXTRAN) from which the typical year can be selected. However, EXTRAN (or other equally complex numerical models) is extremely sensitive to numerical stability conditions and the application of EXTRAN to a complex hydraulic network using an input precipitation time series of 20 years or more is an enormous (and impractical) undertaking. Not only are run-times unreasonable, but *properly* debugging and correcting run-time errors and other problems encountered during the complete simulation period (which can contain between *15 million and 150 million time steps* for a 10-40 year period) is simply not practical.

It must be recognized that any perceived advantage in a detailed EXTRAN model of the combined sewer system is negated if it is used with a statistically invalid representative (e.g. typical year) precipitation time series. And even if a suitable typical year is developed, a continuous EXTRAN model used with that typical year is a cumbersome planning tool, requiring from one million to three million solution time steps for a proper model representation, which creates the associated problems with excessive run-times and extensive debugging efforts. Some modelers have reduced the required number of simulation time steps by eliminating dry periods from the long-term precipitation time series. However, since the modeled system will not immediately return to the pre-event condition, this approach can negate much of the benefit of continuous simulation by neglecting the actual inter-event relationships in the system response. Although application of EXTRAN in continuous mode may be appropriate to support certain modeling objectives involving detailed analysis and preliminary design work, it should be recognized that this approach is generally not appropriate for master planning.

Continuous Simulation to Support CSO Master Planning

The preferred approach in most planning applications is to simulate the hydraulic response of the interceptor network (to uniform loading of runoff) in relatively fine detail and use this information, together with a simpler model of the hydrologic response to generate CSO statistics for the study area. The use of a simple hydrologic model is, from a practical standpoint, only limited in the ability to simulate the response of the more pervious sewersheds to the larger events in the precipitation time series, a limitation that generally acceptable for master planning applications. Given the general absence of a sufficient number of long-term precipitation gauging sites to characterize spatial variation in the long-term precipitation record, detailed hydrologic modeling of the system response to a precipitation time series that assumes uniform spatial distribution is of questionable value. And the advantages of continuous simulation with a simpler model of the physical system more than compensate for its limitations.

The above strategy, which focuses the highest level of detail on the aspect of the system that requires fine detail - the network of interceptor sewers and regulators - while avoiding unnecessary model detail throughout the remainder of the system, has been found to be an effective and efficient approach to CSO analysis. This approach enables the continuous simulation model to support master planning much more effectively than continuous simulation with a detailed model of the entire system, as more modeling resources can be reserved for planning, and alternative control strategies can be tested and evaluated much more quickly. This is an important advantage during plan development, as a broader array of alternatives can typically be evaluated than if the planning process is constrained to evaluation of only a limited set of improvement alternatives that can be simulated with an overly detailed model.

Conclusions

Different physical systems and different modeling objectives require the application of different simulation approaches. The advantages and disadvantages of the various models (SWMM/RUNOFF, SWMM/TRANSPORT, SWMM/EXTRAN and STORM), the different techniques for pre-processing the precipitation time series, and the different approaches to model representation of the physical system must be weighed carefully when selecting a modeling strategy to support CSO analysis and planning. Trade-offs exist between simplifying the hydrologic characterization and abbreviating the precipitation time series. Excessive model detail during the planning phase can severely constrain the ability of the model to support alternatives analysis, whereas a simpler model is often a superior planning tool.

References

Freedman, P. L., M. P. Sullivan and J. K. Marr (1994). "Modeling Requirements for Long Term CSO Control Plans"; Proceedings of the Water Environment Federation Specialty Conference - A Global Perspective for Reducing CSOs: Balancing Technologies, Costs and Water Quality; July 10-13, 1994; Louisville, Kentucky; pp.3-83 through 3-94.

Huber, W. C. and R. E. Dickinson (1988). Storm Water Management Model - Version 4: User's Manual; Cooperative Agreement CR-811607; U.S. EPA; Athens, Georgia.

Hydrologic Engineering Center (1977). Storage, Treatment, Overflow, Runoff Model - STORM: User's Manual; U.S. Army Corps of Engineers; Davis, California.

Roesner, L. A. *et al* (1974). "A Model for Evaluating Runoff Quality in Metropolitan Master Planning"; American Society of Civil Engineers Urban Water Resources Research Program Technical Memorandum No. 23 (NTIS PB-234312); New York, New York.

Roesner, L. A., E. H. Burgess and M. J. TenBroek (1994). "Facility Master Planning for CSO Management in an Uncertain Regulatory Environment"; Proceedings of the Water Environment Federation Specialty Conference - A Global Perspective for Reducing CSOs: Balancing Technologies, Costs and Water Quality; July 10-13, 1994; Louisville, Kentucky; pp.12-25 through 12-35.

CATCH BASIN MANAGEMENT: AN IMPORTANT CSO TECHNIQUE

by

Stephen A. McKelvie & C. Cassandra McKenzie
Parsons Brinckerhoff Gore & Storrie, Inc.
and
Cornelis Geldof, P.E.
Metropolitan District Commission, Hartford, Connecticut

Introduction

Catch basins are an important component of urban runoff conveyance systems. Think about it. Have you ever tried to get from a foundry the capacity curves for a particular grate? Few people who design storm drainage systems consider the number of catch basins that are connected to the pipe system. Do the capacities of the catch basins match with the capacity of the drainage system? In many cases questions of this nature will draw a blank.

This paper discusses a combined sewer overflow (CSO) abatement project in which the catch basins play an important role towards the solution of this nationwide problem.

Background

The Metropolitan District Commission established in 1929 in Hartford, Connecticut provides sewage collection, treatment, water treatment, water distribution, some hydro power, and mapping services to over 400,000 people in eight member communities in the greater Hartford area.

The area of interest in this paper lies in the southern portion of the City of Hartford. Like many cities in the east this area was developed with a combined sewer system. The District has been working virtually full time since the mid sixties to reduce the frequency and volume of CSO. This area known as the Wethersfield Cove CSO Area is tributary to a picturesque cove on the Connecticut River. At the present time the District is under a Consent Order to reduce the frequency of overflows into Wethersfield Cove. The District adopted a CSO Abatement Plan in September 1994 in which catch basin management plays a key role.

The Existing Sewer System

This area of the Hartford has a combined sewer system that has three sources of inflow as follows:

- sanitary sewage
- inflow through rainleader or roof drain connections
- inflow through catch basins

Analysis of the combined sewer system has shown that the combined sewer system can convey the flow which would result from the one year design storm with no overflows providing no surface water runoff entered the system and fifty percent of the existing rainleaders were disconnected. These disconnected rainleaders would then discharge onto the surface where the resultant inflow could be controlled by catch basin management.

The District over the last thirty plus years had installed the skeleton of a separted storm drainage system that was designed to convey the runoff that would result from a 10 year design storm. Therefore the key to getting this system to work is to establish an overland flow regime to convey the runoff to the storm sewer system. Another very important factor is to achieve the rainleader disconnection goals; however rainleader disconnection is not part of the discussion of this paper.

Flowslipping

Flowslipping is the term used to describe the method of restricting the entry of storm water runoff into catch basins while at the same time establishing an overland flow system to convey the runoff to a separated system that is designed to carry this slipped flow.

As a review it is helpful to remember that there are two very distinct drainage systems that are operable during most storm water runoff events. First there is the piped drainage system is usually designed to carry the runoff that results from the more frequent rainfall events. (which typically this system is design to carry the runoff from the 2, 5, or 10 year design storm). This pipe system is sometimes called the "convenience" system as it is installed not necessarily to prevent flooding but to remove water from the roads to allow for easy passage of vehicles during these relatively frequent storms. Secondly there is the overland flow system or the "major" system which will be the natural path that the storm water runoff will take during storms that produce runoff greater than the capacity of the piped system to carry the runoff away. The overland flow system will develop whether it is planned or not. Flowslipping must recognize the existence of both of these two systems.

There are several ways to restrict the inlet capacity of catch basins. Many of these methods are based on providing a small orifice through which the water must flow to enter the piped drainage system. The small orifice in some cases can be installed on the outlet pipe from the catch basin or on the catch basin inlet grating. More sophisticated inlet control methods may use vortex flow controllers.

In the Hartford project there are challenges in acheiving a successful rainleader disconnection program. Therefore in order to prevent overflows from the combined sewer system it became apparent that the flows that could be allowed to enter the combined sewers through the system's catch basins must be strictly controlled. Greater inflow through the catch basins would therefore require a higher the rate of rainleader disconnection necessary to meet the overflow requirements.

A common concern regarding flowslipping is that with the reduced outflow from the catch basin there will be a higher likelihood of the catch basin becoming plugged. This was a serious concern as the maximum allowable catch basin inlet flow rates were low.

As the available methods of catch basin restrictions were considered it was determined that the best method was to restrict inflow into the majority of the system's catch basins altogether. Exceptions would be made for certain catch basins on a case by case basis.

It was understood that the premise of completely blocking any inflow into catch basins would require approval from the City of Hartford. The cooperation of the City of Hartford is an important part of ensuring the success of this program. The District controls the sewer system but it has no authority over surface drainage. Without the cooperation of the City of Hartford the entire concept of flowslipping, the foundation of the Abatement Plan would be difficult to implement.

Field surveys had noted that many of the existing catch basins in the City had a decorative style with relatively small openings for the runoff to enter the grate. The existing grates have a tendency to become plugged with street litter and leaves. Some of the catch basin grates were plugged and if it was not for the curb inlets runoff could not enter the catch basin at all.

The catch basins are also an important interface between City of Hartford and District responsibility. The City is responsible for street cleaning. The District is responsible for catch basin cleaning. Street cleaning removes debris from the street however if the streets are not cleaned regularly the debris can be washed into the catch basin and thus become the responsibility of the District to clean.

In order to convince the City that the proposed concept of flowslipping should be accepted the District organized a full scale demonstration of flowslipping and invited City, State of Connecticut Department of Environmental Protection staff and local residents to attend. The demonstration would simulate actual runoff conditions by opening up adjacent fire hydrants and establishing gutter flow simulating a ten year runoff event. Thus the response of the system could be observed without waiting for a significant rainfall event to occur.

There are two types of catch basins in a flowslipping system. The restricted catch basins and the catchment catch basins. The catchment catch basins must collect the storm water runoff that has bypassed the restricted catch basins. The catchment catch basins in the Hartford system must then direct the captured flow to the separated storm drainage system. In order to improve the convenience to pedestrians the new catchment catch basins are placed upstream of crosswalks, demonstrating that the flowslipped method is in fact an improvement over the existing conditions.

It is also necessary to review street hydraulics when considering a flowslipping system. Due to the shape of the gutter and road cross section it is possible to increase the flow in the street significantly with only a small increase in the depth of flow in the street.

The flowslipping demonstration proved that during major rainfall/runoff events there already was a great deal of flowslipping occurring due to the limited capacity of the existing catch basins. At a low point on Franklin Avenue where there was ponding as a result of the limited capacity of the existing catch basins. After the installation of the flowslipping system the demonstration revealed only a minor increase in gutter flow depths; however, with the new catchment catch basin in place the conditions on Franklin Avenue improved considerably. Most of the water was captured upstream of the cross walk. Thus conditions were improved for both pedestrians and cars. This successful demonstration has allowed the District to proceed with the design and construction of a more extensive full scale flowslipping project incorporating several acres and numerous streets.

Conclusions

This initial phase of the program in Hartford has shown that by understanding both street hydraulics and overland flow systems, and by working in a cooperative way with all those involved, catch basin management has proven to be an important and successful technique in the solution of the City's combined sewer overflow problem.

PHASE VI COMBINED SEWER SEPARATION PROJECT CAMBRIDGE, MA

Paul R. Levine[1], Member

Abstract

While the 1995 ASCE Water Resources Planning and Management National Conference is taking place in Cambridge, MA., the City of Cambridge will be in the midst of constructing Phase VI of its comprehensive Sewer Separation Project. This paper will discuss the technical and funding aspects of the project which started 26 years ago. Since the project is in conformance with the overall plan to clean up Boston Harbor, by reducing Combined Sewer Overflows (CSO's) and wastewater flows to the Massachusetts Water Resources Authority (MWRA) collection and treatment facilities, it is currently being funded by a loan under the State Revolving Loan Fund (SRLF) Program .

Introduction

Twenty six years ago, the City of Cambridge made a conscientious decision to embark on an environmental pollution control plan to reduce CSO wastewater from flowing into the Charles River and Alewife Brook during dry and wet weather conditions. In 1969, because sections of the sewer system were still transporting sanitary and storm water, frequent combined sewer overflows occurred into the above waterways during major rainfall events. At that time, the City engaged Maguire Group Inc. to prepare a Facilities Plan which would establish a prioritized planning program for the City to follow to convert the combined sewer systems into separate sanitary and storm drain systems. Construction of the $44 Million Phase VI Sewer Separation Project-which started in October, 1993-is approximately 25% complete, and is scheduled to be completed in 1999.

Overview of Sewer Separation Phases

Since 1971, the City of Cambridge has expended approximately $20,000,000 for five major phases of construction to reduce CSO's to the Charles River, Alewife Brook, the MWRA Relief Sewers and the MWRA wastewater treatment facility at Deer Island. Each of the previous phases, as well as Phase VI, has been based on the

[1]Vice President, Director of Environmental Engineering; Maguire Group Inc.; 225 Foxborough Boulevard; Foxborough, MA 02035

recommendations of the 1969 Facilities Plan. In addition to the phased construction program, several smaller sewer separation construction contracts have been completed.

Because formal monitoring of the 13 CSO's, located along the Charles River and Alewife Brook, did not exist before completion of the previous five phases, the beneficial impact of the previous construction cannot be quantified. In 1988, the City embarked on a formal CSO monitoring program which will permit future quantification of the impact that the Phase VI, 300 acre, sewer separation project will have on the CSO located at Plympton St., which currently discharges to the Charles River. In addition, storm flows into the MWRA North Charles Relief Sewer System will be significantly reduced and will result in an increase of the capacity of the Relief System.

Project Funding

Regulatory agency funding of the planning, design, construction and construction inspection services for the Cambridge project has played a very major role in assisting the City to implement the project. The proactive attitude of the City in wanting to help improve the environment, by voting to authorize the necessary bonds to fund the project, has facilitated the ability of the City to obtain Grants under the original PL 92-500 federal/state aid program; and now to obtain substantial funding through the Massachusetts State Revolving Fund Program. By working with the City and the State, Maguire Group Inc. has made the obtaining of outside funding a priority in order to financially assist the City by: (1) Preparing a Facilities Plan which adequately addressed the City's needs into the future; (2) Maintaining project continuity by having the same design and construction team working on the project for the last 26 years; (3) Working with the City to address priority areas and issues as they develop; (4) Having the proposed plan approved and adopted by the State because the project fits into the Boston Harbor Clean-up program; and (5) Maintaining continuous dialog between the appropriate City and State officials so that all parties have been kept abreast of existing and proposed funding assistance programs.

Highlights of Phase VI

Completion of The Phase VI project will eliminate the City's largest remaining area that is served by a Combined Sewer System. Approximately 17.4 km (57,000 ft) of new sewers and storm drains will be constructed within the project's 300 acre highly urbanized area, which is located in the central section of the City beginning in DeWolfe St. at Memorial Drive, and ending at Oxford St. and Forest St.. Completion of the Phase VI project will significantly mitigate: CSO to the Charles River; storm flows to the MWRA North Charles Relief Sewer; and the combined wastewater flows to the new MWRA Deer Island wastewater treatment facility.

In order to minimize disruption to traffic and to the commercial, residential and academic sectors within the project area, the project is to be constructed under five (5) major contracts: 1, 1B, 1C, 2 and 3. Contract 1 has been completed, and Contracts 1B & 1C are presently under construction. Contract 1A (Phase VI), which is also complete, has reduced flooding problems along Sherman St. (located in the north western section of the City), and is outside the 300 Acre area which is the focus of this presentation.

- Contract 1: This construction initiated the Phase VI project by the separation of the combined sewer area located west of DeWolfe St., and is generally bounded by Holyoke St. on the west; Mill St. on the south; and Harvard St. on the north. Construction included approximately 10.9 km (3,570 ft) of new storm drain, ranging in sizes from 45.7 cm to 182.9 cm (18" to 72") diameter; and approximately 0.8 km (2,530 ft) of new sanitary sewers ranging in size from 25.4 cm to 61.0 cm (10" to 24") in diameter. This contract also included the construction of a new 182.9 cm (72") diameter storm drain outfall structure into the Charles River; jacking of a 182.9 cm (72") diameter storm drain under Memorial Drive and an existing Harvard University Utility Tunnel. The cost of this construction contract is $2.5 Million.

- Contract 1B: Construction includes the installation of a 182.9 cm (72") diameter reinforced concrete pipe within a tunnel- approximately 2.4 m (8 ft) in diameter- directly underneath the MBTA red line rapid transit subway tunnel in the Quincy Square area. The new storm drain tunnel is to be installed 18.3 m (60 ft) below the ground surface, and the top of the tunnel will be approximately 3.7 m (12 ft) under the concrete base of the subway tunnel. The unique construction procedure requires that the new pipe will be "jacked' and/or tunneled through weathered agilite rock which needs to be stabilized with grout before it can be tunneled. The tunnel/jacking section is approximately 0.4 km (1,350 ft) long; the crossing of the subway tunnel is approximately 15.2 m 50 (ft); and the required pipe tunnel is approximately 41.1 m (135 ft) long. The pipe tunnel is to be installed by drilling and blasting The procedure is subject to significant restrictions due to the proximity to other existing structures and the active subway tunnel. Part of the MBTA requirements include that the blasting must be performed during the hours when the MBTA is not in operation (i.e. between 1:30 am and 3:30 am), and a comprehensive vibration monitoring program is to be conducted during the operations.

- Contract 1C: This contract continues the separation process where Contract 1 ended. It involves the construction of approximately 1.2 km (4,000 ft) of new storm drain, ranging in size from 30.5 cm to 91.4 cm (12" to 36") in diameter; and approximately 0.9 km (2,800 ft) of new sanitary sewers ranging in size from 20.3 cm to 45.7 cm (8" to 18") in diameter. When completed this contract will have separated the area located east of DeWolfe St., and is generally bounded by Harvard St. on the north; Dana St. on the east; and Mt. Auburn St. on the south. The construction cost of this Contract is $2.4 Million.

- Contract 2; This contract will continue the separation process beginning at the end of Contract 1B. It will involve the construction of approximately 1.2 km (3,800 ft) of new storm drain, ranging in size from 30.5 cm to 182.9 cm (12" to 72") diameter; and approximately 2.1 km (7,000 ft) of new sanitary sewers, ranging in size from 20.3 cm to 61.0 cm (8" to 24") in diameter. The project area is located east of Quincy St. between Cambridge St. on the north; Broadway on the south; and Dana St. on the east. Construction of this contract is scheduled for early summer, 1995; with completion by the end of 1997. The estimated construction cost for this Contract is $6.2 Million.

- Contract 3: This the largest of Phase VI construction contracts, and it will begin where Contract 2 will end. It will involve the construction of approximately 4.7 km (15,500 ft) of new storm drain, ranging in size from 30.5 cm to 152.4 cm (12" to 60") in diameter; and 4.1 km (13,500 ft) of new sanitary sewers, ranging in size from 20.3 cm to 61.0 cm (8" to 24") in diameter. Construction of this contract is scheduled to start in July, 1996; with completion by the end of 1999. The estimated construction cost for this contract is $14.3 Million.

Summary

For the past 26 years the City of Cambridge, Maguire Group Inc., the Mass. DEP, and the MWRA have worked together to design, fund and coordinate the construction of a comprehensive Sewer Separation Project to eliminate CSO discharges to the Charles River, Alewife Brook and the MWRA wastewater collection and treatment systems.. Funding for the comprehensive project has been made available because it was designed to augment, and significantly contribute toward, the State's Boston Harbor Clean-up program. The City worked hard to maintain its strong financial status so it could bond the entire amount of the project (a prerequisite of the funding program); and representatives of the City and Maguire Group Inc. never gave up on working with the regulatory agencies toward obtaining financial assistance. In fact, when the Mass. DEP was in the initial stages of developing its low interest SRLF program, Maguire and the City were right there lobbying for the money. The result- Cambridge received one of the first Mass. DEP SRLF moneys for the current Phase VI project. The City's commitment, to implement the 1969 Facilities Plan, eliminates CSO discharges to two urban waterways and reduces combined flow to the MWRA sewer system; which in turn reduces flow to the MWRA wastewater treatment plant now being constructed to clean-up Boston Harbor. The money saved by the potential for reduced wastewater transportation and treatment costs eventually end up in the taxpayers' pockets of all the communities served by the MWRA.

Water Surface Oscillation in A Deep Sewer Tunnel System

Junn-Ling Chao[1], M, ASCE and Dong Mclearie[2]

Abstract:

In order to improve the water quality in the Victoria Harbor, the Government of Hong Kong has undertaken a great endeavor of implementing the stage 1 of Strategic Sewage Disposal Scheme. One of the major components of the Scheme is to construct a collection network of deep drop and rising shafts and tunnels of various diameters up to 3.5m totaling 25 km running under the urban Kowloon and the eastern end of Hong Kong Island. This tunnel system will connect to a main pump station with a capacity of 31 m^3/s capacity to lift the water to a major treatment works. Water surface oscillations in the shafts and its related overflow was simulated numerically resulting from power outage at the main pump station. Pump control to react with flow changes in the system were also modeled to provide design guide.

Introduction:

The rapid population growth in Hong Kong (about 6 million in 1993) and associated commercial and industry expansion has increased the sewerage output tremendously. Presently, Hong Kong produces about 2.0 million cubic meters per day sewerage. As in any coastal cities, those wastewaters are generally discharged into the surrounding water bodies in the past and practiced many places today. Since most of the population and activities in Hong Kong are around the Victoria Harbor, these discharges at the seawalls have great ill impact of the harbor water quality. In 1970 the first comprehensive and integrated sewerage disposal strategy for the harbor was conducted.

Based on the initial study, a new sewerage disposal strategy study was commissioned to protect the territory's watercourses, rivers and marine environment in 1987. Montgomery Watson was the primary consultant to carry out that project known as SSDS (strategic Sewage Disposal Scheme) which has a phased implementation recommendation to clean up the coastal waters. The scheme compromises: Stage 1 - Principal collection and treatment system collect existing screened sewerage in South Kowloon and north east of Hong Kong Island transport it via deep tunnels to Stonecutters Island for primary treatment then discharge via an interim outfall system; Stage 2 - long and deep oceanic outfall some 30 km south from Stonecutters and discharge into the South China Sea; Stage 3 - collection system for North and Southeast Hong Kong Island, treatment at Mount Davis and discharge into deep outfall. The transient hydraulic study is part of the design work for Stage 1 tunnel system.

System Description and Study Objectives:

The tunnel system consists of three subsystems; i.e. the KC-SCI-KT subsystem

---

1. Senior Consultant, Montgomery Watson, Inc., 300 N. Lake Ave., Pasadena CA 91101
2. Director, Montgomery Watson Asia, 311 Gloucester Road, Hong Kong.

(Kwai Chung - Stonecutters Island-Kwan Tong); the CW-KT subsystem (Chai Wan-Kwun Tong) and the TKO-KT subsystems (Tsueng Kwan O-Kwun Tong). In this presentation , we shall only discuss the KC-SCI-KT subsystem. The total length of the tunnel system is about 25 km with diameter ranges from 1200 to 3540 mm. The deepest shaft is about 150m. The system schematic is shown in Figure 1.. Figure 2. shows the typical drop shaft design. Sewerage collected from TKO will be pumped to KT drop shaft. Sewerage from Chai Wan and Shau Kei Wan will be gravitated to KT pump station and then lifted to KT drop shaft. This 7.61 cms flow entering KT drops shaft together with the flow from KTPTW and TKWPTW totaling of 21.8 cms will be gravitated to stonecutter island treatment plant main pump station (SMP). Sewerage collected from Kwai Chung and Tsing Yi will be gravitated to SMP as well with a flow of 9.45 cm. and this gives a total maximum flow of 31.25 cms to the SPM which shall be lifted to the treatment plant headworks.

Objectives of the study are: (1) Simulate the transient conditions under a set of boundary conditions; (2) Ascertain water surface changes in the drop and rising shafts; and (3) Estimate volumes of overflow at those shafts where the water surface rises above the overflow weir elevation and its maximum rate of overflow for a given time span. This will give some estimates on possible pollution impacts to local waters.

Analysis and Results:

As described the operational scheme is to gravity the flow from the drop shafts to the SMP and lifted to the treatment headworks. In order to minimize the pumping cost at the SMP, it is intended to operate the water surface in all drop shafts at high water levels. Under this condition, if power failure takes place at the SMP, it would result of overflows at some of the upstream drop shafts. Water surface oscillations and overflows are sought in the analysis as stated in the objectives. Characteristic method for transient analysis developed by professor Streeter and his colleague at University of Michigan with modified boundary conditions was employed in this investigation.

Starting from steady state flow as initial condition and power failure at SMP as time "**0**", the simulation runs for a number of real time spans. 900 seconds was the longest run. Because it is highly possible that power outage occurs at SMP only, all flows from the PTWS will continue to enter the drop shafts without interruption. The simulations were conducted with the pumps stop operation at SMP but all flows to the individual drop shafts continue until the water surface in the drop shaft reaches its respective overflow weir elevations. At that time, flow will cease to enter the respective drop shafts and starting emergency overflow.

Water surface oscillations and overflows for the drop shafts and SMP rising shafts were obtained from the simulations. Because of the difference on the drop shafts structural configurations, storage volume and overflow Weir crest elevations, except at SMP rising shafts, overflows will occur at all drop shafts starting about 200 seconds to 400 seconds following power failure at SMP. The volumes of overflow at all shafts are not high. However, because of the continuous inflow from the catchment to the PTW which overflow its emergency overflow weir, the total overflow is high unless the power outage at the SMP can be restored promptly. Figure 3 shows the water surface changes during the transient time for the SMP rising shaft and two typical drop shafts

References:

Montgomery Watson: "Technical Note, SSDS Stage 1 Tunnel System Steady State Flow Hydraulics" to DSD, Hong Kong Government, Jan.,1993

Montgomery Watson: 'Technical Note, SSDS Stage 1 tunnel system Transient Hydraulic Analysis" Internal Report, June, 1993

Wylie, E.B. and Streeter, V.L. : Fluid Transients" McGraw-Hill Inc., 1978

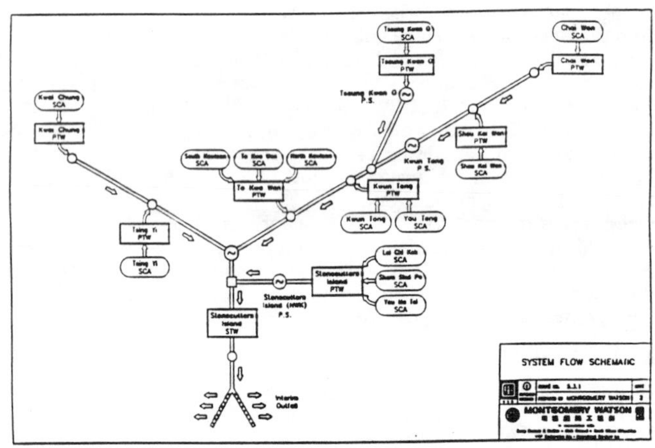

Fig. 1 System Schematic

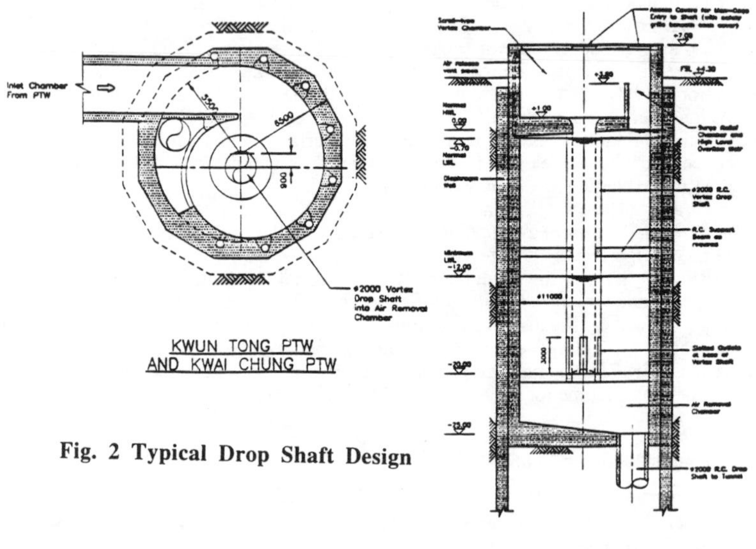

Fig. 2 Typical Drop Shaft Design

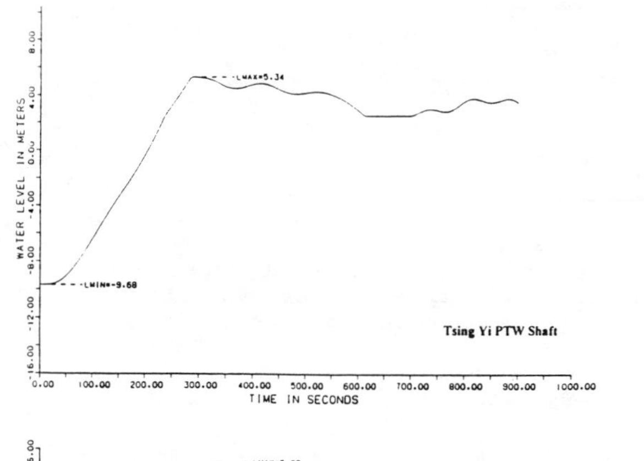

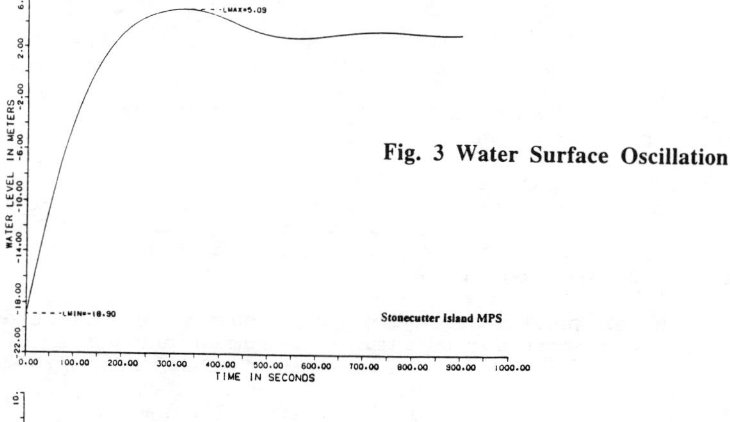

Fig. 3 Water Surface Oscillation

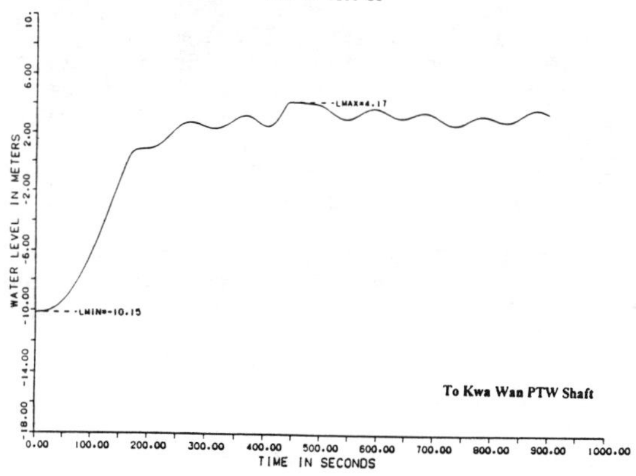

Modeling the Impact of Macrophytes on Instream Dissolved Oxygen Dynamics in the Pennsauken Watershed

James F. Cosgrove, Jr., P.E.[1], Member ASCE
Christopher C. Obropta, P.E.[2], Member ASCE

Abstract

The Pennsauken Watershed is located in the New Jersey counties of Burlington and Camden with a drainage area of approximately 33 square miles. The Pennsauken Creek is tidally influenced and discharges to the Delaware River. Instream total phosphorus concentrations consistently exceed the New Jersey receiving water quality criterion of 0.1 milligrams per liter (mg/l). To determine the effect of reducing the phosphorus point source loads on the receiving waters, Omni Environmental Corporation conducted a monitoring and modeling study within this watershed.

Water quality and hydrodynamic data were collected over a ten month period during the summer and winter of 1992. Although the main channel of the Pennsauken Creek is relatively free of rooted vegetation (macrophytes), large highly vegetated tidal flats are located along the banks. These tidal flats impact the instream dissolved oxygen (DO) during high tides by increasing the instream DO during daylight hours and decreasing DO during nighttime hours. Therefore, a dynamic model was needed to simulate the macrophyte effect on DO as the water depth increased and the tidal flats became submerged.

The Water Analysis Simulation Program (WASP), developed and distributed by the USEPA, was modified by Omni, in conjunction with AScI Corporation, to simulate

----

[1] Associate, Omni Environmental Corp., 211 College Road East, Princeton, NJ 08540-6623.

[2] Associate Project Manager, Omni Environmental Corp., 211 College Road East, Princeton, NJ 08540-6623.

the impacts of macrophytes on dissolved oxygen and incorporate a depth correction factor to account for water depth fluctuations within the tidal flats. This modified WASP model has been named OmniWASP.

Model Theory

Presently, there are no well established mechanistic methods available to predict macrophyte growth and death. Therefore, the approach taken in OmniWASP is quasi-predictive. A maximum productivity rate is specified in the model, which is then modified by available light and nutrients to estimate the specific productivity rate. Macrophyte respiration is computed from the maximum growth rate and a specified ratio of respiration to productivity (AScI, 1993).

In the Pennsauken system, macrophytes are confined to the tidal flats and are covered with water only during periods of high tide (see Figure 1). Simulation of these conditions would typically require the segmentation of the tidal flats separately from the channel areas. However, wetting and drying in the tidal flat segments would make such a model simulation difficult. Alternatively, in OmniWASP, the rate of productivity may be modified by a depth correction term which reflects the submergence of the tidal marshes (AScI, 1993). As the water depth increases above a user specified depth, the impact to instream dissolved oxygen from plant photosynthesis and respiration are simulated by OmniWASP.

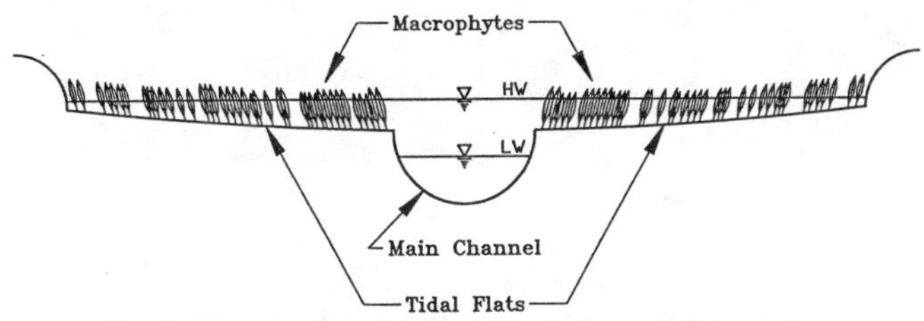

Figure 1: Cross-section of a Tidal Marsh With Macrophytes

Modeling the Pennsauken Watershed

The hydrodynamic program, DYNHYD5, was used to model the hydrodynamics of the Pennsauken Creek. The model was calibrated using the data collected on August 25-27, 1992 during spring tide conditions. The model was then verified using neap tide data collected on September 17-19, 1992 (Omni, 1994a). Manning's roughness coefficient was used as a calibration device.

The output from the DYNHYD5 model is read directly into the water quality model, OmniWASP. Data collected during a two-day (August 26-27, 1992) intensive sampling event were utilized to calibrate nutrient and photosynthesis coefficients in the OmniWASP model of the Pennsauken Creek. The calibrated model was then verified using two other diurnal data sets collected during the 1992 nutrient study (October 1, 1992 and November 17, 1992). As a check, the model was also reverified using the average of all available summer data. The model simulated the observed nutrient and DO data quite well. The addition of macrophytes to the model allowed large diurnal DO swings to be simulated very accurately.

A comparison of typical diurnal DO modeling results using the WASP model before the macrophyte modification and after the modification is shown in Figure 2. This figure clearly illustrates that the impact of macrophytes on DO is required to properly simulate the diurnal DO swing. The modeling depicted in Figure 2 was conducted during the critical conditions when high tide occurred at approximately 14:00 hours and 2:00 hours. Submergence of the tidal flats during mid-afternoon and slightly after midnight created the largest impact on instream dissolved oxygen due to macrophyte photosynthesis and respiration.

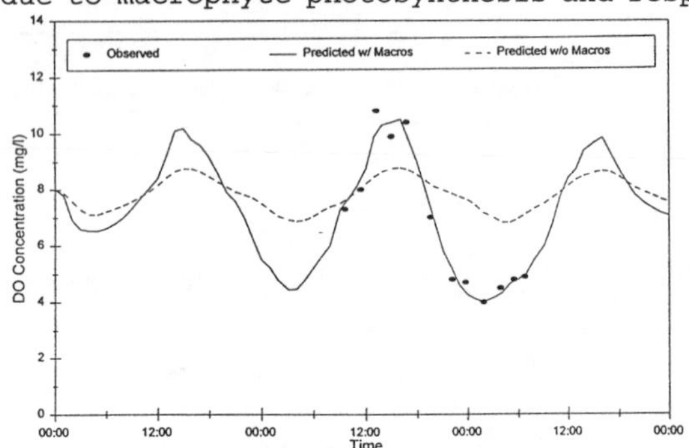

Figure 2: Typical Modeling Results With and Without Macrophyte Impacts

The calibrated and verified model was used to simulate future conditions under critical stream flows for both summer and winter conditions. These simulations demonstrated that a decrease in point source total phosphorus concentration had little effect on the instream DO. As the effluent limitation for total phosphorus approaches the receiving water standard of 0.1 mg/l, the minimum instream DO increased slightly, while the average DO decreased slightly, creating an overall smaller DO swing. This phenomenon is illustrated in

Figure 3. The increase in minimum DO was not substantial enough to require the implementation of strict total phosphorus limitations (Omni 1994b).

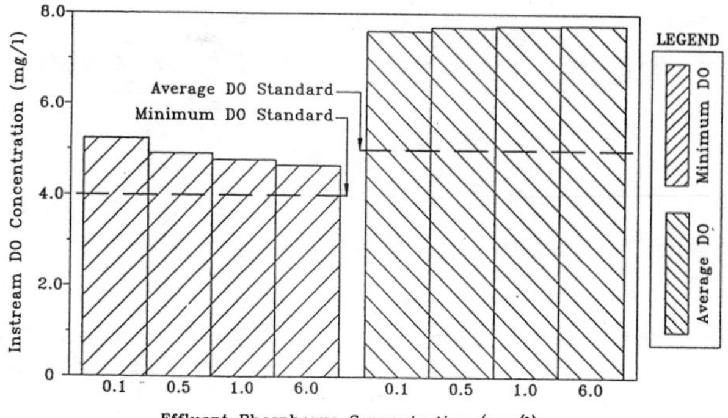

Figure 3: Typical Instream DO Impacts From Varying Point Source Total Phosphorus Concentrations

Conclusion

Since the USEPA developed model, WASP, is incapable of simulating the impacts to instream nutrients and dissolved oxygen from rooted vegetation, it was not capable of accurately simulating water quality in the Pennsauken Creek. However, by modifying the model to incorporate the impacts of macrophytes, OmniWASP can simulate nutrient and dissolved oxygen dynamics well.

Once the OmniWASP model was calibrated and verified with three sets of data collected during 1992, it was used to determine whether a reduction in total phosphorus point source loadings would lead to increased instream dissolved oxygen concentrations. Simulations showed that although a reduction in point source discharges of phosphorus lead to a reduction in instream phosphorus concentrations, the change in instream dissolved oxygen is very minor.

References

ASCI, "MACRO-EUTRO, A Eutrophication Model which includes the Effects of Macrophytes," December 1, 1993.

Omni Environmental Corporation, "Nutrient Study for the Pennsauken Creek," January, 1994.

Omni Environmental Corporation, "Pennsauken Creek Nutrient Study for the Township of Moorestown and Evesham Municipal Utilities Authority," December, 1994.

Impacts of NPS Pollution Control On Water and Sediment Quality Within Salem Sound, Massachusetts

By: Richard M. Baker[1]

Abstract

Nonpoint source pollution residence times predicted using the tidal prism method are compared to those found using transient two dimensional models of Salem Sound. Results suggest that the tidal prism method is not appropriate for estimating residence times in this coastal embayment.

Introduction

Salem Sound is a shallow coastal embayment located approximately 20 miles north of Boston, Massachusetts. The sound is a valuable coastal resource which is heavily utilized for commercial and sport fishing, boating, bathing and a wide variety of recreational activities. Dry and wet weather shoreline monitoring (Massachusetts Audubon, 1990) indicates that significant nonpoint source (NPS) pollution enters the four major harbors within the sound from heavily urbanized watersheds located within the towns of Manchester, Beverly, Peabody, Danvers, Salem and Marblehead. Relative contributions of individual PS and NPS inputs to pollutant levels within the sound's water column and bottom sediments must be determined if beneficial impacts of NPS control strategies are to be adequately quantified. Development of landside watershed and receiving water hydrodynamic and water quality transport models has been initiated. Landside modeling will utilize the Hydrologic Simulation Program - Fortran (HSPF) (Johanson et al.,1984). Receiving water impact analyses are being conducted using the Tidal Embayment Analysis-Non Linear (TEANL) and Eulerian Lagrangian Analysis (ELA) models (Westerink et al., 1985 and Baptista et al., 1984). These transient two-dimensional models are being used to simulate tidal hydrodynamics and water quality transport processes, respectively. TEANL and ELA based models developed previously (Adams, 1989) for simulating Salem Sound and adjacent waters of Massachusetts Bay were adapted for this study.

[1] Principal, NUMERIC Environmental Modeling, R643 Hale, Beverly, MA 01915

Procedure

NPS pollution residence times were developed using a modification of the tidal prism method (Dyer, 1973). This method assumes that initial pollutant mass is instantaneously mixed within an embayment volume at high tide and that the portion of the mass contained within the volume of the tidal prism is lost from an embayment during the ebbing tide. It further assumes that during the flooding tide all mass remaining within an embayment is fully mixed with incoming seawater, which is assumed to be free of pollutant mass. Thus, the tidal prism method predicts an exponential loss of pollutant mass from an embayment and its corresponding residence time, t_{res} , is defined as the time required for the mass of pollutant remaining to decrease to $1/e^{th}$ (approximately 37 %) of the initial mass. Hence $t_{res} = V_h / (V_h - V_l)$, where V_h and V_l are the embayment volume at high and low tide, respectively.

Due to simplifying assumptions, the tidal prism method fails to account for numerous factors and processes likely to influence residence times of NPS pollutants introduced into tidal embayments during storms. These include spatial and temporal variability of tidal and non-tidal (residual) circulation, horizontal diffusion, return of endogenous and exogenous pollutant mass during flooding tides, and the initial spatial distribution of NPS pollution during a storm. To estimate the relative impact of several of the above assumptions on residence time predictions a series of runs were made using the TEANL and ELA models of the sound. These models due not include the simplifying assumptions of the tidal prism method.

TEANL was run in the non-linear mode with stage ocean boundary conditions corresponding to five tidal harmonic constituents known to dominate tidal variations within Massachusetts Bay, including: M_2 , N_2 , S_2 , K_1 and O_1. The "zero tilt" stage ocean boundary condition utilized by Adams (1989) to study circulation and pollutant transport within Massachusetts Bay was adopted as a worse-case assumption. TEANL was also run using these stage boundary conditions plus steady, non-zero normal flux boundary conditions along land boundary segments corresponding to 12 significant river inflow points within the individual harbors of the sound. In leu of detailed catchment runoff modeling to be conducted at a later date, storm event-average river flows were estimated based on subcatchment drainage area data and the rational formula, assuming a runoff coefficient of 0.20 and an event-average precipitation rate of 0.05 inches per hour. The resultant river inflows to each harbor, in units of m^3/s, were 1.6 (Manchester), 7.3 (Beverly), 1.8 (Salem) and 0.5 (Marblehead). TEANL nodal water depths and current velocities, predicted with and without river inflows, were subsequently used in the ELA modeling. In an effort to approximate the initial spatial distribution of NPS pollutants within the sound and its harbors during a rainfall event, ELA was hot-started with shoreline boundary node concentrations set at 1000 mg/l and all other nodal concentrations set to zero. Separate ELA runs were hot-started with non-zero concentrations at nodes located along 1) all shorelines within the sound and its harbors, and 2) only along shorelines within each harbor (one ELA run per harbor). The objective here was to determine residence times of each harbor due to its local NPS inputs, as well as those derived from other harbors and shorelines within the sound.

The two-dimensional model domain was divided into regions contained within each harbor and the sound as a whole. ELA was then modified to generate water volumes and water quality constituent mass contained within each of these regions, during each model time step (11,160 sec). ELA runs were carried out to a duration of one lunar month. Horizontal diffusion coefficients determined by Adams (1989) for the sound ($D_x = D_y = 45\ m^2/s$) were used throughout, except for one ELA run made to determine sensitivity of predicted residence times to a 10% reduction in D_x and D_y.

Results

The tidal prism box model, which was implemented as a spreadsheet program, utilized ELA harbor and sound volumes generated at each time step and the time required to reach $1/e^{th}$ the initial pollutant mass within each region was determined. Results for spring and neap tide conditions are given in Table 1, where it is seen that residence times during neap tide conditions were approximately two times longer than those predicted for spring tide conditions.

Results for each scenario simulated using TEANL and ELA, corresponding to the period around spring tides, are also given in Table 1. For comparative purposes only, residence times determined using ELA predictions correspond to the time required to reach $1/e^{th}$ the initial mass within each region. ELA predicted decreases in harbor and sound pollutant masses over time did not typically follow an exponential curve after approximately one tide cycle, due to return of previously discharged pollutant mass on incoming tides.

Table 1 : NPS Pollutant Residence Times (Days)

	Sound	Manchester	Beverly	Salem	Marblehead
	Tidal Prism Model				
Spring Tide	1.3	0.4	0.6	0.7	0.9
Neap Tide	2.3	1.1	1.1	1.4	1.8
	TEANL/ELA Models				
All Sources	7.0	0.3	2.5	3.0	0.8
Local Sources	-	0.1	1.1	1.0	0.3
All Sources, River Flows	4.8	0.2	1.7	2.0	0.3
All Sources, $0.9{*}D_{x,y}$	8.7	0.7	2.9	3.3	1.1

Residence times predicted for Beverly and Salem harbors and the sound using ELA with all sources were found to be 4 to 5 times longer than those determined using the tidal prism method. These differences are likely attributable to ELA's ability to simulate incomplete initial and subsequent mixing, return of previously discharged pollution mass on flood tide and inter-harbor pollutant transfers. In contrast, ELA predicted residence times for Manchester and

Marblehead harbor due to local sources only were found to be approximately one third those determined using the tidal prism method. This result is likely due to ELA's accounting for dispersive pollutant transport.

Comparison of results for the ELA run with all sources to those including only sources local to each harbor suggests that significant amounts of NPS pollution may be rapidly advected and diffused between Beverly, Salem and Marblehead harbors, due to their close proximity and location approximately along a tidal streamline. In contrast, only a small amount of mass transfer appears to occur between these harbors and Manchester harbor, which is located much further to the east on the opposite shoreline of the sound.

Results of the TEANL/ELA modeling further suggest that harbor and sound residence times are sensitive to rates of horizontal diffusion and river-induced residual circulation.

Conclusions

Modeling results suggest that due to its simplifying assumptions the tidal prism method is not appropriate for estimating NPS pollution residence times in Salem Sound and its harbors. Use of the tidal prism method alone for assessing impacts of NPS pollution and control strategies in other Massachusetts coastal embayments would not appear to be a valid approach.

References

Adams, E.E., 1989. South Essex Sewerage District (SESD) Secondary Treatment Facilities Plan, Receiving Water Modeling Appendix, for CDM, Boston, MA.

Baptista, A.E., E.E. Adams, and K.D. Stolzenbach, 1984. Eulerian-Lagrangian Analysis of Pollutant Transport in Shallow Water (ELA), MIT Energy Laboratory, Report No. MIT-EL-84-008, Cambridge, MA, June 1984.

Dyer, K.R., 1973. Estuaries: A Physical Introduction. John Wiley & Sons, London.

Johanson, R.C., J.C. Imhoff, J.L. Kittle and A.S. Donigian. 1984. Hydrological Simulation Program - Fortran (HSPF), Release 8, Environmental Research Laboratory (ERL), U.S. Environmental Protection Agency (USEPA), Athens, GA, April 1984.

Massachusetts Audubon Society of the North Shore, 1990. Beverly, Marblehead, Salem Harbor Monitoring Report, Wenham, MA., April 1990.

Westerink, J.J., K.D. Stolzenbach, and J.J. Connor, 1985. A Frequency Domain Finite Element Model For Tidal Circulation (TEA), MIT Ralph Parsons Laboratory, Technical Report No. MIT- EL-85-006, Cambridge, MA.

INNOVATIVE DATA MANAGEMENT TECHNIQUES FOR HYDROLOGIC AND HYDRAULIC MODELING IN COMBINED SEWER OVERFLOW ABATEMENT PROGRAMS

By Jennifer L. Angell[1], James T. Smullen, M.ASCE[2], Albert Schneider[3] and Keith Miller[4]

Abstract

The proposed paper describes a comprehensive model support system used by the Allegheny County Sanitary Authority's (ALCOSAN) CSO Permit Compliance Program. This system relies upon the integration of database management software, a geographic information system (GIS), statistical software and a model graphical post-processor. The use of these techniques has resulted in the efficient management of the vast amounts of data required for CSO modeling, facilitating the development of a beneficial CSO control program.

Introduction

The EPA National Combined Sewer Overflow (CSO) Policy requires that CSO dischargers perform a system inventory and hydraulic characterization as part of their CSO compliance programs. Detailed hydraulic and hydrologic models of the sewer system and tributary drainages typically are developed to understand the existing system, to characterize CSO frequency and volumes, and to evaluate the effectiveness of alternative interceptor and wastewater treatment plant operating scenarios. The models are used in later phases of the compliance program to support the development of the Nine Minimum Control documentation and a Long Term CSO Management Control Plan. These complex models require that significant amount of environmental data be stored, managed, analyzed and displayed.

---

1 Environmental Scientist, Camp Dresser & McKee, Inc., Edison, NJ.

2 Principal Engineer, Camp Dresser and McKee, Inc., Edison, NJ.

3 Manager of Construction, Allegheny County Sanitary Authority, Pittsburgh, PA.

4 Environmental Engineer, Camp Dresser and Mckee, Inc., Edison, NJ.

The ALCOSAN CSO Program is using the USEPA Storm Water Management Model (SWMM) Extended Transport (EXTRAN) block for hydraulic evaluations and the USCOE Storage, Overflow, Runoff Model (STORM) to provide continuous simulation of the sewershed hydrology. The use and application of these models in CSO abatement planning is widely documented.

The implementation of hydrologic and hydraulic models on large scale systems requires efficient data management, a method to rapidly develop model input files, and the capability to produce graphic displays (i.e., maps, graphs, etc.) that assist in the interpretation and understanding of facility characteristics and model results. While in the past these management systems may have been considered a desirable feature, today, large-scale modeling projects virtually are dependant on the use of a geographically-based data management system for their efficient and cost-effective implementation. A schematic of the ALCOSAN CSO data management system for modeling support is shown in Figure 1.

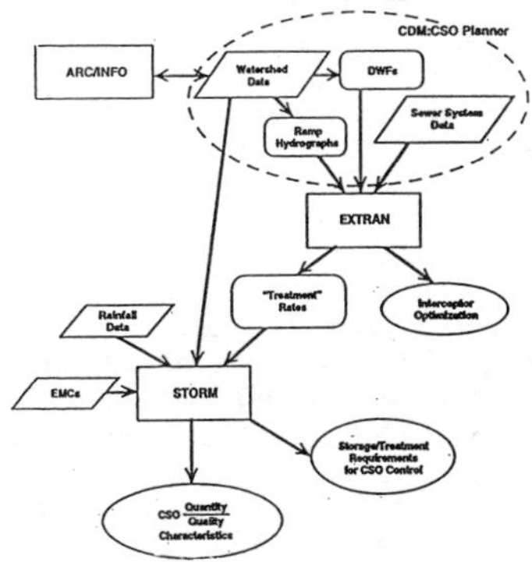

Figure 1 Schematic of the ALCOSAN CSO Program modeling data management system

The principal data management tool for storage and management of hydraulic model input information for the ALCOSAN CSO Program is the CDM:CSO Planner package. The CDM:CSO Planner is a data base management program developed specifically for using the PROGRESS data base application software and the AUTOCAD computer-aided design/drafting application software packages. It includes the following capabilities:

- building and managing large databases via digitization and tabular input;
- evaluating the connectivity of the interceptor and verifying flow directions;
- generating inputs for hydraulic and hydrologic models;

- mapping facility information and modeling results, il-lustrating existing characteristics and future performance.

The ARC/Info software package is the GIS used to store, manage, and display spatial information for the hydrologic modeling of the ALCOSAN service area. The US Census Bureau TIGER/Line files were used to develop a base map of the area. Information in the ALCOSAN CSO GIS includes land use, slopes, population density, sewershed boundaries, sewer billing data, political boundaries, impervious cover estimates, industrial discharger locations and other spatial data. Some of the typical data analyses performed using the GIS are described below.

The GIS was used to combine the 1990 Census data with Alcosan billing records to determine the dry weather flows contributed to the interceptor from households within the service area. The GIS is invaluable in overlaying the individual population and drainage boundary layers to create an interpretative scene that aggregates these data to the watershed/sewershed level.

Using the GIS to provide estimates of the impervious cover characteristics of the sewersheds is another example where the GIS easily facilitated what otherwise would have been an excessively burdensome task. Initially, these estimates were performed using standard percent imperviousness land use characterizations available from the literature. However, ground-truth investigations revealed that the original imperviousness scenes based on literature values were somewhat lower than expected for the central business district and the directly adjacent high-density residential areas. GIS-based analyses of population in the commercial and the three residential land use-classes suggested that alternative approaches to the estimation of impervious land cover should be considered. Two separate population density imperviousness relationships were employed to provide new estimates for percent imperviousness in residential areas. The commercial land use also was reclassified into two land uses, urban commercial and suburban commercial, based on an analysis of the population density in these areas. The combination of high speed computers and the GIS allowed the analysts to perform these evaluations at the Census Block level. Comparisons of results indicate that the new methodology provides superior estimates. This data-intensive effort would have been much too difficult to perform at this level of detail without the GIS. But population-density-based analyses must be performed at this level of detail when relationships between population density and another attribute are nonlinear. Clearly, the GIS facilitates new approaches to supporting models that previously would not or could not be followed.

The effort required to interpret model results is greatly reduced by the use of a graphical post-processor. The CSO Program uses the Model Turbo View-EXTRAN (MTVE) software package as the post-processor for the hydraulic modeling. The MTVE software allows EXTRAN inputs and results to be displayed in multiple formats (e.g., hyetographs, hydro-graphs, hydraulic grade lines, surcharges, and regulator and manhole overflows). Figure 2 shows an MTVE plan view of a interceptor system as it is represented in the EXTRAN modeling network. Today, the use of a highly sophisticated post-processor like MTVE allows the model analyst to elucidate aspects of both the system modeled, and the model itself, that were extremely difficult to discern in the past when model output was mostly limited to print-outs and line printer graphics.

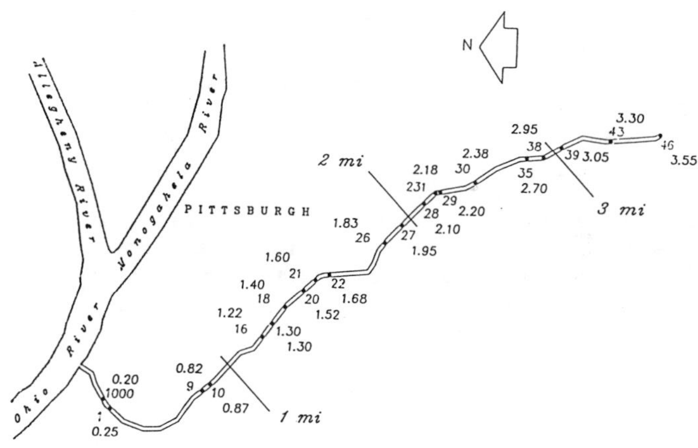

Figure 2 MTVE plan view of intercepting sewer

THE TCHELO DJEGOU WATER-WHEEL

Appropriate Technology Solutions for Water Supplies in Rural West Africa

Randy Brown[1]
Associate Member

ABSTRACT

When the maintenance and repair costs of an out-of-production mechanical hand pump exceeded the financial resources of a remote village in western Niger, the pump was replaced with an affordable appropriate technology water-lifting device.

INTRODUCTION

Over the past two decades many Western governmental agencies and international aid organizations have worked to develop water resources infrastructures in Africa. Large drinking water campaigns were instituted and thousands of deep bore-hole wells and mechanical pumps were installed across the continent. In many regions these programs helped to improve the health and quality of life for large populations. After only two decades, however, much of this infrastructure is in decay. Many of the mechanical pumps installed in small and remote villages are now aged and in need of costly repairs at a time when the manufacturers of these pumps have limited their operations or even discontinued production entirely.

As a result of the low success rates of large-scale development efforts, the practice of appropriate and sustainable technology is being integrated into a new generation of development projects. These projects utilize local materials and technologies, integrate technical training programs, and require a demonstrated financial commitment from the serving population. This paper describes an alternative water-lifting device that incorporates these ideas to provide a reliable water supply for the village of Tchelo Djegou.

---

[1] Civil Engineer, Department of Remediation and Waste Engineering,
ENSR Consulting and Engineering, 35 Nagog Park, Acton, Massachusetts, 01720

BACKGROUND

Tchelo Djegou is located in a remote region of rural Niger, in West Africa, approximately 80 kilometers south of the capital city, Niamey. The village is inhabited by approximately 200 semi-nomadic herders of the Fulani ethnic group who settled the region in search of fields in which to grow the countries primary subsistence staple crop of millet. Located in the southern region of the Sahara Desert, Niger has an extremely hot and dry climate. Most regions of the country receive less than 800 millimeters of annual rainfall, all of which falls during a brief three-month rainy season. The country has suffered several severe droughts during the past 30 years, devastating many of the nomadic populations.

In the region of Tchelo Djegou, the terrain consists of long and barren plateaus cut by fertile river valleys. Traditionally, the region's populations inhabited the valleys due to their close proximity to water and fertile fields. Most of the rivers in this area are temporary, coinciding with the rainy season. However, even in the height of the nine-month dry season, the water table is generally only five to ten meters below the ground surface. During the rainy season, water is collected from the temporary rivers to meet daily needs. As the rivers dry up, unlined, hand-dug wells are constructed in the shallow aquifers along the rivers. As the country's population increased and the limited farm land in the river valleys was allocated, farmers were forced onto the plateaus to clear new land for fields. On these plateaus, aquifers range from twenty to one hundred meters below the ground surface. Because traditional unlined wells cannot be dug to such depths farmers walk long distances, in some cases up to ten kilometers, to collect water for their daily needs.

As in many of the surrounding communities, a 110-meter well and mechanical hand pump were installed in Tchelo Djegou as part of a region-wide international water resources development project in the early 1970's. The new water supply attracted families forced onto the plateau in search of new fields, and the village population increased. For nearly two decades, the mechanical hand pump in Tchelo Djegou continued to meet the water needs of the local population. As the pump aged, however, maintenance costs increased and eventually exceeded the financial resources of the village. The pump was abandoned and the villagers were forced to make an exhausting eight-kilometer walk to a distant river valley with pails atop their heads.

DEVELOPMENT OF AN APPROPRIATE SOLUTION

Between 1991 and 1993, as part of a rural development program implemented by the US Peace Corps, special emphasis was placed on the revitalization of water resources infrastructures in a region of Niger that includes Tchelo Djegou. This program focused on region-wide water resources infrastructure training programs, the substitution of concrete-lined drinking water wells in villages using unlined wells or relying on untreated river water, and the development of sustainable solutions for water supplies in villages located on the region's plateaus.

In early 1991, Tchelo Djegou villagers approached the US Peace Corps requesting financial aid to help repair their pump. When the project was researched it was discovered that their pump was no longer in production, and that used parts were difficult to purchase. Due to the extreme depth of Tchelo Djegou's well, a heavy winch system was required to remove the pump from the well for repairs. This costly exercise required a truck to bring the necessary equipment from the distant capital city, Niamey, placing the village in a position of dependency on non-local resources.

The depth of Tchelo Djegou's aquifer ruled out the option of constructing a concrete drinking water well. However, the village's small population did not justify replacing the pump, as the same financial problems would eventually arise. The operating costs were consistently difficult to meet; moreover, the villagers could never make a significant financial contribution toward the initial cost of a new pump (approximately US $4000).

AN APPROPRIATE TECHNOLOGY DESIGN

In October, 1991 the mechanical pump in Tchelo Djegou failed for a second time that month, and the cost to repair the pump was excessive. As a result, the village elected to have the entire pump assembly removed from the well and to evaluate more sustainable alternatives that could incorporate the existing well. Several preliminary designs were considered, including "animal traction" water lifting systems used locally for agricultural applications. Many of these systems, however, required large amounts of rope to operate, and were consequently rejected due to their high operation and maintenance costs.

Final design criteria for the system included complete in-country fabrication, the use of only locally available materials, and a cost of less than US $1200. In addition, it was essential that the system could be operated safely without requiring large numbers of villagers. After several months of experimentation, a final design was submitted to the United States Embassy Self-Help Fund on January 29, 1992. Figure 1 shows the "human traction" wheel in operation shortly after construction was completed. The basis for this simplistic design is a fabricated wheel three meters in diameter. The wheel acts as a spool onto which 100 meters of nylon cord are wrapped. A modified version of a bailer, designed by Lutheran World Relief engineers for agricultural applications, is attached to the cord, and is centered over the well opening with a frame and pulley. The system is operated by rotating the wheel to raise and lower the bailer into the water table 94 meters below the ground surface.

The 1.5-meter long bailer provides fifteen liters of water per cycle, the volume of a standard Nigerien pail. The system requires three people to provide the traction inside the wheel while a fourth person retrieves and empties the bailer upon its arrival at the surface. One complete bailing cycle takes about three minutes and requires ten full revolutions of the wheel in each direction. The bailers are constructed with 10.5 centimeter PVC, and use a simple check valve. To remove water from the bailer, a six centimeter pin is welded to the valve gate such that when a full bailer is placed inside a pail, the pin's contact with the bottom surface of the pail opens the valve and releases the water.

Figure 1: Operation of the Tchelo Djegou Water Wheel

CONCLUSION

By designing a water system based on appropriate principles, this village was able to remove its defunct mechanical pump from the existing well and install a new system that was both simple to operate and constructed locally with locally available materials. Based on the village's population, and assuming that each person will use approximately ten liters of water per day, the village's total water needs can be met in 5 hours of daily operation. Although not as efficient as the western standards of Tchelo Djegou's original mechanical pump, this system is far more efficient than a three hour walk to the nearest river. Finally, the many problems associated with a sophisticated and imported technology were resolved, while the operation and maintenance costs remained affordable, thus achieving a higher level of infrastructural independence for the villagers of Tchelo Djegou.

FAST-TRACKING EMERGENCY CHLORINATION SYSTEMS FOR YEREVAN, ARMENIA

Frederick J. McNeill[1]

Abstract

This paper describes a fast-track chlorination project in the Republic of Armenia that consisted of systems design, procurement, installation, and operational training services. These services were carried out over a period of 15 months which included five trips to Armenia by various team members and five equipment shipments. This paper details the trials, tribulations and triumphs of working with a foreign utility on a fast-track design and installation project under particularly adverse conditions.

Introduction

Armenia, formerly one of the 15 republics that made up the Union of Soviet Socialist Republics, is now an independent state. The recent political, economic, and social transitions have created many challenges for the people of Armenia. One of the major challenges is an armed conflict with neighboring Azerbaijan over the disputed Nagorno-Karabakh territory. Consequently, Armenia has been under a crippling energy blockade from its neighboring countries for more than four years. The energy blockade has left most of the country with sporadic heat, water, and electric service. Of the many hardships facing Armenia, the safety of its drinking water was identified as a major concern.

In March 1993, a study team sponsored by the United States Agency for International Development (USAID) met with various Armenian officials to assess their critical water and wastewater needs as a result of the energy blockade. The study team identified new chlorination

---

[1] Project Engineer, Camp Dresser & McKee International, Inc. Ten Cambridge Center, Cambridge, MA 02142

equipment for Yerevan's drinking water system as a critical need for protection of public health. Under a technical assistance contract with USAID, a scope of work for a fast-track project was developed to design, procure, transport, install, and provide training in the operation of the chlorination systems. In addition to the chlorination equipment, other critically needed water and wastewater equipment identified by the study team was included in this contract.

Equipment Evaluation and Design

A review was conducted on the recommendations outlined in the USAID field report. Recommendations were prioritized with the emphasis based on the prevention of water borne disease outbreaks. An evaluation was then performed regarding the mechanical, electrical, and general equipment compatibility with existing conditions in Armenia. A technical review was also conducted to ensure that appropriate technology was being utilized and that it was compatible with the existing level of technology in Armenia.

A concentrated effort was required to design new chlorination systems for the eleven drinking water distribution/pump stations serving Yerevan, the capital of Armenia. The Yerevan drinking water distribution system consists of 1,937 kilometers (1,204 miles) of pipeline providing over 993,600 cubic meters per day (262 mgd) of drinking water to a pre-conflict population of 1.5 million people in the greater Yerevan area. The stations are supplied by either mountain streams or artesian well fields. The raw water quality from these sources is generally good, and disinfection with chlorine is the only treatment that the water receives.

The distribution/pump stations were being serviced by failing, antiquated, and dangerous direct pressure feed chlorination systems. The design emphasis for the new chlorination systems, was placed on simple technology, safety, flexibility of installation, ease of operation, and the ability to interchange spare parts. Based on these criteria, vacuum-operated, solution fed, manually adjustable chlorination systems were chosen.

Equipment Procurement

Due to the emergency nature of the project, a fast track procurement procedure was utilized to purchase the equipment for this project. Requests for quotations were sent to a minimum of three vendors for each piece of equipment. Quotations were evaluated on compliance with technical specifications, price, ability to meet project schedule, and the vendor's international experience. All equipment procured was also required to meet USAID's

United States source and origin requirements.

Upon the awarding of the equipment contracts translation of operation and maintenance manuals commenced. Translation services consisted of summaries of key operational information for each piece of equipment. These translated summaries were provided to the end users in Armenia along with the complete English versions of the operation and maintenance manuals.

Equipment Shipments

Armenia being a landlocked country, and currently in a state of conflict with Azerbaijan, presented unique transportation problems. The project contacted the United Armenian Fund (UAF) a nonprofit relief organization that arranges monthly humanitarian flights to Yerevan. The UAF agreed to air lift the equipment, at no charge to the project, from the US to Armenia. All vendors were directed to ship their equipment to a freight forwarder's warehouse in Maryland where it was inventoried and consolidated. The equipment was then flown by UAF to Europe, where the plane was refueled, and then, upon receiving government clearance, flown into Yerevan. A total of five equipment shipments were made during the project.

Chlorination Equipment Installation

Working in conjunction with the Yerevan Water and Sewerage Department (YWSD) a prioritized action plan was developed and implemented to install the new chlorination systems in the 11 stations that make up the Yerevan water distribution system. A team of technicians from the YWSD were chosen to accompany the team on all equipment installations. The YWSD team consisted of electrical and mechanical technicians who would be responsible for the maintenance of the chlorination systems for the YWSD.

A total of 14 new vacuum-operated, solution fed, manually adjustable chlorination systems were installed in the 11 stations. A total of eight systems were fully operational at the end of the first three week assignment in Armenia. The remaining six were installed but required minor electrical modifications by YWSD. The existing chlorination systems were left operational at all of the stations that needed the electrical modifications.

At each installation site the pump station operators, along with the technician team, participated in the actual chlorination system installation. This facilitated the transfer of the appropriate skills needed to operate and maintain these systems. After each installation, classroom training sessions were conducted in which operation and maintenance manuals were reviewed, spare parts examined and labeled, and key components disassembled and reassembled. After one week of installations the YWSD technicians were completing complete system installations and conducting the operator training under the supervision of the team.

Chlorination system spare parts, chlorine residual test kits and self contained breathing apparatus (SCBA) were key components procured for each pump station. Each pump station was equipped with chlorination system spare parts to help maintain self sufficiency. Chlorine residual test kits were supplied to confirm that proper chlorine dosages are reaching all points of the water distribution system. Sufficient chlorine residual reagents were purchased for each test kit to allow for daily testing for a two year period. SCBA units were also provided in case of an emergency situation in which chlorine was leaking and the units had to be shut down in a hazardous environment.

A follow-up visit to Armenia was made during September and October 1993. Each chlorination system was visited and, if required, repairs were made. The follow-up visit confirmed that all eight units that were on-line during the previous visit were still operating. An additional four systems were put on-line during this visit. These follow-up visits also benefitted the YWSD technicians who accompanied the team on all site visits. The technicians were able to learn trouble shooting techniques and general repair procedures.

During the team's fourth visit to Armenia in January and February 1994 follow-up visits were again conducted to all pump stations to inspect the chlorination systems. All twelve systems that were on-line in the fall were found to be in good operating condition. The chlorination systems in the remaining two station were made operational during the team's final visit in May 1994.

Conclusions

During the 15 months of the project, critically needed water, wastewater, and laboratory equipment worth more than $450,000 was provided to strengthen the infrastructures of Armenia. This project has demonstrated an immediate and positive impact on the quality of life in Yerevan. Residents are now assured of safer drinking water as a result of new chlorination systems. This project also represented a fast-track program with the installation of the first chlorination equipment commencing less then two months after the notice to proceed. The success of this project has led to a similar project currently being conducted in the neighboring Republic of Georgia.

The Gulf of Aqaba: a Common Cause in the Middle East Peace Process

by Jonathan A. French, M. ASCE[1]

Abstract

The uniquely beautiful Gulf of Aqaba, an important resource for tourism, commerce, marine research, and coral reef habitat, is environmentally imperilled by several factors. Effective improvement in environmental protection and management, a goal held in common among the four bordering nations, requires international collaboration. It is hoped that such collaboration would be a substrate for peace among these erstwhile warring nations.

Introduction

The Gulf of Aqaba, a small arm of the Red Sea, is bordered by four Middle East nations: Egypt, Israel, Jordan, and Saudi Arabia. The small but growing cities of Aqaba, Jordan and Eilat, Israel share the northern tip of the Gulf, each with a short national shoreline. Each city is a commercial port, with facilities for tourists to enjoy water sports and the coral reefs, and with marine research stations. The long western shore is Egyptian, since the return of the Sinai peninsula to Egyptian control following the Camp David accords. Towns are presently small and few, but huge tourist development is projected. The long eastern shore, in Saudi Arabia, is very sparsely inhabited.

The Gulf of Aqaba has been part of the geography of the region's conflicts throughout history, not least the past half century. The Gulf is also a unique ecological zone, whose great depths and coral reefs harbor a great variety of sea life, some of it found nowhere else in the world. International tourism, as well as international marine science, is attracted by its physical and biological beauty. Threats to the environmental health of this sea include oil spills, inadequately treated wastewater discharges, careless tourists, and poorly managed solid waste.

The Middle East political polarization has hindered marine science, in that scientists from the four bordering countries have not been free to commingle locally, or conduct surveys in each other's waters. It has prevented any measure of international collaboration to control and remediate oil spills or any other accident involving the territory of more than one nation. It has hampered the development of the international tourist industry to the area. All the bordering countries have recognized this. As a contribution to the Multilateral Peace Talks for environmental matters, the U.S. in 1992 sponsored a survey of the environmental resources of

[1]Associate, Camp Dresser & McKee International Inc., 10 Cambridge Center, Cambridge, MA 02142, USA

the Gulf, and the environmental threats to them. Besides any environmental benefits, it was hoped that such collaboration towards a common set of goals would also help to bring these polarized countries closer together.

The survey

Two members of the survey team visited the Gulf of Aqaba from the shores of Egypt and Israel, interviewing scientists and government officials. Two other team members visited Jordan, and conducted a research of the published and unpublished literature regarding the Aqaba coast of Saudi Arabia, which elected not to participate in this study. Upon returning to Washington, the survey team wrote an environmental data survey report (Ref. 1) that cited and described needs--and opportunities--for international collaboration in three particular areas: scientific research, coordinated coastal management, and environmental education that would be markedly advanced by collaboration among the bordering countries. The subject has also been treated extensively by the Environmental Law Institute of Washington, DC (Ref. 2).

Collaboration in Scientific Research

Marine oceanographical research programs are active in all three of the countries visited. Research opportunities identified by these countries' scientists almost all relate to one, and frequently both, of the following needs: (1) Determining environmental baseline conditions in the Gulf of Aqaba and monitoring future discharges against the baseline; (2) Gaining a better understanding of the basic processes of potential global interest that can be studied in the Gulf of Aqaba, considered as a unique, deep-sea laboratory.

Baseline establishment and monitoring. An intial baseline survey, or a series of such surveys covering various topic areas, should be considered. These surveys would collect basic information on water circulation, water quality, sediments, flora, and fauna and would provide the basis for the design of future monitoring programs.

Joint Oceanographic Programs. In 1992, under the auspices of the Intergovernmental Oceanographic Commission and the Regional Organization for the Protection of the Marine Environment, a U. S. research vessel, the R/V Mitchell, hosted 130 scientists from 15 countries for a 100-day cruise for integrated research and exploration of conditons in the Persian/Arabian Gulf following the Gulf War. Likewise, much would be gained from a joint effort to explore and document the physical and biological nature of the Gulf of Aqaba.

Project Forsskal is a proposed international cooperative marine scientific expedition to retrace the route of the first oceanographic expedition in the Gulf of Aqaba, conducted in the 1770s. The hope is that internationally this would be of significant historical as well as scientific interest.

Regional Research Center. An international-class marine laboratory in the Red Sea or the Gulf of Aqaba, collaborating with existing laboratories at Eilat and Aqaba, could attract world-class scholars to the research and monitoring programs needed in this area, and serve not only the Aqaba but the Red Sea as a whole. A marine station was established by Israeli scientists at Naama Bay/Marsa-el-At, by Sharm-el-Sheikh, near the mouth of the Gulf and near some of the world's most spectacular coral reefs. Following the return of the Sinai to Egyptian control, an interlude of dormancy, and a change of site, the station continues under the auspices of Suez Canal University, and is the obvious seed from which such a facility could grow.

Long-Term Research Programs. Specific long-term research opportunities identified by Egyptian, Israeli, and Jordanian scientists include a baseline flora and fauna survey; baseline oceanographic studies; primary productivity measurements; studies of nutrient recycling and food chains; hazardous material spill trajectory modelling; three-dimensional circulation models of the Gulf; longshore sediment transport measurements; exploration for hydrothermal vents; fisheries research; and mariculture.

Coordinated Coastal Management

Integrated coastal resources management plans elsewhere in the world have proven to be useful tools for ensuring that coastal development provides a sustainable economic return. Actually, all four of the Gulf of Aqaba countries already have their own separate coastal management plans, which thus lays the groundwork for coordination of master plans. Topics identified by scientists and government officials in the countries visited identified the following managment issues:

Marine Reserves. A coordinated marine reserves program could protect the Aqaba coral reefs by: zoning human activities; limiting access to the reefs, even establishing no-use zones as needed, and rotating the access to reefs; implementing a reef mooring buoy system; establishing elevated viewing platforms and routes; prohibiting certain activities entirely; implementing a public education program; implementing a resource monitoring and assessment program; and zoning hotel development *away* from coral reefs. *[As of early 1995, efforts are underway to create a binational Israeli-Jordanian Marine Peace Park.]*

Information Exchange and Coordination of Environmental Laws. Each of the four Aqaba countries is developing and strengthening its body of environmental law, and strengthening the agencies that enforce it. It would be useful to provide convenient channels and forums for the key individuals driving the process in each country to share information and to coordinate their goals. *[Indeed, in January 1995, an international conference on coordinated marine research and coastal resource management was held in Eilat, Israel.]*

Marine Debris. With the worldwide advent of plastic wrap and plastic water bottles, marine debris is a problem faced by coastal communities everywhere, including the Gulf of Aqaba. The problem of littering, both from shore and from boats and ships, can be resolved only if all the bordering countries have similar legal constraints against littering, and make coordinated provision for their enforcement at sea.

Geographical Information Systems. All four countries have national planning departments with increasing experience in GIS; it would be mutually beneficial if the systems were compatible.

Coordinated Emergency Response Program. There is an identified need to establish a coordinated international hazardous material spill response capability, particularly in the northern end of the Gulf at Eilat and Aqaba. In an effective network, individual response stations in the separate countries would be in routine (daily) contact; their equipment would be mutually compatible and interchangable; and response capabilities at Eilat and Aqaba would be coordinated and in communication with the response capabilities maintained for oil production platforms in the Gulf of Suez. *[By January 1995, an Upper Gulf of Aqaba oil spill contingency plan has been developed among Egypt, Israel, and Jordan].*

Navigation Systems. The risk of spills of hazardous cargo could be reduced by improving aids to navigation, particularly through the Straits of Tiran, and near the ports of Aqaba and Eilat. A vessel-tracking system comparable to that used in the Gulf of Suez and in the Suez Canal could be considered.

Reducing Bulk Material Dust created by trans-shipment of phosphate and some other materials is a problem at many ports across the world, including Aqaba (Eilat has few such trans-shipments). Techniques such as choke feeders and steps to seal off the loading and storage operations are being considered.

Mooring Buoys for excursion boats to the coral reefs, and in fact for any vessel near coral reefs, eliminates the need to drop anchor, which devastates any coral structure hit.

Fisheries. To effectively protect marine life for the purposes of a sustainable fishery, support of the tourist industry, and as part of a healthy ecosystem generally, it is necessary to have coordinated research, policies, legislation, and enforcement procedures.

Environmental Education

Education of the General Public about the marine environment improves awareness and reduces the frequency of activity that degrades the marine and coastal environment. The bases for such educational programs are well established, with the Marine Science Station in Aqaba and Coral World in Eilat. While the Egyptian Sinai coast is as yet little developed, cautions against reef damage and spearfishing are frequently posted along the coast and in resort areas. Hotel gift shops are well stocked with relevant, attractive diving and coral reef books. Public outreach television programs about coral reef ecology have been broadcast in Egypt.

Environmental Education for Tourists can be undertaken even within the short time available during their visits to the Gulf. Already, posters and signs in hotel rooms and at the beaches warn against breaking coral or spearfishing. There could be closed-circuit educational TV programs or video screenings at visitor centers about coral reef ecology and the behavior required of reef-visiting tourists.

Advanced Education. Cooperative programs, including workshops on site, joint research publications, and joint management initiatives improve international understanding of an ecosystem. Overseas student research programs already exist with European scientific research centers and universities; expansion of these programs would benefit all participants.

References

1. Gulf of Aqaba Environmental Data Survey, prepared for the U.S. Agency for International Development by the Irrigation Support Project for Asia and the Near East (ISPAN), Arlington, VA, October 1992.
2. Protecting the Gulf of Aqaba: A Regional Environmental Challenge, Environmental Law Institute, Washington, DC, 1993.

Integrated Remediation Study for the Guarapiranga Project, São Paulo, Brazil

by Jonathan A. French, M. ASCE[1] and Luiz H. W. Oliveira[2]

Abstract

The Guarapiranga Reservoir suffers increasing nutrient loadings, largely attributable to a growing unsewered urban encroachment in its basin. The resulting algal blooms and periods of anoxia in the lower layers give rise to taste and odor problems, despite heavy use of algicide in the reservoir, and improvement in water treatment plant operation. A five-year program is attacking the problem on many fronts.

Introduction

The Guarapiranga watershed, a 637 sq km subbasin of the Tietê River, lies in and just beyond the southwestern districts of São Paulo, Brazil. The 194 million cubic metre Guarapiranga Reservoir supplies drinking water to over 3 million people, about 20 percent of the population in the São Paulo Metropolitan Area. Presently, the average draft is about 12 m^3/sec; future interbasin transfers being considered would further increase the flow rate of the system. The reservoir is also an important recreational amenity in a beautiful setting, with many yacht clubs and stately residences on its shores.

The rapidly growing metropolitan area has encroached on the lower parts of this watershed, near the northern shore of the reservoir. Much of this encroachment is high-density low-income housing and slums with little or no sewerage. Only 45 percent of the population of the basin is sewered. Blackwater drainage, as well as inadequate collection of solid waste, contributes to excessive loadings of oxygen demand and nutrients to the reservoir.

E. coli counts suggest that much of the reservoir is unfit for swimming. For the past 20 years, algal blooms have reached nuisance proportions. An algicide application program faces a runaway consumption of copper sulphate and hydrogen peroxide, in some months exceeding 100 tons of $CuSO_4$.

---

1 Camp Dresser & McKee International Inc., 10 Cambridge Center, Cambridge MA 02142, USA

2 Companhia Brasileira de Projetos e Empreendimentos, São Paulo 05422-012, Brazil

The Guarapiranga Program

The Government of the State of São Paulo, sensitive to these problems, launched the Guarapiranga Program in 1992. Funded both by local sources and the World Bank, this US$262 million environmental sanitation program has corrective and preventive measures, with permanent investments in watershed management. Its main objectives are to protect and restore water quality in the Guarapiranga Reservoir.

For a five-year period beginning in 1993, COBRAPE (Companhia Brasileira de Projetos e Empreendimentos), a private-sector engineering and planning consulting firm, is supporting the management of the Guarapiranga Waterbasin Environmental Sanitation Program. Camp Dresser & McKee International Inc. (CDM) are technical advisors to COBRAPE for this project, assisting particularly on technical matters relating to water quality and water treatment. The COBRAPE/CDM rôle is both to offer new approaches to watershed and reservoir management, and, as part of a planned mid-term review process, to evaluate the results of the Program's studies conducted to date.

This task is greatly facilitated by the large body of water quality data that has been collected over a period of many years by the state water and sewer agency, SABESP, and by the state environmental sanitation agency, CETESB.

Approach

Through a variety of measures, the Guarapiranga Program has been attacking problems of reservoir and water quality decline on four fronts:

(a) by improved watershed management;
(b) by diversion or treatment of highly polluted tributaries;
(c) by in-reservoir management techniques; and
(d) by upgrading the water treatment plant.

Coordinated consideration of these measures is essential, since the cost and effectiveness of the measures vary greatly. The CDM/COBRAPE team is seeking to identify those measures, and combinations of measures, with the greatest benefit/cost ratio, and those with the greatest immediate benefit, to grant them priority in schedule and funds.

Findings to date

Watershed Management. More than 400 kg/day of phosphorus flows into the reservoir. Much of this influx can be attributed to unsewered residences, whether favela (slum) housing or more substantial housing. Although the largest sub-basin, the Embu Mirim, provides the greatest contribution of phosphorus, more than half of the total phosphorus inflow to the reservoir is produced in a few small sub-basins contiguous to the reservoir. In these heavily developed watersheds, the phosphorus influx is carried by water that is not natural runoff, but municipal sewage. Therefore, the flux of phosphorus is relatively constant, and is only slightly dependent on rainfall patterns.

Reservoir management. A high priority for the Guarapiranga Program is to control and reduce the propensity to algae blooms, and the related heavy usage of algicide. Applied at a rate sometimes exceeding 100 tons of copper sulfate per month, the algicide is costly, and ultimately introduces counterproductive factors, such as:

- The copper sulfate is toxic not only to the algae being targeted, but to parts of the natural ecosystem such as zooplankton grazers that naturally would work to control the bloom; and

- The sulfate component can hasten the depletion of oxygen in lower layers of the water column.

Nutrient control strategy. For algae control, the most important nutrient to manage is phosphorus (P), although silica should be monitored as well. (To try to manage algae blooms by controlling the other principal nutrient, nitrogen, would likely be counterproductive by giving an advantage to the undesirable blue-green algae, which can fix atmospheric nitrogen for their needs.)

It is estimated that an average of 400 kg of P enters the reservoir each day, and that about 40 kg is withdrawn each day. Therefore about 360 kg is precipitated to the bottom sediments each day. The areal distribution of this depositition is not yet mapped.

Most probably the sedimentation of P is not continuous and unidirectional; rather, it is likely that P deposited in sediments when the water is well oxygenated, with a pH of at least 6.0; but that P is released from the sediments on occasions when waters just above the sediments become anoxic. The 360 kg/day of deposition is therefore a net sedimentation rate, with greater and lesser, and even negative values, in some times and places.

The average concentration of P in the reservoir is about 40 mg/m^3. To curtail the propensity for algal blooms, the concentration of P must be reduced to 20 or 25 mg/m^3. Several means are being considered:

- Decreasing the loading rate by 50 percent, from 400 kg to 200 kg of P per day.

- Increasing the net sedimentation rate by about 6 percent, from 360 kg/day to 380 kg/day. The principal means for increasing net sedimentation include nutrient inactivation at the mouths of tributaries by application of $FeCl_3$, and artificial aeration of portions of the reservoir using air bubble columns.

Water Treatment. While the Alto da Boa Vista treatment plant produces water of generally good quality, current problems include episodes of unpleasant taste and odor; and unattractive staining by manganese and iron. Several of the filters are being reconstructed. Ongoing laboratory tests seek to optimize choice and dosages of coagulants.

Recommended Actions for the Immediate Short Term

The COBRAPE/CDM study has identified numerous measures worthy of consideration; these have been categorized as either long-term or short-term measures. Emphasis has been given to the following short-term measures which should be given immediate attention to halt the degradation of the reservoir as quickly as possible:

1) ***Interception of contaminated tributary flows,*** wherein the dry weather flow from the most contaminated streams is intercepted and pumped out of the watershed to treatment and disposal. Such facilities are already on line for two small, heavily contaminated sub-basins. More such facilities are encouraged, with highest priority given to the subbasins with the greatest concentration of P.

2) ***Nutrient inactivation,*** wherein nutrient-laden tributaries that are not intercepted and diverted are dosed with an appropriate flocculant to promote the settling of nutrients to the bottom of the reservoir.

3) ***Aeration,*** wherein air is pumped to one or more points on the reservoir bottom, and released in a continuous stream of rising bubbles to aerate the entire reservoir water column, both directly, by contact with the air bubbles, and indirectly, by induced circulation. Keeping the water column aerobic at all depths, and at all times, keeps nutrients (and iron and manganese) sequestered in the bottom sediments.

4) ***Improvements at the water treatment plant.*** Refinements in the choice and dosage of coagulants; the use of granular activated carbon as a filtering medium; and mechanical flocculation should all improve the quality and reliability of water treatment.

5) ***Algicide application.*** The measures cited above should permit a significant reduction in the amount of algicide needed. Further reduction may be possible by refinements to the dosing process itself, e.g. preemptive dosing at the earliest possible indication of an algal bloom.

Each of these five categories is subdivided into several tasks and ranked according to priority and as to which are prerequisite to others, in order to provide a clear guide to a sequence of steps the Guarapiranga Program can take within the next year, and for several years thereafter.

However, coordinated consideration of these measures is essential, since the cost, and effectiveness, varies widely. The Project seeks to identify those measures, and combinations of measures, with the greatest immediate benefit, to grant them priority in schedule and funds.

A Windows™-based data management and modelling platform has been developed for the Project; it is described in another paper at this conference (see "Development of an Integrated Decision Support System for the Guarapiranga Reservoir" by Jacobs, Agostini, and Oliveira).

An Assessment of Shared Vision Model Effectiveness In Water Resources Planning

Allison M. Keyes[1], Student Member, and Richard N. Palmer[2], Member, ASCE

Introduction

During the National Drought Study the US Army Corps of Engineers sponsored river basin planning efforts at six sites across the country (Institute for Water Resources 1994). These sites included the Cedar and Tolt River basins in Washington, the Green River basin in Washington, the James River basin in Virginia, the Kanawha River basin in West Virginia, the Marais des Cygnes-Osage River basin in Kansas and Missouri, and the Quabbin and Wachusetts River basins in Massachusetts. A common component of these efforts was the use of an object-oriented programming environment for simulation model construction. Two main goals were established for the modeling efforts: 1) To create models that could clearly be characterized as "Shared Vision" models , and 2) to effectively integrate these models into a seven-step paradigm for collaborative planning advocated by the Corps. Because the use of object-oriented programming to create Shared Vision models is relatively new to water planning, the effectiveness of this approach is of potential interest to water managers.

This paper outlines the approach used to assess if Shared Vision models were produced at each site, and the process by which model effectiveness was evaluated. The outcomes at each site are described and influencing factors are noted. The implications of these findings for future planning efforts are also briefly discussed.

Evaluation Approach

It was initially hypothesized that use of object-oriented software would facilitate the creation of Shared Vision models, by allowing potential model users to play a more significant role in the model development process. Furthermore, it was believed that Shared Vision models would be more readily and effectively utilized throughout plan development than those which address a more singular perspective (Palmer et al 1993). To test these hypotheses, the evaluation effort focused on three activities: 1) Assessing the extent to which the object-oriented models could be characterized as Shared Vision models, 2) Assessing model usefulness in the planning activities advocated by the Corps, and 3) Identifying factors which contributed to the outcomes observed. Greater detail regarding these evaluation activities is provided below.

Graduate Research Assistant, and [2] Professor of Civil Engineering, Department of Civil Engineering, University of Washington, Seattle, Washington 98195

Several sources of information were utilized for this assessment. Initial modeling objectives were examined. Models were reviewed and critiqued at different stages in the planning process on the basis of their technical merit, and appropriateness to the planning effort. Numerous planning workshops were attended to observe model applications in these settings. Questionnaires were administered to workshop participants and key members of the planning team were interviewed to assess their level of understanding, trust, and satisfaction with the model, as well as their views on the planning process. Model developers were interviewed to document the model development approach, ascertain their satisfaction with the model development and application, and gain insight regarding the perceived impacts of object oriented programming and other factors to these outcomes. The perspectives of numerous water resources professionals who were involved in the National Drought Study were also sought. Based on this information, case summaries were prepared for each study site.

Shared Vision Model Assessment

Unlike a structured and established analysis tool such as HEC-5, there is no firm prescription for Shared Vision model content, analysis capabilities or output. Rather, the unifying characteristics of Shared Vision models relate to their ability to represent a system in a way that is understood by participants, and to contain information that is relevant to their diverse perspectives. A Shared Vision model must be jointly endorsed by planning participants; a sufficient level of trust must be achieved so that it is viewed as an unbiased and valid source of information in a group decision making context. Finally, a Shared Vision model must be non-proprietary. It must be equally accessible to all groups represented in the planning effort, and a common level of proficiency in model use should be attained.

To assess the extent to which the goals of Shared Vision model development were achieved, the models produced at each site were characterized based on the above attributes. The elements of Shared Vision models that were most frequently attained were examined, as well as those that were most elusive.

Assessment of Model Effectiveness

Models utilized in a planning setting must be appropriate to the problems addressed, and be technically valid, not only from the perspective of planning participants, but also by peers in the field (Loucks 1992). Furthermore, they must be beneficial to the process and outcomes pursued. Site-specific planning objectives, approaches, and participants will help define these general model requirements.

The Corps envisioned that a similar planning process would be followed at five of the six sites. Like many water resources planning efforts, the planning approach advocated by the Corps provided numerous opportunities for model use in activities such as: joint fact-finding, problem identification, assumption and constraint clarification, group brainstorming, alternative screening, trade-off assessment, strategy refinement, communication, and plan maintenance. In general terms, each model's effectiveness was gauged by its usefulness with respect to these tasks, as well as the versatility of support it provided.

To assess model usefulness, the following questions were posed at each site: 1) Was the model utilized in each of the planning activities listed above? 2) Did the model enhance each participants' ability to participate effectively in plan formulation, implementation and maintenance? 3) Did the model influence participants perceptions regarding the quality of management decisions? 4) Did model use enhance the support for planning decisions? Specific metrics were utilized whenever possible to clarify these outcomes.

Assessment of Influencing Factors

Many factors have been noted which can potentially influence the development and effectiveness of Shared Vision model in a planning process (McKinney 1990, Loucks et al. 1985). The factors examined in this study include the interactiveness, transparency, and flexibility of the models; model developer characteristics; the involvement of planning participants in model development; the accessibility of the platform; the types of model training provided; and several aspects of the planning environment. These attributes were characterized for each site. Cross-case comparisons were made to ascertain if there were observable relationships among these factors and study outcomes.

Summary of Model Outcomes and Influencing Factors

Shared Vision Models

The goal of creating a Shared Vision model was attained in each region to varying degrees. The most difficult and elusive goal was attaining a high level of model understanding and sufficient proficiency to allow models to be used by non-model developers. The level of understanding attained was positively correlated with the amount of stakeholder involvement in model development. However, it appears that additional training efforts, wider access to the modeling platform, and greater agency commitment to understand the model were needed.

Planning participants' acceptance of the models were more easily attained. Even though model details were often not well understood, it was nevertheless often trusted and perceived as both relevant and non-biased. The skills of model developers influenced this outcome, as it was common for model developers to assume the role of model translators in group planning activities. Their effectiveness in demonstrating model validity, completeness and relevance was a key prerequisite to model endorsement. The involvement of stakeholders in the model development process also enhanced the likelihood of model endorsement.

The use of object-oriented software greatly facilitated Shared Vision model development efforts. It enhanced the model developers' ability to communicate model content, and to readily incorporate new suggestions into the models. Also, given an appropriate setting, planning participants were able to effectively critique the models because of the transparency afforded by the object-oriented modeling environment.

Model Effectiveness

Model effectiveness varied from site to site. In the most successful case, the model was used in many contexts. In interagency settings it provided an effective means of demonstrating the impacts and trade-offs of different drought management strategies in terms that were relevant to participants. This information greatly facilitated group discussions and helped participants to focus on strategies that were perceived to be most effective and equitable. The model was useful in this role because it met many of the criterion of Shared Vision models: it was trusted by participants and relevant to the various interests represented.

In addition, the primary model users at this site reported that because of the model, a greater variety and number of alternatives were considered and a greater depth of analysis was achieved. The object-oriented environment was cited as the primary source of these advantages, because it was felt to enhance model flexibility and allowed modifications to be easily made.

In three cases, models were developed to facilitate interagency water management activities other than the Corps-sponsored planning effort. Two of the sites were able to attain endorsement of the model for these purposes. Each model development team's ability to establish a shared sense of ownership in the model, to demonstrate model usefulness in relationship to other existing tools, and to identify potential applications for the model outside the Corps environment was extremely important to these successes. However, in one case, the model can not be utilized until a single, key management agency participates in its review.

In two basins, clear modeling objectives were never established, public participation was limited, and model development progress was sporadic. Planning efforts also suffered from a lack of focus and commitment to the process. As a result of these factors, viable models have not been produced.

Conclusions

The National Drought Study has shown that simulation models can enhance water planning efforts if they are appropriately integrated into the process. Adopting a modeling perspective that allows a Shared Vision model to be produced may enhance the likelihood that the model will benefit a collaborative planning effort. The features offered by object-oriented simulation software facilitate both Shared Vision model development and model application in a planning environment.

Development of a Shared Vision model requires extensive coordination, as well as a high level of commitment among model developers and model users. It calls for clearly articulated modeling objectives, a talented model development staff, and a strategy for effectively involving planning participants in model development. Training for both model developers and users may also be required.

The planning environment also plays a significant role in determining the prospects for model success. Shared Vision model development and effective model application in collaborative planning require a setting where mutual commitment to the process is high, planning objectives are well-defined, and each participant has a stake in planning outcomes. If these conditions are not met, the value of model development and use may be seriously undermined.

References

Institute for Water Resources (1994). *National Study of Water Management During Drought: Report to Congress.* U.S. Army Corps of Engineers, IWR Report 94-NDS-13.

Loucks, D. P. (1992). "Water Resource Systems Models: Their Role in Planning". *Journal of Water Resources Planning and Management*, 118(3),214-223.

Loucks, D. P., J. R. Stedinger, and U. Shamir (1985). "Modelling water resource systems: issues and experiences". *Civil Engineering Systems*, 2,223-231.

McKinney, M. J. (1990) State Water Planning: A Forum for Proactively Resolving Water Policy Disputes". *Water Resources Bulletin*, 26(2), 323-331.

Palmer, R. N., A. M. Keyes, and S. M. Fisher (1993). "Empowering Stakeholders through Simulation in Water Resources Planning". *Proceedings of the 20th Annual National Conference, Water Resources Planning and Management Division of ASCE,* Seattle, Washington, 451-454.

UNDERSTANDING WATER SUPPLY SYSTEM BEHAVIOR

Ralph A. Bolognese[1], AM, ASCE
and
Richard M. Vogel[2], M, ASCE

Abstract

There are a myriad of questions that need to be addressed in order to understand the behavior of a water supply system. Some of the relationships which are needed to improve our understanding of the behavior and to properly manage water supply systems are explored in this study using analytic water supply performance indices.

Introduction

Detailed reservoir system studies take months or even years to provide results and usually the resulting model output is extremely complex, making it difficult to see the "big picture". Simple, yet accurate, "back-of-the-envelope" methods such as those described here can provide results quickly and provide insight into the overall behavior of the reservoir system. Such methods are not intended to replace the need for detailed simulation studies. Rather, these "back-of-the-envelope" procedures are needed prior to the application of more complex models to assure an understanding of the general behavior of the reservoir system and most importantly to assure that the proper issues are addressed in more detailed simulation studies.

Application of the following water supply system performance indices requires estimates of parameters of the net annual hydrologic inflow to the system including their serial correlation, ρ, coefficient of variation, C_v, mean annual inflow, μ, and the type of distribution. Vogel et al. (1994) presented a regional approach to estimating these parameters for northeastern U.S. which only require estimates of the drainage area.

---

[1]Associate Project Engineer, Weston & Sampson Engineers, Five Centennial Drive, Peabody, MA 01960-7985

[2]Associate Professor, Department of Civil & Environmental Engineering, Tufts University, Medford, MA 02155

Reservoir System Resilience

In general, two classes of reservoir systems exist, within-year and over-year. Within-year systems generally refill at the end of each year, as opposed to over-year systems, which typically do not. Over-year systems are prone to water supply failures which potentially could extend over several years, whereas water supply failures for within-year systems tend to be short-lived.

In order to determine whether a water supply system is dominated by within-year or over-year behavior, a useful index is

$$m = \frac{1-\alpha}{C_v} \quad (1)$$

where α is the annual system demand as a fraction of the mean annual inflow, μ, and $C_v = \sigma/\mu$ is the coefficient of variation of the annual inflows which is equal to the standard deviation of the annual inflows. The parameter m was introduced by Hazen (1914) and others, and recently used as a measure of water supply system resilience (Vogel and Bolognese, 1995).

Another measure of system resilience is the probability r that a reservoir system will recover from a failure once a drought has set in. Here, we define a failure as the inability of the system to meet its contracted demand. If the inflows are independent and follow a normal distribution:

$$r = \Phi(m) \quad (2)$$

where Φ denotes the cumulative distribution function of a standardized normal random variable and m is defined in (1). The index r can be used to distinguish between systems dominated by over-year (r small) behavior and within-year (r large) behavior.

Reservoir System Reliability

In the U.S., a water supply system's reliability is often defined as the probability p, that the system will meet a given demand without failure over an N-year planning period. This interpretation results from having estimated the "dependable" or "safe" yield based on the drought-of-record, repeatedly, using a stochastic streamflow model (see Vogel, 1987). Elsewhere, the reliability of a reservoir system is usually defined in terms of the steady-state probability of a failure, in any given year, which we term q. Both concepts of reliability are related as follows:

$$p = (1-q)\left[1-r\left(\frac{q}{1-q}\right)\right]^{N-1} \quad (3)$$

where N is the planning horizon and r is defined in (2).

Storage-Reliability-Yield (SRY) Relationships

An understanding of reservoir system behavior requires an evaluation of the effect of storage capacity on system yield, reliability, and resilience. General analytic approximations relate the storage capacity of a reservoir S to moments of the annual inflows μ, σ, and ρ and the demand as a fraction of the mean annual inflow α, for normally distributed inflows are summarized in Vogel and Bolognese (1995) and Figure 1. Figure 1 applies to systems in the northeastern U.S. with demands in excess of 75% of the mean annual inflow μ with $C_v = 0.25$ as was verified by Vogel et al. (1995) for 166 basins in the northeastern U.S. Figure 1 compares these relationships for four water supply systems in the northeastern U.S. New York City (NYC) is a within-year system when the demand is equal to its historical safe yield of 1290 mgd and under those conditions it is a much more resilient system than the other three. However, current NYC demand is closer to 1500 mgd or 0.77 mgd/mi^2, which makes NYC an over-year system. Since all of the systems exhibit high reliabilities when operated at their respective safe yields, their ability to recover from a failure will not be tested very often. Figure 1 also documents that the location of systems on the SRY curve can help guide the direction of future water supply planning. For example, Boston is located on a very steep portion of the SRY curve, which indicates that increases in storage capacity lead only to marginal increases in yield. However, NYC is located on a flat portion of the curve and significant gains in system yield could be achieved by increasing system storage.

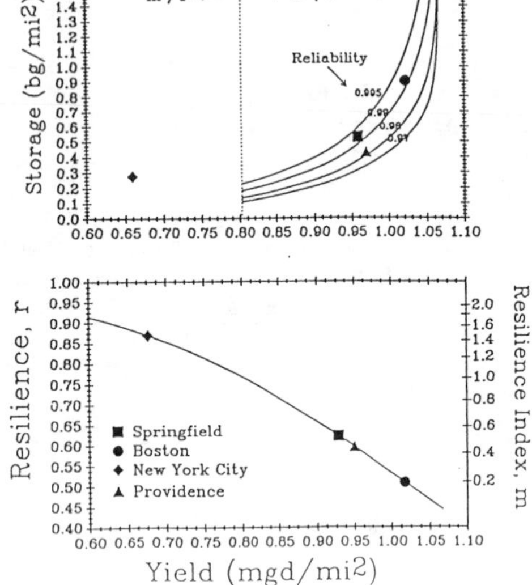

Figure 1 - Storage-Reliability-Resilience-Yield Curves

What is the Relationship Between a System's Resilience and its Reliability?

It is possible to improve the performance of a water supply system by lowering demand thus increasing its resilience, but to what extent can such changes produce significant improvements in overall operations and in particular do increases in resilience always lead to increases in system reliability?

Figure 2 illustrates the relationships between reliability and resilience for systems fed by AR(1) normal inflows. For a fixed storage capacity, (standardized by μ) decreasing the resilience of a system (by increasing yield) lowers the reliability of the system. However, the relationship is not linear. For all systems there is a point where increasing the resilience (decreasing demand) no longer significantly improves the reliability of the system.

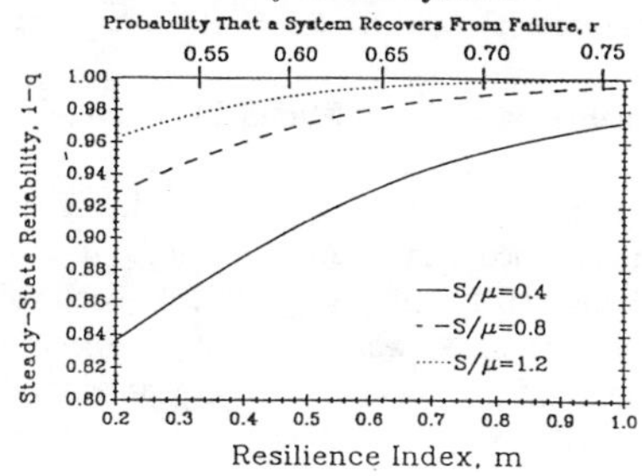

Figure 2 - Resilience and Reliability Relationships

Conclusion

This study demonstrates how recent developments in the theory of storage reservoirs can be applied to explain the detailed behavior of water supply systems which exhibit significant carry-over storage requirements. These procedures can be implemented in a short period of time and yield valuable information about water supply system behavior. They provide a valuable addition to more detailed investigations by focusing upon the overall behavior of the system.

References

Hazen, A., Storage to Be Provided in Impounding Reservoirs for Municipal Water Supply. *Transactions, ASCE*, Vol. 77:p. 1539 (1914).

Vogel, R.M. and R.A. Bolognese, Storage-Reliability-Resilience-Yield Relationships for Over-Year Water Supply Systems, *Water Resources Research*, in press (1995).

Vogel, R.M., N.M. Fennessey, and R.A. Bolognese, Regional Storage-Reliability-Resilience-Yield Relationships for Northeastern United States, *Journal of Water Resources Planning and Management*, ASCE, in press (1995).

Use of Genetic Algorithms for Project Sequencing

Graeme Dandy[1] and Michael Connarty[1]

Abstract. The problem of sizing, sequencing and scheduling of projects for water supply systems has been examined using various methods. In this study the methods of integer linear programming (IP), unit cost (UC) and genetic algorithms (GA) are used to size, sequence and schedule water resources projects for the South-East Queensland Region (Australia). It is found that the GA provides the best solution for the system examined and models used. It has the added advantages over integer linear programming that it can operate on a continuous time scale rather than requiring discrete time intervals and it needs only a fraction of the computer time to run.

Introduction. The simple sequencing and scheduling problem for a water supply system consists of determining the order and timing of future projects for augmentation of the current system. The timing of projects occurs when the expected demand equals the supply capacity. At this stage a project is added to the system to increase the supply. The problem is made more difficult when the sizing of projects is considered. In this case, not only is the order examined but the size of each project also needs to be decided. The sequencing problem has been solved using dynamic and linear programming as well as numerous heuristic methods. In this study integer linear programming (IP), the method of unit cost (UC) and a new method called genetic algorithms (GA) will be used to size, sequence and schedule water resources projects. The South-East Queensland system will be used as a case study for this problem.

Case Study: South-East Queensland System

South-East Queensland (Australia) is undergoing a rapid growth in population. Future options for expanding the water supply include 16 possible dams [3]; 15 of these each have 3 possible sizes and the remaining project has a single possible size. For this study the 16 projects are reduced to the 9 best projects based on the cost and annual

---

[1]**Associate Professor and Postgraduate student (respectively), Department of Civil and Environmental Engineering, University of Adelaide, Adelaide, 5005, Australia.**

yields of the projects. The costs and annual yields for the nine projects used in this study are given in Table 1.

Table 1 Project Cost and Annual Yield for the South East Queensland Region

Project	Cost ($Mill)	Yield (GL/year)	Project	Cost ($Mill)	Yield (GL/year)	Project	Cost ($Mill)	Yield (GL/year)
1/1	107.0	79.71	6/1	309.0	110.1	8/3	88.0	61.34
1/2	123.0	98.60	6/2	327.0	116.4	9/1	83.0	37.58
1/3	139.0	105.5	6/3	350.0	122.3	9/2	94.0	45.60
4/1	114.0	31.30	7/1	60.0	23.20	9/3	104.0	54.27
4/2	120.0	36.40	7/2	99.0	61.50	10/1	97.0	27.29
4/3	127.0	42.40	7/3	119.0	70.05	10/2	110.0	37.13
5/1	83.0	7.000	8/1	54.0	23.53	10/3	126.0	44.60
5/2	109.0	26.20	8/2	61.0	48.69	13/1	72.0	29.38
5/3	134.0	64.40						

There are three likely demand scenarios for a 100 year period from 1990 to 2090. The demand growth rates to be used for the planning period are shown in Table 2.

Table 2 Demand Growth Rates (GL/Year) for Three Demand Forecasts

Demand Forecast	Period 1990-2010	Period 2010-2030	Period 2030-2050	Period 2050-2070	Period 2070-2090
1	5.12	5.36	3.38	2.36	2.23
2	10.62	6.70	4.22	2.96	2.79
3	6.80	6.94	5.80	4.30	4.20

For this study the current system yield and current demand are assumed to equal 500 GL/year and 337.7 GL/year respectively (ie assuming 1990 is the current year).

Integer programming

For the IP model a binary (0,1) decision variable, X_{ikj}, is defined for a project, i, size, k, and timing, j. If the variable is equal to 1 then project, i, of size, k, is implemented in time period, j; if it equals 0 the project is not implemented in that time period. Due to the long computation time of the IP method, only a 55 year planning period is examined and it is assumed that the decision on the capacity expansion is made at discrete 5 year intervals. In addition, only projects 1, 7, 8 and 9 will be used in the IP model. These assumptions help reduce the computation time to a reasonable amount. The project cost and yield are defined as C_{ik} and Z_{ik}. The objective is to minimise the present value of costs (PVC) subject to the constraint that the system yield always is greater than or equal to the demand. The values of X_{ikj} are constrained by the following equation :

$$\left(\sum_{k=1}^{3}\left[\sum_{j=1}^{T}X_{ikj}\right]\right)\leq 1 \qquad \forall i \qquad (1)$$

The IP model is solved using the linear programming optimisation package, LINGO [2].

Genetic Algorithms

The genetic algorithm method is a search procedure based on the mechanics of natural selection and genetics [1]. The simple GA works with a coded string of "bits" (chromosome). The GA process uses a population of these strings. All strings in the population have their fitness or the objective function evaluated, which is used to select suitable parents for the next population. The new population is produced from the previous population by using the operators of reproduction, crossover and mutation. A more detailed discussion on the theory of GAs can be found in Goldberg [1]. For this study a real coded string will be used and thus each bit of a string will be a real number between 0 and 1. An example of a possible string is :

[0.862 0.521 0.923 0.437 0.658]

The position of a number in the string represents the project number. The order of the projects is then evaluated by the magnitude of the real number. The highest value will indicated that the particular project is the first sequenced. This continues until the project with the lowest value is sequenced. For example, the above string will result in the following sequence of projects : 3 1 5 2 4. To incorporate sizing in the coding of the string it is simply a matter of having a real number for every project site and size. Then the decoding of the string to calculate its fitness, should ensure that only one bit per project contributes to the evaluation of the fitness. For the GA, a population size of 100 was used and 100 generations were examined.

The Unit Cost (UC) Method

The UC method is simply the sequencing of projects based on their unit cost. The UC is calculated by dividing the cost of a project by its annual yield . The projects are then ordered from lowest to highest UC to obtain a sequence.

Results

The three methods presented were used to obtain the size and sequence of projects for the South-East Queensland system. Demand forecast 2 and a discount rate of 5 % were used with a 55 year planning horizon. The results are given in Table 3.

Table 3 Optimum Project Sequences for Demand Forecast 2

METHOD	Project Sequencing and Timing						PVC ($Million)
Unit Cost	Project	1/2	8/2	7/2	9/3		
	Year	15	27	34	45		105.924
Integer Programming	Project	1/2	7/2	8/2	9/2		
	Year	15	25	35	45		109.920
Genetic Algorithm	Project	1/2	8/2	7/2	9/2		
	Year	15	27	34	45		104.811

PVC = present value of capital costs

The above solution illustrates that the GA produces the lowest PVC result of all methods. The difference between GA and the UC results is that the latter method sequences a project which is larger than is required for the planning period. This is seen as a problem with the UC method. Other results obtained for longer planning periods demonstrate that the GA is superior to the UC method. An interesting result in the above Table is the alteration in sequence between the IP method and the other methods. This is a result of using discrete time periods in the IP model.

In relation to computer time for these various methods, the IP model required between 40 minutes to 40 hours on a DECstation 5000/240 computer to run (depending on the demand forecast). The GA model took approximately 200 seconds to run and the UC method took approximately 1 second.

Conclusions

The results found in this study favour the GA model. It is a faster method than the IP model and produces better results. In addition it can examine a continuous time period which the IP can not do without taking an impractical amount of time. The GA also outperforms the UC method although it takes longer to obtain a solution. However, the extra time taken by the GA is not considered to be excessive considering the results it obtains. Also the GA will find many near-optimal solutions in obtaining the optimal solution. These may be useful to system managers, if factors other than economic ones are to be considered in the planning of a water supply system.

References

[1] Goldberg, D. E., Genetic Algorithms in Search, Optimisation and Machine Learning, Addison-Wesley Publishing Co., Inc., 1989.
[2] Schrage, L. and K. Cunningham, LINGO Optimisation Modeling Language, Lindo Systems Inc., Chicago, 1991.
[3] Water Supply Sources in South-East Queensland. Volume 2 Main Report, Water Resources Commission, Department of Primary Industries, January 1991.

Dealing With Rapid Growth in The Desert Southwest - Las Vegas, Tucson, and Phoenix: A Case Study

Chris D. Meenan[1]

Abstract

Rapid growth in the cities of Phoenix, Arizona; Tucson, Arizona; and Las Vegas, Nevada, has caused both water capacity and supply problems. In response, utilities have undertaken four types of programs to ensure that the deliverable supply equals demand. These programs can be classified as education and public information, new supplies, micromanagement and price changes.

Introduction

This paper presents a brief summary and comparison of three utilities' responses to the problems caused by rapid population growth and its subsequent impact on water demand. The focus of this paper is Phoenix, Arizona; Tucson, Arizona; and Las Vegas, Nevada. The problems faced by these three communities have been similar. The pattern is characterized by unforeseen rapid economic and population growth followed by short-term water delivery capacity limits and long-term supply limits.

The responses to these problems have been fairly consistent among the utilities. Responses are implemented in stages beginning with those programs which will generate the least public discomfort. As capacity and supply problems persist, progressively more stringent actions are taken. The spectrum begins with voluntary conservation and ends with mandated limitations. In general, utility responses can be categorized into four primary classifications: education and public relations, development of new supplies, micromanagement, and price changes.

---

[1]Senior Resource Analyst, Resources Department, Las Vegas Valley Water District, 1001 South Valley View Boulevard, Las Vegas, Nevada 89153

1. EDUCATION AND PUBLIC INFORMATION

All three communities are very committed to education and information programs. These programs are used to raise personal awareness of conservation techniques, system peaking problems and the need for voluntary reductions in water use. These programs have been shown to have some impact in dealing with short-term drought conditions or when combined with a significant price increase, but have had little success in dealing with long-term problems when utilized as the only demand reduction program.

One problem with these programs that is not generally understood is the question of equity. It is often perceived that current residents are foregoing water use in order to supply this water to developers and new residents at artificially low prices.

2. NEW SUPPLIES

Historically, the expansion of supply has been the primary choice for meeting increasing demands. New supplies were relatively close and inexpensive to obtain, and could often be developed with federal subsidies placing less impact on the end-user. Cost benefit analysis of new supplies and/or capacity expansion also suffered from a demand modelling bias that utilized models which projected demands without accounting for impacts such as rate increases on the demand system.

As local groundwater sources become depleted and current Colorado River allocations reach full utilization, the development of new supplies will become increasingly expensive. As the search for more water continues, potential sources include: fallowing agricultural property and importing the water, time shifting water through recharge programs (banking water underground during periods of surplus to be withdrawn during period of shortage), reuse of grey water, and expanded use of the Colorado River through marketing arrangements.

3. MICROMANAGEMENT

This class of response includes legal limits on turf landscaping, general building codes, retrofits, lawn watering restrictions and waste patrols. These programs have proved popular with policy makers. These programs do not confront consumers with the costs of their usage patterns or force them to change.

The results of these programs are characterized by mixed evidence that suggests the majority of these programs are effective over experimental control areas or for short durations, but little impact is seen at the system level or long term. An example of this process is Phoenix's 1985 emergency retrofit program. The consultants evaluating the retrofit estimated water savings of 9 gallons per day per household for 40,000 homes. Yet, three of the next four years saw GPCD figures in excess of the 249 logged in 1985.

4. PRICE CHANGES

As a demand management tool, price changes are used as little as possible and are usually only instituted to cover increases in operating costs. Historically, rate increases have been disliked by rate payers and policy makers alike. Rate changes cause users inconvenience - they are forced to re-evaluate how they use water as well as how much is being used. Policy makers sometimes dislike rates because of the difficulty in developing a "fair" or equitable rate. Most systems developed seem to favor one group over another. A steep inverted rate structure places a larger burden on large users. A more progressive system, such as the one implemented in Tucson in 1994 that sets rates based on the proportion of summer use to winter use, penalizes the small user who would normally use less water during the winter.

Another drawback from the utilities' perspective has been economists' inability to develop specific price elasticity figures (the percent change in water for a percent change in price). Yet it has been demonstrated that determining the exact elasticity is irrelevant as long as the range that it will fall into can be determined. If the elasticity is greater than one, demand will fall at a faster rate than prices have increased and revenue will drop. If elasticity is less than one, demand will drop more slowly than prices increase and revenue will rise. If it can be shown that the elasticity is less than one, rate changes present a very potent tool for managing demand.

AS THINGS STAND:

Price increases have been shown to be the most effective tool for reducing system-wide demands, yet negative public sentiment, potential inequities in implementation and the inability to estimate price elasticity have limited their use.

Although still heavily pursued, the expansion of supply would require revisions of the current rate structures of all three communities. As communities look for new sources, costs and rates will continue to rise.

Public education utilized in a vacuum has little or no effect on water demand. Yet, combined with a price change as was the case in Tucson and Las Vegas, public information campaigns can improve public acceptance and enhance demand impacts. Public information programs also serve as tools to provide the public with the information they will need to answer critical and expensive questions as to how future demands will be met.

WHERE TO NEXT:

1. Development of price sensitive models allowing planners to determine the price impacts of large projects on user demands.

2. Expanded use of studies which analyze demand management as well as supply augmentation. Already these studies are becoming prevalent, examples which can be seen in the Southern Nevada Water Authorities' (Las Vegas)

Integrated Resource Plan and Phoenix's Resource Plan.

3. Development of coordinated pricing and supply strategies that bring new supplies online after the price structure has been developed to support it. This will test customer willingness to pay for additional water before the utility is locked into a long term capital commitment.

4. Evolution of the public information program. Realization that, if supplies are increased, then price increases are inevitable. Provide users with information as to what the difficult decisions will be and what options there are to address them.

5. Pricing strategy. Separate efficiency issues (how many benefits the system produces) from equity issues (how those benefits are divided amongst stakeholders), and realize that pricing is primarily a mechanism of efficiency. With proper pricing, micromanagement techniques become irrelevant as users themselves determine the most beneficial use of water. Equity issues can be addressed with such things as "lifeline" rates and subsidized education programs.

References

Arizona Department of Water Resources, "Conservation Requirements for the Second Management Period 1990-2000."

Babcock, Tom, and Opitz, Eva M., (1988), "Phoenix Emergency Retrofit Program: Impacts on Water Use and Consumer Behavior", A report submitted to Phoenix Water and Wastewater Department by Planning and Management Consultants, Limited.

City of Phoenix, Water and Wastewater Department, Water Conservation and Resources Division, "Must The Roses Die?", Phoenix Water Needs, Supplies and Strategies - A Summary of the Phoenix Water Resource Plan - 1990.

Geiser, Edward E., P.E., Project Manager (1989), "Tucson Water Resources Plan 1999-2100", CH2M Hill.

Griffin, Adrian H., Ingram, Helen M., Laney, Nancy K., and Martin, William E., (1984), "Resources for the Future - Saving Water in a Desert City", Resources for the Future/Washington, D.C.

Tucson Water Department, "Serving You Today - Planning for Tomorrow....."

Integrated Water Resource Planning in Las Vegas

Susan L. Robinson, A.I.C.P.[1]

Abstract
Integrated Resource Planning (IRP) is the latest buzzword in water planning, although the concept has been used by electric and gas industries for years. Like a number of communities, Las Vegas has jumped on the IRP bandwagon.

Introduction

In recent years the Las Vegas region in southern Nevada has been one of the fastest growing metropolitan areas in the country, with water demands predicted to exceed supplies shortly after the turn of the century. A number of planning efforts are underway to stretch those supplies. The latest stage in Las Vegas' water planning evolution is IRP.

Background

Las Vegas receives almost four inches a year of rainfall. 85% of its water supply comes from the Colorado River, the rest from groundwater. Until the mid-1980s, supplies were expected to last well into the 21st century. Since the tremendous growth that began in 1987, however, annual water usage shot up as high as 13%. If trends continued, Las Vegas would be out of water in less than ten years.

This awareness spurred an intense water supply and demand projection process in 1990 among all the local water and wastewater agencies, complete with computer models and consultants. The bad news was confirmed in early 1991: Supply shortfall by 2007 with conservation, by 1998 without. The good news was the creation in 1991 of the Southern Nevada Water Authority (Authority), a regional water and wastewater agency with members from all seven

1. Southern Nevada Water Authority, 1001 South Valley View, Las Vegas, NV 89153

local water and wastewater agencies and municipalities. The Authority would manage existing supplies, find new supplies and promote conservation. Conservation was instituted in various forms within the member agencies, including increases in water rates, and some more water was acquired by the Authority. Even so, the latest demand projections showed that resources still could not reach 2010.

An even more immediate problem of facility shortages loomed. Las Vegas receives most of its water from the Colorado River through a single "straw" called the Southern Nevada Water System (SNWS). The water is piped from Lake Mead through the River Mountains tunnel and into the valley to the different utilities. Some utilities were facing the inability to meet peak summer demands as soon as 1996, due to lack of facility capacity. While six percent conservation had been achieved by the end of 1993, none was occuring during the summer, when it was needed most.

To buy the community a few more years, plans to expand SNWS as quickly as possible began immediately. However, additional capacity was still needed soon after year 2000. In 1993 the Authority began plans for a second "straw," the Authority's Treatment and Transmission Facility (TTF). Projects of this magnitude normally took as long as twenty years to plan and build; TTF staff had seven years.

At the same time, the concept of IRP in water planning was beginning to surface, and, in early 1994, the Authority began its own IRP effort. Authority consensus was that it did not have many options to meet its resource and facility needs. Colorado River water was the best resource option to pursue, conservation was imperative, and new facilities should have been underway yesterday. Since IRP was supposed to look at the "big picture" of resources, facilities and conservation, however, the Authority wanted to make sure it was going down the right path, before it went much further.

The Authority hired Barakat & Chamberlin, Inc. (BCI) to do a "Phase I" IRP in six months. With existing plans and studies (no new studies) and a technical work group, BCI was to create different strategies that would meet future water demands and then rate the different strategies against various objectives determined by the work group.

IRP for water

The gas and electric utilities have been using IRP for years, as a result of the 1960s' escalating costs of facility construction, accompanying public outcry, and the environmental movement. Its major concepts include: least-cost alternatives, consideration of environmental and social costs, conservation, and public involvement.

Traditional water planning means engineering and cost studies, but it is evolving to include IRP concepts.

Currently, there is no universally accepted, specific procedure for a water IRP, but there are some key documents emerging. In 1993, BCI wrote draft IRP guidelines for the American Water Works Association (AWWA), Wade Miller and Associates was hired by the AWWA Research Foundation to study the various water IRPs that were beginning throughout the country, and Janice Beecher wrote an IRP discussion paper for the Water Industry Technical Action Fund. In the West, a few communities are in various stages of IRP and their experiences will undoubtably help refine the process.

Generally, a water IRP is a planning process: Define the purpose of the process; define objectives (cost, reliability, environmental impacts, etc.) that will help in choosing among different strategies to meeting the purpose; develop objective criteria or measures to rate the different strategies (e.g., probablility of shortages as a measure for the "reliability" objective); determine the various resource, facility and conservation options to consider; combine options into different strategies; rank the different strategies using the objective criteria; use some process to pick the best strategy.

Challenges

Citizen Involvement. Three months into the IRP, Authority staff realized that it had better involve citizens as soon as possible, since important decisions might be made within Phase I IRP. A 21-member IRP Advisory Committee (IRPAC) was created from a cross-section of the community to provide input on which strategies the Authority should pursue. The problem was keeping ahead of IRPAC. Folks wanted to act immediately, before analysis was complete. As one official called it, public involvement is "like herding cats."

Multi-jurisdictional. The IRP was under control of the Authority, but there were other water and wastewater planning efforts in various stages that were not. In such a rapidly growing community, different agencies' efforts could not be put on hold while the IRP churned through a two- or three-year long process. With constant communication, the IRP and other efforts would begin to influence one another in later stages.

Time. The biggest timing issue was the TTF schedule. TTF staff could not put the TTF on hold until results of a lengthy, detailed IRP process would tell them, chances were, that the Authority needed a TTF. So TTF planning continued on a parallel path with the IRP, with no commitment to build, while the IRP process determined whether or not to even build a TTF.

Data. Initially staff thought it had all the data the IRP could possibly need, but soon found it did not. Much had been collected but not as part of a whole picture. People were constantly reminded that this was a first phase of a process and that data would always be changing. In the first phase IRP, there was enough data to see the relative merits of different strategies, even if not the precise merit, and enough on which to make some general decisions.

Uncertainties. The Authority had uncertainties in both its future demands and supplies. The major future supplies would all have to be negotiated. While projections generally showed the community continuing to grow, in a volatile economic climate like Las Vegas', how much and when were anybody's guess. "Low" and "high" demands were analyzed throughout the IRP process, with fingers crossed that the "high" was high enough. It was good that the uncertainties were included in the analysis, because real world decisions must consider uncertainties. However, the uncertainties made the possible strategies that the Authority might follow more complicated. The Authority had to make short-term facility decisions now, before knowing what resources might come available, yet try to avoid eliminating future supply possibilities resulting from current facility decisions. Thus, possible strategies were more like decision trees, rather than specific courses of action. While "trees" best represented reality, they were hard to understand and explain.

Conclusion

Overall, IRP makes good planning sense. If there is a negative side, it is the potential for infinite analysis. Once people's perspectives widen to see connections among more issues, it is tempting to broaden the scope of the project. Before they know it, they are trying to solve world peace. There must be a balance between planning and action. In a rapidly growing community like Las Vegas, where the window of opportunity for action is very small, it is imperative to stick to the agreed-upon project scope, determine the essential data, and act on it. Detailed analysis may simply not be needed for certain decisions.

IRP is not just an event, its a process. The primary goal of an IRP is to show to the public and the decisionmakers, in explicit and consistent language, the tradeoffs among different solutions to a stated problem. Perhaps, though, an equally important thing to come out of IRP might be the process: People with different backgrounds learning to look at possible solutions beyond their own standard ones.

Watershed Planning and Management: Barriers and Opportunities

Jonathan W. Bulkley, F.ASCE[1]

Abstract

Watershed planning and management has been implemented in England and Wales for over twenty years. Systematic planning and management did not take place in the 1972 Clean Water Act. The reauthorization of the Clean Water Act is expected to require watershed planning and management. The evaluation criteria are presented to facilitate effective institutional structures for this purpose.

Introduction

A major institutional change for the provision of comprehensive water services in England and Wales took place on April 1, 1974. On that date, ten regional water authorities based upon specific watersheds took over the provision of water services in England and Wales and effectively replaced more than 1600 separate entities which had been providing these services in the past. The problems which were facing the water industry in England and Wales and which led to the reorganization upon the watershed basis included the following (Bulkley, Gross, 1975):

1. The projected increase in demand for water by the Year 2000 would pose severe difficulties under existing organizational arrangements.
2. Water reuse was anticipated to increase and therefore a much greater concern will be required for treatment provided water after use.
3. There needed to be a sweeping reduction in the number of separate operating units providing sewage disposal and a further reduction in the number of separate units providing water supply.

----

[1]Professor of Natural Resources, Professor of Civil and Environmental Engineering, 2506B Dana, The University of Michigan, Ann Arbor, Michigan 48109-1115.

4. Increasing conflicts of interest existed between the various authorities (local units of government, water supply organization, conservation agencies, etc.) and inadequate mechanisms for resolving these conflicts which included the following:
   a) Inflexibility in the use of existing water resources.
   b) Divided responsibilities for new sources of water.
   c) Difficulties in the promotion of joint or national programs.
   d) Conflicts of interest with regard to water reclamation and water reuse.
5. A need existed to be able to implement plans once agreed upon. Previous management and financial arrangements made implementation most difficult.
6. A need existed to improve planning and coordination.
7. It was seen to be very desirable to have both a five-year capital investment plan for each of the ten regional water authorities as well as a long-term 20-year water plan for each region.

These ten public regional water authorities were privatized by the Conservative Government in 1989. The concept to achieve integrated river-basin management was maintained following privatization. The delivery has been less than expected (Kinnersley, 1994).

The example of England and Wales demonstrates that integrated water management on a watershed basis is feasible. However, implementation of watershed planning and management in the United States must take place within the technical, legal, economic, political, and social realities of this country. This paper will identify certain opportunities and certain barriers associated with the adoption of watershed planning and management in the United States.

Section 208 Area-Wide Water Quality Planning and Management

In the Water Pollution Control Act Amendments of 1972, Section 208 established the requirement for area-wide water quality planning and management. In fact, in the Amendments of 1972, a planning sequence was presented which would have required river basin plans (Section 303) to be followed by area-wide water quality planning (Section 208) to be followed by facility construction (Section 201). In reality, the U.S. EPA proceeded directly with the implementation of Section 201 - facility construction without doing the Section 303 and Section 208 planning prior to facility construction (Metzger et al., 1978). Accordingly, the opportunity for a sequential planning process for water quality control in the United States was lost in part because of the sense of urgency and the need to proceed directly to build waste water treatment plants. It should be noted that the legislative history of the Water Pollution Control Act Amendments of 1972 does not demonstrate significant Congressional debate in Section 208, although Senator Muskie (Maine), a principal author of the Act, endorsed the river basin concept as one possible avenue for Section 208 (Metzger et al., 1978). Institutional conflicts combined with the decision to proceed with facility construction ahead of the mandated planning proved to be fatal blows to the effectiveness of Section 208.

Current Situation

The Clean Water Act Amendments of 1987 brought a number of major changes including a new initiative on nonpoint source pollution. In April, 1991 EPA's Office of Wetlands, Oceans, and Watersheds (OWOW) was created to integrate the protection and management of the nation's watersheds, coastal and marine waters, and wetlands. What had been recognized in this country was the need for integrated holistic approach to plan and manage the nation's critical water resources. Great achievements had been made in surface water quality improvements; yet, the nonpoint source pollution problems remain. A watershed approach is critical to address these issues.

In February 1994, the Administration's Clean Water Initiative proposed a new provision in the Clean Water Act to establish state-wide programs for comprehensive watershed management (U.S. Environmental Protection Agency, 1994). When the Clean Water Act is eventually reauthorized, comprehensive watershed management should be an integral part of the Act. It is sound from an environmental perspective and from an economic efficiency perspective.

Observations

The evolution toward watershed planning and management in this country has followed a difficult pathway. Since the early 1990s many initiatives have been undertaken. The long-term success of watershed planning and management will require the tailored combination of incentives to encourage sub-state federal agencies to undertake effective watershed planning and management. A number of evaluative criteria may be helpful to assess the feasibility of these new partnerships as they evolve. Consider the following:

1. All cost and benefits should accrue within the watershed and should be equitably distributed therein.
2. The watershed unit should have the power and authority to raise adequate capital and the flexibility to select the best means to secure funds and compel performance.
3. The watershed unit should have sufficient authority to resolve conflicts among all stakeholders.

4. The watershed unit should have the legal and administrative authority to perform or caused to be performed the tasks needed in the specific watershed.
5. Lines of communication and the process of coordinating planning and management should be formalized.
6. The watershed unit needs to be accountable to the public including the decision-making processes.
7. The watershed unit should be compatible with the overall governmental structure.
8. There should be sufficient incentives to encourage local governmental units to join into this new partnership organization.
9. The watershed area should be large enough to realize economies of scale.
10. The watershed unit should be able to consider and adjust externalities arising from the system.

References

1. Bulkley, Jonathan W., Gross, Thomas A., "An Innovative Organizational Arrangement for Comprehensive Water Services: The Thames Water Authority as a Model for Complex Urban Areas of the Great Lakes," Research Project Technical Completion Report, OWRT Project No. A-083-MICH, 106 pp., September 1975.

2. Kinnersley, David. Coming Clean: The Politics of Water and the Environment. Penguin (UK), 240 pp, 1994.

3. Metzger, Phillip A., Alter, Craig B., Bulkley, Jonathan W. "Institutional Arrangements for Area-Wide Water Quality Planning and Management: Section 208 Applied to Three Metropolitan Areas of the Great Lakes." Research Project Technical Completion Report, OWRT Project No. A-096-MICH. November 1978.

4. U.S. Environmental Protection Agency. "President Clinton's Clean Water Initiative," Office of Water, EPA 800-R-94-001, Washington, DC, February 1994.

Business Issues in Water Management: Re-Engineering Data Management

Susan K. Lior, P.E., Member A.S.C.E.[1]

Abstract

A great deal of inefficiency creeps into an organization's operations as it quietly adapts to changing technology, administrations, and demands. Periodically, each organization should evaluate its current operations to eliminate redundant and outdated procedures, and to incorporate more effective and efficient practices. This is especially true with water resource management and the evaluation of infrastructure-related information and its computerization. Managers are finding out that good information, in the right form and at the right time is invaluable to saving money and improving productivity.

This paper will address re-engineering data management to optimize the collection and dissemination of information within a water resource organization and to realize greater productivity gains.

Introduction

In the face of escalating costs, it's difficult to justify an increased budget for managing resources and maintaining facilities that are too often taken for granted. Today's managers are expected to do more with less -- less budget, less workforce, less resources. The challenge faced by utility and public works managers is to develop cost effective solutions that are possible with the use of enterprise-wide information management.

The water utility business has experienced rapid growth in the installation of computers and in the use of computer software. There are abundant examples where computer applications have reduced costs and improved the effectiveness

[1] Principal, Susan Lior Engineering, Six Lantern Lane, Media, PA 19063

of an organization. The benefits of billing systems, automated mapping and drawing, maintenance management, leak detection and control, flow analysis, and watershed modeling are well recognized. Unfortunately, the design of each of these applications has typically been quite restricted, addressing only a specific function or narrow problem of the system and often developed on standalone microcomputers. The productivity gains are modest in that the information needs of the entire organization have not been taken into consideration

Methodology

The first step of re-engineering requires an identification of the functions and responsibilities within an organization. Functions represent key areas of activity for a water or public utility. Responsibilities connote the responsibilities that are assigned to the various departments in order to carry out the functions of the utility. Once these are identified, a detailed analysis of activities performed by the utility uncovers specific problem areas that impede efficiency. This analysis also establishes the highest priority activities to improve based upon need.

When the source of the problems associated with these activities are addressed they are normally categorized as relating to four areas:

- Data
- Technology
- Business Practices
- Organization and Skills

Data and technology represent the two largest problem areas. Most of the problems identified with data management issues are usually in conjunction with technology related issues. One of the most oft-cited data issues is that the data is usually obtained in hard copy and that analysis work has to be performed manually or requires substantial re-keying of data. This seriously impacts the ability of staff to carry out detailed analyses and assessments. Problems tend to relate to data format, accessibility and lack of tools required for analysis.

Focusing on data management, the next step of re-engineering identifies significant opportunities to improve the performances related to carrying out activities. Opportunities are assessed in five categories:

- reduction of costs
- collection of greater revenue
- better compliance with regulations and laws
- greater safety for employees or the population at large

- better corporate image

Although the goals of each organization vary depending upon where it sees its greatest need, the common thread is to gather data to improve decision support and to share data among application. To do this, organizations need to implement systems in an integrated environment.

Integrated Data Environment

An analysis of data needs recognizes that often 85% or greater of a utilities data is geographically referenced. To accommodate this, many utilities use a geographic information system (GIS) as a tool to integrate their data. A GIS recognizes the spatial as well as the data component of information.

The following figure presents a data model of an integrated data environment using a GIS as a data integrator and having common standards, integrated technology, and easy access to quality data.

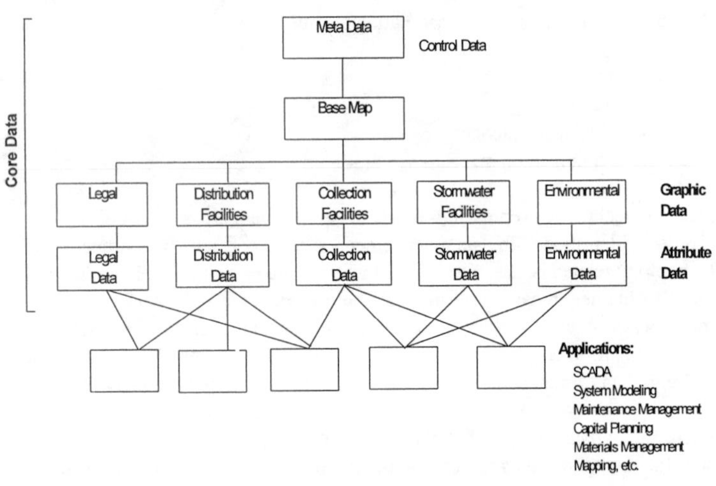

Figure 1 Distributed Data Environment

Each application has the capability of incorporating data from a wide assortment of data sources, eliminating problems related to data format and accessibility. The use of the GIS allows the user to view data on maps; for locating facilities, planning activities, and recognizing trends.

Implications of Developing An Integrated Environment

Developing distributed data in an integrated environment is the more difficult road in the near term requiring greater effort, a structured approach, long term and sustained financial and human resource commitment within the organization. However, the long term productivity gains are well worth the effort. An integrated data environment is a valuable asset that allows utilities to better leverage their resources in the future. Organizations are finding that support for additional functions may be accomplished with little or no additional data acquisition costs after the initial data conversion effort.

Conclusion

By developing *common, distributed databases* and providing a *powerful set of tools* for collecting, sorting, retrieving, evaluating and displaying infrastructure data in the real world, organizations will be able to cost effectively address:

- capital planning (establishing priorities, long range planning, condition assessments, replacement costing);
- maintenance management (tracking and controlling costs, efficient scheduling, status tracking, monitoring productivity);
- preventive maintenance (planning and scheduling, reducing emergency work, ordering materials);
- instrumentation and control (automate and optimize process control, alarming and data logging, monitoring);
- engineering support (project management, estimating, digitized drawings/maps, CADD, modeling); and
- compliance/environmental issues (Right-to-know, MSDS, document management, track employee training, employee trip reduction plans, permitting).

Taking advantage of the "information age" does not have to be a nightmare. End-users today can expect to have innovative, yet practical, solutions that exceed their expectations. The key is acknowledging that information is a valuable resource managed just like any other valuable resource.

Water Availability Evaluation of Five Sources at Portland, Oregon

Tom Davis[1] and Steve Thurin[2]
Montgomery Watson

Abstract

Twenty-seven municipal water providers in the Portland Metropolitan area (Oregon) are currently in the second phase of evaluating the regional water supply outlook regarding long-term demands, water availability, water quality, and system needs. One critical component of the integrated water resources evaluation is the determination of how much water is available for future municipal purposes from each of five sources identified in the first phase of the study; the Columbia, Bull Run, Clackamas, Willamette and Trask rivers. The availability evaluation involved drought year hydrology; planning level reservoir operations models; instream flow considerations; fish species that are listed, or may be listed, under threatened or endangered species criteria; and numerous other political-environmental issues.

Overview of Methods

The hydrology methods were built around the existing information base - which varied considerably for the five sources. Data from flow gages and synthetic data from models were both used in the water availability evaluation. Storage releases from existing, new or expanded reservoirs owned and operated by the municipal water providers were important for the Bull Run and Trask Rivers. Planning level operation models were created for those two systems since they had not been previously modeled in a manner which could provide water availability information for the regional study. The Willamette River is controlled by eleven Corps of Engineers reservoirs and the modeling results available from an existing Corps HEC-5 model were extended for use in the study. Gage data were extended for the Bull Run and Clackamas Rivers to provide a consistent data period.

Regression analyses were done to determine the relationships between data at various flow gages for three of the sources. The analysis methods and time steps used (daily versus monthly) varied from one source to another depending on the needs and constraints.

(1) Principal Engineer, Montgomery Watson, 1800 SW First Avenue, Suite 350, Portland, OR 97201
(2) Principal Engineer, Montgomery Watson, 355 Lennon Lane, Walnut Creek, CA 94598

Bull Run River

USGS gage data for the Bull Run River below Reservoirs No. 1 and No. 2, and reservoir stage data provided by the Portland Water Bureau, were used to estimate the natural river flows for 1928 through 1992. USGS gage data for the Bull Run River near Multnomah Falls, and ratios of precipitation and drainage basin area, were used to develop the inflow record to Reservoir No. 3. A planning level reservoir operations model was developed, supplied with the estimated natural river flow data, and four system scenarios were modeled. The four simulated scenarios were existing storage, existing storage with treatment, future storage, and future storage with treatment. A demand pattern was applied and the yield was determined.

Columbia River

The Columbia River data from The Dalles gage for the period 1928 through 1992 were used, and both monthly and annual flow duration curves were produced. A long term monthly record of continuous available supply was also prepared. The results are monthly flow data for the period of interest and percent exceedance flows.

Willamette River

The COE evaluated the Willamette River system at Salem for a 20-year period from 1967 to 1986 using the HEC-5 model. The results of that study were obtained from the COE and transformed to the Wilsonville area using a regression equation which represents the historical flow differences between Salem and Wilsonville. The Wilsonville data were then extended to the 1928 to 1992 period. Future water rights, river diversions and reservoir operations, as well as instream flow requirements, were evaluated to determine the water available to municipal water providers. The COE model provided the base level of information because it is the best source of realistic future conditions information. The natural flows have been heavily manipulated for a number of decades by the COE reservoirs thus making a "natural flow" analysis complex and unnecessarily time consuming at this study level. The 20 years of COE results were extended to the 1928-1992 study period and these extended storage augmented flows were used in the availability analysis.

Clackamas River

The available Clackamas River streamflow information from an upstream gage was extended to a gage nearer the likely diversion point to produce daily and monthly flows. The differences between mean daily flows and mean monthly flows for the Clackamas can be significant. The two values were compared and a 50 to 100 cfs difference was observed. After flows were prepared for the period of interest, various levels of water rights and instream demands were subtracted. Statistical analyses were performed on the results to produce water availability versus percent exceedance graphs and tables.

Tualatin/Trask System

A Final Environmental Impact Statement (FEIS) was prepared and published in May 1994 by the U.S. Army COE. Inflow data to Barney Reservoir were developed by CH2M Hill for the period of record, 1928 to 1990, and were obtained for this study. These data were extended to include 1991 and 1992. A simple reservoir operations model was developed and applied to the FEIS releases to the Tualatin. Reliability of yield was then calculated.

Instream Flow Considerations

The instream flow picture in the northwest is in a period of rapid change, primarily due to rare and endangered species concerns. Eleven categories of instream flow considerations were identified, ranging from certificated instream water rights to emerging or recent recommendations by resource agencies or environmental organizations. Other factors will affect the degree to which the five sources can be used, including environmental constraints, public perceptions of each source, existing plans and programs of the various agencies involved, and regulatory requirements.

The eleven categories of instream considerations are:

- Certificated instream rights.
- Applications for instream rights.
- Water Resources Commission minimum perennial streamflows.
- Diack flows (court mandated flow).
- Operational criteria for dams.
- Formal agreements between water management agencies and resource agencies.
- Informal agreements/criteria between water management agencies and resource agencies.
- Pre-1909 filings for water rights (the ones of interest are for run-of-river flow).
- Oregon Water Resources Department (OWRD) criteria for threatened/endangered fish species.
- Flows that may be implied by current policies or laws, and which are preferred by an agency or group interested in an instream flow dependent resource.

Results

The results of the study were used in two ways. The water availability information was published in a report and used to help the water providers and the public view and understand the five sources comprehensively. Monthly and daily flow data were also provided for use in an integrated resource planning model which is being developed and applied separately.

The five sources examined vary considerably in the quantities of water physically available and the legal or political constraints and opportunities for using that water for municipal purposes. All five sources involve significant legal or political issues as well as public perceptions that will affect their use. The availability of water is shown on Table 1. The percent exceedance numbers shown are annual for Bull Run and Barney, monthly for the Willamette and Columbia, and daily for the Clackamas. This varying level of detail corresponds to the analytical needs of each source for the purposes of this study.

TABLE 1 SUMMARY OF WATER AVAILABILITY

	Bull Run[1]	Columbia	Willamette	Clackamas	Barney
PERCENT EXCEEDANCE					
95%	434.8 cfs[1]	58,500 cfs[3]	6,456 cfs[5]	777 cfs[8]	39.9 cfs[11]
	280.9 mgd	**37,790 mgd**	**4,170 mgd**	**502 mgd**	**25.8 mgd**
80%	502.3 cfs	75,000 cfs	8,558 cfs	1,066 cfs	47.1 cfs
	324.5 mgd	**48,450 mgd**	**5,530 mgd**	**690 mgd**	**30.4 mgd**
50%	576.2 cfs	99,500 cfs	18,726 cfs	2,884 cfs	53.7 cfs
	372.2 mgd	**64,280 mgd**	**12,100 mgd**	**1,863 mgd**	**34.7 mgd**
NEW SUPPLY YIELD OF SYSTEM FOR PHASE 2 PARTICIPANTS					
95%	280.9 mgd[2]	>600 mgd[4]	165 mgd[6]	152 mgd[9]	25.8 mgd
80%	324.5 mgd	>600 mgd	473 mgd[7]	171 mgd[10]	30.4 mgd
50%	372.2 mgd	>600 mgd	473 mgd	171 mgd	34.7 mgd

1 Average seasonal yield (with Reservoir No. 3) that meets a recent demand pattern 95% of years
2 Average releases per day, June through October
3 95% and 80% exceedance flows occurred during January (low flow month)
4 Available water far exceeds demand by Phase 2 participants
5 Based on extension of 20 years of COE river operations - assuming a similar future operation pattern
6 Existing Phase 2 participants permits total 154 mgd
7 Phase 2 participants applications plus existing permits total 473 mgd
8 Daily average flow in river before diversions
9 Includes existing municipal diversions of 61 mgd
10 Total existing "higher", and "lower" municipal, undiverted rights of Phase 2 participants, i.e. total supply
11 Potential six month releases from an expanded Barney Reservoir to the Tualatin River and the JWC system; reservoir yield maximized

Water Supplies Once Abandoned

Douglas F. Reed, P.E., MEMBER[1]

Abstract

Public water supplies were abandoned in many communities over the last century because better alternatives were available. Today, many of these same communities find that growth and full use of existing sources leaves few new alternatives. Study unveiled the reality that the abandoned and unprotected sources had become hostile to public water supply use. Competing land uses and regulations make it difficult to reactivate these sources.

****

From a water supplier's perspective, today's aquifer and watershed protection programs should ideally include all possible sources of water. However, it is difficult for a source to be recognized as a public water supply unless evaluation and testing has been completed and regulatory support is obtained. Thus, once a source is accepted as a source, it should be maintained as a recognized source even if it is not used in favor of other alternatives. This is important because, once abandoned, restoration can be difficult and perhaps impossible to achieve.

Restoration of an abandoned water supply can be blocked by several factors; conversion of the property to other use, encroachment that is not compatible with water supply use; and environmental impacts of constructing and operating the water supply. A historical perspective of these factors and a review of regulatory policy today suggests the importance of planning for protection of these sources so they will be available when they may be needed.

One means to identify the difficulty of establishing new water sources with respect to competing demands on the

----

[1]Project Manager, Roy F. Weston, Inc., 187 Ballardvale Street, Wilmington, Massachusetts

land, is to examine past efforts. The following generalized examples are based on sources that were abandoned. Because the quality of other sources were better, there were larger sources, or there were less expensive choices. Let us examine what has happened to these sources and reflect upon what might be required to reactivate them in today's business climate.

Case 1. One suburban community gave up its three groundwater supplies between 1920 and 1960. One was adjacent to a flood control reservoir and the other along a river. A much larger source had been located in a coarse gravel deposit beside another impoundment. The new source did not require treatment, the older sources required filtration for iron removal.

Today, the community has lived through years of water use restrictions and economic growth has been curtailed. The older sources are no longer available. On one site is a Little League ball field. The other is a residential area. Attempts to reactivate a source on the site of a ball field might be a very unpopular effort. It would be impractical to remove the homes from the other area.

Case 2. An urban community on the fringe of Boston drew its 4 mgd water supply from a river and used a filtration facility. Industrialization polluted the river and a regional source became available at a lower cost. The source was abandoned and the site of the filtration facility became the town dump. It is now a closed landfill and is incompatible with water supply use.

Case 3. An urban community abandoned its sources for a regional supply and converted the property to a cemetery. Demolition materials from a nearby manufacturing facility were reportedly disposed within the large dug wells. This site has contamination issues and lacks the required radius of protection around the wells.

Case 4. A suburban groundwater supply became surrounded by industrial parks, particularly automobile dealers and service garages. State hazardous waste files document numerous contamination incidents nearby. The wells no longer have the 400 foot radius of protection required by the state and the establishment of land use controls within the zone of contribution would be onerous at best.

Case 5. An aquifer was tested in the 1960's but not developed. Since then, the area was included in an Area of Critical Environmental Concern. This designation increases the level of environmental review. Also, the nearby river is a navigable waterway under the jurisdiction of the Army

Corps of Engineers and a permit is required. Since it is difficult to demonstrate that a well located in a wetland area will not have any impact on the wetlands, this community has not been able to obtain the permit to build the water supply.

Case 6. A surface source was abandoned early in this century. Since that time, waterfront homes with docks have been built, and the pond is used for boating and fishing. Evaluation of the water supply concluded that the water level of the pond would be impacted by withdrawals for public water supply use. The home owners will certainly not want to give up their recreational use of the water.

These examples describe how pristine land that is useful for public water supply, can be made incapable of supporting a supply in only a few generations. Once development occurs, on or near a water supply, the loss is permanent. For some communities, these sites may have been the only choice. The result is that obtaining sources becomes more costly, or the local economy and quality of living become permanently diminished.

There are a variety of competing demands for water supplies. In the last part of this century these forces have become very apparent. Looking into the future, these forces will become more formidable as they become better organized. Thus, it will become more difficult to activate water supplies.

Environmental policy has created a number of entities that control or influence the ability to develop water supplies. These entities can be federal, state or local agencies, private enterprises and citizen groups. A review of the policies that exist identifies this influence and control.

Federal environmental policy influences water supply development by controlling projects that require the discharge of fill material into navigable waters. The Army Corps of Engineers 404 permit is the process and the term navigable waters is broad and covers adjacent marshes and areas where plants are adapted to life in saturated soils.[2] The public review process includes conservation, economics, aesthetics, wetlands, historic properties, fish and wildlife values, flight hazards, flood plain values, land use, shore erosion, recreation, and safety as some of the criteria. Since many water sources impact navigable areas, either by the placement of fill for access roads, drawdown created by aquifer withdrawals, and stream flow

----

[2]Massachusetts Environmental Law, Hinkley Allen Snyder and Federal Publications, Inc. 1992

impacts, approvals can be difficult. State policy also protects wetlands that are not navigable. For instance the Massachusetts Wetlands Protection Act M.G.L. c. 131 & 40 includes all vegetative wetland and landforms such as a bank, freshwater wetland, marsh, or swamp that borders on any creek, river, stream, pond, or lake and land subject to flooding. Again, water supplies are almost always in these areas and performance standards will impact the feasibility of developing a site.

State water supply regulations and water quality regulations influence the land ownership requirements of the source, the recreation access of surface waters impacting the water source, the level of treatment, the quantity of water, and the ability of the water purveyor to acquire land by eminent domain.

The Massachusetts Environmental Policy Act requires a public review for projects of a certain dollar value or increased water use by a threshold value or having certain wetland impacts.

Local policy also excerpts powerful influence on the development of a site. Local zoning bylaws and review boards, land use controls, local wetland bylaws, and neighborhood groups are examples of local influence.

In most of these cases, a formal public review process is required. If a former water source can successfully maintain public recognition as being a public water supply source, then this public process can be less formidable.

We essentially have several choices; preserve what we have, restrict growth, or regionalize. If we limit the development of new water supplies, then we will make a decision to limit growth. Commerce that desires to grow will have to locate elsewhere.

In either case, the theme of preserving sources is valid. This is because the future cannot be reliably predicted. Projections can only be accomplished ten or so years into the future. Thus, securing permanent recognition of potential water supplies is important. Water resources, whether groundwater or surface water, are vital and should be protected for ultimate water supply use.

We need to recognize that, as water suppliers, other parties' interests must be considered in the process of long term water supply management. Thus, we must be in balance with these interest groups. If we plan early enough and recognize that we will need to work with these groups, then we can take the steps to ensure that important sources of public water are kept secure.

Inferring Columbia River System Operating Rules Using Optimization Model Results

Kenneth W. Kirby[1] and Jay R. Lund[2]

Abstract

The U.S. Army Corps of Engineers' North Pacific Division and Hydrologic Engineering Center applied a prescriptive reservoir model to the Columbia River System to suggest system operating rules. This paper presents the methodology applied to formulate preliminary monthly operating rules using an economically derived objective function. The paper also shows a variety of display techniques and statistical methods used to scrutinize prescribed results. The paper compares prescriptive results with existing operations (represented by results from NPD's descriptive model). The two analysis methods produce very similar system wide operations, but significant differences exist for storage allocation among larger reservoirs such as Grand Coulee and Arrow. These similarities and differences are discussed regarding their potential significance.

Introduction

Development and implementation of viable operating rules are extremely important aspects of reservoir system management. One application of the Hydrologic Engineering Center's Prescriptive Reservoir Model (HEC-PRM) is to suggest reservoir system operations optimized explicitly for quantitative statements of system operating objectives. This paper shows how prescriptive models can help reservoir system managers address conflicting demands for water use. Increasingly complex water management problems make prescriptive modeling techniques even more useful. By deriving reservoir system control strategies from operations optimized for specified objectives, prescriptive methods can reveal solution alternatives not intuitive to an analyst through simulation alone. Paired with descriptive models, prescriptive tools

---

[1] Hydrologic Engineering Coop, Hydrologic Engineering Center, U.S. Army Corps of Engineers, 609 Second Street, Davis, CA 95616

[2] Associate Professor, Department of Civil and Environmental Engineering, University of California, Davis, CA 95616

expand the analyst and decision makers' evaluative capability that can produce a larger solution frontier for complex reservoir management problems.

Preliminary strategic operating rules for the Columbia River System were suggested from results of an HEC-PRM model including a 50-year standard inflow hydrology and economically derived penalty functions for each water use considered. (HEC-PRM is a deterministic optimization model which uses a minimum cost network flow structure with an optimal sequential algorithm for incorporating hydropower.) These prescribed results were scrutinized with a variety of statistical and data display techniques and compared with existing operations represented by results from the North Pacific Division's simulation model (HYSSR). The HEC-PRM model suggested operations similar to current operations, with some major exceptions for Grand Coulee, Arrow, and Duncan reservoirs. The preliminary operating rule changes suggested by this analysis require refinement, testing, and evaluation through descriptive modeling studies.

Comparing HEC-PRM Results to HYSSR Results

Solution of the HEC-PRM model of the Columbia River System produces fifty years of monthly time series data for storage and flow at each point of interest. The primary graphical aids used to analyze these results were quartile, exceedance probability, time series, and storage allocation plots. Figure 1 shows an example of monthly storage quartiles (25, 50, and 75%) plotted for HEC-PRM and HYSSR results. The 25% quartile represents that quantity of storage for a given month (i.e. January) for which one quarter of all values occurring for that month are below. The three quartiles plotted for each model in Figure 1 shows the range of storage values for the central 50% of all results for each month. These quartiles give an indication of variation from year to year and also illustrate annual draw down and refill cycles. Figure 1 shows that prescribed Grand Coulee operational patterns are similar to existing operations, but that the prescribed storage results tend to be higher. In other words, operations suggested by HEC-PRM draw down Grand Coulee Storage for fewer months each year and with less departure from full capacity.

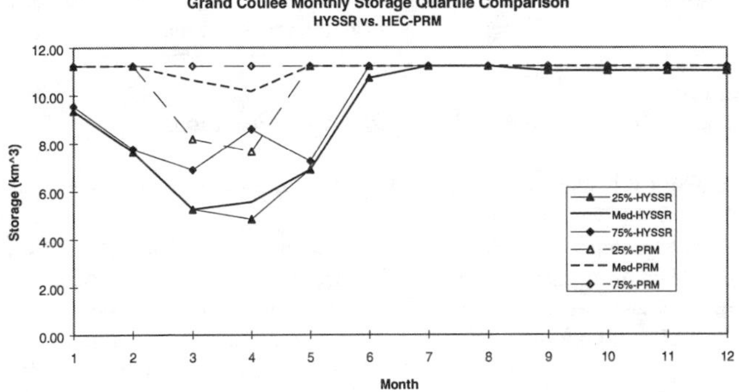

Figure 1 Monthly Storage Quartile Comparison Plot

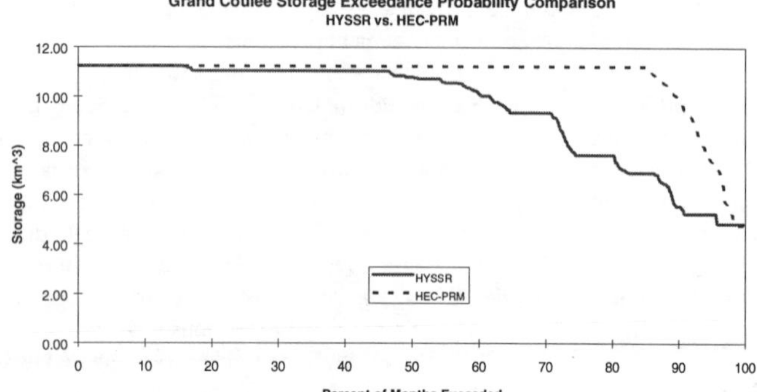

Figure 2 Storage Exceedance Probability Comparison Plot

Another useful plot compares exceedance probabilities for monthly storage. The plots show the percentage of months for which the storage was above a specified quantity. Figure 2 displays exceedance probabilities for Grand Coulee storage from HEC-PRM and HYSSR results. The significant difference for Grand Coulee reservoir is that prescribed results maintain storage at maximum capacity (11.23 km^3) approximately 85% of all months over the 50 year analysis, whereas current operations result in storage at maximum capacity only about 16% of the months.

One technique used to better understand relationships within the system was the storage allocation plot. This approach plots storage results for a given reservoir for one month against storage results for a series of reservoirs (usually those reservoirs on a particular branch of the system) for the same month. Figure 3 illustrates this technique for Grand Coulee Reservoir storage plotted against the total storage contained in the branch of reservoirs including Mica, Arrow and Grand Coulee. This plot illustrates the relationship of operation for Grand Coulee for a given storage level on the branch. Figure 3 shows that HEC-PRM operations keep Grand Coulee Reservoir full as long as is feasible while storage is being depleted from the other two reservoirs on the branch.

Conclusions

The results of this study led to the following conclusions.

1. HEC-PRM results can be used to suggest enhancements to operating rules for the Columbia River System.
2. Potential modifications to current operating rules should be tested, refined, and evaluated using simulation tools. Testing, refining, and evaluating of operating rules based on HEC-PRM results through the use of simulation models (such as HYSSR) is important due to the specialized conditions required to apply a prescriptive model like HEC-PRM. The HEC-PRM model uses a fairly simple representation of the system, a monthly time step, operates with perfect foresight,

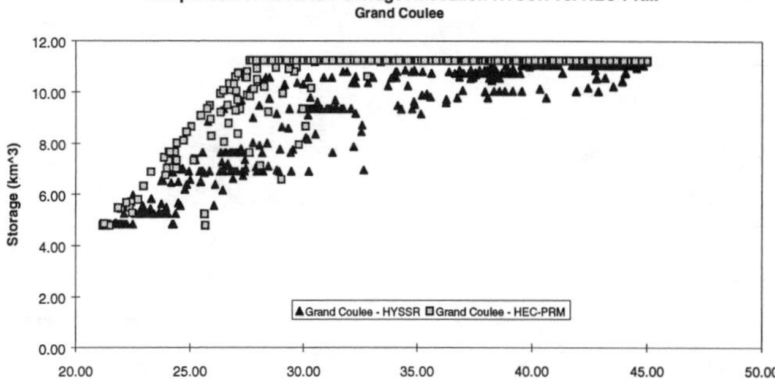

Figure 3 Storage Allocation Plot for Reservoir vs. Branch Storage

and requires that all system objectives be specified as convex linear functions. Results from this simplified representation can be extremely useful, but usually need to be refined.

3. The overall operation strategy suggested by HEC-PRM is similar to current operations represented by HYSSR results. Annual drawdown and refill of system-wide storage under HEC-PRM is very similar to HYSSR results for the 50-year period examined. HEC-PRM operations differ from those of HYSSR mostly in the allocation of total storage within the basin.
4. The most significant suggestions arising from HEC-PRM results are to draw down Grand Coulee less frequently and typically to make smaller drawdowns. This operation entails greater storage variations for Arrow and Duncan. Suggestions for modifying operation of other reservoirs are much less dramatic.
5. Applying HEC-PRM to strategic operating rule development and the screening of planning alternatives is feasible and provides insights and operation justification unavailable from traditional simulation modeling studies alone.

References

USACE (1994). *Preliminary Operating Rules for the Columbia River System from HEC-PRM Results*, Hydrologic Engineering Center, 609 Second Street, Davis, CA, Draft.

Use of MODFLOW-Derived Unit Impulse Functions to Re-Calibrate Return Flow in a Surface Water Planning Model

Steven M. Thurin[1] and Kaylea White[2], Members ASCE.

Abstract

Three separate Environmental Impact Statements are being developed for three separate water resources projects along the Provo River in Heber Valley, Utah. Surface water and groundwater model analyses were needed to assess potential environmental impacts associated with these projects. The key areas of concern were potential impacts to existing wetlands and potential impacts on irrigation return flows which are utilized downstream. Irrigation return flows are of special concern because significant over-irrigation is practiced in the Heber Valley, and downstream users rely heavily on these flows. Unit Impulse Functions were used to calibrate the return flow calculation portion of an existing surface water planning model using simulated results from a groundwater model.

Background

To simulate impacts of the three proposed water resources projects in the Heber Valley, both surface and groundwater needed to be analyzed. Two existing models were to be utilized in combination for analysis of project impacts. A groundwater model of the Heber Valley was developed by the USGS (Roark, et al., 1991). This model utilized the MODFLOW three dimensional groundwater flow finite difference program. A separate surface water model, PROSIM, was developed for operational simulation of the

---

[1]Principal Engineer, Montgomery Watson, 355 Lennon Lane, Walnut Creek, CA 94598

[2]Senior Geologist, Montgomery Watson, Sacramento, CA 95825

complex Provo River surface water and water rights system (Thurin, et al., 1994). PROSIM would be used to establish available diversions, and simulate river flows and reservoir levels. MODFLOW would be used to estimate changes in groundwater levels and to provide outflows from groundwater to the surface water system.

Because groundwater return flows are such a significant portion of the Heber Valley water budget, prior to their use, the two existing planning models needed to be re-calibrated with each other to ensure their close agreement. If not, the simulated diversion produced by PROSIM would not accurately reflect the availability of return flows simulated by MODFLOW. Because the models use completely different methods to estimate return flows, a simple modification of parameters was not a suitable calibration method. PROSIM estimates consumptive use, then uses up to eight different delay patterns for each separate irrigation area, to route unconsumed water back to various points on the stream. MODFLOW accepts unconsumed applied water as an input, then simulates saturated flow through a porous medium to route percolated water. It then estimates discharge flux to streams, drains, and lakes by boundary and stream bed conductance and hydraulic head gradient.

Although the hydrologic process used in MODFLOW is simplified for PROSIM, it can be made fairly representative because the near-surface groundwater flow processes in the Heber Valley are dominated by irrigation recharge and discharge. Furthermore, groundwater levels in the Heber Valley only vary by a few feet from season to season or from year to year, and thus storage effects are minor.

Method

The return flow functions were derived by using MODFLOW to simulate the effects on boundary fluxes of one unit of additional recharge. By subtracting the simulated boundary fluxes without the extra recharge, from the boundary fluxes with it, patterns of discharge or return flow representative of any one unit of recharge were developed. Assuming relatively constant groundwater levels, these patterns may be used to define the PROSIM return flow patterns for all recharge. Each of the 14 irrigation areas in the Valley was analyzed individually to determine the location and timing of resultant return flow. Twenty eight separate return flow patterns were developed. Typical patterns, along with the primary pattern in use prior to recalibration, are shown in Figure 1.

FIGURE 1
Comparison of Return Flow Delay Patterns

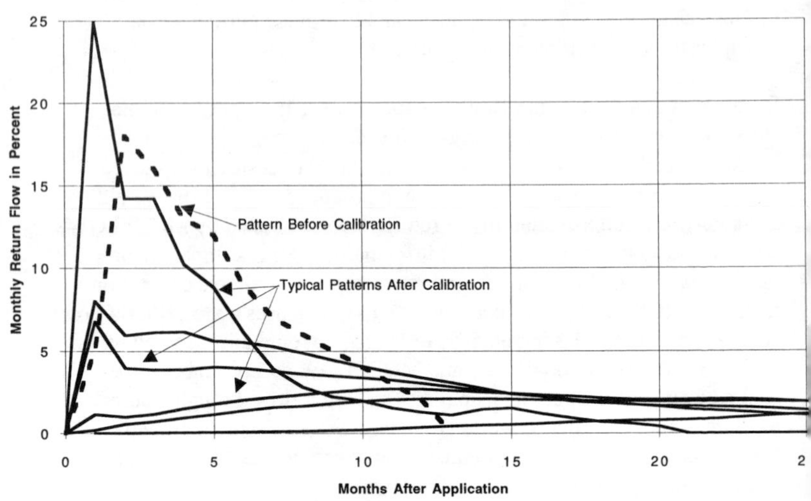

FIGURE 2

Mean Monthly PROSIM Return Flow vs. Mean Monthly MODFLOW Discharge

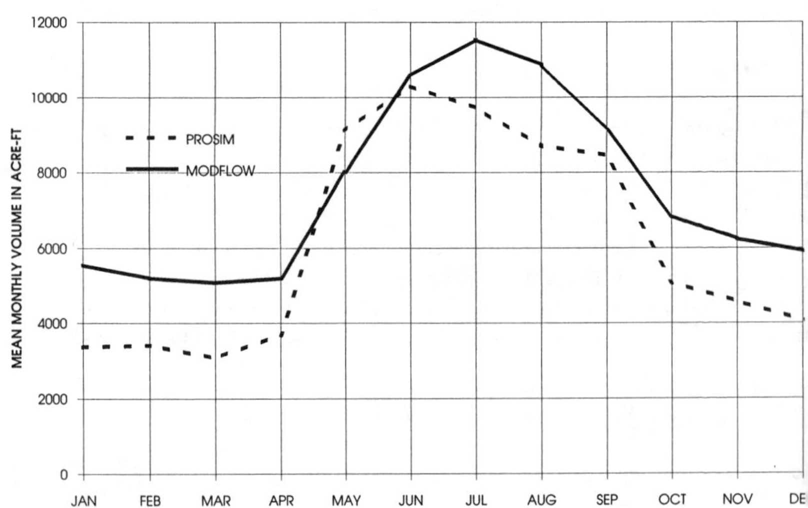

After developing the new return flow patterns, the ungaged PROSIM surface water inflows were adjusted so that the predicted stream flows agreed with the available historic streamgage data. In effect, where return flows were higher, ungaged inflows were adjusted downwards to maintain the total flow in the stream. Likewise, when return flows were lower, ungaged inflows were increased.

As a check on the recalibration process, the total volume of return flow predicted by PROSIM was compared with the total volume of boundary flux simulated with the MODFLOW program. Although there are certain quantities of flow which are included in the boundary flux but not included in the PROSIM return flows, in general the pattern of total flow was observed to be very similar. The total flows are shown in Figure 2.

Results

As displayed in Figure 1, the MODFLOW derived return flow patterns were significantly different from the primary PROSIM pattern in use prior to the recalibration. Based upon the relatively low groundwater transmissivities, MODFLOW indicated that recharge on certain of the more remote irrigated areas would not return to the stream for up to seven years. This contrasts dramatically with the popularly held belief in the Heber Valley that essentially all irrigation water returns in the same season as it was applied. Because the water resources projects being studied involve significant changes to return flow volumes, these longer return flow delays may result in markedly different conclusions with respect to water availability.

References

D. Michael Roark, Walter F. Holmes, and Heidi Shlosar, (1991), "Hydrology of Heber and Round Valleys, Wasatch County, Utah, With Emphasis on Simulation of Ground-Water Flow in Heber Valley", USGS.

Steven M. Thurin, Paul C. Summers, and P. Kirt Carpenter (1994), "PROSIM - A Water Rights-Based Operational Simulation Model of the Provo River", Proceedings of Specialty Conf.:Water Policy and Management, ASCE, Fontane and Torno, ed., New York, N.Y.

CALIFORNIA AQUEDUCT MODELING PACKAGE (CAMP)

J. Giron, M. Morris, J. Willcourt, J. R. Phillips [1] & Sun Liang [2]

Abstract

A model called CAMP (California Aqueduct Modeling Package) has been developed to predict the effect of pump-ins on water quality along the California Aqueduct. CAMP is a stand-alone software package designed to run on an IBM-compatible personal computer system with the Microsoft Window operating environment.

The model has been tested using two tracers, sulfate and chloride, for two historic four month periods. Most of the predicted concentration values were within 10 to 15 percent of those measured along the Aqueduct. Correspondence tended to worsen with increased downstream distance, but was within 25 percent for all cases.

Introduction

The Metropolitan Water District of Southern California, one of the nation's largest wholesalers, is responsible for providing imported and safe water to over 16 million people in Southern California. Metropolitan has two main sources of water: the Colorado River and California's State Water Project. The California State Water Project is a complex system which delivers water from Northern California, covering a 444 mile distance from Lake Oroville, to the San Luis Reservoir and on into the California Aqueduct.

The California Aqueduct runs through the San Joaquin Valley and delivers water to Pyramid and Silverwood Lakes. MWD received water from both lakes. San Joaquin Valley water agencies and agricultural contractor both pump ground water from local aquifers into the Aqueduct and pump out of the Aqueduct. MWD is interested in the effect of this activity on downstream water quality.

Tracer Selection

It was specified that only conservative chemical species should be considered. (A conservative chemical species is one that tends not to dissociate, react, or disappear from solution in appreciable amounts over a given time.) In addition, a second criterion was developed: a water quality constituent should have a significantly different concentration in ground water compared to that in the main Aqueduct flow.

---

1 Department of Engineering, Harvey Mudd College, Claremont, CA 91711
2 Senior Engineer, Metropolitan Water District of Southern California,, 700 Moreno Avenue, La Verne, CA 91750

Two chemical constituents, chloride and sulfate, were determined to be appropriate and were used as tracers for the model. Sulfate was selected in part because concentrations in the pump-ins were nearly an order of magnitude higher than those in the baseflow (an average of 433 mg/L versus 49 mg/L over June-September, 1991). Chloride was selected as a second constituent because its concentration in the groundwater was significantly lower than that in the baseflow-an average of 75 mg/L in a typical well compared to 105 mg/L at the outlet of O'Neill Forebay.

Data Base

Verification, and use in a predictive manner, of CAMP depends to a great extent on the quality of Aqueduct operations data available. Assembling this information was a major task. A summary of the data acquired is given below. (Giron, May 1994).

Type of Data	Best Reporting Frequency	Remarks
Aqueduct Water Quality	Monthly	Consistently available.
Pumped-in Ground Water Quality	Monthly	Sporadic
Pumped-in Flow	Monthly	From unpublished reports
Delivery Flow	Monthly	SWP Monthly Operating Reports
Aqueduct "Base Flow"	Daily	SWP Monthly Operating Reports

Table 1 : Aqueduct Data

Inadequate sampling of water quality and Aqueduct flow data was a problem. Since water quality and flow information were used to set up and verify the model, their quality and sampling frequency determined the accuracy of the model. If the concentration of a particular constituent was not sampled frequently enough, then the historical data may be useless for calibrating the modeling program. Most constituents in the SWP water quality database were sampled (at best) only twice each month. Furthermore, there were only five mileposts between O'Neill Forebay and the terminal reservoirs with consistent data.

Model

The search for a suitable theory to model the Aqueduct was strongly guided by the physical complexity of the Aqueduct and the quality of the available modeling data. A number of simplifying assumptions were made to account for these two factors in the search for an appropriate modeling algorithm.

The first assumption involved neglecting unregulated flow : rain, evaporation, and leakage were ignored in model calculations. Measures of such were not thoroughly quantified in the data provided. Plug (or slug) flow behavior was the second simplifying assumption. That is, a volume of water travels downstream with a definite physical start and endpoint. Any mixing that occurs within the flow happens only within the cross-sectional plane of the Aqueduct. Plug flow behavior also implies that no chemical diffusion occurs normal to the cross-sectional plane of flow. The third assumption is complete lateral mixing within the plugs (a characteristic of plug flow). The fourth assumption involved step input. Inputs to the Aqueduct have constant flow rate and concentration over a given time interval. This is not just a simplification, but a necessity based on the data provided. A constant linear flow velocity rounds out the list of simplifying assumptions. The Aqueduct is designed to maintain an average flow velocity of between three to four feet per second in the open-channel sections, and this was used in the model. (Giron, May 1994).

A mass-balance based modeling concept was then implemented as the slug model. A slug is a volume of water which has a definite start and endpoint as it flows along within the Aqueduct. It does not interact with anything outside those boundaries, although slugs that are coincident spatially within the Aqueduct can combine to obtain composite values for water volume and tracer mass. Within these boundaries, the tracer concentration is assumed to be uniform. Once these slugs enter the Aqueduct, then they can be modeled as step functions. The ease with which step functions can be manipulated and superimposed in a great advantage of the slug model. The assumption of complete lateral mixing within plugs allows the model to superimpose slugs and find composite values for volume and tracer mass. Which the slugs do not interact with anything outside their lengthwise boundaries, they do mix within their cross sections.

Results

Two test periods were selected: June-September, 1991 and January-April, 1992 (Giron, Sept. 1994). Tracer concentrations were predicted and compared to measured values at four mileposts, (172, 245, 293, 303) downstream from the San Luis Reservoir. During these periods, around 800 pump-ins and deliveries had to be taken into account. Pump-in flow ranged from 3 to 41 percent of total Aqueduct flow.

A typical result is presented in Figure 1 which presents predicted vs. observed data for sulfate for the period June-September, 1991 at Milepost 245.

For this period predicted concentration values followed the general trend exhibited by measured sulfate data, with a majority of the data points falling within 10 to 15 percent of actual values. Correspondence between predicted and actual values tended to worsen with increased downstream distance, but were within 25% for all cases. Results were similar for the January to April test period.

Predictions for January - April, 1992 also generally agree with measured chloride concentration samples. Predicted values for early mileposts generally lie within 10 percent of actual sample concentrations, but again the correspondence is somewhat worse for downstream locations. Results were similar for the June to September, 1991 period. CAMP was also tested for sensitivity to changes in input data. Results from doubling concentration and flowrate values for the baseflow and pump-in scenarios were obtained. Four tests were developed to determine CAMP's sensitivity to errors in flowrate and concentration values for both baseflow and pump-ins. The results were as expected.

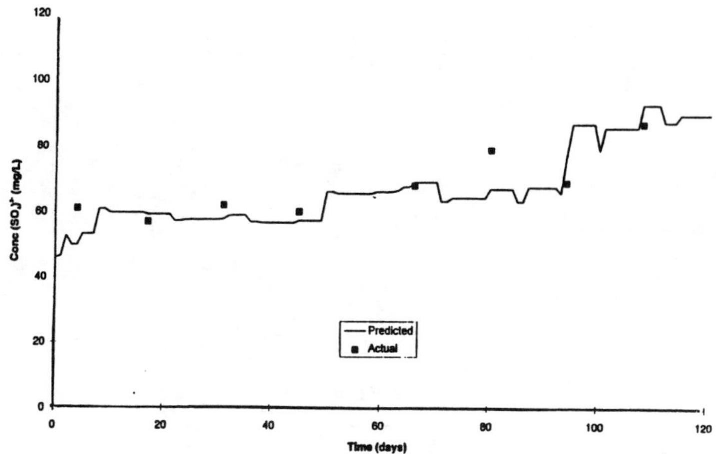

Figure 1. Sulfate at MP 245, 6/91 - 9/91

Conclusions

The California Aqueduct Modeling Package (CAMP) is a plug-flow modeling concept that has been developed to predict the effect of groundwater pump-ins on water quality along the California Aqueduct. Verification of the model was achieved using sulfate and chloride as tracers for two four-month test periods. Most predicted concentration values were within 10 to 15 percent of those measured and all were within 25 percent.

Acknowledgement

This work was funded by the Metropolitan Water District of Southern California, as part of the Harvey Mudd College Engineering Clinic. In the Clinic program, undergraduate student teams conduct sponsored design and development projects. The following undergraduate students also contributed: Brian Auchard, Erik Browne and Lisa Chynoweth. Mary Ann Fingerlos served as MWD Liaison.

References

Giron, Jonathan J. et. al., Final report to Metropolitan Water District of Southern California : Development of a Computer Model to Predict Changes in Water Quality Along the California Aqueduct, Harvey Mudd College, Engineering Clinic, May 1994.

Giron, Jonathan J., Report on Summer Activities : California Aqueduct Modeling Package, Harvey Mudd Engineering Clinic, September 1994.

Applying Decision Analysis To Water Strategies

Joseph Duncan King[1]

Abstract

In a complicated planning environment, a successful water resources strategy must achieve several objectives simultaneously:

- select a strategy from a large number of alternatives
- explicitly address uncertainty and its impacts on the alternatives, and
- comply with multiple regulatory constraints at the least cost.

This paper presents a process for achieving these objectives and a case study of implementing this process based on an actual project. The project involved a wastewater master plan which developed an integrated water resources strategy. Decision analysis was the core decision making process used to develop the strategy. The focus of this paper is how decision analysis could be used to analyze and develop water resources strategies and highlight the benefits of using decision analysis versus more typical planning processes.

Decision Analysis

The project's objectives and constraints would be identified first. For a typical water resources plan, two common objectives could be to:

- balance the future supply of and demand for water resources, and
- minimize the life-cycle costs of the strategy.

Next, the alternatives to be evaluated would be identified. Each of these alternatives would have various options associated with it. These options are mutually exclusive, such that one and only one of the options for each alternative would be chosen as part of the strategy. In addition, these options encompass the

----

[1]Principal, CH2M HILL Strategies Group, 1111 Broadway, Suite 1200, Oakland, CA 94607

full range of options available for an alternative.

Typical alternatives for a water resources strategy which focuses on wastewater issues could be to implement a water conservation program or expand the treatment capacity at the wastewater plant. The water conservation program could be voluntary or regulated, and could include just basic measures, such as commercial conservation, or aggressive efforts that could result in additional water savings from all users. The capacity expansion could be either a trickling filter or activated sludge process. For this example, increments of 2 and 4 million gallons per day (MGD) of capacity will be analyzed. If a trickling filter process is selected, then additional treatment will be required before the wastewater can be discharged to a river. These options include agricultural reuse with either a wet meadows or reservoir during the winter, or construction of an aeration basin or trickling filter for nitrification.

Two important items of information must be analyzed to determine how well these alternatives achieve the objectives: their impact on either the supply of or demand for water and their impact on life-cycle costs. For the sake of brevity, this information is not shown here but is included in the decision analysis model. The alternatives and options are summarized in Figure 1 using a decision tree format. Each alternative is shown as a rectangular decision node with a branch for each option.

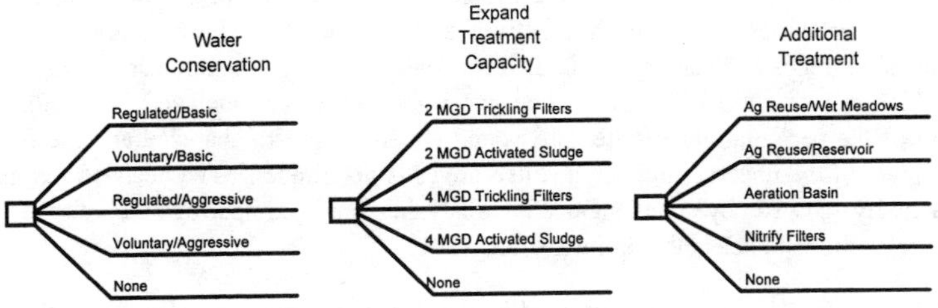

Figure 1. Decision Nodes

The analysis of these alternatives is complicated by three major uncertainties: the effectiveness of the water conservation programs, the rate of growth in future wastewater flows and the possibility of water conservation requirements in the future water delivery and supply contracts. For water conservation programs and rate of growth in future wastewater flows, three possible levels of program effectiveness and future flow increases are analyzed: low, medium, and high. For water conservation requirements, the possible states are low, high or none. Information related to the amount of conservation and rates of flow increase and their probabilities of occurrence are also included in the decision analysis model. These uncertainties are summarized in Figure 2 using a decision analysis format. Each uncertainty is shown as a circular uncertainty node with a branch for each possible state.

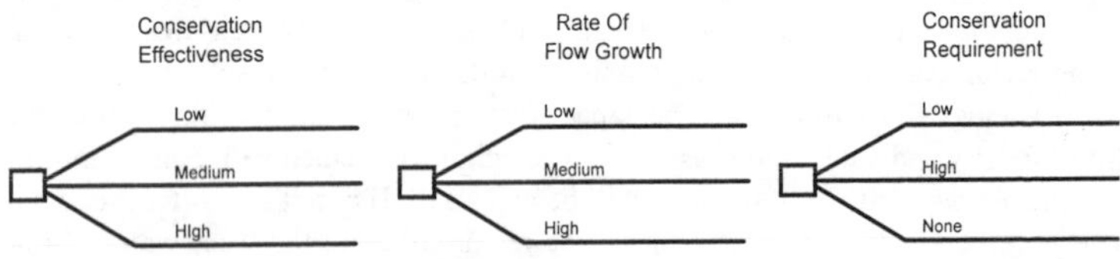

Figure 2. Uncertainty Nodes

This water resources problem has over 1,000 possible permutations of alternative options and uncertainties to be analyzed. To perform a thorough analysis, all of these permutations should be analyzed to determine how well the options achieve the objectives. The best options would then be selected as the optimal strategy. Although it would be difficult to analyze all of these permutations using spreadsheets, decision analysis can easily and quickly analyze these permutations and even solve problems that are orders of magnitude more complicated.

Decision analysis solves this type of problem by constructing a decision tree which combines all of the alternatives and uncertainties shown above. The decision tree is analyzed by methodically moving from node to node and changing which sets of alternative options and uncertainty states are analyzed. Therefore, all of the permutations of alternatives and uncertainties are analyzed and the full impact of the uncertainties on the alternatives is determined. The decision tree is actually solved by mathematically analyzing and comparing all of these permutations to select the optimal strategy.

Based on the information contained in the decision analysis model, the optimal strategy is to implement a voluntary, basic water conservation program, construct 4 MGD of trickling filters, and implement agricultural reuse with a wet meadows system. The expected cost of this strategy is approximately $13.2 million. Decision analysis also calculates the expected cost of every other possible strategy and produces a summary graphic of this information for comparison. Another very useful graphical output of decision analysis is the cumulative risk profile of the optimal strategy. Risk is defined as the amount of uncertainty in determining how well the optimal strategy achieves the objectives. In this example, the primary objective is cost, so the optimal strategy's range of cost that results from the uncertainties in the model is plotted along the x axis. The cumulative probability that the strategy will cost a specific amount or less is plotted along the y axis. For example, the probability that the optimal strategy will cost $15 million or less is 85 percent. This graphic also shows that the cost for the optimal strategy could be as high as $23.5 million. Decision analysis can identify the uncertainties that drive the costs into this upper range, which can show decision makers what issues to focus on in the future. Decision makers can also determine what is the probability of exceeding a specific cost, which provides valuable information for financial and user fee planning.

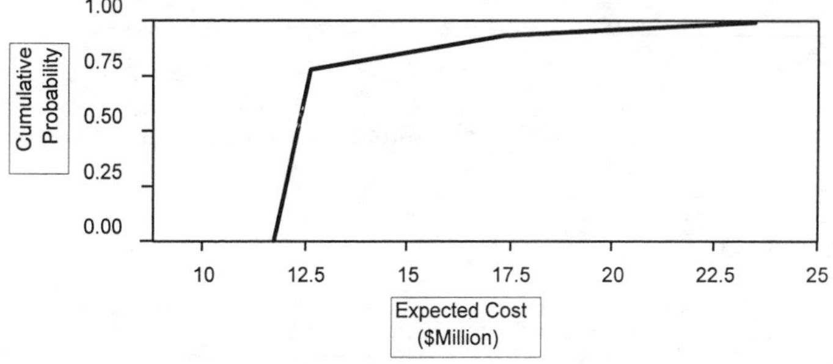

Figure 3. Cumulative Probability

Conclusion

Making a good decision does not guarantee a good outcome. However, using decision analysis to incorporate uncertainty and thoroughly analyze a problem can provide powerful insight that can greatly increase the likelihood of a good outcome for a selected strategy.

This example was derived from a twenty year wastewater master plan. The original wastewater strategy focused on water conservation, activated sludge and biological phosphorus removal. The decision analysis identified an optimal strategy for the overall water resources problem by combining water conservation, trickling filters, agricultural reuse and a natural wet meadows system. The projected costs were less and the projected reuse and conservation were more for this integrated resources plan than for the original strategy.

Decision Analysis Tools for Water Resource Management

Phillip C. Beccue[1]

Introduction

Integrated water resource management is the systematic evaluation of resource options for meeting the needs of utility customers. Integrated planning of water resources is difficult due to the multiple participants involved in the planning process, the conflicting objectives of minimizing rates and environmental impacts as well as maintaining economic growth and shareholder value, the uncertain consequences of technological performance, customer needs, and regulatory requirements, and the many competing alternatives that must be considered.

Decision analysis concepts and tools are uniquely qualified to address many of these challenges. Decision trees are used to incorporate differing perspectives and communicate the basis of the decision to multiple stakeholders. Multiattribute decision analysis allows one to consistently compare water resource options with diverse characteristics. Uncertainty is captured by analyzing several future scenarios simultaneously instead of the more traditional point estimates. Finally, the risks of competing alternatives are communicated clearly through risk profiles.

An example from a recent wastewater planning problem faced by a Western city can show how decision analysis can help address the challenges identified above.

Decision Analysis Wastewater Planning Example

City planners are concerned about meeting wastewater flow requirements over the next several years, while at the same time achieving environmental compliance for standards that may change by the year 2000. They must

[1] Senior Manager, Applied Decision Analysis, Inc.
2710 Sand Hill Road, Menlo Park, CA 94025

act now before flow growth or pollution limits are known precisely. Their goal is to minimize costs, while at the same time avoiding environmental and financial risks.

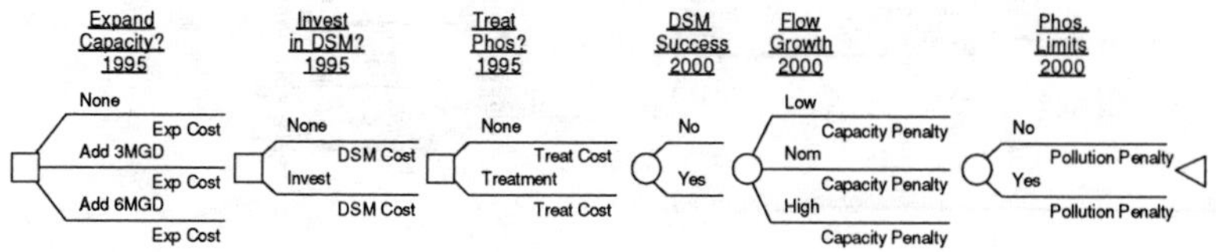

Figure 1. Wastewater Treatment Decision Tree

The illustrative model shows the three major decisions to be made today and three major uncertainties that are resolved only after the decisions are made. As shown in the schematic decision tree of Figure 1, planners can expand existing facilities by adding 3 or 6 million gallons per day of capacity, invest in demand-side management programs, and/or invest in phosphorus treatment facilities now to reduce phosphorus concentrations in anticipation of tighter limits. There are five potential sources of cost: capacity expansion, DSM, phosphorus treatment, penalties for insufficient capacity, and penalties for excess phosphorus. The schematic decision tree, a graphical representation of the decisions, uncertainties, and costs for the wastewater problem, is a compact representation of a 144 path tree.

The decision tree in Figure 1 reflects the fact that the future is uncertain. The demand-side management programs are forecast to have an 80% chance of success, meaning a reduction in flow requirements of 500,000 gal/day. Flow growth, independent of DSM, is projected to be 3%/yr, but could be as low as 2% or as high as 4%/yr. Finally, regulators could limit phosphorus concentrations to 35 lb./million gal. in the year 2000 or leave them at the current limit of 150 lb./million gal.

The optimal decision policy shows the best course of action to take among the combinations of alternatives available, based on minimizing total expected costs. As illustrated in Figure 2, the analysis indicates that planners should expand facilities by 3 MGD, should not make additional investments in DSM programs, and should invest in phosphorus treatment options. The expected costs (shown in brackets) of this plan is just over $13 million. There are 11 other possible combinations of decision alternatives, each of which has a higher expected cost than the optimal policy.

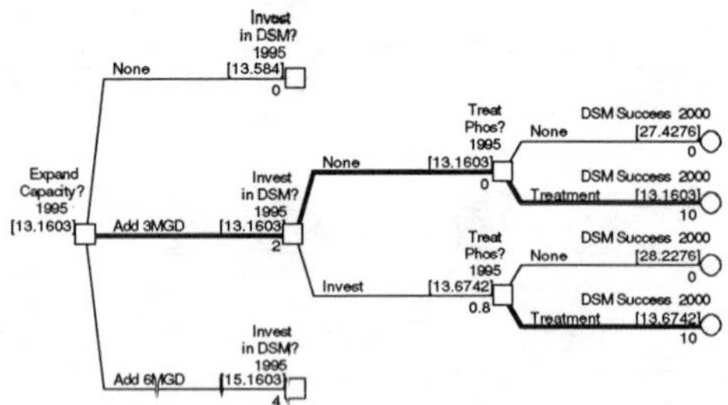

Figure 2. Optimal Policy

What risks do city planners face if they follow the optimal policy? Based on probabilistic data, we can use risk profiles to examine the risks of competing alternatives. The risk profiles provide a graphical representation of the range of possible outcomes and their associated probabilities.

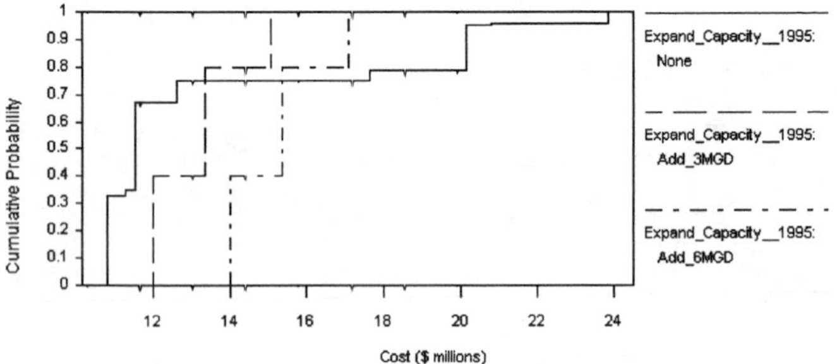

Figure 3. Comparison of Risk Profiles for 3 Alternatives

The optimal policy is the dashed line in the middle of Figure 3. With the optimal policy (add 3 MGD), costs could be as low as $12 million and as high as $15 million. By comparing this policy to one in which no capacity is added (the solid line), the risk of higher costs is easily quantified. The "no capacity" option is less expensive in many cases (low and nominal growth), but is much more expensive when flow growth is high.

Value sensitivity analysis illustrates the sensitivity of the model to input assumptions. For example, a capacity penalty of $10 million/MGD of unserved demand was initially assumed, reflecting the costs of "emergency" measures to meet higher than expected demand. Suppose this penalty was raised or lowered. What impact would that have on both costs and the best course of action?

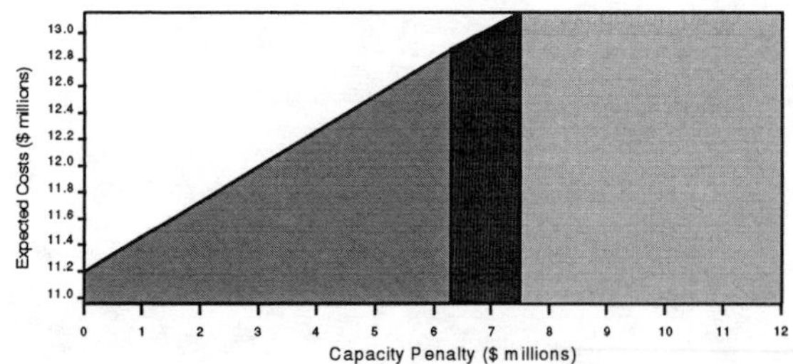

Figure 4. Sensitivity to Capacity Penalty

In performing a value sensitivity analysis, the capacity penalty is changed in increments from $0 to $12 million. The upward sloping line shows the increase in expected costs from $11.2 million to $13.2 million. The shading changes represent a switch in policy. For capacity penalties less than $6.2 million, the best course of action is not to invest in either capacity expansion or DSM. For capacity penalties between $6.2 and $7.5 million, the best course of action is to invest in DSM programs and not invest in expansion facilities. For capacity penalties greater than $7.5 million, the best course of action is to expand capacity by 3 MGD and not invest in DSM programs.

A recent trend in water resource management is to evaluate all the social costs of an alternative, not just those that fall on the ratepayers or the utility. Multiattribute utility analysis (MUA) allows a planner to compare alternatives with diverse environmental and health impacts. A value hierarchy structures the criteria, performance measures define the degree to which each option achieves objectives, and value weights quantify the trade-offs between competing objectives. In the wastewater example, planners discovered that under the optimal policy there is a 20% chance that phosphorus concentrations will surpass the limits, although only by a small amount. Furthermore, the trade-offs between achieving environmental compliance and minimizing other objectives are made explicit through the MUA approach.

Conclusion

Today's water resource professionals face complex interrelated issues in their planning that require the incorporation of uncertainty, balancing DSM and supply, involving public scrutiny, and satisfying societal objectives. This paper has demonstrated with a real example how decision analysis can help meet challenges faced by water resource planners.

USE OF DECISION MODELS FOR PLANNING AND PERMITTING OF WATER RESOURCE PROJECTS

Keith A. Ferguson[1], P.E., Richard C. Pyle[2], P.E., Thomas O. Keller[3], P.E., and Kenneth A. Steele[4], P.E.

Abstract

A decision model has been developed by GEI Consultants, Inc. (GEI) to assist with the permitting of a complex water resources development project called the Emergency Storage Project (ESP) being sponsored by the San Diego County Water Authority (SDCWA). The model has been successful in providing a quantitative comparison of over 32 alternative systems consisting of various combinations of new dams, enlargement of existing dams, re-operation of existing dams, ground water resource development and large interconnecting pipelines and pumping stations. Total estimated project costs range up to $700 million. The alternatives analysis has been used to comply with the requirements of the California Environmental Quality Act (CEQA), the National Environmental Policy Act (NEPA) and Section 404b of the Clean Water Act and has been well received by the Environmental Protection Agency (EPA) and the U.S. Army Corps of Engineers (COE).

Structure of The Decision Model

A rigorous stakeholder involvement process completed during early stages of the environmental compliance process resulted in the identification of 34 independent criteria that were deemed important in the evaluation of project alternatives. The functional hierarchy of the criteria in relation to five overall project goals and related subgoals and objectives is illustrated on Figure 1. This hierarchy was used to rank 15 alternative systems that meet project needs. These 15 systems were selected from the original 32 alternatives using a separate but similar hierarchy.

---

1 M. ASCE, Vice President and Branch Manager, GEI Consultants, Inc., 5660 Greenwood Plaza Blvd., Suite 202, Englewood, CO 80111-2418.

2 M. ASCE, Senior Civil Engineer, San Diego County Water Authority, 3211 Fifth Avenue, San Diego, CA 92103-5718.

3 M. ASCE, Senior Project Manager, GEI Consultants, Inc., 1021 Main Street, Winchester, MA 01890-1970.

4 M. ASCE, Emergency Storage Project Manager, San Diego County Water Authority, 3211 Fifth Avenue, San Diego, CA 92103-5718.

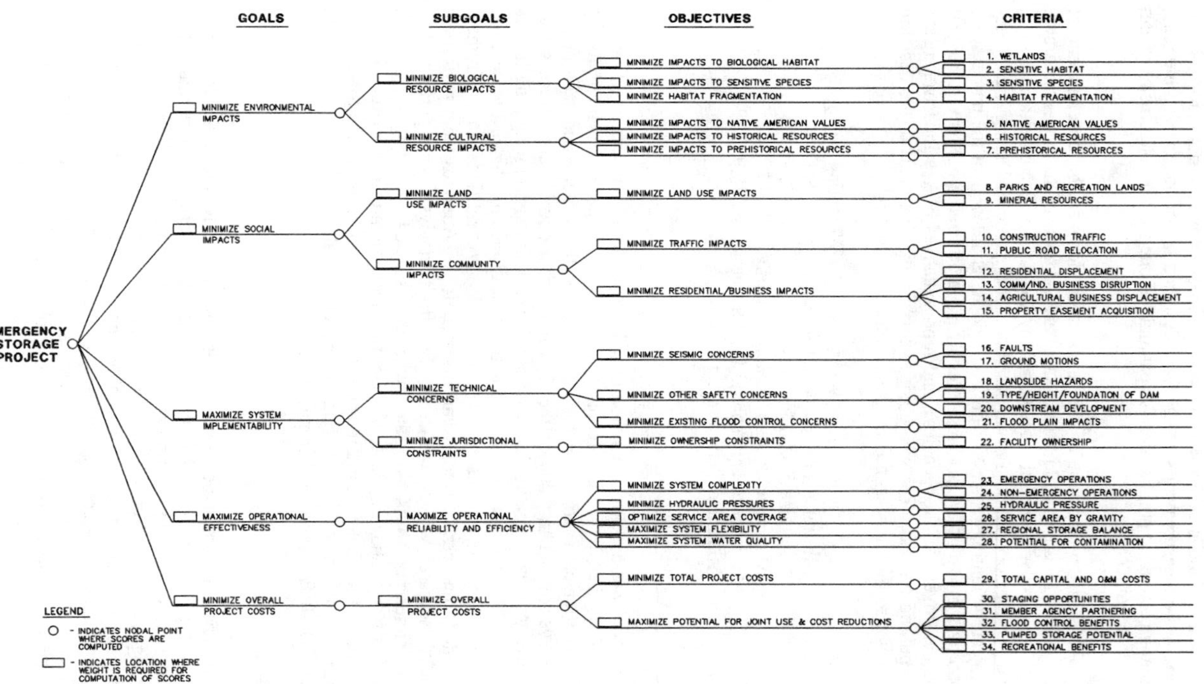

Figure 1.Decision Model Hierarchy

Once decision criteria were identified, a technical committee, comprised of 20 environmental and technical experts, developed measurement systems for each of the criteria. The measurement systems were designed to convert measurable units to a common preference point scale. An example of one such measurement system is shown on Figure 2.

The final step in development of the decision model was the development of weighting factors for criteria, objectives, subgoals and goals of the model hierarchy. Weighting factors were initially developed by the technical committee using a structured participatory polling (group based decision) process. During later stages of the project, weighting factors for the goal and subgoal levels of the model were developed by a group of 27 local residents living in affected areas, environmental groups, local governmental agencies, and representatives of key industries using a similar polling process. The weighting factors for the five goals of the decision model developed by this stakeholder group were as follows:

Goal #	Title	Weighting Factor
1	Minimize Environmental Impacts	24.8
2	Minimize Social Impacts	17.2
3	Maximize System Implementability	14.3
4	Maximize Operational Effectiveness	19.9
5	Minimize Total Project Costs	23.8
	Total	100.0

Results of Alternative Analysis

Once the model was developed, environmental scientists and engineers working for the SDCWA gathered the data required to provide the necessary input to the adopted measurement systems for each of the evaluation criteria and each of the alternative systems. The value of the decision model development process was noted by everyone participating in the project for two reasons: 1) weighting factors helped to identify the criteria that would be most discriminating in the decision process, and 2) the criteria measurement systems clearly identified data requirement and prevented unnecessary studies and investigations. Analyses were performed using the "base case" weighting factors. Sensitivity analyses were performed to determine preference shifts associated with the assigned weighting factors. A typical set of results showing the total score for each system is illustrated on Figure 3. Each bar on this figure shows the contribution of goal scores to the total score. The model was successful in identifying four preferred alternatives for inclusion in the project Draft Environmental Impact Report.

Conclusions

The development and implementation of a decision model for the ESP provided an extremely valuable tool to the SDCWA in its efforts to permit an important water resources project. By using group based decision modeling, the SDCWA was able to bring together the project planning and public involvement efforts. This was an invaluable tool for educating the public about the project and the project proponents about environmental concerns. The modeling process used for the ESP is adaptable to a wide range of other types of projects and should be considered by a wide range of project sponsors to avoid future permitting failures.

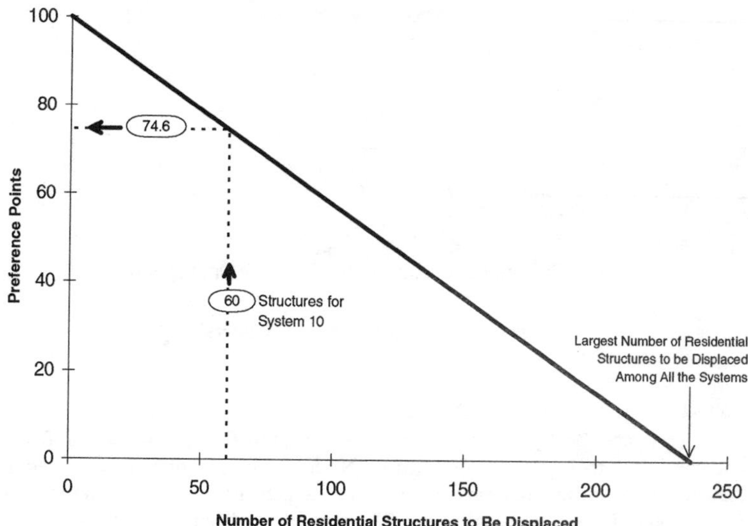

Figure 2. Measurement System for Residential Displacements

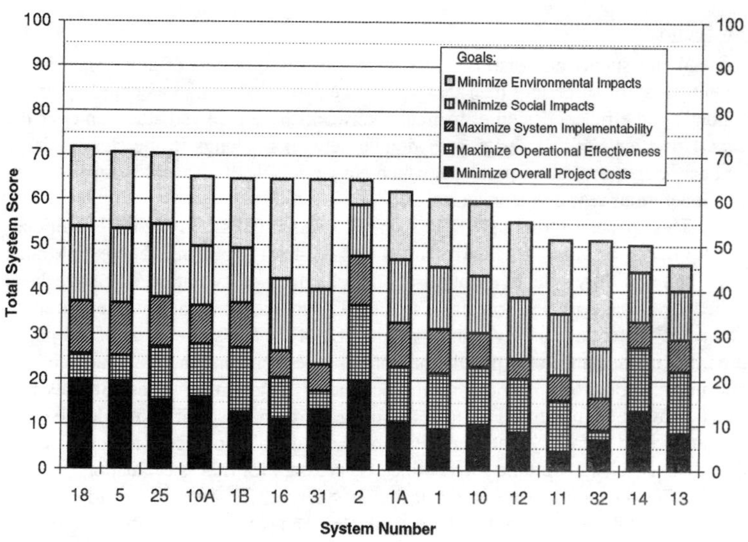

Figure 3. Bar Chart of System Scores for Fine Screening. Base Case Analysis

Place of Evolution Programs in Pipe Network Optimization

Dragan A. Savić and Godfrey A. Walters[1]

Abstract

This paper briefly introduces Evolution Programs (EPs) as general artificial-evolution search tools particularly suited for NP-hard problems often encountered in pipe network optimization. Problems addressed are network layout, pressure regulation and pipe-size selection. Although EPs were allowed only a small number of evaluations per each run (usually a small fraction of the expected number of possible solutions) the solutions identified compare favourably to those obtained using classical mathematical programming techniques or heuristics.

Introduction

Water supply and distribution systems are key elements of any urban development. It is not surprising then that various analyses and design techniques for the economic and efficient design of hydraulic networks have received attention over the years. However, optimal design of hydraulic networks belongs to the class of large combinatorial optimisation problems which are very difficult to handle using conventional operations research techniques.

This paper introduces Evolution Programs (EPs) as general artificial-evolution search tools particularly suited for NP-hard problems. These programs mimic nature's very effective optimisation techniques of evolution and have been applied to various complex domains. Similarly to evolution mechanisms their strength is based on preferential survival and reproduction of the fittest members of the population, the maintenance of a population with diverse members, the inheritance of genetic information from parents, and the occasional mutation of genes. Besides features borrowed from nature EPs take advantage of domain-specific knowledge, e.g., the mature field of steady-state modelling of water distribution networks provides an EP with an efficient hydraulic solver which is necessary for fitness evaluation.

After an introduction to the techniques several EP applications to network optimization will be described. Where available, examples from literature were used for comparison and to illustrate the practicality of the programs.

---

[1] School of Engineering, University of Exeter, Exeter EX4 4QF, United Kingdom.

Evolution Programs

Evolution Programs (EPs) is a collective term for Evolutionary Programming, Evolution Strategies, Classifier Systems and Genetic Programming algorithms which draw their power from principles of biological evolution. However, Genetic Algorithms (GAs) are probably the best known type of EPs. The theory behind GAs was proposed by Holland (1975) and further developed by Goldberg (1989) and others in the 1980's. These algorithms rely on the collective learning process within a ***population*** of individuals, each of which represents a search point in the space of potential solutions. The analogy with nature is established by the creation within a computer of an ***initial population*** of ***individuals*** - ***phenotypes*** represented by ***chromosomes*** - ***genotypes***, which is, in essence, a set of character strings that are analogous to the chromosomes found in DNA (Figure 1). Standard GAs use a binary alphabet (characters may be 0's or 1's) to form chromosomes although non-binary alphabets can be used. The measure of how good the individual solution is in comparison with other members of the population is called the ***fitness*** of the individual. Consequently, the selection of who gets to reproduce is a function of the fitness of the individual. During the reproductive phase of the GA individuals are selected from the population and recombined, producing offspring which will comprise the next generation. This is the ***recombination*** operation, which is generally referred to as ***crossover*** because of the way that genetic material crosses over from one chromosome to another. Crossover takes two individuals and cuts their chromosome strings at some randomly chosen point. The newly created head segments stay in their respective places while the tail segments are then crossed over to produce two new chromosomes. In addition to crossover, another genetic operator called ***mutation*** randomly alters genes (e.g., g0,...,g7) within a chromosome thus providing a small amount of random search.

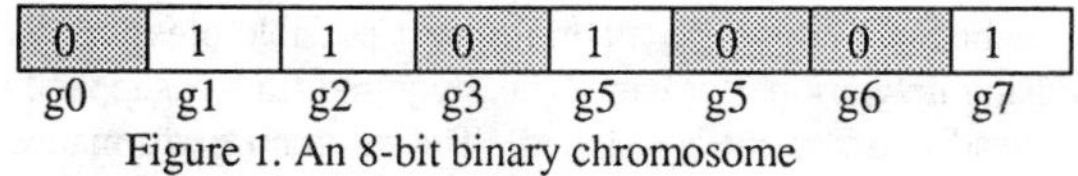

Figure 1. An 8-bit binary chromosome

A variety of GA applications has been presented since they were first introduced. A review by Savić and Walters (1994a) shows that GAs have clearly demonstrated their capability to yield good approximate solutions to difficult problems.

Layout Selection for Tree-like Pipe Networks

Walters and Lohbeck (1993) analyzed the layout problem of tree networks for water, gas and sewer systems. The optimization problem was defined as a search for the best tree layout from a directed base graph of possible pipeline connections. The simplification of using a directed base graph greatly reduces the number of candidate solutions. The authors have shown that for a 25-node grid network (5 × 5) the assumption reduces the solution space from approximately 3.3×10^{13} possible layouts to about 3.4×10^{7} possibilities. Even for this small example the solution space still remains large with a great number of local minima. A Genetic Algorithm is developed and implemented and its performance is compared to that of Dynamic Programming (DP). However, because of the well-known "curse of dimensionally" the possible network size that can be tackled with DP for layout optimisation is very limited. The largest grid example solved by the DP program was a 5 × 6 grid network. It was shown that the best solutions obtained by repeatedly running the GA converge to the global optimum identified by DP. Unfortunately, although GA can solve larger example problems it is not known whether the result obtained is the global or a local optimum.

The above GA is of limited use for distribution networks because of its assumption of predetermined directions. It was noted earlier that if no directions are assumed for the flow along all candidate links in a base graph, the size of the problem increases dramatically. Walters and Smith (1995) developed an EP for the selection of a tree network from a non-directed base graph. The program developed ensures the generation of feasible solutions. This is achieved through innovative coding, recombination and mutation operators. The program was successfully applied to a two-source example problem with 100 nodes and 232 arcs (3.65×10^{54} possible solutions). To reach the best identified solution each run required approximately 30,000 evaluations of network cost, 50 orders of magnitude fewer than for complete enumeration of the problem. When solutions obtained from the directed and undirected base graph problems are compared it was found that the new EP had identified less expensive designs.

Pressure Regulation for Looped Pipe Networks

There are many factors which can affect leakage but pressure is the only one that can be easily controlled once the pipelines have been laid. Savić and Walters (1995) developed an EP for pressure regulation in a looped network using isolating valves. The optimization problem was to find the optimal settings of all isolating valves (i.e., open or closed) to attain the best possible pressure distribution without compromising network performance (i.e., required flow is supplied to each node and minimum head requirement is satisfied). The program performance was tested on a small example (9 nodes and 17 pipes) that allowed complete enumeration. By repeatedly running the EP, the same 'true' optimum was found using far less CPU time than used for enumerating the problem. For larger problems enumeration was impracticable and the 'true' optimum was not known. However, the EP identified several near-optimal solutions whose fitnesses were comparable but the valves identified were different. This may give additional flexibility to a decision maker when deciding which solution to implement. It is also important to note that the EP used only a small number of evaluations per run (usually a fraction of 1% of the expected number of feasible networks).

Optimal Selection of Pipe Sizes

Selection of pipe diameters (especially from a set of commercially available discrete-valued diameters) to constitute a water supply network of least capital cost has been shown to be an NP-hard problem. It is not surprising then that various simplified approaches have been suggested in the past to reduce the complexity of the original problem. Those approaches use simple trial-and-error procedures, enumeration, successive linear approximation or complex non-linear optimization algorithms. However, many of them still treat pipe diameters as continuous variables and/or have limitations to the size of problem they can handle. Savić and Walters (1994b) used a GA to test the sensitivity of the solutions to changes in head-loss equation and compared the GA solutions to the ones found in literature. The problems addressed vary in size from 1.5×10^9 possible designs (a network consisting of 7 nodes, 8 pipes and two loops with 14 discrete diameters considered) to 2.9×10^{26} possible designs (a network consisting of 32 nodes, 34 pipes and 3 loops with 6 discrete diameters considered). The range of GA solutions compare favourably in terms of cost and minimum head requirements to solutions identified by other techniques.

Conclusions

Evolution Programs for pipe network optimization are becoming powerful research tools which will gain wider acceptance among pipe network practitioners. Example problems show the reasons why EPs are preferred over classical optimization techniques: 1) they are easier to implement since basic network modelling (simulation) tools are available to analysts; 2) their performance does not depend on the assumption of the continuity of the search space; and 3) a range of near-optimal solutions identified by an EP may be thoroughly different in terms of the set of selected decision variables which provides a decision maker with a wider range of solutions to choose from. Future work is directed toward development of EPs for parameter estimation (calibration) of network models, optimization of water quality parameters and pump scheduling.

References

Goldberg, D.E., (1989), Genetic Algorithms in Search, Optimization and Machine Learning, Addison-Wesley.

Holland, J.H., (1975), Adaptation in Natural and Artificial Systems, MIT Press.

Savić, D.A. and Walters, G.A., (1994a), Genetic Algorithms and Evolution Programs for Decision Support, Proc. 4th International Symposium: Advances in Logistics Science and Software, J. Knezevic (ed.), pp72-80.

Savić, D.A. and Walters, G.A., (1994b), Sensitivity of Optimal Pipeline System Designs to Changes in Head-loss Equation, Centre for Systems and Control Engineering, University of Exeter, Report No. 94/21, Exeter, UK.

Savić, D.A. and Walters, G.A., (1995), An Evolution Program for Optimal Pressure Regulation in Water Distribution Networks, *Engineering Optimization* (accepted for publication).

Walters, G.A., and Lohbeck, T., (1993), Optimal Layout of Tree Networks Using Genetic Algorithms, *Engineering Optimization*, 22, pp27-48.

Walters, G.A. and Smith, D.K., (1995), Evolutionary Design Algorithm for Optimal Layout of Tree Networks, *Engineering Optimization* (accepted for publication).

Decision Quality in Water Management

John W. Rogers[1]

In today's dynamic workplace, decisions are complex and difficult. Decision making and the implementation process are as important as the quality of service or the quality of a product. Decisions irrevocably commit resources, be it time, money or people. Water utility managers increasingly face the need to do more with less, a higher level of public and media scrutiny, and new regulatory challenges and multiple uncertainties. In the Clean Air Act and the proposed Safe Drinking Water Act and Clean Water Act, environmental managers face increasing levels of risk. Because of what's at stake, water managers must find better ways to do it right the first time. More than 200 senior level water managers recently surveyed by CH2M HILL ranked better decision making as a top priority.

Research shows that decision making requires a balance between the organizational and analytical aspects of a decision process. If you focus too much on content, you may get the right answer but nobody cares. If you focus only on organization and process, you may get consensus based upon ignorance. This paper outlines key organizational and content aspects in decision making for water managers. Its primary focus is on leadership and organizational aspects of a decision process that help ensure a problem-solving decision process and commitment to directed implementation.

Introduction

With increasing exposure of environmental problems and concerns in the media and rising costs at stake for organizations and individuals, (*e.g.*, legal and financial liability, risk management, increased costs, regulatory compliance pressures, etc.) we can benefit from good decision making techniques. At the same time, managers are expected to do more in less time with less ability to sort out useful information from growing piles of data. Today, more and more decisions seem to have a number of these characteristics:

- *Conflicting Objectives*: high quality and low cost
- *Uncertain Consequences*: will this technology meet or exceed future regulatory standards?
- *Multiple Alternatives*: new water supply dams versus conservation measures
- *Multiple participants and interests*: agriculture, environmental groups, community groups, regulators, Native Americans
- *Organizational discord or bias*: no cross-functional teamwork
- *More at risk than money*: business success is on the line

[1] Sr. Vice President, CH2M HILL, 1700 Market Street, Philadelphia PA 19103-3916

When the senior environmental managers were asked in the CH2M HILL survey, "How important is decisionmaking to organizational success on a scale of 1 to 10?", the average score was 9.6. These same managers rated their organizational skills in decisionmaking, on average, at 6.4 and implementation skills at 5.6. There is clearly a gap between current needs and performance.

Why Decisions Fail

Decisions often fail because people "jump to solutions" and rely on advocacy or adversarial processes rather than problem-solving processes. For example, advocacy processes often allow the problem to be defined by others outside the process and hence the wrong problem is addressed. Advocacy processes often overlook good alternatives. Adversarial processes often require overly directive leaders which prevent proper framing of the problem or buy-in by constituents. Advocacy processes often overlook a diversity of opinion in framing a problem. Advocacy and adversarial processes seldom set measures of success or monitor performance and learn from mistakes. Even with a proper problem definition, "group think" can lead to poor communication and wrong conclusions. Finally, managers often view risk from their own personal perspective rather than from that of the organization.

The CH2M HILL survey shows that the "Top Ten" reasons decisions fail are:

- Lack of a clear decision process
- Lack of vision
- Lack of leadership and commitment
- Poor understanding of risk
- Poorly defined problem
- Wrong stakeholders
- Political intervention
- Unclear measures of success
- Poor communication
- Organizational structure

For organizational success, senior managers need assurance that decision techniques, just like accounting practices, are reliable and auditable. Environmental managers and executives must ensure that a systematic, problem solving decision process is used to prevent advocacy or adversarial biases from affecting the integrity of decisions on environmental issues for which they are responsible. Because environmental decisions are increasingly driven by both political and rational considerations, a balance must be struck between analytical focus and organizational process.

Decision Process

Traditional techniques spend very little time clearly defining a need for action and thus lack the support necessary to achieve successful decisions. This often leads to poor problem formulation and buy-in. Leadership is not emphasized and therefore commitment is often hit or miss. Too much time is spent on collecting unnecessary information and diverting resources which could be better used for collecting more useful information or developing alternatives. Finally, without adequate knowledge or tools, very little effort is given to assessing risk and uncertainty. This can lead to risk blindness and risk aversion or overly risky behavior. Worse yet, they may be "betting the ranch" and not know it. The biggest challenge is to build organizational and analytical quality into the process from the start.

Figure 1 shows a six step process to decision making which helps identify and overcome over 45 organizational and content barriers and mistakes in decision making. The process allows individuals and teams to identify the multiple organizational and content factors and catalysts surrounding decisions. By establishing leadership and commitment from the start, future surprises that occur after the decision is reached are greatly reduced. Rather than rely on intuition and generic habits, decision makers can move through a variety of discovery points to clearly frame the problem and identify the stakeholders who need to be involved. A well-developed decision process can build organizational confidence in problem solving and project implementation. This leads to trust, teamwork and alignment.

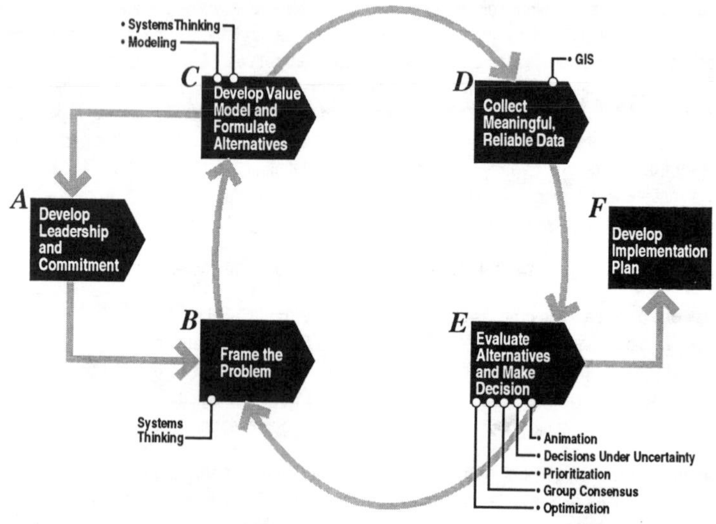

Figure 1. Six Step Decision Process

Decision analysis methodology has been proven technically appropriate as a structured approach to dealing with decision alternatives involving public risk and environmental resources. Decision analysis principles provide a basis for developing and evaluating alternatives in a way that is consistent with the relevant value judgments and information: it aids decision makers in articulating objectives, identifying alternatives, examining the implications of the alternatives, evaluating these implications, and in gaining insights into the nature of preferred alternatives. Computerized tools help decision makers effectively manage complex decisions which deal with uncertainty, prioritization, goal programming, systems thinking, and even group consensus. This is a real help insuring the content aspects of the decision are done to a proper standard.

The Need for Leadership and Commitment in Decision Making

Decisions are made and battles fought not by numbers and computers but by complicated and unpredictable human beings. People often begin a decision process with an expectation that data or science will provide answers. They seldom do. A large part of the answer lies in credibility, leadership, trust, and commitment.

Decisions begin with leadership and are implemented because of trust and commitment. The success of any public or private organization lies in the trust of constituents in their leaders. We call this ***credibility***. Our survey shows that credibility is about what people demand of their leaders and the actions leaders must take to intensify their constituent's commitment to a common cause. The leading descriptions of credibility in our survey were:

- honesty and openness
- leadership/quality of personnel
- trust/integrity
- achievement record
- technical accuracy

Our survey indicated that the traits managers find least effective in leaders are:

- A lack of a clear vision
- Poor communication
- Indecision
- Arrogance or indifference (not involved)

From our survey it was also clear that the barriers to implementation are different than barriers to decisions. Our research shows that implementation is hindered by:

- Unclear need for action and fear of change
- Unclear responsibility
- Lack of clear consensus or commitment
- Timing or lack of teamwork

From the survey results, a review of the literature and from our own experience, the seven habits of effective decision ***leadership*** are to:

1. Provide a vision and motivate teamwork
2. Involve and empower the right stakeholders early in the process
3. Require a systematic decision process and milestones
4. Ensure an open climate for discussion and encourage debate and critical evaluation
5. Ensure the proper level of resources
6. Demonstrate commitment

Trust begins with the understanding that everyone potentially has a stake in decisions that can significantly affect their lives. Bringing the right stakeholders into the process early is critical to good decision making. This is particularly true when dealing in issues which affect the public.

- ***Commitment*** and sponsorship take far more than ideas and rhetoric; they require the ability and willingness to plan, participate and apply the meaningful rewards and pressures that produce the desired results. The idea of commitment means there are clear guidelines to:
- Invest resources (time, energy, money) to insure the desired outcome.
- Establish roles of key participants.
- Engage in symbolic start-ups and celebration of action plans and closure to announced actions and periodic participation in informal meetings.
- Constantly pursue the goal, even when under stress and with the passage of time.
- Apply creativity, ingenuity, and resourcefulness to resolving problems or issues that would otherwise block the achievement of the goal.
- Develop a roadmap for the decision process so people can see how to get from A-Z.

Learning and incorporating the six step decision and implementation process into everyday practice allows the organization and individuals involved to gain and continually expose areas of growth and future benefit. By providing leadership and commitment, a problem-solving process can replace advocacy and adversarial processes. By correctly framing the problem in a problem solving process, the decision maker is able to achieve his/her objective, enhance credibility, facilitate communication and understanding, motivate commitment for action, generate innovative alternatives, use resources wisely, utilize risk to his or her advantage, and provide areas for a learning and improvement.

International Decision Support System for Reservoir Operations

by R. B. Allen (Acres International Limited)

1 Background

Great Lakes Power Limited (GLP) is a private utility, generating, transmitting and distributing power in the Algoma District of northern Ontario, Canada. Total generating capacity is 297 MW from 12 hydroelectric stations in four river systems. All of the plants are under remote operating control. GLP is also linked to the Ontario Hydro grid, which serves most of the balance of the Ontario demand. The hydraulic plants are a mix of run of river/headpond storage and seasonal storage plants. Annual storage at the upstream plants provides significant opportunity to capture spring runoff for use through the balance of the year. Headpond storage at a number of the plants permits daily and weekly retiming of water flows to meet peak demands.

Maximum GLP system demand is approximately 375 MW (winter) and consists of a mix of industrial, commercial and residential load. The annual system load factor is 61 percent. The demands are met from self-generation (constituting 70 to 80 percent of total sales) and the purchase of deficit power from Ontario Hydro. Purchase can be made in various classes of firm and emergency power and at various rates. The rate structure typically has both a capacity and energy component with seasonal and daily peak/off peak pricing. Such a time-of-use rate structure offers considerable incentive for optimal power system management on a long-term (annual) and short-term (hourly) basis.

A number of factors have combined to make a computer-based decision support system for power system operations a financially attractive opportunity for GLP.

- The introduction of a time-of-use rate structure by Ontario Hydro in the sale of their power to GLP makes the optimal decision process for long and short term management much more complex.

- Advanced technologies in both hardware and software now permit a user friendly approach to decision-making, allowing the user to combine computer-based decision support information with operator judgment.

- The retirement of operations staff, who have run the system for years on the basis of experience, has necessitated a more consistent and analytical approach.

- There is a need to adapt optimal system operations quickly to significant changes in the load demand from major customers, the addition of new facilities and ultimately the sale of power out of the system as the generating capability exceeds native demand.

A preliminary assessment of the advantages of a decision support system for GLP operations indicated a saving of approximately $1,000,000 per year in power purchases from Ontario Hydro.

A decision support system (DSS) is a computer based system of programs and routines used to collect, store and process data. The system is designed to be highly interactive in a user friendly manner facilitating optimal long-term resources management and short term dispatching of generating units to meet load demands.
The decision-making process which will achieve an optimal operation of a power utility can consist of a complex network of different decision paths dependent on various specific start and forecast conditions. These decision paths (or variables) may require separate individual assessment before being submitted to an optimization process; while others may consist of multiple constraints or dynamic system data that govern (direct) the analysis procedure. The function of the DSS is to integrate these various constraints, models and procedures, using a common database, to arrive at an optimum use or scheduling of generating facilities. This approach will eliminate the need for file transfers, use of incompatible software, manual data input, slow computer access, and multi-supervision requirements.

An important aspect of a DSS is its requirement for user friendliness. Development of various engineering decision tools and the scheduling of power generation in particular has historically evolved in parallel with the availability and use of computer technology. Unfortunately, at various levels of implementation, the application of computer models has resulted in a strong aversion to the use of computer models by operators, mostly due to lack of provisions for human interaction with the computer systems. Causes for operator aversion to computer models are many — complicated manipulation of data and files, various computer systems with different handling requirements, theoretical rather than practical problem solutions, and the complication of programs often designed to function as black boxes only. A recent survey and study conducted by the author for the Canadian Electrical Association (CEA) has shown that operator satisfaction with computer software is a key aspect for successful DSS implementation.

2 DSS Framework

The components of a DSS can be classified into three main areas of interaction, database management activities, user-activated processes, and a graphical user interface (GUI). All these activities are coordinated and performed on the computer operating system. Database management activities such a data editing, data manipulations, data record/archive functions, and data report generation are all supported. Operator-activated processes such as SCADA data review/updates, forecast modules for load and hydrologic inflows, generation scheduling algorithms, and other operational procedures are also supported. These activities are made available to the operator through the menu displays in the graphical user interface. The scope of these functions will define the specifications for a specific DSS. The development of the DSS for daily operational planning of the GLP power system required the preparation of a design specification and the definition of the overall framework within which various computer programs would function.

A key aspect of the design specification for GLPDSS is the user interface requirements. The user interface not only provides the smooth and easy to use human interaction platform for the operators but also defines the appearance of the graphical user interface to the operator. For each daily activity that the operator will face in making his decisions, a data/program/user interface was designed. Each interface consists of one or more of the following functions:

- user/program dialogue boxes
- user input menu structures
- direct database communications to all SCADA information
- program output displays
- graphical data displays (input and/or output)
- optional program execution menus
- database management information services (MIS) report generation/display.

3 Decision Support Analysis Strategies

The two major decision support strategies developed for GLP are the annual and short-term water resource management decision packages called SCPLAN and SCOPER, respectively. SCPLAN is a long-term operational model developed to maximize the long-term expected operational benefits of the GLP hydropower system. This model addresses such features as annual load forecasts, stochastic hydrology, water resources physical and operational constraints, and long-term energy transactions. The model provides the optimal allocation of the available water resources on a weekly basis and the marginal value of reservoir storage for the present week.

The SCOPER model is a short-term hydropower operations model developed to optimize the following objective: minimize the cost of energy supply to the power grid, accounting for the cost/benefit of energy transactions with neighboring utilities and the marginal value of end-of-period reservoir storage (defined from results from the SCPLAN model). This model addresses such features as hourly load and inflow forecasts, operational and physical constraints, river flow constraints (environmental issues), capacity reserve constraints, and economy sales opportunities. The model determines the optimal unit commitment profile, hourly unit dispatch schedule, hourly water release schedule, and a summary of generation production costs with energy transactions costs/benefits.

4 Model Implementation

The success of implementing a DSS largely depends on its acceptance by system operators. User friendly initiation methods and the introduction of user-directed model evaluation techniques are important aspects of a successful DSS implementation.

Decision-making is a highly practical procedure—the recommended decisions have to be a workable solution. It is therefore important to streamline the review of generation schedules and to provide tools for evaluation, in order to gain acceptance of the DSS as a support system to decision-making. The following techniques have been employed in the GLP DSS.

(i) Practicality of Optimization Results

Several techniques are available for reviewing and testing the theoretical optimum solutions before they are presented to the user. Such techniques are based on comparisons with alternative operating scenarios, subroutines that test the sensitivity of optimum solutions to externally imposed conditions, and internal parameters that condition the optimum solutions to practical solutions. The result of introducing these techniques is that the user is provided with practical guidance that can be interpreted in a confident manner with near-optimal results.

(ii) Data Manipulation Options

To some extent, it is inevitable that calculated optimum scheduling is the result of a black box procedure. Testing of the solution is important for the operators in order to check and understand the sensitivity of the solution and so gain the confidence necessary to execute the recommendations presented by the model. For this purpose, manual data manipulation (spreadsheet models) can be introduced, providing the capability for the user to change generation schedules in order to test the economic sensitivity of the optimal solution to minor variation in load allocation.

(iii) Graphical Displays

A third evaluation technique is the introduction of graphical displays to permit visual examination of the model results. Graphical displays can provide the physical contact needed by the operators to obtain the necessary insight to gain the confidence of applying the recommended solutions.

Development of the DSS for GLP is being undertaken in four phases. Phase I comprises the introduction of operational procedures and the development of the data base system. In Phase II, a planning model is introduced and the data base is linked to the SCADA system. Phase III consists of the development of an operations model, and Phase IV includes the introduction of several model enhancements, mostly related to updated hydrology forecasts, revised unit efficiencies and extended MIS reporting, including graphical displays.

Phase I was implemented in November 1993 and has performed extremely well. Phase II was completed in January 1995. Phase III is scheduled for completion at the end of 1995. The feedback of the implementation and training procedures has been very positive and is considered a very big contribution to the current success of the DSS as a tool to direct system operations.

Boston's Regional Water Supply in 2020

John F. Shawcross[1]
William A. Brutsch[2]

Abstract

Boston's regional water supply is an excellent example of nineteenth and early twentieth century engineering. It will be necessary to invest on the order of $2 billion to provide the level of water supply service that will be required in 2020. Employee training and improved instrumentation and control systems are also important to water quality and improved system reliability.

Introduction

The Massachusetts Water Resources Authority (MWRA), provides water to the City of Boston and to 46 other communities. Significant historical documents describing the growth of the system include reports to the Massachusetts Legislature in 1895(1) and 1937(2), and a 1983 book, Great Waters(3). Some of the decisions made in the past have a direct bearing on the future development of the MWRA water supply system.

Key decisions in the growth of the water system. In 1895, construction of the 246 million cubic meter (65 billion gallon) Wachusett Reservoir 50 kilometers West of Boston provided a pure unfiltered source of water that could supply water by gravity. An alternative considered at that time was to take water from the Merrimack River, filter it, and pump it to the Boston area. This key decision in the 1895 report was discussed as follows: *"The experiments carried on... in Lawrence..., and the filter constructed in connection with the water works of that city, have shown that waters as polluted as those of the Merrimack can be effectually filtered and rendered safe for domestic use; but it is also true that the filtering areas require continuous care on the part of well trained attendants, and that, in a few instances at least, inefficient administration or inherent defects of construction have allowed disease germs to pass through filters which were assumed, by good authority, to be a sufficient protection."* This insightful decision is

----

[1]Director Capital Eng. & Planning, Waterworks Div., MWRA.
[2]Director Waterworks Division, MWRA, Boston MA.

interesting one hundred years later in light of the 1989 Surface Water Treatment Rule requirements for filtration with its neglect of watershed protection, and the 1993 Cryptosporidium outbreak in Milwaukee which resulted from problems at the water filtration plant.

In 1936, construction of the 1.6 billion cubic meter (412 billion gallon) Quabbin Reservoir, began. This reservoir located 105 kilometers West of Boston is also a very high quality source that could supply water to Boston by gravity and without filtration. Transmission of water to Boston is through deep rock tunnels, brick lined unpressurized aqueducts and large diameter steel and concrete pipelines. In the Boston area water is stored in large, well protected but uncovered distribution storage reservoirs.

In 1987, it was decided to pursue a policy of "demand management", instead of seeking additional sources of water. Water use was reduced from 1.3 million cubic meters per day (340 million gallons per day, (mgd)) in 1987 to 950,000 cubic meters per day (250 mgd) in 1994. Safe yield of the system is estimated at 1.14 million cubic meters per day (300 mgd).

Prior to 1986 for at least twenty years, the water system did not receive adequate funding for repair and maintenance. A number of the facilities are in need of renovation. This also is reflected in a labor force that needs to move towards greater use of labor saving technology and information systems.

Needed Improvements

Key improvements relate to the quantity, quality and reliability of the water supply and also to the human resources available for operation and further improvement. These are detailed in an internal MWRA report, 'Twenty Year Waterworks Master Plan, December 1993'.

Quantity. By promoting conservation and demand management, the MWRA has removed the immediate necessity of identifying future sources of water. Further growth of the regional water system is not necessary or desirable. Instead of planning to meet "projected" demands some 20 or 30 years into the future the policy is to control demand to stay within available resources and to develop new sources only as a last resort. The MWRA can focus on how to optimize the system to supply an average quantity of 1.14 million cubic meters per day (300 mgd) with a minimum daily use of 750,000 million cubic meters (200 million gallons) and a maximum daily use of 1.9 million cubic meters (500 million gallons).

Quality. Since the 1895 report(1), it has been assumed that it is better to keep contaminants out of the source water than to attempt to purify polluted water. This policy holds true but additional actions are planned to comply with the various regulations arising from the 1986 Amendments to the Safe Drinking Water Act.

Reliability. The MWRA sells water to community water systems and has not developed a fully looped transmission system in many areas. MWRA has identified those transmission lines which should be looped or paralleled by 2020.

Human Resources. The reliability of the water supply system is also affected by the level of training of the workforce. There is a need in the MWRA for better maintenance of existing facilities, and upgrading many of the employees from manual workers to skilled workers and engineers with access to better equipment, instrumentation and information systems.

Major Projects

Other papers at this conference focus on details of the MWRA water system (eg Track E, Session 10). This paper therefore summarizes the major projects.

- Land is being purchased around Wachusett Reservoir. The percentage of the watershed owned by the State is expected to rise from 8% in 1990 to 25% in 2020. Potential sources of contamination will be identified and eliminated.

- A 1.7 million cubic meters per day (450 mgd) water treatment plant will be constructed to treat all water supplied by MWRA to the Metropolitan area. Currently a plant comprising, dissolved air flotation (DAF), ozonation, deep bed granular activated carbon filtration, corrosion control using lime and carbon dioxide, fluoridation and residual chloramination, has been approved by the State, Department of Environmental Protection. Demonstration scale testing will continue through 1995 and a review of the necessary degree of treatment is expected by July 1998.

- Covered distribution reservoirs will be constructed at up to ten locations within and around the Metropolitan area. The largest of these is a 435,000 cubic meters (115 million gallons) tank currently in preliminary design. Three other reservoirs totalling 246,000 cubic meters (65 million gallons) are in design. The objective is to have covered distribution storage of at least an average day of use and preferably a maximum day of use, by the year 2020.

- o The MetroWest Water supply Tunnel is a $413 million, 29 kilometers long, 60-120 meter (200-400 feet) deep, 4.27 meter (14 feet) finished diameter, concrete lined rock tunnel which will provide redundancy in a key section of the transmission system between the water treatment plant and the major distribution reservoirs serving the Metropolitan area. This project is currently in final design and is expected to be completed around 2002, pending a commitment of State funding assistance.

- o Other projects to improve reliability are being reviewed. These will complete pipe loops or parallel existing major pipes. Unlined iron or steel water pipelines will be replaced or cleaned and lined to reduce water quality deterioration within the water transmission system. These projects will be constructed according to priority, feasibility and financial resources and are expected to continue beyond 2020.

- o MWRA will investigate with State and local agencies how to provide increased public access to facilities placed in reserve status. The old open distribution reservoirs when taken out of service or replaced by covered reservoirs could provide recreational opportunities. Abandoned aqueduct routes could be adapted to bicycling, jogging, walking and cross-country skiing.

Summary A century ago the Boston area regional water supply established a bias towards high quality unfiltered water. Ten years ago it was decided to manage water demand rather than to further expand the system. For the next 25 years at least, the MWRA plans to modernize the current system, improve water quality, improve reliability and retrain its workforce. The total cost is expected to be on the order of $2 billion. In addition to the water supply benefits the region will regain use of some of the land previously reserved for surface aqueducts and large distribution reservoirs.

Appendix References

(1) Metropolitan Water Supply, Report of the Massachusetts State Board of Health. February 1895

(2) Special Report, Chapter 48 of the Resolves of 1936, Commonwealth of Massachusetts. December 1, 1937

(3) Great Waters, A History of Boston's Water Supply, by Fern L. Nessan, Univ. Press of New England, 1983

Building the Case Against Drinking Water Treatment

Cory J. Crebbin, M. ASCE[1]

Abstract

The City of Lacey provides unadulterated groundwater directly from wells to the water distribution system. This paper discusses the various pressures to initiate treatment which are imposed on the water utility and what strategies have been used to avoid treatment.

Introduction

The City of Lacey relies entirely on groundwater for municipal use. The City has 15 active wells, from 34 to 236 meters deep, distributed widely within and outside of the City limits. The City's Water Utility provides over 6.8 billion liters of water to a population of 34,000. The depreciated replacement costs for the entire water utility is $19 million and revenues are approximately $2.8 million per year.

Water is delivered directly from wells into the distribution system without disinfection or other treatment. The untreated nature of the municipal supply is a source of community pride and contributes to pollution prevention efforts.

The water sources currently developed meet the primary drinking water standards established by the Environmental Protection Agency (EPA), but some sources do contain noticeable levels of iron and manganese.

Impact of Installing Treatment

Implementing treatment will be difficult due to the wide geographic dispersion of water sources. Variations in water chemistry between sources also complicates the comparison of treatment options. Concentrations of minerals, pH levels, and turbidity fluctuate at each source and between sources; and chemistry varies unpredictably within the distribution system dependent on location and which sources are active.

---

[1] P.E., Water Resources Manager, City of Lacey, P.O. Box 'B', Lacey, WA 98503

An example of the problems encountered in considering treatment for the Lacey water system was provided during the analysis of a prospective new well. The proposed well produced water with small, but noticeable, quantities of H_2S. Two options considered were air stripping and chlorination. Air stripping was considered too expensive for the amount of water which would be produced. Although much less expensive, the addition of chlorine would probably cause precipitation of minerals introduced by other sources into the distribution system.

The initial costs of installing treatment are also high in relation to development of alternate sources. A preliminary cost estimate of $500,000 was determined for installation of filtration at one well with manganese levels above the secondary Maximum Contaminant Level (MCL).

Additionally, installing treatment will require upgrade of current state-issued operator certificates from "Water Distribution Managers" to "Water Treatment Plant Operators". Such an upgrade will significantly increase the training and experience requirements for water utility staff. The only treatment options which would not require such a personnel upgrade are fluoridation, chlorination, or corrosion control.

Treatment: The Solution to What?

The groundwater available in the Lacey area is safe and of high quality. There is an increasing bias by regulatory authorities to require treatment of groundwater. Lacey and neighboring jurisdictions continue to resist this trend.

a) Disinfection

The Washington Administrative Code (WAC) specifies disinfection as the minimum level of treatment required for public water systems. Waivers may be granted by the Department of Health (DOH) by proving that the aquifer from which water is drawn is "protected" and by demonstrating a satisfactory bacteriological history. The Lacey water system currently operates under such a waiver, but "well susceptibility analyses" required by recent regulations are currently under review by DOH.

b) Lead and Copper

The City has completed sampling for lead and copper. The results were satisfactory for the Main Lacey System, but results for a 350-customer satellite system indicated that a corrosion control study was needed. The City intertied the satellite system to the Main system and persuaded DOH to allow combining Lead and Copper results for both systems, thus avoiding a corrosion control study.

Several new buildings, including residences and schools, have exhibited copper levels in excess of the MCL. One of these new buildings is the Washington State Department of Ecology Headquarters. High copper levels in new construction have resulted in health concerns by occupants, concern about copper plumbing installation methods, and concern about the solutions undertaken by facility managers.

Occupant concerns are generally short-lived as the copper levels decrease rapidly with water use. The DOE Headquarters, considered a worst-case for the City due to long copper pipe runs and relatively low water use, has experienced significantly reduced copper levels over the last year. DOE samples a number of plumbing fixtures in the building. The figure below shows the copper levels for the fixture with the most complete sampling record (3F-WF) and the maximum copper level drawn from any fixture for each round of sampling.

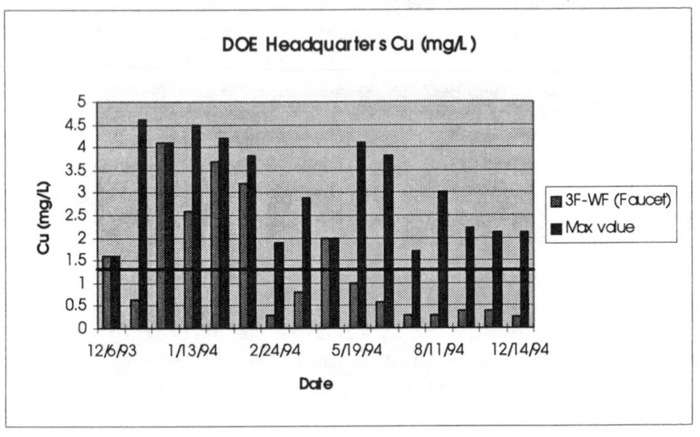

New residential construction demonstrated a wide range of copper levels prior to occupancy. Because the copper levels tended to be consistent within developments, it is suspected that plumber practices, particularly with regard to the application of flux, is a major contributor to high copper levels. All plumbers and builders questioned indicated they paint flux on the outside of pipes rather than dipping pipes flux.

The local school district commissioned a corrosion control study and installed the recommended calcite contactors at three new facilities. This is of concern to the DOH because any party which alters the chemistry of water is deemed a water purveyor according to the WAC. DOH has expressed their desire that the City assume responsibility for these treatment units to avoid establishing the school district as a water purveyor. The City has so far refused to do so since these installations were made without the City's concurrence.

c) Manganese

Current DOH policy is that new water sources must not exceed the EPA's secondary MCL for manganese (Mn), and that existing sources which exceed the secondary standard may continue to be utilized without treatment provided no complaints are received by DOH. The EPA's primary standard for Mn is 0.3 mg/L; the secondary standard is 0.05 mg/L.

The most prolific well in the City system, which provides 20% of the available instantaneous production, produces water with manganese concentrations as high as 0.22 mg/L. This well, designated Well #7, is only utilized to meet peak demands. The average Mn concentration in the distribution system is 0.017 mg/L without Well #7. The average Mn concentration is 0.065 mg/L when Well #7 is operating. This is a low estimate due to the assumption that all other production wells are operating at maximum capacity.

Customers near Well #7 have noted staining of plumbing fixtures and laundry as a result of high Mn levels during the summer peak water demand period. The City has applied for a grant to replace Well #7 due to the threat from three hazardous waste sites within 1000 m of the wellhead. One of these sites has contaminated the upper aquifer with Trichloroethane (PCE) and resulted in the abandonment of a nearby well.

d) Radon

All of the City's wells except one exceed the EPA's proposed MCL of 200 pico-Curies per liter (piC/L) for radon. Some sources have radon levels which exceed 500 piC/L. Indoor air levels of radon gas in the Lacey area have not been definitively studied, but all new houses are provided with radon detectors and there are no known cases of indoor air radon levels exceeding EPA standards. The City contacted members of Congress whose districts include the City and encouraged them to support the MCL of 1000 piC/L proposed in House Bill 3392.

Conclusions

The initial and continuing costs of drinking water treatment are high, and the costs of avoiding treatment are growing. Resources are inefficiently deployed, such as locating a $1.7 million reservoir next to a well with high levels of manganese to provide for dilution, resulting in higher costs with no discernible benefit to water customers.

Customers should be the voice to which water system operators respond. Properly informed, customers can decide whether benefits outweigh the costs of various alternatives. The current regulatory environment has replaced the voice of the customer with guidelines from state and federal bureaucracies. Water system operators must spend resources on innovative ways to avoid treatment while still meeting regulations of questionable benefit, especially with regard to manganese and radon, rather than developing innovative solutions to real operating problems such as pressure zone reconfigurations and reserve sources of supply.

Treatment can be avoided, but water systems must be prepared to identify and commit the resources necessary. In the meantime, water customers will be well served if regulatory agencies change their approach to one in which it is their responsibility to prove that treatment is required as opposed to the present practice of assuming expensive technology is required unless proven otherwise.

Changing the Water Management Criteria of Youghiogheny River Lake

Werner C. Loehlein, P.E., M. ASCE [1]

Abstract

The Municipal Authorities of Westmoreland County (MAWC) and North Fayette, public water supply systems licensed by the Pennsylvania Department of Environmental Resources (PADER), obtain a large proportion of their untreated water from the Youghiogheny River. The operation of these systems is currently regulated by Water Allocation Permits issued by PADER which establish a maximum daily withdrawal based upon minimum acceptable flows which serve to ensure that water quality and low flow augmentation are not compromised. As a result of current and projected increases in potable water demand, and recent regional drought conditions, these water authorities are exceeding the permitted withdrawal amounts. In order to prevent a continuation of these violations, PADER directed these authorities to secure suitable additional storage to provide the needed supplemental releases during periods of low flow / high demand.

Since the Corp of Engineers' Youghiogheny River Lake is located upstream, the water purveyors contacted the Corps to explore the potential for reallocating storage to water supply. The opportunity for this reallocation may exist because recent improvements in water quality in the basin have resulted from the reduction in the area's heavy industry and mining. The region has lost a significant portion of its heavy industry which once dominated the employment base. This loss of heavy industry, as well as a reduction in the severity of acid mine pollution, has reduced the pollution load of the Youghiogheny River. Consequently, the original authorized storage and release schedule for Youghiogheny

----

[1]Supv. Hydraulic Engineer, U.S. Army Corps of Engineers 1000 Liberty Avenue, Pittsburgh, PA 15222-4186

River Lake may be inappropriate since less storage is probably needed to dilute pollution.

This paper will discuss the current storage and release schedule against the backdrop of improved water quality in the watershed, and the needs of the area's competing water users. The paper will discuss any potential for reallocating storage to serve water supply needs and to modify the storage and release schedule to be more responsive to present and anticipated water use in the region.

Youghiogheny River Lake

The dam is located in southwestern Pennsylvania on the Youghiogheny River about 74.2 miles (119.4 km.) above its junction with the Monongahela River at McKeesport, Pa., and 1.2 miles (1.9 km.) above Confluence, Pa. The reservoir is located in Fayette and Somerset Counties, Pa., and Garrett County, Md. The drainage area above the dam is 434 square miles (1,120 km^2.). Youghiogheny River Lake was authorized by the Flood Control Act approved on June 28, 1938, for the purposes of reducing flood stages and providing low-flow water quality control on the Youghiogheny, lower Monongahela, and upper Ohio Rivers. The project was placed in full operation in January 1948.

The Water Resources Development Act of 1988 added downstream and upstream recreation as an authorized project purpose. However, the Act did not authorize the allocation of additional storage or the reallocation of existing storage for recreation. The Act states that the Lake will be "operated in such manner as will protect and enhance recreation associated with such project" and to the extent that recreation is compatible with other project purposes. It should be noted that downstream recreation, especially the whitewater rafting industry which draws about one hundred and fifty thousand visitors every year, is a competing water user with the lake recreation.

In December 1989, commercial hydropower was added to the project. The hydropower that currently is being generated, however, does not require any change in the previous operational procedures. In addition, the water quality of Youghiogheny River Lake currently supports an important two-story warm and cold water reservoir fishery, and a very popular cold water tailrace fishery.

Previous Studies

The reallocation of Youghiogheny River Lake's water

quality storage to water supply was recognized in a 1981 study done by the Pittsburgh District. The study, which was completed under the authority of Section 22 of the Water Resources Development Act of 1974 (Public Law 93-251), supported the supposition that the value and benefits of the project may be enhanced by revising the operation of the lake, particularly changes to the storage and release schedule (Figure 1).

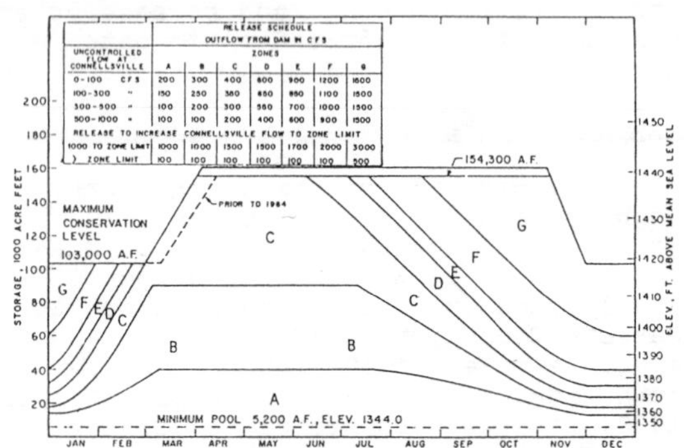

UNCONTROLLED FLOW AT CONNELLSVILLE	RELEASE SCHEDULE OUTFLOW FROM DAM IN CFS — ZONES A	B	C	D	E	F	G
0-100 CFS	200	300	400	800	900	1200	1600
100-300 "	150	250	380	650	880	1100	1500
300-500 "	100	200	300	580	700	1000	1500
500-1000 "	100	100	200	400	600	900	1500
RELEASE TO INCREASE CONNELLSVILLE FLOW TO ZONE LIMIT							
1000 TO ZONE LIMIT	1000	1000	1500	1500	1700	2000	3000
> ZONE LIMIT	100	100	100	100	100	100	500

Figure 1. Youghiogheny River Lake Storage and Release Schedule

The 1981 study demonstrated that there was no surplus water available for water supply utilizing the present storage and release schedules. However, some of the alternatives investigated did produce surplus water, but not without significant trade-offs.

The study concluded that the only apparent viable alternative available to provide water supply from Youghiogheny River Lake would be to increase the maximum summer conservation pool level and physically modify the dam structure and the pertinent features to compensate for the related loss of flood control storage.

The gradual improving trend in water quality in the Youghiogheny River basin that was noted in the 1981 study has continued. With the removal of many of the sources of water quality degradation, for which the original release schedules were established, a unique opportunity now exists to adjust the project's operation.

Several meetings have been held with the water authorities, their engineering representatives, PADER officials, staff members for Congressmen Austin J. Murphy and John P. Murtha, and the Corps. During these meetings, the Corps' study protocol involving an initial assessment, followed by a reconnaissance effort (at

Federal expense), and ultimately a cost shared feasibility study was explained. All participants agreed that utilizing storage within the lake may prove to be a highly effective means of satisfying the changing low flow augmentation requirements.

In November 1993, the District initial assessment was approved. This assessment concluded that the potential to reallocate storage in Youghiogheny River Lake does exist. The district was authorized to begin the reconnaissance study when funding becomes available. The major tasks anticipated for this effort include the following items:

a. Review existing authorizations and operating rules.
b. Examine existing conditions within the watershed.
c. Review water supply issues.
d. Inventory possible alternative water sources to satisfy basin needs.
e. Analyze and quantify the important factors which define the potential of reallocation of storage.
f. Examine the impacts of any reallocation of storage.
g. Evaluate all study analyses and formulate conclusions and recommendations.
h. Estimate future study, engineering and construction costs.
i. Report preparation.
j. Publish and distribute report.

The total cost to complete the reconnaissance study is estimated to be approximately $351,000. However, due to the current emphasis to reduce the nation's budget deficit, funding for this study has been postponed until October 1995.

Conclusion

The opportunity for reallocating storage in the Youghiogheny River Lake exists because of improving water quality in the basin. Moreover, in view of the considerable costs and negative impacts associated with alternative water supply strategies, storage reallocation appears to be the most feasible and cost effective means to alleviate future public water supply shortages or dramatic rate increases. The reconnaissance study will be undertaken under the authority of Section 216 of the Flood Control Act of 1970, which authorizes studies to review the operation of completed Federal projects and recommend project modifications. This will avoid the need for PADER to impose penalties on the water purveyors for exceeding their water allocation permits.

Water Resources Planning for the 20th Century

N. Bruce Hanes[1]
Member

Abstract

No comprehensive conference on Water Resources Planning for the 21st century would be complete without an examination of the planning that occurred as we entered the 20th century. This paper reviews the thinking of our predecessors one hundred years ago in Boston, the site of this meeting, and expresses concern for our future planning efforts.

It is important that we understand conditions and what was known in 1900. In Boston, the sewers were constructed for storm-water runoff and "sanitary sewage" was excluded by ordinance from these drains. Later what was called a "combined sewer" came into use with the addition of "sanitary sewage". The development of the water carriage system for conveying human fecal matter away from places of habitation was itself an advance in the protection of public health because it removed the source of infection from the homes. For example, at this time some vaults and privies in homes were situated on the same or higher levels and their contents oozed through walls into adjacent occupied living quarters. It is important to note that it was not until the mid-nineteenth century that the germ theory of disease was discovered.

In 1889, the Metropolitan Sewerage District (MSD) was created by legislative action. The MSD was charged with the responsibility of providing common action to reduce the discharge of raw sewage into the Mystic, Charles and Neponset Rivers. The MSD was expanded in 1895 adding additional communities along the Charles River to the Boston Main Drainage System which discharged through the Moon Island Outlet. Additional facilities (interceptor

----

[1]Professor Emeritus, Tufts University, 40 Kingston Court, Gibsonville, NC 27249

sewers, pumping stations and outfalls off Deer Island) were constructed to serve twelve northern towns as well as Charlestown and East Boston.

In 1897, the Massachusetts State Board of Health, in a report on the improvement of the Neponset River, noted that "intelligent observers report that these meadows are at times the source of disagreeable odors and the direct cause of much sickness." The report also noted that "One disease has attracted considerable attention in recent years in many portions of this state, malarial fever, and portions of this valley have suffered from it, and severely, when the limited population is taken into account".

A variety of physical information was reported at various stations which included: drainage areas, population, length, slope, width and elevation of the river. In addition, limited chemical analyses were presented from 1875 to 1897. These were reported in parts per 100,000 and included ammonia, total solids, solids loss on ignition, chlorine, color, nitrogen, oxygen consumed and hardness. For example, the average oxygen consumed from river samples taken at Hyde Park were .9548, 1.0003 and 1.0458 parts per 100,000 during 1893, 1894 and 1895. The improvement in our ability to identify and to measure minute quantities of materials in water has vastly improved during the 20th century, however, our ability to intelligently use the results has not kept pace in many cases. It is interesting to observe that, even without a clear knowledge of the cause of malaria, a comprehensive plan was presented that included drainage of the meadows, prevention of further discharge of "domestic and manufacturing sewage", treatment of existing discharges either on site or transporting them to Boston Harbor. The theories in 1897 concerning malaria related the disease with marshy regions which included stagnating water and decaying vegetation. At that time, it was not known that the mosquito was the vector for the protozoal parasite which is the causative agent for malaria. Progress was rapid and by 1904 in a Report on the Mystic River and Alewife Brook the breeding of mosquitoes was related to the cause of malaria.

Boston Harbor also had its problems. In 1899, the State Board of Health was directed by the General Court "to consider the general subject of the discharge of sewage...and to report a plan for an outlet for a high-level, gravity or other sewer for the relief of the Charles and Neponset River valleys". At that time sewage was discharged into Boston Harbor at Deer Island Beacon and on the north side of Moon Island. 50,000,000 gallons per day was discharged through Deer Island Outlet and "while distinctly visible along the northerly edge of the channel for a half mile toward Boston on the incoming tide and toward the sea on the outgoing tide...it completely disappears within a distance of 1 1/4 miles". It was reported that even with population growth "the Board sees no reason to anticipate any trouble from this for many years in the future upon any inhabitable shores, and believes that the only objection that can be raised to the continual discharge of sewage here will be by those sailing through or near to the stream of sewage within a mile of the outlet". The Moon Island Outlet which handled 100,000,000 gallons of sewage a day was another story. Sewage was stored in reservoirs for discharge only on outgoing tides to avoid becoming a nuisance on habitable shores. The storage of sewage in reservoirs between tides made it more offensive and its discharge at a higher rate on the outgoing tide created an "area of a half mile radius much more objectionable than the steady discharge of fresh sewage at Deer Island". It was proposed that the discharge at Moon Island be limited and that new sewage outlets be built in the channel along the northwesterly side of Peddock's Island. "Here the sewage will be discharged about 30 feet below the surface at low tide into a strong and deep current, by which it will be kept well away from inhabited shores until it disappears by commingling with enormous quantities of ever-changing salt water." The third system of outfalls were constructed in 1904 near Nut Island. The overloaded Boston Main Drainage System was relieved by diverting flows from Brookline, Dedham, Hyde Park, Milton, Newton, Quincy, Waltham and Watertown to the Nut Island Outfalls. Dilution was the solution for Boston Harbor as we entered the 20th century.

In the case of water supply Boston obtained its water from Jamaica Pond, Lake Cochituate and the Sudbury River Supply; Charlestown, East Boston, Chelsea, Everett and Somerville were supplied from the upper Mystic Lake; Malden, Medford and Melrose utilized the surface water captured in Spot Pond; Belmont, Milton, Nahant, Newton, Revere, Swampscott, Watertown and Winthrop were supplied from ground water sources; Lexington and Quincy had both reservoirs and wells; Arlington was supplied from a storage reservoir and nearby filter gallery, and Stoneham obtained its water from the Wakefield Water Company. Many of these communities were experiencing difficulties obtaining sufficient water of satisfactory quality from these sources because of encroachment of the watersheds by people and salt water intrusion of the wells. It was reported that the 1888 average daily consumption of water in Boston was 79 gallons per person with a maximum of 151 gallons per person on January 28, 1888. The peak consumption was attributed to allowing the water to flow at night to prevent pipes from freezing.

In 1895, House Bill 500 was passed which established the Metropolitan Water District. Frederick Pike Stearns, while Chief Engineer of the Massachusetts State Department of Public Health, devised the Master Plan for the Water Supply of the Metropolitan Water District. This plan included Wachusett Dam and Reservoir, Quabbin Reservoir and future diversion from the Millers River watershed. A plan which has provided an excellent water supply during the 20th century for the Boston Metropolitan Area.

I now must share my concerns on how a group such as this will view our current effort at the beginning of the 22nd century. Today, it would not be possible to build even Quabbin Reservoir because of our policies and resource allocation. For example, it is now estimated that it will cost $112,000,000 to obtain a license and an additional $65,000,000 for the construction of a low-level radio-active waste repository in North Carolina. When planning and licensing a facility triples the cost of a project, environmental gridlock is not far behind. If we are to respond to today's needs of society, we must return to making decisions based on what is known rather than a fear of the unknown. In the United States, our life expectancy has almost doubled in the 20th century. What will happen to our life expectancy in the 21st century?

References

Massachusetts State Board of Health, Report upon the Improvement of Neponset River, 1897.

Massachusetts State Board of Health, Report upon the Discharge of Sewage into Boston Harbor, April 18, 1900.

Metropolitan Water and Sewerage Board Second Annual Report, January 1, 1903.

Metropolitan Park Commission, Report on the Improvement of the Upper Mystic River and Alewife Brook, September 21, 1904.

Hanes, N.B., "Opening Remarks for Water Supply Workshop", Proceedings of the First New England Conference on Urban Planning for Environmental Health, September 1965.

NATIONAL RESERVOIR DATABASES

by William K. Johnson, M. ASCE[1] and James R. Wallis[2]

Abstract

Two efforts are currently underway to develop databases of the principal reservoirs in the United States: a database of daily storage volumes for the National Drought Atlas and a database of all Corps of Engineers reservoirs. These efforts will allow comprehensive analysis of their physical features, the purposes they serve, their susceptibility to drought, and their place in the ecology of the watershed.

National Drought Atlas

The development of a National Drought Atlas (NDA) by the Institute for Water Resources, Corps of Engineers (1994) has sparked the development of a national database that includes reservoir information maintained by the U.S. Army Corps of Engineers, U.S. Geological Survey (USGS), and U.S. Bureau of Reclamation. Times series data, principally period of record daily storage volumes, are being collected for analysis of drought conditions and verification of the Drought Atlas.

The NDA team had, from its inception, intended to include minimum, maximum, and average daily storage volumes by months for the complete period of record for all the major reservoirs located within the contiguous 48 states. There is still a need for such a file in regional drought alleviation and climate change studies, as well as for flood warning systems resulting from dam failure, and even, as was shown in the 1993 mid-western floods, as a reservoir management tool. What follows is in the nature of a status report

[1] Civil Engineer, Hydrologic Engineering Center, Corps of Engineers, Davis, CA 95616

[2] Research Staff Member, IBM, Thomas J. Watson Research Center, Yorktown Heights, NY 10598

which outlines the progress made, and the difficulties encountered in our attempt to have a national database of reservoir storage volumes.

First, consider the problem of locating and identifying the major reservoirs. The USGS publishes monthly data on a set of reservoirs known as the National Water Conditions Reservoir Index Stations, NWCRIS. In general, this database is distributed at the district office level. Historic volumes, if they are maintained at all, are likely to be only for the last day of the month. The partial files accessible at the USGS NWCRIS office in Reston, VA., proved to be so corrupted that we were finally advised not to use them.

Second, we turned to the National Inventory of Dams, NID, being produced by the Federal Emergency Management Agency (FEMA) contemporaneously with the NDA work. The NID database is still actively being updated and corrected and will eventually result in a very useful product. Our first two retrievals of early versions of the NID database proved to have so many obvious errors as to be unusable for NDA purposes.

There are 74,053 dams listed in the NID inventory (CDROM 1993-1994), of which 41,223 were built in the thirty year period 1950-1979, (future archaeologists may refer to this as "the golden age of giant terrace builders"). Unfortunately, there are still many obvious deficiencies in the NID database, some of which have been summarized in Table 1.

Table 1
Aspects of the 1993-1994 N.I.D.

Total number of dams	74,053
Number of dams built in the 30 year period 1950-1979	41,223
Number of dams having significant to high downstream hazard	27,712
Number of dams with missing latitude	1,462
Number of dams with missing longitude	1,456
Number with incorrect latitude	74
Number with incorrect longitude	61
Number of latitudes given to nearest degree	588
Number of longitudes given to nearest degree	570
Number of dams, normal > maximum capacity	7,144
Number of dams with maximum storage = 0	6,673
Number of dams with normal storage = 0	762
Number of dams with length = 0	1,114
Number of dams with height = 0	48

Accurate location is still a major problem. There are 1400+ sites for which either latitude or longitude are missing. Seventy-four sites have longitude outside their respective state boundaries, two being well into the Pacific Ocean, while others are apparently in Western Europe and on a previously undetected mid-Atlantic ridge. The precision of the locations for many of the other sites is also questionable.

Considerable effort went into acquiring daily volume data starting with NWCRI sites and Corp of Engineers projects identified in their 1992 report. Occasionally we would find an individual at a USGS or Corps of Engineers office who had accurate and complete data, and could make it available over the Internet in any desired format (or alternatively on diskette). However, the more usual case was to ask for volumes and get stages. Duplicate or missing years or months then became apparent, along with obvious errors - for instance, stages sometimes as much as 3000' above the spillway crest. Typically numerous contacts were necessary before a record could be added to the volume database - along with a hope that its major errors had been removed. Complications abounded, as might be expected, data formats changed between reservoirs, but also unexpectedly and inexplicably over time within many individual series.

Hydrologic Engineering Center Database

The Hydrologic Engineering Center (HEC) database provides technical information on 542 federally-owned reservoirs operated by the Corps of Engineers across the nation (U. S. Army Corps of Engineers, 1992;1994). This includes information on project location, authorizing legislation, water control management, hydrologic and structural features, reservoir storage, hydroelectric power facilities, water supply contracts, and project recreation. The information is detailed and comprehensive and is displayed in standard reports selected from a menu. Additional reports can be developed, or the data accessed using standard (SQL) database commands.

The database also provides important links necessary to access other databases and create a database network (Figure 1). For example, it provides for each reservoir, the hydrologic unit number commonly used in U. S. Geological Survey databases, the FIPS (Federal Information Processing Standard) state and county codes used by the U. S. Bureau of Census. At the center, or hub, of the network is a database, NETID, that contains identifying information. Each reservoir has over 20 unique identifiers associated with it. These identifiers include such information as river, state, county, hydrologic unit, latitude and longitude, congressional district, zipcode and so on. Each identifier provides a key piece of information, a link, that can be used to access information in other databases.

Many of the problems encountered in collecting daily volumes for the NDA database were also encountered in developing the HEC database. At present, development of the network illustrated by Figure 1 is only partially complete. The hub developed by HEC has been linked with other Corps databases (those to the left of Figure 1). These are not maintained on-line, but rather the databases have been obtained by HEC from the office responsible and loaded into the network.

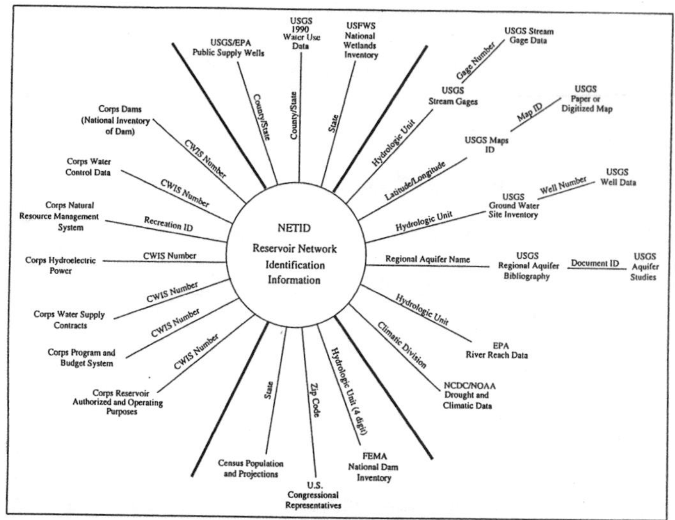

Figure 1 - HEC Reservoir-Database Network

The lessons learned from the development of NDA and HEC databases are as important as the databases themselves. It is common when examining existing databases to find data that are incorrect, missing, or mislabeled. Creating an accurate, complete and properly identified database is a tedious and time consuming task and one which requires finding creative ways to check and cross-check the data.

References

U. S. Federal Emergency Management Agency, (1992). *Water Control Infrastructure, National Inventory of Dams*, Washington D.C.

U. S. Army Corps of Engineers. (1992). *Authorized and Operating Purposes of Corps of Engineers Reservoirs*, Washington D.C.

U. S. Army Corps of Engineers. (1994). *HEC Reservoir-Database Network*, Hydrologic Engineering Center, Davis, California

U. S. Army Corps of Engineers. (1994). *National Drought Atlas*, Institute for Water Resources, Washington D.C.

Water Quality Performance Evaluation
of
Water Distribution Systems

Arun K. Deb[1], Fellow ASCE and Yakir J. Hasit[2], Member ASCE

Abstract

The primary goal of a distribution system manager is to provide customers an adequate quantity of good quality water at adequate pressure. To protect public health, the quality of water provided at the treatment plants should be maintained in the distribution systems. AWWA Research Foundation contracted Roy F. Weston, Inc. to identify distribution system water quality performance measures and to develop methodologies to evaluate the performance of distribution systems regarding water quality.

In this study, a systematic methodology was prepared to develop an action plan to monitor distribution system water quality measures and to evaluate the water quality performance of distribution systems. Primary water quality objectives of providing safe and good quality services by minimizing water quality complaints can be achieved by implementing this action plan.

Introduction

Conventionally, the primary goal of a distribution system manager is to supply an adequate quantity of water at adequate pressure to consumer locations. Water quality in the system has recently become a primary issue because of documented evidence of deterioration of water quality in some distribution systems (Clark et al., 1993). In order to protect public health, the

---

[1] Vice President, Roy F. Weston, Inc., One Weston Way, West Chester, PA 19380

[2] Senior Project Manager, Roy F. Weston, Inc., One Weston Way, West Chester, PA 19380

high quality water that is produced at the treatment plant must be maintained throughout the distribution system. Deterioration of water quality in the distribution system should be prevented.

Primary water quality objectives and management goals for a water distribution system include supplying water of high quality that satisfies all requirements of the federal Safe Drinking Water Act (SDWA) and the state drinking water regulations, as well as providing quality service to customers.

Direct Water Quality Performance Measures

A direct performance measure gauges the actual water quality performance of the distribution system and can be classified into three groups:

- Bacteriological measures (total coliform, HPC, disinfectant residual, bacterial regrowth).
- Aesthetic measures (discolored water, taste and odor, temperature).
- Other measures (corrosion indices, lead and copper concentration, disinfection by-products).

Indirect Water Quality Performance Measures

Indirect performance measures provide information about the sources of water quality problems in the system and include:

Residence Time: Long residence times have significant impacts on concentration of contaminants as they propagate through the system.

System Operation: Pumping water from the source directly to consumers, as opposed to supplying water through storage tanks, reduces residence time and improves water quality.

System Pressure: Maintaining adequate water pressure within the distribution system is a major element in preserving distribution system water quality.

Tuberculation: Tuberculation not only reduces the cross sectional area (and consequently the water-carrying capacity of water mains), but it also provides a greater surface area for bacterial attachment and growth.

Customer Complaints: Customer complaint data provide very good information on the aesthetic quality of water (discolored water, taste, and odor).

Methodology for Performance Evaluation

A flow diagram describing a general methodology for evaluating distribution system quality performance is presented in Figure 1.

Step I - Evaluation of Existing and Historical Water Quality Information — This evaluation should include water quality monitoring data as required by SDWA regulations and customer complaint data. This data review and evaluation should be conducted in order to isolate the historical problem areas and determine the potential causes of the problem. All relevant data should be plotted for each monitoring area, along with other available information. The resulting plot or map should then be analyzed for patterns which may indicate causes and nature of the problems.

Step II - Set Up Annual and Long-Term Water Quality Performance Goals — This plan should be part of the overall water distribution system performance improvement plan. This plan will address the need for long-term and short-term water quality improvement and associated investment for future financial planning. The goal of the plan should be to provide specified quality service to the customers. The quality of service should, at a minimum, meet all SDWA and other state and local regulatory requirements.

Step III - Collect and Load Water Quality Data — Existing regulations in the United States require significant monitoring of water in the distribution system. The requirements, particularly frequency of sampling and water quality parameters, vary with system size. In the future, these regulations will be more stringent and monitoring will play a major role in compliance of these regulations.

Once a proper sampling location plan has been developed, designed, and installed, a comprehensive data collection plan should be developed with the following objectives:

- Collect all data as required by drinking water regulations.
- Collect all data in response to all customer complaints.
- Collect all data to monitor all historical problems identified in Step I.
- Collect all data to monitor impacts of any change in distribution system operation (change in pumping pattern, etc.) on water quality.

Step IV - Analyze Data and Identify Causes of Problems — Step IV involves analysis of historical data and data collected in direct response to a problem. This would include data collected in response to a customer complaint or in response to a water quality rule violation. The objective of this

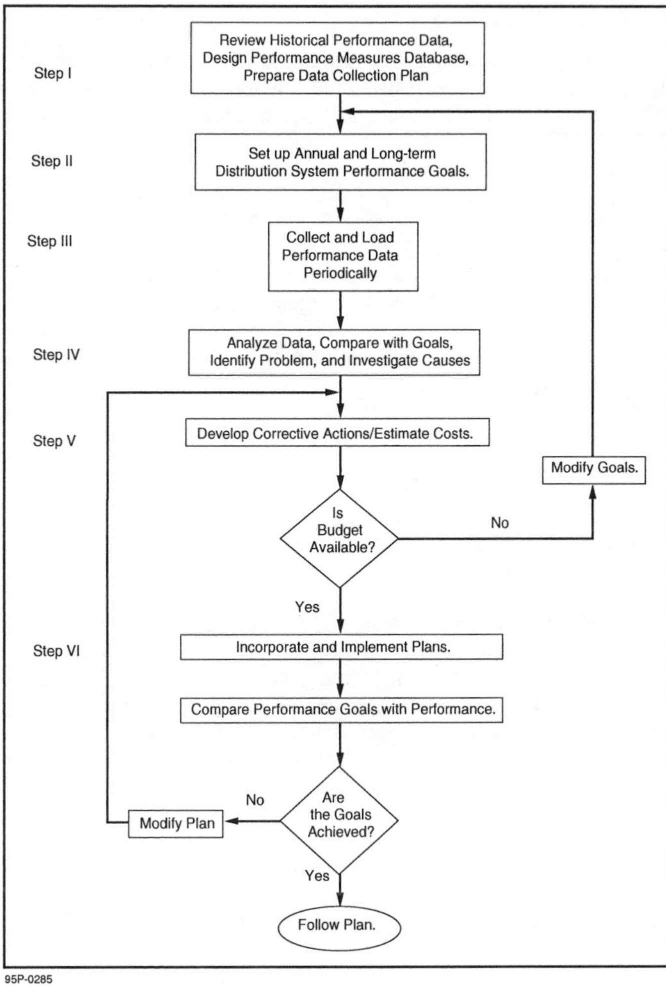

Figure 1. Performance Evaluation Methodology

step is to reassess historical data along with newly collected data in order to identify causes of the problem. After initial analysis of the problem, a plan should be developed to systematically collect any additional data that may be required.

Step V - Develop Corrective Actions — The two main causes of water quality deterioration in the distribution mains are high residence time (stagnation of water) and formation of biofilm on pipe walls which creates high chlorine demands.

The following corrective measures should be considered to improve water quality in the distribution system:

- Design and operate the distribution system to minimize residence time.
- Conduct routine flushing and emergency or corrective flushing.
- Improve disinfection (increase residuals, rechlorinate, use chloramines).
- Prevent tuberculation and corrosion of water mains (cleaning and lining; replacement of water mains).
- Reduce biologically degradable organic carbon at the treatment plant.
- Optimize corrosion control treatment to reduce corrosion rates of water mains and services lines.
- Minimize water main breaks.
- Minimize dead ends in the distribution system. The number of dead ends/mile of distribution system may be considered an indirect design performance measure.

Step VI - Evaluate Performance Goals — The results of the corrective action on the distribution system water quality should be monitored in order to analyze effectiveness of the program. Results achieved should then be compared with the distribution water quality performance goals. This plan should be reviewed annually against actual performance. If short-term performance goals are not achieved, the strategic plan should be modified accordingly.

References

1. Clark, R.M., J.A. Goodrich, and L.J. Wymer. 1993. Effect of the Distribution System on Drinking Water Quality. *J. Water SRT - Aqua,* 42(1):30-38.

Reliability Analysis of Regional Water Supply System

Shu-li Yang[1], Nien-Sheng Hsu[2], Associate Member, ASCE,
Peter W. F. Louie[3], and William W-G. Yeh[4], Fellow, ASCE

Abstract

Two reliability models are presented in this paper. Using the minimum cut-set method, the deterministic model focuses on the source-demand connectivity and helps to identify crucial pipes in the system. Using Monte Carlo simulation, the stochastic model is used to evaluate the supply level that the system can provide under random component failures. A multiobjective optimization model is embedded in the analysis to simulate optimal operation of the system to allow an equal basis of comparison among different system configurations.

Introduction

As the largest water supplier in Southern California, the Metropolitan Water District of Southern California (MWDSC) is confronted with numerous supply management problems. Facility expansion programs of the existing system are underway to secure a more reliable water supply to meet future water demand. However, like many other water distribution systems, performance of the MWDSC system is subject to random failures of its pipelines. A reliability analysis is performed to investigate the impact of various pipe failure patterns on water supply at demand nodes of concern. The results are useful in identifying the facility expansion alternatives. A comprehensive report on reliability analysis of water distribution systems was done by May (1989). However, applications of reliability analysis methods to large-scale water distribution systems have been limited.

Problem Description

The MWDSC system consists of a large number of demand nodes and numerous pipelines connecting the demand nodes to the sources. For planning

[1] Postdoctoral Scholar, Dept. of Civil and Environ. Engrg., UCLA, Los Angeles, CA 90095.

[2] Assoc. Prof., Dept. of Civil Engrg., National Taiwan U., Taipei, Taiwan, ROC.

[3] Senior Engineer, Div. of Planning, Metropolitan Water District of So. California, Los Angeles, CA 90054

[4] Professor, Dept. of Civil and Environ. Engrg., UCLA, Los Angeles, CA 90095.

purposes, the MWDSC system is simplified by grouping some local piping systems into prototype pipes. Represented by a directed graph, the prototype distribution system consists of 214 links and 158 nodes, including 4 source nodes, 38 demand nodes, 9 surface reservoir nodes, 18 groundwater recharge basins, 87 junction nodes, and 2 reservoir spillage nodes. Each of the demand nodes represents a local water agency instead of a single water user. Forecasts of demands and schedules of imported water for the next 25 years are available on a monthly basis.

Performance of the MWDSC system is subject to random failures of its pipelines. However, failures of smaller pipelines may not have significant impact on overall system performance since they can be quickly repaired. Pipe failures of less than one month and pipes shut-down for a few days for regular maintenance are not considered as failures in the monthly operational model. Therefore, 12 pipes are selected as key pipes for the reliability analysis. Only failures of the key pipes are to be examined.

This paper presents two models to resolve the following questions:

1) What are the crucial pipes, the failures of which will completely disrupt water supply to the demand nodes of concern?
2) What is the probability that each of the demand nodes of concern is connected to at least one of its sources?
3) How do random pipe failures affect system performance during the 25-year planning horizon?

Methodology

Two reliability models, deterministic and stochastic, are developed to help resolve the above-mentioned problems.

Deterministic Model

In the deterministic model, system failure is defined as any event in which at least one of the demand nodes of concern is not connected to any source. The system failure probability is determined using the minimum cut-set method. A minimum cut-set is defined as a set of links which, when all links of the set fail, causes system failure, but when any one link in the set has not failed, does not cause system failure. Because computational requirements increase exponentially with an increase in the size of the network, the minimum cut-set method for the reliability analysis of a large-scale water distribution system, such as the MWDSC system, is computationally intensive. A modified algorithm for the identification of the minimum cut-sets for the system is developed, resulting in a significant reduction in the computational effort.

The modified minimum cut-set method consists of the following three stages: 1) identification of the minimum cut-sets for each source-demand pair; 2) identification of the minimum cut-sets for each demand node; and 3) identification of the minimum cut-sets for the complete system. Only the first stage requires running the simulation model. Since simultaneous failure of more than 2 pipes is highly unlikely for the MWDSC system, the minimum cut-sets containing more than 2 pipes are not considered.

Stochastic Model

From a practical perspective, source-demand connectivity is necessary but not sufficient for meeting the water demand. As a complement to the deterministic model, the stochastic model is capable of investigating the impact of component failures on the supply level at each demand node. Since supply and demand on the MWDSC system change with time, and reservoir operations are influential in stabilizing the water supply, multiple-period simulation is needed to incorporate the time-varying supply and demand, and the carry-over effect of reservoir storages.

As pipe failures can be modeled as a random process, Monte Carlo simulation is used in the stochastic model. It has been shown that the time to failure of water pipes follows the exponential probability density function (Quimpo and Shamsi, 1991). An exponential random number generator is employed for constructing system scenario. Operation of the synthetic system is simulated using a multiobjective optimization model, which is efficiently solved by an embedded generalized network algorithm (EMNET) (Sun et al., 1994). The simulation results provide the optimal water distribution among the competing demands in a system with random component failures. However, the simulation results are subject to sampling error. Therefore, a large number of scenarios are created to represent different possible "crippled" system configurations, and to produce a statistically significant simulation result. The use of an optimization model in the simulation process provides an equal basis for comparison, since for each scenario, the system is optimally operated to achieve the same specific objectives. For each simulation, the determination of system state, failed or not failed, is made based on the pre-set required supply level. A chance-constrained criterion is used for defining system success follows

{The percentage of time during the planning horizon that shortage is no greater than $\overline{Q}$% of demand} ≥ $\underline{P}$% at each of the demand nodes of concern.

System performance reliability, defined as the probability that the above criterion is met, is determined after a large number of simulations. Two parameters, $\overline{Q}$ and $\underline{P}$, are involved in the criterion. Constructed by changing the values of $\overline{Q}$ and $\underline{P}$, the reliability curves can provide useful information on the magnitude and probability of system performance failure.

Results

The results from the deterministic model indicate that one demand node in the MWDSC system has the smallest mechanical reliability because only one route connects it to its source and two key pipes are present in the route. Most of the other demand nodes either have multiple routes between demand and source or have more reliable pipes (non-key pipes) in their routes. Twenty-one demand nodes have a reliability value of 1. It has been observed that for nodes having a mechanical reliability of 1, the performance reliabilities may still be subject to random failures of key pipes.

Dimension of the optimization model and the required computation time per simulation run for both deterministic and stochastic analyses are shown in Table 1.

Table 1. Dimensions of the Optimization Model

	Attribute	Deterministic	Stochastic
No. of Variables	Pipe Flow	214	64,200
	Reservoir Storage	54	16,200
	Slack Variable	38	11,400
	Minimum Supply Rate (*p*)	1	1
	Total	307	91,801
No. of Constraints	Continuity Equation	114	34,200
	Definition of *p*	38	11,400
	Total	152	45,600
Computation Time (CPU sec)		0.37	58.36

Conclusion

Two reliability models have been developed for the reliability analysis of the MWDSC system. The deterministic model helps to identify the crucial pipes whose failures will disrupt source-demand connectivity, while the stochastic model determines the impact of pipe failures on the supply level at the demand nodes. Results from both models are valuable for facility expansion planning. Given the size of the MWDSC system, although the analysis is computationally intensive, the simulation processes in both reliability models are accelerated by using EMNET. The developed methodology has been proven to be practical and feasible for the MWDSC system. It is believed that this methodology is also applicable to other large-scale water distribution systems.

Acknowledgment

The research reported herein was supported by funds provided by the Metropolitan Water District of Southern California.

References

Mays, L.W., ed. (1989). *Reliability analysis of water distribution systems*, American Society of Civil Engineers, New York, N.Y.

Quimpo, R.G., and Shamsi, U.M. (1991). "Reliability-based distribution system maintenance." *J. Water Resour. Plng. and Mgmt.*, 117(3), 321-339.

Sun, Y-H., Yeh, W. W-G., Hsu, N-S., and Louie, P. W.F. (1994). "A general network algorithm for water supply system optimization." To be published in *J. Water Resour. Plng. and Mgmt.*, ASCE.

CONSIDERATIONS FOR THE PLANNING AND DESIGN OF IMPROVEMENTS TO EXISTING WATER SUPPLY FACILITIES

David A. Oberlander[1], Associate Member

Abstract

Many communities and water works face the demanding task of supplying clean water to satisfy the water needs of their constituents. Often, existing facilities must be expanded to increase the amount of water available for daily use and for fire protection. Other facilities may require upgrading to meet more stringent water quality regulations. This paper focuses on the planning and design of improvements to a 50-year old water treatment and distribution system located in Beltsville, Maryland. The distribution system serves the United States Department of Agriculture (USDA) Agricultural Research Service's Beltsville Agricultural Research Center--East (BARC-East).

Introduction

BARC was established in 1910 as an experimental farm for animal husbandry and dairy research. The facility grew considerably in the 1930s and 1940s as the research activities expanded. Several agricultural achievements have been reached at the facility. A partial list of accomplishments includes the development of the modern turkey, hog, blueberry, strawberry, and disease resistant potato. Today, research continues on a variety of topics including insect control, animal immunization, parasitology, animal feeds, tissue culture, and human nutrition. A quality water supply and a dependable distribution system are essential to the continued success of the research center. BARC is divided into two subareas known as BARC-East and BARC-West; together they encompass over 28 square kilometers (7,000 acres) of farmland, roadways, wetlands, research facilities, animal quarters,

[1]Senior Engineer, Environmental Division, Maguire Group, Inc., 225 Foxborough Boulevard, Foxborough, Massachusetts 02035-2500.

and other buildings related to the agricultural studies. BARC-East has a completely independent water supply and distribution system. The system's size and complexity are similar to many of the water works serving the smaller municipalities in the United States.

Study

The project began in 1989 with a comprehensive study of the existing water supply, distribution, and treatment system. The study described the existing system conditions and made specific recommendations for repairs, improvements, and renovations to properly operate the water system. The report identified the following deficiencies: low pressure areas, low flow areas, unaccounted-for water, deteriorated storage facilities, inefficient pumps, and a manual disinfection system. The existing average day demand for the system is about 1,700 liters per minute (450 gpm).

Numerous field visits were conducted to identify and evaluate system components. A hydraulic model of the system incorporating CYBERNET® software was developed. Extensive flow tests were conducted to calibrate the model. Once calibrated, the model was incorporated to aid in the development of water system improvements. The recommended improvements included cleaning and lining of existing water mains, installation of new water lines, rehabilitation of three elevated water storage tanks, upgrading of the water treatment plant, and construction of a new 300,000 gallon finished water storage tank.

Existing System

The majority of the existing infrastructure for BARC-East was constructed in the late 1930s and early 1940s by the Civilian Conservation Corps (CCC). The water supply comes from groundwater under the research facility site. At one time, the water system consisted of ten separate gravel packed wells. However, the current operation relies on the continuous use of three of the wells and the intermittent use of two of the wells. A sixth well is reserved for emergency use only. The remaining four wells have been out of service for many years. A raw water transmission system conveys the raw water via six kilometers (4 miles) of 10 cm (4") and 15 cm (6") pipe.

Raw water enters the water treatment facility located near the center of the distribution system. The treatment facility increases the pH and reduces turbidity. The raw water enters an aeration tower at a pH of about 5.3 and a turbidity of 0.1. The purpose of the aeration is to reduce the carbon dioxide concentration and raise the pH to about 6.9. After aeration, chlorine solution is added for disinfection and soda ash is added to further raise the pH to about 8.0. The aerated water then passes over a weir into a 150,000 liter (40,000 gallon) underground tank. Four vertical turbine pumps convey the water from the tank, through six pressure filters and to the distribution system.

The distribution system consists of five elevated water storage tanks, one of the tanks has been abandoned. The four operating tanks have capacities of 95,000, 303,000, and two at 568,000 liters (25,000, 80,000 and two at 150,000 gallons). Distribution piping includes 37 kilometers (23 miles) of 10 cm (4") through 25 cm (10") pipe. The majority of the pipelines are 50± year old cast iron with lead and oakum joints. Some of the pipe was installed in 1976 as part of an expansion of the system; it consists of ductile iron pipe with rubber gaskets.

Deficiencies Identified and Solutions Recommended

Wells: The combined capacity of the operational wells exceeds the average day demand, but falls short of the maximum day demand. Additional flow was determined to be required for the facility. Several existing wells were scheduled for rehabilitation to increase yield. A new well was designed. Additional controls were designed to allow better monitoring and control of the system.

Distribution: Areas of low flow were attributed to two major causes--pipe size and tuberculation. As a result of several trials using the computer model, new, larger diameter pipes were selected for some areas, while others required cleaning and lining to increase the Hazen-Williams C factor. A total of 11 kilometers (6.8 miles) of new pipelines were designed and 0.8 kilometers (0.5 miles) of existing pipes were designated for cleaning and lining.

Storage: The elevated storage tanks were found to be in disrepair. They had not been painted for many years and showed signs of rusting and coating failure. In addition, there were not any level controls for the tanks. Oftentimes, the tanks were filled to the overflow and continued flowing until the pumps at the treatment facility were shut down manually. The recommended rehabilitation of the storage tanks involved the stripping and painting of the interior and exterior of the elevated steel tanks. Specifications require shrouding to confine sandblast residue. Disposal of lead based paint residue will require laboratory testing to determine the leachability of lead in the existing coating. Level controls will be added so that the tank levels can be monitored at the treatment facility.

Water Treatment Facility: The project eventually identified the need for more clearwater storage and larger pumps for the treatment facility. The facility did not have storage for the finished water other than the storage provided by the elevated storage tanks. The earlier study had recommended an increase of 1.13 million liters (300,000 gallons) of storage was required to provide adequate fire protection for emergency conditions. New pumps include two duty pumps at 1,500 liters per minute (400 gpm) each and one high service pump at 5,700 liters per minute (1,500 gpm). The new pumps will draw from a proposed 300,000 gallon concrete storage tank. In addition, an automated disinfection system was designed. The system allows prechlorination which is flow paced and post chlorination which is flow paced and residual trimmed. The controls for the chlorination system as well as the well pumps, process pumps, and tanks will be computerized to provide better control and allow more efficient recording of data for the overall system.

Design Review

Once the design reached the 50% completion point, it was reviewed by another consulting engineer as part of the government's value engineering procedures. Several comments were received at this stage. Many of the comments were addressed by explaining details of the study and preliminary design. A few of the comments resulted in additional research and some redesign of certain portions of the project. For example, the use of one large storage tank versus the four smaller tanks was investigated. The depth of cover over the water mains was decreased from 1.5 meters (5 feet) to 1.2 meters (4 feet). Overall, the review was worthwhile in that it helped clarify the reasons for certain improvements in the design.

Construction Phasing

As part of the design, a construction phasing scheme was developed to allow the continued operation of the existing treatment and distribution system while the improvements were made. The specifications included a section recommending a construction sequence. The purpose of the sequence is to alert the Contractor that construction cannot take place in all areas at the same time. Sufficient time for delays must be included in the schedule. The Contractor will provide a critical path construction schedule to insure the completion of the project in an efficient manner. At present, the USDA is considering breaking the project up into several smaller contracts to allow for additional phasing of the work and perhaps additional competition on the bid prices.

Current Status

Final design of the recommended improvements resumed in November 1994. At the time this paper was submitted, the design had reached the 90% stage. Once the design of the improvements is reviewed and accepted, the project will be finalized and put out to bid. We estimate that the project will be bid in 1995 and completed by the end of the 1998 construction season. The combined efforts of our study, computer modeling analysis, and proposed infrastructure improvements establish a plan to help insure the availability of quality water for drinking, farm use, and fire protection for the BARC-East facility.

Design and Construction of the Puddletown Booster Station

James R. Barsanti[1], Frank J. Ayotte[1]
James D. Bunsey[2], and William B. Powers[1]

Abstract

This paper describes the design and construction of the Puddletown Booster Station for the Metropolitan District Commission in Hartford, Connecticut. Factors affecting the design of the booster pump station include the transmission system modeling and pipeline pressure ratings. Construction issues involved with siting, connection of the booster station piping to the existing transmission system piping and field testing, are presented.

Introduction

The Metropolitan District Commission (MDC) is responsible for the operation of the water supply system servicing Hartford, Connecticut and 10 surrounding communities with a total population of 400,000. Two principal components of the distribution system are the Barkhamsted and Nepaug Reservoirs which were placed in service in the late 1930s and early 1940s. The Barkhamsted Reservoir transfers water by gravity to the Nepaug Reservoir via a 48-inch reinforced concrete cylinder pipe.

----

[1] Montgomery Watson, 40 Broad Street, Suite 800
Boston, MA 02109
[2] Montgomery Watson, 1300 East 9th Street, Suite 700
Cleveland, OH 44144

During drought conditions, the Nepaug Reservoir empties first to meet downstream system demand. Due to the limited capacity of the 48-inch transmission pipeline from Barkhamsted Reservoir, the MDC elected to incorporate a booster pump station along the transmission main. A schematic representation of the reservoir system and booster pump station location is shown in Figure 1:

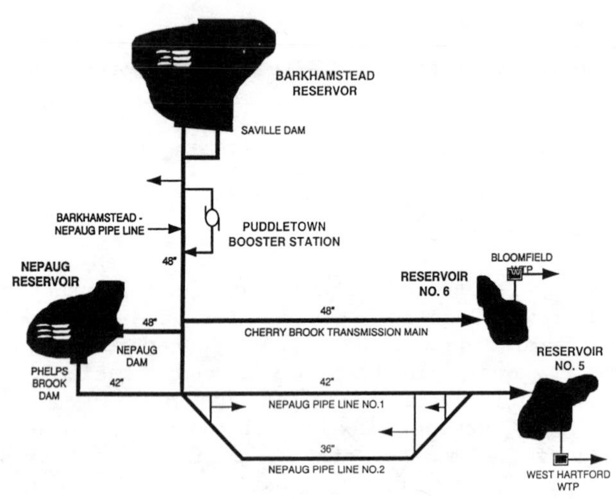

Figure 1. Water Supply System

Design Considerations

To properly design the booster pump station, a thorough understanding of the existing system hydraulics was required. A physical model of the system which incorporated pipe lengths, diameters, venturi meters, fittings, valves and associated elevations was configured using available plans and data supplied by MDC.

The models selected for system analysis were the University of Kentucky KYPIPE and Surge4. The KYPIPE model utilizes the Hardy Cross method to solve system flow, while the Surge4 model utilizes the wave theorem to determine transient conditions within the pipe system.

Once the physical model was configured, MDC staff measured actual flowrates at varying reservoir elevations to develop a data base for use in the model. To calibrate the model, assumed Hazen Williams "C" values were adjusted to fit each pipeline to the measured data. After the model was calibrated, operating curves depicting total dynamic head vs. flowrate were generated for varying operating elevations of each reservoir. These curves were then used to determine the proper operating envelope for the booster pump station.

Another consideration critical to the design was accurate determination of existing pipeline pressure ratings. The maximum hydraulic grade line of the 48-inch transmission main is set by the Barkhamsted Reservoir. Along the pipeline, as the difference between the hydraulic grade line and the pipe elevation increases, the pipe design pressure rating increases. A given pipeline section is constructed from one of twelve pressure classes, ranging from 175 to 278 feet. Using AWWA standard methods, the theoretical working and surge pressures of the pipelines were calculated and compared to the rated pressures along the pipeline. This information provided a design parameter to use in assuring that the booster pump station was designed to prevent over-pressurization of the existing pipeline. Once this was determined, an analysis was performed to identify highest transient pressures caused by the operation of the booster pump station based on various operating scenarios. This procedure was used to determine the optimum control valve closure times for normal pump startup, shutdown and emergency shutdown.

The results of the modeling study indicated that the system was hydraulically limited by suction pressure when the Barkhamsted Reservoir was low and by discharge pressure when Nepaug Reservoir was high.

Table 1 summarizes key design features of the pump:

Table 1

Pump Type	Horizontal Split Case
Drive Type	Direct
Speed (at design load)	509 rpm
Design Capacity	41,700 gpm @ 125 feet
Secondary Rating Point	51,290 gpm @ 95 feet
Suction Diameter	36 inches
Discharge Diameter	30 inches

Construction

During the public participation phase of the project design, the two major concerns raised by local residents were the aesthetics and noise generation of the facility. In view of these concerns, the booster pump station was constructed as an underground facility to mitigate these impacts.

In order to connect the pump station piping to the existing 48-inch transmission main, the MDC shut the transmission main down for a period of one week to allow the contractor to perform his work. The contractor exposed sections of the pipeline where the connections to the booster station piping were to be located and constructed steel H-pile cradles to provide support for the new piping into the station.

The steel cylinder, reinforcing steel and concrete from the pipeline sections that were removed from the transmission main to accommodate the booster station fittings were found to be in excellent condition, both inside and out, showing little signs of corrosion. A core sample was removed and tested per ASTM standards and was found to have a compressive strength of approximately 8000 psi.

Upon completion of construction, field testing data recorded during initial startup and shutdown of the booster station revealed that the actual system transient pressures correlated closely to the transient pressures modeled by the computer analyses.

First Steps in Aqueduct Leak Risk Assessment

David L. Connelly[1] , M. ASCE and Roy J. Perry [2]

Abstract

This paper presents a summary of the findings of the first phase of the Hultman Aqueduct Leak Repair Investigation. Included are the major observations and key recommendations resulting from the analysis. In future phases of work remote leak sensing coupled with utilization of formal risk assessment techniques may be used to properly put into perspective uncertainties, detection costs and costs of failure.

Introduction

The Hultman Aqueduct is the primary conduit conveying water to the Boston metropolitan area from the Quabbin and Wachusett supply reservoirs. Completed in 1942, just before World War II, the Hultman is approximately 28 kilometers (17.3 miles) in length and is 3.5 meters (11.5 feet) in diameter for most of that length with a 2.9 kilometer (1.8 mile) section, having a diameter of 3.8 meters (12.5 feet). In 1991, the Hultman carried an average of nearly 11.2 m^3/s (255 million gallons per day) and has a capacity in excess of 16.4 m^3/s (375-mgd) at a pressure of about 710kPA (103 pounds per square inch).

The Hultman supplies about 85 percent of the demand for an area of more than two million people. No backup for the Hultman will exist until the completion of the Metro West Tunnel early next century, making it impossible to take the aqueduct out of service for maintenance and inspection. Because of its critical role in the area's

----

1 Vice President, Montgomery Watson, 40 Broad Street, Boston, Massachusetts 02109

2 Project Manager, Waterworks Maintenance Engineering Department, Massachusetts Water Resources Authority, 100 First Avenue, Boston, Massachusetts 02129

water supply system, there is concern about the possibility of a failure of the Hultman and the resulting devastating impacts, i.e., a water shortage for days to weeks and economic losses estimated at $60 to $100 million per day.

Twenty-one leaks have been documented along the near surface segments of the Hultman dating back to the 1950's. Since the completion of the report, three additional leaks have been found. While the leaks have been monitored and some steps have been taken to minimize impacts of water leaking from the pipe, there has been minimal formal risk assessment of attempting interim leak repairs versus "doing nothing" until the aqueduct can be removed from service for repairs.

Description of Study

In 1992, Havens and Emerson, Inc., (now Montgomery Watson) was retained by the Massachusetts Water Resources Authority to conduct an initial investigation of the leaks and repair options. The purpose of the study was to collect existing data regarding the leaks, determine what additional data was needed, and observe trends related to physical conditions, change in leakage rates or other criteria.

The focus of the assignment was to identify those factors that may have an impact on leakage and present existing data relative to those factors. The intent was to identify conditions shared by each known leak location that would facilitate a prediction of failure conditions and planning of repairs. Broad categories of information to compare at each leak location included material design, installation condition, and external impacts. For example, external impacts include water pressure/surge, traffic loading, electrical currents, temperature cycles and seismic events. For each broad category, several specific factors were analyzed and addressed along with review of potential repair methods and the risks perceived to be associated with each option.

Observations

The most significant general observation is that over the 50 year history of the aqueduct service there have been no major failures and relatively few reported leaks. The hydrostatic testing upon completion of construction allowed leakage of up to 95 liters (25 gallons) per joint per day at a design operation pressure of 23 meters (75 feet). It is reasonable to assume some increase in leakage over 50 years, although the normal operating pressure of the aqueduct is below the design head for most of its length.

The pipeline design and construction appears to have been very conservative and to high standards. An exposed section of the aqueduct showed the material to be of high quality, the joint mortar to be intact and the bedding conditions (concrete cradle) and backfill to be consistent with good practice.

- The factors which seem to be most likely to result in leakage are as follows:

- Lead gaskets do not have the property of expanding or rebounding in response to joint movement.

- Corrosion of gaskets as a result of chemical or electrical conditions at the joints. This is most likely in locations where protective mortar at joint is cracked or spalled.

- Pipe movement is most likely to occur as a result of poor foundation conditions, laying in a downhill direction, temperature variations, external loads (construction, traffic, seismic), or surge conditions.

- Possible insufficient hand or pneumatic caulking of the lead gasket from within the pipe where direct overhead work was performed.

- Connection of appurtenances to the main line increases the chance for developing a leak due to imperfect welds or differential settlement.

- The application of mortar to the inside upper half of the pipe is a procedure which is difficult to accomplish successfully.

Recommendations

Based on a review of data and the observations described, several recommendations were outlined. These included:

- Conduct an infrared thermography survey of the entire route. This will establish better knowledge of known leaks and identify other potential or existing problem areas. There may be significant leakage occurring at this time which is not visible at the surface due to soil type, groundwater elevation or topography.

- Implement a more vigorous program to identify new leaks and/or changes in appearance along the route.

- Implement a thorough and consistent leak inspection program. Inspections should be made at the same time each month and the leak locations should be physically marked in the field.

- Develop plans to repair the highest priority leaks. Such plans would include a detailed investigation of leak source and quantity, and a repair risk analysis.

- At several locations along the route both near reported leaks and elsewhere, samples of groundwater should be analyzed for fluoride and corrosivity and should be monitored for changes in these characteristics.

- On an annual basis the top of pipe elevations at selected sites to be determined in Phase II should be recorded to monitor settlement.

- A stray current analysis should be undertaken at those points where nearby gas lines are protected with impressed current systems. Several other random locations should be checked also for comparison.

- An analysis of hydraulic transients should be completed for the Hultman, based on the most accurate data available regarding operating procedures.

Conclusion

Because of the high cost of failure, the known leaks should not be ignored. Plans should be developed to repair those which are apparently leaking at the highest rate. Such plans should include a risk analysis which could result in the decision not to pursue repairs unless a leak continues to worsen. Simultaneously with the development of plans to repair the worst leaks, further steps should be taken to monitor the other leaks and the rest of the pipeline to make sure that worsening conditions or new leaks are identified early.

Since the completion of phase I the MWRA has made plans for the Phase II investigation which is anticipated to be underway in February, 1995.

Appendix. References:

Havens and Emerson, Inc., "Hultman Aqueduct Leak Repair Investigation, Phase I" report to Massachusetts Water Resources Authority, December, 1992.

RECENT ADVANCES IN WATER TRANSMISSION AND DISTRIBUTION PIPE MATERIALS AND DESIGN

By

Jey K. Jeyapalan, Ph. D., P.E.[1]

Abstract

The pipe materials used for water transmission are welded steel, ductile iron, pvc, prestressed concrete, pretensioned concrete, and fiberglass. For water distribution, steel, ductile iron, pvc, and hdpe are commonly used. There have been significant advances in the last 20 years in the manufacturing, design, and construction of water projects with these materials. More and more plastics have been introduced at the expense of traditional materials all to provide the user with lower cost materials. The design practice also has changed and although many consensus standards are available within the framework of AWWA to cover the use of above materials, engineers have turned to design tools which are more detailed than these bare minimum standards to complete the design. In underground pipe design, one of the key input parameters controlling the design of flexible pipe materials is the modulus of soil reaction, E', which is not well-understood within the design community. In most cases, this property is generated from information furnished by the pipe manufacturer without adequate attention to the limitations of its use. Failures of numerous miles of water pipe have also been a common occurrence due to the above problems and in some cases led to more stringent AWWA and ASTM standards, for example in the case of prestressed concrete cylinder pipe.

The purpose of this paper is to trace the significant advances which have taken place in these areas, to highlight strengths and weaknesses of each pipe material, to indicate areas which need further improvement and to provide good design guidance.

Engineering Consultant, American Ventures, Inc., 2320 85th Place, NE, Bellevue, WA, USA 98004; Ph 206-462-6261; Fax 206-462-6316; email JEYJEY@AOL.COM

EVALUATION OF PIPE MATERIALS

In water transmission and distribution projects, the primary pipe material used today is welded steel. Ductile iron pipe is used in sizes up to 60 inches. PVC and HDPE are used mostly in the size range of up to 12 or 15 inches. Rarely fiberglass pipe is used due to its history of having problems. Although concrete pressure pipe is a viable material, due to several failures of prestressed concrete pipe, most agencies are not using this type of pipe or other forms of concrete pressure pipe in their projects. Each pipe material has its place in the market. Welded steel comes in a wide range of yield strengths and for projects involving high pressures with little external soil or traffic loads, the welded steel pipe made of high strength steel proves to be best choice provided proper corrosion monitoring and protection systems, coatings, and linings are used. Ductile iron pipe on the other hand sometimes offers a cost advantage over steel pipe in smaller pipe sizes.

o welded steel(WSP)
o ductile iron(DIP)
o concrete pressure(CP)
o fiberglass(FRP)
o high density polyethylene(HPE)
o polyvinyl chloride(PVC)

The selection parameters one should consider in evaluating the above materials are as follows with the desirability of each material:

	WSP	DIP	HPE	FRP	CP	PVC
Corrosion resistance	G	G	E	E	P	E
Handling & Construction	E	E	P	G	P	P
Joints	E	G	G	G	G	G
Quality of Pipe	E	G	G	G	P	G
Flow Characteristics	E	G	G	E	G	E
Structural Capacity	E	E	G	P	G	P

P-Poor, G-Good, E-Excellent, L-Low, H-High

CHARACTERIZATION OF SOIL STIFFNESS

The design engineer should search and locate past geotechnical data in the vicinity of the planned project. All information from regional maps, geological databases, groundwater maps for the area, borings done on projects of the area, all should be reviewed very carefully to plan the locations and depths of new borings. Some borings would be for the purpose of verifying information which were found in past subsurface investigations while other would shed new light on the ground conditions. The methods of subsurface exploration can be a combination of drilling, bore hole testing, laboratory testing, in-situ testing, and geophysical methods. And the following types of data are essential for the success of pipeline projects:

o grain size characteristics

o Atterberg limits
o unified soil classification
o groundwater-shortterm & historical
o permeability-field/lab
o unit weight, moisture content
o blow count
o unconfined compressive strength
o direct shear or triaxial strength
o soil/water contamination levels
o consolidation and compaction properties

Depending on the size, complexity, and financial exposure of the project, some amount of cone penetration testing, seismic and/or ground penetrating radar mapping of the ground conditions are to be part of the subsurface exploration program. Regardless of the size of the budget expended, the extent of the subsurface exploration, nothing could become a substitute for the experience and judgement of a well-qualified pipeline design engineer with a sound background in geotechnical and structural engineering.

The E' used in flexible pipe designs is not a fundamental geotechnical engineering property of the soil. This property cannot be measured either in the laboratory or in the field. This is an empirical soil-pipe system parameter which could be obtained only from back-calculating by knowing the values of other parameters in the modified Iowa equation. An experienced soil-pipe interaction design engineer would expect the pipe-soil stiffness ratio to have an effect on the value one uses for E' in design. It is interesting to note that the range of E' used for stiff ductile iron pipe ranges from 1 MN/m^2(150 psi) to 5 MN/m^2(700 psi), while much softer plastic pipe is designed with values in the range of 7 MN/m^2(1,000 psi) to 20 MN/m^2(3,000 psi).

The value of E' chosen will affect the overall design of the pipeline project and the cost of construction in a major way. Thus, it is important to recognize that the E' is controlled by the following factors:

o depth of soil over the pipe
o size of pipe
o stiffness of the pipe relative to the soil
o trench width, location the water table
o native soil type, compaction density, modulus
o bedding soil type, compaction density, modulus

The following steps are helpful in establishing reasonable E' values:

Step 1: Review soil borings, plans, profiles, and obtain most appropriate blow counts.
Step 2: Calculate total stress, pore water pressure, effective stress.

Step 3: Calculate relative density with some judgement.
Step 4: Estimate Standard Proctor relative compaction density.
Step 5: Select E' from Duncan & Hartley charts (1983).
Step 6: Adjust E' to allow for variations in Pipe-soil stiffness ratio, size, and other factors.
Step 7: Repeat the procedure for Bedding soil to obtain its E'.
Step 8: Select trial trenchwidth.
Step 9: Calculate E' correction factor allowing for its variation from native to bedding.
Step 10: Estimate deign E' for the pipe-soil system and adjust up or down with some judgement.

When fine-grained soils or mixed soils are encountered, the above procedures have to be adjusted based on sound geotechnical input for the area.

DESIGN METHODS

The structural design of the pipe should never be analyzed without proper engineering considerations for its interaction with the soils surrounding the pipe. Also, the pipe-soil system has to be designed to carry design loadings during construction phase and in-service phase. Design criteria should include checking for handling stiffness, internal pressure, bending from external loads, deflection from external loads, combined stresses/strains from internal and external loads, and buckling. Most AWWA standards do not check for all of these with the exception of C-950 for fiberglass pipe. In Europe, design standards are based on engineering concepts and all pipe material designs are usually done within a common standard. If the design engineers were to focus their attention on material strengths and compare these with computed load-induced stresses regardless of the pipe material, most AWWA standards for water pipe could be written into one common standard to avoid confusion among designers.

COATINGS & LININGS

In order to protect metallic pipe materials from internal and external corrosion, the industry has made significant progress in this field. Coal tar epoxy, fusion bonded epoxy, cement mortar, polyethylene tape, and a combination of two or more of these are used today around the world for welded steel pipe. Ductile iron pipe also is designed with some of these protection systems. In most cases, corrosion monitoring and cathodic protection systems are used to ensure that the pipe lasts for its intended service life.

CONCLUSION

If proper procedures are followed for E' interpretation and for the design of soil-pipe systems, significant cost savings could result on water transmission and distribution projects.

Planning for Water Treatment for the Boston Region

Jae R. Kim[1]

Abstract

The Boston region has always had excellent source water supplies. For many years water was delivered to consumers with very little treatment, but the 1986 Amendments to the Safe Drinking Water Act are about to change that. The amendments require the U.S. Environmental Protection Agency to establish water quality standards and water treatment techniques. Several rules have already been promulgated. The draft Disinfectants and Disinfection By Products Rule Stage I and the draft Enhanced Surface Water Treatment Rule have also been published.

Stronger disinfection treatment and additional corrosion control treatment will be necessary to meet the requirements of Surface Water Treatment Rule and the Lead and Copper Rule. Filtration may be required to meet the requirements of Surface Water Treatment Rule, Disinfectants and Disinfection By Products Rule, and Enhanced Surface Water Treatment Rule. However treating source water is only part of the answer. The source water needs to be protected through increased watershed protection to preserve the pristine quality. The treated water quality also needs to be preserved in the transmission and distribution system. This requires providing properly designed enclosed distribution storage reservoirs (the Boston region currently has most of its distribution storage in open reservoirs), and improvements, such as cleaning and cement mortar lining of mains in the distribution system.

The paper describes current water treatment practices, various water treatment processes piloted and tested, and other planning efforts carried out to-date. It also describes what water treatment and storage facilities may be completed by early in the 21st century to comply with the requirements of the above referenced regulations.

----

[1]Manager, Water Quality & Facilities, Waterworks Division Massachusetts Water Resources Authority, Boston, MA 02129

Introduction

The Massachusetts Water Resources Authority (MWRA) supplies water to 47 communities mainly in the metropolitan Boston area. The main sources of water supply are Quabbin Reservoir and Wachusett Reservoir located 105 kilometers and 50 kilometers to the west of the city, respectively. The Wachusett Reservoir is the oldest active water source in the MWRA water supply system. The Wachusett Reservoir serves as the terminal source reservoir prior to distribution of water to the MWRA communities in eastern Massachusetts including the City of Boston. Historically, water withdrawn from the Wachusett Reservoir has been of high quality, requiring only minimal treatment to meet the applicable federal and state water quality standards.

The 1986 Amendments to the Safe Drinking Water Act (SDWA) require the U.S. Environmental Protection Agency to establish water quality standards and water treatment techniques. The Surface Water Treatment Rule (SWTR), Total Coliform Rule (TCR), and Lead and Copper Rule (LCR) have already been promulgated. Draft Disinfectants and Disinfection By Products Rule (DDBPR) Stage I has also been published. Information collected as part of Information Collection Rule (ICR) will set the basis for the Enhanced Surface Water Treatment Rule (ESWTR) and the Disinfectants and Disinfection By Products Rule Stage II which are scheduled to be promulgated over the next several years. These rules will require MWRA to provide additional treatment before Wachusett Reservoir water is delivered to consumers.

Existing Water Quality

Wachusett Reservoir water is of very high quality. Its turbidity rarely exceeds 0.5 NTU and color is low. The natural organic matter content is low, indicating high quality water source. It has low alkalinity and hardness, typical of New England surface waters.

Turbidity(NTU)	0.2 - 0.6
Color(SU)	6 - 17
Alkalinity(mg/l as CaCO3)	5 - 13
Calcium Hardness(mg/l as CaCO3)	9 - 14
pH	6.1 - 7.7
Total Organic Carbon(mg/l)	2.3 - 4.6

Existing Water Treatment

Presently MWRA water is fluoridated, pH adjusted with addition of Sodium Hydroxide (NaOH), and disinfected with chlorine and ammonia (chloramination). The water is rechlorinated and pH is readjusted down stream of open distribution reservoirs within the metropolitan Boston area. No other treatment is currently provided.

Water Treatment Process Evaluation

Studies were carried out to determine the best way to come into compliance with the SDWA regulations for Wachusett Reservoir water. In particular, the purpose of the studies was to determine the best way to comply with SWTR, LCR, DDBPR and ESWTR in the most cost effective manner. Tests were conducted to determine most suitable filtration, disinfection and corrosion control treatment processes.

The studies were prepared as part of the MWRA's dual-track approach to comply with the SDWA requirements. One track includes further study and the implementation of recommendations from the Wachusett Watershed Resources Protection Plan, such that non-filtration options might be viable for MWRA to be in compliance with the law. The second track consists of MWRA proceeding with the planning and design for a filtration plant. The dual-track approach is permitted under a Consent Order signed by Massachusetts Department of Environmental Protection, Metropolitan District Commission and MWRA.

The treatment requirement for the Wachusett Reservoir water, under the new regulations, arises from the following factors:

- The inability of the existing disinfection treatment to inactivate giardia cysts and cryptosporidium oocysts.
- The level of nutrient/organic contents which contribute to taste and odor problems and, more importantly, to the level of disinfection by-products. If disinfection were to be increased sufficiently, or modified to inactivate giardia cysts, cryptosporidium oocysts and other potential pathogens as required by the SWTR and ESWTR, the disinfection by-product level would be elevated to levels which increase carcinogenic risks and other undesirable impacts. These are expected to violate the Disinfectants and Disinfection By-products Rule.
- The low pH and low alkalinity of Wachusett Reservoir water which makes it difficult to meet the requirements of the Lead and Copper Rule.

Treatment processes which are known to be effective in removing organics, reducing disinfectant dosage, reducing disinfection by-products, reducing corrosivity and reducing taste and odor were tested. The ability to meet the requirements of the existing regulations and the anticipated future regulations was used in evaluating the treatment process selection. Also included in the evaluation criteria were their ability to deal with taste and odor issues as well as its operational flexibility to deal with changing raw water qualities and its potential impact on the distribution system water quality.

The two year study of the Pilot Treatment Program indicated that among the feasible treatment processes which were tested at the pilot plant, the treatment train consisting of dissolved air flotation, ozonation and biological activated carbon filtration provides the most appropriate treatment combination for the Wachusett Reservoir water to meet current and future rules at reasonable cost. The study also concluded that use of disinfection only, using chlorine or ozone (non-filtration alternative), may be a viable option. This would be dependent on the MWRA obtaining a waiver from filtration and the source water being further improved to lower the level of disinfection by-product precursors substantially. The study also concluded that using lime and carbon dioxide (CO_2) for alkalinity and pH adjustment and zinc or non-zinc orthophosphate as an inhibitor would provide an optimum corrosion control treatment for the MWRA source water.

Open Distribution Reservoirs

The MWRA water distribution system currently has four active open distribution reservoirs in the metropolitan Boston area: Norumbega and Weston reservoirs in the Town of Weston and Spot Pond and Fells reservoirs in the Town of Stoneham. Planning and design phases are currently under way to take the open distribution reservoirs off line from the water system and provide enclosed distribution storage facilities.

Schedule

The following summarizes current status of projects which are required for MWRA to be in compliance with the 1986 Amendment to the Safe Drinking Water Act and their required completion schedule;

Wachusett Water Treatment Plant - Preliminary Design and EIR phase. Plant is to be constructed by December 2001.

Norumbega Reservoir - Preliminary Design and EIR phase. Enclosed storage is to be constructed by December 2000.

Weston Reservoir - Final Design phase. The reservoir is to be taken off line by December 2001 when MetroWest Water Supply Tunnel is constructed.

Fells Reservoir - Final Design phase. New enclosed reservoir is to be constructed by December 1998.

Spot Pond Reservoir - Construction Phase. The reservoir is to be taken off line by November 1997 when a new pumping station and an associated pipeline are constructed.

To Filter or Not To Filter:
Boston's Two Track Approach to Compliance With the SDWA

Stephen Estes-Smargiassi[1]

Abstract

The Massachusetts Water Resources Authority has undertaken a two track approach to compliance with the Surface Water Treatment Rule under the federal Safe Drinking Water Act. While the agency has been ordered to filter the water leaving its Wachusett Reservoir toward the Boston metropolitan area, it has negotiated a consent order which allows watershed protection to be improved with a second chance to apply for a filtration waiver in 1998 before construction of the filtration plant is to begin.

Introduction

Since early in 1987, the Massachusetts Water Resources Authority, serving over 2 million people in the Boston metropolitan area, has been working towards compliance with the 1986 Amendments to the Safe Drinking Water Act. While early indications were that the 1.1 million m^3 (300 million gallons) per day currently unfiltered Quabbin/Ware/Wachusett Reservoir system might require filtration (CH_2MHill, 1989), the agency has maintained a two track approach: improving protection in the 103,600 hectare (400 square mile) watershed to possibly qualify for a waiver of filtration, and proceeding expeditiously through the piloting, siting and design process to provide filtration if it is required.

Filtration is a sure route to compliance, but will cost about $385 million. A non-filtration approach might save $100 to $200 million, but may not yield as good a

---

[1]Manager of Waterworks Planning, Massachusetts Water Resources Authority, Charlestown Navy Yard, 100 First Avenue, Boston, MA 02129

quality of water, and may not provide long term compliance. Is the improved protection from waterborne disease and taste and odor complaints worth the additional cost? How much risk of making the wrong choice are decision makers willing to take?

To date, it is clear that improved attention to protection can yield dramatic improvements in some measures of water quality, that the costs of keeping options open are small compared to the benefits of having more information and a chance to try things out, and that a prolonged decision making process may yield more educated support for the ultimate decision. As of early 1995, improved protection efforts have succeeded in securing a waiver of filtration for the communities served directly from the 1.6 billion m3 (412 billion gallon) Quabbin Reservoir, and while the MWRA has been ordered to build a 1.7 million m3 (450 mgd) filtration plant for the Boston metropolitan area served by both the Quabbin and Wachusett Reservoirs, it has secured a unique opportunity to have the regulatory agencies re-examine the filtration decision just prior to construction in 1998.

Evaluation Process

Shortly after the SDWA was amended in 1986, the MWRA began a study process to determine what the likely impacts of the amendments would be on the agency's water supply system. Regulatory review, water quality sampling, a pilot scale evaluation of disinfection and filtration technologies, and recommendations for action were within the selected consultant's scope. Early on, it was clear that substantial changes would be needed in treatment including improved disinfection to meet the Ct values, changes in corrosion control to reduce lead levels, covering of open distribution storage, and very likely filtration. The pilot plant results indicated that an innovative approach using ozonation, dissolved air floatation and deep bed biologically active carbon filters, along with changes in pH and alkalinity, would meet the regulatory requirements, but at a substantial cost.

The Surface Water Treatment Rule essentially requires that all surface water supplies be filtered, unless seven criteria can be met (EPA, 1987):

* an effective watershed control program;
* excellent source water quality as measured by fecal and total coliform, and turbidity;
* effective disinfection processes and residual disinfectant in the distribution system;

* no waterborne disease outbreaks;
* annual sanitary surveys;
* distribution system water in compliance with limits on total coliform and THM's; and an
* on-going water quality monitoring program.

The project team felt that several of the criteria were being met, and could continue to be met, but that the watershed control, disinfection byproducts, and source water quality criteria would require substantial improvement to qualify for a waiver.

Initial Two Track Decision

In April of 1989, staff recommended to the Board of Directors that a two track approach be undertaken: staff would simultaneously pursue improved watershed protection and the technical studies necessary for both filtration and non-filtration water treatment plant design (MWRA, 1989). Using the two tracks, would allow the agency to maximize its flexibility without falling behind in compliance planning - whatever the eventual decision, no time would have been wasted. The Board concurred, and work was begun on improved watershed protection plans for the two reservoir systems.

Over the next two years, separate teams of staff worked to move both the treatment and protection tracks forward, and by January 1991 had new materials to bring back to the Board of Directors. Protection Plans (MWRA, 1991) had been completed which called for additional staff resources for the Division of Watershed Management, passage of state legislation limiting development near tributaries, increased land acquisition, remediation of failing septic systems, control of gull flocks near the intake, and several dozen other measures. However, the on-going sampling results indicated that the Wachusett Reservoir, serving 95 percent of the service area, was grossly failing the source water quality criteria. Filtration for that source seemed inevitable, and the Board of Directors choose not to formally request the waiver. The decision seemed to turn on a Board member's statement "If we're spending $4 billion making the harbor safe for fish, why shouldn't we spend $400 million making the water safe to drink for people," and the very high bacteria levels in the reservoir, exceeding even safe swimming criteria.

The Question Reopened

The MWRA believes that watershed protection will always be the first and most important barrier in providing safe drinking water, and so while engineering studies continued, so did implementation of the improved protection plans. Water quality improved, largely due in the short term to harassing flocks of 5000 to 10,000 gulls away from the intake to less sensitive areas. When the initial 18 month period to construct filtration facilities had passed by in mid 1993, and as expected, studies were not yet complete, let alone the plant, the MA Department of Environmental Protection required a binding schedule for design and construction be incorporated into a consent order. Buoyed by the initial improvements in water quality and implementation of protection measures, including the state legislation, the MWRA negotiated to reopen the filtration question.

The DEP would not issue a waiver then, and required a schedule for siting and design of a filtration plant with construction starting in 1999, but was convinced to include provisions in the consent order requiring the official submittal of the protection plans, implementation on an agreed upon schedule, and a date to re-evaluate the evidence - 1998. If by then, the MWRA can demonstrate compliance with all seven criteria, a smaller water treatment may be adequate; if not, then construction of the filtration plant must begin.

Current Status

As of January 1995, source water quality has met the criteria for 10 consecutive 6 month periods, and much of the watershed protection program is being implemented. The two track approach has cost about an additional $1 million so far, and as design for both filtration and non-filtration is undertaken, will cost an additional several million dollars. Changes in regulations or public pressure may ultimately require filtration, but a this relatively low cost, the agency's option have been kept open an additional 5 years. The success of the two track approach to filtration has led the MWRA to look at multi-track approaches to solve other complex regulatory and technical compliance problems.

CH_2MHill (1989). Safe Drinking Water Act Impact Study (for the MWRA), Boston, MA

EPA (1987). Guidance Manual for Compliance with Surface Water Treatment Requirements, Washington DC

MWRA (1989). MWRA Safe Drinking Water Act Compliance Plan, Boston MA

MWRA (1991). Watershed Protection Plans, Boston, MA

2_track.pap

Planning and Design of the MetroWest Water Supply Tunnel for the Boston Region

James Powers[1]

Abstract

Over 85 percent of water supply to the Boston Metropolitan area flows through the High Pressure Aqueduct System. Along a 28km distance, this flow is solely carried by the Hultman Aqueduct. The Hultman was completed in 1941 and planned as a double-barrel surface aqueduct, however only the first barrel was constructed. Over-reliance on this aging facility has placed the system in a precarious condition. Rather than build a second surface aqueduct, a 4.25m diameter deep rock tunnel has been designed which will be constructed from 1996 through 2002.

Introduction

The metropolitan Boston water system serves 47 municipalities with a population of about 2.5 million. Thirty-seven of these, with a population of 2.1 million, surround Boston; the remainder are in the watershed areas. The system wide safe yield of the active supply sources is 13,000 l/s and the Massachusetts Water Resources Authority (the Authority or MWRA) intends to keep average demand below the safe yield into the indefinite future. All transmission facilities are being designed accordingly.

In 1988, Anderson-Nichols & Co. examined the reliability of the MWRA transmission system. The study concluded that the Hultman Aqueduct was the weakest link and its loss would have the most dire consequences to water supply. The study confirmed

----

[1]Manager, Transmission and Distribution Section, Waterworks Division Massachusetts Water Resources Authority, Boston, MA 02129

the validity of the 1936 Master Plan, and recommended rapid construction of the second barrel for the Hultman Aqueduct.

The Anderson-Nichols study examined ten alternative means for carrying out the basic intent of the 1936 plan. They examined the four aqueducts that have been built to convey water to the metropolitan area, as shown below:

Aqueduct	Date Comp.	Length (km)	Diameter (m)	Capacity (l/s)	Grade Line(1) BCBD	Status
Cochituate	1848	24	1.5	1,700	40.9	abandoned
Sudbury	1878	28	2.3	6,000	40.9	standby
Weston	1903	19	3.5	20,150	61	active
Hultman	1941	28	3.5	21,850	83.7	active

(1) Boston City Base Datum (mean low water) Figure 1

Because there was serious concern that failure of one barrel of the Hultman could cause a wash-out which would undermine a second barrel, the A-N study determined that either of two alternatives with equivalent transmission capacity were preferable:

- A new tunnel constructed parallel to the Hultman Aqueduct.
- Reconstruction of the existing Sudbury and Weston Aqueducts as pressure conduits.

Of the recommended alternatives, reconstruction of the older aqueducts has the advantage of incremental construction - but the disadvantage of severe environmental impact associated with linear surface construction. Whereas a sub-surface tunnel could be constructed from a small number of shafts located in areas where environmental impacts could be limited.

A series of public hearings, held early in the environmental review process, indicated a clear preference for the tunnel alternative. Subsequent study of means to bring the water system into compliance with the 1986 Amendments to the Safe Drinking Water Act also showed clear advantages for the tunnel. In particular, the fact that for a small incremental cost, the diameter, and hence capacity, of the tunnel could be increased was a major advantage. This would limit the amount of expensive covered distribution storage needed, would permit this storage to be constructed in an incremental fashion, and would permit the retirement from active service of facilities not in compliance with SDWA.

<u>MetroWest Tunnel</u>

Sverdrup Civil, Inc. was retained to lead a team of firms to design the new tunnel. The MetroWest Tunnel will extend 29km from the terminus of the Cosgrove Tunnel in Marlborough to the start of the City Tunnel in Weston and to the Weston Aqueduct Terminal Chamber, as shown on Figure 2. It will have a capacity of 20,000 l/s, or slightly more than maximum day demand.

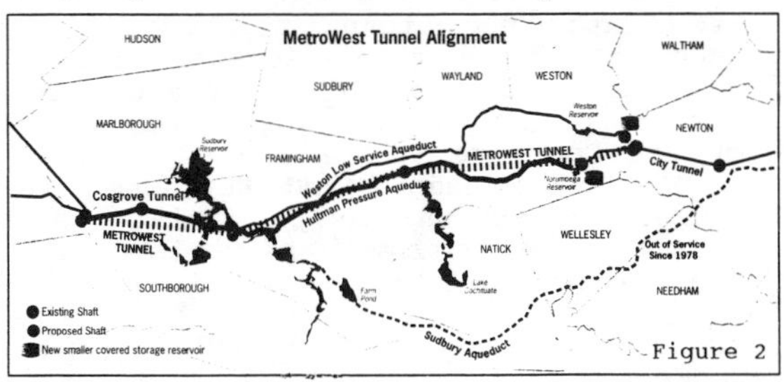

Figure 2

It is to be a deep-rock tunnel, located from 60 to slightly more than 125m below existing ground surface. It is to be capable of operating at the full head of Wachusett Reservoir (120m B.C.B.D.), with pressure generally contained by the weight and strength of the rock through which is passes. Although it traverses some of the most complex geologic formations in the Northeast, the rock is generally hard and of good quality for tunneling. With compressive strengths of up to 250 pascals (36 kips), this will be some of the hardest rock ever excavated by tunnel boring machine. Quartz content of the granite averages 40 percent in the Milford and Dedham granites, raising concerns regarding abrasiveness and, hence, wear of cutter heads.

The western 8km of the tunnel, Contract No. 1, will be advanced from a construction shaft located near the base of Fayville Dam in Southborough. Due to the proximity of existing waterworks structures, concerns about blasting have been raised and appropriate precautions will be undertaken. Inflow to the tunnel during construction must be treated since it will be discharged to a public water supply. Due to the Bloody Bluff fault zone, through which the tunnel will pass, significant water intrusion is anticipated along this stretch. A related concern is that rock wells which supply local residences might be affected by tunnel construction. A contingency plan is being developed to forestall or rapidly react to this situation. Because this construction site is on land owned by the Authority, few local residents are affected by construction activity at the shaft. The Authority originally planned to leave a large percentage of the excavated rock on site, but environmental review led to the decision to truck it off the site to maintain its beauty. Fortunately this site is not far from major highways.

Late into the tunnel design, it was determined that the optimum location for a new water treatment plant would be adjacent to the headworks of the MetroWest Tunnel. The Wachusett Aqueduct and Cosgrove Tunnel end at this point, and the Hultman Aqueduct and MetroWest Tunnel begin here, making this site attractive both hydraulically and from the viewpoint of interconnection to existing facilities.

More than 50 percent of the entire length of the MetroWest tunnel, or 19km, will be constructed from a single shaft. This central portion of the tunnel, designated Contract No. 2, will be constructed from Shaft L, located on the New England Sand and Gravel Property in Framingham, with tunnel headings advanced in both directions. As weight of rock is used to contain water pressure within the tunnel, and as the rock surface dips at Cochituate Valley adjacent to Shaft L, Shaft L is the low-point of the tunnel. Both tunnel headings will be advanced up slope from Shaft L, and water intruding into the tunnel will be pumped from the shaft.

Although the site has been actively used as a sand and gravel quarry for more than 50 years, environmental studies and review indicated numerous measures which were required. Among these were, protection of two white pine forests which buffer the site, an Indian archaeological area, a municipal wellfield, a buffer zone along the Sudbury River, and prevention of a plume of contaminated groundwater from migrating. Over 650,000m^3 of rock excavated at Shaft L will be left on site, forming a hill up to 21m in height.

The eastern portion of Contract No. 2 passes beneath Norumbega Reservoir, the principal balancing reservoir of the high service pressure aqueduct system. A major shaft, designated Shaft

N, will be constructed at the east end of Norumbega Reservoir. Surface connections will be made to the Reservoir, the proposed covered storage reservoir adjacent to Norumbega Reservoir, the Hultman Aqueduct, as well as provisions made for future connections to the planned Northern and Southern Tunnel Loops.

The eastern-most tunnel construction contract, Contract No. 3, consists of only slightly over 1,200m of tunnel. This section traverses the Northern Border Fault, which separates the argillite rock of the Boston Basin from the granite of the Boston-Avalon terrain. This fault zone is complex and poorly understood. There have been a number of exploratory drillings and the Authority has conducted three phases of test borings. Work completed to date has resulted in significant changes to the original plans for the tunnel in this area.

Construction of Contract No. 3 will take place from Shaft 5A adjacent to the existing Shaft 5, where the Hultman Aqueduct ends and the City Tunnel begins. This shaft is located along the bank of the Charles River in the middle of the interchange between the Massachusetts Turnpike (Route I-90) and Route 128 (I-95). The construction shaft itself will be difficult to build as it will pass through over 30m of granular saturated overburden soils and the tunnel will have to be cut through the relatively poor rock quality of the Border Fault. This short tunnel portion, 1,200m in length, will be steel lined. Contract No. 3 also includes construction of a new 75-million liter covered reservoir adjacent to the Weston Aqueduct Terminal Chamber.

Conclusion

Although the MetroWest Tunnel is similar in function to the long-planned second barrel of the Hultman Aqueduct, it is much more. It will form a very strong hydraulic link between the new water treatment plant at its start and the new covered Norumbega Reservoir near its terminus. Taken together, these three closely interrelated projects set hydraulic conditions throughout the transmission system, and form the basis for compliance with the Safe Drinking Water Act. All are anticipated to be on-line by 2002.

The Central Valley Improvement Act

Floyd R. Summers, PE FASCE[1]

Abstract

The federal Central Valley Project (CVP or Project) of California is undergoing substantial statutory, administrative, structural and operational changes to retain fish and wildlife values associated with, and serviced by the Project, and to restore adverse effects of the Project's sixty years of operation. Specific direction and Congressional authority was provided to the Secretary of the Interior by the Central Valley Improvement Act (CVPIA or Act), Section 34 of Public Law 102-575, enacted in 1992.

What is the Central Valley Project?

Extending 400 miles north to south, the Central Valley of California is situated in a basin that is almost completely enveloped by mountain ranges. The valley has two major river systems -- the Sacramento running north to south, and the San Joaquin running south to north. The two streams meet about midway in the valley to form a Delta and then course through natural channels and a series of constructed levies to the Pacific Ocean, at the Golden Gate of San Francisco Bay. The Central Valley Project (CVP) was originally conceived by the State of California to protect the Central Valley from water shortages and floods. The basic concept of the project and many of the initial features were included in the State Water Plan formulated in the 1920's and 30's. However, depression-era economics made it difficult to obtain financing for the massive undertaking, and the plan evolved into State-sponsored Federal authorization, the Rivers and Harbors Act of 1937. The 1937 Act provided that project purposes included river regulation and improvement of flood control, irrigation and domestic uses, and power. Subsequent legislation specifically authorized fish and wildlife and recreation as project purposes. The project later

---

[1]Civil Engineer/ Program Manager, U.S. Bureau of Reclamation, Northern California Area Office, 16349 Shasta Dam Boulevard, Shasta Lake, CA 96019-8400

expanded and evolved to help match the continued growth and developing needs of California.

The CVP consists of some 20 reservoirs with a combined storage capacity of about 12 million acre-feet, 8 powerplants, and 2 pumping-generating plants with a maximum capacity of about 2 million kilowatts, and more that 500 miles of major canals. Most of the project is operated as an integrated system as it was planned and designed to do. CVP water irrigates about 3 million acres of farmland (about 20,000 farms) and provides water to more than 2 million urban residents, in addition to providing water for Federal and State wildlife refuges. Of California's 35 million acre-feet of developed water, about 25 percent comes from the CVP. The CVP supplies about 30 percent of the water used for irrigated agriculture in the state, plus about 13 percent of the urban and industrial water used in California. Urban water deliveries make up about 9 percent of the CVP contractual supply.

Even though most of the Project water supply comes from the Sacramento basin, more than half of it is delivered to agricultural and urban contractors south of the Delta. (Because the CVP and the State Water Project use the Sacramento River and the Delta as common conveyance facilities, reservoir releases and Delta exports must be coordinated to ensure that each project retains its part of the shared water and bears its share of joint obligations to protect in-basin uses, including the water required to maintain Delta water quality standards set by the State Water Resources Control Board. The Coordinated Operations Agreement, which became effective in 1986, defines the rights and responsibilities of the two projects and provides an accounting mechanism to guide operations.

Competing demands for water, prolonged drought, urban growth, and reduced fish and wildlife population are factors that contributed to the passage of the Central Valley Project Improvement Act (CVPIA) on October 1, 1992. This Act, or CVPIA, is Title 34 of P.L. 102-575, and is the most significant legislation affecting the operation of the CVP in over 50 years. It strengthens existing fish and wildlife project purposes by adding fish and wildlife mitigation, protection, and restoration as an authorized purpose of the CVP. Other titles of P.L. 102-575 contain provisions for water resource projects throughout other western states.

What are the Features of the CVPIA?

The Act provides that the use of CVP dams and reservoirs for fish and wildlife purposes is of equal priority with irrigation and municipal and industrial uses. It also revises the statute addressing the renewal of contracts, revises the water pricing system, and expands the scope of water transfers. Reclamation and the U.S. Fish and Wildlife Service are developing interim guidelines for initial efforts to implement Title 34 in the areas of water conservation, water banking, water transfers, contract renewals, funding to non-Federal entities and a restoration fund, and annual water supplies for fish and wildlife. They have also begun work on a plan for doubling the populations of naturally-spawned anadromous fish, primarily four races of Chinook salmon.

Several Major Areas of Change

In 1993 and 1994, significant progress was made on developing administrative procedures for advancing toward, or implementing, on an interim basis:

* Dedicating 800,000 acre-feet of water annually exclusively to fish and wildlife;
* Tierd water pricing applicable to new and renewed contracts;
* Water transfers that under certain conditions will permit sale of CVP water to users outside the CVP service area (such as major municipal and industrial suppliers);
* Reasonable efforts to maintain by 2002 the natural production of anadromous fish at twice the average levels attained during recent years;
* No new water contracts until fish and wildlife goals are achieved; no contract renewals until completion of a Programmatic EIS, and contract durations reduced from 40 to 25 years with subsequent renewal at the discretion of the Secretary of the Interior;
* Numerous specific measures aimed at fish, wildlife, and habitat protection, mitigation, restoration, and enhancement;
* Firm water supplies for Central Valley wildlife refuges;
* Perpetuation of fishery protection measures in the Trinity River (a major tributary of the Klamath River which enter the Pacific Ocean on the north coast of California); and
* Development of a plan to increase the CVP yield.

Most of the CVPIA activities will be funded by a Restoration Fund, acquired from surcharges for water delivered to CVP contractors, and a cost-sharing partnership with the State of California (25 percent for most activities). Some activities already underway will be funded by Congressional Energy and Water Appropriations for Reclamation.

Initial implementation of the CVPIA began in what was the last year of a 6-year drought (1993), which compounded challenges for operating with a minimum amount of water and a limited amount of funds from water users coming into the Restoration Fund. Water year 1994 was somewhat better, and 1995 appears to be at or slightly above a long-term average.

The Act, as expected, has an extraordinary level of interest from many publics, proponents, opponents, and affected stakeholders. As a result, 42 of the 59 public involvement efforts of Reclamation's Mid-Pacific Region in 1993 were directly associated with it.

Highly Visible Activities Underway

Activities underway with a high degree of visibility, most of which were ongoing prior to CVPIA but given new impetus by it, include:

* Shasta Temperature Control Device - Construction is underway on a $63.7 million, 360-feet high by 390-feet wide, and 50-feet deep, steel box structure with 17 adjustable panels, on the upstream face of Shasta Dam, to allow access to deep, cool water for power releases, for salmonid preservation in the upper 60 miles of the Sacramento River;
* Red Bluff Diversion Dam Fish Passage Program - Continuation of evaluation of alternatives to improve upstream and downstream passage for salmon at Red Bluff Diversion Dam, which supplies the Tehama-Colusa Canal (about 3,000 cubic feet per second for 100,000 acres and three wildlife refuges), including completion of construction this year of a 400 cfs Research Pumping Plant to test three low-impact pumps (two 10-foot diameter Archimedes Screw, and one low-speed helical), and provide late spring and early fall water to the canal ($10 million plus);
* Tracy Pumps Mitigation - Develop a mitigation plan for offsetting direct fish losses caused by the Tracy Pumping Plant and Tracy Fish Collection Facility;
* Operation Standards - Operating the CVP to comply with coordinated Operation Agreement provisions, Endangered Species Act requirements, and all flow standards, objectives, and diversion limits required by CVPIA, including pulsed flows, reduced ramp rates, and reservoir carryover storage.
* Delta Barriers and Gates - Installing three seasonal (summer) barriers in Delta channels to direct flows to assist out-migration, operating Delta cross channel gates to enhance direction of smolts through channels and away from pumps, and water-distract testing of acoustical training devices for smolts;
* Screening Diversions - Implementing screening upgrades at the Glenn-Colusa Irrigation District and implementing a Pilot Fish Screen Demonstration Program at other diversions;
* Waterfowl Habitat Incentives Program - Providing funds and water to expand the Sacramento Valley Ricelands/Conjunctive Use Program and other incentives to growers to create winter waterfowl habitat in conjunction with rice stubble decomposition by winter field flooding;
* Trinity River Fishery Flow Evaluation Program - Initiating an Instream Flow Program and associated EIS to determine optimum flow requirements for fish in the Trinity River;
* Investigations - Continuing or initiating investigations for the San Joaquin Basin Resource Recovery Initiative; the Stanislaus River Basin and Calaveras River Water Use Program; temperature control in the Sacramento and San Joaquin Rivers and the Delta; and an evaluation of water augmentation options to offset adverse impacts to users, and to identify potential resources beyond present CVP capability to meet CVPIA needs.

The Status of Water Pollution Problems and Control Initiatives in the Baltics

Igor Runge[1], Erik Petrovskis[2] and Sarma Valtere[3]

Abstract

Years of general disregard of environmental issues has left many of the water resources of the Baltic countries of Latvia and Lithuania in dismal condition. Inadequate sewage treatment, numerous industrial dischargers and poor agricultural practices during the Soviet occupation period have contaminated numerous surface and groundwaters. Especially alarming are the excessive nutrient loads to the Baltic Sea. Financial assistance from abroad will be required to properly address these issues.

Introduction

As part of a National Academy of Sciences (NAS) program to assess water pollution impacts in Eastern Europe, a team of ten engineers and scientists journeyed to the Baltic countries of Latvia and Lithuania during the summer of 1994. The principle task of the mission was to obtain a first hand view of some of the major water pollution issues facing the region, specifically with regard to non point source pollution and agricultural runoff. It is hoped that the future may unfold collaborative research efforts between the U.S. and these Baltic countries.

The Baltic Sea is the major water resource in the region with nearly 80 million people residing in its catchment area. The environmental condition of the Baltic is of paramount concern to the surrounding countries which rely on it for fishing and tourism activity. Environmental pollution of the vulnerable Baltic has reached threatening levels, partly due to long (25-50 years) residence times. Significant

---

[1]Assistant Professor, Dept. of Civil Engineering, Univ. of North Carolina at Charlotte, Charlotte, NC 28223

[2]Ph.D. Candidate, Environmental and Water Resources Engineering, Univ. of Michigan, Ann Arbor, MI 48109

[3]Dean of Postgraduate Education Faculty, Riga Technical University, Riga, Latvia

surface water contributions include inadequately treated sewage, industrial discharges of heavy metals and toxics, and agricultural runoff. Additionally, groundwater resources in the Baltics are threatened by contamination from landfills, petroleum processing and transfer station operations and abandoned former Soviet military bases. During the NAS program, we had the opportunity to assess the impacts of some of these environmental problems.

Surface Water Impacts

Of fundamental concern is the treatment of residential wastewater. Many of the more rural areas of both countries are void of plumbing and rely on the pit privy for personal needs. Their use has been correlated to limited groundwater contamination in some regions. By far the more important issues are the cities. For example, the Riga (Latvia) waste water treatment facility (WWTF), with a design capacity of approximately 400 cubic meters per day, currently experiences flows only half that amount. The facility was completed and became operational in late 1991 and provides tertiary treatment and sludge digestion for approximately half of the population of Riga (1 million inhabitants) due primarily to a lack of pipe network and associated infrastructure. For the remaining areas of Riga, including historic old Riga, sewage is discharged directly into the Daugava River, after mechanical treatment. Estimates show this to be a significant source of bacteria and nutrients to the Daugava, as well as the Baltic Sea. The management of this relatively new WWTF has expressed many concerns. These include fluctuation of final effluent quality due to seasonal temperature swings; ineffective flow equalization; limited pretreatment practices by the industrial dischargers which occasionally introduce significant quantities of toxins that can upset the system; disposal of sludge contaminated with heavy metals; frequent equipment breakdowns; and during wet weather flow bypassing. While not a very important issue presently, stormwater runoff was implicated as a concern which may have to be addressed in the future.

Other large cities in Latvia (Liepaja and Daugavpils) and Lithuania (Vilnius, Kaunas and Klaipeda) provide no effective secondary treatment of wastewater. A visit to the Lithuanian Environmental Ministry reinforced wastewater treatment concerns in that country. Considerable amounts of raw sewage are discharged to the Nemunas which drains into the Curonian Lagoon before entering the Baltic Sea. These discharges are considered to be the major source of nutrient loads in the Lagoon. While the Vilnius WWTF has attempted the implementation of biological treatment, problems persist. Local engineers have estimated that construction of proper WWTFs for the three largest Lithuanian cities along with two smaller towns would decrease nutrient loading into the Baltic by 70%. The Environmental Ministry is unable to make progress on more specialized problems until this fundamental question of sewage treatment is addressed. Funding assistance for the construction of several WWTFs is expected from Western European and Nordic countries, with an anticipated (and hopeful) operation date about 5 years from present (Aug. 1999).

Industrial discharges are another concern in the region. Meat and dairy processing plants have been important point sources of nutrient loads. Many of these plants have initiated biological treatment of wastewater with limited success due to operational difficulties. Because of mainly financial limitations, plant personnel have not been able to follow recommendations proposed by scientists. Paper production in Latvia (Sloka) and in Lithuania has been responsible for significant discharges of chlorinated organic compounds. Pretreatment programs are being developed to minimize their impact.

Agriculture traditional played a very important role in Baltic economies. Under collectivization, intensive agricultural practices severely degraded the water resources. Excessive fertilizer runoff was attributed to poor application technology, improper tillage practices and excessive application. Management of animal wastes remains a problem. Abandoned swine operations that raised tens of thousands of animals still pose a significant source of nutrient enriched runoff. Despite these observations, the impacts of agriculture on surface water resources has recently begun to decrease. With the privatization of huge collective farms and markets open to cheaper excess production from subsidized Western European farmers, agricultural production in some regions has dropped to 1920 levels, hence a reduction in fertilizer application.

Groundwater Impacts

Landfills represent the most prevalent source of groundwater contamination in the region. Without separation of hazardous wastes, without adequate landfill liner installations and without surface water infiltration control, contaminants are relatively easily transported from open dumps to shallow aquifers. Latvia has 600-1000 such dumps. In addition, five to six large chemical plants in Latvia have been responsible for severe groundwater contamination due to dumping of hazardous wastes in open pits. The Republic of Germany is funding a study of groundwater contamination by landfills in Latvia. The study consists of two pilot projects; at the dumps in Daugavpils and in Milgravis. The latter has significant petroleum contamination due to a nearby oil refinery and fuel depot. For each dump, a comprehensive technical study is planned which will include a subsurface transport modeling exercise and future recommendations.

Petroleum contamination of groundwater at refineries and at fuel transfer depots is common. Under former Soviet rule, fuel was inexpensive and leakage during the processing and transfer operations was not a concern. Many of these sites now have more than 1 meter of free product contaminating subsurface soils over areas in excess of several hundred square meters.

Because of their strategic location, the Baltic States had numerous military bases, including communications centers, ballistic missile launch sites and storage facilities, established by the Soviets during the occupation period. The Russian military forces evacuated all such bases in Lithuania during 1993 and were in the process of evacuating remaining bases in Latvia during our visit. These are an especially problematic source of both surface and groundwater contamination since very little is known of the specific types and amounts of contaminants. In Lithuania, military installations occupied 1.3% of the total area with 200 different bases. A limited evaluation of 88 of these sites by local scientists and engineers has resulted in 400 soil and 200 ground and surface water samples. Analyses indicate many of these samples contain high concentrations of petrochemicals, explosives and solvents. At Kedainiai, the third largest military airfield (660 hectares), extensive contamination of all the surface and groundwater resources on the base by improper wastewater and animal manure disposal and by petroleum storage has been confirmed. During our site visit, strong petroleum odors were evident throughout this large former base and rainwater ponds appeared covered with thick, oily films. Significant dumping of unused fuels has cause an estimated 740,000 cubic meters of contaminated soil, with a large layer of free product in the underlying shallow aquifer. The potable water supply of the city of Kedainiai is experiencing some contamination and remains threatened. High concentrations of radioactive waste and cadmium compounds have been detected in the groundwater of nearby communities.

In Latvia, former military operations yield similar problems throughout the country. A visit to the former Soviet Rocket Defense System Headquarters in Zakumuiza (near Riga) included a tour of several large underground bunkers. A most striking observation was the prevalence of vast quantities of large pressured gas cylinders in various conditions. Some of them were listed as containing Freon; the contents of others a mystery. The Latvian officials also expressed concern regarding potential groundwater contamination from underground storage tanks on the site.

Researchers at the Riga Technical University have begun a joint study with the Emergencies Engineering Division of Environment Canada and Gartner-Lee International, Inc. to carry out a demonstration project aimed at site assessment and remediation at a former Soviet military base in Latvia.

Conclusions

With a combined area the size of the State of New York, Latvia and Lithuania appear to have a lions's share of water resource contamination. Inadequate sewage treatment, numerous industrial dischargers and abandoned military installations and poor agricultural practices have led to groundwater and surface water pollution problems. Because of depressed economic activity, the environmental impacts of agricultural runoff and some industrial activity have recently decreased. However, excessive nutrient loading to the Baltic Sea from municipal wastewater treatment remains a most significant issue. Financial assistance will be required to construct wastewater treatment facilities, as well as address the legacy of pollution problems remaining from Soviet rule.

Using Monte Carlo Simulation in Financial Planning

Paul L. Matthews[1]
Gregory H. Tilley[2]

Abstract

As the costs of operating and maintaining water and wastewater utilities continue to increase, a greater need exists for prudent financial planning. The difficulty in fulfilling this need results from the uncertainty that accompanies forecasting. Advances in personal computer hardware and software have made increasingly sophisticated financial planning techniques available to smaller utilities. Computer simulation has become a cost-effective way to measure and account for uncertainty in financial planning. Although the techniques appear to provide great power to the users, unexpected problems can be encountered. This paper addresses how uncertainty can be modeled in the financial plans of small utilities.

Introduction

Monte Carlo simulation was developed during World War II to help in the development of the atomic bomb. For financial planning, Monte Carlo simulation allows the user to assign probability distributions to uncertain variables affecting the financial planning model. When using Monte Carlo simulation, an important distinction should be made between endogenous and exogenous uncertainty. Those variables that are exogenous like interest rates and inflation are especially difficult to assign probabilities. Nonetheless, some reasonable assignment of a probability distribution is better than none at all.

---

[1]Vice President, Integrated Utilities Group Inc., 5445 DTC Parkway, Suite 1035, Greenwood Village, CO 80111.

[2]Utility Economist, Integrated Utilities Group Inc., 5445 DTC Parkway, Suite 1035, Greenwood Village, CO 80111.

A powerful software tool that simplifies Monte Carlo simulation[3] is called @Risk.[4] @Risk assigns random values within a predetermined distribution function to variables the user specifies in his or her spreadsheet. The selected distribution function determines the probability of a value being selected. In other words, those values with higher probabilities are more likely to be selected than those with lower probabilities.[5] Once selected, the spreadsheet is recalculated. Monte Carlo allows the user to run hundreds of iterations with various assumptions to measure the probability of the results. To run an effective simulation, the distribution of variables, dependency between the variables and financial planning principles must be known.

The distribution of variables is determined by the user when he or she builds the spreadsheet. As an add-in to 1-2-3 and Excel, @Risk provides formula functions (in 1-2-3 these are called @functions, hence the name, @Risk) to the user. When specifying a variable with a given distribution, the user selects the distribution function by using the appropriate formula function. The formula functions, in turn, require arguments or inputs, to determine the sampling probabilities. For example, in 1-2-3, to select a variable that has a normal distribution with mean of 160 and standard deviation of 29, the user would enter:

@NORMAL(160,29)

@Risk provides 28 distribution functions. The arguments required in the distribution functions can be entered directly or can be generated by other formulas in the spreadsheet. Dependence among variables must be estimated during simulation so that the dependence is recognized in the sampling. Examples of dependent variables include, housing starts and interest rates, and interest rates and inflation rates.

An Example of Use of Simulation in Financial Planning

An example of using simulation for financial planning involves a water and sanitation district that had filed for bankruptcy protection. To emerge from bankruptcy, the district was required to create a 50-year financial workout plan that met with the bondholders' and court's approval. A major obstacle to the creation of this plan was the

---

[3]The sampling techniques available in @Risk include Latin Hypercube and Monte Carlo. Although the differences between these sampling techniques are technical, Latin Hypercube will generally be more efficient than Monte Carlo. We use Monte Carlo to mean computer simulation, including simulation using Latin Hypercube sampling.

[4]@Risk, by Palisade Corporation, is an add-in for Lotus 1-2-3 and Microsoft Excel.

[5]Palisade Corporation, Risk Analysis and Simulation Add-In for Microsoft Excel or Lotus 1-2-3, March 1, 1994, pg B-5.

uncertainty of assumptions that would affect the plan for the next 50 years. The district had four major concerns it needed to address in its workout plan. The first concern was inflation. Inflation is an important factor in estimating future tax revenues because tax revenues depend on assessed values that change with the general price level. Also, inflation affects the other costs of operating the utility. Part of the workout plan for the district required inflationary increases in user fees and system development charges.

Besides inflation, the rate of development within the district was a critical assumption. The rate of development depends on many factors including net migration within the region. As a result, the financial plan made assumptions about the absorption rate of available land within the district.

It is important to note that not all development is equal. Some development results in higher property tax revenues than others. Also, user charges depend on water and wastewater needs that vary with the type of development. Therefore, to estimate the spending needs and resources of the district, modeling of future development was necessary.

The Results of the Simulation

Policy makers wanted to understand the likelihood that taxes would have to exceed those rates already committed to in the workout plan over the 50-year period. Also important in the negotiations were estimates of the total value of the workout plan to the existing bondholders. To estimate the total value of the workout plan to bondholders, we used @Risk to collect information on the net present value calculated by our spreadsheet. By varying assumptions on inflation, capital needs, operation and maintenance costs, development rates and types, etc., then tracking the net present values, @Risk generates a probability function of the actual net present value. As an example, Figure 1 presents a sample distribution of the average age of a pump before replacement. This variable is one of many variables that influence the net present value to the bondholders. Figure 2 presents the output from @Risk for the net present value.

Beyond providing graphical output, @Risk provides the statistical representations from the simulation. Additionally, the user can vary the number of iterations used in the analysis. Generally, both the accuracy of the results and the time to run the simulation increase with the number of iterations. However, the increase in accuracy diminishes as the number of iterations increases. The financial plan for the district generally used 1,000 iterations.

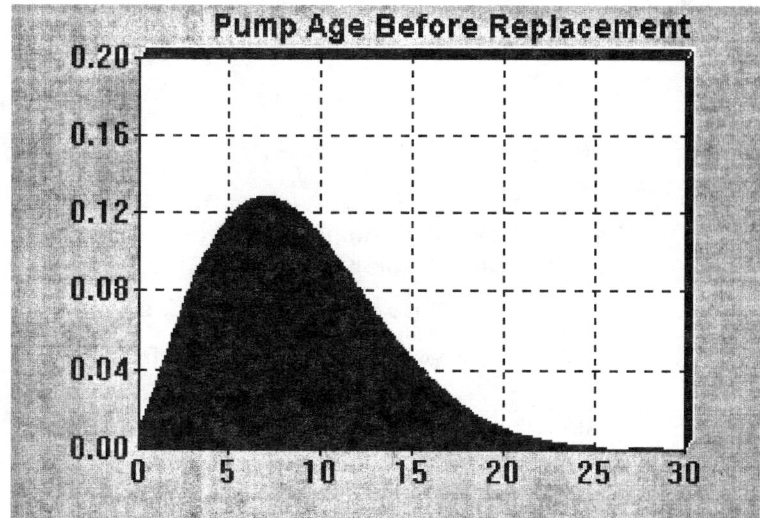

Figure 1. Probability Distribution of Pump Age Before Replacement.

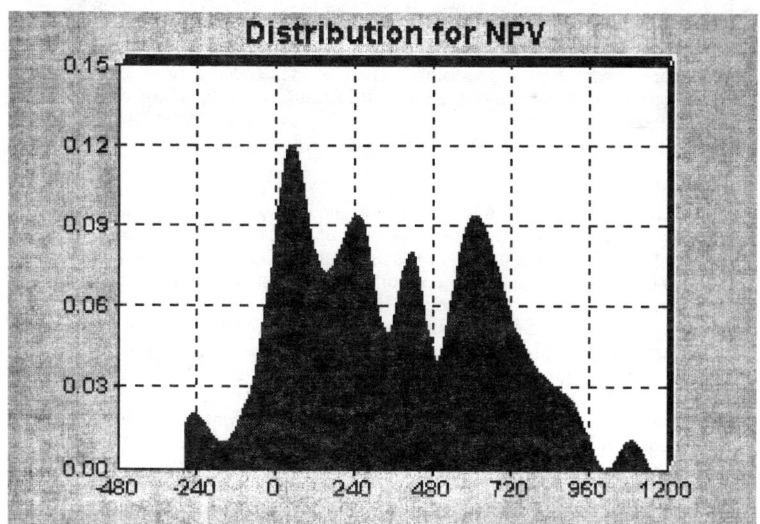

Figure 2. Probability Distribution of Net Present Value of Bondholders' Payments.

Risk-Based Decision Support for Dredging the Lower Mississippi River

Jay R. Lund, Dept. of Civil and Environmental Engineering, University of California, Davis, CA 95616
Vini Vannicola, Apogee Research, Inc., Bethesda, MD
David A. Moser, Institute for Water Resources, Alexandria, VA
Samuel J. Ratick, CENTED, Clark University, Worster, MA
Atul Celly, Apogee Research, Inc., Bethesda, MD
L. Leigh Skaggs, Institute for Water Resources, Alexandria, VA

Abstract

The paper reviews progress of a U.S. Army Corps of Engineers project to develop a prototype risk-based decision support system (DSS) for operational scheduling of dredging on the Lower Mississippi River. The DSS incorporates uncertainty in dredging costs, navigation control depths, and navigation costs avoided in an integrated manner to explore the risks and costs associated with potential operational dredging decisions. The DSS is intended to provide economic and risk-based operational guidance and justification in the scheduling, operation, and movement of maintenance dredges. Several lessons are inferred for risk analysis modeling of maintenance dredging activities, and risk analysis in general.

Introduction

The Lower Mississippi River is one of the largest port and shipping complexes in the world and a depositional environment for sediment. In an operational time-frame, the waterway has uncertain and varying channel bottom elevations (due to sedimentation and scour), river stages, and vessel traffic of various drafts. Therefore, maintenance dredging for this waterway involves significant uncertainty, high costs, and risk. While experienced engineers have successfully overseen dredging on this system for over a century, the economic consequences of their decisions have grown, and so dredging decisions have greater potential for controversy. This has prompted examination of more formal methods for comparing and evaluating alternative maintenance dredging decisions for this system.

Formal economic risk analysis for Lower Mississippi River maintenance dredging decisions requires probabilistic characterization of uncertainties in channel-bottom elevations, river stage, dredge effectiveness, dredging cost, and economic losses to shippers from limitations in navigable channel depth at many locations throughout the system. The characterization of each of these uncertainties is an interesting, and often imperfect, technical task. The formal integration of each uncertainty and consequences and display of results for use by operation engineers are equally important tasks. This brief paper reviews progress on this project to date, and future directions for this work.

Project Objectives

The ultimate objective of this work is to improve quantification of the economic basis for maintenance dredging decisions for the Lower Mississippi River Deep Draft Navigation System. Given the physical, economic, and engineering complexity of the system, as well as the novelty of probabilistic economic analysis of maintenance dredging systems, the more limited and realistic objective for this project is to develop a prototype decision support system (DSS). The decision support system would provide a probabilistic assessment of the risks and costs of alternative short-term maintenance dredging decision alternatives.

The development of the DSS is currently in its third phase. Phase I was the development of an initial conceptual design for a DSS (Moser, et al., 1990). Phase II included the development of a working DSS for one reach of the Lower Mississippi River system, Redeye Crossing (Apogee, 1995). This second phase required the assembly and use of data and expertise and detailed development of the approach for a single-reach system, as a step towards system-wide implementation. Phase III currently involves the extension of Phase II to multiple reaches, but still only a 3-7 reach subset of the system. This third phase is needed to prototype methods for routing dredges in a multi-reach system and examining probabilistic conditions and costs in a multi-reach setting. One task in Phase III is to have a prototype system tested by District personnel during the coming dredging season.

DSS Overview

The DSS consists of several modules. Uncertainty in the physical, engineering, and cost outputs of each module is based on recorded observations of these subjects by the New Orleans District over many years. The modules are (Apogee, 1995):

1. *Controlling Depth Estimation.*

The depth available for navigation varies with both the bottom elevation and river stage, both of which are uncertain in the time-frame of maintenance dredging decisions. This module estimates a probability distribution of controlling depths for each reach and time step. These distributions are based on historical observations of sedimentation rates and river stages, and how they vary about forecast and current conditions (Apogee, 1995). This empirical approach to controlling depth uncertainty replaces a more physically-based control depth modeling concept envisioned in Phase I of the project (Moser, et al., 1990). This empirical approach appears to work well, and has been fully implemented for Redeye Crossing.

2. *Navigation Costs.*

This module estimates a probability distribution of short-term costs to the navigation sector resulting from limited navigation controlling depth. These costs are uncertain, since the vessel traffic desiring to pass a depth-limited crossing on any given day has uncertain draft, cargo, destination, and other characteristics. These vessels have different light-loading, delay, lightering, and diversion options, which incur costs. Probabilistic vessel characteristics and recourse decisions are based on traffic records for the Lower Mississippi River system and Corps of Engineers' deep draft cost estimation methodologies.

3. *Dredging Cost and Productivity.*

The short-term cost of dredging a river reach and the dredge's productivity are not certain. The costs of dredging varies, depending on the dredges available and the amount of dredging required. This module estimates a probability distribution of dredging costs, based on historical dredging cost distributions and discussions with District personnel regarding how dredges are employed and contracted.

4. *Dredging Trigger Rule.*

Over time, the decision to bring in a dredge becomes somewhat regularized, based on conditions in the system. From discussions with District personnel, it became apparent that many dredging decisions for Redeye Crossing could be simulated by a set of rules. These rules were elicited from district personnel and are represented in a Dredging Trigger Rule module. This rule has been tested against the judgment of District experts and has been found to concur with expert judgment over 80% of the time. The incorporation of such rules in the DSS allows conventional maintenance decision-making to be better scrutinized by local experts.

5. *Interface and module coordination.*

A master control module coordinates the user interface and data flow among the different functional modules. This interface also provides a structured format for examining potential dredging decisions.

Risk Analysis Implications

The development of a risk-based decision support system for maintenance dredging has raised some interesting points which might be useful for those working on risk and uncertainty analysis of engineering problems in general. The experience so far with the Lower Mississippi River is shaped by the long duration of engineered management of the waterway and the complexity of the physical (sedimentation and stage) and economic (navigation) aspects of the system.

For problems which are similar in being complex, involving recurrent decision-making, and having been managed over long periods of time:

1. Empirically-based characterization of uncertainty often is more practical than physically-based or derived estimation of probability distributions. For the Mississippi system, the complexity and vast spatial extent of the physical processes involved implied that the use of physically-based models would need to be restricted to greatly simplified representations of physical processes. Even these representations would require daunting quantities of initial and parameter values to be established, either deterministically or probabilistically, and overly burdensome computational effort. Fortunately, the long management of this system by engineers resulted in the existence of large quantities of data regarding the end results of uncertain processes. This, and the recurrent nature of the maintenance decision-making problem, encouraged the use of empirically-based probability distributions for this problem.

2. For a system like the Lower Mississippi River, it is likely that the uncertainty in the mind of operating managers will be somewhat less than that embodied in the DSS, due to the greater knowledge of the expert than most of the statistical relationships embodied in the DSS. The regression and other relationships used to inter-relate the probabilities of different events in the DSS are simpler than the understandings of the system developed by District personnel over the course of

decades of experience and thought. For example, our regressions, based on a few recent years of data, relate sedimentation to river stage. However, District personnel maintain that sedimentation is also significantly related to the present and recent origin of flows in the river. Thus, higher sediment loads are thought to come with higher recent flows from the tributary Missouri River system, relative to Ohio River tributary flows.

3. Managers are well acquainted with uncertainty. A risk-based analysis approach merely provides them with a more quantified and "objective" representation of common sources of uncertainty. Thus, risk-based analysis can provide managers both with information that is more readily communicated to members of the public and resource users which scrutinize management decisions and a standard (albeit somewhat imperfect) on which to scrutinize their own decisions. In these roles, attention must be given to how data is presented and combined. Information must be understood to be useful. Therefore, attention has been given to data display and the orchestration of data entry, decision-making, and results presentation in the master control module of the DSS.

References

Apogee Research (1995), *Risk-Based Analysis of Maintenance Dredging: Development of a Prototype Dredging Decision Support System*, Apogee Research, Bethesda, MD.

Moser, D.A., T. Denes, J.R. Lund, and K.I. Rubin (1990), *A Risk-Based Decision Support System for Maintenance Dredging: Stage 1 - Conceptual Design*, Project completion report for the U.S. Army Corps of Engineers New Orleans District, U.S. Army Corps of Engineers, Institute for Water Resources, Alexandria, VA, May.

A Model for Risk-Based Dredging Decisions

Holly Morehouse Garriga[1], Samuel J. Ratick[1], David A. Moser[2], Wei Du[3], Leigh Skaggs[2], Jay R. Lund[2,4], and Ron Klimberg[1]

Introduction

Planning for channel maintenance dredging activities is a difficult undertaking complicated by various types of uncertainties in hydrologic forecasting, shoaling assessment, and prediction of engineering capacity and economic conditions. Stochastic river conditions make it difficult to maintain the authorized channel dimensions in many waterways and ports of the United States. Historical records for 1970 through 1987 indicate that the 40 foot project depth at the Lower Mississippi River could only be maintained approximately 80% of the time. The economic benefits of projects designed for improving navigation channels can only be realized in so far as the channel is reliably maintained. The lack of reliability in channel depths may alter long-term shipper and vessel operator behavior, further reducing the realized benefits of the project. During the past two decades, a large body of literature dealing with risk and uncertainty in water resources engineering design and planning has been developed. Studies about hydrologic parameter uncertainty and hydrologic model uncertainty can be found in Benjamin and Cornell (1970), Bodo and Unny (1976), Tung and Mays (1980), Lee and Mays (1983), Wood and Rodriguez (1975), Bogardi et al. (1978), Ratick et al. (1988), Ratick et al. (1991), Du (1989), and Du and Ratick (1990), and studies about economic uncertainty in Goicoechea et al. (1982) and Dandy (1985). Specifically, Lund (1990) addresses uncertainty issues for scheduling maintenance dredging on a single reach with uncertainty. Because of these uncertainties, accurately determining annual dredging budgets and

---

[1]Center for Technology, Environment, and Development, The George Perkins Marsh Institute, Clark University, 950 Main Street, Worcester, MA 01610-1477

[2]Institute for Water Resources, Water Resources Support Center, U.S. Army Corps of Engineers, Fort Belvoir, VA 22060

[3]Harvard Design and Mapping Company, Cambridge, MA 02139-2503

[4]Department of Civil Engineering, University of California at Davis, CA

addressing dredging facility allocation over space and time is a difficult task. In this paper we discuss the development and application of a multi-objective dredging decision model, the Reliability Based Dynamic Dredging Decision (RBD^3) model, that explicitly evaluates how dredging costs and channel reliability levels vary under different assumptions for maintaining proper channel depths.

The RBD^3 Base Model

The RBD^3 model (see Ratick, et al., 1992 and 1994, for a more detailed discussion of the model formulation) incorporates factors pertaining to: dredge costs and characteristics, stochastic river conditions and channel dredging requirements, and project objectives and the safe navigation depth in order to estimate a least cost schedule for dredges over time to meet specified channel reliability levels. The model is dynamic in that dredging schedules, costs, and reliability levels are solved for over time. The user can specify the number and length of these time periods (ie., months, weeks, or days). These time periods need not be constant throughout the planning horizon but will depend on the type of data available.

Although it is acknowledged that no model can duplicate reality completely, the RBD^3 model has been designed in a flexible manner, to represent the dredging decision environment as best possible. Various types of dredge costs are included as input to the model: rent (or fixed costs), assessed by length of use; variable costs, assessed by amount of material dredged; and mobilization/demobilization costs, assessed each time a dredge begins or ends operations. Dredge characteristics, such as the stated design capacity of a dredge (dredge capacity), the amount of material it actually handles under current operating conditions (dredge productivity), and its availability at each time period in the planning horizon (dredge availability), are addressed within the RBD^3 framework. Dredging characteristics and costs can be varied by dredge type, time period and reach location. Specific dredges and dredging activity can be restricted by time period and reach location depending on relevant environmental constraints, the type of dredge material at that location, and whether or not a particular dredge is available and fully operable at that time.

The RBD^3 model allows for the scheduling of advanced maintenance dredging which involves dredging extra depth to lower the dredging requirements in following time periods. Advanced maintenance dredging can be an efficient operational strategy to reduce overall project costs by shortening the dredging sequences and lowering the grounding risks for vessels. Extensions to the RBD^3 model include accounting for variations in shoaling and scouring rates due to river conditions and dredging activity and also allowing for the possibility that not all dredging requirements can be met under project constraints. One model extension calculates the unmet dredging requirements and its effect on achievable reliability levels.

The information provided in the output from the RBD^3 model includes:

- Dredging quantity (or dredging depth) for different types of dredges

under different reliability levels at different reaches over different time periods;
- Allocation of dredging facilities, including what types of dredging facilities will be selected, how many dredging facilities will be used, and where and when those facilities will be used;
- Dredging frequency representing how often a reach needs to be dredged;
- Scheduling of advanced maintenance dredging;
- The relationships between total dredging costs and channel reliability levels;
- Sequencing of dredges and allocation of mobilization/demobilization activities;
- Maximum achievable reliability levels within a constrained budget and/or operating conditions; and
- Impact of variations in dredging productivity levels upon the overall project costs and the final scheduling and allocation of dredges.

Reliability Loop Model Extension

The base model formulation for RBD^3 requires that the decision maker select a specified level of reliability and solve the model based on those requirements. The reliability loop model extension allows the model to be solved repeatedly over a range of reliability levels. The basic problem structure as specified by the decision maker remains the same (ie. the data inputs, parameter values, and project assumptions). Within this problem definition a solution is generated across a range of user supplied reliability levels (for example 60%, 70%, 80%, 90%, 91%, 93%, 95%, 97%, 99%). Included in the output from these model runs is a tradeoff curve plotting channel reliability level against total dredging costs.

Variations in Dredge Productivity Model Extension

The base model assumes that the dredge facilities will always perform up to their stated design capacity. Although it may be reasonable to assume that most of the time a dredge will be able to remove an amount of material somewhere close to its design capacity, other factors may come into play that will lower dredge productivity and may have a significant impact upon overall project costs, channel reliability levels, and the scheduling of all project dredges. This model extension can be used to investigate what impact variations in dredge productivity will have on the project solution.

Reliability Maximization Model Within a Constrained Budget

The original formulation of the reliability based dynamic dredging decision model (RBD^3) minimizes the total dredging costs of a channel maintenance project under a specified level of reliability. Many districts try to maintain their projects at high levels of reliability but may be unable to maintain the highest levels because of limited budgets. Instead of minimizing total costs, this extension of the RBD^3 model sets a budget constraint and then, using the allocated funds, maximizes the achievable level of reliability across the locations and time frame of the project. In the base model the objective function was a calculation of total dredging costs for the project. Instead of minimizing project costs, in this model extension project costs are constrained to be within a set budget. The new objective function strives to maximize the minimum achieved reliability level without violating the budget constraint

Appendix

Benjamin and Cornell, *Probability, Statistics, and Decision for Civil Engineers*, McGraw-Hill, New York, 1980.

Bodo, B., and T.E. Unny, "Model uncertainty in flood frequency analysis and frequency-based design," *Water Resources Research*, Vol. 18, pp. 315-330, 1984.

Bogardi, I., L. Duckstein, and F. Szidarovsky, "Multiobjective modeling for regional natural resources management," the paper presented at NOTEC river regional development project task force meeting, Jablonna, Poland, Sept., 1987.

Dandy, G.C., "An approximate method for the analysis of uncertainty in benefit-cost ratios," *Water Resources Research*, Vol. 21(3). pp. 267-271, 1985.

Du, Wei, "Estimating statistical parameters for an N-segment discrete hydrologic data series," *Journal of Hydrology*, New Zealand, Vol. 28(1), pp. 3-17, 1989.

Du, Wei, and Samuel J. Ratick, "An uncertainty analysis for the Log-Pearson Type III probability distribution," *Hydrological Science and Technology*, Vol.6(1-4), pp.7-15, 1990.

Goicoechea, A., et al., "An approach to risk and uncertainty in benefit-cost analysis of water resources projects," *Water Resources Research*, Vol.18(4), pp.791-799, 1982.

Lee, Jan-Lin, and Larry W. Mays, "Improved risk and reliability model for hydraulic structures," *Water Resources Research*, Vol. 19(6), pp. 1415-1422, 1983.

Lund, Jay R., "Scheduling maintenance dredging on a single reach with uncertainty," *Journal of Waterway, Port, Coastal, and Ocean Engineering*, Vol. 116(2), 1990.

Ratick, Samuel J., Wei Du, and Eugene Stakhiv, "Guidelines for evaluating risk and uncertainty in local flood damage protection projects," Draft Technique Report to the U.S. Army Corps of Engineers, 1988.

Ratick, Samuel J., Wei Du, Lin Vallianos, and Eugene Z. Stakhiv, "Uncertainty analysis for sea level rise and coastal flood damage evaluation," a research report to the Institute of Water Resources, the U.S. Army Corps of Engineers, 1991.

Ratick, Samuel J., Wei Du, and David A. Moser, "Development of a Reliability Based Dynamic Dredging Decision Model," *European Journal of Operations Research*, Vol. 58, pp.318-334, 1992.

Ratick, Samuel J., et al., "Development and application of the reliability based dredging decision model (RBD^3)," a draft report to the Institute of Water Resources, U.S. Army Corps of Engineers, October 1994.

Tung, Y. K., and L.W. Mays, "Optimal risk-based design of water resource engineering projects," *Technique. Report: CRWR-171*, Univ, of Texas, Austin, Aug., 1980.

Wood, E.F. and I. Rodriguez-Iturbe, "Bayesian inference and decision making for extreme hydrologic events," *Water Resources Research*, Vol. 11(4), pp.533-542, 1975.

A Probabilistic-based Optimization Model for Stormwater Infrastructure Rehabilitation

Jane T. Ho[1], Timothy L. Jacobs[2], and Miguel A. Medina, Jr.[3]

Introduction and Scope

The deterioration of the nation's infrastructure is understood to be an immediate problem. Cost estimates to rehabilitate the nation's infrastructure are as high as 2 trillion dollars (ASCE, 1984). Grigg (1986) estimates the cost to rehabilitate stormwater drainage systems may exceed $200 billion. With such high costs, rehabilitation activities implemented need to be the best use of typically limited funds.

Most municipalities and utilities use the rational formula to design stormwater networks (Walesh, 1989). The primary attraction of the rational method is its simplicity. However, its drawbacks relate primarily to the fact that the intensity-duration-frequency relationships upon which it is based were originally intended for the calculation of a peak flow rate and not for the calculation of either the runoff hydrographs or flow volumes. Therefore, no discharge hydrographs nor hydro-dynamic components of stormwater are incorporated. The kinematic wave routing method was developed originally for overland flow on a plane surface and has also been applied to flow routing in pipes and channels. The method derives from the assumption that gravity and friction forces dominate all other terms in the momentum equation for gradually varied unsteady flow and, therefore, that the bed slope of the channel or pipe equals the friction slope. This allows use of a uniform flow equation in combination with mass conservation to obtain a solution for the outflow hydrograph leaving a pipe or channel segment (ASCE, 1992).

This paper presents a mixed integer optimization model to design or specify rehabilitation activities for stormwater drainage systems conditioned on the likelihood of exceeding the system's conveyance capacity. The probability that stormwater flows exceed the conveyance capacity of any pipe segment within the system is

[1] Research Assistant, Department of Civil and Environmental Engineering, Duke University, Durham, NC 27708-0287.
[2] Assistant Professor, Department of Civil and Environmental Engineering, Duke University, Durham, NC 27708-0287.
[3] Associate Professor, Department of Civil and Environmental Engineering, Duke University, Durham, NC 27708-0287.

modeled as a chance constraint. The model combines kinematic wave routing and first order-second moment analysis to formulate the chance constraint. Chance constrained models are a specialization of mathematical optimization models in which one or more of the physical constraints is expressed as a probability statement (Houck, 1979). Jacobs and Medina (1994) used a chance constrained model and Monte Carlo simulation to determine optimal rehabilitation stormwater drainage systems.

The results of the multiobjective optimization model define a tradeoff relationship between the cost of the design and the probability of failure. The model results completely specify the system design at a minimum cost and probability of failure. To demonstrate the model, an application is presented utilizing a pipe network located on the East Campus of Duke University.

Model Development

The design of a stormwater drainage system involves conflicting objectives and uncertain decision parameters. Two possible objectives include the minimization of both total design cost and probability of failure. The model directly addresses each of these objectives with a chance constrained mixed-integer program.

To begin, the probability of exceeding the conveyance capacity of any pipe segment is defined by:

$$P_{f_i} = P(q_{cap_i} \leq q_i) \leq \alpha \qquad \forall i \tag{1}$$

where q_{capi}, q_i, and α represent the discharge capacity, the actual stormwater discharge, and the probability of failure of pipe segment i, respectively.

In terms of the safety index (Ang and Tang, 1984), the probability of failure for each pipe i within the network is defined by:

$$P_{f_i} = 1 - \phi(\beta_i) \quad \forall_i \tag{2}$$

where $\phi(\beta)$ is the distribution function of the standard normal variate β. Equation (2) can be restated as a mathematical constraint in which α represents the lower bound for the probability of failure:

$$1 - \phi(\beta_i) \geq \alpha \quad \forall i \tag{3}$$

Assuming the stormwater flows for each pipe are normally distributed and the capacity of the pipe is known and a function of the design decision, the safety index is defined by:

$$\beta_i = \frac{q_{cap_i} - \mu_{q_i}}{\sigma_{q_i}} \quad \forall i \tag{4}$$

where μ_{qi} and σ_{qi} represent the mean and the standard deviation of stormwater flows from pipe segment i, respectively. The values for μ_{qi} and σ_{qi} are determined using Monte Carlo simulations and a kinematic wave routing method. For this formulation, q_{cap} represents the conveyance capacity of pipe i and is a function of the system design. In general, the kinematic wave model for overland flow may be expressed as (Eagleson, 1970):

$$\frac{\partial y}{\partial t} + m\alpha_o y^{m-1}\frac{\partial y}{\partial x} = i - f \tag{5a}$$

$$q = \alpha_o y^m \tag{5b}$$

Equations (3) and (4) are combined to formulate the chance constraint which ensures that the likelihood of exceeding the capacity of pipe i within the stormwater drainage system is realized with a minimum probability:

$$q_{cap_i} - \sigma_{q_i}\phi^{-1}(1-\alpha) \geq \mu_{q_i} \quad \forall i \tag{6}$$

The model objectives considered here include the minimization of total cost and the probability of system failure. The complete model formulation is presented as:

$$Min\ \ Cost = \sum_i \sum_j C_{ij} X_{ij} \tag{7}$$

$$Min\ \ P_f = \alpha \tag{8}$$

Subject to:

$$q_{cap_i} - \sigma_{q_i}\phi^{-1}(1-\alpha) \geq \mu_{q_i} \quad \forall i \tag{6}$$

$$q_{cap_i} = \sum_j q_{cap_{ij}}\ X_{ij} \quad \forall i \tag{9}$$

$$\Sigma X_{ij} = 1 \tag{10}$$

$$X_{ij} \in \{0,1\} \tag{11}$$

X_{ij} is the binary decision variable for selecting available design option j for pipe segment i. C_{ij} is the cost of implementing pipe i with diameter size j. P_f represents the total probability of system failure. Constraints 9 and 10 define the available design options as a function of binary decisions.

Case Study and Discussion

To illustrate the model, a pipe network at Duke University consisting of fifteen drainage pipes serving both residential and commercial areas is used. The mean and standard deviation of the maximum segment flows were determined using 500 Monte Carlo simulations of past storm events and 100 time steps for simulating the kinematic wave progression. The results of the Monte Carlo simulations were used as inputs in the optimization model (Equation 6). The cumulative distribution function ϕ^{-1} (1-α) was incorporated into the optimization model using piecewise linear constraints. For each pipe in the network, six diameters are considered. Figure 1 illustrates the trade-off relationship between total system design cost and the probability of stormwater system failure.

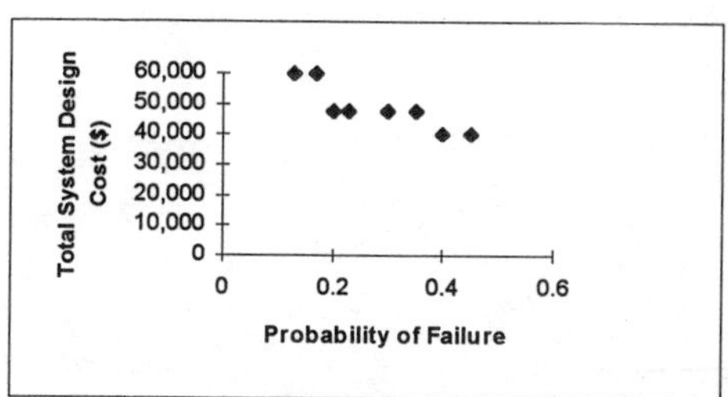

Figure 1. Tradeoff Relationship between Cost and Probability of Failure

Figure 1 shows that the probability of failure can be reduced from forty percent to twenty percent for approximately $8,000, while to reduce the probability of failure further to seventeen percent costs an additional $12,000. The tradeoff relationship shown in Figure 1 is a discrete function due to the discrete nature of the available pipe diameters. All costs are based on Means Building Construction Cost Data (1994).

Conclusion

A chance constrained optimization model has been presented to design stormwater drainage systems conditioned on the probability of exceeding the system's conveyance capacity. The model avoids overdesigning a stormwater system by incorporating the hydrodynamic characteristics of stormwater flows through the mean and standard deviation. This model provides the engineer with a tool for assessing the tradeoffs between system design cost and the probability of failure.

References

American Society of Civil Engineers (ASCE), 1992. Design and Construction of Urban Stormwater Management Systems. American Society of Civil Engineers and the Water Environment Federation, New York, New York.

American Society of Civil Engineers (ASCE), 1984. Research Needs Related to the Nation's Infrastructure. American Society of Civil Engineers, New York, NewYork.

Ang, A. H. and W. H. Tang, 1984. Probability Concepts In Engineering Planning and Design, v.2. Decision, Risk, and Reliability. Wiley, New York, New York.

Eagleson P., 1970. Dynamic Hydrology, McGraw-Hill, New York, New York.

Grigg, N. S., 1986. Urban Stormwater Infrastructure. Wiley and Sons, New York, New York.

Houck, M. H., 1979. A Chance Constrained Optimization Model for Reservoir Design and Operation, *Water Resources Research*, 15(5), 1011-1016.

Jacobs, T. L. and M. A. Medina, 1994. A Chance Constrained Model Using Kinematic Wave Routing for Stormwater Infrastructure Rehabilitation. *In:* Urban Drainage Rehabilitation Programs and Techniques, W. A. Macaltis (Editor). American Society of Civil Engineering, New York, New York.

Means Building Construction Data. 52nd Annual Edition, pp.67-68, 1994.

Walesh, S. G., 1989. Urban Surface Water Management. Wiley and Sons, New York, New York.

Risk-based Analysis for Flood Studies

Michael W. Burnham[1] and Darryl W. Davis[1]

Abstract

The Corps of Engineers now requires risk-based analysis for formulating and evaluating flood damage reduction measures. The adopted approach explicitly incorporates uncertainty of key parameters and functions into project benefit and performance analyses. Monte Carlo simulation is used to assess the impact of the uncertainty in the discharge-probability, elevation-discharge, and elevation-damage functions. This paper summaries the traditional methods, describes the risk-based analysis approach, and presents results of an example application.

Introduction

Planning and design of flood damage reduction projects traditionally applied best estimates of key variables, and other data elements in determining project benefits and performance. Benefit calculations involve discharge-probability, elevation-discharge (or rating), and elevation-damage functions and costs associated with the project configuration over it's life. Historically, inherent errors and imprecisions in these data were acknowledged but not explicitly incorporated into the analysis or considered in the results. Uncertainty was normally addressed through sensitivity analysis, conservative parameter estimates, and addition of extra capacity such as freeboard for levees. Each has limitations in estimating the statistical implications of uncertainty.

Project performance traditionally considered level-of-protection as the primary performance indicator. It is the exceedance probability event that corresponds to the capacity of the project. The importance of this single indicator was often overemphasized, while ignoring other performance information needed to assure proper project comparisons and ultimate selection for implementation.

---

[1]Chief, Planning Analysis Division, and Director, Hydrologic Engineering Center, respectively, U.S. Army Corps of Engineers, 609 Second Street, Davis, CA 95616.

Risk-based Analysis Approach

The Corps of Engineers requires risk-based analysis procedures for formulating and evaluating flood damage reduction measures (Corps of Engineers 1994). The procedures quantify uncertainty in discharge-probability, elevation-discharge, elevation-damage and incorporate it into economic and performance analyses of alternatives. The process applies Monte Carlo simulation (Benjamin et al.), a numerical-analysis procedure that computes the expected value of damage reduced while explicitly accounting for uncertainty.

Discharge-probability derivations depend on data availability. For gaged locations and where analytical methods are applicable, procedures defined by the Interagency Advisory Committee on Water Data (1982) are applied. Uncertainties for discrete probabilities are defined using the associated non-central *T* distribution. For ungaged locations, the cumulative discharge-probability is adopted from applying a variety of approaches (Water Resources Council 1981). The adopted function statistics are computed similar to gaged locations. The equivalent record length is selected based on the reliability of information. Regulated discharge-probability, elevation-probability, and other non-analytical probability functions require different methods. An approach called order statistics (Morgan et al. 1990) is applied for derivation of the cumulative function and associated uncertainty relationships for these situations.

Elevation-discharge functions are defined by observed data or computed water surface profiles. For observations, uncertainty is calculated from derivates of the best fit cumulative rating function. Computed profiles are required for ungaged locations and modified conditions. Where sufficient historic event data exists, profile uncertainty is estimated based on the adequacy of the model calibration to the historic data. Where data are scant, or the hydraulics complex, such as for high velocities, debris, ice, and bulk flow conditions, sensitivity assessments of reasonable upper and lower bound profiles are used to estimate the standard derivation of the uncertainty in stage. For now, the uncertainty distribution is assumed normal (Corps of Engineers 1993).

An elevation-damage function is derived from structure inventory information. They are constructed at damage reach index location along with the discharge-probability and elevation-discharge. Presently, the uncertainty distributions associated with the structure elevation, structure value, and content values are specified and used to produce an aggregated structure elevation-damage-standard derivation of error function for each damage reach.

Chester Creek Example

Chester Creek is a 177 km^2 watershed located near Philadelphia. Formulation and evaluation are performed to determine feasibility of implementing several flood damage reduction plans. This includes comparison of the economic value, performance, and other factors for with- and without-project conditions. Future conditions are projected to be similar as the base year for project implementation. Plans evaluated are 7 and 8 m. high levees, a channel modification configured with 15 m. bottom and 43 m. top widths, and a detention storage project of 5.5 million m^3 capacity.

Without-project condition cumulative discharge-probability is derived using Interagency Advisory Committee (1982) procedures for the data of the stream gage located in the basin. The gage has a 65 year record. Uncertainty is determined from associated confidence limits of the cumulative function. Profile analyses from modeling produce the cumulative rating at the index location. Rating uncertainty, with standard deviation varying from zero at zero discharge to one foot for the 1-percent chance event discharge and beyond, are derived from sensitivity and calibration analysis results. Uncertainty in damage is defined as a standard derivation equal to 10% of the cumulative damage value. Project condition analyses generate revised cumulative functions and associated uncertainties.

Monte Carlo simulations produce flood damage inundation reduction results and performance information for with-and without-project conditions. Summary of economic results are shown on Table 1. The display format is similar to those traditionally used. The results are also similar. Inclusion of other benefits may slightly alter the results. Any project with positive net benefits is acceptable. This excludes detention. The 8 m. high levee is identified as the plan that maximizes national economic development. It also provides the greatest benefits and is the most costly plan.

Table 1
Results of Economic Evaluation

Plan description	Annual with-project residual damage in $1,000	Annual with-project inundation reduction benefit in $1,000	Annual cost in $1,000	Annual net benefit in $1,000
Without-Project	78.1	0.0	0.0	0.0
7 meter levee	50.6	27.5	19.8	7.7
8 meter levee	18.4	59.7	37.1	22.6
Channel	41.2	56.9	25.0	11.9
Detention	44.1	34.0	35.8	-1.8

Performance information is shown on Table 2. Expected annual exceedance probability is similar to the traditional level-of-protection except now uncertainty in the rating is also considered with that of discharge-probability. The probability of exceedance within the 50 year project life is calculated directly using the binomial distribution. Event performance is the conditional probability defined as the chance of the project containing a specific event should it occur. It is a direct result of the risk-based analysis.

Table 2
Results of Performance Evaluation

Plan description	Expected annual exceedance probability	Probability of exceedance in 50 yrs	Event Performance, as %-chance non-exceedance for specified event: 2%-chance event	1%-chance event	0.4%-chance event
Without-Project	0.075	0.92	2.3	0.0	0.0
7 meter levee	0.012	0.46	88.2	48.3	6.6
8 meter levee	0.003	0.14	99.7	97.5	76.3
Channel	0.031	0.79	24.8	1.9	0.0
Detention	0.038	0.86	20.5	4.0	0.3

Inspection of performance results indicate only the 8 m. high levee affords a high level of performance. This is both the expected annual exceedance and event performance through the chance of containing the .4 percent chance event. Since it also provides maximum net benefits it appears to be a logical choice from the federal perspective. Notice, however, it has a 14 percent chance of exceedance during its project life. Since the consequences of capacity exceedance vary for different types of projects it is an important consideration in plan selection. Capacity exceedance for levees may cause sudden deep flooding that results in high risk life and significant damage. Channels and small detention will not normally make matters worse when the capacity is exceeded. These considerations as well as others, such as environmental and social impacts, are requisites for plan evaluation and selection. Economic and performance information derived from risk-based analysis procedures enable better decisions for project selection.

Appendix

Benjamin, J. R., and Cornell, A. C. (1970). *Probability, Statistics, and Decision for Civil Engineers.* McGraw-Hill Book Co., New York, NY.

Corps of Engineers (1994). "Risk-based Analysis for Evaluation of Hydrology/Hydraulics and Economics in Flood Damage Reduction Studies." Engineering Circular.

Interagency Advisory Committee of Water Data (1982). "Guidelines for Determining Flood Flow Frequency." *Bulletin 17B* U. S. Department of the Interior, U. S. Geological Survey, Office of Water Data Coordination, Reston, VA.

Morgan, M. Granger, and Max Hendron (1990). *Uncertainty: A Guide to Dealing with Uncertainty in Quantitative Risk and Policy Analysis. Cambridge University Press.*

Water Resources Council (1981). "Estimating Peak Flow Frequency for National Ungaged Watersheds - A Proposed Nationwide Test." U. S. Government Printing Office, Washington, D.C.

Groundwater Pollution Remediation and Control: The Role of Global Optimizers and Exploitation of Available Information.

G. Cassiani[1], W.H. Liu[1], M.A. Medina[2] and T. L.Jacobs[3]

Abstract

This paper presents a multiobjective optimization model for identifying possible pump and treat strategies for groundwater pollution remediation and control. Two issues greatly influence the effectiveness of groundwater remediation optimization models: 1) the uncertainties associated with the hydrogeologic parameters, and 2) the non-convex nature of the many of the constraints and objective functions. Budgetary limitations and uncertainties about the physical size and characteristics of the affected area often impede investigators efforts to fully understand the groundwater behavior surrounding the site. This characteristic has been referred to as the non-uniqueness or instability of the inverse problem in groundwater modeling. Stochastic techniques involving geostatistical descriptions of the random field, perturbation analysis of the flow equation and the maximum likelihood estimation of unknown geostatistical parameters provide a mechanism for reducing the non-uniqueness of the inverse problem. In addition, the use of "soft" information such as geophysical measurements can lead to more accurate estimates.

The stochastic parameter estimates are used as inputs to a multiobjective optimization model to identify the non-inferior set of possible pump and treat strategies based on cost and likelihood of exceeding the maximum allowable contamination concentration. Due to the highly nonlinear nature of the optimization model, gradient-driven optimization can only be guaranteed to achieve local optimality. The optimization model can be solved using genetic algorithms. The tradeoff relationship provides decision makers with a variety of optimal pump and treat strategies conditioned to the uncertainties inherent in the modeling process.

Geostatistical paradigm and estimation.

The use of geostatistical techniques is established practice in estimating the spatial variability of geological properties. Best linear unbiased estimators (kriging) and stochastic simulators are applied to typically sparse data sets. Spatial variates are treated as random fields (RF), in which the random variables have spatial functional dependence. A RF is fully characterized by its multivariate probability density function (pdf). Assumptions for this process include stationarity and Gaus-

[1]Research Assistants, [2]Assoc. Professor, [3] Asst. Professor; Department of Civil and Environmental Engineering, Duke University, Durham, NC 27708-0287

sian form of the pdf in order to ensure that the RF can be described in terms of mean and spatial covariance only. To begin, an existing model is prescribed for the covariance. Mathematically this is identified by: $C(\bar{\mathbf{x}}_i - \bar{\mathbf{x}}_j) = \sigma_y^2 \exp(-d(\bar{\mathbf{x}}_i, \bar{\mathbf{x}}_j)/\lambda_y)$, where $d(\bar{\mathbf{x}}_i, \bar{\mathbf{x}}_j)$, σ_y^2 and λ_y represent the distance of the two locations $\bar{\mathbf{x}}_i$ and $\bar{\mathbf{x}}_j$, the variance and the correlation length of the RF *y* respectively. If **z** is the vector of measured values of y, the likelihood of **z** is (in the Gaussian hypothesis):

$$l(\vartheta_G|\mathbf{z}) = (2\pi)^{-N/2}|\mathbf{Q}|^{-1/2} \exp\left\{-\frac{1}{2}(\mathbf{z}-\mu_y)^T \mathbf{Q}^{-1}(\mathbf{z}-\mu_y)\right\} \tag{1}$$

where $\mathbf{Q}$ represents the covariance matrix of *y*, μ_y the mean and $\vartheta_G = (\mu_y, \sigma_y^2, \lambda_y)$ the vector of parameters to estimate; a widely applied technique for this latter purpose is the Maximum Likelihood (ML) paradigm with a Gauss-Newton iterative procedure (Kitanidis and Lane, 1985). Once ϑ_G has been evaluated, y can be estimated over the domain using a linear combination of the measured values, the weights of which are determined by the conditions of unbiasedness and minimum estimation variance (kriging). The procedure presented here for a single RF can be extended to include multiple RFs (y_1, y_2, ...) with the goal of gaining additional information about one RF of specific interest (called the primary one) conditional on the measurements from the primary RF and the remaining secondary RFs. The likelihood of **z** follows eq.(1). However $\mathbf{Q}$, μ_y and **z** are augmented to account for the secondary RFs; mathematically this is defined for two RFs as:

$$\mathbf{Q} = \begin{bmatrix} \mathbf{Q}_{11} & \mathbf{Q}_{12} \\ \mathbf{Q}_{21} & \mathbf{Q}_{22} \end{bmatrix}; \quad \mu = \begin{Bmatrix} \mu_{y_1} \\ \mu_{y_2} \end{Bmatrix}; \quad \mathbf{z} = \begin{Bmatrix} \mathbf{z}_1 \\ \mathbf{z}_2 \end{Bmatrix} \tag{2}$$

where $\mathbf{Q}_{11}, \mathbf{Q}_{12}, \mathbf{Q}_{22}$ are the covariance and cross-covariance matrices of the two RFs. Since our interest is to estimate the primary variable y_1, we seek a relationship between y_1 and y_2, which translates into a relationship between covariances and means:

$$\begin{gathered} \mathbf{Q}_{12} = \mathbf{Q}_{12}(\mathbf{Q}_{11}, \vartheta_R) \quad ; \quad \mathbf{Q}_{22} = \mathbf{Q}_{22}(\mathbf{Q}_{11}, \vartheta_R) \\ \mu_{y_2} = \mu_{y_2}(\mu_{y_1}, \vartheta_R) \end{gathered} \tag{3}$$

where ϑ_R represents a set of parameters which affect the relationship between y_1 and y_2. By applying ML, we estimate both ϑ_G and ϑ_R. ϑ_R may or may not be of our interest, in which latter case we refer to it as nuisance parameter. The relationship between RFs can be either statistical (regression) or physical (PDE) in nature. Given the estimate of ϑ_G, the evaluation of the y_1 RF from measurements of y_1 and y_2 is performed through co-kriging (Deutsch and Journel, 1992).

Application to "soft" data utilization and inverse modeling.

In the groundwater flow case the primary RF is given by the logarithm of the aquifer transmissivity T. Many authors (e.g. Freeze, 1975; Hoeksema and Kitanidis, 1985) have found that T is generally log-normally distributed. Therefore the hypothesis that Y=log(T) is Gaussian is valid. The secondary field can be some geophysical quantity such as electrical resistivity or some hydrological variable such as the hydraulic head (inverse problem). For this work these approaches have been tried on a small-scale fractured aquifer threatened by landfill contamination, located in the Duke Forest, Durham, NC.

Geophysical data inclusion.

The electrical resistivity ρ and the thickness b of the water-bearing formation can be evaluated through Vertical Electrical Sounding (V.E.S.) techniques. A power-law regression has been commonly proposed in the literature to relate electrical resistivity and hydraulic conductivity (Mazac et al., 1990). Taking into account the definition of transmissivity as T=bK, where b is the aquifer thickness and K is the hydraulic conductivity, and assuming a log-normal distribution for ρ as well (Mazac et al.,1990), Z = log(ρ), a linear regression can be assumed, such as:

$$Z = \alpha Y + \beta + \log b + \varepsilon \tag{4}$$

where ε is the regression error. Defined the regression variance as $\sigma_Z^2 = \mathrm{var}(\varepsilon)$, the matrix $\mathbf{Q}$ and mean vector μ can be written:

$$\mathbf{Q} = \begin{bmatrix} \mathbf{Q}_{YY} & \alpha\mathbf{Q}_{YY} \\ \alpha\mathbf{Q}_{YY} & \alpha^2\mathbf{Q}_{YY} + \sigma_Z^2\mathbf{I} \end{bmatrix}, \quad \mu = \begin{Bmatrix} \mu_Y \\ \alpha\mu_Y + \beta \end{Bmatrix} \tag{5}$$

Consequently the ML procedure estimates the vector of parameters $\vartheta = \{\mu_Y, \sigma_Y^2, \lambda_Y, \alpha, \beta, \sigma_Z^2\}$.

Inverse modeling.

In this case the relationship between y_1 and y_2 is given by a PDE:

$$\frac{\partial^2 H}{\partial x^2} + \frac{\partial^2 H}{\partial y^2} = -\mathrm{R}e^{-Y} \tag{6}$$

where Y is the logarithm of T, H the hydraulic head and R represents the recharge. In order to find the relationship between the spatial covariances of the Y and H RFs, a small perturbation approach has been proposed (Kitanidis and Vomvoris, 1983): H and Y are written in terms of their expected values $\langle H\rangle, \langle Y\rangle$ and first-order zero-mean oscillations around the means: $Y = \langle Y\rangle + f$; $H = \langle H\rangle + h$. Substituting these relationships into eq.(6) and simplifying yields the stochastic PDE relating the perturbations f and h:

$$\frac{\partial^2 h}{\partial x^2} + \frac{\partial^2 h}{\partial y^2} = -\frac{\partial f}{\partial x}\frac{\partial\langle H\rangle}{\partial x} - \frac{\partial f}{\partial y}\frac{\partial\langle H\rangle}{\partial y} + \mathrm{R}e^{-\langle Y\rangle} f \tag{7}$$

By using a numerical approach such as finite difference (FD) or finite element (FE), the stochastic linearized PDE [eq.(7)] can be written in discrete matricial form as $\mathbf{h} = \mathbf{A}^{-1} \cdot \mathbf{B}(R) \cdot \mathbf{f}$. Both matrices $\mathbf{A}$ and $\mathbf{B}(R)$ are function of the ϑ_G; $\mathbf{B}(R)$ is also a function of recharge R and the boundary conditions. The covariance matrix components yield:

$$\begin{aligned} \mathbf{Q}_{HH} &= \mathbf{V}\left(\mathbf{A}^{-1} \cdot \mathbf{B}(R)\right)\mathbf{Q}_{YY}\left(\mathbf{A}^{-1} \cdot \mathbf{B}(R)\right)^T \mathbf{V}^T \\ \mathbf{Q}_{HY} &= \mathbf{V}\left(\mathbf{A}^{-1} \cdot \mathbf{B}(R)\right)\mathbf{Q}_{YY}\mathbf{U} \end{aligned} \tag{8}$$

where $\mathbf{U}$ and $\mathbf{V}$ are binary matrices which 'sample' the FD or FE matrices for the points where actual measurements are taken. Unfortunately, the inverse problem is physically ill-posed; this leads to ill-conditioned **Q** matrices (Dietrich and Neumann, 1989) and to unstable ML estimations where model errors (discretization errors) and measurement errors are critical. Furthermore, the small perturbation approach is only valid for small values of σ_Y^2 (Gelhar and Axness, 1983).

Optimization of remediation control

The objective of the optimization framework is to minimize the total remediation cost while simultaneously maximize the reliability of the remediation strategy. The uncertainty associated with hydraulic conductivity described by the results of the estimation stage can be used to evaluate the reliability of the remediation system. The reliability is defined as the probability that the concentration at a monitoring well does not exceed a specified maximum contaminant level (MCL). The resulting multiobjective optimization model is solved using a genetic algorithm (GA) which uses an approach analogous to natural evolution to identify possible feasible solution (McKinney and Lin, 1994). The solution completely specifies a set of non-inferior remediation strategies conditioned on the total cost and the reliability of the remediation scheme.

Conclusions

The presented general geostatistical estimation procedure is quite effective at including "soft" information in transmissivity field evaluation. The inverse problem procedure still shows instability problems, but they could be circumvented by including a "soft" RF, such as electrical resistivity, in the inverse procedure itself by augmenting the relevant covariance matrix and mean and measurement vectors. The information from the estimation stage can be used to identify possible efficient remediation strategies under conditions of uncertainty.

References

Deutsch C.V.; Journel, A.G.; 1992; GSLIB Geostatistical Software Library and User's Guide; *Oxford University Press.*

Dietrich C.R.; Newsam, G.N.; 1989; A Stability Analysis of the Geostatistical Approach to Aquifer Transmissivity Identification; *Stochastic Hydrology and Hydraulics*, 3, 293-316.

Freeze, R.A.; 1975; A Stochastic-Conceptual Analysis of the One-Dimensional Groundwater Flow in Nonuniform Homogeneous Media; *Water Resources Research*, 11(5), 725-741.

Gelhar, L.W.; Axness, C.L.; 1983; Three-Dimensional Stochastic Analysis of Macro-Dispersion in Aquifers; *Water Resources Research*, 19(1), 161-180.

Hoeksema, R.J.; Kitanidis, P.K.; 1985; Analysis of Spatial Structure of Properties of Selected Aquifers; *Water Resources Research*, 21(4),563-582.

Kitanidis, P.K.; Lane, R.W.; 1985; Maximum Likelihood Parameter Estimation of Hydrologic Spatial Processes by the Gauss-Newton Method; *Journal of Hydrology*, 79, 53-71.

Kitanidis, P.K.; Vomvoris, E.G.; 1983; A Geostatistical Approach to the Inverse Problem in Groundwater Modeling (Steady State) and One-Dimensional Simulations; *Water Resources Research*, 19(3), 677-690.

Mazac, O.; Cislerova, M.; Kelly, W.E.; Landa, I.; Vehnodova, D.; 1990; Determination of Hydraulic Conductivities by Surface Geoelectrical Methods; in *Geotechnical and Environmental Geophysics*, Ward S.H. Editor, SEG, Vol.2, 125-131.

McKinney, D.C., Lin, M.; 1994; Genetic Algorithm Solution of Groundwater Management Models; *Water Resources Research*, 30(6), 1897-1906.

Wasteload Allocation Using Combined Simulation and Optimization

Andrews K. Takyi[1] and Barbara J. Lence[2]

Abstract

Research directed toward incorporating uncertainty in wasteload allocation models typically concentrates on input information uncertainty. Combined simulation-optimization modelling, SIOP, which is one approach for incorporating uncertainty, is effective for representing the complex mathematical and statistical properties of a stochastic water quality system. However, traditional SIOP may produce management decisions that are inefficient with respect to some of the management objectives, such as cost or reliability. Improved SIOP models and heuristic algorithms have been developed to address this problem, but these approaches require extensive computational resources. This paper presents methods that may be used to reduce the computational burden of one of these heuristic algorithms.

Introduction

The common objectives in a wasteload allocation problem that incorporates uncertainty include minimizing cost and maximizing reliability of maintaining water quality. Although traditional SIOP modelling is often used for analyzing such multi-objective water quality management problems, the identification and evaluation of the non-inferior surface for the objectives can be a complicated task (see, e.g., Fuessle et al., 1987 and Burn, 1989).

---

[1] Graduate Research Assistant, Department of Civil and Geological Engineering, 15 Gillson Street, University of Manitoba, Winnipeg, Manitoba, Canada, R3T 5V6.

[2] Associate Professor, Department of Civil and Geological Engineering, 15 Gillson Street, University of Manitoba, Winnipeg, Manitoba, Canada, R3T 5V6.

Morgan et al. (1993) present an improved SIOP model and a heuristic algorithm that can be used to solve a stochastic multi-objective problem. This algorithm generates several potential sets of system design conditions via Monte Carlo simulation and incorporates these conditions simultaneously in a single optimization model to achieve a solution that maintains the required objectives for all of the potential conditions. To obtain the non-inferior surface for reliability and a cost surrogate, the optimization model is solved several times, each time dropping the set of design conditions that would result in the best improvement in the value of cost objective. The reliability of each of these solutions is represented by the ratio of the number of sets of design conditions included in the optimization model to obtain this solution and the total number of Monte Carlo simulations, S, generated.

Even for small water quality systems, this improved SIOP model and heuristic algorithm may require a large computational effort if an adequate number of simulations, S, is utilized. The CPU time required to solve the heuristic algorithm is determined to a large extent by the number of constraints. However, for each time the model is solved, many of the water quality constraints are non-binding at optimality. Therefore, the CPU time could be drastically reduced without affecting the accuracy of the solution if some of these non-binding constraints could be determined *a-priori* and removed before solving the optimization model. This paper presents an approach for reducing the computational effort for solving the improved SIOP multi-objective model of Morgan et al. (1993) by identifying and removing two groups of potentially non-binding constraints for each level of reliability before obtaining a solution. These two constraint groups are those that represent (i) non-critical water quality checkpoints and (ii) inferior realizations of the Monte Carlo simulation.

Identifying Non-critical Water Quality Checkpoints

The non-critical water quality constraints are identified by employing Monte Carlo simulations prior to solving the wasteload allocation problem. First, S possible realizations of water quality background conditions are generated, then the water quality management model based on each of these realizations is solved and the critical checkpoints for each realization solution are identified. These critical checkpoints are considered as the potential critical checkpoints that must be included in the SIOP multi-objective model. The set of potential critical checkpoints is denoted as Θ_C.

Identifying Inferior Realizations of Water Quality

To identify the inferior realizations from the Monte Carlo simulations that might be removed at each level of reliability, two algorithms are developed, the γ Algorithm and the Ranked Impacts Algorithm. For two realizations of the background water quality conditions, Realization A and Realization B, if the solution to the wasteload allocation model based on Realization A can ensure acceptable water quality for stream background conditions represented by Realization B, but the reverse is not possible, then Realization B is said to be inferior to Realization A. On the other hand, if the wasteload allocation based on either of these realizations of stream background conditions is not able to maintain acceptable water quality levels at the critical locations associated with the other realization, then these two realizations are non-inferior to each other.

The γ Algorithm modifies the improved SIOP model of Morgan et al. (1993). For the improved SIOP model, there is a set of realizations of water quality responses that would result in the best objective function value (i.e., the lowest total cost) and a chosen reliability equal to R, Φ_R. The γ Algorithm determines the parameter γ, such that the worst γS realizations in the set Φ_R, based on the objective function values (i.e., the total cost) from a traditional SIOP solution approach, constitute the non-inferior realizations. There are three steps in this algorithm. In Step 1, N (about 10% of S) realizations of background water quality are generated and the traditional SIOP is solved for each of these realizations. The solutions to the traditional SIOP model are then ranked. In Step 2, the improved SIOP model is solved for all levels of R (between 1 and 0) for the N realizations generated in Step 1 using the heuristic algorithm of Morgan et al. (1993). In Step 3, the smallest proportion γ, of the N realizations, that must be included at all levels of R of the solution process, to obtain the exact results as in Step 2, is determined. The determination of γ is based on the ranking of the objective function values from Step 1, and on an identification of the realizations that have binding constraints for the optimal solution of the improved SIOP at each level of R from Step 2.

The Ranked Impacts Algorithm reduces the number of constraints at each level of reliability R of the improved SIOP model of Morgan et al. (1993). For each realization that has not been dropped by the heuristic algorithm of Morgan et al. (1993) at reliability level R, this algorithm assigns a ranking of each constraint $i \in \Theta_C$ in terms of its importance for determining the realizations that control the improved SIOP model solution at this reliability level. That is, at a given reliability level R, for realization $j \in \Phi_R$, the ranking, W_{ij}, is determined for each constraint $i \in \Theta_C$ based on the following relationship:

$$W_{ij} = S - T_{ij} \qquad \forall\ i \in \Theta_C \quad j \in \Phi_R$$

where T_{ij} = the number of water quality violations at checkpoint i (or constraint i) in the stream under realization j, if the traditional SIOP solutions were implemented under each of the

- 1 other realizations. Once the ranks, W_{ij} , are determined, he algorithm selects a small number of the realizations that orrespond to the highest ranked constraints at each checkpoint $\in \Theta_C$. The selected realizations are then included in the mproved SIOP model, for the given reliability level. This pproach may drastically reduce the size and computational urden of the optimization model solved at each reliability evel, and will produce results that are similar to the onreduced improved SIOP model.

eferences

urn, D. H., Water-quality management through combined imulation-optimization approach, *Jour. of Environ. Engng. Div., SCE, 115*, 1011-1024, 1989.

iessle, R. W., E. D. Brill, Jr., and J. C. Liebman, Air quality lanning: A general chance-constraint model, *Jour. Environ. ngng. Div., ASCE, 113(1)*, 106-123, 1987.

organ, D. R., J. W. Eheart and A. J. Valocchi, Aquifer emediation design under uncertainty using a new chance onstrained programming technique, *Wat. Resour. Res., 29(3)*, 51-561, 1993.

Water Quality Modeling in Distribution Systems : Overview and Methods of Solutions

Avi Ostfeld[1] and Uri Shamir[2], Fellow, ASCE

Abstract

In recent years considerable research is focused on water quality modeling in distribution systems. This is mainly due to the scarcity of water resources, and the concern over quality changes in the distribution system. This paper is aimed at summarizing the state of the art in water quality modeling in distribution systems, with respect to simulation, optimal design, optimal operation, and reliability; and to suggest a new approach for managing water quality in distribution systems.

Introduction

In multiquality water distribution systems (MQWDS) waters of different qualities are taken from sources, possibly treated, conveyed, and supplied to the consumers.

The interest in modeling flow and quality in water distribution systems, stems from three major types of circumstances: (1) Use of waters from sources with different qualities in a single distribution system which serves to mix and convey them. This situation occurs where sources of good quality are limited; (2) concern over quality changes in the distribution system, such as decay of disinfectants and/or growth of organisms; (3) events in which contaminants enter by mishap into the drinking water supply systems and are distributed with flow.

In this paper we summarize the state of the art in water quality modeling in distribution systems, with respect to simulation and management; and suggest a new approach for overcoming existing difficulties in managing water quality in distribution systems.

Simulation of MQWDS

Simulation of MQWDS is the study of the changes of water quality substances, in time and in space, within the distribution system. The most advanced tool for simulating MQWDS is EPANET (Rossman 1993). EPANET models the growth (or loss) of water quality substances through the distribution network, due to first order kinetics within the bulk flow and between the bulk flow and the pipe wall. The method of solution is the Discrete Volume Element Method (DVEM) of Rossman et al. (1993). EPANET does not consider dispersion of the substances within the distribution network (which can become significant for low flow velocities), chemical reactions between the substances in the network and in the reservoirs, and assumes complete and instantaneous mixing in the reservoirs. Consideration of part or all of these phenomena is presumably in the next stage of MQWDS simulation.

---

[1] Research Fellow, Water Research Institute, Technion, Haifa 32000, ISRAEL

[2] Professor, Faculty of Civil Engineering, Technion, Haifa 32000, ISRAEL

Management of MQWDS

The need for management exists whenever the solution to a problem is not unique. In modeling MQWDS this situation occurs in design or operation, but may also appear in simulation in case of insufficient data.

Ostfeld and Shamir (1993a) classified the problems of MQWDS according to the physical laws which are considered explicitly as constraints:

(1) *Discharge-Head (QH) models*: quality is not considered. The network is described only by its hydraulic behavior. (2) *Discharge-Quality (QC) models*: the physics of the system are included only as continuity of water and of pollutant mass at nodes. Quality is described essentially as a transportation problem, in which pollutants are carried in the pipes, and mass conservation is maintained at nodes. Such a model can account for decay of pollutants within the pipes and even chemical reactions, but does not satisfy the continuity of energy law (Kirchoff's Law no. 2), and therefore there is no guarantee of hydraulic feasibility and of maintaining head constraints at nodes. (3) *Discharge-Quality-Head (QCH) models*: quality constraints and the hydraulic laws which govern the system behavior are all considered.

The QH and the QC problems are easier to solve than the full QCH model.

QCH models for design and operation are highly non linear and non-smooth. The non linearities appear both in the objective function and the constraints. The non smoothness is due, for example, to the Hazen Williams head loss formula in which the absolute value or the "sign" function should be used in order to allow the network to be undirected with respect to flow in the pipes. To cope with these difficulties special approximations and/or non-smooth techniques were used (Cohen, 1992 ; Ostfeld and Shamir, 1993a ,1993b)

So far, only a few QCH models have been developed for the optimal design, optimal operation, and reliability of MQWDS. Following is a brief review.

Optimal design of MQWDS

Cohen (1987) proposed a method to obtain the optimal design of a directed MQWDS, based on the LPG method of Alperovits and Shamir (1977). The overall algorithm starts with a given feasible flow distribution which is successively altered using a gradient search type technique. Ostfeld (1994) solved the optimal design problem of a MQWDS by decomposing the problem into an "outer" non-smooth problem, and an "inner" convex quadratic problem. The decision variables of the "outer" problem are the so-called "circular flows" in the system. These are the flows (actually changes of flows) along a set of paths in the systems-all basic loops and open paths between pair of sources. The "outer" and the "inner" problems are connected through the circular flows. The method of solution includes the use of the *r - algorithm* (Shor, 1985) for minimizing the "outer" problem, for which a subgradient of the Lagrangian of the problem with respect to the circular flows is calculated in each iteration. The method allows reversal of flows in pipes, relative to the direction initially assigned.

Optimal operation of MQWDS

Cohen (1992) solved the steady state optimal operation model of an undirected MQWDS by decomposing the QCH problem into QH and QC sub-problems for fixed water flows in the distribution system, and fixed removal ratios in the treatment plants. The QC and QH models are solved first. The combination of their solutions serves for solving the QCH. For the unsteady case a detailed mathematical model has been developed, which has not been implemented. Ostfeld and Shamir (1993a) developed a model for optimal operation of undirected MQWDS under steady state conditions, which has been extended to the unsteady case (Ostfeld and Shamir, 1993b). In the unsteady case, rather than dividing each pipe into segments and tracking the progress of the quality fronts, as is done in simulation, a single equation has been developed which represents the average concentration of the pollutant in the pipe for a specific time period. Both models are solved with GAMS (Brooke et al., 1988) / MINOS

(Murtagh and Saunders, 1982). Mehrez et al. (1992) developed a model which is aimed at addressing the operation of a MQWDS in real-time. The main disadvantage of the model is that it does not consider the unsteady stage of the propagation of the quality fronts in the pipes, and therefore is applicable only for the steady state case. The model is solved using GAMS / MINOS.

Reliability of MQWDS

Reliability is an inherent attribute of any system, which refers to its ability to perform its mission adequately under stated environmental conditions for a prescribed time interval. The only work reported on reliability of MQWDS is that of Ostfeld (1994). Ostfeld (1994) developed a methodology which incorporates in a single framework, optimal design and reliability of MQWDS. The system constructed is able to sustain prescribed failure scenarios, such as any single random component failure, and still maintain a desired level of service in terms of the quantities, qualities and pressures supplied to the consumers.

Managing MQWDS using Hybrid Techniques

Previous work on managing MQWDS was limited with respect to both the physical processes modeled, such as the explicit inclusion of the unsteady stage of the propagation of the quality fronts in the pipes, and with respect to the optimal solution found which is dependent on a good starting point and is at most local. This is a result of the highly non-linear and non-smooth properties of the models, and the large number of constraints and decision variables.

A possible approach to overcome these difficulties is to combine previous methods for managing multiquality water distribution systems, with one or a combination of Simulated Annealing, Genetic Algorithms, and Tabu Search -- yielding hybrid optimization techniques.

Simulated Annealing (SA) (Kirkpatrick et al., 1982), Genetic Algorithms (GA) (Holland, 1975), and Tabu Search (TS) (Glover,1986), are heuristic discrete optimization techniques that invoke some implicit law of nature. Simulated Annealing is based on physical processes in metallurgy, Genetic Algorithms seek to imitate the biological phenomenon of evolutionary reproduction, and Tabu Search is based on selected concepts from artificial intelligence.

Following is an example of a hybrid optimization framework for the optimal design of MQWDS. The hybridization is among the method of Ostfeld (1994) (as the management model), and GA (as the heuristic discrete optimization search technique):

1. Initialization

1.1 Set the generation counter, t : = 0 ;

1.2 Generate an initial population, G(0) ;

I. Solve the optimal design problem of a multiquality water distribution system using the formulation of Ostfeld (1994), from several starting points. Within this formulation the objective function is the minimization of system cost (capital and operation); the constraints are on: continuity of flow and energy; pressure heads at consumption nodes; length of each pipeline; power of pumping stations, and threshold concentrations at consumption nodes. The decision variables are: the vector of flows in all pipes for each loading condition; pumping heads for each pumping station and loading condition; the pipe segment lengths; the maximum power of each pumping station, and the treatment facilities volumes and removal ratios.

II. Code the decision variables of each solution into a string in which each bit position takes on a value of 0 or 1. The bundle of strings comprise the initial population. *Note that this stage is contrary to the traditional GA in which the initial population is generated randomly. The hybridization is the result of this stage.*

1.3 Evaluate, G(0).

Each coded string in G(0) is set a fitness value which is a function of system cost. The fitness value is selected such that the probability of the best string to survive will be the highest.

2. Main scheme

Repeat

2.1 Set : t = t + 1 ;

2.2 Generate G(t) using G(t - 1) ;

The operations of Reproduction, Crossover, and Mutation are applied to G(t - 1), to generate G(t).

2.3 Evaluate G(t). G(t) is evaluated by decoding each string to its corresponding decision variables, computing system cost, and setting a fitness value to each string.

Until stopping conditions are met.

Stopping conditions may be some logical combination of: t is greater than the maximum number of generations allowed; for some sequential generations there is no improvement to system cost; a limit of the maximum number of generations since improvement to system cost was reached.

Acknowledgment

This research was supported by the Fund for the Promotion of Research at the Technion.

References

Alperovits E. and Shamir U. (1977). "Design of Optimal Water Distribution Systems". *Water Resources Research,* Vol. 13, No. 6, pp. 885 - 900.

Brooke A., Kendrick D., and Meeraus A. (1988). "GAMS : A User's Guide". *Scientific Press U.S.,* 289p

Cohen D. (1987). "Optimal Design of Multiquality Networks". *M.Sc Thesis,* Faculty of Agricultural Engineering, *Technion - Israel (In Hebrew)*, 114 p.

Cohen D. (1992). "Optimal Operation of Multiquality Networks". *D.Sc Thesis,* Faculty of Agricultural Engineering, *Technion - Israel (In Hebrew)*, 400 p.

Glover F. (1986). "Future paths for integer programming and links to artificial intelligence", *Comp. Oper. Res.,* Vol. 13, pp. 533 - 549.

Holland J. H. (1975). "Adaptation in natural and artificial systems". *Ann Arbor : The University of Michigan Press.*

Kirkpatrick S., Gelatt C. D., and Vecchi M. P. (1982). "Optimization by simulated annealing". *IBM Research Report RC 9355.*

Mehrez A., Percia C., and Oron G. (1992). " Optimal Operation of a Multisource and Multiquality Regional Water System". *Water Resources Research,* Vol. 28, No. 5, pp. 1199 - 1206.

Murtagh B. A. and Saunders M. A. (1982). " A projected Lagrangian Algorithm and its Implementation for Sparse Nonlinear Constraints". *Mathematical Programming Sudy,* Vol. 16, pp. 84 - 117.

Ostfeld A. and Shamir U. (1993a). "Optimal Operation of Multiquality Distribution Systems : Steady State Conditions". *Journal of Water Resources Planning and Management Division, ASCE,* Vol. 119, No. 6, pp. 645 - 662.

Ostfeld A. and Shamir U.(1993b). "Optimal Operation of Multiquality Distribution Systems : Unsteady Conditions". *Journal of Water Resources Planning and Management Division, ASCE,* Vol. 119, No. 6, pp. 663 - 684.

Ostfeld A. (1994). "Optimal Design of Reliable Multiquality Water Supply Systems". *D.Sc. Thesis,* Engineering and Management of Water Resources Program, *Technion Israel (In English),* 164p.

Rossman l. a. (1993). "EPANET Users Manual". *Cincinnati : Drinking Water Research Division, Risk Reduction Engineering Laboratory, US EPA.* 107p.

Rossman L. A., Boulos P. F., and Altman T. (1993). "The Discrete Volume Element Method for Modeling Water Quality in Pipe Networks". *Journal of Water Resources Planning and Management Division, ASCE,* Vol. 119, No. 5, pp. 505 - 517.

Shor N.Z. (1985). "Minimization Methods for Non - Differentiable Functions". *Springer - Verlag,* 159 p.

Fort Peck Reservoir Monte Carlo Simulation Model

William Doan[1]

Abstract

Cylinder gates controlling flow through flood control tunnels in Fort Peck Dam have become unreliable. Use of the tunnels has resulted in operational problems and high operational costs with potential for major damage. In order to evaluate the economic effectiveness of a major gate rehabilitation, a Monte Carlo risk based analysis was conducted on the dam to determine the probability distribution of potential reservoir disruptions and the consequences of disruptions.

Introduction

Fort Peck Dam and Reservoir are located on the Missouri River in northeast Montana. Fort Peck Dam is the fourth largest dam in the United States. The project consists of a hydraulic filled dam, two power plants, two flood control tunnels with outlet works, and a spillway. Flow through the tunnels or outlet works is controlled with cylinder gates which may be unreliable. As part of the risk based analysis of the overall Fort Peck Project, the frequency and consequences of reservoir operation disruptions were estimated to determine the economic effectiveness of any cylindrical gate rehabilitation measures. In order to estimate and combine the frequency of reservoir operations disruptions with the associated consequences of disruptions at Fort Peck Reservoirs, a Monte Carlo simulation of Fort Peck Reservoir was developed. This simulation was required to determine the probability distribution of reservoir damage costs given the probability distribution of variable parameters (reservoir pool levels, reservoir inflows, reservoir operating uncertainties, etc.). The simulation model developed for Fort Peck Reservoir allows for both deterministic and stochastic elements of the reservoir system to be analyzed to determine if repairing the cylindrical gates would be cost effective.

---

[1]Hydraulic Engineer, U.S. Army Corps of Engineers, Omaha District

Monte Carlo Simulation Model

The Fort Peck simulation model can be used to perform the following tasks: (1) generate synthetic annual maximum pool levels using a simple autoregressive time dependent model, (2) generate synthetic annual peak inflows using the probability integral transform, (3) determine the required release rates for reservoir regulation rule curves for given pool levels and inflows, (4) utilize event tree to determine which release mechanisms (power plant, outlet works, spillway) will be used, consequences of use, and damage costs associated with consequence, (5) simulates specified number of years of simulation and determines present value damage costs and benefit-cost ratio for the rehabilitation measure.

Reservoir Pool Levels- The first step in the Monte Carlo simulation was to generate the time series of annual maximum pool levels for Ft Peck. Due to carry-over storage in the reservoir, the pool levels in Fort Peck are not independent. Because of the annual dependency of pool levels, a simple first order autoregressive model was used to simulate the time series of pool levels.

Twenty-five years of data (1968-1992) since the Mainstem Missouri River System has filled and operated as a complete system were used in the development of the autoregressive model. A correction factor was applied to the generated pool levels to insure that the generated pool level probability curve would match the historic pool probability curve. Figure 1 shows a typical time-series of reservoir pool levels for a period of 1000 trials.

Reservoir Inflows- The simulation model was used to generate annual maximum daily inflows into Fort Peck by combining a log-Pearson III analysis with the probability integral transform (Benjamin and Cornell, 1970). An uniform random number between 0 and 1 was selected to represent the cumulative probability distribution quantile value. These randomly selected numbers were applied to the inflow-probability curve to generate the maximum daily inflows. A typical time-series of reservoir inflows is shown in Figure 1.

Reservoir Releases- The simulated releases were based on application of the Emergency Regulation Curves - Late Spring Flood Season for given pool levels and inflows.

Event Tree- For the given required release rate for each year, the model was programmed to select the release system (power plant, outlet works, spillway, or combination). The releases are prioritized in that for the given release rate, the model will first use the power plants if they are available. If the power plants are not available, the model will go directly to the spillway or outlet works, depending on pool elevations and user specified release preference. If the required release rate exceeds the power plant release capacity for the particular pool level, the residual release rate exceeding the capacity of the powerplants is sent to either the spillway or outlet works depending on the pool level and user specified priority. If the residual release rate is sent to the outlet works first, the residual release rate exceeding the outlet works capacity for the particular pool level is then sent to the spillway.

After the release mechanisms have been chosen, the model runs through an event tree to determine possible consequences of use. The event tree consists of possible consequences of using the power plant, outlet works, and spillway. Consequences include minor and major damages of the cylinder gates, failures of the gates, minor and major erosion damage to the spillway, etc. The consequences of use are defined by probability distributions which are randomly sampled with the Monte Carlo simulations. The consequences of uses are converted to damage costs.

Economics- The model repeats for a specified number of years of simulation. The specified period of simulation is divided into fifty-year periods. The fifty-year periods simulate possible scenarios that may possible occur during the life of the Fort Peck project. Within each fifty-year period, the damage costs were converted to present value. The average present value damage cost is calculated by averaging the present value damage costs for all 50-year periods. The model was used to simulate the existing or base condition and with-project conditions. The reduction in average present value damages between the base condition and with-project conditions was divided by the estimated project cost to determine the benefit-cost ratio.

Model Verification

Due to the complexity of the simulation model, verification of the model results were based on the different levels listed below.

Time Series Plotting- The simulation model automatically plots the generated time series of pool levels, inflows, releases, and damage costs as a quick means of examining the model results.

<u>Generated Data Statistical Analysis-</u> The simulation model automatically was used to calculate and plot pool-probability curves and inflow-probability curves for the generated data. The simulated data probability curves were plotted alongside the historic data probability curves to insure that the generated data had the same statistical properties as the parent populations.

<u>Reservoir Operations-</u> The simulation model accounts for all release mechanisms and combinations of release mechanisms that were used in the analysis for comparison to release combinations historically used.

Conclusion

The analysis proved to provide a good method to incorporate stochastic elements of the reservoir along with operating uncertainties associated with the Fort Peck Project. The analysis provided a systematic approach to evaluate the economic effectiveness of a major gate rehabilitation project.

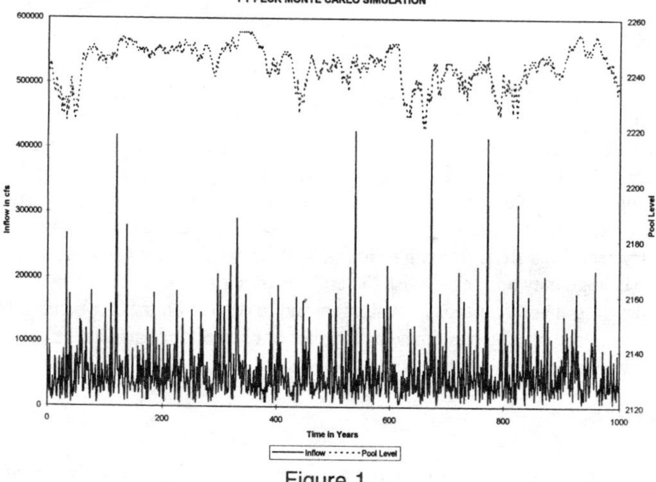

Figure 1

References

Benjamin, J.R., and C.A. Cornell, *Probability, Statistics, and Decision for Civil Engineers*, McGraw-Hill Book Company, 1970.

Salas, J.D., J.R. Delleur, V. Yevjevich, and W.L. Lane, *Applied Modeling of Hydrologic Time Series*, Water Resources Publications, 1980.

Using Probabilistic Temperature Forecasts
to Support Operation of TVA's Power System,
The First Year's Experience

Chris Walker[1] and Janis Dintsch[2]

Abstract

The utility industry is rapidly changing, spurred by the Energy Policy Act of 1992 which gives utilities the right to equitably use the transmission lines of other utilities. Presently, most customers do not have an option on who serves their power needs. Under open access, customers ultimately will have the option of choosing their supplier of electricity. This will increase competition for customers and decrease profit margins to utilities. The effects of open access on the natural gas industry serve as a stern warning to utilities: those who do not adequately prepare for the change may no longer exist. Therefore, utilities must quickly prepare themselves to meet the needs of customers and to zero in on making better informed decisions in order to be positioned as supplier of choice. Probabilistic temperature forecasting is one step on the path to this goal.

Introduction

This new open operating environment has drawn many new players into the electric utility environment. The impending birth of futures and options markets in electricity has captured the interest of companies such as ENRON, Louis Dreyfus, and Morgan Stanley. This interest comes because the dollar volume for electricity options will be several times the physical electricity market, which is $200 billion per year, if the electricity market unfolds as the natural gas market did.

---

[1]Power Supply Engineer, Electric System Operations, Tennessee Valley Authority, 1101 Market Street, Chattanooga, TN 37402
[2]Power Supply Analyst, Electric System Operations, Tennessee Valley Authority, 1101 Market Street, Chattanooga, TN 37402

Open access and the electricity markets force utilities to better analyze the increased risks in their decision-making processes. The weekly bulk power market is an area rich in complexity due to the many factors affecting the decision to buy or sell power for the week. Having identified this as an area with many uncertainties, Electric Systems Operations (ESO) at TVA saw the need for a methodology which would give insight into the risks involved in bulk power market transaction decisions, rather than one which provides a "best estimate". ESO adopted the discipline of Decision Analysis as the methodology of choice to provide that insight.

Application

Many uncertainties affect bulk power market transaction decisions, such as temperature, unit status, rainfall, spot market demand, likelihood of bid acceptance, and load forecasting model accuracy. Decision Analysis shows where to focus attention to improve forecasts. By using decision analysis modeling, ESO can see how sensitive these decisions are to the aforementioned uncertainties. Temperature uncertainty has proven to be the unknown which most often has the greatest impact on which alternative should be chosen, therefore ESO focused its attention on improving temperature forecasting.

Probabilistic Forecasting

Decision Analysis requires explicit encoding of the probabilities of the occurrence of certain events. For example, not only is the weatherman's "best estimate" needed, but also how much lower or higher the temperature may turn out to be.

Therefore, in the spring of 1994, TVA's two operating centers, ESO and the Reservoir Forecasting Center, contracted with two weather forecasting services to provide hourly temperature and quantitative precipitation forecasts for the Tennessee Valley. The two companies entered into a one year weather forecast competition which would determine which of the two companies could provide the more accurate forecasting services for the Tennessee Valley.

A team was formed to establish criteria for temperature forecasts from the weather forecasting vendors. Out of the team meetings came the requirement for the 10-50-90 percentile temperature forecasts. The 10 percentile (low) forecast stipulates that in the long run 10 percent of the actual temperatures should fall below the forecast. Likewise, the 50 percentile (medium or expected) forecast stipulates that 50 percent of the actuals should fall below the forecast, and the 90 percentile (high) forecast stipulates that 90 percent of the actuals should fall below the forecast. Therefore, 80 percent of the actuals should fall between the low and high forecasts. When more than 80 percent of the actuals fall between the low and high forecasts, then forecast uncertainty was assessed more than truly exists. When less than 80 percent of the actuals fall between the low and high forecasts, then forecast uncertainty was assessed less than truly exists.

Analysis of the Probabilistic Forecasts

The terms of the contracts with the two weather forecasters stipulates that the "weather rodeo" is to run from June 1994 through May 1995. It further stipulates that an analysis of the two companies' forecasts will take place on a monthly basis and will be performed by TVA.

There are two components of the analysis; the 48 hour forecast which begins with day 1 (current day) of the eleven day forecast and moves through 48 hours, and the ten day outlook which uses forecasts from days two (next day) through eleven. The 48 hour component consists of two analyses; the average absolute error between actual and expected temperatures and the distribution of actual temperature in relation to the low forecast and the high forecast.

The hourly low, medium, and high temperature forecasts along with other information for a period spanning eleven days are electronically transmitted three times daily in an ASCII format. The first forecast is made at 5:30 A.M. with updates occurring at 11 A.M. and 2:30 P.M. Only the 5:30 A.M. forecast is used in the analysis. Temperature forecasts are made for six points in the Tennessee Valley, five in Tennessee and one in Alabama. The Tennessee locations are Memphis, Nashville, Chattanooga, Knoxville, and Tri-Cities (Bristol). The forecast point in Alabama is Huntsville. The actual hourly temperatures for the six points for the same period are furnished by the National Oceanic Atmospheric Administration and are used as the basis for the monthly analysis. The six-city hourly temperatures in each case are averaged to form one hourly value for both forecasted and actual temperatures.

Analysis Results

As one would expect, the error in the temperature forecast increases with time. That is, more error is expected in the forecast for Day 10 than for Day 1. From July to September, the average error increased by only 2 degrees from Day 2 to Day 10. However, from October to December the average error increased by 4-5 degrees from Day 2 to Day 10. The range of error experienced from October to December was much greater than the range of error experienced from July to September. The average error for the first 48 hours was 2.0 degrees Farenheight for one company and 2.4 degrees (F) for the other.

The distributions of actual temperatures provided some mixed results. From July to September, one company had 96% of the actual temperatures fall between the low and high forecasts, leading to the aforementioned misperception of risk. During this same time period, the other company was skewed to the high side in their forecasts. From October to December, as the temperature forecasting became more uncertain, the distributions for both companies began improving, which gave ESO a more accurate picture of the net system load uncertainty.

Conclusions

Better 10-50-90 temperature forecasts has helped ESO make better bulk power market transaction decisions. This experience helps give a clearer picture of the risks involved in transactions, as well as other decision-making areas. Working with the weather vendors has been beneficial because the vendors understand our needs more clearly and have given unsolicited information which has proven useful. ESO views this new temperature forecasting methodology and relationship with the weather vendor as an important step in determining risks and ultimately becoming a supplier of choice when "open access" arrives.

TVA's Electric System Operations and Reservoir Operations have chosen to use Probabilistic Weather Forecasts

Susan L. Morris[1]

Abstract

Three of TVA's quality task teams have identified reliable weather forecasting as a critical success factor to important processes. These teams are the Spill Management Task Team, Short Term Load Forecasting Task Team, and the Economic Dispatch of Hydro Units Task Team. The objectives of these teams were related, such that they were attempting to determine if greater value could be realized from TVA's existing assets and processes. The increasingly competitive environment in the utility industry today demands that decisions are made which explicitly consider operating uncertainties. In response, more and more new utility models on the market today can receive probabilistic information as input and based on that information give an output range which allows utilities to manage risk and assign dollar amounts to contingencies. Therefore TVA set out on a process to find the best weather forecasting service available.

Strategy

Select the top two bidders using a multi-criteria matrix selection process. A notice was published in the Commerce Business Daily in December 1993. As a result eighteen requests were received for bid packages that included the specifications. Six bid proposals were received. A Weather Forecasting Service Selection Team was formed to consider the bids. Team members represented both Reservoir Operations and Electric System Operations. The team also used a TVA meteorologist as a technical consultant and a Contracting Officer to implement the selection process and the final contracts. Before the bid packages were opened, the selection team developed a detailed criteria matrix to use in evaluating bids. After performance of a pre-screening process to determine compliance with bid specifications, the team decided that site visits should be made to the four remaining

[1]Electrical Engineer, Electric System Operations, Tennessee Valley Authority, MR BA, 1101 Market Street, Chattanooga, Tennessee 37402-2801

companies in support of the quantitative ranking procedure. Two of the companies had been visited previously. After visiting the two remaining companies, clear distinctions could be made, and the quantitative ranking procedure determined the top two bidders. Those two companies were notified in April 1994.

In May 1994, TVA met with the two companies to set up file formats and protocols to send the weather forecast files via modem to Reservoir Operations in Knoxville and Electric System Operations in Chattanooga. The meeting also resolved questions concerning the official benchmark process. It was decided to begin the service on June 1, 1994 and begin benchmark analysis on July 1, 1994. Both the contracts and the benchmark process are scheduled to end June 1, 1995. In addition, the benchmark analysis will only be performed on one of the daily (5:30am CPT (central prevailing time)) forecast. A total of three forecasts are received on a daily basis from each company. The forecasts give the system weather outlook for ten days into the future. It was agreed that a monthly verification report will be prepared and circulated to all parties and that each vendor will be shown both vendors' results.

Analysis of the data consists of verification of Temperature and Quantitative Precipitation Forecasts. Reservoir Operations conducts the benchmark on precipitation on the 5:30am forecast for the 48 hour precipitation forecast. For precipitation the benchmark is performed for the Tennessee and Cumberland river basins. The calculation will also be done for the Tennessee River basin alone. The TVA rain gauge system is being used to calculate the precipitation in each sub basin using the Theissen weighting coefficients which were previously distributed to each of the contractors. The forecast is probabilistic in nature, consisting of low, medium, and high values known as A, B, and C. The absolute mean error compared to the B value (most likely) is calculated for each sub basin and the actual calculated precipitation is compared with the A, B, and C value. Specifications requested that 80% of the values will occur between the A and C values. In addition, 10% of the actual calculated precipitation values should be below the A value and 10% should be above the C value.

Electric System Operations performs the benchmark on temperatures. The 5:30am forecast is used for the next 48 hours of hourly temperature forecast for six cities in the TVA region. The average of the six cities is calculated for the low, medium, and high values known as X, Y, and Z values for each hour. The absolute mean error between the average of the Y (most likely) values and the actual values is calculated. Next, the average of the actual values is compared to the forecasted average hourly values of the X and Z numbers. Again, specifications requested that the average actual values will fall between the X and Z values 80% of the time and that 10% of the time the average actual value will be below the X value and 10 % of the time the average actual value will be above the Z value.

An important goal for TVA was to determine and establish the partnering potential with the forecasters during the benchmark process. The intent was to enter into a close partnership for communication with each of the forecasters to determine which one could convey meaningful weather information in the clearest manner for non-meteorologist users. It was important that the forecaster understand the critical decisions, based on different weather information affecting the TVA power and reservoir systems. It was also important that the forecaster understand the financial implications these decisions have on TVA and the Valley residents. The evaluation of the degree of communication and other value adds along with the accuracy of the different forecasters will be essential in selecting renewal contract options.

Evaluating potential by establishing Value Add estimates in four different areas is another important goal. The first area of documentation is information useful in making decisions in instances when weather forecasts were critical, such as unit commitment and bulk power transaction decisions and regaining flood storage. The second area concerns early communication so that essential personnel can be available during critical operational activities. A third area is the improved quality of information used for scheduling energy-limited generation both weekly and seasonally. The fourth area concerns the adequacy of probabilistic weather information in support of the new generation of operational decision analysis support models. These models are in development and being used operationally and incorporate formal decision analysis processes to explicitly assess risks and rewards of complex decisions.

Progress Report

The following list details the progressive steps taken and processes established for the project thus far and the progressive steps planned during the remaining contract period.

- Contract specifications were received and finalized in November 1993.
- The Request for Proposals were advertised in the Commerce Business Daily December 1993.
- Bid Packages were mailed to 18 companies during February 1994.
- Six proposals were received and evaluated during March 1994.
- Made recommendation to management and finalized funding during early April 1994.
- Awarded contract by May 1, 1994.
- Met with each vendor during May, 1994 to finalize forecast formats and communication protocol.
- Implemented forecasting service on June 1, 1994.
- Began benchmark process on July 1, 1994.
- Provide monthly feedback until June 1, 1995.
- Meet every three months to discuss improvements.

- Celebrate success and learn and improve continuously.
- Make modifications if necessary.
- Evaluate Value Add criteria for entire contract period.
- Assess improved partnering opportunities.
- Evaluate future forecasting needs beyond June 1, 1995.

Using Probabilistic Quantitative Precipitation Forecasts to Support Operation of TVA's Reservoir System, The First Year's Experience

Bill C. Arnwine, PE, M.ASCE(1)
and Arland W. Whitlock, PE, M.ASCE(2)

Abstract
The Tennessee Valley Authority (TVA) has used observed and predicted rainfall throughout its 63-year history to forecast streamflow, and thus inflow, for reservoirs in the TVA system. Until last year, the predictions were obtained from a single vendor for the most likely values for 7 subareas comprising the TVA and Cumberland River areas. As new tools are developed to provide quantitative risk assessments of the uncertainty in scheduling reservoir operations, more detailed QPF data is required to weather variable. Thus, TVA decided to expand its standard QPF forecast to a more probabilistic approach in examining future operating scenarios.

TVA entered into a one-year contract with two forecasting services beginning July 1, 1994, to obtain probabilistic QPF forecasts. These forecasts were requested for the 90 percent exceedence, the most likely, and the 10 percent exceedence values amounts for the next 48 hours, the next 3 to 5 days, and the next 6 to 10 day periods. The 7 subareas were reduced to 5 based on input from potential vendors that this would adequately describe the variation over the Tennessee and Cumberland Valleys. The evaluation of these forecasts focus on quantifying the uncertainty and the accuracy of precipitation predictions.

Introduction
As new models are developed that allow evaluation of uncertainty associated with operation of TVA's reservoir system, data for input

(1) Civil Engineer, Reservoir Operations Planning and Development, Tennessee Valley Authority, 400 W. Summit Hill Drive, WT10-B, Knoxville, Tennessee 37902.

(2) Operations Specialist, Reservoir Operations Planning and Development, Tennessee Valley Authority, 400 W. Summit Hill Drive, WT10-B, Knoxville, Tennessee 37902.

must also be risk-based. TVA has decided to start looking at the distribution of risk associated with rainfall forecasts and to work with vendors toward improvement. Thus, new contracts were issued for weather vendors to obtain probabilistic QPF predictions.

TVA's criteria for the 10-day probabilistic QPF forecast was for the 90 percent exceedence, the most likely, and the 10 percent exceedence amounts, expressed as 6-hour values for the first 48 hours, and 24-hour values for the next 8 days for 5 sub-basins.

These sub-basins are divided along watershed boundaries according to general topographical features to provide homogenity within the areas. These divisions allow forecasters and reservoir planners to account for spatial and timing differences in rainfall events and provide for more accurate forecasts and analysis in each area.

Logistical Issues

Concurrent with the selection of two vendors to provide probabilistic QPF forecasts, logistical issues associated with this major change were being identified and addressed. These issues included standardized formats for compiling and transmitting data in computer sensible form, development of data bases and interfaces with existing operational software for storage and analysis, and determining which set of values, or combination of values and which vendor to use for actual reservoir operating plans. In addition, a process to provide verification of the QPF forecasts to compare the results and provide feedback to vendors for process improvement was developed. An additional logistical issue was how to implement two additional scenarios without increasing scheduling staff needs since corporate downsizing was resulting in a reduction in both the number and experience of the staff.

Because no data was available to support the use of one vendor's predictions over the other, the most likely value predictions from both vendors were averaged for analysis as input into actual operations. In addition, the software was developed to allow the scheduler flexibility to choose data from either vendor or the average of both vendors for the three predicted amounts - the 10 percent, the most likely, and the 90 percent values.

Displaying of the probabilistic QPF data in an easy to understand format utilized graphical displays of the forecasts. This provided scheduling staff a visual summary of both the predicted values and the variation of predictions between weather vendors (Figure 1).

Verification of the QPF predicted values required the computing and storing of rainfall averages for 5 sub-basins utilizing 6-hour rainfall amounts from a 300 gage network. In addition, the data needed to be stored on-line long-term to allow monthly, seasonal, annual, and perhaps multi-year comparisons for comparison, benchmarking, and process improvement.

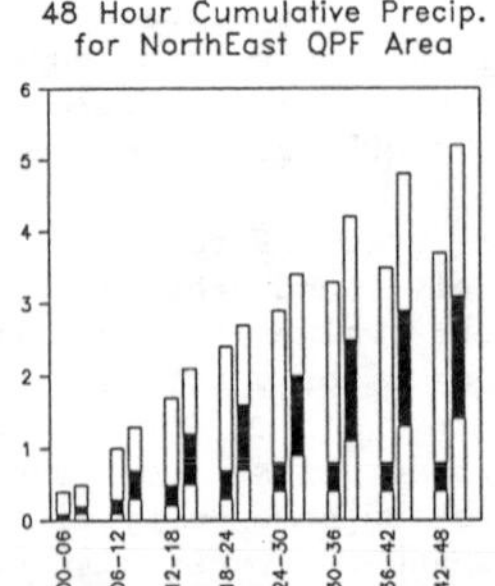

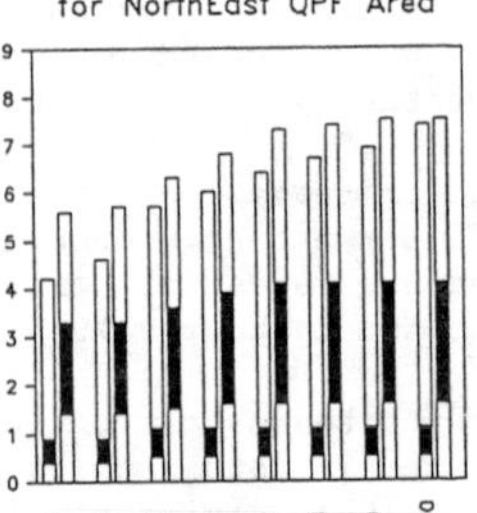

Figure 1

Two bars are shown for each forecast period. The first bar represents vendor 1, the second vendor 2. The top of the bars is the 10 percent exceedence value, the top of the black bar the most likely value, and the bottom of the black bar the 90 percent exceedence value for the January 14, 1995, forecast.

Statistical Evaluations

Six months of data are currently available. Although statistical computations have been made and are shown in Figure 2 for illustration, the data set is much too small for extensive evaluation. This section is to define processes only, and is not intended as a benchmark of the industry standard.

Evaluations included computations of percentiles for observed values versus predicted values for 5 categories - above 90 percent, between 90 and 10 percent, below 10 percent, above the most likely value, and below the most likely value. These computations cover three periods, days 1-2, days 3-5, and days 6-10 for both vendors. In addition, computations are made for the historical, or climatic, data (Figure 2). Comparisons were also made for the three time periods of relative accuracies of the most likely QPF predicted values utilizing forecast ranges that represent increasingly complex decisions by reservoir schedulers.

Lessons Learned

A search of literature and conversations with several of the larger forecast service vendors indicates that no reservoir system operator has undertaken the use of QPF forecasts in their daily scheduling to the extent that TVA is currently doing. In addition, the products that TVA desired were not commercially available. Both TVA staff and the vendors had to develop basic skills and technology to produce and utilize the desired products.

Changes for the Future

TVA and the suppliers have been working very closely in developing feedback methodology and dialog during the last few months. If these partnerships lead to improved accuracy over the traditional 10-day forecasting skills, TVA will evaluate utilizing predicted rainfall in actual operations. New models that are developed will likely require a probabilistic input data set.

Examples of Value-Add

An evaluation of operation to provide the most benefits in meeting the multipurpose demands of the TVA system utilizing the 10 and 90 percentile forecasts will provide the range of decision-making and flexibility that exists in the system. Coupled with an improved scheduling model, the range of possibilities can be computed and evaluated with less time and experience of staff than the standard evaluation of the most likely value that is now used. These additional analyses on a daily basis may allow the system to be operated in the future with a smaller safety cushion. This could be worth millions of dollars in hydropower increase while maintaining or improving benefits obtained for the other statutory purposes.

PERCENTILE COMPARISON

Supplier	FCST Period	ACT<=90	90<ACT<10	ACT>=10	ACT<=50	ACT>50
Perfect Score		10	80	10	50	50
Vendor 1	Days 1-2	54	44	1	90	10
Vendor 2		52	44	3	90	10
Climatic		64	19	17	77	23
Vendor 1	Days 3-5	43	55	2	63	37
Vendor 2		49	38	13	71	29
Climatic		57	25	18	77	23
Vendor 1	Days 6-10	24	74	2	39	61
Vendor 2		33	48	19	61	39
Climatic		47	36	17	72	28

Figure 2
period July 1, 1994 - December 31, 1994

MOST LIKELY VALUE COMPARISON

Forecast Range		0 -.5	.5 - 1	1 - 1.5	1.5 - 2	2 - 3	3 - 4
		average observed amount					
Vendor 1	Days 1-2	.27	.51	.81	.68	.59	.00
Vendor 2		.28	.48	.75	.61	.25	.47
Climatic	2-day total (.26)						
Vendor 1	Days 3-5	.45	.34	.00	.00	.00	.00
Vendor 2		.39	.56	.71	.06	.00	.00
Climatic	3-day total (.39)						
Vendor 1	Days 6-10	.64	.33	.00	.00	.00	.00
Vendor 2		.68	.54	.56	.61	.18	.00
Climatic	5-day total (.65)						

Figure 3
period July 1, 1994 - December 31, 1994

Probabilistic Temperature and Precipitation Forecasts to Support a Customer's Needs: A Vendor's Perspective

H. Thomas Bulluck[1]

Abstract

Weather Services Corporation (WSC) was one of many private weather forecasting companies approached by the Tennessee Valley Authority (TVA) approximately 2 years ago to review the very special and complex needs of TVA for a 10 day probabilistic temperature and quantitative precipitation forecast. WSC responded to a request for bids and was selected as one of two weather suppliers for a period of 1 year. Although probabilities have been used in weather forecasting for a number of years, the length of projection and the amount of detail requested by TVA was a new concept and quite extensive.

This paper will provide an insight to WSC's approach to TVA's requirements for 10 day probabilistic temperature and precipitation forecasts. A brief review of the methodology used in preparing the weather forecasts for Power and Reservoir Operations at TVA will be provided as well as the need to structure a new approach to accomplish these goals.

Background

WSC was founded in 1946 to provide specialized weather information for government and industry. During the past 49 years, WSC has grown into one of the largest and most respected sources of worldwide commercial weather information. WSC's global forecast center in Bedford, Massachusetts, USA is located on the famed 128 technology highway just outside of Boston. The center operates around the clock 24 hours per day, 7 days per week. It is supervised by Certified Consulting Meteorologists (the highest designation awarded by the American Meteorological Society), and is continuously staffed by dozens of experienced consulting meteorologists and support personnel. WSC provides customized weather information services to a variety of industries including electric and gas utilities, energy, agribusiness, media, and government.

---

[1] Utility Projects Manager, Weather Services Corporation, 131A The Great Road, Bedford, MA 01730

Probabilities are used in weather forecasting to convey the likelihood of the occurrence of an event. There usually is no quantified value associated with these types of probabilities. Probabilistic temperature and precipitation forecasting has never been provided by commercial weather providers on the scale TVA receives. The National Weather Service (NWS) in just the past few years has been performing experimental probabilistic precipitation forecasts for use as input into hydrologic models to better forecast river stages. The request from TVA for a detailed probabilistic temperature and precipitation forecast required WSC to develop new forecasting procedures to meet this requirement.

TVA's Probabilistic Forecast Requirements

The probabilistic forecast is a new and innovative method of forecasting because it allows probabilities to be placed on different scenarios of weather events. Assigning probabilities allows the meteorologist to better define what events may occur and provides the end user with more quantified information. TVA's probabilistic temperature forecast includes a very specific 10 day, hourly forecast for 6 weather stations located within TVA's service area. The probabilistic precipitation forecast incorporates a detailed 10 day forecast divided into 8, 6 hour forecasts for all 5 of TVA's basins for the first 48 hours, followed by a 24 hour projection through day 10.

Each hourly temperature forecast consists of three values denoted by "X", "Y", "Z". The "X" value is the lowest forecast temperature expected to occur and the "Z" value is the highest forecast temperature expected to occur. On average, the goal is to have the actual temperatures fall below the "X" value or above the "Z" value only 10% of the time. This implies that the "X" value has a 90% certainty of occurrence and "Z" value has a 10% certainty of occurrence. The "Y" value is the mostly likely forecast temperature to occur and 80% of the time, this temperature should fall between the "X" and "Z" values. A perfect forecast distribution would yield over time a X=10%, Y=80%, and Z=10% distribution of actual occurrence versus forecast during the 10 day forecast period.

The precipitation forecast requirement is identical to the temperature forecast requirements, except the three values are denoted by "A", "B", "C". The "A" value is the minimum precipitation event expected, the "C" value is the maximum precipitation event expected, and the "B" value is the most likely precipitation event that will occur.

WSC's Approach To TVA's Probabilistic Forecast Requirements

Technical and Operations personnel from WSC met with TVA Reservoir Operation and Power Supply personnel in Tennessee to review TVA's operational requirements and the probabilistic forecast goals. After a consensus was reached among both parties on the required parameters and forecast periods, WSC altered its conventional forecasting methodology to meet TVA's objectives.

The probabilistic forecast is unique because the meteorologist is required to provide three forecast scenarios quantifying probability of occurrence for each scenario. Furthermore, forecasting precipitation in a time series forced WSC's meteorologists to re-evaluate how precipitation forecasts are made. In the past, there has been a tendency to over estimate precipitation amounts because the meteorologist tends to increase precipitation amounts whenever confidence is low. The lack of numerous reporting stations to confirm precipitation events makes it difficult to verify forecasts.

To meet the unique requirements of TVA, WSC developed computer programs to manipulate forecast data from WSC's database while incorporating the new probabilistic forecast data for temperature and precipitation. Proprietary programs allow the meteorologist to input necessary forecast temperature and precipitation ranges while at the same time, taking advantage of WSC's computer system to output the forecast information. Due to the coverage of TVA's service area, WSC's most experienced operational meteorologists in the Southeast and Tennessee Valley region were assigned to this project. Using their experience and resources available at WSC, the meteorologists are able to develop the probabilistic forecast elements.

Creating the Probabilistic Forecast

Probabilities in the past have been used to forecast the likelihood of weather events occurring. The probabilistic temperature and precipitation forecasts raises the art of weather forecasting to a higher level and quantifies the uncertainty of differing weather events. More detail is provided but more importantly, specific information is supplied quantifying the likelihood of occurrence and providing a specific value for each probability. WSC's forecasters incorporate past experience while developing new skills to produce the probabilistic forecast.

The meteorologist analyzes current weather data to provide a solid foundation to create the probabilistic forecast. The NWS runs several short and medium range models that produce statistical numerical guidance for use in operational forecasting. Past model performance, as well as model bias based on recent and historical experience, is reviewed and a determination is made as to the likelihood of the success of the numerical guidance. If the numerical guidance agrees with the meteorologist's assessment of the evolving weather scenario, then the meteorologist will likely have a higher confidence. On the other hand, if the numerical guidance is inconsistent or disagrees widely with the meteorologist's opinion, then the meteorologist will have lower confidence in the forecast scenario.

After the meteorologist has decided on a forecast scenario, a projection of the most likely temperature and precipitation event is made ("B" and "Y" values). Next, the meteorologist determines the 10% tails to develop the minimum and maximum forecast values which also yields the 80% occurrence scenario. Finally, local climatic effects on weather are assessed for the overall forecast. Fine tuning is performed on the forecast values and timing of events to finalize the forecast.

Probabilistic forecasting is a new concept and will require experience to improve the accuracy of the distribution curves. The probabilistic temperature forecast should prove to be more accurate initially since temperature is usually more continuos from point to point where precipitation can vary widely from location to location. Hourly temperature data is available to track the performance of a forecast which allows for necessary adjustments when actual weather events deviate significantly from the forecast scenario. Probabilistic precipitation forecasting for an areal average is more difficult since limited information (point verifications) is available for verification of precipitation events. TVA's weighted basin averages have been made available to WSC's meteorologists to verify precipitation forecast accuracy and to improve forecaster skill.

The Consultation Factor

A key asset for TVA decision makers is unlimited consultations with operational meteorologists for collaboration about the forecast or weather events. WSC monitors TVA's service area 24 hours a day, 7 days a week, and issues updated forecasts when conditions deviate enough to require a forecast adjustment. WSC notifies TVA when weather conditions develop that likely will result in significant damage or flooding. This allows TVA to be in state of preparedness while at the same time, mitigating potential liability.

Statistically, climatology will likely beat the weather forecaster on average in the longer range of the forecast period if there is no variable weather or if weather conditions conform to climatological normals. The meteorologist, unlike climatology, will alert TVA to rapid weather changes with ample notification to prepare for impending weather events. Climatology is ineffective in predicting major weather changes or alerting key decision makers to impending adverse weather that may occur. Since the art of weather forecasting is not 100% accurate, consultation with the meteorologist proves to be the single most valuable aspect when critical decisions have to be made.

Summary

TVA is the first utility to commercially employ the use of probabilistic forecasting. The new forecasting methodology requires the meteorologist to quantify and associate probabilities of occurrence with weather variables, something which has never been done in the past. The probabilistic temperature forecast will likely have the best accuracy while precipitation forecasts will be somewhat less accurate, simply because precipitation events are more variable in nature. The probabilistic forecast incorporates forecaster confidence, which gives the end user greater detail about impending weather events than ever before. The probabilistic forecast combined with consultations gives TVA an advantage in making informed decisions that can result in significant economic rewards.

Natural Resource Damage Assessment for Groundwater Resources

Robert E. Unsworth and Timothy B. Petersen[1]

Abstract

Under the Comprehensive Environmental Response, Compensation, and Liability Act (CERCLA), government agencies, acting as trustees on behalf of the public, can claim economic damages resulting from injuries to natural resources associated with the release of hazardous substances to the environment. One of the fastest growing categories of claims brought by state and federal trustee agencies involves contaminated groundwater resources. The purpose of this paper is to review the principal economic, legal and policy issues associated with these claims. A recent groundwater damage assessment is described.

Introduction

CERCLA allows trustee agencies to recover damages for injury to, destruction of, or loss of natural resources, including the reasonable costs of the assessment. Damages include both the cost of restoring the resource as well as interim lost use and nonuse values (i.e., those economic damages that occur from the point when injury first occurs until restoration is complete). These damages are separate and distinct from the cost of site remediation. For example, remedial actions under CERCLA are intended to reduce to an acceptable level the risks associated with the release of hazardous substances to the environment. Remediation, however, may not result in the restoration of injured natural resources to "baseline" conditions (i.e., conditions that would have existed had the release of hazardous substances not occurred). Restoration activities

---

[1] Industrial Economics, Incorporated, 2067 Massachusetts Avenue, Cambridge, MA 02140.

accomplished through the damage assessment process are intended to do just that. Thus, the purpose of the natural resource damage assessment process is the restoration of all injured resources and compensation of the public for losses suffered prior to full restoration of the injured resource.

In order to aid trustee agencies in the development of damage claims, the U.S. Department of Interior has developed rules for the conduct of damage assessments. The process, as defined by Interior, can be divided into three components: documentation of the release of hazardous substances to the environment; documentation of injury to natural resources from such a release; and estimation of damages due to the documented injury. Damages include the cost of restoring the resource, as well as the compensable losses that occur prior to full restoration of the resource. Natural resource injury, as defined in the Interior rules, is a measurable adverse change, either long or short term, in the chemical or physical quality or viability of a natural resource.

Groundwater Damage Assessment

Groundwater is a natural resource under the Interior rules, as are surface waters, biota, geologic resources and air. Injury to groundwater is defined to include exceedences of state or federal drinking water criteria, exceedances of water quality criteria for committed public water supplies, or the presence of contaminants in groundwater sufficient to cause injury to other resources (e.g., the discharge of contaminated groundwater to a surface water body which results in injury to fish). For a full definition of injury to a groundwater resource, see 43 CFR Section 11.62.

Given the large number of hazardous waste sites and other sources of groundwater contamination, trustee agencies are pursuing an increasing number of groundwater claims. Many of the challenges faced in bringing claims for damages to surface water bodies, biota, soils and sediments are not encountered in groundwater claims. For example, there is often little question as to the identity of the responsible party in groundwater cases. In addition, documentation of injury to groundwater generally requires application of commonly used sampling and testing techniques, while efforts to document injury to biota can require complex laboratory and field studies. In many cases the data required to document injury to groundwater already exist. For these and other reasons, the number of claims for economic damage to groundwater resources will likely increase in the future.

The first step in the development of a groundwater damage claim involves documentation of a continuing release to an aquifer. The definition of continuing release has been broadly interpreted by state and federal trustees to include the continued migration of contamination within an aquifer, as well as more obvious ongoing discharges of hazardous materials to an aquifer. If such a continuing release exists, the next step involves determining the extent and nature of the contaminated plume. This step includes documenting the aerial extent of the plume, based on current and expected future conditions, and the concentration of contaminants in the plume (current and expected). In addition, this step generally includes documenting the extent to which the plume has resulted in well closures or is expected to result in well closures. The result of these two steps is a record of injury to the resource; for example, it might be determined that a given aquifer has been contaminated with a hazardous substance over the federal drinking water standard, and that this contamination will ultimately result in the closure of a municipal well.

Once the injury has been documented, damages can be estimated. The most commonly applied method to estimate economic damage to groundwater resources is the added cost of providing a safe drinking water source to affected users of the resource. For example, the release of contaminants to a groundwater resource may result in the closure of a municipal well field. The cost to the affected community to develop an alternative source of drinking water represents one source of economic damage to the resource. In addition, in some cases the presence of contaminated groundwater may result in "passive use" losses. Such losses represent the willingness to pay on the part of the public to protect groundwater, even if they do not currently use the resource.

Below we outline the techniques used and issues encountered in the conduct of a groundwater damage assessment at a state Superfund site.

Example

The Charles George Landfill is located in Tyngsboro, Massachusetts. As a result of the disposal of a variety of commercial, industrial and municipal wastes, the aquifer underlying the Charles George Landfill has become contaminated. Given this contamination, present and future water users who could have utilized this aquifer will be forced to develop more expensive water systems. In this case, all of these water users were households with private wells.

Economists estimate the damage associated with the loss of the use of an aquifer to be the loss in consumer and producer surplus due to the higher cost of meeting the demand for water from an alternative source. Under certain circumstances, the cost of restoring a resource or replacing its lost services can be used as an approximation of the surplus loss. CERCLA and the Interior rules for damage assessment sanction the use of replacement cost as a measure of damages.

In this case we estimated the replacement cost (and thus, damages) by comparing the costs of an alternative water supply (i.e., a central supply and distribution system) for the affected homes with the cost of water to these homes in the absence of contamination. Several factors were considered in comparing these costs. For example, municipal systems provide a fire fighting function, thus resulting in reduced insurance premiums and higher property values. This factor was taken into account in estimating the net economic damage of groundwater contamination. In addition, the study area was expected to experience a growth in housing; thus, our estimate accounted for the added cost to future homeowners who would be affected by this problem.

One consideration in many damage assessment cases is whether the cost of the proposed restoration option is "grossly disproportionate" to the value of the resource. This term has been drawn from the footnote of a judicial review of Interior's regulations for damage assessment, as they were originally proposed. The principal behind such a consideration is that the cost of the restoration action may outweigh the benefit of accomplishing it. Replacement cost is a valid measure of damage only if people value the resource at least as much as the cost of replacement (or at least the costs should not be "grossly disproportionate" to the benefits).

In the Charles George case we implemented a grossly disproportionate test by comparing the cost of replacement with an estimate of the loss in consumer surplus that consumers would experience if the price of water were to increase sufficiently to cover the cost of purchasing water from the next best alternative to the contaminated aquifer. In this case, we found that the estimated lost use value was slightly less than the estimated replacement cost. If they were forced to bear the replacement cost, consumers would likely choose to continue to purchase water, albeit at smaller quantities. Since the replacement and lost use values were similar, however, we determined that the benefits of the proposed restoration action was not grossly disproportionate to the cost.

Substitution Between Natural Spawning and Hatchery Stocking in the Restoration of Atlantic Salmon

Kevin J. Boyle[1]

Abstract

The goal of the Atlantic salmon restoration program on the Penobscot River, Maine is an annual run of 6,000 salmon supported by natural spawning. The run peaked in 1986 with 4,529 salmon returning to the river, but only 1,049 salmon returned in 1994. In a survey of Atlantic salmon anglers we found current salmon anglers prefer natural spawning to hatchery stocking, but that stocking a large number of salmon can compensate for the lack of a natural run. These anglers are unwilling to give up the opportunity to chose to keep a salmon.

Introduction

By any measure, the Penobscot River is the premier Atlantic salmon (Salmo salar) river in the United States. The river is the recipient of the largest Atlantic salmon stocking effort, has the largest annual run of Atlantic salmon, and receives the heaviest Atlantic salmon fishing effort in the United States. The restoration effort is not without controversy; restoration cost over $22 million during the 1980s, a new dam is proposed for the river, and environmental groups advocate listing Atlantic salmon as an endangered species.

The major constituency supporting the restoration effort has been salmon anglers. In 1990 there were 3,299 anglers who held a Maine Atlantic salmon fishing license, with 88% percent of the licenses being held by Maine residents. In addition to providing political support for the restoration effort, angler responses to fishing regulations can influence the ultimate effectiveness of these management actions.

---

[1] Associate Professor, Department of Resource Economics and Policy, 200A Winslow Hall, University of Maine, Orono, ME 04469

We conducted a mail survey of 894 Maine residents holding a 1990 Atlantic salmon fishing license. The purpose of the survey was to examine anglers preferences for the current restoration program and to elicit their preferences for alternative management programs.

Methods

Angler preferences were elicited using conjoint analysis, which is a marketing tool used to elicit consumer preferences for multi-attribute commodities (Green and Srinivasan, 1990). The attributes of salmon fishing considered were the type of fishery (natural spawning or hatchery), size of the annual run (3,000, 6,000 or 10,000), fishing method allowed (fly fishing or flies and lures), catch restrictions (1 salmon and 4 grisle, 5 grisle, catch and release, or any 5 salmon), restoration efforts on other Atlantic salmon rivers in Maine (constant or increased), and cost per day of salmon fishing. (Grisle are male salmon that return to the river prior to sexual maturity.) Current runs of Atlantic salmon on the Penobscot River are supported by hatchery stocking, a 1994 run of 1,049 salmon, fly fishing only, keeping 1 grisle, and constant restoration efforts on other rivers. The goal of the restoration program is to achieve an annual run of at least 6,000 salmon supported by natural spawning (Baum, 1983).

Anglers evaluated different combinations of these variables. Each combination comprised a different management program in the survey. Anglers were asked to rate programs on a scale ranging from 1 (poor) to 10 (excellent). All anglers were asked to rate four randomly selected management programs. Responses to the conjoint question were analyzed using the conjoint rating as the dependent variable and the independent variables were the characteristics used to describe the management programs.

The sample was stratified into two groups. Salmon anglers who currently fish the Penobscot River for Atlantic salmon and salmon anglers who fish other Maine rivers for Atlantic salmon (Penobscot and non-Penobscot anglers here after). Penobscot anglers are the individuals who are most familiar with the fishery. Non-Penobscot anglers have a demonstrated interest in Atlantic salmon fishing and are the anglers who may be most likely to enter the fishery if fishing conditions improve.

Results

The survey was conducted in the fall of 1991. Eighty-two percent of the deliverable surveys were completed and returned. Separate equations were estimated to analyze responses from Penobscot and Non-Penobscot anglers. The null hypothesis of no difference in the estimated vectors of coefficients is rejected (Chi-square=20.7, df=8). Thus, Penobscot and Non-Penobscot anglers provide different evaluations of salmon fishing on the Penobscot River and the empirical results for these two groupings of anglers are reported separately. Complete equations are not reported for each subsample of respondents to economize on space. (Readers seeking more information about the study are referred to Teisl, Boyle and Roe (1995).)

For Penobscot anglers, six variables had significant coefficients with signs that are consistent with prior expectations. Conjoint ratings increase significantly with a fishery supported by natural spawning, a larger annual run, and increased restoration on other rivers. The ratings decrease if lure fishing is allowed, catch and release is required, or if anglers are allowed to take any 5 salmon. These last two results indicate salmon anglers are not willing to give up the opportunity to keep a salmon, nor do they support an angler keeping 5 adult salmon that would otherwise be available to spawn.

Only two variables are significant for Non-Penobscot anglers. Conjoint ratings increase significantly with the size of the run and decrease if lure fishing is allowed. Non-Penobscot anglers are expressing the general disdain of salmon anglers toward lure fishing. Run size is the only other variable with a significant coefficient, which indicates Non-Penobscot anglers are only likely to enter the Penobscot River salmon fishery if the run size increase substantially.

Coefficient estimates are used to predict conjoint ratings for management programs (Table 1). Program 1 received average ratings of 4.6 and 4.9. The catch restriction was 1 salmon and 4 grisle at the time the study was conducted. If the catch restriction was changed to catch and release (Program 2), ratings drop to 3.7 and 4.4. In contrast, increasing the run size to 6,000 salmon with natural spawning improves ratings to 5.7 and 5.9. If the increase is supported by hatchery-stocked salmon the ratings only increase to 5.0 and 5.7. With a hatchery-stocked run of 10,000 the highest ratings are attained, 6.4 and 7.0.

Discussion

Restoration of Atlantic salmon is an expensive undertaking supported by a very small number of anglers who appear unwilling, as a group, to give up their opportunity to have the choice of keeping a salmon they catch. Our results also suggest that new anglers will only enter the Penobscot River fishery if there is a substantial increase in the annual run of Atlantic salmon. The predicted conjoint ratings reveal that while anglers prefer natural spawning to hatchery stacking, they would give a higher rating to a hatchery stocked fishery if the run were substantially above what would be supported by natural spawning, e.g., 6,000 versus 10,000.

Table 1. Predicted conjoint ratings of alternative Atlantic salmon management programs for the Penobscot River.

	Penobscot Anglers	Non-Penobscot Anglers
1. Natural spawning, run of 3,000, fly fishing, 1 salmon and 4 grisle, current restoration	4.6	4.9
2. Natural spawning, run of 6,000, fly fishing, catch & release, current restoration	3.7	4.4
3. Natural spawning, run of 6,000 fly fishing, 1 salmon and 4 grisle, current restoration	5.7	5.9
4. Hatchery-stocked, run of 6,000 fly fishing, 1 salmon and 4 grisle, current restoration	5.0	5.7
5. Hatchery-stocked, run of 10,000, fly fishing, 1 salmon and 4 grisle, current restoration	6.4	7.0

Although our study deals with salmon anglers, there are individuals who do not fish for Atlantic salmon who support restoration. The voices of these individuals have been largely silent until the recent push to list Atlantic salmon as endangered. If Atlantic salmon are listed as endangered, even a catch and release regulation may be deemed unacceptable. Thus, while the efforts to protect salmon are strengthen, the cohesive support of the traditional constituency may wain.

References

Baum, Edward T. 1983. "The Penobscot River: An Atlantic Salmon River Management Report." Maine Atlantic Sea-Run Salmon Commission, Bangor, ME.

Green, Paul E., and V. Srinivasan. 1990. "Conjoint Analysis in Marketing: New Developments with Implications for Research and Practice." Journal of Marketing 54: 3-19.

Teisl, Mario F., Kevin J. Boyle, and Brian Roe. 1995. "Angler Evaluations of Atlantic Salmon Restoration on the Penobscot River, Maine (U.S.A.)." Submitted to the North American Journal of Fisheries Management.

Policy Instruments for Improving the Efficiency of Bureau of Reclamation Supplied Water

Benjamin M. Simon[1]

Abstract: A number of policy instruments are available to promote more efficient use of Bureau of Reclamation supplied irrigation water. These include pricing, facilitating water transfers, conservation incentives, and water quantity restrictions. Different instruments may be appropriate in different institutional and hydrologic situations.

Introduction

Over the past several years the Bureau of Reclamation has begun to redefine its mission to one of resource management, water conservation, and environmental protection. One of the principle questions facing policy makers is the choice of policy instruments that might be used to promote these new objectives. Policy instruments potentially available to improve the efficiency of federally supplied irrigation water include water pricing modifications, facilitating water transfers, conservation incentives, and water quantity restrictions.

Legal Setting for Reclamation Project Water Pricing

The arrangements governing repayment are typically contained in 40 year contracts between irrigation districts and the U.S. In general, capital costs allocated to irrigation are repaid on an interest free basis; costs allocated to power and municipal and industrial uses are repaid with interest. By law, once a project is complete and operational, the overall repayment obligation allocated by the Secretary of the Interior to project irrigators for repayment cannot be changed without Congressional approval. However, the Secretary has broad discretion to establish terms for new or renewed contracts.

---

[1]Economist, Office of Policy Analysis, U.S. Department of the Interior, 1849 C St., NW, Washington, D.C., 20240. The views in the paper represent the views of the author not the Department of the Interior.

Choice of Policy Instrument

Factors that influence instrument choice include implementation costs, the flexibility and reliability of each instrument, the magnitude of potential efficiency gains (or deadweight losses) associated with each, and distributional effects (Roberts, 1976). While similar instruments might be used to improve efficiency and raise revenue, there may be situations where policy instruments achieve one goal at the expense of another, or where unintended results occur. Site specific conditions, existing institutional arrangements, and the underlying distribution of property rights all need to be considered in choosing the appropriate policy tools. For example, unless the price of water can easily adjust to reflect varying hydrologic conditions, an inflexible price instrument may result in smaller efficiency gains compared to what might result if a more flexible policy of facilitating transfers were pursued.

Policy Alternatives

Price Increases: One frequently suggested approach to improve the efficiency with which Bureau water supplies are used is to increase contract prices so that they reflected the marginal value of water for each district. However, determining this price is not straight forward because water is not traded freely in markets, and the value of water (and thus the price level required to induce behavior effects) will vary from location to location depending on conditions that effect demand and supply--climatic and hydrologic conditions, cropping patterns, internal district pricing arrangements, etc. In addition, price increases raise difficult implementation issues: the need to identify the desired effects sought by the price change; the magnitude of the price change required to induce these effects; the methods used to determine the "right" price; the nature of the rate structure that might be adopted; and the extent to which Federal price increases are passed on by water retailers to their customers.

The value of most uses of water in irrigation is probably less than $50 per acre-foot (Young,1984; Gibbons, 1986). While relatively small price increases might be required to cause behavioral effects on some projects, Reclamation contract rates for irrigation water might have to be increased by substantial amounts (at least from 30% to 100% and perhaps more) on many projects to reflect the value of the water supply in its current uses in agriculture. These values would vary from district to district.

While price increases may be attractive to policy makers because the increases could reduce the magnitude of the subsidies involved in repayment of irrigation construction expenses, they are likely to be strongly resisted by water users. On older projects, many of the lands have been resold, so that the present owner has already paid the values resulting from the subsidized water rates to the original land owner. In addition, the differential responses across irrigation districts--due to differences in the marginal value of water as well as to districts' internal pricing arrangements--may make it difficult to achieve targets for obtaining water for environmental or other purposes.

Water Transfers: Removing any disincentives to engage in voluntary water transfers created by Reclamation administrative and contacting practices and law is another means by which more efficient use of federally supplied water could be promoted. Increasing the flexibility of the terms governing the use and distribution of Reclamation supplied water and encouraging short and long term water markets to develop is essentially another means of establishing a price for Bureau supplied water. There is an extensive literature on potential efficiency gains from water transfers. See, for example, Wahl (1989).

Facilitating the ability of water right holders to engage in voluntary transfer activities may be more acceptable to Bureau contractors, easier to implement, and will result in the same outcome as increasing water prices to more accurately reflect the opportunity cost of water. Voluntary transfers would also necessitate considerably less government intervention in terms of administratively determining prices and lend a degree of flexibility that may not be available through pricing policies.

Water Conservation: Yet another means of promoting more efficient water use is through regulations requiring Reclamation contractors to implement water conservation measures and practices, or meet certain water conservation standards. Currently, Bureau contractors subject to the Reclamation Reform Act are required to develop and implement water conservation plans. In addition, the Bureau plans to establish criteria for evaluating water conservation plans modeled on the criteria recently established for implementing the CVPIA. Alternatively, a policy could be developed to provide water users with subsidies for conserving water. To be most effective, the subsidies would need to be provided on the basis of the amount of water conserved, rather than subsidizing the purchase or installation of water conserving technologies.

Neither of these approaches take into account differential district or individual costs for implementing conservation improvements. Establishing conservation requirements will be difficult and require costly monitoring and enforcement activities. While subsidizing conservation activities might find supporters among water users, it has several drawbacks: it is not the least costly means of achieving a given conservation objective and it may raise concerns about providing additional subsidies to water users. In addition, the subsidies would have to be large enough to induce conservation efforts, a problem similar to using price increases. However, conservation requirements (such as establishing criteria to evaluate district conservation plans) do have the advantage of being able to be applied to virtually all Reclamation contractors immediately. This avoids the difficulties associated with attempting to alter existing contracts.

Water Quantity Restrictions: Where legally possible, contracts could be altered to reduce the amount of water delivered to a Reclamation contractor. The difference between the water quantity in the reduced contract and the original contract might then be available for reallocation to higher valued uses. The problems associated with implementing such a policy include determining the extent to which each contractor's water quantities can be reduced, the extent to which the Bureau had the ability to reallocate this "saved" water, and determining the process by which reallocations are made. Another obvious disadvantage of this approach is that it would be difficult to adjust contract quantities except when contracts were being renewed.

Choosing to reduce existing contract supplies might be achieved if the water being provided by the Bureau was not being used in accordance with relevant state laws. For example, if a contractor was not putting water to beneficial use, or was using the water on a parcel of land not associated with the water right, the Bureau might have reason to adjust contract quantities.

Conclusions

A number of policy instruments are available to promote more efficient uses of Reclamation supplied water. Focusing solely on increasing prices raises questions about whether similar results could be achieved in a more cost effective manner using other policy instruments, such as facilitating voluntary transfers of water. Given the differences in locations, climatic conditions, state laws and internal district organization, etc., pursuing a mixed strategy may offer more opportunities to increase efficiency than might be available through a single policy instrument. Other policy instruments--such as facilitating water transfers--may be easier to implement and encounter less resistance from water users. In addition, strengthening its role as a facilitator of voluntary transfers would assist in moving Reclamation forward to meet the challenge of its new mission as a water manager.

References

Gibbons, D. 1986. The Economic Value of Water, Resources for the Future, Washington, D.C.

Wahl, Richard, 1989. Markets for Federal Water, Resources for the Future.

Young, R., 1984. *Direct and Indirect Regional Impacts of Competition for Irrigation Water*, in Water and Scarcity: Impacts of Western Agriculture, E. Engelbert, ed., 1984.

Economic Effects of Climate Change on U.S. Water Resources

Paul Kirshen[1]
Brian Hurd[2]
Mac Callaway[2]
Joel Smith[2]

Study Objectives

The objective of this study is to develop a comprehensive and credible estimate of welfare changes to U.S. water users under alternative projections of climate change. The goal is not to arrive at a single number; rather, it is to determine how the economic value of water resources are affected by simulated changes in runoff and demand parameters that vary across alternative scenarios of climate change. This is accomplished by:

1) Developing Spatial Equilibrium economic models for several large water resource regions across the U.S. -- specifically the Colorado, Missouri, Delaware, and Apalachicola-Chattahoochee-Flint rivers,

2) Using climate sensitive results from simulated runoff models as inputs to the models,

3) Comparing model results (e.g., sector allocations, economic welfare levels, reservoir storage conditions, implicit water prices) across alternative climate scenarios,

4) Extrapolating economic welfare results to remaining U.S. water resource regions, and estimating the aggregate impacts of climate change.

Spatial Equilibrium Modeling

The class of models known as spatial equilibrium models was initially developed in a theoretical framework to rationalize inter-regional trade. Today, spatial equilibrium models are widely used to characterize market behavior in natural resource sectors, such as agriculture and forestry, and in electricity supply and distribution planning.

---

[1]Manager, Water Program, Tellus Institute, 11 Arlington Street, Boston, MA 02116.

[2]RCG/Hagler, Bailly, Inc., P.O. Drawer 0, Boulder, CO 80306-1906

In the context of representing water resources and allocations, the models are designed as non-linear, mathematical programming models with the following elements:

- An objective function that enumerates the various economic values captured by the model. For example, measures of consumer and producer surplus associated with the provision and use of water by various economic sectors in the model.

- A set of constraints that describe the spatial patterns of runoff, storage, and flow, in the basin. These constraints serve to define where and how much water "enters" the basin, where it flows, and where it is stored.

- A set of constraints that describe the spatial patterns of diversion, use, and return flows. These constraints reflect the allocation and use of water along the various reaches of the river.

- Additional constraints, as necessary, to reflect institutional and other economic interests. For example, compact or treaty obligations requiring certain deliveries of water.

Decision variables in the models are chosen to maximize the net economic welfare of society while satisfying physical and institutional constraints. The primary decision variables are:

1) Temporal and spatial allocations of water use across economic sectors,
2) Reservoir releases and storage qualities in each time period,
3) Flow volumes in each river segment for each time period,
4) Economic welfare for consumptive and non-consumptive users.

The models are deterministic and solve for *all* decision variables within a single optimization step (i.e., perfect foresight is assumed). The models are further assumed to simulate steady-state conditions that are expected to exist under changed climate conditions projected to exist in 2060. The models also incorporate projected changes in population and income that are expected to affect energy, municipal, and industrial water demands. Variations in water demand by agriculture due to climate change are also reflected in the models.

Missouri River Basin Model

As an example, the structure and components of the Missouri River basin model are presented in this section. The model uses a seasonal time-step to allocate use, flows, and storage quantities. This allows for greater temporal definition of many of the economic sectors with seasonal variations (e.g., agriculture, flooding, and navigation). Seasonal uses are modeled in three periods of four months each, January-April, May-August, September-December.

Major features and components of the model are derived from major physical features and uses of the Missouri River and its tributaries (the River). Draining a vast region of the interior U.S., the River affects the economies of many states, including: Wyoming, Montana, North Dakota, South Dakota, Nebraska, Colorado, Kansas, and Missouri.

The River is an important source of water for many economic uses, including:

- irrigation and livestock

- municipal and industrial
- electric power generation
- pollution assimilation
- flood control
- navigation
- recreation
- ecosystem and habitat

In developing a spatial equilibrium model of the River, we have focused attention on the first six of these economic uses. These uses are readily identified and relate generally to marketed goods and services with identifiable prices. Prices are a key indicator of the value of the good or service provided by the water. Where possible, we have used observed prices and quantities to identify underlying demand and supply relationships. Similar types of data and information are not available for recreation and ecological services provided by the River. Available time and resources do not permit us to include these important sectors in our analysis. The inclusion of these sectors would unarguably add to the competition for the River's resources. Greater competition would likely result in higher reported opportunity costs and greater welfare losses from reductions in water resources.

The model simplifies the physical system of the River to its most basic elements. These elements are defined by the location of the River's most significant users, tributaries, and reservoirs. Information on water use and storage is aggregated to preserve the fundamental allocations even within the large regions represented in the model.

Groundwater is included as an additional source of water at each of the modeled withdrawal points. For modeling purposes, we have assumed that water supplied from the river is the marginal supply and that groundwater is pumped and used in quantities that are held constant at levels reported by the USGS. This simplification abstracts from the short-run view that groundwater pumping would increase during periods of drought and low flows. However, much of the groundwater in this region (e.g., Ogallala aquifer) has very slow recharge rates and current pumping rates are not sustainable over the long-run conditions that we our modeling.

Consumptive Sectors

In this model, the consumptive sectors are defined as:

- agricultural
- municipal and industrial
- thermoelectric power

Linear specifications are used in these models to represent competing demands. Linear demand functions are estimated from observations on: a) the quantity of water withdrawn; b) an estimate of the users willingness to pay for water; and c) an estimate of the price elasticity of demand in the model for each of the 18 consumptive users in the model.

Non-Consumptive Sectors

Several valuable services are provided by the flow of water left in the River. The value of the River to these non-consumptive sectors, forms an important component of the model. Non-consumptive sectors in the model include:

- hydroelectric power generation
- flood damages (an economic cost)
- navigation
- wastewater assimilation (secondary and tertiary effluent)
- waste heat assimilation from thermoelectric power plants

Hydropower

The Missouri River basin has several hydroelectric power stations. The majority of these power stations are associated with reservoirs constructed by the Bureau of Reclamation.

We have made several simplifying assumptions to model the generation of hydroelectric power. These assumptions include:

- Constant generation head.
- Average generation efficiency of 85%.
- Generated electricity is valued at $54 MWh.

Navigation

Value functions for commercial navigation were developed by the U.S. Army Corps of Engineers. These monthly and flow based values were derived from the costs of alternative methods of transportation that would be necessary during periods in which water flows were insufficient for commercial navigation.

The value of navigation -- as a function of the level of flow in the River -- is conceptually similar to the S-shaped curve known as the logistic function. At minimal flows, navigation is limited to relatively small vessels and is, perhaps, limited to the most downstream reaches of the river. These flows produce minimal benefits. As flow increases, more and larger vessels can be used. Benefits begin to increase at an increasing rate. As flows continue to rise, limitations on the number and size of vessels decrease, and marginal benefits begin to decline. At a certain flow, there are no longer restrictions on the number and size of vessels, and the maximum value for navigation is achieved. This value, we assume, remains constant for all greater flows.

Flood Control

Flood damage functions were developed using data from the U.S. Army Corps of Engineers (1993) for eight reaches in the model. These functions, where appropriate, reflect losses in urban, agricultural, and navigational sectors for each specific reach during the month of June. (Flooding during only one season was modeled to prevent an area being flooded more than once per year.)

Total flood damages for different levels of flow were calculated for each of the reaches of the River. Total damages were calculated by summing damages across each of the different sectors at the various flow levels. These total damages were then plotted against flows to: a) identify the appropriate threshold flow; and b) estimate a functional relationship between damage and flow.

Secondary Wastewater Assimilation

The remaining economic categories relate to the value of water quality. This value reflects the assimilative capacity of the river to accommodate pollutants without violating established standards.

Secondary waste-water treatment is presently employed by municipalities discharging into the river. To account for the value of assimilative capacity of wastewater effluent, we use inflation-adjusted values from a previous study that estimates the value of water for diluting and assimilating BOD in several river basins. These values reflect the costs of pretreating the effluent.

Tertiary Wastewater Assimilation

In the event that dissolved oxygen standards are not adequately met, tertiary treatment of municipal wastewater may be necessary in the future. To account for this possibility, minimum flow levels were identified below which it would be necessary to construct and operate tertiary treatment facilities. Costs are associated with the decrease in the dilution capacity of the flow.

Waste Heat Assimilation (Once-through-cooling plants)

Waste heat from once-through-cooling (OTC) thermal power plants primarily affect near-field river temperatures. During periods of low flow, discharges of heated water are reduced from plants such that temperature tolerances in the river are maintained. At higher flows, the river has greater assimilative capacity and plants can be operated at higher capacities. Once flows reach a specified level, existing plants can operate at full capacity.

An exponential function is used to model the value of river flow for power production at OTC plants. We assume that power production ceases at zero flow levels, and rises at a decreasing rate approaching full capacity at a certain flow. The slope parameter is determined by assuming that 90% of the maximum power is attained at a specified flow level.

Economic Analysis of Optimal Water Policy

Duane J. Rosa, Ph.D., Member ASCE[1]

Introduction

Efficient water policies can be regarded as a set of decision rules in a multistage decision process. In this process, a sequence of decisions extend over time and space in an open system. The first step in analyzing an efficient water policy is to identify the system or systems in the control of which decision rules are sought. Specifically, the characteristics of the water resource system must be evaluated, and then the most efficient water policies determined to achieve an efficient allocation of the resource.

The purpose of this paper will be to analyze the issues involved in defining an effective water resource system policy. Most water resource systems are a mixture of a groundwater-surface water system. In these systems groundwater use can be quantitatively at least as significant as surface water use, and the integration of the two raises some of the most important issues for water policy. Also, the design of these systems does not necessarily involve large engineering structures. The appropriate institutional structures, on the other hand, are a necessary and frequently sufficient condition for their functioning. Such institutions relate to water law which influences water development, water allocation and water quality. They also consist of water district law controlling the establishment, organization and operation of public water districts.

A final and equally important concept in the determination of an efficient water policy is the issue of sustainability. A sustainable policy is one that can persist over generations, is far-seeing enough, flexible enough, and wise enough not to undermine either its physical or its social systems of support. A sustainable water resource policy is thus one that meets the needs of the present without compromising the ability of future generations to meet their own needs.

---

[1]Associate Professor of Economics, West Texas A&M University, Canyon, Texas, 79016.

Water Policy and Decision Levels

In order to find the decision rules necessary for an integrated water resource system, it is first necessary to differentiate among three levels of decision making. On the first or lowest level, the decision making process relates directly to the control of inputs, outputs, and other quantitative characteristics of the water resource system. These characteristics may be either deterministic or stochastic; they may be in physical or in value terms. Decision making at the second level controls the institutional framework of the decision making process on the first level. Finally, on the third level the framework of the decision making process on the second level is the subject of decisions.

Each of these decision levels may also be conceived as optimizing levels. On each level, decision rules are sought for making the best decision on that level. Although decision making processes differ from level to level, they are interrelated because the effects of each decision can be traced through all the lower levels. From the observation of these effects on the water resource system, decision makers can learn how to make improvements in the decision making process at all levels. In this way feedback is provided for future planning purposes.

The majority of all integrated water resource systems contain both public and private operating sectors. Thus, decision making on this level constitutes the first decision level in the hierarchy. The common decision rules for both private and public operating sectors specify maximization of an objective function (ie profit or welfare), under the constraints regarding institutions, technology and resource availability. Maximization is accomplished by fulfilling the necessary and sufficient conditions given by the first and second derivative of the objective function. Time may also be factored into this analysis.

Developing an optimal water resource policy at this level, however, poses many operational limitations. It is often difficult to evaluate and quantify all of the constraints involved in the optimization process. This is especially true for institutions. In social decision making, however, institutions tend to behave more as independent and dependant variables than as constraints. Thus, their actions and reactions are often difficult to analyze and even respond counter to the overall objective process. In the decision making process the issue of achieving a sustainable allocation of water resources at this first level depends heavily on quantifying resource availability and use. Accurate data about total water supplies in a given integrated water resource system is often inaccurate, especially in estimating the amount of groundwater available. Recent technological advances, however, using GIS techniques offer a possible way to categorize the multitude of natural and artificial factors that collectively constitute an integrated water resource system.

On the second level of the decision making process the purpose is not aimed at the control of inputs, outputs and other quantitative aspects of the system, but rather its purpose is to maintain and to increase welfare by continuously influencing decision making on the lower level. This is done under sometimes constantly changing conditions that from any point in time cannot be projected, and they are always uncertain with respect to actual occurrence. Here, each water institution may be regarded as a decision making system that functions as a whole with a particular pattern of change over time. The system is modified through the political process at the federal, state or local levels with the state level being the most important for water institutions.

Performance of such a decision making system at this second level can be evaluated by viewing the system as it functions over time under various economic conditions. The criteria for evaluating the performance of this level must be determined under the unique conditions that are present in the political process. For the second level of the decision making process criteria need to be identified that can serve as conceptually and operationally meaningful proxies for the institutional goal of maximizing social welfare.

The third and final level of the decision making hierarchy for integrated water resource systems in the United States relates to the constitutional organization of the United States. However, all countries and regions of the world do not have the same structure of governmental organization. So we can regard this third level conceptually as a social decision making process that defines the way people within a society are governed.

Criteria for an Optimal Water Policy

Having developed the three levels in an integrated water resource system, the next step is to specify criteria for judging the performance of water institutions. Decision making on the first level, as we discussed previously, is usually a process of maximizing an objective function subject to certain resource, technical and institutional constraints. The overall goal for this process is usually maximization of profits or social welfare. The criteria for judging performance at this level should involve evaluating the specification of the objective function and constraints. Multidimensionality of the objective function has long been a major conceptual and operational difficulty in quantitative optimizing. The constraints must be critically evaluated as to their relevance and operational possibility of fulfilling the necessary and sufficient conditions of the problem. Finally, it is important to keep in mind the differences between the decision making process on different levels. The political process should be brought in at the second and third levels, and not relied upon to solve technical difficulties that arise in the quantitative optimizing process.

Decision processes on the second and third levels are part of the political process. The criteria for evaluating an optimal water policy at this level involves defining the meaning of political agreement and a specification

of the means of bringing it about. Since quantitative precision and optimizing are not involved on this level of decision making, the overall goal should relate to not just making feasible decisions, but rather decisions that are appropriate for the common welfare of society.

Long-Run Sustainability of Water Resource Systems

The last issue in developing an effective water resource system policy that is often overlooked in evaluating water institutions is the aspect of long-run sustainability. For an integrated water resource system to be sustainable, water resources should be used in a manner that respects the needs of future generations. The presence of long-run sustainability should be a critical part of any optimal water policy. However, when water policies are evaluated in terms of efficiency, only the costs and benefits for the current generation of users are usually considered. It is difficult to completely evaluate resource needs for future generations due to inherent uncertainty of determining their water needs.

If we are going to ensure sustainability in water resource systems, three principles should be in place. First, all water resources should be valued using a full-cost principle. By this principle, all users of the resource should pay the full-cost for the use of the resource. Second, a cost-effectiveness principle must be included that insures that a water policy is cost effective if it achieves the policy objective at the lowest possible cost. Cost effectiveness is an important characteristic because it can diminish political backlash by limiting wasteful expenditures. The final principle that needs to be in place is a property-rights principle. Part of the loss of efficiency in water resource systems involves misspecified property rights, which can create perverse incentives. Property rights to all water resources must be clearly defined and specified. Adopting these three principles will go a long way toward insuring a long-run sustainable and optimal allocation of water resources.

Conclusion

This paper has focused on some of the issues involved in defining an effective sustainable water-resource system policy. The approach has not been to establish criteria for economic optimizing of water resource systems, but rather to offer a basis for water policy involving successive stages of decision making.

Sharing the Challenge of Effective Floodplain Management

Shannon E. Cunniff[1] and
Brigadier General Gerald E. Galloway, Jr.[2]

Abstract

Today the nation faces three major problems in floodplain management: people and property remain at risk; loss of habitat; and, attention to these issues varies widely among and within federal, state, tribal and local governments. A national effort must be organized to conduct effective and efficient floodplain management with responsibility and accountability for accomplishing floodplain management shared among all sectors of society. This paper summarizes the recommendations of the Interagency Floodplain Management Review Committee for an integrated floodplain management strategy.

Introduction

In 1993 the Midwest was hit by disastrous flooding. That flooding, which caused approximately $12-16 billion in damages and the loss of at least 38 lives, was one of the most costly flood events in our nation's history and led the Clinton Administration and others to question how the nation manages its floodplains. As this and subsequent floods have clearly demonstrated, people and property remain at risk throughout the nation. Many of those at risk neither fully understand the nature and

[1] Assistant Executive Director, Floodplain Management Review Committee; Assistant to the Commissioner, Bureau of Reclamation, 1849 C St., NW, Washington, DC 20240

[2] Executive Director, Floodplain Management Review Committee; Dean of the Academic Board, US Military Academy, West Point, NY 10996

the potential consequences of that risk nor share fully in the fiscal implications of bearing that risk. Consequently, the Clinton Administration convened an Interagency Floodplain Management Review Committee (FMRC) to independently examine the causes and consequences of the Midwest flood of 1993 and evaluate the performance of existing floodplain management strategies. Their report, *Sharing the Challenge: Floodplain Management into the 21st Century* recommends changes in current policies, programs, and procedures to lessen the vulnerability of the nation to flood damages (FMRC, 1994).

Floodplain management deals with the appropriate use of the floodplains. Modifications from upstream and upland activities can exacerbate flooding conditions downstream, therefore floodplain management must always be considered in the context of each watershed and the realities of the flooding that will occur. The FMRC proposes a better way to manage the nation's floodplains. Its fundamental precept is that all levels of government, all businesses, and all citizens interested in the floodplain should have a stake in properly managing its resources and all those who contribute to flood risk, either directly or indirectly, must also share in the management and the cost of reducing that risk. The FMRC maintains that the federal government must lead by example, state and local governments must manage the floodplains, and individual citizens must adjust their actions to the risk they face. The tools, authorities, and programs are available to move toward accomplishment of these goals -- the issue is not one of lack of understanding of how to manage floodplains and their associated watersheds, but is an issue of determination and organization. The FMRC proposes modifications of many aspects of current programs to unify and streamline the national approach and meet the goals of floodplain management.

The FMRC supports an approach to floodplain management that replaces the focus on structural solutions with a sequential strategy of avoidance, minimization and mitigation. In many cases by controlling runoff, managing ecosystems for all their benefits, land use planning, and identifying those areas at risk, flood hazards can be avoided. Where risk cannot be avoided, damage minimization approaches, such as elevation and relocation of buildings should be pursued. Construction of reservoirs or flood protection structures should integrated into an overall systems approach to flood damage reduction in the basin. When floods occur, damages to individuals and communities should be mitigated with a flood insurance program that obtains its support from those who are at risk. Full disaster support for those in the floodplain should be contingent on participation in self-help mitigation programs. By internalizing these risks, the moral hazard associated with full government support is reduced.

Committee Findings

The division of responsibility for floodplain management activities among federal, state, tribal and local governments needs to be clearly defined. Without a fiscal stake in floodplain management, state and local governments have few incentives to be fully involved in floodplain management. The FMRC recommends legislation to develop and fund a national floodplain management program with principal responsibility and accountability at the state level. It also proposes revitalization of a mechanism to better coordinate federal activities building on the strengths and achievements of the former Water Resources Council.

The FMRC recommends establishing environmental quality and national economic development as objectives of water resources projects and calls for a review of the federal guidelines (Water Resources Council, 1983) to accommodate the new objectives, current priorities and activities of federal water resource development programs, and to ensure full consideration of nonstructural alternatives.

The linkage between floodplain protection, habitat protection, and disaster programs could be strengthened to the advantage of both goals. Specific recommendations are made to: increase attention to the environment in federal operation and maintenance and disaster recovery activities; pursue programmatic acquisition of needed lands from willing sellers; and, to increase post-disaster flexibility in the execution of the land acquisition programs.

Federal pre-disaster response recovery and mitigation programs need streamlining but are making marked progress. To enhance coordination of project development to address multi-objective planing and to increase customer service, collaborative partnerships between federal, state, tribal, and local agencies, and land owners should be sought. To ensure continuing state, tribal and local interest in floodplain management success, the cost-sharing in pre-disaster, recovery, response and mitigation should be the norm.

The National Flood Insurance Program (NFIP) needs

improvement: penetration of the flood insurance into the target market is poor (20-30%); repetitive payments on properties is a significant drain on NFIP resources; insurance can be purchased with a flood looming; the perception that the government will make flood victims financially whole persists and can be perpetuated by government agency actions. The FMRC identified an array of specific problems and solutions. Public Law 103-325 addresses mitigation insurance, extension of the waiting period, and limits the amount of repetitive payments to a property is consistent with the FMRC recommendations.

Opportunities exist to use science and technology to full advantage in gathering and dissemination critical water resources management information to better plan the use of the floodplain and to operate during crisis conditions. An information clearinghouse to provide information gathered by the state and federal governments; state of the art modeling; and, the development of decision support systems are needed for floodplain activities.

Next Steps

Although floods cannot be predicted nor stopped, adoption of a new approach to floodplain management will lessen the vulnerability of our nation to the costly damages and expenses that occur during and following floods. Congress held a hearing on the report during summer of 1994 and is expected to continue its interest. Many of the FMRC recommendations have already been implemented by the Administration. Further Administration decisions should be forthcoming. What remains to be seen is whether the conviction exists across the nation to affect the changes needed. Only through working together with a common vision, can the nation accomplish this mission.

Appendix

Interagency Floodplain Management Review Committee. 1994. *Sharing the Challenge: Floodplain Management into the 21st Century*. Report to the Administration Floodplain Management Review Committee. Washington, D.C.

Water Resources Council. 1983. *Economic and Environmental Principles and Guidelines for Water and Related Land Resources Implementation Studies*. Washington, D.C.

Case Studies of Multi-Objective Management in New England

Kevin Merli, P.E.[1] and Michael Goetz[2]

Since 1978 FEMA has spent almost $5 million per year to purchase flood damaged properties and demolish or remove structures to a non floodprone area under Section 1362 of the National Flood Insurance Program. Under the program FEMA purchased thirteen properties in Scituate, MA and three properties in Falmouth, MA following two coastal floods that hit Massachusetts in August and October of 1991. The project costs, benefits and impacts are provided in References 1 and 2 at the end of this paper.

To qualify for participation in the Section 1362 buy-out program each property must have been insured under the National Flood Insurance Program (NFIP) and have been damaged by a recent flood. In the interest of cost efficiency it is FEMA's goal to buy multiple properties in proximity of each other.

Once the individual properties meet specified flood damage criteria, it must be demonstrated that the economic, social and environmental benefits of the package of properties must exceed the detriments of the project. We have often found that the purchase of flood damaged properties has rendered tremendous benefits to the community and the local, state and federal governments. The economic benefits have included:

- Avoidance of repair costs incurred by the owners
- Reduced flood insurance claims payments made by the National Flood Insurance Fund
- Reduced disaster benefits given to the home owners for damage repairs and living expenses following a major flood event. This assistance is paid by FEMA out of taxpayer funds or by non profit organizations.
- Reduced disaster assistance given to the communities

---

[1] Deputy Director, Mitigation Division, FEMA Region I

[2] Team Leader, Hazard Identification and Risk Assessment Team, FEMA Region I

for debris clearance and public facility repairs.

In addition to the flood loss reduction benefits, other purposes are served by the acquisition program. The creation of recreational areas that are ideal for the floodplain environment. Recreational activities include swimming, boating, sunbathing, walking and picnicking. The quality of the water could be improved by the removal of the residential structures along with the associated substandard septic systems. In addition, wildlife areas could be created.

Following the coastal floods of August and October, 1991, many of the homes of the southeastern Massachusetts area incurred damages exceeding 50% or more of their fair market value. In accordance with the Massachusetts State Building Code (SBC) and the NFIP regulations, these substantially damaged buildings were required to be elevated above the 100 year flood elevation and to meet other flood hazard reduction requirements.

The development of coastal areas often exacerbates erosion rates by limiting the natural movement of sand and beach materials. Foundations, staircases, paved driveways, patios, walls and revetments associated with such development interfere with the natural processes and often increase erosion rates. Communities have a better opportunity to control erosion once the beach properties are removed. This can be done through the construction of sand dunes and by allowing the natural coastal processes to restore normal cycles of erosion and accretion.

Falmouth, Massachusetts

FEMA purchased three properties located of Surf Drive in Falmouth which runs along a barrier beach fronting a salt marsh. The properties, which consist of small beach bungalows serving as second homes, are located on the ocean side of the road and face southeastward overlooking Vineyard Sound and the Atlantic Ocean.

A private non-profit land trust group, the 300 Committee, coordinated the effort to acquire the properties and the Town of Falmouth supported that effort. Following the purchase by FEMA the town took ownership of the properties and will maintain them for open space purposes in perpetuity.

This area is subject to coastal flooding resulting primarily from tropical storms or hurricanes. The most severe floods of the century in this area were caused by hurricanes of 1938, 1944 and 1954. On August 19, 1991, Hurricane Bob struck the New England coastline east of Falmouth. The hurricane was determined to have a 15 to 20-year return frequency. Although this event was small relative to other hurricanes of the twentieth century, it caused considerable damages to low lying properties.

Fortunately, most of the properties carried flood insurance coverage because of their high vulnerability to coastal flooding.

The 300 Committee reached agreements with three of the owners to purchase their properties. In doing so, the owners were pleased to know that the properties would be turned over for open space purposes, i.e., they will be used by the town for a much needed extension to the public recreational beach.

As a result of these purchases, the water quality in the area should improve because of the removal of the homes and their substandard septic systems. Today the vacated properties are being used by the public for swimming, sunbathing, walking, boating and even for educational activities such as field trips for elementary school and day care children. The property owners and federal and state governmental agencies will benefit from the reduced expenditures needed for damage repairs. The Town and its residents will all benefit by the extension of the public beach and the improved water quality.

Scituate, Massachusetts

The second case study is another Section 1362 acquisition project located in Scituate, Massachusetts. The project consisted of the purchase of thirteen homes located on Peggotty Beach which faces northeasterly toward Cape Cod Bay. These homes have been damaged by ten different flood events in a fourteen year period. Like the properties on Surf Drive in Falmouth, the Peggotty Beach properties are located on a barrier beach which fronts a saltwater marsh area. This barrier, however, is not subject to hurricanes but instead to extra-tropical winter storms. These storms, known locally as "nor'easters", occur several times each year during the late fall, winter and early spring. Normally about one event a year does significant damage to the buildings and/or roads on Peggotty Beach.

A major concern in the area has been the loss of the front beach as a result of erosion which has accelerated in recent years due to several factors including the rising sea level and the reduced volume of the littoral sand supply carried along the shore. Much of the sand supply has been trapped by man-made devices such as seawalls, revetments, groins and jetties located to the north and south of the area. The private development of homes and associated appurtenances may have exacerbated the erosion and their removal may help reduce the rate of erosion.

To help further retard the erosion the Town of Scituate plans to construct sand dunes on back portion of the barrier beach on the vacated properties. Without the above effort, the barrier beach could continue to erode and eventually would be breached, thus allowing the storm surge and ocean waves to enter the marsh behind the barrier. This would allow for significant wave action to attack the mainland including the downtown village area resulting in considerable damage.

The former Scituate Building Commissioner Ralph Crossen coordinated the acquisition of the properties and the Town Administrator oversaw the project.

Like Falmouth, the vacated properties will be used as a public recreation beach to support such activities as swimming, sunbathing, boating, fishing, picnicking and walking. In addition, the sand dune area is expected to help support a bird habitat. The water quality is also expected to improve once the buildings and their septic systems are removed.

About the National Flood Insurance Program

The National Flood Insurance Program is a self sustaining program in the average loss year. Since 1986 all premiums that have been collected by the program have exceeded the claims payments. In the last three years the flood insurance fund which is supported by the flood insurance policy premiums has paid for all program expenses including the FEMA and state flood staff salaries, mapping, special studies and the Section 1362 acquisition program.

A program for acquisition and other mitigation actions is necessary to sustain a strong flood insurance fund by reducing or eliminating the most severe flood risks.

In September 1994, President Clinton signed into law the National Flood Insurance Reform Act of 1994. This law has made some sweeping changes to the program aimed at achieving greater flood loss reduction and increasing the sale of flood insurance policies. A key change in the program is the elimination of the Section 1362 acquisition program and its replacement with a Mitigation Grant Program. Under this new grant program communities can receive funds from the NFIP to undertake a variety of flood mitigation options which could include acquisition and relocation, elevation of structures and/or utilities, floodproofing, etc. Criteria for participating in the grant program will be made available when the interim regulations are published by FEMA in March 1995.

References:
1. FEMA Region I. ***Environmental Assessment, Section 1362 Acquisition Project for Scituate, MA***, July 28, 1992
2. FEMA Region I. ***Environmental Assessment, Section 1362 Acquisition Project for Falmouth, MA***, October 8, 1992

Nonstructural Flood Mitigation Projects in New England

John R. Kennelly[1]

Abstract

The Corps of Engineers has constructed a number of flood control projects in New England to address the region's flooding problems. Included are two nonstructural projects, Charles River Natural Valley Storage, Massachusetts and Belmont Park, Rhode Island, which demonstrate the Corps' use of floodplain management approaches advocated in the Galloway Report. These projects are examples of the complementary use of structural and nonstructural measures.

Introduction

The New England region has experienced several major flood events in the last half century, including a devastating flood event in 1955. Torrential rainfall accompanying hurricane Diane resulted in record flooding in New England, leaving 90 people dead and $458 million in flood damages. The Corps of Engineers (Corps) responded to these floods by constructing a regional system of 35 flood control reservoirs and 58 local protection projects at an aggregate cost of approximately $496 million. The total value of flood damages prevented by these projects has been estimated at $2.3 billion. The Corps also constructed two large-scale nonstructural projects to mitigate flood damages in New England. These projects illustrate the Corps' commitment to nonstructural approaches when they can be economically justified within existing Federal regulations, and demonstrate how structural and nonstructural projects can complement each other in solving a region's flooding problems.

In June 1994, the Interagency Floodplain Management Review Committee to the Administration Floodplain Management Task Force completed a report on the Midwest Flood of 1993. This report is commonly known as the "Galloway

[1]Civil Engineer, US Army Corps of Engineers, New England Division, 424 Trapelo Road, Waltham, Massachusetts 02254

Report". The principal recommendation of the Galloway Report is the need for the nation to look at flooding from a comprehensive watershed perspective, considering both structural and nonstructural measures, and evaluating solutions within the context of all social and environmental issues. The report further recommends that the nation establish as a goal the preservation and enhancement of the natural resources and functions of floodplains. The two nonstructural projects completed by the Corps in New England are excellent examples of the floodplain management approaches advocated in the Galloway Report.

Charles River Natural Valley Storage Project, Massachusetts

The Charles River watershed is located in eastern Massachusetts, extending inland from Boston Harbor southwesterly toward the Massachusetts-Rhode Island border. The watershed is hour-glass shaped, approximately 31 miles long by 16 miles wide, with a drainage area of approximately 310 square miles. The upper portion of the watershed is characterized by a meandering river channel and extensive contiguous marshes, swamps and meadows. The lower portion of the watershed is highly urbanized with significant development in the floodplains. The 1955 flood caused $5.5 million in damages in the Charles River watershed. During the flood event, it was observed that peak flood stages in the lower, urban portion of the river occurred within hours of the rains while the natural storage areas in the upper portion attenuated the peak discharges by as much as four days.

In June 1965, Congress authorized the Corps of Engineers to study the flooding problems of the Charles River watershed. The Corps developed a comprehensive solution for the watershed which incorporated two very different approaches. The first study, completed in 1968, recommended a structural solution for the lower portion of the river consisting of a flood control dam, pumps and navigation improvements. The project was completed in 1978 at a cost of approximately $61 million. The second study, completed in August 1972, recognized the importance of the natural storage in reducing flood peaks and the need to preserve these areas to prevent larger floods in the growing metropolitan Boston region. The plan presented in the report, known as the Charles River Natural Valley Storage (NVS) Project, concluded that the preservation of existing natural storage areas was superior to any other solution and recommended the Federal acquisition of 17 parcels of land containing over 8,000 acres of wetlands. The wetlands were acquired between 1977 and 1983 at a cost of approximately $9 million. The report also recommended that the state and local governments provide protection to the additional 12,500 acres of wetlands in parcels that were too small to be advantageously acquired by the Federal government.

The authorized purpose of the Charles River NVS Project is flood control, however, the federally preserved NVS lands also provide benefits to the public in the form of recreation opportunities and fish and wildlife management.

Recreation development consists of the layout and marking of trails, designation of canoe launching areas and provisions for limited parking in upland areas. The Corps encouraged an active fish and wildlife management program to preserve and enhance these resources. The Commonwealth of Massachusetts leases approximately 2,500 acres for this purpose.

The Charles River NVS Project remains a unique project for the Corps of Engineers. Numerous attempts have been made to justify similar projects, but the Corps has been unable to develop a proposal which is economically justified solely on the basis of reduced flood damages. In 1991, the New England Division completed a study for the Commonwealth of Massachusetts to determine the factors that enabled the economic justification of the Charles River NVS Project. The study determined that there were two primary reasons the project was justified. The first reason was the high rate of future NVS loss assumed due to development pressures associated with the construction of a major highway and the state's emerging high-tech boom. The project was formulated prior to the implementation of many current wetlands and floodplain laws which now regulate development in these areas. The second major factor was the assumption of large increased floodplain occupancy as a result of the same development pressures and the resulting increases in estimated flood damages.

The Corps' report concluded that the Charles River NVS Project would not be justified today based on a current economic evaluation. While the Corps' evaluation regulations have undergone only minor modifications since the project was approved, the assumptions regarding wetlands loss and floodplain occupancy would be very difficult to support and would not permit the project to be incrementally justified on the basis of reduced flood damage.

Belmont Park - Warwick, Rhode Island

In June 1982, the New England Division of the Corps completed an investigation of flooding problems on the Pawtuxet River in Warwick, Rhode Island. The study identified a 38-acre residential area of Warwick, known as Belmont Park, which was prone to frequent and severe flooding. In January 1979, a combination of above normal temperatures and precipitation caused extensive flooding of about 30 acres of Belmont Park causing significant flood damage and the temporary evacuation of most homes. The Corps considered several structural solutions consisting of dikes and channel improvements as well as nonstructural solutions. The recommended plan, which was authorized by the Corps and implemented from September 1982 to July 1985, was the purchase or relocation of 61 homes; the purchase of 19 privately owned vacant lots; and the construction of 12 above ground utility room additions to residences that experienced less severe flooding. The Corps also participated with the National Weather Service in the development of a flood warning system to further protect those homes which could be evacuated and where the structures' contents could be protected. The cost of the project was estimated at $4 million. The purchase of the 38 acres of land along the Pawtuxet River effectively prevents any further development in the floodplain and restored the area back to its natural state. The Belmont Park Project was one of the Corps of Engineers' first floodplain evacuation and flood proofing projects.

Conclusions

The Corps' traditional response to flooding in New England is similar to that of the nation, comprised mostly of flood control reservoirs, walls, dikes, and other structural improvements. By 1955, development in the floodplains of the region's major rivers was significant. Wholesale evacuation would not have been practical nor economically prudent. Structural solutions were the logical means of protecting the public. The flood control works constructed by the Corps provide a high degree of protection, prevent millions of dollars in losses and, undoubtedly, save many lives. In spite of their economic achievements, these projects have had some negative impacts on the region. While floodplain management practices are best used in areas of lower development, their inclusion as a non-Federal requirement would have complemented the protection afforded by the Corps project and, as well, would have protected the natural and social aspects of the floodplains.

The evolution of the Corps' role in addressing the nation's flooding problems has been hampered by the nation's difficulty in defining the appropriate responsibilities of the Federal government. The Corps' involvement initially took the form of structural measures because of their association with maintaining the nation's navigation infrastructure and because they protected the public from the dangers of flooding. In doing so, it contributed to the national economy. Floodplain management approaches such as land-use regulations were viewed as local concerns. Because the Federal government would fund structural solutions, local governments supported these measures in lieu of land-use restrictions imposed at their expense. Floodplain management did not find public acceptance until it was recognized that the risk of flood damage was rising faster than the Federal government could fund its solution. The Corps responded to these concerns, broadening its approach to include nonstructural projects and stress the implementation of floodplain management.

The Galloway Report presents a possible next step in the evolution of the Federal flood control program. In order to implement the findings, existing regulations used to evaluate Federal projects need to be updated to reflect the values the public now places on protecting the environmental and social aspects of the floodplains. Efforts such as the Charles River NVS Project in Massachusetts and the Belmont Park Project in Rhode Island demonstrate that the Corps can successfully implement nonstructural projects to address the nation's flood control problems and protect or restore the natural floodplains.

Engineer-Economic Approach to Better Floodplain Usage

by Glenn D. Lloyd JR[1]

Abstract

The Flood of 1993 in the upper Mississippi and Missouri river systems has given new impetus to analyze the interaction of man and nature. The flood has caused us as a nation to again ask the age-old question of the optimum mix of man and his environment. This research looks at the economic/engineer interface and points to some non-engineer reasons for the continued problems and proposes ways to better utilize our limited economic and natural resources.

Introduction

As long as man exists on this planet, two things are certain, there will be rain events that cause floods and man kind will want to reside in proximity to the water sources that are the absolute necessity of life. The river courses that carry water are in a constant state of flux, however the change is measured in a way much different manner than the way man does. We can construct and build structures in months, while nature changes the rivers over decades. Regardless of who is doing the changing, flood events will occur and cause river modifications as well as damage, and the goal of mankind must be to minimize the damages by working with, rather than against, the forces of nature. Like a well-run business, we needed to dissect the various components of the damage costs and then analyze them to see which can and can not be decreased in a cost effective manner.

Flooding/Investment

Excess rainfall events result in runoff that gathers

----

[1] Student Member, Teaching Fellow, Department of Civil Engineering, University of Missouri-Rolla, Rolla, MO 65401

and eventually become the swollen rivers which result in floods of various magnitudes depending on the rainfall event. Mitigation of the peak and volume of runoff can be accomplished naturally with features such as wetlands and natural lakes or thru man made enhancements such as reservoirs, channel improvements that increase flow capacity for the same cross sectional area, levees that increase the cross sectional flow area, and diversion channels that take the water around the effected area. The estimates are that the nation has invested approximately $25 billion in man made flood protection systems and just in the 93 Flood alone, these features prevented an estimated $19 billion in damages which can be added to the savings already totaled from all previous floods that occurred since the system has been in place. The case can easily be made that the cost of the system has more that paid for itself, but that does not address the question of how to further reduce the damages that did occur.

Cost of 93 Flood

The estimated damages for the flood range from $15 to $20 billion dollars with the most reliable number being $15.6 billion. Closer inspection of this number shows that it can be further broken down into sub categories and that when this is done over half the damages are related to agricultural costs. Other areas of damage expense include communications systems, utilities, urban areas and damage to flood protection facilities(levees). The cost reimbursements are not yet final because they have not all been quantified, but, the most recent subtotal is $10.8 and still rising. While much has been said about the negative effects of river improvements, much of the agricultural damage was due to upland flooding away from the river, which was not affected by the rivers. Because all the dollar amounts must be obtained from various agencies and repair work etc. is still on going, the final cost comparisons are not yet available, however, there is a good indication that a large portion of the cost of this flood event is NOT related to the channel improvements and levees that have been built. What is yet to be done, is breaking down of the costs by the categories of transportation systems, utilities, large cities, small cities, rural areas and open land (agricultural and natural), and comparing these to the reimbursements using the same categories. Once done, we can begin to assess if in fact flood control measures such as flood zoning, etc. have worked to reduce costs to man occupied regions and our problem is one of flooding in other than urban areas. Another aspect that is also being looked at is the damage, by category, and the benefits returned to the effected people. No one agency really sums all the inputs of funds put back into a region as a result of floods, and there is a theory that the net cost vs regional benefits may be a net positive which rather than cause an economic hardship creates an economic boom. If this is the case, then this net increase of revenue eliminates the incentive to change the way we do business. The issue of flood relief benefits exceeding costs is a key area that no one appears to have tackled. The costs of the flood appear to be reasonably well documented, but the relief benefits are not summarized by any one agency, to include FEMA. Issues like payments of workman unemployment compensation, contributions from a host of volunteer agencies, federal and state highway dollars, farm subsidies, SBA loans, free FEMA provided housing for up to 18 months and the like are not rolled up into one package for future use and comparison.

Other Systems

Of interest also, is a look at river systems where a single agency has been able to create a comprehensive flood prevention plan by using all the tools available. Systems like the Tennessee River Valley, the Columbia River, the Delaware River and even the lower Mississippi River have had little damage since their systems have been put in place. By contrast, the upper Mississippi and Missouri Rivers systems are not under the control or oversight of any one agency and no comprehensive plan has ever been developed for this system. Even an inexperienced hydrologic engineer can see system gaps in this region. Additionally, the myth that many levees failed must be corrected. Only a few levees failed, most worked exactly as designed and were over topped when the design capacity was exceeded, just as any structural facility will fail if the allowable load is exceeded.

Conclusions

The systematic or holistic approach is being taken in review of this flood and comparison made of all the costs and benefits. The task under way is the assembly of all cost data, engineer, human, agricultural etc., and the breaking of it down by category. Against this, the benefits received to the region and the people residing there are being assembled in like categories. The necessary cost indexing of this event against other events, by category will begin to show where the real costs are rising and/or falling by area affected. Clearly some measure such as hardening key facilities, like water treatment plants, need to be done as well as building in failure points on levees to minimize damage. There is also a need to compare this area to other more sophisticated river systems and their damage costs over time. Finally, the question of the proper balance of activities such as cities, towns, agriculture and natural land will be dictated by the economic judgement based on sound engineer practice.

References

FEMA After action Report, Kansas City Office, The Great Flood of '93, dated 11 December 1993, Jefferson City Field Office

U.S. Army Corps Of Engineers, North Central Division, The Great Flood of 1993 Post-Flood Report, September 1994.

U.S. Department of Commerce, NOAA Natural Disaster Survey Report, The Great Flood of 1993, February 1994.

Report of the Interagency Floodplain Management Review Committee, Sharing the Challenge: Floodplain Management into the 21st Century, June 1994 Washington D.C.

ASCE Magazine, January 1994 VOL. 64 Number 1, Page 38, " When the Levee Breaks" by James Denning

Integrating Community Needs with Engineering Design

Deborah A. Foley, P.E., Member ASCE[1]
and Ferris W. Chamberlin, P.E.[2]

Abstract

This paper describes how the Corps of Engineers and a local community worked together to design and construct a flood control project that integrated hydraulic, geotechnical, and structural design requirements with landscape architecture, recreational planning, and environmental sensitivity.

Introduction

In 1854, settlers attracted to the fertile farm land of southeastern Minnesota staked claims along the South Fork Zumbro River. Despite a major flood in 1855 and a disastrous tornado in 1883, the city of Rochester flourished. The 1883 disaster led to the establishment of the world-renowned Mayo Clinic, a major contributor to Rochester's economy.

Steep channel gradients and rapid runoff in the South Fork Zumbro basin provide the necessary ingredients for flash flooding. Rochester has experienced major flooding more than twenty times since it was founded. The flood of July 1978 was the most devastating, claiming five lives and causing damages in excess of $58 million.

Project Description

Nearly sixty years after it was first authorized for study by the landmark Flood Control Act of 1936, the Rochester, Minnesota flood control project is nearing

----

[1]Project Manager, St. Paul District, U.S. Army Corps of Engineers, 190 Fifth St., St. Paul, MN 55101-1638
[2]Hydraulic Engineer, St. Paul District, U.S. Army Corps of Engineers, 190 Fifth St., St. Paul, MN 55101-1638

completion. Together with a system of small upstream reservoirs designed and constructed by the Natural Resources Conservation Service, the Rochester flood control project will protect the community from approximately the 200-year flood.

The project was a cooperative venture between the St. Paul District, U.S. Army Corps of Engineers and the City of Rochester. It consists of more than nine miles of channel modifications--widening and deepening the existing channel and bank stabilization--on the South Fork Zumbro River and two of its tributaries, Bear and Cascade Creeks. Its structural features also include:

- reconstruction of an existing low-head dam.
- 1.3 miles of levee.
- three grade control structures.
- replacement of one railroad and two road bridges.
- replacement of nine pedestrian bridges.

A non-structural stage of the project involves acquiring and demolishing homes in the floodplain.

Project Planning and Team Building

Project planning, which began in the mid-1970s, included numerous meetings with local officials and citizen groups to solicit input for project design. During the transition from project planning to preconstruction design in the early 1980s, however, much of the early conceptual design that resulted from the local input was lost or given little attention.

Final project design began following congressional authorization in the Water Resources Development Act of 1986. Design was accomplished on a fast-track, beginning with the downstream stages.

As the final design began on the downstream stages, some city representatives who had been involved with the project since the 1970s, began to question the Corps' responsiveness to local needs. Three strategies were devised by the design team to address these concerns.

First, close coordination of project design was established between the Corps and the city through daily contact and frequent site visits. This coordination was aided by the Corps' new project management system, which focussed on customer responsiveness. City staff actively participated in project review. The city also retained a consultant to assist in developing recommendations for landscape design and in design review.

Second, a radical change was made in the culture and operation of the project design team. In the past, the landscape architect typically played a minor role in flood control design--mainly locating trees and plantings as a final design step. The landscape architect was elevated to full-fledged membership on the Rochester design team, with a lead role in the final project design. Aesthetic, recreational, and environmental needs were given equal weight with hydraulic, geotechnical and structural design requirements.

The design team's third strategy was to encourage innovative thinking throughout the design process to create a project that integrated high quality engineering with the needs of an urban environment.

Design Innovations

Riprap had been specified in the preliminary design for bank erosion protection throughout much of the project. Weed growth in riprap can be a visual detriment in an urban environment or in a park setting. The rock provides a source of "ammunition" for would-be vandals and a habitat for undesirable rodents. A number of innovative approaches were used by the design team to address the negative impacts caused by riprap.

Through Mayo Park, the upper part of the riprap slope was replaced with architecturally treated concrete retaining walls. In several park areas, riprap was replaced by interlocking concrete blocks. Voids in the blocks were filled with topsoil and seeded, providing a pleasing visual effect. The blocks also provide pedestrian access to the water's edge.

In several other locations, the upper portion of riprap was buried under 12 inches of sand and topsoil, and sodded, softening the visual impact of riprap. In Soldier's Field municipal golf course, this technique was used together with modular block walls to maintain course playability where holes crossed the river channel.

Through the downtown area, channel construction was constricted by commercial and industrial development. To minimize the channel width, concrete retaining walls were constructed. An elastomeric form liner was used to create a visually-pleasing finish on these walls, as well as on other walls and grade control structures throughout the project. Decorative handrail was used to create visual interest.

A "water feature" was created in the downtown government center area from an existing storm water outlet. A fountain was constructed from fabricated rock. On the walls adjacent to the fountain, a mural depicting the history of the city was cast in the concrete. The water feature is visible from a pedestrian crossing immediately downstream, from a plaza at the government center across the channel, and from pedestrian trails on both sides of the channel.

Local concern about the visual and safety impacts of the grade control structures led to the use of new design concepts developed by the Corps' Waterways Experimental Station. The design for two structures located in densely populated areas incorporated steps into the weir, providing a cascade of flow over the structure and a means of escape should a person fall from the structure.

Innovative teamwork also led to the redesign of an entire stage of the project on Cascade Creek. Through this redesign, channel improvements were restricted to three short reaches while causing only a minimal reduction in the level of flood protection. In addition, a segment of levee construction was replaced by a non-structural flood control plan, consisting of acquiring and demolishing twenty-one homes in the floodplain. The non-structural work, which is now in progress, is estimated to provide $1 million in savings over the cost of the proposed levee.

Recreational Design

Through extensive coordination with city staff, a comprehensive recreational plan was developed for the project, consisting of over nine miles of hiking and bicycling trails, bridge underpasses, pedestrian bridges and plazas, picnic shelters, river accesses, and a canoe launch area. The trail system provides an urban link into the Minnesota Statewide Recreational trail system.

Summary

Construction of the $98 million project began in 1987 and will be completed in September 1995. The costs of the project's flood control features are shared by the city of Rochester and the federal government as 75 percent federal and 25 percent local. Recreational features, such as the bicycle and pedestrian trails, are shared on a 50-50 basis. Amenities such as the water feature were constructed at 100 percent City cost.

As the Rochester flood control project nears completion, the community's delight with its flood control and recreational improvements demonstrates the success of close Corps-city coordination, an emphasis on landscape architecture, and innovative design techniques.

FLOOD CONTROL AND FLOODPLAIN PLANNING FOR THE HAVASUPAI INDIAN RESERVATION

Peter D. Waugh, P.E.[1], Leo M. Eisel, Ph.D., P.E.[2], and Mark R. Bonner[3]

Abstract

Severe flooding of Havasu Creek on the Havasupai Indian Reservation in northern Arizona in 1990 and again in 1993 demonstrated the need for floodplain management on the reservation. The tribe utilized a three-phased approach to floodplain management: (1) repairing flood-damaged facilities with a new emphasis on flood protection; (2) mapping of areas that would be inundated during major floods; and (3) initial planning of a new community away from Havasu Creek to ease land use pressures within the floodplain.

Introduction

The Havasupai Indian Reservation is located in north-central Arizona adjacent to Grand Canyon National Park. The 600 tribal members live in the village of Supai which is situated 2,000 feet below the rim of the canyon on the banks of Havasu Creek, a tributary of the Colorado River. Supai is eight miles from the closest road and accessible only by foot, horse, or helicopter.

Major floods occurred on Havasu Creek in September 1990 (20,300 cubic feet per second [cfs]) and February 1993 (13,800 cfs). The September 1990 flood, although larger in magnitude, was caused by a brief thunderstorm and therefore had a short flood peak and caused only relatively minor damage. The February 1993 flood was caused by six days of continuous rain. The flood peak lasted several hours and caused severe erosion and relocation of the creek channel.

---

1 *Senior Water Resources Engineer, Wright Water Engineers, Inc., 2490 West 26th Avenue, Suite 100A, Denver, Colorado 80211.*

2 *Former Vice President, Wright Water Engineers, Inc., 2490 West 26th Avenue, Suite 100A, Denver, Colorado 80211.*

3 *Civil Engineer, Wright Water Engineers, Inc., 2490 West 26th Avenue, Suite 100A, Denver, Colorado 80211.*

Numerous Tribal facilities were damaged including the 500-year old irrigation system, pedestrian/equestrian bridges and trails, and water, sewer, and electric facilities.

A three-phased approach was taken to address the floodplain management problem on the reservation:

1. Flood-damaged facilities were repaired and bank stabilization was performed with a new emphasis on flood protection;

2. Floodplain mapping was prepared for a variety of flood recurrence intervals for use as a planning tool for the tribal government; and

3. Initial steps were taken to develop a new community away from Havasu Creek to ease land use pressures in Supai that contributed to building within flood prone areas.

Flood Inundation Mapping

The Havasu Creek drainage basin is 2,950 square miles in size. The upper portion of the basin has desert plateau topography. The lower portion of the watershed is narrow canyon with vertical rock walls.

A hydrologic analysis using gaged watersheds similar to Havasu Creek was carried out to determine the peak flow rates of the 2-, 10-, 50-, and 100-year recurrence interval floods. The U.S. Geological Survey (USGS) gage on Havasu Creek near the mouth had information regarding the 1990 and 1993 floods, but with only five years of record, had insufficient data upon which to base the hydrologic analysis. Fourteen large, gaged watersheds in northern Arizona were identified as potential watersheds to be used in the regional hydrologic analysis. Two of these watersheds have similar topographic and climatic characteristics as Havasu Creek. They also have a long period of record. Data from these watersheds were analyzed using a Log Pearson Type III distribution. The Havasu Creek flood flows were interpolated from the two gaged watersheds based on watershed area. The peak flows for the 2-, 10-, 50-, and 100-year return frequencies were estimated to be 5,300 cfs, 11,900 cfs, 18,600 cfs, and 21,700 cfs, respectively.

The U.S. Army Corps of Engineers HEC-2 computer program was used to determine sub-critical water surface profiles for Havasu Creek for flows of 5,300 cfs (estimated 2-year event), 14,000 cfs (estimated February 1993 flood, approximate 25-year event), and 21,000 cfs (estimated September 1990 "historic flood," approximate 100-year flood). Channel and floodplain geometry were obtained from aerial mapping conducted specifically for this purpose in July 1993. Base mapping was produced on a scale of 1 inch = 100 feet with 2 foot contour intervals. Twenty (20) cross-sections were used in the model along 7,300 feet of channel length resulting in an average of one cross-section for every 330 feet of channel.

Since normal base flow for Havasu Creek is approximately 80 to 100 cfs, all three flows modeled produced substantial overbank flooding. With this in mind, Manning's "n" values, channel and overbank reach lengths, and other hydraulic parameters were selected to best represent severe flood conditions. The following summarizes the hydraulic parameters used in the model:

Manning's "n"		**Transition Loss Coefficients**		
Left Overbank	Right Overbank	Channel	Expansion	Contraction
0.045 - 0.05	0.045 - 0.05	0.03	0.3 - 0.5	0.1 - 0.3

To calibrate the model and validate the model results, a preliminary topographic map was produced with the extent of flooding (floodlines) plotted for both the 1990 (21,000 cfs) and 1993 (14,000 cfs) floods. Residents of Supai were asked to accurately identify the exact location of flooding near their homes for both floods on blank topographic maps. The firsthand accounts were then compared with the model results indicating comparison in every case. Good calibration existed between the HEC-2 model and the actual 1990 and 1993 flood conditions.

With the model calibrated by firsthand accounts, a floodplain map was produced for the September 1990 "historic flood" of 21,000 cfs. The floodplain spanned from wall to wall within the canyon except near the center part of Supai. A large number of homes and roads were located within the defined floodplain. However, no homes were lost in either flood event and only two homes were in serious jeopardy of being lost.

Because a large number of homes are located within the floodplain, additional land is not available, and the purpose of this project was for future tribal planning and development, an alternative to the standard engineering practice of defining the floodway and flood fringe within the floodplain was selected. The floodplain was separated into the following three zones based largely on depth and velocity:

1. **Low Hazard Flood Zone**
   Depths are shallow (0 to 2 feet).
   Velocities are low (0 to 4 feet per second [fps]).
   Risk to human life is relatively low.
   New structures should be elevated a minimum of 1 foot above the 1990 flood stage.

2. **Intermediate Hazard Flood Zone**
   Depths are moderate (2 to 4 feet).
   Velocities are high (3 to 6 fps).
   Risk to human life is relatively high.

New building is not recommended; existing structures should be relocated where possible.

3. **High Hazard Flood Zone**
Depths are extremely deep (4 to 15 feet).
Velocities are extremely high (5 to 20 fps).
Risk to human life is extremely high.
No permanent structures allowed.

Repair of Flood Damaged Facilities

The February 1993 flood caused damage to a number of tribal facilities including: (1) rendering two fifty-foot span pedestrian/equestrian bridges unsafe; (2) destruction of approximately 400 linear feet of 3-inch diameter potable water pipeline; (3) exposing 200 linear feet of eight-inch diameter gravity sewer line; (4) destruction of 2,000 linear feet of foot/horse trail; (5) wash-out of 1,500 linear feet of irrigation ditches; and (6) relocation of approximately 1,000 linear feet of the Havasu Creek channel by cutting the original ground surface down by about eight feet.

The reconstruction efforts were guided by several constraints. There is no road access to the site. All construction equipment and materials were transported to the site using horse or helicopter. There was little heavy equipment available for construction works. The community is located in an environmentally pristine setting and therefore all construction work had to take this into account. The usable land within the community is extremely limited and there is no room for relocation of homes or community facilities.

The water, sewer, and electric facilities, as well as the irrigation ditches, were repaired or replaced in their original locations because there was no room for relocation out of flood hazard areas. The damaged portions of trail were relocated where possible. Where sacred tribal sites limited the trail to its present location, it was rock-armored to withstand low frequency flood events. There was insufficient funding available to construct bridges out of the flood hazard area. The bridge abutments were designed to resist flooding while the wide flange steel girders can wash off the abutments during severe flood events.

Channelization of the creek through Supai was considered as a means of avoiding future flood damage to the community. Numerous obstacles to this solution (including lack of readily available construction equipment and materials) prevented its implementation. As an alternative, groins constructed of native rock were placed in four locations to train the creek and keep it in its historic channel. Additionally, riprap bank protection was provided to channel banks in areas where HEC-2 modelling showed high velocities.

The Alberto Flood of 1994
Water Management and Flood Forecasting Aspects

Edmund B. Burkett and Cheryl L. Struble[1]

Abstract

In the aftermath of Tropical Storm Alberto, heavy convective rains of ten inches or greater fell across a large area of central Georgia and southeastern Alabama. Record or near record flooding occurred at numerous sites along the rivers and streams in these areas. The fact that there had been many years since any comparable floods had occurred made this a particularly challenging flood to forecast and to manage.

Storm Background

On July 1, 1994 an area of low pressure near Cuba strengthened to become the first tropical storm in the Atlantic hurricane season. The storm drifted almost due north and was close to hurricane strength as it made landfall on the morning of July 3rd near Ft. Walton Beach, Florida. The storm produced only minor wind and storm surge damage along the coast and initial indications were that this storm was typical of the three of four tropical weather systems that affect the central Gulf Coast every Summer and Fall. However, because of high pressure to the north which resulted in stagnation of upper air steering currents, the storm stalled between 32° and 33°, approximately mid-way along the Georgia-Alabama border.

Over the next four days the storm maintained a distinct circulation and meandered across four river basins producing very heavy rainfall in west-central and southwestern Georgia and in southeastern Alabama. The storm track is shown in Figure 1. Storm rainfalls of over ten inches occurred over a large area. The greatest amount of rain and, coincidentally the greatest loss of life, occurred in Americus, Georgia. About 635 mm (25 inches) fell in the five days of the storm. Figure 2 shows the storm rainfall totals across the region.

---

[1]Chief and Civil Engineer, respectively, Water Management Section, U. S. Army Corps of Engineers, Mobile District, P. O. Box 2288, Mobile, AL 36628-0001

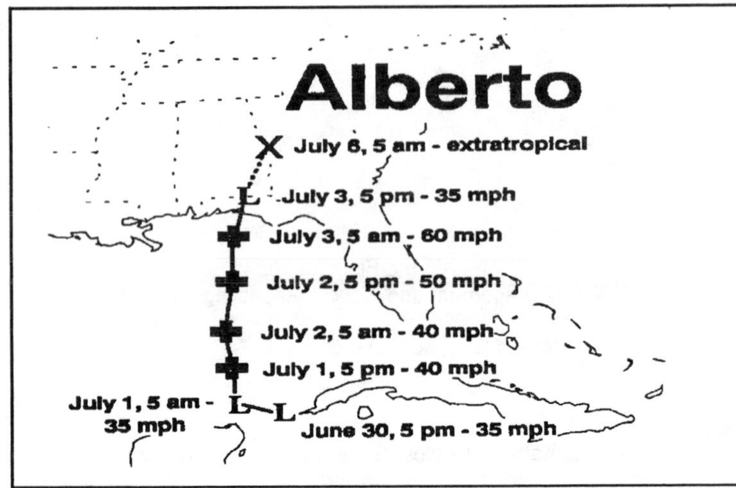

Figure 1

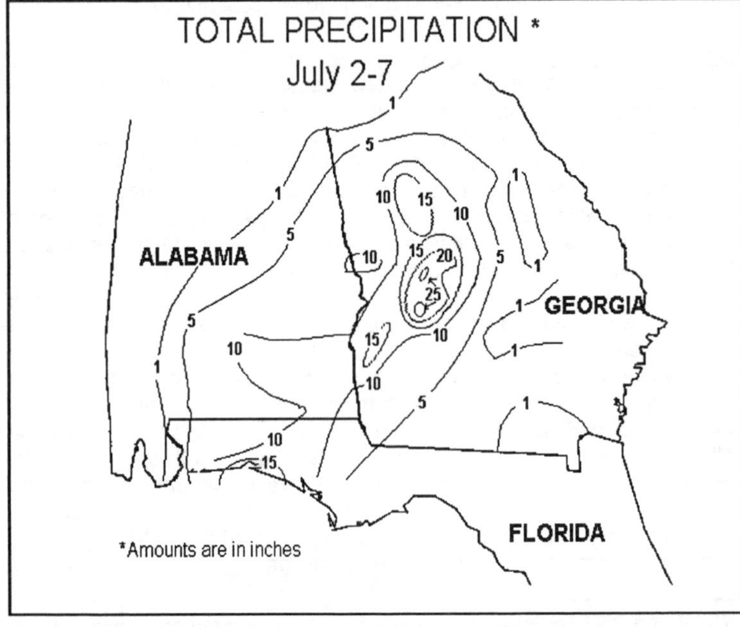

Figure 2

Flood Description

For weather forecasters and water managers a perplexing aspect of this storm was the lack of predictable movement. The National Weather Service issued Quantitative Precipitation Forecasts based on the storm gradually drifting northeast up the Appalachians. However, the storm stalled about 121 km (75 miles) southwest of Atlanta for a day and then drifted back into south-central Alabama before dissipating on July 7th.

The storm produced greater than 100-year floods in many streams. Table 1 illustrates a log-Person Type III frequency analysis for flows at selected sites. Table 1 also shows the flow for July of 1994. Table 2 illustrates the magnitude of flooding at selected sites in the affected basins.

Table 1 - Frequency Analysis

	Return Interval (years)					
Station	**500**	**200**	**100**	**50**	**25**	**July 1994**
Chattahoochee, Fl.	325,000	284,000	253,000	224,000	195,000	209,000
Blountstown, Fl.	320,000	281,000	252,000	224,000	196,000	210,000
Albany, Ga.	130,000	112,000	99,100	86,700	74,700	120,000

Table 2 - Flooding Data

River	Station	Flood Stage (ft)	Previous Highest Stage (ft)	Date	July Crest (ft)	Date
Chattahoochee	W.F. George TW	134	158.5	3-17-29	149.9	7-6-94
Flint	Albany	20	37.8	1-21-25	43.0	7-11-94
Flint	Bainbridge	25	40.9	1-24-25	37.2	7-14-94
Ocmulgee	Macon	18	29.8	3-19-90	35.3	7-7-94
Chattahoochee	Andrews TW	113	122.1	3-19-90	124.3	7-7-94
Apalachicola	Woodruff TW	66	74.2	3-21-90	76.3	7-10-94
Apalachicola	Blountstown	15	28.6	3-21-29	27.4	7-10-94
Choctawhatchee	Caryville	12	27.1	3-17-29	23.9	7-9-94

Reservoir Activities

Although the flood affected several local flood projects, the flood originated downstream of most major flood control storage projects. The one project which did modify the flood hydrograph slightly is the Walter F.

George Reservoir on the Chattahoochee River. On July 6 the daily peak inflow to the reservoir was approximately 4420 cms (156,000 cfs). The flood control plan produced a daily discharge of 2950 cms (104,000 cfs).

A major concern in this event was flooding at and downstream of the junction of the Flint and Chattahoochee Rivers. As Figure 3 indicates, the peak flows for the two rivers arrived about 5 days apart. Figure 3 is an illustration of the stages at Andrews and Bainbridge, Georgia which are located at points above the junction of the two rivers on the Chattahoochee River and the Flint River, respectively. The greater of the two peaks came from the Chattahoochee and exceeded the gated-spillway capacity of about 5950 cms (210,000 cfs). However, auxiliary and emergency spillways did not come into play. The peak discharge was about 6570 cms (232,000 cfs).

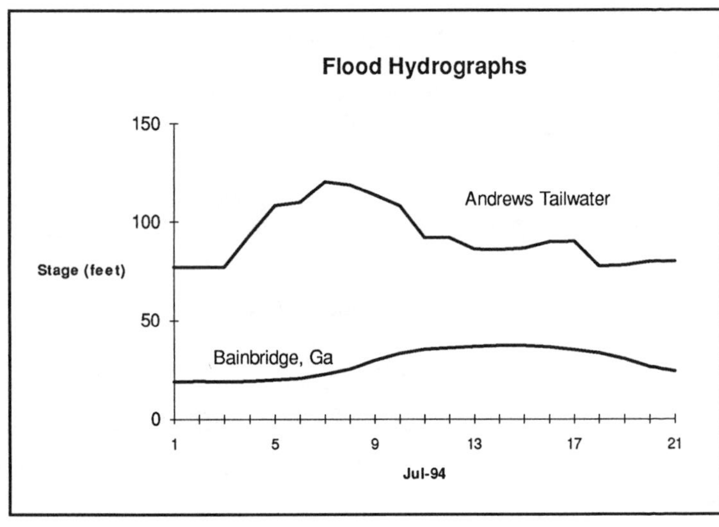

Figure 3

Summary

The Alberto Flood was a major catastrophic flood event equaling or exceeded the food-of-record on many stream. Although flood storage projects were not a major factor in managing the flood, the event provided water managers valuable lessons in flood fighting and flood management.

Water Resources Management
with the NWS
Water Resources Forecasting System (WARFS)

John J. Ingram[1], M. ASCE

Abstract

The Water Resources Forecasting System (WARFS) is managed within the Department of Commerce, National Oceanic and Atmospheric Administration (NOAA), National Weather Service (NWS), Office of Hydrology. WARFS is an operational program providing for the modernization of hydrologic forecasting services across the Nation. Preparation for WARFS national implementation activities have begun within the upper Mississippi River basin.

Introduction

Deaths and economic losses resulting from The Great Flood of 1993 and our Nation's floods and droughts of 1994 have heightened the need for improved predictions to support flood/drought management and damage mitigation (NWS, 1994). Furthermore, the allocation of water among competing demands (e.g. fisheries, irrigation, hydropower and municipalities) looms as a national problem that requires improved water quantity forecasts for sustainable development. For these needs, the National Weather Service (NWS) is preparing for national implementation of the advanced Water Resources Forecasting System (WARFS).

The NWS has the mission to provide river and flood forecasts and warnings for the protection of life and property, and to provide hydrologic forecast information for economic and environmental well being. In support of this mission, the NWS operates thirteen

[1] WARFS Program Manager, Office of Hydrology, National Weather Service, 1325 East-West Highway, Silver Spring, MD 20910

regionally based River Forecast Centers (RFCs) across the Nation. The RFCs typically issue stage forecasts for only 1, 2, and 3 days into the future at most of its 4000 forecast points and crest forecasts out to about 1 week for a few selected forecast points. WARFS, including Extended Streamflow Prediction (ESP) enhancements, will provide for analyses of streamflow trace ensembles within specified future time windows, objectively couple meteorological/climatological forecasts in the ensemble analysis, provide for a variety of probabilistic analyses of ensembles, and package probabilistic streamflow forecast products with extended lead times (out to several months). Thereby, WARFS will provide river forecasts which not only account for precipitation already on the ground but also will probabilistically account for estimates of future precipitation.

Forecasting System Advancements

WARFS provides an advanced hydrologic prediction system as it is coupled with: 1) NOAA's current scientific and operational infrastructure, including the NWS River Forecast System (NWSRFS, Fread, et. al., 1991); 2) National Weather Service (NWS) modernization technologies (Friday, 1994), especially NEXRAD (NEXt Generation Weather RADar - WSR-88D) and AWIPS (Advanced Weather Interactive Processing System); and 3) cooperative and supportive partnerships with other government agencies, universities, and the private sector. WARFS, including ESP enhancements, will take advantage of all these program relationships by building on the technological and information framework they provide.

WARFS is an integrated real-time modeling and data management/analysis system which includes provisions for the use of historical hydrologic/hydrometeorologic data and meteorological/climatological forecasts for input to ESP simulations (Figure 1). As indicated above, implementation of WARFS services builds upon the NWSRFS and NWS modernization technologies and is divided into three interdependent functional requirement areas: Integrated Data Management and Analysis, Advanced Hydrologic/ Hydraulic Modeling, and Advanced Product Packaging/ Dissemination. ESP is the portion of NWSRFS which produces probabilistic forecasts out to several months. WARFS probabilistic forecasts not only will greatly improve the capability of emergency managers and water facility managers to take timely and effective actions to mitigate the

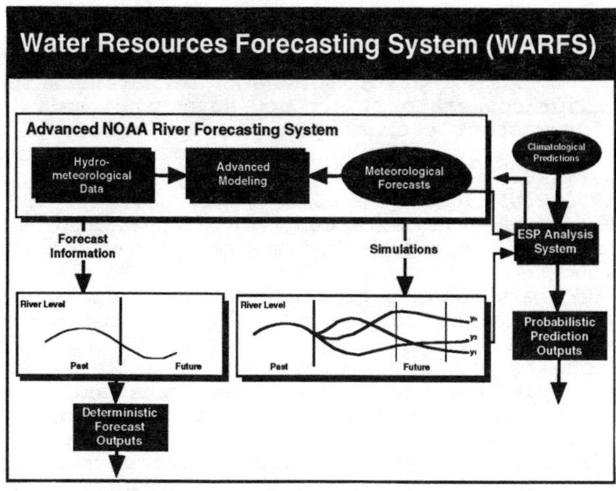

Figure 1. Schematic of the Water Resources Forecasting System (WARFS) integrated real-time modeling and data management/analysis system.

impact of major flood and drought situations, they also will provide better support for overall water resources management (e.g. better management of competing water demands between irrigation, fisheries, and hydropower). For example, the application of ESP on the Nile River has increased forecast lead time of the annual flood from two weeks to over three months with acceptable accuracy.

Current and Potential Implementation Activities

Preparation for WARFS National implementation activities have begun within the upper Mississippi River basin at the NWS North Central River Forecast Center, located in Minneapolis, Minnesota. Further implementation will expand to other parts of the Mississippi Basin, and begin in additional basins. These will include basins which are of critical economic and environmental importance. As these technologies are advanced, they may be shared with other countries through memoranda of agreement or through other technological exchange mechanisms as appropriate.

Summary

The Department of Commerce's National Oceanic and Atmospheric Administration (NOAA), in partnership with other major cooperators, are now advancing their hydrologic forecasting technologies through the development and implementation of the Water Resources Forecasting System (WARFS). This effort is a key component in the *NOAA 1995-2005 Strategic Plan* (Baker, et.al., 1993) to improve prediction services for the Nation and to enhance NOAA's role in environmental prediction.

WARFS extended forecast lead times (up to several months) will allow for: more effective mitigation of extreme events (e.g. floods and droughts), improved operations of water resource facilities (e.g. irrigation and hydropower facilities), and enhanced ecosystem management (e.g. fisheries and wetlands management). As these advanced technologies are developed, DOC/NOAA/NWS can exchange them with other governments in order to help meet sustainable development needs. These exchanges may be executed through memoranda of agreements or through other technological exchange mechanisms as appropriate.

References

Baker, D. James; Douglas K. Hall, Diana H. Josephson, and Kathryn D. Sullivan, July 1993, *National Oceanic and Atmospheric Administration, 1995 - 2005 Strategic Plan*, NOAA, Washington D.C.

Fread, Danny L., George F. Smith, and Gerald N. Day, January 1991, "The Role of Real-Time Interactive Processing in Support of Water Resources Forecasting in the Modernized Weather Service," *Seventh International Conference on Interactive Information and Processing System for Meteorology, Hydrology, and Oceanography*, New Orleans, La. pages 294-298.

Friday, Elbert W. Jr., January 1994, "The Modernization and Associated Restructuring of the National Weather Service: An Overview," *Bulletin of the American Meteorological Society*, pgs. 43-52, Vol. 75, No. 1.

NWS, February 1994, *Natural Disaster Survey Report, The Great Flood of 1993*, U.S. Department of Commerce, National Oceanic and Atmospheric Administration, National Weather Service, Washington, D.C.

Lessons Learned From The Great Flood of 1993

Eugene A. Stallings[1], M.ASCE

Abstract

The Great Flood of 1993 in the Upper Mississippi Basin devastated over 500 counties in nine states and was unprecedented in modern history. Immediately following this event, a National Oceanic and Atmospheric Administration (NOAA) disaster survey team was formed to identify opportunities to improve NOAA's weather and flood forecast and warning systems, not only for the nine-state area but also throughout the Nation. The NOAA report, issued in May 1994, summarizes 106 findings from the survey team's investigation as well as the associated recommendations for improvement where deficiencies were found. The consensus was clearly that the National Weather Service (NWS) provided exceptionally good services during this unique event.

Introduction

The Great Flood of 1993 had its origin in an extended wet period starting 9-10 months prior to the onset of major flooding. The wet period moistened soils to near saturation and raised many stream levels to bankfull or flood levels. This set the stage for rapid runoff and record flooding that followed excessive June and July rainfall. In terms of precipitation amounts, record river stages, areal extent of flooding, persons displaced, crop and property damage, and flood duration, this event (or sequence of events) surpassed all floods in the United States during modern times. Initial economic impact assessments of The Great Flood of 1993 indicate that losses range between $15 and $20 billion.

---

[1]Chief, Hydrologic Services Branch, Office of Hydrology, National Weather Service, 1325 East-West Highway, Silver Spring, Maryland 20910

NOAA Disaster Survey

The NWS is widely known as the Federal agency in charge of weather forecasting and warning for the Nation. Many people, however, are not aware that the NWS also is charged by law with the responsibility of issuing forecasts and warnings of floods. Whenever a natural disaster of unusual magnitude occurs, a disaster survey team, headed by a high-ranking NOAA official, is established. The focus of the survey is to assess NOAA's effectiveness in the warning process, evaluate how NOAA worked with the numerous organizations of the hazards community to ensure a proper public response, and identify opportunities to improve forecasting services. The survey team also attempts to estimate the economic and life-saving significance achieved because of the warning and to determine any potential for increased effectiveness.

The Deputy Under Secretary for Oceans and Atmosphere was selected to be the Team Leader for the NOAA Disaster Survey. The National Operational Hydrologic Remote Sensing Branch Chief served as Technical Leader, while the Director of the NWS Office of Hydrology served as Team Coordinator. The shear magnitude of the flood necessitated the formation of a larger-than-normal survey team. This team of 16 members (normal is 6) split into two groups in order to interview as many agencies and individuals as possible without extending the length of the field visits to an unacceptable limit. The survey team interviewed more than 120 individuals representing more than 61 Federal, state, and private organizations across the flood-stricken region.

NOAA Disaster Survey Report

The NOAA Natural Disaster Survey Report was completed in February 1994 (NWS 1994). The Report contains over 100 findings and recommendations. Obviously, it is not possible to list all of these lessons learned but emphasis is given to the most important. Most of the deficiencies identified by the NOAA survey team resulted from inadequate technological capabilities within the current forecast and warning system. Those identified weaknesses can mostly be corrected through implementation of more advanced hydrologic prediction capabilities. A substantial number can also be corrected by the NWS modernization and restructuring (MAR) efforts.

A major recommendation of the NOAA survey team is that MAR must be maintained on schedule or accelerated, if possible. This finding resulted from the fact that the NWS modernization had progressed in 1993 to the point that some limited technological advances were readily visible during The Great Flood of 1993. The installed base of NWS Weather Surveillance Radar 88 Doppler (WSR-88D) system within MAR provided major benefits during The Great Flood of 1993. Maximum use of WSR-88D data for hydrologic forecast and warnings awaits completion of the WSR-88D network. In addition, the Advanced Weather Interactive Processing System, under MAR, is needed at the River Forecast Centers to process and mosaic information from multiple radars in their areas of forecast responsibility. Such capabilities are currently being used at several River Forecast Centers.

Of equal importance to the technological enhancements are the advances in human resources also associated with the NWS modernization. These include (1) training and (2) increased hydrometeorological capabilities, particularly to facilitate the coupling of meteorological and hydrological operations required to effectively include quantitative precipitation forecasts and climate information in the hydrologic prediction models.

There were other lessons learned during The Great Flood of 1993 which are not associated with modernization of the NWS. For example, there is a definite need to make substantial advances in NOAA's abilities to model and to predict the complex hydrologic and hydraulic conditions experienced in the Upper Midwest during The Great Flood of 1993. Prediction of streamflow conditions on the Missouri and Mississippi River Basins and their tributaries during the event requires the best possible physical representation of all phases of the water cycle. Hundreds of levees failed during The Great Flood which complicated the determination of accurate stage discharge relationships needed to make flood forecasts. Additionally, discharge measurements were difficult to obtain because of the shear magnitude of the event; and, equipment failures were not uncommon due to the tremendous volumes of water. Other factors affecting streamflow predictions are proper accounting for soil moisture conditions and transport of water through complex river channels, reservoirs, and locks and dams.

Finally, in the future, hydrologic forecasts will require greater quantification. That includes bracketed confidence limits, or probabilistics, which provide the likelihood of occurrences for a range of specific stage forecasts. These more specific and timely forecasts would enable emergency managers and water facility operators to make more accurate, precise, and informal decisions required to effectively carry out routine operations and emergency flood mitigation actions.

Advanced Hydrologic Prediction System

The disaster survey team, through a survey of forecast product users, identified ways to improve all components of the current forecast system. The basic components of an advanced hydrologic prediction system are (1) NWS modernization technologies, (2) the current NWS River Forecast System operation foundation, (3) NOAA partnerships, and (4) advanced Water Resources Forecasting Systems (WARFS) responsibilities. The first two components have been addressed in the previous section. The third component involves the NWS hydrologic services program, which is critically dependent on major collaborative efforts with many of NOAA's partners at the Federal, state, and local levels as well as academia and the private sector. Although communication was generally good during this event, one recommendation in the disaster survey report was to increase efforts at all levels (national, regional, and local) to ensure maximum coordination and cooperation among all agencies involved in disaster mitigation. The fourth component, WARFS, offers the opportunity to increase lead-times, from the present practice of days, to weeks. Federal, state, and local groups also experienced a need for a range of forecast stages with associated probabilities of occurrence. The surveyed users indicated great interest in improving hydrologic prediction to help mitigate the impact of future floods. Improvements in the Nation's capabilities to more accurately predict the hydrologic extremes of droughts and floods, as well as to provide day-to-day information for improved water management decisions, will translate into enormous economic and environmental benefits for the Nation.

Summary

The performance of NWS employees during The Great Flood of 1993 was superb as well as their outstanding devotion to high-quality services and protection of life and property. In many cases, human judgment and expertise compensated for serious deficiencies in the current technological capabilities of the forecast and warning system. The identified deficiencies can, for the most part, be corrected through implementation of a more advanced hydrologic prediction system. The Report also emphasized the need to take systematic action to capitalize on the ongoing NWS modernization and restructuring.

Reference

NWS, February 1994, *Natural Disaster Survey Report The Great Flood of 1993,* NOAA, U.S. Department of Commerce, Washington, DC.

FLOOD WARNING SYSTEM PRODUCT USAGE PATTERNS

Lynn E. Johnson[1]

Abstract

Flood warning system risk reduction research provides a basis for understanding how new-generation forecasting tools are used in the operational environment. Radar-rainfall remote sensing, and other real-time data on precipitation and flood levels were used by National Weather Service forecasters to formulate flood warning and other messages. A prototype realization of an advanced interactive workstation and real-time data system was placed at the Norman, OK forecast office to support risk reduction assessments. One research aspect addressed herein concerned analysis of workstation product usage during flood periods. The analyses indicate that the radar-reflectivity and radar-rainfall products are the most popular with the forecasters as these data provide detail on storm magnitude and movement. Further, these data provide the temporal and spatial detail reflected in the issued flash flood warning messages.

Introduction

A responsibility of a National Weather Service Forecast Office (WSFO) is to provide observations, forecasts and warnings of weather conditions to the public and to provide timely warnings of severe weather events. Risk reduction activities at the Norman, Oklahoma WSFO were to test new workstations, data sets, and operational procedures in an actual working environment before they are made available nationwide later this decade. The Norman WSFO is responsible for county zone and aviation forecasts and issuing severe weather watches and warnings for the western two-thirds of Oklahoma and a portion of northern Texas.

----

[1]Assoc. Prof., Dept. Civil Engineering, Univ. of Colorado at Denver, and Research Hydrologist, NOAA Forecast Systems Laboratory, Boulder, CO 80303.

By 1993, a wide variety of advanced datasets and information were made available to Norman forecasters, so they could utilize their new workstations. Norman received satellite information from the National Environmental Satellite Data and Information Service (NESDIS) and weather modeling results from the National Meteorological Center (NMC). Doppler radar data was received from the Twin Lakes and Frederick WSR-88D sites and surface products from the Mesoscale Analysis and Prediction System (MAPS) were obtained from the NOAA Forecast Systems Laboratory (FSL) in Boulder, CO. Also, Norman forecasters had access to the Oklahoma State Mesonet, which has been providing 5-minute averages of surface parameters from 111 surface observation sites deployed across the State of Oklahoma.

Forecasters' use of the advanced hydrologic functions during April through July 1993 were examined through analysis of workstation product usage logs. A subset of 408 hydrologic workstation products were selected from the more than 27,000 callable products. These products were grouped into function-based categories for frequency analysis. Frequency of products calls for these categories were calculated on a daily basis for the four month period and on an hourly basis for the May 7-13 flood event.

Results

There were 27,898 hydrologic product calls or "usages" made over the 122 day flood season of 1993. Table 1 lists the total products calls for the fifteen hydrologic groups. Radar Reflectivity and velocity products (R) accounted for the most product usage with a group total frequency of 22,541 or 80.8% of the total product usage. The group with the next highest frequency was the Radar Precipitation (RP) products with 1,790 calls or 6.4% of the total product usage. The two dominate individual Radar Precipitation products used were the Storm Total Accumulation from the two available radar sources, comprising 51.1% of the RP group, and Rainfall Accumulation (1hr and 3hr from both radars) totaling 39.7% of the RP group.

The land surface-based hydrological groups were not used much. The Flash Flood Guidance (FFG) group reported only one product with a count of 64. The Quantitative Precipitation Forecast (QPF) group had a frequency of 144, but River Stage Plot (RSP), containing the OK WFO river stage plot, was only counted 24 times. Displays from the River Basin group (RB) were not used.

The 7 day period from the May 7th to the 13th was used to characterize total product usage for the May 1993 flood event. During this period the Norman WSFO issued 20 flash flood warnings, 2 flash flood watches, and 2 flash flood statements. A total of 3,086 hydrologic product calls were counted for the 7-day flood event period. Again, Radar Reflectivity and Velocity (R) dominated, accounted for 2,617 counts or

84.8% of all products used. The Radar Precipitation group was called 297 times (9.4 ercent); this is a significant increase compared to compared to 6.4 percent during the April to July period. Also, 29 percent of the river stage plot calls made during the 1993 storm season were recorded during the flood event.

Table 2. HYDROLOGIC PRODUCT GROUPS

		April - July, 1993		May 7 - 13, 1993	
Group	**Name**	**No. Calls**	**Percent**	**No. Calls**	**Percent**
AVN	Aviation Products	83	0.30%	7	0.23%
FFG	Flash Flood Guidance	64	0.23%	5	0.16%
maps	map backgrounds	262	0.94%	27	0.87%
mapsH	map backgrounds, Hydrologic	118	0.42%	5	0.16%
MOS	Model Output Statistics	198	0.71%	3	0.10%
MRF	Medium Range Forecast	0	0.00%	0	0.00%
P	Precipitation Gage Data Displays	543	1.95%	20	0.65%
QPF	Quantitative Precipitation Forecast	144	0.52%	5	0.16%
RAFS	Regional Analysis & Forecasting System	167	0.60%	14	0.45%
R	Radar Reflectivity & Velocity Displays	22541	80.80%	2617	84.80%
RP	Radar Precipitation Displays	1790	6.42%	297	9.62%
RSP	River Stage Plot	24	0.09%	7	0.23%
SFC	Surface Data Displays	387	1.39%	11	0.36%
MAPS	Mesoscale Analysis & Prediction System	1577	5.65%	68	2.20%
RB	River Basin Data Displays	0	0.00%	0	0.00%
	Total =	27898	100.00%	3086	100.00%

References

Kucera, Patrice C. and W. F. Roberts, 1994. Warm Season Product Usage Patterns Form the DARE Workstations at the Denver and Norman WSFOs. NOAA Technical Report (in press), NOAA Environmental Research Laboratories, Boulder, CO.

Acknowledgments

The cooperation of the Norman WSFO forecasters, management and Risk Reduction Team are gratefully acknowledged. The hydrologic product evaluation activities were conducted with oversight by Lee Larson, Regional Hydrologist and Chair of the Norman Evaluation Committee - Hydrology. Gary Grice lead the overall Norman Risk Reduction effort. The efforts of FSL staff in developing the workstation hydrologic functions are also gratefully acknowledged. Other members of the FSL Evaluation Team, including William Roberts, Cynthia Lusk, Patrice Kucera helped design the approach and provided review comments.

Site-Specific Probable Maximum Precipitation (PMP) Great Miami River Drainage in Southwestern Ohio

Ed Tomlinson[1], George Wilkerson[2], Tom Wiscomb[3], James L. Rozelle[4]

Probable Maximum Precipitation (PMP) is defined by the American Meteorological Society's Glossary as:

> "The theoretically greatest depth of precipitation, for a given duration, that is physically possible over a particular drainage area at a certain time of year."

Since at least the mid-1940's several government agencies have been developing methods to calculate PMP for various regions of the United States. These agencies include the National Weather Service (formally the U.S. Weather Bureau), Department of the Army Corps of Engineers, and the Bureau of Reclamation. The PMP estimates derived form these reports are used to calculate the Probable Maximum Flood (PMF).

Generalized PMP studies are published for use in the conterminous United States. Hydrometeorological Report #51 (HMR 51) provides generalized PMP values for various area sizes and durations for U.S. locations east of the Rocky Mountains. This study focused on providing a site-

---

[1] Ph.D., Chief Scientist, North American Weather Consult. 1293 West 2200 South, Salt Lake City, UT 84119.

[2] Senior Meteorologist, North American Weather Consult.

[3] Staff Meteorologist, North American Weather Consult.

[4] The Miami Conservancy District, Dayton, Ohio.

specific determination of PMP values for a much smaller region in southwestern Ohio.

The study continued the basic approach used in prior studies to estimate PMP. This methodology includes the following:

- Identifying major flood producing storms which have occurred over and adjacent to the study region.
- Maximizing each storm's precipitation.
- Moving the storm's precipitation pattern over the study region, adjusting for geographic and meteorological variations (this procedure is called transpositioning).
- Analyzing the largest precipitation amounts by enveloping the rainfall amounts.

This methodology was updated with computer based applications and refinements to the moisture parameterization. The goal was to maintain as much continuity as possible with previous studies while providing reliable site-specific PMP values for the study region.

The study domain was the Great Miami River drainage in southwestern Ohio. The topography is characterized by low rolling hills covered by forest or farmland. The topography is not significant enough in elevation to affect precipitation patterns.

The largest storms that have occurred within the study domain and the surrounding area provided the data base for the development of the PMP estimates. Storms from surrounding areas with similar meteorological controls may be transpositioned into the study domain when it can be established that these storms, with appropriate adjustments considering topography and available precipitable water, could have occurred over the study domain.

Analysis of precipitation data identified the largest storms which have occurred within the storm search area. Thirty storms were identified. Of these, 22 occurred during the warm season. The remaining eight storms were cool season storms, i.e., storms which could occur during the times of year when the ground could be snow covered and/or frozen.

Depth-area-duration (D-A-D) analyses provide the depth and areal extent of precipitation over various temporal intervals for a particular storm event. The D-A-D analysis provides the hydrologist and/or engineer with storm precipitation data which can be uniformly compared with similar events and may be used for various hydrologic scenarios. In the study, the methods used in past D-A-D analyses were kept intact, but the procedures used to derive the final product were refined.

The HMR 51 procedure for storm maximization uses a representative storm dewpoint as the parameter to represent available moisture for a storm. The storm precipitation amounts are maximized using the ratio of precipitable water based on the storm representative dewpoint to a maximum amount of precipitable water based on a maximum dewpoint climatology. The HMR 51 procedure was followed in this study with several refinements in determining dewpoint values. Average dewpoint values were used instead of persisting and various temporal periods (i.e., 6, 12 and 25 hour) based on precipitation durations were used.

Maximization and transpositioning provide an indication of the maximum total amount of precipitation that a particular storm could have produced over the region of interest. These values alone do not provide assurances that PMP values are provided for the particular basin sizes and durations since some of the maximized values may be less than the PMP values should be. By enveloping the values resulting form the maximization and transpositioning of the storm D-A-D amounts from all of the storms, values indicative of the PMP magnitude will result (WMO, 1986).

Enveloping is the process of selecting the largest value from a set of data. This technique provides continuous smooth curves based on the largest precipitation values from the set of maximized and transpositioned storm rainfall data values. The largest precipitation amounts provide guidance for constructing the curves.

This enveloping procedure addresses the possibility that for certain sizes and durations, no significantly large storms have been observed which provide large enough values after being maximized and transpositioned. The result of this procedure is a set of smooth curves which maintain continuity among temporal periods and areal sizes.

PMP maps were produced for southwestern Ohio for area sizes ranging from 100 to 10,000 square miles and for time periods of 6, 12, 24, 48 and 72 hours. A separate set of PMP maps were produced for cool season storms. Figure 1 shows the PMP maps for the warm and cool seasons for the 1000 square mile area for 24 hours.

The site-specific PMP values resulting from this study vary from those provided in HMR 51. The differences vary from approximately 20% smaller values for the smaller area sizes and shorter durations to approximately a 2% decrease for the larger area sizes and longer durations. These differences are primarily the result of the more detailed treatment of the moisture maximization.

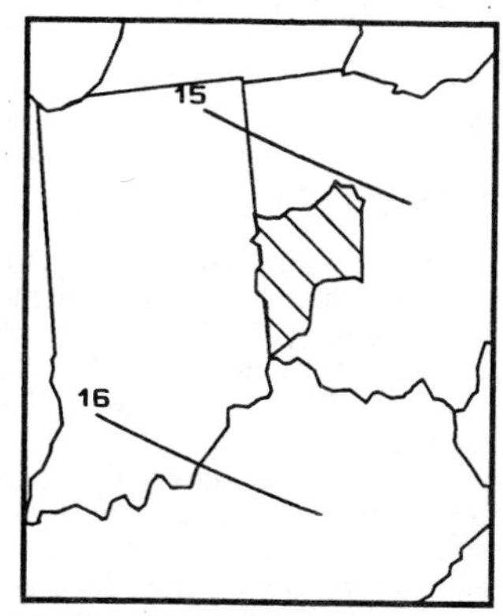

Warm Season

Cool Season

Figure 1. PMP values for 48 hours and 1,000 square miles.

<u>References</u>

American Meteorological Society, 1959: Glossary of Meteorology, Boston, MA, 638 pp.

Schreiner, L.C., J. T. Riedel, 1978: Probable maximum precipitation estimates, United States East of the 105th Meridian. <u>NOAA Hydrometeorlogical Report</u>, No., 51 prepared by NWS, Washington, D.C., June, 1978.

World Meteorological Organization, 1986: Manual for estimation of probable maximum precipitation (2nd Ed.), WMO, No. 332, Geneva, Switzerland, 250 pp. 22.

An Expert System for Evaluating Scour Potential and Stream Stability at Bridges

Lea Adams[1], Student Member, Richard Palmer[2], Member ASCE, and George Turkiyyah[3]

Introduction

Scour is the erosion of bed and bank material by flowing water; scour poses a threat to bridge security by undermining piers and abutments over time. In fact, "the most common cause of bridge failures is floods with the scouring of bridge foundations being the most common cause of flood damage" (Richardson et al., 1993). While regular inspections and maintenance reduce the risk of a scour-related failure, identification of conditions indicative of scour can be difficult. Good field judgments require significant practical experience as well as formal training. This paper describes an expert system that aids bridge inspectors with the scour portion of regular bridge inspections; this system results in more consistent and accurate recommendations, and provides the inspector with information on scour processes. In addition, the system functions as a screening tool to identify the severity of scour and stream stability problems at a wide variety of sites. This paper discusses the scour inspection process and the benefits of applying an expert system. Elements of the system development addressed in this paper include the scour inspection process, objectives and approach, system variables and recommendations, and initial testing. Proposed future refinements and modifications are also discussed.

Inspection Process

A survey of scour inspection practices at twenty-one state Departments of Transportation (DOTs) provided information on data collection, data analysis, and the mechanics of inspections. Bridges over water generally require a scour inspection every two years (Pagan, pers. comm., 1994); the level of detail considered during these inspections varies from state to state, however. The most commonly performed inspection action among the surveyed DOTs is the investigation of scour holes around pier foundations by either direct observation or probing. General site characteristics are also frequently noted. Many DOTs produce streambed profiles of a cross-section at the bridge site, which can be used to determine trends in the bed elevation over time. The data collected during a scour inspection is utilized in two ways: it is recorded to provide a history of the site, and it is analyzed to screen the bridge for severity of scour problems. After a bridge is inspected, a code reflecting the status of scour is assigned and reported to the Federal Highway Administration.

---

[1]Graduate Research Assistant, [2]Professor, and [3]Assistant Professor, Department of Civil Engineering, University of Washington, Seattle, Washington 98195.

Other factors impacting scour inspections were also surveyed, including inspector experience and inspection resource limitations. Virtually all of the DOTs reported performing scour and structural bridge inspections jointly; sites with scour problems are typically flagged for further investigation by more qualified personnel. This indicates that the majority of inspectors investigating scour are primarily knowledgeable about the structural aspects of bridges. Both temporal and economic limitations also restrict scour inspections. The survey revealed that the number of bridges inspected each year varies depending upon the state; some states may review as few as 700 and others as many as 8,000. The large number of bridges limits the amount of time available for each inspection. Scour budgets range from approximately $300 to $1200 per bridge. Again, this limits the amount of time and equipment available for scour inspections.

The results of the survey indicate that a system to aid scour inspection must be designed for individuals with limited background in scour processes and river mechanics. In addition, the system must support an inspection process that is typically conducted in less than thirty minutes. By providing a system that meets these requirements, it is expected that the quality of scour inspections will improve and inspectors will become more knowledgeable about scour processes.

Objectives and Approach

There are two main objectives of the expert system: to aid bridge inspectors with the scour portion of regular bridge inspections and to provide a screening tool. The system software assists inspectors by improving the consistency of conclusions at similar sites, improving overall accuracy of conclusions, and presenting information about scour processes. Because interpreting the signs of scour is not an exact science, there can be significant variability in the conclusions reached by different inspectors at the same site, or by one inspector at different sites.

By encoding expert knowledge on scour, this system can also aid inspectors by increasing the accuracy of site recommendations. Such recommendations include requesting maintenance, calling in an engineer for further site investigation, and identifying situations which require immediate action. Improving the accuracy of these conclusions enables inspectors to recognize site problems at an early stage, preventing the development of larger problems or even catastrophic failures later on. Increased accuracy of the site recommendations can also improve the efficiency with which temporal and economic resources are applied.

Because this tool is intended to aid bridge inspectors not formally trained in hydraulics and river mechanics, one of the design objectives is to provide information on scour and river processes. To achieve this objective, additional information is made available to the user in the form of help screens. This allows a new user to access more detailed information on particular questions and answers, while more experienced users can operate the system without scrolling through unnecessary information. Information within the help screens is arranged hierarchically.

The second objective of the system is to provide inspectors with a scour screening tool that identifies three categories of bridges: 1) bridges at the extremes of the scour continuum (from severe scour problems to those with no scour problems), 2) bridges that require some attention, but are not threatened by scour, and 3) bridges that require analysis beyond the screening capabilities of the expert system.

System Development

Once the objectives of the system were established, the development of the expert system logic was initiated. The development process can be organized into three primary activities: discussions with intended system users, interaction with scour experts, and iterations of the logic. The survey of the DOTs revealed the preferences of the intended users, contributing information about the relative importance of the site factors and possible logic formats. Later review of the complete system by state DOT personnel also suggested methods of improvement.

Consultants from Northwest Hydraulic Consultants, Inc. (NHC) provided the scour-related knowledge encoded in the system. Meetings with the scour experts established a working knowledge base, organized the data collection format and developed system recommendations. Expert involvement also contributed insight on the interaction of the site and office information; this interconnection of factors affecting scour formed the core of the expert system logic. In later stages of development, experts from NHC worked with the research team to correct the system logic through iterations of review and modification. Both specific variables and the overall logic structure were evaluated in detail; suggestions for improvement were then incorporated into the system.

System Variables and Recommendations

A group of variables relevant for assessing scour and stream stability was defined through discussions with both state inspection personnel and consultants. These variables were arranged into two categories: site factors and office information. Table 1 displays examples of both types of variables. More time is required to gather office information than site information, but the data is collected only once and saved electronically. This and any site data files are recalled at the beginning of each system run. The site factors necessary for a scour analysis require little time to collect as they are generally observable at the bridge site. Both types of variables are used in the logic to evaluate the state of scour and stream stability at each bridge site.

Table 1. Typical System Information

Office Information	Site Information
deepest historical scour elevation	observed scour elevation
deepest predicted scour elevation	thalweg (main channel of stream)
pier countermeasures	- location
- type	- elevation of deepest part
bank countermeasures	type of levees present along stream
- type	amount of vegetation on banks
- location	presence of channel bar formations
foundation type	debris amount
floodplain width	bank material composition
subsurface bed material composition	surface bed material composition

The results of the system are presented as recommended actions. These actions are classified into several main categories of recommendations: further engineering analysis, site monitoring, underwater inspections, and immediate action. After the collected data is analyzed, the system results are organized into general site recommendations and pier-specific recommendations; general recommendations involve factors such as debris accumulation, bank erosion and bank countermeasures, while pier-specific recommendations reflect observed scour holes at each pier and pier countermeasures.

Programming Environment

The expert system software was developed in two parts: an interface and an inference engine. The interface was programmed in Microsoft Visual Basic which provided the software with a standard Windows look and operation. Because the interface is a function of the system data collection requirements, it was organized into an office module and a field module. This allows the inspector to complete all necessary data entry in the office before proceeding to the field. The interface saves office information and field information in separate files. The inference engine was developed in EXSYS, an expert system programming platform. Several factors influenced the choice of this software, including ease of use, built-in de-bugging tools and external connection capabilities. Several programming formats were evaluated utilizing CLIPS, C and EXSYS's rule-based language; the eventual use of the rule-based programming language greatly simplified the software development.

Testing

The prototype expert system was tested to determine its completeness, sensistivity and reliability, and the results evaluated for possible system improvements. Completeness was measured by the presence of a system response to various case studies. A complete system generates a recommendation for each site, while an incomplete system may fail to recommend an action. System sensitivity is reflected by the degree of change in the final recommendations as compared to the degree of change in the entered data. This measure is important for evaluating system response to unknown or unavailable information. The final measure of system performance is the appropriateness of the final recommendations (Pollon, 1994).

The prototype was tested on six case studies provided by Northwest Hydraulic Consultants, Inc. and the New York State Department of Transportation. The results showed that system completeness ranged from inadequate to good. The most significant flaw in the system at this stage was incompleteness; the knowledge lacking in each case was readily identified, however, and the logic necessary for the appropriate recommendations added. System sensitivity as tested with the case studies was either robust or sensitive. Again, the sensitivity reflected gaps in the knowledge base which were addressed in a later version of the system. Evaluation of the system reliability indicated generally appropriate site recommendations.

Future Efforts

This prototype expert system shows promise as a decision support tool for field use by bridge inspectors for detecting scour. The system is designed to provide advice on the action to be taken at a site, including noting when more detailed analysis is needed. Work is continuing on the system, with emphasis on increased field testing by seven targeted state Departments of Transportation. It is anticipated that the breadth of problems that can be successfully identified by the system will increase as the logic is improved with the addition of more sophisticated rules based on evaluations of more sites.

References

Pagan, J., (1994), Hydraulic Engineer, Federal Highway Administration, Washington D.C. Personal communication, December.

Pollon, S., (1994) "Knowledge Formalization for the Development of a Bridge Scour and Stream Stability Expert System," Master's Thesis, Univ. of Washington.

Richardson, E.V., L.J. Harrison, J.R. Richardson, and S.R. Davis, (1993) Evaluating Scour at Bridges, Federal Highway Administration Report, Hydraulic Engineering Circular No. 18.

Is The National Economic Development Plan the Best Plan?

By Roy G. Huffman and Darryl W. Davis[1]

Abstract

The policy of the Federal Government is to select the alternative flood damage reduction plan that maximizes national economic development (NED) benefits. Benefit calculations are based on average annual values. Flooding is a random process so that the average benefit will not occur on a recurring annual basis. Basing project selection on average benefits obscures the true random nature of floods and places equal weight to protecting against small frequent floods as protecting against large, catastrophic floods. The shortcomings of NED plan selection and alternative considerations are presented and discussed.

Introduction

The planning process employed by the Federal Government for flood damage reduction project investigations is based on legislation and policy that evolved over a number of years and is documented in the Planning Guidance Notebook (USACE 1990a). The process includes defining the flood problem; formulating alterative solutions; evaluating economic, environmental, and other impacts of the alternatives; and selecting the plan to be recommended for implementation. Unless there are compelling reasons for selecting an alternative plan, the plan that maximizes NED benefits is selected.

The accepted approach to estimating flood damage reduction benefits is to compute expected annual damage and damage reduction benefits. Because floods occur at random, flood damage reduction benefits in any year may be more or less than the planned-for benefit. Furthermore, due to the random nature of flooding and the short economic lives of flood damage reduction projects, the actual benefit over the life of the project may be more or less than the expected benefit. The computed net benefit (or benefit cost ratio) for the economic life of a flood damage reduction

---

[1]Respectively, Hydrologic Engineering Consultant, Vienna, Virginia, and Director, U.S. Army Corps of Engineers Hydrologic Engineering Center, Davis, California.

project is therefore a random number. Thus, maximizing the expected annual net benefit may not be the best single criteria for plan selection.

Flood Damage Analysis

Flood damage analysis involves inventory of properties subject to flood threat and hydrologic engineering analysis to determine the flow, stage and exceedance frequency descriptors of flooding. The data are formed into a damage-frequency function that is subsequently integrated to determine the expected value of annual damage, often referred to as "average annual damage". A consequence of performing flood damage analysis in this manner is that the variability and magnitude of the flooding and consequent flood damage is obscured. Damage potential is associated with a probability but visual, graphic evidence of the extremely variable nature of flooding is lost. Data for Chester Creek, Pennsylvania are used to illustrate the issues. Figure 1 is the damage probability function for Chester Creek. Damage from the 1% chance flood would be $3.8 million and the expected annual damage is $78,000. Figures 2, 3, and 4 show flood damage that would occur from likely alternative flooding sequences that were derived from Monte Carlo simulation. A levee that would provide a level of protection with 1% chance of being exceeded would eliminate all damage in Figure 3 but damage would occur from one flood in Figure 2 and 3 floods in Figure 4. The expected annual damage reduction for the levee is $21,000.

Several items about the figures are noteworthy. A few large events dominate; the event damage is typically many times larger than the expected annual value; increasing the levee height by small increments would not likely alter the computed damage reduction for these sequences; there are many other scenarios that are likely, this is simply a sample set;

Decision Making and Averages

Several items about making decisions based on averages are troublesome. One is that the variability of the flooding, and thus flood damage reduction, is obscured, as the figures demonstrate. Damage potential is associated with a probability but the true nature of flooding and project performance is lost. As demonstrated, Chester Creek project benefits would accrue from protecting against a few moderate sized floods. However, the area would remain vulnerable to large floods. Optimizing the plan by computing the expected NED benefits for increments of levee size increase/decrease in such a case is questionable, but is the current standard approach. Another troublesome item, not unrelated to the above, is that the computation of expected damage and benefits for the above situation involves weighing very high damage consequences for large catastrophic floods with very small probabilities. The value of the resulting product (the increment of expected annual damage for events in that range) is often about the same as the increment from

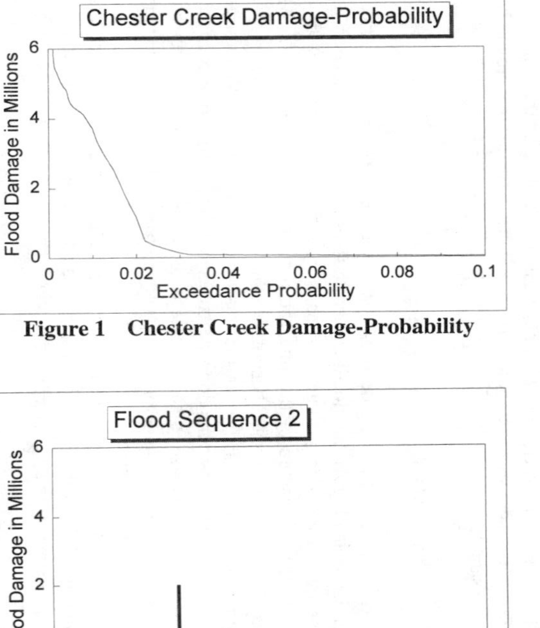

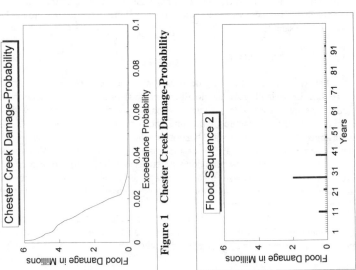

Figure 1 Chester Creek Damage-Probability

Figure 3 Flood Sequence 2

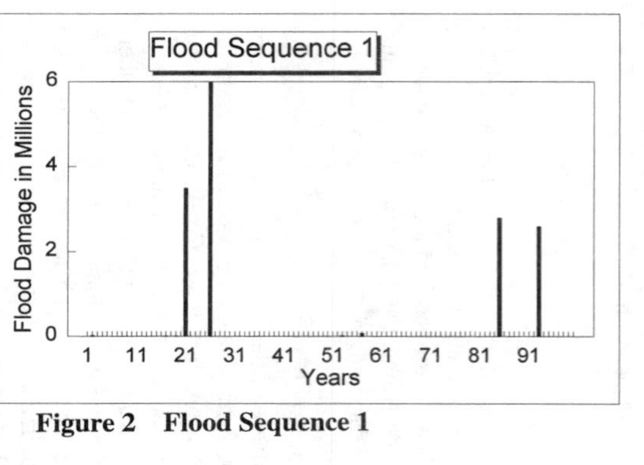

Figure 2 Flood Sequence 1

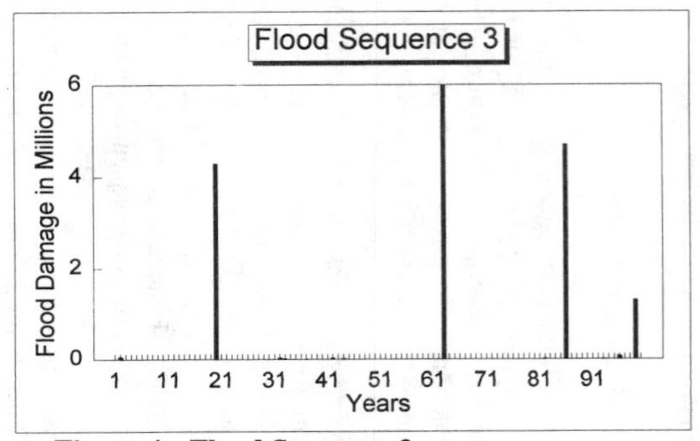

Figure 4 Flood Sequence 3

a more frequent but less damaging flood. As a consequence, to decision makers, small and large floods may appear to have nearly the same consequences. We know that damage from a single catastrophic event can be devastating to an urban community whereas damage from smaller events may be within the capacity of the community to absorb without critically crippling the area. While NED based planning is said to implicitly incorporate such issues as performance and reliability, averages again obscure these issues in the real world of random occurrences of the large, rare floods of interest.

Some Thoughts

Present Federal policy requires that flood damage reduction plans be consistent with environmental and other legislation and policies but does not encourage consideration of plans other that the NED plan for recommendation. As a result, other impacts are given minimum attention. Selection of flood damage reduction plans should be an informed decision making process that considers the full range of issues. Along with the traditional NED analysis and environmental impact assessment, project performance and reliability should be meticulously analyzed and explicitly displayed. Project performance includes such issues as the impacts on flow rates, flood duration, and likely impacts in the event of a flood exceeding project capacity. A pamphlet is available (USACE 1990b) that describes in lay terms the generic performance of various types of flood damage reduction measures. Reliability encompasses issues ranging from the time required/available to operate closure structures in a levee to how sediment and debris during a flood event could influence project capacity. A local sponsor's ability to maintain and operate a plan may also be a reliability consideration.

The authors recognize the obstacles confronting the full implementation of the principles expressed herein. Consistent methods are needed to measure and display performance and reliability of plans to enable the selection of flood damage reduction plans based on a full range of issues besides average NED benefits. Improving methods for analysis and display of performance and reliability of flood damage reduction measures should be priority issues.

References

U.S. Army Corps of Engineers, Regulation No. 1105-2-100, Guidance for Conducting Civil Works Planning Studies, December 1990.

U.S. Army Corps of Engineers, Hydrologic Engineering Center, Measures to Reduce Flood Damage - Pamphlet, March 1990.

Issues in the Estimation of IDF Curves

Paolo Bartolini[1] and **Juan B. Valdés**[2] F. ASCE

Introduction

Several methods of estimation of IDF curves and its application to the development of designs storms are widely reported in the hydrologic literature, e.g. Chow et al, 1988. The estimation starts with the fitting historical values of annual maximum precipitation depths at a site, for different durations, is the starting point in most procedures aimed at evaluating the design storm at that site. A common procedure is to fit the different sets of historical depths, for different durations, with the same function of τ (duration) and T (return period), giving rise to Depth-Duration-Frequency (DDF) curves, $h(\tau, T)$. A well known form of the DDF is, for example,

$$h(\tau, T) = \frac{cT^m \tau}{\tau^e + f} \tag{1}$$

If a Least Squares procedure is then followed, the values of the parameters of the DDF curves (c, m, e, f in the previous example) may be derived by minimizing the sum of the squared deviations E between the observed and theoretical frequencies of the historical data. Once the constrained optimization procedure has been performed [where the constraints regard the invertibility of the DDF curves, or the functions $h = h(\tau, T)$, $T = T(h, \tau)$ and $\tau = \tau(h, T)$ must be single-valued) and the DDF curves have been derived, the possibility arises of defining a set of "equivalent storms", from which to choose the design storm, with a given return period. The final goal may be the use of a Rainfall-Runoff Model (RRM) for processing the design storm to yield the corresponding discharge. The hyetographs of the equivalent storms may be defined in different ways. One possibility is to consider the very simple hyetograph

$$i(t, \tau, T) = \frac{h(\tau, T)}{\tau} \qquad for\ t \in 0, \tau \tag{2}$$

and zero elsewhere. Thus, the non-zero values of the hyetographs coincide with the values of the Intensity Duration Frequency curves (IDF), defined as

$$i(\tau, T) = \frac{h(\tau, T)}{\tau} \tag{3}$$

For the example above

$$i(\tau, T) = \frac{cT^m}{\tau^e + f} \tag{4}$$

[1]Institute of Hydraulics, University of Genova, Via Montallegro 1, 16145 Genova, Italy.

[2]Department of Civil Engineering and Climate System Research Program, Texas A&M University, College Station, TX 77843-3136; (409)845-1340.

The shortcoming of this simple procedure is that an infinite set of hyetographs (with different durations) corresponds to a given return period T. In the common practice this multivariety is cutt off by considering the unique hyetograph which, for a given return period, has a specified value of the duration τ. In view of the ultimate goal of the DDF analysis, i.e. the estimation of the design flood at the outlet of a basin, the choice of the duration depends on the choice of the RRM. If a linear RRM is used, the common practice is to choose the duration of the hyetograph equal to the time of concentration of the basin. Even if this procedure may be accepted in the case of linear RRM, the same cannot be done in the case of non-linear RRM, for which there is no possibility of defining an unique time of concentration. Bartolini and Valdes (1994) have shown a procedure for address this problem, by using Monte Carlo simulation. The procedure has the following steps:

1. fit the DDF curves to historical data, and derive the corresponding IDF curves;
2. calibrate a non-linear RRM model for the basin under consideration;
3. for an appropriate set of different durations (theoretically, the infinit set of all durations, in practice the set of durations multiple of $\Delta\tau$, up to a proper maximum), find the corresponding (constant) intensities able to give rise (through the RRM) to a selected maximum discharge Q_o, and calculate the corresponding return periods from the IDF curves. These Return Periods (RP) must be viewed as "*marginal*" RP for each duration;
4. derive the RP of Q_o by a Monte Carlo generation of sequences of annual maximum depths for the selected set of durations, and by enumerating the years in which at least one out of the generated intensities overtakes the limiting value for producing the discharge Q_o. At this first step, the generation of the intensities follows a semi-independent procedure, i.e. a procedure which constraints the probability function (PDF) for the next duration to be equal to the bi-truncated form of the corresponding marginal PDF, with the lower and upper truncation points given as function of the generated preceding intensity $i(\tau - \Delta\tau)$ by:

$$i_l(\tau) = i(\tau - \Delta\tau) frac\tau - \Delta\tau\tau \quad (5)$$

$$i_u(\tau) = i(\tau - \Delta\tau) \quad (6)$$

5. during the Monte Carlo generation, derive the empirical marginal periods of the given intensities (obtained at step 3), and compare them with the theoretical marginal periods of the same intensities;
6. in case of significant deviations between the empirical and marginal periods (step 5), modify the generation procedure of the annual maximum intensities for different durations: in the referenced paper, a distortion has been introduced in the generation of annual maxima for different durations, such that values of the intensities closer to those of the previous duration have been privileged;
7. the dependence mechanism between intensities of growing durations is validate when an acceptable concordance between the empirical and theoretical return periods of the selected intensities has been achieved.

The outlined procedure is aimed at finding a way for generating conditional annual depths for each duration, in view of the fact that the DDF do not contain any information about the dependence structure of the former. The key point is the definition of the constraints between annual maximum intensities for different durations, able to preserve their marginal return periods. An alternative approach to the problem is to determine from the IDF curves a set of different design storms, each with its own return period, such that the selected RRM produces different maximum discharges, for each different storm. This procedure is widely used in hydrologic practice (Chow et al., 1989). To this kind of approach belong the alternate block method and the Chicago method. If the alternate block method is used, an arbitrary value must be taken for the time step Δt, and the constant value of the intensity $i(k\Delta t, T)$ during the interval $[(k-1)\Delta t, k\Delta t]$ is given by

$$i(k\Delta t, T) = \{h[2(k-1)\Delta t, T] - h[(2k-3)\Delta t, T]\}/\Delta t \tag{7}$$

while, for the interval $[-k\Delta t, -(k-1)\Delta t]$, the constant intensity is given by

$$i(-k\Delta t, T) = \{h[(2k+1)\Delta t, T] - h[2k\Delta t, T]\}/\Delta t \tag{8}$$

Note that, in this case, the hyetograph does not depend on a particular value of the duration τ. The arbitrariety in the choice of Δt may be removed by letting $\Delta t \to 0$. In this case, the final form of the hyetograph will be

$$i(t,T) = i(-t,T) = \frac{dh(\tau,T)}{d\tau}]_{\tau=2t} \tag{9}$$

and the method yields a symmetric hyetograph. In the case of the previous example, the hyetograph has the form (for $t > 0$)

$$i(t,T) = cT^m \frac{[(2t)^e(1-e)+f]}{[(2t)^e+f]^2} \tag{10}$$

As discussed above, the single hyetograph obtained with the alternate block method has the disadvantage of a not always realistic symmetric form. A way to circumvent this problem has been proposed in the triangular method, by introducing a 'shape coefficient r', wich allows a skewed form for the hyetograph. The justification of the method is that a single-maximum hyetograph (like the one wich is produced by the alternate block method) has the property that the corresponding values of the annual maximum depths for different durations, defined by

$$h(\tau,T) = max(t)[\int_{t-\tau}^{t} i(x,T)dx] \tag{11}$$

are shared by all the hyetographs which have in common the function $j(\tau,T)$, defined by the conditions

$$j(\tau,T) = i[t(\tau),T] = i[t(\tau)-\tau,T], \tag{12}$$

where $t(\tau)$ is the time at which the hyetograph $i(t,T)$ takes the same value as at the time $t-\tau$. In fact the annual maximum depths are also defined as

$$h(\tau,T) = \int_0^{\tau} j(x,T)dx \tag{13}$$

It follows that, given the single-maximum hyetograph $i_o(t,T)$, an infinite number of new hyetographs with the same maximum depths for different durations may be obtained by using the function $j_o(\tau,T)$ corresponding to $i_o(t,T)$. In the Chicago method the whole set is obtained as

$$i(t,T) = j_o[t/r,T] \qquad for\ t>0 \tag{14}$$

$$i(t,T) = j_o[t/(1-r),T] \qquad for\ t>0 \tag{15}$$

where the shape factor r may take on any value between 0 and 1. The arbitrariety in the choice of r may be cutt off by examining typical form of the observed hyetographs, and a cumbersome procedure is needed, unless informations about the common value of r in a given region are at disposal. Otherwise, some assumption about the common likelihood of different hyetographs (for different values of r) must be done, and some empirical procedure (like the one proposed by Bartolini and Valdes) must be followed.

Direct Derivation of the Design Storm from Annual Maximum Depths

The already outlined procedures have the common feature to derive the hyetographs from the DDF (or the related IDF). A different approach may be to make directly use of the selected form of the hyetograph of the design storm even in the fitting of the historical maximum annual depths. It means that, instead of assuming an analytical form for the DDF (or the IDF), a form for the design hyetograph is assumed, and the DDF and IDF curves may be obtained as a consequence. In fact, it is easy verified that the alternate block and the Chicago methods may be followed by expliciting the form of the hyetograph, and by directly fitting the historical annual maximum depths. In the alternate block method, if the limiting case is considered (with $\Delta t \rightarrow 0$) the theoretical maximum annual depths, for different durations τ and Return Periods T, are given by:

$$h(\tau, T) = 2\int_0^{\tau/2} cT^m \frac{[(2t)^e(1-e)+f]}{[(2t)^e+f]^2} \qquad (16)$$

and the Least Squares Method (LSM) may be followed. But, it is also easy verified that if the hyetograph has a unique maximum, with a form dependent on a shape factor, the direct fitting of the historical maximum depths (e.g. with the LSM) fails in finding the solution (the parameter of the hyetograph). The reason is shown above, and lies in the multivariety of single maximum hyetographs of a given parametric form which have the same DDF (in the Chicago method this implies that the shape coefficient r cannot be determined).

A simple assumption can then be made of that the hyetograph of the design storm has more than one maximum. In this case, the already proposed procedure, i.e. to direct fit the historical maximum depths with the selected hyetograph, may be followed and will give, instead of the parameters of the DDF curves, the parameters of the hyetograph. It is important to note that very often the real hyetographs (from which the annual maximum depths are obtained) show more then one maximum. In fact, most of the model aimed at reproducing the real world characteristics of storms take into account the possibility of different phenomena, each able to give rise to more than a local maximum in the hyetograph. The succession of rain cells is one out of these phenomena. Again, the succession of Cluster Potential Centers (CPC), with their ability to produce clustered groups of cells, gives rise to a variety of local maxima in the rain intensity. In our work the hyetograph of the design storm was given the simple form

$$i(t,T) = j_o(T)\{exp(b_2 t) + k\ exp[b_2(t-t_1)]\} \qquad for\ t < 0 \qquad (17)$$

$$i(t,T) = j_o(T)\{exp(-b_1 t) + k\ exp[b_2(t-t_1)]\} \qquad for\ t \in 0, t_1 \qquad (18)$$

$$i(t,T) = j_o(T)\{exp(-b_1 t) + k\ exp[-b_1(t-t_1)]\} \qquad for\ t > t_1 \qquad (19)$$

where $j_o(T)$ is the maximum intensity of the first CPC, for a given return period, and $k \leq 1$ is the ratio between the maximum intensities of the second and the first CPC. The time t_1 of the second maximum, and the coefficients b_1 and b_2, together with the parameters enbedded in $j_o(T)$, complete the set of parameters of the hyetograph. The proposed form of the hyetograph of the design storm belongs to the set of functions

$$i(t,T) = j_o(T) f(t) \qquad (20)$$

with $f(t) = 1$ for $t = 0$. It is easy verified that the hyetograph of the alternate block method belongs to the same set. As an example of the previous statement about the multivariety of solutions obtained by fitting the annual maximum depths with a single maximum hyetograph, let us consider the case of the single maximum version of the proposed model, i.e. a model with k set to zero. In this case:

$$h(\tau,T) = \int_{t(\tau)-\tau}^{0} j_o(T)exp(b_2x)dx + \int_{0}^{t(\tau)} j_o exp(-b_1x)dx \qquad (21)$$

with $t(\tau)$ such that $exp[b_2(t(\tau)-\tau] = exp[-b_1t(\tau)]$ or $t(\tau) = \frac{b_2}{b_1+b_2}\tau$ and it follows that $h(\tau,T) = j_o(T)\frac{b_1+b_2}{b_1b_2}[1-exp(-\frac{b_1b_2}{b_1+b_2}\tau]$ The theoretical value of $h(\tau,T)$ does not depend on individual values of the parameters b_1 andb_2, but on the value of the combination

$$\eta = \frac{b_1b_2}{b_1+b_2} \qquad (22)$$

At the end of the estimation procedure, a value for η will be estimated, but not the individual values of b_1 and b_2.

One versus Two-Maximum Hyetographs

A comparison between the performances of the models with one and two-maximum hyetograph has been made, with regard to 58 raingage stations in Liguria (northern Italy), with data ranging from 30 to 63 years. The fitting of historical annual maximum depths has been performed by using MLS, by limiting the analysis to the durations of 1, 3, 6 hours. This restriction is due to the fact that our model is supposed to ressemble some of the characteristic of one out of the two physical mechanism which produce the maximum annual depths, i.e. the convective storms. The possibility that the annual maxima are produced by the other phenomenum (i.e. frontal storms) is growing with growing durations, and it is possible to rely on the fact that the annual maximum depths for the selected durations are produced only by convective storms. Later, a mixed model will be introduced to allow the use also of the greater durations (12 and 24 hours) for which data are at disposal.
The structure of the function $j-o(T)$ has been derived by assuming that the annual maximum intensity for the first CPC is distributed as a Gumbel variate. Thus, the number of parameters is 3 for the first model (α_0, U_o, η) and 6 for the second model $(\alpha_o, U_o, b_1, b_2, k, t_1)$. As a performance index, the average value of the error 1000 $(E)^{1/2}$ of equation (1) has been used, and the results were $E_1 = 61.397$ and $E_2 = 36.058$ respectively. The values of the parameters, for selected stations of the 11 stations in Northern Italy analyzed and with at least 50 years of records, for the two-maxima case, are listed in Table 1.

Acknowledgments

Parts of this paper were written with the support of the Italian Ministry of University and Scientific Research and the U.S. National Aeronautics and Space Administration (NASA). The support of the NATO Scientific Affairs Division allowed visits and the collaboration between the authors. All contributions are gratefully acknowledged.

Table 1: Parameters of the Two-Maxima Model.

station	N_d	α_o	U_o	b_1	b_2	k	t_1	MSE
Isoverde	56	19.39	57.85	1.81	1.99	0.90	5.57	40.54
Mignanego	58	22.52	53.87	2.32	2.07	0.87	5.63	29.96
GE-Università	59	29.31	59.84	1.27	5.79	0.47	5.43	39.92
Tigliolo	59	18.60	51.52	1.10	4.96	0.99	5.66	39.96
S.Michele	58	20.99	49.68	1.28	4.37	0.73	5.58	28.85

PROMOTING ACCURACY FOR TIME OF CONCENTRATION IN DAM SAFETY ANALYSIS

S. Samuel Lin[1], Duncan McGregor[1], Joseph S. Haugh[2]

Abstract

The purpose of this paper is to promote the accuracy of the t_c estimate for dam hydrologic safety analysis. The use of more physically based factors to determine values of parameters related to t_c can reduce uncertainties involved in estimates. Hence, velocity methods are generally better than empirical formulas. A more scientific approach based on physical factors and specific site measurements/ observations is recommended over subjective judgment methods. A practical example is provided for a study in dam upgrade to accommodate the spillway design flood. The Hughes's formula is also applied with some constraints for mountain flow travel time estimate. The opinions/views expressed here are solely those of the authors and do not necessarily represent the policy of the State.

Introduction

A flood hydrograph shows the variation of the flow discharge with time and represents the hydrologic nature of a flood at a specific site. The SCS synthetic unit hydrograph is generally used for determination of flood hydrographs for small ungaged watersheds. The surface-runoff time of concentration (t_c) is a key factor to construct the unit hydrograph. As an idealized concept, t_c is the time when the entire drainage basin is contributing simultaneously to flow at the outlet where runoff is assumed to reach a peak. At a dam site the t_c is the travel time of the flow that starts from the hydraulically most remote point in the drainage area to the upstream end of the reservoir. It is a major input parameter in modeling the reservoir inflow hydrograph and affects the predicted direct runoff peaks and time distributions of hydrographs for reservoir and river routings. The accuracy of t_c estimates can therefore affect the results of dam safety evaluations, upgrade requirements, or inundation steadies at a given downstream location.

---

[1]P.E., Environmental Engineer Senior; [2]P.E., Chief of Dam Safety Program; Virginia Department of Conservation and Recreation, Division of Soil & Water Conservation

However, the t_c estimate in hydrologic modeling is often not given the necessary attention when compared with other important parameters for prediction of a flood peak. In addition, it is often estimated by empirical methods involving applicabilities and uncertainties which require extensive engineering judgment. Therefore, t_c is the hydrologic variable most subject to error and can significantly impact the results. Empirical formulas and velocity methods generally used to estimate t_c are discussed below.

Velocity Methods

The travel time of flow in a watershed usually reflects the storage and hydraulic characteristics of runoff-contributing areas and flow channels. Velocity methods are focused herein for t_c estimates in an unaged small watershed (upper limits of 10 mi^2 considered) The estimated t_c at a dam site is equal to the summation of the travel times for each flow regime (type). The principal flow path should be divided into segments (reaches) at points where the slope or other hydraulic conditions significantly changes in a profile of the flow path.

Upland Flow: Upland flow, including sheet and shallow concentrated flows, is usually concentrated into terrace channels within less than 1,000 feet of its origin. Sheet flow over rough terrain is assumed to usually exist in the upper reaches of the (hydraulic) flow path of the runoff. Shallow, concentrated flow usually begins where sheet flow converges to form small rills or gullies before runoff concentrates in an open channel. The velocity of upland flow in a forest area can be determined for conditions of each site from the SCS (1985) figure entitled "Velocities for Upland Method of Estimating t_c". The upland method is more accurate than the SCS curve number method.

Open Channel Flow: An open channel flow is assumed to begin where either the stream is colored blue on USGS quadrangle sheets, the channel is visible on aerial photographs, or a well-defined cross-section channel is verified by field investigation. Manning's equation can be used to estimate average flow velocities in each relatively uniform reach subdivided along the watercourse for more accurate prediction of hydraulic performance. Use of the channel bankfull discharges or 2-year frequency discharge with valley lengths is a compromise that gives a travel time for average floods. (SCS, 1985) The Manning's roughness coefficient (**n**), reflecting channel boundary resistance, can be more accurately assigned by determining the channel base value and the value of each roughness-causing factor and combining those values. (FHA, 1984; McCuen, 1989) Typically mountain channels tend to follow a stair step descent of riffles and pools. The increased boundary resistance and turbulence dissipates sufficient kinetic energy to maintain a relatively constant velocity regardless of channel slope. (Hughes, 1993) A natural regulation which maintains a subcritical flow condition appears in mountain streams. The value of **n** was found directly related to water surface slope (S_w), and the bankfull flow velocity in gravel and rock lined natural channels never exceeded 10 fps and was always greater than 5 fps (Jarrett, 1985). An empirical relationship between Manning parameter ($\mathbf{R}^{2/3}/\mathbf{n}$; **R** - hydraulic radius) and S_w of a bankfull flow, which can be directly determined from topographic maps, was developed from 42 historical data points (correlation coefficient = 0.94) for high gradient (S_w > 0.002) stream channels with natural beds and banks of large gravel, cobbles and boulders. It can be used for estimating travel time for average floods when channel slopes are averaged over longer spans on the order of 0.5 mile. The average velocity is inversely related to $0.33S_w^{0.1}$. (Hughes, 1993)

Empirical Formula Method

A variety of empirical formulas are available for t_c estimates for different flow types with little or no knowledge of its accuracy. (McCuen, 1989) When an empirical method is applied, its origin for an adequate flow type should be considered and applicabilities of its formula for particular situations need to be investigated. Applicability of most of these formulas is constrained because they have been based on very limited data in some specific watersheds. Constraints for using these formulas are therefore required to avoid misapplication of methodology and divergence of estimated results.

Summary and Recommendations

The velocity methods for t_c estimates are based on more physical factors and specific site measurements/ observations. So these methods do not require as much subjective judgement and experience to apply as empirical formula methods which depend heavily upon the person applying the methods and t_c estimates could vary considerably. Also, t_c can be more accurately estimated for channels by the Chezy or Manning formulas in terrain where a contributing watershed has complex, irregular, and diverse drainage patterns. The velocity methods are recommended.

Flood runoff in mountain basins is really a spatially varied unsteady flow. To simplify hydraulic computations involving irregularities of channels, a uniform bankfull flow is assumed herein using velocity methods to estimate t_c. Within some constraints Hughes's formula can be applied for estimating travel time in mountain basin streams instead of through a costly site investigation and field survey for obtaining sufficient information to estimate **n** and **R** values.

Example Case

The watershed is about 0.69 square miles of mostly woodland with slopes ranging from 10% to 50%. Watercourse slopes range from 40% in the upper reaches to 8% near the reservoir. The 78-foot high earth dam with a volume of 280 acre-feet is classified as high potential hazard requiring a full PMF as its spillway design flood (SDF). The primary spillway is a single concrete type chute with a box inlet weir 62 feet long, capable of accommodating 69% PMF at the top of the dam elevation 3292.5. The velocity methods are applied to estimate t_c for developing a SCS synthetic unit hydrograph. The travel time is computed by dividing the length of the flow path by the average velocity of flow along the path. Manning's formula is used to compute flow velocity in stream channels. While Hughes's formula is used for a reach when its channel slopes are averaged over longer spans on the order of about 0.5 mile. Otherwise, the **n** value exceeds the upper bound of 0.07 (McCuen, 1989) for using Manning's formula to maintain a subcritical flow on a theoretical basis. The hydraulic characteristics and travel times (t_i) are shown in Table 1 for each of the two subbasins. One minute was used as the computation interval for HEC-1 hydrograph modelings. Table 2 shows the effect of subdividing the watershed on the peak flow at the dam and how it affects the spillway upgrade. A major merit of subdividing into two subbasins is a relatively uniform distribution of runoff on each subbasin to better define t_c. Results in Tables 1 & 2 reflect homogeneous hydraulic characteristics of drainage areas. Peak outflow discharges and their occurrence times are close due to the relatively short t_c.

Table 1. Hydraulic Characteristics of Channel Reaches

Subbasin	Reach	Flow Type	S(%)	n	L(ft.)	V(fps)	t_i(hr)
WEST BASIN	1^1	^{1}upland flow	15	--	200	1.0	0.06
	2^1	^{2}open channel flow	40	--	650	1.6	0.11
	3^2		22	0.070	450	9.6	0.01
	4^2		14	0.065	1,450	8.9	0.05
	5^2		8	--	3,350	5.7	0.16
EAST BASIN	1^1	^{1}upland flow	20	--	100	1.2	0.02
	2^1	^{2}open channel flow	33	--	600	1.6	0.10
	3^2		10	0.070	1,000	6.4	0.04
	4^2		19	--	2,500	5.2	0.13

Table 2. Comparison of Effects of Subdividing Watershed on Spillway Upgrade

Water-shed	Drain. Area (mi)^2	t_c (hrs)	Peak Flow at Reservoir (cfs)			Spillway Upgrade to Width of 250 ft.		
			Single	Total	Time (hrs)	Outflow (cfs)	Overtop Depth (ft)	Overtop Time (hrs)
Whole	0.69	0.4	9,630	--	1.7	9,030	0.03	0.02
West Basin	0.04	0.4	5,580					
West Basin + East Basin				9,930	1.7	9,320	0.12	0.05
East Basin	0.29	0.3	4,350					

Discussion

1. Overestimating or underestimating t_c is not unusual due to uncertainties in analysis processes. Although the above approaches are not a precise science, the increased accuracy of t_c can nevertheless improve the prediction of runoff for dam safety

hydrologic studies. The economic consequences of t_c reductions or increases could significantly affect the costs of modifications and improvements to dams.

2. Upland flow velocity on a forest land cover can be estimated using the SCS empirical chart more accurately than using the kinematic wave equation which is good for impervious areas. However, the upland component of travel time becomes essentially insignificant as basin size exceeds a square mile in area. (Hughes, 1993)

3. Topographic maps, photos and any other available information can be used for comparison with field situations to help verify selected physical factor values. For example, a reliable **n** value can be assigned by use of an analytical approach involving a series of decisions addressed in the FHA guide (1984) instead of engineering judgment for open channel flows.

4. The above site-specific study recommends that t_c estimate techniques should incorporate appropriate concepts and data bases of hydraulic characteristics, such as stream slopes, and cross sections. Properly subdividing a watershed into subbasins and estimating t_c for each subbasin can improve the prediction accuracy of routed or combined flood hydrographs from each subdivision.

5. The t_c is held constant for all storms if it is justified as a lumped parameter which ignores effect of spatial variations throughout an entire homogeneous area. (Viessman, 1977) However, watershed characteristics may change over time and time-varying interactions may act as a dynamic model such as land use changes. These changes should be reviewed with the periodic dam safety assessment to ensure rational decisions regarding the adequacy of spillway capacities.

6. To improve accuracy of predicted hydrographs in lieu of a synthetic unit hydrograph, a regional unit hydrograph developed from watersheds with gage stations that have similar watershed characteristics and meteorological conditions should be applied to a study dam site.

References

Federal Highway Administration (FHA) (1984). Guide for Selecting Manning's Roughness Coefficient for Natural Channels and Flood Plains. Springfield, Va.

Hughes, William (1993). "Travel Time in Mountain Basins". Engineering Hydrology Proceedings. ASCE Hydraulics Conference, San Francisco, CA.

Jarrett, R.D. (1985). "Determination of Roughness Coefficient for Streams in Colorado", Water-Resources Investigation Report 85-4004. USGS, Washington D.C.

McCuen, Richard H. (1989). Hydrologic Analysis and Design. Prentice-Hall, Inc. Englewood Cliffs, New Jersey.

Soil Conservation Service (SCS) (1985). Hydrology. National Engineering Handbook Section 4, Washington, D.C.

Viessman, Warren, Jr. and John W. Knapp etc. (1977). Introduction to Hydrology, Harper & Row, Publishers, Inc., New York, New York.

Conveyance Distribution to Define Flow Boundaries at Culverts

Stephen R. Sands, P.E.[1]
James A Harned, P.E.[2]
Mark A. Sites, P.E.[3]
James R. Stahl, E.I.T.[4]

Abstract

This paper presents a method to quantitatively analyze the location of the effective flow areas at culvert/bridge openings using a one-dimensional flow model based on the conveyance distribution in the cross-section contiguous to the culvert/bridge. The conveyance distribution method can be applied to a full spectrum of storm events and associated peak flows to result in independent water surface elevation results and effective flow boundary locations for a particular storm event.

A comparison of standard one-dimensional model, one-dimensional model with effective flow boundaries defined using the conveyance distribution method, and two-dimensional hydraulic model results of an actual complex bridge opening is performed to verify the conveyance distribution method of effective flow boundary definition.

Introduction

A typical one-dimensional hydraulic model applies step backwater computations assuming the flow characteristics vary longitudinally in the direction of flow. Standard modeling techniques specified for one-dimensional hydraulic models in flow transition areas adjacent to culvert/bridge openings are based on empirical

----

[1] Branch Manager, Ogden Environmental and Engineering Services, Inc., 9800 West Kincey Avenue, Suite 190, Huntersville, NC 28078
[2] Senior Water Resources Engineer, Ogden Environmental and Energy Services, Inc., 690 Commonwealth Center, 11003 Bluegrass Parkway, Louisville, KY 40299
[3] Development Review Engineer, Louisville and Jefferson County Metropolitan Sewer District, 400 South Sixth Street, Louisville, KY 40202
[4] Water Resources Engineer Ogden Environmental and Energy Services, Inc., 690 Commonwealth Center, 11003 Bluegrass Parkway, Louisville, KY 40299

relationships. For example, the HEC-2 model documentation recommends the model user contract the effective flow area upstream from bridge openings at a 1:1 ratio and expand the effective flow area downstream from bridge openings at a 4:1 ratio. Standard practice for typical bridge openings center the effective flow areas over the culvert centerline extended in a longitudinal direction perpendicular to the opening.

However, in many floodplains analyses the conveyance distribution across a section of the main channel and adjacent overbank areas at a bridge opening are asymmetrical in relation to the opening. Therefore if standard modeling techniques are applied, the effective flow areas will be located in a region of the floodplain that is not, in reality, providing significant conveyance. With experience, model users are able to subjectively locate the effective flow areas centered over the majority of the conveyance for the storm of interest. For example, an experienced model user may locate the effective flow areas over the centerline of the channel during the 2-year storm event, and centered over the regulatory floodway during the 100-year storm event.

The stream reach that was studied exhibits several characteristics which inhibit the physical occurrence of standard flow expansion. The study reach includes an interstate bridge at the upstream end followed by a secondary road crossing. The upstream bridge was modeled as an opening with a bottom width of 23 meters and depth of 3.5 meters. The channel downstream from the bridge opening was surveyed to be approximately 2 meters deep with bottom width of 8 meters and top width of 14 meters. The channel meanders 90 degrees to the left immediately downstream of the outlet of the bridge and back 90 degrees to the right 75 meters downstream to be parallel with the centerline extended from the bridge opening. At that location the channel flows through two arch culverts at a secondary roadway. Floodplain topwidths were determined from existing hydraulic model output to vary from 300 to 700 meters. The study discharge was the 100-year storm event of 421 cms.

Methodology

Three methods were used to analyze the stream reach for water surface profiles, HEC-2 with standard 4:1 expansion, HEC-2 with weighted conveyance 4:1 expansion, and FESWMS-2DH, a two-dimensional flow model. Standard HEC-2 modeling techniques dictated that the conveyance downstream from the structure was parallel with the centerline of the bridge, and therefore was not centered over the channel and extended into the right overbank. The modeler felt that due to the low top-of-bank elevations and the meander of the channel that the momentum of the flow would force a majority of the discharge out of the channel into the overbank. Flow would eventually follow the contouring of the overbank area and re-center over the main channel several hundred feet downstream.

The second method used to determine the water surface profiles in the stream reach was the "weighted conveyance method" using the HEC-2 model. This method was developed in order to account for the off-center floodplain and channel geometry in the expansion area of the subject bridge. The basis of this methodology was that the expansion area downstream from bridges should be centered over the centerline of the conveyance of the floodplain. The method and theoretical assumptions were as follows:

- The existing base HEC-2 model was executed without any assumed ineffective flow areas due to the bridge restriction. The model output control option was set to provide a minor trace so that discharge, velocity, depth and percentage of total discharge was reported for intermediate segments of each cross-section within the reach.

- The original centerline of the direction of flow was superimposed onto the topographic maps to provide a reference line. This centerline is parallel with the centerline of the bridge based on standard HEC-2 modeling practices. The original base model 4:1 expansion areas had been set in relation to this line.

- The percentage of conveyance on each side of this line and for each cross-section was determined. The theory was that the expansion may follow a curved path based on the distribution of conveyance in each cross-section. However, the maximum increase of the 4:1 expansion rates on either side of the channel could not be exceeded due to the inability of flows to expand at a greater rate.

- The expansion rate was revised based on the percentage of conveyance on each side of the centerline. For example, the conveyance distribution for the cross-section downstream from the outlet of the bridge was 17.6 percent of the conveyance on the right side and 82.4 percent of the conveyance on the left side. Therefore, the total expansion of two - 4:1 expansions was adjusted based on these rates.

- The distance to the next upstream cross-section was determined to be 49 meters. Based on the two - 4:1 expansion rates, the top width increased from 23 meters at the bridge outlet by 26 meters to 49 meters. Therefore, 82.4 percent of the increase was applied to the left side and 17.6 percent of the increase was applied to the right side.

- All subsequent downstream cross-sections were revised with the same method.

The third method used was a two-dimensional flow model FESWMS-2DH which applies a finite element method to solve a set of differential equations accounting for the effects of bed friction, wind-induced stress, turbulence, and the effect of the Earth's rotation. To evaluate only the two-dimensional properties of channel flow, the model ignores the variation of all parameters in the vertical dimension. A grid pattern of nodes and elements was constructed over the entire stream reach defining the overall geometry and channel roughness. By solving the set of differential equations for this grid pattern with finite element analysis, the model internally accounted for the asymmetrical properties of the floodplain.

Water Surface Elevations (meters)				
HEC-2 Model Cross-section	No Culvert In-effective Flow Areas	Standard HEC-2 Expansion	Conveyance Weighted Expansion	Two-dimensional Model
8.727	158.53	158.53	158.53	158.61
8.813	158.72	158.73	158.73	158.73
8.849	159.20	159.15	159.17	159.19
8.862	159.23	159.61	159.22	159.41
8.877	159.54	159.97	159.57	159.45
8.914	159.71	160.23	159.81	159.55
9.017 (Bridge)	159.84	160.65	160.58	159.91
9.071	160.46	160.95	160.89	160.11
9.187	160.65	161.10	161.04	160.66

Conclusion

As shown in the above results, the weighted conveyance method can be used to approximate two-dimensional flow situations downstream of an atypical bridge opening more accurately than relying solely on standard one-dimensional modeling techniques. However, the engineer must have a complete understanding of the limitations of the weighted conveyance method and the physical situation to which it will be applied. The results also indicate that standard flow expansion assumptions are not valid for this particular study area and account for much of the discrepancy between the results of the one-dimensional and two-dimensional models.

Standard engineering judgement concludes one-dimensional flow modeling should be applied only to stream reaches where the assumption of one-dimensional flow is not violated. However, in an actual stream, one-dimensional flow is rarely present over the entire study reach. Therefore, the engineer must use sound judgement to determine when the system requires more detailed modeling techniques. These more detailed modeling techniques such as two-dimensional modeling using finite elements are laborious and time consuming. Therefore, for many stream reaches the modeler may be able to avoid the additional effort associated with a complete finite element analysis by using the weighted-conveyance method of locating bridge flow expansion areas.

Management of Regional Flood Damages to Railroads

Gary L. Lewis,[1] Member, ASCE

Abstract

Railroads throughout the U.S. deliver goods over thousands of miles of tracks and hundreds of yards that were constructed in the worst possible location, along river channels and in major floodplains. Advances in the business of railroading have increased the demand on these systems to perform during regional, catastrophic flood occurrences such as the general floods that occurred this winter in California, in the south last summer, and in the midwest during the summer of 1993. This paper examines the 1993 floods in the upper Mississippi and lower Missouri, describes how railroads cooperated in meeting the challenges, and looks at lessons learned and the proposed national response to identify possibilities for an integrated railroad approach to surviving catastrophic regional floods.

1993 Flood Damages

The 1993 midwest floods, based on the statistics complied by the Interagency Floodplain Management Review Committee (IFMRC), also called the Galloway Committee, resulted in the following damages:

- Total damages of $12 to $16 billion
- National response and recovery costs of $6 billion
- 100,000 homes damaged
- 50,000 homes damaged from sewer backup
- Agricultural damages of $1.4 billion
- Road, railroad, and bridge damages of $250 million
- Over 13,300 FIA insurance claims for $297 million

---

[1]Senior Water Resources Engineer, CH2M HILL, P.O. Box 22508, Denver, CO 80222-0508

The flood breached or broke 2,000 miles of levees, damaging almost 1,000 levees in the upper Mississippi and lower Missouri Rivers.

Railroad System Damages

The 1993 floods have been deemed the "worst disaster ever to befall the nation's railroads." Statistical summaries suggest over $200 million in damages to railroads; over 2,000 trains detoured; and up to 4,000 miles of service track impacted. Hubs at St. Louis and Kansas City were immobilized. Flooding was continuous over the summer.

Floods such as these result in numerous, obvious direct impacts to railroads. Not as obvious, but just as important, are indirect losses due to: additional train miles, detour expenses, crew expenses, overtime, deadheading, held aways, transporting and lodging added crews, congestion in yards, congestion along open corridors, interrupted cycle times, missed connections, security costs, acquiring additional trains, fuel, slower average speeds, and repairs of locomotives and freight cars,

Federal Response

Railroads were not completely neglected by the federal flood relief program. Federal aid to railroads, however, was limited to $21 million through the Emergency Flood Relief Bill, and provided assistance only to short-line railroads. Kansas alone received $3.8 million, for example, spread among the Northeast Kansas and Missouri Railroad, the Kyle Railroad, the Central Kansas Railway, and the Southeast Kansas Railway.

Orders from Washington following the 1993 floods resulted in official refusal by the Corps to fix over 350 private levees that did not meet Corps inspection and maintenance requirements. Environmentalists applauded the action as good fiscal management, and saw it as federal support of conversion of farmland to wetlands. The segment of track between St. Louis and St. Joseph is crucial to railroad traffic. It contains most of these private levees and remains a "weak spot" in the overall federal, state, local, and private response.

Environmental Organization Response

USA Today provided front page coverage to the *American Rivers Association* opinion that the Corps levees, dams, and channelizing made things worse in the 1993 midwest floods. Their remedy includes rebuilding levees only at greater setback distances, allowing no new levees for the protection of farmland, and rewarding communities for prevention of flood plain developments. This type of proposal is not without political opposition. Overheard after the 1994 floods in the south, "The 1994 flood in Georgia should not make good farmers leave their land" (Bill Clinton).

Other environmental groups are calling for re-regulation of locks and dams to create improved habitat for threatened and endangered species. One plan calls for deliberately raising Missouri River levels in the spring to improve habitat. By all reports, this action will increase the risk of flooding and decrease barge traffic. Some have estimated that water surface levels could increase as much as 4 feet.

Railroad Response

All major U.S. railroads except Conrail and CSX took a hit from the 1993 flood. A few of the reported ways the railroads coped in 1993 are numerous traffic diversions were allowed without prior detour agreements, One railroad carried 140 trains for seven other railroads, cost concerns over sharing of facilities were deferred until after the crisis, Amtrak chartered buses (and one airplane) and made arrangements with United and TWA for reduced fares, Santa Fe's transcontinental main line was "out" July 10 and back "in" July 25 when a bridge 90 miles east of Kansas City disappeared but was quickly repaired, BN was significantly impacted with a high percentage of their track under water; they detoured over a half dozen other carriers lines, and SP had minor problems until the Missouri flows were regulated, causing a closing the yard at Kansas City.

Fast repairs were another important individual and corporate response by railroads. Repairs to key railroad facilities were often accomplished within a fraction of normal times because of the importance of revenues. "The strength of the American railroad system lies in its ability to continue to operate, even in the face of natural disasters such as the 1993 Midwest flood" (Association of American Railroads).

National Response

Following the 1993 floods, programs were recommended by the IFMRC to accomplish floodplain management goals. They were:

- Consider permanent evacuation of flood prone areas
- Reduce the vulnerability of urban, industrial, and agricultural areas
- Avoid new development in floodplains
- Relocate people and property as appropriate
- Purchase marginal farmland and return it to its natural use
- Locate infrastructure outside of floodplains
- Lessen vulnerability of existing structures

The IFMRC report (USGS, Reston Va, June 1994, 703-648-4792) contains a long list of strategies in each of four programs.

Opportunities for Integrated Railroad Management

As a first step in management, railroads should consider making permanent the temporary actions taken during regional floods. Other possible railroad actions, arising from the ashes of lessons learned in 1993, are:

- Write pre-flood detour agreements with other carriers
- Develop new standards for flood-resistant design and construction of railroad facilities
- Seek incorporation of private levees into Corps jurisdiction
- Look for means to accomplish repairs within short time periods
- Establish legislative language for tax relief or low-interest emergency loans to railroads to offset emergency repairs
- Stockpile riprap, ties, rail, bridge, or culvert construction materials at strategic locations
- Strive for increasing flood-protection requirements to increasingly-important, national-interest structures
- Become involved in seeking ways of regulating upstream development that results in increased flows
- Stress the need to recognize railroads as having the same status as interstate highways
- Improve emergency communications through a single federal flood forecast and information system, somewhat like the electronic links that currently relay earthquake information
- Support an up-to-date floodplain mapping system
- Improve accuracy of floodplains maps
- Establish loan and grant programs for damaged infrastructure
- Establish new flood elevation standards

The Mission Creek Flood Control Project

Rudolf E. Ohlemutz M.S. D.Eng.[1], Member ASCE

Abstract

A flood control project for Mission Creek in Santa Barbara, California, has been pursued for many years by the U.S. Army Corps of Engineers and local interests. Urban encroachment and sedimentation problems complicate the design. After a highly controversial Corps proposal was abandoned, local interests formed a committee to identify a feasible and acceptable flood control project. The identified project is currently under investigation by the Corps in a new reconnaissance study.

Introduction

Mission Creek has a watershed of about 30 square kilometers (11.5 square miles) and a length of about 13 kilometers (8 miles). It originates in the Santa Ynez Mountains at an elevation of about 1150 meters (3750 feet) and discharges into the Pacific Ocean at Santa Barbara, California. Within the City of Santa Barbara, Mission Creek is crossed by more than 20 bridges and urban development has encroached in some locations to the top of the bank.

Because of the well documented flood threat to downtown Santa Barbara, the U.S. Army Corps of Engineers (Corps) proposed the Lower Mission Creek Project which was authorized by Congress in the 1988 Water Resources Development Act (WRDA). The project called for a rectangular concrete channel which would be able to convey a 100-year peak flow of 220 cubic meters per second (cms) or 7,800 cubic feet per second (cfs) in a supercritical regime.

The Corps' design attempted to preserve most of the existing bridges and developments along the creek. Consequently, the channel width was left largely unchanged and additional capacity was derived from lowering the roughness by

[1] Senior Engineer, Kennedy/Jenks Consultants, 1000 Hill Road, Suite 200, Ventura, CA 93003

concrete lining and increasing the slope and cross-sectional area by deepening. In the zone of tidal influence, the Corps' calculations showed that at peak discharge the hydraulic jump occurred on the beach and downstream of the last bridge crossing.

Upon closer investigation, the Corps discovered that at rising discharges, when the hydraulic jump was assumed to migrate downstream through the tidal zone, sediment would fill up much of the channel downstream of the hydraulic jump. This sediment deposition would not be fully re-mobilized at peak flows and render the last reach of the channel unable to convey the design discharge. Upon discovery of this problem, the Corps abandoned the project in the summer of 1993.

Public Participation

The demise of the Corps' project received a mixed welcome from the people of Santa Barbara. In the eyes of some the creek was a flood threat that required attention, others saw a degraded urban creek that needed rehabilitation. While the Corps' project had addressed the first concern, it was quite inadequate in the areas of aesthetics and habitat preservation. In fact, the opposition to the Corps project was so vehement that legal challenges on the basis of endangered species and wetland issues may likely have brought the project to a halt.

The proponents of a flood control project for Mission Creek decided on a consensus-based approach to come up with an alternative project and convened a committee consisting of City Council members, County Supervisors, City Public Works personnel, County Flood Control personnel, Corps representatives, affected landowners and representatives of various environmental groups.

Despite considerable odds and even ridicule, the committee met its objectives by identifying an achievable, acceptable, and economic project that preserves and enhances the values of the creek. In the process, valuable lessons were learned both in the technical and the consensus-building areas.

Technical Aspects

The committee employed several consultants to assist with the analysis of the technical aspects of the project. Important findings were derived in the areas of hydrology, sediment transport and water surface profile modeling.

A review of the hydrology revealed that the Corps had used "expected probability" rather than "computed probability" to determine the design discharge. Expected probability takes into account the period of record available for the statistical analysis. For a given frequency or return period, expected probability will yield a higher discharge than computed probability. The difference between the discharges is inversely related to the period of record available. In the case of Mission Creek, the use of computed probability reduced the design discharge from 220 cms to 164 cms (5800 cfs). It should be noted that the Federal Emergency Management Agency (FEMA) uses computed probability. Therefor, a 100-year design flow eliminates the need for flood insurance whether it is based on computed or expected probability.

Sediment yields from the mountainous watersheds of Southern California are very large. In the case of Mission Creek, controlling the amount of sediment expected from a 100-year event at the apex of the alluvial fan would require a basin of significant size. The sediment transport capacity downstream, however, is dramatically smaller than the sediment yield and decreasing towards the mouth. Therefor, sediment control facilities at any point of the creek need to be sized only for the sediment transport capacity of the upstream reach, accumulated over the flood hydrograph.

Water surface profile modeling of near-critical flow with bridges in short intervals stresses the capabilities of HEC-2. It was determined that reliable results could only be achieved if the split flow option was used and careful manual analyses were carried out to determine the amount of flow leaving the creek at each bridge and the point where the flow returns to the creek.

Consensus-building

The committee which was expected to come to a consensus on the Mission Creek project had members with very diverse interests. The spectrum ranged from those who wanted the original Corps design resurrected to those who wanted to see a creek restoration and naturalization project. The fact that the committee was able to come to a consensus was due to the intensity and openness of the process, the participation of elected officials, and the commitment of staff and consultants to provide unbiased answers to difficult questions.

To be effective and maintain its momentum, the committee had, at times, weekly meetings and long agendas. A generally accepted "mission statement" was of great value as was a facilitator who guided the committee temporarily.

Meetings were open to the public. Since the committee itself had a limited number of members, it was important to created a sense of representation as causes introduced from the outside were championed by committee members.

The presence of elected officials had a very positive effect. It signaled that the issue was important, that a resolution was expected, and that some of the committee members would be a part of the next higher decision-making body to move the project forward. The presence of elected officials also had positive impacts on discipline and goal-orientation in the committee's work.

Finally, the committee was backed up by staff and consultants who provided unbiased and comprehensive information required for the decisions by the committee.

The Project

The project which was developed during the public participation process represented a true compromise.

The concrete lining was abandoned in favor of gabions, stepped walls, and reinforced earth banks and an unlined invert. This assured access to the creek and an opportunity for vegetation to be established on the banks.

The depth of the creek was not increased so as not to disturb the existing sediment transport equilibrium in the tidal zone. Sediment transport capacity was shown to decrease gradually in a downstream direction. The differences from reach to reach, however, were not large enough to suggest sediment depositions which could adversely affect the performance of the channel.

The channel was widened to the maximum extent possible. Although some bridges would have to be removed and some properties would have to be purchased and their residents relocated, the majority of creekside properties would loose a narrow strip along the creek but end up with a more attractive and safer bank.

The design discharge was largely governed by the maximum possible widening. The recommended project has a 25-year capacity (90 cms), which still represents a significant improvement over existing conditions. The FEMA 100-year flood plain will be narrowed by the project.

The project was returned to the Corps in October of 1994. The design was changed too significantly for the Corps to continue working under the 1988 WRDA project authorization. However, the Corps has started working on a reconnaissance level. Since the project is expected to cost about $10 million (as compared to $15 million for the previous Corps project), it will be funded under the Corps' "Continuing Authorization Program" (federal share not to exceed $5 million) rather than go through another congressional reauthorization process.

On January 10, 1995, Mission Creek caused significant flooding in downtown Santa Barbara. Details were not yet available at the time of writing.

Confidence Limits For Hydroelectric Generation

Robert P. Armstrong and Ronald L. Rossmiller, M. ASCE[1]

Abstract

Estimates of average annual energy production for proposed run-of-river hydroelectric projects are often developed using historic or simulated historic average daily flows associated with a potential project site. These average annual energy estimates are then used to forecast the expected long-term average energy production at the site. The forecasted energy estimate is generally a primary piece of data used to determine the economic feasibility of the proposed hydropower development. However, these estimates are often based on relatively short (less than 30 years) periods of record. Determination of confidence intervals to assess the uncertainty associated with these energy estimates is accomplished by utilizing recent developments in the interpretation of flow-duration curves (FDCs). These recent developments may also be applied to energy-duration curves (EDCs), which present the relationship between the frequency and the magnitude of energy generation for a particular hydroelectric project.

Introduction

Flow-duration curves (FDCs) have been utilized in hydrologic analyses for many years to provide a graphical representation of the relationship between the magnitude and the frequency of daily (or other time period) flows for a particular stream. FDCs are generally a plot of flow on the ordinate, versus the percentage of time that flow is equaled or exceeded, on the abscissa. Guidelines for the construction of FDCs can be found in Searcy (1959). Analysis of FDCs on an annual basis rather than the traditional period-of-record analysis enables one to establish confidence limits about a particular stream's FDC. A detailed explanation of this procedure along with technological advances in the construction of FDCs can be found in Vogel and Fennessey (1994).

---

[1]Water Resources Project Manager and National Program Director - Stormwater, respectively, HDR Engineering, Inc., 500 - 108th Avenue N.E., Suite 1200, Bellevue, WA 98004-5538

In the same fashion as traditional FDCs, energy-duration curves (EDCs) can be constructed that present the relationship between the magnitude and frequency of daily energy production for a particular hydroelectric project. Like FDCs, EDCs can be interpreted on an annual basis (rather than a period of record analysis) to facilitate the development of confidence limits. The main difference between FDCs and EDCs is that while FDCs are bounded only by a lower limit (zero flow), EDCs are bounded by a lower and an upper limit created by the minimum and maximum operating ranges of a given hydroelectric project.

Typically, forecasts of energy production estimates for a proposed run-of-river hydroelectric development are determined by simulating energy production over a particular period of record for the project site. These energy simulations are usually based on the estimated average daily flow available for the project; the turbine/generator/switchyard efficiencies and operating characteristics; and the available net head for the project.

Method of Analysis

Analysis of a hypothetical hydroelectric project on the Snoqualmie River near Snoqualmie, Washington, is used to demonstrate this method of analysis. Average daily discharges recorded by a United States Geological Survey streamflow gage are available for this site for a continuous period from water year (WY) 1959 to WY1993. The average daily discharge for this 970 square kilometer (375 sq. mile) contributing drainage basin is 73 m^3/sec. (2,579 ft^3/sec.). Utilizing the recorded daily streamflow information, daily energy production estimates were computed for the period from WY1959 to WY1993. These energy estimates were a function of the actual daily streamflow records and hypothetical minimum instream flows, discharge/head relationships, and turbine/generator characteristics. The resulting average annual energy production estimate for the site was approximately 97.54 terajoules (270.9 gigawatt-hours).

Using the daily energy simulation results, annual EDCs were then developed for each year for the period WY1959 to WY1993. To develop the annual EDCs, twenty-eight energy class intervals were used. The percentage of time that daily energy simulation values equaled or exceeded each of these twenty-eight energy class intervals was computed for each of the thirty-five years of record. Mean annual percent exceedance values for each of these twenty-eight class intervals were then computed as

$$\overline{C}_j = \sum_{i=1}^{n} C_j \tag{1}$$

where n = number of years (1 to 35), C_j = percent exceedance value of class interval j where j = class interval number from 1 to 28. The resulting mean annual EDC for the complete period of record is shown graphically in Figure 1. Integration of the area under an FDC yields volume of flow. Likewise, integration of a mean annual EDC yields total mean annual energy production. To verify that the selected EDC class intervals provided an accurate representation of the relationship between the magnitude and frequency of average daily energy estimates, Figure 1 was integrated to compute the mean annual energy production. This integration resulted in a value of 97.52 terajoules (270.9 gigawatt-hours), within 0.02 percent of the result as with computing the average annual production based upon daily simulated values. Therefore, it is concluded that the chosen class intervals adequately describe the relationship between the magnitude and frequency of energy production for this hypothetical project.

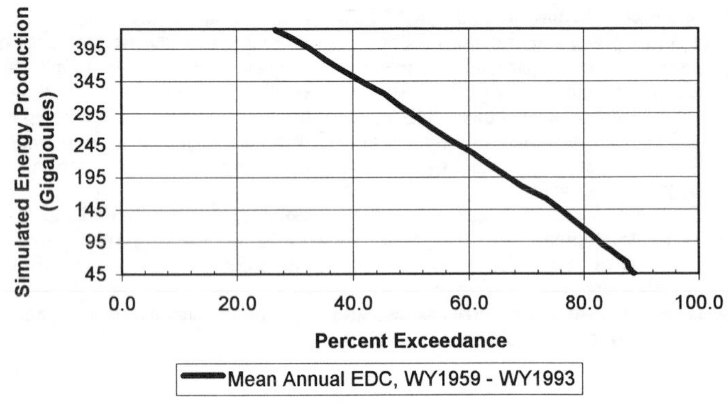

Figure 1. Mean Annual EDC

With the mean annual EDC computed from the twenty-eight annual EDCs, confidence limits about the mean annual EDC can be computed. To estimate confidence limits, the standard deviation of the annual percent exceedance values must be computed for each of the twenty-eight class intervals. These standard deviations were computed as

$$s_j = \left(\sum_{i=1}^{n}(C_j - \overline{C}_j)^2/(n-1)\right)^{0.5} \tag{2}$$

where s_j = standard deviation of class interval j. 95% confidence limits were then computed for each of the twenty-eight class intervals as

$$\overline{C}_j \pm 1.96\, s_j/\sqrt{n} \tag{3}$$

The mean annual EDC along with the estimated confidence limits are shown in Fig. 2.

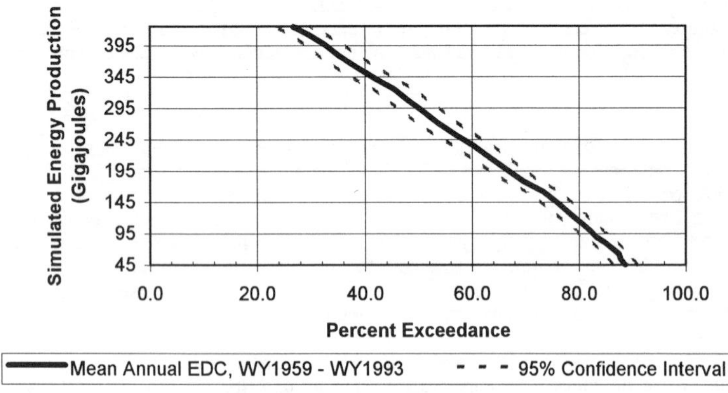

Figure 2. Mean Annual EDC With 95 % Confidence Intervals

Integration of the upper and lower 95% confidence limit EDCs resulted in annual energy estimates of 102.47 and 92.57 terajoules (284.6 and 257.1 gigawatt-hours), respectively. Therefore, one would conclude that, assuming hydrologic and meteorlogic conditions within the project's drainage basin remain unchanged, the forecasted average annual energy production is 97.52 terajoules (270.9 gigawatt-hours). Furthermore, there is a 95% probability that the long term energy production will be between 102.47 and 92.57 terajoules (284.6 and 257.1 gigawatt-hours), or approximately 5 percent of the annual average estimate.

The forecasted mean annual energy estimate is often a primary piece of the data used in assessing the benefit/cost ratio for a project. If the projected benefit/cost ratio for a project is close to a developer's threshold of viability, the 95% (or other) confidence interval energy estimate becomes a valuable source of information. This information will aid the developer in assessing the risk of investment in a particular project.

Summary

Recent advances in the interpretation of traditional flow-duration curves are applicable in the interpretation of energy-duration curves. Proper selection of EDC class intervals will facilitate an accurate determination of confidence limits. Determination of these confidence limits provides information that enables an assessment of the uncertainty associated with forecasting energy production estimates from daily streamflow records. Thereby, benefiting project developers by supplying potentially critical information needed when assessing a marginal project.

References

Searcy, J. K. (1959). "Flow-Duration Curves." *Water Supply Paper 1542-A*, U.S. Geological Survey, Reston, Virginia.

Vogel, R. M., and Fennessey, N. M. (1994). "Flow-Duration Curves. I: New Interpretation And Confidence Intervals." *J. Water Resour. Plng. and Mgmt.*, 120(4), 485-503.

Self-Optimization in Water Resource Systems

Jay R. Lund, Associate Professor
Department of Civil and Environmental Engineering
University of California, Davis, CA 95616
E-mail: jrlund@ucdavis.edu

Introduction

A regular operation of a water resource system, if adhered to long enough, often tends to become optimal. This paper develops examples and explanation to support this statement and suggests some implications of this phenomenon for water resources management. Typically, the overall economy and society surrounding a water resource system is relatively insensitive to its operation. Recreation and water supply demands are affected relatively little by small changes in water availability compared to equally large changes in regional population, employment, income, and other factors. Water-related infrastructure to support these demands adapts to whatever water resources infrastructure and operating rules exist. These adaptations include establishment of land use patterns, water withdrawal intake elevations, dock and boat ramp elevations, and other aspects of infrastructure affecting the optimality of water resource system operations. The accumulation of such water-related infrastructure and land use over time in response to relatively constant reservoir operating policies can result in those policies being increasingly relied upon to support these valued economic activities. Thus, any given set of water system operations becomes increasingly expensive to change and so can become more optimal over time.

Some Examples of Self-Optimization

Self-optimization in water resource systems arises from the investment of the surrounding economy in long-lived capital projects which rely on an established regime of water control. A regular operation of a water resources system, no matter how misguided initially, becomes more desirable to maintain because the economy's long-lived and difficult to replace investments have become accustomed to this regular operation. Some examples of this phenomenon are provided below.

Docks and boat ramps set to accustomed water levels

In many reservoirs with recreation facilities, these facilities are designed based on accustomed water levels which result from a reservoir system's established operating policy. Long adherence to an operating policy increases number of facilities dependent on continuation of this operating policy, increasing recreation losses from deviating from current operations.

Intakes set to accustomed water levels

The operation of the Missouri River system in the summer and fall was initially established to accommodate what is now relatively minor navigation use of the lower river. Consistent operation for these navigation flows since the late 1960s has led to an expectation of consistent low flow river stages. Water withdrawal intake elevations for municipal and thermal electric water supplies have been set according to these stage expectations. In the optimization of reservoir operations for this system, the great economic importance of these water withdrawals tends to prevent great reduction from traditional low-flow system operations. Changing these withdrawal elevations is a long-term process.

A similar problem arises when a region accustomed to a regular sustained yield of groundwater desires to develop the use of an aquifer for greater over-year storage as part of a conjunctive use scheme. Here, well elevations are likely to be established as a level too shallow to allow significant groundwater withdrawals during prolonged droughts (Blomquist, 1992). This problem is likely to be especially problematic where the groundwater is currently shallow. The costs of modifying well elevations based on historical water resource operations tends to make current operations closer to optimal.

Hydropower and the Power Grid

For regional energy systems with a major hydropower component, such as the U.S. Pacific Northwest, the entire energy system, including the structure of energy demands, can become dependent on the regular generation of hydropower, arising from regular reservoir operation. Power plants have been built and sized based on expectations of hydropower availability. In the Pacific Northwest, large nuclear power plants are the region's major thermal power source, and are attractive for their large baseload capacity where extensive hydropower can be relied on for seasonal and daily peaking. Similarly, large power users, such as Aluminum smelting, have moved to the region, anticipating regular hydropower availability from established reservoir system operations. The use of hydropower for winter seasonal peaking encouraged electric heating investments in homes throughout the Pacific Northwest.

Re-operation of this system in ways which modify the seasonal availability of hydropower or its overall availability would likely impose larger costs now than if they had been made decades ago. The "water budget" operation of the lower Columbia and Snake Rivers diminishes hydropower production during the winter peak, but provides additional hydropower during the spring and summer, when power demands are less.

Environmental Preservation

Numerous occasions have developed where environmental species have become dependent on a habitat provided by an established regime for operating reservoir systems. Fisheries develop which depend on the flow, temperature, and sediment regimes established by the construction and operation of reservoirs. Often replacing devastated natural aquatic ecosystems, these new fisheries quickly develop a recreational clientele and a degree of environmental protection when changed operations are sought.

On a larger regional scale, some rivers are often selected as "wild and scenic", developing a recreational and environmental clientele to the exclusion of reservoirs and abutting land uses that would jeopardize these uses. Other rivers in the same region develop different uses less compatible with wildland recreation and environmental uses. These rivers become more completely used for hydropower, water supply, irrigation, and waste assimilation. Land uses grow up adjacent to these rivers which are incompatible with "wild and scenic" activities, resulting in often substantial pollution and destruction of riparian habitat. In California, the Eel River in the north of the state is a "wild and scenic" river, while the San Joaquin river has become used almost exclusively for irrigation and other water supply and drainage water disposal, with relative destruction of its environmental uses. Recent decades have seen attempts to develop the Eel River for water supply and change use of the San Joaquin River for environmental uses. Both changes have been strongly resisted by the established water uses on these streams. As a classic non-convexity, it is often economically optimal to concentrate different types of uses in different regional river basins (Mar, 1981). This selection having been made, more often by historical than planning processes, it becomes difficult to change river usage, due to the establishment of infrastructure and habitats on these rivers. The historical selection tends to become optimal.

Development of Metropolitan Water Markets

Cities have frequently constructed reservoir systems to supply mistakenly overestimated future water demands for their jurisdictions (Lund, 1988). While classical planning would hold this to be a great mistake, many cities have found this to be a great advantage regionally, using the "surplus" water supply to foster annexation, sell to neighboring jurisdictions to lower the cost of water within the city, and as a source of regional cohesion in often politically divided metropolitan areas. Indeed, many cities actively modify their operations and expand their systems to continue supplying these initially unintended customers.

Planning and Management with Self-Optimization

Adaptations in regional economies, land use, and human behavior in response to the construction and operation of water resource facilities have long recognized roles in determining the success or failure of water resource systems. Gilbert White's (1957) work of floodplain encroachment, responding partially to upstream flood control projects, provides an example of self-defeating water resource management. We also should recognize that our facilities and operations can also become optimal in a sense, despite our mistakes and limitations as planners and engineers. While many of our successes derive from our efforts, some operations apparently are successful because the larger economy, society, or environment have adapted well to what we have done.

For an established water resources system, with an established physical, social, and political infrastructure adapted to a long-standing set of operations, it is likely that the most feasible, and perhaps the most desirable improvements will be small and incremental. Such situations encourage water managers to "muddle through" in seeking to improve water resource management and operations (Lindblom, 1979). Managers often tend to focus, probably very rationally, on improving or maintaining day-to-day operations, with less attention to strategic changes in operations.

The Limits of Self-Optimization

Many small local adaptations in an economy and society tend toward self-optimization in water resource systems. However, the accumulation of exogenous changes in the society and economy can limit or reverse this tendency. Recent decades have seen changes in economic demands for water, with growth in some sectors and declines in others, or simply growth in all sectors with resulting economic competition for water. Social changes accumulate as well. Changes in societal demands for environmental protection, recreation, and aesthetics have significantly changed the "objective functions" for water management. Some amount of change can be accommodated within an existing structure of water resource operations. Adaptive, or self-optimizing, economic and social activities tend first to take advantage of water-related opportunities within existing operations (Palanisami and Easter, 1984), avoiding the more expensive and institutionally difficult rearrangement of water-related infrastructure. However, large changes in economic water demands often provide sufficient impetus to overcome the costs of escaping the current "local optimum" created by self-optimization processes for a new, and perhaps structurally different solution (Figure 1). Self-optimization is not absolute, but an important tendency.

Figure 1: Changes in Penalty Surface for Operation over Time

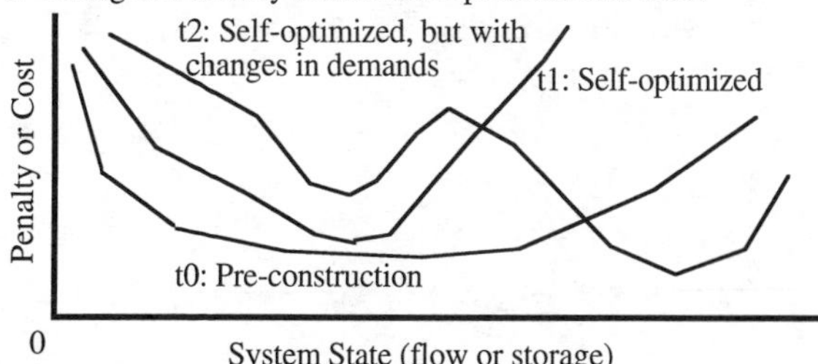

Conclusions

While long-term planning is important, there are many economic, social, and environmental aspects of water resource system performance that we will be unable to accurately evaluate or predict. Among the processes making long-term performance assessment difficult is the tendency of systems to "self-optimize," or adapt to what would otherwise be non-optimal operations. The existence of self-optimization should not discourage planners, engineers, and modelers, but rather should caution us in our work. Some practical implications of self-optimization for water management include:

1. Water management of long-established water resource systems will typically involve small changes to existing operations, since large operational changes typically require too large and rapid changes in economic infrastructure and environmental species which have become habituated to a current operating policy.
2. To adapt to exogenous changes in water demand, management of long-established water resource systems should pursue actions to increase the flexibility of operations in the long-term. Actions to increase flexibility could include long-term efforts to widen the range of design elevations for intake structures and recreational facilities. New, replacement, and rehabilitated infrastructure should be designed to function under a wider range of water elevations, for example.
3. Managers of newly-established water resource systems should try to maintain the flexibility of their system's operations. Relatively small costs for maintaining a wide range of design elevations for intake and recreational infrastructure can reduce the costs incurred if changes in management objectives or drought require changes in operation.
4. Perhaps above all, managers and students of water resource systems should feel some humility and caution in the planning and management of water resource systems. We planners, economists, and engineers have an important, but ultimately small role in water management, particularly where systems have been existing for some time.

References

Blomquist, W. (1992), *Dividing the Waters*, ICS Press, San Francisco, CA.

Lindblom, C.E. (1979), "Still Muddling, Not Yet Through'," *Public Administration Review*, Vol. 39, No. 6, pp. 517-526.

Lund, J.R. (1988), "Metropolitan Water Market Development: Seattle, Washington, 1887-1987,"*J. of Water Res.Plan. and Man.*, V.114, No.2, pp.223-240.

Mar, B.W. (1981), "Dead is Dead - An Alternative Strategy for Urban Water Management," *Urban Ecology*, Vol. 5, pp. 103-112.

Palanisami, K. and K.W. Easter (1984),"Ex Post Evaluation of Flood Control Investments: A Case Study in North Dakota,"*Water Res. Res.*,V.20, pp.1785-1790.

White, G., et al. (1957), *Changes in Urban Occupancy of Flood Plains in the United States*, Res. Paper No. 57, Dept. of Geog., Univ. of Chicago, Chicago, IL.

Hydroelectric Project Relicensing: Federal Regulation of Water Use

Dick Westmore, P.E.[1] and Dick Petzke[2]

Introduction

The Federal Energy Regulatory Commission (FERC) regulates non-federal hydroelectric power projects in the United States. Licenses to operate projects are issued by FERC, giving terms and conditions for operating these hydroelectric projects. In 1986, the Electric Consumers Protection Act (ECPA) was enacted. It mandates that the licensing process must include formal step-by-step consultation with natural resource agencies and the public to identify and resolve issues relating to hydroelectric project construction, operation and maintenance. This process is mandated for projects seeking initial operating licenses, as well as existing hydroelectric projects for which new operating licenses are required.

The Licensing Process

The process of relicensing an existing hydroelectric project requires about five to eight years from filing a notice of intent to relicense a project through completion of National Environmental Policy Act (NEPA) compliance and issuing by FERC of the new license with its operating terms and conditions. The process of balancing "developmental" values with "non-developmental" values is the heart of the FERC licensing process. The process brings the project owner into close contact with federal and state resource agencies and FERC personnel to "negotiate" a plan for future operation of the project that, hopefully, balances these values to the satisfaction of all parties involved.

Studies to document hydroelectric project impacts and enhancement potentials are time-consuming, technically challenging and expensive. This process typically leads to a change in operation to benefit environmental values and in reductions of energy

---

1 M. ASCE Water Resources Division Manager, GEI Consultants, Inc., 5660 Greenwood Plaza Blvd., Suite 202, Englewood, CO 80111-2418

2 Life Management/Project Licensing Specialist, Public Service Company of Colorado, 5900 E. 39th Avenue, Denver, CO 80207

production. The relicensing process, which is described in this paper for a small hydroelectric project in Colorado, owned by Public Service Company of Colorado (PSCo), demonstrates federal policy with respect to regulations that affect the hydropower industry and other users of our nation's water resources.

PSCo's Projects

Federal licenses to operate the Georgetown and Salida Hydroelectric Projects expired on December 31, 1993. Approximately five years earlier, Public Service Company of Colorado (PSCo) began the arduous and expensive process of obtaining new operating licenses from the FERC. As of January 18, 1995, neither project had received a new 30-year license, and both continue to operate on interim licenses granted annually by the FERC.

The Georgetown and Salida hydro projects were constructed in the early 1900s and once produced energy that spurred economic development in the small towns of Georgetown, located 30 miles west of Denver, and Salida, located 150 miles southwest of Denver. The two projects, with capacities of 1440 kW and 1350 kW, respectively, are now small parts of PSCo's integrated energy system. Both projects operate on a run-of-river basis. Their generation totals about 13.6 million kWh per year. Despite their relative size, the projects are important parts of PSCo's system because they produce low-cost energy from renewable sources, and because the water diversion and storage rights associated with each project have significant economic value to PSCo.

Key Issues at the Georgetown Project

This paper deals with the efforts of PSCo to obtain a new license for the Georgetown Hydroelectric Project. Collectively, these efforts are termed the relicensing process. The project (Figure 1) includes a 5400-foot long penstock and a powerhouse containing two 720 kW horizontal-shaft Francis turbine/generator units.

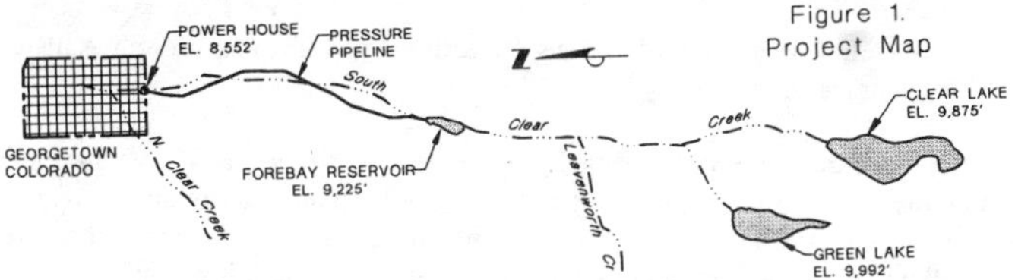

Figure 1.
Project Map

ECPA requires that hydro project owners, in consultation with natural resource agencies and the public, evaluate the effects of hydro operations on the environment and identify measures to enhance environmental values. At Georgetown, the primary environmental impacts of project operation occur as the result of diversions of water from South Clear Creek. These diversions impact a 5400-foot reach between the small Forebay Dam on South Clear Creek and the tailrace of the Georgetown Powerhouse. The effects on project operation on streamflows in South Clear Creek are depicted in Figure 2.

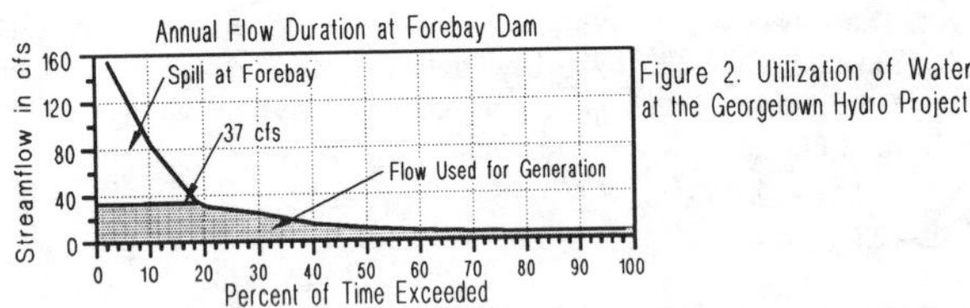

Figure 2. Utilization of Water at the Georgetown Hydro Project

Hydroelectric operations have reduced streamflows and the amount of aquatic habitat. Theoretically, the affected reach of stream could support about 188 pounds of fish biomass (mainly rainbow and brook trout). With current hydroelectric operations, only about 80 pounds of biomass are believed to exist in the affected stream reach, based on an electrofishing survey conducted in September 1990. The Colorado Division of Wildlife (CDOW), the U.S. Forest Service (FS) and U.S. Fish and Wildlife Service (FWS) expressed concerns about the impact of operations on aquatic habitat and fish life in South Clear Creek and, during the course of relicensing, suggested that PSCo modify operations to increase streamflows and aquatic habitat.

The Balancing Process

Initially, PSCo proposed the following modifications and enhancements for the Georgetown Project:

- Minimum streamflows below the Forebay of 2 cfs from September through April and 4 cfs from May through August;
- Modified release pattern from Clear Lake;
- Cooperation with CDOW to create a recreational fishery between Clear Lake and the Forebay; and
- Implementation of various recreational and historical resource management measures.

If these measures were implemented, impacts to PSCo in terms of cost and operational difficulties would occur. The lost generation, due to minimum streamflow requirements, would cost about $34,000 per year (1993 dollars). The resource agencies, however, were not willing to accept these enhancements and requested the following:

- Minimum streamflows below the Forebay of 3.25 cfs from September through April and 7.5 cfs from May through August;
- The same minimum flows (*i.e.,* 3.25 and 7.5 cfs) between Clear Lake and the Forebay;
- Allowing non-motorized boating on Clear Lake (which is currently closed to boating and swimming); and

- Implementation of significant recreational enhancement measures beyond those proposed by PSCo.

The additional instream flows requested by the agencies below the Forebay would reduce generation by 17% and would have required PSCo to shut down the hydro plant about every two years during the winter due to inadequate natural flows.

In their project review, FERC staff recommended that instream flows be maintained as suggested by PSCo between the Forebay and the Powerhouse, and as suggested by the agencies between Clear Lake and the Forebay. This would reduce economic costs to PSCo (assuming that enhanced instream flow must be provided), while providing aquatic enhancements in the stream reach between Clear Lake and the Forebay. The resource agencies, however, have not yet concurred with FERC staff's recommendations.

The main hurdle for negotiations between PSCo and the resource agencies was the issue of instream flows. PSCo argued that the Habitat Quality Index (HQI) was reasonable for assessing project impacts and for identifying and evaluating enhancement options. The agencies, although initially supportive of HQI, fell back on the results of simpler analytical tools. Their instream flow recommendations were based on the R2-Cross Method and the Tenant Method. These methods do not rely on assessing habitat quality. They provide guidelines on the amount of flow needed to achieve a desired flow depth, which then is equated to aquatic habitat. HQI suggested that the 2/4 cfs minimum flow pattern proposed by PSCo would provide habitat to support an increase in fish biomass of 74% over present operations. When HQI was applied to the current flow patterns, it predicted 52 pounds of trout biomass per 1,000 feet of stream, which closely matched results of actual surveys (80 pounds). Despite this finding, FERC staff did not support PSCo in the use of HQI. Fortunately, FERC did concur with PSCo's flow recommendation for the stream reach between the Forebay and the Powerhouse.

Conclusion

PSCo expects that final license conditions will be as recommended by FERC and is prepared to modify operations accordingly. FERC staff have indicated that energy reductions at relicensed hydro projects will result in an average 15% reduction in historic energy production on a nationwide basis. Modified operations at Georgetown will reduce energy production by 8%, somewhat better than the national average. Other enhancements will be provided by PSCo at additional cost to rate payers. However, by virtue of the relicensing process, FERC and the Federal and State resource agencies have implemented regulations on the historic and legally permitted use of water at the Georgetown Hydroelectric Project.

A Stochastic Multi-Reservoir Hydroelectric Scheduling Model

J. Jacobs[1], G. Freeman[2], J. Grygier[1], D. Morton[3],
G. Schultz[1], K. Staschus[4], and J. Stedinger[5] M. ASCE

Abstract

Pacific Gas and Electric Company (PG&E) obtains a significant fraction of its electric energy and capacity from its own hydrogeneration facilities. Hydropower is very flexibility, but is constrained by energy limitations. Optimal scheduling of hydrogeneration, in coordination with other energy sources and streamflow forecasts, is a stochastic problem of great significance to PG&E. This paper describes the SOCRATES system for the optimal scheduling of hydropower generation over a one- to two-year horizon. Possible hydrologic inputs are described by an event tree and the resulting stochastic optimization problem is formulated as a network and solved using Benders decomposition. The network model includes a description of operations within each period. A streamflow forecasting model and data handling capabilities are important components of the system.

Introduction

PG&E is the major electric utility company for northern California. It obtains a significant fraction of its electric energy from hydropower systems in the Sierra Nevada mountains of California. The utility is interested in employing available spring snowmelt-season runoff forecasts and their uncertainty in the derivation of optimal release schedules for its Sierra reservoirs. Several basins contain 2-3 seasonal storage reservoirs, in addition to other facilities with forebays that allow regulation of daily flows. The SOCRATES scheduling model calculates weekly and monthly reservoir release targets that maximize the value of generated hydroelectric energy. Hydropower systems in different basins are modelled using a network formulation with either weekly or monthly time steps that reflect whether releases

[1]Electric Supply Systems, Pacific Gas and Electric Co., Mail Code T25A, P.O Box 770000, San Francisco, CA 94177.

[2]Hydro Generation Dept., Pacific Gas and Electric Co., Mail Code P10A, San Francisco.

[3]Naval Postgraduate School, Monterey, California.

[4]Deutsche Verbundgesellschaft, Heidelberg, Hauptstraße, Germany.

[5]School of Civil and Envir. Engineering, Cornell University, Ithaca, NY 14853-3501.

generate electricity during four subperiods, corresponding to weekday and weekend, on-peak and off-peak hours, within each weekly or monthly period. The model is used to develop weekly release targets for reservoir operators, and forecasts of monthly energy generation for resource planning.

Literature Review

A challenge is the derivation of efficient operating decisions for reservoir systems employing realistic forecasts with consideration of uncertainty. Available methodologies include the use of streamflow forecasts in simulation models with heuristic policies (Johnson et al., 1991) and deterministic optimization models (Ikura et al., 1986; Mizyed et al., 1992). By employing an appropriately selected critical (wet or dry) inflow quantile, Randall and McCrodden (1992) developed operating decisions that hedged against adversity at the selected probability level. Pereira and Pinto (1985) used multiple-stage decision trees (Bertsekas, 1976) to model successfully a Brazilian system, as we do. Howard (1992) described ongoing experience with an LP model that uses multiple inflow forecasts corresponding to specified quantiles of the cumulative inflow distribution; the model determines optimal generation in the first period and a distribution for subsequent operation.

Dynamic programs are another and a powerful way to capture the sequential nature of decisions and the joint distribution of inflows and forecasts (Stedinger et al., 1984; Vasiliadis and Karamouz, 1994; Kelman et al., 1990). Pereira and Pinto (1992) obtain a stochastic DP model as an extension of a multi-stage analysis such as that employed here. Stedinger et al. (1992) use stochastic DP models with different hydrologic state variables to show that the value of hydrologic forecasts in California depends upon the levels of energy and water targets and the severity of shortage penalities. Johnson et al. (1991) and Kelman et al. (1990) also considered the value of forecasts in California reservoir system operation.

Event Tree Description of Hydrologic Uncertainty

Most precipitation in PG&E's watersheds in the Sierra Nevada mountains falls between Oct. and May, and that above PG&E reservoirs is generally snow on relatively shallow soils or bare rock. The timing of runoff is largely dependent on temperatures during the precipitation event and later during the spring melt season.

Streamflow forecasting is done in two steps: estimating the total flow in a water year and projecting the timing of that flow. The timing in turn depends on the total flow forecast – deep snowpacks melt later than shallow ones – and on temperatures. Hydrologic statistical models use measured snowpack and possible precipitation to develop snowmelt-season streamflow forecasts which they disaggregate into weekly and monthly flows for each scenario (Grygier et al., 1993).

System operation is modelled using an event tree that describes possible basin-wide inflows for up to 2 years. For example, inflows for the first few weeks are assumed to be known with certainty, after which 5 different inflow sequences might begin; these can then split three ways to allow for 15 possible inflow sequences during the late spring and summer; a three-way branch in Sept. would yield 45 possible inflow scenarios in the following water year (Jacobs et al., 1995).

Network Model and Modelling of Subperiod Operations

Networks make convenient models of hydropower systems because they are easily solved and required assumptions are not too restrictive. Because the value of energy varies significantly within each period (week or month), periods were divided into four *subperiods* representing weekday-peak, weekday-off-peak, weekend-peak and weekend-off-peak hours. A "weekday-peak" arc includes all flow in weekday-peak hours during the given period (perhaps 20 separate time intervals). Individual powerhouses are modelled to approximate possible flow control which is constrained by the timing of inflows and the size of forebays and afterbays that allow regulation.

Benders Decomposition for Stochastic Optimization

A Benders decomposition approximates the future value of water and optimally combines network solutions for different deterministic streamflow segments to identify the optimal stochastic reservoir schedule for a basin over the entire scheduling horizon. The following table compares solution times on an HP 9000/750 workstation using CPLEX to solve deterministic equivalents of the entire decision tree as a single problem, with Benders decomposition using the NETSIDE solver for min-cost network flow problems with side constraints; for the four cases, the size of the deterministic equivalent problem were 1,200x4,650, 9,500x3,600, 22,000x85,000, and 36,000x140,000 constraints by columns.

Performance of Algorithms on Mokelumne Basin

Number of scenarios	Benders Iterations	Benders Time (sec)	Simplex Time (sec)	Speedup (ratio)
1	10	17	7	0.4
9	13	85	540	6.3
27	9	130	3660	25
45	10	220	10,500	50

These results include subproblem "warm starts" from closest previous solution, which improved performance by a factor of approximately 15, and multiple single-scenario Benders cuts rather than expected-value cuts, which improved performance by a factor of approximately 1.5. The use of an improved tree traversal strategy in the Benders algorithm gave an additional factor of 2 speedup (Jacobs et al., 1995).

A master program is being developed to combine predicted generation in individual basins, and from other energy resources, with a probabilistic thermal-system production costing model. It will identify the expected system-wide marginal thermal production cost for all four subperiods in each modelled period.

Results and Conclusions

SOCRATES is an ambitious attempt to integrate medium-term operations planning and uncertainty for a large, complex electric utility. It incorporated many ideas. We have demonstrated the feasibility and advantages of using four subperiods to describe system operation and the value of power at different times within a week; this was not available with the previous planning model (Ikura et al., 1986).

The benefits of the stochastic optimization capability is being tested.

Efficient data handling and forecast generation is a necessary condition for user acceptance of the scheduling software because the stochastic model forces them to deal with much more data than before.

The Benders Decomposition as a computational approach works well and is many times faster than solving deterministic equivalents of even moderate-sized problems. The value of enhancements to the Benders algorithm has been demonstrated so that

SOCRATES optimizes individual river basins under uncertainty in a matter of minutes. Decomposition of the overall coordination problem into resource-specific subproblems, such as the hydro scheduling module, allows submodels to be "owned" by the responsible departments, and makes the software easier to maintain. Work remains to be done on coordination and costing submodules. Uncertainties in loads and gas prices (both partly functions of temperature) are also important.

SOCRATES is gaining user acceptance within PG&E, but it is a slow process.

Acknowledgments

We appreciate the support of personnel in PG&E's Hydro Generation, Power Control, and R&D departments, and many of our colleagues who have worked in Electric Supply Systems, including Zhiming Wang. The advice of Roger Wets and Mario Pereira is gratefully acknowledged.

Appendix -- References

Bertsekas, D.P., *Dynamic Programming and Stochastic Control,* Academic Press, NY, 1976.

Grygier, J., J.R. Stedinger, H. Yin, and G. Freeman, Disaggregation Models of Seasonal Streamflow Forecasts, *Proceedings of the Fiftieth Annual Eastern Snow Conference,* pp. 283-289, Quebec City, Quebec, July 8-10, 1993.

Howard, C., Experience with probabilistic forecasts for cumulative stochastic optimization, Am. Water Resour. Assoc. Sym., Reno, Nevada, Nov. 1992.

Ikura, Y., G. Gross, and G. Sand Hall, PGandE's state-of-the-art scheduling tool for hydro systems, Interfaces, 16(1), 65-82, 1986.

Jacobs, J., G. Freeman, J. Grygier, D. Morton, G. Schultz, K. Staschus, and J. Stedinger, SOCRATES: A system for scheduling hydroelectric generation under uncertainty, to appear *Annals of Operations Research,* 1995.

Johnson, S.A., J.R. Stedinger, and K. Staschus, Heuristic operating policies for reservoir system simulation, *Water Resour. Res. 27*(5), 673-685, 1991.

Kelman, J., J. R. Stedinger, L. A. Cooper, E. Hsu, and S. Yuan, Sampling stochastic dynamic programming applied to reservoir operation, *Water Resour. Res. 26*(3), 447-454, 1990.

Mizyed, N.R., J.C. Loftis, D.G. Fontane, Operation of large multireservoir systems using optimal-control theory, *J. Water Resour. Plng. and Mgmt., 118*(4), 371-387, 1992.

Pereira, M.V.F., and L.M.V.G. Pinto, Stochastic optimization of a multireservoir hydroelectric system-a decomposition approach, *Water Resour. Res., 21*(6), 1985.

Pereira, M.V.F., and L.M.V.G. Pinto, Multi-stage stochastic optimization applied to energy planning, *Math. Programming, 52*(2), 359-375, 1991.

Randall, D., and B.J. McCrodden, Modeling the Savannah River system for improved operations, 4th Operations Management Workshop, Am. Soc. of Civil Engineers, Mobile, AL, March 1992.

Stedinger, J.R., B.F. Sule, and D.P. Loucks, Stochastic dynamic programming models for reservoir operation optimization, *Water Resour. Res., 20*(11), 1499-1505, 1984.

Stedinger, J.R., J.A. Tejada-Guibert, and S.A. Johnson, Performance of hydropower systems with optimal operating policies employing different hydrologic information, 4th Operations Management Workshop, Am. Soc. of Civil Engineers, Mobile, AL, March 1992.

Vasiliadis, H.V. and M. Karamouz, Demand-driven operation of reservoirs using uncertainty-based optimal operating policies, *J. Water Resour. Plng. and Mgmt.,* 120(1), 101-114, 1994.

Stochastic Finite Element Model For Transient Advective-Dispersive Transport In An Uncertain Hydrodynamic Environment

Kuo-Ching Lin*, Timothy L. Jacobs† Miguel A. Medina‡

Abstract

A finite element formulation that incorporates a first-order second-moment(FOSM) algorithm is developed for the uncertainty analysis of groundwater solute transport in saturated, isotropic media. The contaminant concentration is the state random variable resulting from two major uncertain hydrodynamic parameters: hydraulic conductivity and dispersivity. Applying a Taylor's series expansion, the mean and standard deviation of solute concentration are calculated directly. This algorithm has the flexibility of accommodating steady or unsteady solute transport with any type of uncertain hydraulic parameter. A set of two-dimensional examples with one and two uncertain parameters in homogeneous and heterogeneous media is presented to demonstrate the applicability of proposed method. These examples are compared to the exact solution from a convolution integral approach and show that the proposed FOSM method is far more cost-effective than Monte Carlo simulation. The numerical error resulting from the FOSM method and potential methods for reducing the error are also discussed in this paper.

Introduction

The objective of this study is to develop a methodology to quantify the reliability of groundwater quality predictions. Many researchers have focused on the estimation of hydrodynamic parameters in deterministic models. A general review of the literature is presented by Yeh[1986]. However, several studies show the difficulty of calibrating two or three dimensional models (Martin [1986], Konikow and Bredehoeft [1992], Maloszewski and Zuber [1993]). Using two general cases, Martin [1986] showed that the 'optimal ' parameters selected on the basis of their sampling properties in parameter space can leads to nonoptimal decisions relative to planning objectives. Konikow and Bredehoeft [1992] claim that flow models cannot be validated and the term 'validation' should be abandoned. However, Maloszewski and Zuber [1993] state that validation is usually obtainable only with respect to some quantities or parameters and should be regarded as a partial validation. Eventually, due to incomplete information about the system, the uncertainty associated with the physical parameters lead to errors in simulating the

*Research Associate, Department of Civil Engineering, Duke University, Durham, NC27708
†Assistant Professor, Department of Civil Engineering, Duke University, Durham, NC27708
‡Associate Professor , Department of Civil Engineering, Duke University, Durham, NC27708

groundwater transport. This paper uses the first-order second-moment method (FOSM) to analyze the influence of parameter uncertainty on contaminant plume predictions.

Mathematical Development

The FOSM method is defined as the statistic analysis of state variables in a random function based on its first order Taylor series expansion. Let c be a generic nonlinear function of some random variable p. The function is expanded in a Taylor series about the mean value of its independent variables:

$$c(p) = c(\mu_p) + \sum_{i=1}^{n} \frac{\partial c}{\partial p_i}(p_i - \mu_{p_i}) +$$

$$0.5 \sum_{i=1}^{n} \sum_{j=1}^{n} \frac{\partial^2 c}{\partial p_i \partial p_j}(p_i - \mu_{p_i})(p_j - \mu_{p_j}) + O\{(p_i - \mu_{p_i})^3\} \tag{1}$$

where p_i, p_j are the ith and j^{th} member of the set p, and μ_p is the set of means for p. The values of μ_{p_i} and μ_{p_j} represent the ith and j^{th} member of the set μ_p. Ignoring higher order terms, the mean and variance of c with respect to p are:

$$\mu(c) = \int_{-\infty}^{\infty} c(p) f_x(p) dp \approx c(\mu_p) \tag{2}$$

$$Var(c) = \int_{-\infty}^{\infty} (c(p) - \mu(c))^2 f_x(p) dp \approx \sum_{i=1}^{n} \sum_{j=1}^{n} \frac{\partial c}{\partial p_i} Cov(p_i, p_j) \frac{\partial c}{\partial p_j} \tag{3}$$

where dp represents $dp_1 dp_2 ... dp_n$, for all values of n. The derivative $\frac{\partial c}{\partial p}$ represents the sensitivity of the state variable, c, to changes in the model parameter, p.

In the FOSM formulations, the sensitivity of state variables with respect to parameters is determined indirectly via finite elements. An interesting point is that with the exception of the right hand side, the formulation of sensitivity equations is identical to groundwater governing equations. This means the numerical scheme used in solving groundwater flow and mass transport can be applied to evaluate the sensitivities. Possible errors that result from this method include truncation error from the FOSM formulation, numerical error from discretization of the continuous function of the groundwater model and discretization error of the random field. However, the errors can be greatly reduced by reducing the variance-covariance of the hydrodynamic parameters, by obtaining more field data and applying Bayesian statistical updating techniques.

Numerical illustration

A two-dimensional advection-dispersion transport problem for a homogeneous medium in a uniform flow field, with Dirichlet boundary conditions, is presented. The governing equation for this example is defined as:

$$D_l \frac{\partial^2 C}{\partial x^2} + D_t \frac{\partial^2 C}{\partial y^2} - v_x \frac{\partial C}{\partial x} = \frac{\partial C}{\partial t} \tag{4}$$

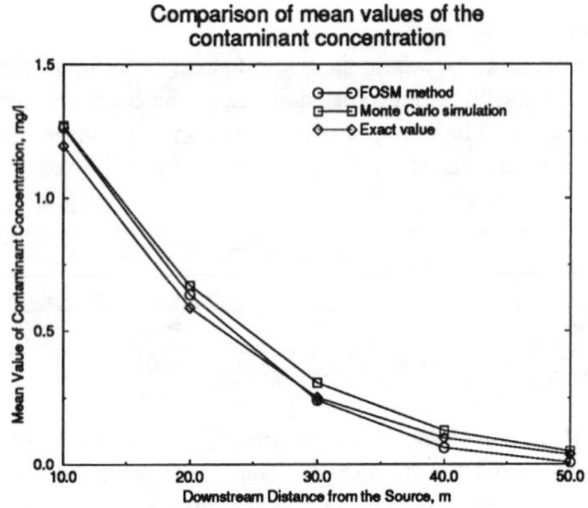

Figure 1: Comparison of FOSM and Monte-Carlo method with exact solution, elapsed time = 500 days with 100 time steps, $\mu_{ln(k)} = -3$, $\sigma_{ln(k)} = 0.5$, $\mu_a = 60$, $\sigma_a = 10$.

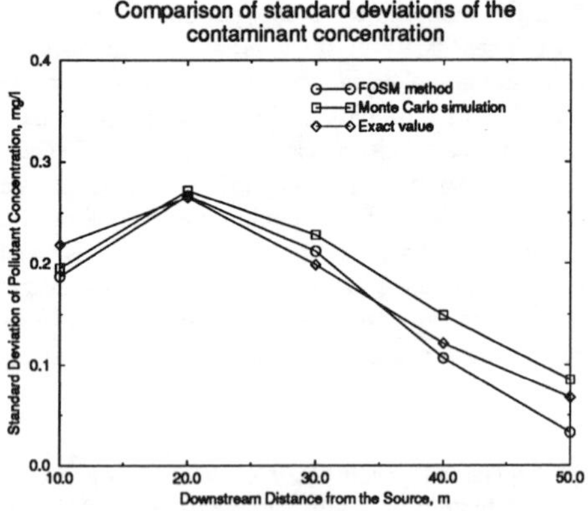

Figure 2: Comparison of FOSM and Monte-Carlo method with exact solution, elapsed time = 500 days with 100 time steps, $\mu_{ln(k)} = -3$, $\sigma_{ln(k)} = 0.5$, $\mu_a = 60$, $\sigma_a = 10$.

where the groundwater flow follows Darcy's law (i.e. $v_x = ki$), and dispersion coefficients are assumed linear with respect to v_x (i.e. $D_l = a_l v_x, D_t = a_t v_x$). This example considers hydraulic conductivity k, and the dispersion coefficients, a_l and a_t, as independent random variables with log-normal and normal probability distributions, respectively.

Using the FOSM method, in the case where both hydraulic conductivity and dispersivity are modeled as random variables, the mean and variance of the contaminant concentration are defined as:

$$\mu_c \simeq C(\mu_a, \mu_{ln(k)}) \tag{5}$$

and

$$Var(C) \simeq (\frac{\partial C}{\partial a})^2 Var(a) + (\frac{\partial C}{\partial [ln(k)]})^2 Var(ln(k)) \tag{6}$$

Figures 1 and 2 compare results using an analytical, Monte Carlo and FOSM solution for the two-dimensional advective-dispersive transport problem.

Conclusion

The results presented in this paper show that the FOSM method adequately estimates the mean and variance-covariance properties of the contaminant concentration predictions using both analytical solutions and numerical solutions for advective-dispersive transport problems. An extension of this method was presented by Lin [1994] and Lin, et. al. [1995] for large scale two dimensional coupled groundwater flow and solute transport models with pump-and-treat remediation schemes to predict the probabilistic characteristics of the pollutant concentration due to uncertainties in hydrodynamic flow field.

The results illustrate that a significant tradeoff exists in the amount of computational effort required in determining the cumulative distribution of the contaminant concentrations. These results can provide decision-makers in real time with useful information concerning the probabilistic nature of contaminant transport for planning possible remediation strategies.

References

[1] Konikow, L.F. and J.D. Bredehoeft, Groundwater models cannot be validated. *Advances in Water Resources*, Vol 15, No. 1, 75-83, 1992.

[2] Lin, K.-C., *Estimating the reliability of groundwater contaminant remediation strategies in an uncertain hydrodynamic environment,* Ph.D. dissertation, Duke Univ., 1994.

[3] Lin, K.-C., T.L. Jacobs, and M.A. Medina, Reliability analysis in stochastic groundwater flow and transport modeling, submitted to *J. Water Resour. Plan. Manag.*, 1995.

[4] Maloszewski, P. and A. Zuber, Principles and practice of calibration and validation of mathematical models for the interpretation of environmental tracer data in aquifers. *Advances in Water Resources*, Vol. 16, No.3, 173-190, 1993.

[5] Martin, C.M., Parameter estimation in water resources planning and management: Optimal actions or optimal parameters?, *Water Resour. Res.*, Vol 22, No. 3, pp 353-360, March, 1986.

[6] Yeh, W. W-G., Review of parameter identification procedures in groundwater hydrology: The inverse problem, *Water Resour. Res.*, 22(2), 95-108, 1986.

Regional Scale Effects of Aquifer Parameters on Exposure

Susan D. Pelmulder[1], A.M.ASCE, William W-G. Yeh[2], F.ASCE, and William E. Kastenberg[3]

Abstract

A framework of linked models was used to simulate the human exposure resulting from residual concentrations of contaminants at restoration sites migrating to wells. The framework includes horizontal two dimensional flow and transport in an aquifer, fate and transport of irrigation water in an environmental compartment model, and multiple pathway exposure assessment. The models in the framework are appropriate for regional scale simulations. A case study using the framework determined the sensitivity of exposure estimates to changes in several parameters.

Introduction

With ground water supplies scarce, aquifers with contamination must be managed to restore and retain use of the resource. After most aquifer restoration efforts, there remains a residual concentration of contaminant in the aquifer. This research estimates the health exposure associated with these residual plumes for use in regional scale aquifer management models. If the final objective in managing the aquifer is to protect public health, then it is advantageous to use a

---

[1] Lecturer, Dept. of Civil and Environmental Engineering, University of California, Los Angeles CA 90095-1593
[2] Professor, Dept. of Civil and Environmental Engineering, University of California, Los Angeles CA 90095-1593
[3] Chair, Dept. of Nuclear Engineering, University of California, Berkeley CA 94720

management model which has human exposure to the contaminant as a decision variable.

There have been several studies related to this research. In Massmann et al. (1991) a risk-cost-benefit framework is applied to ground water remediation, however, risk is associated with financial loss, rather than human exposure. Kaunas and Haimes (1985) also addressed the issue of multiobjective optimization of ground water contamination and included the objective of minimizing the fraction of time that the well water concentration was above the regulatory limit. In Reichard and Evans (1989), human health risk modeling was integrated with fate and transport modeling to estimate the value of minimizing uncertainty in transport and exposure parameters.

In previous studies, either a surrogate has been used for human exposure and risk assessment, or the method has only been designed and used on a local scale. This research includes human exposure with other models appropriate for regional scale simulation of aquifer management of problems.

Approach

In this research, a linked framework of ground water flow and contaminant transport in the horizontal plane, irrigation water use in an environmental compartment model, and multiple pathway human exposure were used in a study of the sensitivity of exposure to changes in aquifer, initial plume, and water supply related parameters. Eight quantifications of exposure, representing different approaches to protecting public health, were used. These included calculating individual risk, societal risk with three assumptions regarding population growth. For each of these four types of exposure, the 200 year average and maximum 30 year average were calculated.

The aquifer modeled is rectangular, homogeneous and isotropic, with dimensions of 8000m in the x direction (general flow direction), 4000 m in the y direction (crossflow direction), and 100m in thickness. The ground water head along the x=0 m boundary is held fixed at 200 m above the base of the aquifer, while all other boundaries have no flow. There are three rows of wells in the y direction located x= 2500, 4000, and 5500 m from the inflow boundary.

A base case of parameters was set to match values typically found in aquifers in the field. The base case plume is assumed to be 350x700 m with a uniform concentration of 10 ppb of tetrachloroethene. Several parameters were chosen for the sensitivity study and each was varied individually. Simulations were carried out for 200 yrs and a nondimensional sensitivity coefficient was calculated for each of the eight quantifications of exposure.

Results

The parameters used in the sensitivity analysis were ranked according to their absolute value, with a ranking of 1 associated with the highest sensitivity coefficients. These rankings are contained in Table I. The nondimensional sensitivity coefficients for individual and no growth societal exposures are the same, and therefore, have the same rankings. These rankings are for a particular region, so care must be taken in transferring the results to other areas. There are, however, several observations in comparing these coefficients which may be of general interest. It should also be noted that all parameters except population have been held constant over the 200 year simulation period, including the fraction of water supplied by ground water.

The rankings indicate that exposure is most sensitive to changes in contaminant concentration in the water supply. This implies that reducing the concentration in the water supply is the most direct method of reducing exposure, although no consideration has been given to cost efficiency or feasibility. For a homogeneous aquifer, the exposure is about as sensitive to hydraulic conductivity as longitudinal dispersivity. However, real aquifers are heterogeneous and the result could be significantly different than the homogeneous case. When comparing nondimensional sensitivity coefficients, it is also useful to consider the magnitude of variation in the parameter. Since hydraulic conductivity can vary over several orders of magnitude in a single aquifer, large differences in exposure can result from relatively small sensitivity coefficients.

The location and boundaries of the plume can also have a significant effect on exposure. This is evidenced by the relatively large sensitivity coefficients and high rankings for the distance to the wells and area of the plume across all methods of quantifying exposure. These result are consistent with those of Massmann et al.(1991) in their paper on optimum design of remedial measures for contaminated ground water.

The final aquifer parameter for which the sensitivity coefficient was calculated is the retardation factor. The sensitivity coefficients indicate that R has a significant impact on exposure from residual plumes only if societal exposure is considered. The use of societal exposure, or total number of added cancers, introduces the effects of time delay in exposure.

Another observation regarding the use of societal exposure is that modeling of the entire region, rather than the contamination site and nearest wells, becomes important. When societal exposure was used as the management parameter, the exposure associated with the plume reaching the second row of wells became more important than that associated with the first row of wells.

Table I. Rankings of absolute value of sensitivity parameters.

Parameter	Individual and No Growth		Linear Growth		Exponential Growth	
	200 Yr Average	30 Year Average	200 Yr Average	30 Year Average	200 Yr Average	30 Year Average
Initial Concentration	1	1	1	1	1	1
GW Fraction	1	1	1	1	1	1
Plume Area (x 3.8)	2	2	2	3	2	3
Plume Area (x 2)	3	4	3	4	3	4
Distance to Wells	4	3	6	5	6	5
Hydraulic Conductivity	5	6	4	7	5	6
Retardation Factor	6	7	5	2	5	2
Longitudinal Dispersivity	7	5	7	6	7	7
Transverse Dispersivity	8	8	8	8	8	8

Conclusion

A study of the sensitivity of exposure to aquifer, plume, and exposure parameters was conducted for a hypothetical region. The computational framework demonstrated with this study includes spatially distributed ground water transport and fate simulation and multiple pathway exposure modeling. A compartment model for agricultural use of contaminated well water is also included. It is believed that this is the first aquifer parameter sensitivity study to use multiple pathway exposure as the dependent variable. The methods used to quantify exposure are also the first to demonstrate the effect of the definition of exposed population on sensitivity results. This type of modeling provides information for the discussions of what population should be protected and how decisions may affect future populations.

References

Kaunas, J.R., and Haimes, Y.Y. (1985) "Risk management of groundwater contamination in a multiobjective framework," *Water Resour. Res.* 21(11):1721-1730.

Massmann, J., Freeze, R.A., Smith, L., Sperling, T., and James, B. (1991) "Hydrogeological decision analysis: 2. Applications to groundwater contamination," *Water Resour. Res.* 29(3):536-548.

Reichard, E.G., and Evans, J.S.(1989) "Assessing the value of hydrologic information for risk-based remedial action decision," *Water Resour. Res.* 25(7):1451-1460.

Development of Dynamic Groundwater Remediation Strategies for Variable Aquifer Configurations

Changlin Huang[1] and Alex S. Mayer[1]

Abstract

Discrete-time optimal control theory is used to develop dynamic strategies for groundwater pump-and-treat remediation. The genetic algorithm is applied as an alternative optimization approach for solving the optimal control problem. A nonlinear term for water treatment costs is included in the model. The model is applied to a set of layered aquifer combinations to study the impact of aquifer configurations on remediation costs and strategies.

Introduction

Dynamic strategies for pump-and-treat (PAT) groundwater remediation have been suggested for resolving the problems posed by aquifer heterogeneity, the presence of trapped non-aqueous phase liquids, or rate-limited mass transfer processes [Keely, 1989]. These strategies include varying pumping rates, pumping locations, pumping well configurations, or the number of wells from one remediation management period to the other. The concept of dynamic remediation strategies has been addressed in mathematical optimization models by the use of discrete-time optimal control theory [Andricevic and Kitanidis, 1990; Culver and Shoemaker, 1993; Whiffen and Shoemaker, 1993]. Gradient-based numerical approaches have been used to solve these models to determine optimal, time-varying pumping rates and numbers of wells. Genetic algorithms have been applied to optimize static remediation strategies [McKinney and Lin, 1994]. In this work, we apply genetic algorithms to determine optimal, dynamic management strategies for PAT remediation. A nonlinear treatment cost term is introduced into the objective function of the mathematical optimization model. The impact of aquifer configurations on remediation strategies is investigated by applying the model layered aquifer systems with variable hydraulic conductivities.

---

[1]Department of Geological Engineering, Michigan Technological University, Houghton, Michigan

Numerical Model

The problem of finding an optimal, dynamic remediation strategy can be represented as a minimization problem for a set of discrete time intervals, or management periods. A set of constraints for hydraulic and contaminant characteristics, such as maximum pumping rates, maximum drawdowns, and aquifer remediation goals, is imposed. In order to implement the numerical approach, the constrained optimal control model is converted to an unconstrained problem by using penalty techniques. A performance index, or cost function, is expressed as the sum of the pumping and treatment costs at each management period and a terminal cost. Treatment costs usually are formulated as a linear function of pumping rate [e.g. Whiffen and Shoemaker, 1993]. However, a nonlinear treatment cost term that is a function of pumping rate and solute concentrations in the extracted water may provide a more accurate description of treatment costs for many treatment processes. Granular activated carbon (GAC) was selected as the treatment process in this case. A nonlinear cost function for GAC treatment from Crittenden et al., [1987] was used to replace the linear treatment cost in the performance index. The state equations in the mathematical control model are governed by groundwater flow and solute transport equations. The simulation models that are used within the discrete-time optimal control framework are the solute transport model MT3D [Zheng, 1992]. and the flow model MODFLOW [McDonald and Harbaugh, 1983].

Application

An example problem was chosen to investigate the effectiveness of the optimization model and the impact of different aquifer configurations on remediation strategies and cost. The example site was adopted from the field application example given in the user's guide for MT3D [Zheng, 1992]. The site is a 750 m by 430 m by 25 m confined/unconfined aquifer system consisting of two or three distinct hydraulic conductivity zones. A plume of contaminants is found in an area at the center of the site, with the majority of the plume contained in the middle layer. The details of the initial concentration distribution may be found in the MT3D user's guide [Zheng, 1992].

The domain is divided into 31 rows, 20 columns and 3 layers with a variable finite-difference mesh in the vertical direction and horizonal direction. No-flow boundaries are applied on the north and south sides and the bottom of the domain; constant-head boundaries are imposed on the east and west sides. Initial conditions for the flow consisted of an east-to-west gradient. The flow field is simulated using the MODFLOW model assuming steady-state conditions for each management period. Table 1 lists the properties used for all of the aquifer configurations. Table 2 indicates the parameters used in the genetic algorithm scheme. The optimization model was applied to three different configurations of the aquifer system. The hydraulic conductivity (K) distribution for each case is given in Table 3.

The remediation strategies consisted of a single well at a fixed location. The pumping rate for this well was optimized for each aquifer configuration. The total remediation time was fixed at 500 days and consisted of three equal-length management periods. The contaminant concentration constraints were set such that (1) during the first two management periods, the contaminant concentration goal should be met at the extraction wells and at monitoring points down-gradient of the plume and (2) during the last management period, the goal should be met throughout the aquifer. A penalty cost is assessed at the end of the last management period if the contaminant concentration remains above the contaminant concentration goal.

Table 1 Aquifer properties and input parameters

Porosity	0.3	Bulk density (g/cm^3)	1.7
Longitudinal dispersivity (m)	3.3	Transverse dispersivity (m)	0.7
Distribution coeff. (cm^3/g)	1.0	Diffusion coeff. (cm^2/s)	0.0
Contaminant conc'n. goal (ppb)		5	
Freundlich adsorption parameters		K_{ab}=37.9 1/n=0.83	

Table 2 Genetic algorithm parameters

Max. Population Size	35	Generation size	100
Probability of best agent	0.2	Probability of crossover	0.6
Probability of mutation	0.00333	Length of binary string	32

Table 3 Aquifer configurations and optimization results

	Case A	Case B	Case C
K in layer 1 (m/d)	1.8	1.8	180
K in layer 2 (m/d)	1.8	18.2	180
K in layer 3 (m/d)	18.2	1.8	1820
total cost ($1,000)	287	148	4
total # of generations	22	25	22
Q for period 1 (m^3/d)	3685 2089	8469	
Q for period 2 (m^3/d)	1062	912	60
Q for period 3 (m^3/d)	12	710	61

Results and Discussion

The results of the remediation strategy optimizations are given in Table 3 in the form of total remediation costs and optimal pumping rates, Q, for each management period. Optimization of remediation strategies for all the cases results in higher pumping rates in the first management period. As the size of the contaminant plume shrinks due to the remediation, lower pumping rates are required for plume capture. The remediation cost for case B is lower than the cost for case A because the majority of the plume initially resides in layer 2, and the high conductivity in the middle layer in case B results in a reduction in the amount of pumped water, which reduces both pumping and treatment costs. The costs for case C are lower than the costs for case A even though the overall pumping rate is higher in case C. This result is due to the very high penalty costs that are assessed at the end of the remediation in Case A, because there is still a relatively high concentration of contaminant remaining in the aquifer. These results indicate that dependence of remediation costs on aquifer configurations may be highly nonlinear. The number of generations required for convergence also is indicated in Table 3. All of the cases require approximately the same number of generations to converge. However, different convergence rates may result when the optimization model is applied to a wider range of heterogeneous configurations.

References

Andricevic, R., and P. Kitanidis, Optimization of the Pumping Schedule in Aquifer Remediation under Uncertainty, *Water Resour. Res.,* 26, 875-885, 1990.

Crittenden, J. C., D. W. Hand, H. Arora, and B. W. Lykins Jr., Design Considerations for Granular Activated Carbon Treatment of Organic Chemicals, *J. A*, 79. 74-81, 1987.

Culver, T. B., and C. A. Shoemaker, Optimal Control for Groundwater Remediation by Differential Dynamic Programming with Quasi-Newton Approximations, *Water Resour. Res.* , 29, 823-831, 1993.

Keely, J. F., Performance Evaluation of Pump-and-Treat Remediation, U.S. Environ. Protect. Agency, *Rep. EPA/540/4-89/005*, Washington, D. C., 1989

McDonald, M.G. and A. W. Harbaugh, A Modular Three-Dimensional Finite-Difference Ground Water Flow Model, *USGS TWRI Chapter A1, Open-File Report 83-875*, 1983.

McKinney, D. C and M. Lin, Genetic Algorithms Solution of Groundwater Management Models, *Water Resour. Res.,* 30, 1897-1906, 1994

Whiffen, G. J. and C. A. Shoemaker, Nonlinear Weighted Feedback Control of Groundwater Remediation under Uncertainty, *Water Resour. Res*., 29, 3277-3289, 1993.

Zheng, C., MT3D: A Modular Three-Dimensional Transport Model for Simulation of Advection, Dispersion and Chemical Reactions of Contaminants in Ground-Water Systems, *US. EPA, R.S. Kerr Environmental Research Laboratory, Ada, Oklahoma*, 1992.

**Protecting Groundwater from Leaking Underground Tanks-
An Innovative Approach**

Peter T. Silbermann, Member[1]
Douglas R. DeNatale[2]
Frederick C. Conley[3]

Abstract

The groundwater supply of the Town of Natick was threatened by leaking home heating fuel from hundreds of residential underground tanks buried in close proximity to the Town's watershed drinking supply wells. The Town undertook, at no cost to the taxpayers, a program to remove the tanks and petroleum contaminated soils from the private sites and remediate the contaminated soils at a permitted facility. The innovative features of the program included:

- First comprehensive townwide program to deal with residential underground fuel storage tanks in Massachusetts
- Funded by a prototype EOCD discretionary grant
- First major project to be performed under the newly revised DEP regulations
- Special legislation promulgated for assessment of betterment charges to homeowners specifically for tank removals and soil remediation
- Obtained waiver on requirement for use of prevailing wage rates
- No cost to town
- Program can serve as prototype for other towns
- Program may be used as a basis for the Legislature to establish a fund to help finance future similar townwide projects

[1]Peter T. Silbermann, P.E., Vice President, Whitman & Howard, Inc., 45 William Street, Wellesley, MA 02181

[2]Licensed Site Professional, Hydrogeologist, Whitman & Howard, Inc., 45 William Street, Wellesley, MA 02181

[3]Town Administrator, Town of Natick, Town Hall, 13 East Central Street, Natick, MA 01760

Introduction

Town officials in Natick became aware of the magnitude of the problem when some homeowners decided to have the tanks and contaminated soils removed from their yards at a cost, in some cases, in the $50,000 to $60,000 range. The consequences facing homeowners were heightened when banks and mortgage companies stopped approving mortgage or equity loans on homes with underground tanks.

Because of the imminent threat to the Town's water supply and the potentially catastrophic financial impact and liability to individual homeowners, Natick decided to explore how to solve the problem. They realized that protection of the town's water supply was critical and they also realized that consolidation of all planning and construction activity would likely be substantially less costly for each individual owner if the work could somehow be done as one project.

At the request of the Town, in 1992, the State Legislature enacted a Home Rule Petition providing Natick with authority to borrow funds for the purpose of removing residential underground fuel storage tanks and contaminated soil and the authority to charge back to the residents the cost of removal and remediation through "betterment" assessments. The Town then applied for and received a $1 million Small Cities Grant from HUD to be administered by the state Economic Office of Community Development (EOCD). With the EOCD grant money available, the Town worked out a formula for financing the work. Homeowners were asked to pay the first $5,000 of tank removal and cleanup cost, through the betterments approach; all costs in excess of the $5,000 cap would be paid through the grant.

An Innovative Program

Whitman & Howard developed a comprehensive plan for tank removal, soil remediation and project scheduling and worked with the town to provide services in the following areas:

- Administering the EOCD grant
- Developing a Geographic Information System (GIS) database to maintain the enormous quantity of data being generated in connection with each individual property
- Preparing technical specifications and bidding documents for the tank removal and soil remediation work
- Ensuring compliance with State regulations and obtaining necessary permits
- Overseeing of all construction work by Licensed Site Professionals.

In early February 1994, the specifications for the construction work were advertised for competitive bids and a great deal of contractor interest was expressed. There were 14 qualified bidders on the project and very competitive prices were obtained.

Phase I Tank Removals - In Phase I, 320 tanks were removed over a three month period. In general, the tanks were found to be under two to three feet of soil cover and approximately within two feet of the house. As expected, almost all of the tanks had a capacity of 500 gallons, or less.

Because of the rapid pace of work, a system was devised to aid the field staff in deciding whether the tank was leaking. Five indicators were considered in making this determination: 1) Oil staining in the soil; 2) Oil odors in the soil or excavation; 3) Condition of the tank (visible holes); 4) A headspace reading of soil; and 5) Laboratory testing of soil. Normally, these five indicators corroborated one another. However, on occasion, conflicts arose and a sample had to be retested, or the property was revisited in Phase II.

Phase II Soil Cleanup - Soil cleanup activities took place on 144 properties and consisted of excavating accessible soils, and transporting the soils to Environmental Soil Management, Inc. (ESMI) of Loudon, New Hampshire for thermal treatment. When it was encountered, contaminated groundwater was also removed from excavations, drummed on site, and transported to a licensed Treatment/Storage/Disposal Facility (TSDF).

Project Results and Observations

Soils and Groundwater - The soils encountered in Natick are glacially derived soils, typical of New England, that can be classified into two broad categories: till and stratified drift. As Phase II progressed, we made another interesting observation about till contaminated with fuel oil. The till soils appeared to produce "falsely" high headspace readings and stronger odors than sandy or gravelly soils with comparable levels of TPH. The explanation for this may also relate to soil permeability. Most of the tank leakage occurred at the base of the tank, roughly four to nine feet below the ground surface. Leaking fuel migrated downward and laterally, as discussed above. The till soils, because of their low permeability, had a tendency to trap fuel vapors, not allowing them to escape upward through the soil and into the atmosphere. In contrast, the sandy or gravelly soils were highly permeable, allowing fuel vapors to escape more readily.

In Phase I, groundwater was not encountered, except on rare occasions. This was due primarily to the fact that the excavations to remove the tanks did not generally extend below a depth of about nine feet. During Phase II, groundwater was encountered more frequently because the excavations to remove contaminated soil were extended to depths ranging from eight to 19 feet. Groundwater was encountered more commonly in excavations in till soils due to their low permeability.

Contaminated Soil - The volume of contaminated soil excavated from any one property ranged from five to 400 cubic yards (cy), about the size of a typical single story slab-on-grade home in Natick. The total volume of contaminated soil excavated from 144 properties in Natick was 9,500 cy, averaging 66 cy per home. TPH concentrations in the contaminated soil ranged from less than 100 ppm to nearly 60,000 ppm (directly below the tank before any contaminated soils had been removed). The average TPH concentration in the excavated petroleum contaminated soils was between 3,000 and 5,000 ppm.

The best indicators of "severe problems" were sight and smell. In most cases, when the tank was found to be ruptured or to have numerous holes and the soil was noted to be heavily oil-stained and have heavy oil odors, more than 100 cy of PCS was excavated. These sites represented a relatively small percentage of the total number of leaking tanks: 17 out of 144.

Contaminated Groundwater - One of the principal reasons the Town of Natick undertook this project was because of the Town's concern for its drinking water supply wells. Fortunately, we found no significant amount of groundwater contamination associated with the leaking tanks. Floating oil was not found in any excavation. In only three cases were dissolved concentrations of TPH or BTEX above reportable concentrations, and only two of these were in the zone of contribution of the Town wells. To Town officials, who were fearing the worst, this result was extremely comforting. Certainly, had the Town not acted when it did and had homeowners had waited another five or ten years to remove their tanks, both soil and groundwater contamination would have been more severe given the high number of leaking tanks that were discovered and removed, most of which were still in active use at the start of the program.

Project Status

In all, 326 homeowners participated in the project, which represents 98 per cent of homeowners who are known to have had underground fuel tanks before the project began. 45 per cent of the individual sites had leaking tanks, which agrees with the rate observed by the Natick Board of Health for tanks that were removed before this project began. To date, 294 sites are "closed cases", that is, the tank has been removed, and soil and groundwater have been cleaned up. The remaining 32 sites are unresolved ("open cases"). Of these, 26 sites have soil contamination beneath the home (sub-slab). A Method 3 Risk Assessment, recently performed by Alceon Corporation, indicated that for No. 2 fuel oil TPH levels of up to 4,000 ppm on residential property do not pose a significant risk to human health or the environment. This would have the effect, with the concurrence of DEP, of eliminating 13 of the 26 unresolved sub-slab sites, and will allow these 13 sites to be "closed" under the MCP. The remaining 13 sub-slab contamination sites will require continued cleanup (Phase III) or an Activity and Use Limitation (AUL). In addition, there are 6 sites, without sub-slab contamination, which require further investigation and action to deal with groundwater or other problems, such as contamination from non-heating fuel oils.

Optimal Dynamic Groundwater Remediation with Heuristic Pumping Constraints

Teresa B. Culver, A.M. ASCE
Hong Dai[1]

Abstract

Inefficient handling of constraints has reduced the performance of optimal control algorithms for groundwater remediation design. A simple heuristic check was added to differential dynamic programming (DDP) algorithms with hyperbolic penalty functions to prevent violations of the pumping constraints. With the exception of the most dynamic case, the successive approximation linear quadratic regulator method with the heuristic was found to perform better than the quasi-Newton DDP algorithms which require significantly more computational memory.

Introduction

Control theory algorithms have been combined with numerical modeling of groundwater flow and transport to determine optimal dynamic pump-and-treat groundwater remediation schemes. For groundwater reclamation, dynamic policies have been shown to be more cost-effective than the best static policies (Ahlfeld 1990, Chang et al. 1992, Culver and Shoemaker 1992). Algorithms that have been successfully applied include differential dynamic programming or DDP (Whiffen and Shoemaker, 1993) and its variants, the successive approximation linear quadratic regulator (SALQR) method (Culver and Shoemaker, 1992) and quasi-Newton DDP (Culver and Shoemaker, 1993). When applied to groundwater remediation, constraints on the control (pumping) and the state (hydraulic heads and concentrations) must be imposed. Previous applications have utilized penalty function techniques, thus changing the constrained problem into an equivalent unconstrained problem.

To solve the nonlinear optimization problem of groundwater remediation, the DDP algorithm iteratively converges on an optimal policy. In each iteration, an improved pumping policy is found in two steps. First, the vector-space direction with the greatest decrease in costs is determined based on local derivative information. Then a line search is performed to determine how large of a change in the control, along that direction is appropriate. A large change in pumping rates may result in a violation of constraints, causing an increase in costs due to the penalties.

[1]Assistant Professor and Graduate Research Associate, Civil Engineering and Applied Mechanics, University of Virginia, Charlottesville, VA 22903-2442

In such a case, the line search, which requires an additional run of the numerical simulation model, must be repeated using a smaller change in the pumping rates. (The entire algorithm is described in Culver and Shoemaker, 1992.) In our application, we utilized the hyperbolic penalty function of Lin (1990), which avoids the oscillatory behavior that can occur with quadratic penalty functions (Georgakakos, 1989). Nevertheless, our algorithm typically requires 4 to 5 line searches per iteration to find an appropriate change in the pumping policy. The inefficiency of the penalty function and line searches, given the complexity of the simulation models used for contaminant transport, contributes to the large computational requirements of the DDP algorithms.

This work explores using an heuristic adjustment to improve the performance of the hyperbolic penalty function for constraints on the controls. Constraints on the states are handled as before.

Approach

A constraint on the control is a constraint with respect to the vector of pumping rates in time period k ($\mathbf{U}_k$) only. In our application the constraint on pumping alone is that only extraction is permitted ($U_{ik} \geq 0$, for all k and i, where i indicates a particular well). Let $\bar{\mathbf{U}}_k$ indicate the suggested new pumping policy based on the standard DDP update method. To speed the performance of the line search and penalty function, we added the following heuristic check after the calculation of $\bar{\mathbf{U}}_k$:

$$\text{if } \bar{U}_{ik} < 0, \text{ then } U_{ik} = 0$$

$$\text{else } U_{ik} = \bar{U}_{ik}$$

Then the vector $\mathbf{U}_k$ is applied in the next line search. This simple check does not increase the computational time per iteration or the memory requirements. This heuristic is similar to the truncation algorithm used in the feedback policy of Whiffen and Shoemaker (1993). They applied this policy only during the feedback step, while we are applying the heuristic check throughout the search. The heuristic check avoids unnecessary violations of the pumping constraints. In addition, the heuristic control only alters the pumping rates of those wells in violation of the extraction constraint, while wells in compliance are not adjusted unless the subsequent line search indicates that an adjustment is necessary. This is significantly different than the standard DDP algorithm, which reduces the step size of each element of the vector U_k proportionally during a line search. Although proportional adjustment maintains the same vector-space direction of change, it implicitly forces the magnitude of the change in all pumping rates to be limited by the most tightly constrained well.

Example Results

A hypothetical groundwater reclamation test problem, which was described in detail in Culver and Shoemaker (1993) was used as the basis for comparison. The test case described a homogeneous, confined aquifer that has been contaminated. There were 77 finite element nodes, 18 potential wells and 13 observation wells. All parameter values remain the same, except the convergence criterion, θ_{min}, was relaxed from $\theta_{min} = 0.0001$ to $\theta_{min} = 0.01$. Two algorithms, the SALQR algorithm and the QNDDP algorithm, were each run with and without the heuristic check, for a total of four DDP variations. Each of the four optimization approaches was used to determine the optimal 5-year management plan given 20, 10, 5, 4, 2, and 1 management period(s) within which the pumping rates are held constant. All runs were performed on an IBM RS/6000 with 64 megabytes of memory provided through the Information, Technology and Communications division of the University of Virginia.

As expected, relaxing the convergence criterion had little impact on the quality of the optimal policy. In earlier works, a stringent criterion was used to test the algorithms. The larger convergence value is more realistic for practical purposes. In all cases with the larger convergence criterion, the percent increase in the objective was less than 1%, while the average CPU time saved was over 40% for the SALQR algorithm. At this convergence criterion, the advantage of QNDDP over SALQR is reduced (Figure 1). When compared to SALQR, QNDDP provides an average reduction in iterations over all management period of only 10.4%. This reduced performance of QNDDP would be expected, since QNDDP was most effective at fine tuning the policy in the neighbor nearest the optimal policy. Furthermore, the QNDDP algorithm requires significantly more computational memory than the SALQR technique since additional second derivative information must be stored.

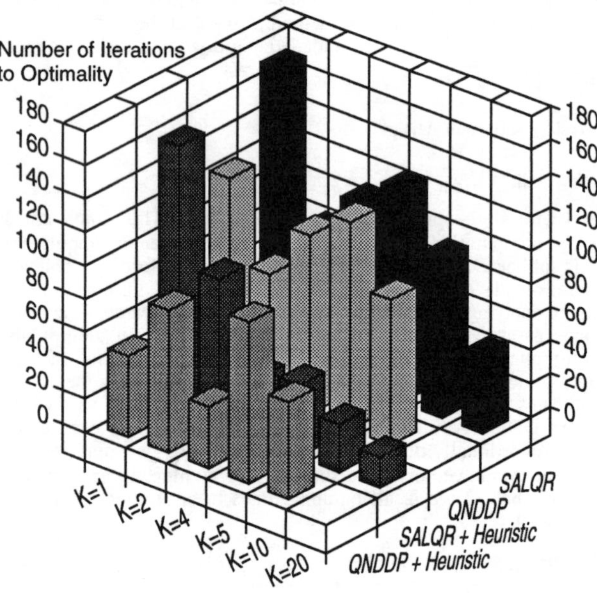

Figure 1. Number of iterations required for the four optimization algorithms to reach an optimal policy for the test problem. K is the number of time steps per management period, given a total of 20 time steps. QNDDP has no impact for steady-state policies (when K=20).

Addition of the heuristic check reduced the number of tries per iteration by only 16%, but it significantly reduced the overall number of iterations to optimality (Figure 1). When compared to SALQR alone, the addition of the heuristic check to the SALQR algorithm reduced the number of iterations on average by 44.5%, while QNDDP with the heuristic reduced the number of iterations by 40.5% on average. The average reduction in CPU time is slightly higher due to the combined effect of reducing both the number of line searches within an iteration and total number of iterations; the average reduction in CPU time was 48% for SALQR with the heuristic check and 44.5% for QNDDP with the heuristic check. Thus the heuristic which requires no additional memory, performed better on average than either QNDDP algorithm which require extensive memory. However, SALQR with the heuristic did not significantly impact the number of iterations for the most computationally intensive cases (management periods of 1 and 2 simulation time steps). Only the QNDDP with the heuristic check performed well on the 20 management period problem (k=1), reaching optimality in 48 iterations compared to 172 required by SALQR and requiring less than a third of the CPU time needed for SALQR (1900 seconds versus 6850 seconds). There was little difference between the performances of the four algorithms for the case with 2 simulation time steps per management period.

Conclusion

A simple heuristic check significantly improved the performance of the optimal control algorithms when applied to groundwater remediation. The heuristic check adds engineering common sense to the algorithm, avoiding unnecessary violations of constraints and allowing larger adjustments in the policy within any particular iteration. The overall effect is an increased convergence rate. The SALQR algorithm with heuristic is the most effective algorithm, except for when the policy is allowed to change often. In those cases, the QNDDP with the heuristic performed the best. Further algorithmic development may determine other variations of the heuristic check that would provide effective performance for the SALQR algorithm for the most dynamic policies and avoid the large memory requirements of QNDDP.

References

Ahlfeld, D.P., Two-stage ground-water remediation design, *J. Water Resour. Planning and Management*, 116(4), 517-529, 1990.

Chang, L.-C., C.A. Shoemaker, and P. L-F. Liu, Optimal time-varying pumping rates for groundwater remediation: Application of a constrained optimal control algorithm, *Water Resour. Res.*, 28(12), 3157-3171, 1992.

Culver, T. B. and C. A. Shoemaker, Dynamic optimal control for groundwater remediation with flexible management periods, *Water Resour. Res.*, 28(3), 629-641, 1992.

Culver, T. B. and C. A. Shoemaker, Optimal control for groundwater remediation by differential dynamic programming with quasi-Newton approximations, *Water Resour. Res.*, 29(4), 823-831, 1993.

Georgakakos, A.P., Extended linear quadratic Gaussian control: Further extensions, *Water Resour. Res.*, 25(2), 191-201, 1989.

Lin, T.W., Well behaved penalty functions for constrained optimization, *J. Chin. Inst. Eng.*, 13(2), 157-166, 1990.

Whiffen, G. J. and C. A. Shoemaker, Nonlinear weighted feedback control of groundwater remediation under uncertainty, *Water Resour. Res.*, 29(9), 3277-3289, 1993.

Comparison of Continuous- and Discrete-time Optimal Control Applied to Groundwater Remediation Design

Gregory J. Whiffen and Christine A. Shoemaker[1]

Abstract

The purpose of this paper is to improve the feasibility of computing cost efficient solutions to groundwater remediation design problems by exploring the computational efficiencies of two alternative nonlinear optimal control approaches (*Jacobson and Mayne*, 1970): Continuous-time and Discrete-time Differential Dynamic Programming (cDDP and dDDP). Both algorithms are applied to the same 2-D (third-dimension averaged) transient flow and transport model to identify minimum pumping remediation strategies. Arithmetic operation counts and memory usage are compared for hypothetical applications using a workstation. Both implicit and explicit time stepping are used in the flow and transport model.

Introduction

This paper focuses on the optimal control approach for pump and treat groundwater remediation design with time varying policies. An advantage to time varying policies is pumping rates can change intensity and location during the remediation - thus adapting to the reduction and movement of the plume over time. This adaptation can result in significantly more efficient policies than the steady state approach (*Culver and Shoemaker*, 1992). Despite DDP's numerical superiority to nonlinear programming for time-varying problems, it has been limited to problems involving only small groundwater model meshes due to computational requirements.

We compare dDDP and the cDDP for remediation design. Memory usage and operation counts are compared as a function of mesh size. cDDP has never before been applied to groundwater remediation design and has several properties which make it easier to apply than the dDDP algorithm. *Lin and Arora*, [1994] showed that cDDP is more computationally efficient for structural control problems than dDDP.

The problem description used to compare the cDDP and dDDP approaches is as follows: minimize the sum of the squares of extraction rates at flow and transport model nodes subject to the constraints a) at the end of the simulation period all model nodes must have contamination concentration levels less than a given standard; and b) allow only extraction. cDDP and dDDP require different formulations of the groundwater remediation problem: one tailored to each algorithm. We attempted to make both formulations as similar as possible to study the similarities and differences of cDDP and dDDP for this application.

---

1. School of Civil and Environmental Eng., Cornell University, Ithaca, New York 14850

Discrete-time Differential Dynamic Programming Formulation

We choose the following problem formulation to apply dDDP:

$$J = \min_{\vec{u}} \sum_{t=1}^{N} a\Delta t((\vec{u}^t \cdot \vec{u}^t)) \quad ,$$

where $\vec{u}^t \in \Re^m$ is a vector containing pumping rates at each remediation well at time t; Δt is the flow and transport model time step, and N is the number of decision time periods. The constant a is a scaling parameter for the objective. The minimization is constrained by the model's prediction of the state of the aquifer at a future time step given the control and state at the current time step. In the context of dDDP we must represent the flow and transport model as a discrete-in-time recursion:

$$x^{t+1} = T(x^t, u^t) \qquad t = (1, 2, \ldots, N-1) \qquad x^1 = x_o \qquad (1)$$

where $x^t \in \Re^n$ is a vector containing the head and concentration at each active node of the mesh at time step t. Further constraints were added to ensure only extraction is allowed (this is optional) and to ensure a water quality standard C_{standard} is met at all nodes by the final time step:

$$u^t \geq 0: \ t = (1, 2, \ldots, N); \qquad c(x^t) \leq C_{\text{standard}}: \ t = N$$

The function $c(x^N)$ represents the concentrations at all model nodes at the final time step, N. We extended both cDDP and dDDP to handle nonlinear constraints using a penalty function method similar to *Chang et al.*, [1992].

Continuous-time Differential Dynamic Programming Formulation

The following continuous-time formulation of the groundwater remediation problem was constructed to be as similar to the dDDP formulation as possible:

$$J^* = \min_{u(t)} \int_{t^0}^{t^f} a(\vec{u}(\tau) \cdot \vec{u}(\tau))\, d\tau$$

Subject to the continuous-time representation of the flow and transport model:

$$\frac{d\vec{x}(t)}{dt} = f(\vec{x}(t), \vec{u}(t), t) \quad \vec{x}(t^0) = \vec{x}^{\text{initial}} \qquad (2).$$

In this formulation each component of $\vec{u}(t)$ represents the pumping rate at a well site as a function of time. Analogous to the constraints for the dDDP case we require:

$$\vec{u}(t) \geq 0: \ \forall t \in (t^0, t^f); \qquad c(\vec{x}(t)) \leq C_{\text{standard}}: \ t = t^f$$

Continuous- and Discrete-time transition function representations

A substantial difference between cDDP and dDDP is how the transition function is represented. Both algorithms require derivatives of the respective transition functions with respect to $\vec{x}$ and $\vec{u}$. dDDP requires derivatives of (1) whereas cDDP requires derivatives of (2). This difference is large from a computational standpoint when an implicit time stepping method is used by the flow and transport model. To understand the difference, consider the transient flow model formulation used in the numerical examples. The same qualitative results follow for the transport model. Transient flow for a 2-D confined aquifer with storage can be expressed as

$$\nabla \cdot (bK\nabla\bar{h}) + \sum_{i \in Q} \bar{q}^i \delta(x^i, y^i) - bS^s \frac{\partial \bar{h}}{\partial t} = 0 \tag{3}$$

Where b is the saturated vertical thickness, $\delta(x^i, y^i)$ is a 2-D Dirac delta function, $\bar{h}$ is the vertically averaged hydraulic head, K is the hydraulic conductivity tensor, $\bar{q}^i$ is the pumping rate for well i located at (x^i, y^i), and S^s is the specific storage. We employed a centered (second order accurate) finite difference method to approximate spatial derivatives in (3) to obtain the ordinary differential equation:

$$\frac{d\bar{h}}{dt} = \frac{M}{\Delta x^2}\bar{h} + \frac{\vec{q}}{bS^s\Delta x^2} - V^b \qquad \bar{h} \in \Re^{N_{nodes}} \tag{4}$$

where Δx is the spatial discretization length, $M \in \Re^{N_{nodes} \times N_{nodes}}$ is a constant coefficient (Mass) matrix, $\vec{q}$ is a vector of pumping rates at all active nodes, and V^b is vector of boundary condition forcing terms. The variable N_{nodes} is the number of active nodes in the model mesh.

Notice that (4) has the form of the transition equation we need to apply cDDP (compare (4) to (2).) Thus we only need to approximate the spatial derivatives of the governing equations, before cDDP can be applied. This is important because cDDP acts independently of the time discretization of the model's governing equations.

To apply dDDP, we must approximate the time derivatives in the governing equations before we can obtain the derivatives we need to solve the optimal control problem. The general vector difference equations we used to approximate the time derivative in (4) is as follows:

$$\frac{\bar{h}^{t+1} - \bar{h}^t}{\Delta t} = -V^b + \varepsilon\left(M\bar{h}^{t+1} + \frac{q^{t+1}}{bS^s\Delta x^2}\right) + (1-\varepsilon)\left(M\bar{h}^t + \frac{q^t}{bS^s\Delta x^2}\right) \tag{5}$$

The implicit parameter ε is chosen from the interval [0,1] where a value of 1 is fully implicit and a value of 0 is fully explicit. Notice that (5) has the proper form of the transition equation we need to apply dDDP (compare (5) to (1).) Two cases of interest occur: first is when explicit time stepping is used, and second when implicit time stepping is used. In the first case it easy to verify that all the derivatives of $\bar{h}^{t+1}$ with respect to q^t and $\bar{h}^t$ involve at most matrix-vector multiplication operations (if the matrices have narrow band width, the work is $O(N_{nodes})$.) However, evaluating these same derivatives when an implicit time step is used requires the inversion (or factoring) of matrices (if the matrices have a narrow band width, the work is $O(N_{nodes})^2$.) This is a considerable work for large applications.

From this analysis one would expect the explicit approach will be superior to the implicit approach if derivative calculations dominate the overall work load in solving the optimal control problem. However, there is a trade-off when one uses an explicit method because more time steps are necessary. Individual applications will vary.

Numerical Investigations

In the test cases we allowed all of the active nodes in the finite difference mesh to be potential extraction sites. We considered cases where N_{nodes} = 77, 160, 216, 280, 320, and 414. The largest mesh allowed 378 wells. The modeled aquifers were assumed to have the same parameters as used by *Chang et al.*, [1992].

The explicit (implicit) time step must be $\leq 1.5\,(90)$ days in order to achieve numerical stability. We choose a clean up duration of 30 months. All cases were run on a SPARCstation 10. The sparsity detection and sparse storage capability of MATLAB was used to reduce the operation count and memory use needed for the sparse matrices produced by the flow and transport model.

Table 1: Observed Asymptotic Complexity

Algorithm-Time step	FLOPS/iteration	Memory use (Megabytes)
dDDP-explicit 1.5days	$\sim 2340\,(N_{nodes})^{3.22}$	$\sim 1851\,(N_{nodes})^{2.25}$
cDDP-explicit 1.5days	$\sim 436\,(N_{nodes})^{3.33}$	$\sim 1791\,(N_{nodes})^{2.26}$
dDDP-implicit 90 days	$\sim 48\,(N_{nodes})^{3.38}$	$\sim 79.73\,(N_{nodes})^{2.20}$

Conclusion

Table 1 indicates cDDP and dDDP/explicit approaches become computationally intractable for large meshes more rapidly than the dDDP/implicit approach. We can conclude that the advantages of the sparse derivatives which occur when using the explicit model do not overcome the work associated with the short explicit time step for this application. Given the same size explicit time step, cDDP outperforms dDDP which agrees with the work of *Lin and Arora*, [1994].

In general, an analysis of the characteristics of the OCP of interest will aid in the decision of which optimal control algorithm will be best. It is reasonable to expect that for some systems with highly nonlinear dynamics the cDDP approach will be better due to accuracy rather than stability constraints on the largest possible time step for the simulation of the dynamics.

Acknowledgments

G. J. W. is supported by a fellowship awarded by the D.O.E. Computational Science Graduate Fellowship Program. Support for C. A. S. was from NSF (ASC 8915326). We also acknowledge many useful conversations with James A. Liggitt.

References

Chang, L.-C., C.A. Shoemaker, and P. L.-F. Liu, Application of a constrained optimal control algorithm to groundwater remediation, *Water Resour. Res.*, 28(12), 3157-3173, 1992.

Culver, T., and C.A. Shoemaker, Dynamic Optimal Control for Groundwater Remediation with Flexible Management Periods, *Water Resour. Res.*, 28(3), 629-641, 1992.

Jacobson D. H., and D.Q. Mayne, *Differential Dynamic Programming*, Elsevier Scientific, New York, NY, 1970.

Lin, T. C. and J. S. Arora, Simultaneous Design of Control and Structural Systems, *J. Optimal Control: Applications and Methods*, 15, 77-100, 1994.

Optimizing Ground Water Remediation Design Using Linear-Quadratic Functions

Charles S. Sawyer *

Abstract

A general modeling framework to help in the design of a groundwater remediation sytem using extraction wells is presented. The uncertainty in the modeling parameters (e.g. hydraulic conductivity) is of importance in the design and is considered in the formulation of the model.

Introduction

Ground water remediation by extraction at pumping wells is a widely used method in aquifer remediation. Modeling ground water flow and contaminant transport in an aquifer can help in deciding remediation scenarios by pumping, for a contaminated aquifer. The aquifer properties play a very important role in the modeling procedure since they determine input parameters to models. The output from the models depend on these input parameters and can give erroneous and misleading results if these parameters are not determined accurately.

Many models use to simulate an aquifer treat the model parameters as deterministic. For most field situations, these parameters are uncertain. This is because most field tests performed to determine these parameters cannot completely capture the spatial distribution of these parameters and the tests themselves are not very accurate. Taking into account these uncertainties will lead to a more accurate model and hence more reliable results. However, this can lead to a more complex model that may be difficult to solve.

In this paper, a model that combines groundwater flow and contaminant transport with optimization methods is developed. The model should enable us to

*Assistant Professor of Civil Engineering, Dept. of Civil Engineering, University of Connecticut, Storrs, CT 06269

design remediation scenarios by pumping, that takes into account uncertainty in modeling parameters. Determination of the optimum number of wells, their location and pumping rates for remediating an aquifer subject to constraints on hydraulic gradients and upper bounds on concentration at specified locations within the aquifer can be addressed by the model.

The Optimization Problem

The objective function is formulated as minimizing an objective function subject to certain constraints. The nature of the objective function and the constraints formulation follows.

Constraints Formulation

Suppose we have a contaminated aquifer to be remediated by groundwater extraction, and the extent of spreading of the contaminant plume is known precisely. As part of our remediation design, we may want to prevent further spreading of the contaminant plume. This can be achieved by hydraulic gradient control. The constraints will be such that an inward gradient is achieved around the plume periphery.

To formulate this constraint in a stochastic framework, assuming that the hydraulic conductivity is the only uncertain parameter, we will need to generate a set of realizations for the hydraulic conductivity distribution. Let each realization be represented by $\omega = 1, 2, \ldots, \Omega$. By solving a flow model for each ω, we can assemble a constraint coefficient matrix for each ω corresponding to a set of potential pumping well and constraints locations. Taking uncertainty into account, the hydraulic gradient control constraints can be written as

$$h_{l1,w}(q) - h_{l2,w}(q) \geq 0 \quad l \in L \quad \omega \in \Omega \tag{1}$$

where

$h_{l1,w}(q)$ = head at control node $l1$ due to pump rate q for realization ω
$h_{l2,w}(q)$ = head at control node $l2$ due to pump rate q for realization ω

$h_{l1,w}(q)$ and $h_{l2,w}(q)$ form an inward gradient in the direction from 1 to 2. Each realization ω will yield a different value of head at location 1 and 2. These heads are a linear function of pumping for situations where drawdown is not excessive.

To monitor the progress of contaminant removal, minimum concentration standards can be set at certain monitoring positions in the aquifer. Imposing constraints on concentration as a function of pumping at specified locations for a similar set of realizations result in a non–linear relationship. These constraints can be written as:

$$c_{k,\omega}(q) \leq c_k^* \quad k \in K \quad \omega \in \Omega \tag{2}$$

where

$c_{k,\omega}(q)$ is the contaminant concentration at monitoring location k in a realization ω

c_k^* is the desired concentration limit.

Both the groundwater flow and contaminant transport models are needed to formulate the concentration constraints. To incorporate these constraints in the optimization model, the Jacobians that relate concentration at control location i due to pumping at location j need to be computed. This can be done using a non–linear optimization algorithm such as MINOS [Murtagh and Saunders, 1987].

The hydraulic head gradient and concentration constraints as written do not take into account parameter uncertainty. In solving the optimization problem taking into account all the realizations, we may not be able to satisfy all the head gradients and concentration constraints in the model. However, we can device a method to come up with acceptable solutions if we can distinguish between situations that barely violates the desired constraints and those that deviate substantially from the desired. This is done by introducing a penalty term in the objective function and accounting for the deviation from the desired goals. The head gradient and concentration constraints can therefore be written as:

$$h_{l1,\omega}(q) - h_{l2,\omega}(q) = v_{l,\omega} \quad l \in L \quad \omega \in \Omega \tag{3}$$

$$c_{k,\omega}(q) - c_k^* = u_{l,\omega} \quad k \in K \quad \omega \in \Omega \tag{4}$$

where

$v_{l,\omega}$ and $u_{l,\omega}$ represent the magnitude of the violation of head gradients and concentration gradients respectively.

The Objective Function

The objective function is to minimize pumping when taken in a deterministic sense. Because of violations that may occur due to the multiple realizations, a penalty term is incorporated into the objective function. The objective function can therefore be written as

Minimize

$$\sum_{\omega \in \Omega} \pi_\omega \left(\sum_{j=1}^{n} c^t q_j + \sum_{l \in L} \rho(v_{l,\omega}) + \sum_{k \in K} \rho(u_{k,\omega}) \right) \tag{5}$$

where

$\rho(v_{l,\omega})$ and $\rho(u_{k,\omega})$ are penalty terms.

c_j^t are unit costs of pumping.

q_j is pumping rate at candidate well location j.

π_ω is the probability of occurence of each realization

These penalty terms can be written as quadratic function terms as follows [Dembo and King].

$$\rho(v_{l,\omega}) = \begin{cases} 0 & \text{if } h_{l1,w}(q) \geq h_{l2,w}(q) \\ \frac{1}{2p_l}(h_{l1,\omega}(q) - h_{l2,\omega}(q))^2 & \text{if } h_{l2,w}(q) \leq h_{l1,\omega}(q) + p_l v_l \\ v_l(h_{l1,\omega}(q) - h_{l2,\omega}(q)) - \frac{1}{2}p_l s_l^2 & \text{if } h_{l2,\omega}(q) \geq h_1(q) + p_l v_l \end{cases} \tag{6}$$

$$\rho(u_{k,\omega}) = \begin{cases} 0 & \text{if } c_{k,\omega}(q) \leq c_k^* \\ \frac{1}{2p_l}(c_{k,\omega}(q) - c_k(q))^2 & \text{if } c_k^* \leq c_{k,\omega}(q) \leq c_k^* + p_l u_l \\ u_l(c_{k,\omega}(q) - c_k^*) - \frac{1}{2}p_l u_l^2 & \text{if } c_{k,w}(q) \geq c_k^* + u_l s_l \end{cases} \tag{7}$$

Varying the relative sizes of the scaling parameters will lead to slight variant of the solutions obtained.

Conclusion

The remediation design model has been formulated. The resulting formulation leads to a quadratic objective function and a set of linear and non–linear constraints. A stochastic problem of this type can be solved with a modified version of the MINOS package. Work is in progress to apply this model to an example problem.

References

[1] Murtagh, B.A., and M.A. Saunders, Minos 5.1 Users Guide, *Technical Report SOL 83-20R* Jan., 1987, Stanford University.

[2] Denbo R.S. and A.J. King, Linear-Quadratic Tracking Models For Choices under uncertainty, *Unpublished*

Optimizing In Situ Bioremediation of Groundwater

Barbara Spang Minsker, ASCE Student Member
Christine A. Shoemaker, ASCE Member[1]

Abstract

A nonlinear optimization model for improving the design of in situ bioremediation of groundwater is presented. The model selects injection and extraction rates at specified pumping wells in each time period to minimize the cost of the cleanup. Numerical issues associated with the model are discussed and results are presented for three hypothetical cases.

Introduction

Previous researchers have demonstrated that optimization models coupled with groundwater models can significantly reduce the cost of pump-and-treat groundwater remediation. In this paper, we present a nonlinear optimal control model for designing cost-effective aerobic in situ bioremediation and discuss numerical issues that arise due to the highly nonlinear nature of bioremediation processes. To our knowledge, this research is the first application of an optimization algorithm to in situ bioremediation design.

Optimization Formulation

The optimization model uses differential dynamic programming to select injection and extraction well sites and pumping rates. The model considers the dynamic nature of in situ bioremediation by selecting pumping strategies that change at user-specified time periods (called management periods) over the course of the cleanup. As presented in Culver and Shoemaker (1992), the optimization formulation is as follows:

---

[1]School of Civil and Environmental Engineering, Hollister Hall, Cornell University, Ithaca, NY 14853

$$\min_{U_1 \ldots U_k} J(U) = \sum_{k=1}^{K} G_k(X_k, U_k, k)$$

subject to:

$$X_{k+1} = Y(X_k, U_k, k), \qquad k = 1, \ldots, K$$

$$L(X_k, U_k, k) \leq 0, \qquad k = 1, \ldots, K$$

where J is the total cost of a pumping strategy (the objective function); U_k is the control vector of pumping rates during management period k; X_k is the state vector at the beginning of management period k; L is the set of r constraints on the control and state vectors; G_k is the cost of a strategy during management period k; Y is the transition equation describing the change in X from one management period to the next, given U; and K is the number of management periods.

The state vector consists of hydraulic heads and concentrations of the contaminant (substrate), oxygen, and biomass at each node in the finite element mesh. The transition equation is a nonlinear, two-dimensional, vertically averaged finite-element simulation model called Bio2D (Taylor, 1993), which assesses the effects of pumping strategies on the state vector. The simulation model includes the Haldane variant of Monod kinetics, which allows inhibition of biodegradation to occur at high contaminant concentrations when appropriate.

For bioremediation, the set of constraints in L are:

$$L(X_k, U_k, k) = \begin{cases} c_{s,K}(j) - c_{\max}(j) \leq 0, \quad \forall j \in \Phi \\ c_{s,k}(l) - c_{\max}(l) \leq 0, \quad k = 1, \ldots, K, \ \forall l \in \Psi \\ U_k^{ext} \leq 0, \quad k = 1, \ldots, K \\ U_k^{inj} \geq 0, \quad k = 1, \ldots, K \end{cases}$$

where $c_{max}(j)$ is the water quality goal at node j [mg/L], $c_{s,k}(j)$ is the substrate concentration in time period k at node j [mg/L], Φ is the set of monitoring wells where water quality compliance must be attained by the end of the cleanup, Ψ is the set of monitoring wells at the downgradient end of the solution domain where water quality compliance must be maintained in each management period, U_k^{ext} is the extraction rate vector in period k [m^3/day], U_k^{inj} is the injection rate vector in period k [m^3/day], and the other variables are defined as previously.

Numerical Issues

The biological processes of in situ bioremediation are highly nonlinear; hence, solution of the optimization model is considerably more difficult than solution of a pump-and-treat remediation. A finer mesh and a smaller time step must be used, which increases the computational effort.

The optimization algorithm requires that the derivatives of the transition equation be computed with respect to the state and control vector in each time step within each iteration of the optimization. The number of state variables in the bioremediation problem (heads and concentrations of contaminant, biomass, and oxygen) is twice the number in the pump-and-treat case (heads and contaminant concentrations). Ignoring sparsity considerations, the computational effort required to compute the derivatives is $O(n^3)$, where n is the number of state variables. Thus the computational effort of computing the derivatives is eight times larger in the bioremediation case than in the pump-and-treat case. Additionally, each derivative evaluated is more complex due to the nonlinear biodegradation kinetics terms in the transition equation.

Model Application

To demonstrate the capabilities of the model, this section presents an application to a hypothetical site. Initial contaminant concentrations range from 0 to 40 mg/l at the center of the plume. Initial biomass concentrations are assumed to be 10^{-3} mg/l at all nodes. Oxygen concentrations are assumed to be depleted at the center of the plume initially and to be 3 mg/l surrounding the plume. The water quality goal for the contaminant is 2 mg/l.

The finite elements are 10 by 10 meters, with a total site area of 7,200 m^2. The optimization model selects pumping rates for nine injection wells. The aquifer is assumed to be homogeneous and isotropic. The contaminant was assumed to be toluene, hence all biological parameters were selected from the literature for toluene. The injection water was assumed to be aerated to a concentration of 8 mg/l. A biomass retardation factor of 20 assumes that 95 percent of the biomass in the aquifer is sorbed onto the solid particles. The groundwater velocity prior to pumping can be computed from the heads and hydraulic conductivity to be 10^{-6} m/s.

Using this site, we compare the optimized costs and strategies of three case studies. For each case, an initial, time-invariant pumping strategy was chosen using trial-and-error simulation runs. This initial strategy was input to the optimization model and an optimal pumping strategy was found. In all cases presented below, the optimal costs are relative costs and no units are given.

For the first case, called the time-varying strategy, pumping rates are allowed to change twice a month. The simulation time steps are 0.5 days, so there are 28 simulation steps per management period. The relative cost of the optimal time-varying strategy was 54. The optimal injection rates increase in each period until the fourth period, when the pumping peaks at 3.2 m^3/hour at the center wells, and then tapers off to the end of the cleanup.

The second case, called the time-invariant strategy, illustrates the potential cost savings associated with time-varying pumping strategies. In this second case, pumping rates are required to be fixed at one value over the entire cleanup. This case is easier to solve computationally and requires less adjustment in the field, but the optimal time-invariant cleanup costs are 50 percent larger than the time-varying strategy. The relative cost of the optimal time-invariant strategy was 81.

The third case provides an example of one of the many issues that could be considered using the model, the sensitivity of the model to inhibition. Inhibition causes the degradation rate to decrease at high contaminant concentrations due to toxicity effects of the contaminant on the microorganisms. The inhibition coefficient was decreased from 10^{20} mg/l to 0.1 mg/l, allowing the effects of high levels of inhibition to be examined. The optimal time-varying pumping strategy for high inhibition costs 125, more than twice the relative cost of the first case, which has essentially no inhibition. The strategy selected by the model is also quite different from the first case. Injection of oxygen into areas of high contaminant concentration where inhibition would occur is inefficient. Hence the optimal pumping begins further downgradient where lower contaminant concentrations allow more efficient degradation. Significant pumping occurs at the upgradient wells only after the third management period when advection, diffusion, and flushing effects have decreased the contaminant concentrations in the center of the plume to a level where inhibition effects are less significant.

Appendix. References

Taylor, S. W., Modeling enhanced in-situ biodegradation in groundwater: Model response to biological parameter uncertainty, *Proceedings: 1993 Groundwater Modeling Conference*, International Ground Water Modeling Center, Golden, Colorado, June 1993.

Culver, T. B., and C. A. Shoemaker, Dynamic optimal control for groundwater remediation with flexible management periods, *Water Resour. Res.*, *28*(3), 629-641, 1992.

Optimization of the Surfactant Enhanced Pump-and-Treat Remediation Systems

Min-Der Lin[1] and Daene C. McKinney[2] AM. ASCE

Abstract

The optimization of surfactant enhanced pump-and-treat (SEPT) remediation designs are presented in this paper for the restoration of nonaqueous phase liquids (NAPLs) contaminated aquifers. A three-dimensional, finite difference, multicomponent, multiphase compositional model (UTCHEM) is used to simulate the SEPT process. A hypothetical aquifer, whose characteristics are based on the aquifer at the Canadian Forces base in Borden, Ontario, was employed for the model application. Nonlinear programming was applied to find optimal SEPT designs which provide efficient restoration at the lowest cost. Optimal operating strategies for 2D SEPT designs are presented. The methodology developed in this research is equally applicable to the design of other enhanced pump-and-treat remediation systems (e.g., polymer or cosolvents flushing) for restoration of NAPL contaminated vadose zone or saturated zone.

Introduction

Among the contaminants found in aquifers, organic contaminants are more difficult to remediate than most others due to their low solubilities and high interfacial tensions. Many organic contaminants are relatively insoluble in water and may infiltrate through the subsurface environment as nonaqueous-phase liquids (NAPLs). As the NAPL proceeds through the medium, a portion of the organic liquid leaves behind blobs trapped in the pores because of surface tension effects, resulting in a residual saturation of NAPL. Although the aqueous solubilities of NAPL contaminants are low, they are still high enough to seriously degrade water quality and can persist as contaminant sources for passing groundwater for long periods.

Conventional pump-and-treat remediation technologies have proved to be ineffective for cleaning up NAPL contaminated sites. Surfactants are one of the most promising enhanced aquifer remediation techniques proposed to efficiently remove NAPL contaminants. Due to the high costs of surfactants, it is important to optimize the remediation strategies to provide the most efficient restoration at the lowest cost.

Model Development

A model linking a multiphase simulator to optimization algorithms was developed

[1] Grad. Res. Asst., Dept. of Civil Engrg., Univ.of Texas, Austin, TX 78712.
[2] Asst. Prof., Dept. of Civil Engrg., Univ. of Texas, Austin, TX 78712, (512) 471-8772.

Model Development

A model linking a multiphase simulator to optimization algorithms was developed to investigate the design of cost-efficient SEPT remediation systems. UTCHEM, a multiphase, compositional simulator developed for applications to enhanced oil recovery, was adapted to simulate the SEPT process. The objective of the optimization model is to minimize the total amount of injected surfactant required to cleanup a NAPL contaminated aquifer in a timely manner. The optimization model constraints include: (1) a reduction of the NAPL concentration at designated compliance points to less than some regulatory standard, (2) balanced production, and (3) bounds on extraction and injection rates.

Model Application

The optimization model was used to design a SEPT system to restore a hypothetical aquifer contaminated with the tetrachloroethylene (PCE). The hypothetical aquifer, whose characteristics are based on the aquifer at the Canadian Forces base in Borden, Ontario, was initially developed and presented by *Brown et al.* [1994]. It is assumed to be 49 m long, 15 m wide and 12 m thick. Impervious top and bottom boundaries, and a natural horizontal hydraulic gradient of 0.0043 are assumed. Several assumptions have been made to keep the study within a reasonable scope: (1) the aquifer properties and extent of NAPL contamination are known, (2) all simulations are restricted to the saturated zone, (3) only one surfactant mixture, one contaminant, and one groundwater electrolyte concentration are considered, and (4) local equilibrium conditions are achieved at all times during remediation.

A 25 by 24 rectangular grid was used to represent a two-dimensional vertical cross-sectional model. A contamination event was created by a hypothetical NAPL spill scenario to simulate initial conditions for the SEPT remediation [*Brown et al.*, 1994]. A point source of PCE was introduced in the central grid block at the top of the saturated formation. A 28 m^3 spill of PCE was released at a constant rate of 0.94 m^3/day for 30 days. The PCE source was then removed and the PCE allowed to redistribute and create a contaminant plume under the natural hydraulic gradient for 60 days. Two injection wells located on either side of the spill and one production well located at the spill site. The injection wells were assumed to have the same surfactant solution injection rate which is the decision variable in the optimization model. The production rate is assumed to be equal to the total injection rate. The objective is to minimize the injection rate which reduces the PCE effluent concentration at the production well to less than 5 ppb in one year. Three-dimensional models and more detailed information can be found in *Lin and McKinney* [1995].

Results

Results of solving five cases of 25x24 grid block model are summarized in Table 1. Case 1.1 uses the Winsor Type I system (NAPL-swollen micelles in the aqueous phase) to model the phase behavior between water, microemulsion and PCE. Case 1.2 uses a Winsor Type III system (middle-phase microemulsion) to model the water/microemulsion/PCE phase behavior. Mobilization of PCE through interfacial tension (IFT) reduction effects is not included in Cases 1.1 or 1.2. Mobilization caused by IFT reduction is included in Cases 1.3 and 1.4. There the residual saturation of each phase is treated as a function of capillary number, which is a function of IFT. Phase behaviors in Cases 1.3 and 1.4 are modeled as Winsor Type I and Type III systems, respectively. The hydraulic gradient in case 1.5 (Winsor Type I

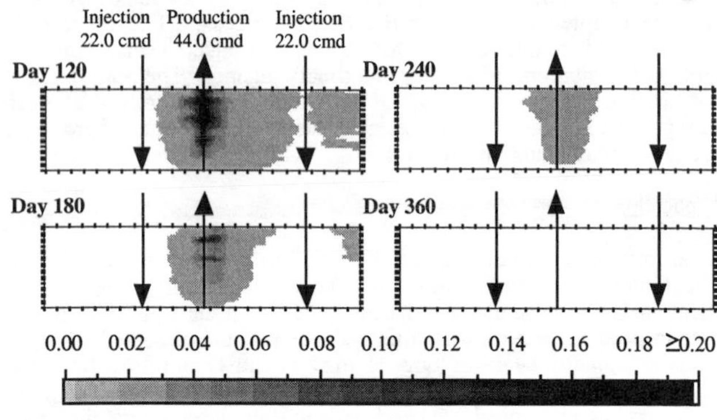

Figure 1. PCE concentration distribution SEPT Case 1.1. (unit: vol./vol.)

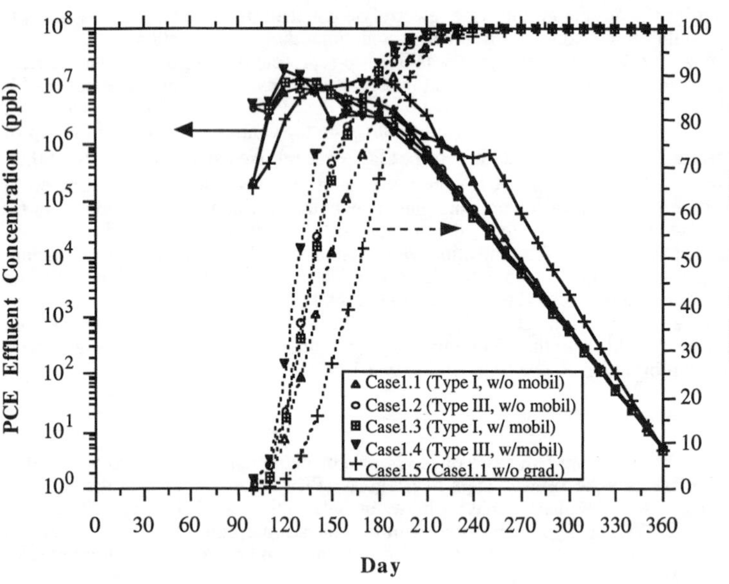

Figure 2. Comparison of PCE recovery performance of SEPT Cases 1.1 to 1.5

system without PCE mobilization effect) is assumed to be zero, instead of 0.0043 as in the previous cases.

A reduction of surfactant injection rate is observed in the cases including mobilization effect. However, the impact is minimal (less than 3%) in reducing the PCE effluent concentration to the level of 5 ppb, which is equivalent to nearly 100% of PCE recovery. This is due to the fact that the PCE in the lower-permeability sand lenses is hard to completely mobilized and achieve 100% recovery.

Phase behavior is an important factor affecting surfactant enhanced recovery processes. The results indicate that the Type III systems recover more PCE with a smaller amount of surfactant within the same period than the Type I systems. But the improvement is less than 3%. The higher efficiency of Type III systems is more apparent during the early stage of the remediation period before about 90% of the PCE is recovered.

The hydraplic gradient significantly affects the volume of aquifer contaminated with NAPL during the spill event, and the dilution of the surfactant solution during remediation. The presence of hydraulic gradient makes it difficult to contain the NAPL plume and the surfactant solution within the treatment area hence results in a lower sweep efficiency of the surfactant solution. Taking advantage of less dilution of the surfactant solution and a more concentrated distribution of PCE in the aquifer due to the lack of hydraulic gradient, the total injection rate of case 1.5 (32.8 m^3/day) is 25% less than case 1.1 (44 m^3/day).

Acknowledgments

This research was funded by the Department of Energy, DOE Grand No. XXXX. Thinking Machine, Inc., provided the parallel computer resources (CM5 machine) which was used to solve genetic algorithm models in this research.

APPENDIX. References

Brown, C.L., G.A. Pope, L.M. Abriola, and K. Sepehrnoori, "Simulation of Surfactant Enhanced Aquifer Remediation," *Water Resour. Res.*, **30**(11), 2959-2977, 1994.

Lin, M-D. and D. C. McKinney, "Optimization of the Design of NAPL Remediation Systems: Surfactant Enhanced Pump-and-Treat Systems," submitted to *Water Resour. Res.*, 1995

Table 1. Results of 2D SEPT Remediation Design Model.

	Case 1.1 Type I w/o mobilization	Case 1.2 Type III w/o mobilization	Case 1.3 Type I w/ mobilization	Case 1.4 Type III w mobilization	Case 1.5 Type I w/o hyd. gradient
Function Calls	32	83	89	98	78
Total Injection (m^3/day)	44 .0	43.4	43.0	43.0	32.8
Total Surfactant (m^3)	475.2	469.4	465.3	463.5	353.8
PCE concentration (ppb)	5.00	4.92	4.98	4.98	4.98
CPU time (hrs)	3.9	11.4	11.0	17.1	8.6

Use of Numerical Sparsity in a Groundwater Quality Management Model

Christopher M. Mansfield, ASCE Student Member
Dr. Christine A. Shoemaker, ASCE Member[1]
Dr. Li-Zhi Liao[2]

Abstract

The use of numerical sparsity in a non-linear optimal control model as it is applied to the pu and treat remediation of groundwater problems is developed. The control model selects the le cost combination of pumping rates and well locations to achieve specified cleanup goals at a speci time horizon, based on the predictions of a finite element simulation model; this is a computation intensive problem. This research focused on the use of the numerical sparsity arising from the fi element model within the optimization procedure to reduce the overall computational demand solving the problem.

Introduction

It has been demonstrated in several published papers that optimization models combined v groundwater simulation models are capable of significantly reducing the cost of pump and tı contamination remediation plans. Linear, nonlinear, neural network, and simulated annealing me ods have all been employed for this purpose; our paper looks only at nonlinear optimal con methods. The solution of this problem is numerically intense, which has limited application of method. However, when consideration of the numerical sparsity of the simulation model is ta in the optimization procedure, the computational difficulty of solving the problem is reduced by order of n, where n is the number of grid points used in the simulation model, enabling the solut of much larger problems.

Problem Formulation

The optimal control model uses a modified form of differential dynamic programming (SALC to select well sites and pumping rates that will give the least cost solution to a pump and tı remediation problem. The control model computes a time-varying pumping policy which has b demonstrated in Culver and Shoemaker (1992) to provide significant savings over a steady st pumping policy. The optimization formulation employed is as follows (as presented in Chang, Sh maker, and Liu, 1992):

$$minimize\ J' = \sum_{t=1}^{N} \left[g^t(x_t, u_t) + \sum_{j \in r_t} Y_j^t(x_t, u_t, w_j) \right],$$

subject to:

$$x_{t+1} = T(x_t, u_t, t), \qquad t = 1, ..., N-1$$
$$L(x_t, u_t, t) \leq 0, \qquad t = 1, ..., N$$

where J' is the total cost for the pumping strategy (i.e., the objective function for the optimizati which is equal to the cost function g plus a penalty function Y, summed over the time horizon of project N. The vector $u_t \in \mathcal{R}^m$ is the set of pumping rates applied in time t, there are m poten

---

[1] School of Civil and Environmental Engineering, Cornell University, Ithaca, NY 14843.
[2] Department of Mathematics, Hong Kong Baptist College, Kowloon, Hong Kong.

pumping well locations; u_t is called the control vector. The vector $x_t \in \mathcal{R}^{2n}$ is the set of non-Dirichlet nodal head and concentration values from the simulation model at time t; this term is called the state vector; n is the number of non-Dirichlet head or concentration nodes in the model. T is the transition equation (i.e., the simulation model) describing the changes of the head and concentration in the aquifer from one time step to the next. Evaluation of the transition equation is achieved with the ISOQUAD finite element flow and transport model (Pinder and Gray, 1977), which is a two dimensional vertically averaged single contaminant model. L is the set of r_t constraints applied in time step t; these constraints are enforced by the penalty function Y with weight w_j. The set of constraints L is:

$$L(x_t, u_t, t) = \begin{cases} u_t \leq 0 & t = 1, ..., N, \\ c_{x_{t=N}}(s) \leq c_{standard} & \forall s \in \Phi \end{cases} \tag{4}$$

which specify that there be no injection occurring at any of the pumping wells, and that the aquifer be clean at the end of the remediation at the s wells in the set Φ of observation wells. The transition equation can be more fully described by the matrix equations (with notation adapted from Chang et al., 1992):

$$([A] + [B])\{h_{t+1}\} = [B]\{h_t\} - \{F_h\} + [L_h]\{u_t\} \tag{5}$$

$$([N(h_t, u_t] + [M])\{c_{t+1}\} = [M]\{c_t\} - \{F_c\} - \sum_{i=1}^{m} u_{t,i}\left(c_{t+1,i} - c_i'\right)[P^i] \tag{6}$$

where h and c are the sets of head and concentration values at the node points in the model, u is the set of pumping rates, F represent boundary conditions, and $[A]$, $[B]$, $[N]$, $[M]$, $[L]$, $[P]$ represent coefficient matrices assembled from the finite element simulation model.

Optimization Procedure

The SALQR optimal control algorithm employs an iterative process of improving from an arbitrary starting policy to an optimal policy. The attainment of cleanup objectives is ensured by progressively increasing the weights on penalties for constraint violations; the loss associated with constraint violation penalties is added to the cost function which also considers pumping and treatment costs (as in Chang et al., 1992). The algorithm employs derivative information obtained from simulation of the current policy u^k to determine a policy increment leading to a new policy $u^{k+1} = u^k + \epsilon\delta u$ that will yield an anticipated quadratic decrease in the cost function. The size of the policy increment taken is scaled by the factor ϵ to ensure that the actual decrease in the cost is the same as the anticipated decrease; this is necessary because the objective and control spaces are not quadratic over large intervals. The iterative process will be repeated until the constraints are met and the anticipated decrease in cost from another iteration is small.

The derivatives of the simulation model (i.e., the derivatives of equations (5) and (6), $< \frac{\partial h_{t+1}}{\partial h_t} >$, $< \frac{\partial c_{t+1}}{\partial h_t} >$, $< \frac{\partial c_{t+1}}{\partial c_t} >$, $< \frac{\partial h_{t+1}}{\partial u_t} >$, $< \frac{\partial c_{t+1}}{\partial u_t} >$) are computed analytically. A method of employing the sparsity structure of the finite element coefficient matrices on which these derivative equations are based was developed. A coefficient matrix from a finite element model will exhibit a sparse, banded structure because of the grid used to discretize the space described by the model. The coefficient matrices will only have non-zero entries at points corresponding to points in the model grid that have direct physical interaction. Through a careful ordering of the computations involved in obtaining the derivatives of the simulation model, the overall complexity of computation is reduced from the order of n^3 to the order of n^2, where n is the number of finite element grid points. Matrix operations such as multiplication or solution of a linear system will produce dense (i.e., non-zero containing) resulting matrices, even if the systems being multiplied or solved are sparse. However, the actual multiplication or solution will be an order n^2 rather than an order n^3 operation if at least one of the matrix terms involved in the operation is sparse. Hence, even though these derivative terms are dense, through a careful ordering of operations in evaluating derivative terms the sparsity structure is used at all steps in the evaluation of the derivatives because at least one of the matrix terms being operated on will be sparse.

Description of Computational Results

To demonstrate the improved performance possible from the algorithm using sparsity as oppos to the algorithm without using sparsity, various problems of increasing state dimension were ru This was done to investigate the actual relative performance of the algorithms, which due to t order of n reduction in the leading term of the computational complexity, was expected to be batt for the sparse algorithm at higher state dimensions. Each case was run through several iterations the optimization several times, to obtain average CPU per iteration values for each of the algorith for the range of state dimensions tested. At the largest size problem successfully tested ($n = 57$ we obtained a ratio of CPU required of about 1 : 32 (about 6.75 minutes per iteration as opposed about 219.5 minutes per iteration) for the case with sparsity as opposed to that without. The relati improvement in CPU per iteration corresponds well to the projected relative differences between t two algorithms based on a theoretical flop analysis.

A secondary advantage of employing sparsity within the optimization computations is that there a large reduction in the computer RAM storage required. This reduction is because it is unnecessa to store the information from coefficient matrix locations that contain entries known to be zer When the banded structure of the coefficient matrices is considered, altering the algorithm to ta advantage of this reduces the leading term in the memory use computation from $(\hat{C} + N)n^2$ stora units to $\frac{\hat{C}}{4}n^2$ storage units, where N is the number of time steps in the remediation horizon and is an integer constant approximately equal to 40 in our case (the exact value of $\hat{C}$ is dependent the formulation of the simulation model). A storage unit is taken to be the amount of RAM need to hold a single floating point number; in the case of double precision Fortran as used in this work, storage unit is equal to 16 bytes. The reduction in memory requirements allowed us to run problen of over 1000 nodes on our IBM RS6000 workstations when using sparsity, while the same problen could not be run on a supercomputer at the Cornell Theory Center when not taking advantage the sparsity of the problem, due to the large size of the RAM required.

Appendix: References

1. Culver, T.B., and C.A. Shoemaker, Dynamic Optimal Control for Groundwater Remediati with Flexible Management Periods, *Water Resour. Res.*, *28*, (1992) **3**, 629-641.

2. Chang, L.-C., C.A. Shoemaker, and P. L.-F. Liu,Application of a Constrained Optimal Contr Algorithm to Groundwater Remediation, *Water Resour. Res.*, *28*, (1992) **12**, 3157-3173.

3. Pinder, G.F., and W.G. Gray, *Finite Element Simulation in Surface and Subsurface Hydrolog* Academic Press, Orlando, FL, 1977.

Reliability Under Uncertainty of Optimal Hydraulic Control Solutions

Joan V. Dannenhoffer David P. Ahlfeld *

Abstract

The robustness of optimal hydraulic control solutions developed using an average hydraulic conductivity is examined. A base case optimal pumping strategy with constant hydraulic conductivity was developed with a groundwater flow simulation-optimization model. A monte carlo approach, using the same well locations and head constraints as the base case, was used to determine the feasibility of simulations with heterogeneous conductivity fields. Heterogeneous conductivity fields were generated using the turning bands method. Results are presented which indicate the robustness of the optimal solutions under varied scenarios.

Introduction

The design of hydraulic control strategies is enhanced when optimization methods are combined with groundwater simulation models. Flow regimes which produce capture or containment of a specified portion of an aquifer can be designed using explicit hydraulic head and gradient constraints. The resulting optimization problem is a linear program for cases, such as confined aquifers, where transmissivity is not a function of hydraulic heads.

The optimal hydraulic containment approach to groundwater management has been in the literature since the early 1970's. It is increasingly used in field applications for both management of groundwater levels and for remediation. An extensive literature review can be found in *Gorelick* [1983] or *Ahlfeld and Heidari* [1994] for a more recent review. Some work has been done to incorporate uncertainty of parameters into simulation models. Recent literature reviews on

*Graduate Research Assistant and Associate Professor of Civil Engineering, Dept. of Civil Engineering, University of Connecticut, Storrs, CT 06226

uncertainty modeling can be found in *Tiedeman and Gorelick* [1993] and in *Chan* [1994].

The uncertainty of parameters is the major problem in obtaining results from models which equate to the actual response of an aquifer to hydraulic controls. Extensive and expensive field testing can not always provide an adequate level of detail. As a result, averages for parameters such as hydraulic conductivity are often used. Even assuming the simulation model is an excellent representation of the physical model and accurate results can be obtained by modeling a continuous process with a discrete model, it is not known how well the hydraulic control system will work, or if it will work, given a heterogeneous conductivity field.

Using an optimization-simulation model in field applications, well location and pumping rates are selected. Since the wells represent a significant capital investment, the question of whether the designed hydraulic control system can be adjusted, without additional wells, to achieve the desired results when the physical parameters differ from the assumed model parameters is paramount. This work examines the robustness of the optimal solution with uncertain hydraulic conductivity under various scenarios.

Problem and Background Formulation

The aquifer domain size is 1690m by 1080m. The numerical discretization consists of 4680 nodes. In the area of study, the node spacing is 10m in the x and y direction, increasing outside of this area, to a maximum of 100m.

A confined and isotropic system was chosen for this experiment. The deterministic physical characteristics of the hypothetical aquifer are elemental longitudinal and transverse dispersivities of 15.0m and 6.0m, respectively; elemental porosity of 0.25; and elemental saturated thickness of 35.0m. An elemental hydraulic conductivity, K, of 4.0 m/day was used in the base case.

A simulation model, using linear quadrilateral finite elements [*Bredehoeft and Pinder*, 1973], which describes vertically averaged, two-dimensional, confined, steady-state flow was used. The simulation model is linked to a linear optimization model [*Ahlfeld and Heidari*, 1994] through the head constraints.

The initial candidate well set of 267 wells included all of the nodes within the area of the generated plume boundary. The plume boundary was defined as the 50 concentration units contour which resulted from simulating contaminant transport of a single source for 6.25 years with a convective-dispersive equation [*Bredehoeft and Pinder*, 1973]. Forty seven head gradient constraints of 4cm/10m were placed symmetrically around the outside of the plume at every third node pair. This ensures capture by requiring that flow has a component approximately perpendicular to the concentration contour. The result of solving this linear optimization problem was a set of 43 wells located at nodes around the inside of the plume, with pumping rates for each individual well, and a total pumping rate of $1128 m^3/day$. The wells and head gradient constraints are depicted in Figure 1 as stars and arrows, respectively.

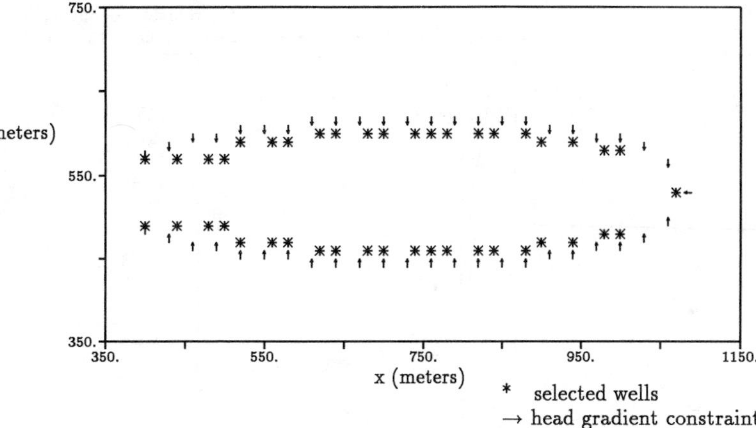

Figure 1: Model Constraints

The results from the optimal homogeneous conductivity design was then used as the basis for comparison. Heterogeneous hydraulic conductivity fields were generated using the turning bands method [*Tompson et al.*, 1989]. Monte carlo simulations were performed, using 100 simulations each, for ten standard deviations of K, $\sigma_{\ln K}$, between 0.2 and 2.0, at five correlation lengths, λ, between 50m and 250m. A solution was considered feasible if, with the heterogeneous conductivity field, an optimal solution could be achieved using base case parameters and wells with maximum pumping constrained to twice the base case pumping at each well.

Results and Conclusions

As shown in Figure 2, the solutions are robust at small deviations. However, the percent feasible solutions declines rapidly as the standard deviation of the hydraulic conductivity fields increase above 0.2. At a standard deviation of 0.4 the percent feasible solutions ranges from 78 percent down to 7 percent , with correlation lengths of 250m and 50m, respectively. This result shows that optimal solutions designed with average conductivity are only robust with small variations in the conductivity field.

Further analysis of the head gradient constraint violations showed that by allowing a smaller head gradient, of .005 for example, the percent feasible solutions increases significantly. For the worst case, $\lambda = 50$m, the percent feasible increased from 7 to 92 percent at $\sigma_{\ln K} = 0.4$ and from zero to 31 percent $\sigma_{\ln K} = 0.6$. The implication is that designing the system with a higher head gradient constraint than required is a practical way of obtaining more reliable results.

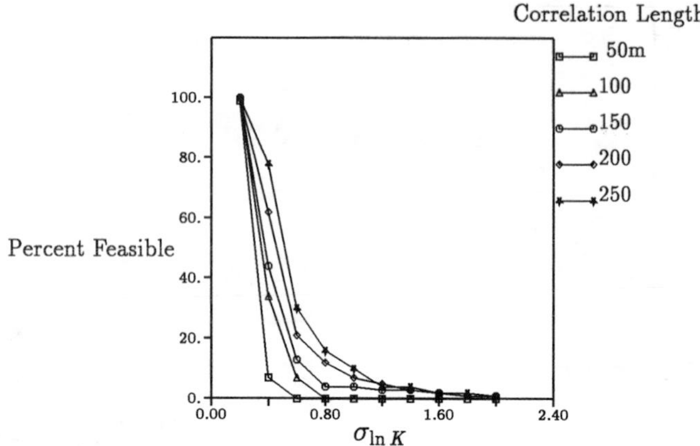

Figure 2: Percent Feasible Solutions versus Standard Deviation

References

[1] Ahlfeld, D.P.and Heidari, Applications of Hydraulic Control to Groundwater Systems, *ASCE Journal of Water Resources Planning and Management*, v. 120, pp. 350-65,1994.

[2] Bredehoeft, J.D. and G.F Pinder, Hydraulic Gradient Control For Groundwater Contaminant Removal, *Journal of Hydrology* v. 76, pp. 86-106, 1973.

[3] Chan, N., Partial Infeasibility Method For Chance-Constrained Aquifer Management, *Journal of Water Resources Planning and Management*, v. 120, pp. 70-89, 1994.

[4] Gorelick, S.M., A Review of Distributed Parameter Groundwater Management Modeling Methods, *Water Resources Research*, v. 19, no. 2, pp. 305-319, 1983.

[5] Tiedeman, C., and S.M. Gorelick, Analysis of Uncertainty in Optimal Groundwater Contaminant Capture Design, *Water Resources Research*, v. 29, no. 7, pp. 2139-2153, 1993.

[6] Tompson, A.F.B., Ababou, R., Gelhar, L.W., Implementation of the Three-Dimensional Turning Bands Random Field Generator, *Water Resources Research*, v. 25, no. 10, pp. 2227-2243, 1989.

Comparison of Different Optimization Techniques for the Solution of a Groundwater Quality Management Problem.

George P. Karatzas
George F. Pinder
College of Engineering & Mathematics, University of Vermont, Burlington, VT 05405, USA

ABSTRACT

In the recent developments, for the solution of the groundwater quality management problem, the main objective is to improve the robustness and the speed of performance of the existing optimization techniques. In this work, three different optimization techniques are applied to solve a groundwater remediation scenario for the Woburn Aquifer (MA). Comparison of the results and the computational effort allows us to evaluate the three techniques.

1 BACKGROUND INFORMATION

In the early '70's, groundwater flow were combined with optimization techniques, to evaluate aquifer system management alternatives related to groundwater 'quantity' or 'quality'. Thereafter, this pioneer work rapidly extended to several methodologies and techniques including the classical linear/nonlinear programming methods (Gorelick et al., 1984, Ahlfeld et al., 1988), simulated annealing (Dougherty et al., 1991), dynamic programming (Culver et al., 1992), an outer approximation method based on combinatorials (Karatzas et al., 1993), neural network and genetic algorithms (Rogers et al., 1994, McKinney et al., 1994) and cutting plane techniques combined with line search directions called the primal method (Tucciarelli et al., 1994). All of these methods aim to achieve robustness and speed of performance.

This work compares the results of one remediation scenario applied to the Woburn Aquifer (MA); the same problem was solved using three different optimization algorithms and the same numerical simulator. The first algorithm is based on the outer approximation method presented by Karatzas and Pinder (1993, 1994). The objective is to minimize a concave function over a closed set of constraints. The outer approximation is one of the 'cutting plane' techniques for global optimization in which the original problem is approximated by a sequence of problems and the optimal solution of the approximated problem is approaching a global optimal solution of the original problem. The second algorithm is a combination of the cutting plane technique and a primal method. In this method, the concept of successive approximation is used. In addition a line search is performed in each optimization step such that the obtained solution is located in the feasible region (Tucciarelli et al., 1994). The third algorithm is MINOS 5.1 (Murtagh et al.,1987), a linear/nonlinear optimization model widely used by several researchers in the area of groundwater management.

A GROUNDWATER MANAGEMENT PROBLEM FOR THE WOBURN (MA) AQUIFER

The Woburn aquifer was used as an application of the method to a groundwater quality management problem. The aquifer has been studied in the past by Ahlfeld (1987), and the current problem formulation was based on his work. For a description of the aquifer, the numerical and physical parameters and the boundary and initial conditions see Ahlfeld (1988). For details of the optimization formulation of the problem see Karatzas et al. (1994). It should be noted that the outer approximation method includes the well installation cost into the objective function (Karatzas and Pinder, 1993).

The solution of the problem, using three different algorithms is shown in Fiqure.1 (the height of the bars is propotional to the pumping rates) and the results are summarized in Table 1. The extraction wells are indicated by a '-' sign and the injection wells by a '+' sign.

DISCUSSION AND CONCLUSION

At the optimal solution only six wells, from the 40 potential wells, were activated using the outer approximation method and the primal method, and three using MINOS. Regarding total pumping (Table 1), the optimal solution of the outer approximation method was 4% less than the optimal solution of the primal method and 16.6% less than MINOS' solution. The reported computational time (user time) was 5.08 hrs for the outer approximation method, 2.6 hrs for the primal Method and 76.5 hrs for MINOS. (MINOS had to be restarted three times in order to obtain the optimal solution.) All runs were performed on a 24 specfp92 Silicon Graphics PC4600 Indy with 96 Megabytes of memory. Comparison of the results suggests both the outer approximation method and the primal method reduce the computational effort and reach a better optimal solution than MINOS. This is attributed to the fact that the former methods can better handle the nonconvex problem that arises in groundwater management problems.

REFERENCES

1. Ahlfeld, D.P., J.M. Mulvey, G.F. Pinder and E.F Wood (1988), 'Contaminated Groundwater Remediation Design Using Simulation, Optimization, and Sensitivity Theory. 2. Analysis of a field site ', *Water Resources Research*, vol.24(3), 443-452.
2. Culver, T.B., and Shoemaker, C.A. (1992), 'Dynamic Optimal Control Groundwater Remediation with Flexible Management Periods.', *Water Resources Research*, vol.28(3), 629-641.

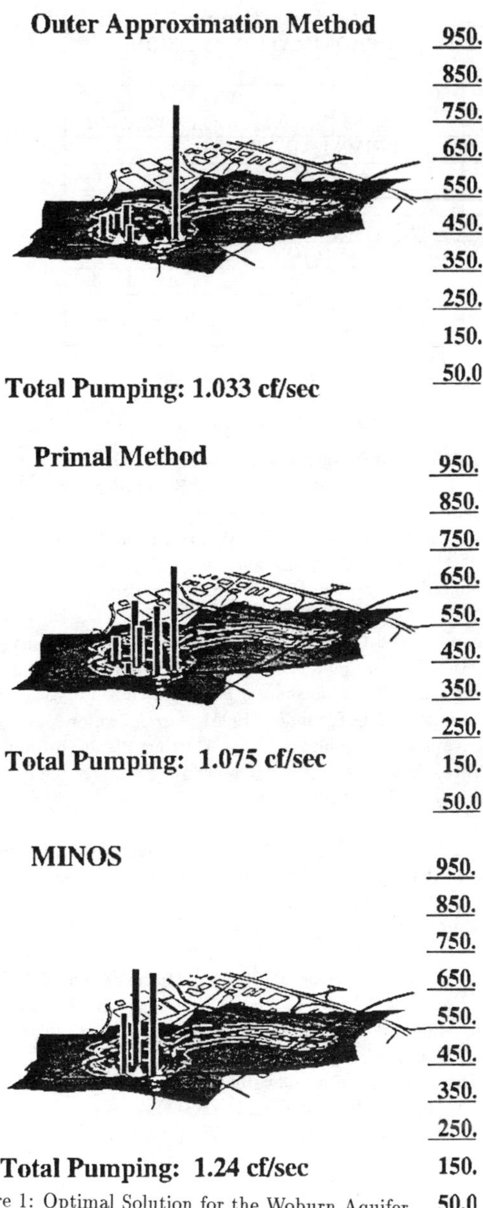

Figure 1: Optimal Solution for the Woburn Aquifer

Pumping rates (ft^3/sec)							Total Pumping ft^3/sec
OUTER APPROXIMATION MATHOD							
Well #	1	2	3	4	5	6	
Node #	90	120	71	106	188	227	
	-.071	-.082	-.100	-0.123	-.023	-.634	1.033
PRIMAL METHOD							
Well #	1	2	3	4	5	6	
Node #	90	120	123	139	188	227	
	-.114	+.05	-.157	-.194	-.156	-.404	1.075
MINOS							
Well #	1	2	3	4	5	6	
Node #	120	139	188				
	-.290	-.459	-.492				1.241

Table 1: Optimal Solution for the Woburn Aquifer.

3. Dougherty, D.E., and Marryott, R.A. (1991), 'Optimal Groundwater Management. 1. Simulated Annealing.', *Water Resources Research*, vol.27(10), 2493- 2509.

4. Gorelick, S.M., C.I. Voss, P.E. Gill, W. Murray, M.A. Saunders and M.H. Wright (1984), 'Aquifer Reclamation Design: The Use of Contaminant Transport Simulation Combined with Nonlinear Programming', *Water Resources Research*, vol.20(4), 415-427.

5. Karatzas, G.P. and G.F Pinder (1993), 'Groundwater Management Using Numerical Simulation and the Outer Approximation Method for Global Optimization', *Water Resources Research*, vol.29(10),3371-3378

6. Karatzas, G.P. and G.F Pinder (1994), 'Groundwater Management Using a 3-D Numerical Simulator and a Cutting Plane Optimization Technique', Proceedings of the X Conference on Computational Methods in Water Resources, Heidleberg, Germany, Kluwer Academic Publishers, vol.2, 841-848.

7. McKinney,D.C and M. Lin (1994), 'Genetic Algorithms Solution of Groundwater Management Models, *Water Resources Research*, vol. 30(6), 1897-1906.

8. Murtagh, B.A., and M.A. Saunders (1987), 'MINOS 5.1 Users Guide.' Technical Report SOL 83-20R, Department of Operations Research, Stanford Univ., CA.

9. Rogers, L.L. and F.U. Dowla (1994), 'Optimization of Groundwater Remediation Using Artificial Neural Networks with Parallel Solute Transport Modeling', *Water Resources Research*, vol.30(2),457-481.

10. Tucciarelli,T.,G.P.Pinder and G.P.Karatzas (1994), 'A Primal Method for the Solution of the Groundwater Management Problem', *Water Resources Planning and Management*, in review.

Robust Optimization for Groundwater Quality Management Under Uncertainty

David W. Watkins, Jr.[1] SM. ASCE and Daene C. McKinney[2] AM. ASCE

Abstract

A robust optimization model for containing a groundwater contaminant plume through the installation and operation of pumping wells is developed. The model explicitly considers uncertainty in hydraulic conductivity and allows decision makers to evaluate the trade-offs among the expected cost, the risk of cost overruns, and the penalty incurred if the plume is not contained completely. The model includes a two-stage decision process in which the investment decision of how many and where to locate wells is made first, and then the pumping rates of the wells are determined assuming that information has been gained during installation. The cost function includes both the fixed installation cost and the variable pumping cost, resulting in a mixed-integer nonlinear stochastic programming problem. This problem is solved using generalized Benders decomposition, and results show that robust optimization can be useful in evaluating the trade-offs faced by environmental decision makers.

Introduction

Robust optimization (RO) [*Mulvey et al.*, 1993] attempts to find solutions which are nearly optimal under most conditions and at least reasonably safe under the worst conditions. Essentially, RO is an extension of classical stochastic programming--in which only an expected value of the objective is considered--which incorporates decision makers' preferences towards risk and allows the consideration of "nearly feasible" solutions in a multiple objective framework. RO thus allows decision makers to search for solutions with varying degrees of two types of robustness: (1) Solution robustness, meaning the solution remains close to optimal for any realization of the random input data, and (2) Model robustness, meaning the solution remains nearly feasible for any realization of the random input data.

The groundwater plume containment model formulated in this paper is similar to the stochastic programming model formulated by *Wagner et al.* [1992], except that the objective function is extended to include the risk and infeasibility penalty terms which characterize RO. Also, fixed costs are included in the model through the use of binary (0-1) variables, and second-stage decisions can be made to minimize recourse costs.

Model Formulation

The problem considered is to contain an area of groundwater contamination by

[1] Grad. Res. Asst., Dept. of Civil Engrg., Univ. of Texas, Austin, TX 78712.

[2] Asst. Prof., Dept. of Civil Engrg., Univ. of Texas, Austin, TX 78712, (512) 471-8772.

maintaining an inward hydraulic gradient across a capture curve. Using a finite-difference representation, steady-flow is modeled in a heterogeneous, confined aquifer in which the hydraulic conductivity is unknown. The decision variables in the model are the location and pumping rates of wells used to contain the contaminant plume. The finite-difference grid, the aquifer boundary conditions, the capture curve, and the potential well locations are shown in Figure 1.

The objective function of the robust optimization model includes three terms: (1) the expected cost of installing and operating the wells, (2) the sum of positive deviations from this expected cost over all realizations of hydraulic conductivity, and (3) a penalty for not maintaining the required hydraulic gradients. This objective can be written as

$$\text{Minimize } Z = \lambda_1\left[IC + \frac{1}{N_s}\sum_{s\in S} OC_s\right] + \lambda_2\left[\frac{1}{N_s}\sum_{s\in S} V_s\right] + \lambda_3\left[\frac{1}{N_s}\sum_{s\in S}\sum_{i\in I} z_{is}\right] \quad (1)$$

where

$$\text{Investment cost:} \qquad IC = \sum_{w\in\Omega} X_w C^{inv} \quad (2)$$

$$\text{Operating cost under realization } s\text{:} \quad OC_s = \sum_{w\in\Omega} C^{op} * \{Q_{ws}(E_w - H_{ws})\} \quad \forall s \quad (3)$$

$$\text{Positive deviation from expected cost:} \quad V_s = max\left\{0,\ OC_s - \frac{1}{N_s}\sum_{s'\in S} OC_{s'}\right\} \quad \forall s \quad (4)$$

$$\text{Violation of gradient constraint } i\text{:} \quad z_{is} = max\left\{0,\ H_{is}^{in} - H_{is}^{out}\right\} \quad \forall i, \forall s \quad (5)$$

X_w is a (0,1) variable which is 1 if well w is installed; C^{inv} and C^{op} are the investment and unit operating costs; Q_{ws} and H_{ws} are the pumping rate and aquifer head at well w under realization s; E_w is the elevation of the well casing; N_s is the number of realizations; and λ_1, λ_2, and λ_3 are multiple objective programming weights to be chosen by the decision maker. The first term in the objective function (1), the expected cost, is identical to the objective of most stochastic programming models. The second term, though, adds a measure of financial risk. The third term prevents the overemphasis of feasibility in the model by penalizing "soft" constraint violations rather than imposing hard constraints. The other constraints in the model are the finite difference groundwater equations and the following pumping capacity limits:

$$0 \le Q_{ws} \le X_w CAP \qquad \forall w, \forall s \quad (6)$$

where CAP is the pump capacity.

A matrix decomposition algorithm [*Yang*, 1990] was used to generate 10 realizations of a heterogeneous, lognormally-distributed hydraulic conductivity field, with a mean of 1.16×10^{-4} m/s, a standard deviation of the underlying normal distribution of 1.45, and a correlation length of 100 meters in all directions. Each of the realizations is assumed equally likely to exist in reality.

Solution Method and Results

The MINLP problem is solved with generalized Benders decomposition [*Geoffrion*, 1972], which involves the successive solution of a relaxed mixed-integer linear programming (MILP) master problem and a parameterized nonlinear

Figure 1-Schematic of the Aquifer.

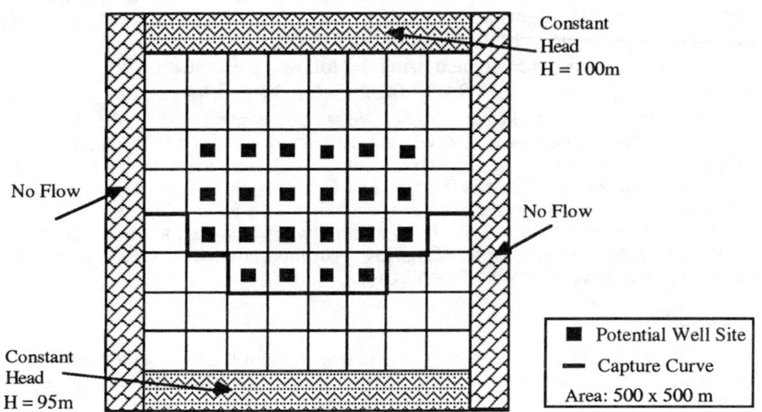

Table 1-Some Interesting Solutions (λ_1 = 1.0 in call cases).

λ_2	λ_3	Expected cost ($1000)	Pos. dev. from exp. cost ($1000)	Expected sum of infeasibilities (m)	No. wells
any	< 10	0	0	7.657	0
0	100	229.5	15.2	0.462	3
0	1000	319.9	97.3	0.057	5
0	5000	508.6	447.3	0.026	7
1.0	100	227.6	11.9	0.049	3
1.0	1000	318.9	19.9	0.059	5
1.0	5000	798.2	29.3	0.01	22*
10	100	230.8	0.00	0.673	3
10	1000	364.9	0.15	0.068	5

* A feasible solution, but not necessarily optimal for the given values of the weights.

programming (NLP) subproblem. The algorithm begins by fixing the binary variables (representing the investment decisions) to an initial guess and solving the resulting NLP subproblem (representing the operating decision problem). Solution of the subproblem provides an upper bound on the objective value, as well as Lagrange multipliers which are used to form the MILP master problem. Solution of the master problem provides a lower bound on the objective value and new values of the binary variables to be used in the next NLP subproblem. Iterations proceed until the upper and lower bounds on the objective value converge to within the desired tolerance. In this work, the GBD algorithm is implemented in GAMS [*Brooke et al.*, 1992] following the procedures of *Paules and Floudas* [1989].

Numerical results are obtained using the following data: C^{inv} =\$25,000; C^{op} = \$.584 /m^4/day (based on \$0.10 /kWh, .0032 kWh to lift 1 m^3 water 1 m, and a five-year operating period); and CAP = 5000 m^3/day. Some interesting results are shown in Table 1. The expected cost is given for various values of the solution and model robustness weights, λ_2 and λ_3, respectively, along with the corresponding investment decisions. The expected cost is 0 when $\lambda_3 = 0$, corresponding to the "do nothing" solution. The highest expected costs are seen when $\lambda_3 = 5000$, when at least 7 wells are installed for each value of λ_2. Even with all wells installed, however, all of the constraints cannot be met for all scenarios. An important aspect of RO is that the model is not declared "infeasible" in this case.

Conclusions

The results obtained from this simple plume containment model clearly demonstrate the potential of robust optimization in environmental and water resources decision-support. Though the two-stage decision model is a great simplification of the real world, in which perfect information is never acquired, RO can help decision makers find solutions which hedge against the risks which remain.

Acknowledgments

This research was funded in part by a National Science Foundation graduate fellowship. The authors also thank C. Floudas for providing them with an example model.

APPENDIX. References

Brooke, A.,D. Kendrick, and A. Meeraus,*GAMS: A User's Guide, Release 2.25,* The Scientific Press, San Fran., 1992.

Geoffrion, A.M. (1972). Generalized Benders decomposition. *J. of Opt. Theory and Appls., 10*(4): 237-260.

Mulvey, J.M., R.J. Vanderbei, and S.A. Zenios. (1993). Robust optimization of large scale systems, Report SOR-91-13, Princeton Univ., Princeton, N.J.

Paules IV, G.E., and C.A. Floudas. (1989). APROS: Algorithmic development methodology for discrete-continuous optimization problems, *Ops. Res., 37*(6), 902-915.

Wagner, J.M., U. Shamir, and H.R. Nemati. (1992). Groundwater quality management under uncertainty: Stochastic programming approaches and the value of information. *Water Resour. Res., 28*(5): 1233-1246.

Yang, A.P. (1990). Stochastic Heterogeneity and Dispersion. Ph.D. Dissertation, Univ. of Texas, Austin, TX.

Mixed Integer Nonlinear Optimization of Soil Vapor Extraction Systems

Yung-Hsin Sun[1] and William W-G. Yeh[2], Fellow, ASCE

Abstract

A mixed integer nonlinear programming (MINLP) approach is proposed for the design of a soil vapor extraction system (SVES), which includes the determination of extraction well locations and the corresponding pumping schedules. A *k*-change local search algorithm is employed in the MINLP model to avoid the pitfall of sequential decision rules in the combinatorial optimization of an SVES design. Results of the test problem indicate that the MINLP approach is computationally efficient and logically correct in the design of a multi-well SVES.

Introduction

Soil vapor extraction is a cost-effective technique for the removal of volatile organic compounds (VOCs) immobilized in the unsaturated zone and can be used conjunctively with other remediation techniques for cleaning up the contaminated soil. However, few guidelines are available for the design of an SVES. The multi-phase vapor flow in the unsaturated zone presents a highly nonlinear problem, and the existence of the mutual interference among well operations further complicates the design problem. Existing approaches for SVES design are mainly based on experiences, simplified analytical approximation or limited simulations (Pedersen and Curtis 1991). An optimization-based design was first introduced by Davert and Yeh (1994), in which a systematic reduction method (SRM) was employed for the design of an SVES.

Given a budgetary constraint or a cleaning up requirement, the design of an SVES is to decide where to pump and how to pump in order to control the contaminant movement effectively and efficiently. The design of an SVES can be

---

[1] Postdoctoral scholar, Dept. of Civil and Environmental Engineering, UCLA, Los Angeles, CA 90095.

[2] Professor, Dept. of Civil and Environmental Engineering, UCLA, Los Angeles, CA 90095; Phone: (310)825-2300; Fax: (310)206-2222.

decomposed into well location optimization and schedule optimization, because schedule optimization is required only if well location is selected. In addition, a numerical simulation model for a 2-D multi-phase multi-component transport in the unsaturated zone (Rathfelder et al. 1991) was linked to the MINLP model for performing system evaluation. The externalization of the simulation model provides flexibility in the selection of various alternative designs.

Problem Formulation

The design problem in this study can be formulated as follows.

$$Max \quad \sum_{l \in \Omega} (C_l(\mathbf{Y},0) - C_l(\mathbf{Y},T)) \tag{1}$$

$$s.t. \quad \sum_i (\alpha_i x_i + \sum_{j \in J_i} \sum_{k \in K} \beta_{i,j,k} y_{i,j,k} q_{i,j,k}) \leq B \tag{2}$$

$$\sum_{j \in J_i} y_{i,j,k} = x_i \qquad k \in K, i \in I \tag{3}$$

$$x_i, y_{i,j,k} \text{ Binary} \qquad j \in J_i, k \in K, i \in I \tag{4}$$

in which α_i is the capital cost of well installation at well location i; $\beta_{i,j,k}$ is the operational cost of pumping rate j at well location i in time period k; Ω is the contaminated area; B is the total budget; C_l is the contaminant concentration at location l; I is the set of potential well locations; J_i is the set of possible pumping rates at well location i; K is the set of time periods; $q_{i,j,k}$ is the extraction of pumping rate j at well location i in time period k; T is the final time period; x_i is the indicator of well locations i; $y_{i,j,k}$ is the indicator of pumping rate j at well location i in time period k; $\mathbf{Y}$ is the matrix of $y_{i,j,k}$'s.

In this formulation, pumping rate is discretized, and the resulting schedule optimization becomes a combinatorial problem similar to location optimization.

Methodology

Sequential decision rules based on the branch-and-bound method, such as the systematic reduction method (SRM) (Davert and Yeh 1994) and the implicit enumeration (Garfinkel and Nemhauser 1972), have been used to solve nonlinear combinatorial problems. To reduce the computational efforts in the nonlinear combinatorial problem, tests are often included to decide the most profitable branching. The implicit enumeration starts with the optimal single-well system, and includes additional wells one by one in the most cost-effective manner. Contrarily, the SRM starts with the full set of potential wells, and sequentially identifies one inferior well for removal from the current well set. The logic of the implicit enumeration implies that the best (n+1)-well system evolves from the best n-well system, and the logic of the SRM implies that the best (n-1)-well system is a degenerate of the best n-well system. These two implications are often incorrect in a multi-well system due to the existence of mutual interference among well operations.

In this study, an intuitive approximation method, the *k*-change local search algorithm (Papadimitriou and Steiglitz 1982), is employed to solve the design problem. In each iteration of the *k*-change local search algorithm, only *k* out of the total *n* variables are allowed to change their values. Because the local search algorithm always keeps the desired number of wells during the entire optimization, the mutual interference among well operations is automatically incorporated. Besides, the user-defined search neighborhoods and heuristics of the *k*-change local search algorithm provide flexibility in its application. In the location optimization, the search neighborhoods are defined by the relative distances between wells; and the neighborhoods in schedule optimization are defined similarly to the nonlinear optimization with controlled search direction.

With the *k*-change local search algorithm, no presumption is imposed on the optimal design and the optimization process. A well location once eliminated from the current consideration may still be reconsidered as the optimization proceeds. A distinct advantage of the proposed MINLP model is that it avoids the possible pitfall of the sequential decision rules.

A Benchmark Problem

A benchmark problem is used to validate the proposed MINLP model. Because the same problem was used for the validation of SRM (Davert and Yeh, 1994), and the complete solution has been generated, results of the MINLP model and those of the SRM can be compared directly.

The benchmark problem uses a 48m×48m homogeneous benzene-contaminated site, in which there are eight possible locations for extraction well installation. The design problem herein is to find a two-well system design that maximizes the total contamiant mass removal within 20 days. Constant pumping rates are assumed for vapor extraction in the entire planning horizon, and the rate can vary from 0.0 to 226.5 *l*/min (8.0 ft^3/min) with an increment of 2.83 *l*/min (0.1 ft^3/min). In addition, the budgetary constraint requires the aggregate pumping rate of all active wells to be under 226.5 *l*/min (8.0 ft^3/min).

In the SRM approach, three initial policies with all eight wells and a schedule of 226.5 *l*/min (8.0 ft^3/min) aggregate pumping rate were used. The SRM converged to a near-optimal design in all three cases, and the numbers of simulation evaluations required for convergence with three different initial policies are 151, 229 and 208, respectively. In addition, the optimal design by SRM contains the location of the optimal single-well system. Based on the available solution of the benchmark problem, the implicit enumeration would give the same optimal two-well design. Therefore, the optimal designs of these two sequential decision rules are consistent, although the directions of approach are opposite.

Two initial well-pairs were used to examine the robustness of the MINLP model. One is a good initial estimate and another is an inferior one, judging from their locations and the potential interference induced by the operation of well-pairs. The MINLP model resulted in the same optimal design when two different initial estimates were used. The complete solution indicates that the resulting system design is the global optimum of the benchmark problem. The convergence took place after 52 simulations when using the good initial estimate, and required 107 simulations when the inferior initial estimate was used. Based on the number of simulations required for convergence, the proposed MINLP approach is two to four times faster than the SRM.

Conclusions

In the design of a multi-well SVES, the proposed MINLP model is logically more accurate than the sequential decision rules. Not only is the computational efficiency of the MINLP model highly competitive, the flexibility of the *k*-change local search algorithm gives rise to the possibility of a wide spectrum of applications.

Acknowledgements

This research was supported by the University of California Systemwide Toxic Substances Research and Teaching Program.

References

Davert, M. W., and Yeh, W. W-G. (1994). "Optimization of the design and operation of soil vapor extraction systems," *Proceedings of the 21st annual conference of the Water Resources Planning and Management Div.*, ASCE, Denver, CO., 181-184.

Garfinkel, R. S., and Nemhauser, G. L. (1972). *Integer programming*, Wiley, New York, NY.

Papadimitriou, C. H., and Steiglitz, K. (1982). *Combinatorial optimization: Algorithms and complexity*, Prentice-Hall, Englewood Cliff, NJ.

Pedersen, T. A., and Curtis, J. T. (1991). *Soil vapor extraction technology*, Noyes Data Corporation, Park Ridge, NJ.

Rathfelder, K., Yeh, W. W-G., and Mackay, D. (1991). "Mathematical simulation of soil vapor extraction systems: Model development and numerical examples," *J. of Contaminant Hydrology*, 8, 263-297.

Optimal Measurement Location for the Calibration of Groundwater Management Models

Antonio Criminisi[1], Tullio Tucciarelli[2]

Abstract

A new criterion is proposed for the location of transmissivity measurements aimed to the calibration of a groundwater management model. According to a random perspective of the transmissivity, a series of management problem, each one defined by a single, conditioned realization of the transmissivity field, is first solved. The cost variation coefficient, defined as the ratio between the minimum cost variance and the minimum expected cost, is then evaluated. The best location is assumed to be the one that minimizes the variation coefficient.

Introduction

The solution of the flow problem, inside a groundwater or a groundwater quality management problem, is function of the transmissivity values inside the aquifer. These values can be thought as the realization of a random field, conditioned to assume given values in the measure locations. Because these measures are very expensive, it is worthwhile to find a good criterion to choose the location of each new transmissivity measure. To take into account the value of information of existing measures, we assumes that transmissivities follow a conditional lognormal joint probability distribution.

Stochastic optimization problems can be numerically solved by means of Monte Carlo simulations. Given the vector of **m** known measures located at known points

[1]Ph.D. student, University of Palermo, Italy.
[2]Associate Professor, University of Reggio Calabria, Italy.

$x_1^*, x_2^*, \ldots x_m^*$, a certain number N of conditional lognormal transmissivity fields are generated by using the theory proposed by Journel(1974). For each realization an optimal cost and an optimal strategy can be found. The statistics of this hystogram fully characterize the stochastic output.

Stochastic structure of the transmissivity field

Statistical properties of log-transmissivity conditional random field are described by means of the joint conditional normal distribution function. With reference to a finite element scheme, assuming constant transmissivity over each element, each value of transmissivity can be referred to the center of each one. We divide the centres of all the n_e elements into s simulation points $\hat{x}_1; \hat{x}_2; \ldots \hat{x}_s$ and m measure points $x_1^*, x_2^*, \ldots x_m^*$ ($s+m=n_e$).

We shall refer to simulated variables and their moments using the symbol ^ and to measure variables and their moments using the symbol *. We define a vector of transmissivities in all the s+m points:

$$\mathbf{T} = (\mathrm{T}(\hat{x}_1); \mathrm{T}(\hat{x}_2); \ldots\ldots \mathrm{T}(\hat{x}_s); \mathrm{T}(x_1^*); \ldots\ldots \mathrm{T}(x_m^*))^{\mathrm{T}} = \begin{pmatrix} \hat{\mathbf{T}} \\ \mathbf{T}^* \end{pmatrix} \tag{1}$$

The corresponding vector of log-transmissivities is given by:

$$\mathbf{Y} = \ln \mathbf{T} = \begin{pmatrix} \ln \hat{\mathbf{T}} \\ \ln \mathbf{T}^* \end{pmatrix} = \begin{pmatrix} \hat{\mathbf{Y}} \\ \mathbf{Y}^* \end{pmatrix} \tag{2}$$

Correlation structure of $\mathbf{Y}$ is described by means of a positive definite, exponential-type covariance matrix $\mathbf{C_Y}$, given by:

$$\mathrm{Cov}[\mathrm{Y_i}; \mathrm{Y_j}] = \sigma_{\mathrm{Y}}^2 \cdot e^{-\mathrm{h}/\lambda} \tag{3}$$

where σ_Y^2 is a positive scalar, $h=|x_i - x_j|$ is the euclidean norm and λ is the correlation scale or lenght. It can be proved that the conditional probability $f_{\mathrm{Y}}(\hat{\mathbf{Y}}|\mathbf{Y}^*)$ is s-variate normal (Dagan,1982) and given by :

$$f_{\mathrm{Y}}(\hat{\mathbf{Y}}|\mathbf{Y}^*) = ((2\pi)^s \cdot |\mathbf{C}_{\hat{\mathbf{Y}}|\mathbf{Y}^*}|)^{-1/2} \cdot exp((\hat{\mathbf{Y}} - \mathrm{E}[\hat{\mathbf{Y}}|\mathbf{Y}^*])^{\mathrm{T}} \cdot \mathbf{C}_{\hat{\mathbf{Y}}|\mathbf{Y}^*}{}^{-1} \cdot (\hat{\mathbf{Y}} - \mathrm{E}[\hat{\mathbf{Y}}|\mathbf{Y}^*])) \tag{4}$$

where $\mathbf{C}_{\hat{\mathbf{Y}}|\mathbf{Y}*}$ is the conditional covariance matrix (s×s). Moments of pdf (4) are related to moments of the unconditional pdf by (Dagan,1982):

$$E[\hat{Y}_i|Y*] = E[\hat{Y}_i] + \sum_{j=1}^{m} \lambda_{ij}(Y_j^* - E[Y(x_j^*)]) \qquad i = 1,2,3...s \qquad (5)$$

$$Cov[\hat{Y}_i, \hat{Y}_j|Y^*] = Cov[\hat{Y}_i, \hat{Y}_j] - \sum_{k=1}^{m} \sigma_Y^2 \cdot \rho(|\hat{\boldsymbol{x}}_j - \boldsymbol{x}*_k|) \cdot \lambda_{ik} \qquad i,j = 1,2,3...s \qquad (6)$$

where $Y(\boldsymbol{x}_j^*)$ is the log-transmissivity, in the point $\boldsymbol{x}_j^*$, of the unconditioned random field and each vector $\boldsymbol{\lambda}_i$ is the solution of the following linear system (Dagan, 1982):

$$\sum_{j=1}^{m} \lambda_{ij}\rho(|\hat{\boldsymbol{x}}_j - \boldsymbol{x}*_k|) = \rho(|\hat{\boldsymbol{x}}_i - \boldsymbol{x}*_k|) \qquad k=1,2,...m \qquad (7)$$

Moments (5) and (6) form respectively a vector and a matrix. We can relate these moments with moments of transmissivity $\mathbf{T}(\boldsymbol{x})$ using the moment generating function (Johnson and Kotz, 1976):

$$\mu_{r_1,r_2,...,r_s} = exp(\mathbf{r}^T E[\hat{\mathbf{Y}}|\mathbf{Y}*] + \tfrac{1}{2}\mathbf{r}^T \mathbf{C}_{\hat{\mathbf{Y}}|\mathbf{Y}*}\mathbf{r}) \qquad (8)$$

where μ is a generic moment and $\mathbf{r}$ is an auxiliar vector. For r_i=1 and all the other r's equal to zero, we find the i-th mean:

$$E[\hat{T}_i|\mathbf{T}*] = exp(E[\hat{Y}_i|\mathbf{Y}*] + \tfrac{1}{2} Var[\hat{Y}_i|\mathbf{Y}*]) \qquad (9)$$

for r_i=1, r_j=1 and all the other r's equal to zero, we find the covariance:

$$Cov[\hat{T}_i, \hat{T}_j|\mathbf{T}*] = \{exp(Cov[\hat{Y}_i, \hat{Y}_j|\mathbf{Y}*]) - 1\} \cdot$$
$$exp(E[\hat{Y}_i|\mathbf{Y}*] + E[\hat{Y}_j|\mathbf{Y}*] + \tfrac{1}{2} Var[\hat{Y}_i|\mathbf{Y}*] + \tfrac{1}{2} Var[\hat{Y}_j|\mathbf{Y}*]) \qquad (10)$$

Finally, from (9) and (10) we find that the conditional variation coefficient of transmissivity at location $\hat{\boldsymbol{x}}_i$ is given by:

$$\alpha_{\hat{T}_i} = \frac{Var^{\frac{1}{2}}[\hat{T}_i|\mathbf{T}*]}{E[\hat{T}_i|\mathbf{T}*]} = [exp(Var[\hat{Y}_i|\mathbf{Y}*]) - 1]^{\frac{1}{2}} \qquad (11)$$

since the conditional variance $Var[\hat{Y}_i|\mathbf{Y}*]$ does not depend on measured values (see eq. 7) we obtain that the variation coefficient of each transmissivity does not depend on the measured values but only on their location. This property is fundamental in order to validate the criterion that will be proposed in the next section.

Proposed criterion to choose the best measure location

The main idea is to choose the new measure location which gives the maximum reduction of the pumping cost uncertainty. To characterize the uncertainty about the management strategy we solve a "distribution problem" (Wagner, 1992) and we choose the variation coefficient of the optimal cost distribution (α_{cost^*}) as cost uncertainty indicator:

$$\alpha_{cost^*} = \frac{\mathrm{Var}^{\frac{1}{2}}[\mathrm{cost}^*]}{\mathrm{E}[\mathrm{cost}^*]} \tag{12}$$

The most important property of (12) is its strong correlation with the variation coefficient (11) and, therefore, the low sensitivity of (12) with respect to the actual value of the new measure. Because of this property the kriging value in the potential new measure location is used in order to generate the conditional random log-transmissivity field. The exact proportionality between (11) and (12) can be proved only in some particular cases, but numerical tests (not documented here) suggest that the probability to obtain the optimal location is quite large. The results obtained with the use of a synthetic example are going to be published in the next future.

References

Dagan G. (1982). "Stochastic Modelling of groundwater Flow by Unconditional and Conditional Probabilities: 1. Conditional Simulation and the Direct Problem.", *Water Resour. Res.*, 18(4), 813-833.

Johnson, N. L., and S. Kotz. (1976). *Distribution in Statistics: Continuos Multivariate Distribution*, John Wiley & Son Inc..

Journel, A. G. (1974). "Geostatistics for Conditional Simulation of Ore Bodies.", *Economic Geology*, Vol. 69, 673-687.

Wagner J. M., Shamir U., Nemati H. R. (1992). "Groundwater Quality Management under Uncertainty. Stochastic Programming Approaches and the Value of Information", *Water Resour. Res., 28*(5), 1233-1246.

Development of Methodology for the Optimal Operation of Soil Aquifer Treatment Systems

Zongwu Tang[1], Guihua Li[1], Larry W. Mays[2], M. ASCE, and Peter Fox[3]

Abstract

A new methodology is being developed for determining the optimal operation of soil aquifer treatment (SAT) systems. The new methodology is based upon mathematically describing the problem as a large-scale nonlinear discrete-time optimal control problem. The problem is solved by interfacing a nonlinear programming optimizer, a variably saturated flow model, and a water quality (transport) model. Because this discrete-time optimal control problem is nonlinear, both from the view point of the objective function and the simulator (transition) equations, differential dynamic programming (DDP), specifically a successive approximation linear quadratic regulator (SALQR) approach, is being interfaced with the simulator. In order to describe the unsaturated flow and transport mechanism, the HYDRUS model *(Kool and van Genuchten,* 1991) has been modified to incorporate the transport mechanisms to form the simulator. Once developed, the methodology will be applied to proposed SAT systems in Arizona. One application will be the proposed SAT site for the City of Phoenix 91st Avenue wastewater treatment plant.

Mathematical Formulation of Optimization Model

The purpose of the operation model is to determine optimal *control variables* values: pumping rate (wastewater application rate, Q), application time (T_A), and drying time (T_R) in order to maximize the infiltration. The simulator determines the state of the SAT system for these decisions. *State variables* are defined as maximum surface water

[1] Graduate Research Assistant, Department of Civil Engineering, Arizona State University, Tempe, Arizona 85287

[2] Professor and Chair, Department of Civil Engineering, Arizona State University, Tempe, Arizona 85287

[3] Assistant Professor, Department of Civil Engineering, Arizona State University, Tempe, Arizona 85287

ponding depth (D_{MAX}), draining period (T_D), water content (ω) averaged along the soil column, and water quality through the soil column (in this case, total nitrogen TN and organics TOC). The overall optimization model mathematical formulation is as follows:

$$Maximize \quad I = \frac{1}{CT} \int_0^{CT} q_s(t)dt \tag{1}$$

Subject to

1. The general unsaturated flow equations are defined as

$$g_1(\omega, h, q_s, z, t) = 0 \tag{2}$$

where CT is the cycle time, h is the pressure head, q_s is the water flux at the soil surface, z is the positive downward soil column depth, and t is time

2. The general solute transport equations are defined as

$$g_2(TN, TOC, z, t) = 0 \tag{3}$$

3. Mass balance for water in the pond is

$$\frac{dD(t)}{dt} = Q(t) - E(t) - q_s(t) \tag{4}$$

where E(t) is the evaporation rate. The initial and boundary conditions are

$$D(t)\Big|_{t=0,\, t \geq T_A + T_R} = 0 \tag{5}$$

and

$$Q(t)\Big|_{t=T_A} = 0 \tag{6}$$

4. Maximum ponding water depth

$$D(t) \leq D_{MAX} \tag{7}$$

5. The cycle time defined by the operation time constraint

$$T_A + T_R + T_D = CT \tag{8}$$

6. Infiltration potential recovery constraint

$$\omega(L, CT) \leq \omega^* \tag{9}$$

where ω^* indicates the limit below which infiltration potential insufficiently recovered by the drying process, and

7. Upper limits on concentration constraints

$$TN(t) \le TN^* \tag{10}$$

and

$$TOC(t) \le TOC^* \tag{11}$$

where TN* and TOC* indicate water quality standards for bulk liquid concentrations of total nitrogen and total organic carbon.

Mathematical Statement of the Simulator

The unsaturated flow simulator, $g_1(\omega, h, q_s, z, t) = 0$, and the transport simulator, $g_2(TN, TOC, z, t) = 0$, are described as follows:

1. Water flow in unsaturated soil

The basic governing equation for unsaturated flow is the one dimensional Richard's equation

$$C\frac{\partial h}{\partial t} = \frac{\partial}{\partial z}\left(K\frac{\partial h}{\partial z} - K\right) - s(z,t) \tag{12}$$

where z is the aquifer depth positive downward, h is the pressure head, and K is the hydraulic conductivity. The upper boundary conditions are

$$q_0 = \begin{cases} -K\dfrac{\partial h}{\partial z} + K & 0 \le t \le t_W, z = 0 \\ E & t > t_W, z = 0 \end{cases} \tag{13}$$

where E is the evaporation rate during the drying period. The lower boundary conditions are

$$\left.\frac{\partial h}{\partial z}\right|_{z=L} = 0 \tag{14}$$

The hydraulic characteristics are described according to *Kool* and *van Genuchten*, (1991) by

$$\omega(h) = \omega_r + \frac{\omega_s - \omega_r}{(1 + |\alpha h|^{\beta})^{\gamma}} \tag{15}$$

and

$$K(S_e) = K_s S_e^{1/2}(1 - (1 - S_e^{1/\gamma})^{\gamma})^2 * f_{clog} \tag{16}$$

where ω_s and ω_r are the saturated and residual water content, S_e is the degree of saturation, and α, β, and γ are curve fitting parameters. The factor f_{clog} expresses the effect of surface clogging

$$f_{clog}(0,t) = \exp(-rst) \tag{17}$$

and the effect of biological clogging is described according to *Taylor* and *Jaffe* (1990) by

$$f_{clog}(z,t) = \begin{cases} \exp(aB + bB^2) & B \leq B_0 \\ f_0 & B > B_0 \end{cases} \tag{18}$$

where r, a, b, f_0 and B_0 are empirical values, B is the biomass (M/L^3).

2. Solute transport

The transport model describing the advection, dispersion, and transformation of solutes in the SAT system has the mass balance form

$$\frac{\partial}{\partial t}(\omega[C_i]R_i) = \underbrace{-\frac{\partial}{\partial z}(q[C_i])}_{\cdots advection} + \underbrace{\frac{\partial}{\partial z}(\omega E_i \frac{\partial [C_i]}{\partial z})}_{\cdots dispersion} + \underbrace{(\frac{dC_i}{dt})_{bio}}_{\cdots transformation} \tag{19}$$

where R_i is the retardation factor considering the effect of adsorption, q is Darcy's flux which is solved from Richard's equation, E_i is the coefficient of dispersion, and C_i denotes solute concentration, notations of solute components $\{C_i\}$ are: ammonium nitrogen [NH]; nitrite and nitrate nitrogen [NO]; dissolved oxygen [DO]; organic carbon [SC]; and biomass [B]. The $(dC_i/dt)_{bio}$ refers to the microbiochemical reactions.

References

Kool J.B., and M. Th. van Genuchten, 1991, *HYDRUS: One-Dimensional Variably Saturated Flow and Transport Model, Including Hysteresis and Root Uptake*, v3.31, U.S. Salinity Laboratory, U.S. Dept. of Agriculture, Agriculture Service, Pineside, CA.

Taylor, S.W., and P.R. Jaffe, 1990, Subsurface and biomass transport in a porus medium, *Water Resources Research*, 26(9):2181-2194.

A Unified Approach for Stochastic Parameter Estimation, Experimental Design, and Reliability Analysis in Groundwater Modeling

Angelina Y.S. Tong[1], and William W-G. Yeh[2], Fellow, ASCE

Abstract

A methodology for unifying stochastic parameter estimation, experimental design, and reliability analysis is developed to advance the modeling of groundwater flow and contaminant transport. Observations of transport under transient conditions are incorporated with observations of transient head and point estimates of hydraulic conductivity. The adjoint state method efficiently determines the Jacobian for the Maximum Likelihood Estimation of the mean, variance and correlation scale of the parameters. Cokriging then gives an unbiased and minimum error variance estimate of the parameters. A prediction-equivalent identifiability criterion, assuming that model parameters are represented by a random field, is used to optimize the experimental design (sampling strategies) for parameter estimation. Consequently, the relationship between the quantity and quality of data and the reliability of the identified stochastic model can be evaluated.

Problem Statement

Effective control and remediation of groundwater is dependent on accurate and reliable models to predict groundwater flow and contaminant transport. However, there are many sources of error, as the key model parameters cannot be directly measured from the physical point of view, and in practice must be estimated by an inverse procedure. Additionally, the heterogeneity of many aquifer systems requires

---

[1] Graduate Student, Department of Civil and Environmental Engineering, UCLA, Los Angeles, CA 90095.

[2] Professor, Department of Civil and Environmental Engineering, UCLA, Los Angeles, CA 90095; Phone: (310) 825-2300; Fax: (310) 206-2222.

model parameters to be represented by random fields. Thus an experimental design determines the data input for estimating the stochastic parameters and is an important consideration in model reliability.

A major objective of this study is to develop a methodology to optimally determine model parameters from sparse point measurements of parameters as well as observations of transient head and concentration.

Methodology

Assuming that the fluid density and viscosity are constant, the groundwater flow and mass transport equations may be solved independently. The groundwater flow and transport model to be used is two dimensional, subject to appropriate initial and boundary conditions and governed by

$$S\frac{\partial h}{\partial t} = \nabla \cdot (Km\nabla h) + W \quad (1)$$

$$\frac{\partial (nC)}{\partial t} = \nabla \cdot (n\underline{D}\nabla C) - \nabla \cdot (nVC) + M \quad (2)$$

where h is head, C is solute concentration, m is thickness of the aquifer, W and M are sink or source rates to unit aquifer area, V is average velocity, S is storage coefficient, K is hydraulic conductivity, n is porosity, and $\underline{D}$ is the hydrodynamic dispersion tensor.

Statistical Parameter Estimation

The first step in statistical parameter estimation is to determine the statistical structure of the parameter to be estimated. It is assumed that the parameter to be estimated is hydraulic conductivity K and that it is log-normally distributed so that Y=log$K$ is normally distributed. Also, first order stationarity and an isotropic, exponential covariance is assumed. K can then be described and determined by θ, a vector of three statistical parameters (Sun and Yeh, 1992): constant mean of Y, μ_Y; variance of Y, σ_Y^2; and correlation scale of Y, l_Y.

Maximum Likelihood Estimation (MLE) is used to determine θ. Assuming that the observations, z, are from a normal distribution, the objective of the MLE method to find the θ that most likely have produced the TM number of observations z is (Kitanidis and Vomvoris, 1983):

$$\min L(\theta) = \tfrac{TM}{2}\ln(2\pi) + \tfrac{1}{2}\ln|Q| + \tfrac{1}{2}(z-\mu)^T Q^{-1}(z-\mu) \quad (3)$$

where Q is the measurement covariance matrix. Taking the first order approximation, Q can be determined from Jacobians of head and concentration data with respect to the parameter of interest, as determined by the adjoint state method.

The adjoint state method is used to determine the Jacobians as required by the MLE above. The following adjoint state equations are developed for the stochastic partial differential equation (SPDE) for the perturbations based on a first order approximation:

$$S\frac{\partial \psi_1}{\partial t} = -e^{\bar{Y}}\nabla\cdot\left(m\nabla\psi_1\right) - e^{\bar{Y}}\nabla\cdot\left(\bar{C}\nabla\psi_2\right) + \frac{\partial R}{\partial h} \tag{4}$$

$$n\frac{\partial \psi_2}{\partial t} = e^{\bar{Y}}\left(\nabla\bar{h}\cdot\nabla\psi_2\right) - \nabla\cdot\left(Dn\nabla\psi_2\right) + \frac{\partial R}{\partial C} \tag{5}$$

where $\bar{h}$ is expected head, $\bar{C}$ is expected concentration, $\bar{Y}$ is expected log-hydraulic conductivity, R is performance function (e.g., head or concentration at a given observation location j at time t_k), and ψ_1, ψ_2 are adjoint state variables. The Jacobian, change in R due to a unit change in the parameter p, can then be calculated by integrating over time and Ω_j, the exclusive subdomain of node j:

$$\begin{aligned}\frac{\partial R(x_i,t_k)}{\partial p_j} = \int_0^{t_k}\iint_{\Omega_j}\Big[\, & e^{\bar{Y}}m\nabla\cdot\left(\psi_1\nabla\bar{h}\right) - S\psi_1\frac{\partial\bar{h}}{\partial t} + Q\psi_1 \\ & - e^{\bar{Y}}\bar{C}\nabla\cdot\left(\psi_2\nabla\bar{h}\right) + \psi_2\nabla\bar{C}\cdot\nabla\bar{h} + \psi_2\bar{C}\cdot\nabla^2\bar{h} + \frac{\partial R}{\partial p}\,\Big]\partial\Omega\,\partial t\end{aligned} \tag{6}$$

After the statistical parameters are determined, cokriging, based on linear minimum variance estimation theory, gives the estimator of the true Y (log *K*) field:

$$\hat{Y}_o = \sum_{m=1}^{M}\lambda_m Y_m + \sum_{k=1}^{K}\sum_{l=1}^{L}\rho_{l,k}\left(h_{l,k} - \bar{h}_{l,k}\right) + \sum_{k=1}^{S}\sum_{l=1}^{P}\tau_{l,k}\left(C_{l,k} - \bar{C}_{l,k}\right) \tag{7}$$

where $\hat{Y}_o$ is estimated parameter at x_o, K index is number of observation times, L index is number of observation wells, M index is number of Y measurements, $h_{l,k}$ is head observation at t_k and x_l, $\bar{h}_{l,k}$ is mean value of head at t_k and x_l, $C_{l,k}$ is concentration observation at t_k and x_l, $\bar{C}_{l,k}$ is mean value of concentration at t_k and x_l, Y_m is log$K$ measurement at the mth measurement point, and $\mu_{l,k}, \lambda_m, \tau_{l,k}$ are cokriging coefficients. The cokriging coefficients are determined by solving a set of linear equations derived by minimizing the variance of the estimate under the constraint of finding an unbiased estimator. With ν as the Lagrange multiplier for the unbiased estimator constraint and Y_o^* as the true parameter at x_o, the variance of cokriging estimate is:

$$\begin{aligned}Var\left[\hat{Y}_o - Y_o^*\right] = \sigma_Y^2 & - \sum_{m=1}^{M}\lambda_m Cov\left[Y_m, Y_o\right] - \sum_{k=1}^{K}\sum_{l=1}^{L}\rho_{l,k}Cov\left[h_l\left(t_k\right), Y_o\right] \\ & - \sum_{k=1}^{S}\sum_{l=1}^{P}\tau_{l,k}Cov\left[C_l\left(t_k\right), Y_o\right] - \nu\end{aligned} \tag{8}$$

Reliability Analysis

Applying the first order approximation for the combined system, the estimation of the variance of prediction errors is estimated by:

$$\text{Var}\left[\phi(p^*) - \phi(p)\right] = \sum_{i=1}^{N}\sum_{j=1}^{N}\frac{\partial\phi}{\partial p_i}\frac{\partial\phi}{\partial p_j}E\left[\left(p_i^* - p_i\right)\left(p_j^* - p_j\right)\right] \tag{9}$$

where ϕ is a model output corresponding to a given set of prediction objectives, p is the parameter to be estimated, and p^* is the true value of the parameter. The sensitivities $\frac{\partial\phi}{\partial p_i}$ *and* $\frac{\partial\phi}{\partial p_j}$ can be calculated using the adjoint state method and the term $E\left[\left(p_i^* - p_i\right)\left(p_j^* - p_j\right)\right]$ is given by the cokriging estimate variance.

Experimental Design

Using prediction–equivalent identifiability (PEI) as a criterion, a design D is sufficient if for a set of realizations M of parameter p that generate the same observations as that in D,

$$\left\|\phi(p^r) - \phi(p^s)\right\| < \varepsilon \text{ for any } p^r \in M \text{ and } p^s \in M \tag{10}$$

Using the variance norm, the criterion is thus a measure of reliability. In order to determine the best number and locations of new observations, an efficient combinatorial optimization algorithm is required.

Conclusion

Adjoint state equations relating transient head, contaminant concentration, and hydraulic conductivity perturbation were derived and applied in the developed stochastic inverse procedure for the coupled groundwater flow and contaminant transport system to estimate hydraulic conductivity. An estimate for the variance of prediction errors for stochastic models of groundwater flow and contaminant transport is also developed as a criterion for reliability analysis. Continuation of this research involves development of experimental design procedure and the estimation of other parameters such as dispersivity and retardation factor.

Acknowledgment

The research reported herein is supported by funds provided by the National Science Foundation under Award No. MSS-9213963.

References

Kitanidis, P.K., and Vomvoris, E.G. (1983). "A geostatistical approach to the inverse problem in groundwater modeling (steady state) and one-dimensional simulations," *Water Resources Research,* 19(3), 677-690.

Sun, N-Z., and Yeh, W.W-G. (1992). "A stochastic inverse solution for transient groundwater flow: parameter identification and reliability analysis," *Water Resources Research*, 28(12), 3269-3280.

Conjunctive Use and Groundwater Management

Babs Makinde-Odusola[1], P.E.

Abstract

In recent past, conjunctive use projects seek to maximize the combined yield of aquifer-stream systems. Now, many conjunctive use projects, especially in California, are multiobjective in scope, and seek to improve water quality, wildlife habitat, and safe yield. Water used to replenish aquifers now comes from many sources, including: storm flow, urban runoff, river base flow, reclaimed water, and imported water from outside the watershed. Many conjunctive use projects, especially in effluent dominated streams of semi-arid areas now include wastewater reclamation, such as denitrification.

Introduction

Conjunctive use is the coordinated and planned management of both surface and groundwater supplies to meet water requirements while conserving water. The required water may be for municipal, industrial, and agricultural use. This paper focuses more on conjunctive use projects that emphasize municipal and industrial water supply. It is prudent to examine the environmental impacts of conjunctive use projects. Resource managers must also consider uncertainty, stakeholders, and societal issues. Integrated Resources Planning (IRP) addresses those complex interrelated issues. Conjunctive use is part of IRP for water utilities.

Forms of Conjunctive use projects

The California State Water Resources Control Board (SWRCB) defines two forms of conjunctive use - seasonal storage and long term storage. Seasonal storage use requires that the recharged water be extracted on an annual basis. Long-term storage use implies banking large quantities of water during times of surplus. The

[1]Senior Engineer, Riverside Public Utilities, 3900 Main Street, Riverside CA 92522.

banked water may alleviate overdraft conditions during droughts. Another form of conjunctive use is the so-called "In-lieu" in which surface water is used instead of groundwater, or water is exchanged, because of quality, for more appropriate use. For example, a domestic water purveyor may exchange non-potable water for potable water from an agricultural user.

Traditional stream-aquifer conjunctive use relies on hydraulically contiguous stream and aquifer. Environmental concerns, and regulations may reduce the attractiveness of conjunctive use projects within stream-aquifer systems. Many conjunctive use projects, such as aquifer storage recovery (ASR) projects, now involve noncontiguous aquifers that may be located a substantial distance from the stream-aquifer system.

Miller et al. (1993) reported that ASR can be used for long-term storage of potable water. Some ASR projects use available aquifer capacity to store potable water, even in aquifers containing non-potable water. Des Moines, Iowa may store treated surface water within a 610-meter deep aquifer (*Engineering News-Record,* Sept. 26, 1994). The stored potable water could be recovered during emergencies, such as the flood of 1993, which shut down the City's treatment plant for 19 days. There is concern that future or more stringent drinking water standards may impact long-term stored potable water in aquifers. Miller et al. (1993) reported on the fate of some disinfection by-products (DBP) contained in potable water during aquifer storage.

Conjunctive Use Modeling

The Santa Ana Regional Water Quality Control Board used a model based on *QUAL2E* (James M. Montgomery, 1990) to upgrade the basin plan objectives for the Santa Ana River. Nitrate concentrations within the river and some groundwater basins exceed standard, partially because of agriculture. Cunha et al. (1993) and O'Mara and Duloy (1984) modeled conjunctive use projects that included agricultural production. Young et al. (1986), and Muller and Male (1993) presented conjunctive use from legal and regulatory perspectives. El-Kadi (1989) identified the limitations of using existing watershed models to model conjunctive use. One such limitation is that some models emphasize surface flow more than groundwater flow. This limitation may not apply to conjunctive use models, in semi-arid areas, which emphasize groundwater flow.

Conjunctive use Projects in California

The motivation for developing conjunctive use projects in California include: the 1987-92 drought, cutbacks in quantity of imported water, increasing costs of imported water, the need to maintain stable water rates, stringent wastewater discharge requirements, adoption of state's guidelines for wastewater reuse, reduced number of available sites for surface reservoirs, the huge costs of developing new surface reservoirs, and environmental obstacles in getting approvals for new reservoirs.

Conjunctive use projects are being proposed in California to: increase sustainable yield from aquifer-stream systems, meet peaking demand, improve water quality, manage plume migration, manage fresh/salt water interface, reduce the need for expensive wellhead treatment facilities, reduce costs of pumping, and improve wildlife habitat. The SWRCB promotes conjunctive use as a means of diverting surface water and storing excess water in aquifers closer to point of use. The stored water can be used, during prolonged drought, in-lieu of water export from the Sacramento-San Joaquin Delta. SWRCB promotes long-term conjunctive projects so that the state can survive a drought of 10 to 15 years.

Examples of conjunctive use projects in California include: Orange County Water District (OCWD) Recharge Projects, Kern Groundwater Banking Projects, and the Monterey-Peninsula/Carmel River Projects. OCWD uses sand levees and inflatable rubber dams in the Santa Ana River to increase aquifer recharge. In 1993, OCWD replenished 17 million cubic meters of water worth an estimated $2 million (Markus et al., 1994). OCWD also injects reclaimed water to create a hydraulic barrier to salt water intrusion.

Institutional and Regulatory Constraints

Implementing conjunctive use projects sometimes requires that some existing institutional constraints be relaxed. New institutions may have to be created to fund and administer conjunctive use projects. For example, some regional and state agencies subsidize conjunctive use projects. Local agencies and regulators must cooperate to find innovative means for realizing the full benefits of conjunctive use.

California Water Code Sections 13550 and 13551 prohibit the use of potable water for non-potable uses if adequate reclaimed water is available at "reasonable cost." The code recognizes that the use of reclaimed water must not cause loss or diminution of existing water rights. However, in California, the water right to reclaimed water is still unsettled. Water rights permitees and licenses are now required to report periodically on the potential to use wastewater for all or part of their water needs. California Water Code Section 10750, otherwise known as AB 3030 allows an existing water agency to develop a groundwater management plan (GMP). An AB 3030 GMP can include control of salt water intrusion, replenishment of groundwater, wellhead protection, and the management of recharge areas.

During the 1991 and 1992 droughts, California established a Drought Water Bank. Many farmers pumped underlying groundwater to replace surface water transferred to the Bank. The Water Code now prohibits replacing transferred surface water with groundwater unless the use is consistent with an AB 3030 GMP, and will not result in long term overdraft of the basin. There are areas where existing environmental regulations conflict. For example, the United States Environmental Protection Agency rejected California's adoption of site specific standards for effluent dominated streams.

Some groups opposed components of some conjunctive use projects because of water quality. Tucson, Arizona elected to rely exclusively on groundwater because of taste problems associated with substitute imported surface water. The Upper San Gabriel Valley Municipal Water District proposed to annually recharge the aquifer underlying the Miller Brewing Company's plant in Irwindale, California with 20 million cubic meters of tertiary treated effluent. The Company sued and petitioned against the District.

Summary and Conclusions

Conjunctive use models will be used more extensively to increase reliability and yield of water supply projects. However, the feasibility of some projects may be affected by more stringent and forthcoming environmental regulations. For example, the pending Groundwater Disinfection Rule (GDR) may limit further development of conjunctive use projects that rely on aquifer-stream systems for domestic water supply. If the GDR is very stringent, future aquifer-stream system conjunctive models may incorporate viral and solute transport modeling to estimate biological and chemical purity of the aquifer.

APPENDIX - REFERENCES

Cunha, M.C.M.O., P. Hubert, and D. Tyteca. 1993. "Optimal management of a groundwater system for seasonally varying agricultural production." *Water Resources Research* 29(7):2415-2426.

El-Kadi, A.I., 1989. "Watershed models and their applicability to conjunctive use management." *Water Resources Bulletin* 25(1):125-137.

James M. Montgomery Inc. 1990. Nitrogen and TDS Studies, Santa Ana River Watershed.

Markus, M.R., C. A. Thompson, and M. Ulukaya, 1995. "Aquifer recharge enhanced with rubber dam installations." *Water Engineering Management*, 142(1),37-40.

Miller, C.J., L.G. Wilson, G.L. Amy, and Kay Brothers, 1993. "Fate of organochlorine compounds during aquifer storage and recovery: The Las Vegas experience." *Ground Water* 31(3):410-416.

Mueller, F.A. and J.W. Male, 1993. "A management model for the specification of groundwater withdrawal permits." *Water Resources Research* 29(5):1359-1368.

O'Mara, G.T. and J.H. Duloy, 1984. "Modeling efficient water allocation in a conjunctive use regime: The Indus Basin of Pakistan." *Water Resources Research* 20(11):1489-1498.

Young, R. A., J. T. Daubert, and H.J. Morel-Seytoux, 1986. "Evaluating institutional alternatives for managing an interrelated stream-aquifer system." *American Journal of Agricultural Economics*, 68(4), 787-797.

Selection of Modeling Tools for a Conjunctive Use Decision Support System

John C. Tracy, Assoc. Member ASCE[1]

Introduction

A variety of models have been developed to simulate many hydrologic and hydraulic systems. Typically, these models tend to focus on one specific element in the hydrologic cycle to simulate one or several problem variables. However, when conjunctive use systems must be simulated for planning or management studies, multiple hydrologic, hydraulic and sometimes management systems must be simultaneously simulated to similar degrees of accuracy to determine the state of the entire system. Few stand alone models have been created to simulate conjunctive use systems, thus simulations must be performed using selected models that can simulate each of the hydrologic components comprising the entire system. These results must then be integrated together to predict the response of the conjunctive use system to management decisions related to water use in the area. A critical phase in a project of this type is the selection of the models that will be used to simulate each hydrologic component. The purpose of this paper is to evaluate procedures as to how these modeling components may be selected.

Model Selection Criteria

Many criteria can be employed in the selection of models to be used in simulating a conjunctive use system. In general, these criteria can be categorized into addressing one of four concerns, these being: (1) Prediction accuracy; (2) model detail; (3) defendability; and (4) usability. Prediction accuracy is both a function of what a model predicts and how well the model predicts it. An increase in the **prediction accuracy** of a model entails a model that more specifically focuses on predicting the state variable of interest and predicts the state variable more accurately. This typically means that more general models that simulate a variety of state variables tend to produce a lower

---

[1]Assistant Professor, Department of Civil and Environmental Engineering, Crothers Engineering hall Box 2219, South Dakota State University, Brookings, SD 57007, (605) 688-5316.

prediction accuracy, while highly specific models tend to produce a higher prediction accuracy. **Model detail** relates the ability of a model to incorporate detailed information on a variety of physical processes in the model simulations. An increase in a model's detail leads many people to assume that the model's prediction accuracy will increase. However, this is only true if the increased level of model detail is directly related to the state variables of interest, and information on the physical processes being modeled can be obtained within the project's budget constraints. **Model usability** refers to the user friendliness of a modeling package. The user friendliness of a modeling package relates both to the ease with which necessary information can be used to develop the computer model and the ease with which model output can be synthesized into useful decision making information. Models that require extensive editing of input and output data to accomplish these tasks can be considered to have a low usability factor, and models that provide on screen digitization of input geometry and graphical output of results can be considered to have a high usability factor. **Model defendability** is related to the model documentation, history, frequency of use, and acceptability of use within the modeling profession. Typically, for a model to be highly defendable requires that it have an extensive history of use in the profession and have a high level of user support and maintenance. Ground water models such as MODFLOW (McDonald and Harbaugh 1988) would be considered highly defendable, while new research codes would be considered minimally defendable.

Personnel Influencing Model Selection

The criteria listed in the introduction seems straight forward enough when used as the basis for selecting a model or set of models to simulate the state of a surface-ground water system. However, what has to be kept in mind when using these criteria is that different personnel involved in the selection process can place a different value on each criteria. A variety of personnel are typically involved in any large scale modelling project. For the most part the personnel can be classified into four separate groups, these being: (1) **technical personnel**, which includes scientists and engineers who do the field and computer modeling work necessary to develop the model; (2) **end users**, who are the people that will use the model most frequently once it has been fully developed; (3) **interest groups**, comprised of people that are directly or indirectly affected by the decisions made from usage of the model; and (4) **decision makers**, comprised of people who are the overall model project managers and quite often represent the agency that is funding the modeling project.

Each of these groups tends to put a greater emphasis on different model selection criteria, and what follows is a subjective assessment of the ranking of the importance of each criteria for each group based on past experience. The most important criteria that the chosen models must possess for technical personnel is typically a fairly even weight between model prediction accuracy and model usability, with slightly more importance given to model accuracy. The third most important factor is the model detail and last is model defendability. This ranking should be somewhat expected since technical personnel feel more comfortable in the actual model development exercise, and once the model is developed they feel that the model is defendable, since they know that it is accurate. This makes model defendability a part of the model development excercise and not a separate model selection criteria, and hence a non-issue. The model detail is considered somewhat important, but not nearly as important as usability and accuracy since more model detail does not necessarily lead to more accurate model predictions, and many model details that only indirectly affect the state variable to be predicted can be approximated using simpler physical or empirical representations. The most important criteria for end users tends to be modeling detail, followed by defendability, usability and accuracy. This may appear somewhat surprising since one would assume that an end user would be more concerned with the ease with which a model could be used. However, it is possible that usability of the recently developed models are just assumed to be good, so other issues receive more concern. The placing of such emphasis on the modeling detail and defendability seems to result from the end user's understanding that decisions resulting from their model use could be questioned. Thus it would be very important for them to justify these decisions by showing they incorporated the highest possible degree of detail into the model and also used the most accepted models available. For interest groups, the most important criteria typically seems to be the modeling details, followed by defendability, prediction accuracy and model usability. The ranking appears to be quite similar to that of the end users for much the same reason as the end users. That is, the interest groups seem to think that including the highest level of model detail in widely accepted models would produce the most accurate results. Typically, for the decision making personnel the most important and overriding model selection criteria is its defendability, followed by almost equal weights of model accuracy, model detail and model usability. This ranking seems to result from the decision makers need to justify the selected models to an extremely wide audience, who view their decisions in a highly political atmosphere. Thus, their main focus appears to be assuring all interests that the most current and widely accepted modeling packages are being used, reemphasizing their objectivity in the model selection process.

Placing a quantitative value on each criteria for each group in the selection process is a somewhat subjective procedure. However, based on the above assessments and using a scale between 0 (low importance) and 10 (high importance), and the sum of the criteria ratings for each group must equal 10, the relative value of each selection criteria for each group can be summarized in Table 1.

Relationship Amongst Groups

In the model selection phase of a project, there are a variety of ways the groups involved in the model selection process could interact. Many different approaches have been explored and prcedures have been developed using tools such as STELLA (e.g. see Palmer and Keyes, 1992) to help in resolving water related issues. However, for many new modeling project supported on a state or regional level the project interaction seems to be centered primarily around the decision makers. That is the decision makers have a strong interaction with all groups, while the technical personnel have a weak interaction

Table 1. Relative Ratings of Selection Criteria Importance.

Criteria	Technical Personnel	End Users	Interest Groups	Decision Makers
Accuracy	4.0	0.5	1.0	1.0
Detail	1.5	5.0	5.0	1.0
Usability	4.0	1.5	1.0	1.0
Defendability	0.5	3.0	3.0	7.0

with the end users, and very little interaction with interest groups. Note, this is only a subjective assessment by the author (a technical person) based on previous discussions with technical and decision personnel. Personnel involved in specific project that can be classified under the remaining groups may have a different perspective on these relationships. None the less, for a process where the decision makers are the focal point of the model development discussions, they would obviously have the most influence in the model selection process. Using a subjective weight to represent each groups influence on the selection process, the decision personnel would receive about a 0.4 (since all information flows through them) the technical and end user personnel would receive about 0.25 each (since there is some degree of independent interaction between them) and the interest groups would receive about a 0.1. To determine the relative overall importance of each criteria to be used in the final model selection, a weighted average for each criteria could be developed, and the resulting average importance of each criteria for a model selection process as described above would be: Prediction Accuracy = 1.62; Model Detail = 2.52; Model Usability = 1.88; and Model Defendability = 3.98.

Summary

For this type of model selection process, the above rankings would end up placing defendability and modeling detail much higher than predictive capability and usability. This would result in the models selected for simulating a conjunctive use system producing a highly detailed and defendable modeling system, sacrificing model usability and to some degree predictive accuracy. Use of different model selection processes could result in changes in the model selection influence factors. This in turn could lead to a completely different ranking of the final model selection criteria, and hence producing a different conjunctive use modeling system.

References

Keyes, A. M. and Palmer, R. N., 1992. "Use of interactive simulation environments for the development of negotiation tools," Proceedings of the 1992 Annual Water Resources Planning and Management Conference, ASCE, pgs. 68-73.

McDonald, M. G. and A. W. Harbaugh, 1988. "A modular three-dimensional finite-difference ground-water flow model," *Techniques of Water--Resources Investigations*, 06-A1, U.S. Geological Survey.

Pilot Program To Demonstrate Cost-Effective Solution To Groundwater Overdrafting In Ventura County, California

Patrick J. Reeves[1], Member, ASCE; Frederick J. Gientke[2], Member, ASCE.

Abstract

To combat groundwater overdrafting and seawater intrusion on the Oxnard Plain, the United Water Conservation District has embarked on a Pilot Program to demonstrate the cost-effective use of gravel mine pits for use as recharge basins and storage reservoirs.

Introduction

The Oxnard Plain, in Ventura County, California is suffering from groundwater overdrafting. The Oxnard Plain is one of the most fertile and productive agricultural areas in the nation and also provides a home to over 200,000 residents. A major portion of this area is located within the boundaries of the United Water Conservation District (District). The District is comprised of 86,000 hectares (212,000 acres), of which approximately 33,000 hectares are irrigated. The District boundaries and vicinity are shown in Figure 1.

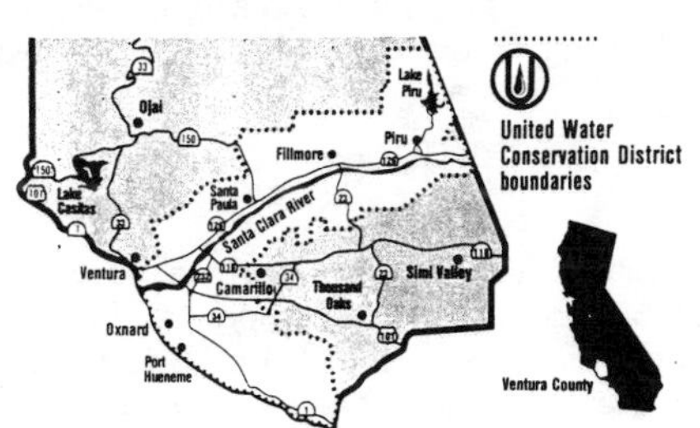

Figure 1 - District Boundary and Vicinity

----

[1] United Water Conservation District, 725 E. Main St., #301, Santa Paula, CA, 93061

[2] Penfield & Smith, 2530 Financial Square Drive, #110, Oxnard, CA, 93030

The Oxnard Plain is underlain by the Fox Canyon Aquifer which is being annually overdrafted by over 37 million m^3 (30,000 acre-feet (A.F.)) per year. These excess extractions which exceed the safe yield of the groundwater basin cause numerous problems including; a) increased pumping costs, b) dewatering of aquifers, c) water well abandonment or expensive lowering costs and d) seawater intrusion.

To resolve these problems the State required that a Groundwater Management Agency (GMA) be formed to monitor all groundwater extractions and to develop and implement a plan to reduce the basin overdrafting. Baseline groundwater extractions were developed by averaging the extractions between 1985 and 1989. The GMA then established goals within an ordinance to reduce groundwater extractions by 5% increments every 5 years. By the year 2010, there should be 25% less water extracted than the baseline, or approximately 48 million m^3 (39,000 A.F.) per year. If a groundwater pumper exceeds their extraction goal, they can be fined up to $162 per thousand m^3 ($200 per A.F).

Alternative Water Supplies

In order to combat the groundwater overdrafting and seawater intrusion, water purveyors in the region have been working together to develop alternative water supplies. The County of Ventura has maintained an entitlement of 25 million m^3 (20,000 A.F.) per year from the State of California's "State Water Project". Unfortunately the state facilities are located many miles away. To implement the project, it has been estimated will cost between $713 to $972 per thousand m^3 ($879 to $1,199 per A.F.) delivered. In addition, during the recent California drought, it was demonstrated that the State project cannot always deliver it's contract entitlements, therefore the project reliability was questioned by the general population. Several studies have also been completed regarding the feasibility of seawater desalination. These reports cite that this supply would cost on the order of $1,702 per thousand m^3 ($2,100 per A.F). These two alternative supply projects are both extremely costly to implement, meanwhile the groundwater overdrafting continues.

Project Concept

Along the Santa Clara River there are many acres of mineral resources which are extracted by sand and gravel companies. The County of Ventura allows the extraction down to a level of 1.5 m (5 feet) above historic high groundwater. After the sand and gravel is extracted, a layer of top soil is placed in the mined area or pit, and agricultural production is established.

The District has been conserving water resources for over 60 years and has developed many facilities to combat groundwater overdrafting and seawater intrusion on the Oxnard Plain. The District's major facilities include a 111 million m^3 (90,000 A.F.) reservoir, three spreading grounds with a combined acreage of 113 hectares (280 acres) and a recharge capacity of 1.1 million m^3 (900 A.F.) per day, the Freeman Diversion Structure in the Santa Clara River, a surface water delivery system, and the Oxnard-Hueneme delivery system.

The Santa Clara River flows through Ventura County and discharges to the ocean between the cities of San Buenaventura and Oxnard. The District has constructed a river diversion facility and has the right to continuously divert 10.62 m^3 (375 cubic feet) per second provided that the total diverted volume does not exceed 147 million m^3 (119,000 A.F.) per year. Currently, the District rejects inflow from this facility whenever; a) the river flow exceed 566 m^3 (20,000 C.F.S.), b) the suspended solids in the river exceed 10,000 part per million, c) the fish screens become plugged, or d) there is inadequate capacity in the recharge facilities to accept the excess flow not being delivered into the pipeline delivery systems.

Due to the proximity of the gravel pits to the District's facilities, the District commissioned an engineering feasibility report prepared by Penfield & Smith in 1992. The report found that these pits totaled over 202 hectares (500 acres) and could be developed with a maximum storage capacity of 23.4 million m^3 (19,000 acre-feet). The pits could act as either recharge basins or reservoir storage and could harvest 18.5 to 24.7 million m^3 (15,000 to 20,000 A.F.) per year of additional supply from the Santa Clara River. The anticipated costs for developing this additional supply will range between \$56.75 and \$97.28 per thousand m^3 (\$70 and \$120 per A.F.) depending on the cost of money (interest) and right-of-way acquisition costs. The flow schematic of the proposed ultimate project is shown in Figure 2. However, prior to committing significant dollars, the District developed a pilot program to validate this concept.

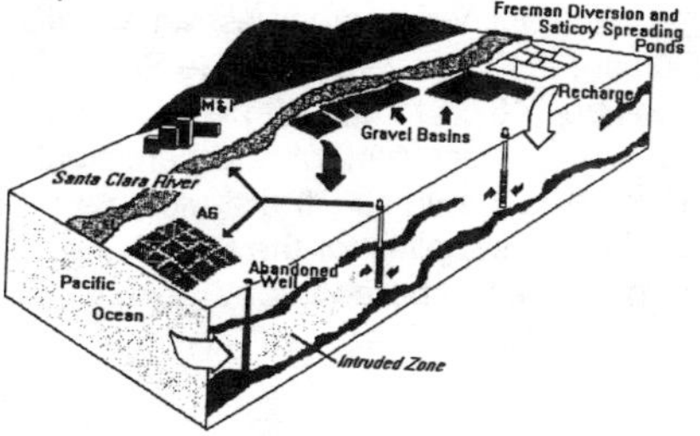

Figure 2 - Oblique Flow Schematic

Pilot Program

The Fox Canyon Seawater Intrusion Abatement Project Pilot Program was constructed in late 1994. The construction cost was approximately $1.6 million and the land acquisition costs were approximately $1.8 million. The project includes the construction of new and improved canals to convey 10.67 m^3 per second (375 C.F.S.) through the existing Saticoy Spreading Grounds. The canal included new control structures for eleven of the ponds within the spreading grounds. Flow meters were constructed to monitor the flow to both the spreading grounds and the Noble Pit. The Noble Pit consists of 53.4 hectares (132 acres), and will have a maximum storage capacity of 1.5 million m^3 (1,200 acre-feet). The Noble Pit is connected to the Saticoy Spreading Grounds via a triple 1.22 m (48-inch) diameter pipe crossing which had to be bored and jacked under State Highway 118 (Los Angeles Avenue).

The Pilot Program could not be installed without complying with the close scrutiny of the California Environmental Quality Act provisions. In addition, significant coordination and cooperation with the gravel mining company who owned the pit, and the adjacent Rose Pit was also required. Monitoring groundwater wells are being constructed around the periphery of the project to document the efficiency and impacts of the Pilot Program.

By having the additional capacity of the Noble Pit, the District can convey the remainder of their full annual river flow entitlement to this facility. It is anticipated that the Noble Pit will recharge a minimum of .91m (3 feet) per day for the first several years of operations, and eventually begin to seal. At that time it will be converted into a reservoir, and the water can be conveyed to future gravel basins, or into another part of the District's surface water delivery system.

Summary

The Pilot Program is anticipated to harvest an additional 37 to 49 million m^3 (3,000 to 4,000 acre-feet) per year from the Santa Clara River. The unit cost of this water will be $48.64 to $64.86 per thousand m^3 ($60 to $80 per acre-foot). If the pilot project performs as anticipated, the District will proceed with additional EIR studies. If the ultimate project is successful, significant additional water supplies for the County will be obtained and the seawater intrusion problem will be abated.

Implementing an Artificial Recharge Program in an Alluvial Ground Water Basin in Southern California

John S. Hurlburt, P.E., M. ASCE[1] and
Keith A. London, P.E., M. ASCE[2]

Abstract

As part of an ongoing ground water management program, Rancho California Water District (RCWD) is planning to artificially recharge up to 37 million cubic meters (30,000 acre-feet) of groundwater annually through a series of recharge basins located in the Valle de los Caballos (VDC) area of the upper Pauba Valley.

The program concept is to purchase untreated, imported water from the Metropolitan Water District of Southern California (MWD) during the winter months for artificial recharge and subsequent recovery during the summer months. The system will also be able to percolate natural runoff collected upstream in Vail Lake. RCWD anticipates the artificial recharge program will provide up to 25% of the District's ultimate water supply requirements.

Design of the facilities included geohydrological explorations, extensive ground water modeling, HEC-2 modeling, sediment transport analyses, hydraulic analyses, and various soils analyses. Ground water modeling and engineering were coordinated to produce the final design. Initial modeling was used to generate gross design criteria and a basic design, which led to more detailed modeling efforts and detailed design. Modeling was conducted in three phases with each phase based on more refined engineering and providing more detailed design criteria. This paper focuses on how the modeling was used to develop the engineered design.

---

[1]Planning & Capital Projects Manager, Rancho California Water District, 42135 Winchester Road, Post Office Box 9017, Temecula, CA 92589-9017

[2]Principal, Camp Dresser & McKee Inc., 1925 Palomar Oaks Way, Suite 300, Carlsbad, CA 92008

Introduction

The Rancho California Water District (RCWD) is located near the town of Temecula in Riverside County, California. Significant geographic features in the area are the Santa Margarita River, Temecula Creek and Murrieta Creek. The two creeks merge to form the Santa Margarita River at the Temecula Gorge. Temecula Creek flows through the Pauba Valley and the Valle de los Caballos (VDC) is located in the upstream area of the Pauba Valley. Vail Lake is located on Temecula Creek, upstream of the VDC area, and captures surface runoff from 829 square kilometers (320 square miles).

RCWD overlies and manages one of the largest ground water basins in Southern California. The basin covers over 230 square kilometers (90 square miles) of area. Ground water is produced from saturated alluvial deposits of Pliocene to Halocene age which reach typical depths of 300 meters (1,000 feet). The Pleistocene Pauba formation comprises the shallow aquifer which overlies the pliocene Temecula Arkose. The ground water basin is surrounded and underlain by non-water bearing basement complex rocks of Pliocene through Triassic age. Ground water occurs throughout the basin with movement towards the basin's primary outlet, the Temecula Gorge. The basin provides an annual sustained yield of approximately 37 million cubic meters (30,000 acre-feet). Development of ground water as a resource has been ongoing since the mid-1800s. The focus of the development then and now is the Pauba Valley which contains the coarsest alluvial deposits and consequently the highest yielding aquifers. Within the Pauba Valley, the VDC area has always been recognized as the best location for artificial recharge of the basin's aquifers (see Figure 1).

The VDC area was identified on maps as early as 1917 as "spreading grounds" and was used continuously for that purpose without improvements until the mid-1970s. In 1978, RCWD acquired the VDC property for a conjunctive use project utilizing recharge from both Vail Lake as well as imported water through the MWD system. Vail Lake lies upstream of the VDC area and is the historic source of water for the spreading grounds. Temporary basins were constructed in 1979 and were used to percolate Vail Lake water until 1992 when engineered recharge facilities totaling 53 hectares (131 acres) were constructed. The recharge system also includes 11 kilometers (7 miles) of 1.1 meter (42-inch) transmission main and five recovery/conditioning production wells.

Design Process

While property had been secured for the project years previously, limited physical percolation testing had been performed resulting in only crude estimations of the expected recharge performance. As described in the

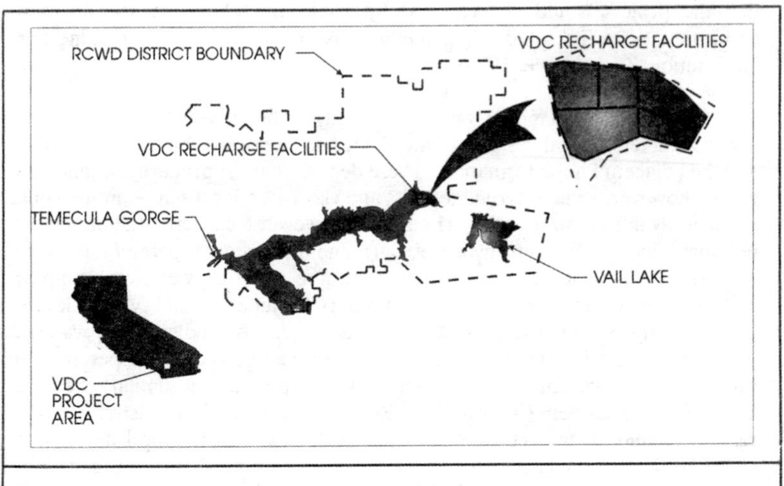

LOCATION OF VDC RECHARGE PROJECT

FIGURE 1

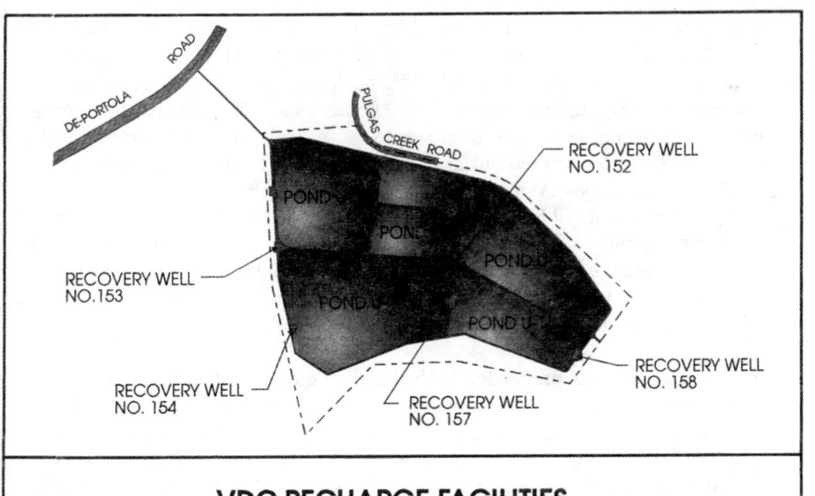

VDC RECHARGE FACILITIES

FIGURE 2

introduction, significant facilities (about $12M in construction) would be required to deliver and recharge water so that it was critical 1) to demonstrate that the project would be economically viable and 2) to size the facilities accurately. The first phase of modeling was directed towards providing this information (Geoscience, 1989).

The VDC Recharge model was a single layer model which simulated the unconsolidated alluvial deposits of the Temecula Creek (Pauba Valley) channel and the adjacent Pauba formation. These deposits consist primarily of sand and gravel; however, some interbedded silts and clay cause local semi-confinement, particularly in the lower areas. The model, however, considered the aquifer to be unconfined. Various hydrogeologic and operational parameters were necessary to simulate artificial recharge and the effect on well field pumping and ground water flow. The parameters included initial water levels, transmissivity, storativity, pumping and recharge. As indicated previously, percolation tests had been performed in the recharge areas and percolation rates were known for several other recharge projects in similar southern California alluvial deposits. RCWD also operated several production wells in close proximity to the recharge sites and had constructed several monitoring wells within the recharge sites. The model parameters were estimated or derived from the available physical data. Operationally, the recharge water was assumed to be available during the winter, thus recharge would occur over the eight-month period from October to May. Existing water production wells located within the model limits were operated per their normal schedule.

The percolation tests completed for the VDC area had indicated that percolation rates of 0.46 meters per day (1.5 feet per day) could be expected. However, the maximum potential recharge is subject to both percolation rate and the available aquifer storage. The Phase I modeling verified the storage issue. The recharge basin was shown to be severely limited by ground water mounds rising to the land surface, creating a water logging condition. The model predicted, however, that if depth to water levels could be lowered by 12 to 15 meters (40 to 50 feet) prior to the start of the recharge cycle, the annual recharge volume could exceed the target volume of 37 million cubic meters (30,000 acre-feet). The modeling verified the recharge goals but identified the need for a number of dewatering wells to "condition" the basin before the recharge cycle began in the Fall.

The design engineer proceeded to design the delivery system and to develop a basic layout of the percolation basins which optimized recharge areas (Figure 2). At this point a second phase of modeling was required to design the distribution and control facilities to each basin (Geoscience, 1990). The 53 hectares (131 acres) of recharge area were comprised of seven separate basins of differing area and soil conditions. As the water distribution of 1.7 to 2.3 cubic meters per second (60-80 cfs) was to be controlled automatically via telemetry, it was essential to accurately estimate recharge per basin. The modeling was completed using the same model as Phase I, except that more refined criteria was utilized. Internal basins were evaluated based on soils, percolation rates (estimated), proximity to impermeable boundaries, and probable problems once recharge began. Recharge problems (reduced percolation rates over time) had been observed in other recharge projects in Southern California so the Phase II modeling included allowances for these problems. Finally, the Phase II modeling was used to simulate basic operational strategies and verify the effectiveness.

The third phase of modeling (Geoscience, 1991) involved locating the dewatering wells to provide the best dewatering capacity and further evaluation of operational scenarios. Initial expectations for water production from the wells were on the order of 190 to 250 liters per second (3,000 to 4,000 gallons per minute). However, this proved to be over-optimistic. The areal extent of the semi-confining layers of silt and clay proved to be much further upstream than anticipated. The actual dewatering rates averaged about 125 liters per second (2,000 gallons per minute). The reduced flows caused the number of dewatering wells to increase from three to five.

Conclusions and Operational History

The VDC Recharge Project is an excellent example of integrating geohydrologic modeling with conventional civil engineering to produce a successful design. Subsequent to the construction of the recharge basins, RCWD was required to complete a pilot scale project to demonstrate compliance with the Surface Water Treatment Rule. The study involved percolating Vail Lake water in a 2 hectare (five-acre) sub-basin over a period of three months. The percolation rates and mounding effects mirrored the model predictions. RCWD expects to place the facility in full operation in 1995.

References

GEOSCIENCE Support Services, Inc. October 1989. *Present and Potential Artificial Recharge Capability Pauba Valley Facilities*. Rancho California Water District

GEOSCIENCE Support Services, Inc. August 1990. *VDC Recharge Project Preliminary Design Report*. Rancho California Water District

GEOSCIENCE Support Services, Inc. April 1991. *Location of VDC Recovery Wells*. Rancho California Water District

Use of Capture Zone Analyses in Delineation of Wellhead Protection Areas

Francis K. Cheung, M. ASCE, Sujit K. Bhowmik, M. ASCE and W. Winston Chen, M. ASCE[1]

Abstract

This paper summarizes a case study on delineation of wellhead protection area for a proposed wellfield in Florida. Capture zone analyses for the proposed wellfield were performed using the USGS MODFLOW model and the EPA WHPA model based on a 10-year travel period. Important hydrogeological factors such as aquifer flow gradient and direction, storage coefficient, transmissivity, vertical recharge and interference with nearby pumping wells that affect the capture zones of individual wells in the wellfield are discussed. Finally, buffer zone requirements for delineation of wellhead protection areas and limitations of capture zone analyses in establishing wellhead protection areas are discussed.

Introduction

Because of ever increasing population growth and diminishing rural areas in Florida, adequate protection of groundwater resources to ensure a safe, economical and long term supply of drinking water is a major concern. Proper planning and design of wellfields have become a key issue in regional development. Determination of setback distances of new wellfields from potential pollution sources is an important consideration in siting and planning of any wellfield. Setback distances stipulated in regulatory guidelines are often based on fixed radii and, in many cases, ignore site-specific hydrogeological factors. Based on a case study and using capture zone analyses, this paper addresses some site-specific hydrogeological factors that could influence the siting and design of a wellfield.

Wellfield Description and Aquifer Characteristics

The wellfield discussed in this paper is a proposed wellfield that covers an area of over 10,000 acres and consists of 27 wells with a combined withdrawal rate of 24 mgd from an underlying semi-confined aquifer. The average design pumping rate from each

---

[1] Senior Project Engineer, Project Engineer and Project Engineer, respectively, Ardaman & Associates, Inc., 8008 South Orange Avenue, Orlando, Florida 32809.

well is approximately 0.89 mgd. Because the proposed wellfield is located adjacent to an industrial facility, there was concern that potentially contaminated groundwater could migrate toward the wellfield. To protect the water supply for the wellfield, the local authority would like to identify a wellhead protection area for the wellfield within which any industrial activity would be prohibited or regulated.

The semi-confined aquifer that supplies water for the proposed wellfield has a thickness of 800 feet, a transmissivity of 43,200 square feet per day, and a storage coefficient of 0.05. The aquifer receives recharge from the surficial aquifer above at a rate of about 6 inches per year. The vertical leakance of the confining layer was determined to be 6.8×10^{-4} gpd/ft^3 or 0.00009/day. Both the direction and gradient of natural groundwater flow in the semi-confined aquifer vary across the proposed wellfield site and from dry to wet seasons. Generally, the natural aquifer flow occurs from west to east under an average gradient of approximately 0.0005.

Capture Zone Analyses

Although the fixed radius approach specified in many regulatory guidelines provides a simple and inexpensive method for protection of a water supply source, this type of method could result in overprotection of some areas that could otherwise be used for economic development. On the other end of the spectrum, this type of approach could result in underprotection of groundwater resources in some highly vulnerable aquifers. In recent years, land use planning has been leaning towards a resource protection approach that recognizes site-specific characteristics and focuses on providing protection for groundwater that contributes to a wellfield. Under this approach, the areal extent of the aquifer that would yield water to a wellfield for a specified travel period, typically in the range of 10 to 20 years, would be identified, and the land surface area within that extent plus a buffer zone would be designated as the wellhead protection area within which any industrial development or activity would be prohibited or regulated.

For this case study, the United States Geological Survey MODFLOW model and the Environmental Protection Agency WHPA model were used in conjunction to delineate the groundwater capture zones of withdrawal wells in the proposed wellfield. The MODFLOW model was used to determine water level drawdowns during wellfield operation and to produce head input data for the WHPA model. The WHPA model was used to generate pathlines and capture zone for each withdrawal well based on a 10-year travel period. Without the head data produced by the MODFLOW model, the WHPA model could only be used to analyze a homogeneous and uniform aquifer. As indicated previously, natural aquifer flow varies in both direction and gradient across the subject wellfield site, and from dry to wet seasons.

Based on the aquifer properties indicated previously, the natural potentiometric surface elevation in the project area and the magnitude of drawdown predicted using the MODFLOW model, the capture zones of the wells in the proposed wellfield for the wet and dry seasons are superimposed together on Figure 1. The capture zones shown on the figure are the combined effects of many site-specific hydrogeological factors, including aquifer flow direction and gradient, transmissivity, storage coefficient, time of travel, and groundwater withdrawals in the site vicinity.

The orientation of the capture zone of a withdrawal well is primarily governed by the direction of natural aquifer flow. If there is no aquifer flow and the aquifer is isotropic, the boundary of the capture zone would be circular. In cases where the natural aquifer flow gradient is significant, the capture zone of a withdrawal well will have an elongated shape with the tail of the capture zone located on the upgradient side of aquifer flow. On the downgradient side, the capture zone will extend a shorter distance because it is more difficult to capture groundwater against the direction of natural aquifer flow.

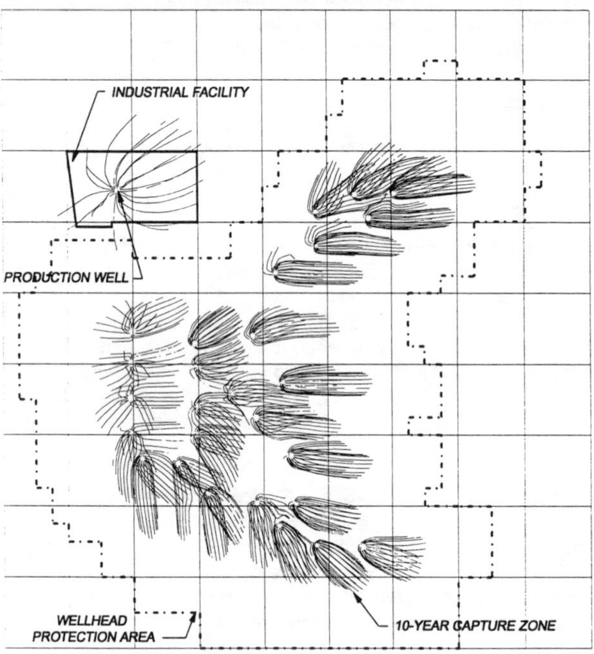

Figure 1. Capture Zones and Wellhead Protection Area

For a fixed setback distance, a pollution source that is located on the upgradient side of aquifer flow would pose a greater threat than one that is located on the downgradient side. This effect becomes more pronounced with increasing aquifer flow gradient. The length of the capture zone on the upgradient side increases with increasing aquifer flow gradient. The width and downgradient length of the capture zone, however, decrease with increasing aquifer flow gradient.

The transmissivity and storage coefficient of an aquifer also play important roles in determining the size of the capture zone. The length of the capture zone on the upgradient side increases with increasing aquifer transmissivity and decreasing storage coefficient. Conversely, the width and downgradient length of the capture zone decrease with increasing aquifer transmissivity and decreasing storage coefficient.

The size of the capture zone is also a function of travel time. The longer the time, the longer the upgradient capture zone will be. The width of the capture zone and its length on the downgradient side, however, are independent of the time of interest.

Groundwater pumping in the vicinity of the wellfield, interference among withdrawal wells in the wellfield and hydrogeological conditions of the surrounding areas could also have a significant influence on the size and shape of the capture zone of a withdrawal well. Nearby pumping could significantly alter aquifer flow in the vicinity of a withdrawal well. Hydrogeological features such as groundwater divides or surface waters could limit the size of the capture zone of a withdrawal well.

Effects of Production Well Pumping and Vertical Recharge

As indicated previously, an industrial facility is located north of the proposed wellfield. Groundwater has been withdrawn from an on-site production well at an average rate of 6 mgd to supply the water need for the facility. To determine if the production well should be included in the capture zone analyses, the magnitude of drawdown at the three nearest gaging stations on the potentiometric surface map were analyzed. Based on the drawdown analyses, it was concluded that the production well pumping had no effect on the potentiometric surface contour maps that display regional aquifer flow. Although the production well has minimal influence on regional aquifer flow, it could have a significant local impact on the capture zones of withdrawal wells located closest to the industrial facility. Accordingly, the production well was included in the wellhead protection modeling to determine the capture zones for the proposed wellfield. The analyses indicated that the production well actually prevents movement of potentially contaminated groundwater from migrating to the wellfield.

For this case study and under most situations, vertical recharge from the surficial aquifer does not affect the capture zone of a withdrawal well significantly unless the time of interest is so long that the capture zone extends close to the boundary of the cone of influence of the withdrawal well. Although vertical recharge rate could affect the magnitude of drawdown substantially, the velocity of water at distances close to a withdrawal well remain relatively constant for varying recharge rates.

Delineation of Wellhead Protection Area and Buffer Zone Requirements

To account for variabilities in aquifer characteristics, a buffer zone was added to the boundaries of the capture zones to obtain the boundary of the wellhead protection area. For this case study, the local authority has adopted the following buffer zone requirements: (i) a 1.0-mile buffer zone on the upgradient side of the capture zones; (ii) a 0.25-mile buffer zone on the downgradient side of the capture zones; (iii) a 0.5-mile buffer zone on the two sides of the capture zones. Using these buffer zone requirements, the delineated wellhead protection area boundary for the proposed wellfield is shown in Figure 1.

Although adoption of a rigid set of buffer zone requirements is convenient, such requirements may not be applicable for all situations. As an example, if the aquifer transmissivity is the predominant unknown factor in the capture zone analyses and a large buffer zone was used on the upgradient side of the capture zone to account for the possibility that the aquifer transmissivity could be greater than expected, use of a correspondingly large buffer zone on the two sides and on the downgradient side of the capture zone would be inappropriate as increased transmissivity would actually decrease these dimensions. Similarly, if time of travel is the only element of uncertainty, a buffer zone on the two sides and on the downgradient side of the capture zone would not be necessary.

Limitations of Capture Zone Analyses

Although capture zone analyses provide a rational approach in protecting a water supply source, they do have their limitations in delineation of wellhead protection areas. First, such analyses simulate only time-related groundwater movement to a withdrawal well. Movements of contaminants, which depends on other variables such as dilution and geochemical attenuation, are not considered. Second, for the case of semi-confined aquifers, capture zone analyses assume that any possible contaminated surface pollution source would take no time to reach the deep aquifer, while in reality water from a potential contamination source on land surface must first travel through the surficial aquifer and confining unit before reaching underlying semi-confined aquifers. Third, a well-designed containment system for a significant potential pollution source located within a well capture zone could present little threat whereas a poorly-designed facility located outside the capture zone may eventually impact the wellfield.

Conclusions

The capture zone analyses presented herein provided a rational approach in delineation of the wellhead protection area for this case study, and indicated that the existing nearby industrial facility can co-exist with the proposed wellfield.

Site-specific hydrogeological factors often dictate the behavior of a wellfield. Capture zone analyses provide a valuable tool in delineation of wellhead protection areas. Good data should be collected and sensitivity analyses should be performed using the probable ranges of aquifer parameters to identify the capture zones and to avoid an overly conservative buffer zone. The limitations of capture zone analyses should be kept in mind while delineating wellhead protection areas.

Conjunctive Use Planning in the Tanshui River Basin, Taiwan

Nien-Sheng Hsu[1], Associate Member, ASCE, Ming-Jame Horng[2], Chian Min Wu[3], Affiliate Member, ASCE, William W-G. Yeh[4], Fellow, ASCE, and Jan-Tai Kuo[5], Member, ASCE

Abstract

In this paper, a simulation model is developed for conjunctive use planning for the Tanshui river basin. The overall simulation model consists of a surface water model and a groundwater model. The surface water model is basically a water-budget-accounting type of model with a special function for water right simulation and is used in the upper portion of the basin. The groundwater model, a two-layered flow model, is used in the lower portion of the basin, i.e., the Taipei basin. The groundwater model computes water level changes due to pumping and recharge as estimated independently and/or by the surface water model for various management alternatives. The developed simulation model has been used to investigate the effects of the construction of Feitsui Reservoir on streamflows in the Tanshui river and the groundwater levels in the Taipei basin..

Introduction

Over the last forty years, growing industrial and urban developments in the Taipei basin, a major aquifer in the Tanshui river basin, have placed severe demands on local water supplies. Since the construction of Shihmen Reservoir in 1964, shortage in water supply has been alleviated. However, over pumping of groundwater occurred prior to 1970's, and, as a consequence, groundwater level dropped by as much as 40 meters at certain locations. This has resulted in serious land subsidence problems as well as deterioration of groundwater quality. In 1974, a regulation was enforced by the government to completely stop pumping of

---

[1] Assoc. Prof., Dept. of Civ. Engrg., Nat. Taiwan Univ., Taipei, Taiwan, ROC.

[2] Engineer, Water Resour. Planning Commission, MOEA, Taipei, Taiwan, ROC.

[3] Chairman, Water Resour. Planning Commission, MOEA, Taipei, Taiwan, ROC.

[4] Professor, Dept. of Civ. and Environ. Engrg., UCLA, Los Angeles, CA 90095.

[5] Professor, Dept. of Civ. Engrg., Nat. Taiwan Univ., Taipei, Taiwan, ROC.

groundwater, and the groundwater level has been rising gradually since then. In an effort to remedy the adverse effect on the underground engineering construction due to rising water level and to efficiently utilize the total water resources, the Water Resources Planning Commission (WRPC), a governmental agency, has been investigating various water resources management alternatives, including conjunctive use of surface water and groundwater. As a prerequisite to evaluating any proposed alternative, a set of mathematical models has been developed to assist in decision-making.

Model Structure

The Tanshui river basin in the northern part of Taiwan is a major political and population center in Taiwan, the Republic of China, and has a drainage area of 2,762 km^2 and a population of 7,344 million in 1989. In the Tanshui river basin, the major aquifer (the Taipei basin) is located in the lower portion of the basin. Based upon its geographical and hydrogeological conditions, the Tanshui river basin is divided into an upper portion and a lower portion. A flow model for groundwater level simulation in the Taipei basin has been developed by Hsu and Lin (1988) and Hsu et al. (1989, 1990, 1991 and 1993). On the other hand, a water-budget-accounting type of model has been developed by Hsu (1992) for surface water-right simulation. By coupling the surface water model and the groundwater model, the overall simulation model can be used for conjunctive use planning of surface and groundwater water. Features of the surface water model and the groundwater model are summarized and briefly described below.

Surface Water Model

A water right simulation subroutine is added to the water budget accounting model previously developed by WRPC (1991). In processing water rights, a direct method is used. This method is more efficient than the commonly-used iterative approach.

Groundwater Model

Based upon the results from a detailed analysis of the geological data available, the Taipei aquifer was conceptualized as a two-layered system, consisting of an upper unconfined aquifer and a lower semi-confined aquifer. The two aquifers are separated by an aquitard. Vertical leakage takes place between the two aquifers due to difference in head. The upper layer can be divided into three subregions. Two of the three subregions are considered as unconfined aquifers which receive most of the natural recharge, and the third subregion is treated as a semi-unconfined aquifer which receives a limited amount of recharge.

The governing equations of the conceptual model of the Taipei aquifer system consist of an equation for each of the upper and lower layers. The upper layer is assumed to be an inhomogeneous, isotropic, and leaky unconfined aquifer. On the other hand, the lower layer is assumed to be an inhomogeneous, isotropic, and leaky confined aquifer. The governing equations are solved numerically using the Galerkin finite element formulation and the Crank-Nicholson scheme.

Model Calibration

The overall simulation model is calibrated separately by comparing model outputs with historical measurements of streamflows and groundwater head values.

Surface Water Model

Computed downstream flows were compared with actual (measured) flows at the four downstream locations corresponding to the desired flow nodes: Wutu, Hsiulung, Chihtse and Sanying for the water years 1986 and 1987. For these two years, the computed results for Wutu are almost exactly equal to the measured values. This can be explained by the fact that no reservoir is present in the Keelung system, so that the model to a large extent merely reproduces the measured inflows which were provided as input data to the model. This is not the case for the Hsiulung, Chihtse and Sanying stations, which are downstream of Feitsui and Shihmen reservoirs, respectively. The differences which were already observed between simulated and measured reservoir operations directly translate into the differences observed in the downstream flow. Given this explanation, the computed results seem to confirm the general flow characteristics of the river systems within reasonable accuracy limits.

Groundwater Model

The developed two-layered groundwater model was calibrated by a systematic trial-and-error procedure to obtain the parameter values in the model. The comparison between the computed and measured head values at 16 observation wells shows an acceptable match. Model validation was also performed by comparing the water budget obtained from the model calibration and that from the system identification (Hsu et al., 1993).

Model Application

The developed simulation model has been used to investigate the effects of the construction of Feitsui reservoir, i.e., the current condition, on the streamflows at the three stations in the Tanshui river basin and the groundwater levels at 16 observation wells in the Taipei basin. The hydrology of relatively dry period, from 1973 to 1982, was selected and assumed to be repeated at the beginning of 1991. Water shortages were analyzed under various conditions for the two cases: with Feitsui reservoir in operation and without Feitsui in operation. The corresponding streamflows at Hsiulung station and the contours of head values for both cases were investigated. For the case with Feitsui reservoir in operation, the results show: (1) less shortage in water supply; (2) higher flows in low flow seasons and lower flows in high flow seasons; and (3) higher groundwater levels in the Gingmei formation. These results are evident because of the carry-over effects of the reservoir.

Concluding Remarks

An overall simulation model has been developed for conjunctive use planning for the Tanshui river basin. The model consists of a surface water model and a groundwater flow model. The simulation model was calibrated individually. The results have shown a reasonable match between the observed and simulated streamflows at three gaging stations and groundwater head values at 16 well locations. Validation of the groundwater model was also performed by comparing the water budget obtained from model calibration and that from the system identification.

The calibrated simulation model has also been used to investigate the effects of the construction of Feitsui reservoir on streamflows in the Tanshui river and the groundwater levels in the Taipei basin. The results clearly demonstrated the importance of reservoir operation.

References

Hsu, N.S., and Lin, K.F. (1988). "A study on the management of groundwater quality in Taipei Aquifer (I)." *Project completion Rept.,* Prepared for the Water Resour. Planning Commission, Dept. of Civ. Engrg., Nat. Taiwan Univ., Taipei, Taiwan, ROC (in Chinese).

Hsu, N.S., Yeh, W.W-G., Kuo, J.-T., and Jeng, M.C. (1989). "A study on the management of groundwater quality in Taipei Aquifer (II)." *Project completion Rept.,* Prepared for the Water Resour. Planning Commission, Dept. of Civ. Engrg., Nat. Taiwan Univ., Taipei, Taiwan, ROC (in Chinese).

Hsu, N.S., Yeh, W.W-G., Kuo, J.-T., and Lee, L.W. (1990). "A study on the transfer and application of techniques for groundwater quality management in Taipei Aquifer." *Project completion Rept.,* Prepared for the Water Resour. Planning Commission, Dept. of Civ. Engrg., Nat. Taiwan Univ., Taipei, Taiwan, ROC (in Chinese).

Hsu, N.S. (1992). "Development of a model for conjunctive use planning of surface and ground water." *Project completion Rept.,* Prepared for the Nation Science Council, Dept. of Civ. Engrg., Nat. Taiwan Univ., Taipei, Taiwan, ROC (in Chinese).

Hsu, N.S., Yeh, W.W-G., Kuo, J.-T., and Lee, L.W. (1991). "A study on optimal conjunctive use planning in the Tanshui river basin (I) Development of simulation model." *Project completion Rept.,* Prepared for the Water Resour. Planning Commission, Dept. of Civ. Engrg., Nat. Taiwan Univ., Taipei, Taiwan, ROC (in Chinese).

Hsu, N.S., Horng, M.J., Wu, J.M., and Yeh, W.W-G. (1993). "Development of a two-layered groundwater model for the Taipei basin." *Proceedings of the 1993 National Conference on Hydraulic Engineering*, ASCE, San Francisco, California, 305-310.

Water Resources Planning Commission (1991). " National master plan for water resources management, Final report." Taipei, Taiwan, R.O.C.

Optimization of Conjunctive Use of Surface Water and Groundwater in a Coastal Zone

Pamela G. Emch[1] and William W-G. Yeh[2], Fellow, ASCE

Abstract

A nonlinear multiobjective optimization model is developed for managing the conjunctive use of surface water and groundwater supplies under conditions of potential saltwater intrusion. Two conflicting management objectives are considered: cost-effective allocation of available surface water and groundwater supplies, and minimization of saltwater intrusion due to groundwater overdraft. Optimal control of the system is examined by studying the response of these objectives to changes in the magnitude of groundwater pumping, and the transfer of surface water supplies between sources and users. System constraints including economic, operational, and institutional criteria are met.

Introduction

Coastal aquifers are an important groundwater resource. Under undeveloped conditions an equilibrium gradient exists within each aquifer of a multi-layered system, with excess freshwater discharging to the sea. Within each layer a wedge of denser saltwater develops beneath the lighter freshwater. A persistent reduction in freshwater flow towards the sea reduces the equilibrium gradient, inducing intrusion of saltwater into the aquifers as the interface, or boundary between the freshwater and saltwater zones, moves inland. Therefore, although the development of groundwater supplies tends to be cost-effective in the short term, the sustained overdraft of groundwater can affect the natural equilibrium, leading to long term problems. In aquifers in general, groundwater overdraft has produced declining water levels and well discharges, and has contributed to land subsidence. In coastal aquifers the problem is compounded as a result of saltwater intrusion.

---

[1] Graduate Student, Dept. of Civil and Environmental Engrg., UCLA, Los Angeles, CA 90095.
[2] Professor, Dept. of Civil and Environmental Engrg., UCLA, Los Angeles, CA 90095; Phone: (310)825-2300; Fax: (310)206-2222.

Both linear and nonlinear optimization models have been developed for the management of groundwater systems. Saltwater intrusion problems are often included in the latter category due to the complexity of the governing equations. The methodology developed here combines optimization of the conflicting objectives of minimizing water supply cost and saltwater intrusion with the management of the conjunctive use, treatment, and disposal of both surface water and groundwater supplies. It is assumed that the optimization is to be carried out on a regional scale. This allows use of the simplifying assumption that the freshwater and saltwater are considered immiscible fluids separated by a sharp interface (Freeze and Cherry, 1979). The computer code used to simulate the groundwater flow is described below. The optimization model is solved using the reduced-gradient method in Program MINOS (Murtagh and Saunders, 1982). The necessary linkage between the optimization procedure and the simulation model is created by incorporating the underlying simulation equations into each optimization algorithm. The objectives and the flow equations of the aquifer system are formulated as nonlinear functions of the freshwater and saltwater heads and the pumping schedules of the coastal region. Nonlinear constraints are also incorporated. The optimization algorithm then uses finite difference techniques to determine the objective and constraint gradients of the planning problem.

Simulation Model

The management model incorporates the quasi-three-dimensional sharp interface model available in Program SHARP (Essaid, 1990). This model allows for regional-scale simulation of coastal groundwater flow in a multi-layer aquifer system and includes both saltwater and freshwater dynamics. A pair of two-dimensional, vertically integrated, partial differential equations were developed which represent the flow in each layer of the aquifer system. The equations are coupled via the boundary condition at the interface. The flow is modeled as horizontal within each aquifer layer, and the layers are coupled through the terms in the governing equations which define the recharge occurring in each layer due to natural recharge and vertical leakage from adjacent aquitards. These equations are solved simultaneously for the freshwater and saltwater heads within each aquifer layer using an implicit finite difference scheme. The elevation of the interface within each aquifer follows from continuity of fluid pressure at the interface.

Multiobjective Optimization Model

This methodology incorporates a trade-off analysis between two conflicting management objectives: minimize the cost of supplying water, and minimize the intrusion. The mathematical formulation of this multiobjective optimization problem can be stated as

$$\min_{x \in X} Z = \left(Z_1(\bar{x}), Z_2(\bar{x})\right) \tag{1}$$

subject to:

$$g_i(\overline{x}) \le 0, \quad i = 1, 2, \ldots, m$$
$$x_j \ge 0, \quad j = 1, 2, \ldots, n$$

where $Z_k(\overline{x})$ is a mathematical function representing objective k, $\overline{x}$ is the vector of n decision variables for which optimal values are desired, X is the bounded set of decision variables, and $g_i(\overline{x})$ is a mathematical function representing the ith of m constraints.

A non-inferior solution to this problem, referred to as the "Pareto Optimal" alternative, is the solution for which any increase in the optimum level of satisfaction of one of the objectives results in a corresponding decrease in the optimum level of the other objective. The set of non-inferior solutions of a multiobjective problem is obtainable only if comparisons between the two objectives are possible. There are a number of ways of approaching this problem. The method chosen here is known as the "Constraint Method", in which one objective is sequentially optimized while parametrically varying the second objective from a lower bound to an upper bound in the form of a constraint. This traces out a trade-off curve representing the set of non-inferior solutions. A manager can then utilize the trade-off curve to select that combination of objectives which best meets his/her needs.

The two objectives of this problem are formulated as follows:

Water Supply Objective

$$\min Z_1 = \sum_{j=1}^{\tau}\left[\sum_{i\in\Omega_1} Q_{i,j}\cdot C_i\cdot\left(L_i - h_{i,j}\right) + \sum_{i\in\Omega_2} Q_{i,j}\cdot C_i\right] \tag{2}$$

Saltwater Intrusion Objective

$$\min Z_2 = \sum_{l\in\Gamma l}\left[\iint z_I^l(x,y)\,dx\,dy\right] \tag{3}$$

where $Q_{i,j}$ is the pumping or surface water supply rate for source i in period j [L^3/T], $h_{i,j}$ is the freshwater head at well i in period j [L], C_i is the unit cost of either groundwater extraction per height of required lift [$\$/L^3/L$] or of surface water supply for source i [$\$/L^3$], L_i is the elevation of the ground surface above the datum at well i [L], and z_I^l is the elevation of the interface in layer l [L].

Each of these objectives is subject to the same set of constraints which may include but is not limited to: demand constraints, well capacity constraints, supply source upper bounds, and drawdown costraints. It is this last which acts as a nonlinear constraint, since it may be formulated as:

$$h_{i,j} \geq h_i^{\min} \qquad \text{for all wells } i, \text{ an all periods } j \tag{4}$$

where $h_i^{\min}$ is the minimum allowable head for well i [L].

Case Study

In order to test the methodology described here, a case study based on the Waialae aquifer of Oahu, Hawaii was developed and examined. Reasonable surface water sources and system demands and constraints were devised. Sources included two wells, imported water, and treated wastewater. Water was used for municipal and agricultural purposes, was sent to a treatment plant and was exported. The operational horizon of 3 years was divided into 12 seasonal time periods. The problem comprised 60 decision variables and 96 constraints, of which 24 were nonlinear. The methodology provided a trade-off curve consisting of pairs of feasible, locally optimal solutions for the two objective functions.

Summary and Conclusions

A methodology has been developed that manages the conjunctive use of surface water and groundwater in a coastal region while minimizing the cost of water supply and the amount of intrusion. The methodology can accomodate user-defined system constraints, whether linear or nonlinear. The final product is a trade-off curve with which to assess the relative importance of each objective.

Acknowledgements

The research reported herein was partially supported under a fellowship from TRW, Inc., and by the National Science Foundation under Award No. MSS-9213963.

References

Essaid, H.I. (1990). The computer model SHARP, a quasi-three-dimensional finite-difference model to simulate freshwater and saltwater flow in layered coastal aquifer systems, *U.S. Geological Survey Water-Resouces Investigations Report 90-4130*, 181 pp.

Freeze, R.A., and J.A. Cherry (1979). *Groundwater*, Prentice-Hall, Englewood Cliffs, New Jersey, 604 pp.

Murtagh, B.A., and M.A. Saunders (1987). MINOS 5.1 user's guide, *Technical Report 83-20R*, Systems Optimization Laboratory, Department of Operations Research, Stanford University, Stanford, California, 118 pp.

IMPERIAL COUNTY GROUNDWATER EVALUATION: MODEL APPLICATIONS

M. Najmus Saquib[1], Larry C. Davis[2], Abdul Q. Khan[1], and S. Alireza Taghavi[1],

Abstract

An Integrated Groundwater and Surface Water Model (IGSM) was developed for the Imperial County in southern California (Taghavi, et. al., 1995). The calibrated model was used to simulate the response of the groundwater basin to three potential management scenarios: 1) an aquifer storage and recover project; 2) a canal lining project; and 3) a canal seepage recovery project. The model results for the aquifer storage and recovery project indicate that the longer the extraction period, the less is the recharge recovery. The results for the canal lining project show that the groundwater level near the previously unlined canal stabilizes to a level that is about 25 feet lower than the historic water levels. The canal seepage recovery by pumping near unlined canal includes seepage that nearly equals to the pumped water, and hence is not effective in recovering seepage water.

Introduction

The groundwater management activities in the Imperial County are largely permit oriented and/or restrictive measures that provide protection to the groundwater resources of the County. In 1991, the Board of Supervisors of Imperial County passed a resolution emphasizing a moratorium on the permitting, construction and operation of all new groundwater wells and the appropriation of all groundwater. Exceptions to the moratorium were made for domestic water wells proposing to use less than two acre-feet per year and non-domestic wells proposing to use less than five acre-feet per year. Application in excess of these limits can be considered if the applicant can provide satisfactory evidence that the requested water well will not adversely affect the underlying groundwater basin. The moratorium exempts

[1, 2] Respectively, Senior Engineers and Principal Engineer, Montgomery Watson, 777 Campus Commons Dr., Suite 250, Sacramento, CA. 95825

Imperial Irrigation District (IID) water seepage/recovery wells less than 30 feet deep within 100 feet of an unlined canal, water and geothermal test wells, and the four wells along the All-American Canal associated with the Bureau of Reclamation (USBR) sponsored Lower Colorado River Water Supply Project. The moratorium is to remain in place until the Board of Supervisors adopts a comprehensive Groundwater Management Plan or for a period of two years, whichever comes first. The moratorium was extended for another two years in July, 1993. The draft Groundwater Management Act and Ordinance developed in 1991 provides the framework for the regulation and control of groundwater throughout the Imperial County, allowing the implementation of measures to manage and allocate available groundwater.

Groundwater Development Proposals

Groundwater development proposals in the Imperial County range from new urban development in subareas outside the IID to proposals to recover seepage from unlined canals. Any additional groundwater pumping raises a concern that it could result in overdraft and/or migration of poor quality groundwater into areas where groundwater is suitable for domestic and agricultural purposes. It should also be noted that the California State law encourages voluntary transfer of conserved water and allows conserved water to be sold, leased, exchanged or otherwise transferred for use within or outside the District. The Imperial County IGSM (Taghavi et. al., 1995) was developed and calibrated with historic data to provide a quantitative tool for evaluation of groundwater development proposals in the County.

Aquifer Recharge and Storage Project

In 1988, a recharge demonstration project was conducted in a portion of the old Coachella Canal that has been replaced by a lined canal. That demonstration project involved the recharge of 17,000 acre-feet of water over a three month period. The Imperial County IGSM was used to simulate the basin response to groundwater withdrawal near the demonstration project site over various periods of time after recharge: 17,000 acre-feet of water was recharged in three months, the withdrawal began nine months later, and the withdrawal of 17,000 acre-feet of water took place over a period of one year (Alt 1-A), five years (Alt 1-B), and ten years (Alt 1-C). The objective of this project was to answer the question of how much of the recharge amount could be recovered and over what period of time, without significantly affecting the groundwater storage in the pre-recharge condition. Figure 1 shows the average water level near the project site. The baseline condition is no-action scenario with 40 year hydrology developed by repeating 1970-1990 historic hydrology. The impact of the length of the withdrawal period on the recharge recovery is demonstrated by 12,750 acre-feet, 11,600 acre-feet, and 10,200 acre-feet of recovery respectively for one year, five year, and 10 year extraction period. The water level rise in the proximity of the recharge location can be as high as 12 feet during the recharge period, but it reduces down to two feet within a year.

All American Canal Lining Project

All American Canal (AAC) is a major canal that carries about 2.5 million acre-feet of water every year for use in the Imperial County. The seepage loss from the unlined canal amounts to about 25,000 acre-feet per year. The amount of water savings that can be achieved by lining AAC has been a hotly debated topic because of water transfer interests. For this evaluation, a part of the AAC in the upper reaches is considered to be lined for model simulation and its impact was investigated. The model results show that in the vicinity of the lined canal, water level drops as much as 25 feet before stabilizing to this lower level after 30 years (Figure 2).

The groundwater flow direction does not change, but the gradient from the canal to Mexico is flattened; this reduces the annual outflow to Mexico from 43,600 acre-feet under baseline scenario to 22,750 acre-feet under the lined canal scenario.

Canal Seepage Recovery Project

This project consists of activating two existing wells to pump 8,400 acre-feet of groundwater near the AAC for delivery to the IID. The purpose of this study was to determine whether pumping in the vicinity of the canal induced increased seepage that would offset the seepage recovery by pumping. The model results indicate that due to 8,400 acre feet per year pumping near the unlined canal, the seepage loss increases by about 7,000 acre-feet per year, indicating that this project is not effective in recovering seepage water.

Conclusions

Imperial County IGSM was used to evaluate the response of the groundwater basin under three potential management alternatives: an aquifer storage and recovery project, a canal lining project, and a canal seepage recovery project. The results for the aquifer storage and recovery project indicate that the longer the extraction period, the less is the recharge recovery. The results for the canal lining project show that the groundwater level near the previously unlined canal stabilizes to a level that is about 25 feet lower than the historic water levels. The canal seepage recovery by pumping near unlined canal includes seepage that nearly equals to the pumped water, and hence is not effective in recovering seepage water.

Acknowledgments
The development of Imperial County IGSM was the result of a study funded jointly by the Imperial County and Imperial Irrigation District.

References

1. S. A. Taghavi, M.N. Saquib, A.Q. Khan, and L.C. Davis, 1995. Imperial County Groundwater Evaluation: Model Development and Calibration. In Proceedings of 22nd Water Resources Planning and Management Specialty Conference Division, American Society of Civil Engineers. Boston, MA.

2. Montgomery Watson, 1995. Imperial County Groundwater Study. Draft Report submitted to County of Imperial, January 1995.

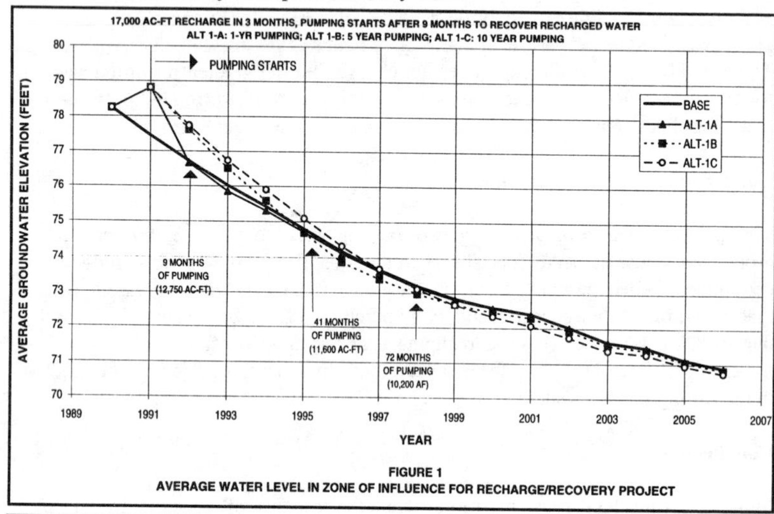

FIGURE 1
AVERAGE WATER LEVEL IN ZONE OF INFLUENCE FOR RECHARGE/RECOVERY PROJECT

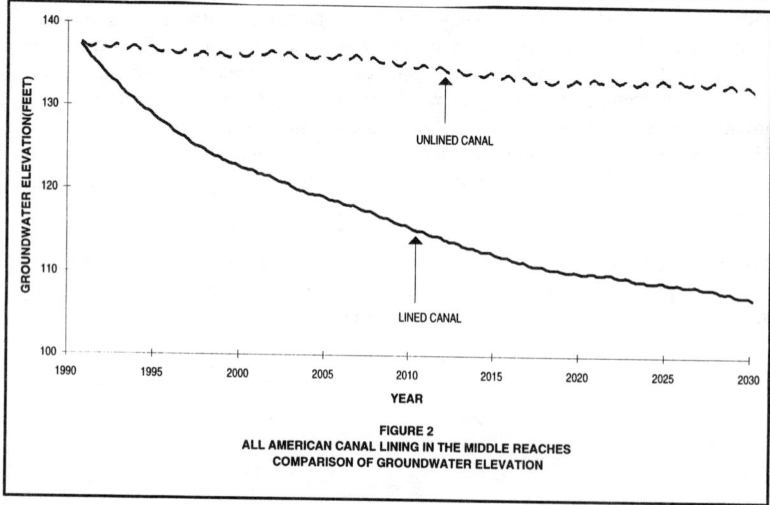

FIGURE 2
ALL AMERICAN CANAL LINING IN THE MIDDLE REACHES
COMPARISON OF GROUNDWATER ELEVATION

IMPERIAL COUNTY GROUNDWATER EVALUATION: MODEL DEVELOPMENT AND CALIBRATION

S. Alireza Taghavi[1], M. Najmus Saquib[1], Abdul Q. Khan[1], and Larry C. Davis[2]

Abstract

An Integrated Groundwater and Surface Water Model (IGSM) has been developed for the Imperial County, located in southern California. This largely agricultural area is supplied by Colorado River water, which conveys to the different points of use by extensive canal networks. Since many of these canals are unlined, substantial amounts of canal seepage takes place within the basin. The current groundwater use is small, but is expected to increase in the future as a result of anticipated developments. In recent times, canal seepage to groundwater has been targeted by water transfer patrons because of potential water savings by lining major canals. Groundwater represents a small fraction of Imperial County's water supply; however, it plays a significant role in the overall water balance in the area. Local groundwater is the only source of supply outside the Imperial Irrigation District (IID) service area. There is, therefore, a great need for an analytical tool that can be used to evaluate the groundwater conditions of the basin. The primary purpose of developing the Imperial County IGSM is to satisfy that need. The monthly simulation model has been calibrated with historic data for the period between 1970 and 1990. The calibrated model will serve as a useful tool for evaluating the impact of land and water use changes resulting from new developments in the county. As part of this study, a groundwater quality model was also developed. This model can potentially be used to analyze the impacts of various alternatives on the quality of groundwater in the basin.

Introduction

Imperial County in southeastern California, covers about 4,600 square miles, and borders the state of Baja California, Mexico. The Colorado River marks the eastern boundary of Imperial County, and is the state boundary between California and Arizona. Despite its naturally arid environment, Imperial County is one of the richest agricultural regions in California, owing to the diversion of water from the Colorado River to the fertile soils of the Imperial Valley.

Groundwater represents a small but very important component of Imperial County's water supply. Outside of the Imperial Irrigation District (IID) local groundwater is the only source of supply. The major source of water within the IID service area is the imported Colorado

[1, 2] Respectively, Senior Engineers and Principal Engineer, Montgomery Watson, 777 Campus Commons Dr., Suite 250, Sacramento, CA. 95825

River water via vast transfer canal systems, from the last reaches of the River in the United States. The quality of groundwater is highly variable throughout the County. Generally, the quality of groundwater underlying the IID area is high in TDS and is unsuitable for drinking water purposes and irrigated agriculture. The increasing demand on groundwater in the area, potential water savings, and water transfers from lining of major canal systems in the County have led to the need for better understanding of the groundwater basin in the County. In July 1991, the County recognized that groundwater resources are limited, and passed a moratorium preventing the drilling of new wells in Imperial County until a groundwater management plan is adopted. Responding to these concerns, the Imperial IGSM was developed for the County, to be used for technical evaluations in this groundwater management plan, and other groundwater studies in the County.

Basin Description

Imperial County has a typical desert climate, with extreme summer and mild winter temperatures, low humidities, and little precipitation. Rainfall generally ranges between 3 and 4 inches per year and, in higher elevations, occurs most often between November and April. Summer thunderstorms are characteristic for the lower desert valley regions.

The limited availability of good quality water from either local surface or groundwater supplies has been a dominant factor in the demarcation of human activity in the County. The Colorado River, flowing along the eastern boundary of the County is the only source of good quality surface water in the County. In late nineteen and early twentieth century, the Colorado River water was diverted to the Imperial valley for irrigation of fertile lands in the area. This diversion was later increased to an average annual flow of 2.5 million acre-feet, when the All-American Canal (AAC) was constructed (Figure 1). However, the importation of Colorado River water to the interior of the valley has significantly impacted the natural system of intermittent and/or polluted surface waters draining to the Salton Sea, as well as groundwater basin underlying the valley. The AAC and other distribution system canals totaling approximately 1,600 miles in the IID act as the main source of recharge to the groundwater system. Salton Sea occupies the deepest part of the valley, whose lowest point is nearly 280 feet below sea level, and drains most of the natural runoff in the valley, as well as, the irrigated lands in the area.

Model Development

As part of the technical analysis required for initiating the Groundwater Management Plan, a County-wide groundwater model was developed. The conceived model was required to simulate the groundwater flow in the basin, including effects of seepage losses from the major canal systems, recharge from irrigated and natural lands, and interaction between groundwater and the major river systems in the area. IGSM (Montgomery Watson, 1993) met the requirements of such a model, and was therefore selected for this application. This comprehensive hydrologic model simulates all the components of the hydrologic cycle of a groundwater basin including precipitation, runoff, groundwater recharge, evaporation, consumptive use, pumping, and subsurface inflows and outflows. IGSM can simulate complex surface hydrologic, multiple aquifer systems, and the interaction of flows between streams and groundwater aquifers, all in accordance with mass balance and water budget accounting procedures. The groundwater simulation module is based on quasi-three dimensional Galerkin finite element method for saturated confined and unconfined aquifer systems. The unsaturated flow is simulated based on one-dimensional unsaturated flow equations. IGSM requires a comprehensive set of geologic, hydrologic, and land and water use input data. Key output of the model are streamflow, groundwater levels, as well as hydrologic budgets for soil, surface use, groundwater, and stream systems.

The Imperial County IGSM is designed to reflect the regional nature of water management in the context of conjunctive use. In this regards, the 4017 square miles of model area is discretized to 1432 finite elements of triangular and quadrilateral shapes with an average size of 2.8 square miles, and 1419 nodes (Montgomery Watson, 1994). The groundwater basin is assumed to be stratified into an unconfined and a confined aquifer with a relatively thick semi-permeable clay layer sandwiched in between, acting as an aquitard. The average thickness of this aquitard is 60 feet, and the maximum is 280 feet. The aquifers are mostly non-marine deposits of late Tertiary and Quaternary age, and have average thickness of 200 and 380 feet, with maximum of 450 feet and 1500 feet, respectively. Eight streams and rivers, and 7 major unlined canals are included in the model to simulate the stream-aquifer interaction dynamically. Historic monthly data from ten precipitation stations are used to provide the rainfall patterns in the model. The Soil Conservation Service hydrologic soil classification is used to determine the runoff and recharge components of rainfall. Model boundary conditions are simulated as constant flux over each time step. The surface and subsurface flux into the main basin from 36 tributary watersheds peripheral to the model area are also simulated. The model area covered by the Salton Sea has a fixed head boundary condition.

In order to evaluate the hydrologic budget in the study area, the model area is subdivided into eight subregions. Calibration period for this model is selected to be September 1970 to October 1990. The calibration of such complex model begins with the water budget analysis within each system. Once the water budget looks reasonable, the hydrogeologic parameters of the aquifer are adjusted to ensure reasonable agreement between model simulated groundwater elevations and observed ones in each of the selected monitoring wells. In addition, the model parameters are fine-tuned to attain reasonable agreements between simulated and observed streamflows at selected stream gaging stations. The error analysis performed on the model calibration results indicates that over 75% of model simulated groundwater levels are within ±5 feet, and over 95% are within ±10 feet of observed values, which is very reasonable tolerance. Figure 2 shows the model calibration results for streamflows.

Conclusions

The Imperial County IGSM, developed as part of this study, is a valuable analytical tool for evaluation of existing groundwater conditions in the County. In addition, the model can be used to evaluate the impacts of changes in land and water use, surface and groundwater operations, and facilities planning and design. The model is currently used to evaluate the response of the groundwater basin to a set of operations that have been under study. These include the Colorado River recharge and recovery project, AAC lining, and recovery of AAC seepage. The context and model results from these and other scenarios are discussed in Saquib et al. (1995).

Acknowledgments

The development of Imperial County IGSM was the result of a study funded jointly by the Imperial County and Imperial Irrigation District.

References

1. Montgomery Watson, 1993. Documentation and User's Manual for the Integrated Ground and Surface water Model (IGSM), Sacramento, CA.
2. Montgomery Watson, 1995. Imperial County Groundwater Study. Draft Report submitted to County of Imperial, January 1995.
3. Saquib, M.N., L.C. Davis, A.Q. Khan, and S.A. Taghavi, 1995. Imperial County Groundwater Evaluation: Model Applications. In Proceedings of 22nd Water Resources Planning and Management Specialty Conference Division, American Society of Civil Engineers. Boston, MA.

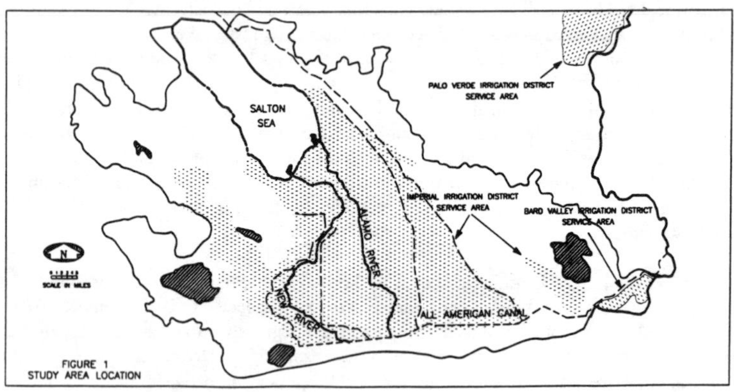

FIGURE 1
STUDY AREA LOCATION

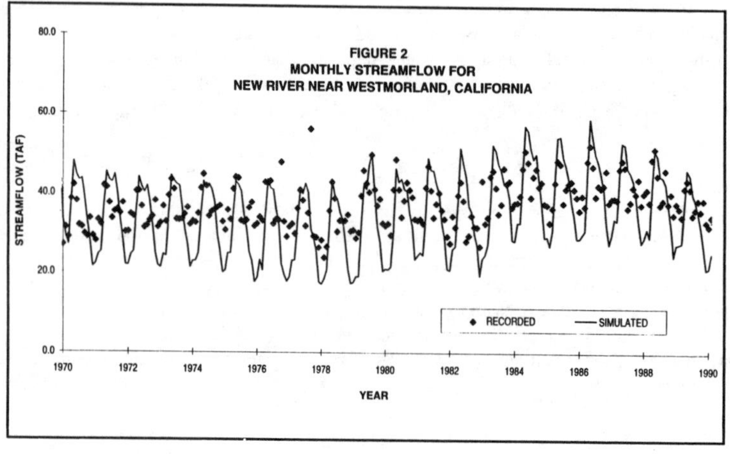

FIGURE 2
MONTHLY STREAMFLOW FOR
NEW RIVER NEAR WESTMORLAND, CALIFORNIA

Qué pasa with Water Resources Management in the Americas?

Vinio Floris [1], Associate Member ASCE

Abstract

North and South America have an immense quantity of water resources, a large portion of which is found in the Latin America and Caribbean Region. Water has been one of the most important and limiting factors in almost all human activities and the environment. However, after several decades of mismanagement, the quality of human life, its economics and the ecosystem have been affected seriously. Governments have the critical and urgent mission of redefining their policies, laws and administration. They must develop a participatory and efficient system to manage water -for quality and quantity- in the context of sustainable development. If no assertive and timely actions are taken, clean and reliable freshwater may soon become an endangered resource.

Introduction

The Latin American and Caribbean Region has a vast territory of 20.4 million km^2, which includes a vast water resources potential. The Amazon basin alone has an area of more than 7 million km^2 producing a runoff of around 3,767 km^3/year (Spediel and Agnew, 1988), making the Amazon River the largest river discharge in the world and even larger than the next five largest river systems all together. Jointly, North and South America have almost 45% of all the freshwater releases of the planet.

Since pre-Columbian times, most human activities in the Region have been closely related to water resources. This is evidenced by the magnificent hydraulic infrastructure built by ancient civilizations in Central and South America. During the 20th century, attempts have been made to improve the use of water. Large investments of time and money and great efforts have been expended, but often with unfortunate negative outcomes. As a result of decades of ineffective tactics, the quality of life and the environment of the Region has been seriously affected. However, changes in the economic order throughout the world are aiding most countries in the implementation of new measures to radically modify the old system and develop a more efficient one.

This paper reviews historical information, events, and actions taken in water resources

---

[1] Planning Department, South Florida Water Management District, P.O. Box 24680, West Palm Beach, Florida 33406

projects in the Region, and their socio-economic and environmental relationships and impacts. Findings and discussion are grouped by topics in order to facilitate the analysis.

The Agricultural Sector

There is approximately 700 million ha of potentially arable land and some 570 million Ha of natural grassland in Latin America. Farm exports (1989 data) totaled $ 317.2 billion (US) while imports were $ 128.5 billion US (IDB, 1994). Although this may sound positive, most countries have serious agricultural production and -especially- productivity problems, water being one of the most limiting factors.

Negative balance of payments has been a chronic problem in the Region. In order to strengthen their economies, countries have invested in large number of reservoirs and water distribution networks. However, the production and productivity levels still remain low. Water has been misused, dramatically increasing the drainage problems downstream basins and affecting the water quality of streams and groundwater. Also the agricultural frontier has expanded into lands of limited potential, seriously altering the fragile ecosystems.

The Energy Sector

Hydropower generation has been one of the most important energy sources in Latin America (64% of the total). However, while the Region has an enormous hydropower potential (around 22% of the world) representing 700,000 megawatts, the installed capacity is only 22% of the total (153,500 Megawatts). Over the last 30 years the production of electric energy increased nearly eight times but the region's average consumption per capita is low (1,150 kwh per person), which is one-tenth of US per-capita consumption and below the world average (IDB, 1994). The energy demand, however, is estimated to double in the next 13 years.

The Environment and Natural Resources

Latin America is relatively rich in natural resources. It is the repository of the planet's greatest biodiversity and of very rich fishery resources. It contains more than 800 million ha of forests and a water runoff which is almost 27% of the world total (IDB, 1994). However, the environment has suffered mismanagement of several human activities: mining, agriculture, and industry, among others. Several relatively new pollution problems have appeared most recently in the Region. These problems are dangerously affecting the environment. Some of them are related to the increase of mining and oil extraction activities in several countries and contamination of groundwater with nitrates and pesticides from agriculture. It is important to mention that groundwater supplies almost 50% of the populations freshwater needs in the Region.

Because of the serious problems mentioned above, there is growing concern in preserving the environment. Dourojeanni (ECLAC, 1994) has estimated that 50% of all environmental problems could be reduced by improving water quality at the watershed level. Governments are producing legislation and rearranging institutions to protect and restore unique ecosystems.

Health and Sanitation

Latin America and the Caribbean have insufficient and generally low-quality services for water, sewerage and solid waste. During the 1980s, the situation worsened in light of high rates of population growth and lower levels of public investment and maintenance. Data collected in 1988 shows that 79% of urban areas had water connections, 9% with some access to them and 12% with no services at all. In rural areas only 55% of the population had water services. For sewerage and excreta disposal, in urban areas 49% of the total population had house connections, 32% had some other solutions (septic tanks, latrines, etc.) and 19% no services at all. In the rural areas only 32% of the population has access to sewerage and excreta disposal. The amount of urban and rural population is around 291.6 and 124 million, respectively. A problem of major concern is the establishment of urban areas on the periphery of existing cities without any planning of infrastructure and services. These marginal regions already harbor 40% of the urban population and are expected to absorb 80% of the population growth during the 1990s. Also, less than 10% of the sewerage systems have treatment plants and between 5% and 10% of the collected wastewater receives treatment which is often inadequate. It is estimated that in 1990 a total of 350 m^3/second of untreated wastewater were disposed improperly (PAHO, 1992)

Water pollution is a serious problem that has a significant and accelerated degradation of fresh water resources and an increase in the incidence and prevalence of associated diseases. This problem is augmented by industrial releases such as heavy metals and chemical substances coming from pulp/paper, and steel/iron industries -two of the Region's biggest polluters-. These industries are growing twice as fast as the economy as a whole (Postel, 1988) and give off by products which are more harmful that those found in domestic wastes.

Institutional Aspects

Traditionally in Latin America, the role of managing water was assumed by the governments. This responsibility resided in public organizations which rarely had the political authority or necessary human and/or financial resources. This created a concern of efficiency and efficacy, and the need for users to participate and be involved in decision-making. There is also a role for the private sector which historically has been left out of the system.

The lack of a water administrative system generates great difficulty when there is a need to launch massive projects with user participation. The existing system only works for formal users, well educated and with an entrepreneur's attitude. However, the great majority is left out physically and legally (ECLAC, 1994). The environmental protection level of developed countries can not be reached without having an organization which embodies public and private interests.

Legal and Water Policy Issues

Most countries in the Region have sectoral water legislation which is complemented by international water law and treaties. In some countries legislation is very general, not updated, and superficial. In others it is more complete and modern. It is not uncommon to have laws and codes without any water quality or environmental considerations. This greatly affects water policy implementation and legislation enforcement. However, there is a strong trend towards modifying water laws and introducing more holistic concepts. This will contribute to translate public opinion on natural resources into political will and specific actions. Effective results can not be obtained by the mere promulgation of decrees. Education and the dissemination of information are therefore essential (ECLAC, 1994).

Conclusions

As was discussed above, the supply of freshwater in the Region -in quantity and quality- continues to diminish per habitant and the environment. This will continue to give rise to conflicts derived from the management system, but also provides a big challenge to all providers and users to find an efficient, participatory water management system that can achieve a true sustainable development. It is important for governments to recognize that the role of the State and the role of individuals should be defined as clearly as possible, with regards to water management policy. There is a need to conceive investment projects at the user level. Another need is to achieve decentralization by delegating decision-making at river basin level. These bodies have an important role to play by administrating water resources (ECLAC, 1994) and promoting water markets.

Currently, governments in the Region are seeking active participation of the private sector by privatizing water utilities. This may be a successful decision but only if the governments are aware about the creation of monopolies. It is essential to uphold the regulatory and normative function of the State in order to avoid monopolies and promote competition. Such regulation should take effect before privatization (ECLAC, 1994).

Water is just a component of a large and complex body and it should be managed as an integrated system with the participation of users and suppliers. Governments should recognize the importance of water resources planning and clean technologies, and make provisions for environmental monitoring. This is a challenge which requires immediate action.

Appendix. References

Inter-American Development Bank - IDB (1994). IDB Projects. Washington, D.C.

Pan American Health Organization- PAHO (1992). Consultative Meeting on Excreta and Wastewater Disposal in Latin America and the Caribbean. Washington, D.C.

Postel, S. (1988). "The Consequences of Mismanagement." Perspectives on Water, Uses and Abuses, D. Speidel, *et al*, ed. Oxford University Press, New York, N. Y., 307-325.

Speidel, D. and Agnew A. "The World Water Budget." Perspectives on Water, Uses and Abuses, D. Speidel, *et al*, ed. Oxford University Press, New York, N. Y., 27-36.

United Nations Economic Commission for Latin America and the Caribbean - ECLAC (1994). Report of the Group of Experts on the Effects of Agenda 21 on Integrated Water Resources Management of Latin America and the Caribbean. Santiago, Chile.

Water Supply and Sanitation in the Hemisphere: Issues and Options

Neil S. Grigg[1]

bstract

he paper outlines issues of water supply and sanitation system development the Hemisphere with emphasis on emerging problems of economic tegration. Topics include: links between water supply, public health, social evelopment and economic progress; need to relate water supply development rational patterns of urban and rural development; need for effective planning f systems; financing systems, including systems of charges; institutional ructures for planning, management and regulation; and need for effective perations and maintenance.

troduction

ur Hemisphere includes a vast array of cultures and lands, ranging from those ove the Artic Circle to those in the steamy jungles of Amazonia. In between e examples of living in mountains, plains, and coastal areas; and living in ega-cities, industrial towns, and rural villages. While the needs for water ipply and sanitation in these areas vary in many ways, the Hemisphere shares e fundamental need to provide adequate, safe, and affordable water and nitation systems to the people.

ne issues were summed up by Chapter 18 of Agenda 21, which adopted ven program areas for the freshwater sector. Four of these relate directly to ater and sanitation infrastructure: drinking-water supply and sanitation, water nd sustainable urban development, protection of water resources, and water r rural development. The remaining themes also relate, because "integrated ater resources development and management" must include water and astewater infrastructure.

ternational Drinking Water and Sanitation Decade

a paper prepared for the United Nations Water Conference, the Intermediate echnology Development Group concluded that there were just over 1000 illion people in the world's rural areas who lacked access to a safe water ipply (Pacey, 1977). As a result of these findings, and others that were just dramatic, the United Nations proclaimed the 1980's as the International rinking Water Supply and Sanitation Decade. The Decade was conceived at

---

lead, Department of Civil Engineering, Colorado State University, Ft Collins, O 80523.

the UN Water Conference in 1977 at Mar del Plata, Argentina, and endorsed the General Assembly of the UN in November 1980. The goal was to supply of the world's population with safe drinking water and sanitation by 1990.

Canemark (1989), the World Bank division chief for the water supply sector, reported in 1989 that the coverage had improved, but that the greatest achievement had been the communication, awareness and priority-setting tha had occurred to deal with the problem.

In 1977, the world economy was on a relative upswing. However, the 1980' were a time of economic and social decline. More than half of the developing countries experienced negative economic growth, and their debt burdens increased. Levels of investment in water and sanitation did not keep pace wi population growth or rates of urbanization.

Anatomy of the problem

Okun (1991) summarized his theories in a 1991 paper for the Abel Wolman Distinguished Lecture, and how appropriate the topic was, given the enormou contributions of Abel Wolman in this field. Okun reported, to his regret, that his 50 years of work on the water and sanitation problems in urban areas, the had gotten worse. The reasons were an inadequate supply of water in the cities attributable to limited water resources, and/or poor facilities for treating and distributing the water compounded by an absence of proper sewerage.

What happens is that intermittent supplies of water create opportunities for infiltration of heavily contaminated water into the distribution systems when the pressure is off. Water-borne infectious agents then can reach taps, even when the water is safe as it leaves the treatment plants.

Okun's paper emphasizes the need for water resources capacity-building to create a favorable policy climate and appropriate institutional development which would include establishing sound management systems, incentive structures, and the human resources development needed for sustainable development of water-related programs.

Responding to the Decade, the US Agency for International Development organized a special project called the Water and Sanitation for Health (WASH) project, and in 1990 reported the lessons learned (US Agency for Internationa Development, Lessons Learned from the WASH Project, USAID, Water and Sanitation for Health Project, Washington, 1990). WASH organized their lessons learned into four principles:

- Technical assistance is most successful when it helps people learn to do things for themselves in the long run.
- Water supply and sanitation development proceeds most effectively when its various elements are linked at all levels.

The basic measure for success of both the national system for development and the community management systems it creates is sustainability, the ability to perform effectively and indefinitely after donor assistance has been terminated.

Sustainable development is more likely to occur if each of the key participants recognizes and assumes its appropriate role and shoulders its share of the responsibility.

'ater and Sanitation Systems in the Hemisphere

s do social and environmental conditions, access to modern water and initation systems varies widely in the Hemisphere. Although the problems in tin America are formidable, a great deal of progress has been made since out 1959 when the Inter-American Bank was organized. In fact, the Bank's st loan was to expand a water supply and sewerage system in Peru (Inter-nerican Development Bank, 1992). Problems continue to increase due to pulation growth, urbanization, and industrialization. Although the percentage people served has increased, the total number without service has actually creased. Another problem is that because safe drinking water is so important d has received priority, sewage systems have been neglected, at the expense basic sanitation.

hile attention to the problems has intensified, investment has not. While the 980's were designated the International Drinking Water and Sanitation ecade, it turned out that this decade also introduced the worse economic isis of the century in Latin America. A worldwide economic problem also duced aid to developing nations. Unfortunately, in Latin America, a gnificant proportion of all disease is still attributed to polluted drinking water d untreated sewage, and water-related diarrheal diseases continue as the ading cause of infant mortality in many countries.

llaboration to Improve Water and Wastewater Infrastructure

e range of water and wastewater problems facing the nations of the misphere is too large to address, but let me summarize a few to focus on w we might all benefit from collaboration:

cess. Disparity in access to safe drinking water and adequate sanitation rvices, and in levels of service, need attention throughout the Hemisphere. is is a worldwide problem as evidenced by the International Drinking Water d Sanitation Decade.

chnology. Modern technologies are not available to those nations and gions lacking investment capital. Appropriate technologies offer help, but ey require training, expertise, and local capacity-building.

Management. Improving management efficiency, especially at local levels, is critical issue. One of the most urgent issues is obtaining qualified and trained personnel.

Finance. All nations in the Hemisphere are stuggling with issues related to financial capacity. As economic integration proceeds, ways are needed to upgrade and equalize water and sanitation as a basic issue in trade and politi cooperation.

Institutions. In all nations, institutional factors are the most important in upgrading and equalizing water and sanitation services.

How can the water and sanitation sectors in nations and institutions in the Hemisphere collaborate to improve our joint approaches to solving these problems? Possible areas of collaboration include technology transfer, improving access to information and innovations, and formation of alliances. This might be facilitated by mechanisms of cooperation such as a collaborativ network of research and training institutes. While there is a wide variety in t nature of the problems faced at local levels, there certainly exists a potential help each other solve problems in the Hemisphere through cooperation and sharing of experiences and knowledge.

References

Canemark, Curt, The Decade and After: Lessons from the 80s for the 90s an Beyond, World Water 89, London, November 14, 1989.

Grigg, Neil S., Water and Sanitation In the Hemisphere: Issues and Options, Interamerican Dialog for Water Management, Miami, October, 1993.

Inter-American Development Bank, Water and Sanitation, June 1992.

Okun, Daniel A., Meeting the Need For Water and Sanitation For Urban Populations, The Abel Wolman Distinguished Lecture, National Research Council, May 1991, Washington.

Pacey, Arnold, ed., Water for the Thousand Millions, Pergamon Press, Oxford 1977.

United Nations, Global Consultation on Safe Water and Sanitation for the 1990's, New Delhi, 1991

US Agency for International Development, Lessons Learned from the WASH Project, USAID, Water and Sanitation for Health Project, Washington, 1990.

World Bank, World Bank Atlas, 25th Anniversary Edition.

The Role of Wetland Conservation in Sustainable Development

Gonzalo Castro[1]

Abstract

Wetlands, or freshwater ecosystems, provide a variety of important services to society, including water supply and purification, flood control, sustenance for fisheries, protection from natural forces, sediment retention, nutrient and toxicant removal, and others. In addition, wetlands provide numerous natural products, and opportunities for recreation, fishing, and hunting. The combined economic and ecological value of these services is enormous. Wetlands, however, are rapidly disappearing because of a lack of understanding about their importance. This loss is widespread and in most cases irreversible.

Any rational strategy for the long-term, sustainable use of water resources must incorporate wetlands as one of its crucial elements. This paper discusses specific examples of wetlands services, and presents a case study from northern Perú demonstrating sustainable utilization of wetland resources.

Wetlands

Aquatic ecosystems, or wetlands, are an integral component of the water cycle and are thus inextricably linked with the management of water resources. Given the variety and importance of the services that wetlands provide to society, their conservation must be a central

[1] Senior-Vice President for Policy and Development, Wetlands for the Americas, c/o World Wildlife Fund, 1250 24th Street, NW, Washington, DC 20037

component of any rational strategy for the long-term utilization of water.

Traditionally, wetlands have been considered useless ecosystems. Widespread ignorance about the important benefits that wetlands provide to human societies has contributed to this notion, promoting the destruction and degradation of wetlands throughout the world. The conterminous U.S., for example, has lost an estimated 53-55 percent of its original wetlands. By the 1980's, twenty-two states had lost 50 percent or more of the wetland areas that were contained within state borders (Dahl 1990). This loss is equivalent to an area larger than the state of California, and translates into a loss of one acre every single minute for the last two hundred years. Although reliable figures are not available for other countries in the hemisphere, evidence indicates that wetlands are rapidly being destroyed and degraded everywhere.

Wetland Benefits

Wetlands provide a great variety of benefits to society. These benefits are often referred to as functions, uses, values and attributes, features, and goods or services, and are defined as any of these terms which may have a value to people, wildlife, natural systems or natural processes (Claridge 1991). According to Adamus and Stockwell (1983), there are about 75 such characteristics of wetlands that can be considered beneficial. These include water supply, flood regulation, protection from natural forces, nutrient retention and toxicant removal, micro-climate stabilization, water transport, global carbon intake, tourism opportunities, socio-cultural values, and many others. The economic and social magnitude of these benefits is enormous.

Public policy requires economic valuation of the public benefits of wetland conservation. Since most wetlands provide several of these benefits simultaneously, the total value of a wetland thus cannot be accurately estimated unless all functions, products, and attributes are incorporated into the calculations (James 1991). Quantification at the local scale for the harvestable products such as food and fuel is straightforward. Recreational and aesthetic values of wetlands, and their diversity of plants, fish, and wildlife can also be approached with conventional economic methods based on the businesses supported by recreational experiences, or based on willingness to pay for the recreational experience (Castro et al. 1994).

Higher ecological values are external to the market system because the benefits are accrued by society as a whole. Attempts have been made to assign economic value to these functions based on what it would cost to replace the function, or by depreciating the natural capital that is eroded when a natural resource is depleted (Solórzano et al. 1991). Regardless of the method utilized, it is clear that the very valuable benefits that wetlands provide to society must be somehow incorporated within national accounting schemes in order to change the perception that wetlands are useless ecosystems, and to promote the conservation of their valuable benefits.

The Huanchaco Extractive Reserve in Perú: An Example of Sustainable Use

Many ancient civilizations were organized along the coast of Perú to benefit from coastal wetlands. These wetlands were managed to obtain a variety of products including birds, eggs, fish and mollusks, peat (as fuel), and cattail (*Typha*) fibers. Many of these products were utilized for the manufacturing of tools, containers, housing, and fishing vessels. After the Spanish conquest, these traditions were abandoned in most coastal areas, except for Huanchaco in northern Perú. Today, more than 500 people, mostly fishermen, directly depend upon the totora extracted from Huanchaco for survival.

The main goal of this project was to rescue the ancient techniques of coastal wetland management in coastal Perú, by developing a demonstration project in Huanchaco. Specific objectives are to document the ancient techniques of coastal wetland management and study their adaptation within a contemporary context; to refine these techniques to maximize their values for biodiversity conservation; to explore the development of alternative markets for the wetland products generated; and to promote the utilization of these techniques in additional wetlands along the coast of Perú.

The results of the project will rescue techniques that can be used for the sustainable management of wetlands, while providing important habitats for biodiversity. In addition, it will help conserve this important site, its biological and cultural values, and the 500 people that depend upon it. It will serve as an international showcase demonstrating that wetlands can be sustainably managed to benefit both people and biodiversity. Culturally, this project will help conserve the last location where a cultural tradition has been continuously utilized for more than 2,000 years.

References

Adamus, P.R., and L.T. Stockwell. 1983. A Method for Wetland Functional Assessment. Vol 1: Critical Review and Evaluation Concepts. U.S. Department of Transportation, FHWA-IP- 82-83. Washington, DC. 178 Pp.

Castro, G., et al. 1994. Wetland and bird conservation in North America. American Ornithologists' Union Wetland Conservation Sub-Committee. Ms. in preparation.

Claridge, G.F. 1991. An Overview of Wetland Values: A Necessary Preliminary to Wise Use. PHPA/AWB Sumatra Wetland Project Report No. 7, AWB, Bogor.

Dahl, T.E. 1990. Wetland losses in the United States 1780's to 1980's. U.S. Department of the Interior, Fish and Wildlife Service, Washington, D.C. 13 pp.

James, R.F. 1991. The Valuation of Wetlands: Approaches, Methods, and Issues. PHPA/AWB Sumatra Wetland project Report No. 29, Bogor. 95 Pp.

Solórzano, R., R. de Camino, R. Woodward, J. Tosi, V. Watson, A. Vásquez, C. Villalobos, J. Jiménez, R. Repetto, and W. Cruz. 1991. Accounts Overdue: Natural Resource Depreciation in Costa Rica. World Resources Institute, and Tropical Science Center. Washington, DC, and San José. 110 Pp.

Financing Water Resource Projects in
Latin America and the Caribbean

Eduardo A. Figueroa, Ph.D. 1/

The period under study, 1989 to 1994, corresponds to the seventh capital replenishment of the Inter-American Development Bank (IDB). This period closes a very critical decade for Latin America and the Caribbean during which the worst economic performance has been registered since the great depression of the 1930's. During the decade, regional GDP dropped 5%, and on average, per capita income levels fell to those of the previous decade. However, the 1989-94 period marked the beginning of an important era of economic recovery for the region, with the IDB as the main source of external multilateral financing.

The Seventh Replenishment represented a capital increase of US$ 26.5 billion, leading to record lending levels totalling US$ 29 billion (see table below). Special priority was given to water resources projects, a total of 30 projects were approved, accounting for US$ 4.3 billion, approximately 15% of the total lending.

TABLE 1.
(US $ million)
IDB PROJECT APPROVALS 1989 - 1994

YEAR	1989	1990	1991	1992	1993	1994	TOTAL
OTHERS	2,471	3,189	4,874	5,143	4,530	4,484	24,691
WATER	82	614	496	966	1,448	747	4,353
TOTAL	2,553	3,803	5,370	6,109	5,978	5,231	29,044

Water resource management problems in the region are not significantly different from the ones found in more developed countries: competition for scarce resources; water shortages, very critical during the dry seasons; lack of adequate water markets and an increasing number of problems related to water quality. Water management must face complex dilemmas involving ecological,

1/ Senior Environmental Engineer, Inter-American Development Bank (IDB), 1300 New York Avenue, N.W. Washington D.C. 20577.
Findings, interpretations and conclusions expressed in this article are entirely those of the author and should not be attributed to the IDB.

political, legal and institutional issues. Some of these inter-related issues involve: (a) competitive consumptive uses (domestic, industrial, agricultural); (b) pollution control and treatment for water reuse; (c) non-consumptive use (energy, transportation); (d) flood control, and; (e) other uses, such as tourism, recreation, and habitat conservation. Their solution has been further jeopardized by the decision of many governments to assign this responsibility to institutions with neither a clear mandate nor adequate financial and human resources.

At the IDB, water resource project design underwent significant changes during the Seventh Replenishment period. Projects traditionally had been designed with a sub-sectoral focus (potable water, sewage collection, irrigation and hydropower). However, there has been a gradual move toward integrating related activities from different sub-sectors, such as water supply with sewage collection and treatment. By 1990, new environmental procedures were introduced, resulting in a more complete operational analysis during project preparation, including environmental impact evaluation. The environmental impact analysis considers the interactions existing within the geographical units of planning. This has led to an improved understanding of environmental problems which resulted in a new approach to water projects within the context of watershed or basin integrated management.

One of the most important operational policies related to pricing and cost recovery relates to the financial management of public enterprises. This policy recognizes that economic marginal costs will determine tariff structures which favor efficient resource allocation. As a minimum, the policy states that upon loan approval, costs for operation and maintenance of public services for residential water and sewerage systems should be covered by the rates charged. For commercial services such as electricity, surpluses must exist to cover future investments. It is also recommended that depreciation costs, debt service, system expansion and a profitability margin be included in public enterprise determination of water and sewerage rates.

During project design, special attention is now given to organizational and management efficiency, integrated resource management mechanisms, fair pricing systems and investment cost recovery. New principles in the IDB for efficient demand planning include: (a) adequate prices through efficient rate structures; (b) laws and regulations which influence demand and incorporate norms for effluents; (c) education and information programs; (d) more flexibility and liberalization of the water market, and; (e) efficient operational control.

As an example, the recently approved IDB project, "Pollution Control of the Tiete River" in Sao Paulo, Brazil totalling US $450 millon loan, will expand the sewerage system. Activities to begin the process of decontamination involve construction of sewage treatment plants, industrial pollution control, strengthening institutional capacity, environmental education, and pollution monitoring. Another example of IDB action in water resource management is an environmental sanitation and flood control project for the Reconquista River in Argentina to be financed with a US$160 million. Project activities include construction of flood control structures, implementation of an action plan to control industrial pollution, sanitary education and environmental awareness. Some similar projects are the Environmental Sanitation of the Guanabara Bay in Brazil and the Basic sanitation of the Medellin River in Colombia.

A project reflecting the IDB's integrated watershed management approach is "Environmental Management and Recovery of the Guaiba River Basin" in the state of Rio Grande Do Sul, Brazil. The Guaiba basin covers a surface of 85,959 Km^2, which represents more than 30% of the total area of the State. Most of the state's economic activity is concentrated in this basin, where approximately 6 million people live.

Environmental degradation of the basin can be attributed to: (i) deforestation, soil erosion and inefficient use of agro-chemicals in the rural areas; (ii) inadequate system support for protected areas; (iii) water pollution caused by domestic and industrial effluents, particularly in the metropolitan Porto Alegre area.

This environmental degradation is reflected in downstream areas, such as Guaiba Lake. It is estimated that only 5% of sewage is treated. Municipal and industrial wastewater flows directly into the river systems and lake. As a result, the level of "fecal coliforms" is above 100,000 org/100 ml in long stretches of the Cachoeriha and Gravatai rivers and even in areas located several kilometers downstream of the city of Porto Alegre on Guaiba Lake. Some of these highly contaminated areas are found upstream of water supply systems for urban areas, including Porto Alegre and some popular public beaches areas, representing a serious public health threat.

The general objective of the proposed US$ 220 million program is to improve the environmental quality of the Guaiba River Basin by reducing the level of rural and urban contamination, and by preserving the natural resources. The specific objectives are: (i) control and reduce contamination caused by industrial and domestic pollution; (ii) introduce soil conservation practices and improve the management of agro-chemicals in priority micro-watersheds; (iii) support the consolidation of Parks and Protected Areas (iv) support environmental education and awareness at a formal and informal level; (v) strengthen the environmental control capabilities at the state and local levels.

Toward these objectives, a series of activities and projects have been proposed. Sewerage systems for the Port Alegre Metropolitan Area will be expanded in coverage, including construction of 3 sewage treatment plants. Pollution control in the Guaiba Lake and its tributaries will be improved through the development of appropriate monitoring systems. Soil, agro-toxic and protected area management will be upgraded through extension services, institutional strengthening and training.

The goals of the program will result in the improvement of the environmental conditions and the quality of life of the inhabitants of the Guaiba River basin. The proposed sewerage system will benefit approximately 400,000 people in the MAPA (Metropolitan area of Puerto Alegre), eliminating open sewers and treating sewage before it flows into the natural systems. Industrial pollution control will be also improved by decreasing the organic load in approximately 50% in the coming 4 years. A total of 7,820, low income rural families will be able to increase their farming productivity, at the same time, the soil will be preserved and the use of agro-chemical products will be reduced. Biodiversity will be protected by habitat preservation and important species, some of them in danger extinction; and alternative recreational areas for the population will be established.

Regarding water quality of the Guaiba lake, predictive models, QUAL2E and HAR03, were run for different scenarios, with different assumptions for river flows, sewage treatment levels, location of treatment plants and final disposal sites. The solutions were studied in stages for a period to the year 2015. The first stage would cover the period between 1995 and 2002. The results of the predictive model show a comparison between situations with and without project. These results indicate the significant degradation that could be produced by the year 2025 if the project is not implemented and the substantial improvement in water quality if works are completed as planned.

The main objective of the economic analysis was to determine if the beneficiaries willingness to pay is higher than the cost of the project. The analysis was carried out considering economic efficiency prices assuming existing economic distortions. The selected alternative corresponds to the least cost technically feasible alternative. The economic feasibility of including the sewage treatment plants was verified when the environmental benefits calculated over the situation without project were higher than the costs. In order to estimate the benefits, a dichotomous choice contingent valuation method was used, with a follow-up question in the survey. The results present an internal rate of return of 28% for the program in Porto Alegre and approximately 18% for the down and up-stream systems.

The financial analysis showed the need to adjust the rate system of the sanitary companies of the State and Municipality of Porto Alegre. The purpose was to generate enough revenues to cover the operation and maintenance of the system as well as to meet the financial responsibilities and generate surplus for additional investments. This tariff revision was accomplished during the preparation of the operation.

For coordinating and executing purposes, the State established an Executive Secretariat with both public and private sector participation. An institutional strengthening program was proposed which includes equipment, training, advisory services and the installation of a State-wide data bank and Geographic Information System with interactive laboratories at partner institutions. The project is scheduled to start during 1995, a close monitoring from the Bank is proposed to follow up its implementation.

Looking at the future, the Bank has an even bigger challenge, since it has received a new capital increase of US$40 billion with the commitment that most of these new resources will be used in operations directed to the social sectors. This policy supports the national efforts throughout the region toward poverty reduction and social and economic reform. The projects related to water resources will continue to be an important part of the project portfolio as they relate to poverty reduction, environmental improvement and sustainable development of natural resources in Latin America and the Caribbean.

Integrated river basin management in Latin America

Axel Dourojeanni[1] and Andrei Jouravlev[2]

Unlike the United States and Western Europe which have traditionally relied on the comprehensive multi-purpose river basin planning approach (Rogers, 1992), Latin American and Caribbean countries have not widely applied it and there are few river basin management agencies in the region (Lee, 1990). Although there have been sporadic and isolated applications of the concept in several countries, few, if any, successful examples of integrated river basin management exist. Even the application of the concept in water planning has not been common.

In most countries which did attempt to apply the concept, the interest in river basin development emerged following the success achieved by the Tennessee Valley Authority (TVA) created in 1933 in the United States and the experience of other similar agencies created there in the late 1950s. The interest was reinforced by the influence of the international financial and development institutions founded following the Second World War. Bilateral technical assistance was also very influential, e.g. the United States in the training of professionals assisted the diffusion of the integrated river basin management concepts in the region.

Administrative approaches to river basin management adopted in the region varied from those in which the departmental set-up was preserved and the programs were managed by the central government departments to the cases in which, following the TVA model, largely autonomous agencies were created. Early examples are afforded by Brazil (1948) and Mexico (1950). During the next two decades, several river basin management agencies were established. Some of them were created

[1]Director, Water Resources and Energy Division, United Nations, Economic Commission for Latin America and the Caribbean (ECLAC), CEPAL, Casilla 179-D, Santiago, Chile. Phone: (562)-210-2248/2262. Fax: (562)-208-0252/1946.
[2]Associate Economic Affairs Officer, Water Resources and Energy Division, ECLAC.

to meet temporary emergency situations (Stöhr, 1975).

The period following the Second World War was marked by a complete reversal of the historical tendency regarding the role of the public sector in water resources management in the region. Latin America and Caribbean countries began to use public intervention in water resources management as a means to achieve the goals of economic and social development (Lee, 1990). This had important implications for the newly founded river basin agencies in that even where they were initially charged with similar functions to those of TVA, with the passage of time their responsibilities have been gradually reduced to those of a regional agency for the development of irrigation or hydroelectricity.

The boom at the end of 1970s marked the climax in the expansion of public economic activities typified by the undertaking of a number of grandiose water-related projects (UN/ECLAC, 1991). In 1975 there were 18 river basin authorities in existence affecting 10 countries, two of them related to international rivers (UN/ECLAC, 1979). They operated within a water management system which was highly centralized within the public sector and in the central government and dominated by national single-purpose agencies. In general, the private sector was excluded from participation in water resources management, though with some notable exceptions (irrigation) (UN/ECLAC, 1994a).

The 1980s - the "lost decade" - were marked by the most serious economic recession since the 1930s. Overstretched government budgets and the growing dissatisfaction with the performance of the public sector provided a fertile soil for an intellectual revival of faith in free markets and marked a region-wide movement towards privatization, deregulation and decentralization. The water-related activities have not been immune to these changes which provided an incentive for the general adoption of institutional arrangements based on a concept of resource-oriented, integrated river basin water resources management.

In most countries of the region, the role of the state in water resources management (and the economy as a whole) has been radically revised. Although in many countries these reforms are not yet complete, it is possible to discern some general tendencies: 1) water resources management is being privatized and decentralized in most countries, in some cases this involves the creation of water markets through the assignment of property rights to the water and permitting the holders to trade them freely (existing in Chile and proposed in Peru); 2) a definite move towards the self-financing of water services; 3) an aspiration to establish a global set of rules for the integral and sustainable management of water resources; and 4) a realization of the growing importance of the environment (this is in this area where there is least experience), including an acceptance of the need for the water services to finance the externalities associated with their provision (UN/ECLAC, 1994b).

These tendencies imply a need for a new water resources management system which should: 1) provide an enabling environment for the decentralization and private sector participation with a view to ensure efficient delivery of water-based services which remain woefully inadequate; and 2) be capable of addressing the largely ignored environmental problems and the challenges presented by the intensification of water use, including the increasing frequency of multiple, successive and mutually interfering uses of the water courses.

As a result, there is considerable interest in river basin-based water resources management which is increasingly being considered in the region the most appropriate way of internalizing the externalities of the water system (UN/ECLAC, 1994b). Chile, where the implicit water policy is largely to leave all management decisions related to irrigation in the hands of the autonomous user bodies, developed a decentralized management system in which these entities are responsible for regulating and administering the water resources and related infrastructure under their jurisdiction (UN/ECLAC, 1994a). However, a river basin is not used as a basic management unit (those are irrigation canals and sections thereof) and the user bodies' jurisdiction is limited to matters related to water quantity. The proposed reforms to the water law provide for the creation of autonomous private, but non-profit, river basin authorities in which both government and users would participate. In Peru, a new water law, currently under consideration, provides for a water management system similar to that found in Chile, but with the important addition of river basin regional authorities and river basin directorates. In Brazil, there are proposals in both the federal and state governments to establish a management system for water resources based on integrated river basin management. Some of the states have already enacted legislation (e.g. the state of Sao Paulo has established a new water management system based on the concept of integrated river basin management with considerable public participation; a similar system has been established in the state of Ceara). In Colombia, which is considered to have the most successful river basin agencies in the region, Autonomous Regional Corporations have been strengthened and assigned new environmental responsibilities.

In Mexico, the new water law provides for the establishment of river basin councils as advisory bodies for coordinating and facilitating cooperation between the National Water Commission, other agencies and representatives of the users to formulate and execute programmes and activities for improved water management, the development of hydraulic infrastructure and related services, and for the river basin conservation (most examples are from UN/ECLAC, 1994c).

These and other examples suggest that there are some indications of a reconsideration of the applicability of the river basin concept, even river basin planning, in water management in the region. However, there has been little tangible progress towards a river basin management system capable of challenges of the future and there remains a strong emphasis on the study of the physical components of the systems, on sectoral activities and investments. Much remains to be done (e.g. regarding the role and applicability of water markets; developing principles, standards and regulatory mechanisms; defining the roles of the public and private sectors; the policy paths, organizational forms and sources of financing of river basin entities). A change of this magnitude is a very complex matter and much will depend on the local political support and the ability to develop local representative entities.

Appendix

Lee, Terence Richard (1990), ***Water resources management in Latin America and the Caribbean***, Studies in Water Policy and Management, Nº 16, Westview Press, ISBN 0-8133-7999-7.

Rogers, Peter (1992), ***Comprehensive water resources management. A concept paper***, The World Bank, Policy Research Working Papers, WPS 879, Washington, D.C.

Stöhr, Walter (1975), ***Regional development experiences and prospects in Latin America***, the United Nations Research Institute for Social Development, the Hague, ISBN 90-279-7661-9.

UN/ECLAC (United Nations/Economic Commission for Latin America and the Caribbean) (1979), ***Water management and environment in Latin America***, Water Development, Supply and Management, Volume 12, New York, Pergamon Press, ISBN 0-08-024457-2.

___ (1991), ***The administration of water resources in Latin America and the Caribbean***, LC/G.1694, Santiago, Chile.

___ (1994a), ***Sharing responsibility for river basin management***, LC/R.1365, Santiago, Chile.

___ (1994b), ***Agenda 21 and integrated water resource management in Latin America and the Caribbean***, LC/R.1316(Sem.76/3), Santiago, Chile.

___ (1994c), ***A guide to water resources administration in the countries of Latin America and the Caribbean***, LC/R.1471, Santiago, Chile.

Toward An Inter-American Water Resources Network

Cathleen C. Vogel [1]

Abstract

The 1992 United Nations Conference on Environment and Development (UNCED), better known as the Rio Earth Summit, and more recently, the Summit of the Americas, focused tremendous attention on the need to develop strategies for sustainable development wherein thriving economies and healthy environments co-exist. No other resource is as critical to successful implementation of sustainable development as water. Today, water presents a unique challenge worldwide but especially within the Western Hemisphere. Here, disparities between developed and developing nations with respect to management and protection of water resources confront us with both crises and opportunities. In 1993, the South Florida Water Management District hosted the Inter-american Dialogue on Water Management which brought together more than 400 water resource professionals from 19 countries throughout North America, Latin America and the Caribbean. The goal of the Dialogue was to establish an Inter-American Water Resources Network (IWRN) for improved communications, technology transfer and information-sharing among practitioners throughout the hemisphere.

Background

While global warming has yet to claim its first human victim, water pollution and water shortages continue to claim thousands of lives in Central and South America and the Caribbean. 98% of sewage goes untreated in Latin America. But even in the face of these water crises, many Latin American countries are moving forward to revolutionize their approach to water. While some countries are decentralizing water utilities and management systems, others are adopting national water laws. Massive projects such as the Hidrovia are dictating new, environmentally-sound approaches to water projects at the same time that the World Bank's adoption of a Water Policy requires borrowing countries to enact national water management strategies.

[1]Deputy Director, Office of Government & Public Affairs, South Florida Water Management District, PO Box 24680, West Palm Beach, FL 33416-4680

In the United States, water resource professionals may not be struggling with standard of living issues, but we are spending billions to rebuild and retrofit deteriorating infrastructure or restore ecosystems that have been brought to the brink of collapse by shortsighted developmental practices that epitomized unsustainable development.

These situations represent tremendous opportunities for linking the water resource "resources" of the Americas. As borders and trade opportunities open, water resource professional are increasingly working at "nuts and bolts" levels to share expertise and know-how in water management and protection of natural systems.

Inter-American Dialogue on Water Management

The Inter-American Dialogue on Water Management was an experiment in international relationship-building based on a demonstrated need and demand. Throughout the 1980's numerous delegations of Latin American water resource interests visited the South Florida Water Management District[1] to exchange information and concerns about shared problems: aquatic weed control, drought management, lake eutrophication, wetlands preservation, irrigation technologies - even environmental education. This "bottom-up" phenomenon wherein water resource professionals were coming together - outside of any formal exchange program - prompted the water management district to organize the Inter-American Dialogue in order to draw attention to the need for interactive, collaborative and networking opportunities.

The Dialogue was both well attended and well supported. It was jointly sponsored by the five water management districts of Florida, the Florida Department of Environmental Protection, and the Interstate Council on Water Policy. The Global Tomorrow Coalition - a private, non-profit organization dedicated to international education for sustainable development - was a key organizer of the conference. Beyond these sponsoring organizations, a number of stellar financial contributors made it possible to underwrite the participation of many Latin American attendees. These included the John D. and Catherine T. MacArthur Foundation, the Organization of American States, the World Bank, the Tennessee Valley Authority, the University of Miami North-South Center, the City of Miami, the National Geographic Society, Law Companies, Ecology and Environment, Inc. and Blockbuster Corporation - to name a few. The sheer number and diversity of financial supporters - from international banking institutions, to charitable foundations, to academic institutions, to consulting firms - underscored the fact that the Dialogue was an undertaking whose time had come. And beyond financiers, the array of groups and organizations that got involved was astounding - the Army Corps of Engineers, Florida Power & Light Company, the

[1] South Florida Water Management District is a regional agency responsible for flood control, water supply protection, water quality protection and environmental enhancement.

US Environmental Protection Agency, the National Audubon Society - in fact, one of the first entities to take an active interest in the Dialogue was the International Section of the American Society of Civil Engineers! In short, it was an effort that made sense, held promise, and, in an era when the nation's focus is shifting from an east-west orientation to a north-south orientation, the Dialogue was a great conduit for sharpening that focus.

The Dialogue generated a variety of interactions: an NGO forum was held; an Inter-American Survey of Water Resources Professionals was conducted; a subregional discourse was held between the countries of Costa Rica and Nicaragua on the San Juan River Talks; two case studies were presented - one on the Everglades and the Pantanal, and one on water supply and sanitation infrastructure in the hemisphere. But the heart of the Dialogue was the convening of three Roundtable discussions: Management of Aquatic Ecosystems, Infrastructure for Water Supply and Sanitation in the Context of Sustainable Development, and Water Governance and Policy. And if the heart of the event was the Roundtables, the soul of the Dialogue was the "Statement of Miami" [1] - a summary document produced collaboratively by the conferees which outlined a call for action, the key to which was the establishment of an Inter-American Water Resources Network which would facilitate communication, cooperation and interaction among groups with a shared commitment to the sound management of water resources in the Americas.

Toward An Inter-American Water Resources Network

One of the key factors in the success of the Dialogue and its evolution to an Inter-American Network was the formation of a multi-national advisory council comprised of public and private sector representatives which met before, during and since the Dialogue to provide guidance, counsel and continuity to the effort. The Advisory Council created a sense of ongoing purpose and ownership in this grass roots project which furthered the notion of inclusion and direct involvement among many diverse interests. Through the work of the Advisory Council, the Organization of American States (OAS) was enlisted and agreed to house the Technical Secretariat of the IWRN.

In reality, the IWRN is a network of networks, operating out of the Department of Regional Development and Environment within the OAS. Its purpose is to create and strengthen water resources partnerships in the Western Hemisphere that address public health issues, water supplies, sanitation, environmentally sound management of water resources, and the conservation of ecosystems.

[1] Copies of the Statement of Miami and Dialogue proceedings are available from the South Florida Water Management District.

Barely a year old, the IWRN is by all accounts a fledgling organization, but it is already making progress and taking on several initiatives. The OAS has employed two consultants to manage the Technical Secretariat and the United Nations Environment Programme has provided first year funding. The Technical Secretariat is concentrating its present efforts on fund raising, program development, and formalizing links between the OAS and institutions and agencies that will serve as nodes on the network.

The building blocks of the IWRN will be its nodes - locations of activities that are sponsored by or affiliated with the IWRN. A long-term goal will be to link all nodes electronically, but it is recognized that in the meantime, newsletters, telefax, telephone and mail will be important connecting tools.

While the nodes will serve as key linkages in the IWRN, the users of the network will be a fluid and limitless group - government agencies, academic and research institutions, professional associations, consulting firms, NGOs, corporations, the media - literally anyone with an interest in inter-american water resource matters. The IWRN will serve its users with information on specific water issues such as conservation, tropical hydrology, floodplain management, etc. And the IWRN will also assist with geographic-specific areas, mainly international river basins, where there are explicit water management objectives and a commonality of interests. The first such river basin, the Titicaca-Desaguadero-Poopo-Salar system of Bolivia and Peru will use the network to facilitate communication between the local water community and geographically similar areas in other parts of the Americas. OAS water project leaders will employ the network to exchange information about water resources in transboundary watersheds.

There is no doubt that the goals of the IWRN are ambitious. But its game plan is to start small and build on success. The Technical Secretariat has articulated a two-year action agenda that will focus on 1) Expanding political support for the IWRN among OAS member countries and from international organizations; 2) Holding periodic meetings of the Advisory Council - one which will be held in Washington in the Spring of 1995, and a second scheduled in conjunction with the annual conference of the American Water Resources Association in Houston in November 1995; 3) Compiling directories of water agencies, water networks, and sources of funding for water partnerships; 4) Establishing national IWRN nodes; 5) Preparing a periodic newsletter; 6) Implementing an international river basin project; and 7) Cooperating with the country of Argentina which has offered to host the Second Inter-American Dialogue on Water Management.

See you in Buenos Aires in 1996!

The Water Resources Management System of the State of São Paulo, Brazil

Rubem L. Porto[1]
Benedito P.F. Braga Jr.[1]
Flavio T. Barth[1]

Abstract

Brazil is a large developing country characterized by regional diversities in social, economical and political aspects. Water availability, quality and demand greatly differ from region to region. Clearly, the country needs appropriate strategies to tackle the water resources problems that arise from such complex system. This paper gives a general overview of the Brazilian situation in general and of the situation of the State of São Paulo in particular, and describes the efforts the authorities are doing toward the organization of a National Water Resources Management System. Simultaneously, several states are organizing their own systems to deal with different regional conditions.

Introduction

Brazil is a continent country. Its large territory with 8.5 million Km^2, its population of 150 million and significant diversity in physiographic, economic, hydrologic, cultural and environmental terms poses a complex problem of management. The average annual flow of the country is estimated at 251,000 m^3/s, of which 202,000 m^3/s alone come from the Amazon basin. In a significant portion of the territory, including the Northeast and the São Francisco River basin, the climate is semi-arid and water is a limiting factor to the socioeconomic development. In the south and southeast,

[1]Engineer, DAEE-CTH/USP, Av. Prof. Lucio M. Rodrigues, 120, São Paulo, SP, 05508-900, Brazil

regardless of the great availability of water, large urban centers and the industry produced such a pollution load that there are several watersheds with shortage of good quality water.

The Brazilian Water Resources Association-ABRH, in the Declaration of Rio (1981), states the following directives as the appropriate management for the Brazilian waters and the environment: 1) sustainable development of the Amazon, observing the economic-ecologic zoning and regional development plans, existing or to be formulated in the near future, with adequate water resources projects included, 2) social and economical development of the northeast and the São Francisco River basin, based on irrigation, harmonized with hydropower production and environmental protection, 3) social and economical development of the north and central west savannah region with irrigation, multiple use of water with special attention to hydropower, navigation and environmental protection, 4) in the southeast critical watersheds, the management should be integrated, decentralized with public participation and harmonious with regional development plans, 5) in urban and industrial centers, water management should be integrated with land use to prevent flooding, landslides and pollution of selected watersheds for water supply purposes, 6) rational development, protection and conservation of coastal areas through appropriate management, and 7) preservation and conservation of the Pantanal, through the protection of fragile ecosystems and the use of flood warning systems to manage agriculture and cattle production.

Water Resources Management of the State of São Paulo

The State of São Paulo, southeastern Brazil, with an area of 250,000 Km^2 and a population of 31 million, is the most developed and industrialized state. Large urban and industrial areas, specially in the Alto Tiete River basin, impose water shortages due to severe water pollution. To cope with this situation, a new water resources policy, based on a state law, is being implemented. This policy is producing important changes in government attitudes as well as the society behavior, with respect to the development, control, protection, conservation and restoration of the water resources.

This state law defines the objective of the policy as "to assure that water, an essential resource necessary to economic growth and social welfare, may be used, under satisfactory quality standards, by all users and guaranteed to future generations, throughout the state territory". Among the principles of the State of São Paulo Water Resources Policy, it can be cited:

- integrated, participatory and decentralized management;

- adoption of the watershed as the planning unit;

- recognition of the water as a public good, which utilization must be charged, maintained all aspects of quantity and quality;

- harmonizing water resources management with regional development and environmental protection.

The basic elements for implementing this policy are water charging and the polluter-pay principle, with institutional arrangements similar to those introduced in France in 1964, such as the watershed commissions and watershed agencies.

The User/Polluter-Pay Principle in the Legal System

Water charging was a hypothesis already envisioned in the Brazilian legal system as back as 1934 with the Water Code and also in the 1981 National Environmental Policy Act. More recently, water charging is being considered in the constitution of several other states, besides São Paulo. A federal law, in debate since 1991, proposes that the same principle be applied to all federal rivers.

Water charging has shown, in those countries where it has been applied, an effective measure in solving water shortage difficulties, allowing the protection and restoration of water bodies. The main objectives to be achieved through water charging in the State of São Paulo are:

- to redistribute costs in a more equitable way;

- to manage the increasing demand in order to achieve efficiency in water use;

- to constitute a fund, in order to provide means of implementing regional plans;

- to stimulate integrated regional development, especially in its social and environmental aspects.

To charge for the water is expected not to be another formula to collect funds for government investment, since society would be contrary to. The level of public participation in the State of São Paulo has increased significantly in the past years, probably due to the effort of promoting awareness made by some non-governmental organizations.

Water charging will have an immediate effect and a direct impact over:

- decreasing significantly the water demands, with losses reduction and consequent abatement of capital investments;
- introducing rationality in the decisions over the implementation of large scale developments, which may have great impact on maintaining water quality;
- financial and economic feasibility of environmental restoration;
- reducing the use of fertilizers, pesticides and energy in the irrigated agriculture.

In order to assure an efficient water charging system, the following mechanisms will be used:

- equitable participation of all government levels and society representatives in state boards and watershed commissions;
- preparation of a 4-year water resources master plan, to be approved both by the State Congress and the watershed commission;
- publication of a yearly report on the water resources status, to be shared with the society.

References

Barth, F.T., et al., 1987, *Modelos para Gerenciamento de Recursos Hídricos*, ABRH/Nobel, São Paulo.

Barth, F.T., 1994, "Alternativas propostas para o Sistema Nacional de Gerenciamento de Recursos Hidricos", *Boletim ABRH*, no. 50, Abr./Jun. 1994.

Governo do Estado de São Paulo, *Lei no. 7663, 31 de dezembro de 1991*.

Educational and Research Issues on the Road to Integrated Water Resources Management

Hector R. Fuentes[1] and Vassilios A. Tsihrintzis[1]

Abstract

The 1992 United Nations Conference on Environment and Development formalized an international environmental agenda, referred to as Agenda 21, which outlined an integrated approach to the management of water resources and associated ecosystems. This Agenda offers to the hemisphere's academic community the opportunity to develop educational curricula and research programs in accordance with its goals. Overall, the academic sector should move forward with innovative approaches, based on new and integrating concepts such as sustainability and ecosystem management. These approaches must provide future engineers with the knowledge to effectively implement the growing number of international agreements and national plans that are directing the world toward an environmentally sustainable future.

Introduction

The search for sustainable patterns of economic development by the nations of the hemisphere demands a careful evaluation of university educational curricula and supporting research efforts. Results of this evaluation are expected to ensure that both curricula and research effectively respond to the spirit of the Rio Declaration on Environment and Development and the contents of Agenda 21 (United Nations 1992). Twenty-seven articles of the Declaration form the political framework for **seven priority actions** and **seven essential means**, whose timely implementation are expected to halt and reverse the effects of environmental degradation, and promote sustainable and environmentally sound development in all countries.

---

[1]Department of Civil & Environmental Engineering, Florida International University, University Park, VH-160, Miami, FL 33199, USA.

All **priority actions** include specific themes that directly point at a university's mission for education and research in the areas of water resources, including:

- urban water supplies;
- urban pollution and health;
- protection of the quality and supply of freshwater resources;
- protection of the oceans and all kinds of seas, including enclosed and semi-enclosed seas; and
- managing fragile ecosystems.

Various aspects of the above mentioned themes have traditionally been a prime responsibility of engineering disciplines, such as civil, environmental, water resources, coastal, agricultural, and public health engineering, that typically offer water resources undergraduate and graduate programs. *The challenge today for these programs is to link environmental protection within sustainable patterns of development.*

Additionally, at least three of the seven **essential means** strongly connect to a university's mission in education and research, namely:

- science for sustainable development;
- transfer of environmentally sound technology, cooperation and capacity-building;
- information for decision-making;

These essential means directly relate to the university's mission as a provider of educated human resources that will assure the continuous sharing and expansion of general knowledge, technical knowledge and knowledge building (James 1993). In fact, it is known that universities prepare students for careers that may span decades, when individuals contribute with sound technological applications, sound decision-making, and acquisition of new knowledge. *As a result, universities are key players in the development of those essential means that will lead to the realization of the priority actions described by Agenda 21.*

Conclusively, the need is to evaluate and enhance university educational curricula and supporting research to ensure that future graduates will have the fundamental abilities to manage water resources within an integrated context driven by sustainability goals.

Current Situation

In the last decades, university programs in water resources have continuously searched for the best approaches to the incorporation of relevant environmental elements into their educational curricula and supporting research activities. Today, over one hundred water resources programs are offered in the USA. A number of authors have written historical and future perspectives of these programs addressing the issue either holistically (Luthy et al. 1991; Grigg 1992; James 1993) or specifically (Öziş 1992; Edwards et al. 1992). There are a number of common considerations among these different perspectives:

a) The education of engineers in water resources, regardless of the discipline (e.g., civil, environmental, agricultural engineering), has been a continuous concern, with changes driven mostly by emerging problems, societal attitudes and technological advancement.

b) The focus has gradually shifted from local and regional, to include national and global issues, as an acknowledgement of the scale and extent of environmental problems.

c) There has been a permanent search for the best package of academic coursework and experiences that will endow water resources students with the required abilities to successfully contribute to the management of water resources.

d) A trend of specialization with incorporation of multidisciplinary elements from environmental science, engineering and management areas (e.g., ecology, chemical engineering, and economics) has recognized the political, legal, financial, scientific and technological characteristics of water resources problems.

e) Research efforts have primarily supported the decision-making needs of industry and government (at all federal, state and local levels).

Today, the most relevant issue is how water resources programs should address conceptual integration of the trilogy of water resources, environment and economic development, called for by Agenda 21. *The new driving descriptor is sustainability and the question is **how** to incorporate it into educational curricula and research endeavors.*

It is important to consider that pressure is building from the various levels of governmental action. For instance, the State of Florida has already passed legislation that mandates ecosystem management. These two concepts force policy, regulatory, scientific, technical, and socio-economic actions to approach water resources within a holistic context, in contrast with common practice.

Hemispheric Status

Although the analysis, herein presented, focuses on the USA situation, the academic curriculum question is quite comparable in the rest of the american context. Water resources programs in other nations, e.g., Mexico and Brazil (OMS 1991), have been somewhat modeled after USA approaches with various levels of influence from the UNESCO initiative (Gilbrich 1991). This initiative successfully sponsored studies of the hydrological cycle that began in the mid-sixties. Programs in the USA and Canada particularly differ from those in other countries in the limited number of graduate programs (OMS 1991), which is significantly critical at the doctoral degree level. This difference is aggravated due to the very limited supporting research infrastructure and efforts. A common denominator throughout the continent is that Agenda 21 should be addressed and effectively articulated by all university water resources programs in universities around the hemisphere.

Recommendations

Of utmost priority for academia is understanding sustainability. This concept, with its associated agendas (e.g., Agenda 21), must reach the water resources academic sector, so that awareness, debate and exploration in curricula and research can actively continue within the newly proposed framework. Academic efforts should be complemented by initiatives from other groups, including research sponsors, governmental agencies, professional organizations, and accreditation boards. These groups and their counterparts in each country can provide a framework for debate through their publications, conferences and workshops.

Appendix. References

James, L. D., 1993. "An Historical Perspective on Water Resources Education," The Universities Council on Water Resources, Issue No. 91, Carbondale, Illinois.

Luthy, R. G., D. A. Bella, J. R. Hunt, J. H. Johnson, D. F. Lawler, C R. O'Melia, and F. G. Pohland, 1991. "Future Concerns in Environmental Engineering Graduate Education," AEEP Conference on Environmental Engineering Education for the Year 2000, Oregon State University, Corvallis, Oregon.

Edwards, H. W., D. W. Hendricks, F. A. Kulacki, J. C. Loftis, V. G. Murphy, T. G. Sanders, and R. C. Ward, 1992. "Environmental Education in Engineering," Water Resources and Environment: Education, Training and Research Conference, Colorado State University, Fort Collins, Colorado, July 13-17.

Gilbrich, W. H., 1991. "25 Years of UNESCO's Programme in Hydrological Education under IHD/IHP," UNESCO, Paris, France.

Grigg, N. S., 1992. "Water Resources Management: A Challenge for Educators," in Hydrology and Water Resources Education, Training and Development (ed. J. A. Raynal), Water Resources Publications, Littleton, Colorado.

OMS, 1991. Directorio de Programas de Formación en Ingenieria Sanitaria y Ambiental en América Latina y el Caribe, Asociación Interamericana de Ingeniería Sanitaria y Ambiental, Organización Panamericana de la Salud, Washington, D. C.

Öziş, Ü., 1992. "Towards Establishing Water Resources Engineering as an Undergraduate Engineering-degree Program," in Hydrology and Water Resources Education, Training and Development (ed. J. A. Raynal), Water Resources Publications, Littleton, Colorado.

United Nations, 1992. "The Global Partnership for Environment and Development: A Guide to Agenda 21," New York, New York.

Evaluating Reuse Feasibility with Predictive Models

David K. Ammerman, P.E.[1]

Abstract

Evaluating the interaction of supply and demand for water reclamation systems is a critical design element. Anticipation of high demand periods is required to ensure adequate facilities are available to handle peak flows. When reclaimed water demands are less than supplies, provisions for storage and/or alternative disposal are required to obtain regulatory approval. Traditional water balance models using a monthly time step were compared to a daily water balance model. The results indicate that monthly time steps and synthetic rainfall distributions may be inappropriate and are poor permitting criteria.

Discussion

One of the major challenges for the design of an irrigation based reuse system is to provide the water when it is needed. Designing storage simply to meet the median irrigation requirements results in a system, that by design, fails to provide the needed water half of the time. Seasonal storage may be required during periods of low demand to meet later peak demands, or to avoid an un-permitted discharge. Often storage facilities have a substantial impact on the capital cost of the system. Thirteen states (Payne, 1993) have adopted regulations pertaining to the amount of storage required to limit discharges. These vary considerably by locale. Most states require that a water balance be developed for an abnormally high annual rainfall total.

Most of the water balances contained in the regulations of various states are derivatives of those published by the Environmental Protection Agency (EPA) (USEPA, 1981), or the State of California (PettyGrove and Asano, 1985). These methods generally use a monthly time step, and determine the demand for reclaimed water as the difference between potential evapotranspiration (PET) and rainfall. PET is determined using either the Blaney-Criddle or Thornthwaite equations. In the EPA methodology, synthetic monthly rainfall totals are derived by distributing the annual

----

[1]Project Manager, Camp Dresser & McKee Inc., 1950 Summit Park Drive, Suite 300, Orlando, FL 32810

rainfall total with a 10-year recurrence interval against the normal distribution.

There are several limitations to both the synthetic distribution and the monthly time step. Analysis of rainfall records suggests that the years of abnormally high rainfall are not always due to proportional monthly increases, but often are the result of a few high monthly totals. Further, specific months of abnormally high rainfall may be the result of intense but brief storm events that effect irrigation demands for only a limited period. On the other hand, well distributed rainfall equal to or less than the average may have a significant impact on irrigation demands.

These issues, and other problems associated with monthly water balance estimates suggest the need to perform a daily water balance in order to fully estimate the risk of not having the correct balance of supply, demand, and storage. The use of daily water balance has been suggested as appropriate for Florida (Payne *et al.*, 1994) and Mississippi (Pote and Wax, 1995). EPA (USEPA, 1981) recommends the use of daily water balance models (EPA 1, EPA 2, or EPA 3 as appropriate) developed by the National Climatic Center to estimate storage days required for land application systems. These programs incorporate an available soil moisture routine, and a daily depletion rate to account for gravitational loss of excess rainfall to the soil, but do not appear to simulate groundwater directly (Whiting, 1976).

Figure 1 illustrates the relationship between supply and demand based on actual recorded irrigation use for a golf course in southwest Florida.

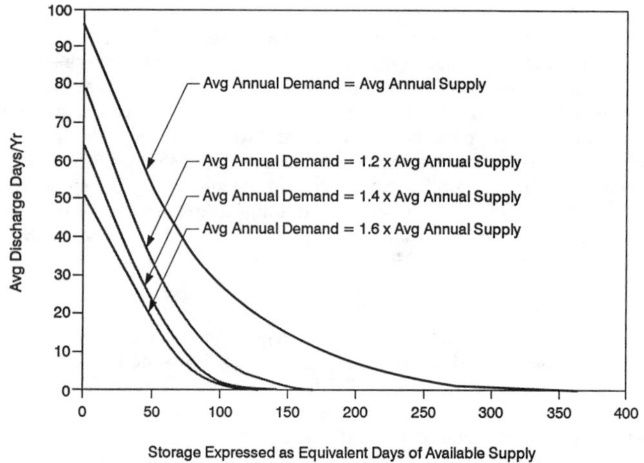

Figure 1: Relationship of Supply and Demand Based on Long Term Irrigation Records

For the purposes of developing this example, irrigation demands were converted into demand factors by dividing actual recorded monthly irrigation usage by the average monthly irrigation usage over the period of record (12 years). A dimensionless supply of 1 unit of reclaimed water was assumed to be available to meet irrigation demands. The reclaimed water demand was then applied to the available supply over the available period of record as demand varied from 1 times available supply, to 1.6 times available supply.

Referring to the curve where it has been assumed that supply is equal to demand on a long term basis, it is interesting to note that storage volumes in excess of 100 days of equivalent supply would be required to reduce average discharges below 30 days per year. This compares well with the results of a daily water balance simulation for this area which predicted a storage volume of 120 days of equivalent supply in order to prevent discharge in the 1-in-10 year rainfall event. Monthly water balances run for the same area predict the storage volumes of less than 20 days will be required to obtain the same results.

As an alternative to providing massive volumes of storage, Figure 1 also illustrates the use of expanding reclaimed water application sites such that the average annual demand for reclaimed water will exceed the average annual supply. While it can be seen that this strategy will decrease discharge days associated with the given volume of storage, 50 days of storage would be required to reduce effluent discharges below 30 days per year, even as reclaimed water demands represent 140 percent of the available supply. Further, the strategy of increasing average reclaimed water demands beyond that of the supply, will create periodic shortfalls to the reclaimed water customer. Where reclaimed water is intended to be used for beneficial purpose (i.e., urban or agricultural irrigation), chronic shortfall cannot be tolerated because: (1) it will devalue the benefits afforded to the reclaimed water customer, and (2) it may potentially inspire the reclaimed water customer to seek alternative sources during shortage, thus creating potential cross-connections between the reclaimed water system and other water supplies.

Conclusions

The daily distribution pattern of rainfall has more of an effect on irrigation requirements than total annual rainfall, and thus annual rainfall totals are poor criteria for use in permitting. While monthly water balance models are appropriate for preliminary design considerations, long-term daily simulations are recommended for final design. The components of the water balance must reflect all of the sources of water available to the crop, and even daily simulation results should be carefully scrutinized and compared to actual irrigation practices wherever possible.

References

Payne, J.F., 1993. Summary of Water Reuse Regulations and Guidelines in the United States. Proceedings of the Water Environment Federation 66th Annual Conference. Anaheim, CA. October 3-7, 1993.

Payne, J.F., Heyl, M.G., and D.K. Ammerman. Modeling Reclaimed Water Uses. American Water Works Association Water Environment Federation. 1994 Water Reuse Symposium Proceedings, Dallas, Texas.

PettyGrove, G.S. and T. Asano, ed. 1985. Irrigation with Reclaimed Municipal Wastewater - A Guidance Manual. Lewis Publishers. Chelsea, MI.

Pote, J.W. and C.L. Wax. Climatic Criteria for Land Application of Municipal Wastewater Effluent. Water Resources, Volume 29, No. 1, pp 323-328, 1995.

USEPA, 1981. Process Design Manual. Land Treatment of Municipal Wastewater. EPA 625/1-81-013.

Whiting, D.M., 1976. Use of Climatic Data in Estimating Storage Days for Soil Treatment Systems. USEPA Office of Research and Development. EPA 600/2-76-250.

EFFLUENT REUSE OPTIONS IN THE DESERT ENVIRONMENT

Michael Gritzuk[1], Paul Kinshella[2], Andrew W. Richardson[3], Member, ASCE, and Michael P. Valentine[4], Member, ASCE

Abstract

Future State Water Quality Standards for Navigable Waters and associated National Pollutant Discharge Elimination System permits are expected to become ever increasingly more stringent. As a result, the City of Phoenix, Arizona, is studying the technical feasibility and financial viability of various treatment alternatives that can meet these anticipated standards. Treatment alternatives include providing an upgrade of conventional equipment at the plant with continued discharge to the Salt River, constructing wetlands with wildlife habitats, or developing underground storage and recovery with soil aquifer treatment. A combination of these alternatives may be selected.

Introduction

The City of Phoenix, Arizona, operates the 91st Avenue Wastewater Treatment Plant (WWTP) on behalf of the Multi-Cities Sub-Regional Operating Group (SROG). SROG consists of the cities of Glendale, Mesa, Phoenix, Scottsdale, Tempe and Youngtown. Currently, the 91st Avenue WWTP treats approximately 5.68EE5 m^3/day (150 MGD) equating to 2.07EE4 ha-m/year (168,000 acre-feet/year). Treated wastewater can be discharged into the adjacent Salt River and also diverted to Palo Verde Nuclear Generating Station. Palo Verde Nuclear Generating Station has a contract allowing them to obtain up to 1.30EE4 ha-m/year (105,000 acre-feet/year) of 91st Avenue WWTP effluent. Approximately seven miles downstream of the WWTP, the Buckeye Irrigation Company is allowed to divert up to 0.52EE4 ha-m/year (42,000 acre-feet/year)

[1]Director of Water Services, City of Phoenix, 200 W. Washington Street, Phoenix, AZ 85003

[2]Wastewater Engineering Superintendent, City of Phoenix, 200 W. Washington Street, Phoenix, AZ 85003

[3]Partner, Greeley and Hansen Engineers, 426 N. 44th Street, Phoenix, AZ 85008

[4]Engineer, Greeley and Hansen Engineers, 426 N. 44th Street, Phoenix, AZ 85008

of Salt River flow for seasonal agricultural use. Salt River flow upstream of the WWTP typically occurs only in the rainy season.

Triennial review of the Arizona Water Quality Standards for Navigable Waters (WQSNW) and associated National Pollutant Discharge Elimination System (NPDES) permit requirements has promulgated SROG into studying various alternatives that maximize the ability of 91st Avenue WWTP reclaimed water to continually meet the WQSNW and minimize negative impact on the public. The various alternatives studied include:

- Continued discharge of reclaimed water into the Salt River by upgrading the 91st Avenue WWTP with conventional equipment
- Tres Rios constructed wetlands for further treatment of the water with development of natural habitats
- Underground storage of reclaimed water via percolation ponds for Soil Aquifer Treatment (SAT) and future recovery for various uses

Continued Discharge of Reclaimed Water into the Salt River

For SROG to continue discharging the 91st Avenue WWTP reclaimed water into the Salt River and meet the anticipated continually stricter WQSNW, an upgrade to WWTP with conventional equipment will be required. The extent of the upgrade will be determined by the severity of the adopted standards. It was estimated in 1992 that the upgrade could cost as much as $368 million based on the proposed WQSNW.

The effluent from 91st Avenue WWTP discharged into the Salt River has always been of high quality, and as a result, SROG has won numerous awards. Continual discharging of this high quality, treated water into the Salt River without any means of recovery and reuse, seems wasteful. Consequently, SROG questioned why they should treat the water, essentially to drinking water standards, just to release it into the Salt River where it is lost and cannot be reused.

Tres Rios Constructed Wetlands

Downstream of the 91st Avenue WWTP, the Salt River merges with the Gila and Agua Fria Rivers. It is at the confluence of these "three rivers - tres rios" that a full-scale constructed wetlands is being considered. Tres Rios, a free water surface wetlands, would receive flow from the WWTP. The wetlands would also provide water treatment, wildlife habitats, public recreation and education, and flood protection.

To facilitate the design, construction and operation of the potential full-scale wetlands project, a wetlands demonstration project has been initiated. The demonstration project consists of three separate test sites. The Riparian site, located in a hayfield adjacent to the WWTP, the Cobble site, located on a cobble gravel bar in the river bed adjacent to the WWTP, and the research cells near the chlorine contact tanks, in an off-line, solar drying sludge lagoon. The cobble site best represents the conditions existing at the proposed full-scale Tres Rios wetlands project. The basins will be formed by grading river rock into embankments that contain the reclaimed water to flow in an ox-bow shape direction, but not limit the water's movement into the subsurface.

The Tres Rios demonstration wetlands entails a process stream flowing first through a marsh and then into a deeper quiescent pool. The flow exits the deeper pool and enters into an additional marsh before treatment is complete. The first marsh provides initial wetlands treatment to the influent. The deep, quiescent pool allows for plug flow distribution into the second marsh. The second marsh provides the final treatment.

Traditionally, constructed wetlands have incorporated cattails, reeds and bulrushes in the marsh for treatment purposes. Treatment of the water at the Riparian and Cobble sites demonstration projects will be facilitated primarily through the employment of bulrushes. In addition to the bulrushes, submersed aquatic species will be planted in the deeper pools. The submersed plant species compete against algae species, provide migrating waterfowl a source of food, and provide supplemental treatment of the water.

Underground Storage of Reclaimed Water with Future Recovery

Water in the arid southwest is a natural and financial resource. Accordingly, an alternative considering the underground storage and recovery of reclaimed water, after Soil Aquifer Treatment (SAT), is presently being studied as well. The public, regulatory, technical, and financial aspects of implementing a Total Reuse Facility (TRF) with SAT are addressed within this portion of the overall study.

SAT research is being conducted by Arizona State University, the University of Arizona, the University of Colorado, and Greeley and Hansen Engineers. The research is supported by the American Water Works Association Research Foundation (AWWARF) and the Water Environment Research Foundation (WERF). The objectives of the research are to verify previous research findings that indicate the primary drinking water standards can be achieved and to verify the sustainability of a full-scale SAT system.

The proposed location of the TRF recharge site has been identified near the confluence of the Agua Fria and New River. The percolation basins could be constructed either inside or outside of the Agua Fria floodway. The proposed recharge site is located approximately 16 km (10 miles) northwest of the WWTP.

SAT is facilitated through the employment of percolation basins above a permeable vadose zone and transmissive aquifer. To determine in-situ infiltration rates, two pilot percolation test basins were constructed in July 1993 and operated for approximately six months. The basins were both within the proposed full-scale TRF recharge area, outside of the floodway, but separated by approximately one mile. Well water was used during the testing to represent the 91st Avenue reclaimed water. Both sites produced clean water infiltration rates that averaged approximately 0.76 m/day (2.5 feet/day). Operation of the basins was typically 14 days of standing water followed by 14 days of solar drying.

Assuming equal wet and dry cycle times, and applying a conversion factor for well water infiltration to reclaimed water infiltration, it has been estimated that a full-scale TRF can sustain a long-term infiltration rate of 0.23 m/day (0.75 feet/day) of reclaimed water. For the conceptual TRF maximum design capacity of 7.57EE5 m^3/day (200 MGD), approximately 425 ha (1050 acres) of percolation basins are required.

The TRF has been conceptually designed to manage the WWTP's maximum potential flow and eliminate discharges to the Salt River. SROG's TRF consists of the following:

- 9.84EE5 m^3/day (260 MGD) Low Lift Pumping Facilities to lift reclaimed water into the proposed equalization basins
- 9.1EE4 m^3 (24 MG) Equalization Basins to equalize the maximum anticipated diurnal flow volume
- 7.57EE5 m^3 (200 MGD) High Lift Pumping Facilities to pump the reclaimed water from the Equalization Basins to the Recharge Site
- 2.4 m (96-inch) diameter, 16 km (10 miles) transmission pipe from the WWTP to the recharge site
- Percolation Basin Recharge Facilities with SAT to treat the reclaimed water and recharge the aquifer
- Recovery Facilities and Distribution System to recover the groundwater from the aquifer and provide it to various customers

The TRF has been estimated in 1994 to cost between $180 million and $201 million. The costs of recovered water range from $1,610 per ha-m ($199 per acre-foot) to $1,810 per ha-m ($223 per acre-foot), which indicates that underground storage and recovery is a financial and viable water reuse alternative.

Conclusion

Analysis of the results from both the wetlands demonstration study and the reclaimed water study may indicate that the best solution available to SROG is to develop both a wetlands project and a scaled-back recharge and recovery project.

Wastewater Reclamation and Recharge: A Potential Water Resource Management Strategy for Albuquerque

Paul J. Gorder[1]
Robert J. Brunswick[2]

Abstract

Similar to other major communities in the arid Southwest, Albuquerque, New Mexico has both a growing demand for water and a limited long-term supply. Currently supplied by a groundwater-based system of finite capacity, Albuquerque's options to meet future demands will all be complex. One of the most promising scenarios for extending the life of the groundwater resource is the indirect potable reuse of treated wastewater, by further treatment of the flow to drinking water standards and direct injection into the City's groundwater supply. Studies have been conducted by CDM to determine the best treatment techniques and injection schemes, toward development of a potential recharge program for Albuquerque. The City will then evaluate this scenario along with other long-term water management options.

Introduction

The City of Albuquerque, New Mexico is characterized by an arid climate with an annual rainfall of about 8 inches. Like many growing southwestern cities, Albuquerque consumes more water than is naturally replenished to its supply. The City relies exclusively on groundwater, currently pumping about 120,000 acre-ft. per year from the Albuquerque Basin aquifer, accomplished with about 100 active wells located throughout the metropolitan area. Natural recharge of the groundwater aquifer is only about 30,000 acre-ft./year, and the groundwater withdrawals also have a depletion effect on flow in the Rio Grande. This depletion is institutionally limited to 20,000 acre-ft. annually. Indications are that current depletions far exceed this amount, but the difference is offset by the return of treated wastewater to the river. Albuquerque also owns a total of 48,200 acre-ft./yr. of surface water in the San Juan-Chama diversion system, most of which is currently leased or used to maintain stream flows in conjunction with wastewater discharge requirements.

---

[1]Vice President, Camp Dresser & McKee Inc., 1331 17th St., Denver, CO 80202
[2]Environmental Engineer, Camp Dresser & McKee Inc., 2400 Louisiana Blvd. N.E., Ste. 740, Albuquerque, NM 87110

Indirect Potable Reuse Options

To accomplish indirect potable reuse, reclaimed wastewater could be injected directly into the saturated zone, or introduced at the ground surface with spreading basins or in the vadose zone with shallow wells. Direct injection requires a higher degree of pretreatment than does surface spreading, since no opportunity is provided for soil-aquifer treatment prior to introduction of the flow into the water supply. However, it was determined that the most favorable approach for Albuquerque would be direct injection via deep wells into the saturated zone, since relatively little available land is available for spreading basins, known local geological obstacles would impede infiltration of water applied at the surface, and concerns exist regarding the potential evaporative water losses from surface spreading systems.

Recharge System Capacities

The Phase 1 evaluations were conducted for two different recharge capacities: 76 million gallons per day (mgd), which is the design capacity of the SWRP, and would result in zero-discharge to the Rio Grande; and 30 mgd, which is the amount of effluent that could presently be recharged to the aquifer without any impact on the administration of the City's current water rights. Surface water from the City's San Juan-Chama supply would be released to offset the reduced discharge of treated wastewater to the river.

Constraints for Reuse System Design

Albuquerque's situation is not ideal for application of high-tech treatment methods toward attainment of potable water from sewage. Relatively high natural silica levels in the groundwater greatly reduce recovery rates for membrane treatment methods, requiring low-pressure, less-selective processes (nanofiltration) to be considered. High-lime pretreatment for a plant of this size would generate hundreds of tons of chemical sludge; and microfiltration is ineffective for silica reduction. There is no economical means for disposal of liquid concentrate or residuals, nor is it desirable to waste water in this arid region. Treating only with activated carbon will not achieve adequate TOC reduction (see following discussion), thus requiring blending with treated surface water prior to recharge. However, blending would require additional water rights, which are expensive and not readily available. Therefore, system design for Albuquerque must seek to achieve adequate quality to avoid blending, while minimizing the production of liquid and solid waste residuals.

Water Quality Considerations for Indirect Potable Reuse

No regulations for intentional recharge of treated wastewater into potable supplies exist anywhere in the USA, although the EPA and some states have developed guidelines, and the state of California is in the process of developing specific regulations for this practice.

The lack of knowledge about the fate and long-term health effects of contaminants in reclaimed wastewater dictates conservative water quality standards for groundwater recharge where the injected water will later be extracted for potable uses. Attainment of drinking water standards has been used as a minimum guideline for the few such reclamation systems that exist.

In establishing specific treatment requirements for Albuquerque, both the SDWA primary standards and the requirements of the New Mexico Groundwater Protection Standards have been used. Other criteria from the draft California regulations have also been incorporated, including organics control (as measured by TOC concentration), underground retention time for the injected water, and spacing distances between injection and extraction wells. It is anticipated that the known and unknown trace organics will be reduced to acceptable levels by the treatment methods to be provided. TOC is to be limited to 1-2 mg/l. As would be expected, high reductions are necessary for pathogens, microbials, and turbidity. Disinfection and pathogen barrier processes have been designed for:

- 6-Log reduction
- Total coliforms to <1CFU/100 ml (SDWA standard)
- Fecal Coliforms to <1CFU/100 ml (CA proposed GW recharge standards)
- Virus, Cryptosporidium, Giardia, and Legionella to below detection limit (SDWA standard)
- Turbidity to 0.5 NTU (SDWA standard)

Treatment Technologies

The study initially considered a wide range of methodologies for achieving the treatment requirements, including membranes, activated carbon adsorption, chemical precipitation, oxidation, ion exchange, air stripping and normal wastewater biological and disinfection processes. An initial screening was conducted, resulting in eleven initial alternatives comprised of various combinations of reverse osmosis, nanofiltration, lime precipitation, granular activated carbon, and microfiltration.

In identifying final alternatives for detailed evaluations, major emphasis was placed on the design constraints previously discussed. The need to minimize waste residual generation while also attaining efficient TOC reduction (to minimize the need for blending prior to recharge) was a critical issue in process selection. None of the alternatives are ideal in all aspects, nor are they all equal in terms of treatment capability.

The recommended treatment method is nanofiltration (NF) and GAC treatment in parallel. This alternative combines the use of NF with GAC to provide a relatively high degree of TOC removal while reducing the residual generation to levels lower than those associated with purely membrane options. The relative percentages of GAC and NF treatment will be dictated by the treatment goals, although it is anticipated that the GAC portion would treat 25 to 35 percent of the load, with the remainder being treated by the NF train. TOC levels of 2 mg/l can be achieved, with waste residual in the 3 to 5 mgd range for a 76 mgd system. The preferred method of disinfection is ozone, but in order to maintain a residual into the recharge well fields and reduce biofouling on and near well screens, a small dosage of chlorine is also recommended. Membrane concentrate disposal is proposed to be accomplished with distillation concentration followed by evaporation ponds for the concentrated residual.

Geohydrological Aspects

The key elements of the geohydrologic work accomplished in this study were: 1) definition of potential areas where injection wells would most feasibly be located; 2) refinement of the known characteristics of the basic geohydrologic, geochemical and water quality conditions, including aquifer transmissivity and flow rates, spatial and temporal distribution of supply well pumping, and geologic characteristics; 3) further evaluation of the aquifer system using simple flow modeling to identify the best potential recharge areas in terms of cones of depression/impression, flowfield effects of injection and extraction, annual travel distances and zones of influence, potential injection rates and capacities, and sensitivity to artificial recharge variables; 4) review of information relative to groundwater geochemical characteristics and sensitivity to varying injected water quality conditions; and 5) selection of the best zones in the Albuquerque area for groundwater recharge. The three best recharge zones were identified, along with their hydrologic characteristics and other design criteria.

Recommendations and Costs

To provide a range of what the initial costs for a reclamation and recharge program might be, preliminary estimates for two potential system layouts were requested by the City: one for a 30 mgd capacity system and one for a 76 mgd system. The 30 mgd facilities include an advanced wastewater treatment system constructed at the wastewater treatment plant utilizing activated carbon and nanofiltration technologies, distillation concentrators and evaporation ponds for resulting waste streams, on-site treated effluent storage, two pumping stations to convey the effluent to well fields, 10 miles of large-diameter transmission piping, approximately 12 miles of small-diameter well and tank service lines, 5 holding reservoirs near the injection sites, and about 26 injection wells with pumps. The recharge wells would be sited in only one of the three identified zones. Capital costs for this system would be from $250 to $300 million, with annual operation costs in the $20-25 million range. The facilities for 76 mgd would include an additional 46 mgd of advanced treatment process capacity, 39 more recharge wells, 10 additional miles of transmission piping, 18 more miles of tank and well service lines, another booster station and 7 additional holding reservoirs, along with additional pumps, clear well capacity, and support equipment. Recharge would occur in two of the three identified zones. Total capital costs for the 76 mgd system are estimated from $600 to 700 million, with annual operating costs in the vicinity of $50 million.

THE ROLE OF WATER REUSE IN THE LAS VEGAS VALLEY

Andrew W. Richardson[1], Member, ASCE, Gary Grinnell[2], Member, ASCE, Mark T. Owens[3] and Ram G. Janga[4]

Abstract

Rapid growth in the northwest portion of the City of Las Vegas has placed a stress on the existing water distribution system and wastewater collection system. Included in this growth are the development of golf courses and other turf areas. The water demands for these areas could be satisfied with water of lesser quality than potable water. A feasibility study was conducted to determine if water reclamation facilities could help relieve some of the stress on the existing water and wastewater systems. The feasibility study consisted of developing and evaluating several water reclamation system alternatives from direct-reuse to indirect-reuse. A unique aspect of this study was the public-private partnership arrangement to conduct the study. The study was performed for the City of Las Vegas, the Las Vegas Valley Water District and the Summerlin Development.

Introduction

The City of Las Vegas is one of the fast growing cities in the United States. The northwest portion of the City's growth is in an area known as the Summerlin Development. A distinguishing feature of the northwest portion of the City is that it has gradually rising elevations to the mountains that create the western boundary of the City. In addition, the northwest portion of the City is at a considerable distance from Lake Mead, the Las Vegas Valley's principle source of water supply. Historically, the majority of all wastewater generated in the City was returned to Lake Mead for "return-flow-credits". In essence, this is a large-scale form of water reclamation.

---

[1]Partner, Greeley and Hansen Engineers, 426 N. 44th Street, Phoenix, AZ 85008

[2]Senior Civil Engineer, Las Vegas Valley Water District, 3700 W. Charleston Blvd., Las Vegas, NV 89153

[3]Engineering Supervisor, City of Las Vegas, 400 E. Stewart Avenue, Las Vegas, NV 89101

[4]Engineer, Greeley and Hansen Engineers, 4680 S. Polaris Avenue, Las Vegas, NV 89103

The City of Las Vegas has responsibility for wastewater collection and treatment in the study area. The Las Vegas Valley Water District has responsibility for water distribution in the study area.

Several golf courses were either constructed or planned to be constructed in the northwest growth portion of the City. This presents an opportunity to take the golf course demands off of the potable water system and satisfy those demands with reclaimed water. To determine if water reclamation should be implemented in the northwest portion of the City, a feasibility study was conducted. To conduct the feasibility, a public-private arrangement was made between the City of Las Vegas, the Las Vegas Valley Water District and the Summerlin Development. The feasibility study consisted of a phased approach to develop and evaluate different types of water reclamation systems for implementation in the northwest portion of the City. The phased approach included:

- Compilation of northwest area characteristics
- Development of reclaimed water supply and demand
- Development of water reclamation alternatives
- Evaluation of water reclamation alternatives

Compilation of Northwest Area Characteristics

The first step to determine if water reclamation was feasible for the northwest portion of the City was to compile land use, population projections and wastewater generation through the year 2040. In addition to the compilation of planning data, a review of Nevada's regulatory considerations relevant to wastewater collection, treatment, reclamation, recharge, storage and reuse was conducted under this phase of the study.

Development of Reclaimed Water Supply and Demands

A key aspect to the viability of water reclamation for an area is the balance between reclaimed water available and reclaimed water demands. A reclaimed water balance was developed based on the planning information. The results of the reclaimed water balance showed that by the year 2010, there would be 43,080 m^3/day (11.38 mgd) of reclaimed water available with 41,380 m^3/day (10.93 mgd) of annual reclaimed water demands. The reclaimed water demands were for existing and planned golf courses in the northwest portion of the City.

Development of Water Reclamation Alternatives

Several water reclamation system alternatives were considered including:

- Direct reuse with surface storage
- Indirect reuse with shallow vadose zone recharge wells

- Indirect reuse with deep vadose zone recharge wells
- Prioritized distribution with and without direct and indirect reuse
- Prioritized distribution with potable water recharge and recovery wells

For a water reclamation system to be economical, there has to be a means to economically handle the seasonal fluctuation in reclaimed water demand. The above alternatives were developed to determine the most economical means of managing excess reclaimed water storage during the low demand winter months.

The type of seasonal storage dictates the level of reclaimed water treatment. For this study, two levels of reclaimed water treatment were developed. The two reclaimed water quality requirements were for direct and indirect reuse. Direct reuse treatment is for surface storage and indirect reuse treatment was for groundwater recharge. Reclaimed water treatment process trains were developed for each alternative based on either direct or indirect reuse.

Evaluation of Water Reclamation Alternative

The evaluation of the reclaimed water system alternatives included the development of both monetary and non-monetary criteria. An evaluation criteria workshop was held with potential project stakeholders to develop consensus during project development. Evaluation criteria were developed in three areas: technical criteria, intangible criteria and financial criteria. For each area, the criteria developed were as follows:

- Technical Criteria
  - Impact on existing water pollution control facility
  - Impact on return flow credits
  - Availability of reclaimed water versus demand
  - Impact of reclaimed water recharge on groundwater quality and quantity
  - Impact on potable water treatment and distribution
  - Speed of implementation
  - Water reclamation system process and complexity
  - Impact on Lake Mead nutrients discharges

- Financial Criteria
  - Impact on existing sewer customers
  - Reclaimed water costs versus increased potable water requirements
  - Capital costs
  - Annual costs
  - Stakeholder costs

    - Avoidance costs
    - Equity of sewer rates versus water rate impacts

- Intangible Criteria
    - Sensitivity to neighbors of the water reclamation facility
    - Assignment/crediting of potable water "freed-up" by reclaimed water
    - Allocation of recovered reclaimed water from groundwater
    - Public acceptance including public education
    - Institutional arrangements to implement

The above criteria were placed in a matrix and the alternatives were evaluated. Based on that evaluation, two alternatives were selected for a more detailed benefit/cost analysis. The detailed benefit costs analysis consisted of identifying the cost savings associated with water reclamation versus staying with conventional wastewater collection and treatment as well as conventional potable water treatment and distribution.

Summary

The detailed benefit/cost analysis showed that the implementation of water reclamation in the northwest growth portion of the City has the potential to accrue benefits to both the wastewater treatment/collection system as well as the water treatment/distribution system.

Leaking Plumbing Fixtures and Water Conservation

Andrew H. Hildick-Smith, ASCE Member[1]

Abstract

Historically, leaking plumbing has not been a significant component of water conservation. A residential conservation program in the metropolitan Boston area showed that 22% of all housing units, 15% of single family units, and 32% of multi-family units had at least one leaking plumbing fixture. An estimated lower limit of the total leakage from plumbing fixtures in this study area is 2% of residential usage.

Historical Perspective

In 1873, shortly after the Civil War, unacceptably high night time flow rates in Boston led to a comprehensive leak survey of street mains and fittings (Fitzgerald, 1876). After all repairs were made and flows remained high, Boston Water Works staff embarked on a house to house search for leaks. Approximately 4000 leaks were discovered including water-closet and tap leaks. Leak repair orders were issued and the Water Works staff waited for a drop in night time flows. It never happened. They concluded that leaks were not a significant part of daily water consumption.

Current Perspective

One hundred and twenty years later, the Massachusetts Water Resources Authority completed a residential water conservation program that installed over a million water saving fixtures in 349,000 households in the metropolitan Boston area. Approximately one third of the installations were in multi-

---

[1] Emergency Response Planner, Massachusetts Water Resources Authority, 100 First Avenue, Boston, MA 02129

family units, which were defined as those with greater than 4 units per building. The remaining two thirds of the installations were made in single family units, which were defined as those with 4 or fewer units per building. The program offered free installation of faucet aerators, toilet dams, and low flow showerheads to homeowners, renters, and landlords.

As part of the educational component of the program, the field installers performed a leak detection survey of plumbing fixtures. The survey included a visual inspection of valves, bath fixtures, faucets, and toilets. In addition, dye tabs were used to detect toilet leaks. All leaks were recorded and classified as slow, medium, and fast.

A representative sample of 127,000 (36% of the total) installation records were used for analyzing leak trends. Of these installations, a total of 22% had at least one leaking fixture. A total of 32% of the multi-family installations had at least one leak. A total of 15% of the single family installations had at least one leak. By comparison, a single family audit program in San Diego found that approximately 6% of households had leaks (Bamezai, 1994). The following table shows a break down of leaking fixtures by housing type as found in the representative sample:

Fixture	Housing type	# of fixtures present *	# of fixtures leaking	Percent leaking
Valve	Single family	93,073	1,756	1.9
	Multi-family	53,466	2,955	5.5
	Combined	146,539	4,711	3.2
Bath (shower & tub)	Single family	93,073	2,328	2.5
	Multi-family	53,466	4,193	7.8
	Combined	146,539	6,521	4.5
Faucet	Single family	197,894	7,824	4.0
	Multi-family	106,834	10,345	9.7
	Combined	304,724	18,169	6.0
Toilet	Single family	113,797	2,740	2.4
	Multi-family	56,282	3,413	6.1
	Combined	170,079	6,153	3.6

* *The showerhead count was used to approximate both the number of valve fixtures and the number of bath fixtures present.*

Leakage Rate

In order to estimate a reasonable lower limit for leakage, flow rates were experimentally derived to match the field definitions for slow, medium, and fast leaks. Valve, bath, and faucet leaks were defined as: slow when they released 1 drop every 2 seconds, medium when they flowed at 2 drips per second, and fast when they ran as a small stream. Toilet leaks were defined as: slow when the dye appeared in the bowl in 5 to 15 minutes, medium when the dye appeared in 1 to 5 minutes, and fast when either the dye appeared in less than 1 minute or there was an overflow leak in the toilet.

Measuring faucet flows at the above leak rates resulted in a slow rate of 11 l/d (3 gpd), a medium rate of 45 l/d (12 gpd), and a fast rate of 102 l/d (27 gpd). Toilet leaks were estimated using the drip regulator of a medical IV bag that was paced to delivery appropriately concentrated dye to the toilet bowl via the overflow tube. Using this approach, toilet leaks were identified with a slow rate of 2 l/d (0.5 gpd), a medium rate of 19 l/d (5 gpd), and an estimated fast rate of 189 l/d (50 gpd). While the above slow and medium leak rates are probably good estimates, the fast rates used for all fixtures are very conservative. For example, a New York City study found leaking toilets averaging 746 l/d (197 gpd) (Ostrega, 1994). The following table estimates the volume of water lost per day by the representative sample to leakage:

Fixture	Leak classification	# of fixtures leaking	Estimated leak rate l/d (gpd)	Total leak rate m^3/d (mgd)
Valve	Slow	2,562	11 (3)	189 (0.05)
	Medium	1,293	45 (12)	
	Fast	856	102 (27)	
Bath	Slow	3,848	11 (3)	227 (0.06)
	Medium	1,624	45 (12)	
	Fast	1,049	102 (27)	
Faucet	Slow	12,503	11 (3)	530 (0.14)
	Medium	3,687	45 (12)	
	Fast	1,979	102 (27)	
Toilet	Slow	2,934	2 (0.5)	265 (0.07)
	Medium	1,978	19 (5)	
	Fast	1,241	189 (50)	
Total				1,211 (0.32)

Applying this leakage rate to the estimated 747,000 housing units in the retrofit program's coverage area, yields a total leakage rate of 7,192.3 m^3/d (1.9 mgd). In 1993, this represented approximately 1% of total system use, and approximately 2% of total residential use. If the fast leakage rates presented above are increased by a factor of four, as suggested by the New York City study, plumbing leakage in the Boston metropolitan area would represent approximately 4% of total residential use. To put this into perspective, the following single family savings were achieved by audit and retrofit programs: 4.6% in North Marin, California (Nelson, 1992), 6% in Concord, California (Bruvold, 1993), and 5.7% in the Boston metropolitan area (Kempe, 1994).

Conclusion

While plumbing fixture leakage is small when compared to system water use, it takes on significance when viewed from two water conservation perspectives. First of all, the relatively high percent of homes with leaks is important. If approximately one fifth of all residents live with leaking fixtures in their homes, how receptive will this group be to messages promoting water conservation? Secondly, the leakage rate is an appreciable component of the potential water savings that is available from residential retrofit or water audit programs. In an environment where achievable single family water savings are approximately 5% to 6%, a 2% or more residential leakage rate is important. Therefore, while plumbing fixture leakage may be a small part of the big picture, it is significant from a water conservation point of view.

References

Bamezai, A., T.W. Chesnutt, J. Wiedmann, M. Steirer. How Persistent are Residential Audit Water Savings? Evidence from San Diego's Single Family Audit Program. Presented at the 1994 AWWA National Conference in New York. p.4.

Bruvold, William H., P. R. Mitchell. Evaluating the Effect of Residential Water Audits. Journal AWWA, v.85, n.8, p.79 (Aug. 1993).

Fitzgerald, Desmond. History of the Boston Water Works, from 1868 to 1876. Rockwell and Churchill, City Printers, Boston. 1876. pp.35-37.

Kempe, Marcis, B. Lahage, J. Mendez-Isenburg. Unpublished staff summary presented to the MWRA Board of Directors on March 3, 1994. p.1.

Nelson, John Olaf. Water Audit Encourages Residents to Reduce Consumption. Journal AWWA, v.84, n.10, p.59 (Oct. 1992).

Ostrega, Steven F. New York City: Where Conservation, Rate Relief and Environmental Policy Meet. Presented at the 1994 AWWA National Conference in New York. p.4.

Water Conservation: A Key Component of Integrated Resource Planning

Harold (Skip) Schick, Gary Fiske, Jennifer Stout, Anh Dong[1]

Abstract

Integrated Resource Planning (IRP) seeks to develop water resource strategies with an appropriate mix of supply-side and demand-side (conservation) resources for meeting increasing water demands. This paper describes a comprehensive approach developed by Barakat and Chamberlin to assessing conservation's potential contribution to the resource mix. Treating savings from conservation activities in the same manner as future water supply options is fundamental to this analysis.

Introduction

This paper describes a systematic, step-by-step approach to examining water conservation on a "level playing field" with supply-side options in Integrated Resource Planning. The approach combines thorough demand forecasting, economic analysis, and market assessment in a series of well-defined steps described below.

Step 1: Development of a Demand Forecast

Forecasting future water needs is crucial to integrated resource planning because it anticipates the expected magnitude and type of demand growth. A important component of the forecast is "naturally occurring" conservation expected to take place over time. Naturally occurring conservation results from code and legislative changes requiring water-efficient technologies in all new installations.

---

[1]Harold (Skip) Schick and Gary Fiske are Principals, Anh Dong is an Associate, and Jennifer Stout is an Analyst with Barakat & Chamberlin, Inc., 620 SW Fifth Avenue, Portland, Oregon 97204.

Step 2: Identify the Universe of Conservation Measures

Next, a comprehensive list of water saving "measures" is developed. A "measure" is defined as a conservation technology or management practice. Examples of conservation measures are low-flow showerheads and water-efficient landscaping.

Step 3: Apply Qualitative Screen

The qualitative screen eliminates candidate measures that are clearly unsuitable for implementation in the particular region under study following criteria are applied:

- *Better Measure Available.* Another, clearly more appropriate, measure exists.
- *Technological or Market Maturity.* The technology is either not commercially available or not supported by the necessary service industry.
- *Poor Utility Match.* The technology is not applicable to the climate, building stock, or equipment typical in the service territory.
- *Poor Customer Acceptance.* Customers may be unwilling to implement the measure.
- *Environmental and Health Concerns.* The measure raises unacceptable concerns regarding health, safety, or environmental impacts.

Step 4: Develop "Technology Profiles"

After applying the qualitative screen to identify appropriate measures, information on their characteristics is compiled into "technology profiles." The profiles contain the following information: measure description, savings impact, measure life, cost, and expected changes in measure cost and market penetration.

Step 5: Apply Economic Screen

Measures that pass the qualitative screen are subjected to an economic screen to eliminate measures that are clearly not cost-effective. The economic screen compares the cost of water savings from individual conservation measures with the cost of water from viable future water supply options.

Step 6: Package Passing Measures Into Programs

Measures that passed both screens are "bundled" into programs that effectively target particular "markets" critical to achieving water savings. To assure that all program options are considered, Barakat & Chamberlin has developed two "matrices" as tools to facilitate the program planning process. One matrix is for residential programs, and the other is for commercial, industrial, and institutional customers.

On the vertical axis of each matrix, the various "delivery mechanisms" are listed in order of their "aggressiveness." A "delivery mechanism" is the means by which conservation information and measures are delivered to customers, or the means by which customers are encouraged to take action. Aggressiveness is defined as the degree of provider action or financial incentive involved. The various target markets are laid out on the horizontal axis. Target markets consist of well-defined water end uses or practices for particular customer classes. A mock-up of the commercial, industrial, and institutional (CI&I) matrix is provided in Table 1:

TABLE 1. CI&I Matrix

CI&I	INDOOR Plumbing Fixtures	HVAC Equipment	Equipment & Applnces	Indust'l Processes	OUTDOOR Irrigation System	Irrigation Scheduling	Land-scaping
Education/ Awareness							
Technical Assistance							
Financial Incentives							
Direct Install							
Regulation							

Next, program concepts must be formulated by combining matrix cells to create comprehensive program concepts. Table 2 on page 4 identifies concepts selected in a recent Barakat & Chamberlin study. All program details are then developed, the most critical being the participation, cost, and savings estimates.

Step 8: Pass Conservation Programs to Integration Element

In this phase, programs are grouped into three increasingly aggressive "scenarios" or levels. The costs and savings for all programs at each level are aggregated and evaluated against several supply-side options.

Level I programs take an educational and informational approach, relying on customers' initiative to achieve savings rather than on provider actions or incentives. In Level II programs, the providers are more directly involved, offering customers on-site audits, technical assistance, and financial incentives. Level III programs involve full incentives, direct installation, and ordinances. See Table 2, page 4.

TABLE 2. Regional Conservation Programs to be Incorporated into Integration Phase

	INDOOR		OUTDOOR	
	Residential	Commercial, Industrial, Institutional	Residential	Commercial, Industrial, Institutional
Level I	Public education and awareness	Commercial plumbing and appliances education HVAC equipment education Industrial processes workshop	Public Education and Awareness Customer workshops	CI&I education and awareness Customer and trade ally workshops
Level II	Appliance incentives & equipment tagging Indoor audit (combined with outdoor)	Commercial plumbing incentives HVAC financial incentives Commercial and industrial audits New industrial processes incentives	Outdoor audits Incentives for new efficient landscaping and irrigations systems	CI&I outdoor audits Large landscape audits Incentives for new efficient landscaping & irrigation systems Landscaping & irrig. plan review
Level III	Ultra low-flush toilet rebate Plumbing retrofit kit	Direct installation and financial incentives for replacing plumbing fixtures Commercial plumbing regulation Incentives for early retirement of single-pass cooling	Landscaping ordinance Incentives for landscaping and irrigation system retrofit	Landscaping ordinance

WATER USE MEASUREMENT AS A WATER MANAGEMENT TOOL

Jeff Adkins[1] and David Still[2]

Abstract
The St. Johns River Water Management District's (SJRWMD) water conservation rule sets mandatory conservation criteria for water users. The criteria include a requirement that permittees measure the quantity of water they use. Water use data collected is enabling SJRWMD to evaluate resource availability and encourage effective water conservation measures.

Introduction
Florida has witnessed incredible population growth in this century. The state was blessed with an extensive aquifer system and many surface lakes and rivers. In many areas of Florida, however, the sustainable use of potable water sources has reached its limits; competition for water has forced adoption of serious water conservation efforts. SJRWMD covers the St. Johns River basin in northeast and east central Florida, about 21 percent of the state's land area. SJRWMD was created to manage the area's water resources. SJRWMD programs include hydrologic conditions monitoring, resource-related regulation, public education, research and environmental restoration.

SJRWMD issues individual water use permits for withdrawals of water above a threshold volume from surface and underground sources. To support the water use permit program, SJRWMD introduced comprehensive water conservation rules. The rules place restrictions on all forms of irrigation, require that water withdrawals be measured, require water users to implement all feasible water conservation measures and encourage use of reclaimed water.

These rules require that water use from each permitted well and surface water withdrawal be measured and reported to SJRWMD every six months. Water use measurement is required

---

[1]Supervising Professional Engineer, St. Johns River Water Management District, Palatka, Florida 32178
[2]Agricultural Engineer, Suwannee River Water Management District, Live Oak, Florida 32060

for all categories of water users (e.g., agricultural, golf course irrigation, public supply, commercial, industrial). New uses after July 1991 are required to install flow meters. Users prior to this time can use either a flow meter or an alternative flow measurement method, provided the proposal is accurate and verifiable.

Water Use Reporting Program

The Water Use Reporting program was created by SJRWMD in March 1993 to implement this requirement. The elements of the program are described below:

Flow Meter Distribution

In one area of the District, water use by a concentration of commercial fern growers raised concerns about surface water and groundwater availability for irrigation and public supply. In response to these concerns, SJRWMD instituted a special data collection effort targeting in this 400-square mile "Delineated Area", including the distribution of flow meters at SJRWMD cost to established permittees. To date, over 800 totalizing flow meters have been distributed to permitted sites in this area. In addition, flow meters are provided to selected permittees in other parts of the District. SJRWMD staff deliver the meters, which are to be installed by the permittees within 90 days. SJRWMD staff also verify installation and calibration of the meters.

A Geographic Information System (GIS) is being employed to track the flow meter distribution in the Delineated Area. The GIS will also contain information about other flow-metered and alternative method sites throughout the 19 counties of the District.

To encourage the use of flow meters District-wide, not just within the Delineated Area, SJRWMD arranged with our flow meter vendor for all existing permittees to purchase meters directly at SJRWMD's contract cost. Other venders have responded with competitively-priced flow meters.

Technical Guidance on Flow Meter Installation

During the flow meter delivery process, SJRWMD staff visit Delineated Area sites to assist permittees in determining the appropriate location for flow meters at each wellhead and surface water pump. Staff also provide guidance on the proper use and maintenance of these meters. Several workshops on flow meter installation have been held in the Delineated Area. Technical assistance is also provided District-wide to permittees upon request.

Alternative Flow Measurement Methods

SJRWMD personnel assist permittees in evaluating appropriate alternative water use measurement methods.

Staff review the submittals using a objective set of minimum criteria for alternative methods. An applicant incorporating these minimum criteria would have the presumption of fulfilling SJRWMD requirements. Approximately 15% of the proposals received to date have not met the criteria, and others that do meet the criteria may be less than satisfactory in practice. After review, permittees will be notified either accepting their proposed alternative method or requesting modification of the proposal to meet SJRWMD guidelines. A random sampling of the alternative method sites will be checked annually for accuracy. Where a reliable alternative cannot be identified, the permittee will be required to install a flow meter.

The typical alternative method proposals pursue a "time-volume" methodology. Records are maintained of the time each pump is operated, along with a verifiable flow rate determination for that pump. Unfortunately, private firms have not come forward with flow rate calibration services, so the availability of reliable and verifiable pump flow rates by alternative means is limited. SJRWMD staff will identify site locations for testing of percent accuracy of each classification of alternative method. Through this research, the minimum criteria may be modified and improved.

Public Education and Encouraging Compliance

During the program's first year, SJRWMD directed its primary compliance efforts among Delineated Area permittees. Those eligible for meters from SJRWMD received direct mailings about the program. In an effort to increase awareness of the program District-wide, Staff have solicited support from the agriculture trade associations, the Natural Resources Conservation Service and agriculture extension services. Workshops are held throughout the District to help permittees understand the compliance forms.

Database Development and Management

The value of permittee compliance, meter distribution and alternative method accuracy is minimal if the data is not collected into a usable format by SJRWMD. Reporting forms are to be submitted during March and September for water use during the previous six months. Improvements in reporting forms, instruction sheets and permittee training seek to improve the percentage of forms completed correctly and accurately.

Water use data is received on computer-scannable forms, on computer disks and on handwritten forms. Staff scan the forms and print out reference summaries. Quality assurance procedures include checks for reporting form completeness, data entry errors and technical content. Further technical quality assurance is being implemented to increase the level of confidence in the summary reports. The summary output is in database format with semi-annual reports prepared for SJRWMD staff.

Applying the Water Use Reports

Knowledge of the amount of water used, both by individuals and SJRWMD, is expected to encourage better management of the resource. For example, irrigators who can quantify their applications have a tool to adopt more consistent and conservative practices. Public supply users can detect leaks more quickly, and commercial users may be encouraged to identify candidate processes for reuse, recirculation or flow rate reduction.

The database can also be used by SJRWMD as a tool for resource management, compliance and public education. As the period of record increases, the large quantity of water use data accumulated by the District will be made available to SJRWMD engineers. Staff will use the data to model the changes in the Floridan aquifer resulting from water use and hydrogeologic patterns. This is an improvement over past efforts where SJRWMD has relied on estimates of water use to identify areas of special concern with regard to future water availability.

Semi-annual reports include water use summaries by permit, summary statistics by county, use type and source type, and listings of permittees who have not complied with the water use reporting requirement. The database will identify permittees whose water use exceeds their allocation. Based on this data, SJRWMD may increase the allocation or determine additional appropriate water conservation practices to keep water use beneath allocated levels. Finally, the data can be formatted and distributed to the public, to convey current water use practices and areas of concern.

Summary

The SJRWMD Water Use Reporting program is a tangible effort to have the right tools in place for the tough water management decisions to be faced in the near future. Accurate measurement and reporting of water use will require extensive education and support of the permittees by SJRWMD. Care must be taken at every step - data collection, reporting, and database management - to ensure a meaningful record of water use. Though it will be several years before the SJRWMD water use database is mature, these actions are intended to ensure its value for water resource management, permit compliance and public education.

Estimating Demand Variability

Richard N. Palmer[1], Member, ASCE, Joan M. Kersnar[2], Member, ASCE, and Dora Choi[3]

Introduction

In the US, summer demand for water varies considerably from year to year. This variability is often ignored in long-term forecasts used in yield models, where "average" weather conditions are used to define long-term demand with a typical seasonal variation. However, the variability between demand based on summer weather conditions may be larger than the uncertainty associated with other explanatory variables in long-term forecasts.

This paper describes the process of generating a "historic" demand trace that might have occurred had the population currently served by Seattle Water existed during past meteorological events. Several mathematical models that define water demand as a function of average annual demand, summer peaking, temperature, and rainfall are described. The process for selecting the most appropriate model is presented. A historic demand trace was generated and used to determine the impacts of using variable demand on the yield of a supply system.

Approach

To evaluate the reliability of a water supply system, it is necessary to compare the availability of water with the demand for water. This is often determined by using a computer simulation of the water supply system with historic streamflows and anticipated demands. In the past, year-to-year demand variability has been modeled as a function of time of year. Monthly demand factors have been calculated to incorporate the average seasonal variability into evaluations of water supply. This has been accomplished by examining the yearly demand and determining the relative demand for each month. Use of average monthly demand factors typically underestimated the peak demands and overestimated the minimum demands.

This study models short-term demand variations by regressing monthly demand data against weather data for Seattle Water's service area. The demand data includes water supplied from all sources. The weather data used in the analyses were the average of daily maximum temperatures and total rainfall for each month.

---

[1]Professor of Civil Engineering, Department of Civil Engineering, University of Washington, Seattle, Washington 98195, and [2]Senior Civil Engineer and [3]Student Intern, Seattle Water, Seattle, Washington 98105.

The 1970-1985 period was selected for model calibration because operations were relatively normal. The weather data used to predict demand were needed for the entire period over which the yield analyses were to be performed, 1929 through 1993.

Modeling of variable demand was limited to the peak-use season (May through September). Demand during the off-peak season was assumed to be independent of weather and was estimated using average monthly factors.

Model Alternatives
Several model alternatives, differing in mathematical forms, were analyzed as potential candidates for the relationship between demand and weather. The alternative equation forms were as follows:

A Dm,y = Fm * Dy
B Dm,y = exp[X1*ln(Dy*Fm) + X2*ln(Rm,y) + X3*ln(Tm,y) + X4]
C Dm,y = (Dy*Fm) * (Rm,y)^X1 * (Tm,y)^X2 * exp[X3]
D Dm,y = (Dy*Fm) * (Rm,y/Rm)^X1 * (Tm,y/Tm)^X2 * exp[X3]

where:

Dm,y	=	Demand for month m, in year y (million gallons per day, mgd)
Dy	=	Demand for year y (mgd)
Fm	=	Factor for month m, averaged for all years
Rm,y	=	Rainfall in month m, in year y (inches)
Tm,y	=	Average maximum daily temperature in month m, in year y (°F)
Rm	=	Average rainfall for month m (inches)
Tm	=	Average maximum daily temperature for month m (°F)
ln( )	=	natural logarithm of the value in the parentheses
exp[ ]	=	exponent of the value in the brackets
Xn	=	regression coefficients

Method A represents the conventional method for predicting monthly demands using average monthly factors. The proposed methods, B through D, use rainfall and temperature to calculate or predict monthly demand. The model parameters for each of the proposed methods, denoted here by Xn's, were determined using linear regression methods (with transformations to natural logarithms) and data from the calibration period (1970-85).

Model Selection
For the proposed methods, several regression techniques were explored. These regression techniques included the following:

(1) Aggregate all months during the peak season (May, June, July, August and September) to derive one set of regression coefficients.

(2) Analyze each month separately to derive monthly regression coefficients (May, June, July, August and September).

(3) Include the previous month's weather data in the equation to derive one set of regression coefficients for all months.

The above techniques did not perform equally well. Approach (2) produced parameters such that rainfall and temperature impacted demand in an inconsistent manner. That is, low rainfall and high temperatures would produce lower demands in some months than high rainfall and low temperatures in other months. The added complexity of approach (3) did not produce significantly better results and was abandoned. Approach (1) produced adequate results and was used in model selection.

Selection of the demand variability model was based on goodness-of-fit statistics, R^2 and adjusted R^2. Method D had an R^2 of 0.59 and an adjusted R^2 of 0.58, whereas Method C had an R^2 of 0.31 and an adjusted R^2 of 0.29. Method B produced the best fit with the data as indicated by an R^2 and adjusted R^2 of 0.85. These statistics were not computed for the conventional approach.

The selected method was compared to the conventional approach (Method A) using other goodness-of-fit statistics. These statistics were calculated for peak season demands only since the two methods do not differ in the non-peak months. As shown by the following summary of these statistics, Method B outperforms the conventional method in predicting monthly demands.

Statistic	Method A Conventional Method	Method B Variable Demand
Root of the Mean Squared Error (RMSE)	9.9 mgd	6.5 mgd
Mean Error	0 mgd	0.9 mgd
Mean Percent Error	-0.2	0.3
Mean Absolute Deviation	7.9 mgd	5.5 mgd
Mean Absolute Percent Error	4.3	2.9

Method B was used to generate a demand trace with the rainfall and temperature data for 1929 through 1993 and an annual average demand of 60 mgd. Peak month demand ranged from 75 mgd in August of 1932 to 108 mgd in July of 1958. For a comparable demand trace using the conventional method with average monthly factors, the peak month value would be (1.395 * 60 mgd =) 84 mgd in July of each year. One result of this study was that demand during the summer was more sentitive to the average maximum daily temperature in month than to rainfall during the month.

Impact on Yield

Analyses were performed to determine the impact of demand variability on the yield of the supply system. A simulation model of the South Fork Tolt system was created to determine system yield using the variable demand method and the conventional demand method. The simulation occurred over the analysis period of 1929-1993.

To make the comparison, a simplified approach was used. System yield was defined as the maximum amount of water that could be supplied on an average annual basis without shortages, whereas Seattle Water uses a 98% reliability standard. The ability to switch to lower instream flow requirements was not incorpated into the model, which is a mangement option that is practiced by the Seattle Water Department during a period of real low flows. Also, the full storage capacity of the reservoir was utilized and was not constrained by water quality concerns nor pipeline capacity.

The results of the yield analyses showed that both demand methods result in the same average annual yield of 60 mgd. However, although the annual demand delivered in the simulation with the conventional demand method was 60 mgd for each year, the annual demand delivered with the variable demand method ranged from 52 mgd to 64 mgd. The reason that both demand methods produced the same yield was that the critical year that constrains system deliveries is 1941, and weather conditions were such that summer demands in 1941 were close to average values. That is, the annual demand for 1941 was 60 mgd for both the conventional demand method and the variable demand method.

Thus, use of variable demands may not impact system yield if drought conditions at the supply source do not correspond to hot, dry summers in the service area. This condition could occur in areas similar to Seattle where supply sources are located far from the service area and winter snowpack plays an important role in determining the subsequent summer's water supply.

Summary

In the past, water demands for planning purposes have often been estimated using average monthly demand factors. Estimates of water demands during the peak season can be improved by using weather data. Several alternative methods that use rainfall and temperature data to predict demands were evaluated. A method for estimating demands was selected based on its overall ability to predict peak season demands. The selected method for calculating short-term demand variability can be used to estimate monthly water demands when the long-term or base demand is known. Use of variable demands, instead of conventional average monthly demands, may not impact system yield if periods of high demands associated with hot, dry summers do not occur in the same periods of low inflows at the supply source. However, if

Acknowledgements

The authors wish to thank the members of the Supply Reliability Task Force, a joint working committee comprised of staff of the Seattle Water Department and regional purveyors, for their participation in this study.

Managing Water Supplies During Drought, The Search for Triggers

Selene M. Fisher[1], Member, ASCE, and Richard N. Palmer[2], Member, ASCE

Introduction

Water management in the US no longer focuses primarily on the construction of large reservoirs to supply water for growing demands. Limitations on acceptable new reservoir sites, rising development costs, and growing public concern over environmental impacts has placed renewed interest in better management of existing supplies. Both strategic and tactical drought planning is increasingly focused on creative water management, both in more effectively operating supplies and carefully influencing demands. This paper explores the search for an indicator that identifies the onset or increased severity of a municipal drought and can function as a triggering mechanism for the implementation of drought response measures. The index, Days of Supply Remaining (DSR), is proposed and its effectiveness examined.

Approach

Two steps are required to determine a water supply index: establishing the index levels at which response activities are triggered and evaluating the use of the index as a decision tool. The first step establishes trial triggers and compares system behavior using a perfect forecasting ability. The second step evaluates the index under less than perfect forecasts. With competing interests, it is unlikely that a single set of triggers will minimize all appropriate system performance measures. A multi-objective analysis is used to evaluate the impact of trigger levels on performance. The Seattle water supply system is used as a case study for this paper, although the techniques are equally applicable to any setting.

Days of supply remaining is calculated by predicting future inflows and demands and determining when supply is no longer adequate to meet demands. Table 1 illustrates this calculation. The DSR is calculated at the beginning of each time step, when forecasts are made of the subsequent weeks' inflows and demands. Inflows are added and demands subtracted for each successive week until the remaining supply is inadequate to meet demand (negative).

The DSR index is evaluated in a simulation model of the Seattle system (Fisher, 1993) which receives its water supply mainly from the Cedar and South Fork Tolt Rivers. The Seattle Water Department (SWD) currently supplies water to about 1.2 million residents through both direct and purveyor sales. The mostly residential demand peaks in late spring and summer, primarily due to increased outdoor water

[1] 2741 NE 143 St., Seattle, WA 98125, (selene@eskimo.com), and [2]Department of Civil Engineering, FX-10, University of Washington, Seattle, WA 98195, (palmer@u.washington.edu)

Table 1 Sample Calculation of DSR

Week	Beginning Storage (volume)	Inflow (volume per week)	Demand (volume per week)	Ending Storage (volume)	Demand Met?	DSR (days)
Begin	100	40	40	100	√	7
+ 1	100	60	40	120	√	14
+ 2	120	20	40	100	√	21
+ 3	100	20	40	80	√	28
+ 4	80	0	40	40	√	35
+ 5	40	20	40	20	√	42
+ 6	20	10	40	-10	no	
+ 7	-10	60	40	10		

use and coinciding with reduced precipitation. In addition to water supply, the Cedar/Tolt system also provides water for instream fish flows, lake level maintenance, and lock flushing. The model simulates the system with operation based on current streamflows and estimates of future inflows and demands. The index is calculated based on predicted inflows and compared to triggers to determine water supply status. If a drought is indicated, appropriate response activities are instituted; in this study these are demand reduction as a function of time of year and drought stage. Performance is evaluated as the system state evolves with the actual hydrologic inflows (approximately sixty years of weekly flows). System performance measures include the number and magnitude of municipal supply shortfalls, deviations from instream fishflow targets, and the number of days of demand restrictions (at the various restriction, or drought stage, levels).

Evaluation of the DSR Index

Since future flows can not be forecast perfectly, predicting the DSR incorporates an estimate of expected inflows. It is anticipated that detrimental impacts will increase since inaccurate estimates of inflow prevent perfect operation. Several simple predicted inflows (mean and 10th through 50th percentile calculated from the historical record) are used. Mean flows provide insight into average system operations, however, they obviously overestimate actual inflows during drought episodes. Ten-percentile flows are used as relatively conservative estimates. These flows generally under-predict inflows and shortfalls from lack of water are reduced, but episodes of unnecessary restrictions are instituted. Perfect information illustrates how well the index could perform if forecasts were accurate and incorporated into the decision process. Less than perfect information illustrates how forecast error deteriorates the effectiveness of the DSR index as a decision aid. A status quo condition is defined as the system with no institutionalized restrictions to cutback demand. Full demands are met without hedging against future shortages. This results in fewer, but larger, shortfalls. Four stages of restrictions are considered, ranging from low impact to mandatory curtailments.

By operating the system with the triggers selected in the first step and different inflow sequences, the effects of increasingly less conservative streamflow estimates can be assessed. During periods of potential shortfalls in Seattle, July through November, streamflows are driven by rainfall (rather than snowmelt). Extended forecasts for periods greater than two weeks during this period are typically poor. The authors recognize that the inflows used are not true forecasts, that is, they are not based upon watershed conditions and conditional situations. In the absence of reliable rainfall predictions (as in the case in reality), percentile flows represent a reasonable substitute.

Results and Conclusions

The system simulation calculates the DSR at each time step and compares that to predefined levels of severity to 'trigger' drought response. A "set of triggers" are the levels of DSR associated with each drought stage. To find the trigger levels resulting in the smallest impacts, 320 sets of triggers were evaluated, ranging from 25-175 days for Stage 1 to 0-145 days for Stage 4. All sets resulting in unreasonably large municipal or instream flow shortfalls were eliminated. The range for each stage for the remaining sets of triggers is shown in Figure 1. A weighting scheme assigned priorities for water allocation and shortfalls, and a 'best' set of triggers was found for each objective. With the conflicting demands for water, no single set of triggers minimizes all parameters. Figure 1 shows a selection of sets of triggers for different objectives: minimize municipal impacts (Muni), minimize instream flow impacts (IFR), minimize days spent in restrictions (Days), and a combination assigning first priority to instream flows, but also including municipal concerns (IFR->Muni).

Figure 1 Triggers for Successive Drought Stages

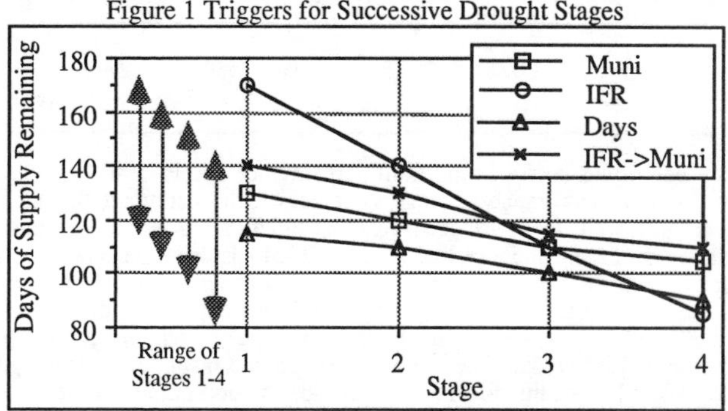

These results show that municipal impacts are minimized with moderately high triggers that progress rapidly from Stage 1 to 4 (130, 120, 110, 105 days). Time spent under restrictions is minimized by triggering drought stages as late as possible. To minimize instream flow impacts, it is best to restrict municipal demands as early as possible, thus the high Stage 1 trigger. However, since Stage 3 and 4 declarations can cause the instream flows to be reduced, it is best to delay these as long as possible; hence the comparatively low Stage 4 trigger. The combined objective is a compromise between the instream and municipal results. Once a drought occurs, the least impacts are achieved when demands are reduced quickly, but not too early, avoiding long periods of possibly unnecessarily restricted demands.

Figure 2 provides a summary of the relative performance of the system with the 'Muni' triggers for the different predicted inflows. The less conservative predicted inflows (mean and 30th-50th percentile flows) resulted in at least one day of 100% municipal shortfall. The 10th and 20th percentiles had lower shortfalls, although still quite high. An examination of the weekly output identified the maximum municipal shortfalls as occurring in the fall, when instream flow requirements increase quickly as municipal demand drops off slowly and remaining storage and inflows are insufficient to meet both needs. In real droughts, the instream flows would likely be returned to the higher fall levels more slowly, until inflows are replenished by autumn rains.

Figure 2 Measures of System Performance

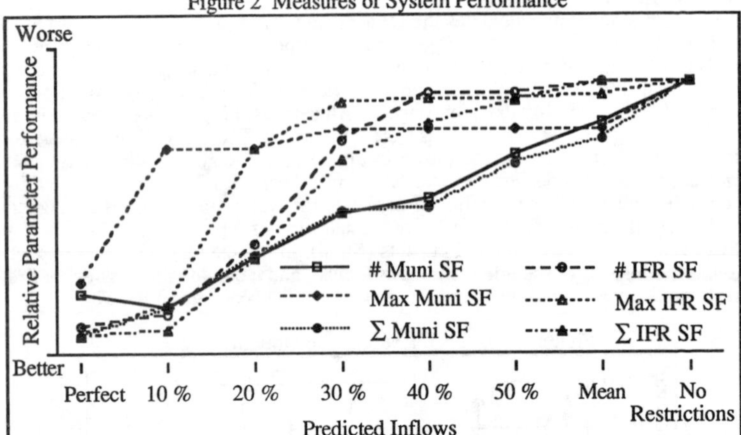

Screening techniques such as minimum performance levels or weighted sums can help identify the general range of triggers against which to compare the index. Since it is difficult to find 'perfect' trigger levels that satisfy all users, the emphasis should be placed on developing an index that is acceptable to decision makers and on considering a range of trigger levels in concert with other observations, such as snowpack or cumulative precipitation. Different trigger levels depending on season may be appropriate. When a system is dependent on highly variable, poorly predicted inflows (precipitation) during the drought susceptible season, using some prediction method correlated to the historical record provides a good guide to inflows. Figure 2 shows it is better to be conservative with these predictions in a municipal supply system.

Summary

The results show that using the DSR index can improve the management of water. The index triggers provide valuable insight on when to initiate hedging against prolonged shortages to minimize specific impacts. This evaluation assumes that loss functions are convex and many small shortfalls (demand reduction) have less impact than a few large ones. However, the DSR index alone, in the absence of factors correlating with physical system status, may be insufficient. Using the index, which incorporates system operating rules, along with hydrologic or meteorological parameters which factor in antecedent conditions may provide better results. The suggested approach requires iterative simulations rather than a single optimization. Incorporation of the nuances of real time management currently preclude either linear or dynamic programming formulations of this complex problem.

References

Fisher, S.M., 1993, Days-of-Supply-Remaining as an Indicator of Drought Severity in Water Supply Planning and Management, Masters Thesis, University of Washington, Seattle, Washington, December 1993.

Seattle Water Department, 1986, 1985 COMPLAN (Seattle Comprehensive Regional Water Plan), Vol. I, Summary, Vol. VI, Conservation Plan, City of Seattle, Washington, 1986.

Practicing for Droughts, Guidelines for Virtual Drought Exercise

Allison M. Keyes[1], Student Member, Richard N. Palmer[2], Member, ASCE, and William Werick[3]

Introduction

The effectiveness of drought planning is inherently limited by its episodic nature. Once developed, plans typically lie dormant until hydrologic events or changing conditions mandate their review. Limited plan maintenance diminishes the benefits that may be accrued when they are most needed. The concept of Virtual Drought Exercises was developed and tested during the National Drought Study (Werick et al. 1994) to maintain the benefits of drought planning during non-drought periods. Virtual Droughts contain many elements that are similar to other drought exercises, but are more strongly linked to the DPS planning methodology and philosophy. Virtual Droughts require the participation of a broad spectrum of key water managers and stakeholders in drought response. They are conducted to achieve clearly specified goals of joint interest to its participants. Moreover, they encourage participants to adopt a collaborative drought management approach. To support this collaboration Shared Vision Models are used. This paper discusses the benefits offered by Virtual Droughts to water managers, and describes the primary activities and components of the process. Considerations for Virtual Drought design and implementation are presented.

Virtual Drought Exercises

A Virtual Drought Exercise is a process in which water managers guide a region through a simulated drought to maintain and improve drought preparedness. These exercises allow the testing of existing plans or proposed drought strategies, and can improve drought response coordination among managers and stakeholders in a region. Virtual Droughts are an extension of the concepts employed in the Potomac River Basin (Sheer et al. 1989) and other areas. They are also an outgrowth of other natural hazard preparedness exercises. Although many approaches may be effective in conducting such exercises, this paper proposes important considerations that help ensure the effectiveness and success of any such exercise.

Conducting the Virtual Drought Exercise

Typically, the Virtual Drought Exercise is conducted in three stages: briefing, drought response simulation, and debriefing. A briefing assures that a common

---

[1]Graduate Research Assistant, and [2]Professor of Civil Engineering, Depart. of Civil Engineering, University of Washington, Seattle, WA 98195, and [3]Policy Analyst, Institute of Water Resources, 7701 Telegraph Road, Ft. Belvoir, VA, 22060-5586.

understanding of the objectives and procedural logistics of the process is established prior to initiating the simulation. The required scope of the briefing effort will depend on participants' familiarity with the Virtual Drought concept, the Shared Vision Model, and their roles in the exercise. The briefing should leave the participants with a clear vision of what is expected of them, the benefits that they should derive from the day's activities, and the long-term benefits of the exercise.

In the simulation, attendees are given specific goals to achieve, subject to system constraints and an unknown sequence of future inflows. This requires understanding of planning guidance, and effective interagency dialogue and decision making. Shared Vision Models are used to analyze proposed management strategies. Other supplemental materials such as runoff forecasts are provided emulate the types of information which participants would utilize in a drought response setting.

Management decisions are implemented by making appropriate changes in a separate Shared Vision Model, whose purpose is to track the status of the virtual scenario. The model is run for a specified number of time steps to provide an updated status report on system conditions. At this point, participants must initiate a new round of assessment and negotiation. The simulation proceeds iteratively in this manner until its conclusion.

Debriefing is the evaluative portion of the Virtual Drought exercise. Normally, participants and observers are encouraged to share their observations on three topics: 1) the quality of operational decisions made during the exercise, 2) institutional constraints to optimal resource management, and 3) the strengths and weaknesses of the exercise, including the Shared Vision Model used. Debriefing can be accomplished through a combination several approaches including facilitated group discussion, interviews and questionnaires.

Shared Vision Models

A key ingredient in the success of a Virtual Drought Exercise is the use of a Shared Vision Model. The term "Shared Vision" implies a highly interactive computer model that is developed jointly by stakeholders in a region, under the direction of those experienced in their construction, that contains a collective view of a water resources system and how it is operated. Because of stakeholder involvement in the development of these model, they better represent true system operating policies and constraints, and are typically more trusted by those engaged in the Virtual Drought Exercise. The authors have used the simulation tool STELLA® II at eight sites throughout the US to facilitate Shared Vision Model construction (Palmer et al. 1993).

Design Considerations

Careful consideration must be given to the goals and format of a Virtual Drought Exercise. It is extremely important that it be run effectively, as those attending are likely to be key decision makers in the region. An interagency review committee, comprised of water managers, stakeholders, and other impacted parties, should be formed at the start of the design process. Committee input will help to assure that the Virtual Drought will be sufficiently realistic to be of value to attendees, offer management challenges, and test the most pertinent aspects of drought response. Essential design considerations include: 1) exercise objectives, 2) design constraints, 3) attendees, 4) exercise format and rules, 5) drought scenario, 7) model use and interaction, and 8) documentation.

Objective and constraint definition will influence the design of the all other aspects of the exercise. The primary objective in conducting a Virtual Drought Exercise is to enhance a region's ability to respond to a drought. This may be achieved in a variety of ways. An exercise's design can improve communication between individuals or clarify an agency's role in drought response. Another objective might be to better understand lines of authority within an agency or across a region. An exercise could explore potential inefficiencies in information flow within or across agencies. Another typical objective is to improve the general understanding of system operation and to explore the effects of alternate operating policies.

Proper representation must be obtained if the Virtual Drought is to reflect reality and be of greatest benefit. Key individuals in the drought response process must be identified and encouraged to participate. This includes, where applicable, representatives of state, federal, and local water managers, the staff of elected officials, environmental groups, recreational vendors, irrigators, Indian tribes, resource agencies, and other stakeholder groups. Representatives of local media, should be invited to attend as they are often the primary venue for distributing information about drought to the general public. If key players are unable to participate, appropriate surrogates must be identified.

To define the general structure of the exercise, the time allocated to briefing, simulation and debriefing should be established, as well as the general activities of each phase. The anatomy of each decision making iteration within the simulation phase should also be outlined. Greater detail must be provided through rule and role definition. The rules established for the Virtual Drought Exercise typically define factors such as the allowable modes of participant interaction, the time available for specific tasks, the extent that information will be shared, protocols for decision making, and allowable management actions. Rules should be consistent with the overall objectives of the exercise and reflect awareness of time and resource constraints.

The drought scenario created by designers will play an central role in influencing the events that transpire during the simulation. The scenario should be sufficiently realistic to be of interest and value to participants, providing all with an incentive to actively participate in the process. Definition of the drought scenario requires consideration of: the objectives of the exercise, the types of conflicts that should be examined, the lessons that should be emphasized, and the time constraints of the process. In addition to selecting an appropriate hydrologic data for the exercise, the information requirements of participants must be defined. These requirements will determine the types of information provided to participants throughout the exercise, both in the creation of initial conditions, and in ongoing status reports.

As discussed previously, Shared Vision Models are used in two contexts during a Virtual Drought, requiring that two separate versions of the model be maintained. A "reality" model is needed to act as an intermediary between participants and the virtual scenario, allowing implementation of management decisions, and providing system status updates during the simulation. An "analysis model" provides a tool for joint inquiry during negotiation of management decisions. Designers should carefully define the functional requirements of each version of the model, and make necessary modifications made for the purposes of the exercise.

Often an important goal of the Virtual Drought is to communicate its outcome to a broader audience, particularly those with the ability to implement exercise recommendations. Documentation can be accomplished through several approaches including exercise minutes, interview summaries, and videotape.

In addition to the general considerations listed above, other factors should be considered in Virtual Drought design. A facilitator is typically required to introduce the objectives and format of the exercise, manage time and activities, and guide discussions among the participants. This facilitator should have the respect of the attendees, be knowledgeable about water resource issues, and be independent of any political issues that might be raised by the exercise. Model technicians, well versed in the structure of the model and capable of modifying the model during the exercise, are essential. Independent observers are also valuable to provide an impartial perspective to the events. Like the facilitator, these observers should be knowledgeable and unbiased.

Measures of Exercise Success

After a Virtual Drought Exercise is concluded, it is important to determine the degree to which the goals of the exercise were met. It is instructive to answer questions such as: 1) To what extent did the exercise help improve communication among the participants?, 2) To what extent did the exercise improve participant understanding of the system?, 3) Did the appropriate stakeholders in the region participate in the exercise?, 4) Did the exercise help improve decision making during drought?, and 5) Did the exercise detect shortcomings in management that can be effectively addressed?.

A revealing measure of the value of a Virtual Drought Exercise is the perception of need for continuing such exercises in the future. As the purpose of a Virtual Drought is to better prepare for and manage a real drought, one such exercise is unlikely to provide all of the possible benefits. Repeated exercises are valuable to train new water resource managers and stakeholders, and to maintain a high level of interest and ability among all of the key participants in a region.

Conclusions

Although Virtual Drought Exercises require considerable preparation and the involvement of key stakeholders in a region, they offer an excellent venue to test the drought readiness of a region, to improve communication between agencies, and to explore alternate operating policies without the risk of actual financial losses or public embarrassment. No precise rules exist for performing drought exercises, but the recommendations in this paper serve as a guide and checklist for approaches and steps that have proven successful in the past. With increased competition for dwindling water management resources, Virtual Drought Exercises represent a new and potentially intriguing approach to improve drought preparedness.

References

Werick, W.J., Keyes, A.M., Palmer, R.N., and Lund, J., (1995) "Virtual Droughts, and Shared Visions," in *Proceedings of the Conference on Drought Management in a Changing West, A New Direction for Water Policy*, May 10-13, edited by D.A. Wilhite, and D.A. Wood, pp. 165-177.

Palmer, R.N., A.M. Keyes, and S.M. Fisher, Empowering stakeholders through simulation in water resources planning, *Proceedings of the 20th Annual National Conference, Water Resources Planning and Management Division of ASCE*, Seattle, Washington, April 1993, pp. 451-454.

Sheer, D. P., J. Wright, M.L. Baeck, (1989). "The Computer As Negotiator." *AWWA Journal*, February, 81(2), 68-73.

Assessing the Hydrologic Impacts of Climate Change on the Colorado River Basin

Erik V. Mas, Student Member ASCE[1],
Eric F. Wood[2], Dennis P. Lettenmaier[3], Bart Nijssen[4]

Abstract

A grid-based modeling approach consisting of a series of linked, conceptual-type models is developed to assess the hydrologic effects of climate change on large-scale watersheds. The National Weather Service snow accumulation and ablation model is used in conjunction with a 2-layer Variable Infiltration Capacity model (VIC-2L) to simulate historical naturalized streamflow at various nodes within the Colorado River Basin. The model is then run with adjusted meteorological forcings to determine the sensitivity of streamflow and evaporation to potential climate change scenarios. The output of two general circulation models (GCMs) is also used as input to the hydrologic simulation models to test a more realistic spatial pattern of precipitation and temperature changes. Results indicate a reduction in mean annual flow volume of between 10% and 30% for temperature increases of 1.5C to 5.0C at most locations, with up to a 50% flow reduction for the most extreme scenario. A shift in the average hydrograph peak from June to May also occurs for the increased temperature scenarios due to a smaller snow pack and earlier melt. The GCM results are somewhat contradictory, pointing out the uncertainty in climate change related research.

Introduction

The planning, design, and operation of water supply and flood control systems is based on the assumption that future trends in precipitation, temperature, and streamflow will repeat those which have occurred in the past. However, climate change research strongly suggests that increased atmospheric levels of greenhouse gases will result in higher global temperatures and altered precipitation patterns. In addition, climate change related hydrologic research conducted over the last decade has shown that the regional effects of anticipated precipitation and temperature changes may include decreased runoff and increased evaporation. Consequently, serious water shortages could result in areas of the Western U.S. such as the Colorado River Basin, where water demand is near or exceeds water supply.

Although previous hydrologic studies have shown significant sensitivities of streamflow and water supply to changes in precipitation and temperature, the studies

[1] Hydrologist, National Weather Service, Northeast River Forecast Center, Taunton, MA 02780
[2] Professor, Department of Civil Engineering and Operations Research, Princeton University, Princeton, NJ 08544
[3] Professor, [4] Graduate Student, Department of Civil Engineering, University of Washington, Seattle, WA 98195

utilizing conceptual type models have focused on medium or small size catchments in specific climatic and physiographic settings. In this study, a method for assessing the regional hydrologic effects of climate change is developed for *continental-scale* river basins. The emphasis of this approach is on providing a modeling framework suitable for macroscale catchments on the order of 1,000,000 km^2, which also maintains a realistic description of the hydrologic processes governing streamflow response. The method is applied to the Colorado Basin to examine the effects of climate change scenarios on streamflow at 20 different inflow locations for large-scale water resources impact assessment purposes.

Model Description

The modeling framework consists of the National Weather Service snow ablation and accumulation model (Anderson 1973), a macroscale hydrologic model based on the Variable Infiltration Capacity (VIC) formulation, and a simplistic routing model to account for the lateral transport of water through a river network. The snow model is a temperature-index model which requires precipitation and temperature as its only inputs. The land surface hydrology model (Liang et al. 1994), which was developed for use within coupled land-atmosphere GCMs, uses the VIC spatial distribution of soil moisture capacities (Wood et al. 1992), 2 soil layers, and multiple vegetation types. The snowmelt and hydrologic models are applied separately over 1° latitude × 1° longitude grid cells within a river basin, with the output of the snowmelt model used as the input to the soil moisture accounting model. A river network connectivity is defined to link the 1 degree grid cells in the routing component, which utilizes a unit hydrograph and the impulse response function for a linear reservoir. Both the transport of water *within* and *between* 1 degree grid cells is accounted for by the routing scheme.

Model Application

The models were applied to the Colorado River basin to simulate daily streamflow at important water supply nodes for 40 years of historical input data. Time series of mean areal precipitation and temperature were estimated for each grid cell from the hydroclimatological data prepared by Wallis et al. (1991). Spatially distributed climatological PET estimates at 10 km resolution were obtained from Marks (1993) and averaged up to 1 degree resolution. Initial estimates of soil water holding capacity and vegetation parameters were obtained from spatially distributed global data sets (Patterson 1990, Willmott and Klink 1986). Final values of the snow and VIC model parameters were determined through manual calibration using naturalized streamflow data for the period 1949 - 1960. Model performance was validated by comparing simulated and naturalized streamflow for the period 1961 - 1982. A current climate scenario was then generated for the entire 40 year historical data record (1949 - 1987). Climate change scenarios were generated by adjusting the historical precipitation, temperature, and PET time series by prescribed amounts and by using GCM output interpolated to a 1 degree grid. Simulations were performed for 15 combinations of fixed precipitation (P-10%, P+7%, P+15%) and temperature (T+1.5C, T+2.5C, T+5.0C) changes, and for the Canadian Climate Center and Geophysical Fluid Dynamics Laboratory GCMs. The spatial distribution of changes in snowmelt, runoff, evaporation, and streamflow over the entire basin were investigated. Of particular interest for water supply purposes are the resulting changes in the magnitude and timing of streamflow at reservoir inflow locations throughout the Colorado.

Current Climate Simulation Results

A current climate streamflow simulation was performed for 40 years of historical meteorological forcings using the model parameter values determined through calibration. Figure 1 shows plots of average monthly naturalized and simulated flows of the Green River near Greendale, Utah. The hydrographs at this location exhibit a strong snowmelt peak during June, which is typical of most of the Upper Colorado sites. The simulated and observed hydrographs show close agreement throughout most of the year (r^2=0.89), although the peak flow is slightly underpredicted while winter flows are slightly overestimated. On an annual basis, the simulated flow volume is within 1% of the naturalized observed flow volume for this location.

Climate Change Sensitivity Analysis Results

Four types of climate change scenarios were examined in this study: 1) a basin-wide prescribed precipitation change 2) a basin-wide prescribed temperature change 3) combinations of basin-wide precipitation and temperature changes, and 4) spatially varying GCM derived precipitation and temperature changes.

A change in precipitation without any change in temperature results in a corresponding change in annual flow volume and the magnitude of the hydrograph peak, but virtually no change in hydrograph timing. A temperature increase ranging from 1.5C to 5.0C without any change in precipitation tends to decrease streamflow volume due to increased evaporation and a smaller snow pack, while the hydrograph peak occurs earlier because of an earlier onset of the snowmelt season. Combinations of fixed precipitation and temperature changes are shown in Figure 2 for the Green River near Greendale. Only the extreme scenarios, along with the base case, are shown since all other combinations fall within this bounded region. A precipitation increase of 15% combined with a temperature increase of 1.5C causes mean annual streamflow volume to increase by about 13% in the Upper and Lower Colorado. Increased evaporation due to warmer temperatures is offset by the larger overall streamflow volume in the late spring and early summer. However, at the opposite extreme, when precipitation is reduced by 10% and temperature increases by 5.0C, streamflow is dramatically reduced by as much as 43% in the upper basin and nearly 50% in the lower basin, and the hydrograph peak is shifted from June to May. The two GCM scenarios agree in their shifting of the hydrograph peak from June to May, although the Canadian Center Climate Model reduces mean annual streamflow by 23% while the GFDL model increases mean annual flow by 15%.

Overall, the climate change sensitivity results are in general agreement with previous studies conducted by Nash and Gleick (1991) and Lettenmaier et al. (1988), which predict decreased flows for most scenarios. However, the strength of this approach is its use of simplified conceptual-type models to simulate streamflow across large areas, while still maintaining a sufficiently realistic hydrologic description for use in regional water resources studies. This modeling strategy is being applied concurrently to other continental-scale catchments including the Columbia, Missouri, and Delaware, and has the potential to provide a consistent assessment of the climate change sensitivity of water resources for large regions of the U.S.

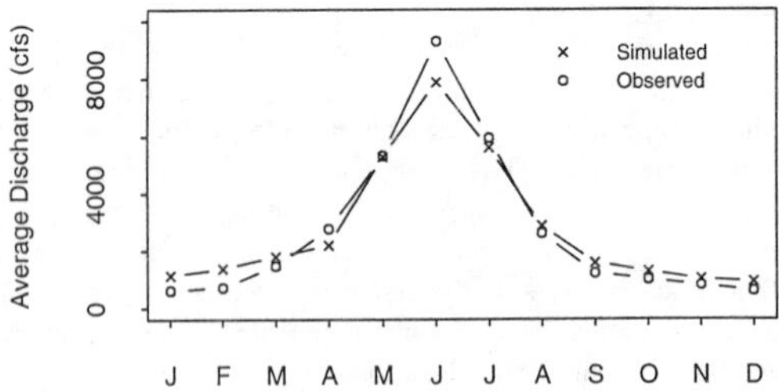

Figure 1. Average simulated and naturalized streamflow hydrographs of the Green River at Greendale, UT (1949 - 1987) (1 cfs = 1 ft.^{3}/s = 0.028 m^3/s).

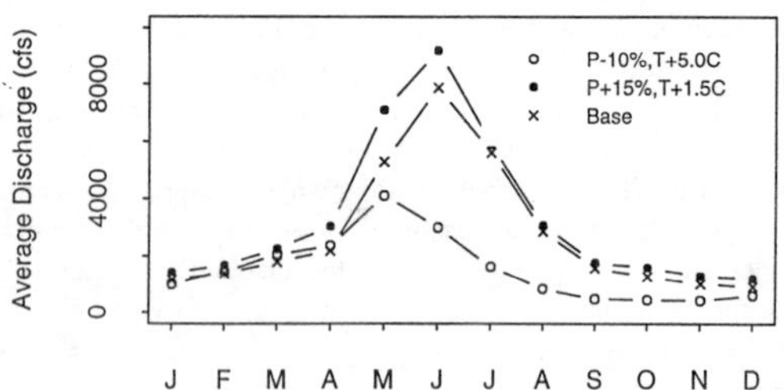

Figure 2. Effect of precipitation and temperature changes on the average hydrograph of the Green River at Greendale, UT (1 cfs = 1 ft.^{3}/s = 0.028 m^3/s).

References Cited

Anderson, E.A., National Weather Service River Forecast System--Snow Accumulation and Ablation Model, NOAA Technical Memorandum NWS HYDRO-17, U.S. Department of Commerce, Silver Spring, Maryland, 1973.

Marks, D., Personal Communication, 1993.

Liang, X., Lettenmaier, D.P., Wood, E.F., and S.J. Burgess, A simple hydrologically based model of land surface water and energy fluxes for GCMs, *Journal of Climate*, in press 1994.

Wood, E.F., Lettenmaier, D.P., and V.G. Zartarian, A land-surface parameterization with subgrid variability for general circulation models, *Journal of Geophysical Research*, 97(D3), pp. 2717-2728, 1992.

Wallis, J.R., Lettenmaier, D.P., and E.F. Wood, A daily hydroclimatological data set for the continental United States, Water Resources Research, 27(7), pp.1657-1663, 1991.

Patterson, 1990, Master of Science Thesis, University of Delaware, 1990.

Willmott, C.J. and K. Klink, Representation of the terrestrial biosphere for use in global climate studies, Proc. ISLSCP Conference, Rome, Italy, December 1985, ESA SP-248, May 1986.

Nash, L. and P.H. Gleick, The Colorado River Basin and climate change: the sensitivity of streamflow and water supply to variations in temperature and precipitation, U.S. EPA, Report Number EPA 230-R-93-009, December, 1993.

Lettenmaier, D.P., Gan,T.Y., and D.R. Dawdy, Interpretation of hydrologic effects of climate change in the Sacramento-San Joaquin River basin, California, *Water Resour. Series Tech. Reports No. 110*, Dept. of Civil Engrg., University of Washington, Seattle, Washington, 1988.

A REGIONAL MODEL OF POTENTIAL EVAPORATION FOR THE NORTHEAST UNITED STATES

Neil M. Fennessey[1] and Richard M. Vogel[2]

Abstract

A regional regression model is presented which approximates the mean monthly free water surface evaporation rate that would be estimated using a modified Penman equation. The model, which requires only the twelve annual values of mean monthly air temperature, site elevation and site longitude, is valid for all locations in New England, New York, New Jersey and Pennsylvania.

Introduction

The modified Penman equation (see Van Bavel, 1966; or Eagleson, 1972) describes potential evaporation, E_p: the atmosphere's near-earth surface demand for water vapor above a water body. The primary disadvantage of the modified Penman equation is its data requirements which include: net solar radiation; windspeed, dewpoint temperature, and air temperature. These data are generally only available at NOAA First Order observatories which are staffed by trained weather observers. In contrast, local average monthly temperature data are far more readily available from other Federal or State supported networks (over 1000 NOAA Summary-of-the-Day versus 26 First Order stations in New England, New York, New Jersey and Penn-sylvania). Since Penman equation derived estimates of E_p can only readily be estimated at First Order observa -tories, a method to estimate $E_p(\tau)$, τ=1,12, the mean monthly potential evaporation rate for any location in the northeastern United States would be useful in a broad range of water resources studies.

Model Development

Using monthly mean values of 1951-1980 climate data

[1]Senior Hydrologist, AM. ASCE, Massachusetts Department of Env. Protection, One Winter St., Boston, MA. 02108.
[2]Assoc. Professor, M. ASCE, Department of Civil and Env. Engineering, Tufts University, Medford, MA. 02155.

observed at 34 NOAA First Order stations, estimates of the average monthly potential evaporation, $E_p(\tau)$, using the modified Penman equation were obtained using pro-cedures described by Fennessey (1994) and Fennessey and Kirshen (1994).

A two harmonic Fourier series function, denoted as $E_{p,f}(\tau)$, was found to provide an excellent approximation to $E_p(\tau)$ and is shown as Eq. 1.

$$E_{p,f}(\tau) = E_{pa} + \sum_{k=1}^{2} a_k \cos\left[\frac{\pi k \tau}{6}\right] + b_k \sin\left[\frac{\pi k \tau}{6}\right] \quad (1)$$

The five $E_{p,f}(\tau)$ Fourier coefficients are: a_1, a_2 b_1, b_2 and E_{pa} where the latter is the annual average daily potential evaporation rate.

Regional regression equations for the five $E_{p,f}(\tau)$ Fourier coefficients were developed using the 34 NOAA First Order station estimates of E_{pa}, a_1, a_2, b_1, and b_2 as the dependent variables. These equations are

$$\hat{E}_{pa} = 2.95 - 0.0187\text{Long} + 0.000431\text{Elev} + 0.136T_a \quad (2a)$$

$$\hat{a}_1 = -1.13 - 0.0144\text{Long} + 0.000323\text{Elev} - 0.0401T_a + 0.0988c_1 - 0.129d_1 - 0.869c_2 \quad (2b)$$

$$\hat{b}_1 = 0.819 - 0.00817\text{Long} - 0.0190T_a - 0.0496c_1 + 0.172d_1 \quad (2c)$$

$$\hat{a}_2 = 0.101\text{Long} - 0.000190\text{Elev} - 0.0145T_a - 0.0459c_1 + 0.101d_1 + 0.500c_2 \quad (2d)$$

$$\hat{b}_2 = -0.491 + 0.00922\text{Long} - 0.000241\text{Elev} - 0.0158T_a - 0.0305c_1 + 0.245c_2 + 0.153d_2 \quad (2e)$$

where Long equals the site longitude in decimal degrees; Elev is the site elevation (meters); and T_a, c_1, c_2, b_1 and b_2 are estimated using the procedure discussed below. The station latitude was found to be highly correlated with the $E_{p,f}(\tau)$ Fourier coefficient d_1 (r=0.920), hence, latitude is not an independent variable in any of the regional equations.

All model coefficients were significantly different from zero at the 5% level and all five model residuals were found to be approximately normally distributed, using a 5% level normal probability plot correlation coefficient hypothesis test (See Fennessey, 1994).

T_a, c_1, c_2, b_1 and b_2, which are independent variables required by Eq. 2, are Fourier coefficients estimated from each NOAA station's twelve monthly values of mean air temperature. These coefficients are

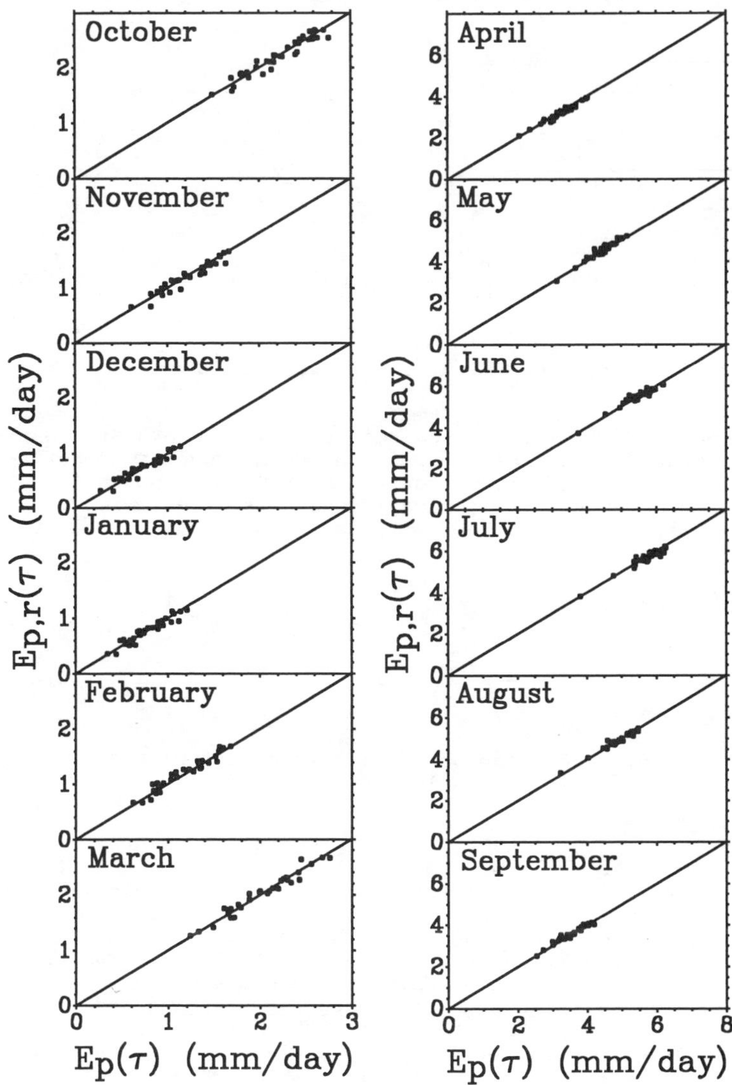

Figure 1. Comparison between modified Penman estimated $E_p(\tau)$ and regional regression model $E_{p,r}(\tau)$ potential evaporation.

$$T_a = \frac{1}{12}\sum_{\tau=1}^{12} T(\tau) \tag{3a}$$

$$c_1 = \frac{1}{6}\sum_{\tau=1}^{12}\left[T(\tau) - T_a\right]\cos\left[\frac{\pi\tau}{6}\right] \tag{3b}$$

$$d_1 = \frac{1}{6}\sum_{\tau=1}^{12}\left[T(\tau) - T_a\right]\sin\left[\frac{\pi\tau}{6}\right] \tag{3c}$$

$$c_2 = \frac{1}{6}\sum_{\tau=1}^{12}\left[T(\tau) - T_a\right]\cos\left[\frac{\pi\tau}{3}\right] \tag{3d}$$

$$d_2 = \frac{1}{6}\sum_{\tau=1}^{12}\left[T(\tau) - T_a\right]\sin\left[\frac{\pi\tau}{3}\right] \tag{3e}$$

where $T(\tau)$, τ=1,12 is the mean monthly temperature in degrees Celsius and $T_a(\tau)$ is the mean annual temperature.

The regional model for the mean monthly potential evaporation $E_{p,r}(\tau)$ (mm/day), which is an estimate of $E_{p,f}(\tau)$ and therefore $E_p(\tau)$, is shown as Eq. 4.

$$E_{p,r}(\tau) = \hat{E}_{pa} + \hat{a}_1\cos\left[\frac{\pi\tau}{6}\right] + \hat{b}_1\sin\left[\frac{\pi\tau}{6}\right] + \hat{a}_2\cos\left[\frac{\pi\tau}{3}\right] + \hat{b}_2\sin\left[\frac{\pi\tau}{3}\right] \tag{4}$$

The overall goodness-of-fit associated with the regional $E_{p,r}(\tau)$ model is best summarized in Figure 1.

References

Eagleson, P.S., (1970), Dynamic Hydrology, McGraw-Hill Book Company, New York, N.Y.

Fennessey, Neil M., (1994), A Hydro-climatological Model of Daily Streamflow for the Northeast United States", Ph.D. dissertation, Tufts University.

Fennessey, N.M. and P.H. Kirshen, (1994), Evaporation and Evapotranspiration Under Climate Change in New England. ASCE J. Water Resources Planning and Management, 120(1), pp. 48-69.

Van Bavel, C.H.M., (1966), "Potential Evaporation: The Combination Concept and Its Experimental Verification", Water Resources Research, 2(3), pp. 455-467.

CONSERVATION VERSUS TRADITIONAL SUPPLY MANAGEMENT

James P. Scott, P.E.[1]
ASCE Member

BACKGROUND:

New Jersey-American Water Company is the largest investor owned water utility in New Jersey and part of the American Water Works Company's nation-wide system of water companies. The American system has a corporate policy supporting proper water resource management. New Jersey-American has a comprehensive conservation plan in place supporting both supply side and demand side management strategies.

NEED:

New Jersey-American's sister company, Connecticut-American Water Company, completed a three year state mandated program of retrofit kit distribution. Before New Jersey-American started a similar program in response to regulation or in lieu of traditional supply argumentation, it wanted to determine 1) the effectiveness of retrofit devices for it's customers, 2) what delivery methods yielded the best results for the least cost, 3) what penetration and retention rates could be expected and 4) the actual savings that could be expected from a company-wide program. Since there was no information available on similar programs for New Jersey, the company initiated it's own retrofit pilot study using kits with low flow shower heads, faucet aerators and toilet devices.

PLAN:

New Jersey-American Water Company hired an economic consulting firm to assist in the design and evaluation of a retrofit pilot study. The company surveyed 3500 select residential customers and chose two methods of kit distribution: 1) direct mail (low-retrofit) and 2) contractor installed (high-retrofit). The study design targeted 400 hundred residential customers for each method of distribution and a total of 800 customers for a control group.

----

[1]Manager of Engineering, New Jersey-American Water Company, 500 Grove Street, Haddon Heights, New Jersey 08035

IMPLEMENTATION:

Of the 3500 customers selected to receive an initial questionnaire, 42% indicated they wanted free conservation devices. This was very encouraging, since a 10% to 20% response rate is typically expected. The customers chose which distribution method they wanted which the company expected would increase the satisfaction levels of the participants.

In the Spring of 1992, kits were mailed to 400 customers and a contractor installed kits for 276 customers. After the customers received their retrofit kits, the company conducted a satisfaction survey. The results of this were somewhat disappointing. After two mailings, only 58% of the low-retrofit customers and 72% of the high-retrofit customers returned the questionnaire. This may reflect the customer's waning interest in conservation after they see what is involved.

SAVINGS ANALYSIS RESULTS:

The consultant performed a billing analysis to determine the quantity of water that could be expected to be saved through a full scale retrofit program. They used two methods. First, a winter comparison of pre-study and post-study consumption with a control group and second, a state-of -the-art statistical analysis.

Price and income elasticity studies show that declining income or increasing rates could effect water consumption. Control groups were used to account for these and other contributing factors through time.

The first methodology compares winter usage for customers who participated in the study during the pre-conservation period with the post-conservation period. For comparison, a control group of non-participants was utilized. The method was performed as follows:

1. calculate the percent change for the pre- and post-periods for the participants.
2. calculate the percent change for the pre- and post-periods for the control group.
3. subtract the control group change from the participant percent change.

The second methodology involves the use of specialized statistical techniques to normalize for the effect of weather, such as precipitation and cooling degree days. These techniques can pool data effectively across participants and the control group to determine the effect of conservation measures. The technique also can allow for changes such as price, income, and other economic effects that a simple comparison may not be able to do.

A "fixed estimator" statistical method was used to estimate the conservation effects through time across individuals. Temperature and precipitation data were utilized from three NOAA weather stations in New Jersey.

Three years of billing data was collected from both participants and the control group. Participants were identified as those who indicated that they installed and did not remove the retrofit devices.

Reconciling both the pre- and post-conservation method of analysis and the statistical method of analysis, the participants in the low-retrofit program reduced their winter consumption by an average of 3.75%. The high-retrofit customers reduced their winter consumption by an average of 6.60%. The greater reductions for the high-retrofit group can be attributed to additional devices installed by the contractor at the customer's request. The combined average reduction in winter water consumption was 5.18%.

PROJECTED COMPANY-WIDE SAVINGS:

From the initial survey, 42% of New Jersey-American's customers were willing to install conservation devices in their home. The final installation rates for both high and low retrofit programs ranged from 20% to 29%. This translates into a projected total savings of 1.04% of residential winter usage if a company-wide low-retrofit program were implemented. If a company-wide high-retrofit program were implemented, a 1.40% savings of residential winter use would be expected.

Residential usage accounts for 49% of New Jersey-American's water sales and summer usage is 144% of winter usage. Therefore, a full scale low-retrofit conservation program for residential customers should yield approximately a 0.46% annual reduction in demand on the system. Similarly, a full scale high-retrofit program would yield an annual reduction in water consumption of 0.63%.

DECISION: CONSERVATION VS. TRADITIONAL SOURCE OF SUPPLY

Now that New Jersey-American Water Company has completed this retrofit pilot study, how can the results be used in its planning process? The first consideration is to quantify the deficit in source of supply. How much water is needed? Is there a maximum day demand deficit or an annual safe yield deficit? Are there alternative sources of supply available? Will "stacking" several different options be the most cost effective solution? How long will each option satisfy the demands before additional sources are required? What is the life expectancy of each alternative? If conservation is chosen, will an additional source of supply still be required to make up the deficit?

The results of the pilot study suggest that if there is more than a 0.6% deficit in annual demand, residential retrofit devices alone may not be enough. Although increased levels of outreach, marketing, and kit distribution might yield higher levels of demand reduction, this would also add to the cost of the program. Also, residential retrofit devices typically last five to ten years before needing replacement.

Are alternative sources of supply available? If there are, then a cost analysis will be necessary. The levelized cost of the high-retrofit program was \$1.52 per 3.8 m^3 (1,000 gallons) saved. The lėvelized cost of the low retrofit program was \$0.80 per 3.8 m^3 (1,000 gallons) saved. These figures do not include any administrative costs of the program. A 3,800 m^3 per day (1-MGD) well that costs \$1,000,000 to construct costs \$0.30 per 3.8 m^3 (1,000 gallons). Surface supplies could cost from \$0.60 per 3.8 m^3 (1,000 gallons) for a 3,800 m^3 per day (1-MGD) \$2,000,000 plant expansion to \$1.50 per 3.8 m^3 (1,000 gallons) for a 3,800 m^3 per day (1-MGD) \$5,000,000 new source.

If residential retrofit devices are determined to be the most cost effective means of making up the source of supply deficit, then the length of time that they will satisfy the deficit must be considered. If demand is increasing by 1.0% per year and residential retrofit devices will reduce demand by 0.6%, an additional source of supply will be required in about 7 months. If the next source of supply being considered will be adequate for many years of growth, then residential retrofit devices will only add additional costs and should be considered later.

CONCLUSIONS:

1. For New Jersey-American Water Company, residential retrofit devices can reduce demand by up to 0.6% and therefore may be appropriate for areas with little to no growth in demand.

2. Residential retrofit devices can be utilized as a source of supply where other sources are not available.

3. The cost of implementing a residential retrofit device program may be greater than traditional source of supply development.

4. A residential retrofit device program needs to consider the short life span of the devices, customer participation and cost to implement.

Potential for Irrigation on the Delmarva Peninsula

William F. Ritter   Robert W. Scarborough[1]
ASCE Member

Abstract

Corn and vegetables are the two largest irrigated crops on the Delmarva Peninsula. Total ground-water use on the peninsula is 0.74×10^6 m^3/day with 35% used by agriculture. High nitrate concentrations in the water-table aquifer are a concern on the peninsula. Nitrate concentrations ranged from 0.48 to 48 mg/L N in a USGS study with 33% of the samples exceeding 10mg/L N. There is a great potential to significantly increase the irrigated corn area because of an abundant ground-water supply.

Introduction

The Delmarva Peninsula is one of the largest agricultural production areas along the east coast. It consists of three counties in Delaware, nine counties east of the Chesapeake Bay in Maryland and two counties in Virginia. The area, approximately 16,800 km^2, is part of the Atlantic Coastal Plain that extends from Long Island, NY to the Gulf of Mexico. It is the most concentrated broiler production area in the U.S. Currently there are approximately 605 million broilers a year produced on the Delmarva Peninsula. The broiler industry employs over 22,000 people and is a $1.3 billion industry. Corn, soybeans and vegetables are the major crops grown on the Delmarva Peninsula.

Irrigation Trends

The total irrigated area for Delaware, Maryland and Virginia for 1993 is

[1] Professor and Research Associate, Agricultural Engineering Department, University of Delaware, Newark, DE 19717.

29900, 28400 and 38000 ha, respectively (Irrigation Association, 1994). There were large increases in irrigation in all three states from 1978 to 1987 but since that time the increase in irrigation has been relatively small. In all three states vegetables and corn are the two largest irrigated crops. The 1992 agricultural census indicates that in the nine counties in Maryland on the Delmarva Peninsula, 10% of the corn (7644 ha) and 73% of the vegetables (4169 ha) are irrigated (USDC, 1994b). In Delaware, 13% of the corn (7900 ha) and 59% of the vegetables (9330 ha) were irrigated (USDC, 1994a). The irrigated area listed in the census for all three states in 1992 is somewhat lower than the Irrigation Association survey. Delaware has a lower percentage of fresh market vegetables than Maryland which may account for the lower vegetable area irrigated. The two Virginia counties had 5160 ha of irrigation in 1992 (USDC, 1994c) with vegetables and potatoes as the only two crops irrigated. Soybeans, wheat and vegetables are the three largest crops in the Virginia counties. There is very little corn grown in the two counties in Virginia.

In all three states, sprinkler irrigation is the major irrigation method with most of the systems being either center pivot or travelling gun.

Water Supply and Geology

The peninsula is underlain by unconsolidated sediments that range in thickness from a featheredge at the Fall Line (the northern boundary of the peninsula) to as much as 2440 m along parts of the Atlantic coast. Precipitation averages 109 cm/yr and streamflow averages 38 cm/yr. About 16.5 cm of the average streamflow is overland flow, and the remaining 21.5 cm is ground-water discharge. The streams are generally perennial, however, the 7-day, 20 year return period, low flow ranges from 0 to 0.085 m^3/km^2 (Cushing et al, 1973).

Ten regional aquifers furnish nearly all the water used on the peninsula. Hydraulic characteristics vary from aquifer to aquifer and from place to place within each aquifer. The Quaternary aquifer has the highest transmissivity and is the most productive water-bearing unit on the peninsula with an estimated perennial yield of 3.79×10^6 m^3/day (67% of the estimated 5.68×10^6 m^3/day potential perennial fresh-water yield).

Annual water use on the peninsula in 1970 was about 0.52 $m^3 \times 10^6$/day (Cushing et al, 1973). By the year 2010 it is estimated to be 0.98×10^6 m^3/day. The estimated irrigation water use and total ground-water use for the three states on the Delmarva Peninsula for 1994 are presented in Table 1. The total ground-water use is approximately 0.74×10^6 m^3/day. Agriculture accounts for approximately 35% of the ground-water use with the largest amount being used for irrigation (0.17×10^6 m^3/day).

Ground-Water Quality

There have been numerous ground-water quality studies conducted on the Delmarva Peninsula (Hamilton et al, 1993; Ritter and Chirnside, 1982). The Delmarva Peninsula was one of seven pilot project areas in the U.S. Geological Survey National Water Quality Assessment (NAQA) started in 1986. Natural ground water on the Delmarva Peninsula is moderately acidic (pH is about 5.8) and dissolved constituents are low (median specific conductance value of 115 μsiemens/cm at 25°C) (Hamilton et al, 1993). Concentrations of nitrates in 185 water samples collected in agricultural areas ranged from 0.48 to 48 mg/L N. Thirty-three percent of the samples exceeded the EPA drinking water standard of 10 mg/L N. Concentrations of calcium and magnesium were also high in agricultural areas because of liming. Nitrate and other inorganic constituents in the water-table aquifer differed among hydrogeomorphic regions of the Delmarva Peninsula. Highest nitrate concentrations were found in ground water from the well-drained uplands (median concentration 8.9 mg/L) and lowest nitrate concentrations were found in the poorly drained low lands (0.1 mg/L N). The confined aquifers had a median nitrate concentration of 1.1 mg/L N. In an earlier study in Kent and Sussex counties, Delaware, Ritter and Chirnside (1983) found that 32% of the wells sampled in Sussex County had nitrate concentrations above 10 mg/L N.

As part of the NAQA study more than 100 wells in the water-table aquifer were analyzed for about 40 pesticides. Atrazine and/or alachlor were detected in more than 20 samples (Hamilton and Shedlock, 1992). Ninety-four percent of the water samples with detectable concentrations of pesticides were less than the EPA maximum contaminant and health advisory levels for drinking water. Water samples where pesticides were detected were generally collected from no more than 6.1 m below the water table.

Ground-Water Quantity and Quality Problems

Water levels have declined in the confined Piney Point aquifer in some parts of Delaware and Maryland. There is also the potential for well interference if large quantities of water are withdrawn, such as the case in irrigation wells. The principle solution to this problem rests in the judicious location of new wells.

Salt water intrusion is a problem in a number of areas on the Delmarva Peninsula. Most of the problems occur close to the bays because of large industrial users. Widespread nitrate contamination of the water-table aquifer and the difficulties in reducing it is also a major concern.

Table 1. Irrigation and Ground Water use on the Delmarva Peninsula for 1994.

State	Irrigation Use		Total Ground-Water Use
	Ground Water	Surface Water	
		(m^3/day x 10^3)	
Delaware	102.2	11.0	401
Maryland	53.7	29.3	302
Virginia	14.4	0	39

Future Irrigation Potential

There is a great potential to increase irrigation area in Sussex County, Delaware and Wicomico, Worcester, Dorchester, Sommerset and Caroline counties, Maryland. The Quaternary aquifer has the potential to supply large irrigation wells in many areas of these counties. In Sussex County, the estimated long-term water available from the Quaternary aquifer and the subcropping Manokin aquifer is 9.8×10^5 m^3/day. The confined aquifers in these counties would offer little potential for increased irrigation water pumpage.

From 1986 to 1993 there has been an average of 50.8 x 10^7 kg of corn imported to the Delmarva Peninsula for the broiler industry. By increasing the irrigated area the amount of corn imported could be reduced. This reduction in imported corn could help solve the ground-water nitrate problem and also may help reduce some of the disease problems caused by aflatoxins in the imported corn. If the irrigated corn area was increased by 40500 ha on the peninsula, the need for imported corn would be reduced in half. Also, because of the increased nitrogen demand, almost 25% of the poultry manure produced could be used to meet the increased nitrogen requirements for irrigated corn.

Appendix - References

Cushing, E. M., J. H. Kantrowitz and K. R. Taylor. (1973). "Water resources of the Delmarva peninsula". USGS, Washington, DC. Professional Paper 822.

Hamilton, P. A. and R. J. Shedlock. (1992). "Are fertilizers and pesticides in the ground water?" USGS, Washington, DC. Circular 1080.

Hamilton, P. A., J. M. Denver, P. J. Phillips and R. J. Shedlock. (1993). "Water-quality assessment of the Delmarva peninsula, Delaware, Maryland and Virginia - effects of agricultural activities on, and distribution of, nitrate and other inorganic constituents in the surfical aquifer." USGS, Washington, DC. Open File Report '93-40.

Irrigation Association. (1994). "1993 irrigation survey." Irrigation Journal 44(1):26-41.

Ritter, W. F. and A. E. M. Chirnside. (1982). "Ground-water quality in selected areas of Kent and Sussex counties, Delaware." Agricultural Engineering Department, University of Delaware, Newark, DE. Technical Report.

U. S. Department of Commerce. (1994a). "1992 census of Agriculture, part 8 Delaware." Department of Commerce, Washington, DC.

U. S. Department of Commerce. (1994b). "1992 census of agriculture, part 20 Maryland." Department of Commerce, Washington, DC.

U. S. Department of Commerce. (1994c). "1992 census of agriculture, part 46 Virginia." Department of Commerce, Washington, DC.

Impacts of Potential Water Use Restrictions on Row-Crop Irrigation in the ACF Basin

Daniel L. Thomas and James E. Hook[1]

Abstract

Crop model predictions of agricultural water consumption in the Flint River portion of the ACF basin in Georgia can exceed 4.0 million m^3/day during peak use periods in drought years. Potential emergency water restrictions could save between 25 and 150 million m^3 annually in the Flint River basin, based on selected 30- or 60-day restrictions on withdrawals. However, these water savings would cause economic losses to growers of $10 to $144 million, depending on the timing and duration of restrictions. Potential water management strategies must be developed which reduce irrigation demands in drought years, but also decrease the risk of economic losses.

Introduction

The Appalachicola-Chattahoochee-Flint River basins are a part of the continuing discussion on water use between Alabama, Florida and Georgia. Requests by metropolitan Atlanta for increased water allocation sparked litigation and debate over water resources associated with all three states. The Flint River basin is primarily agricultural and is not directly accessible to Atlanta. However, the Flint and Chattahoochee rivers do feed directly into the Appalachicola River and the Appalachicola Bay. In addition, the Flint River has unique hydrologic characteristics by the direct connection of ground water resources to the river. Agricultural withdrawals from ground and surface water resources within the basin can directly impact the water levels in the Flint River and downstream (Hicks et al., 1987). One of the primary concerns of Florida is to maintain consistent flow volumes into the Appalachicola Bay for oyster production.

To help resolve the water rights case, the U.S. Corps of Engineers initiated comprehensive studies of water use in the ACF and the adjacent Alabama-Coosa-Tallapoosa River basins. These studies included evaluations of

[1]Associate Professor and Research Leader, Bio. & Ag. Engineering Dept.; and Associate Professor, Crop and Soil Science Dept., University of Georgia, Coastal Plain Experiment Station, Tifton, GA 31793-0748

agricultural water uses and projections of potential water consumption for the next 50 years (Soil Conservation Service, 1994). The average (or most expected projection) is that agricultural water use will increase about 40% in the next 50 years, but could increase as much as 146%. Since the Flint River basin contains the majority of the current agricultural irrigation, these potential withdrawals will require effective water management strategies. One concern is how water management strategies can or will be implemented.

This paper briefly describes parts of the SCS report (1994, pg. 98-105). In this analysis, irrigation water demands for the major irrigated crops were evaluated with arbitrary irrigation restrictions. The primary emphasis is on potential water saved and the economic impacts of these water restrictions.

Methods

The methods used in this study are presented in the Soil Conservation Service Report (1994) and by Hook and Thomas (1995). Additional data was developed which is not represented in this paper. The current distribution of the four main irrigated row crops (corn, soybean, peanut, and cotton) in the Flint River basin were used to calculate agricultural water use. Irrigated areas have remained somewhat stable in the past 10 years. Recent increases in cotton irrigation have been at the expense of soybean irrigated areas.

Predictions of crop yield and irrigation timing and amount were performed with three crop models; CERES-maize, SOYGRO, and PNUTGRO; using procedures indicated by Hook (1994). Cotton water use was assumed proportional to soybean water use since no validated cotton model was available. Climatological data was from the Tifton, Georgia meteorological station which included 57 years of rainfall, temperature and pan evaporation data and 20 years of solar radiation data. Tifton is within 32 km of the edge of the Flint River basin. Troup/Lakeland deep sand, Tifton/Dothan loamy sand, and Orangeburg sandy loam represent ~95% of the agricultural soils in the lower Flint River basin (where most irrigation is) and were the soils used.

The analysis was performed on all years and on drought years. Agricultural drought years were the lowest quartile of non-irrigated yield compared to no-stress yields. These 14 "drought years" were not the same year for each crop because of different irrigation seasons.

Water use restriction options were assumed as the most feasibly implemented potential scenarios. Alternative 30- and 60-day no-irrigation periods were selected as regulatory restrictions. With the yield and irrigation amounts, direct economic impacts to growers was calculated for each restriction scenario. This discussion is based on potential water savings and cost of water savings. Water savings were the difference between full and restricted irrigation amounts times the total irrigated area for that crop. The loss of revenue was the value of full irrigation yield - restricted irrigation yield -

savings for less water applied divided by the water savings. Fixed irrigation system costs are not included in the analysis.

Results and Discussion

Figure 1 illustrates the mean daily irrigation water demand for each 10-day period for the Flint River basin during the 14 drought years. Corn irrigation is most critical from May to July, peanut from July to September, and cotton from August to October. The crops grown in August and September can result in water needs over 4.0 million m^3 of water/day.

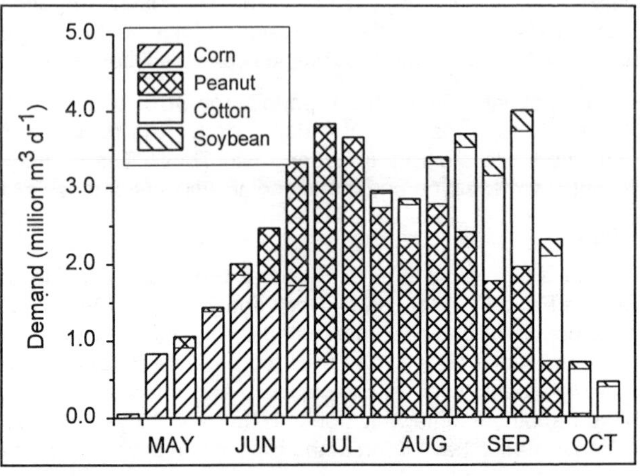

Table 1 illustrates the potential revenue lost by growers and the resulting water savings. For each crop, if the monthly water restrictions were imposed during the period of greatest water need (i.e., June for corn, September for soybean ...), the net farm receipts would be negative. With the two month restrictions, the net losses would be even greater. These economic losses are difficult to manage since restrictions are imposed after planting and after most fixed and variable costs have been incurred. If water use restriction periods and durations were known prior to planting, cropping scenarios, production levels, and irrigated areas could be adjusted to minimize risk of economic loss. In any event, new, economically viable crop alternatives that require water at different times, irrigation methods that minimize losses, and improved scheduling techniques are necessary to optimize irrigation water use under any irrigation reduction plan.

The potential water saved by these restriction scenarios may be less than expected. For example, peanut received an average 6.2 cm of water during August with full-season irrigation. When irrigation was restricted during August, seasonal water savings were only 3.1 cm. Extra water was applied in September to recover the soil water depletion in August. Obviously, techniques which have been used in the west, including deficit irrigation, can be applied to humid areas. Since humid area irrigation is supplemental to rainfall, options to minimize inputs in anticipation of rainfall may be appropriate. in any case, more thought, research, and planning is necessary to minimize negative economic impacts of water management and restriction alternatives.

Table 1. Grower Revenue Losses and Water Savings in the Flint River Basin from Water Restrictions that Eliminate Irrigation.

Water Restriction Period	Crop				All Crops	
	Corn	Peanut	Soybean	Cotton	Loss Total	Water Savings
	------------------------ Million $ ----------------------					Mil. m^3
May	9.9	-	-	-	9.9	24.6
Jun	24.8	-	-	-	24.8	50.3
Jul	0.5	33.4	0.0	-	33.9	68.9
Aug	-	34.8	0.3	-	35.1	42.0
Sep	-	9.8	4.6	9.3	14.4	81.0
May-Jun	37.0	-	-	-	37.0	82.5
Jun-Jul	25.7	-	-	-	25.7	60.6
Jul-Aug	-	143.2	0.5	-	143.7	130.6
Aug-Sep	-	79.6	5.2	9.3	94.1	153.7
No-Irr	37.3	241.5	5.7	26.8	311.3	418.5

References

Hicks, D. W., H. E. Gill, and S. A. Longsworth. 1987. Hydrology, chemical quality, and availability in the upper Floridan aquifer, Albany Area, Georgia. U.S. Geological Survey. Water Res. Invest. Rpt. 87-4145. 52p.

Hook, J. E. 1994. Using crop models to plan water withdrawals from irrigation in drought years. Agric. Systems 45:271-289.

Hook, J. E. and D. L. Thomas. 1995. Projected seasonal water consumption and water restrictions for row crop irrigation in the AC basin. in Proceedings of the 1995 Georgia Water Resources Conference, April 11-12, 1995. Kathryn J. Hatcher, Editor, Inst. of Natural Resources, University of Georgia, Athens.

Soil Conservation Service. 1994. ACT/ACF river basins comprehensive study: Agricultural water demand. Tech. Rpt., Aug., 1994. USDA, Soil Conservation Service, Auburn, AL.

AGRICULTURAL WATER SUPPLY AND CONSERVATION ISSUES IN NORTHEAST STATES

Leland Hardy, Carl Gustafson, David Nelson[1/]

Abstract

In Northeast States, the agricultural community competes with urban, industry, wildlife and other uses for water. Annual precipitation is adequate to grow most crops. However to meet plant water needs, individual rainfall events are often untimely, especially during July and August. Irrigation is beneficial to crops and the environment.

Except for isolated hot spots, agriculture is a small contributor to real (or perceived) water quality and quantity problems in Northeast States. Many water users are using state-of-the-art tools to help make water management decisions. Conservation practices help minimize water moving below the plant root zone and as surface runoff. This presentation will discuss what agricultural water users are doing to properly use available water, including irrigation methods used, crops grown, irrigation scheduling, and potential impacts on surface and ground water quality.

Background

Over 200,000 hectares (500,000 acres) of high value crops are being irrigated in Northeast States. Close proximity to markets, productive soils and available water supplies strengthen irrigated agriculture. Competing land uses include urban, industry and wildlife. In the last five years total irrigated land in the northeast increased 8%, while irrigation in Maine, New Hampshire, and Vermont decreased over 20%.

1/ All authors are USDA Natural Resources Conservation Service engineers, 451 West Street, Amherst MA 01002.

Annual precipitation is adequate to grow most crops. However, individual rainfall events are often untimely for providing daily plant water needs, especially during critical bloom or fruiting stages. When plant functions are limited by a lack of water, even for a few hours, the plant begins to enter into a survival mode, closing down less critical functions, thus reducing potential yield or biomass. In addition as the soil continues to dry, soil microorganism activity slows, and if the soil is dry enough, activity ceases. Soil microorganism activity is necessary for most user applied and natural occurring chemicals to become available for plant use. Sources of natural occurring chemicals include: residue decomposition (nitrification), acid rain (nitrogen oxides and sulfates), and phosphorous deposition in surface waters.

IRRIGATED CROPS GROWN IN NORTHEAST STATES 1/

CROP	CT	DE	ME	MD	MA	NH	NJ	NY	PA	RI	VT	VA	WV
Alfalfa						*					*	*	
Barley		*											
Beans												*	
Corn	*	*		*		*		*				*	*
Cotton												*	
Cranberries			*		*		*						
Grapes					*							*	
Melons			*	*								*	
Nursery	*			*	*		*			*		*	
Pasture/Hay	*		*	*	*		*			*	*	*	
Peanuts												*	
Potatoes	*	*	*		*			*	*	*		*	
Sm Fruit/Nuts		*	*		*	*	*	*	*			*	
Sod/Turf	*			*			*			*		*	*
Sorghum												*	
Soybeans		*		*								*	
Sweet Potatoes												*	
Tobacco				*	*							*	*
Tree Fruits		*		*	*		*	*	*	*		*	*
Vegetables		*	*	*	*	*	*	*	*	*		*	*
Wheat		*		*									

Most farmers like clean water for human consumption, irrigation, fishing, and recreation; and they must live in the community. Irrigators in Northeast States generally want to protect the environment by practicing good water management on their fields. This action is mostly personal, but also to help avoid a group

1/ 1993 Irrigation Association Survey

perception that irrigators are major contributors to degraded surface and ground water. Except for a few hot spots, this is generally true. Irrigators in the northeast are above average.

Irrigation Methods

Irrigation methods used to apply water in the northeast states are: sprinkle (92%), micro (8%), and surface (<1%). Center pivot, linear move, traveling gun, side roll (wheel line) and hand move sprinkler systems are most common. Micro irrigation (primarily drip emitters, micro spray and line source) is used to irrigate orchards, vineyards and vegetables. In addition to supplying plant water needs, other water uses such as frost control and crop quality control can influence irrigation method selection.

With proper water management, sprinkle, micro and surface irrigation systems can apply water uniformly across the field, in the proper amount at the proper time. Excess irrigation and precipitation results in surface water runoff or water moving through the soil below plant roots to ground water.

Waste Disposal

The soil is perhaps the best media for waste disposal and improved water quality. Soil microbes in the presence of moisture can metabolize pollutants into harmless chemical compounds. With proper management, most agricultural wastes and many municipal and industrial waste products can be safely disposed on agricultural land as a plant nutrient source. Some metabolites and heavy metals can be harmful.

Irrigation Scheduling

State-of-the-art tools to help the irrigation decision maker determine "when and how much water to apply" include computer software, and soil or plant moisture monitoring devices. SCS SCHEDULER and several other University computer software programs are available. These software programs use crop, soil and daily climatic data from a weather station to calculate potential plant water use for each crop. Results are field specific.

Some irrigation scheduling methods simulate plant water use. Two of these methods, calibrated evaporation pans and atmometers are placed in the field at approximately the plant canopy height. The crop water stress index (CWSI) method uses a hand-held gun to measure plant canopy temperature, ambient air temperature, solar radiation and relative humidity. Periodic soil moisture monitoring is used to calibrate each method with actual soil water used by plants.

Frequent soil moisture monitoring can be used directly to schedule irrigation applications. Most common of these methods is the "feel and appearance" method, where a sample of soil is removed from a preselected soil depth with a shovel, auger or probe. The hand and fingers are used to manipulate the soil sample where soil wetting patterns on the hand, color, and strength can be observed. With experience the observer can estimate soil moisture within 5% of actual. Indirect soil moisture measurements include electrical resistance blocks and tensiometers. These devices are placed in the soil at preselected depths. Electrical current resistance within the block as affected by moisture, or soil water suction is measured. These devices are most accurate in medium and course textured soils.

The accuracy of any irrigation scheduling tool need only be sufficient to help make the decision "Do I irrigate or do I wait?" Subsequent soil or plant water measurements compensate for over or under estimation of actual water available for plant use. Periodic rainfall events can mask an otherwise clear decision. Rainfall events can also salvage a less than adequate irrigation schedule. In all cases, applied irrigation water is supplemental to precipitation events.

Negative impacts on the environment resulting from poor water management include reduced crop yield and biomass, reduced water availability, nutrient-laden surface runoff and deep percolation water, algae bloom in water bodies, and sediment in streams. Proper irrigation water management has very little, if any, negative impacts. Positive impacts include: increased plant biomass, clean stabilized water bodies, typically reduced water use; and, assuming humans are part of the environment, improved quality of life, job stability from year to year, stabilized community economy and improved community aesthetics. Conservation practices, including good irrigation water management, crop residue use and reduced tillage are used to minimize negative impacts on the environment. Thus, the irrigation decision maker who practices good water management not only improves the farm economically, but also improves the quality of life of others and the general environment in which he/she lives.

Untapped Water Conservation Opportunities in New England

Amy Vickers[1]

Abstract

Over the past 10 years there has been a virtual explosion in the field of water conservation technology and policy. Interestingly, several of these initiatives were made in New England despite its "abundant water" image. This paper will review one recent conservation policy initiative which, if adopted by New England (and other) utilities, would realize increased efficiency in water usage in a demand category that is all too often ignored: outdoor landscape and irrigation management.

Introduction

Water conservation is typically associated with California and the arid southwestern regions of the United States. At the same time, while a number of water utilities there have implemented conservation programs over the past 10 to 15 years, some of the most aggressive efforts to adopt and implement innovative conservation technologies in recent years were actually pioneered in New England. For example, Massachusetts became the first state in the country to require low-volume, 1.6 gallons per flush (gpf) or less toilets in 1988 through an amendment to the Massachusetts Plumbing Code. That initiative later resulted in federal legislation, the Energy Policy Act of 1992, which established for the first time national water efficiency standards for plumbing fixturesConnecticut also became the first state to require all water utilities to implement plumbing fixture retrofit programs targeted to all residential customers. In addition, the Massachusetts Water Resources Authority's

[1] President, Amy Vickers & Associates, Inc., Water Planning, Policy, and Management, 100 Boylston Street, Suite 1015, Boston, MA 02116-4610

conservation program has reduced systemwide demand since the mid-1980s by approximately 20 percent, a figure equaled by no other major U.S. water utility to date.[2],[3],[4]

What's interesting about these two examples is that the technology for low-volume fixtures and the conservation program design for distributing retrofit kits in the U.S. was largely "born" in California. And yet, they weren't fully realized - implemented on a broad scale - until they were established by New England water utilities. If there's a pattern here perhaps it is this: conservation innovations may be dreamt in California, but New England can make them happen in a big way.

Efficient Landscape And Irrigation Standards

There have been a number of improvements in water conservation technology and policy within the last 5 years that offer new and exciting opportunities to increase water use efficiency, particularly in the area of outdoor water usage.

Despite the 40 to 50 inches of annual rainfall that they receive, New England homeowners, commercial businesses, golf courses and others demand considerable amounts of water to irrigate their turf and landscaped areas. For example, the MWRA's 1994 peak summer month demand of approximately 325 million gallons a day (mgd) was nearly 1.5 times the average indoor winter monthly usage. Clearly, some of this discretionary water demand could be trimmed if efficient landscape and irrigation standards were established on a permanent basis.

A number of states and communities - Florida, Tucson, AZ, California, and others - have established Xeriscape-type policies and ordinances. Case studies have shown that efficiently designed and managed turf and landscaped areas can reduce water demand by 40 to 75 percent, if not more. Establishing standards to promote the installation of more water efficient plants and irrigation systems could be adopted in New England as well.

An example of an efficient landscape and irrigation policy is the California Model Landscape Ordinance law.[5] This

---

[2] Vickers, Amy. New Massachusetts Toilet Standard Sets Water Conservation Precedent. *Journal AWWA*. (March 1989)
[3] Vickers, Amy. The Energy Policy Act: Assessing Its Impact on Utilities. *Journal AWWA*. (August 1993)
[4] Ruzicka, Denise. Recent Water Conservation Initiatives By The State of Connecticut. *Proceedings of Conserv90*, American Water Works Association. (1990)

legislation requires all cities and counties to either establish their own landscape standards or adopt the State's. Specific provisions of the law include: irrigation schedules, quantitative standards for plant water requirements and irrigation efficiency (which are not to exceed the maximum applied water allowance -- "MAWA," approximately half the amount of water needed to irrigate 4- to 7-inch cool season grass at an irrigation efficiency of 62.5 percent), prohibition on runoff, mandatory irrigation audits for existing landscaped areas over one acre, and other features.

Conclusions and Recommendations

This paper briefly presents one but significant and untapped opportunity to realize permanent reductions in outdoor water use -- the establishment of efficient landscape and irrigation standards -- which have yet to be fully developed by utility conservation programs in New England. While such measures originated in more arid regions of the U.S., they are just as applicable and needed in the New England environoment and present additional opportunities to permanently reduce both average and peak day water demand in the future.

---

[5] California State Ordinance AB.325, 1990.

Living Within Our Means:
A Successful Demand-side Water Resources Plan for Boston

Stephen A Estes-Smargiassi[1]

Abstract

A thoughtful, well funded program of demand management measures from the reservoir to the tap was successful at reducing long-term year round demand in the Boston metropolitan area by over 25 percent in less than 7 years. This reduction eliminated the need to develop new supplies saving hundreds of millions of dollars.

Introduction

From 1652 to 1986, water resource planning for the Boston metropolitan area meant developing new sources. In 1986, the Massachusetts Water Resources Authority made a radical change in policy and instituted demand-side planning. Major new sources were out, efficient use of available resources was in. In the past 7 years, base level demand has been reduced by 0.3 million m^3 (85 million gallons) per day, more than 25 percent, and the system is operating within its safe yield for the first time in 20 years. No new source development is planned for the indefinite future.

The MWRA effort was designed to result in long-term, year-round reductions in use, rather than seasonal or occasional changes. It examined each component of water use, every step of the way from the reservoir to the tap, and attempted to produce measurable improvements in efficiency. The program involved leak detection and improved metering at both the wholesale and retail level; conservation projects for residential, commercial,

----

[1]Manager Waterworks Planning, Massachusetts Water Resources Authority, Charlestown Navy Yard, 100 First Avenue, Boston, MA 02129

industrial, and municipal users; more efficient use of both major surface supplies and local groundwater sources; protection of vulnerable supplies; and a program evaluation component which allowed for mid-course changes in direction.

Moment of Decision

In 1986, the newly appointed Board of Directors of the newly created MWRA was faced with almost 50 volumes of technical reports, several volumes of citizens' detailed point by point responses to the reports, and a series of heated public meetings. At issue was whether more water was needed by the Boston area, and if so where to get it. It seemed clear that the current use was more than 10 percent higher than the 1.1 million m^3/day (300 mgd) safe yield of the sources, but also clear that a portion of this demand left the reservoir to serve not people, but leaks and inefficient water sue practices. The predicted potential of a 50 percent demand increase by 2010 was there, but so was the potential for a significant reduction in demand. (MWRA, 1986) (ESRG, 1986)

The Board directed to develop a three year trial of the demand management alternative, and to essentially re-evaluate the core assumptions of traditional water supply planning. How much demand reduction was possible? How much would it cost? Must existing supplies inexorably be contaminated and lost as supplies? Could the MWRA meet the legitimate needs of its users indefinitely without a major new supply? Only if the demand management alternative failed were they prepared to resume their look at the traditional alternatives.

The Program

The program (MWRA, 1987) had six principal areas, with initial or pilot efforts to be underway during the three year trial:

* Leak detection, repair and metering to eliminate waste and better account for use;
* Conservation and demand management to help customers use water more efficiently;
* Improved use of sources so every gallon of yield could be used;
* Water supply protection so no existing source would need to be replaced with a new one;
* Management and planning for the future to ensure a reliable, safe supply; and
* Outreach and reporting to build and maintain support of the programs.

Each area was structured to test out its effectiveness, with frequent reports on individual efforts and the entire program to the Board, and mid-course revisions as needed to achieve the desired outcome.

Leak detection and repair programs were designed to find and repair leaks on the 426 km (265 m) of MWRA owned major pipes and the 10,140 km (6300 m) of community owned pipes. The effort included a once through survey of every mile, funded and managed by the MWRA, with communities required to promptly repair any leaks found. This was followed up with new regulations requiring leak surveys every other year. Several thousand leaks were found initially and over 0.1 million m^3/day (30 mgd) of waste found, about 8 percent of total demand. (MWRA, 1990) Subsequent surveys have typically found over a thousand leaks per year. This type of program is needed to show that the utility is serious about getting its own "house in order" as well as resulting in significant savings. Our goal is to reduce total unaccounted for from its high of about one third to less than 15 percent. We are more than half there.

Demand management programs are designed to provide a comprehensive approach to reducing water use over the long term, with coordinated efforts in each sector of use. Water use in homes was reduced by a aggressive water saving device retrofit program. A pilot program researched how to best get the low use aerators, showerheads and toilet devices into peoples homes. We found that actually having a contractor go in and install the devices was the most cost effective way, on a dollar per gallon basis. Over 1.3 million devices were installed in about 363,000 houses (about half the service area) with estimated water savings of 0.02 million m^3/day (5 mgd). Water use in business was reduced by funding audits of industrial, commercial and institutional customers, and then using the results not just in the audited facility, but in other similar locations. This area has a large potential with 20 to 30 percent reductions possible with economic paybacks of less than several years. A school education effort with its focus on teacher training and curriculum development will keep these issues alive for the next generation.

Water supply protection is key to our long term success. Much of the historical, and the previously projected, increases in demand was from communities losing their sources to contamination and turning to the MWRA. These programs looked at the potential pollution threats to sources in both communities partially served by the MWRA and those adjacent to the service area. Over 50 ground and surface water supplies were evaluated and protection efforts recommended. Failure to succeed could represent almost 0.4 million m^3/day (100 mgd) of additional demand.

A new way of approaching planning was begun, where planning, rather than being an every decade or so massive episodic effort, would be a continuous process with monitoring and triggering points established. (USACOE, 1994) Significant players both inside and outside the agency were involved in the process to increase effectiveness and consensus on the results.

All the good intentions for changing our way of thinking would not have been possible without the support of significant outside interest groups. A formal process of consultation and reporting on our programs and progress resulted in continuous feedback and critical support at key moments. A funded citizens advisory group, with their own staff, looked over every detail. This separate voice lent credibility to our efforts.

Conclusions and prognosis

Our success indicates that a well thought-out program, with high level support from within the agency, interest from decision makers and momentum driven from citizen pressure can result in significant long-term reductions in demand. Our Board of Directors made it clear that demand management was to be taken as seriously as any engineering project, and funded and oversaw the program appropriately. Demand management was less expensive on a life cycle basis than new sources, more environmentally acceptable, and appear to be at least a dependable as any new source would be. We foresee no new sources in our future.

Energy Systems Research Group (1986). Closing the Gap: A Preliminary Assessment of a Conservation Scenario for Meeting the Long Term Water Needs of the MWRA Communities, Boston, MA

MWRA (1986). Summary Report - Long Range Water Supply Study and EIR-2020, Boston, MA

MWRA (1987). Long Range Water Supply Program, Boston MA

MWRA (1990). Long Range Water Supply Program - Program Briefing and Recommendations, Boston MA

MWRA (1993). LRWSP - Update, Boston MA

USACOE (1994). The New England Drought Study: Trigger Planning, in the National Study of Water Management During Drought, Waltham, MA

means.pap

Demand Management Strategies for Providence Water Supply Board

Arun K. Deb, Ph.D., P.E., Fellow, ASCE[1]
Frank Grablutz, P.E., Associate Member, ASCE[2]
Paul Gadoury, P.E.[3]

Abstract

The Providence Water Supply Board (PWSB) supplies water in and around the City of Providence, Rhode Island. The PWSB sponsored a demand management study in recognition of the importance of conserving a valuable natural resource while still maintaining adequate supplies of public water. Projections of service area water demand were made through the year 2010 based on existing water use patterns. A series of demand management strategies were developed, and the effectiveness of these strategies in reducing water demand were estimated using a computer model. The demand management study later became a part of a state-mandated water supply management plan.

Introduction

The City of Providence, Rhode Island, obtains its water from the Scituate Reservoir complex, located on the North Branch of the Pawtuxet River. The Providence Water Supply Board (PWSB) supplies water to about 66% of Rhode Island's population through retail water service to customers in

[1]Vice President, Roy F. Weston, Inc., One Weston Way, West Chester, PA 19380

[2]Project Engineer, Roy F. Weston, Inc., One Weston Way, West Chester, PA 19380

[3]Director of Operations, Providence Water Supply Board, 552 Academy Avenue, Providence, RI 02908

the City of Providence and three surrounding towns, and through wholesale supply to ten other surrounding water utilities.

The demand management study began with a major data collection effort and the development of databases of water utilities' customer billing data and associated tax assessor data for the study area municipalities. In compiling accurate water consumption data by water use classification, a database was developed by superimposing water utility billing databases with tax assessors' databases. The new database provides water use classifications and water consumption information for all customers. This database is used in the development and analysis of water demand management strategies that would be effective in reducing water demand over the 20-year planning period.

In 1991, the Rhode Island legislature passed the Water Supply Management Act (WSMA) requiring all Rhode Island water systems that deliver more than 50 million gallons per year to prepare and file a detailed water supply management plan. The plan was to be developed in accordance with a guidance manual prepared by the Rhode Island Department of Environmental Management (DEM). The plan was to include a review of the existing water system; an identification of problem areas; an assessment of existing water supply yields and an estimate of future water demand; and the development of supply, demand, emergency, and financial management plans. The demand management study described here satisfied many of the water supply management plan requirements.

Study Area

The study area includes 16 cities and towns that receive water from the Scituate Reservoir complex. Eleven water utilities provide public water supplies within these municipalities. PWSB provides retail service to customers in Providence, North Providence, Cranston, and Johnston in addition to selling water on a wholesale basis to the other utilities. The Kent County Water Authority (KCWA) supplies a portion of its water supply from Authority-owned wells in addition to wholesale purchases from PWSB.

Demand Projections

The average water supply for all the utilities during 1990 was 76.91 million gallons per day (mgd). This includes the water distributed by the PWSB (retail and wholesale) as well as lesser amounts of water supplied from other utilities' sources. Water supply represents all water that is treated and distributed and includes water that is lost to leaks and other unmetered uses.

Three separate projections of service area population and corresponding water demand projections were made for each water utility in the study area by considering a range of possible development patterns (i.e., Low, Most Likely, and High growth). These projections represent the unrestricted water demand that would be expected to occur without the implementation of any water conservation measures.

Demand Management Strategies

In developing demand management strategies, the ongoing conservation activities in each utility were considered. The major existing conservation measure is the Rhode Island plumbing code requiring 1.6 gallons per flush for toilets and 2.5 gallons per minute (gpm) for showerheads. Other existing conservation measures include public education programs (ranging from water conservation literature included with water bills to comprehensive school curriculum programs), water-saving device retrofit programs, leak detection and repair programs, and conservation-oriented water price structures.

Three demand management strategies were analyzed for each utility. The baseline strategy considered only those conservation measures currently active in a utility. The moderate strategy includes the baseline measures and other proven water-saving measures. An aggressive strategy included the same basic measures as the moderate but assumes a higher degree of implementation by the utilities.

The effectiveness of the three demand management strategies was calculated using a computer model that was developed as a part of a U.S. Army Corps of Engineers project (Richards et al., 1984). The results show that the unrestricted water demand in the year 2010 can be reduced by 7.7%, 14.4%, and 24.3% under the baseline, moderate, and aggressive strategies, respectively. These results are summarized in Figure 1.

The total effectiveness of the demand management program is the result of reductions in water use from various conservation measures. The effectiveness of a particular conservation measure will vary from utility to utility as a result of different water-use patterns. Therefore, an important consideration in designing a demand management strategy for a specific water utility is to consider which conservation measure (or measures) provides the greatest probability of implementation and produces the desired savings in water use.

Acknowledgment

The authors would like to acknowledge Mr. Richard Rafanovic, Chief Engineer, Providence Water Supply Board for his valuable guidance in conducting this study.

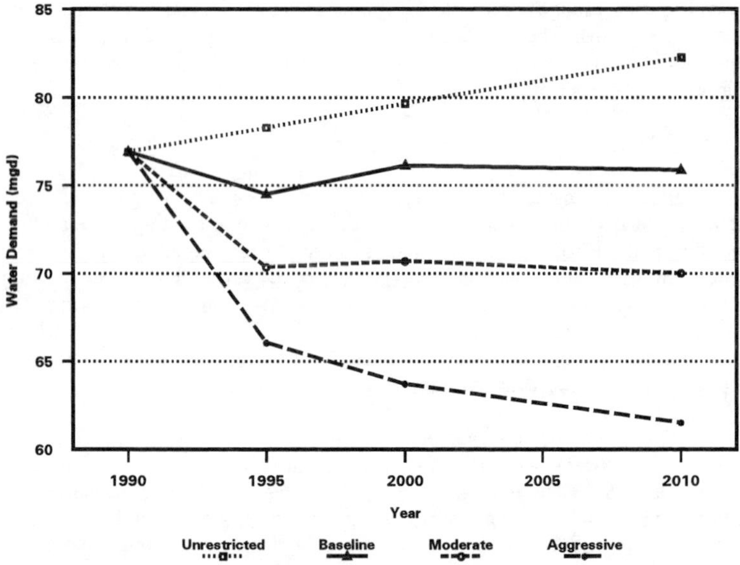

Note: Water demands shown are based on the Most Likely growth scenario.

Figure 1. Water Demand Under Demand Management Strategies

Reference

1. Richards, W.G., D.J. McCall, and A.K. Deb. 1984. *Algorithm for Determining the Effectiveness of Water Conservation Measures*, Technical Report EL-84-3, prepared by Roy F. Weston, Inc., West Chester, PA, for the U.S. Army Corps of Engineers Waterways Experiment Station, Vicksburg, MS.

AN ANIMATION/SIMULATION MODEL FOR MULTIPLE RESERVOIR OPERATIONS IN A WATER SUPPLY SYSTEM

Vladimir Shnaydman[1] and Richard M. Vogel[2]

Abstract

Simulation/animation software is introduced for the purpose of enhancing multiple reservoir decision support systems. The combined use of simulation and animation software provides a visual assessment of complex reservoir systems. Animation/simulation software can also enhance marketing and educational opportunities. A simulation model of a water supply system normally incorporates balance equations and an operations algorithm for the individual and joint operations of the reservoir system. In this study, rule curves are specified for the operation of each individual reservoir and a joint reservoir operations algorithm termed a 'compensation streamflow regulation' is introduced. The algorithm consists of a set of heuristic rules around which the water surpluses and deficits in the various reservoirs are balanced. The animation portion of the model includes visualization of the water supply system layout including different dynamic features such as releases, storage levels and conditions downstream of each reservoir.

Introduction

We examine the utility of using two dimensional color animation for the purpose of simulating multiple reservoir operations. A literature review reveals that exploiting animation models in a simulation environment can lead to benefits both during the model creation phase and the model implementation phase. During the model creation phase animation can: (1) show gross errors in the logic of the system; (2) uncover design flaws inherent in the layout; (3) make complex interactions between elements of the system more apparent and understandable; and (4) involve other decision-makers in addition to the simulation analysts in the development and verification process. Once the model has been created and simulation studies are conducted, the results need to be communicated to managers responsible for the decision-making. Such decision-makers often have little or no simulation background. Hence during the model implentation phase, animation can (1) simplify the presentation of results to managers and engineers; (2) serve as a catalyst for the creative resolution of

---

[1] Advanced Visual Data, Inc., Waltham, MA, 617-736-0860

[2] Department of Civil and Environmental Engineering, Tufts University, Medford, MA, 02155, Tel - 617-628-5000 Ext. 4260, Email - rvogel@pearl.tufts.edu

decision-making negotiations and conflicts, (3) provide engineers and operators with valuable visual information to help schedule and control the system; (4) serve as a training and educational aid; and (5) provide environmental organizations concerned about environmental impacts of water projects with detailed visual information.

An Algorithm for the Joint Operation Multiple Reservoir Systems

We envision the operation of each reservoir in a multiple reservoir system to be dictated by a rule curve. Each reservoir is considered to have a target draft zone, a reduced draft zone, and an excess zone. The rule curve method has several peculiarities for the coordinated control of reservoir operations. The yield from each reservoir depends not only on the amount of water in storage, but also on previous period inflows, and the storage, yields and losses from other reservoirs.

The target zone for each reservoir can be divided into: (1) a compensation zone where additional water can be released downstream; and (2) a critical zone where only the target draft can be provided. The dividing line between a compensation zone and the critical zone can be obtained from simulation experiments [Shnaydman, 1992]. Each individual reservoir can be either an upstream donor reservoir or a deficit zone reservoir. An upstream donor reservoir is one which has abundant water and its storage level is in the compensation zone, hence it can release additional water downstream. A deficit zone reservoir has a storage level below the compensation zone hence it cannot donate water to downstream users, rather, it requires donations from an upstream donor reservoir. For parallel reservoirs, donations can be dispersed in proportion to priorities or current storage levels.

The joint operations of multiple reservoirs is implemented by first executing the simulation model to determine which reservoirs are potential upstream donors and to determine those potential donations. Individual reservoir storage, release and losses are computed and these signals form the current output for the system. A control algorithm recalculates releases from each reservoir to balance the donations and deficits according to a prespecified priorities and balancing scheme. Again releases are adjusted for every reservoir and the control algorithm is implemented again to balance the donations and deficits according to prespecified priorities. These control adjustments are all made for an individual time-step, without advancing to the next time-step until the control adjustments are completed.

Animation/Simulation Modeling for Multiple Reservoir Operations.

A variety of software packages are available for simulating multiple reservoir systems. This study employs a simulation model based on the GPSS simulation language (Shriber, 1991) because effective animation software (PROOF) is available using the GPSS language (Earle and Henriksen, 1993). PROOF is a PC-based, "post-processing" animation software package (i.e. animation is seen after simulation is run). Two ASCII files are required to run PROOF: a layout file and a trace file. The layout file describes the geometric details of the background over which objects move, provides geometric definitions and properties for such moving objects, and provides logical paths along which these objects move.

An animation of a water resource system requires the creation of a sequence of animation screens. The first screen contains a multicolor layout of a watershed including reservoirs, rivers, canals, aqueducts, etc. Several dynamic features can be shown using bargraphs, for example, to illustrate reservoir volume surpluses and deficits, inflows and releases from reservoirs, water demands and water consumption for various users, etc. Colors may be used to enhance the relationships among rule curve zones and reservoir operating data. A second screen illustrates the application of the two-step control algorithm for multiple reservoir operations. The general information on this screen includes a schematic layout of the system. Dynamic features include: (1) information signals from reservoirs to the control device and (2) control signals from the control device to reservoirs. The third screen contains a more detailed explanation of the reservoir operations algorithm. The information on this screen includes: the layout of reservoirs and several dynamic features including (1) a set of rule curves for operating each reservoir for the given time interval; (2) an animation of reservoir volumes and releases; (3) illustration of all operating zones in the reservoir; and (4) animated inflows, actual releases, releases requirements from downstream reservoirs and donations from upper reservoirs.

Every dynamic feature animates with objects. Every object is representative of a CLASS. Three classes are considered: (1) CURVE is intended for the animation of operating curves; (2) SLICE is intended for the animation of the impoundment and drawdown of reservoirs; and (3) FLOW is intended for the animation of numerical information regarding reservoir inflows and releases. Each class contains several subclasses. For example, the class CURVE includes a number of subclasses equal to the number of rule curves for every reservoir. The class SLICE consists of a number of subclasses equal to the number of slices into which a zone of a reservoir is divided. The class FLOW includes subclasses which animate detailed information regarding water balances in every reservoir. Other output screens may include plots, tables, and bargraphs related to various input and output data.

Case Study

The animation/simulation technology was implemented to assist in the development of operating rules for reservoirs in the Eastern Massachusetts Water Supply System. A detailed description of this system is provided by (Vogel and Hellstorm, 1988). We consider a slightly simplified version of the system which includes monthly operations of the Quabbin and Wachusett reservoir systems. One important problem relating to this system is the efficient operation of the reservoirs under all hydrologic conditions ranging from flood to drought conditions.

These two reservoirs are connected by a tunnel that passes under the Ware River and whose flow can be diverted to either reservoir. Inflow to Quabbin consists of local inflow from its own watershed plus possible diversions from the Ware river. Inflow to Wachusett consists of the local inflow from its own watershed and diversions from Quabbin through the aqueduct. In practice, water is always diverted from the Ware River to Quabbin using the tunnel. The Ware River receives significant natural purification due to the long detention period of the Quabbin Reservoir. Obviously, such transfers are not allowed to the Wachusett Reservoir during diversions to Quabbin. Water transfers from the Ware River to Wachusett Reservoir is possible, but undesirable, because it can seriously deteriorate water quality in the Wachusett reservoir. A schematic diagram of the reservoir system is presented in Figure 1. A simulation model of this system was developed using a monthly time interval.

Summary

Animation is a powerful engineering tool which can enhance the verification, validation and presentation of complex simulation results. Multiple purpose reservoir operations involve complex interactions and trade-offs between various water resource system functions. Animation software, when integrated in a simulation environment can be attractive for explaining the behavior of a water supply system to both engineers and managers and can provide an appropriate instrument for implementing the complex negotiations which often take place to arrive at coordinated operations.

References

Earle, N., and J. Henriken, (1993), Proof Animation: Better Animation For Your Simulation, Proceedings of the 1993 Winter Simulation Conference, IEEE, pp. 172-178.

Schriber, T.J., (1991), An Introduction to Simulation Using GPSS/H, John Wiley & Sons, New York.

Shnaydman, V., (1992), The Application of the Aggregative Approach in Simulation Modeling of Water Resource Systems, Water Resources Management, Vol. 6, 135-148.

Vogel, R.M. and D. Hellstrom, (1988), Long Range Surface Water Supply Planning, Civil Engineering Practice, pp. 7-25.

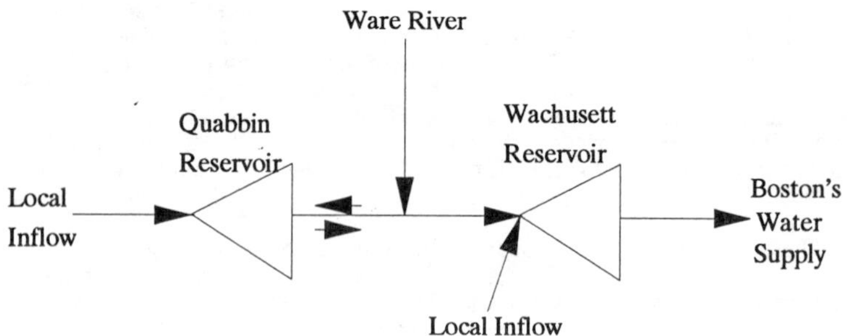

Figure 1 - Layout of the Eastern Massachusetts Water Supply System

CRITICAL PERIOD YIELD VERSUS PROBABILISTIC YIELD FOR WATER SUPPLY SYSTEMS

Richard M. Vogel[1], G. Kenneth Young[2], Stuart Stein[2], Peter Rogers[3], and Christopher Bell[1]

Introduction

In the U.S., it customary to design water supply systems to provide a yield which could be sustained, without failure, if the historical streamflows were to repeat themselves in the future. The reliability of this "safe", "critical period" or "drought-of-record" yield, which we term historic yield, is a matter of speculation. The historic yield is the maximum yield that could have been released from the water supply system over an n-year historical period with absolutely no water shortages. We explore the reliability of systems designed to withstand the drought-of-record. The water supply systems which service the cities of Boston and Springfield, Massachusetts; Providence, Rhode Island; and New York City (NYC) are used as case studies to demonstrate the concepts introduced. These four systems illustrate a wide range of storage capacities, yields and reliabilities and serve as examples of how the graphs presented can lead to a detailed understanding of the behavior and sensitivity of water supply systems to some operational assumptions.

The Average Return Period of the Drought-of-Record

There is a tendency for water supply engineers to treat sequences of droughts in much the same way they deal with floods, assigning an average return period T, to the worst drought in an n-year period using T=n, where n is the planning period or length of historical record. One cannot assign the average recurrence interval using T=n because sequences of water supply system surpluses and failures are no longer independent, and droughts tend to last for several years, hence the return period is no longer a geometric random

---

[1] Department of Civil and Environmental Engineering, Tufts University, Medford, MA, 02155, Tel. - 617-628-5000 Ext. 4260, Email - rvogel@pearl.tufts.edu

[2] GKY & Associates, Inc. 5411-E Backlick Road Springfield, VA, 22151, Tel - 703-642-5080

[3] Harvard University, Division of Applied Sciences, Pierce Hall, 29 Oxford Street, Cambridge, MA, 02138, Tel-671-495-2025

variable as is the case for flood events (see Vogel 1987). Using a simple two-state Markov model, one can show that the average return period of a reservoir system failure T can be approximated using $T \approx 1.44\ R_a$ where R_a is the steady-state annual reliability of the system. So a water supply system which was designed to provide a no-failure yield using an n-year period of record, should experience failures, on average, every 1.44n years. This is because most water supply systems have annual reliabilities R_a, very close to unity.

Is a Water Supply System Subject to Carry-over Storage Requirements?

In order to draw general conclusions about the behavior of water supply systems, it is necessary to simplify the problem, hence we concentrate on the carryover or overyear storage requirements of water supply systems. Systems dominated by overyear storage requirements do not usually refill at the end of each year; such systems are prone to water supply failures (empty reservoirs) during periods of drought which extend over several years. Systems dominated by carryover storage requirements tend to have values of the index $m = (1-\alpha)/C_v$ in the range [0,1], where α is the demand level as a fraction of the mean annual streamflow, and C_v is the coefficient of variation of the annual inflows to the water supply system. Overyear storage requirements increase as the index m decreases. For example, Vogel et al. (1995) show that the index m is equal to 0.2, 0.44, 0.53 and 0.96 for the Boston, Springfield (MA), Providence and NYC systems, hence they are dominated by overyear storage requirements, with the exception of NYC which is includes both overyear and within-year behavior.

Study Assumptions

This study assumes that annual inflows are normally distributed and follow a lag-one autoregressive process with serial correlation 0.2. Vogel et al. (1995) found that annual flows in 166 basins in northeastern U.S. were well approximated by a normal distribution with a fixed lag-one correlation coefficient of $\rho = 0.2$ and a fixed coefficient of variation $C_v = 0.25$. The figures shown here were developed using analytic expressions derived by Vogel (1987) and Vogel and Bolognese (1995).

Reliability of the Critical Period or No-failure Yield

When a stochastic streamflow model is used in combination with Rippl's mass curve or the sequent peak algorithm, one obtains the probability distribution of no-failure yields over an n-year period, R_n. The n-year reliability R_n, denotes the likelihood the system will provide its yield, without failure, over the next n years. The annual reliability R_a, denotes the likelihood that the system will deliver its yield without failure in **any** particular year. Figure 1 uses dotted lines to illustrate the relationship between the n-year reliability R_n and the steady-state annual reliability R_a for systems in the northeastern U.S. For comparison, the solid lines denote the relationship $R_n = R_a{}^n$ which assumes that failure events are independent. Figure 1 illustrates the dramatically different interpretations associated with the two reliabilities R_a and R_n. Naturally, as n approaches unity, R_n will approach R_a.

Probabilistic Yield of a Water Supply System

We define a probabilistic yield as the yield corresponding to annual reliability R_a. Figure 2 illustrates the storage-yield relationship for water supply systems in northeastern U.S. dominated by carry-over storage requirements based on Vogel et al. (1995). The dotted lines represent the storage-yield curves for annual reliabilities of R_a = 0.95, 0.98, 0.99 and 0.995. The solid line represents the storage-yield curve for systems designs based upon the drought-of-record (R_n=0.5) using an n=50-year historical record. Here storage S and yield Y are expressed as a fraction of the mean annual inflow μ.

Figures 1 and 2 document that designing systems to withstand the drought-of-record (R_n=0.5) based upon an n=50 year record, leads to systems with an annual reliability of roughly R_a=0.98=(1-(1/50)), as expected. Figure 2 also illustrates the storage-yield behavior of the four water supply systems Providence, Boston, Springfield and NYC. The historic yields for the Providence, Boston and Springfield systems are based on monthly simulations which include the 1961-67 "drought-of-record". The NYC yield in Figure 2 assumes the current demand of 1550 mgd. Using pooled precipitation series, Russell et al. (1970) estimated the 1961-67 drought to have an average return period of roughly 140 years which corresponds to an annual reliability R_a = (1 - (1/140)) = 0.993. Interestingly, in Figure 2 all four systems appear to have annual reliabilities which are very close to 0.993. The Springfield, Boston and Providence historic yields are based on n = 30, 50 and 71 year historical streamflow records, respectively, including the 1961-67 drought, leading to annual reliabilities of roughly R_a = (1 - (1/n)) = 0.97, 0.98 and 0.99. The reason Figure 2 documents these three systems as having much higher (theoretical) reliabilities is probably because the 1961-67 drought was a more severe drought than one would have expected during a "typical" 30, 50 or 71 year period. The NYC system appears to have a (theoretical) reliability in excess of 0.995. The actual reliability of the NYC system is lower due to within-year spills which are not accounted for by the annual model and due to required conservation releases during low-flow periods.

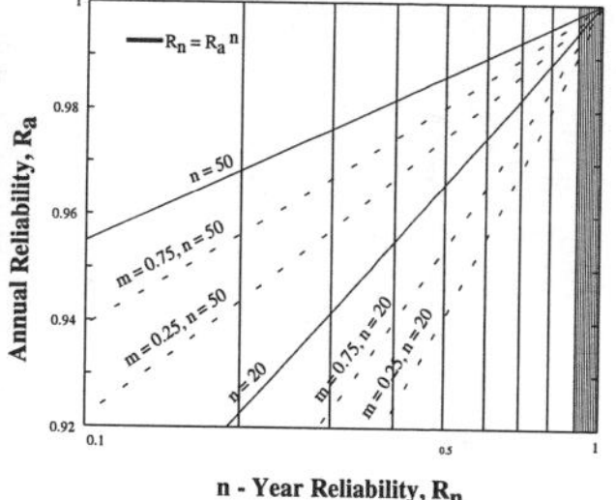

Figure 1

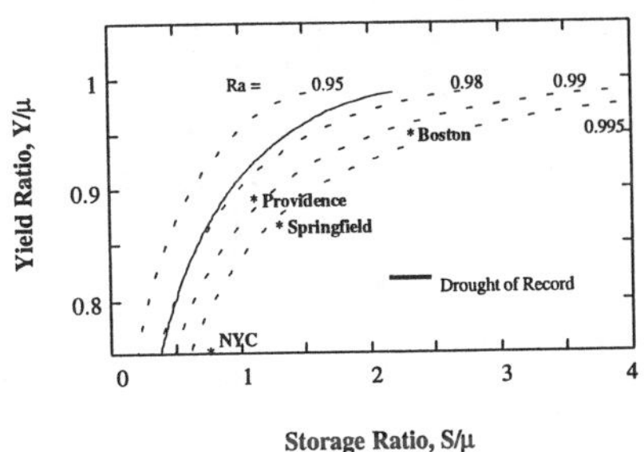

Figure 2

The Sensitivity of Yield to (Reserve) Storage Capacity

Most reservoir systems contain some "reserve" or "dead" storage, which is assumed to be unavailable. Such reserve storage may be allocated due to: water quality restrictions, potential reservoir sedimentation, flood control storage or other requirements. We derive the sensitivity or elasticity of water supply system yield to assumptions about reserve storage. The storage elasticity is defined as the sensitivity of water supply system yield to changes in storage. Mathematically storage elasticity is $\epsilon_S = (dY/Y)/(dS/S) = (S/Y)(dY/dS)$ where S is storage capacity and Y is yield. The elasticity is dimensionless and is generally positive since an increase in storage capacity causes an increase in yield, so that $dY/dS > 0$. A storage elasticity of $\epsilon_S = 0.5$ indicates that a 10% increase in storage will only lead to a 5% increase in yield. Figure 3 illustrates the relationships among storage elasticity, yield ratio and storage ratio for water supply systems in the northeastern U.S. Figure 3 documents that yield improvements due to additions to storage will be greatest for systems with low yields, high storage ratios and low reliabilities.

Figure 3 can be used to approximate the impact of assumptions regarding reserve storage on the yield of a water supply system. For example, assuming the Boston system has an annual reliability of 99%, a storage ratio of $S/\mu = 2.3$ and a yield ratio of $\alpha = 0.95$, leads to a storage elasticity of $\epsilon_S = 0.1$. A 40% reserve storage capacity for the Boston system would only produce a (0.1)(0.4) = 0.04, or a 4% increase in yield if it was exploited.

References

Russell, C.S., D.G. Arey and R.W. Kates, Drought and Water Supply, The Johns Hopkins Press, Baltimore, pp. 42-47, 1970.

Vogel, R.M., Reliability Indices for Water Supply Systems, Journal of Water Resou Planning and Management, ASCE, Vol. 113, No. 4, pp. 563-579, 1987.

Vogel, R.M., N.M. Fennessey, and R.A. Bolognese, Storage-Reliability-Yield Relat for the Northeastern U.S., Journal of Water Resources Planning and Manage ASCE, Vol. 121, No. 5, (in press), 1995.

Vogel, R.M., and R.A. Bolognese, Storage-Reliability-Resilience-Yield Relationshi for Overyear Water Supply Systems, Water Resources Research, (in press), 1

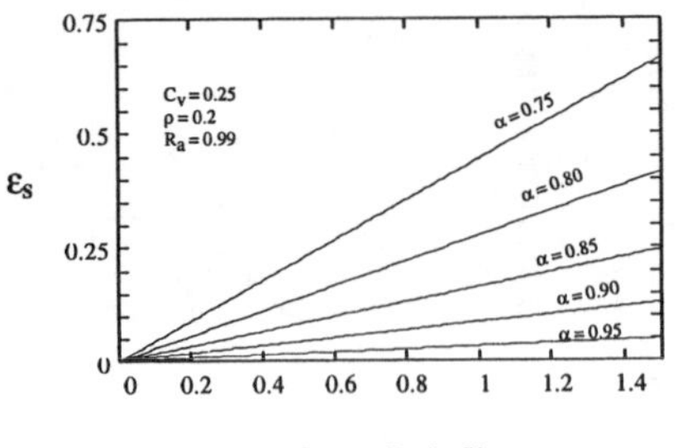

Figure 3 - Storage Elasticity as a Function of the Storage and Yield Ratios

Modeling the Río Colorado in 1970: How Modeling Has Changed in the Past 25 Years

Walter M. Grayman[1], M.ASCE, Juan B. Valdés[2], M.ASCE, Brendan M. Harley[3], M.ASCE

Abstract

A suite of five interacting simulation and optimization models were developed and applied as part of a study of the development of the Río Colorado in Argentina in 1970-1972. The characteristics of these models and their interaction are presented. Changes that have occurred in the past 25 years in modeling are discussed.

Introduction

A systematic study of the Río Colorado was undertaken in 1970 as a joint project by the State Secretariat (later Subsecretariat) for Water Resources of the Republic of Argentina and the Water Resources Division of the Department of Civil Engineering at the Massachusetts Institute of Technology (MIT). There were three primary objectives: to adapt modern water resource systems planning techniques to Argentina; to train selected Argentine professionals in the use of these techniques; and to apply the techniques to a particular river basin, the Río Colorado. MIT (with influence from their Cambridge neighbors at Harvard University) had long been a bastion of the development and use of mathematical models in water resources and as a natural consequence, mathematical modeling became the focus of this study. The models that were developed and applied in the study are described briefly in this paper. A more thorough description of the models is presented in *Applied Water Resource Systems Planning* (Major and Lenton, 1979). Finally, in a retrospective mode, the models are examined from a 1995 perspective and compared to what is available today.

Descriptions of Models

The planning process that transpired during the 2-year life of the study was a truly dynamic one. Models evolved as the specific needs of the study were identified. The dynamics were also influenced by the academic setting of the study in which research and thesis requirements played a part in guiding the work. It was evident from the start

[1] Owner, W.M. Grayman Consulting Engineer, Cincinnati, Ohio 45229

[2] Professor, Dept. of Civil Engineering and Climate System Research Program, Texas A&M University, College Station, TX 77843

[3] Vice President, Camp Dresser & McKee, Cambridge, MA 02142

of the project that a single model was not feasible, but rather, several interacting models should be developed. The result was a suite of five models that utilized a common data base and were used in tandem to identify and evaluate alternative development plans for the Río Colorado. These models are described below:

Screening Model - The screening model was a mathematical programming model used to select initial alternatives. Three season, steady state, mean hydrologic inputs were used with implementation assumed to occur at the start of a 50-year planning period. The model was formulated as a mixed-integer problem in order to constrain solutions for dams and some other facilities to discrete, technically reasonable solutions. The formulation included continuity constraints, reservoir constraints, irrigation constraints, hydroelectric energy constraints, import/export constraints, conditionality constraints (requirement that certain projects be built in tandem), policy constraints, and the objective function. The form of the objective function that was generally used was the maximization of weighted discounted net national and regional income benefits.

Basic Simulation Model - The basic simulation model was a simplified simulation model of the overall system behavior. It was designed to estimate the contributions to objectives that would result from the implementation of a set of management and development measures. The river was represented by a series of nodes and arcs with nodes representing starts, reservoirs, confluences, irrigation sites, import/exports, continuations, and termini, with arcs connecting the nodes. It operated on a 4-month period in order to provide consistency with the screening model and was run for a 51 year record. A combination of historical and synthetic streamflow records were used as input. The model included various reservoir operating rules and water allocation rules for handling periods when a water deficit existed.

Sequencing Model - The sequencing model was a mathematical programming model used to determine when a series of previously selected (and sized) projects should be constructed. The model was formulated as a mixed integer problem. As applied to the Río Colorado, it was assumed that projects could be constructed during one of four 10 year periods and the useful life of a project was 40 years. The model included a budgetary constraint to limit the total capital expenditure by period and a population constraint which constrained site development by the availability of farm population.

Detailed Simulation Model - The detailed simulation model was developed to simulate the hydrologic and development processes within the basin. It was a modular program including: hydrologic modules for snowmelt runoff, channel routing, overland flow, infiltration and evapotranspiration, groundwater flow, and salt balance; and development modules representing the behavior of dams, irrigation areas, hydroelectric plants, diversions, exports and imports. Another module was used to generate synthetic streamflow series which could be used as an alternative to the snowmelt runoff model for upstream flows. The model simulates behavior on a component by component basis with outputs from each component stored in a data bank for later display, analysis or use as input to downstream components. Time steps varied from hours to weeks and could differ between components.

Migration Model - A model simulating farmer migration to newly developed irrigation areas was built as an adjunct to the study. The model was a Systems Dynamics simulation model written in the DYNAMO language comprising a demographic sector, an economic sector and a water sector,and representing the interdependence of demographic phenomena such as migration with investment strategies in irrigation infrastructure.

Model Interactions

The previously described models were used together to address the issue of how best to develop the Río Colorado Basin. Interactions between the models are described below and shown schematically in Figure 1. Each of the models operated off of a common general data base. The range of alternatives included 8 potential reservoirs, 13 potential hydro power plants, 17 potential irrigation sites, and 3 major exports or imports.

The screening model served to investigate a wide range of alternatives and to identify the configurations that resulted in the best solutions based on the specified objective functions. Since the screening model used a steady state hydrologic pattern and, by nature, required certain simplifications in the physical representation, the selected alternatives were further tested and refined using the basic simulation model. The refined alternatives were then scheduled using the sequencing model. In this step, the timing for constructing the selected components of the alternatives were determined.

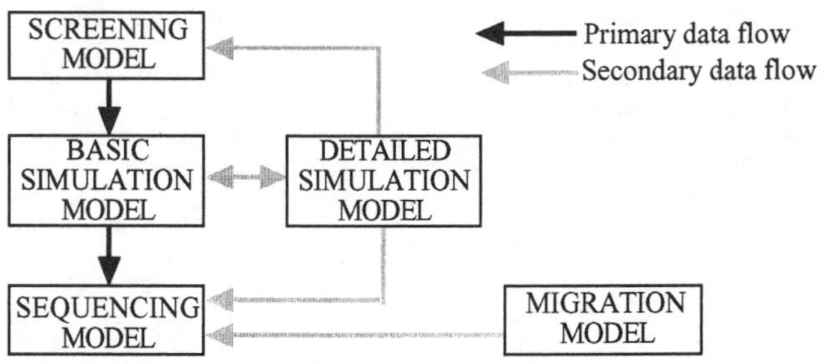

Figure 1. Model Interaction

As illustrated in Figure 1, the detailed simulation model and the migration models functioned as "off-line" advisory models in the process. The detailed model, originally planned to fill the role taken by the basic simulation model, proved to be too complex and computer intensive. As a result, rather, it was used for specific tasks such as generating streamflow patterns for use in the other models, parameterizing the other models, and confirming the validity of the basic simulation model. The migration model provided an enhanced understanding of the migration process and, in a minor way, helped establish the population constraint used in the sequencing model.

A Retrospective Look

In the 25-years since the Río Colorado study commenced, there has been a revolution in the computer market. The room sized behemoths used in that project have been largely replaced by desk top systems of comparable speed at less than 1% of the cost. The whole computer graphical arena has exploded and altered the face of computer programs. How would these changes and advances in the water resources technical area affected the modeling process used 25 years ago? If a project like the Río Colorado commenced today, what would the modeling look like? The answers to these questions are, of course, conjecture; but, ones that can be supported by looking at what is being done today.

Without too much disagreement, it is likely that the modeling study would have been performed on a networked work station. It would have featured a user-friendly graphical user interface in order to simplify usage and, undoubtably, would have included a geographic information system (GIS) in conjunction with a relational data base as a central unit. The result would have been the ability to easily display the characteristics of the existing system and proposed alternatives. Programming would have been performed in a structured, object oriented language such as C++ replacing FORTRAN, the largely outmoded but still loved by many engineers language. A full turnkey hardware-software system would have been presented to Argentina at the end of the project, thus, encouraging its continued use.

As engineers, we are good (but sometimes a little slow) at using these technical advances to our advantage. However, what has happened in the water resources technical arena that would have resulted in a better modeling effort? The answer to this question is far more discouraging. If we examine each of the models used in the study, we find that the advances have been limited.

Mathematical programming models such as the screening model and sequencing model are still used today, with a little more sophistication in non-linear and integer programming, but essentially minimal technical change. In fact, the passionate use of such models in the mid-60's to early 70's has cooled considerably. If you examine the 'guts' of simulation models of today, you see relatively minor changes in the technical methods. The interest in groundwater contamination has spurred significant improvements in groundwater modeling, but surface water modeling is using largely the same techniques employed 25 years ago. Many of the concepts that were new in the early 1970's and utilized in the simulation models such as central data bases (e.g., data centered modeling) and modular systems have become commonplace today. The Systems Dynamics modeling used in the migration model is still around today in forms such as the popular STELLA™program.

One of the identified shortcomings of the original study was the lack of active interaction between the research group at MIT and the sponsoring Argentine agency (Major and Lenton, 1979). There has been much discussion in the past several years about bringing modeling into the decision process by encouraging more of a real-time interactive process between the modeler and the decision maker. However, when one examines most modeling studies of today, one finds really little improvement in the interaction.

Overall, it is clear that if the Río Colorado were conducted today, the modeling effort would have a drastically different and improved 'look and feel' and, hopefully, there would be some improvements in the interaction between the researchers and the 'decision makers'. However, there is little to suggest that the actual water-related (hydrologic, hydraulic) methods would have been substantially improved. Does this mean that we, as engineers and researchers have advanced this field to its ultimate limit or does it mean that over the past 25 years we have ignored the basics. Unfortunately, it appears that the latter is the case and that improvements in the computational and display methods and data availability have not been matched by commensurate improvements in the underlying water resources methodologies.

References

Major, D.C. and Lenton, R.L. (1979), "Applied Water Resource Systems Planning", Prentice-Hall, Inc., Englewood Cliffs, NJ.

MIT - ARGENTINA PROGRAM : WHAT WENT WRONG?

Juan Dalbagni, Fellow, ASCE[1]

ABSTRACT

25 years ago, the Argentina Government sought to achieve the following goals:

. develop techniques for river basin planning;
. train a selected interdisciplinary group of Argentine professionals for the implementation of these techniques;
. test them on a case study in the Rio Colorado basin; and
. apply them to the other 98 Argentine river basins.

To attain these goals, the Argentine Government associated with the Massachusetts Institute of Technology, in a multi-year joint research effort which has been called the "MIT-Argentina Program".

Today the techniques are applied everywhere **outside** the borders of Argentina. The trained professionals and their disciples made remarkable careers **abroad.**
The only remains within Argentina of this ambitious Program are the results of the Rio Colorado's case study.

What went wrong?

---

1 Former Co-Director MIT-Argentina Program.
INGARTE Consultores Técnicos, Buenos Aires, Argentina.
Universidad Austral, Buenos Aires, Argentina.

1. INTRODUCTION

In 1969, Argentina affiliated with the Massachusetts Institute of Technology (MIT) in a multi-year joint research effort which sought to develop techniques for river basin planning, for their standardized application to the 99 Argentina basins; to train a group of Argentine professionals for their implementation; and test them on a case study in the Rio Colorado basin.

Although many models were developed during this program, its emphasis was on adapting existing optimization, simulation, and other techniques for river planning into an integrated methodology of water resources development. This became fully operational by 1976.

The Program ended with only one significant application in Argentina: the solution of the case study.
On the other hand, both developed and developing countries have used the Program's approach of comprehensive river basin planning, based on this case study. This later applications included the Vardar/Axios River (Yugoslavia and Greece), the Lower Nile (Sudan and Egypt), the Kavlinge River (Sweden), the Kalu Ganga (Sri Lanka), the Yellow River (China), the Brazos and Colorado Rivers (Texas, USA), the Lower Mekong (Southeast Asia), and many other.

The Argentina Program's multi-model approach has also been used in teaching in U. S. universities as well as in Egypt, India, Israel, Sweden and other countries. On major book has been published .

As the Program's applications and participants emigrated from Argentina, so too did the know how. Consider, for example, the hydro-economic simulation model (MITSIM) that - even though its main part remains as developed by the MIT - Argentina Program - was modified to address more diverse international river basins. MITSIM was them adopted by the United Nations' World Meteorological Organization (WMO), who recognized it as a valuable tool for river basin planning and included it as a Hydrologic Operational Management System (HOMS) component for distribution to developing countries.
Through this route the model has been applied worldwide and its popularity continues. Private organizations have also adopted Argentina Program models such as the MIT-TAMS.

As a conclusion, it may be stated that the initial Argentine goal of developing an integrated and modern water resources planning technique was achieved.

So too was the application in Argentina to a case study.

Training of Argentine professionals in its use was successful. However, Argentina Program was not a permanent success.

2. WHAT WENT WRONG

If this or a similar Program is to be rooted in Argentina and other countries, the causes of its failure must be identified.

One problem lied within Argentina itself; in its capacity to implement, create or extend the Program's components. Another problem is people. Those who had the know how no longer formed a team. Some no longer live in Argentina. Others are now involved in other business and lack motivation. Hence, Argentina has lost her people and with them has lost the Program.

One might say that it was due to the inherent inability of Argentina's public sector to retain key personnel. In this case there were several factors : job instability, limits to public employment, insufficient infrastructure, lack of proper recognition, political discontinuity, economic incentives and loss of contact.

The first factor was job instability. Team members were only granted a maximum of a one-year contract, with renewal always uncertain.

Second, there were and still exist severe limitations on public employment; which are likely to get even tighter in the future.

Third, the support infrastructure was inefficient. Today, though computers are no longer a limitation.

The fourth factor, lack of proper recognition, reflected the fact that the far-reaching potential of the Program was ignored. Government never sought to use these methods beyond the solution of the politically urgent Rio Colorado case study. In this manner its orderly application to the rest of national basins was neglected.
No effort was made either to consider its potential for horizontal cooperation with other countries, such as Argentina's co-riparian neighbors.

The fifth factor was political discontinuity. Argentina's tradition was that public officials had short lives. With political shifts, priorities changed. In these circumstances, river basin planning is hard to achieve or implement. Plans are abridged or halted altogether causing the loss of enthusiasm and professional motivation. Argentina seems firmly determined to amend this situation. It may take a significant amount of time though to modify what has become practically a mental constraint.

Economics was a sixth factor. Economic considerations make it difficult for the public sector to retain highly skilled professionals. If other research or consulting opportunities exist in the private sector, the temptation of a higher income may provoke shifts in employment, if emigration does not occur.

A final factor was the abrupt loss of contacts. After termination of the Program, there was a loss of contact with MIT. This precluded further developments of the Program or additional implementation in Buenos Aires. Most important was the loss of a locus of interest which made the task of the Argentineans interested in continuing the work more difficult.

3. WHAT'S NEXT

Is it worth reviving the Program a quarter century later?
Considering the increasing awareness on environmental problems and the tremendous technical progress occurred, the answer is undoubtly affirmative.
What went wrong before though should be either eliminated or avoided and new developments should be contemplated.

In this context, a strong emphasis will have to be placed on the environmental aspects.
Regional economic integration should orientate Argentina to illustrate the Program's approach not only as an example but also to provide some sort of horizontal cooperation to other countries in Latin America and beyond.

Finally, the strong privatization process of Argentina points to a private organization such as a University or a Foundation as an appropriate partner for the new MIT - Argentina Program.

Next time it can't go wrong.

Integral Resource Management of the Río Colorado (Argentina)

Juan Enrique Perl [1]

Abstract

This paper presents a list of surveys and research that has been done on the watershed of the Río Colorado (Argentina) related to the integral management of the water resource between the five riverside provinces. It summarizes the activity of COIRCO and presents a description of the current situation along with a list of special studies not foreseen in the original program of 1976.

Introduction

The Río Colorado which flows into the Atlantic runs from the Andes. Its main tributaries are the Grande and Barrancas Rivers. The course of these rivers affect five argentine provinces: Mendoza, Neuquén, La Pampa, Río Negro and Buenos Aires. The module of the Río Colorado is 147 m^3/s; its extension is 1.300 km and the area of the basin is 34.200 km^2.

Forty years ago, long before development and use of the river resources had begun, government representatives initiated action to find a plan for efficient and equitable use of the water.

After conducting a series of initial surveys, and transcendental meetings of the first committees, a Water Management Model was created with the help of the MIT (Massachusetts Institute of Technology). On October 26, 1976 the riverside provinces signed the "Unified Program for Irrigation and Water Distribution of the Río Colorado" *(Programa Unico de Habilitación de Areas de Riego y Distribución de Caudales del Río Colorado).*

----

[1] Civil Engineer, Manager, COIRCO, Comité Interjurisdiccional del Río Colorado, Belgrano 366, 8000 Bahía Blanca, República Argentina.

The fundamental objectives of this program are: a) Efficient use of the water of the Río Colorado, b) Priority is given to human consumption and irrigation over other possible uses of the water, c) All activities and developments should contribute to obtain Territorial Integration between the provinces.

This program set the framework for a series of significant projects, initially allowing the potential development of more than 3,500 km^2 of irrigable land, different flood control earthworks, and construction of hydroelectric plants.

In 1976 COIRCO (*Comité Interjurisdiccional del Río Colorado*) was created. The primary responsibility of this organism is to control the correct use of the water each province is assigned according to the Unified Program and to act as a centralized information center of all studies and surveys carried out on the watershed of the river. Another important task is handling situations not initially foreseen in the Program.

Present Situation

Water Supply to Towns and Cities: Water is supplied to all urban areas, cities, towns and farms on the banks of the Río Colorado. In some cases the water is provided by aqueducts and canals to cities that are not located directly on the margins of the river.

Irrigation areas: According to the latest agricultural statistics, the river supplies water to 945 km^2 of farmland.

Diversion Dams: There are two dams that supply water to agricultural lands of provinces Río Negro and La Pampa:

Dique Punto Unido: Supplies water to both provinces with a potential irrigation area of 623 km^2 and small hydroelectric plants.

Dique Salto Andersen: Supplies irrigation water to Río Negro and could possibly supply La Pampa too. Presently a small hydroelectric plant is under construction.

Other irrigation areas are supplied by water derived directly by canals located on the banks of the river.

Flow Regulation and Flood Control: The complete development of all potential irrigation areas requires the construction of storage dams along the river course.

Presa Embalse Casa de Piedra: This dam is the first major regulator construction to be built on the Río Colorado. Its importance is based on its multiple uses and its regional impact. It has the capacity to regulate the water flow of the river, control river swelling, provide irrigation water to large areas and has a hydroelectric energy plant. COIRCO gave priority to this project over other alternatives. Presently the dam and hydroelectric plant are almost completed.

Control and Execution of the Water Management Program

The first surveys undertaken by COIRCO were to obtain more information about certain parameters that had been estimated when the model was designed. Water salinity, infiltration and evaporation water loss, quality and quantity of water returned from irrigation areas, sediment origin and quantity were determined with better precision in areas were the information was scarce or non-existent. Other studies included the river Barrancas watershed, river topology and preliminary studies of water transfers from the Río Negro river to the Colorado.

It is very important to point out that the success of the Water Management Program is based on its flexibility to adapt itself as more information about the river becomes available.

Other positive aspects derived from the integral Program and centralized Committee are the possibilities to analyze environmental impacts of any significant development plans for the river's watershed and it provides an excellent forum for each province to protect its rights in reference to the use of the resource.

This last point is crucial in efficient planning and use of the river's water. By establishing priorities and keeping control of the major parameters defined in the Program, the Committee can resolve conflicting interests of the provinces. The Committee authorizes and monitors the volume and quality of water that each province uses, adjusting the authorized quantities according to low or high water situations.

COIRCO centralizes the water requests of each province for every year, with a complete description of its intended uses. At the end of each period it verifies the effective supply and returns.

COIRCO receives snow-melt forecasts and maintains ten riverside stations to obtain statistical information about the river. These stations measure water flow, conductivity, dissolved and suspended solids and take water samples for chemical analysis. These stations produce the two most important variables used to control the Integral Program: annual volume and the salinity of the water delivered to each province.

Unpredictable events that were not foreseen in the original program, some caused by Nature, others by construction of hydroelectric or hydraulic projects, and others by activities not directly related to the use of the Río Colorado water (oil, mining, etc.), were handled by special studies and sometimes produced modifications in the internal statutes of COIRCO so it could extend its capabilities.

The following extended capabilities can be mentioned: project control, construction and operation of hydroelectric plants in those aspects referred to in the Unified Program. Detection and control of pollutants in the water. Surveys of natural and induced ecosystems. Environment impact of different projects. Definition of guidelines for borderline determination.

The following actions produced additional studies and research of the Colorado watershed:

a) *Actions caused by Nature*: A series of rich hydrological years that occurred in the 80's reactivated the Desaguadero-Salado-Chadileuvú-Curacó river network for the first time in the last century. These waters are noted for their high salt content. The sudden incorporation of a new affluent impeded irrigation downstream and required emergency action to handle the contingency. This water flow had not been considered in the original program. Presently this river system no longer discharges its waters into the Colorado and long term plans are being considered to find a permanent solution to the problem.

b) *Action caused by construction of dams and dikes*: The building of the dam Casa de Piedra required environmental surveys to determine the impact this site would cause in the region. Special studies evaluated the modification of the river bed, sedimentation in the lake and reservoir, water quality, modifications to the regions fauna and flora, fish population dynamics and climate changes.

c) *Action cause by oil installations*: The higher Colorado and its tributaries are now under intense oil prospecting, drilling and extraction. COIRCO controls and monitors the preservation of the water resource that is modified by waste waters and accidents. Frequent inspections of the installations check the chemicals dumped into the river and injected into the wells along with their impact on the river and subterranean aquifer. Contingency plans and policies have been designed in collaboration with the oil companies to handle major spills or accidents.

Conclusions

This paper presents a descriptive account of the handling of an integrated resource management program for the Río Colorado. It has proven to be a positive and effective Program.

Each province has had the ability to plan the use of their assigned resources and has had guarantee that the water it receives are of the quality and quantity that were previously agreed upon. (i.e. they are not affected by uses of the resource that occurs upstream and are not in their jurisdiction).

The most important aspects of the Program are centralized planning, adaptability to changes in the watershed of the river (caused by natural or human action) and handling of emergency situations.

COMPARISON OF DISTRIBUTIVE VS. LUMPED RAINFALL-RUNOFF MODELS ON GOODWIN CREEK WATERSHED

Billy E. Johnson[1], Roger H. Smith[2], and Jerry L. Anderson[2]

Introduction

In recent times, Geographical Information Systems (GIS) have gained considerable attention in solving a variety of problems. Included among these are using GIS to help in hydrologic modeling. In recent studies a new two dimensional (2-D) distributive hydrology model, (CASC2D) developed at Colorado State University by Dr. Pierre Julien, Dr. Bahram Saghafian, and Dr. Fred Ogden (University of Connecticut),"CASC2D User's Manual" and "de St-Venant Channel Routing in r.hydro.CASC2D", has been used to simulate rainfall-runoff in rural areas.

The focus of this paper will be to describe the distributive model in its present state, and to discuss briefly a study conducted to evaluate CASC2D versus the traditional Lumped Model approach.

CASC2D Methodology

Overland Flow Routing. Overland flow is generally a two-dimensional process which is controlled by spatial variations in slope, surface roughness, excess rainfall, and other parameters. As the overland flow drains into stream channels, one dimensional flow prevails. The diffusive wave equation for channel flow can predict the possible backwater effects in main channels and tributaries. As in the other watershed processes, the spatial variations in channel parameters must be accounted for in the model.

---

[1]Research Hydraulic Engineer, Waterways Experiment Station, 3909 Halls Ferry Road, Vicksburg, MS. 39180.
[2]Associate Professor, University of Memphis, Department of Civil Engineering, Memphis, TN. 38152.

Rainfall Distribution. In CASC2D, rainfall can be analyzed using an interpolation scheme based on the inverse distances squared or the Thiessen polygon method. Recent improvements in the model allow for the input of rainfall radar raster maps. If no raingage or radar data is available, rainfall is assumed to be uniform over the watershed.

Precipitation Loss. An infiltration scheme must accommodate both spatial variations due to soil texture changes, and temporal variations due to the time-variant nature of both rainfall and soil infiltration capacity. Additionally, the fact that rainfall history affects the infiltration rate at the present time has to be accounted for in the infiltration scheme. Ideally, the scheme should also rely on physically measurable soil infiltration parameters. The Green-Ampt infiltration equation adequately satisfies these requirements and is therefore well-suited for distributed watershed modeling.

Channel Routing. There are two one-dimensional channel routing techniques incorporated into CASC2D. The first is based on the diffusive wave equation, similar in principal to the two-dimensional overland flow routing technique (explicit routing scheme). The second is based on the Holly-Priessman channel routing routine (implicit routing scheme). For each time step, infiltration and overland flow routing are processed. This determines the net rate of overland flow pouring into the channel elements.

In developing the channel network, a link map and a node map must be generated. The channel cross section properties for each node within a channel link can be approximated using either a trapezoidal cross-section or a hydraulic properties table (ie., area, conveyance, top width).

Backwater effects can be properly handled, even at the junctions of tributary channels. In the model formulation, the channel cross section is not subject to infiltration, which is likely to be negligible compared to the flow rate in the channel. However, any overland flow running toward the channel, channel overflow, and any specified detention storage all remain subject to infiltration.

Previous Study Methodology

Goodwin Creek Watershed is located in North Mississippi approximately 60 miles south of Memphis Tennessee. The watershed area is 8.4 square miles and is extensively gaged by the Agricultural Research Service. Five rainfall events were simulated using 17 rainfall gages and 6 discharge gages. Two HEC-1 models, SCS Unit Hydrograph and Snyder Unit Hydrograph, and CASC2D were used to simulate rainfall runoff for the purpose of comparing them to observed flow records at the six discharge gage locations. In this analysis, all models used the Green-Ampt infiltration routine. The HEC-1 models used the Muskingum-Cunge channel routing routine while CASC2D used a one-dimensional diffusive wave routing routine. From the output, the peak flow, time to peak, volume of runoff, and hydrograph variance parameters were summarized for all three models.

Conclusions. Based upon the results of the observed and hypothetical storm events simulated for the Goodwin Creek Watershed, the following conclusions were made:

1.) In the case where there is accurate spatial data representation of the watershed variability in soils and landuse, a distributed model will simulate more closely the true shape, rate of rise, and volume of the streamflow runoff hydrograph than the lumped unit hydrograph methods;

2.) In the case where there is sufficient sub-basin stream gage data available for calibration purposes, the lumped unit hydrograph models such as HEC-1 can reproduce the observed hydrograph reasonably well;

3.) The lumped models rely heavily on sub-basin stream gage data in order to adequately simulate the observed hydrograph, however CASC2D can simulate adequately as long as accurate spatial data is available. If accurate spatial data and sub-basin stream gage data are both lacking, then both models (i.e., lumped or distributed) may produce questionable results;

4.) Since the distributive model CASC2D consistently produced more realistic results in terms of hydrograph shape and volume of runoff, it offers more flexibility, when performing sediment studies, than the lumped unit hydrograph models. This will be especially true when evaluating the effects of specific landuse changes or best agricultural management practices on erosion and sediment control within the watershed.

Recommendations. The principal objective of this previous study was to evaluate the watershed hydrology model, CASC2D, for purposes of application to ungaged watersheds. The simulation results from this study showed that the CASC2D model would produce adequate results for design purposes with a limited amount of gage data. However, the following recommendations were made in order to make CASC2D a more robust hydrologic routing model:

1.) The channel routing component of the CASC2D model should be revised as soon as possible to more realistically represent the channel cross-sections in order to improve the timing of the simulated runoff hydrographs.

2.) The channel routing component should be uncoupled or separated from the overbank routing component for modeling overbank flows. This would allow other numerical channel routing techniques to be evaluated and perhaps eliminate the stability problems caused by too long of time steps.

3.) For design purposes of erosion control measures, the model must be able to handle high intensity, short duration storm events.

4.) In the near future, CASC2D should be enhanced by adding sediment yield and transport subroutines for both the overland flow and channel routing components. This will allow the evaluation of planned watershed best management practices and erosion or sediment control structures.

Present Study Methodology

Again, Goodwin Creek watershed, as described above, is being used to see if improvements made in the CASC2D model after completion of the previous study will improve the study results. The ability to use a trapezoidal cross-section instead of a rectangular cross-section and the implementation of the Holly-Priessman channel routing scheme should improve the model results. At present, the study is on-going, but it is anticipated that by the time of the conference, revised results using the improved model should be ready to present.

Acknowledgements

Special thanks are given to the colleagues of the authors who helped in obtaining the required data and in developing the model. Thanks also goes to the project sponsors for their review of the model procedures and study findings. The opinions stated in this paper are the author's. The study described herein, unless otherwise noted, was performed from research sponsored by the US Army Engineer District of Vicksburg and the Improvements of Operation and Management Technology Research Program and conducted by WES, Vicksburg MS. Permission was granted by the Chief of Engineers to publish this information.

GRASS INTEGRATED SYNTHETIC SEDIMENT ROUTING MODEL (GISSRM)

R.H. Smith[1], S.N. Sahoo[2], S.S. Sunkara[2], and L.W. Moore[1]

ABSTRACT

For most watersheds, extensive rainfall, streamflow, and sediment discharge data are not available. Without such data, it is difficult to accurately calibrate and verify watershed hydrologic models for the purpose of predicting sediment yield. However, land use, topographic, edaphic, and meteorological information are generally available. Spatial and physical attributes of a watershed that affect water quality and sediment yield can be assembled into a geographic information system (GIS) data base. This information can then be used to generate synthetic water quality data or estimate sediment yield by a simplistic approach. An important problem, particularly in the watersheds located in North-Central Mississippi that are part of the Demonstration Erosion Control (DEC) Project, is how to estimate the reduction in sediment yield from the watershed due to the construction of grade control drop structures. This paper focuses on the development and testing of a synthetic watershed sediment routing model that can be integrated with the GRASS GIS. It is anticipated that this model can also be used to assess the effectiveness of the drop structures.

Introduction

Erosion and sedimentation by water embody the processes of detachment, transportation, and deposition of soil particles (sediment) by the erosion and transport mechanisms of raindrop impact and runoff over the soil surface. The physical processes incorporated in the

----

[1] Associate Professor, Civil Engineering Dept., Univ. of Memphis, Memphis, TN 38152

[2] Graduate Research Assistant, Civil Engineering Dept., Univ. of Memphis, Memphis, TN 38152

current version of the simplified model (GISSRM) include: (a) Surface runoff arising from the rainfall and causing soil surface detachment and sediment transport in a sub-basin; (b) Sediment delivery due to the sheet and rill erosion to the outlet of a sub-basin (i.e., edge of stream loading); and (c) Deposition of suspended sediment that arises from the amount of sediment delivered to the stream reach.

In quantifying these processes, the following methodologies were adopted: (a) Computation of runoff from rainfall by the Soil Conservation Service (SCS) curve number method; (b) Construction of the sub-basin storm runoff hydrographs by the incomplete gamma distribution method; (c) Channel routing by the Muskingum method for ungaged channels and reservoir routing by the improved Euler method with parabolic interpolation; (d) The modified universal soil loss equation (MUSLE) for computing sediment delivery due to sheet and rill erosion at the outlet of a sub-basin; and (e) The Yang-Stall method for computing sediment transport in the main channel reaches.

The GISSRM model as described above was previously applied (Smith et al., 1992) to North Reelfoot Creek Watershed (approximately 144 sq. Km.) located in the northwest corner of Tennessee. The watershed had only one gaging station with recorded data. Synthetic runoff values (cms) were between +11% to -20% of the recorded values on a monthly and annual comparison basis. Computed annual sediment loads were between +17% to -31% of the recorded values. Stream channel bed and gully erosion processes were not included in the model, which could explain some of the differences noted above, especially during the more extreme storm events.

Current University of Memphis Study

Research is currently underway involving the modification and enhancement of the GISSRM model and its application to the Goodwin Creek watershed (approx. 22 sq. km.) located in Panola County, North Mississippi. This watershed has been extensively monitored (14 or more stations) and rainfall, streamflow, and sediment data that has been recorded over a period of 10 years or more are available for analysis. Also, as part of the DEC Project, a GIS database using a square grid cell (1.62 hectare) has been created for Goodwin Creek. The database contains landuse, soil type, elevation, SCS Curve Number, and other data. Landuse consists of forest, cropland, pasture, and small ponds. Over the last two decades, this watershed has experienced streambank instability and sedimentation problems due to change in land use. Loam, sandy loam, and silt loam are the primary soil types in the watershed with silt loam being predominant. Spatial and physical attributes of each sub-basin within the watershed were assembled as necessary for the GISSRM model using the GRASS GIS and the GeoShed Computer models.

Preliminary Simulation Results

The GISSRM model is currently being run for the same major storm events used in a previous study (Johnson, 1994) for comparison of lumped vs. distributive rainfall models. A comparison of the runoff hydrographs simulated for the Oct 17, 1981 rainfall event is shown in Figure 1. The simulated runoff peak value of 39.41 cms is comparable to the CASC2D model (39.5 cms) and observed flow peak of 39.77 cms.

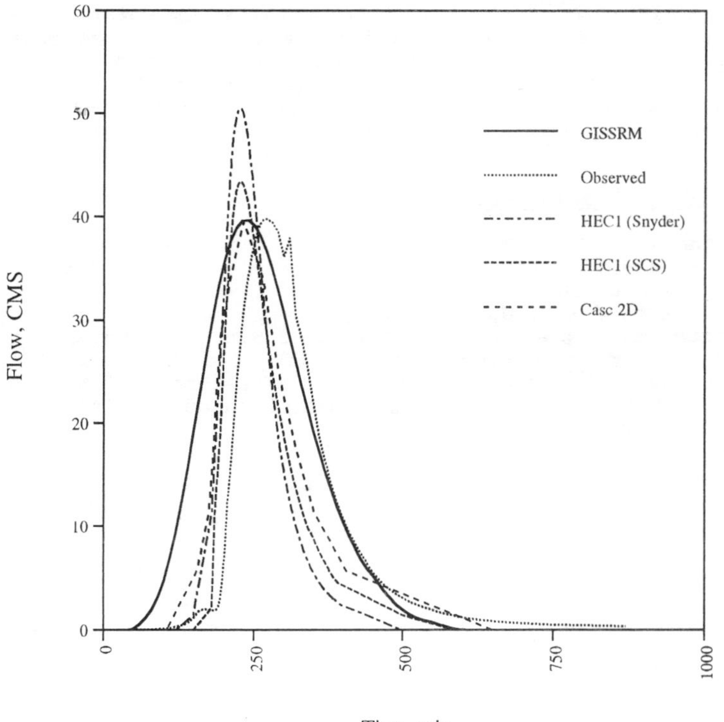

Figure 1. Simulated Runoff Hydrographs (Oct 17, 1981)

The sediment yield analysis portion of GISSRM study is still underway with no definitive results to be reported at this time.

Summary

The use of the GRASS GIS and Geoshed models to develop data input to event-driven models such as HEC-1, GISSRM, and CASC2D was found to be very beneficial in enhancing the overall modeling efforts as well as enhancing the performance of each model. In addition to improving streamflow and sediment simulation results, the GIS integrated approach minimizes time requirements and human error, especially when accurate digitized data are available.

The present version of GISSRM uses the gamma distribution to fit a lumped storm hydrograph rather than the unit hydrograph. Therefore, the time variation of the rainfall as it occurs within the total storm duration is not simulated and the resulting time of simulated peak flows will not match the observed values, as can be seen in Figure 1. However, this should not greatly affect long term computed averages of sediment yield. Sediment yield analysis should be completed in the near future and the results will be presented at the conference.

Appendix - References

Johnson, B. E., "Comparison of Distributive versus Lumped Rainfall -Runoff Modeling Techniques" DEC Project Monitoring Program, Fiscal year 1993 Report, Vol. V, Appendix D, July 1994.

Smith, R. H., et al., "A GIS Based Synthetic Watershed Sediment Routing Model" ASCE, Proceedings of the Water Resources Planning and Management Division (Water Forum 92), Edited by M. Karamouz, Baltimore, Maryland, Aug 2-6, 1992.

Field Reconnaissance for Stream Stability Analysis

Robert W. McCarley[1], M. ASCE and
Colin R. Thorne[2], Affil. M. ASCE

Abstract

Stream reconnaissance is an important means for compiling a broad overview of a fluvial system for engineering and geomorphic analyses. Field reconnaissance work includes a combination of qualitative observations and quantitative measurements of indicative channel and flow parameters. However, data collection and measurements must be relatively quick and easy if site reconnaissance is to be cost-effective. The paper mentions the types of data that should be collected in the field, lists recommended tools and instruments, and suggests typical applications for stability analysis.

Introduction

Increasingly, engineers designing channel stability projects are required to work in harmony with the natural processes, forms, and features of the channel. However, significant fluvial features, signs of instability, and sedimentation problems can be difficult to identify and record in the field. Since millimeter accuracy may not be required, the use of relatively simple techniques, instruments, and tools should be considered.

Basic Approach and Types of Data to be Collected

It is very important to plan a field trip carefully with the objectives, information to be obtained, and time constraints in mind. Examples of the required types of information and data are briefly considered in the following paragraphs.

---

[1]Research Hydraulic Engineer, U.S. Army Engineer Waterways Experiment Station, Hydraulics Laboratory, 3909 Halls Ferry Road, Vicksburg, MS 39180-6199.

[2]Professor, Department of Geography, University of Nottingham, UK NR72RD.

Basin and stream geometry (topography and hydrography)

Noteworthy basin geometry data, much of which can be obtained from maps and air photos or satellite images backed up by a few sample field observations and measurements for verification, include (1) the general relief of the region and stream valley, (2) general planform characteristics of the channel, and (3) the existence of control points. Maintenance activities in the study stream or its tributaries can also be investigated using existing records. Streambank and bed geometry include cross sections, bed slopes, meander planform parameters, and present bank line locations.

Geology and soils

A preliminary geological and soils investigation should identify surface and subsurface geological deposits in the catchment, river valley, and in the channel banks, bars, and bed. Sediment types, size distributions, and depths can be quickly estimated from representative measurements and samples taken in the field.

Hydraulic data

Ideally, hydraulic measurements should be made in different characteristic reaches. Surface velocity can be measured by tracking floats over a known reach length. Observing and recording stages, surface-flow directions, and velocities at relatively high flows give invaluable insights into stream hydraulics and the fluvial and sedimentary processes. Low-stage measurements are less relevant to geomorphological studies but have a crucial bearing on in-stream habitats and the ecological value of the stream. Wave action from wind or boat traffic, ice conditions, and sediment conditions may also be recorded.

Audio-visual records (photographs, videos, audio recordings)

A variety of excellent cameras, camcorders, and hand-held voice recorders is available for stream reconnaissance. The area around the valley, the valley and its sides, and the floodplain, including features such as terrain, surface geology, land use, and vegetation, together with channel features (signs of instability, significant failures and scour, sediment problems, structures, or bed and bank materials) should be photographed and/or videotaped with verbal descriptions recorded.

Components of the Field Equipment Pack and Their Uses

A stream reconnaissance backpack has been assembled to contain the instruments and tools needed. Contents may be tailored to the specific purpose of a particular trip. The following items make up a complete equipment set:

Instruments for measuring basin and stream geometry -- rangefinder; string-operated pedometer; open-reel measuring tape (100 ft); chaining pins and vinyl flagging tape; clinometer; hand level (5X magnification); "pocket rod."

Equipment for soils and geology reconnaissance -- foldable shovel; thick plastic bags and felt markers; geologist pick; probe rod; rock identification charts; gravelometer.

Field gear for hydraulic measurements -- stop watch; float (an unpeeled orange works well).

Audio-visual equipment for recording site inspection findings -- 35-mm camera with carrying case; 8-mm camcorder; hand-held voice recorder.

Miscellaneous auxiliary gear and reference material -- safety gear (first aid kit, canteen, flashlight, matches, sunscreen, insect repellent, snake-bite kit, life jackets); miscellaneous supplies (clipboard, compass, architect and engineer scales, protractor, calculator, magnifying lens, hammer/hatchet, rope, brush knife, rain suit, umbrella); reference material (checklists, forms, plans, maps, aerial photographs).

Engineering Applications

Field reconnaissance of alluvial streams will provide for the (1) collection of input data for stable channel design and (2) assessment, modeling, and control of bank retreat. A framework for the collection of field data using the field pack and a new type of record sheet has been developed at the U.S. Army Engineer Waterways Experiment Station (USAEWES), Vicksburg, MS (Thorne 1991 and 1993).

Collection of input data for stable channel design

The field pack and record sheets can be used in gathering the basic data necessary to characterize existing channels, identify key flow and sediment processes and mechanisms, and estimate the severity of any flow or sediment-related problems. This is an important first step in the design of engineering works to improve channel stability. Only after these steps have been taken is it possible to determine the true causes of the problems and make sound remedial recommendations.

New and innovative approaches to stable channel design, such as the SAM package currently being developed at the USAEWES, require input data of this type (Thomas 1990). Accurate field data are essential in selecting the most appropriate quantitative equations for flow resistance, sediment transport, and one-dimensional modeling.

Assessment, modeling, and control of bank retreat

The explanation, prediction, and stabilization of bank retreat are all aided by the application of new approaches to the analysis of soil erosion processes and geotechnical failures (Thorne and Osman 1988; Hagerty 1989). The input data and qualitative information necessary to apply such methods may be collected in the course of stream reconnaissance using efficient techniques and equipment.

Conclusions

Stream reconnaissance is composed of a combination of qualitative observations and quantitative measurements of indicative channel and flow parameters. Data collection should be quick and require only one or two field hands if it is to be cost-effective. The required instruments and tools are not new, but they may not be routinely available to many practicing engineers. In response to the need to support sound and efficient field reconnaissance, a field pack and record sheets have been designed. The tools and instruments are sufficiently versatile to deal with most situations, while being compact and light enough for the field worker to carry in a single backpack.

References

Hagerty, D.J. (1989). "Geotechnical Aspects of River Bank Erosion," Hydraulic Engineering, M.A. Ports (ed.), Proceedings of the National Conference on Hydraulic Engineering, American Society of Civil Engineers, New Orleans, LA, August 1989, pp 118-123.

Thomas, W.A. (1990). "Example of the Stable Channel Design Approach in Hydraulic Engineering," H.H. Chang and J.C. Hill, ed., Proceedings of the 1990 National Conference, American Society of Civil Engineers, San Diego, CA, pp 175-180.

Thorne, C.R., and Osman, A.M. (1988). "Riverbank Stability Analysis. II: Applications," Journal of Hydraulic Engineering, American Society of Civil Engineers, Vol. 114, No. 2, pp 151-172.

Thorne, C.R. (1991). "Field Assessment Techniques for Bank Erosion Modelling," Final Report to the U.S. Army European Research Office, Contract Number R&D 6560-EN-09, Department of Geography, University of Nottingham, UK.

Thorne, C.R. (1993). "Guidelines for the Use of Stream Reconnaissance Record Sheets in the Field," Contract Report HL-93-2, U.S. Army Engineer Waterways Experiment Station, Vicksburg, MS.

Trend Analyses of Sediment Data for the DEC Project

by Richard Allen Rebich[1]

Abstract

Daily stream discharge, suspended-sediment concentration, and suspended-sediment discharge data were collected at eight sites in six watersheds of the Demonstration Erosion Control project in the Yazoo River Basin in north-central Mississippi during the period July 1985 through September 1991. The project is part of an ongoing interagency program of planning, design, construction, monitoring, and evaluation to alleviate flooding, erosion, sedimentation, and water-quality problems for watersheds located in the bluff hills upstream of the Mississippi River alluvial plain. This paper presents preliminary results of trend analyses for stream discharge and sediment data for the eight project sites. More than 550 stream discharge measurements and 20,000 suspended-sediment samples have been collected at the eight sites since 1985.

Introduction

In 1984, Congress directed the U.S. Army Corps of Engineers and the U.S. Department of Agriculture, Soil Conservation Service, to establish demonstration watersheds to address critical erosion and sedimentation problems. The Demonstration Erosion Control (DEC) Project is in the Yazoo River Basin in north-central Mississippi. In July 1985, the U.S. Geological Survey (USGS) began collecting sediment data for the Yazoo River Basin DEC project. Data were to be collected prior to, during, and after watershed-conservation and channel-stability measures were implemented in the study area. This paper presents preliminary results of trend analyses for stream discharge, and sediment data for eight USGS DEC project sites for 6 years of data collection, specifically, the period July 1985 through September 1991.

Study Area, Sampling Sites, and Data-Collection Activities

Sediment data-collection activities were conducted at eight sites in six watersheds in north-central Mississippi for the study period (fig. 1). In downstream order, the site names are: 1. Hotopha Creek near Batesville (Panola County), 2. Otoucalofa Creek Canal near Water Valley (Yalobusha County), 3. Peters Creek near Pope (Yalobusha County), 4. Hickahala Creek near Senatobia (Tate County), 5. Senatobia Creek near Senatobia (Tate County), 6. Batupan Bogue at Grenada (Grenada County), 7. Fannegusha Creek near Howard (Holmes County), and 8. Harland Creek near Howard (Holmes County).

Stream discharge is routinely measured by personnel of the USGS once every 6 weeks and during selected storms. For the study period, about 550 stream discharge measurements from the eight study sites were analyzed, reviewed, and stored in USGS computer files. Suspended-sediment samples were collected by observers, automatic point samplers, and personnel of the USGS. Sampling procedures include single, vertically integrated samples and multiple, vertically integrated samples taken at several sections across the stream. From July 1985 through September 1991, about 20,000 suspended-sediment

---

[1]Hydrologist, U.S. Geological Survey, Water Resources Division, 100 West Capitol Street, Suite 710, Jackson, Mississippi, 39269

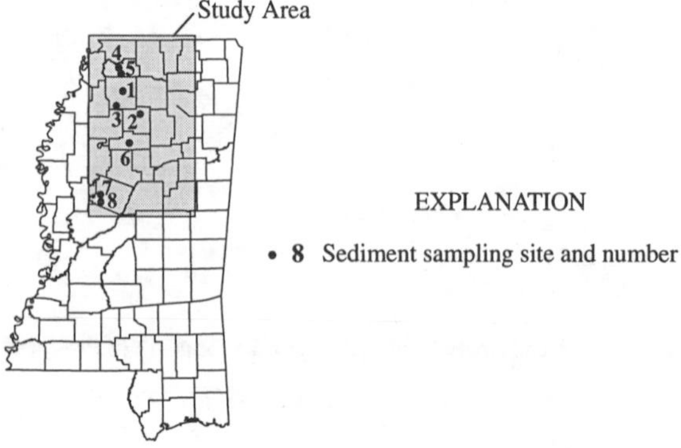

Figure 1. -- Location of study area and sediment sampling sites.

samples from the eight study sites were analyzed and reviewed, and data were stored in USGS computer files. Statistical summaries of stream discharge, suspended-sediment concentration, and sediment discharge data are presented in, "Preliminary Summaries and Trend Analyses of Stream Discharge and Sediment Data for the Yazoo River Basin Demonstration Erosion Control Project, North-Central Mississippi, July 1985 through September 1991" (Rebich, 1993).

Preliminary Trend Analyses of Stream Discharge and Sediment Data

The Seasonal Kendall test for trends was selected to detect trends in stream discharge, suspended-sediment concentration, and sediment discharge data for the eight sites. In addition, the Seasonal Kendall Slope Estimator was used to indicate magnitude and direction of detected trends, and flow-adjustment procedures were used to attempt to remove stream discharge as a source of variance in suspended-sediment concentration and sediment discharge data. Complete explanations of the Seasonal Kendall test, the Seasonal Kendall Slope Estimator, and flow adjustment are presented in "Statistical Methods in Water Resources" by Helsel and Hirsch (1992).

Requirements for the Seasonal Kendall test include an adequate period of record, number of seasons, and pre-selected level of significance. First, a minimum of 5 to 10 years of record is considered adequate to conduct trend analyses (Schertz, 1990). For the eight DEC project sites, four sites had more than 5 years of record, two had nearly 5 years of record, and two sites had less than 4 years of record. Trend analyses were conducted at all sites except the two sites with less than 4 years of record (sites 5 and 7 as previously listed). Next, the number of seasons used to perform the Seasonal Kendall test was selected to represent the range of values in the sediment data for a year of record; however, a large number of seasons could cause potential problems with the trend test, such as eliminating independence in the test data. Because of the large number of stream discharge and sediment data, trend analyses were performed on subsets of data based on 12 seasons per year or one set of values per month. Replicate subsets were formed to support an overall trend result at a particular site. Finally, a level of significance was pre-selected to indicate whether the results of the Seasonal Kendall test conducted on a particular subset is considered statistically significant. A "p-value" associated with the results of the test is the probablility that a trend resulted from a chance arrangement of the data rather than an actual change in the data (Schertz, 1990). If the p-value of the trend test is less than the pre-selected level of significance, then the result can be considered statistically significant. The author chose a pre-selected level of significance of 0.1, the same as that used by Smith and others (1982).

After using the Seasonal Kendall test to identify trends in the sediment data, the Seasonal Kendall Slope Estimator is used to estimate the magnitude and direction of the trend. The magnitude is expressed as a slope (value per unit time), although linearity is not implied in the trend. This slope estimate is the median of the differences (expressed as slopes) of the ordered pairs of data compared in the trend test. The median of differences is the change per year due to the trend (Smith and others, 1982). Because the median of the differences is used, this slope estimate is resistant to extreme values (or outliers) and to seasonal variation (Helsel and Hirsch, 1992). A positive value of the slope estimate indicates an increasing trend, and a negative value indicates a decreasing trend. The Seasonal Kendall Slope Estimator was computed for all trend analyses at the six sites analyzed.

Suspended-sediment concentration and sediment discharge are strongly correlated with stream discharge. Suspended-sediment concentration and sediment discharge generally increase as stream discharge increases because of the transport of particulates within stormwater runoff (Schertz, 1990). If the variability due to stream discharge is removed, trend testing would have greater probability of detecting a trend when one exists, and the trend would not be an artifact of the history of stream discharge at a particular site (Schertz, 1990). Therefore, a statistically significant trend would indicate changes in the factors that contribute to sedimentation and erosion at a particular site.

The technique used to remove the effects of stream discharge on suspended-sediment concentration or sediment discharge is to compute a time series of flow-adjusted concentrations (FAC's) and test this time series for trend. The FAC is defined in this report as a residual computed by subtracting a predicted value from an actual value of suspended-sediment concentration or sediment discharge. Predicted values are computed from a mathematical expression that describes the relation between stream discharge and either suspended-sediment concentration or sediment discharge. Many expressions that describe the relation between stream discharge and either suspended-sediment concentration or sediment discharge at a particular site were considered. Such expressions included linear regression, multiple regression (quadratic polynomial regression), and locally weighted scatterplot smooths (LOWESS) [see Helsel and Hirsch (1992) for a more detailed explanation of LOWESS].

Trend Analyses Results

Trends were detected in stream discharge for all subsets of data at each site analyzed, except for one subset of stream discharges at Batupan Bogue at Grenada (site number 6). In addition, slope estimates indicated an increase in stream discharge for the study period where trends in stream discharge were detected. The increase in stream discharge at the six DEC sites was supported by the annual mean stream discharge extremes generally having lowest annual means in the 1988 water year and highest annual means in the 1989 and 1991 water years, which is consistent with rainfall conditions in the study area. Annual mean stream discharge extremes are discussed in the report by Rebich (1993).

Trends were detected in flow-adjusted suspended-sediment concentration at Hotopha Creek near Batesville (site number 1). For all replicate subsets of flow-adjusted suspended-sediment concentration data, p-values were less than 0.1, indicating a statistically significant trend at Hotopha Creek. Slope estimates associated with these tests were all negative, which may indicate a decrease in the factors that contribute to sedimentation and erosion at this site.

An example of the relation of flow-adjusted suspended-sediment concentration and time is plotted in figure 2 for a subset of data at this site. Each data point (log transformed) represents a residual defined earlier as the difference between actual and predicted values of suspended-sediment concentration. The trend line in figure 2 is based on an equation derived during trend computations and is provided for purposes of illustration only. Although the trend line is depicted as linear in figure 2, a linear trend for this data subset (which represents the entire data set) is not assumed. Rather, the trend line represents the relative decrease in flow-adjusted, log-transformed suspended-sediment concentration per year at this site.

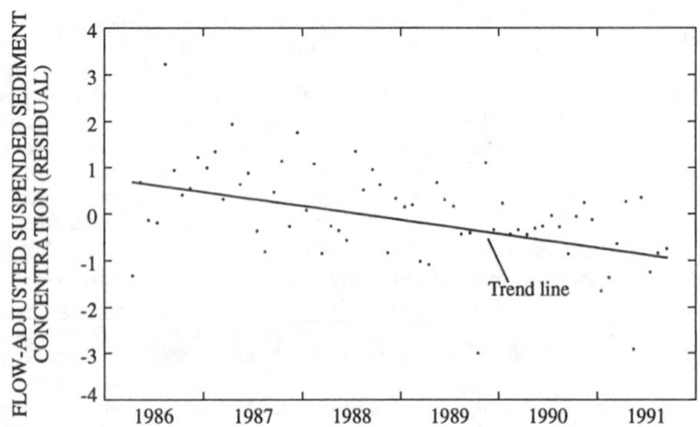

Figure 2. -- Flow-adjusted suspended sediment concentration for Hotopha Creek near Batesville (site number 1).

Similar trends were detected at Otoucalofa Creek Canal near Batesville (site number 2) for flow-adjusted suspended-sediment concentration and flow-adjusted sediment discharge. For subsets of flow-adjusted suspended-sediment concentration and sediment discharge data, p-values were less than 0.1, indicating a statistically significant trend. Slope estimates associated with both sets of trend results were all negative, which may indicate a decrease in the factors that contribute to sedimentation and erosion at this site.

Summary

This paper presents preliminary results of trend analyses of stream discharge and sediment data for six of the eight sites in the Yazoo River Basin Demonstration Erosion Control project in north-central Mississippi for the study period July 1985 through September 1991. The Seasonal Kendall trend test was used to detect trends in stream discharge, suspended-sediment concentration, and sediment discharge data for the DEC project sites. Trend analyses were also conducted for flow-adjusted suspended-sediment concentration and sediment discharge.

Trends were detected in stream discharge at each site indicating an increase in stream discharge for the study period, which is consistent with rainfall conditions in the study area. Trends were detected in flow-adjusted suspended-sediment concentration at Hotopha Creek near Batesville and in flow-adjusted suspended-sediment concentration and flow-adjusted sediment discharge at Otoucalofa Creek Canal near Water Valley indicating a possible decrease in the factors that contribute to sedimentation and erosion at these two sites.

Appendix

Helsel, D.R., and Hirsch, R.M., 1992, Statistical methods in water research: New York, Elsevier, 522 p.

Rebich, R.A., 1993, Preliminary summaries and trend analyses of stream discharge and sediment data for the Yazoo River Basin demonstration erosion control project, north-central Mississippi, July 1985 through September 1991: U.S. Geological Survey Water-Resources Investigations Report 93-4068, 80 p.

Schertz, T.L., 1990, Trends in water-quality data in Texas: U.S. Geological Survey Water-Resources Investigations Report 89-4178, 49 p.

Smith, R.A, Hirsch, R.M., and Slack, J.R., 1982, A study of trends in total phosphorous measurements at NASQAN stations: U.S. Geological Survey Water-Supply Paper 2190, 34 p.

MONITORING OF DEC GRADE CONTROL STRUCTURES

by

Chester C. Watson [1], F.ASCE, Steven R. Abt [2], F.ASCE, and Charles Little [3]

BSTRACT

Field monitoring of grade control structures was conducted as part of the Demonstration rosion Control (DEC) monitoring program. Monitoring of the structures was conducted to inforce previous physical modelling studies and analyses conducted by Waterways Experiment tation (WES) and Colorado State University (CSU). Six common problem were observed in the ıspection of approximately 60 grade control structures, and recommendations were made for ıaintenance and restoration activities.

NTRODUCTION

The Demonstration Erosion Control (DEC) Project provides for the development of a ystem for control of sediment, erosion, and flooding in the foothills area of the Yazoo Basin, lississippi. Structural features that are used in developing rehabilitation plans for the DEC 'atersheds include: high-drop grade control structures similar to the SCS Type-C structure, and w-drop grade control structures similar to the ARS low-drop structure.

Watson, et al. (1988) reported on the evaluation of channel response for several Yazoo asin low drop structures. This evaluation was based on comparison of then-current structure ırveys with available historic information. Although this data base was limited, several commendations were reached: design procedures for the low-drop structure must incorporate ilwater definition, upstream aggradation was limited due to the lack of constriction at the weir, ttle test data existed for operation of the structures at high submergence, and structures should e planned based on a comprehensive watershed stabilization plan. Some of these commendations have resulted in additional physical modeling to supplement the work by Little nd Murphey (1981, 1982). A series of hydraulic model tests has been conducted at Colorado tate University by Abt, et al. (1990,1992) to evaluate the low-drop structure under conditions f flow that were not considered by Little and Murphey (1981, 1982). Results from the CSU tests ıdicated that riprap stability in many existing structures was poor, and field confirmation of the prap instability has been documented by Lenzotti and Fullerton (1990).

[1] Associate Professor, Dept. of Civil Engr., Colorado State University, Ft. Collins, CO 80523
[2] Prof. and Director, Hydraulics Lab, Dept of Civil Engr., Colorado State University, Ft. Collins, CO 80523
[3] Hydraulic Engineer, Vicksburg District, U.S. Army , Corps of Engineers, Vicksburg, MS 39180

As a result of the studies cited, additional model testing by the Waterways Experime Station (WES), and continuing efforts to improve the structure design and application b Vicksburg District personnel, the drop structure design has continously evolved. Each change ha improved performance for the subsequent structures. However, resources and personnel hav not previously been available to make a field evaluation of each structure and to make restoratic recommendations that may be required for the older structures.

The objective of this investigation is to document the condition of U.S. Army, Corps o Engineers high-drop and low-drop grade control structures constructed in the Yazoo Basin as part of the Demonstration Erosion Control (DEC) program, and to make recommendations fo restoration of these structures if necessary. A general objective of this investigation is t contribute to the development of improvements in the general design of grade control structure and to compare the various types and ages of structures to develop a database that may usef in predicting restoration needs and design improvements for similar structures.

MONITORING DATA

A structure evaluation form for each structure was prepared based on field inspection of eac site by Watson and Abt (1993). Field inspections were made in June, 1993. Photograph documentation of each structure was provided by a series of slides which were cataloged an referenced by film roll number and slide number given on each field evaluation form. Approximate five hours of video tape have been narrated and annotated showing the 1993 field conditions.

RESULTS & DISCUSSION

Although each structure operates under unique conditions, many similarities exist and son of the problems that were observed can be summarized. The six most common problems observe are listed and the percentage given denotes the percentage of total structures evaluated in which t problem occurred:

1. Riprap is displaced from the face of the weir. (41%)
2. The channel bank upstream or downstream of the structure is failing. (37%)
3. Bank erosion or piping beneath the riprap is occurring caused by overba drainage. (24%)
4. Riprap is launching at the upstream or downstream apron. (28%)
5. Severe headcutting is migrating into the basin. (17%)
6. Woody vegetation has become established in the upstream or downstream apron, and is impairing the conveyance or the weir unit discharge of the structure. (19%)

Recommendations are made for restoration that may be required at each structure to solve hese or other less common problems. The priority of for restoration of each site has been divided nto three categories, as follows:

- * Category 1 structures are under an imminent threat of loss of function (5 structures);
- * Category 2 structures have problems that should be resolved (32 structures);
- * Category 3 structures have no significant problems (18 structures).

Several issues that are beyond the scope of this field evaluation pertaining to the structures re:

a.) access to structure sites, both for inspection and for maintenance;
b.) establishment of a regularly scheduled maintenance program;
c.) consideration of improved overbank drainage-caused erosion control;
d.) consideration of a vegetation removal practice.

Access to the structures for inspection and for any recommended maintenance is essential. resently, access is limited to the goodwill of the land owner, and while no problems with access for nspection were encountered, maintenance will require heavy equipment and materials that will often equire a construction road.

Some of the situations requiring maintenance developed relatively quickly, in less than one ear. Regular inspection and timely attention to problem situations can result in lessening costs and everity of structure damage.

As noted in the previous section, overbank drainage-caused erosion problems are difficult to revent. Solutions to this problem could be closer attention to detail in the final inspection of the onstruction site, and consideration of the use of some of the numerous manufactured erosion revention products and methods. Many products and methods have been developed for highway onstruction such as jute mats, plastic and fiber mats, and mat and geo-grid combinations. These ame methods and materials could also be used to minimize erosion and sediment delivery to the tream during structure construction. Testing of these materials may also be considered to be within he overall mission.

Removal of woody vegetation from the structures could have ecological consequences. Also, he impact of varying amounts of woody vegetation in the structures is unknown. The approach to he complexities of this issue should be considered.

ECOMMENDATIONS

Based on the evaluation of the structures, the following recommendations are made:

1. Restore the Category 1 structures in accordance with the recommendations given on each structure evaluation form as soon as possible.

2. Develop a regular maintenance program to restore the Category 2 structur resolving the problems identified in the 1993 field evaluation.
3. Establish a regular annual field evaluation program of all structures: Categc 1, Category, 2, Category 3, and new structures.
4. Establish access to all sites for restoration and evaluation.
5. Develop a policy for maintenance of woody vegetation within the structures.
6. Develop a program for investigating and testing of erosion control methc for the structure construction site and for overbank drainage followi construction.

REFERENCES

Abt, S.R., Peterson, M.R., Combs, P.G., and Watson, C.C. (1990). "Evaluation of the A low drop grade control structure." Hydraulic Engineering , Proc. of the 1990 N Conf., Vol. 1, U.S. Army Corps of Engrs., Vicksburg, Miss, 342-347.

Abt, S.R., Peterson, M.R., Watson, C.C., and Hogan, S.A. (1992). "Analysis of ARS lo drop grade-control structure." J. of Hyd. Engr., ASCE, 118(10), 1424-1434.

Lenzotti, J. and Fullerton, W., (1990). "DEC grade control structure evaluation". a rep submitted to Vicksburg District, Corps of Engineers, Vicksburg, Miss.

Little, W.C. and Murphey, J.B., (1981). "Stream channel stability, model study of the low dr grade control structures." USDA Sedimentation Lab., Oxford, Miss.

Little, W.C. and Murphey, J.B. (1982). "Model study of low drop grade control structure J. Hydr. Div., ASCE, 108(10), 1131-1146.

Watson, C.C., Harvey, M.D., Biedenharn, D.S., and Combs, P.G., (1988). " Geotechnical a hydraulic stability numbers for channel rehabilitation: Part I, The approach:" Steven Abt and Johannes Gessler (eds), Proc. ASCE Hyd Div. 1988 National Conf. P. 120-12

Watson, C.C. and Abt, S.R., (1993). Inspection of Low-Drop Grade Control Structures, prepar for Waterways Experiment Station, U.S. Army Corps of Engineers, Contract DACW39-93- 0028.

Measurement and Prediction of Bed Material Load on Goodwin Creek, A DEC Subwatershed

Roger A. Kuhnle[1] and Jurgen Garbrecht[2]

Introduction

The Goodwin Creek Research Watershed was established to study the impact of land use and watershed processes on the stability of the channels and on the movement of sediment through the watershed. The 21.3 km^2 watershed is highly instrumented and serves as a testing ground for sampling and predictive techniques of flow and sediment movement which will be used to evaluate erosion control and channel rectification measures on the Demonstration Erosion Control (DEC) watersheds in northern Mississippi.

An important aspect in the evaluation of channel stability is the accurate prediction of the rate and size of the bed material load. Any long term program of channel stabilization must take into account the transport of sediment through the watershed. An imbalance between supply and transport capacity of sediment will cause channel adjustment or instability to occur. If sediment supply is modified at one location of the watershed a compensatory adjustment must be made to the downstream sediment transport environment to reflect the new conditions. In streams which contain an appreciable percentage of gravel in the bed material, as in Goodwin Creek, determination of the transport rate of the different size fractions is important to determine whether bed surface armoring occurs with time. The objective of this paper is to compare measurements of bed material load from Goodwin Creek to calculated values from a predictive technique which consists of four combined transport equations.

---

[1]Research Hydraulic Engineer, USDA-ARS, National Sedimentation Laboratory, P.O. Box 1157, Oxford, Mississippi, 38655

[2]Research Hydraulic Engineer, USDA-ARS, National Agricultural Water Quality Laboratory, P.O. Box 1430, Durant, Oklahoma 74702

Data Collection

A bed material sampling program was conducted near the downstream end of Goodwin Creek from 1984-1988. Approximately 1000 sand load and 500 gravel load samples were collected during this period. Samples of sand in transport were collected using DH-48 samplers at several locations across the flow area through the entire depth of flow (Willis et al., 1986). Samples of gravel in transport were collected using modified Helley-Smith samplers (Kuhnle, 1992) from foot bridges at several locations across the supercritical flow structure.

Sediment Transport Algorithm

The sediment transport algorithm calculates the rates of 9 size groups from silt to gravel (Garbrecht et al., 1995). Four different equations are used: Laursen (1958)- silt to fine sand (3 groups) 0.010 - 0.250 mm, Yang (1973)- medium to coarse sand (2 groups) 0.250 - 2.000 mm, Yang (1984)- gravel (2 groups) 2.000 - 8.000 mm, and Meyer-Peter Mueller (1948)- gravel 8.000 - 50.000 mm. The sediment transport algorithm has been validated using extensive laboratory and field data collected under equilibrium conditions for sediments with narrow size distributions.

Originally the transport rate of each sediment size group was calculated independently of each other. For cases in which the bed material has a wide distribution of sizes, the flow strength at which the bed material begins to move is affected by the entire size distribution. Therefore, the grain size used to calculate the critical flow strength for each sediment size group was calculated as

$$D_{ci} = D_i \left(\frac{D_i}{D_m} \right)^{-x} \qquad (1)$$

where D_i and D_{ci} are mean and critical sediment diameter for size fraction i, respectively; D_m is the mean size of the bed material sediment; and x is a constant. For x=1 the D_m of the sediment is the critical diameter for all size fractions and all fractions begin to move at the same flow strength. For x=0 each size group behaves independently of the others and the D_i for each size fraction is used to calculate the flow strength that will cause that size fraction to begin to move. Values of x of 0.5 and 1.0 were used in this study for sediment mixtures with bimodal and unimodal size distributions, respectively. This pattern of initiation of movement of the sediment grains has been shown to closely predict both laboratory and field transport data for mixed size gravel sediments (Wilcock, 1993).

Comparison to Transport Data

Comparisons were made between the transport rates calculated using the sediment transport algorithm, outlined above, and measured transport rates collected in a laboratory flume at the National Sedimentation Laboratory Hydraulic Laboratory (Kuhnle, 1993). Figure 1 shows that estimates of the total transport rate and the size distribution of the sediment in transport match closely.

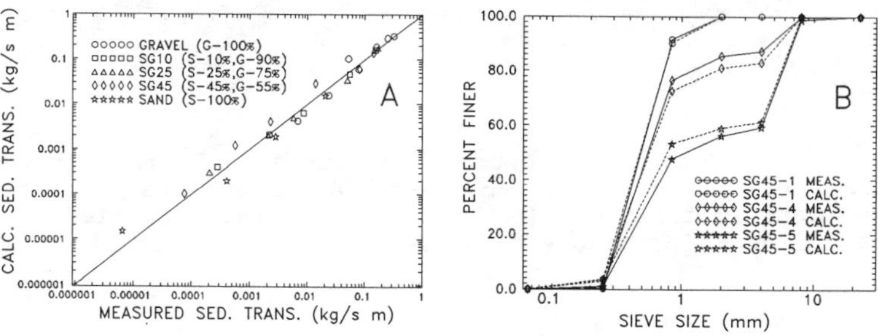

Fig. 1. Comparison between measured and calculated transport rates (A), and size distributions of the sediment in transport (B) for laboratory experiments. For clarity, data from intermediate experiments were not plotted in (B).

Comparisons were also made between calculated and measured sediment transport data from Goodwin Creek (Fig. 2). The correspondence between the calculated and measured data is not as good for Goodwin Creek as it is for the laboratory data. This is to be expected given the greater level of uncertainty in the measured field data compared to measurements collected under controlled laboratory conditions. The ratio of calculated to measured transport rate for Goodwin Creek was greater than 0.5 and less than 1 in all cases except one in which the value was 0.26. Total transport rates closer to the measured rates were obtained by using a value of $x = 0$ in equation (1), however, the calculated size distributions of the sediment in transport contained virtually none of the coarser sediment sizes of the bed material. In long term model studies of watersheds the increase in accuracy of the predicted size distribution of the sediment in transport was determined to be of greater importance than more accurate estimates of the total rate of sediment in transport.

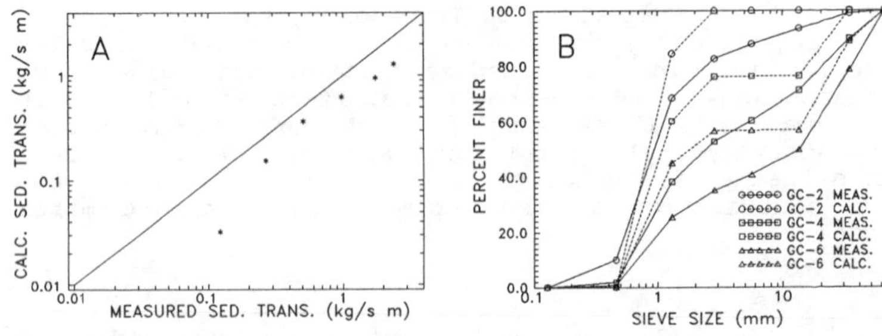

Fig. 2. Comparison between measured and calculated transport rates (A), and size distributions of the sediment in transport (B) on Goodwin Creek. For clarity, data from intermediate flows were not plotted in (B).

References

Garbrecht, J., Kuhnle, R. A., and Alonso, C. V. (1995). "Sediment transport formulation for large channel networks", International Symposium on Water Research and Management in Semiarid Environments (in press).

Kuhnle, R. A. (1992). "Fractional transport rates of bedload on Goodwin Creek", in Dynamics of Gravel-bed Rivers, P. Billi, R. D. Hey, C. R. Thorne, and P. Tacconi, eds., John Wiley and Sons, Chichester, U.K., 141-155.

Kuhnle, R. A. (1993). "Incipient motion of sand-gravel sediment mixtures", ASCE Journal Hydraulic Engineering 119(12), 1400-1415.

Wilcock, P. R. (1993). Critical shear stress of natural sediments", ASCE Journal Hydraulic Engineering 119(4), 491-505.

Willis, J. C., Darden, R. W., and Bowie, A. J. (1986). Sediment transport in Goodwin Creek", Proceedings of 4th Federal Interagency Sedimentation Conference, Las Vegas, Nevada, 4-30 - 4-39.

Channel Stabilization Methods
In The Demonstration Erosion Control Project

Raymond O. Wilson, Jr.[1]
Charles D. Little, Jr., P.E.[1]
Kelly B. Mendrop, P.E.[1]
John B. Smith[1]
Charles A. Montague, Jr.[2]

Abstract

Erosion is a problem that is faced everyday throughout the United States. Besides the obvious loss of productive topsoil, erosion produces indirect effects such as poor water quality, loss of aquatic and terrestrial habitat, damage to the transportation infrastructure, and increased frequency and duration of flooding due to excessive sedimentation. The Demonstration Erosion Control (DEC) project is an intensive program designed to demonstrate the effectiveness of a watershed or systems approach to reducing erosion, sedimentation, and flooding problems. Various types of erosion control structures are being used as part of the DEC project including high drop grade control structures, low drop grade control structures, box culverts, riser pipes, bank stabilization methods, land treatment methods, flood water retarding structures, debris basins, intermediate dams, channels, levees, and pumping stations. This paper focuses on DEC grade control structures which include the first four structure types in the above list. The magnitude and location of the erosion problems dictate the type of grade control structure best suited in each case.

----

[1]Hydraulic Engineer, U.S. Army Corps of Engineers, Vicksburg District, 2101 North Frontage Road, Vicksburg, MS 39180-5191

[2]Civil Engineering Technician, U.S. Army Corps of Engineers, Vicksburg District, 2101 North Frontage Road, Vicksburg, MS 39180-5191

Introduction

The DEC project was first authorized in 1983 as a cooperative effort through the U.S. Department of Agriculture (USDA) Soil Conservation Service with the design and construction of the project's structural features being accomplished as a joint effort by the Corps of Engineers, Vicksburg District and the Soil Conservation Service. The project is located within the Yazoo River basin in northeastern Mississippi. It consists of 15 subwatersheds of the Yazoo River, all of which are located in the foothills of the Yazoo basin. The total drainage area of the Yazoo basin is approximately 34,706 sq. km. (13,400 sq. mi.). The fifteen DEC watersheds encompass 14,640 ha (1,463,970 acres) or 5,923 sq. km. (2287 sq. mi.).

High Drop Grade Control Structures

High drop grade control structures are used in deeply incised channels to intercept head cuts with vertical drops greater than 1.83 meters (6 feet) and to decrease channel slopes and raise channel bottom elevations to improve bank stability. High drop structures have been used in the DEC project for up to 4.27 meters (14 feet) of vertical drop. The deeply incised streams which are common in the DEC area have unstable banks that are in most cases still adjusting toward stable side slopes. The natural bank adjustment, if allowed to continue in an uncontrolled manner, will result in the loss of a significant area of productive land, increased sediment yield to receiving streams, and degeneration of riparian habitat. The primary design considerations for a high drop structure are (a) proper location of the structure along the stream's course for maximum benefit; (b) proper alignment of the structure to prevent upstream flanking of the structure and downstream bank erosion problems; (c) proper sizing of the weir to provide the cross-sectional flow area which results in the desired effects; and (d) adequate sizing of the riprap to be located in the upstream approach to the structure. High drop structures are much more expensive than low drop structures, in some cases twice as expensive, so their applications are limited. Figure 1 is a simplified sectional view of a typical high drop grade control structure along the longitudinal centerline of the structure.

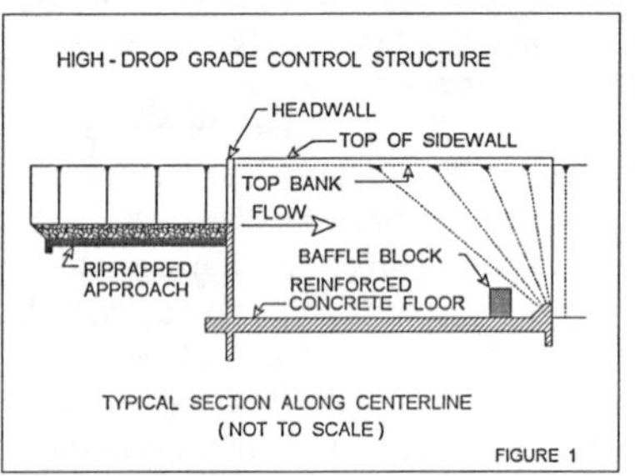

FIGURE 1

Low Drop Grade Control Structures

Low drop grade control structures are used to intercept upstream migrating head cuts where a vertical drop of no greater than 1.83 meters (6 feet) is required. All of the DEC low drop structures consist of sheet pile weirs (most with concrete caps), riprap lined approach channels, riprap lined stilling basins, and energy dissipation baffles located in the stilling basins. The primary design considerations are (a) proper location of the structure along the stream's course for maximum benefit; (b) proper alignment of the structure to prevent upstream flanking of the structure and downstream bank erosion problems; (c) proper sizing of the weir to provide the cross-sectional flow area which results in the desired effects, considering both upstream impact and required stilling basin design; (d) adequate sizing of the riprap to be located in the most turbulent zones of the structure; and (e) proper location and dimensions for the energy dissipation baffle. Two basic structure configurations have been used, vertical drop structures and sloped drop structures. Drawings showing sectional views along the longitudinal centerlines of typical structures for each type are given in Figures 2 and 3.

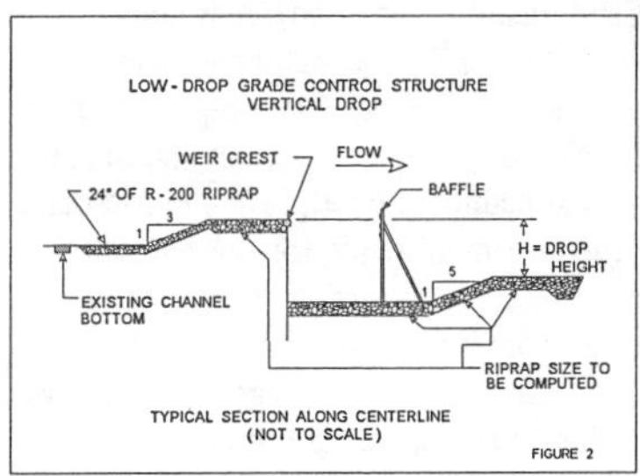

FIGURE 2

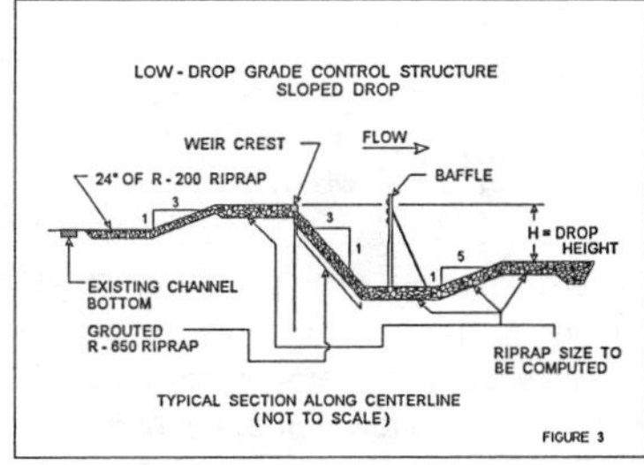

FIGURE 3

Box Culvert Grade Control Structures

Box culvert grade control structures are being used in the DEC project to intercept migrating head cuts which are approaching aging road crossings. In those special cases they have provided a means of addressing two problems at the cost of a single structure. Box culverts are being designed and constructed to provide road crossings which meet modern specifications while providing excellent control of stream degradation. They have been constructed with vertical drops ranging from 1.5 meters (5 feet) to 3.05 meters (10 feet). Two basic box culvert designs have been used in the DEC project with the main difference being the inlet configuration as indicated in

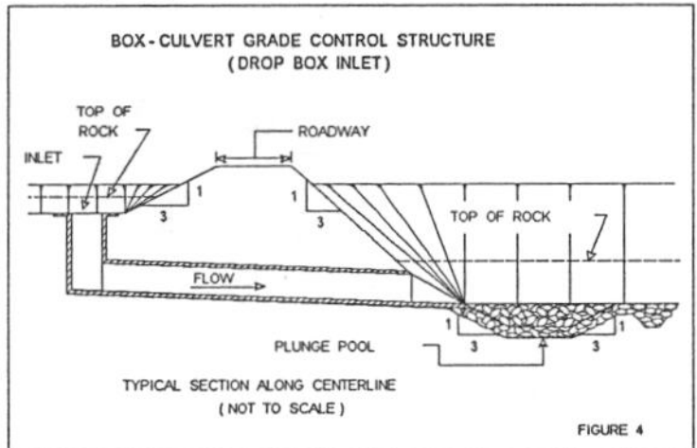

FIGURE 4

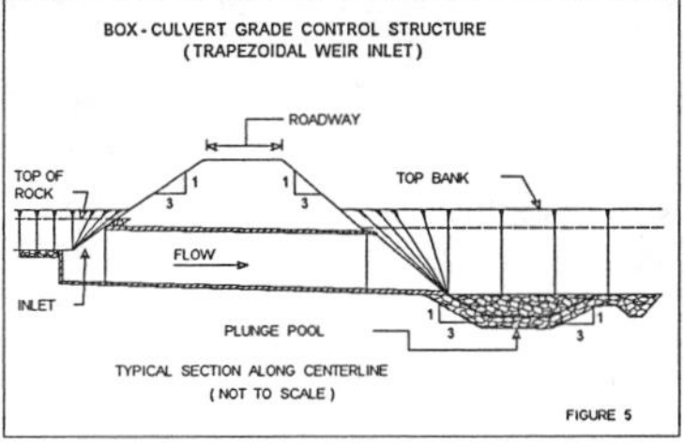

FIGURE 5

Figures 4 and 5. The primary considerations in the design of these structures are (a) the amount of vertical drop required to control the stream degradation; (b) the best orientation of the structure relative to the stream and road; (c) the flow capacity required by the structure; and (d) proper sizing of the structure components (inlet weir and or vertical drop section, conduit, and stilling basin) to provide for the required hydraulic performance.

Riser Pipe Grade Control Structures

Riser pipe grade control structures are utilized in cases where stream degradation has induced localized erosion where lateral drains enter a stream. Gullies typically develop when a stream bottom has been lowered at the point where a lateral drain enters. Development of the gullies is from the stream, landward, resulting in the loss of significant areas of adjacent land. Active gullies can be significant point sources of sediment. Riser pipe structures act as erosion control points, and in some cases as sediment traps, near the top bank of the stream allowing for the recovery of the lost land area. The structures can be designed as non-storage structures or as a temporary storage structures. The primary design considerations are (a) accurate run-off calculations; (b) proper elevation of the riser crest, dam, and emergency spillway; and (c) proper elevation of the outlet invert. Figure 6 is a basic diagram of a typical DEC riser pipe structure.

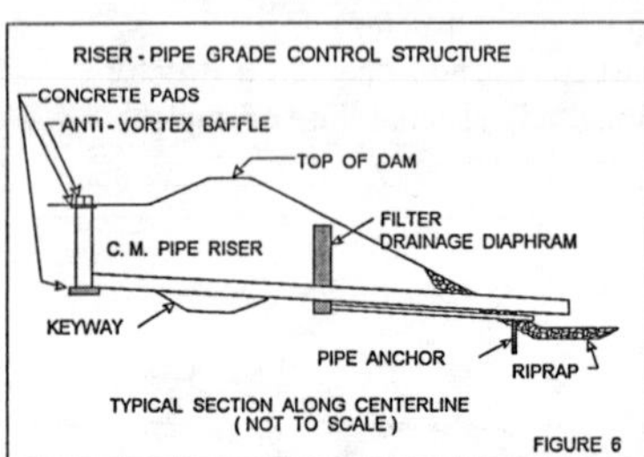

FIGURE 6

Appendix I - References

Little, W. C., and J. B. Murphy, 1982 "Model Study of Low Drop Grade Control Structures", Journal of Hydraulic Division, ASCE, Vol. 108, No. HY10, October, 1982, pp. 1132-1146.

U.S. Army Corps of Engineers, "Yazoo Basin, Mississippi, Demonstration Erosion Control Project", General Design Memorandum No. 54 (Reduced Scope)

Appendix II - Unit Conversion Factors

Multiply SI Unit	By	To Obtain English Unit
Meters	3.280840	Feet
Square Kilometers	0.386102	Square Miles
Hectares	2.471054	Acres

Channel Stabilization Using Box Culvert Grade Control Structures

Kelly B. Mendrop, P.E.[1]
Timothy J. Hubbard[2]
Michael D. Weiland, P.E.[3]

Abstract

The Demonstration Erosion Control (DEC) Project was initiated to address flooding and major sediment and erosion problems in areas of the Yazoo Basin in northwest Mississippi. A major watershed problem is bed and bank instability resulting from past and present channel degradation. This degradation has caused very incised channels in the middle and upper reaches of the watershed and the loss of channel capacity due to sedimentation in the lower reaches of the watershed. One technique used to address this channel instability aspect of the project is the use of Box Culvert Grade Control Structures. These features are a combination road culvert/grade control structure used when active headcuts are near existing road crossings. Up to 2.7 meters (9 ft) of vertical grade control has been established at various sites. The Box Culvert Grade Control Structures provide channel stability by alleviating channel degradation near the road crossing. These structures also provide bank stability upstream of the structure by arresting headcutting and reducing bank heights by raising the channel bed. These structures have been successful in reducing channel slope and providing bank stability in channel reaches in which they have been used.

----

[1]Hydraulic Engineer, U.S. Army Corps of Engineers, Vicksburg District, 2101 North Frontage Road, Vicksburg, MS 39180-5191

[2]Civil Engineering Technician, U.S. Army Corps of Engineers, Vicksburg District, 2101 North Frontage Road, Vicksburg, MS 39180-5191

[3]Civil Engineer, U.S. Army Corps of Engineers, Vicksburg District, 2101 North Frontage Road, Vicksburg, Ms 39180-5191

Hydraulic Design

The hydraulic design for these Box-Culvert Grade Control Structures are based on The U.S. Department of Agriculture's Handbook No. 301 titled Hydraulic Design of the Box-Inlet Drop Spillway. The box culverts referred to in this paper are considered a trapezoidal weir box inlet as referred to by Handbook No. 301. and presented in figure 1.

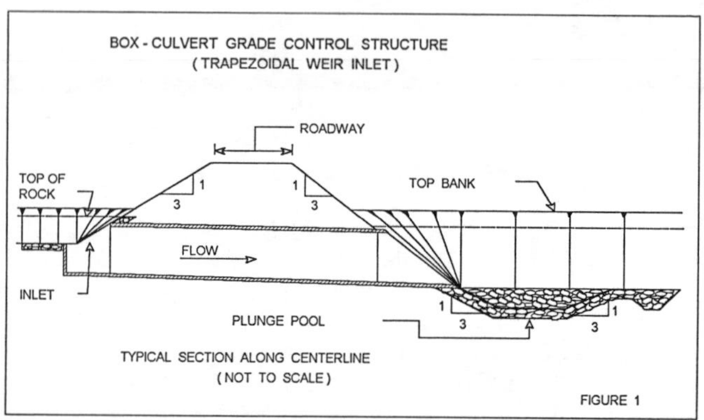

FIGURE 1

In the design of an individual structure or a series of structures, a HEC-2 deck is compiled to analyze the existing hydraulic conditions that comprise the site or reach. When the preliminary drop height is determined, a rating curve for the structure is calculated and input into the HEC-2 deck at the location of the structure. The structures dimensions can then be adjusted to maximize it's performance for channel and bank stability. This is done by analyzing the impact that the structure has on the hydraulic conditions of the unstable reach.

Case Study

In the DEC Project, five of the trapezoidal weir box inlet type grade control structures have been constructed as of January 1995. These culverts have ranged in size from a single cell 2.4 m x 2.4 m (8 ft x 8 ft) to a double cell 3.1 m x 3.7 m (10 ft x 12 ft). The drop height of these structures have ranged from 1.5 m to 2.7 m (5 ft to 9 ft) in height. The Beartail Creek Tributary is a very incised stream located in the headwaters of the Beartail Creek Watershed in North Mississippi. Some watershed and stream characteristics are presented in table 1.

Table 1
Watershed and Stream Characteristics

Drainage Area	5.9 sq. kilometers (2.3 sq. miles)
SCS Curve Num.	77
2-yr. Flow	19.8 cms (700 cfs)
100-yr. Flow	62.3 cms (2200 cfs)
Bank Height	6.1 meters (20 ft)
Channel Slope	.005 meter/meter

This incised stream had very active oversteepened zones (headcutting) occurring near two bridges in a reach approximately 1829 m (6000 ft) in length. These oversteepened reaches have channel slopes in the active headcutting areas as great as .008 (Figure 2). To alleviate this problem, two box culvert grade control structures were built in 1992/1993 at the road crossings where existing bridges were located. The two box culverts were both 3.6 m x 3.6 m (12 ft x 12 ft) single cell structures and have drop heights of 2.6 m and 2.1 m(8.5 ft and 7 ft).

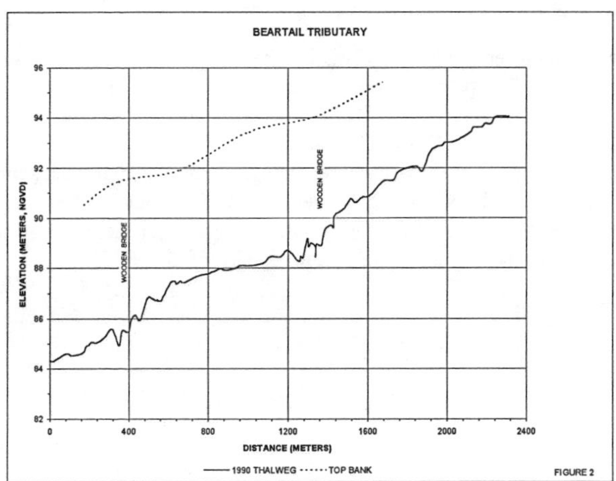

FIGURE 2

These two culverts have taken out 4.7 m (15 ft) of grade along with providing a new road crossing (figure 3). If these oversteepened zones would not have been in the vicinity of road crossings, other types of grade control would have been constructed. By building these two grade controls, the channel slope is expected to be reduced to a range between .002 to .003 which has been determined to be the range of stable slopes for this channel reach. Also, bank stability will be improved due to the reduction of bank heights in this area.

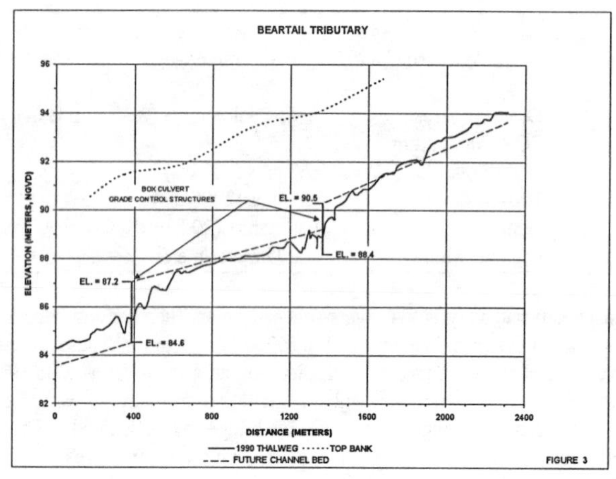

Conclusions

Many bridges and culverts in North Mississippi have been affected by past and present channel instabilities. Trapezoidal weir box grade control structures are being used throughout North Mississippi in the DEC Project to address this problem of channel stability. Channel bed and banks are being stabilized with grade control at road crossings as well as providing major infrastructure improvements. An intensive monitoring program is in place by the Vicksburg District Corps of Engineers to monitor channel responses to these structures and the entire system responce for these and other channel and watershed stability measures which are a part of the DEC Project in North Mississippi.

References

U.S. Department of Agriculture, "Hydraulic Design of the Box-Inlet Drop Spillway, Agriculture Handbook No 301."

U.S. Army Corps of Engineers, "Yazoo Basin, Mississippi, Demonstration Erosion Control Project", Coldwater River Watershed, Supplement H to General Design Memorandum No. 54.

STREAM CLASSIFICATION

by

Chester C. Watson [1], F. ASCE, Brian L. Van Zanten [2],M.ASCE, and Steven R. Abt, F.ASCE [3]

ABSTRACT

Stream classification is an essential element in transferring knowledge and experience pertaining to channel design from location to location. A computer program has been developed to record a comprehensive data set for a watershed and for channel sites, and to present alternative classification of each based on three classification systems: Schumm (1977), Rosgen (1994), or Buffington and Montgomery (1993). A goal of the program is to develop understanding between groups who are most familiar with only one or two of the classification systems compared. Examples of the application of the program will be presented from a variety of locations.

INTRODUCTION

Stream classification can be defined as a systematic arrangement of stream types into groups according to an established criteria. The criteria may include habitat value, channel order, channel pattern, geomorphic processes active within the channel, sediment type, and many other factors, depending on the importance of particular attributes to the person developing the classification system. To the Colorado cattle rancher, the most important stream attribute may be the determination of whether the stream is wet or dry, which may be of little value to the transportation engineer who must design a bridge for long-term stability. Unfortunately, differences in selecting which stream attribute should be the most important to individuals of varying backgrounds can lead to misunderstanding.

The primary objective of this investigation was to attempt to develop a cross-reference system for three channel classification systems: Schumm (1963 and 1977), Rosgen (1993) and Montgomery and Buffington (1993). Thorne (1993) developed a set of stream reconnaissance data sheets which, when properly completed, provide a thorough description of a stream and watershed based on geomorphic characteristics. Using the data sheets as a basis for data input and the three stream classifications to delineate stream types, a program was written (Van Zanten, 1995) in Visual Basic to facilitate data input, analyses, and objective channel classification.

---

1. Associate Professor, Dept. of Civil Engr., Colorado State University, Ft. Collins, CO 80523
2. Civil Engineer, Ayres Associates, Fort Collins, CO 80527
3. Prof. and Director, Hydraulics Lab, Dept of Civil Engr., Colorado State University, Ft. Collins, CO 80523

STREAM CLASSIFICATION SYSTEM

Schumm (1977) developed a classification system based primarily on the modes of sediment transport, divided into three classes: suspended load, mixed load, and bed load. This system was primarily developed from field work on the Great Plains of the western United States. Most of the data is for relatively flat streams, less than 2%, with sand bed material. Rosgen (1993) developed a classification based on morphological features that can be measured in the field. Although widely applied through the U.S., much of his early work was in the western, mountainous forest region. The Rosgen system includes a wide range of slope and materials. The Rosgen stream classes are combinations of alpha and numeric symbols. Montgomery and Buffington (1993) developed a complex and comprehensive process-based classification system. The criteria used to define channel types in each of the three channel classification systems analyzed are presented in Table 1.

Table 1. Stream Classification System Criteria

System	Criteria
Schumm (1963, 1977)	Width/Depth Ratio
	Sinuosity
	Slope
	Channel Perimeter % Silt/Clay (M)
	Bedload Percentage
Rosgen (1985, 1993)	Width/Depth Ratio
	Sinuosity
	Slope
	Entrenchment Ratio
	Dominant Channel Bed Material
Montgomery and Buffington (1993)	Typical Bed Material
	Dominant Roughness Elements
	Dominant Sediment Sources
	Slope
	Confinement
	Pool Spacing

Schumm's system can be used to define three channel types (suspended load, mixed load, bed load) and experienced geomorphologists can assess stability of these three classes. Montgomery and Buffington define eight channel types. Rosgen's system, by contrast, defines 41 channel types and has been recently been expanded into 94 separate classifications.

The criteria used by Montgomery and Buffington are sophisticated; for example, the dominant oughness element can be grain roughness, any one of several alluvial bed forms, large woody debris, r channel planform roughness. Determination of the dominant roughness element is difficult to bserve in any stream of significant depth, and changes in the bed form are a function of bed material ze, flow velocity, depth, and other factors. Considerable analysis would be required for use of the ystem, and, although the system has great potential, only a few of the criteria have been sufficiently uantified for use in an objective program. Criteria for the Schumm and Rosgen systems are uantified and can be used without subjective consideration. Schumm and Rosgen use the same three f the five variables required for classification. Slope and bed material size are both in the Rosgen nd the Montgomery and Buffington systems.

NALYSIS AND RESULTS

Seven stream reaches were selected and data for these reaches were compiled. The computer rogram, CLASSIFY, (Van Zanten, 1995) was set to select a stream classification for each stream each if 80% or more of the criteria defined in each classification system could be met. If more than ne stream classification scored equally at the 80% level, then all classifications of equal score were elected. If no stream classification scored at least 80% in a classification system, then the program elected NEC (...not enough criteria). An 80% level was chosen because many streams do not meet 00% of the criteria, and an 80% match represents a realistic degree of application accuracy. Table provides the results of the classification tests.

Table 2. Results of Stream Reach Classification Tests

tream Classification:	Schumm	Rosgen	Mont & Buff
ee Creek	suspended load	A5a+, A5, G5c	NEC
ed Banks Creek	mixed load	A5a+, A5, G5c	NEC
biaca	mixed load	B5c, C5, DA5,F1, F2, F3, F4, F5, F5b,F6, G5c	NEC
oila	mixed load	DA5, F5	NEC
oose Creek (natural reach)	NEC	B3	NEC
oose Creek (restored reach)	NEC	F3	NEC
oose Creek (degraded reach)	mixed load	F3	NEC

Unfortunately, none of the seven reaches met 80% of the criteria to be classified in th Montgomery and Buffington system. The Schumm classification system worked well for th primarily sand reaches of Lee, Red Banks, Abiaca, and Coila Creeks, which are located in northe Mississippi, and met 80% of the criteria for the Goose Creek degraded reach that is affected by fin grain mining debris. The Rosgen system gives clear, single type classification for the reaches c Goose Creek, which is a trout stream in southern Colorado with gravel and cobble bed material. Th multitude of classification types that met 80% of the Rosgen criteria for the sand bed reaches is tc confusing, and is caused by the overlapping of class definition for a single criteria. For example, slo of 2% to 4% fits approximately 44% of the total classification types.

CONCLUSIONS AND RECOMMENDATIONS

Additional effort to quantify the Montgomery and Buffington criteria is needed, ar modification to the CLASSIFY program is required to incorporate those criteria. The Schum criteria have been suitably quantified and the classification system works well for sand bed stream which are the streams from which the system was originally developed. The Rosgen system wor well for the mountainous, cobble and gravel streams, which are the streams for which the system w originally developed. CLASSIFY has been demonstrated to cross reference stream classification

With either 41 channel types of the earlier Rosgen classification or the 90+ types of Rosg (1993), perhaps consideration should be given to reducing the number of channel types. For exampl the A5 channel appears to be a sand bed stream of 10% or greater slope, which is unlikely to occu certainly not for any significant period of time. The Rosgen system is comprehensive, easi quantified, and is based on measurements easily made in the field. Therefore, the Rosgen system a pragmatic choice for additional development using the CLASSIFY program. Relating Rosg classification types to the fluvial process dominating the channel system would raise the value of tl Rosgen system to the value of the Schumm system for sand bed streams.

REFERENCES

Montgomery, D.R., and Buffington, J.M., (1993). "Channel Classification, Prediction of Chann Response, and Assessment of Channel Condition", Report prepared for the SHAM committee of the Washington State Timber/Fish/Wildlife Agreement, 107 pp.

Rosgen, David L., (1993). "A Classification of Natural Rivers", Short Course on Strea Classification and Applications, Wildland Hydrology, Pagosa Springs, Colorado, pp. C C118.

Schumm, Stanley A., (1963). "A Tentative Classification of Alluvial River Channels", U. Geological Survey Circular 477, 10 p.

Schumm, Stanley A., (1977). *The Fluvial System,* Wiley and Sons, New York, 338 p.

Van Zanten, Brian L., (1995). "An Expert System for Channel Classification", M.S. Thesi Colorado State University, Ft. Collins, CO.

Sediment Yield Reduction Study--Hotophia Creek Watershed

R.H. Smith[1], B.E. Johnson[2], S.S. Sunkara[3], and R.L. Hunt[3]

Abstract

The Demonstration Erosion Control (DEC) Project provides for the development of a system for control of sediment, erosion, and flooding in the foothills area of the Yazoo Basin, Mississippi. Structural features used in developing rehabilitation plans for the DEC watersheds include high-drop grade control structures similar to the Soil Conservation Service (SCS) Type-C structure, and low-drop grade control structures similar to the Agricultural Research Service (ARS) low-drop structure. To assist in the evaluation of the performance of these drop structures, the Hydraulics Laboratory (HL) of the U.S. Army Engineers Waterways Experiment Station (WES) initiated a comprehensive monitoring program in July, 1991 for selected sites representative of the different DEC watersheds. The purpose of this study on Hotophia Creek watershed is to develop and compare various hydrologic and hydraulic modeling techniques for estimating runoff and sediment yield from rainfall events.

Introduction

Hotophia Creek is located east of Batesville, and west of Oxford, Mississippi. One of the selected sites (No. 13) by WES encompasses about two miles between the confluence of Marcum and Deer Creek tributaries. A low-drop structure on Hotophia Creek is at the downstream extent, two low-drop structures are on Deer Creek, a low-drop structure is located on Marcum Creek, and a high-drop structure is located on Hotophia Creek just downstream of

----

[1]Associate Professor and [3]Graduate Research Assistant, respectively, Civil Engineering Department, University of Memphis, Memphis, TN 38152.
[2]Research Hydraulic Engineer, U.S. Army Engineer Waterways Experiment Station, Hydraulics Laboratory, 3909 Halls Ferry Road, Vicksburg, MS 39180-6199

the Marcum Creek confluence. This monitoring site is important due to the complexity of the various constructed elements and the need to document and evaluate channel response to the high-drop structure. The high-drop structure was the first one constructed by the Corps of Engineers in the DEC Program and was placed into operation in the fall of 1991. It was designed to pass the estimated 100-year flow of 210 cms. It has been instrumented with recording water-surface gages upstream, downstream, and at the downstream end of the stilling basin well. Channel thalweg and cross section (at 0.8 kilometer interval) surveys are available for Hotophia Creek for 1977, 1985, and 1992. Semiannual detailed surveys for the selected DEC monitoring sites began December 1991. Two additional high-drops, one within the selected monitoring reach and one upstream of the reach, were under construction at the time of the 1993 surveys. The watershed area at the downstream end of Site No. 13 is approximately 44 square kilometers. A USGS gaging station (watershed about 91 square kilometers) is located at Highway 35 near the mouth of Hotophia Creek.

Hydrologic Models

Three different rainfall-runoff computer models are being utilized in this study for comparison of lumped vs. distributed hydrologic modeling techniques. These included a new two-dimensional and spatially distributed model (CASC2D) which is based upon square grid cells and integrated to use GRASS GIS data. The Corps of Engineers' HEC-1 model and the University of Memphis' synthetic sediment routing model (GISSRM) are traditional lumped sub-basin parameter models based upon unit hydrograph techniques. Each of these three models have channel routing routines that are based upon a composite or representative cross section for the reach. Results of previous application of these three models to Goodwin Creek Watershed, a nearby basin extensively gaged for measurement of rainfall, streamflow, and sediment load, are also being presented in companion papers (Johnson, et al 1995) (Smith, et al 1995) at this conference.

Sediment Models

The GISSRM model results will be compared to two other concurrent study results: one by Colorado State University (CSU) using the Corps of Engineer model (SAM), and one by Mobile Boundary Hydraulics Consulting firm using the Corps of Engineers model (HEC-6). These three hydraulic models are currently being used to estimate the sediment yield from the watershed with the selected hydrology. Efforts to calculate the amount of sediment reduction due to the drop structures at the monitoring site will be made by setting up the watershed hydrology and channel hydraulic computer models with surveyed cross section data representative of before and after conditions.

Colorado State University (CSU) Study

A preliminary analysis (Watson, 1995) was conducted by CSU to estimate the sediment yield at the mouth of Hotophia Creek. Using the Corps' computer program for Stable Channel Design(SAM), composite cross section data were developed for each of the three surveys. The HEC-2 backwater program was used to get slope, velocity, width, and depth for each cross section. A sediment rating curve was then developed for the composite sections for each of the three surveys. Flow-duration relationships were developed using the USGS gaged data for both mean daily and 15-minute discharges. The relationships show the mean daily relationship under-predicts the higher discharges, which can result in under-prediction of the sediment yield and the effective discharge. Because the DEC streams typically have hydrographs that respond quickly, sometimes rising from normal conditions to peak discharge in only a few hours, the sediment yield was based upon the 15-minute discharge data and was calculated as 290 tonnes/day. By comparison, the sediment discharge based upon a 2-year discharge was computed to be about 60 times greater or approximately 17,470 tonnes/day. However, this 2-year discharge only occurs about 0.12% of the time, or 10 hours per year. Therefore, monitoring the performance of stabilization measures solely on the 2-year discharge was shown to be in error and underscored the necessity for developing more accurate and dependable watershed hydrology.

Although not conclusive, the preliminary results available indicate that the high-drop structures result in lower sediment yield (median = 173 tonnes/meter/day) than the low-drop structures (median = 865 tonnes/meter/day). The better performance may be because the low-drop structures generally lack hydraulic control for the effective discharges and many of the early ones were employed only to arrest advancing headcuts and no consideration was made for reducing the energy gradient.

Mobile Boundary Hydraulic Study

Sediment delivery from the Hotophia Creek Watershed into the Little Tallahatchie River is currently being analyzed by Thomas (Thomas 1995)using the HEC-6 model along with a 6.5 year streamflow records (1986-1992) at the USGS gaging station. Results of this study will be presented in companion paper at this conference. A preliminary analysis showed the bed material load at the mouth of the creek to be 512,000 tonnes (approx. 215 tonnes/day) with existing DEC projects. When the no-project condition was simulated, the yield was calculated as 595,000 tonnes (approx. 245 tonnes/day). At channel station 6,100 meters, which is downstream from all significant tributaries entering the main channel of Hotophia Creek, the bed material load reduction due to the DEC Project was estimated at 23 percent.

University of Memphis Study

At the present time, all hydrologic models have been developed and some preliminary analysis has been made. The models are being adjusted or calibrated based upon the streamflow and sediment data measured at the USGS gaging station and rainfall data from two nearby NWS gages with 15 minute intervals. Both the HEC-1 and GISSRM lumped models consist of 26 sub-basins while the CASC2D distributed model is based upon 150 meter grids. Channel surveys dated 1985 and 1992 are being used as a base for before and after conditions representing the DEC Grade Control Structures Project.

Summary

Definitive conclusion as to comparison of the accuracy and the relative ease of usage of each model can not be made at this time. However, preliminary comparison of the individual model results with the historical rainfall-runoff event data used for calibration purposes are favorable.

The GISSRM model will be used to estimate sediment yield for individual storm events, and after calibration is completed the various rainfall frequency events will be simulated to obtain an annual estimate of sediment delivery for conditions with and without the DEC Project. Comparisons will be made with the other two concurrent modeling studies using SAM and HEC-6.

Appendix-References

Johnson, B.E. et al, "Comparison of Distributive vs. Lumped Rainfall-Runoff Models on Goodwin Creek Watershed"

Smith, R.H., et al, "Grass Integrated Synthetic Sediment Routing Model (GISSRM)"

Thomas, W.A., "A Systematic Method for Spacing Grade Control Structures"

Watson, C.C., et al "Monitoring of DEC Drop Structures"

(All four papers listed above are concurrently being presented at this conference.)

HARLAND CREEK BENDWAY WEIR/WILLOW POST BANK STABILIZATION DEMONSTRATION PROJECT

David L. Derrick

Abstract

State-of-the art Bendway Weir and willow post method bank protection technology was applied to 14 eroding bends of a stream as part of the U. S. Army Corps of Engineers Demonstration Erosion Control program.

Introduction and Site Description

This paper presents the planning, design, construction, costs, and first year results obtained from a comprehensive bank stabilization \ habitat improvement project for an 3,570-meter (11,705 ft) reach of Harland Creek (a tributary of Black Creek) near Tchula, MS located within the Yazoo River drainage basin.

The test reach is approximately 29 m (95 ft) wide and average depth is approximately 2 m (6 ft) with wide variance on these dimensions. Bed and bank material ranges from silts and clays to gravel. The stream is of a "flashy" nature (i. e. stages rise and fall quickly during rainfall events). Surrounding land is forested. The stream is deeply incised, with the outer banks of all 14 bends in the reach being over-steepened and actively eroding. Bank height in the bends ranges from 1.5 to 13.4 m (5 to 44 ft), with most banks in the 3- to 6-m (10- to 20-ft) range. Existing vegetative cover on the banks ranged from bare soil to fairly dense vegetation (grasses, shrubs, small trees, etc.). Clay lenses, silts, sands, and gravel were observed in eroded the eroded banks. Bend radii ranged from 46 to 194 m (150 to 635 ft) while degree of curvature ranged from 0.52 to 3.11 radians (30 to 178 degrees).

Research Hydraulic Engineer, Waterways Division, Hydraulics Laboratory, US Army COE Waterways Experiment Station, 3909 Halls Ferry Road, Vicksburg, MS 39180

Aquatic Habitat Restoration

Stream corridor habitat enhancement/restoration was a top priority of this project. It was felt that the types of bank protection planned for this study would present a unique opportunity to demonstrate new channel stability methods while at the same time enhancing aquatic and terrestrial stream corridor habitat. The following habitat enhancing features were identified and, whenever possible, incorporated into the project: occasional deep pools, stable scour holes, stable habitat, diversity of habitat, solid substrate for invertebrates, well sorted stone size gradation, canopy cover, and woody debris.

History, Theory, and Design of Bendway Weirs

Since 1988 Bendway Weirs have been successfully tested in eight movable-bed models at WES and over fifty weirs have been built in 6 bends of the middle Mississippi River. The weirs are submerged upstream-angled stone sills (2,270 kg (5,000 lb) maximum weight rock) located within the navigation channel of the bend. Model and prototype results suggested that several observed benefits from the weirs could be beneficial to small-stream bank protection.

The Bendway Weirs employed in this study were designed to reduce erosion on the outer banks of the bends by reducing near bank velocities, reducing concentration of currents on the outer bank of the bend, moving the thalweg in the bend from the toe of the eroding bank to the stream end of the weirs, and producing an overall better current alignment through the bends and crossings. A number of complex factors went into determining the design parameters and layout of the Bendway Weirs. The following guidelines were used: all sets (fields) of Bendway Weirs were built in an upstream to downstream progression, all weirs were keyed into the bank using the standard Corps hardpoint key design, spacing was set at either 23 or 30.5 m (75 or 100 ft), weirs were sloped from an elevation 0.6 m (2 ft) above the bed at the stream end to 1.2 m (4-ft) above the bed at the bank end, crest width was set at 0.6 m (2-ft), side slopes were specified as the natural angle of repose (1 on 1 1/2), and all weirs and weir roots were constructed of R-650 stone (295 kg (650 pound) maximum weight stone). Maximum weir angle was set at 0.35 radians (20 degrees). In most cases the angles of the weirs at the upper and lower ends of the bends were reduced so that flow would be well aligned when entering or exiting the bend. This was based both on experience gained from the movable-bed model studies and consideration of the geometry (small-radius bends) of the test reach. Weir length was determined by engineering judgement, largely based on the anticipated relocation of the thalweg and the expected reshaping of the point bar.

The Willow Post Method and Design Guidelines

The modern willow post method, developed and used by Mr. Don Roseboom of the Illinois State Water Survey since 1986, is a means of controlling streambank erosion through the systematic vertical installation of large native willow cuttings to stabilize eroding streambanks. The stabilization process is multi-faceted. Willow foliage lowers floodwater velocities on and near the eroding bank; the root system of the willow binds the soil together; the post itself helps to stabilize the bank soil; transpiration aids in lowering the moisture content of the bank; and the resulting stable bank allows volunteer plant growth. Some of the advantages of the willow post method are: low costs, both in terms of materials and installation; maintenance costs are usually low; bank protection is long-term; and improved aquatic and terrestrial habitat is provided.

The design guidelines used in this study are a slightly modified version of the design advocated by Mr. Roseboom. They are: spacing set on a 1 m (3 ft) grid, i. e. 1 m (3 ft) between rows and 1 m (3 ft) between posts in each row; minimum post diameter was set at 76 mm (3 inches) at the butt end; minimum willow post length was 3.3 meters (10 ft); posts were planted at least 2.44 m (8 ft) deep using an 203 mm (8 inch) diameter auger with no more than 1.2 m (4 ft) of the post showing above the ground. Native willows in good condition were used with the elapsed time between cutting and planting not to exceed 48 hours. Willows were kept wet after cutting with the tops of the posts marked to insure posts were planted upright. **Willow posts must be planted when dormant.** Dormancy is defined as after the leaves have dropped and before the leaf buds have appeared (usually between December 1st and March 1st, dependant on location and weather). Willows were planted for the entire length of the outer bank of each bend with no willows planted above top bank or in the water.

Construction, Costs and First Year Results

The construction contract was awarded to Procon Inc. of Brandon, MS., a general contractor with some streambank protection experience but no experience with the protection methods employed in this study. The total contract amount was for $322,845.00. The average cost per meter over the entire reach was $90.43 (contract cost of $322,845.00 divided by the total length of protected bank, 3,570 m). This figure is approximately half of the cost of a typical DEC project. The construction contract bid was broken down as follows: $75,000.00 for willow post planting (divided by 9,383 posts for a per post cost of $7.99); $161,832.00 for construction of 54 Bendway Weirs and keys; $70,913.00 for 3 sections of longitudinal peaked stone toe protection and tiebacks, and miscellaneous costs of $15,100.00 (mobilization and demobilization, debris removal, erosion control).

The weirs and longitudinal peaked stone toe were constructed in 1993, while the willows were planted while dormant (Spring 1994). One bend consisted of willows behind longitudinal peaked stone toe, 4 bends were willow posts only, 6 bends were Bendway Weirs only, 2 bends were willows with Bendway Weirs, and one bend employed all three protection methods.

A comprehensive project evaluation was carried out in the fall of 1994. Since March 1994 the project has been subjected to the two year recurrence interval discharge about once every two months. None the less, initial results have been satisfactory, with most areas of the project appearing stable and maturing quickly.

The Bendway Weir reaches have appeared especially stable. As expected with any discontinuous bank protection some scalloping has occurred between weirs, but the bank slopes have readjusted and benched and native plants have quickly taken root. Included in these plants are a large amount of native willow trees. The biggest problem noted was some bank erosion and caving between the last two weirs and downstream of the last weir in all bends. This problem has cured itself in all but two of the bends where there is still some bank erosion downstream of the last weir. Thirty meters of longitudinal peaked stone toe protection downstream of the last weir would cure this problem, but at this time no fix is planned. All bends experienced the expected amount of erosion on the point bar with the channel thalweg moving to the stream ends of the weirs. A large storm event in January 1995 that overtopped the county bridges had a minimal effect on the project with most bends experiencing sand deposition on the outer banks of the bends. Within the weir fields many fairly deep stable pools have been formed. A most important habitat feature is the stability of woody debris trapped within the weir fields. Even with repeated out-of-bank events this debris has stayed in place.

Most willow post areas appear stable, but overall survival rates have proved to be less than expected. In most areas the row of willows closest to the water have been drowned or lost. According to Mr. Roseboom this is fairly typical. However, in some areas more than one row has been lost. The survival rate in June 1994 was 80%, but had dropped to 42% by October 1994. The reasons for the low survival rate are still being actively investigated. Several factors probably come into play, possibly including soil conditions, the fact that due to contract language the auger holes were not properly backfilled, and a large flood event three weeks after planting broke most branches of the initial flush. However, many of the willow posts that have appeared dead for months have recently sent up live shoots.

Capstone Course with a Writing Twist

by Arthur C. Miller[1]

Abstract

A unique combination is achieved by integrating the capstone course with the writing-across the curriculum requirement passed by the Penn State Faculty Senate. Instead of one capstone course incorporating many of the principals of civil engineering, which was initially contemplated, because of limited resources, seven capstone courses will be offered to encompass each of the civil engineering disciplines. The courses will be limited in enrollment, and the preliminary prognosis is one of success.

Introduction

The Penn State Civil and Environmental Engineering Department has chosen to have several capstone courses available for senior-level electives. Because of the large undergraduate enrollment, the Department graduates approximately 200 seniors throughout the academic year. The faculty determined that given the limited resources in the Department (approximately 22 full-time equivalent (FTE) facutly available to teach) offering a single capstone course would not be feasible. Therefore, each specialty area such as structures and hydrosystems, would offer its own capstone course. The example presented in this paper is the Capstone Course that is currently offered in the hydrosystems specialty area in the Spring Semester.

Background

On April 18, 1989, The Pennsylvania State University Faculty Senate established a Writing-Across-the-Curriculum graduation requirement, effective Summer Session 1990. This requirement applies to all students entering the

---

[1] Professor of Civil Engineering, The Pennsylvania State University, University Park, PA 16802

University as baccalaureate degree candidates. Students are required to complete at least three semester credits of writing-intensive courses selected from W courses that are offered within their major or college of enrollment. For engineering majors, this is in addition to the required Technical Writing Course. Writing-intensive instruction, as proposed by the Faculty Senate, could either be in courses that are designed to be taught only as writing-intensive courses, or in courses or sections that include intensive writing for for a portion of a particular semester.

Courses were proposed by the various departments and the contents of the writing-intensive courses and grading policies were reviewed by the University Writing Subcommittee. This committee made recommendations to the Senate Committee on Curricular Affairs as to whether a course met the Senate requirement.

Faculty in the Department of Civil and Environmental Engineering Department voted in 1991 to include the writing component into each of the capstone courses. Therefore, each capstone course offered in the Department has been designated with a 'W'. This identifies to the students as well as prospective employers that a significant writing component is required in that course.

Hydrosystems Capstone Course

The capstone course in hydrosystems is a design course for seniors in the Civil Engineering major who have an interest in water resources. The course fulfills three semester credits of technical electives as required by the major and the emphasis that is placed on technical writing satisfies the University's "writing-across-the-curriculum" requirement. The content also satisfies three credits of ABET design criteria.

The established Senate guidelines require that the writing assignments be integrated into the course in such a manner that they are related to the overall course objectives. That is, writing assignments are used as instruments for learning the subject matter. Within the hydrosystems area, the capstone course builds upon the material presented in other undergraduate courses: Fluid Mechanics, Hydrology, Applied Hydraulics, and Water and Wastewater Transport Systems.

The capstone course deals with the hydrologic and hydraulic design of a water supply reservoir and integrates the writing assignments into a design project. The course requires writing assignments that give the students the opportunity to practice writing and receive responses from the instructor throughout the course. This means multiple assignments and a series of preparatory writings for the final project. The student writing assignments are clearly outlined and the type of writing to be used is identified in the course syllabus. The syllabus allows for some course time for such activities as writing proposals for major projects, writing and revising drafts, and peer critique student drafts.

The writing assignments are evaluated by the instructor, and the writing quality is a significant portion of the class grade. For the capstone course that I have been teaching, I have hired an English teacher with five years experience as a technical writer for a large civil engineering firm to assist me in grading.

The English Technical Writing course, which is required by the College of Engineering, is a prerequisite for the capstone course. The writing assignments in the capstone course include:

(1) A proposal letter soliciting the project (Solicitation letter to borough council for consulting work)

(2) A written memorandum to document progress of the project to date, including the preliminary requirements for water supply storage and addressing the hydrologic safety criteria for the impoundment reservoir (summary of the hydrologic portion of the project)

(3) A report on the preliminary design of the hydraulic structure. (This phase of the project includes changes made to the hydrologic analysis and a resubmittal of the hydrologic phase incorporating any corrections.)

(4) Documentation of one of the computer programs written as part of the design project. Several computer programs are written throughout the semester and one of these must be documented as part of the writing assignment. This is resubmitted with the final project.

(5) A final report incorporating comments from all of the preliminary drafts. (Final designs will be completed and all pertinent corrections are made to this report.)

Project Description

The project is divided into two parts: Part I is the hydrologic portion and Part II is the hydraulic design portion. Part I begins with a water supply demand study. This includes the population prediction for the water authority's distribution area for the design year, and the estimated water demand for the predicted population. Next, the firm yield of the watershed must be determined. A regional analysis is for rainfall-runoff simulation and the firm yield of the watershed is determined from that study. The storage capability of the reservoir site is then determined and the minimum reservoir elevation to contain the needed supply is resolved. Low flow augmentation for stream habitat is also considered as part of this portion. The students are then introduced to a hydrologic model, in this case, the U.S. Army Corps of Engineers Hydrologic Engineering Center's HEC-1 model. Watershed characteristics are determined and the model is quasi-calibrated. Sensitivity analysis is conducted to determine which watershed characteristics have the greatest effect on the flood flows. The probable maximum precipitation is then determined and the probable maximum flood is predicted: this is later used to size the hydraulic conveyance system.

Part II of the project is the actual design of the hydraulic structures. First, an ogee spillway is designed based on the minimum pool elevation that is needed for storage. Next, a computer model is written to compute spatially varied flow in a side channel spillway. To accomplish the backwater computations of the downstream transitions, the U.S. Army Corps of Engineers Hydrologic Engineering Center's Water Surface Profile Program, HEC-2, is utilized. The flow is passed through the transition section to a chute spillway. Again, a computer program is written to solve for gradually varied supercritical flow. At this point, the HEC-2 program is once again utilized. Tailwater computations are performed to determine downstream effects of the stream channel on any type of energy dissipator that would be needed at the toe of the dam. A stilling basin is then designed based on the water depth from the tailwater computations and the depth of flow calculated from the chute spillway program.

Last, but not least, the National Weather Service's DAMBRK model is used for simulating a dam breach analysis. The failure time of the breach is estimated, as well as the breach opening. The downstream floodwave is calculated.

Integration of Design in the Civil Engineering Curriculum

Michael E. Mulvihill P.E., M. ASCE
Joseph C. Reichenberger[1] P.E., M.ASCE

Abstract

The focus of the Department of Civil Engineering at Loyola Marymount University for the past two years has been on the integration of design through the curriculum. This integration commences in the freshman creative problem solving course, continues with design experiences incorporated in sophomore & junior courses and culminates with comprehensive design projects in senior required and elective courses. A design methodology and description of design experiences are presented.

Introduction

Entry level civil engineers must be educated to enter the civil engineering profession in a variety of positions. They will work for regulatory and public works agencies, educational institutions, private engineering consultants, and industry. These different career paths all have elements of design. Regulators must be able to evaluate designs and suggest alternatives. Researchers need to be creative in designing pilot and bench scale equipment and tests. Those in the public sector review designs. Knowledge of the design process is key to understanding the evolution of the submitted design. Engineers in private practice must be flexible and creative to ensure that their client is afforded a reliable, cost effective solution. Engineers involved in design must be able to deal with open-ended problems with minimal constraints and propose alternative methods of solution.

Civil engineers must be able to perform economic analyses and identify the environmental, political and social factors that control the project and impact its acceptability and implementability. They must be able to organize their calculations to come to sound conclusions and be able to present the findings in written and oral form.

---

[1] Professor and Chairman, and Associate Professor respectively, Department of Civil Engineering and Environmental Science, Loyola Marymount University,Loyola Blvd. @ W. 80th St., Los Angeles, CA 90045-2699.

Once the project concepts are accepted and the solution is selected, the civil engineer must execute the detailed design, complete with plans, specifications, cost estimate, operation and maintenance manuals, and other documentation.

Based on the above description, the Civil Engineering Department at Loyola Marymount University (LMU) developed a model for civil engineering design. (See Figure 1.) This model forms the framework for the design experiences in our curriculum.

Design Experiences in the First Three Years

Our freshman introductory engineering course includes creative problem solving and mini-design projects. In two open-ended group projects students design and construct something that performs a task. The first is to make something "useful" out of materials given to each group. This tests their creativity. The second project requires design, construction and testing of a simple device. Each project requires a formal presentation by each group member.

In the second semester, groups of three or four students evaluate the future engineering needs for Southern California. This open-ended project allows the students to use their creativity, explore and evaluate alternative solutions and expand their view of the engineering profession. Each group develops a schedule, including milestones for submittal of draft reports for review and comment. The project concludes with formal presentations and a final report.

In the sophomore year, students in Statics design a roof truss using RISA-2D (a frame modeling computer program). Students, given the size, materials and methods of calculating loads, configure and optimize a roof truss. Students make a model of the truss, give an oral presentation and prepare a written report.

In Surveying students work in groups designing a horizontal highway curve. With the use of computer software, they test several possible solutions to arrive at an optimal design.

Students in the Dynamics class design and create a device, system, or instrument for directly or indirectly measuring the components of acceleration of a body in curvilinear motion. The device may be actual operating hardware or simulated hardware using the software "Working Model."

In the junior year, design projects are required in several classes. In Thermodynamics, students use computer-aided design software for design optimization of a prototype device considering economics and aesthetics, then construct and competitively test it.

The detailed design of a 150 foot long bridge is assigned in Mechanics of Materials. Students design the bridge trusses, floor beams, and stringers and produce a structural report. RISA-2D, along with manual calculations for verification, are used.

Figure 1

Civil Engineering Design Methodology Steps and Constraints

STEP	CONSTRAINTS
1. Identification of client REQUIREMENTS.	Societal and environmental impact, client master plans, client operator sophistication and capabilities, costs, regulations and codes, schedule.
2. RESEARCH existing information.	Availability of records and operating reports.
3. Establish DESIGN CRITERIA.	Technical feasibility, client preferences, regulations and codes, budget.
4. Develop ALTERNATIVES.	The process of generating alternatives should not be constrained, except for the design criteria.
5. Secure Client CONCURRENCE.	Client's willingness to move forward and commit funding for evaluation.
6. Select and Analyze FEASIBLE ALTERNATIVES.	Societal and environmental impact, reliability, costs, institutional implementability, public acceptance, schedule, permits and approvals.
7. Select the BEST ALTERNATIVE.	Ditto.
8. ANALYZE and OPTIMIZE the design.	Ditto.
9. Prepare DETAILED DESIGN.	Materials and equipment availability, safety, codes and standards, reliability, maintainability, operability, durability, efficiency, functionality, cost.
10. CONSTRUCT facility.	Constructor experience, labor and material availability, competition, schedule, regulatory.
11. Facility START-UP AND OPERATION.	Operator experience and training, maintenance budget, management support, equipment performance.
12. Continued technical SUPPORT.	Client requirements and budget.

The detailed design of a moment resistant frame is assigned in Structural Theory. Students, given a two-dimensional building frame and design loads, develop an efficient design for the frame. A finite element structural analysis program, SAP90, is used for the design.

In Fluid Mechanics II students design a pipe-pump system including development of system curves and select pumps based on an economic evaluation.

Design Experiences in the Senior Year

By the senior year students have developed an appreciation of the engineering design process. Senior level courses build upon this foundation through a number of different design projects.

Students in Environmental Engineering are required to design a sanitary sewer system for a small city, given a street layout, topography, land use information and hydraulic constraints. They work in groups and start with the initial flow estimates and pipeline route selections finally culminating in a formal design report.

Students in Analytical Methods II determine the least cost flood control reservoir spillway and downstream channel given the design flood, storage curves, outflow criteria and cost information. Several open-ended problems which require the development of computer programs are also assigned. These include discharge under a falling head with and without inflow, seepage beneath a dam, and development of a probabilistic model to determine an inflow forecast for the real time operation of a dam for hydroelectric power and water supply.

In Soil Mechanics students investigate, research, evaluate and discuss the impacts of water on the failure of engineering structures and embankments. This provides the students with a perspective on the uncertainty in designs and the potential consequences of those uncertainties. In the laboratory students work in small project teams analyzing different soils to determine their suitability for a specific project.

Students in Hydraulics I may work in pairs or on their own designing a flood control system. Students have the option to solve an existing flood problem or to develop the solution to a hypothetical problem using HYDOC, a knowledge-based computer system. A formal design report including calculations, hydraulics characteristics of each channel reach and hydraulic plan and profile is submitted.

In Water Resources Planning and Design students integrate fluid mechanics, hydraulics, statistics and economics into three design projects involving a stormwater retention basin using using unit hydrograph procedures, an urban storm drainage system using the rational method and local intensity-duration curves, and a water supply and distribution system for a small community using industry-standard computer models.

Students in Reinforced Concrete Design initially prepare a design of a reinforced concrete beam, given the beam length, a structural (and resulting cost) efficiency requirement, and information on design loads. Students then fabricate the beam and load it to failure in the laboratory. Later students develop a design for a reinforced concrete retaining wall combining their knowledge of the design of one-way slabs and wall footings with soil mechanics.

Conclusion

The objective of the civil engineering design sequence at LMU is to produce entry-level engineers who can transition quickly and smoothly into the particular design framework of the organizations for which they work. The entire civil engineering curriculum is formulated to develop graduates with the breadth of understanding, motivation and vision to become highly productive professional engineers capable of becoming leaders in their profession.

A Decade of Capstone Design

Orville E. Wheeler, Fellow, A.S.C.E.[1]
Jerry L. Anderson, Member, A.S.C.E.

Abstract

The University of Memphis introduced a senior design course in the Civil Engineering Department in the fall 1985, semester. That the basic form and intent of the course has stood the test of time reflects the foresight and planning that went into its original implementation. The course's reception by students, and their follow-up comments on course evaluations, have helped validate its importance in the curriculum. A wide range of projects, both individual and team, have been selected and performed by students. Consideration of ethical matters and professionalism are primary components of the course.

Introduction

A required senior design course was added to the curriculum of the Civil Engineering Department for the fall semester of 1985. The course was modelled, in some measure, on an earlier course that was taught as a construction management course in the Department. The earlier course was operated as a construction company office, with bidding, scheduling, etc., simulated by the students using design drawings from local firms. The construction course was felt to be very successful and a decision was made to broaden the experience and make it a required course for all Civil Engineering majors. The course was reviewed by the E.A.C. of A.B.E.T. in 1986 and again in 1992, as part of the periodic accreditation visits. It has grown in recognition and importance by both faculty and students as a measure of overall preparation for the job market.

The original instructor for the course was the (then) Department Chairman who introduced it. He developed the syllabus and established the general pattern for the course that still makes up most of the curriculum today. After teaching the

---

[1]Herff Professor of Structural Mechanics and Associate Professor of Civil Engineering, Civil Engineering Department, The University of Memphis, Memphis, TN 38152

course twice a year for seven years, he suggested that a change would be good for all and the first author took over the course. It is anticipated that a cycle of no more than 6 years will become standard for instructors for the course. Initially, students were allowed to register for the course prior to their final semester but experience has shown this to be a bad idea, consequently, the enrollment is now rigorously limited to students in their final semester.

Purpose

The purpose of the course is stated to the students as follows:

This course is a "capstone" course required to be taken in the final semester of the student's degree program and intended to be an experience which will unify some of the diverse subjects studied in the course of a Civil Engineering undergraduate curriculum. It is aimed at helping the student apply engineering fundamentals, science, design principles, and common sense to practical design problems. It is also intended to be a very small exposure to the environment awaiting the graduate entering the engineering work force. Students will be expected to exercise judgment in choosing between alternatives as they apply quantitative as well as qualitative decision criteria. Team design will be required as part of one of the projects.

Students successfully completing this course will be able to perform analyses of alternate solutions to a design problem, prepare a detailed design of the recommended alternate, present economic justification for the recommended design, and present both a written and oral summary of the design process and results in sufficient detail to be convincing to a knowledgeable audience. Use of the computer as a tool in these design efforts is expected.

This course is considered a "writing intensive" course under the General Education requirements at The University of Memphis. Students are expected to submit all non-problem assignments in typewritten form, preferably using a word processor. All assignments will be graded for style, grammar, spelling, clarity, etc., in addition to engineering content. Occasionally, it may be necessary to require re-submission of assignments.

All students will be evaluated by the instructor as indicated in the syllabus. In addition, a jury of practicing engineers, faculty and other students will be invited to participate in the oral presentations to assist in evaluation.

Students will be required to complete both an individual design project and a portion of a team design project. The individual design project will be proposed by the student and approved by the instructor. A broad range of design topics (within the civil engineering field) is encouraged.

Syllabus

The course presentation consists of 26 one hour lecture sessions, 14 two hour lab sessions, 2 one hour quizzes, and a 2 hour final exam. Mixed into the lecture and lab sessions, each student is required to speak in front of the class (and sometimes others) eight times, presenting progress and final reports. In addition there are eleven required written submittals, including the progress and final reports on both individual and team projects. The only text used in the course is A.S.C.E. Manuls and Reports on Engineering Practice No. 73, "Quality in the Constructed Project." Each student is given a copy of this manual by the department.

The lecture sessions are divided between discussion of design paradigms, optimization, design goals, etc., on the one hand and ethics and professionalism on the other. The six hours spent on ethics include distinguishing between personal and professional ethics. We want to encourage serious thought and study of the former while emphasizing that the latter are externally defined and even in some cases, such as the Tennessee Rules of Professional Conduct, have the force of law. The very fine N.S.P.E. video "Gilbane Gold" is used as a vehicle for discussion of professional ethics and responsibilities. It is cast in such an open ended manner that students are exposed to the difficulties of making ethical decisions even when a published code of professional ethics is at hand.

Several hours are devoted to discussing the work place and conditions the students will encounter there. Many of our students already work or have work experience but these discussions have still been very well received by them. Observations in industry (2 decades of experience) suggest that the most difficult idea for new engineers to understand is that making a good effort is not enough, work must be completed and completed correctly in a timely fashion. The course is run on this basis, no late work is accepted, no written material is accepted expect that printed on a laser printer in a good format with a professional appearence, etc. We try to instill some of the behavior that will have to become second nature in a professional environment.

Projects

The most visible result of the course is, of course, the student's design projects. The students are required to propose both an individual and team project. The proposals are reviewed and either approved or returned to the students for re-writing. The range of topics is interesting, some of the titles for projects completed in the last three years are:

"Redesign of the Poplar-Goodlett-Central Intersection"
"Feasibility of Converting a Grist Mill to Produce Hydroelectric Power"
"Design of a Private Grass Airfield"
"Design of a Small Zoo Aquarium"

"Industrial Waste Water Pre-Treatment System"
"Artificial Wetland as a Polishing Unit"
"Design of a Hydroelectric Dam"
"Crane Bay Bridge"
"Leachate Treatment System"
"Pirogue Design"
"Residential Drainage Improvements"
"Water at My Door: A Preventive Measure to Flooding"
"Design of a Wastewater Effluent Conveyance System for Millington, TN"
"Design of a Timber Pavilion"
"Temporary Roof Design"

Most of these projects deal with an existing problem, perhaps 10% are fictional. Approximately 45% of the projects deal with water or wastewater, 30% with structures and the balance with traffic engineering, site planning, or other topics.

Students are required to complete an individual project to expose them to all of the steps in recognizing a problem, conceiving of a solution, and carrying out the design to effect that solution.

The team (or group) projects are more extensive in scope and intended to provide an experience in cooperative design. The class is divided into groups (usually two) that then form themselves into companies for the duration of the project with an internal structure they select themselves. Our recent practice has been to have a competition between groups to add motivation, the winning group receives a letter grade increment to their group project grade. Recent project titles are:

"Shady Grove Road at Sweetbrier Creek Bridge Replacement"
"Culvert Replacement at Appling and Southern Way in Barlett, TN"
"Design of a Wastewater Treatment Plant for Chickasaw, TN"
"The Development of a Commercial Site Plan"
"Drainage Structure Rehabilitation"
"Feasibility Study and Preliminary Design for a Pedestrian Bridge on the Memphis State Campus"
"Proposed Design of a Small Dam and Lake"

Conclusion

The Department is pleased with the senior design course and its results, most faculty attend all of the open presentations as do many other students. A possible improvement for the course that is being discussed is team teaching. It is felt that the more view points the students can be exposed to, the better. Student comments on the course include a recommendation to make this a two semester course, but thus far, we have not found a way to make room in the curriculum for the expansion. At present, we plan to continue it as it is.

Planning and Managing a Multi-Disciplinary Capstone Project

Brian L. Kruchoski[1]

Abstract

This paper describes a senior capstone design project conducted by the Civil and Environmental Engineering Department at Villanova University. In this six credit hour, two semester course, teams of students addressed a multi-disciplinary design problem.

Introduction

Recent changes to the ABET criteria for accrediting civil engineering programs prompted a change at Villanova University from individual senior projects to a team-oriented capstone project. ABET specifies that the curriculum "must include a meaningful, major engineering design experience", and highly recommends team design projects (ABET 1994). The civil and environmental engineering faculty agreed that a project which draws on two or three of the specialty disciplines within civil engineering would offer the greatest benefit to the forty-five seniors in the department. The project selected was a dam for water supply and flood control on the Brandywine River in suburban Philadelphia.

Faculty and Student Staffing

Three full-time professors were assigned to the course to act as faculty supervisors. They received credit for a three hour load, and were assigned for both semesters to provide continuity. These faculty represented the fields of water resources, environmental engineering, and structures. The specialty areas of available faculty members dictated the choice of the project topic. Other faculty in

[1]Assistant Professor, Dept. of Civil & Environ. Engineering, Villanova University, 800 Lancaster Ave, Villanova, PA 19085-1681

the department were available on an as-need basis during their regular office hours.

Students enrolled in the department select design electives in their junior and senior years from the above specialty areas, as well as from geotechnical and transportation engineering. Students enrolled in the project were assigned to teams so that each team had at least one member from each of the specialty areas chosen for the project.

For the first semester, teams of four students were formed. Since the second semester involved more intensive technical work, the teams were reformed with five students in each. These new teams were staffed so that no member had previously worked with any other member. Each team elected a leader at the beginning of the term; a new team leader was elected at midterm.

Course Elements

In the first semester, each team addressed the same issues at the same level of detail. At midterm, teams submitted a written proposal to conduct a feasibility study. The proposal included technical discussions, time and cost estimates for engineering services, and Gantt charts. The feasibility study submitted at the end of the semester addressed siting and sizing of the dam, local geology, hydrologic aspects, water quality, benefit-cost ratio, and other topics.

In the second semester, each team performed a detailed preliminary design focused on two of the three specialty areas. The two areas assigned to a given team were chosen to reflect the course backgrounds of the members. Second semester tasks related to water resources included: floodplain and stormwater management calculations, reservoir design, analysis of downstream effects of a dam failure, sediment control calculations, and design of water distribution systems. Environmental engineering topics included: public water supply projections, process selection, plant layout, site selection, process sizing, and preparing an environmental impact statement. The structures component of the project included design of the following: concrete spillway, outlet control tower, treatment plant building, pumphouses, and treatment basins. Other topics addressed by all the design teams included: geological studies, removal/relocation of highways, railroads, and structures, construction cost estimate, and benefit-cost ratio.

At the conclusion of the project, teams submitted an extensive design report, calculations, plans, and specifications. Each team made a 45 minute oral presentation to the faculty, fellow students, and the interested public. Each team member participated equally in the preparation and execution of the presentation.

In keeping with the intent of the ABET requirements, both semesters of the capstone project also included significant exposure to teamwork, project management, open-ended/iterative problems, communication skills, and operating under multiple constraints.

Course Organization

Both semesters of the project course were scheduled for two meetings per week: one hour on Tuesday and two hours on Friday afternoon. Although an unpopular time with both students and faculty, Friday afternoon was chosen to accommodate field trips and the work schedules of some outside guest speakers.

Tuesday sessions were used in both semesters for formal lectures by the three faculty supervisors. Faculty presentations in the first semester centered on the design process, managing technical people, and estimating costs and benefits. Lectures were also given on selected topics in the three specialty areas of the project. In the following semester, design specialty lectures and question/answer meetings were held concurrently for the three areas; teams sent a representative to each of these sessions.

Tuesday class meetings were also used by guest speakers, as were some Friday meetings. Guest lecturers were from government and private industry, and spoke on such topics as: ethics, professional liability, dam design, geology, wetlands, environmental impact assessment, regulatory issues, and project management.

Each team had a biweekly progress meeting with their assigned faculty supervisor. At these Friday meetings, the team presented a written and oral report; responsibility for the oral presentation was rotated among team members. Technical problems were also resolved at these meetings. Alternate Fridays were used for question/answer sessions with the assembled teams. Teams also had access to the three supervisors during their regular office hours for problems that needed to be addressed immediately.

Two organized field trips were conducted in the first semester. Visits were made to a flood control dam and to a water treatment plant. The treatment plant was revisited in the following semester, to allow the students to study details relevant to their project. Design teams also visited potential dam sites on their own.

Grading

Course components which were graded in the first semester consisted of the written and oral progress reports (biweekly), the proposal (submitted at midterm), the feasibility study (submitted at end of semester), and project notebooks (submitted at midterm and at end of semester). In the following semester, the design calculations, plans, specifications, and design report were graded at the end of the course; the final oral presentation was also graded. As in the previous semester, biweekly progress reports and project notebooks contributed to the final grade. All members of a team were initially assigned the same grade. Each student conducted a written peer evaluation for all team members; this information, along with attendance and professional conduct, was used to adjust individual grades upward or downward.

Other Considerations

Because of the style and range of the project, and because of the extensive faculty involvement required, many schools will find that it is not feasible to offer a "trailer" section of a capstone project course. Enrollment in the course is contingent upon satisfying several prerequisite and corequisite design courses. It is therefore essential that students are aware of these constraints, and have planned their schedules accordingly.

Some data and materials must be made available to the students by the department (e.g., quadrangle maps, stream gage data, population data); these need to be ordered well before they are needed. Also, arrangements for guest speakers and field trips must be made far in advance, and contingency plans should be in place.

A final consideration concerns faculty effort. Each faculty supervisor received credit for a three hour course load; however, the effort expended easily exceeded the level required for a traditional three hour lecture course. This is especially true the first time the capstone project is offered. Extensive, cooperative, advanced planning is essential to the success of the course.

Conclusions

A capstone design course in civil engineering must be well planned and carefully executed if it is to meet the needs of the students, satisfy the requirements of ABET, utilize the talents of the faculty, and recognize the constraints involved. The course conducted at Villanova University and described in this paper seems to have fulfilled these objectives.

References

Criteria for Accrediting Programs in Engineering in the United States. (1994). Engineering Accreditation Commission Accreditation Board for Engineering Technology, Inc., New York, N.Y.

The Capstone CE Design Course - an Undergraduate College Experience

James R. Groves[1]

Abstract

Several years of experience in planning and teaching the required capstone design course in a small undergraduate college is described. The recommended model is three or four separate design problems involving real-world experiences of the professors. Students work in groups in environmental, structural, and land-use sections, and present both oral and written design reports.

Introduction

The Virginia Military Institute (VMI) is an undergraduate college for 1300 men in rural southwest Virginia. About one-third are engineering majors, with about 200 students in the ABET-accredited four-year civil engineering (CE) program. About thirty-five students are enrolled each spring in the 3 credit-hour capstone design course. The nine full-time CE faculty are all registered Professional Engineers: such registration is required for promotion and tenure. About one-third of the VMI graduates are commissioned in one of the military services upon graduation, but graduate surveys indicate that less than one-sixth go on to military careers.

VMI developed a capstone design course in the 1980s when the requirement first appeared in the ABET criteria. Several models have been tried and revised over the years. The current course is well received by the students, and was praised by the visitor in a recent (September 1994) ABET visit.

---

[1]Professor and Head, Department of Civil and Environmental Engineering, Virginia Military Institute, Lexington, VA 24450

Integrated Design Teams

The first VMI capstone course in 1988 permitted students to designate themselves as environmental, structural, or site planning specialists. Instruction was accomplished in one section for each specialty, with design teams composed of one student from each section. The problems using this organizational model for the first few years were small buildings, such as layout and design of a printing plant or a fast-food facility. The design teams were assigned the same problem, and oral presentations were given within each section, and then to the other sections.

Practitioner involvement was obtained by soliciting a nearby consulting engineering firm to provide the design problem, make presentations to each section, and assist in evaluating the student work. While this process was conceptually sound, scheduling problems and the relative remoteness of the VMI campus prevented the practitioners from extensive participation.

The three-man "interdisciplinary" design team organization was not altogether successful, largely due to sequential effort required in the project planning. The structural and environmental members of the team could accomplish little until the land use work was completed. The instructors were also not always completely comfortable in evaluating work within specialties other than their own.

One Project in Specialty Teams

In 1990 the organization of the course was changed. The same three sections were retained, but cadets worked within their own section on one major specialty-specific design effort for the entire semester. As before, both oral and written reports brought the course to closure.

In many respects this approach was found to be an improvement, but interest and enthusiasm lagged over the long semester. Students required considerable prodding to keep them from procrastinating and waiting to do all their work at the last minute. Some students had little innate interest in the particular project selected.

Multiple Projects

The next, and current, revision of the course resulted from the desire to provide additional variety to the design effort. Students complete three or four smaller problems in discipline-specific three-man teams. Leadership of the teams rotates to ensure that each student contributes.

This model has served to maintain active student participation. By spreading the deadlines for graded work over the semester, there has been less reliance upon last minute effort, particularly as the end of the semester and graduation approach.

Professors, each of whom have experience in practice, have required project reports in the form of technical memoranda similar to those used in many consulting firms. As in the earlier course models, emphasis is placed on group effort and oral and written communications skills.

Practitioner involvement, strongly recommended by the accreditation criteria (ABET 1994), is obtained by using real- world problems, often introduced by "clients" or practicing professionals, who also participate in evaluating the student efforts. Students have designed facilities for several college departments and local public service agencies.

Conclusion

The current VMI capstone design course has evolved through the efforts of the faculty to provide the best possible practice-oriented design experience for their students. A model of several short design projects has proven to be both successful and popular with the small (10 to 16 students) section sizes employed at VMI, and should be successful at other small colleges.

Reference

Accreditation Board for Engineering and Technology (1994), General Criteria for Accrediting Programs in Engineering in the United States, 1994-95 Cycle.

Appendix- Extract of accreditation criteria (ABET 1994)

IV.C.#.d.(3)(d) Each educational program must include a meaningful, major engineering design experience that builds upon the fundamental concepts of mathematics, basic sciences, the humanities and social sciences, engineering topics, and communications skills. The scope of the design experience within the program should match the requirements of practice within that discipline. The major design experience should be taught in section sizes that are small enough to allow interaction between teacher and student. This does not imply that all design work must be dome in isolation by individual students; team efforts are encouraged where appropriate. Design cannot be taught in one course; it is an experience that must grow with the

student's development. A meaningful, major design experience means that, at some point when the student's academic development is nearly complete, there should be a design experience that both focuses the student's attention on professional practice and is drawn from past course work. Inevitably, this means a course, or a project, or a thesis that focuses upon design. "Meaningful" implies that the design experience is significant within the student's major and that it draws upon previous course work, but not necessarily upon every course taken by the student.
****

Program Criteria for Civil and Similarly Named Engineering Programs

2.b. ...A majority of those faculty teaching courses which are primarily design in content must be registered Professional Engineers.

3.b. ...Since the civil engineering design process generally involves a team approach, team design projects are highly recommended. The final design experience should include practitioner involvement whenever appropriate and possible. Student reports and presentations should be an integrated part of the final design experience.

Electronic Journals: ASCE's Present and Future Plans

John Pape[1] and Carol Reese[2]

Abstract

ASCE publishes 22 technical and professional journals which appear in print and electronic formats. This is the first year that ASCE journals are offered in CD ROM format. The research, development, vision, features and advantages of **ASCE Journals on CD ROM** are examined with particular emphasis on the corporate and university library viewpoint and implications of technological developments. ASCE determined that information specialists held a positive view of CD ROM products if they provided value-added capabilities. Rapidly changing computer technology has enabled ASCE to create a product with many requested features.

The American Society of Civil Engineers enjoys one of the oldest and most prestigious reputations for technical publishing. ASCE publications serve as a clearinghouse or forum for the exchange of up-to-date technical and professional knowledge, issues and ideas. In 1867, then-President James P. Kirkwood stressed the importance of the dissemination of papers as essential to ASCE's "continuous and successful existence." ASCE's publications program began that year with the initial appearance of **Transactions**. In 1896, the monthly **Proceedings** (the forerunner of ASCE journals) began publication.

In 1995, the ASCE publications program includes 22 technical and professional journals, the monthly magazine **Civil Engineering**, the newsletter **Emerging Technology**, several information retrieval publications, an on-line bibliographic database, the monthly newspaper **ASCE News**, and 60-70 new books each year. From the first seven papers published in 1867, the program has grown to include over 5,000 papers, articles, books, and news items in 1994.

In 1992, increasing recognition of the importance of electronic formats in the publishing community led ASCE staff to study the relative merits of the CD ROM format and its potential in relation to ASCE's currrent publications. The format was examined for technical character, historical and current usage, suitability for technical publications, market acceptability, and the desirability of developing a proprietary ASCE product.

---

[1] Manager, Marketing Services, American Society of Civil Engineers
[2] Manager, Information Products, American Society of Civil Engineers, 345 E. 47th St., NY, NY 10017-2398

Staff compiled and reviewed information on products available from other sci/tech publishers and societies, results of other industry marketing studies for such products, cost and income implications associated with various types of CD ROM products; executed a survey of the nonmember/library civil engineering marketplace, conducted a focus group on the topic with current customers; attended educational and commercial events, and made inquiries to knowledgeable industry figures.

From general sources of information, a broad picture of the CD ROM field emerged. The advantages to the publisher and librarian include the ability to repackage, store and manipulate a vast amount of information -- one disc can hold over 600 megabytes of information. To the librarian, space, control, and ease of use are premium assets, with obvious advantages. Many practicing engineers utilize advanced computer programs and many students prefer searching literature using the latest technology. Publishers can select sophisticated software to limit information manipulation or permit a multitude of custom approaches -- adding value to material. Unlike on-line electronic information, discs can be shipped, stored and monitored with ease. From the librarian's perspective, the CD ROM can be used over and over without incurring unforeseen on-line charges. It retains many of the advantages of print publications without requiring the same level of manual effort to index, store, monitor and retrieve. According to published sources, the sales numbers of CD ROM readers (hardware) took almost ten years to reach 1,000,000 in the US (1991). By the end of 1992, the number had doubled to 2,000,000 and industry estimates are that 40% (or more) of all new PCs sold in 1995 will be equipped with readers. Other reports indicate that information centers, business users and scientists in Japan prefer this form to other types of electronic delivery. In the US, current price of a CD ROM reader is well within the budget of most organizations. Although industry reports indicate that librarians view CD ROM technology with "guarded enthusiasm," college and university libraries reported large increases in spending for CD ROM products. This attests to the strength of the mandate to rapidly transform the traditional academic library to a 21st century information center.

In order to gather specific information related to civil engineering, ASCE developed a survey and mailed it in April 1993, to 1,566 domestic and foreign recipients of which 744 were members of the engineering and transportation divisions of the Special Libraries Association and the remainder were ASCE member and nonmember subscribers to All ASCE journals. The 182 respondents indicated that ASCE's university and corporate market is heavily involved -- 2/3 of respondents have access -- and those that aren't involved are converting at a rapid rate. 50% of respondents anticipated adding additional CD ROM equipment within the next year. About 50% of those responding devote more than 5% of current budgets to CD ROM and those with current CD ROM budgets plan to increase budget allocations, one by as much as 25% in one year. Larger organizations appear to be the most committed to CD ROM usage with about 70% of respondents reporting more than 25 users. Individual usage appears to be fairly heavy with about 70% using separate titles more than 5 times per week. Ironically, usage of on-line databases has generally decreased in those organizations with CD ROM titles while requests for document delivery have increased. Few available engineering titles restricted collection; 97 out of 105 respondents hold fewer than 10 engineering titles and 82 of 104 respondents would consider adding to their holdings if product was available. As further support of the findings, a June, 1993 computer usage study by Hagen Marketing Research surveyed 1,000 principals of civil engineering firms for ASCE. Of the 59% who responded, 22.6% indicated that their firms currently have CD ROM capability with an additional 19.7% planning to purchase or replace CD ROM drives.

A focus group of civil engineering librarians which met June 7, 1993 at the Special Libraries Association Convention in Cincinnati helped to confirm the responses to the written survey. In addition, the information specialists' insight helped us formulate a vision of an ASCE product, specifically the potential of journals. A product which simply presents papers as they appear in the publications will not be sufficient for the needs of the user. The added value must be in the ability to search through bibliographic or text information to locate the desired reference and then recall the exact information in the paper text. An ideal system should be sophisticated for "expert" use, but simple so that it can be easily understood and manipulated by the novice. Search protocols which are now familiar to engineers and librarians should be used if possible.

As first envisioned, the ideal ASCE CD ROM product uses search strategies to locate desired information for the publications contained on the disc; can bring to the screen the exact image of the printed document; can hot link to access the first page and all subsequent pages of a document which is the result of a search; can be menu driven or command driven; can lock out access to nonsubscribers; has print and download capability; can zoom in and zoom out; defaults to a Table of Contents screen.

In early 1994, Online Computer Systems, Inc. (a Reed/Elsevier company) was selected to develop **ASCE Journals on CD ROM**. Computer technology moves rapidly, resulting in several changes to the original plan. Almost overnight, windows-based programs became the standard for software applications. After several years, the Standard Generalized Markup Language (SGML) became the standard for coding text for digital electronic applications. Hypertext applications enhance the ability to rapidly access charts, graphs and illustrations.

ASCE Journals on CD ROM

Features

Full-text Searching	Windows-based
Hypertext Links	Software Included
Mathematics	Flexible Ordering
All Graphics	Quarterly Updates
Zoom In/Out	Postscript Printing

The resulting **ASCE Journals on CD ROM** represent the best current use of the format for the nature of the technical information encompassed by the journals. Discs, issued quarterly, utilize the familiar window-based screen format. All journals' contents appear on every disc, but access is through a special customer code number which allows the subscriber access only to those journals which have been authorized. Through the subscription year, information will be cumulative with each disc superseding the previous disc. The final disc of the year contains the full contents of the year.

The sophisticated search engine, developed by Online Computer Systems with advanced OPTI-WARE software technology, executes precise word searches, boolean logic operations, and truncation. Search by keyword, author, title, or subject. Cull terms from an index of all words. Scope a search to any combination of journals, issues, or papers. A bookmark feature tags important papers, sections of text, or illustrations. Browse through the table of content or go directly to graphics. Enlarge an illustration by 200% or print the full paper, text and graphics. Possible uses and applications are limited only by the imagination of the reader.

ASCE Journals on CD ROM position ASCE for future electronic developments. With the information digitally coded, any electronic format, which evolves from CD ROM or on-line deliverabilty, becomes possible. With user acceptability, CD ROM could prove to be a desirable alternative format for other ASCE publications such as proceedings, **Civil Engineering** magazine, **ASCE News**, or **Transactions**. Projects designed specifically for the format, such as multimedia texts, could be developed. Internet could provide additional access, promotional opportunities or new complementary electronic formats.

With continued electronic developments, the nature, organization, and application of the information can dictate the best available presentation. Whatever the format, ASCE will continue its role as the foremost disseminator of professional and technical information to the civil engineering community.

Resources:

"CD-ROM Commentaries: The Myth and Reality of Costliness, Or How Much Is Too Much," Information Today, September, 1993
"The CD-ROM Imperative," Folio, March 15, 1994
The CD-ROM Market in Canadian Libraries, (Report) Meckler, 1990
"CD-ROM Marketing: A Case Study," Scholarly Publishing, January, 1994
Construction Engineering Computerization, (Report), 1994
"The Changing Scene in Journal Publishing," Publishers Weekly, May 31, 1993
"Copyright and Licensing in the Electronic Environment," Scholarly Publishing Today, March/April 1993
The Faxon Planning Report, 1993
"GEOROM--57 Years of Geophysics on CD-ROM," The Leading Edge, March, 1993
"How to Market Subs to a Magazine on CD-ROM," Circulation Management, July/August, 1993
"Market Testing – Tips for Successful CD-ROM Product Development," CD-ROM Professional, May, 1993
"The Online Journal Of Current Clinical Trials: An Innovation in Electronic Journal Publishing," Database, February, 1993
"Powering Up," SLA Research Series #5, 1990
"Profit and Loss in CD Publishing," CD-ROM World, October, 1993
"Publishers Deliver Reams of Data on CDs," The Wall Street Journal, Feb. 22, 1993

Developing an Information Server: UWIN

Faye Anderson[1] and Greg A. Wade[2]

Abstract

This paper addresses the development of a particular information server available on the Internet, the Universities Water Information Network (UWIN). It also addresses several key points to consider when developing any publicly accessible information server.

Introduction

The explosive growth of the Internet has transformed the way we think about information and its dissemination. We are in the midst of reconsidering how we can best communicate with our colleagues, share data and research findings, distribute our journals, manage our professional organizations, and advance our areas of expertise. Shrinking both space and time, the Internet is opening up access to tremendous amounts of information. Who has access to the Internet and the quality of its information resources are emerging concerns.

Historical Background

The idea for UWIN came from the Board of Directors of the Universities Council on Water Resources (UCOWR) and was spearheaded by Gen. Gerald G. Galloway. The United States Geological Survey (USGS) funded the pilot project in 1992 as part of its then National Water Information Clearinghouse. The USGS supplied the initial hardware and began providing annual funding for the network. UWIN was housed at UCOWR Headquarters on the campus of Southern Illinois

---

[1]Water Information Specialist, UWIN, 4543 Faner Hall, Southern Illinois University, Carbondale, IL 62901-4526

[2]Systems Administrator, UWIN, 4543 Faner Hall, Southern Illinois University, Carbondale, IL 62901-4526

University at Carbondale. Two half-time staff, a computer scientist and a water information specialist, were hired. Initially, UWIN was accessible to potential users by telnet or gopher clients on the Internet, as well as by modem. UWIN's mission was seen as bringing water resources to the information superhighway and the superhighway to water resources professionals.

UWIN's Informational Resources

UWIN went on-line in August, 1993. Since that time it has grown both in response to its stated goals and to the needs of its users. UWIN's resources include:

USGS WRSIC Research Abstracts. The USGS provided their WRSIC database for inclusion on UWIN. This database contains abstracts of water resources research since 1967. UWIN staff indexed this database and it is fully searchable using user-provided keywords.

Expert Directory. UCOWR previously maintained a small listing of water resources experts for its own use. UWIN staff updated and expanded this database from a few hundred names to over 2500. Each expert listing includes contact information and their individual areas of expertise. Particular attention has been paid to collecting international water experts for the database.

Water Organizations. This file incorporates organizational information from various water groups. The National Institutes for Water Resources (NIWR) listing includes their publications directory, public information directory, and state institute expert listings. This resource also provides UCOWR's organizational news and a directory of graduate education programs in water resources.

Calendar of Water Events. The Calendar gives information on water-related meetings, conferences, workshops, etc.

WaterWiser. This is the Water Efficiency Clearinghouse administered by the American Water Works Association. It disseminates information on all the various aspects of water conservation and water efficiency.

Employment Opportunities. Jobs, assistantships, fellowships, etc. are listed here to help centralize the exchange of information related to career development in water resources.

Press Releases. Any water-related group can send their press releases and other timely announcements to UWIN for posting.

WaterTalk, UWIN's Bulletin Board System

Over the course of the first eight months on-line, several UWIN users suggested the need for a bulletin board system. Some water related bulletin board systems existed, but most were only accessible via modem. WaterTalk went on-line in April, 1994. WaterTalk is reached by a telnet link from UWIN. It is a series of boards that discuss various water topics. Currently, we have discussions on Hydrology, GIS, International water issues, Education, Groundwater quality, and Water policy. Users can suggest new boards if these do not suit their needs.

WaterTalk has many features including e-mail, private talk, on-line conferencing, etc. All postings to WaterTalk are archived, with full searching capabilities, on UWIN.

Technical Aspects of UWIN

UWIN supplies information via two primary servers, a Data General Aviion and a Sun Microsystems SPARC classic. These machines both run versions of the Unix operating system. Because of its stability, Unix is an ideal platform for providing Internet services such as gopher and the World Wide Web (WWW).

These services are provided at UWIN by utilizing gopher server software created at the University of Minnesota and WWW server software created by the National Center for Supercomputing Applications. Both of these servers are then interfaced through C and Perl programming to a Wide Area Information Server (WAIS) to provide searching over UWIN's databases. In addition to WAIS, the WRSIC database uses a custom package of routines created by the UWIN staff to efficiently index the database and to retrieve single abstracts for on-line presentation from over 400 megabytes of data.

Many other programs were either created or modified at UWIN to provide extra features to our users, and to aid the UWIN staff in maintaining our data. Examples of this include, the WWW interface to WaterTalk, and a dBase IV application used in our office to maintain the expertise directory.

To see this technology in action, UWIN can be accessed by: pointing your gopher client at gopher.uwin.siu.edu; accessing http://www.uwin.siu.edu with your WWW browser; or dialing either (618) 453-3324 or (618) 453-3090 with your modem. Modem users should set their communications parameters to 8N1 and be able to emulate a vt100 or ANSI terminal.

Some Considerations in Developing an Information Server

Clearly identify your purpose. Your purpose in constructing and maintaining a server on the Internet should be well thought out. Make sure to "surf" the Internet to investigate what others are doing related to your purposes. Develop your niche carefully. The proliferation of servers is frustrating to users if they contain little in the way of useful information.

Identify your potential users. What audience will be interested in your server and how will you reach them? Consider the types of information that these users might find helpful and consider if you can meet these needs. Also, consider the technical sophistication of this group - will they need educating about the Internet. UWIN's purpose was well-served by utilizing the various water professional organizations to reach its targeted audience. The water institutes (NIWR) were also very helpful.

Build on other's efforts. The Internet is a collaborative place. When developing your system, see what is already out there that is suited to your needs or can be modified. This eliminates wasted effort. Regularly look around the Internet for new ideas and features that other servers have. We built WaterTalk by modifying software that was freely available from an anonymous ftp site. Similarly, we found the "UWIN Top Ten List" idea and software on another server.

Information and Data Collection Efforts. It can take considerable time to collect information for your directories (and perhaps a good deal of word processing as well). Are these efforts best done in-house or can they be done by an external group? Carefully considering what information, how it will be presented, and how it will be updated helps eliminate wasted effort in data collection.

System Testing. Make the effort to thoroughly test your system and any new features you add. The UWIN staff has found it helpful to ask our users to aid us in this process. The multitude of software and hardware in use makes it difficult to know if things work as well for others as they do for us.

Technological Advances. A difficult, yet interesting, task is trying to stay on top of all the technological advances occurring. These advances influence both what you can provide to your users and how your users interact with your services. For example, the WRSIC database on UWIN appeared to work flawlessly with the many gopher clients we have in our office. After having the database on-line for almost a year, we had to make minor changes to our programs to allow searching from some of the new gopher+ clients.

Importance of User Feedback. We cannot stress the importance of user feedback in maintaining a server. The "UWIN Comment Box" provides a direct opportunity for users to ask questions, make suggestions, and tell us what we could be doing better. These have been invaluable in finding "bugs" in the system and letting us know what users think as they are utilizing the network.

Conclusion

This paper has provided an overview of our development of the UWIN system and some of the lessons we have learned since going on-line. We hope these are useful and provide some issues to consider in undertaking such a project. It is an exciting time to provide on-line services and the rate of change is phenomenal. Careful attention to these informational issues will help improve the quality of all our Internet experiences.

WEAP: A Tool for Sustainable Water Resources Planning in the Border Region

Paul Kirshen, Paul Raskin, and Evan Hansen[1]

Abstract

WEAP, the Water Evaluation and Planning System, is a menu-driven microcomputer program designed to assist policy makers in evaluating water supply policies and developing sustainable water resource plans. WEAP provides an integrated picture of a supply and demand system, which can include rivers, reservoirs, and groundwater as water supply sources, and offstream and instream flow requirements as water demands. WEAP operates on the basic principle of water balance accounting. By modeling, in an integrated framework, present and proposed actions on the supply side and demand side, the analyst can study the behavior of the total system. WEAP can be applied to single or multiple interconnected river systems at the city, regional or national level to implement most water development evaluation tasks. WEAP is a planning laboratory for exploring development futures.

Background

Under conditions of economic development, limited natural resources, and environmental pressure, many countries and regions are facing formidable resource management challenges. In particular, the allocation of limited water resources has become an issue of increasing concern. Conventional supply-oriented water

---

[1] Water Program Manager, President, and Research Associate at Tellus Institute, 11 Arlington Street, Boston, MA 02116-3411.

development strategies are not always adequate. A new approach to water development has emerged in the last decade, emphasizing resource conservation, demand management, water use efficiency, and social, cultural, and environmental impacts of water resources development. WEAP aims to incorporate these emerging values into a practical planning tool.

Scenario analysis is central to WEAP. Scenarios are used to ask a broad range of "what if" questions, such as: What if the population or economy grows faster than expected? What if cropping patterns, irrigation efficiency or acreage changes? What if a new canal lining project is built? What if a reservoir is constructed or operated differently? What if groundwater is more fully utilized? What if environmental requirements are altered?

The Stockholm Environment Institute provided primary support for the development of WEAP. The Hydrologic Engineering Center of the US Army Corps of Engineers funded significant enhancements, and a number of agencies, including the World Bank, USAID, and the Global Infrastructure Fund of Japan have provided project support. Though new, WEAP has already been applied in water assessments in the United States, Mexico, China, Central Asia, and India.

The WEAP Approach

The analyst represents the system in terms of its various surface and groundwater sources; withdrawal, transmission, and wastewater treatment facilities; and water demands. WEAP uses an end-use, demand-driven approach to model these demands. To the degree that data permits, it builds up water requirements from a set of final uses. This requires disaggregating water demand into specific end-uses or "water services" in different economic sectors. For example, the agricultural sector could be broken down by crop type, irrigation district, and irrigation technique. An urban sector could be organized by county, city, and water district. The data structure and level of detail may be easily customized to meet the requirements of a particular analysis, and to reflect the limits imposed by restricted data. This approach places development objectives -- providing end-use goods and services -- at the foundation of water analysis.

WEAP operates on the basic principle of water balance accounting by tracking water flows through system components on a monthly basis. To account for hydrologic fluctuations, the analyst can enter gauged inflow data for a selected historical sequence of years, or can construct hypothetical scenarios to assess issues such as climate change.

WEAP applications generally include several steps, as illustrated in Figure 1. The *study definition* sets up the time frame, spatial boundary, system components, and configuration of the problem. The *current accounts* provide a snapshot of actual water demand, resources, and supplies for the system. Alternative sets of *future assumptions* are based on policies, costs, and factors which affect water demand, supply, and hydrology. The *base case* projection serves as a reference. *Scenarios* are constructed consisting of alternative sets of assumptions or policies. Finally, the scenarios are *evaluated* with regard to water sufficiency, costs, and benefits; compatibility with environmental targets; and sensitivity to uncertainty in key variables.

Modular Structure

WEAP consists of five linked programs, as shown in Figure 2. The **Setup** program characterizes the problem under study by defining the study area, time horizon, physical components of the system, and their spatial relationships. The **Demand** program accepts demographic, economic, and water requirement data and assumptions to generate current and future annual water demands for each demand site. An extremely flexible structure allows the user to define the level of demand detail and the projection methodology. The **Distribution** program converts annual water demands to monthly demands, characterizes the

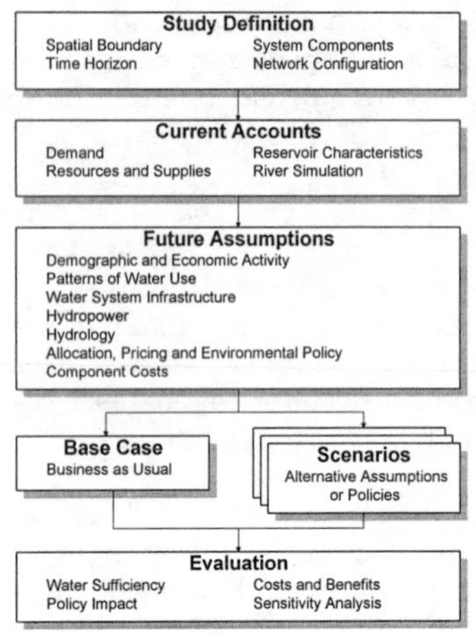

Figure 1: The WEAP Approach

distribution losses within each demand site, and defines the transmission characteristics between supply sources, demand sites, and wastewater treatment plants. The **Supply** program allocates water from supply sources to demand sites, and from demand sites to wastewater treatment plants and other return flow locations. A river simulation mode can track streamflows along a main river and its tributaries. The **Evaluation** program evaluates and compares scenarios in terms of physical demand and supply, environmental requirements, and economic costs and benefits.

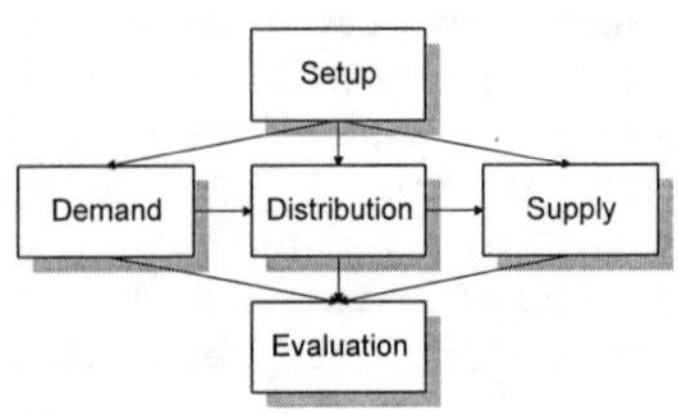

Figure 2: WEAP's Modular Structure

Water Quality Management in Mexico: Actual and Future Policies

Blanca Jiménez[1]

Abstract

In order to know the availability of water resources one must know the characterization of water bodies, the use given to the water and the quantity ecologically available, all three items in actual and future terms. This information permits to define the needs for infrastructure and policies for management and preservation. The project presented herein (Jiménez, 1995) is intended to establish a basis for an approach to the administration of water quality and is divided in three items: quantity, uses and quality of water in Mexico; water quality monitoring, and mexican legislation on water quality.

Quantity, Uses and Quality

At present Mexico has 5,125 m^3 of renewable water/inhabitant/year. According to the Scarcity Index (Falkenmark, 1989), the country is on a jeopardized availability level because of its temporal and spatial distribution and pollution problems. Not considering hydropower generation, the total annual extraction is 2,290 m^3/s, of which 77% is used in irrigation, 13% in industries and 10% in municipal uses. The ground water extraction amounts a 26% of the total, and it is used in irrigation (71%) and industry (23%).

Starting from a developed index, named Use Potential (UP), which indicates how many times the parameters of a water exceed the criteria for a given use, the water quality of the country was evaluated and maps were established on uses for irrigation, drinking water supply and protection for aquatic life. This UP index determines

---

[1]Titular Researcher, Instituto de Ingeniería, UNAM, Apdo. Postal 70-472, Coyoacán, México, D.F., 04510, México

which are the parameters that surpass the norms, what is their importance, what is the feasibility of control and also, the validity of the norm itself.

According with this information, the main source of pollution is the industrial sector, with a registered total of 382 m^3/s of waste waters discharged, that contributes with a variety of pollutants such as solids, organic matter, heavy metals, aliphatic and aromatic compounds, phenols, etc. Municipal discharges amount a total of 160 m^3/s and the main contribution to pollution is faecal coliforms, followed by organic matter and solids.

In order to correct this situation, the government agency in control of water quality, the National Water Commission or CNA for its initials in spanish, has established a national program on "Clean Water", by means of which the whole drinking water supply is being at least disinfected. Aside from this, there are 289 potabilization plants in operation and 18 under construction, which assure a 30% of municipal supply being appropriately treated.

With respect to municipal and industrial discharges, there are 952 waste water treatment plants in the country, 778 (82%) for domestic wastes and 174 (18%) for industrial wastes. With this infrastructure only a 57% of the total waste waters discharged are treated, and 62% of BOD, 59% of COD and 70% of TSS are removed from the domestic discharges, and 64% of BOD, 53% of COD and 62% of TSS are removed from the industrial discharges. Taking into account these three parameters, the information was classified according with level of treatment, and actual and required infrastructure in order to obtain maps that show this information in a comprehensive manner.

It is important to mention that in Mexico there is a wide experience in the field of reusing waste water for irrigation purposes, and nowadays this is being done under a strict control program to minimize the impact on the health of field workers and consumers.

Water Quality Monitoring

The National Monitoring Network (NMN) is in operation since 1974; has 793 monitoring sites and performs yearly 4,024 samplings per site and 114,523 analysis. However, actually is under restructuration, due to a lack of sufficient covering, and that the size, stratification and variability of each water body is not taken into account; also, some parameters measured do not correspond with the actual problems of the region monitored. The aims of this restructuration are to establish more precisely the quality of the water bodies of the country, to develop adequate policies for its improvement and to back feed and evaluate the sanitation programs and also the mexican legislation, under the principles of operation such as a design with a concept of "client oriented", and the use of parameters that are simple, good indicators of pollution effects, few but representative.

Mexican Legislation on Water Quality

In Mexico the legislation related with water quality is stated in 5 Laws, 33 Norms and 2 Regulations, and there are 3 government ministries directly involved in this matter (Secretariat of Agriculture and Hydraulic Resources, Secretariat of Fisheries and Environment, and Secretariat of Health) and 2 government agencies (National Water Commission and Federal Procuratory for the Environment). To this legislation it should be added all the international treatises and conventions and the agreements derived from the NAFTA.

The main problems related with the mexican legislation are: the lack of a defined goal on water quality for each water body of the country; the little importance given to the aquifers even though they are a very important supply of good quality water; the pollution due to non-punctual sources is practically not considered; the problem of pollution due to rainfall is not considered and the dilemma of using combined sewerage is not approached; the norms related to the use of waste waters in agriculture must be revised, for this is a common practice in our country; there are a number of agencies involved in water administration and this causes a great deal of inefficiency in dealing with this matter; there is no way of measuring the efectiveness of the legislation.

In order to solve these problems, the legislation is being revised and evaluated taking into account these basic principles: the instruments developed must be means for the government to evaluate, measure and establish in a quantitative basis the quality of the water bodies, and also how the management policies improve it; the legislation should be "measurable" in terms of analytical techniques, laboratory instruments and trained personnel to assure its compliance; it should guarantee the gradual improvement of water quality in a reasonable term; it should be easy to adequate according with the scientific and technological developments and advances so it can become gradually stringent; to establish a group of "basic" parameters, which should be obligatory to measure and comply with in the national level and a second group called "guidelines", which should express concentrations in terms of tolerance ranges and which should be measured and comply only in some regions of the country; to establish a more appropriate classification of water uses and to define the quality required for each one including realistic values for each parameter; this last item should be in accordance with the conditions and characteristics of the country and not based only in a bibliographic revision of foreign norms.

Conclusions

Without doubt, the most important item concerning water quality, monitoring and legislation, is to find an adequate procedure for surveillance, effective compliance of the legislation, and the evaluation of its impact, in order to back feed how is the quality of water being administered.

Lastly, the acquired experiences in our country, along with a wide number of analytical data, should permit us to establish an adequate framework for a realistic outlining of the administration of water resources in Mexico, and to effectively improve the water quality in the near future.

Appendix

Falkenmark, M. (1989) "The massive water scarcity now threatening Africa- Why isn't it being addressed?", *Ambio*, **18**(12):112-118.

Jiménez, B. (1995) "Bases para el manejo integral de la cantidad y calidad del agua en México ". Elaborado para la Comisión Nacional del Agua. Instituto de Ingeniería, UNAM, 120 pp.

Impacts of NAFTA on the Practice and Licensing of Civil Engineers

Mark W. Killgore, M. ASCE[1]

Abstract

This paper addresses the professional licensing and practice developments under the North American Free Trade Agreement (NAFTA). The negotiations for the transnational licensing of engineers under NAFTA are part of a rapidly evolving process which will likely conclude in 1995. Civil engineers who are licensed Professional Engineers (PE) in Canada, Mexico or the United States may be eligible for a temporary license to practice in either of the neighboring countries, provided that educational, experience and other requirements are met. Citizenship and/or residency requirements for the PE can no longer to be required under NAFTA. This paper summarizes the history and the on-going process of the negotiations and the status of the Mutual Recognition Statement of Professional/Licensed Engineers.

History of the Negotiations

Canada and the United States began negotiating principles for the mutual recognition of registered Professional Engineers and issued a statement to that effect on February 4, 1992. The basis for these discussions was the Canada/United States Free Trade Agreement (FTA) which preceded the NAFTA agreements and included provisions for the free exchange of services. The Canadian Council of Professional Engineers (CCPE) and United States Council for International Engineering Practice (USCIEP) were assigned the task of developing procedures to implement the FTA provisions regarding engineering services(1).

Extensive efforts were undertaken to document the current requirements in each nation for licensure and to identify differences. As progress continued with the North American Free Trade Agreement negotiations, preliminary meetings were held in Lubbock, Texas in June of 1992(2). Various Mexican and U.S. professional engineering societies participated in this first round table discussion. Subsequently, in November of 1992, at a second round table in Cuernavaca, discussions between

[1] Principal Engineer, Raytheon Infrastructure Services Incorporated, 10900 NE 8th Street, Bellevue, WA 98004-4405.

the Mexican College of Mechanical and Electrical Engineers and the Canadian Council of Professional Engineers established the concept of the NAFTA Forum on Engineering Registration and Practice (NFERP). Each nation would have six representatives on the Forum. Several Mexican associations formed the Mexican Council for International Engineering Practice (MCIEP) on April 21, 1993. At the first formal gathering of the Forum in June of 1993 in Austin, Texas, a memorandum of understanding was signed by the participants(3). The U.S. Council for International Engineering Practice (USCIEP) represents the United States and includes members of the Accreditation Board for Engineering and Technology(ABET), National Council of Examiners for Engineering and Surveying(NCEES), and National Society of Professional Engineers(NSPE).

The Memorandum of Agreement (MOA) summarizes current licensing practices in the three nations and commits the three councils to undertake consultation for the purpose of developing recommendations on standards and criteria for licensing or certification including education (accreditation of schools and academic programs), examinations, experience requirements for licensing and conduct, and ethics. In addition professional development, scope of practice, requirements for local knowledge and consumer protection issues are included.

The councils have agreed to develop recommendations by December 31, 1995. Recommendations will address procedures for temporary licensing of engineers, engineering specialties to be included in the procedures and other matters related to temporary licensing. The councils will determine a process for removing citizenship and permanent residency requirements and seek to encourage the competent authorities in all three nations to implement the recommendations within a mutually agreed time. Since the signing of the memorandum, several additional meetings have been held. Significant progress has been made by the parties towards greater uniformity. For example Mexico has established its Council for the Accreditation of the Teaching of Engineering, commonly referred to by its Spanish acronym, CACEI. Schools of higher education in Mexico can request accreditation by CACEI of their engineering curricula and syllabi. The General Assembly of CACEI includes primary representation by engineering associations and engineering schools. The Mexican Government and private sector have a much smaller role(2). Canada accredits its programs through the Canadian Engineering Accreditation Board of CCPE and the U.S. with ABET.

Requirements for Temporary P.E. License

The requirements for temporary licensing will be more stringent than for obtaining the original PE License. A record of progressively responsible experience beyond the PE may be incorporated in the final criteria to obtain a temporary license without further examination. Applicants will be required to demonstrate knowledge of codes, standards, laws and rules and regulations relating to the practice of engineering in jurisdictions where they apply. Competency in the language of commerce will also be required and applicants must be of good character (4). Several additional requirements and other details are included in issue papers which are updated on an ongoing basis.

Progress is also underway towards establishing a uniform code of ethics for engineers in the three nations and a special committee was formed in January 1994 for that purpose (2). At the same January 1994 meeting held in Cancun, the NFERP issued an interim statement on mutual recognition of registered engineers (5). The statement summarizes current procedures in the three nations and outlines the subsequent steps that need to occur to "establish principles whereby registered/licensed engineers will be able to move freely between Canada, USA and Mexico." The principles are similar to those in the initial MOA.

The NFERP held three meetings in 1994 and has scheduled three meetings for 1995. They have produced a glossary of Spanish and English terms related to the licensing of engineers and are close to closure on the experience requirements to be included in future accords.

Visa Requirements

A synopsis of immigration laws related to the practice of engineering was developed by Michael Charles of ASCE(6). U.S. laws will still require Mexican professionals to obtain a visa to work in the U.S. Canadian engineers are not required to obtain a visa. Two classes of visas are open to Mexican nationals. First, they may apply for one of 65,000 visas issued world-wide on an annual basis for business professionals wishing to work in the U.S. on a temporary basis. Secondly, NAFTA authorizes the issuance of up to 5,500 visas for Mexican professionals who wish to practice a NAFTA regulated profession in the U.S. These visas are termed "Trade NAFTA" (TN) visas. The rules were published in the December 30, 1993 issue of the Federal Register. The potential employer must request the visa on behalf on the professional and make a bonafide job offer.

U.S. and Canadian professionals must obtain a "FMN" permit and present proper credentials to work in Mexico. The permit is good for 30 days. The professional must then obtain a "FM3" special NAFTA work permit which is valid for one year and can be renewed up to five additional years. The permit costs $130 U.S. and is payable in pesos.

Mexican and U.S. professionals do not require visas to work temporarily in Canada. A Canadian employer must make a bonafide job offer and proper professional credentials must be presented in addition to proof of citizenship. Furthermore, an "Application for Employment Authorization" must be completed which includes a fee of $95 U.S. The application can either be completed at the border or obtained from a Canadian consulate. The Authorization, once granted, must be renewed each year.

Recent Developments

On November 8, 1994 a draft statement entitled "Mutual Recognition of Registered/Licensed Engineers by Jurisdictions of Canada, Mexico and the United States to Facilitate Mobility in Accordance with the North American Free Trade Agreement" was circulated to USCIEP Members and other interested parties. Under this draft, an engineer who graduates from an accredited engineering curriculum of four years or more and with 19 years of progressively responsible engineering practice including 15 years of registered practice shall be registered for international practice without further examination. The actual years of practice requirements are still under negotiation and specialized local requirements such as the California seismic exam would still be required. Also, under the draft Mutual Recognition document, a temporary license could be obtained with twelve years of acceptable engineering experience including eight years following registration. Additional requirements would include knowledge of local regulations, codes and laws regarding the practice of engineering, fluency in the language of commerce, consumer protection requirements and a "willingness to accept cross-border discipline and enforcement and any fines, restrictions or sanctions imposed in the case of unprofessional practice and/or violations of local laws, rules and regulations(7)." The responsible engineering organizations in all three countries have agreed to work with local licensing authorities to promulgate the above-mentioned requirements. These negotiations represent an important first step towards the globalization of engineering practice for North Americans. Additional details will be published in a forthcoming work entitled "The NAFTA Handbook for Water Resources Engineers and Managers.

Appendix: References

(1) E. Walter LeFevre, "Activities of the U.S. Council for International Engineering Practice", October, 1993.

(2) Comite Mexicano Para La Practica Internacional de La Ingenieria, "Informe de Actividades", June, 1994.

(3) NAFTA Forum on Engineering Registration and Practice, "Memorandum of Understanding Concerning Negotiations of Procedures for the Mutual Recognition of Registered Engineers by State, Provincial and Territorial Authorities in Canada, The United States and Mexico to Facilitate Mobility in Accordance with the North American Free Trade Agreement", June 12, 1993.

(4) E. Walter LeFevre, P.E., Ing. Fernando Ocampo, and Ken Williams, P. Eng. "NAFTA Forum on Engineering Registration and Practice Issues Paper", September 24, 1994.

(5) NAFTA Forum on Engineering Registration and Practice, "Interim Statement on the Mutual Recognition of Registered/Licensed Engineers by National, State, Provincial and Territorial Authorities to Facilitate Mobility in Accordance with the North American Free Trade Agreement", January 23, 1994.

(6) Charles, Michael, "NAFTA and Civil Engineers: The Temporary Entry of Business Professionals under the North American Free Trade Agreement", Unpublished work presented in Denver, Colorado, June 7, 1994.

(7) Kimberling, Charles, draft of "Mutual Recognition of Registered/Licensed Engineers by Jurisdictions of Canada, Mexico and the United States to Facilitate Mobility in Accordance with the North American Free Trade Agreement", November 8, 1994.

Impacts of NAFTA on Canadian Water Policy

Isobel W. Heathcote[1]

Abstract

Under the North American Free Trade Agreement, water is considered a "good" subject to trade agreements and challenges. Canadian critics of the Agreement charge that it will facilitate major water diversions and hydroelectric projects involving major negative environmental impacts. Although Canada appears to command a significant portion of the world's fresh water supplies, many rivers flow north to the Arctic Ocean. In the border region, many lakes and rivers are already adversely affected by industrial and municipal pollution. Under NAFTA, there will be increasing pressure to divert the remaining pristine waters to water-starved regions of the United States. While the legal and political implications of NAFTA are still emerging, the debate is prompting Canadian policy analysts to a long-overdue re-examination of domestic water use and regulatory efficiency.

Introduction

Sustainable development of natural resources has become the cornerstone of modern environmental management. One of the most essential of those resources is fresh water. Canada is one of the most water-rich countries in the world, with almost 9 percent of the world's renewable water supply serving less than 1 percent of the world's population. Little of this abundance is available to the urbanized border areas, however. Most northern rivers flow away from the major population centres to James Bay and the Winnipeg River. In the border area, some municipalities are experiencing shortages in water supply, while in others, particularly areas of urban and resource development, there has been

---

[1]Associate Professor, Environmental Engineering, School of Engineering, University of Guelph, Guelph, Ontario Canada N1G 2W1, and President, Canadian Institute for Environmental Law and Policy, 517 College St., Toronto, Ontario Canada M6G 4A2.

considerable degradation of water quality. Yet Canadians remain wasteful water users, with a per capita consumption almost three times that of many water-poor countries and water prices among the lowest in the world.

Over the last decade, Canadians have become more aware of the precariousness of their water supplies, and more protective of those supplies against international diversion or sale. In 1991, the Ontario MISA Advisory Committee wrote:

> The traditional assumption of water is there is always enough, it's always clean and it's always free. Today's reality is that there is not enough, it is not always clean, and it will never again be free.

The advent of the Canada-U.S. Free Trade Agreement, and later the North American Free Trade Agreement, considerably escalated public and private concerns about the conservation of Canadian water resources. Indeed, many public interest groups used this issue as their primary argument against NAFTA. This paper will argue that NAFTA indeed has many important implications for Canadian water resources policy, some potentially negative, but that it also creates new and exciting opportunities for trilateral collaboration on complex water management issues.

The Distribution of Power in Canada

Canada's Constitution was written in 1867 and reflects that era's emphasis on resource extraction and property rights. Jurisdiction over water management is not clearly articulated, with the result that water management responsibility historically has been split between the federal and provincial levels of government. The provinces (with their municipalities) traditionally have had primary responsibility for water supply and pollution control through constitutional rights over their own natural resources. In recent years the federal government has taken an increasingly active role in regulation writing and enforcement, but this role may again be reduced under proposed "harmonization" and "regulatory efficiency" initiatives, begun December 1994.

The Legal Framework

Canadian water laws have evolved over at least the last hundred years, primarily from public health legislation such as the *Public Health Act* and related instruments dating from the last quarter of the 19th Century. They reflect a paradigm typical of that era, but perhaps less appropriate for modern society, including assumptions that pollution is the inevitable consequence of development; that the environment can absorb some amount of every pollutant; and that it is the task of government to find that "acceptable" level of pollution and set standards at that level. In some provinces, including Ontario, the discharge permitting system reflects this attitude in granting lifetime discharge permits for each facility, provided that the facility is not expanded or its process changed.

Today, Canadian water policy is a labyrinth of statutes, regulations, policies, and guidelines, administered by dozens of agencies at four levels of government. Like many other jurisdictions world-wide, Canada and the provinces are under pressure to streamline this complex system and improve equity and consistency across the country. NAFTA has in a sense brought this "dirty laundry" into the sunlight and, implicitly rather than explicitly, increases the pressure on Canada and the provinces to rationalize cumbersome legal and administrative systems.

Under Canadian water law, there is no requirement to manage water resources on a watershed basis. Historically, cross-jurisdictional water management (for instance, mercury pollution in the Wabigoon River and management of binational channels in the Great Lakes System) has relied on federal-provincial or binational agreements. Although these have usually been adequate for the task at hand, they are necessarily short-lived and inconsistent from problem to problem. One of the most important implications of NAFTA for Canadian water policy is therefore that it forces a national consideration of binational and trinational responsibilities. No ad hoc working agreements will now be possible without examination of NAFTA rights and responsibilities; the result will likely be more consistency and better repeatability in cross-jurisdictional water management.

NAFTA and Water Exports

The NAFTA provisions most of concern for Canadian water policy analysts relate to the designation of water as a tradeable "good" under the agreement. Although Canadian federal analysts have argued that water is not covered by NAFTA or the environmental side agreement, private analysts (e.g. Linton and Holm 1993) conclude that the agreements cover "all natural water other than sea water" and not merely bottled water, as has sometimes been suggested. While certain items are specifically exempted, water is not among them: "ice, snow and potable water not containing sugar or sweetener" are considered tariff items in the agreement.

Authors such as Makuch and Sinclair (1993) note that the agreement does not bind the provinces and thus could not be used to enforce provincial environmental laws, which are currently the backbone of environmental protection in Canada. Furthermore, the definition of "environment" in the agreement excludes laws regarding natural resource management, so no enforcement of such laws is possible under NAFTA. Canadian policy analysts therefore fear that the agreement will encourage accelerated extraction of natural resources, including water.

Canadians' greatest fear is that NAFTA will facilitate major water diversions such as the GRAND Canal scheme, a northern diversion/hydroelectric project whose ultimate goal is to generate electricity for sale in the U.S. Accelerated extraction of groundwater for bottled water sale is another major concern. Conservative Canadian analysts, including those at the Fraser Institute in British Columbia, justify water exports to the U.S. on the grounds that the water will be used to irrigate vegetables grown for export back to Canada. Other authors are less optimistic that Canada will reap benefits proportional to the costs of such diversions.

At present, there is consensus that major water diversions are an unlikely prospect, if only because of their high cost and complex environmental impacts. Yet it appears clear that if water were to be diverted from a river and reserved for the use of Canadians only NAFTA would allow the United States to launch a trade challenge to obtain a proportional share of the resource.

Opportunities for Water Management Under NAFTA

Although Canadians are justifiably nervous about the implications of NAFTA for water management, they are also enthusiastic about the opportunities the agreement will provide for binational and trinational collaboration. Canada has a long history of integrated watershed management, and many observers believe that NAFTA creates both enhanced opportunities and stronger incentives for collaborative decision-making and technology transfer. To date, binational initiatives have encountered obstacles of language, data compatibility, and political will; under NAFTA, stronger incentives exist to resolve these differences.

References

MISA Advisory Committee, 1991. ***Water Conservation in Ontario Municipalities: Implementing the User Pay System to Finance a Cleaner Environment.*** Technical Report. Ontario Ministry of the Environment.

Linton, J. and W. Holm. 1993. NAFTA and water exports. Canadian Environmental Law Association special report. CELA, Toronto, Canada.

Makuch, Z. and S. Sinclair. 1993. The NAFTA environmental side agreement: implications for Canadian environmental management. Canadian Environmental Law Association special report. CELA, Toronto, Canada.

Applying U.S. Technology to Water Supply Problems in México

Emilia Salcido[1], Lynn Lovell[2], Quentin Williams[3]

Abstract: This paper describes the applications of remote sensing, a geographical information system (GIS), and a water accounting computer model to a study of water supply and demand in northern Mexico. The WEAP (Water Evaluation and Planning System) model was used to quantify and evaluate the current and future water requirements and availability for the San Juan River watershed. Municipal, commercial, industrial, and agricultural demand conditions were quantified using WEAP. Supply sources modeled with WEAP included the river-reservoir network, all aquifers, spring sources, and sources of interbasin transfers. Legal and physical links between demand sites and supply sources were defined and prioritized according to the current system of water rights. The interpretation and classification of LANDSAT imagery facilitated the estimate of runoff coefficients to calculate watershed yield. Two GIS platforms, ARC-INFO and Microstation, were used to manipulate the data collected and generated in this study. Major findings of the study included identification of areas of inconsistent reporting and areas of inadequate permits for reported and estimated use. Proposed alternatives to reduce the supply deficit included rate increases as a means of demand management, increased reuse in industry, decreased loss rates, and artificial recharge of an overexploited aquifer.

Introduction. In an effort to insure an adequate water supply, the Mexican Water Commission (CNA) initiated a comprehensive water resources planning study to examine current and future supply and demand conditions for the San Juan River basin. The methodology and computer models developed in this study will be applied to other basins in Mexico. The primary tools used to develop, manage, and analyze the system were remote sensing, a geographical information system, and a comprehensive water accounting computer model WEAP (Water Evaluation and Planning System) (Tellus, 94).

The Río San Juan basin spans three northeastern states in Mexico: Coahuila, Nuevo León, and Tamaulipas. The Metropolitan Area of Monterrey (AMM) and Saltillo are the principal urban centers. Major rivers include the San Juan, Pesquería, Salinas, Santa Catarina, Pilón, Garrapatas, Blanquillo, and Ayancual. Four large reservoirs supply the basin, those being El Cuchillo (1123 Mm^3) in China, Nuevo Leon; Marte R. Gómez (1026 Mm^3) in Camargo and Miguel Aleman, Tamaulipas; La Boca (41 Mm^3) in Santiago, Nuevo León; and Cerro Prieto (326 Mm^3). Cerro Prieto is located outside the watershed but was included in the model because it supplies the AMM. Average annual rainfall in the basin varies from 300 mm in the arid west to 800 mm in the Sierra Madre Mountains.

[1].Hydraulic Engineer. Halff Associates, 4000 Fossil Cr. Blvd, Fort Worth, TX 76137
[2].Vice President. Halff Associates, 4000 Fossil Cr. Blvd, Fort Worth, TX 76137
[3].Engineer. HARZA Engineering, 233 South Wacker Drive, Chicago, IL 60606-6392

Remote sensing. Estimates of hectares of agricultural land use and overall basin runoff-yield volumes for this study were developed from the interpretation of 1993 LANDSAT imagery. The LANDSAT image for the 32,600 km^2 basin is a composite of 4 scenes taken during the summer of 1993, selected to capture optimum agricultural land use. The image was classified using ELAS (Earth Resources Laboratory Software by NASA) into 12 categories (woodlands, grasslands, improved pasture, barren, irrigated agriculture, non-irrigated agriculture, urban areas, transportation, water bodies, and high, medium and low density brush) by selecting a range of spectrum signature for each.

Geographical Information System. A geographical information system (GIS) was used to manipulate, analyze, and retrieve information needed for the different components of the San Juan River Basin study. Two GIS platforms were used for this study. The first one, used to manipulate raster images, was developed using ARC-INFO. Land use data obtained from the analysis and classification of the LANDSAT imagery was imported into ARC-INFO and superimposed on maps of soils, sub-basins, and *municipios* (counties) digitized from available maps. ARC-INFO generated reports of land use area per municipio as well as land use area per soil type per basin. This information was used to estimate each basin's runoff coefficient and irrigated hectares per municipio.

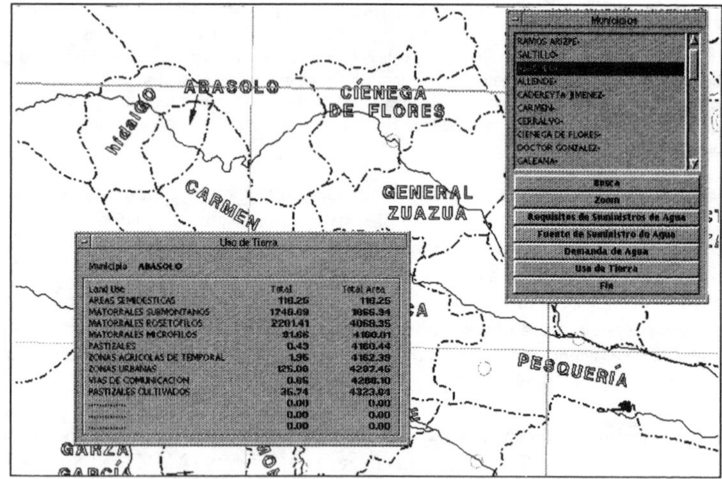

Figure 1. Geographic Information System GIS- *Municipios* Module

A second platform was developed in Microstation 4.0 and Dbase IV using MDL's (Microstation Development Language). Results displayed with this GIS by municipio included a breakdown of current and projected point of use demand, point of withdrawal water requirement, and supply sources as shown in Figure 1. Rain, stream gage, and water quality gages were also included. Other physical parameters available through the GIS include: political boundaries, hydrologic watersheds, roads, railroads, urban areas, and water bodies.

WEAP model. Since most of the available demand-related data was at the *municipio* level, in most cases the municipio was chosen as a demand site. One exception was AMM (Monterrey) where its seven municipios were modeled as one, since all supplies go to a common distribution ring. Three demand sites, however, were defined to represent the AMM's water use to allow more detailed reporting and analysis, and to allow the use of different monthly demand distributions.

The WEAP model included five demand sectors: municipal, commercial, public, industrial, and agricultural. Legal and physical links between demand points and supply sources were made and prioritized according to the local system of water rights to simulate actual conditions of water allocation. Link capacities were based on physical capacity of the transmission lines and permitted withdrawals issued to individuals, corporations, or operating authorities. Losses included distribution, clandestine connections, and unpaid bills among others. Estimated losses were 40% for AMM, 30% for AMM industrial sector, and as high as 50 % for demand sites in Coahuila.

Supply sources included the river-reservoir network, all aquifers, spring sources, and sources of interbasin transfers. Maximum storage capacity, user-specified initial storage level, and monthly inflows for aquifers and reservoirs were estimated from several sources, most of them obtained from local representatives of CNA.

Surface runoff, used as inflow at main river nodes, was estimated for 48 sub-basins using a locally developed computer program, Hidrolo, which uses a direct runoff equation. The runoff coefficient, the key component of the Hidrolo runoff equation, varies annually and considers 3 main factors: annual precipitation, hydrologic soil types, and land cover. Land cover-soil complexes were calculated using ARC-INFO. The average annual runoff, resulting from the Hidrolo program, was distributed monthly based upon monthly rainfall.

Comparisons of supply and demand were made at discrete intervals (1990, 1995, 2005, 2015) and under normal or drought conditions. A normal hydrologic year was based on rainfall for all available gages, averaged through the available period of record. A drought year was estimated as a percent deviation of the normal, at the entire basin level, for the historical drought period of August 1979-July 1980. It was estimated that under 1979-1980 historical drought conditions, annual water availability would be 77% of normal.

Alternatives for Reducing Supply Deficit. Results of the WEAP supply-demand balance model highlighted areas and conditions of supply deficit. Various proposed management strategies were tested to estimate the effect on the reduction of supply deficit. The projected supply deficit for the Nuevo León portion of the watershed could be reduced by as much as 600 Mm^3 (over 60%) while the Coahuila portion could be reduced from over 100 Mm^3 down to less than 10 Mm^3 through a combination of small-scale infrastructure improvements and water management strategies. These include rate increases for selected water use sectors, reduced loss rates, increased reuse in industry, greater reliance on underutilized groundwater reserves, and artificial recharge of the Saltillo-Ramos Arizpe aquifer (Halff, 94).

Major Findings of Study. Results of the study show no consistent or complete reporting on industrial water use in the Rio San Juan Watershed. Groundwater permits for industrial withdrawals in the Nuevo León portion of the watershed appear to be grossly inadequate. The calculated 1990 industrial demand is 389 Mm^3, while industrial permits for groundwater withdrawals total 56.1 Mm^3, of which 18 Mm^3 are assigned to the municipio of García. The annual withdrawal from the aquifer below García (Durazno) is estimated at 8.6 Mm^3, and the annual recharge is estimated at 3.1 Mm^3 (Halff,94).

A number of exceedingly large, unrealistic permits exist for surface water withdrawals. One example is two permits in Rayones to Rio Casillas for 67 Mm^3 each. In other cases, permits for surface water withdrawals and groundwater extractions are insufficient to sustain current levels of activity. For example, Pesquería has significant irrigated agriculture believed to be supported by surface water withdrawals. Model results show that surface water is available but surface water permit totals for the municipio are significantly less than the withdrawals required in the model. Marin is another example of a municipio with restrictive permit levels (Halff, 94).

A number of aquifers in the Rio San Juan watershed may be experiencing damaging levels of overdraft. These include the Saltillo-Ramos Arizpe aquifer, the Monterrey Aquifer system, the Mina aquifer, the Durazno aquifer, and the Buenos Aires aquifer. Aquifers underlying the municipios of Abasolo, Los Aldamas, Allende, Cadereyta Jiménez, El Carmen, Doctor Coss, Hidalgo, Juárez, Marin, Pesquería, Rayones, and Salinas Victoria have not been analyzed with respect to their water yielding capacity despite significant withdrawals from groundwater resources.

Results of this study indicate that Cerro Prieto may be experiencing significant losses by infiltration, Marte Gómez inflow depends considerably on the return flows from Monterrey, and the projected withdrawals from El Cuchillo for 2005 exceed the inflow in an average hydrologic year.

References

Halff Associates, July, 1994. *Estudio de Demanda de Agua de la Cuenca del Río San Juan (Monterrey) y su Aplicación para el Diseño de Tarifas por Concepto de Derechos de Agua*, Forth Worth, TX.

Tellus Institute- Stockholm Environment Institute, June 1994. *WEAP Water Evaluation And Planning System USER GUIDE*, Boston, MA.

THE RÍO SAN JUAN PILOT STUDY- A WEAP APPLICATION

Q.R. Williams

Abstract: The Río San Juan watershed covers 32,667 km^2 and includes the city of Monterrey in Northern Mexico. A comprehensive water balance was carried out using the WEAP (Water Evaluation and Planning) model to assess the future adequacy of the watershed's water resource system through the year 2015. Industrial, municipal, agricultural, commercial, and public demand were estimated for 1990, and totaled 1380 million cubic meters (Mm^3). Point of use water requirements were derived by accounting for loss rates and reuse rates. Surface and groundwater sources were located and quantified for an "average" hydrologic year and a historical drought year. Total water resources under average conditions were estimated to be 1820 Mm^3, and 1400 Mm^3 under drought conditions. Physical and legal restrictions were modeled in comparing supply with demand for 1990, 1995, 2005, and 2015 under both "average" and drought hydrologic conditions. Based on a scenario of unrestricted demand growth, model results show a supply deficit of over 1125 Mm^3 for 2015 under average hydrologic conditions. Proposed water management strategies could reduce the supply deficit by as much as 700 Mm^3 without major infrastructure additions.

I. Introduction. The Río San Juan watershed covers 32,667 km^2, spans the three states of Coahuila, Nuevo León, and Tamaulipas, and includes the principal cities of Monterrey, Nuevo León, and Saltillo, Coahuila. Principal rivers include the San Juan, Pesquería, Salinas, Santa Catarina, Pilón, and Blanquillo. The recently completed reservoir El Cuchillo in the watershed is designed to meet the growing demand of the *Area Metropolitana de Monterrey* (AMM) through the year 2006. The cost of El Cuchillo and the scarcity of potential new supplies led the *Comisión Nacional del Agua* (CNA) to initiate a comprehensive water resource planning study to examine current and future supply and demand conditions, and investigate the water management alternatives for insuring a safe and adequate water supply through the year 2015. The Water Evaluation and Planning System (WEAP), a commercially available software package, was used to carry out this study (Tellus Institute, 1994).

II. Demand in the Watershed. Current demand was categorized and quantified, and projected through the year 2015. Main categories of demand were industrial, municipal, agricultural, commercial, and public, in accordance with the local institutions' classification systems (Halff Assoc., 1994).

Demand in the Río San Juan Watershed

Category		1990		2015	
		Demand (Mm^3)		Demand (Mm^3)	
Industrial	*(AMM)*	389	*(302)*	629	*(487)*
Municipal	*(AMM)*	290	*(221)*	567	*(419)*
Agricultural		674		907	
District 026		440		440	
Commercial	*(AMM)*	24	*(19.0)*	54	*(21.2)*
Public	*(AMM)*	3.9	*(1.6)*	6.9	*(2.3)*
Total	***(AMM)***	**1380**	***(544)***	**2160**	***(930)***

Harza Engineering Co., Chicago, IL, 60611

Industrial Demand. Industrial demand included that from self-supplied industry and from industries connected to the municipal distribution network. The primary method used to calculate self-supplied industrial demand related number of employees to water use by type of industry. Only industrial establishments employing at least 100 persons and having net sales in excess of 1115 times the minimum wage were considered self-supplied. This category includes 159,341 employees, making up 71% of industrial employment. It was assumed that smaller industries rely on the municipal distribution systems for their water supply. Industrial demand for these smaller companies averaged about 3% of annual industrial demand, and was calculated using the distribution agencies' reported number of connections and an average consumption rate per connection (Halff Assoc., 1994). An annual growth rate of 2% was applied to industrial demand.

Within the industrial sector, a relatively small number of industry types were responsible for the majority of demand. The paper and cardboard industry (28.7%), an oil refinery (16.2%), the plastic industry (13.0%), and the chemical fiber industry (12.5%) represent about 70% of 1990 industrial demand. Because of the dominance of a few industries, the sector demand as a whole is quite sensitive to errors in estimating demand for these few industries. A potential source of error is a possible disparity between the U.S. and Mexico in employment levels for similar processes or industries. Sensitivity analysis on the derived water use coefficients shows that if coefficients for these industries are overestimated by 50%, the self-supplied 1990 industrial demand is still significant at about 180 Mm^3.

Municipal Demand. Municipal demand was calculated using 1990 census data and per capita water use rates. These rates vary according to the urban and rural subsectors, and within the urban subsector according to level of service. Urban residents with a direct connection to the municipal network were assigned a per capita water use rate of 250 liters per person per day (lppd) in 1990, increasing by roughly 30% by the year 2000 to reflect an anticipated improvement in standard of living. The urban population without a direct connection to the municipal network (relying on public taps or trucked delivery) was estimated to have a per capita use rate of 125 lppd. Direct service for all urban residents was projected by the year 2000. In accordance with local water authorities' reports, a rural per capita rate of 190 lppd was adopted (Halff Assoc., 1994). Projections of municipal demand were based on the above noted changes in water use rate coupled with area-specific projected population growth rates.

Agricultural Demand. Agricultural demand was based on irrigated agriculture and livestock water requirements. Irrigated hectares per municipio are based on government reports and results of the LANDSAT imagery analysis (Halff Assoc., 1994). Projections of agricultural demand outside of District 026 were based on current trends in population growth. Demand in District 026 was held constant.

Where data were available, the water use rate per hectare of irrigated land was derived based on hectares of each crop, and monthly water requirements for each crop type adjusted for precipitation. Monthly totals were computed, translated into total annual demand, and then divided by the total number of hectares to give an average annual water use rate per hectare.

For the watershed's primary irrigation district, District 026, all of the district's 75,000 hectares were used to calculate demand. Only about 16% of the district is within the San Juan watershed, but the other 84% is potentially irrigated by water originating from the reservoir Marte R. Gómez, the district's primary water source. Where no data were available, irrigated hectares were assigned the net water use rate of 48.4 m^3 per hectare per year. This water use rate was calculated from the average of 6 crops' (cotton, early maize, late maize, early sorghum, late sorghum, and wheat) consumptive use in District 26.

Livestock demand was based on number of head of cattle, swine, horses, goats, and sheep per municipio and an annual water requirement per animal.

Commercial and Public Demand. The commercial sector covered all service industries, small establishments such as restaurants, bars, etc, that all rely on their respective municipal networks. The public sector included all public buildings and schools. The primary method used to calculate demand for both these sectors was based on number of connections and an average water use rate per connection. Projections of commercial and public demand were based on the continuation of current trends in commercial and population growth.

III. Supply Requirements and Link Capacities. The effects of losses and reuse were accounted for in the study. The results reported above reflect the quantity of water that is needed at the point of use and do not include these influences. Results shown in this section represent the quantity of water that must be withdrawn at the point of supply to satisfy the eventual demand at the point of use.

Loss rates vary from a low of 30% for self-supplied industry to 58% for smaller urban areas, with the average running at about 50%. Reuse is most prevalent within the industrial sector in the AMM, and was estimated to be 19% based on reported reuse quantities.

The computed 1990 supply requirement for the Río San Juan watershed and District 026 is 1868 Mm^3 and is estimated to increase to 2919 Mm^3 by 2015. Not including District 026, the supply requirement was 1242 Mm^3 in 1990 and 2293 Mm^3 by 2015.

The supply requirement for all water uses in the AMM totaled approximately 704 Mm^3 in 1990, representing 38% of the watershed's supply requirement, or 57% of the requirement not including District 026. By 2015, the requirement was estimated to be 1256 Mm^3.

Maximum transmission capacities on links between demand and supply sources were specified according to lists of official permits. Physical restrictions such as pipeline and canal capacities were also used. Where data were lacking, link capacities were set according to historical use rates plus a percentage to accomodate demand growth.

IV. Supply. Total annual water resources originating the Río San Juan watershed under average conditions were estimated to be 1820 Mm^3, and 1400 Mm^3 under drought conditions. The Cerro Prieto Reservoir is outside the watershed, but it exclusively supplies the AMM. As a source for transfer into the San Juan watershed, inflow to Cerro Prieto was estimated at 405 Mm^3 in a normal hydrologic year, and 172 Mm^3 in a drought year.

Based on a scenario of unrestricted demand growth, model results show a supply deficit of over 1125 Mm^3 for the year 2015 under average hydrologic conditions. The supply deficit for the AMM was projected to be over 500 Mm^3 by 2015 under normal hydrologic conditions, not allowing for aquifer overdraft.

V. Observations and Potential Solutions. More comprehensive reporting of industrial and agricultural water use and further study of the watershed's aquifers will improve results from future analyses of this type.

The supply-demand relationships and comparisons developed in the model clarified a number of previously indefinite correlations. The location of El Cuchillo significantly reduces the runoff to Marte R. Gómez, and makes the inflow to the latter strongly dependent on return flows from the AMM. In addition, the AMM's effluent quality will strongly influence water quality in Marte R. Gómez. Increased water reuse and conservation in the AMM is likely to directly reduce irrigation from Marte R. Gómez.

Cerro Prieto Reservoir was shown to have almost twice the supply capacity currently utilized. In a normal year, model results show that over 130 Mm^3 must be released from Cerro Prieto to avoid overtopping. In reality, releases have been made from Cerro Prieto only during Hurricane Gilbert in 1988. If 50% of the unaccounted water arriving at Cerro Prieto were recovered, an additional 65 Mm^3 could be provided to the Río San Juan watershed.

Implementing a combination of water management measures including rate increases, reduced loss rates, artificial recharge, and increased reliance on underutilized aquifers could further reduce the projected supply deficit of about 1125 Mm^3 by 640 Mm^3.

References

Halff Associates, 1994. *Estudio de Demanda de Agua de la Cuenca del Río San Juan (Monterrey) y Su Aplicación por Concepto de Derechos de Agua.* Fort Worth, TX.

Tellus Institute-Stockholm Environmental Institute, 1994. *WEAP Water Evaluation and Planning System USER GUIDE.* Boston, MA.

Subject Index

Page number refers to first page of paper

Author Index

Page number refers to the first page of paper

Date Due

IL 7807458			
8/15/03			
9487130	11/18/04		

BRODART, CO. Cat. No. 23-233-003 Printed in U.S.A.